International Conference on Mechanical Engineering, Automation and Control Systems 2015 (MEACS 2015)

IOP Conference Series: Materials Science and Engineering Volume 124

Tomsk, Russia
1 – 4 December 2015

ISBN: 978-1-5108-2412-6
ISSN: 1757-8981

Printed from e-media with permission by:

Curran Associates, Inc.
57 Morehouse Lane
Red Hook, NY 12571

Some format issues inherent in the e-media version may also appear in this print version.

Printed by Curran Associates, Inc. (2016)

For permission requests, please contact the Institute of Physics at the address below.

Institute of Physics
Dirac House, Temple Back
Bristol BS1 6BE UK

Phone: 44 1 17 929 7481
Fax: 44 1 17 920 0979

techtracking@iop.org

Additional copies of this publication are available from:

Curran Associates, Inc.
57 Morehouse Lane
Red Hook, NY 12571 USA
Phone: 845-758-0400
Fax: 845-758-2633
Email: curran@proceedings.com
Web: www.proceedings.com

TABLE OF CONTENTS

INTEGRATED COMPUTER CONTROL SYSTEMS IN MECHANICAL ENGINEERING

NUMERICAL SIMULATION OF APPLIED PROBLEMS

MATERIALS SCIENCE IN MECHANICAL ENGINEERING

MECHANICAL ENGINEERING PROCESSES AND METAL TREATMENT

IOP Conf. Series: Materials Science and Engineering **124** (2016) 011001 doi:10.1088/1757-899X/124/1/011001

Preface

The organizing committee of the conference: 'MEACS 2015' expresses its gratitude to the Russian Foundation for Basic Research for the financial support of the event. In addition, the organizing committee is grateful to Tomsk Polytechnic University, on which basis the conference was held. A special gratitude is expressed to the employee of Tomsk Polytechnic University, Sharkov Ivan, for the assistance in publication of the proceedings of the conference in the journal.

Peer review statement

All papers published in this volume of *IOP Conference Series: Materials Science and Engineering* have been peer reviewed through processes administered by the proceedings Editors. Reviews were conducted by expert referees to the professional and scientific standards expected of a proceedings journal published by IOP Publishing.

IOP Conf. Series: Materials Science and Engineering **124** (2016) 012001 doi:10.1088/1757-899X/124/1/012001

Temperature evaluation of a hyper-rapid plasma jet by the method of high-speed video recording

A E Rif[1], V V Cherevko[2], A S Ivashutenko[2], N V Martyushev[2] and N Ye Nikonova[3]

[1] Municipal budgetary comprehensive establishment of lyceum attached to TPU, 4a. Ivanova st., Tomsk, 634028, Russia

[2] National Research Tomsk Polytechnic University, 30, Lenina ave., Tomsk, 634050, Russia

[3] National Research Tomsk State University, 36, Lenina ave., Tomsk, 634050, Russia

E-mail: ivashutenko@tpu.ru

Abstract. In this paper the procedure of comparative evaluation of plasma temperature using high-speed video filming of fast processes is presented. It has been established that the maximum plasma temperature reaches the value exceeding 30 000 K for the hypervelocity electric-discharge plasma, generated by a coaxial magnetoplasma accelerator with the use of the 'ImageJ' software.

1. Introduction

Plasma is a basic state of a substance in the universe. It is divided into low-temperature (1 000 000 °K) and high-temperature (above 1 000 000 °K). Different substances transform to a plasma state at different temperatures, which accounts for the structure of the outer electron shells of the atoms of the substance: the easier the atom releases its electron, the lower is the temperature of transition into a plasma state.

Plasma behaviour is characterized by a sufficiently vast set of phenomena, so that in complexity its research is comparable to the study of complex biological systems, therefore it is necessary to be able to diagnose it. The diagnostics itself has a sufficiently important meaning for the practical purposes, and first of all, in obtainment of the controlled thermonuclear reaction. On the other hand, the understanding of the processes flowing in the production facility allows the intentional improvement of its operating characteristics. The difficulty in the measurement of plasma temperature consists in the fact that any device located in the medium with the temperature of million degrees should either cool or burn it. Therefore it is necessary to judge about the plasma parameters at a distance and a 'thermometer' should undoubtedly be noncontact. At present there are different methods of plasma temperature measurement, most of them function on the basis of optics law, such as:

- Thomson scattering. The range of the measured temperatures varies from the tens of electron volts in the near-wall region up to 10–20 keV on the axis of a plasma column. The measurement error in electron temperature does not exceed 10 % in the temperature range 0.3–20 keV;
- Laser-induced fluorescence. The domain of applicability of the method is from 200 to 3000 K. The temperature measurement error of gas or low-temperature plasma nearby 2000 K does not exceed 5 %.

IOP Conf. Series: Materials Science and Engineering **124** (2016) 012001 doi:10.1088/1757-899X/124/1/012001

- Doppler broadening of absorption spectral lines [1]. The error of temperature measurements equals ± 5 %. And others.

The analysis of the existing methods of plasma temperature measurement has allowed us to conclude that in spite of sufficiently low error in the measurements they have their own drawbacks, in particular the necessity of using quite complicated complex engineering solutions and a subsequent mathematical treatment [2-4]. Besides, the above-mentioned methods function preliminary with stationary plasma. At this the problem of recording of fast nonstationary processes remains poorly investigated. In this paper we suggest filling this gap by our method based on high-speed video recording of phenomena.

2. Materials and methods

The essence of the method is in the shooting of the entities with the known temperature (standard) and subsequent treatment (calibration curve plotting) and evaluation of plasma temperature of the real process. Hypersonic plasma generated by a coaxial magnetoplasma accelerator [5-7] and recorded by a high-speed camera 'Photron Fastcam SA1.1.' has been chosen as an object of the research. The typical photograms of plasma effusion are presented in Figure 1.

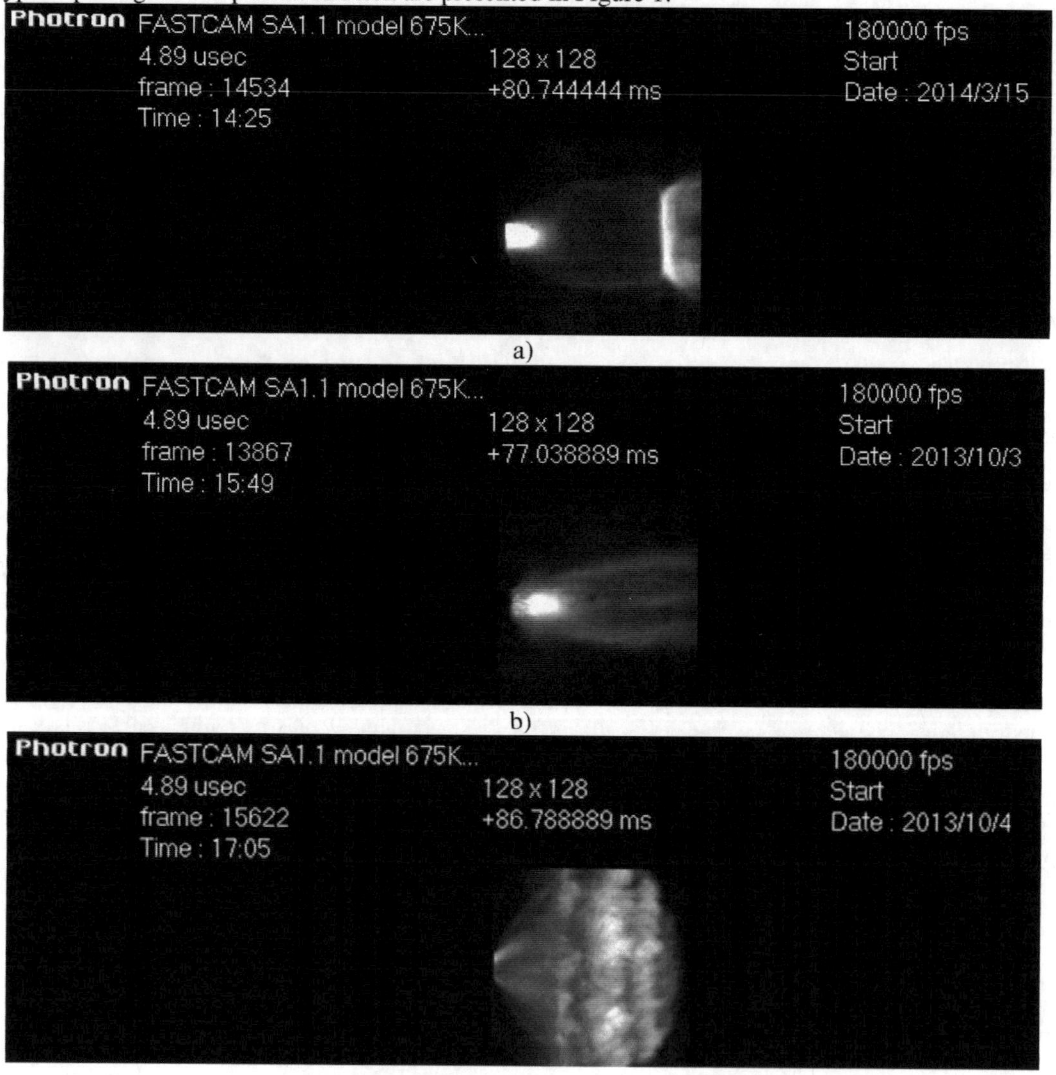

c)

Figure 1. Photograms of plasma effusion.

For the experiment a glowing filament of an incandescent lamp of 100Wt power has been chosen as a 'standard'. For the temperature evaluation of the glowing filament a pyrometer produced by 'Raytek' Company (Germany) with the range of the temperature record from 600 to 3000 °C has been used. Plasma shooting has been conducted by means of the monochrome high-speed camera 'Photron SA1.1.' equipped with the installed lens 'Helios 44-2' in a shooting mode: diaphragm is 2.0 with the use of two light filters 'ND64' and 'ND32'. The use of addition light filters provides the record of the 'non-exposed' picture of plasma effusion.

3. Results and discussion

The procedure of plasma temperature evaluation consists in the following: by means of the laboratory autotransformer we change the power of the incandescent lamp (and accordingly the glowing intensity of the filament itself), the temperature of the filament is simultaneously evaluated using the pyrometer, at this the images are recorded with the use of the high-speed camera. As a result of this work we have obtained a series of images of the glowing filament with different level of luminous intensity at known temperatures. The results of the conducted work are presented in the table.

Table 1 – The parameters of glowing filament with different level of luminous intensity at known temperatures.

№	U, B	I, A	T, °C	Power, W	R_t, Om
1	94	0.256	1204	24.1	367
2	128	0.301	1420	38.5	425
3	173	0.353	1664	61.1	490
4	216	0.397	1866	85.8	544
5	246	0.425	1996	104.6	578

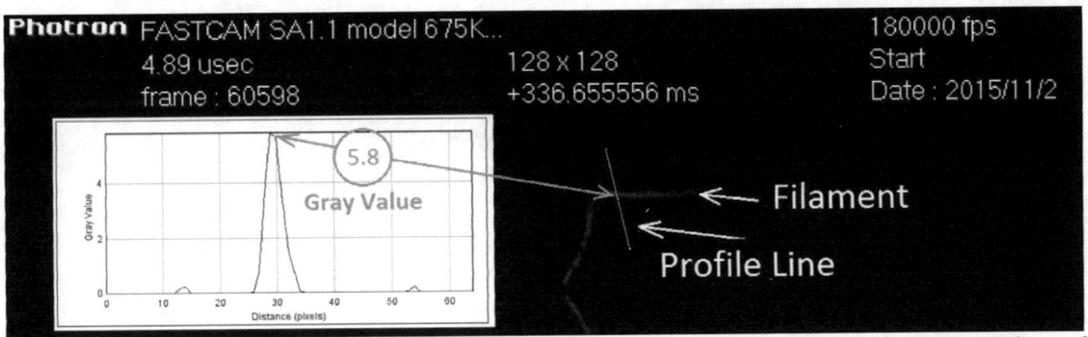

Figure 2. The example of the evaluation of luminous intensity of glowing filament in the gradations of grey.

The level of luminous intensity of the glowing filament (in the gradation of grey) for all obtained images has been evaluated by means of the software product '***ImageJ***'. In order to evaluate the maximum value of filament luminosity in gradations of grey in the photos a line was drawn across the area, which is of great interest. Using a system function of the software '***Plot Profile***' the programme creates automatically an array data with indication of the value in the gradation of grey for each pixel (from 0 to 255). The example of the image treatment is shown in Figure 2.

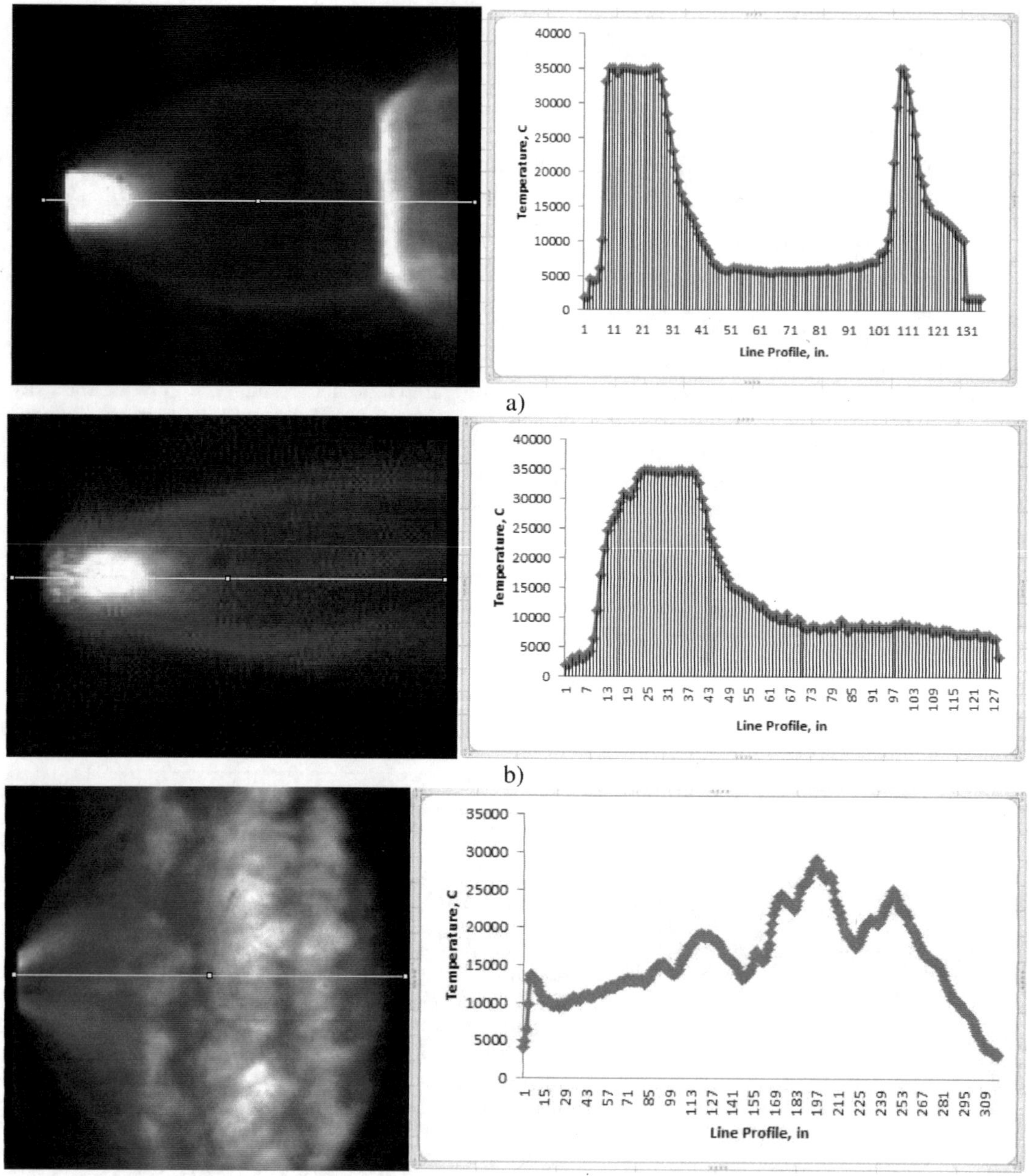

Figure 3. The photograms of plasma effusion after treatment and temperature analysis.

Having conducted the analysis of the photos with different level of luminous intensity we have obtained the array data, the processing of which allowed revealing a describing function – a calibrating curve. The resulting analysis of photograms of hypersonic plasma effusion is presented in Figure 3.

4. Conclusion

On the basis of the conducted work the following peculiarities of such approach to plasma temperature evaluation or, in other words, pros and cons have been found. The advantages of the method are the following: first, this is sufficiently simple technical realisation; second, it is possible to measure the temperature in a semiautomatic mode (realisation of batch processing).

The disadvantages are a higher measurement error (about 20 %) relatively other methods described earlier and impossibility of evaluating the maximum temperature in case of 'exposure' of photo-video shooting or separate elements on it.

The indicated disadvantages are inessential and can be eliminated completely in case of selection and use of more powerful light filters and conducting of the preliminary calibration of the system on the entities with known temperature similar to the analysed one.

5. Acknowledgments

This work was financially supported by the Russian President Grant (SP-1179.2015.1).

References

[1] Magunov A 2010 *Photonica.* **3** 43–47
[2] Ivashutenko A S, Martyushev N V, Vidayev I G 2014 *Advanced Materials Res.* **1040** 819–823
[3] Shevtsova L I, Sameyshcheva T S, Munkueva D D 2014 *Appl. Mechan. and Mater.* **682** 188–191
[4] Shevtsova L I, Korchagin M A, Thommes A, Mali V I, Anisimov A G, Nagavkin S Yu 2014 *Advanced Materials Research* **1040** 772–777
[5] Sivkov A A, Ivashutenko A S, Lomakina A A, Shanenkov I I 2015 *Applied Mechanics and Materials.* **756** 325–328
[6] Shanenkov I I, Sivkov A A, Pak A, Kolganova Y L 2014 *Advanced Mater. Res.* **1040** 813–818
[7] Sivkov A, Pak A, Shanenkov I, Kolganova Y, Prosvirin I 2014 *Advanc. Mater. Res.* **880** 36–41

Milling complex surfaces with cutting edge displacement towards the cut surface

S Ambrosimov[1] **and A Zhirkov**[2]
[1] Lipetsk State Technical University, 30, Moskovskaya st., Lipetsk, 398600, Russia
[2] Bryansk State Technical University, 7, 50 years of October st., Bryansk, 241035, Russia

E-mail: ambsk@ mail.ru

Abstract. The article presents the prospective milling methods with non-linear correlation of feed motions enabling sliding of the cutting edge towards the cut surface, temperature decrease in the cut area and increase of the milling cutter wear resistance.

1. Introduction

Milling problems of complex surfaces, such as dies and molds, arise due to a large volume of chipped material and, as a result, significant expenditures on expensive end and disc radial milling cutters having an accelerated wear.

The most common method of milling the internal surfaces, such as notches and pockets, as well as external profiled surfaces, for instance, punches, cams or turbine blades is cutting with ball-nose end milling cutters made of high-speed steel or solid hard-alloy steel, as well as modular milling cutters with two petal-type plates or one plate with two sided CoroMill-type cutting edges. The latter are rotated by 10° – 15° to vertical to increase the cutting speed. The tool path relative the work piece shall correspond to the radial location of the machined surface, which significantly simplifies the development of control software. In the course of process development, a milling cutter diameter (d_m) is chosen first, which, according to Sandvik Coromant, shall not exceed 1.5…1.0 times a spherical radius of pocket profile R_p, where the milling width is $0.2d_m$, enabling an embrace angle optimal range of 100°to 80° for purposes of angle vibration minimization. However, it raises the cut off layer thickness due to the accelerated feed rate, which is inevitable while milling with a programmed radius. To provide authentic conditions during machining of linear and radial pocket sections the feed rate is decreased while moving along the programmed area to obtain equistability of tooling. The decrease of the milling cutter diameter below R_p unreasonably reduces the machining capacity. Pockets milling starts with the processing of the hole in the center of the pocket, so that the depth of milling does not exceed a double diameter of the milling cutter. In these conditions machining is performed spirally from the center toward outside with the programmed radial areas of tracing radius $r_П = R_П - 0{,}5 \cdot d_Ф$, with feed rate decreased.

Over recent years many new milling methods have appeared, for example, a highly efficient method of pocket plunger milling with one axial feed movement, which contradicts the definition of a milling cutter and a milling process formulated by GOST 25721-83: ‘A milling cutter is an edge tool for machining performed by the main rotational cutting movement of the tool without a possibility to change the radius of this motion and with at least one feed motion direction not corresponding to the rotational axis’.

While machining flat surfaces with face milling cutters, it is reasonable to set the curved cutting enabling minimal thickness of the layer to be cut off by the tooth when leaving the machining area. First of all, this reduces the length of the tool path. Secondly, it raises significantly the resistance of hard-alloy cutting teeth. The decrease of the cut off layer thickness results in the mitigation of the unloading impact occurring after machining by the cutting edge because of the sharp fall in the chip supporting force at the final point of cutting. This causes chip bending, that is, changes the flow pattern of force on the front surface and eventually changes the stress pattern on the front surface of the tooth from compression to elongation. In case of linear cut the thickness of the cut off layer when the tooth is at the end of the cut area remains high until the moment of milling cutter penetration into the work piece by $0.5d_m$.

According to the experimental research carried out by Sandvik Coromant, curved cutting and tracing of angles increase the resistance of cutting hard solid plates 2 up to 16 times while machining steel work pieces.

Therefore, many new machining methods have appeared over recent years, the specifics of these methods being a new kinematic forming diagram characterized by the complex correlation of forming motions with a variable feed value and direction, aimed at an increase in machining resistance and capacity.

Rolling complexly correlated feed motions result in higher milling cutter resistance due to smaller stripe dimensions. Curved cutting with complex correlation at the moment of face milling cutter penetration into the work piece reduces the negative influence of unloading impact on the cutting edge. A gradual decrease of the feed rate along with the cutting edge embraces an angle increase in the pocket angle, tracing points lessen the machining vibrations.

One of the prospectives and new fields of investigation can be presented by methods enabling continuous displacement of the cutting edge toward the cut surface. Displacement of the cutting edge, ensuring non-linear generating of the machined surface as well as generating with sliding, helps to continually involve new areas of the cutting edge not operative before and to reduce heat density at the top of the cutting edge by high-speed sliding. Besides, it increases the contact and general length of the cutting edges and also contributes to even wear of the cutting edges owing to their uniform displacement.

2. Non-linear form-generating milling method

One of the most universal non-linear form-generating methods of the machined surface is a method of complex curved surfaces machining (patent No. 2167746 of the RF), performed by a tool with two taper or one toroidal surfaces (Figure 1), as it enables machining of the profiles intersecting at any angle.

Tool linear generators are machined at an α angle, with the value equal or below minimum angle β_{min} between the tangent lines to the opposite sides of the machining surface profile at the points of interference with the profile concave areas.

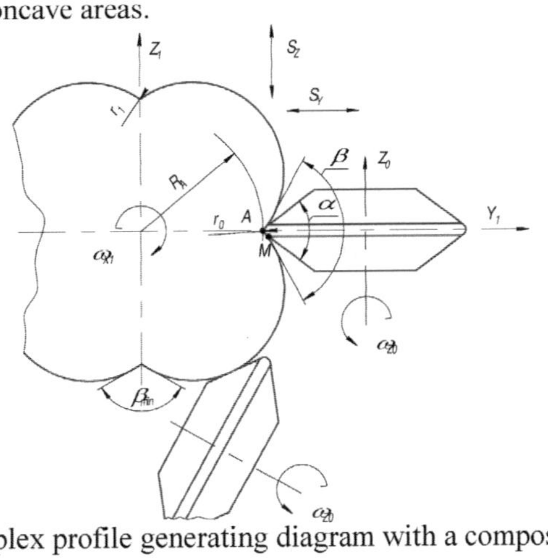

Figure 1. Complex profile generating diagram with a composite profile tool.

Machining of the profile convex side is performed by the linear corresponding side of the tool profile, while the part profile concave area is machined by the radial toroidal surface of the tool. Machining is performed by three simultaneous forming motions with non-linear correlation lying in the same profiling plane. One of them, rotational ω_{X1}, is effected so that the linear generator of the tool is serial to the tangent line in each point of the machined profile. Two other motions, S_Z and S_Y, correlate with rotational ω_{X1}, so that the linear generators can roll over the machined surface while machining the profile areas. The developed method of complex surface machining contributes to a better accuracy owing to part profile generating by linear tool generators. Furthermore, this method is universal and highly-efficient.

3. Milling with a screw feed motion
For machining profiled concave surfaces with variable profile such as grooves, a new method (patent No. 22008502 of the RF) which is performed by a tool in the form of a rotary body with the toroidal generating surface, is developed (Figure 2.). In the course of machining, the tool (work piece) is subject to rotation with ω_{OY} feed motion and two progressive motions, S_Y and S_Z , which are performed simultaneously and lie in the plane perpendicular to the plane of rotational movement. Progressive motions are performed normal (S_Y) and parallel (S_Z) to the base plane and have a non-linear correlation with rotational motion ω_Y. Rotational feed motion is effected until a periodical and simultaneous contact of the tool is realised within the machined area at two points M and M_1 (M' and M_1') on the opposite sides of the profile. At the contact moments the rotational feed motion is reversed. The developed method significantly adds to the improved cutting plate resistance and plate wear evenness as the cutting top is displaced along the cutting edge left side to the right once per each turn of the cutting edge, while the work piece is rotated around OY axis and vice versa. This method also enhances work efficiency and smoothness due to the increased number of simultaneously working teeth.

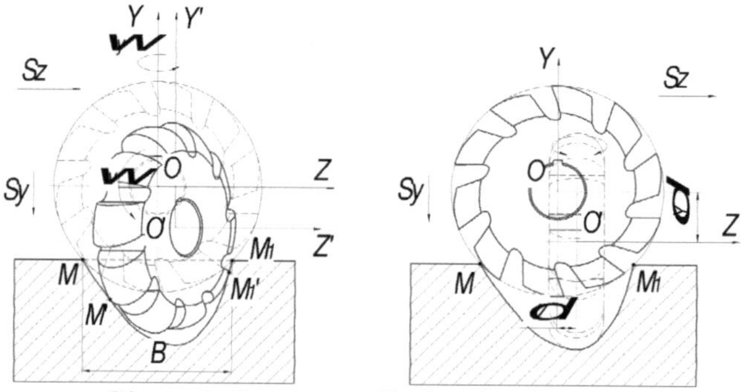

Figure 2. Groove profile forming diagram.

4. Swinging motion milling
A new machining method with swinging motion of the milling cutter is particularly interesting (patent No. 2497636 of the RF). It enables a continuous displacement of the cutting edge top toward the cut surface, a decrease of the cutting edge heat density, its more even wear and, as a result, a better tool resistance.

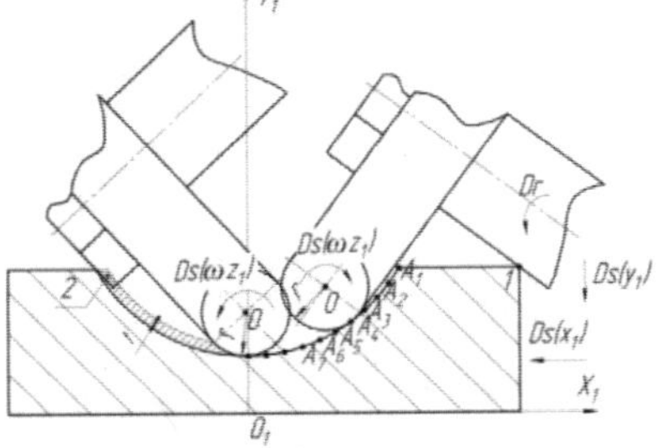

Figure 3. Profiled surface machining diagram with the possible gouging: (1) contact point; (2) penetration area

Machining is performed by a progressive contact of the tool with generated machining surface, so that the toroidal surface of the tool machines convex, flat and concave areas of the machined profile; in this case, the tool performs swinging feed motions with non-linear sliding of the cutting edge top toward the cut surface.

During forming, the tool and the work piece perform four feed motions, the main motion is *Dr* (Figure 3), then the tool is moved to the work piece via X_1 and Y_1 axis displacement so that the tool performs radial penetration till the toroidal profile of the tool reaches initial point A_1 of the machined surface profile, that is, to the motion start point. Then, the work piece is subjected to three simultaneously correlated feed motions $Ds(\omega z_1)$; $Ds(x_1)$; $Ds(y_1)$, so that the tool progressively touches the machined profile at points A_2 and A_3, simultaneously rolling counterclockwise around the center of the tool toroidal surface profile section (O) through angle ψ, which is defined on the one hand, by normal cutting conditions, that is, all the cutting edge length is involved if it is possible, and with consideration of cutting depth *t*, on the other hand, by the conditions of tooling block non-penetration into the work piece.

At point A_3 rotational motion $Ds(\omega z_1)$ is reversed with simultaneous correlation of feed motions $Ds(x_1)$ and $Ds(y_1)$ with the profile, so that the tool progressively contacts the machined profile at the subsequent points with the above conditions observed. Thus, the machining is effected with a relative motion of the tool toward the profile at points A_4, A_5 etc. and swinging feed motions $Ds(\omega z_1)$, which enables both conditions.

5. Temperature and wear resistance investigation

Figures 4 and Figures 5 feature the results of various theoretical investigations related to the tool front surface temperature for swinging motion and line-by-line milling methods when one of the parameters is variable and others are constant.

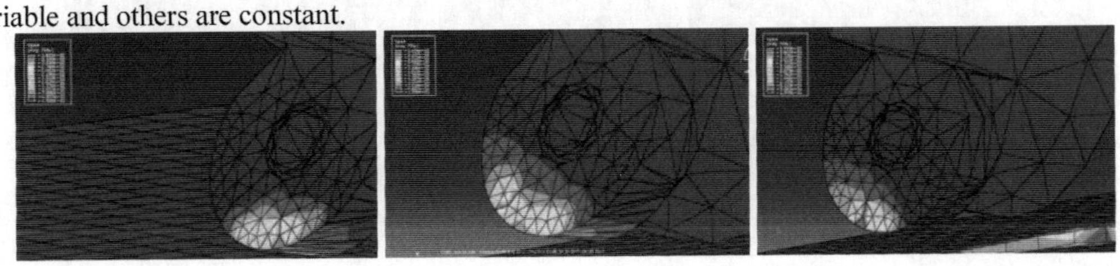

a b c

Figure 4. Distribution of the cutting temperature of the tooth front surface with variable swinging motion rate V_{SW}; $V = 150$ m/min; $Sz = 0.12$ m/min; 9XC steel: (a) $V_{SW} = 0.16$ m/min; (b) $V_{SW} = 1.6$ m/min; c – $V_{SW} = 3.2$ m/min.

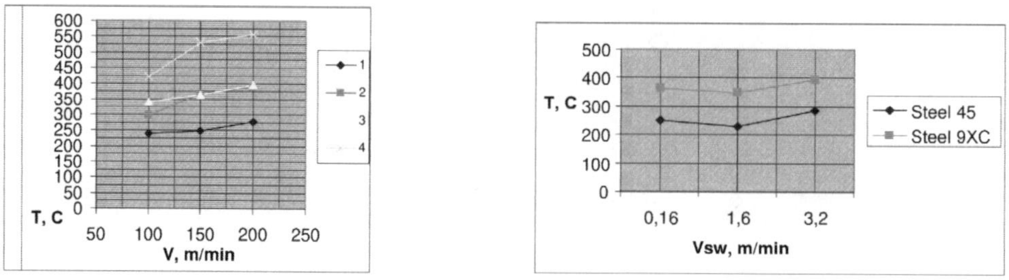

a b

Figure 5. Tooth front surface temperature change: (a) with variable cutting rate V, 45 steel: (1) swinging feed motion milling, (2) line-by-line milling; 9XC steel: (3) swinging feed motion milling, (4) line-by-line milling; (b) with variable swinging feed motion rate V_{SW}.

The experimental research of cutting plate wear resistance (Figure 6, 7) and temperatures in the cut area confirm the temperature behavior of the theoretic process model.

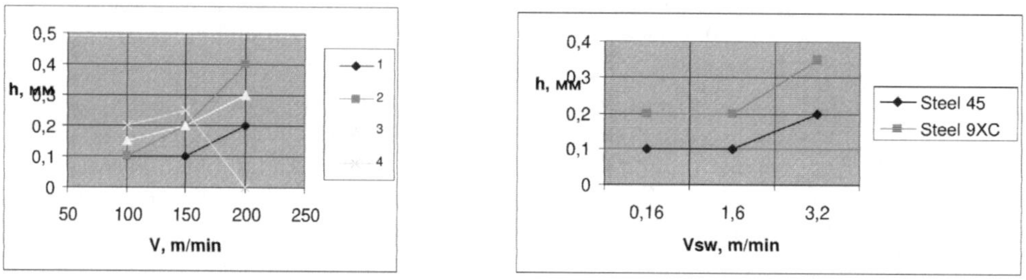

a b

Figure 6. Rear-surface tool wear as a function of: (a) cutting rate V, 45 steel: (1) swinging feed motion milling, (2) line-by-line milling, 9XC steel: (3) swinging feed motion milling, (4) line-by-line milling; (b) swinging feed motion rates.

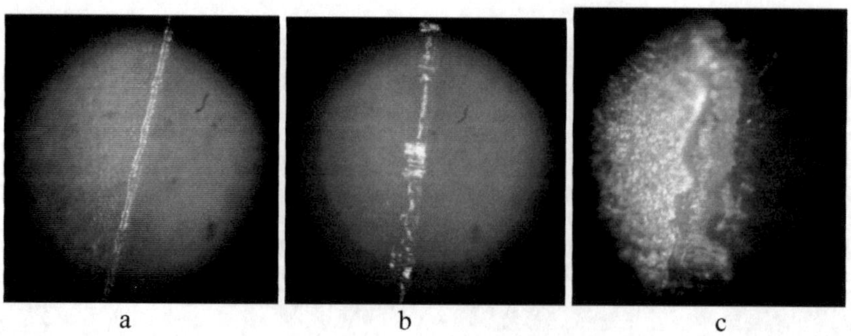

a b c

Figure 7. Rear surface tool wear (a), swinging feed motion milling (b); line-by-line milling (c).

6. Conclusion

The theoretical model of heat flow distribution in the cutting wedge developed with the aid of Abaqus 6.12 software and applied for two milling methods enabled a decrease of the cutting edge temperature to 330.2...395.5 °C, that is, by 80.6...181.6 °C for 9XC steel and to 193.8...285 °C, that is, to 56.6...120.2 °C for 45 steel in comparison with line-by-line milling. It is found that the contact length of the cutting edges is increased by 2.4 times and in these conditions extreme temperature values are localized in the middle part, like in case of line-by-line milling.

The experiment shows that the tool wear at the cutting rate of $V = 100...200$ m/min, tooth feed of $Sz = 0.08...0.16$ mm/tooth, swinging motion rates of $V_{SW} = 0.16...3.2$ m/min is, on the average, 1.7 times less for 45 steel machining and 1.34 times less for 9XC steel machining than in case of line-by-line milling [3-5].

References

[1] Mokritskii B Ya, Kirichek A V, Shpilev A M, Pustovalov D A, Sablin P A 2013 *Russian Engineering Research* **33(2)** 74–78
[2] Vereshchaka A S, Grigor'ev S N, Kim V A, Mokritskii B Ya, Kirichek A V 2014 *Russian Engineering Research* **34(8)** 516–521
[3] Kirichek A V, Soloviev D L, Altuhov A Yu 2014 *Jour. of Nano and Electr. Physics* **6(3)** 03069
[4] Kirichek A V, Soloviev D L, Altuhov A Yu 2014 *Jour. of Nano and Electr. Physics* **6(3)** 03070
[5] Kirichek A V, Soloviev D L 2013 *Jour. of Nano and Electr. Physics* **5(4)** 04009

IOP Conf. Series: Materials Science and Engineering **124** (2016) 012003 doi:10.1088/1757-899X/124/1/012003

Aspects of geometric calculation of the planetary gear train with intermediate rollers. Part 1

An I-Kan[1], A S Il'in[2] and A V Lazurkevich[1]
[1] Department of Theoretical and Applied Mechanics of Tomsk Polytechnic University, 30, Lenina ave., Tomsk, 634050, Russia
[2] Siberian Machine Building Company, 28/2, Vusotskogo st., Tomsk, 634040, Russia

E-mail: igwan@sibmail.com.

Abstract. The paper presents the geometric calculation of the planetary gear train with intermediate rollers, where the number of intermediate rollers is greater or less than that of the teeth in the annular gear by one, the profile surface of the annular gear teeth being the same. The gear ratio is changed by both a value and a sign.

1. Introduction

Planetary gear trains with intermediate rollers are currently widely used in automotive machinery and robotic machines (Figure 1) [1-3]. They possess a lot of advantages, namely: large gear ratios with small total dimensions; high loading capacity due to the multiple contact between their members; smooth running; alignment of the carrier and driven members; arrangement of the movable part in the carrier space. In some ways, this planetary gear transmission is better than that of the Fine Cyclo Japanese company 'Sumitomo' [4].

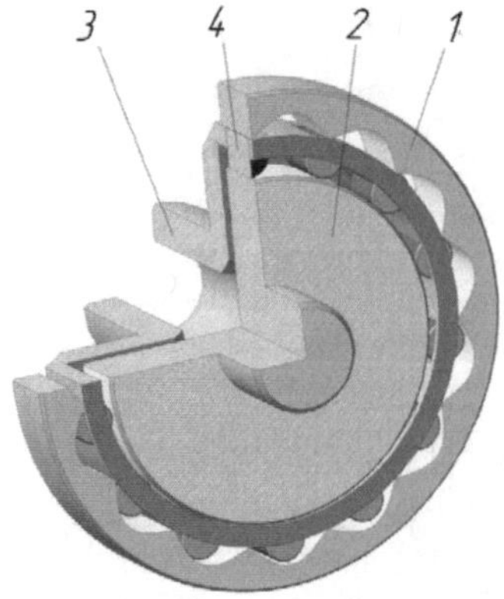

Figure 1. A planetary gear train with intermediate rollers: *1* – annular gear; *2* – carrier; *3* – separator plate; *4* – intermediate roller.

Despite a wide use of these types of planetary gear trains in different industries, especially in petroleum industry (JSC Tomsk Electric Drive Plant, LLC Sibirskii mashinostroitel', RPE Tomsk Electronic Company, LLC Siberian Machine Building Company) and a lot of publications, there are still many unsolved problems, in particular, the dependence between kinematic and geometric parameters with a certain number of intermediate rollers.

2. Calculation of the gear ratio

Figure 2 presents the schematics of the planetary rotation system. If the assemblage of intermediate rollers undergoes planetary (wave) motion by rotating about the common axis of the annular gear and the carrier, this planetary rotation system can be considered as a conventional system of a *K-H-V* type. In this type of the system, planet gears are represented by the intermediate rollers that roll in the slots of the separator plate.

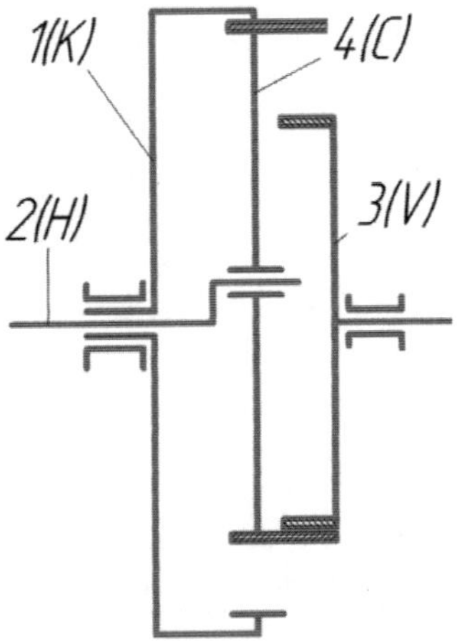

Figure 2. Schematics of the conventional gear train system of a *K-H-V* type: *1* – annular gear; *2* – carrier; *3* – separator plate; *4* – intermediate roller.

Let us consider the example of a stationary annular gear. The gear ratio between the carrier and the separation plate can be obtained using the Willis method [5] for the reversed mechanism:

$$i_{23} = \frac{Z_4}{Z_4 - Z_1},$$

where Z_1, Z_4 are the number of the teeth annular gear and intermediate rollers, respectively.

The analysis of the obtained equation is carried out in relation to the design variants, namely: $Z_1 = Z_4 + 1$ (the number of intermediate rollers is 1 less than that of the teeth annular gear) and $Z_1 = Z_4 - 1$ (the number of intermediate rollers is greater than that of the teeth annular gear by one).

In the first case, the gear ratio is

$$i_{23} = -Z_4,$$

and directions of the member rotation are opposite. In the second case the gear ratio is

$$i_{23} = Z_4.$$

Let us consider the example of a stationary separator plate. The gear ratio between the carrier and the annular gear can be obtained using the Willis method for the reversed mechanism:

$$i_{21} = \frac{Z_1}{Z_1 - Z_4}.$$

Let us suppose that $Z_1 = Z_4 + 1$; then

$$i_{21} = Z_1. \quad (1)$$

Now let us suppose that $Z_1 = Z_4 - 1$; then

$$i_{21} = -Z_1. \quad (2)$$

In the latter case, the members rotate in different directions.

3. Calculation of the profile surface teeth annular gear

Let us consider the planetary gearing design with dependence (2). The schematic layout and design parameters of gearing are shown in Figure 3. Two coordinate systems are used: fixed *S*(*Oxy*) and moving $S1(O1x1y1)$ rigidly connected with the annular gear.

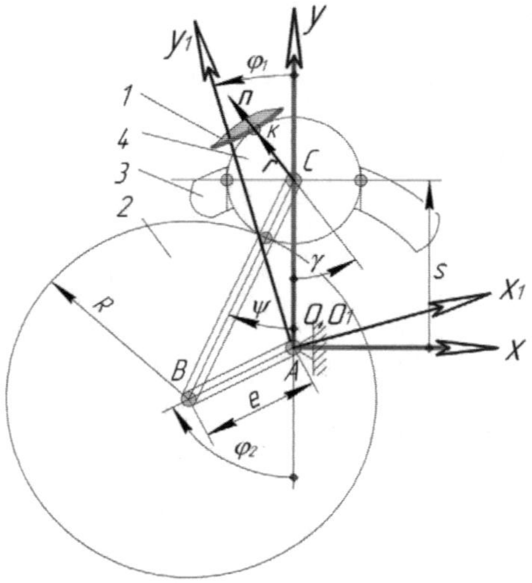

Figure 3. A schematic layout of gearing with dependence (2): *1* – annular gear; *2* – carrier; *3* – separator plate; *4* – intermediate roller.

Conformal rotations of the annular gear and the carrier are determined by directional angles φ_1 and φ_2 connected by (2):

$$i_{21} = \frac{\varphi_2}{\varphi_1} = -Z_1 \cdot \tag{3}$$

As shown in Figure 3, common contact normal ***n*** to point *K* is a perpendicular to vector $V^{(41)}$ of the relative velocity of the intermediate roller and the teeth annular gear at point *C*.

Fixed coordinate system *S*(*Oxy*) is used for these calculations. From triangle *ΔABC* it follows that the law of central displacement in the intermediate roller relative to the common axis of the annular gear and the carrier has the following form:

$$s = l\cos\psi - e\cos\varphi_2 \ . \tag{4}$$

Let us note that

$$\cos\psi = \sqrt{1-\sin^2\psi}\ ; \tag{5}$$

$$\sin\psi = \frac{e}{l}\sin\varphi_2\ , \tag{6}$$

where *e* is the eccentricity.

$$l = R + r\ , \tag{7}$$

where *R* and *r* are the radii of the carrier and the intermediate roller, respectively.

Now we introduce unit vectors ***i*** and ***j*** for axes of fixed coordinate system *S*(*Oxy*). The velocity of point *C* of the intermediate roller can be obtained from

$$\boldsymbol{V}^{(4)} = \frac{ds}{d\varphi_2}\boldsymbol{j} = e\left(\sin\varphi_2 - \mathrm{tg}\,\psi\cos\varphi_2\right)\boldsymbol{j}\ . \tag{8}$$

The velocity of point *C* of the annular gear can be defined as follows:

$$\boldsymbol{V}^{(1)} = -\frac{s}{Z_1}\boldsymbol{i}\ . \tag{9}$$

The relative velocity vector can be found as the difference between velocities obtained in (8) and (9).

$$\boldsymbol{V}^{(41)} = e\left(\sin\varphi_2 - \mathrm{tg}\,\psi\cos\varphi_2\right)\boldsymbol{j} + \frac{s}{Z_1}\boldsymbol{i}\,.$$

Common contact normal $\boldsymbol{n}$ is obtained using $\boldsymbol{k} = \boldsymbol{i}\times\boldsymbol{j}$.

$$\boldsymbol{n} = \boldsymbol{k}\times\boldsymbol{V}^{(41)} = -e\left(\sin\varphi_2 - \mathrm{tg}\,\psi\cos\varphi_2\right)\boldsymbol{i} + \frac{s}{Z_1}\boldsymbol{j}\,. \tag{10}$$

The radius-vector of the profile surface at point K can be represented in the following form:

$$\boldsymbol{r}^{(K)} = s\boldsymbol{j} + \boldsymbol{r} = -r\sin\gamma\boldsymbol{i} + \left(s + r\cos\gamma\right)\boldsymbol{j}\,. \tag{11}$$

Using (10), we obtain the components included in (11):

$$\cos\gamma = \frac{\boldsymbol{n}\cdot\boldsymbol{j}}{n} = \frac{s}{Z_1\sqrt{e^2\left(\sin\varphi_2 - \mathrm{tg}\,\psi\cos\varphi_2\right)^2 + \left(\frac{s}{Z_1}\right)^2}}, \tag{12}$$

$$\sin\gamma = \sqrt{1-\cos^2\gamma}\,. \tag{13}$$

In order to obtain the profile surface of the annular gear teeth, (11) should be transformed to the moving coordinate system $S1(O1x1y1)$. The matrix technique can be used for this purpose [6].

Without going into details, let us show the final system of equations for the profile surface of the annular gear teeth:

$$\left.\begin{aligned} X_1 &= -r\sin\gamma\cos\varphi_1 + (s + r\cos\gamma)\sin\varphi_1\,, \\ Y_1 &= r\cos\gamma\sin\varphi_1 + (s + r\cos\gamma)\cos\varphi_1. \end{aligned}\right\} \tag{14}$$

For the calculation of the profile surface coordinates, the system of equations (13) should be combined with (3)-(7), (12) and (13).

Let us further consider the planetary gearing design with dependence (1). The schematic layout and design parameters of gearing are presented in Figure 4.

Iterating the operations given above and considering the change of the directional angle φ_1 (1), we obtain

$$\left.\begin{aligned} X_1 &= r\sin\gamma\cos\varphi_1 - (s + r\cos\gamma)\sin\varphi_1\,, \\ Y_1 &= r\cos\gamma\sin\varphi_1 + (s + r\cos\gamma)\cos\varphi_1. \end{aligned}\right\} \tag{15}$$

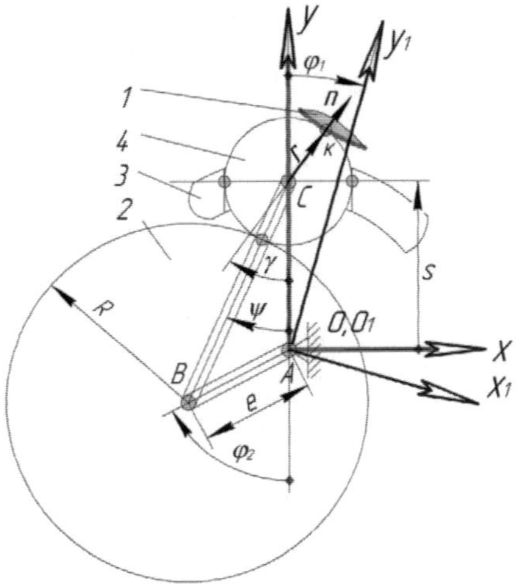

Figure 4. A schematic layout of gearing with dependence (1): *1* – annular gear; *2* – carrier; *3* – separator plate; *4* – intermediate roller.

Note that (14) and (15) describe the teeth profile surface opposite to the symmetry axis of the gear tooth. Hence, we now prove that the profile surface is similar in both cases.

4. A calculation example

Figure 5 depicts the planetary gear train with intermediate rollers designed by the obtained parameters, namely: $z_1 = 16$; $r = 5.16$; $R = 45$; $e = 2.58$.

Under the condition of the stationary annular gear and the driven separator plate, the gear ratio as shown in Figure 5, is $i_{23}^{(1)} = -15$ and $i_{23}^{(1)} = 17$ according to *a*) and *b*) design variants, respectively.

It should be noted that, unlike the separator plate that has the different number of intermediate rollers and slots for them, the change in the separator plate parameters has no effect on the parameters of the annular gear and the carrier.

5. Conclusions

1. The planetary gear train with intermediate rollers may incorporate the number of intermediate rollers, which is greater or less by 1 as compared to the number of the teeth annular gear.
2. The teeth profile surfaces of the annular gear are similar in both cases described herein.
3. A change of the number of intermediate rollers provides a change of the gear ratio and the direction of rotation of the driven member.

References

[1] Yangulov V S 2008 Gears of high accuracy and durability (Tomsk: Publishing house TPU) p 136
[2] Samsonovich S L, Konstantinov S A, Stepanov V S 2005 *Aerospace Instrument* **5** 28–33
[3] Korotkov V S, Razumov S V 2014 *Engineering magazine* 9 58–61
[4] Ivanov A S, Fomin M V, Ermolaev M M 2012 *Russian Engineer. Res.* 2012 **32(11–12)** 707–710
[5] Frolov K V, Popov S A, Musatov A K 1987 Theory of mechanisms and machines (Moscow: Publishing High School) p. 496
[6] Faydor L, Litvin A F 2004 Gear Geometry and Applied Theory (Second Edition: Cambridge University Press) p. 801

IOP Conf. Series: Materials Science and Engineering **124** (2016) 012004 doi:10.1088/1757-899X/124/1/012004

Load analysis of the planetary gear train with intermediate rollers. Part 2

An I-Kan[1], A S Il'in[2] and A V Lazurkevich[3]
[1] Department of Theoretical and Applied Mechanics of Tomsk Polytechnic University, 30, Lenina ave., Tomsk, 634050, Russia
[2] Siberian Machine Building Company, 28/2, Vusotskogo st., Tomsk, 634040, Russia

E-mail: igwan@sibmail.com.

Abstract. The paper presents the load analysis of the planetary gear train with intermediate rollers based on solving a statically indeterminate problem including contact area problems.

1. Introduction

The planetary gear train with intermediate rollers possess a high loading capacity due to the multiple contact between their members. In order to estimate the loading capacity, the force distribution in the system members and, thereby, the statical indeterminateness of the system should be measured.

The dependencies between the geometry and linkage parameters of the planetary gearing system are described in work [1].

2. Effective forces in members of the planetary gearing with intermediate rollers of a unit length

Figure 1 presents the interaction between the planetary gearing members under the carrier loading conditions induced by load moment M_2. Let us calculate the forces acting on the intermediate roller in the form of a circular cylinder. First, we shall consider the planetary gearing design shown in Figure 1,*a*.

For simplification, let M_2 be 1. Then force N_2 (a response of the carrier to the intermediate roller) can be obtained from

$$N_2 = \frac{M_2}{h} = \frac{M_2}{s\sin\psi} = \frac{1}{s\sin\psi}.$$

From the condition of projection equality

$$N_1\cos\gamma = N_2\cos\psi\ ,$$

response N_1 of the annular gear is given by

$$N_1 = \frac{1}{s\cos\gamma\,\mathrm{tg}\psi}.$$

The response of a separator plate equals the sum of force projections:

$$N_3 = N_1\sin\gamma - N_2\sin\psi = \frac{\mathrm{tg}\gamma}{s\,\mathrm{tg}\psi} - \frac{1}{s}.$$

The iteration of all the operations given above results in forces applied to the intermediate roller for the planetary gearing design shown in Figure 1b:

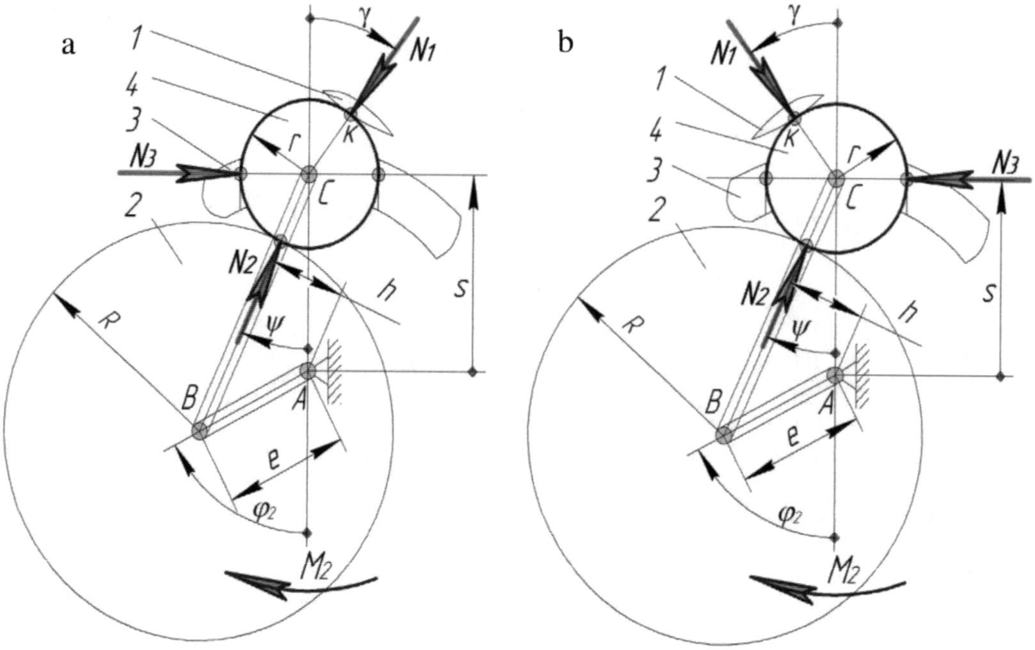

Figure 1. Schematics of loading conditions for the intermediate roller: *a* – the number of intermediate rollers is less than that of teeth by one; *b* – the number of intermediate rollers is greater than that of teeth by one; *1* – the annular gear; *2* – the carrier; *3* – the separator plate; *4* – the intermediate roller.

$$N_2 = \frac{M_2}{h} = \frac{M_2}{s \sin\psi} = \frac{1}{s \sin\psi};$$

$$N_1 = \frac{1}{s \cos\gamma \, \mathrm{tg}\psi};$$

$$N_3 = N_1 \sin\gamma + N_2 \sin\psi = \frac{\mathrm{tg}\gamma}{s \, \mathrm{tg}\psi} + \frac{1}{s}.$$

3. Statically indeterminate problem

In case the annular gear and the separator plate are stationary, load moment M_2 of the carrier, as shown in Figure 2, results in plastic deformation at the place of engagements that, in turn, leads to increment $\delta\varphi_2$ of the rotation angle. The angle increments are common for all members (an annular gear, a carrier, a separator plate, an intermediate roller) participating in the multiple contact in the planetary gearing system. Taking this principle into account, the force distribution can be detected between the members of the system.

Let us calculate increment $\delta\varphi_2$ of the rotation angle of the carrier. Figure 2 shows the positions of the carrier and the intermediate roller before and after loading for two abovementioned design variants of the planetary gearing system.

Let us consider the planetary gear train shown in Figure 2,*a*.

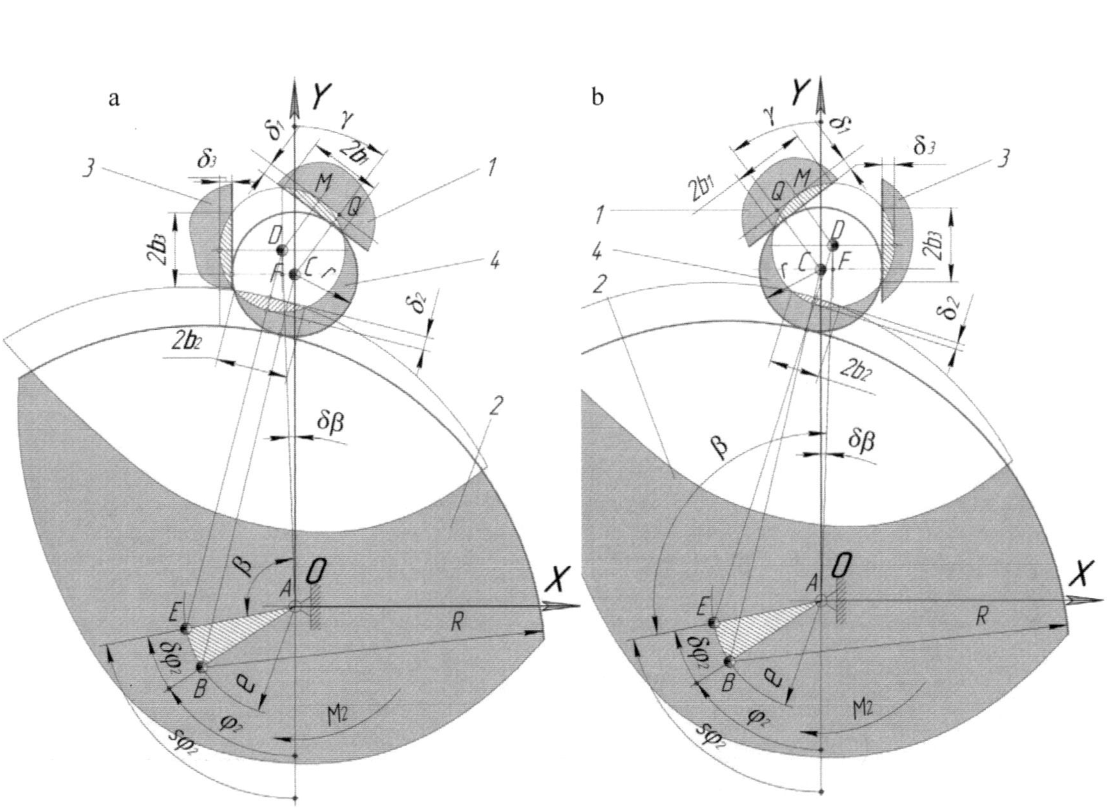

Figure 2. Schematics of loading conditions for system members in two design variants: B, E – positions of the carrier geometrical centre before and after loading, respectively; C, D – positions of the intermediate roller before and after loading, respectively; δ_1, b_1 – engagement and half-width of the contact area, respectively, between the intermediate roller and the annular gear tooth; δ_2, b_2 – engagement and half-width of the contact area, respectively, between the intermediate roller and the carrier; δ_3, b_3 – engagement and half-width of the contact area, respectively, between the intermediate roller and the separator plate.

3.1. A new position of centre D of the intermediate roller

The vector equality has form

$$\overrightarrow{CF} + \overrightarrow{FD} = \overrightarrow{CQ} + \overrightarrow{QM} + \overrightarrow{MD}, \tag{1}$$

the components of which are expressed by basis vectors $\boldsymbol{i}$ and $\boldsymbol{j}$ in the $S(Oxy)$ coordinate system and equal:

$$\overrightarrow{CF} = -\delta_3 \boldsymbol{i}; \tag{2}$$

$$\overrightarrow{FD} = FD\boldsymbol{j}; \tag{3}$$

$$\overrightarrow{CQ} = (R+\delta_1)(\sin\gamma\boldsymbol{i} + \cos\gamma\boldsymbol{j}); \tag{4}$$

$$\overrightarrow{QM} = QM(-\cos\gamma\boldsymbol{i} + \sin\gamma\boldsymbol{j}); \tag{5}$$

$$\overrightarrow{MD} = -R(\sin\gamma\boldsymbol{i} + \cos\gamma\boldsymbol{j}). \tag{6}$$

Transformations of the set of equations (1)-(6) yields

$$FD = \delta_1\cos\gamma + (\delta_1\sin\gamma + \delta_3)\mathrm{tg}\gamma. \tag{7}$$

As shown in Figure 2a the radius-vector of new centre D in the accepted coordinate system is

$$\overrightarrow{AD} = -\delta_3\boldsymbol{i} + (FD+S)\boldsymbol{j}. \tag{8}$$

3.2. *Increment of the rotation angle of the carrier*

According to Figure 2a,

$$\overline{AE} + \overline{ED} = \overline{AD}. \quad (9)$$

Angle β can be obtained from (9):

$$\beta = \arccos\left(\frac{AD^2 + AE^2 - ED^2}{2AD \cdot AE}\right), \quad (10)$$

and

$$ED = R + r - \delta_2; \quad (11)$$

$$AE = e; \quad (12)$$

$$AD = \sqrt{\delta_3^2 + \left[S + \delta_1\cos\gamma + (\delta_1\sin\gamma + \delta_3)\mathrm{tg}\gamma\right]^2}. \quad (13)$$

The latter equation is based on (7) and (8).

Based on dependence

$$\overrightarrow{AD} \cdot j = AD\cos\delta\beta,$$

Let us obtain $\delta\beta$:

$$\delta\beta = \mathrm{arctg}\frac{\delta_3}{S + \delta_1\cos\gamma + (\delta_1\sin\gamma + \delta_3)\mathrm{tg}\gamma}. \quad (14)$$

At the same time, the new position of the carrier is defined by angle

$$s\varphi_2 = \pi - \beta - \delta\beta, \quad (15)$$

while the angle increment of the carrier rotation is

$$\delta\varphi_2 = s\varphi_2 - \varphi_2. \quad (16)$$

Thus, to detect angle inclement $\delta\varphi_2$, it is expedient to regard (16) in combination with (10)-(15). The parameters of S and γ were obtained in work [1].

Note that calculations for the design variant presented in Figure 2b differ from those presented in Figure 2a only by the angle $s\varphi_2$ value:

$$s\varphi_2 = \pi - \beta + \delta\beta.$$

Calculations of engagement parameters $\delta_1, \delta_2, \delta_3$ are somewhat difficult. Design formulas given in [2-4], are not universal and intended rather for particular cases. Thus, for example, in case the radius of one of the profile surface approaches infinity (a cylinder/plane transfer), these formulas produce erroneous outcomes.

Moreover, the profile surface curvature of the annular gear (Figures 1 and 2) ranges from one sign to another in different sections. Therefore, to simplify the problem solving, formulas given in [4] are used for the schematics of cylinder/plane engagement with substituting radius R by that of relative curvature R_n with a positive sign:

$$\delta = 0{,}5795\frac{q}{E}\left(0{,}41 + \ln\frac{2R_n}{b}\right),$$

where q is the specific force due to pressure; E is the Young's modulus similar to both of engaged members; b is the half-width of the contact area.

The above-stated equation also implies the equality of Poisson numbers, namely: $\mu = 0.3$.

The half-width of the contact area has form

$$b = 1{,}523\sqrt{\frac{qR_n}{E}},$$

while the highest contact stress is

$$\sigma_{\max} = 0{,}418\sqrt{\frac{qE}{R_n}}.$$

The main difficulties of the problem solving are connected with cumbersome calculations and impossibility of the analytical study of the profile surface curvature. In this connection, the computer program has been developed on the Rad Studio XE8 environment basis for the numerical implementation of this problem. The curvature radius of the teeth profile surface can be obtained using two chords between three points on it and the sine theorem [5, 6]. The developed computer program allows solving a wide range of problems. The simplest problem algorithm is suggested so as to determine the total rotation moment of the carrier accounting for the condition of restricting the highest contact stress of the given geometry and properties of materials of the planetary gear train.

The problem solving is based on the method of successive approximations, which includes:

– assignment of a random variable of angle increment $\delta\varphi_2$;

– definition of load moment M_2 in all engaged members; maximum contact stresses, i.e. $\sigma_{max}^{(1-4)}$ between the surfaces of teeth and the intermediate roller; $\sigma_{max}^{(2-4)}$ between the surfaces of the carrier and the intermediate roller; $\sigma_{max}^{(3-4)}$ between the surfaces of the separator plate and the intermediate roller;

– control of the condition of restricting highest contact stress $\sigma_{max} \leq [\sigma]$. In satisfying this condition, the total rotation moment of the carrier is obtained, which equals to the sum of all moments obtained in the previous paragraph, and also the sum of forces having effect on each member. Otherwise, the calculations must be done anew taking into account the improved value of angle increment $\delta\varphi_2$.

Note that the obtained results depend on the member arrangement in the planetary gearing system, i.e. angle φ_2 (engagement phases). To avoid this dependence, the averaged values of parameters should be determined:

$$SM_2 = \frac{\sum_{i=1}^{n} M_{2i} \cdot Z}{2n},$$

where SM_2 is the averaged total moment of the carrier; n is the number of calculations at $\varphi_2 \in (0, \pi)$; M_{2i} is the moment at the i-th step of calculations; Z is the number of intermediate rollers;

$$S\sigma_{max} = \frac{\sum_{i=1}^{n} \sigma_{max\,i}}{n},$$

where $S\sigma_{max}$ is the average value of the maximum stress between the contact areas (indices are omitted); σ_{maxi} is the contact stress at the i-th step of calculations;

$$SN_x = \frac{\sum_{i=1}^{n} N_{xi} \cdot Z}{2n},$$

$$SN_y = \frac{\sum_{i=1}^{n} N_{yi} \cdot Z}{2n},$$

where SN_x and SN_y are the components of the total force on axes Ox and Oy, respectively; N_{xi} and N_{xy} are the force components on axes Ox and Oy at the i-th step of calculations.

4. Example of load analysis

The initial data: R = 104 mm; r = 5 mm; e = 2.5 mm; Z = 31; Z-1 is the number of teeth in the annular gear; E = 200000 MPa; μ = 0.3; 1 mm link height; M_2 = 5600 N·mm.

Results of calculations obtained using the developed computer program are as follows. The moment of rotation of the separator plate is 173458.1 N·mm.

The average contact stress between the intermediate roller and

– the annular gear is 1500 MPa;

– the carrier is 1195 MPa;

– the separator plate is 871.5 MPa.

The force acting on:

– the carrier is 2239.9 N;

– the annular gear is 2590.5 N;

– the separator plate is 1329.6 N.

5. Conclusions

1. The results of multiple calculations have showed that the most load-subjected member is an intermediate roller/tooth of the annular gear.
2. The suggested calculation methodology and the computer program have been successfully used in designing drivers of different functions at Tomsk industrial enterprises.

References

[1] An I-Kan, Il'in A S, Lazurkevich A V 2016 *IOP Conference Series: Mechanics and Engeneering.* (*Part 1*)
[2] Pisarenko G S, Yakovlev A P, Matveev V V 1988 Handbook of resistance of materials (Kiev: Naukova Dumka) p 736
[3] Birger I A, Schorr B F, Iosilevich G B 1993 *Mechanical Engineering* 640
[4] Aleksandrov A Y, Ambartsumian S A, Biederman V L 1968 *Mechanical Engineering* **2** 463
[5] An I-Kan 2000 *Engineering Journal* **11** 39–41
[6] Novoselov S I 1967 Special course of trigonometry (Moscow: Publishing High School) p 536

The design of a die with a vacuum thermal insulation

A G Baginski, O M Utyev and Y M Kondratyeva
National Research Tomsk Polytechnic University, 30, Lenina ave., Tomsk, 634050, Russia

E-mail: utievv@mail.ru

Abstract. A new design of a die for the polymer underwater granulation plants has been developed. It differs from similar plants in the design of heat-insulating elements. Vacuum hollows are used as a heat insulator. A vacuumization process does not require a separate operation, and it is conducted simultaneously with connection of all the elements of the die by a brazing method.

1. Introduction

Modern production consumes a large amount of structural materials developed on a polymer base. An increased consumption of polymeric compositions requires a constant increase in production of polymer raw materials from manufacturers. The main product that manufacturing plants consume is a primary polymeric granulate. It is used as a material for product forming on automatic plastic machinery and as a raw material for a secondary granulation with addition of fillers and dyes.

The peak of production chains in the granulate manufacture is obtaining an extruder. The formation of the polymer granules is carried out due to the crushing of a polymer mass, passing through the holes in the die grid. Equipment manufacturers use various schemes of granulation: in the liquid state (underwater granulation), in the solid state (strang granulation). But at any scheme the quality of the product (size, shape, uniformity of granules) depends on many simultaneously affecting factors. They are: a wear degree of the die grid and granulating knives, temperatures of the polymer mass and dies, and etc. Therefore, in order to obtain a stable granulation process, the manufacturers use various technological solutions in the design of pelletizing heads of their machines both for maintaining a heat balance and for reduction of the wear and tear of the granulation unit.

2. Results and Discussion

The present paper is dedicated to the increase of the thermal stability of the die. Manufacturers use a variety of schemes of heating in order to control its thermal balance. SANDVIK Company applies oil or steam-heating (Figure 1) to maintain the heat balance of the granulation process. The heating channels (Figure 1, p.3) are maximally approximated to the surface of the die, and provide the warm-up of the granulation zone to avert the danger of melt 'sticking'. In order to concentrate the heat flux, exiting on the working surface of the die, an internal heat-insulating layer (Figure 1, p.2) is installed around the holes (Figure 1, p.4). That is, the energy capacity of the die heating elements after launching of the granulation process into the operation mode should provide a stable temperature of the die working surface, which is constantly cooled by the incoming polymer mass.

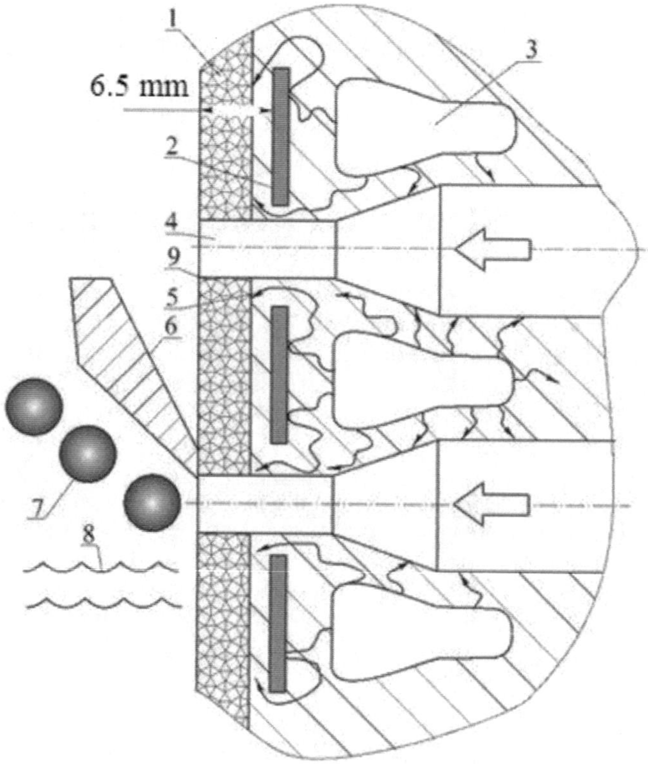

Figure 1. Draft of a working part of a die manufactured by SANDVIK Company: *1* – working surface, *2* – internal heat-insulating layer, *3* – heating channels (oil, *T*=200 °C), *4* – channels for polymer exit, *5* – heat fluxes, *6* – knife, *7* – polymer granules, *8* – water ring, *9* – cutting edge.

It is not always possible to solve this problem by a common increase of the power of the heaters (Figure 2, p.7–8) or the number and the flow cross section of oil channels (Figure 1, p.3) structurally because of the limited sizes of the die. Therefore, some manufacturers have taken a different road. For example, the manufacturers of TRUSIOMA Company, along with die array heating (Figure 2, p.7–8), simultaneously introduce additional structural elements into constructions. They isolate the channels, which form the polymer mass, from the die body, thus, protecting the die from undercooling. The working plane temperature is more than 200 °C and therefore the choice of materials for thermal insulators is limited.

Most often the porous ceramics (manufactured by 'TRUSIOMA') is used as heat-reflecting bushes. This material has a good thermal conductivity coefficient, but is not very convenient technologically in manufacture and installation as it has high hardness and brittleness. The most serious disadvantage of ceramics is a small linear expansion coefficient compared to steel. Therefore, coupling sizes of the heat insulator and conjugated metal parts of the die during assembly and operation differ significantly, what adversely affects the durability and reliable perfromance of the entire structure.

After the analysis of the existing constructions of dies and disadvantages identified in the process of the pelletizer operation, at the MTM Department in TPU a modernized die has been designed for the TRUSIOMA pelletizer.

Earlier the department has already carried out repair works on worn-out dies. The technology of repair of their work surface was developed [1]. A highly wear-resistant cast-alloy was used as a new working surface based on high-speed steel, which is additionally alloyed with chromium and vanadium. The favorable influence of these elements on the morphology of carbides in the alloys with the cast structure and a significant increase in wear-resistance of the material with the same a chemical composition was shown in papers [3] and [4].

The same material was used for manufacture of a working granulating surface of new dies.

The proposed design has a working surface in the form of a solid ring made of a highly wear-resistant alloy. To fix the ring a soldering technology in the high-temperature vacuum, which is developed at the Department (patent № RU 2297307 C1) is used [2]. It provides obtainment of a solid weld seam (Figure 2, p.2) that has high strength. Since vacuumization is one of the most effective ways to reduce the heat transfer coefficient, this opportunity was used to replace the ceramic heat insulators with vacuum hollows. Bushes feeding the polymer (Figure 2, p.6) are separated from the body of the die by air space. To maintain the vacuum, which is created in the hollows when the die is in the furnace, at the top of each bush an annular groove is provided, in which copper solder is laid during the die assemblage. The conjugation of the bush with the body at the outlet from the die is sealed with a layer of solder (Figure 2, p.2) and a retaining ring (Figure 2, p.1).

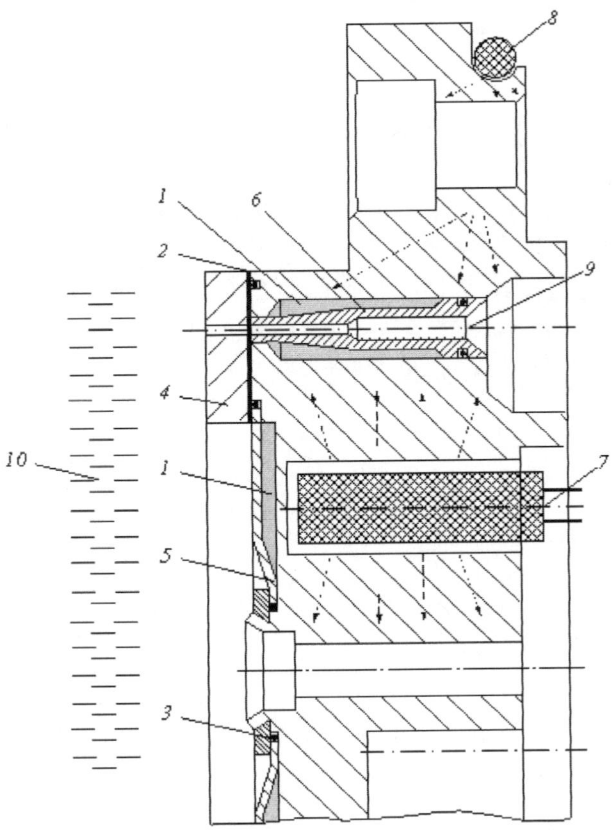

Figure 2. Draft of a die with a vacuum thermal insulation: *1* – vacuum heat insulating chambers, *2* – main solder layer, *3* – ring insertions of solder for sealing the vacuum chambers, *4* – working ring, *5* – vacuum chamber cover, *6* – bush guides, *7*, *8* – electrical heaters, *9* – polymer channels, *10* – process water.

For more efficient operation of heaters (Figure 2, p.7), a protecting cover (Figure 2, p.5) has been introduced into the die configuration. During heating the cover is also sealed with the solder (Figure 2, p.3), positioned in the form of two rings laid along the perimeter of the cover, to create an efficient vacuum layer. To prevent deformation of the cover under atmospheric pressure it is implemented in the form of a non-flat concave profile. The solder, which is put into every connections of the die, has the same melting point (about 1200 °C); therefore, during formation of the main seam, the simultaneous meltdown and sealing of all the cover guides occur. The die cooling after completion of the soldering process occurs under vacuum, thus, exhaustion (about $0.0193 \cdot 10^{-2} \ldots 0.0193 \cdot 10^{-3}$psi) at all sealed hollows is securely preserved. This was confirmed by control unsealing of several vacuum hollows under the water layer.

3. Conclusion

The tested design of the die has showed good results while at operation. It can be quickly heated to the operating temperature. Besides, this design has a sufficient range of temperature adjustment of the die working plane during the setting-up of the granulation process for various polymer compositions.

The production tests were held on the basis of Angarsk polymer plant. After successful completion of the tests, a batch of dies with vacuum heat insulation has been manufactured and shipped to the enterprise.

References

[1] Baginski A, Egorov Y and Utyev O 2000 Restoring dies for 'Bershtorff' and 'Werner Pflyayderer' granulators *Mechanics and Mechanical Engineering* (Tomsk: Tomsk Polytechnic university Publishing) pp 47–52

[2] Baginski A, Egorov Y, Melnikov A and Utyev O 2007 *Russian Federation Patent 2297307.*

[3] Filipovic M, Kamberovic Z, Korac M and Jordovic B 2013 Effect of niobium and vanadium additions on the as-cast microstructure and properties of hypoeutectic Fe–Cr–C alloy *ISIJ International* **53(12)** 2160

[4] Kawalec M and Fras E 2008 Structure, mechanical properties and wear resistance of high-vanadium cast iron *ISIJ International* **48(4)** 518

Software and hardware user authentication methods in the information and control systems based on biometrics

Z Boriev, A Nyrkov, S Sokolov and S Chernyi

Admiral Makarov State University of Maritime and Island Shipping, 5/7, Dvinskaya st., Saint-Petersburg, 198035, Russia

E-mail: za_mir_b@mail.ru

Abstract. The possibility of using biometric information technologies in management is considered in this article. A brief review of access control and time attendance is made as well. Biometric methods and systems for user identification are analyzed. Recommendations on the use of various systems depending on the specific tasks are given in the article.

1. Introduction

Nowadays the problem of using of biometric identification systems is becoming more urgent. However, in fact, traditional methods of personal identification and authorization based on the use of passwords or physical media (in the form of a pass, a passport, a driving license, electronic keys and cards) do not meet modern requirements for reliability in the determination of the individual. A password can be forgotten or intercepted, a material carrier can be copied, lost or given to another person. All of the above mentioned does not allow the traditional access control systems to ensure an adequate level of reliability. As a result, companies are looking for more efficient ways of ensuring security. One way to solve this issue is to move to biometric identification systems [1].

2. Information and Control Systems

Effective management requires constant monitoring and planning of all processes in the organization. Planning activities are carried out taking into account the fact that they will be followed by the monitoring function. It is necessary for the control to have timely, accurate information that will be effective with respect to the cost to develop it. To implement these requirements it is necessary to enter information and to control the system.

Management Information System (MIS) is a set of hardware and software that allows the company management to obtain timely information which is necessary for making effective decisions.

The administration of the enterprise daily deals with the problems of payroll, time tracking, providing access to certain resources of the staff, whose number is constantly changing. The automation of management processes in the company is not a big problem itself. The real difficulties arise due to the fact that the components of any organization are constantly changing, upgraded or even can be deleted. The environment and the structure are changing; technological processes of the enterprise are being modernized. If the administration of the organization is deprived of the opportunity to react promptly to these changes, the consequences for the enterprise can take a serious nature [2].

3. Access control systems

Access control systems (ACS) is an essential component of the information management system of any business or office building. ACS is a very complicated hardware-software system that is responsible for the registration and the restriction of entry and exit of people and vehicles in a certain area through check points (gates, doors). The main objectives of the ACS are to determine the time of admission to the territory and identify the objects that are eligible for admission. In addition, the access control system can perform a range of additional features, which include:

- keeping records of working time.
- conducting employees' payroll (integrated if possible with the accounting system).
- maintaining a database of visitors and employees.

Any ACS has an email id (key), which is used to determine the rights of the owner. The following arrangements can be used as a key in Access control systems: barcode, magnetic or smart (smart cards) Plastic cards; 'Electronic tablets' (Touch Memory); Wiegand cards; remote reading cards (proximity); PIN-code. Moreover, the biometric parameters of a person can be used as an identifier.

'Electronic tablets' (Touch Memory) are microchips arranged in a sturdy metal case. The code information is recorded in the memory of the circuit. It is necessary to attach the tablet to the reader to identify the user. The Reading Speed is 0.1 second. Some models allow one to enter some information about the user. The advantages are compactness, high resistance to mechanical damages, corrosion, temperature extremes and a low cost (comparable to the cost of magnetic stripe cards).

A card with a bar code is a plate coated with black stripes (strokes). The code information is contained in the variable width of strokes and the distance between them. The code of the card is read by the optical reader. The code information of a magnetic card is recorded on the magnetic strip. The bar code can simply be soiled with mud. A magnetic card can easily be damaged in daily use.

A perforated card is a plate (plastic or metal). The code information to a perforated card is applied as the holes arranged in a specific order. The code card is read by mechanical or optical readers.

The code information on a Wiegand card is contained on the thin metal wires arranged in a certain way and glued with special glue. An electromagnetic reader reads the information from a card.

Proximity technologies have a lot of advantages, which are much greater compared with other methods of identification: reliability and durability, lack of power (in passive cards). A proximity reader constantly sends a radio signal. When being in contact with the zone of the reader, a card takes its radiation and sends a signal containing a recorded microchip code. The distance between the reader and the card depends on the power of the reader and varies from 5 cm to several meters. The lack of the mechanical contact during operation allows producing identifiers of an arbitrary form, the identifier can be implanted into the object of the identification. The advantages of proximity technologies are the complexity of falsification, the possibility of using encryption algorithms.

A smart card ('smart card') is a plastic card with an integrated microcontroller and all its attributes (a processor, RAM, nonvolatile memory with a file system, input-output means, additional coprocessors). The main advantages of smart cards are a large memory capacity and high information security which protects against attempts of modifying and duplication. The drawback is the high cost. A smart card is the standard equipment of appropriate automated systems and it can be simply implemented in any arbitrary system.

PIN-code. The memory of a man is the code information carrier. A user independently types the code on the keyboard and it gives the signal to the actuating device [3].

4. Biometric identification systems. Legislative regulation

In the Russian Federation in the field of rights protection of personal data a special attention is paid to biometric data. Article 11 of the Federal Law 'On Personal Data' gives the following definition of biometric data - 'data that characterize the physiological and biological characteristics of the person, according to which it is possible to establish his or her identity (biometric personal data), are used by the operator to identify the subject of personal data. It is necessary to have a written consent of a person to process the biometric personal data, except as provided by paragraph 2 of this Article'.

Part 2 of Article 11 of the Federal Law 'On Personal Data' states that the processing of biometric personal data can be carried out without the consent of the subject of personal data in connection with the implementation of international treaties of the Russian Federation on readmission, in connection with the administration of justice and all that is connected with it.

Besides, the law provides specific requirements for material carriers of biometric personal data and storage technologies of such data if personal data are outside of information systems.

A material carrier must provide:

• protection against unauthorized addition and re-recording of information after removing it from the information system of personal data;

• access to biometric personal data recorded on a material carrier carried out by an operator and persons authorized to work with biometric PD in accordance with the legislation of the Russian Federation;

• ability to identify the personal data information system, in which a biometric record of PD was implemented, as well as the operator, who carried out such record;

• preventing from unauthorized access to biometric personal data contained on a material carrier.

An operator of biometric data, used outside of personal data information systems, must:

• maintain the registration of the number of copies of material carriers;

• provide an appropriate material carrier with a unique identification number that allows the operator to be accurately determined.

5. Biometric identification systems. Hardware and software

Biometric identification systems are the most effective ones because they do not recognize the physical media, but the unique physical characteristics of a human. Access and data protection systems, based on such technologies, are not only the most reliable but also the most user-friendly nowadays. All biometric devices have specific requirements for hardware and software. In any identification system users must be registered first. A lot of biometric systems allow users to do it themselves.

Biometric access control systems allow organizing an efficient, reliable and cost-effective system of access control. The use of biometric technologies avoids such problems as loss, re-release and transfer of electronic passes to other persons.

Timesheet is no less important issue for the management of companies. Time tracking system supplies reporting data, corresponding to the time of staff arrival and departure at the workplace, delays at the workplace, overworking and underworking. In its turn, the information availability allows managing the distribution of wages, bonuses, penalties more flexibly, increases discipline in the enterprise and, of course, results into economic benefits. It is important that the time tracking system, constructed on the basis of biometric identification, allows eliminating instances of intentional violation of the regime related to the lack of binding - e.g., the magnetic card to a specific person, so that the card can easily be transferred from one person to another.

There are several ways to provide access to the necessary resources:

1. Verification based on biometrics and a unique identifier that identifies a particular person (e.g., ID number), that is, this method is based on a combination of identification and authentication techniques.
2. Identification based only on the biometric measurements. The measured parameters are compared to all records from the database of registered users, rather than with one of them, selected on the basis of any ID.

Generally, both methods are used equally. The selection of the authentication method is carried out directly on the basis of security requirements, as well as the funds allocated for these purposes. But the reliability of the access system depends not only on the way of authentication. An important factor in this regard is the selection of the biometric parameters based on the type of access, which is going to be implemented.

6. Biometric parameters

There are a number of biometric parameters, on which base the identification and authentication of a personality are carried out. They are the iris of the eye, the retina of the eye, a hand vein pattern, a fingerprint, a handprint, a hand geometry, a face geometry, 2D / 3D face models, a signature, a voice, a gait, a shape of ears.

Every year more and more biometric parameters, which can identify and authenticate a personality, appear. But it takes a lot of time and money to study them and find methods and algorithms of processing. Therefore, the scrutiny degree of a biometric parameter directly affects the extent of its using.

A fingerprint is the most-used parameter today. This is due to the fact that fingerprinting got its spread in the late 19th and early 20th century. Every day the cost of fingerprint scanners is more and more reduced, but the number of additional parameters, such as finger temperature, pressing force and the like ones, increases. We can say that scanner manufacturers will reach the limit of opportunities in this area in the nearest future.

The most statistically reliable and forgery resistant access parameters are the iris of the eye and a hand vein pattern. But the process of collecting and processing such parameters takes a bit longer time than the processing of a fingerprint or a face shape. The indicators of reliability are such features as a false rejection and a false access. A false denial of access for biometric data tends to 0%, and a false access does not exceed 1-2%.

Despite all the differences between the biometric parameters, it was decided to use the False Acceptance Rate (FAR) and False Rejection Rate (FRR) as the basic characteristics of biometric systems. Obviously, the smaller these ratios, the more reliable one or the other parameter. These factors help to decide whether to use a particular method of identification and authentication, depending on the task. Table 1 shows the average values for the various biometric systems[4]:

Table 1. Mean values for different biometric systems.

	Fingerprints	Face Geometry	Iris
FAR, %	0.001 %	0.1 %	0.00001 %
FRR, %	0.8 %	7 %	0.10 %

7. Structure of biometric systems of access control

For the most part ACS and the working time control systems are network distributed systems with the user access rights, stackable if necessary, open to the integration with other manufacturers' equipment. The controllers connected to the control PC or the server via USB or LAN Ethernet are set at the points of passage [7, 8].

The user registration is performed by means of special software. A control reader, connected via the USB port of a personal computer, is applied to register fingerprints. Users can also be registered independently at the terminals of the system using specific procedures.

As we can see in Figure 1, the ACS has a modular structure. It allows the system to be scalable and more stable. In the event of failure of one of the elements, the system continues working, of course, in a limited function mode. In most cases the software consists of two modules – a 'working time control system' and an 'access control system'. The time tracking system works with a staff, shifts, schedules, events, obtained automatically or listed in the manual mode. It allows making reports. The access control program is responsible for connecting and configuring devices, setting the rights of passage and time zones, synchronization of data between the database and terminals [9].

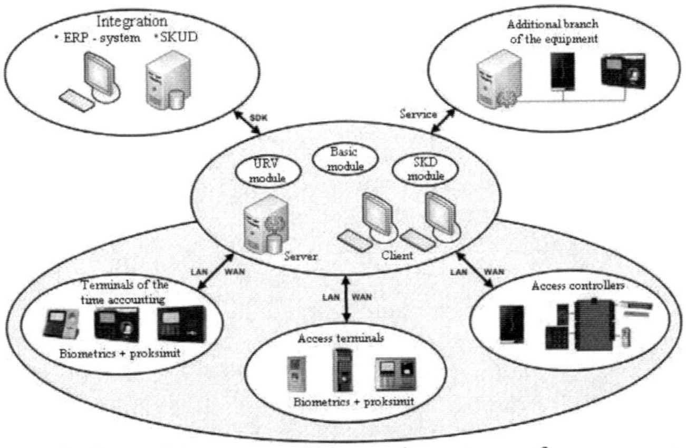

Figure 1. General structure of biometric systems of access control.

8. Comparative evaluation of biometric systems

Let us select several empirical characteristics for assessing the quality of the system:

1. Forgery resistance.
2. Environmental resistance – a characteristic estimating system stability under various environmental conditions, such as changing the lighting or the room temperature.
3. Ease of using - shows how difficult it is to use a biometric scanner, whether it is possible to identify 'on the fly'.
4. Operation speed.
5. The cost of the system

You can see a comparative evaluation of biometric systems based on the ten-point scale.

Table 2. Comparative evaluation of biometric systems.

	Forgery resistance	Environmental resistance	Ease of using	Speed	Cost	Stability of characteristic over time
Iris	10	9	8	7	10	10
Fingerprinting	6	10	9	10	10	9
2D face	4	6	6	10	10	8
3D face	9	8	10	5	7	10
Arm veins	10	7	9	7	8	7
Retina	10	10	6	3	6	9

It can be concluded that forgery resistance of one or the other biometric parameter is inversely proportional to the speed of processing it by the system.

9. Conclusion

The choice in favour of using a certain parameter when organizing access or working hours accounting depends directly on the system requirements. The most statistically reliable and forgery resistant access systems are the admission systems based on the iris of the eye and on a hand vein pattern. The least expensive and easiest to use, but having good statistics are the fingerprints admission systems. The admission on 2D face is convenient and cheap, but it has a limited range of applications due to poor statistics.

In addition to combining the used biometric parameters, it is possible to combine the parts of the body, from which the parameters are removed. For example, for the iris of the eye we can increase the accuracy of the system almost quadratically without time-consuming. We can complicate the system by making it for two eyes. As for the fingerprint method, it is possible to do it by combining several fingers; as for the vein pattern recognition, it may be done by combining two hands, but in this case, the improvement is only possible by increasing the time of human interaction with extractors.

Thus, summarizing the results of the methods it can be said that the iris biometric access and possibly the vein pattern recognition should be used for medium and large objects as well as for the objects with maximum safety requirements. The fingerprint access control will be optimal for the objects with the number of staff to several hundred people. 2D face recognition systems are very specific. They may be required in cases when the detection requires the absence of a physical contact, but it is impossible to supply the control system on the iris of the eye. For example, if identification of a person is required without his or her participation, with a hidden camera or with an outdoor motion detection camera, which is possible only for a small number of subjects in the database and for a small stream of people shot on a camera.

References

[1] Boreysho A 2005 *J. Security of Business Information* **59** 40–43
[2] Aleshin A 2008 *Dashkov & Co. Moscow*
[3] Balle R, Pankanti S, Ratha N, Senior A 2004 *Springer Science+Business Media. New York*
[4] Kharitonov A 2013 *J. Cybernetics and programming* **2** 12–19
[5] Korobeynikov A, Gatchin Y 2014 *World Applied Sciences Journal* **30** 165–169
[6] Nyrkov A P, Belousov A S, Sokolov S S 2015 *Modern Applied Science* **10(5)** 162–176
[7] Chernyi S, Zhilenkov A, Sokolov S, Titov L 2015 *Metallurg. and Mining Industry* **7(3)** 66–73
[8] Chernyi S 2015 *Metallurgical and Mining Industry* **3** 60–65
[9] Nyrkov A, Sokolov S, Chernyi S, Mamunts D 2015 *Metall. and Mining Industry* **7(6)** 521–524

Determination of Radius of Curvature for Teeth With Cycloid Profile

E V Shatalov, E A Efremenkov and K G Shibinskiy
Tomsk Polytechnic University, 30, Lenina ave., Tomsk, 634050, Russia

E-mail: ephrea@mail.ru

Abstract. In the article the geometric determination of curvature radius is considered for teeth with cycloid profile. The equations are obtained for the determination of a radius of curvature with point coordinates of a cycloid profile. The conditions of convexo–concavity of a teeth profile are defined for transmission with intermediate rollers.

1. Introduction

Now strict requirements are imposed at balance of wait – dimensions - power for modern mechanisms and technical unit more-and-more. On the base on transmissions with intermediate rollers (TIR) reducers has optimum ratio of technical data. Half of rollers number of TIR hold loading (sometimes all rollers hold loading in some scheme of the transmissions). So, the rollers transfer loading by the angle of 180°. It allows transferring torque more than 5-10 times and repeated shot-time overload in comparison usual toothed gear transmissions. One of representatives of TIR is coaxial flat transmission with intermediate rollers, which is showed on Figure 1.

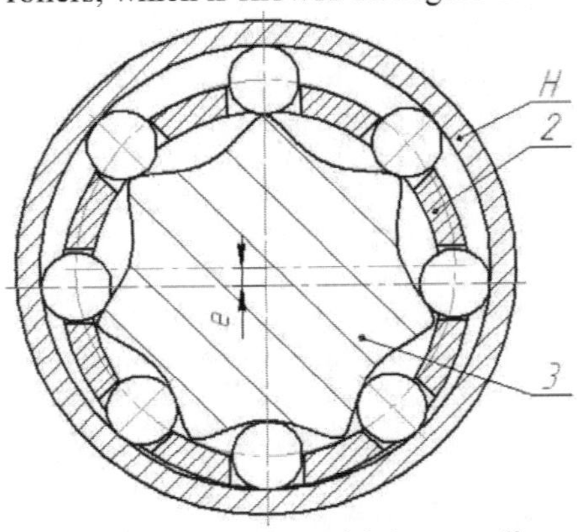

Figure 1. Cross section of flat radial transmission with intermediate roller: *H* – generator (planet carrier); *2* – separator with rollers; *3* – central gear; *e* – eccentricity of the transmission.

The transmission works following way: motor rotates plant carrier *H* and the planet carrier made rollers to move radial in grooves of separator 2, as shows on Figure 1. The separator 2 rigidly fixed in housing of the transmission. The rollers contact with cycloid profile of central gear 3. By the radial movement rollers made to the central gear 3 rotated. The central gear 3 is jointed rigidly with output shaft of the transmission.

It is known that contact durability depends from the specified curvature radius volume of cycloid profile. This shows following equation for calculation of contact stress [2]:

$$\sigma_{max} = \left(\frac{F(R_2 \pm R_1)}{\pi L R_1 R_2 \left(\frac{1-\varepsilon_1^2}{E_1} + \frac{1-\varepsilon_2^2}{E_2} \right)} \right)^{1/2}, \tag{1}$$

where F – normal force to surfaces of contacting bodies, H;

R_1, R_2 – curvature radii of contacting bodies, mm;

L – contact length of roller and profile, mm;

E_1, E_2 – elasticity modulus of first and second contacting bodies respectively, MPa;

ε_1, ε_2 – Poisson's ratios for first and second contacting bodies respectively.

Thus, radii of curvature need for determination of stress-strained state in transmission with intermediate rollers. Besides profile surface radius of curvature needs for technological process of cycloid gear production. Therefor determination of radius of curvature for teeth with cycloid profile is up-to-date.

2. Determination of curvature radius

There are contact between central gear, groove wall of separator, rollers and generator. For roller and generator contact radii of curvature are constant and they equal radius of roller and radius of generator respectively. This contact is convexo-concave.

In roller and groove wall of separator contact is constant. There is contact between radius of roller and flat. Here contact equation will change a little bit.

Let us consider contact between roller and central cycloid gear. Here a radius of curvature of the central gear is variable in difference from both previous cases. Besides, in during this contact roller is moving from convexo-convex contact to convexo-concave contact along cycloid profile of gear and vice versa per one revolution of generator.

Some works are devoted to determination of radiuses of curvature of cycloid profile [3, 4]. There the task be solved by following ways: differential method, Euler-Savari method. Both methods are precise and convenient, but have the difficulties. First method is complex very much, because it demands the capture of the first and second derivative on a generator angle of rotation in the equations of a profile of tooth. These may lead to rather difficult mathematical transformations of the equations. The Bobillier construction is possible at a known a pitch point, but it is unknown for this transmission.

Thus, we need to define method of determination of radius of curvature simpler, but the same effective, as well as considered above. In our case, the method is more suitable for determination of curvature radius by geometric method (Figure 2). For chords are choosing, we need now a coordinates of points around contact point of cycloid profile. The points can be choose by special computer program. Then we create the massif from the points of a profile is found and we define a angle between the two chords.

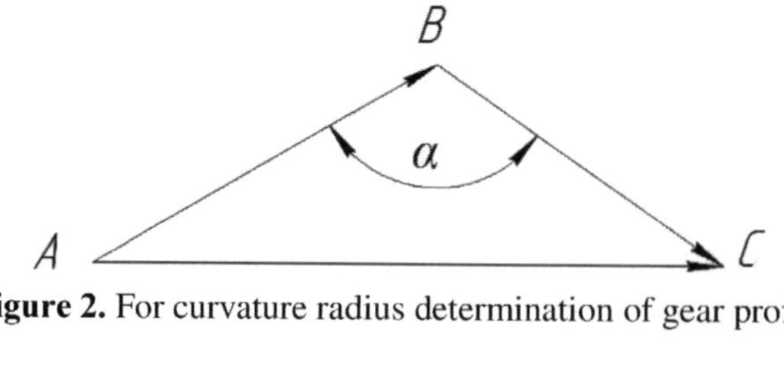

Figure 2. For curvature radius determination of gear profile

Let us consider the points (Figure 2) of the cycloid profile $A(x_1, y_1)$, $B(x_2, y_2)$, $C(x_3, y_3)$. We connected these points by the lines and constructed through them three vectors $\overrightarrow{AB}$, $\overrightarrow{BC}$, $\overrightarrow{AC}$.

Let us determine the coordinates of vectors through the coordinates of pointes A, B, C:

$$\begin{aligned} \overrightarrow{AB} &= \{x_2 - x_1, y_2 - y_1\} \\ \overrightarrow{BC} &= \{x_3 - x_2, y_3 - y_2\} \\ \overrightarrow{AC} &= \{x_3 - x_1, y_3 - y_1\} \end{aligned} \tag{2}$$

Scalar product of vectors $\overrightarrow{AB}$ и $\overrightarrow{BC}$:

$$\overrightarrow{AB} \cdot \overrightarrow{BC} = (x_2 - x_1)\cdot(x_3 - x_2) + (y_2 - y_1)\cdot(y_3 - y_2) \tag{3}$$

Lengths of vectors:

$$\begin{aligned} \left|\overrightarrow{AB}\right| &= \left((x_2 - x_1)^2 + (y_2 - y_1)^2\right)^{1/2} \\ \left|\overrightarrow{BC}\right| &= \left((x_3 - x_2)^2 + (y_3 - y_2)^2\right)^{1/2} \\ \left|\overrightarrow{AC}\right| &= \left((x_3 - x_1)^2 + (y_3 - y_1)^2\right)^{1/2} \end{aligned} \tag{4}$$

The Cosine of the angle between vectors $\overrightarrow{AB}$ и $\overrightarrow{BC}$ are determined by a formula:

$$\cos(\alpha) = \frac{\overrightarrow{AB} \cdot \overrightarrow{BC}}{\left|\overrightarrow{AB}\right| \cdot \left|\overrightarrow{BC}\right|} \tag{5}$$

Define radius of a circumscribed circle according to the theorem of sine [5]:

$$2R = \frac{\left|\overrightarrow{AC}\right|}{\sin(\alpha)}, \tag{6}$$

Transform (6) taking into account that:

$$\sin(\alpha) = \left(1 - \cos^2(\alpha)\right)^{1/2} \tag{7}$$

Then

$$R = \frac{\left|\overrightarrow{AC}\right|}{2 \cdot \sin(\alpha)} = \frac{\left|\overrightarrow{AC}\right|}{2 \cdot \left(1 - \cos^2(\alpha)\right)^{1/2}} \tag{8}$$

Substituting (5) in (8) received:

$$R = \frac{\left|\overrightarrow{AC}\right|}{2 \cdot \left(1 - \left(\dfrac{\overrightarrow{AB} \cdot \overrightarrow{BC}}{\left|\overrightarrow{AB}\right| \cdot \left|\overrightarrow{BC}\right|}\right)^2\right)^{1/2}} \tag{9}$$

3. Determination concave – convex contact

The offered option allows determining curvature radius any curve in engagement. The equation (9) gives a radius which always positive, but curvature of a profile changes from convex to the concave. As the profile of tooth located in the first or fourth quarter of the Cartesian coordinate system and the roller is above this profile, it is possible to define convexo-concave considered Figure 3.

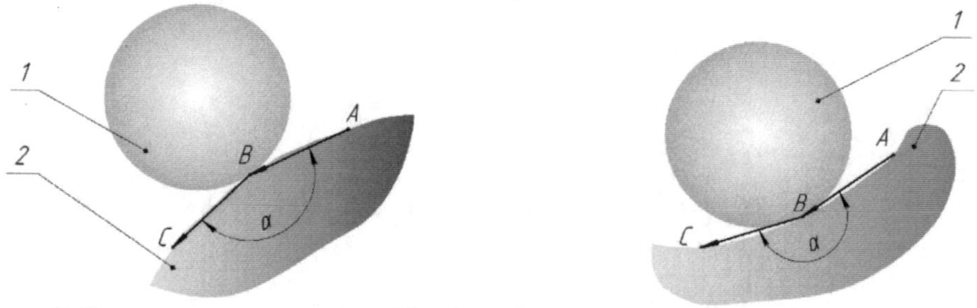

Figure 3. For convexo-concave determination of gear profile: *1* – roller; *2* – profile of the teeth.

It is possible to define profile convexity by angle α:
if $\alpha<180^{\circ}$ – the profile is convex;
if $\alpha>180^{\circ}$ – the profile is concave.

It is possible to determine the angle α through the directing cosines of vectors $\overrightarrow{AB}$ and $\overrightarrow{BC}$, which is showed on Figure 4:

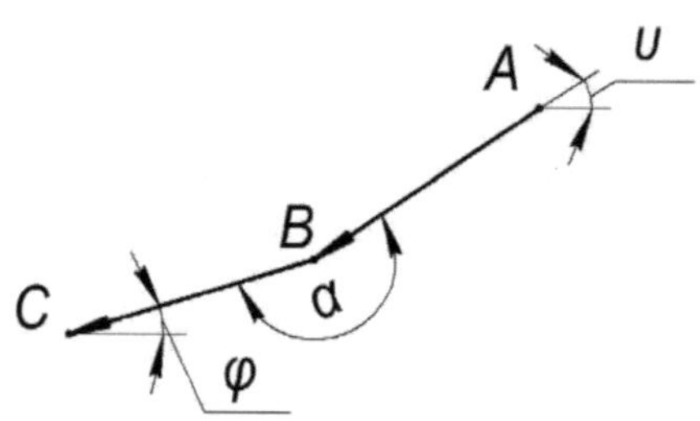

Figure 4. The directing cosines of vectors $\overrightarrow{AB}$ and $\overrightarrow{BC}$

For calculation of an angle α the equation is following:

$$\alpha = 180 - \varphi + \upsilon \tag{10}$$

For determine of an angles υ and φ equations is using [4]:

$$\upsilon = \arccos\left(\frac{x_2 - x_1}{\left|\overrightarrow{AB}\right|}\right) \tag{11}$$

$$\varphi = \arccos\left(\frac{x_3 - x_2}{\left|\overrightarrow{BC}\right|}\right) \tag{12}$$

So, necessary angles having defined from expressions (10) – (12) it is possible to know radius is convex or concave, it found on the equation (9).

4. Example

Let us consider example of calculation by the technique for flat radial transmission with intermediate rollers. We will use equations (4), (9) – (12) to determine the radius of curvature of cycloid profile for a transmission with the following initial parameters: r_2=22 mm – centrode radius; z_2=10 – number of rollers; r_{tk}=3 mm – roller radius; i=11 – transmission ratio.

Profile equation is taken from [1] and calculate a profile points by special program for Matlab-system.

After calculation we obtain the diagram:

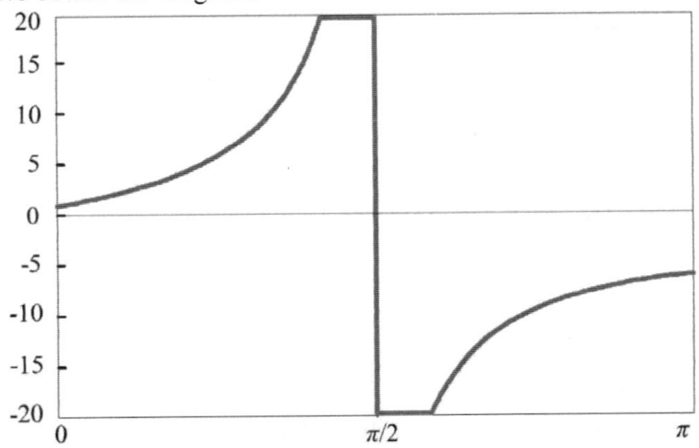

Figure 5. Dependence of curvature radius of a tooth profile on a rotation angle of an input link

On abscissa axis, (Figure 5) the angle of input link is presented. On ordinate axis, values of curvature radius are presented. On graphics the straight section is shown (near an angle π/2) for which points weren't set. The technique of dependence of the directing cosines and curvature radius was used [5] by preparation of the program for Matlab. On graphics, we can see in positive area radiuses of a convex part of the profile, and in negative area radiuses of a concave part of the profile.

Thus, this dependence can be used in definition of a sign of radius of curvature.

It does not matter to consider any position of the tooth profile in space, but it is better to start calculation from top point of profile and move along profile to tooth space.

Conclusions

So, we considered geometric determination of curvature radius for teeth with cycloid profile. The equation was obtained for the determination of a radius of curvature. The conditions of convexo–concavity of a teeth profile were defined for transmission with intermediate rollers.

References:

[1] Shibinskiy K G, Efremenkov E A and An I-Kan 2008 *Izevsk: ISTU. Theory and practice of toothed gearing and reducer building* 188-192

[2] Shibinskiy K G and Efremenkov E A 2004 *Izevsk: ISTU. Theory and practice of toothed gearing* 166-171

[3] Kudryavcev V N 1966 *Moscow: Mashinostroenie*

[4] Efremenkov E A and An I-Kan 2010 *Russian Engineering Research* **30(10)** 1001-1004

[5] Privalov I I 2003 Analytical geometry: textbook (SPb.: Lanj)

The stabilization interval system of a tethered descent underwater vehicle

S A Gayvoronskiy, T Ezangina, I Khozhaev and S V Efimov
Department of Automatics and Computer Systems, Tomsk Polytechnic University, 30, Lenina ave., Tomsk, 634050, Russia

E-mail: eza-tanya@yandex.ru

Abstract. To damp the vertical oscillations of a descent submersible caused by dusting the control system utilizing a shock-absorbing hoist located on the submersible was developed. A robust proportional-plus-integral action controller was included in the control loop to ensure acceptable dynamic properties of the system by interval variations of the module mass, the rope length, the equivalent value of stiffness of a spring linkage and the equivalent value of damping factor of the spring linkage. A parametric synthesis of the controller was carried out on the basis of the robust expansion of the coefficient method of the quality rating estimation. The system operability was confirmed by the results of the digital simulation parameters

1. Introduction

At present, there is an active development of the great oceans by stand-alone unmanned submersibles. However, according to [1-3], some real-world problems are still urgent, which are reasonable to solve by means of a tethered descent underwater vehicle (DUV) connected with a ship-carrier by a cable-rope. In accordance with [1-3], engineering, exploration oceanographic and another marine research are conducted by DUV.

The descent submersible during its descent and ascent and by its location close to the sea-bed undergoes some vertical oscillations under the influence of dusting and can become unable to carry out some undersea explorations (DUV can hit the bottom). The possibility of the resonance oscillation due to frequency coincidence of the longitudinal oscillations in the cable with the sea roughness is the most dangerous phenomenon.

The problem of DUV control becomes more complicated since some physical parameters of the control system are precisely unknown or tend to change within certain limits in the process of operation in accordance with laws that are unknown beforehand (the cable parameters, hoist drive parameters, DUV mass).

Therefore, the development of a speed control system of DUV able to damp the oscillation of DUV in the dusting environment and the interval uncertainty of parameters is very urgent. Therefore, the solution of the following problems is of great importance:

- the development of the robust stabilization system structure of DUV with interval parameters;
- synthesis of the robust controller (regulator) capable to maintain system operability under any variations of the cable length and DUV mass;
- DUV control system simulation for the analysis of its dynamic properties.

2. Block diagram of the stabilization system of a submersible

In order to solve the assigned tasks the block diagram of the stabilization system of a submersible, presented in Figure 1 is offered. The system includes two hoists: a boat hoist (BH) located on the ship and intended for DUV descent and ascent and a shock-absorbing hoist (SAH) installed on DUV used to damp its vertical oscillations. The following symbols are introduced in Figure 1: SCTD-sensor of cable tension deviation; CS – comparison-summator, CON – control block, which implements the selected law of control, SS – speed selector.

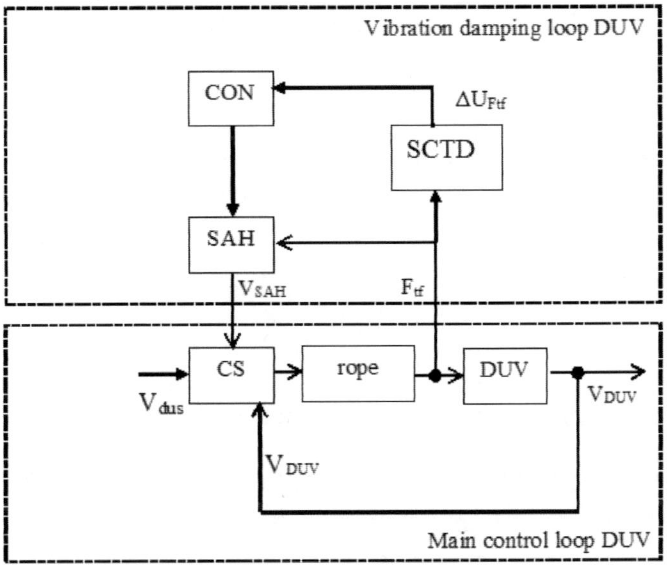

Figure 1. Functional diagram of the DUV stabilization system

The system has a disturbance input (signal V_{dus} – the speed of the vertical vehicle movement under the influence of dusting. An output signal is V_{DUV} – speed of DUV. The intrinsic coordinates of the system are lettered in Figure 1 as follows: V_{sah} –linear speed of SAH; F_{tf} – tension force in the rope. SAH installed at a tethered descent underwater vehicle is used to compensate for the sea oscillating motions.

In this mode the tension deviation from the set value corresponding to the DUV weight is measured by a sensor of cable tension deviation. The output signal of the sensor passes to the SUH controller.

Let us write down the equations of separate system components. The block diagram is arranged on the basis of these equations. The equation of the vertical movement of the tethered submersible vehicle is as follows:

$$m\frac{dV_{DUV}}{dt} = F_{tf} ,$$

where m is mass of the tethered submersible vehicle. The inclusion of the resilient member of quasi-zero stiffness in the rope linkage between the ship and the tethered submersible vehicle allows considering the stiffness of the resilient member as equivalent stiffness. The damping of the resilient member will be determined in the same way as the damping of spring linkage. The equation of spring linkage is obtained on the basis of Hooke's law.

$$F_{tf} = C((x_{dus} + x_{bh}) - (x_{sah} + x_{tsv})) + \chi\frac{d((x_{dus}) - (x_{sah} + x_{tsv}))}{dt},$$

where C is an equivalent value of stiffness of the spring linkage, which is equal to the stiffness of the resilient member, χ is an equivalent value of the damping factor of the spring linkage, which is equal to the damping of the resilient member, x_{dus} is a vertical movement of the ship, x_{SAH} is a movement of the rope on the drum of SAH, x_{tsv} is the vertical movement of the tethered submersible vehicle. The electric drive of SAH is described by the following equation:

$$J_2 \frac{d\omega_{sah}}{dt} = M_d + M_{tf},$$

where J_2 is a moment of SAH inertia, ω_{sah} is angular velocity of the drum rotation of SAH, M_d is a controlling torque of the SAH drive, $M_н$ is a torque produced on SAH by the rope tension force. Thus,

$$M_d = k_{m2}(U_c - U_e),$$

where k_{m2} is the transfer constant of SAH drive per torque, U_c is the output voltage of the SAH controller, $U_e = k_{e2}\omega$ is a reverse electromotive force (EMF) voltage of the SAH drive, k_{e2} is a reverse EMF coefficient of the SAH drive, R_2 is a drum radius of SAH. The mathematical description of a BH electric drive is similar to the description of the SAH drive. The analysis of the transfer function of the speed feed forward control system of DUV under the steady-state conditions with different W_{con} has shown that it is reasonable to use a controller which ensures the astaticism of the first order in the system. Its transfer function can be presented as follows:

$W_{con}(s) = (k_1 + k_2 s)/s.$

The given controller has two setting parameters k_1, k_2, determining the quality of the transient processes in the system. Thus, the block diagram of the system obtained on the basis of the above mentioned mathematical description is illustrated in Figure 2. The system has the following constant parameters [1]: K_{m2}=0,3 (Nm/A); K_{e2}=1 (Vs/rad); J_2=0,5 (kgm^2); R_2=0,1 (m) and interval parameters χ =[35;200] (Ns/m); $C = [800; 4.25\times10^3]$](N/m); $m_{DUV} = [4.47\times10^3; 4.65\times10^3]$ (kg); l=[2;10] (m).

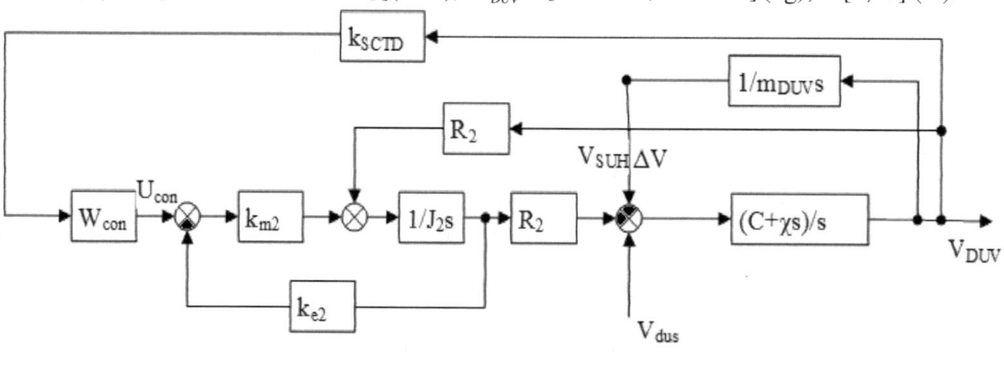

Figure 2. Structure diagram of the DUV stabilization system

The differential equations corresponding to the block diagram shown in Figure 2 are as follows:

$$A(s) = [a_{33}]s^3 + [a_{23}]s^2 + [a_{13}]s^1 + [a_{03}], \quad (1)$$

where

$[a_0] = [C]k_{m2}(k_{e2} + [m^*_{DUV}]R_2 k_{SCTD} k_1);$

$[a_1] = [C]R_2[m^*_{DUV}](k_{m2}k_{SCTD}k_2 + R_2) + [C]J_2 + [\chi]k_{m2}(k_{e2} + R_2[m^*_{DUV}]k_{SCTD}k_1);$

$[a_2] = [m^*_{DUV}]k_{m2}k_{e2} + [\chi]J_2 + R_2[m_{DUV}][\chi](R_2 + k_2 k_{SCTD} k_{m2});$

$[a_3] = J_2[m^*_{DUV}]$.

As it is shown in (1), the polynomial coefficient includes linearly the controller settings and multilinearly the interval system parameters. Due to availability of the interval-undetermined parameters in the system it is necessary to impart this system some robust properties ensuring the preservation of a permissible performance quality at any possible variations of unstable parameters [4]. It is suggested that one should apply a robust approach by the parametric synthesis of PI-controller in the control loop of SAH ($k_1 = 0.078$, $k_2 = 0.079$).

3. Control process simulation

To test the operability of the robust stabilization system of DUV with a synthesized proportional - integral (PI) –controller, let us carry out its mathematical model in Matlab by means of Simulink appendix. In the simulation we used the model of dusting, which is maximally close to real conditions. A block diagram of this model is shown in Figure 3.

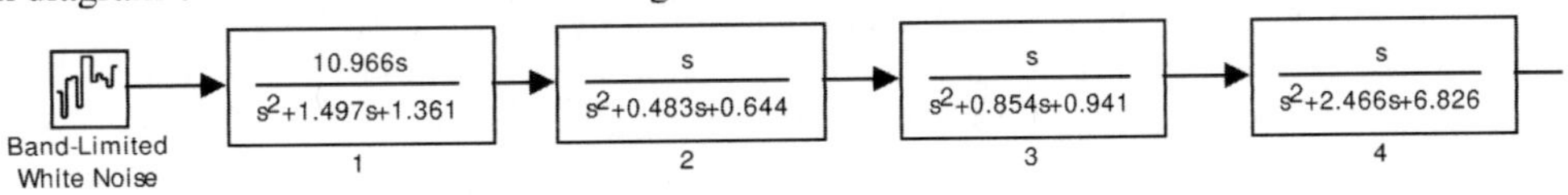

Figure 3. Block diagram of a dusting model

The analysis of the robust stabilization system of DUV has shown that by any possible values of the interval parameters of the system the amplitude of speed fluctuation V_{DUV} is insignificant under the impact of the dusting, and its maximum is 4×10^{-3} m/s (Figure 4, a). We can conclude from Figure 4 (b) that if the stabilization system of DUV is not utilized, the amplitude of speed fluctuation V_{DUV} is significant and is equal to the amplitude of vertical movement of the watercraft in the dusting environment.

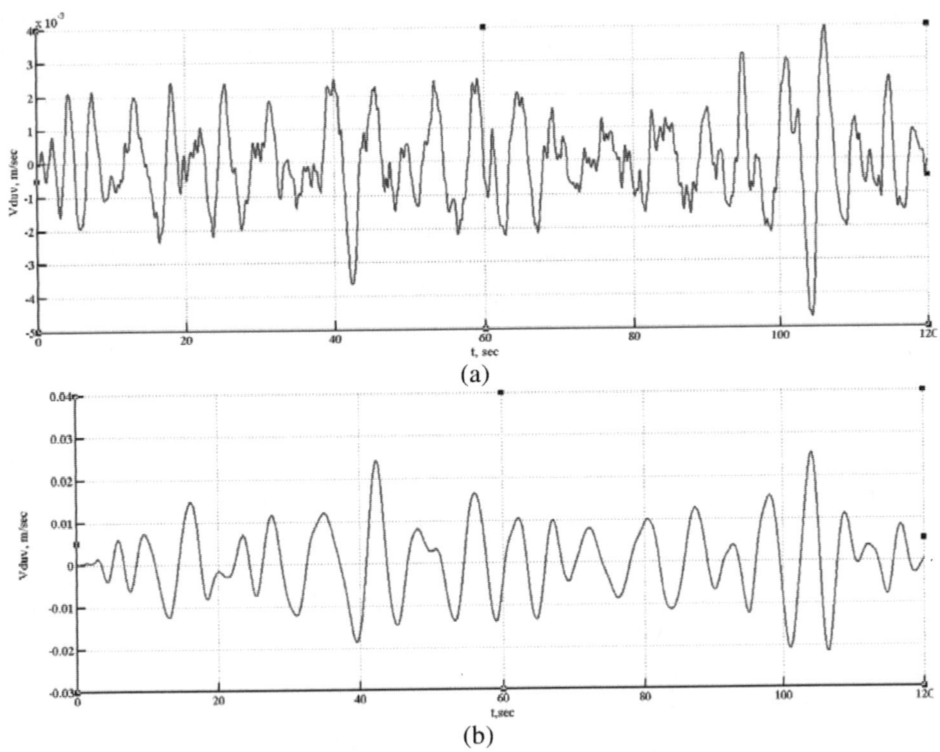

Figure 4. Diagrams of the speed fluctuation of the DUV vertical movement in the descent mode (a) the robust stabilization system is utilized; (b) the stabilization system of DUV is not utilized

Thus, the suggested block diagram of the robust stabilization system of DUV and the technique of the parametric synthesis of the robust PI – controller settings enable a solution of the problem associated with oscillation damping of DUV in the marine roughness environment and variations of the system parameters.

4. Conclusion

The robust stabilization system of DUV that enables compensation of the influence of the ship-carrier rocking by descent and ascent of DUV as well as passing near the sea-bed was developed in the given article. To adjust the interval parameters of the PI- controller used in the system, the interval expansion of the coefficient method and the criterion of a maximum robust degree of stability were applied. The operating efficiency of the designed system was confirmed in Matlab by the results of its simulation under different modes of the operation corresponding to the boundary values of the interval parameters of the system.

Acknowledgments

The reported study was partially supported by RFBR, research project No 14-08-31031.

References

[1] Nam B W, Hong S Y and Kim Y S 2013 *International Journal of Offshore and Polar Engineering* **23(1)** 33–37
[2] Rowe S, Mackenzie B and Snell R 2001 *Proceedings of the 10th Offshore Symposium* 1–9
[3] Neupert J, Mahl T, Haessig B, Sawodny O and Schneider K 2008 *American Control Conference* 538–543
[4] Ezangina T A and Gayvoronskiy S A 2014 *Advanced Materials Research* **1016** 394–399

Development of power supply devices for limitations of short circuit on the ship's hull

R S Fediuk and A M Cherneev
Far Eastern Federal University, 8, Sukhanova st., Vladivostok, 690091, Russia

E-mail: roman44@yandex.ru

Abstract. The authors of the paper have analysed the reasons and consequences of single-phase ground faults (hull faults). For all varieties of devices limiting the current single-phase ground faults, the most effective devices were found to be the arc-suppression coils with different switching circuits. In this case the measurement of circuit capacitance takes on a great importance. A number of variants of capacitance measurement is presented in the paper. The authors have had a detailed look at a device, limiting the single-phase short-circuit current. This device was developed on the basis of the Far Eastern Federal University under the direction of Dr. G.E. Kuvshinov. The device is provided with power supply that converts alternating current (AC) to direct current (DC), and is realised due to the use of semiconductor devices - transistors and diodes - in a bridge circuit. The technical outcome of this power supply application consists in the reduction of size and weight parameters (compared to the closest analogues) in order to connect the controlled voltage rectifier to the alternating voltage source, including mass and size of the capacitor bank of the current limiting circuit, and enhancing the dynamic parameters of the stage of uncontrolled charge of the output capacitor of the controlled voltage rectifier.

1. Introduction

Single-phase ground faults are a predominant type of damage in the power grid of medium voltage, which is 6-10 kV (75 to 90% of the total number of electrical damage). Single-phase ground faults are accompanied by an electric arc that can cause a fire and destroy the current conductors at a spot of short-circuit, which is why single-phase short circuit can be coverted into a two- or three-phase one [1-3].

The vast majority of violations of the normal operation of power grids with an isolated neutral wire is connected with local insulation damage. In such grids, even at high insulation resistance heavy current can pass through the short circuit (SC) spot. This occurs due to the large value of the total capacitance of the power grid with respect to the ground surface. The greater the single-phase short-circuit current, the greater the value of the mentioned total capacitance and the higher the voltage in the power grid [4-6].

To ensure the maximum possible reliability of the grid it is necessary that the short-circuit current was so small that for a sufficiently long time (the time required for search and elimination of the damage) it would be possible to dispensed with disconnection of consumers from the power grid. According to the rules of technical maintenance of power plants and grids, the triggering current capacities that do not require an immediate shutdown of such damaged connections are considered to be the currents: in the overhead networks of 6-20 kV on concrete or metal poles and in all networks at a voltage of 35 kV - 10 A; in the overhead networks without metal or concrete supports, at a voltage of

6 kV - no more than 30 A, at a voltage of 10 kV - no more than 20 A, at a voltage of 15-20 kV - no more than 15 A. If the currents exceed the mentioned values, a compensation of the capacitive earth fault current is required.

The effective value of short-circuit current decreases rapidly up to a small steady-state value, if the neutral point of the power grid is grounded to earth through the arc suppression coil (ASC). The inductive reactance of ASC at a frequency of the voltage source, which supplies the electrical network, is equal to the resistance at this frequency and the total capacity of all three phases of the power grid relatively the earth [7-9].

This capacitance is often changed not only in case of a single-phase circuit, but also during changing of the grid configuration, as well as due to changes in the length of any line during manipulation when networking, or under the influence of temperature, icing, frequency oscillations and for other reasons. Therefore, the inductance of the arc suppression coil must be constantly maintained at a value that resonates with the capacity of the grid. Because of this, the coil is executed in two parts: an adjustable load bearing element and a control unit. The latter affects the load bearing element so that its inductance changed right after the change in the total capacity of the power grid [10-12].

The objective of this work is to develop such power source for the device which will limit the current of the single-phase short ground fault in the power grid.

2. Results and Discussions

In the Far Eastern Federal University under the guidance of prof. G.E. Kuvshinov a device that limits the current of the single-phase ground fault (hull fault) was invented [11, 13].

The electrical block diagram of device 1 for limiting the current of the single-phase ground fault in the power grid with floating neutral 2 is shown in Figure 1. Unit 3 represents the source of three-phase voltage. Unit 9 of capacitors containing capacitances emulates the capacitive susceptance of phases relatively the earth. Resistor R simulates the resistance of the circuit of the single-phase ground fault for phase C. An inductive component of device 1 is unit 4, which includes unit 6 - an electrical filter, unit 7 – a multiplier unit 8 - a voltage-controlled current source, unit 5 - control block.

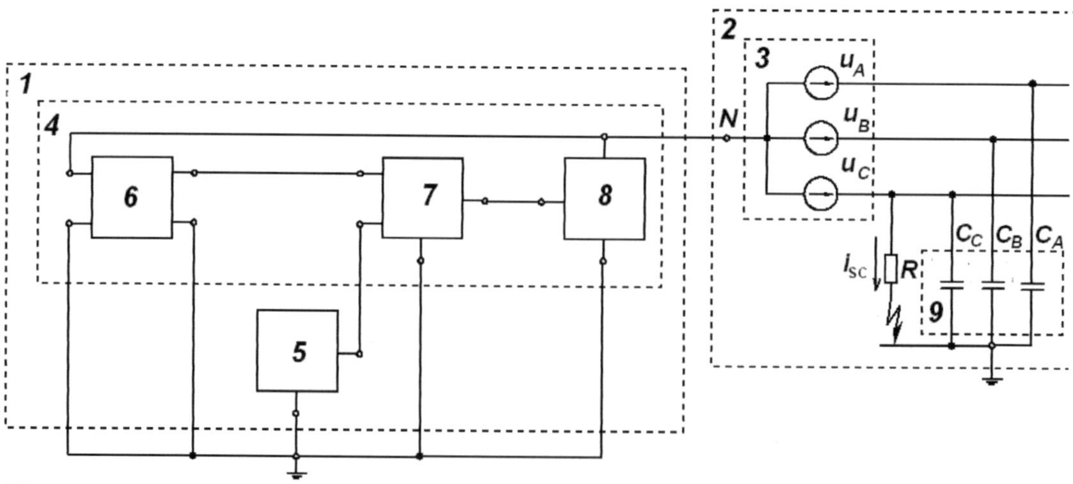

Figure 1. Schematic block diagram of a device limiting the current of the single phase short-circuit

When the SC occurs, voltage appears on the neutral wire of the power grid. This voltage enters the input of electric filter 6. A schematic diagram of the filter realised as a low-pass filter is shown in Figure 2.

The transfer function of the filter shown in Figure 2 has the form:

$$W_{RC}(S) = \frac{U_{F.OUT}}{U_N(S)} = \frac{1}{1 + \tau \cdot s}, \quad (1)$$

where $\tau = R_F C_F$ - the time constant of the filter. This time constant is many times greater than the value, which inverse relatively angular frequency ω_1 of source 3. Therefore, at angular frequencies satisfying condition $\omega \geq \omega_1$, the amplitude and phase frequency characteristics of filter 6, which are obtained after substitution of $s=j\omega_1$ into the transfer function of Eq. 1, differ little from the similar characteristics of the ideal integrator with transfer function $1/(\tau \cdot s)$.

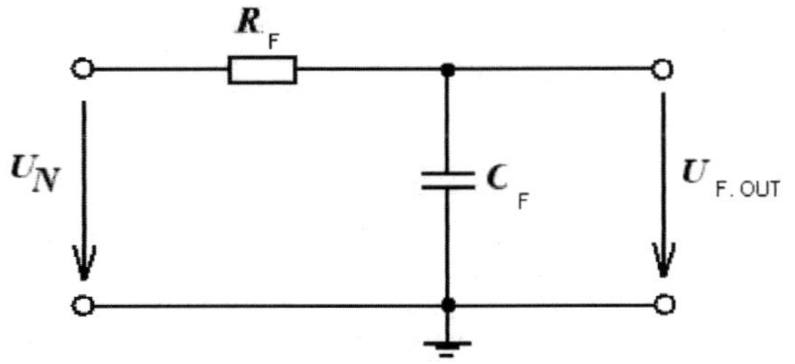

Figure 2. The low pass filter

The output voltage of the filter represented in a complex form:

$$\overline{U}_{F.OUT} = \frac{\overline{U}_N}{1 + j\omega_1 \tau} \quad (2)$$

where $\overline{U}_N$ - the voltage on the neutral wire, ω_1- circular frequency.

The instantaneous value of the voltage is supplied to the second input of multiplier 7, on the first input terminal of which the required value of transmissive conductivity G_{TC} is served, which is controlled by current source voltage 8. In this case the output current of the current source is described by:

$$\overline{I}_{TC} = \overline{U}_{F.OUT} \cdot G_{TC} = \frac{\overline{U}_N \cdot G_{TC}}{1 + j\omega_1 \tau}. \quad (3)$$

From Eq. 3 we can deduce the relation of $\overline{U}_N$ to $\overline{I}_{TC}$ that determines the complex impedance, which is equivalent to inductive component 4:

$$\underline{Z}_{IC} = \frac{1}{G_{TC}} + j\frac{\tau \cdot \omega_1}{G_{TC}}. \quad (4)$$

Imaginary component $\underline{Z}_{IC}$ is equivalent to the inductive reactance of the inductive component and is equal to:

$$X_{IC} = \frac{\tau \cdot \omega_1}{G_{TC}}. \quad (5)$$

The inductance of the inductive component is:

$$L_{IC} = \frac{\tau}{G_{TC}}. \quad (6)$$

Real component $\underline{Z}_{IC}$ is equivalent to the active resistance of the inductive component and is equal to:

$$R_{IC} = \frac{1}{G_{TC}} \tag{7}$$

The Q-factor of the inductive component is:

$$q_{IC} = \frac{X_{IC}}{R_{IC}} = \frac{\omega_1 \cdot L_{IC}}{R_{IC}} = \omega_1 \cdot \tau \tag{8}$$

Q-factor q_{IC} of the inductance component exceeds the quality factor of the arc suppression coil greatly, since time constant τ can reach one or more seconds. Consequently, the quality factor of the inductive component is 6 or more times higher than the quality factor of the arc suppression coil.

When replacing inductive component 4 with equivalent complex impedance $\underline{Z}_{IC}$, consisting of resistance R_{IC} and inductance L_{IC}, the equivalent circuit (Figure 3) of the system with the earthed neutral wire through the inductive component results. The capacitor with a total capacity of three phases relatively ground (3C) is connected in parallel to the inductive component. The circuit consisting of two parallel connected branches (3C and $\underline{Z}_{IC}$), can be conFigured for current resonance.

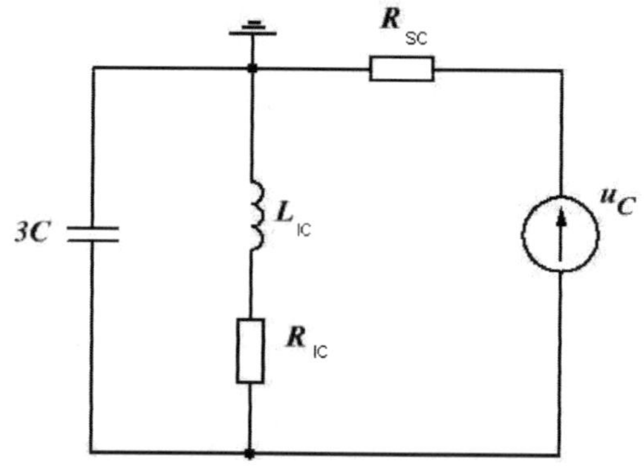

Figure 3. A scheme of replacement of system with the earthed neutral via an inductive component

Then the sum of currents of these branches becomes minimum. This resonance is achieved when susceptance of the indicated parallel connection is equal to zero, that is, condition $X_{IC} = X_{C\Sigma}$ has to be met. From it one derives a formula for the required value of the transfer conduction, which is controlled by the voltage of current source 8:

$$G_{TC} = \tau \cdot \omega_1^2 \cdot C_\Sigma, \tag{9}$$

where C_Σ - the total capacity of phases relatively the earth.

Before connecting inductive component 4 to power supply 2, it is necessary to set the value of transfer conductivity G_{TC} from control block 5, which depends, according to Eq. 9, on the total capacity of all phases relatively the earth. A corresponding signal about value G_{TC} from the output terminal of control block 5 is supplied to the first input terminal of multiplier 7.

Using the electron coverter-source of current it is possible to generate the current at the input marked by the bridge rectifier, which is sinusoidal and is in phase with the input voltage.

The choice of the electronic converter is based on the following assumptions:

- generally, electrical isolation between the input of the load and the output of the power electronic system is unnecessary (e.g., in AC and DC engines) or can be realised provided that the second level converter is applied in the same way as in the mode switch of DC power devices;

- in many cases it is desirable to stabilise DC voltage U_d with a bit of margin of the maximum input AC voltage;
- the input current must be as perfect as it is possible as in the case of a power factor equal to one, so that the interface of the power electronics resistor could emulate a resistor, represented by the source load. This also implies that the electric current is always unidirectional - from the power source to the electrical equipment.
- price, losses and dimensions of the source current should be minimized [14,15].

Based on these rules, the isolation of the linear-frequency transducer is excluded. Also, if it is acceptable, then U_d> U_s, where U_s - peak AC input voltage. Thus, the obvious choice for the power supply is a DC-DC upconverter.

This converter is shown in Figure 4, where capacitor C_d is used to minimize fluctuations U_d and to meet the requirements of energy storage systems of the power electronics. Constant current I_{load}, is the power supplied to the rest of the system (a high-frequency component in the output current is actually filtered C_d). For simplicity, the internal inductance *Ls* of the source is not included into this Figure.

The input current is advisable to supply with sinusoidal and co-phase ones with *Us*. In practice, the losses in the bridge rectifier and in the DC-DC upconverter are quite small, and in a number of cases they can be neglected. Considering that, $\hat{U}_S = \sqrt{2}U_S$, $\hat{I}_S = \sqrt{2}I_S$ is the input power of AC source:

$$p_{in}(t) = \hat{U}_S|\sin\omega t|\hat{I}_S|\sin\omega t| = U_S I_S - U_S I_S \cos 2\omega t$$

Because of relatively large capacity C_d, voltage U_d can initially be considered permanent: U_d (t) = U_d. Therefore, output power P_d (t) = $U_d I_d$ (t),
where, according to Figure 4:

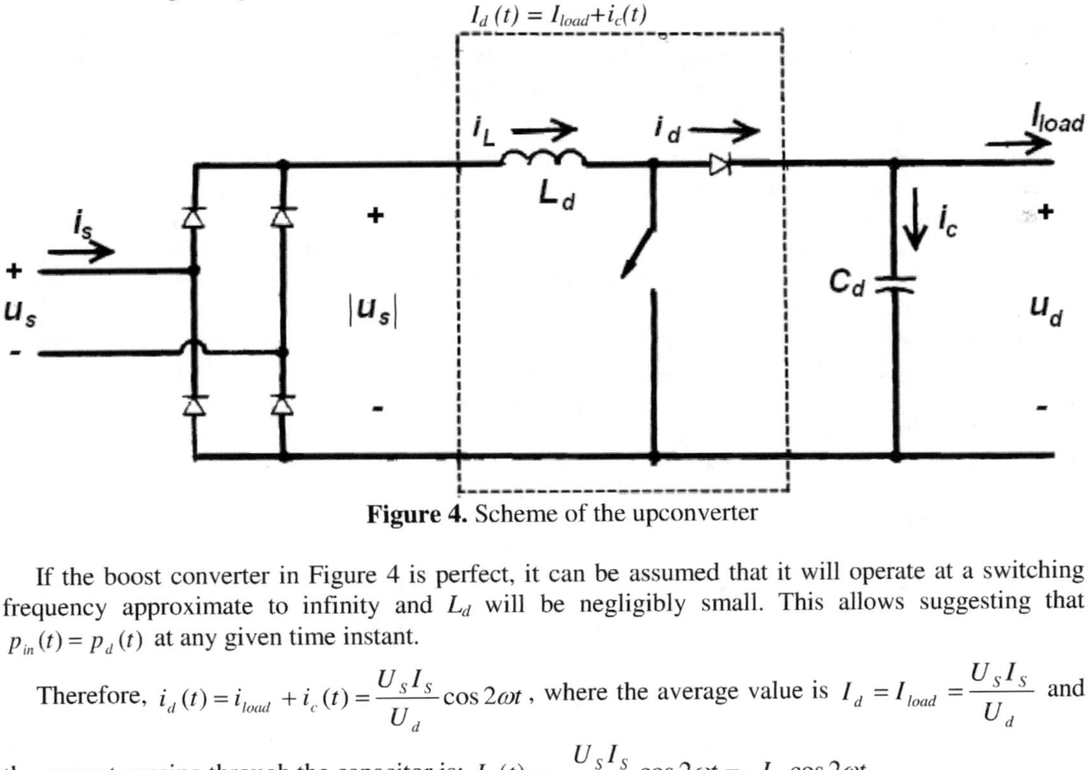

Figure 4. Scheme of the upconverter

If the boost converter in Figure 4 is perfect, it can be assumed that it will operate at a switching frequency approximate to infinity and L_d will be negligibly small. This allows suggesting that $p_{in}(t) = p_d(t)$ at any given time instant.

Therefore, $i_d(t) = i_{load} + i_c(t) = \frac{U_S I_S}{U_d}\cos 2\omega t$, where the average value is $I_d = I_{load} = \frac{U_S I_S}{U_d}$ and the current passing through the capacitor is: $I_c(t) = -\frac{U_S I_S}{U_d}\cos 2\omega t = -I_d \cos 2\omega t$.

Even if this assumption is satisfied, the capacitor will have DC voltage virtually without any pulsation, and pulsation U_d can be calculated as: $V_{d,ripple}(t) \approx \frac{1}{C_d}\int i_c dt = -\frac{I_d}{2\omega C_d}\sin 2\omega t$, which can be kept at a low level having selected accordingly large value C_d. A series of LC-filters tuned into a doubled AC frequency can be connected in parallel to C_d to minimize the pulsation of DC voltage.

It should be noted that the frequency components of i_d currents and high frequency components of the load currents will also pass through C_d. Since the input current of the upconverter has a certain shape, the upconverter functions in a current control mode, due to conversion from DC to AC.

The feedback control is shown in the block diagram (Figure 5), where i^*_L - the predetermined or desired current value.

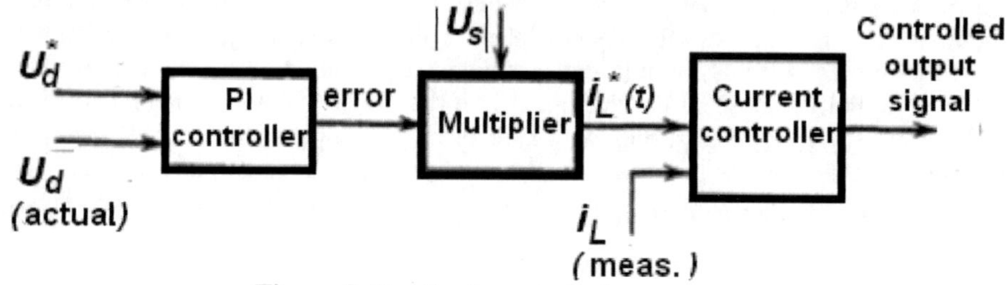

Figure 5. Feedback control of upconverter

Here, i^*_L has the same shape as $|U_S|$ does. The amplitude of i^*_L should be at certain level so as to maintain the output voltage at a predetermined or reference level U^*_d, despite the changes at load and the mains voltage deviation from the nominal value. Waveform of the i^*_L-signal was obtained by measuring $|U_S|$ using resistive voltage divider and having multiplied it with the increased error between initial value U^*_d and actually measured value U_d. Actual current i_L is usually determined by measuring the voltage across a small resistor installed in the direction opposite to i_L. The condition of the switch in the upconverter is controlled by the process of comparing actual current i_L and i^*_L.

Instantaneous value of i^*_L and i_L is determined by various methods of the running control of the mode from the upconverter. These four methods are presented below, where f_s - switching frequency and I_{rip} - peak pulsation of i_L within one period of time of the switching frequency. Determination of the constant frequency control is described in more detail.

1. Controlling the constant frequency. In this case, switching frequency f_s remains constant. When i_L reaches i^*_L, the switch of the upconverter is disconnected. The switch is on for a certain time at fixed frequency f_s, which leads to value i_L, as it is shown in Figure 6, a. The compensation of acceleration tilt is to be performed; otherwise i_L will be uneven in the ratio of switch - debt in excess of 0.5 of normal I_{rip} as shown in Figure 6.

2. The constant of the access to a range of adjustment. Here, current i_L is controlled in a such way that peak pulsation I_{rip} remains constant. With pre-selected values of I_{rip}, value i_L is forced to be within tolerances $i_L^* + \frac{1}{2}I_{rip}$ and $i_L^* - \frac{1}{2}I_{rip}$, and should be adjusted by the position of the switch.

3. Switching to a manual control. In this case, peak pulsation I_{rip} increases in proportion to instantaneous value $|U_S|$.

Otherwise, this approach is similar to the control of the constant of the tolerance range.

4. Controlling intermittent current. In this scheme, the switch is off when i_L reaches doubled value i^*_L.

The key is kept locked until i_L reaches zero. At this moment, the switch is on again. This can be considered as a special case of the variable of the control range allowance.

During the period of the switching frequency it is assumed that the output voltage is constant and input voltage U_d of the upconverter is considered to be constant at this moment of time; I_{rip} - peak pulsation of the current during one period of time of switching frequency.

The following formulas can be derived from Figure 4, during interval t_{on} and interval of switch locking t_{off}.

$$t_{on} = \frac{L_d I_{rip}}{|U_S|} \tag{10}$$

$$t_{off} = \frac{L_d I_{rip}}{U_d - |U_S|} \tag{11}$$

Switching frequency f_s:

$$f_S = \frac{1}{t_{on} + t_{off}} = \frac{(U_d - |U_S|)|U_S|}{L_d I_{rip} U_d} \tag{12}$$

Constant frequency f_s, controlled by the circuit, is constant in this formula and, therefore:

$$I_{rip} = \frac{(U_d - |U_S|)|U_S|}{f_S L_d U_d} \tag{13}$$

Figure 6,*b* shows a graph of normal I_{rip} as a function of $|U_S|/U_d$, taking into account that ratio $|U_S|/U_d$ in the upconverter should be ≤ 1.

The maximum ripple current is set as:

$$I_{rip,\max} = \frac{U_d}{4 f_S L_d} \tag{14}$$

where $|U_S| = \frac{1}{2} V_d$.

In the active loop of shaping current with the use of the DC upconverter the following additional comments have been made:

- Output voltage U_d across capacitor C_d contains 120 Hz of pulsations, which is 2 times higher than the frequency of the grid. A feedback control scheme is used to control U_d at a certain level, and can not compensate this pulsating voltage without distorting the current flow.
- If switching of pulsation frequency i_L is maintained with a small amplitude, it is possible to use a steel core of the inductor coil, which will be smaller in size because of its higher density of saturation flux as compared with high-frequency ferrite materials.
- Higher switching frequency allows for lower value L_d and increase the ease of filtering of high-frequency pulsation. However, the switching frequency is selected as a compromise between the above-mentioned advantages and increasing switching losses.
- U_d voltage exceeding the nominal voltage by more than 10% after reaching the peak of input AC voltage $\hat{U}_S$ will cause a decrease in efficiency.
- To limit the starting current one can use a current-limiting resistor in series with L_d. After the initial transient process, a resistor is bypassed by the contactor or a thyristor connected in parallel to the current-limiting resistor.
- The upconverter topology is well suited for the current input signal, since when the switch is off, the input current directly (via the diode) feeds the output stage. In the current control of the constant frequency, the switch of duty cycle *d* is taken as an example and is shown in Figure 6, *c*, as a function of ωt. It is known that a step-up transformer has input voltage $|U_S|$ and output voltage U_d:

$$(|U_S|/U_d) = 1 - d,$$

therefore

$$d = 1 - \frac{|U_S|}{U_d}.$$

From Figure 6, *c* one can see that *d* - the lowest peak of i^*_L. Thus, the maximum value is passing through the switch only during a small part of the switching period.

- A small filter capacitor must be installed at the output of the diode of the bridge rectifier to prevent pulsation i_L. The filter of electromagnetic interferences (EMI) at the input is still required, as well as in the conventional circuit without active current signal.

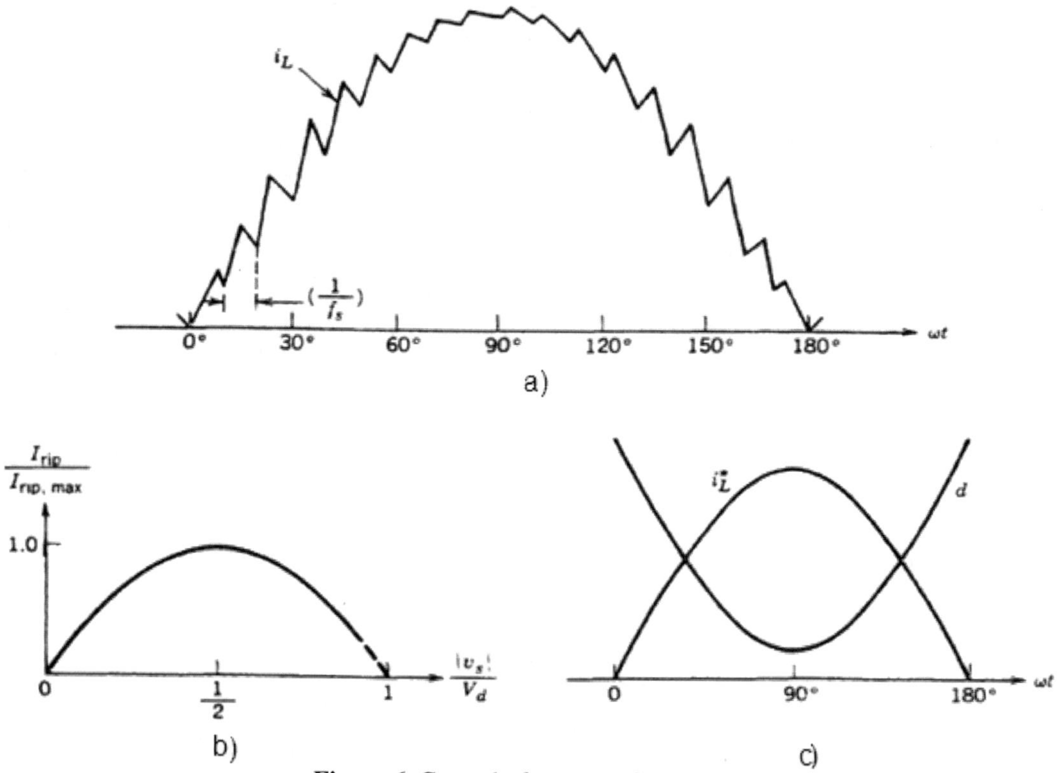

Figure 6. Control of constant frequency
a) "saw-tooth" form of i_L at constant switching frequency f_s; b) steady-state value of peak pulsation I_{rip}; c) dependence of *d* on *ωt*

3. Conclusion

Thus, the described electrical device can convert AC to DC, and it is realised using semiconductor devices - transistors and diodes - in a bridge circuit. The technical result of the application of this device is to reduce the weight - size parameters (compared to the closest analogues) to connect the controlled rectifier of the voltage to AC power supply, including mass and size of the capacitor bank of the current-limiting circuits, and improving the dynamic parameters of the stage of the uncontrolled charge of the output capacitor of the voltage controlled rectifier.

References

[1] Fediuk R 2014 *Proceedings of 2014 International Conference on Mechanical Engineering, Automation and Control Systems, MEACS* 6986928

[2] Andrianova L P and Baiburin E R 2006 *Proceedings of the International scientific and technical conference* **1** 302-305

[3] Chazov J O, Perminov I A and Kochetkov N P 2013 *Agricultural Research journal* **7** 56-59

[4] Osipov D S, Kiselev B J, Plankov A A and Janiszewskaya A G 2014 *Modern problems of science and education* **6** 274

[5] ChenVey Hsien and Chen Ho 2009 *Electricity* **1** 54-57
[6] B Xu and X Cai 2014 *Power System Technology* **38** 107-112
[7] Komlev A V and Kuvshinov G E 2013 *Electro. Electrical power generation, electrical engineering* **3** 13-18
[8] Kostinsky S S, Mikhailov V V, Alekseeva D S and Lopansky V V 2014 *Elektromehanika* **3** 51-54
[9] Li T, Miao X-P and Li X-B 2014 *Power System Protection and Control* **9** 86-90
[10] Shkrabets F P and Ostapchuk A V 2004 *Mountain information-analytical bulletin* **6** 320-322
[11] Kuvshinov G E, Rybalev A N and Skobelev D P 2015 *RU Patent* 537973C1
[12] Cai B 2014 *Applied Mechanics and Materials* **532** 595-602
[13] Kuvshinov G E, Sebto Yu G, Kraskovsky M V, Kravtsov P A and Kravtsova A V 2015 *RU Patent* 2521613C1
[14] Mohan N, Undeland T M and Robbins W P 2005 *Power electronics*
[15] Wang J, Hu X and Chen S 2014 *Transactions of China Electrotechnical Society* **29** 239-245

Diagnostics of the management system of the machine-building plant based on strategic management tools

T M Gerashhenkova and O A Shvecova
Bryansk State Technical University, 7, 50 years of October st., Bryansk, 241035, Russia

E-mail: gerash-tatyana@yandex.ru

Abstract. The research of the paper approaches to the assessment of the capabilities of enterprise management system engineering based on profile building environment. The the long term aassessment of management effectiveness is proposed for the use of the normative structures of performance.

1. Introduction

In the structure of the industrial production of Bryansk Region the engineering production is in the leading position producing the third part of all output products in the region.

At present, when engineering enterprises operate in the conditions of quick-change multifaceted external environment, increasing volume of production is possible only with ordered strategic management [1, 2].

2. Results and discussion

A strategic approach to the diagnostics of the management system of the machine-building plant first of all intends to use tools of management system for creating and providing rationalization for directions of the plant long-term development. A starting point of such diagnostics from the perspective of strategic management is a process of problems definitions and their identification, first of all, in external environment. The main goal of external environment diagnostics is to reveal potential opportunities and neutralize threats for the plant that external environment provides.

Therefore, the first base of the analysis of the external environment is to highlight its 'slices' (both in macro-environment and in the structure of business environment of the enterprise, including branch and competitive slices) that are important for successful functionality and development of mechanical facilities.

The next stage is estimation of the highlighted factors of the external environment using such tool of strategic management as SWOT-analysis. According to the procedure of the SWOT-analysis the list of threats and opportunities of the external environment as well as strengths and weaknesses of the analyzed object is determined. After making a definite list of the SWOT factors the relations between these factors should be established, for which purpose it is advisable to use a procedure of a correlation SWOT-analysis (Table 1) [3].

Table 1. Procedure of a correlation SWOT-analysis

Internal environment	OPPORTUNITIES	THREATS
STRENGTHS	Strength and opportunities	Strength and threats
WEAKNESSES	Weaknesses and opportunities	Weaknesses and threats

For the successful use of the SWOT-analysis methodology during the process of the management system diagnostics of the machine-building plant, a procedure of estimation of the highlighted external environment factors from the perspective of the importance of their consideration in the strategy of the plant behavior presents a milestone. For realization of this stage, one should use the matrix method of the opportunities positioning (Table 2)

Table 2. Matrix of opportunities positioning of the external environment

Probability of making use of possibilities	Effect of opportunities		
	Strong	Medium	Weak
High	Box 'HS'	Box 'HM'	Box 'HW'
Medium	Box 'MS'	Box 'MM'	Box 'MW'
Low	Box 'LS'	Box 'LM'	Box 'LW'

The plant administration should pay attention to the opportunities in boxes 'HS', 'HM', 'MS' of the matrix during creating (correcting) the strategy of the machine-building plant development – the highlighted factors of external environment should be used by all means.

The matrix method is needed for positioning of the treats. (Table 3).

Table 3. Matrix of threats positioning of the external environment

Probability of threats realization	Influence degree of threats			
	Catastrophic	Critical	Medium	Weak
High	Box 'HCa'	Box 'HCr'	Box 'HM'	Box 'HW'
Medium	Box 'MCa'	Box 'MCr'	Box 'MM'	Box 'MW'
Low	Box 'LCa'	Box 'LCr'	Box 'LM'	Box 'LW'

It should be mentioned that in aggregate of all identified factors during the first stage of the diagnostics procedure of the external environment factors, it is advisable to identify the key factors that have the dominant influence on functioning and development of the plant. For realization of this stage, it is recommended to use the expert method of paired comparisons (Table 4). [4]

Table 4. Example of the expert analysis of the most important factors of the external environment, that influence the activity of the machine-building plant

	1	2	3	4	5	$\sum k$	$\sum k^*$	r
1	-	0	0	2	2	4	0,2	2
2	2	-	2	0	2	6	0,3	1
3	2	0	-	1	0	3	0,15	4
4	0	2	1	-	2	5	0,25	3
5	0	0	2	0	-	2	0,1	5
Total	X	X	X	X	X	20	1	X

The ranking procedure of the external environment factors allows creating a profile of the external environment of themachine-building plant that is listed in Table 5. (initial data are listed in Table 4).

Table 5. Profile of the external environment of the machine-building plant

Key factors	Expert evaluation				
	1	2	3	4	5
1st factor		•			
2st factor	•				
3st factor				•	
4st factor			•		
5st factor					•

One of the most important moments of diagnostics of the management system is a research of the conformity of the machine-building plant's strategy with the factors of the external environment for which purpose the quality of the strategy under formation is analyzed first of all.

The research of the quality of the formed strategies of development implies the presence of economical aspect as the most informative one, within the framework of which the advisability and effectiveness of strategic measures are considered as an integrated characteristic of sustainable development of the machine-building plant in the long run [5].

For strategic aspects of the activity that implies a prior the excessive duration of the process itself the trackingof the definite tendencies in dynamics is of great importance. So the analysis of the external environment of the machine building plant should be madeboth from the perspective of current situation and dynamically, whichimplies evaluation of tendencies of changing environment factors of plant functionality and research of the influence of these tendencies on the management system (Figure 1).

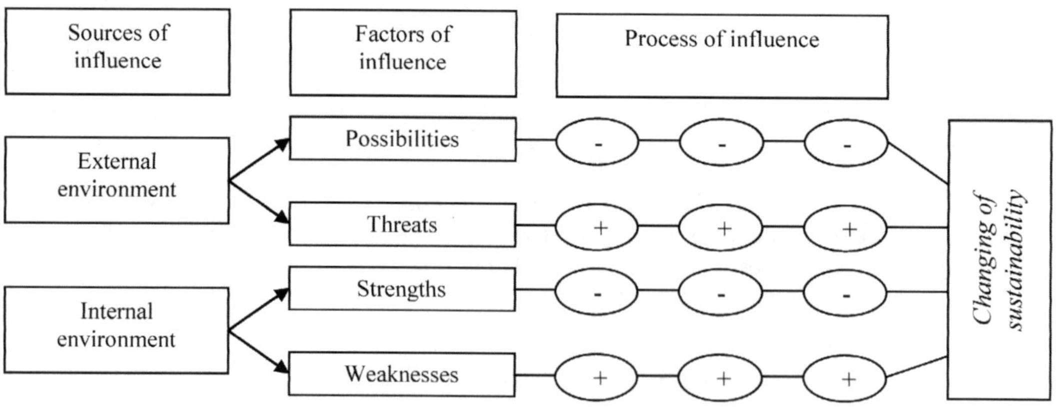

Figure 1. Mechanism of influence of threats and opportunities of the business environment on sustainability of the machine building plant.

(-) - external environment phenomenon, facilitatinga decrease of the sustainability level due to decreasing the influence of the corresponding factor.
(+) - external environment phenomenon, facilitatinga decrease of the sustainability level due to increasinginfluence of the corresponding factor

The characteristic of strengths and weaknesses of the plant, possibilities and threats of the external environment must be evaluated in a convenient form for the plant administration [6]. The most demanded quantitative indices are thosethat allow determining the following data during regular monitoring:

- instantaneous indices;
- rate of increase of indices;
- rate of increase of some indices relatively others.

These are the quantitative indices that allow forming the general concept of performance evaluation of the machine-building plant activity.Consequently, an estimation of the development strategy and an estimation of strategy accordance to the external environment factors can be made. Using a multilevel approach for describing the object of the research, the growth of competitiveness can be displayed as a consistent change of indices that characterizes the internal environment of the plant (Figure 2).

The analysis of effectiveness of the plant development strategy must be done from the perspective of complexity and agility of processes taking place during its realization. Due to this fact a realization of two stages is necessary: measurement and estimation of results characterizing the currenteconomical processes obtained using methods of the economic analysis; calculation of indices, providing identification of rates and directivity of their changing and also the ratio of growth of some indices in comparison with others. Calculated indices also provide information about total instantaneous indices.

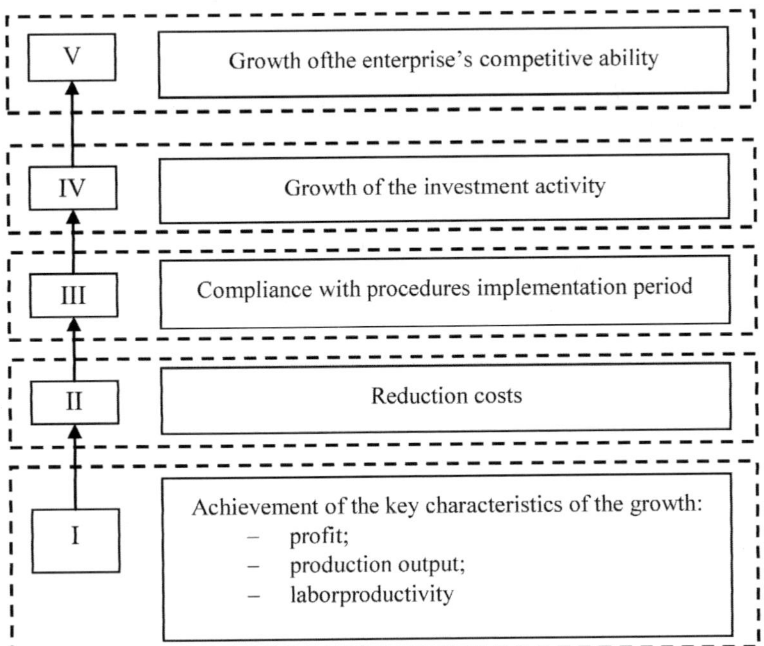

Figure 2. A hierarchic structure of the indices showing the effectiveness of the development strategy and itsconformity with the factors of the external environment.

The research on effectiveness of the development strategy implementation of the machine-building plant using the method of creating performance standards is advisable to present in the following sequence:

The number of indices in a forming sequenceis estimated on the basis of aims of the estimation, availability of information and precision of necessary results. For estimation of the development strategy of the plant on the whole, a group of indices, which is system-wide,should be considered. The estimation of the strategy targeted at intensification of the innovative and investment activity implies using a specific detailed structure of indices. It is conditioned by the fact that in the situation of innovation introduction the value of the balance will significantly grow dynamically, which connected with activation of the investment activity, an implementation of new facilities and so on. However, one should not allow this tendency to effect negatively the characteristic of efficiency of the strategy because regarding the growth of investment it is possible to estimate the growth of the investment activity.

Let us present an array of normative indices that are comparable to a previously established priority and show the presence of an innovative-investment component in processes of achievement of desired goals by machine-building plants (formula) [7]:

$$GR_w < GR_{mc} < GR_Q < GR_{inv} < GR_{pr}$$

This consecution is based on the fact that the aim of any commercial enterprise is profit earning, so it is logical to conclude that the desired function during evaluation of effectiveness of the development strategy is an increment in profit that outruns any other indices. The normative structure of indices should include the next increment of indices of the volume:

- profit growth rate (GR_{pr}) – desired function;
- investment growth rate in the capital asset (GR_{inv}) as a condition of activation and provision of the first stage of realization of innovative-investment development strategy;
- growth rate of production volume (GR_Q) (their possible lagging from the investment growth rate, that is conditioned by the specificity of the strategy);
- manufacturing charges growth rate (GR_{mc});
- growth rate of the average listed number of manual workers (GR_w) is the last index of the normative array that shows positive changes in the performance level when the volume of the labor production growth advances.

As part of the accepted normative structure during the analysis, specification of the quantity of the chosen indices can change within the limits that allow saving the consecution of the normative array. In order to establish that fact of development the normative array of indices must be non-decreasing.

2. The quality of conformance of the true structure of indices belonging to the order scale, formed normative quantitative tightness of the grade correlation shows the effectiveness of strategy realization in the method that we use. To determine the quantitative tightness of the grade correlation, Spearman correlation analysis is used in mathematical statistics.

3. The higher the measure of concordance of true and normative structures of economic indices, the higher is the strategy of effectiveness and harmonious plant development. In case of full conformity of the actual indices array with normative one or the tendency of their dynamic convergence, the involved effectiveness of the management influence and controllability of development factors leads the business activity to achievement of the strategic goal.

In case of non-conformity of these structures, one can make suggestions about incorrect methods of goal achievement and therefore about a decrease of strategic management quality. The worstvariant is a detected reciprocal sequence of indices of the true structure compared with the normative structure.

- Apart from the mentioned directions of application of the presented approach,it is possible to use it for:
- revealing of a “key link” in industrial and economic processes of the plant accenting the control action on it;
- expansion of the sphere of application and usage of new methods of the plant’s activity analysis.
- business forecasting and risk level of occurrence of negative tendencies in case of a negative growth of certain industrial and economic indices at an early stage;
- provision of systematicalness during the work of production andfunctional units as well as the goal orientation and activity for realization of long-term development plans.

3. Conclusion

Specified approaches to the analysis of the business environment of the plant have revealed tendencies and presented methods of evaluation of the effectiveness of development strategy realization. These allow making a conclusion about the implemented policy, a system of cooperation with business partners, competitors, financial structures, infrastructural complex and regulatory bodies. A positive dynamics of the estimated Figure allows relying on improvement of the state of the economic entity and machine-building industrial together.

References

[1] Zotov F P and Muzipov R M 2012 *Economy of Region* **4(32)** 191-196
[2] Gershman M 2013 *Foresight-Russia* **7(1)** 28-43
[3] Belkin V N, Belkina N A and Vladykina L B 2015 *Economy of Region* **1(41)** 144-155
[4] Zaichenko S, Kuznetsova T and Roud V 2014 *Foresight-Russia*, vol. **8(1)** 6-23
[5] Shvecova O A 2011 *Kursk: Publishing office Kursk. State. Agriculture* 272
[6] Goedertier S, Vanthienen J and Caron F 2015 *Enterprise Information Systems* **9(2)** 161-185
[7] Gerashhenkova T M 2014 *Scholarly notes of Pertozavodsk State University* **1(138)** 94-99

Geometrical product specifications: A structure of linear dimensions tolerances

V I Glukhov, O Y Zlatkina and I A Ivleva
Omsk State Technical University, 11, Mira ave., Omsk, 644050, Russia

E-mail: mips@omgtu.ru

Abstract. The paper reveals the relationship of linear dimensions tolerances with other geometrical specifications: coordinates, deviations of position and deviations of surface configuration of component geometric features. The dependence of the precision structure of the linear dimensions on the functional purpose of features. The basic concept of the paper is informativeness of the feature, which represents the total number of linear and angular degrees of freedom, constrained by the feature in the component at operation. The very informativeness of the feature determines the two-dimensional linear-angular composition of the structure of the linear size, its tolerance and standard grade. This paper discusses only the linear dimensions of the features involved in couplings and forming fits.

1. Introduction

Two indicators characterize the precision of the component feature: a dimensional precision and a geometric one. The dimensional precision is a precision of dimensions of the feature, whereas the geometric precision is a total precision of deviations of position and surface configuration of the feature or geometric deviations. Dimension tolerance is an overall characteristic of dimensional and geometric precision.

Geometric precision of the component feature is determined by the deviations of the surface configuration and the deviations of position, which are included into the tolerance structure of the feature dimension. In Russian national standardization, the basic norms of interchangeability of components by geometric precision are divided into three levels A, B, and C. Level A of normal geometric precision means that 60 % of the tolerance zone of the dimension is occupied by the total tolerance zone of geometric deviations in the diametric expression. Level B of the extended geometric precision occupies 40 % and level C of high geometric precision occupies 25 % of the tolerance of the feature dimension. Since the geometric precision determines the difference of dimensions of each feature, two dimensions of one feature different from each other by 25…60 % of the dimension tolerance is an objective reality, which must be considered at normalization of precision [1]. Consequently, the share of dimensional accuracy of the feature accounts for only 40…75 % of the dimension tolerance [2].

2. Setting goals

In order to optimize the normalization of precision during designing of technical products based on registration of informativeness of geometric features of components the authors of the paper have set and solved the following tasks:

IOP Conf. Series: Materials Science and Engineering **124** (2016) 012011 doi:10.1088/1757-899X/124/1/012011

- the informativeness of prismatic and cylindrical features of different functional purpose has been determined;
- a precision structure of dimensions and tolerances for two-, four-, six-sided orthogonal prismatic features has been discovered;
- a precision structure of sizes and tolerances of diameters of cylindrical features of different informativeness has been revealed.

3. Features informativeness

A geometric feature of a component involved in coupling and fitting, serves as a constructional datum depriving the attached component of several degrees of freedom (from one to five). Only three of them prevent from linear movement along three mutually perpendicular axes or prevent from three angular rotations around these axes [1]. The number of degrees of freedom constrained by the feature in datum settings can serve as a quantitative characteristic of its informativeness. If a component feature functions as an actuating feature then, in case of contact loss with a coupling component, its informativeness drops to zero (0). Prismatic features with pairwise parallel flat planes can have informativeness equal to 3, 2, 1, 0, while cylindrical ones – 4, 2 or 0. This difference in feature informativeness impacts the structure of dimension tolerance, and consequently, the feature precision, tolerance numeric value and precision grade.

4. The sizes and tolerances structure for prismatic features

In prismatic features the dimensions form two nominally parallel flat planes, one of which represents the datum depriving the component of 3, 2 or 1 degrees of freedom and determining informativeness of the prismatic feature on the whole, and the second one is the actuator surface with zero informativeness. This prismatic feature peculiarity specifies the structure of the tolerance zone, which consists of two pieces: T′ depends on datum deviations and T″ – on the executive surface deviations (Figure 1-4).

If datum informativeness equals 3, the height of prismatic feature is *h3* (for the shaft) or *H3* (for the hole) (Fig.1). The structure of the first piece of tolerance includes only datum flatness deviation EFE1, while the structure of the second one encompasses the deviation of the centre coordinate of actuator surface Ezc (Ez′c), the deviation from parallelism *EPA* of the executive surface relatively the assembly base and the deviation from flatness EFE_2 of the executive surface:

$$\begin{aligned} &\text{for the shaft } Th3 = T'h3 + T''h3, && \text{for the hole } TH3 = T'H3 + T''H3, \\ &T'h3 = EFE1; && T'H3 = EFE1, \\ &T''h3 = Ezc + EPA + EFE2; && T''H3 = Ez'c + EPA + EFE2 \end{aligned} \quad (1)$$

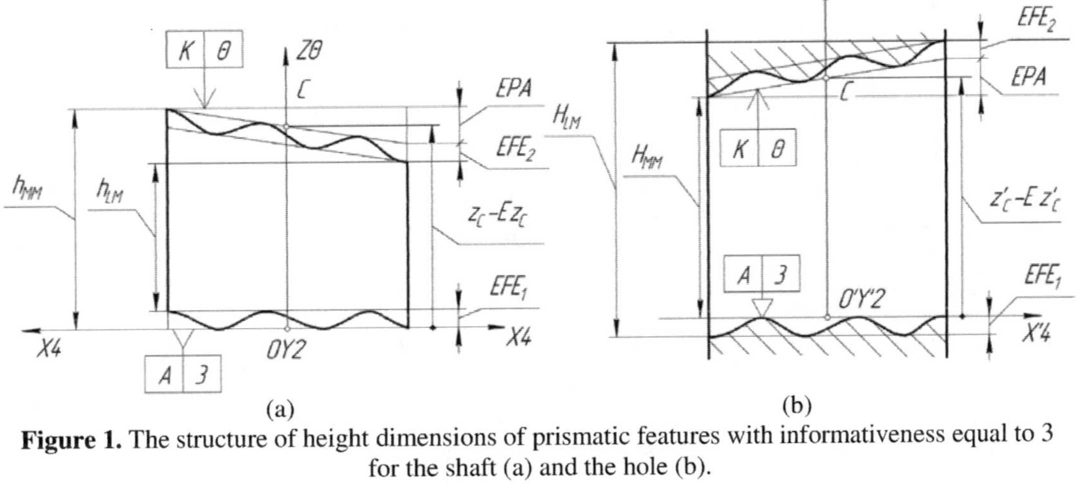

(a) (b)

Figure 1. The structure of height dimensions of prismatic features with informativeness equal to 3 for the shaft (a) and the hole (b).

In the course of base informativeness reduction, the structure of the tolerance of prismatic feature dimension becomes more complex. Width w2(W2) of the feature results from the base with informativeness 2 (Figure 2). The structure of the first piece of tolerance T′w2 (T′W2), along with the deviation from flatness of base EFE1, includes the deviation from perpendicularity EPE1 of base B2 with respect to base A3 with informativeness 3. In the structure of second piece T′w2 (T′W2), as compared to the structure of the zone of height tolerance T″h3(T″H3) the deviations from parallelism EPA is replaced by the deviation from the perpendicularity of executive surface EPE2 with respect to base A3 and the deviation of coordinate Ezc (Ez′c) is replaced by deviation Exc (Ex′c):

for the shaft Tw2= T′w2+ T″w2, for the hole TW2=T′ W2+T″ W2,
T′w2= EPE1+ EFE1; T′ W2= EPE2+EFE1, (2)
T″w2= Exc+ EPE2+ EFE2; T″ W2= Ex′c. + EPE2+ EFE2

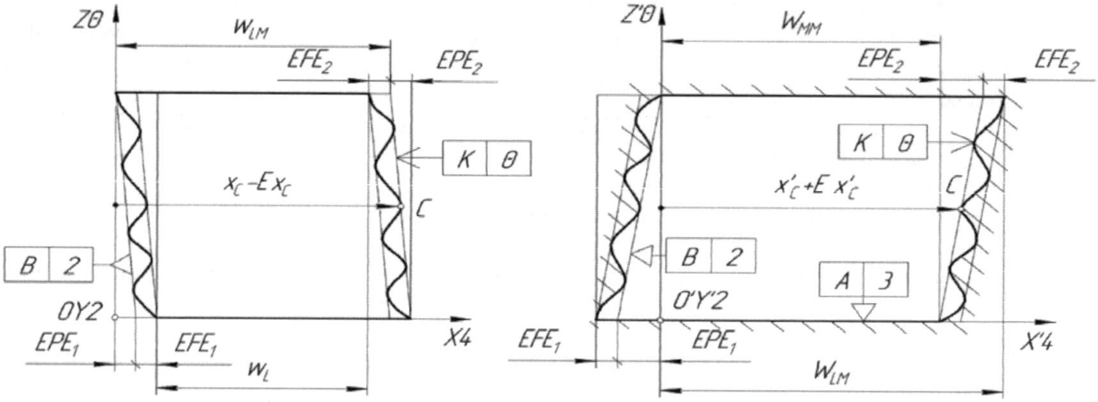

Figure 2. The structure of width dimensions of prismatic features with informativeness 2 for the shaft (a) and for the hole (b).

The makeup of the tolerance of length dimension l1(L1) calculated from datum C1 of the prismatic feature with informativeness 1 (Figure 3) is supplemented with deviations from perpendicularity EPE3 of datum C1 and from perpendicularity EPE4 of executive surface *KΘ* relatively the datum of component B2 with informativeness 2:

for the shaft Tl1= T′ l1+ T″ l1; for the hole T L1=T′ L1+T″ L1;
T′ l1= EPE1+ EPE3+ EFE1; T′ L1= EPE1+ EPE3+EFE1; (3)
T″ l1= Exc+ EPE2+ EPE4+ EFE2; T″ L1= Ex′c. + EPE2+ EPE4+ EFE2.

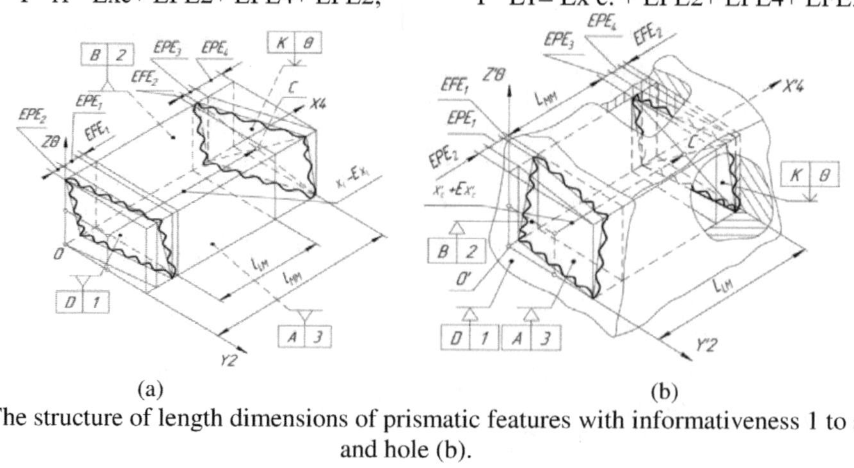

Figure 3. The structure of length dimensions of prismatic features with informativeness 1 to shaft (a) and hole (b).

Finally, in case both planes of prismatic feature function as executive surfaces the feature becomes functionally symmetrical (Figure 4). It forms an overall dimension of height g(G), the structure of which tolerance Tg (TG) include: four deviations Ezc (Ez′c) from nominal coordinate zc of centre C of the plane of feature symmetry, doubled deviation from perpendicularity EPE of the plane of the feature symmetry relatively axis Z4 of generalized coordinate system of the component, represented base A4, deviation of proper height Egc (EGc) from the nominal value in the middle cross-section of the feature and two deviations from the plane of executive surfaces EFE:

for the shaft Tg0=4Ezc +4 EPE+ Egc+2 EPA+2 EFE;

for the hole TG0=4 Ez′c+4 EPE+ EGc+2 EPA+2 EFE. (4)

It stands to reason that such complex tolerance framework does not allow a precise coupling for the executive surfaces of connecting overall dimensions of prismatic features of the components, for instance, in an automobile door both at height and in width.

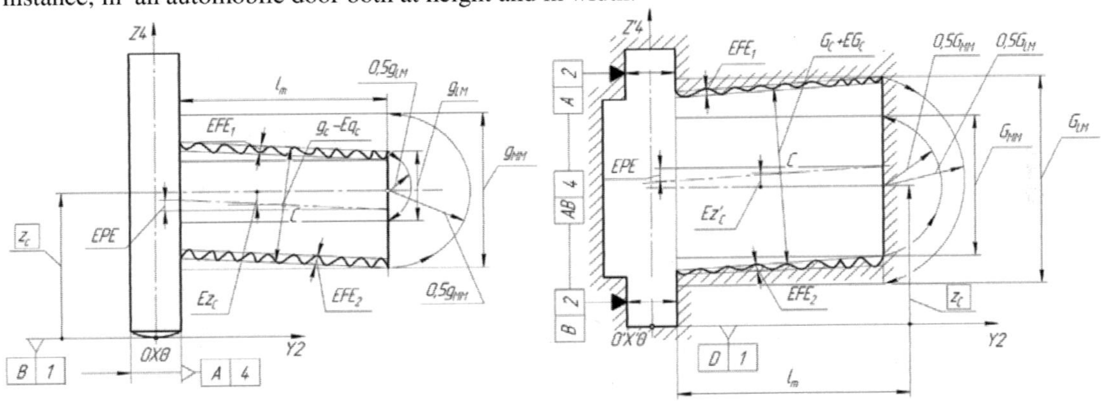

(a) (b)

Figure 4. The structure of dimensions of the length of prismatic features with informativeness *Θ* (zero) for shaft (a) and for hole (b).

5. The dimension and tolerance structure for cylindrical features

The tolerance zone with diameter size Td4 (TD4) of cylindrical element serving as a component of main or auxiliary datum with informativeness 4 (Figure 5) represents a circular area in the component material included between two cylinders. These cylinders are coaxial with adjoining cylinder ZE, the axis of which represents an axis of the generalized or auxiliary coordinate system of the component. The dimensions of two cylinders of the tolerance zone are the limits of maximum dlMM (Dlmm) and minimum dlMM (Dlmm) of the material. The thickness of the circular tolerance zone is equal to half of the tolerance for diameter 0.5Td (0.5TD). The tolerance structure of the feature of geometric deviations includes only the deviations in configuration of cylinder surface EFZ (Figure 6). They include conicity *EFC*, because of the angular distortion of the cylinder feature in the linear expression, which in essence is a deviation from parallelity of the features, that is, angular deviation of position, and deviation of longitudinal profile EFP (barrel or bow), and deviation from the straightness of axis EFL, and deviation from circularity of ovality-type EFV and faceting EFH. The dimensional constituent of the tolerance is a doubled deviation of radius Er(ER) of the cylinder in the midsection from nominal value.

Finally, at informativeness 4 of the cylinder feature we have:

for the shaft Td4=2 Er+2EFZ;

for the hole TD4=2ER+2EFZ;

where EFZ= EFC+ EFP+ EFL +EFV+ EFH. (5)

The overall diameter tolerance zone cannot be exclusively represented by form of deviations because otherwise it would be necessary to make the cylinder radius equal to the nominal one, that is, with zero deviation, which is impractical. Moreover, safe deviations of the radius dimension generally determine the numerical value of the tolerance necessary for compensation of all constituents of machining errors, namely locating, static and dynamic setting, power and temperature-induced deformations.

In the conventional pictorial symbols of tolerance zones of dimensions accepted by the Unified standard system of tolerances and fits [2] in the form of rectangles located relatively a zero line corresponding to nominal diameter d_0 (D_0) (Figure 5b), a part of the tolerance zone occupied by shape deviations should be shown in diametrical expression *TFd (TFD)* and be equal to a doubled standardized tolerance of the shape in radius expression *2TFZ*.

Diametrical tolerance zone of the shape is adjacent to minimum limit of the material as it determines the structure of the dimension of the minimum of the material. However it in essence is movable within the tolerance zone of dimension as it limits the allowable difference of the maximum and minimum dimensions of one and the same feature.

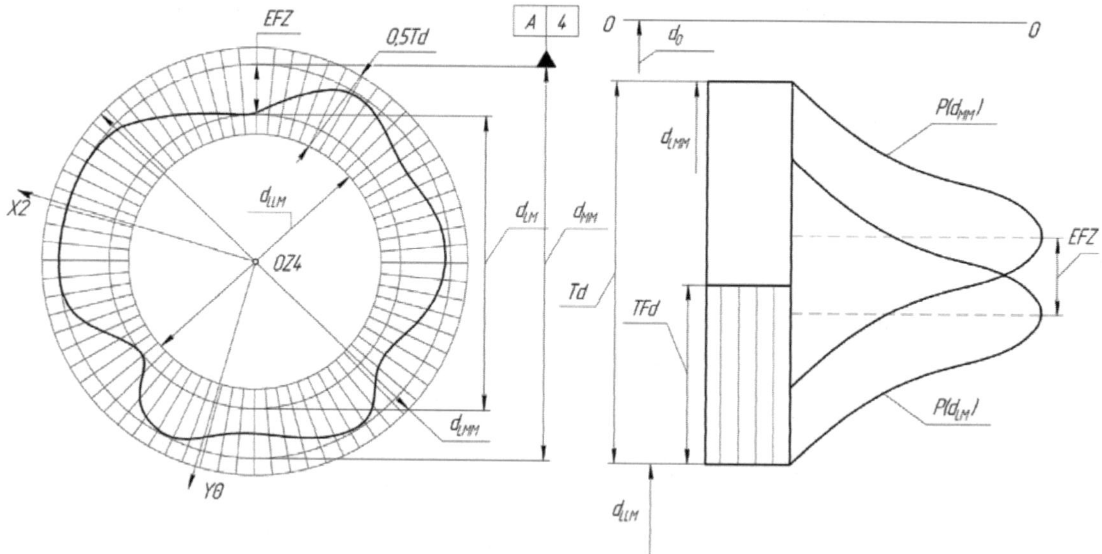

(a) (b)

Figure 5. The sizes and tolerance structure diameter of cylindrical features with informativeness 4 in cross-section (a) and size distribution laws (b).

The material dimensions of maximum *dmm* and *Dmm* (maximum is for the shaft and minimum for the hole) and minimum d_{LM} and D_{LM} (minimum is for the shaft and maximum is for the hole) [4] serve different purposes. Maximum material dimensions play a key role since they determine the nature of fittings between coupled components and materialize the axis of the coordinate system of the component by the axis of the adjacent feature cylinder. Thus, the maximum dimension deviations of the feature material relatively nominal dimension are in fact critical, and therefore they should serve as a basis for the whole system of tolerances and fits. Nevertheless, in the standardized system [2] the primary deviations of tolerance zones are the deviations that are closest to a zero line, which is not scientifically substantiated.

Maximum and minimum feature dimensions of the material comply with their own laws of distribution (Figure 5b) that are displaced relatively each other by the mean value of shape deviations EFZ taken for a component lot. The structure of shape deviation depends on coupling length lm, on which the feature serves its purpose (Figure 6).

As cylindrical element informativeness decreases from 4 to 2 (in case it serves as an assembly base with informativeness 2) (Figure 7) the structure of the tolerance zone of the diameter dimension expands only due to geometrical angular deviations AE of the axes of two coaxial bases relatively common axis Z4, which is the axis of the generalized coordinate system of the component (Figure 7a) or axis Z′4 of the auxiliary coordinate system (Figure 7b). Linearly expressed angular displacements are limited to deviations from coaxiality relatively the common axis in diametrical expression EPC. These deviations are included in the structure of tolerance zone of diameter dimension as doubled value 2 EPC, which is typical only for symmetrical features:

for the shaft Td2=2Er+2EPC+2EFZ;

for the hole TD2=2ER+2EPC+2EFZ. (6)

Despite the angular misalignment of axes, tolerance zones TD of features diameters are arranged in circular zones nominally relatively coordinate axes Z4 и Z′4 represented by two common axes of two bases. No matter to what extent the generators of the features are distorted, but if they fit in their tolerance zones along with shape deviations, the diameters of such features – maximum and minimum – will not go beyond the tolerance zones limits. Hence, the component will be fit and will serve its purpose. In case of a similar length of coupling cylinder features with informativeness 2 are less accurate than those with informativeness 4 due to more complex tolerance structure.

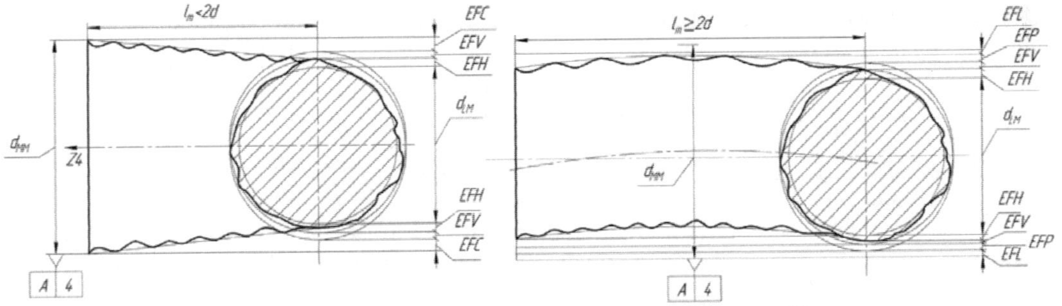

(a) (b)

Figure 6. The tolerance structure of diameter dimensions of a cylindrical feature with informativeness 4 in the longitudinal section, depending on the length of the coupling (a) and (b).

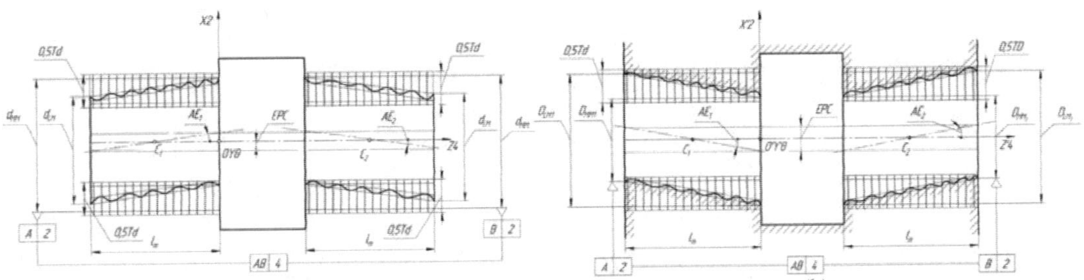

(a) (b)

Figure 7. The tolerance structure of diameter dimensions of a cylindrical feature with informativeness 2 for the shaft (a) and for the hole (b).

As the informativeness of the cylindrical feature drops to zero (which is true, when it operates in assembly in the coupling without direct contact with the coupling component (Figure 8)), tolerance structure Td (TD) additionally increases by four eccentricities EE of the axis of the executive surface with respect to the axis of the coaxial cylindrical datum with informativeness 4. However, there is no provision for standardization of eccentricity by the International Standard [3]. That is why the eccentricity of the axis of the executive surface is subjected to limitation in tandem with linearly expressed angular misalignment AE in terms of standardized deviation from coaxiality relatively the reference axis in diametrical expression EPC. The part of tolerance zone of diameter, which will be occupied by eccentricity and misalignment, is equal to doubled deviation form coaxiality in diametrical expression 2EPC:

$$\text{for the shaft } Td\Theta=2Er+2EPC+2EFZ;$$
$$\text{for the hole } 2D\Theta=2ER+2EPC+2EFZ. \qquad (7)$$

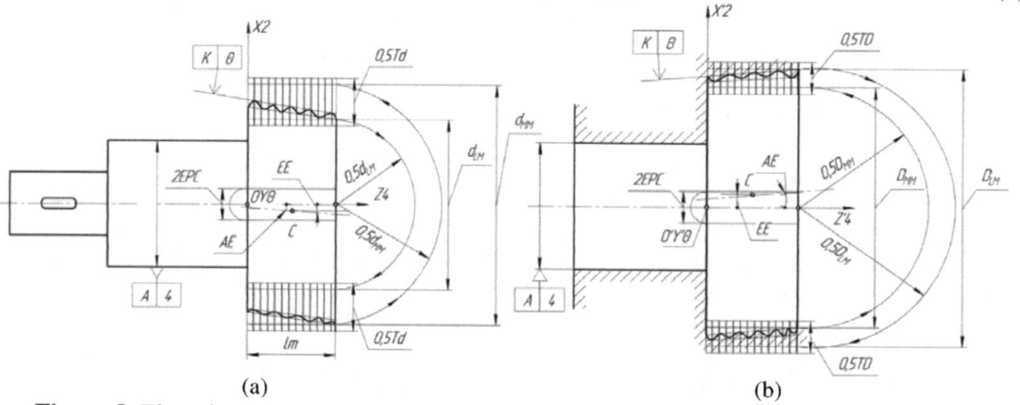

Figure 8. The tolerance structure of the diameter dimensions of the cylindrical feature with informativeness Θ for the shaft (a) and for the hole (b).

$$\text{for the shaft } Td0=2Er+2EPC+2EFZ;$$
$$\text{for the hole } 2D0=2ER+2EPC+2EFZ. \qquad (7)$$

Taking into consideration that such significant tolerance expansion occurs both in the encompassed and in the encompassing components, the precision of executive surfaces fitting cannot be high. If a cylindrical executive surface of the hole with informativity Θ (zero) is placed in the generalized coordinate system, the component represented by a set of three flat datum (Figure 9) the structure of tolerance zone TD0 of its complex diameter will include the doubled tolerances of position deviation in diametric expression 2TPO of executive surface axis relatively three flat data. These tolerances encompass both sides of tolerance zone TDc of proper diameter Dc, of the executive surfaceand thereby constituting the tolerance zone of complex diameter TD0:

$$TD0 = TDc + 2TP0. \qquad (8)$$

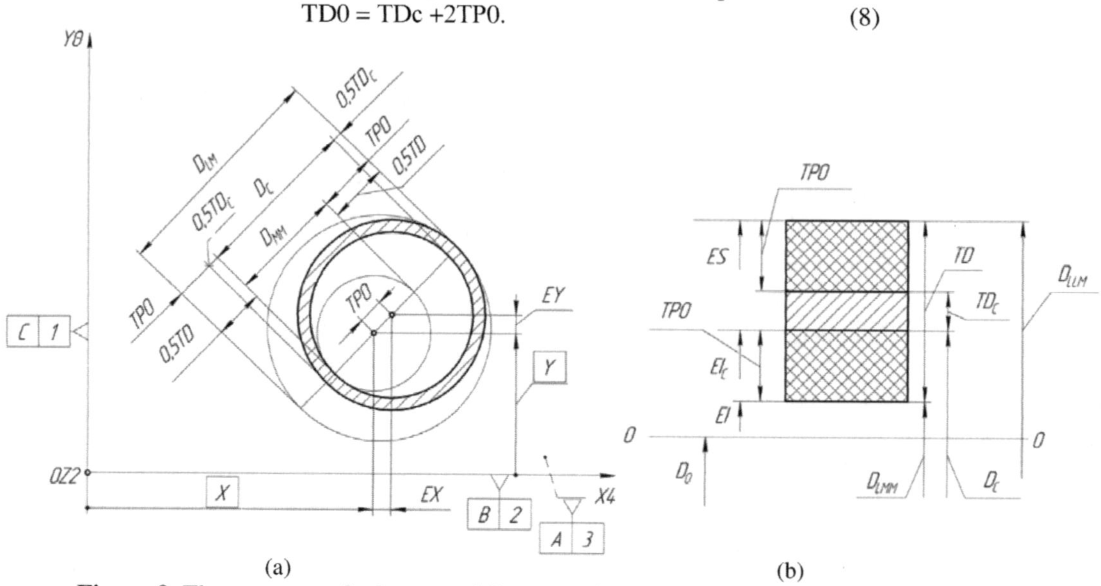

Figure 9. The structure of tolerance of diameter dimensions of the cylindrical hole with informativeness Θ for the model (a) and for the tolerance zone (b).

6. Discussion of scientific results

The following scientific results are submitted for a discussion:

The precision of linear dimensions is characterized by tolerance, which consists of two parts - dimensional precision and geometric precision. Dimensional precision determines the precision of the maximum material dimension or envelope size. Dimensional precision is 40…70 % of the dimensional tolerance. Geometric precision is defined by deviations from the shape and location of feature surfaces. Geometric accuracy amounts to 25…60 % of the dimensional tolerance.

The precision structure of the linear dimensions of the features includes two kinds of components: linear and angular, that is, it is two-dimensional. The composition of constituents in the structure of a linear dimension depends on the informativeness of the feature.

A valid feature dimension is the maximum material dimension, serving for defining the nature of the fitting in coupling. Therefore, the maximum material dimension should determine the position of the underlying deviation in the system of tolerances on the linear dimensions.

The geometric deviations of a single feature are systematic functional distortions that create the variability of feature dimension and form two boundary values: maximum material dimension and minimum material dimension.

The maximum and minimum material dimensions in the kit of parts have different distribution laws, shifted in relation to each other on the average value of the total geometrical deviations in the diametric expression.

References

[1] Glukhov V I 2014 *International Scientific and technical Conference on in Dynamics of Systems, Mechanisms and Machines* 1–9

[2] ISO 286-1-2010 Geometrical product specifications (GPS) – ISO code system for tolerances on linear sizes – Part 1: Basis of tolerances, deviations and fits

[3] ISO 1101-2011 Geometrical product specifications (GPS) – Geometrical tolerancing – Tolerances of form, orientation, location and run-out

[4] ISO 8015-2011 Geometrical product specifications (GPS) –Fundamentals – Concepts, principles and rules

Development of Management Methodology for Engineering Production Quality

O Gorlenko, V Miroshnikov and N Borbatc
Bryansk State Technical University, 7, 50 years of October st., Bryansk, 241035, Russia

E-mail: goa-bgtu@mail.ru

Abstract. The authors of the paper propose four directions of the methodology developing the quality management of engineering products that implement the requirements of new international standard ISO 9001:2015: the analysis of arrangement context taking into account stakeholders, the use of risk management, management of in-house knowledge, assessment of the enterprise activity according to the criteria of effectiveness

1. Introduction

The quality of engineering products largely determines the effectiveness of a country industry because engineering creates a technological basis of all industries [1]. A successful management of a machine-building enterprise requires that it is carried out systematically and in an apparent way. The best world experience shows that success in business can be achieved only through implementation and maintenance of the quality management system (QMS) that confirms the requirements of international standard ISO 9001. Up to date, there have been published four versions of this standard: in 1987, 1994, 2000 and in 2008.

In September, 2015, a new version of standard ISO 9001: 2015 was enforced, which includes new conceptual requirements, compared with the previous version: the analysis of the organization context, understanding the needs and expectations of stakeholders, risk management, knowledge management, performance assessment, etc. The implementation of these requirements at a machine-building enterprise faces methodological difficulties [2]. Therefore, the authors of this article offer the directions for methodology development that provide implementation of the requirements of quality management standard ISO 9001: 2015.

2. Analysis of the Organization Context Taking into Account Stakeholders

ISO 9001 : 2015 includes a section ' Organisation Context' which requires from the Organisation to understand the needs and expectations of stakeholders in its functioning. For this purpose the Organisation should identify the stakeholders relevant to the QMS, define their requirements, monitor and analyse the information to meet their requirements.

The quotient of satisfaction of all stakeholders is becoming a global criterion of perfect (optimal) activities of any organisation in economically developed countries. This quotient should be measured (assessed) quantitatively. Measurement techniques of this indicator are currently still underdeveloped. In order to solve this problem, it is proposed to use a technique [3], in which the assessment of the satisfaction degree of all stakeholders in an organisation is done using avcomplex criterion, representing a weighted average amount of local criteria of the satisfaction degree of stakeholders:

consumers, employees, owners, partners, suppliers, society. There are constructed hierarchy schemes of local criteria interconnection with weights of importance. The assessment of local criteria at a lower level in the hierarchy is made by experts using fuzzy sets.

After the initial assessment of stakeholder satisfaction it is necessary to solve the following task of QMS optimization [4]: 'QMS will be optimal if there is a set of regulatory requirements to key processes of the QMS (management responsibility (leadership), resource management, product life-cycle processes, measurements, analyses, improvements), which gives harmonic integrity of degrees (indexes) of stakeholder satisfaction, that presents the balance of values of local stakeholder satisfaction criteria among themselves, taking into account their subordination and respective proportions, when none of the requirements for the processes cannot be changed (increased or decreased) without violating the integrity of an optimal composition". A minimax algorithm is used as a search method for effective solutions (the compromise).

3. Application of Risk Management

A new requirement, which is found in many sections of ISO 9001:2015 is the need for application of risk management in QMS (the concept of risk-oriented thinking), which replaces the warning action. To meet the requirements of this international standard, the Organisation should plan and implement actions in response to risks and opportunities. Responding to risks and opportunities, creates the basis for increasing QMS effectiveness, achieving better results and preventing negative effects.

The risk management process consists of the following steps (ISO 31000:2009): 1. Information exchange and consulting with internal and external stakeholders. 2. Definition of the situation. 3. Risk identification. 4. Risk analysis. 5. Risk evaluation. 6. Impact on risk. 7. Monitoring. ISO/IEC 31010:2009 describes more than 30 methods for risk assessment. According to the authors of this article a well-known FMEA-Potential Failure Mode and Effects Analysis is the most suitable for risk analysis of quality management processes in engineering, because it allows identifying and evaluating QMS processes that may not perform the intended functions and do not meet the established requirements. However, FMEA has a substantial drawback. The matter is that this risk analysis is to be carried out by a group of experts (FMEA team) that identify risks, assess relevance score S, mark emergence O and discovery point (D) of each risk. Then they calculate the priority number of risks (PNR) by multiplying the points. However, as we know in Metrology, points are only fixed points on the rank scale, so they should not be used for any mathematical operations. As a consequence, the definition of PNR by multiplying the points is incorrect. To solve this problem the authors offer to apply the theory of fuzzy sets [5] to conduct FMEA and subsequent interpretation of the analysis results for quality management. Despite the apparent complexity of calculations, the proposed approach to defining FMEA is easily implemented in practice by using the simplest software.

4. Managing Company Knowledge

A new conceptual requirement of ISO 9001:2015 is introduction of knowledge management in QMS. The Organisation should identify the knowledge necessary for the operation of its processes and how it will acquire or receive access to additional knowledge and necessary updating of its knowledge. The development of the knowledge management system (KMS) is one of the priorities in the field of quality management. The establishment of KMS is a new and weakly developed problem [6]. To solve this problem, the authors suggest using a KMS structural model, which defines the following principles for its creation.

The core of KMS at an engineering enterprise is a multivariate corporate storage of knowledge in quality. The storage should provide a quick and easy search and also allows the storage of diverse information resources.

In its functioning KMS should form two flows of knowledge. Firstly, it enriches the storage, extracting knowledge from different sources. Secondly, it gives knowledge to users at their request or 'on its own initiative', identifying the need for available knowledge by indirect signs.

One of the most difficult practical problems when building KMS in a company is forming the base of knowledge. In this paper, it is proposed to accumulate knowledge (experience) as a result of task solution by various professionals using funds provided by the subsystem of work with frames. Other means of KMS should provide access to existing knowledge. Besides it is preferable not to search knowledge at the user's request, but using the task automatic provision of knowledge found through ontologies.

Knowledge acquisition by the system is performed using the search engine, retrieving, structuring and sorting out knowledge. This subsystem modules are designed to analyse sources and identify various models for knowledge representation.

Information resources in the storage are classified into three types: data, information and knowledge. There is no universally accepted definition of these concepts, so in this work, they are treated as follows:

Data – fact (indicator) which value (usually digital) is uniquely identified by a fixed set of measurements. Information is a formalised text in a human language, which describes properties and characteristics of analyzed objects. Knowledge is the rules, procedures and methods for solving problems in the field of quality, the need for which occurs in a company.

The concept of a frame of knowledge is used as a basic structural part of knowledge in the storage, which in a generalized form can be represented by a tuple of elements: a hierarchy of objectives, sub-objectives (criteria) and alternatives; a formalised description of the relative importance of the hierarchy elements; mapping of the set of names of slots and factors in a variety of formal descriptions that make up the information in KMS.

A frame from the 'framework' is transformed into specific knowledge by symbols (names) and 'attribution' slots, i.e. by assigning their elements-factors to particular values, connecting them with varying information from the storage (which may be replenished).

When using the knowledge frame the hypercube of knowledge of multivariate KMS storage is defined by six basic axes of measurement (participants, organisations, products, processes, time and human resources), which have been determined earlier. Each cell of the hypercube, resulting from the intersection of the basic axes, presents a set of previously solved tasks in the form of the vector of frames. In other words, the cell of knowledge hypercube represents a 'book' of 'pages' strung 'on the axle of a problem.

Within KMS there have been developed frame models to solve the tasks of strategic planning in the Organisation's QMS, as well as solution models for the challenges of QMS processes improvement using KMS.

5. Evaluation of Enterprise Performance Based on Criteria of Effectiveness

In accordance with the requirements of section 'Performance Evaluation' of ISO 9001:2015 the Organisation should evaluate indicators of functioning and effectiveness of QMS. However, the existing methods have the problem of quantifying the value of deviations (change) of QMS effectiveness, which is undeveloped. Thus, this article proposes to apply the methodology for assessing the impact of changes on the basis of cluster analysis [7].

Evaluation of QMS according to this method is carried out by analysing the performance of the basic processes of quality management based on actual and planned values of group performance indicators. QMS processes are determined depending on the specific character of the organisation activities, which will apply the proposed methodology. QMS which will be certified in compliance with ISO 9001: 2015, will include the following processes: management responsibility (leadership), resource management, product life cycle, QMS analysis, management of inconsistencies, QMS improvement.

To calculate the performance of processes included in QMS, we should determine the performance criteria and significance of these indicators. As a criterion we should understand an indication or a rule, on the basis of which evaluation, definition or classification are carried out. Single performance indicators serve as these criteria. Group indicators are calculated as a sum of single indicators.

Evaluation of QMS performance begins with the definition of single performance indicators of QMS processes. To determine single performance indicators of QMS processes the following procedure should be done:

- Define the process customers.
- Define requirements and expectations of the process customers.
- Formulate single performance indicators and formulas for their quantification on the basis of this information.

After defining single performance indicators of QMS processes the information should be gathered and calculation of actual and planned values defined as indicators needed to evaluate the impact of each process included in QMS. Then, according to the formulas for calculation of group indicators there should be determined the actual and planned values of performance of each process in the system.

After determining the effectiveness of all processes the value of changing QMS performance over the planning period is calculated. To determine the changes of QMS performance we propose to use one of the two elements of the cluster analysis. The first element is related to the definition of an extent of objects similarity, the second – to an extent of objects proximity in Euclidean space.

On the basis of the calculated indicators the conclusion on the degree of QMS effectiveness is made. If the analysis proves that the system functions efficiently, the measures to improve the system functioning are taken in order to increase its effectiveness. If the results of the analysis reveal that operating system is ineffective, it is necessary to identify inconsistencies and their causes, as well as to develop and implement a set of corrective measures.

Application of the technique described above and quantification of changes in the effectiveness of QMS processes creates the possibility of efficient (monthly, quarterly and annual) planning and recording QMS performance in the departments of an engineering enterprise.

6. Conclusion

In general, a new version of ISO 9001 : 2015 is certainly a significant step forward in the development of quality management. However, the implementation of its key provisions, such as the analysis of the company context, risk-oriented thinking, leadership, knowledge management and other new concepts at engineering enterprises will cause certain difficulties. The implementation of the proposed in this article directions of quality management methodology development in order to ensure a balance between the needs and expectations of all stakeholders of the organisation, risk management introduction, knowledge management, evaluation of effectiveness of quality management processes, will allow the company to restructure its QMS in time (three years), established by ISO 9001: 2015.

Reference

[1] Cane M et al 2010 *Moscow: Mashinostroenie* 416
[2] Gorlenko O and Miroshnikov V 2015 *High Tech in Mechanical Engineering* **8** 44–48
[3] Miroshnikov V and Borbatc N 2007 *Information Technologies* **3** 63–69
[4] Mashunin K and Mashunin Yu 2013 *Journal Computer and Systems Sciences International* **52(4)** 519–534
[5] Miroshnikov V et al 2014 *Bulletin of Bryansk State Technical University* **4** 153–163
[6] Karatkevich S et al 2011 *Journal Computer and Systems Sciences International* **50(2)** 250–292
[7] Miroshnikov V et al 2014 *Bulletin of Bryansk State Technical University* **4** 164–170

The Process Approach to Management of Enterprise Human Resources

O Gorlenko and T Mozhayeva
Bryansk State Technical University, 7, 50 years of October st., Bryansk, 241035, Russia

E-mail: goa-bgtu@mail.ru

Abstract. The paper describes the approach to human resources management in the quality management system of an enterprise on the basis of their dual nature. The expediency of this approach application is analysed to ensure harmony between the interests of the company and its personnel and to improve the quality of labour.

1. Introduction

In modern conditions of the Russian economy development the problem of effective management of resources is of special urgency in the quality management system (QMS) of an enterprise, allowing the improvement of the quality of its products and, consequently, competitiveness on the market of commodity producers. A new version of ISO 9001 : 2015 approved in September, 2015, assigns a special role to human resources (HR), raising them to the level of the most important resources of the organisation.

However, experience of introduction and certification of QMS for compliance with ISO 9001 shows that many companies have not achieved significant benefits in improving the quality and competitiveness of their products. This fact is explained, in particular, by the formal approach to the enterprise human resources, without taking into account their specificity, and insufficient development of personnel processes. In this connection we propose a process approach to the enterprise personnel management that meets the requirements of ISO 9001: 2015.

2. Analysis of Recent Researches and Publications

Currently personnel management for developing and implementing personnel policies and strategies widely uses the following concepts [1]: the use of human resources, human resource management (conventionally attributed to a resource-based approach); human resource management, management of humans (conditionally attributed to a humanistic approach). In its turn, these concepts define the direction of development towards implementation models and principles of personnel management in the organisation, which only to a certain degree as the object of management consider a person as a specific resource (goods, raw materials), for example, the model of Internal marketing [2]; a person having a high potential and needs realised in employment, for example, the model of Personnel Management in Corporation 'MatsushitaDenki' [3].

The analysis of scientific literature and personnel practices in this field has shown that today for most models of QMS (TQM, ISO 9000, a model of business excellence, etc.), the terminology of the concept of 'human resource management' (HRM) is used to describe the personnel processes. This concept appeared in the West in the early 70-ies of XX century, is characterised by the following shortcomings:

- the lack in the scientific community (Harvard school, Michigan school, etc.) of unity of views on the nature of human resources and, as a consequence, free interpretation of essence of a human resource management concept or its substitution by other conceptual approaches in the personnel management of the organisation while preserving the existing terminology that leads to ambiguity in determining the object, level, and results of this management activity, etc.;
- effectiveness and efficiency indicators of HR processes are connected with formal indicators (competence, qualifications, length of service, etc.) of human resources and doo not deal with the quality of labour and products (human resource management in the management system should, in our view, be associated with a product quality). Moreover, practice shows that the personnel having these formal indicators does not always lead to manufacturing quality products;
- the absence of formalised models of mutual harmonizing personal and organisational requirements and monitoring of the extent of their implementation that does not allow the company to provide itself with quality human resources, and at the same time ensures employees of the expected reward system that reduces the level of loyalty;
- the lack in personnel practices of predictive models of risk management related to human resources - the only type of resource that can distort the control effect, which leads to an increase in the level of uncertainty in the quality of HR processes.

3. Problem Statement

In order to improve human resource management in the quality management system (QMS) of the organisation it is necessary to identify the parameters of HR processes; their proper development will reduce the level of uncertainty in their implementation and eliminate the revealed discrepancies, in particular:

- identify suppliers and consumers of HR processes;
- establish their input and output parameters, as well as indicators of effectiveness and efficiency;
- identify resource provision;
- develop regulatory documentation.

The idea "quality processes make quality products", extrapolated on human processes, can bring its positive results. However, it is obvious that it is impossible to control processes efficiently without determining the object and the purpose of control and parameters to achieve these goals.

In this context, researches aimed at defining the nature of human resources and development of the process model, taking into account the specific character of the organization QMS, are of particular importance.

4. Nature of Human Resources

Today, most discussions of this country and foreign scientists are based on the differences in opinions on the object, level, results of the human resource management.

The problem of determining the object of activity in the field of 'human resource management' is ambiguous, but it is important because it reveals the very reason for incipient contradictions. Some scholars [4] consider people themselves (organization staff), motivated on the basis of organisation capabilities to qualitative work as a control object.

However, other scientists [5, 6] consider a specific resource of the organisation – staff as a management object, which despite the number of peculiarities (the ability to transform the control action, to increase their value during functioning, to change their properties and behaviour in time, etc.), has the characteristics typical to other resource types, which causes a need for its reasonable use and activation.

Within this view there are two approaches to human resource management – rigid and soft [5, 6]. The rigid approach is based on the idea of matching human resources to business strategy, which regards employees as a resource to be managed as efficiently as any other one to maximize profit. In contrast to this view, the soft approach, based on the idea of commitment, considers employees a valuable asset and a source of competitive advantage due to their commitment to the organisation, adaptability, high level of skills and quality of work.

These debates suggest that there is no unity of views on the object of management, and, in its turn, this causes difficulties in formation of effectiveness and efficiency indicators of HR processes. In this regard, it seems appropriate to define the nature of human resources as dualistic, where the staff is considered, on the one hand, as the most important specific resource that defines the competitiveness of the organisation, and on the other hand, as an internal consumer of the system of awards (salary and benefits, promotion, training programs, work significance, etc.) offered to employees for qualitative work.

The prerequisites of this approach can be found in a number of QMS models, in particular international standards ISO 9000. In standards ISO 9000, versions of 2011 and 2015, it is noted that the consumer may be internal or external to the organisation. In this case, the organisation should determine the needs and expectations of its employees in recognition of their work, job satisfaction, and in their development. Such attention helps to ensure the strongest motivation of employees. The organisation should conduct a survey of employees' satisfaction, achieving their results and implementation of appropriate actions. At the same time, top management should determine and provide the resources necessary to implement the strategy and achieve the goals of the organisation.

Based on this approach with the generic name 'human resource management' it is useful to consider two interrelated, integrated, however, principally independent processes – 'human resource management' and 'management of staff motivation' with their own suppliers, customers, control objects and performance indicators that are not always or not fully defined in the framework of the traditional approach.

Orientation on satisfaction of personnel requirements with the career, even in 'soft' models of human resource management is a priori based on the assumption that workers share the interests of the employer, even if they do not match their own. In other words, the employer motivates staff by the system of awards for the qualitative and productive work, based on the capabilities of the organisation. Thus, poorly developed or the lack of harmonisation of personal and organisational requirements in HR management leads to lower level of mutual satisfaction between the employer and employee.

It should be noted that in the development of HR processes in QMS, most organisations also face the problem of determining their level of control. The most preferable for the development of HR processes is a point of view that considers human resources management as operational activities of managers with the staff to implement the strategy of organisation development at the level of departments. The activity of human resources management provides working out the company's development strategy and includes personnel management as a subprocess.

Summing up all mentioned above, human resource management can generally be defined as the direction of administrative activity, in which the human component of the organisation is one of the most important resources used to achieve strategic goals. Its maximization is possible on the basis of a balance of personal and organisational values.

5. Process Model of Human Resource Management

Identifying the dual nature of human resources allows one to develop personnel processes properly and integrate them into QMS of the organisation.

In accordance with this approach we proposed a process model of human resource management in QMS of Bryansk machine-building enterprise. To implement this model it is necessary to solve the following problems:

- Development of a descriptive model of integrated human resources processes of the enterprise, based on the dual nature of human resources, allows for a clear identification of the users and

suppliers (personnel and enterprise), management object (human resources, and system of awards offered to the staff by the enterprise), resource provision, normative and methodological documents and effectiveness and efficiency indicators;
- the formation of a typical registry of personnel processes of the organisation, taking into account the level of tasks (strategic and operational) and the nature of human resources;
- developing processes of harmonisation of personal and organisational requirements;
- formation of a system effectiveness and efficiency indicators of the processes of human resource management in accordance with their targets.

Process model 'Human Resource Management' provides a clear decomposition of personnel processes into integrated processes 'Personnel Management', whose customer is an enterprise, and 'Management of Motivational System,' whose consumer is the personnel, formation and management of the balance of the employee and company interests through the process 'Harmonisation of Personal and Organisational Requirements' and designing processes 'Staffing of Qualitative Work', allowing a connection of the performance indicators testifying the satisfaction of the enterprise resource expressed in the personnel competence and insurance of a high level of quality and productivity and staff satisfaction with the system of awards by the enterprise.

Taking into account the specificity of the identified personnel management a model register of the process 'Human Resource Management Organisation' has been developed (Table. 1). Integrating HR processes into the proposed register, based on the dual nature of human resources, allows:
- eliminating the existing shortcomings in personnel management - the lack of a clear separation of operational and strategic HR processes in the organisation QMS; processes mixed with different users (organisation and staff) and management objects (human resource and a system of awards), which leads to discrepancy of final scheduled performance indicators to the process content;
- to harmonise the interests and expectations of the staff and the employer, thereby maximising the potential of human resources in the organisation that will enhance the efficiency and effectiveness of its QMS.

Provision of quality personnel processes causes the need to balance the requirements of the employee and the company imposed by each other [11]. Harmonisation of these requirements involves determining the closeness of the relationship between the dominant organisational and personal values based on the correlation of their weights.

For statistical significance of the correlation coefficient organisational and personal values are considered as harmonised in favour of declared corporate values. If there is no statistical significance of the correlation coefficient it is necessary to conduct correction of personal values in accordance with dominant organisational values of the declared type of culture.

Effectiveness and efficiency indicators of HR processes have to be directly related to their targets in the organisation QMS. HR process targets should be designed to achieve the integration of the interests of the employer and employee, in particular: ensuring the organisation with human resources of required quantity and quality, capable to realise a corporate strategy and to support and share organisation values; improving the efficiency of employees by meeting their requirements and expectations set on the organisation and given system of awards.

A number of HR processes parameters are simply monitored, as they have a formalised (or reference) performance indicators (employee turnover, absenteeism ratio, etc.), but for other parameters the descriptive (or vague) characteristics (competent specialist, employee's satisfaction, etc.) are generally typical.

Moreover, some parameters of the staff cannot be subjected to consistent monitoring, as these actions are difficult or not economically sound. In particular, the competence of professional adaptability and mobility demanded on the labour market may occur only during the period of employment, which increases the risks of the businessman when hiring such a specialist. In the current situation, using the methods of expert evaluations is reasonable to monitor uncertain effectiveness parameters for a number of HR processes.

IOP Conf. Series: Materials Science and Engineering **124** (2016) 012013 doi:10.1088/1757-899X/124/1/012013

6. Conclusion

Thus, the management of human resources is considered through the prism of the quality management system. Taking into account the dual nature and integratedness of HR processes focused on organisation satisfaction with a resource, expressed by staff competence and ensuring a high level of quality and productivity and the staff satisfaction with the system of awards allow ensuring achievement of efficiency and competitiveness of the organisation as a whole.

Table 1. The Register of the Process 'Human Resources Management of the Enterprise'

1.	**Human resources management**
1.1.	*Development of human resource management policy and strategy*
1.1.1.	Development of personnel policy
1.1.1.1.	Definition of organisation values
1.1.1.2.	Formation of organisational behaviour model
1.1.2.	Development of human resources strategy of the enterprise
1.1.2.1.	Analysis of the labor market and human potential of the enterprise
1.1.2.2.	Cost planning of human resources
1.1.2.3.	Human resource requirement planning
1.2.	*Recruiting*
1.2.1.	Recruitment
1.2.2.	Personnel selection
1.2.3.	Personnel hiring and release
1.3.	*Personnel adaptation and career guidance*
1.3.1.	Personnel adaptation
1.3.2.	Personnel career guidance
1.4.	*Personnel development*
1.4.1.	Personnel training
1.4.2.	Personnel career planning
1.5.	***Personnel management***
1.5.1.	Facilitating the formation of organisation departments
1.5.2.	The definition of pay and benefits
1.5.3.	Management of registration activities
1.5.4.	Advising line managers
2.	**Motivation system management**
2.1.	*Identification of personal values*
2.2.	*Development of motivation programmes*
2.2.1.	Designing the system of awards and material values
2.2.2.	Designing the training system
2.2.3.	Designing the system of promotion
2.2.4.	Designing labour conditions and work importance
2.2.5.	Documentation of motivation programs
2.3.	*Personnel Motivating*
2.3.1.	Implementation of motivation programs
2.3.2.	Monitoring of motivation programs
3.	**Harmonisation of personal and organisation requirements**
3.1.	*Analysis of enterprise personnel satisfaction (certification)*
3.2.	*Analysis of employee satisfaction with the enterprise*
3.3.	*Harmonisation of personal and organisation values*

4.	**Human resources for quality and productivity**
4.1.	*Identification of performance and labour quality parameters*
4.2.	*Personnel management of parameters and quality*
4.3.	*Evaluation of the performance and labour quality indicators*

Reference

[1] Egorshin A P 2008 *Personnel management* (Nizhny Novgorod: NIMB) 607.
[2] Novatorov B E 2013 *Materialy IX Miedzynarodowie naukowi-praktycznej konferencji 'Wschodnie partnerstwo - 2013'* **3** 3-13
[3] Pshennikov B 2007 *Moscow: Japan today* 325
[4] Torrington D *et al* 2004 *Moscow: Business and service* 752
[5] Shadrin A 2013 *Standards and Quality* **4** 78–83
[6] Antonov S and Antonova I 2013 *Standards and Quality*, **4** 82–84
[7] Mozhayeva T and Erohina V 2012 *Bulletin of Bryansk State Technical University* **2** 132–139

The estimation technique of the airframe design for manufacturability

A Govorkov and A Zhilyaev
Institute of Aircraft Construction, Mechanical Engineering and Transport, National Research Irkutsk State Technical University, 83, Lermontova st., Irkutsk, 664074, Russia

E-mail: anton@el.istu.edu

Abstract. This paper discusses the method of quantitative estimation of a design for manufacturability of the parts of the airframe. The method is based on the interaction of individual indicators considering the weighting factor. The authors of the paper introduce the algorithm of the design for manufacturability of parts based on its 3D model

1. Introduction

There is a high variety of technical publications and guides that describe designs for manufacturability (DFM) for the aircraft industry [1, 3], particularly, the methodical basis of its support and evaluation.

Nowadays, it is reasonable to consider DFM evaluation from early design stages when using CAD/CAM/CAE/PDM solutions. Therefore, to reduce the period of documentation preparation and to hasten the product launch, design engineers should do the following tasks:

- choose modern design solutions, an optimal fabrication variant and a structural configuration;
- choose part's design depending on its functions;
- use standard library features when designing the airframe (for example in Siemens NX);
- consider typical manufacturing processes for airframe parts;
- decrease the complexity of parts manufacture.

During the DFM process for an airframe, design engineers should take into account the scale of the manufacture, the production type, and the specifics of the plant to be sure that every part is manufacturable. In the pre-production stages, it is hard to evaluate DFM [2, 4] due to the lack of information (time rates, equipment list). To assure high DFM levels, one can make a qualitative evaluation using DFM properties and a quantitative evaluation by the indicators.

Final DFM evaluations on pilot airframe manufacture are described below and were done by comparison of the author's and manufacturer parts. While there is no analog for the part, the complex indicator must be set above 0.5. It is the value, which is determined by expert evaluations on an unfavorable combination of factors.

This paper proposed the method of DFM evaluation in the aircraft industry that focused on formalization of the process and customization of DFM indicators.

2. The method of quantitative estimation of design for manufacturability

Initial data for calculation of technological design parts are listed below (in Table 1).

Figure 1. Diaphragm of the airplane's brake flap

Table 1. Initial data for DFM evaluation

№	Data title	Value
1	Data for geometric parameters	Model parts, specification
2	Amount of issue production per year	0-100 units.
3	Details' weight	Less than 0,100 kg
4	Amount of element part	More than 3
5	Contacting with a theoretical contour	yes
6	Material	D16AT
7	Admission to the aerodynamic contour	+-1,0

The quantitative DFM evaluation of the technological parts of the 'aperture' in the design phase and a comparison of technology options were realized using complex indicator [2]:

$$p=\frac{\sum_{i=1}^{n} k_i\varphi_i}{\sum_{i=1}^{n} \varphi_i}, \quad (1)$$

where k_i – value of i – st quantitative DFM indicator;
φ_i – weight rate i – st quantitative DFM indicator;
n – number of indicators defined by the expert, shall be more than 4

Table 2. DFM calculating summary

№	DFM private indicator	Value	Weight, φ_i	$k_i\varphi_i$
1	Amount of issue production per year, k_{year}	0,30	0,06	0,018
2	DFM for split operation, k_{tm}	1,00	0,12	0,12
3	DFM for shaping operation, k_{frm}	0,90	0,13	0,117
4	Repeatability of products' element of construction, k_{el}	0,90	0,12	0,108
5	Weight, k_M	1,00	0,07	0,07
6	Size, k_{tm}	0,80	0,08	0,064
7	Shape of the contours, $k_{f.ob}$	0,75	0,06	0,045
8	Shape of the circuit, $k_{f.con}$	0,60	0,05	0,03
9	Curvature level, $k_{lv.cr.}$	0,9	0,11	0,099
10	Contacting with a theoretical contour, $k_{aer.}$	0,40	0,05	0,02
11	Contacting with a perimeter, k_{TK}	0,5	0,08	0,04
12	Details location in relation to conditional plane parts, k_N	0,8	0,07	0,056
Total			1,00	0,787

Moreover, the values of DFM indicators are conditional.

The amount of issue production per year (k_{year}) is described in chart 3 and evaluates the possibility of mechanization and automation of the parts manufacture process. The higher the amount, the more efficient mechanization and automation implementation, and thus reducing complexity will reduce the production price.

Table 3. Amount of airframe production per year

№	Amount of airframe production per year	k_{year}
1	1-10	0,1
2	10-50	0,2
3	50-100	0,3
4	100-200	0,4
5	200-300	0,5
6	300-400	0,6
7	400-600	0,7
8	600-800	0,8
9	800-1000	0,9
10	1000 and more	1,0

DFM indicator depends on the type of ($k_{\text{tms.}}$) split operation for the part manufacture process (Table 4) and on the technological cost of a blank.

Table 4. DFM indicator depending on the type of split operation

№	DFM for split operation	$k_{\text{tms.}}$
1	On milling machines	1,0
2	On a milling machine from a template	0,8
3	The stamp	0,6
4	Guillotine scissors	0,4
5	Roller scissors	0,2

The DFM value depends on the type of the molding (k_{frm}) operation in the production of shaped parts from sheets metal and on DFM properties: technological cost of a blank, technological simplicity.

While there are many types of molding operations for the 1-st part, the average value is determined by formula

$$k_{\text{frm}_{\text{cp}}} = \frac{\sum_{i=1}^{n} k_{\text{frmi}} p_i}{p}, \tag{2}$$

where p – total number of molding operations for details manufacture;

k_{frmi} – DFM indicator value by i – st parameter;

p_i – number of i – st type operations.

The indicators describing repeatability of the part' elements of construction (k_{el}), are taken from the DFM guides.

When the amount of elements of the construction is more than 40, the indicator value is considered within the range of 0,40…0,10.

Indicator k_M depends on the weight of the part and can be determined in the DFM guides. The indicator takes into account DFM properties suitable only for airframe parts.

The indicator describing size of the parts k_{tm}can be determined in the DFM guides.

Indicator $k_{f.\text{ob}}$ takes into account the shape of the contours and depends on the purpose of whole part or its elements. The indicator is described in the DFM guides.

Indicator $k_{f.\text{con}}$ takes into account the shape of the circuit and depends on the purpose of whole part or its elements. The indicator is described the DFM guides.

Indicator $k_{\text{lv.cr.}}$ takes into account the level of part curvature. The indicator is described the DFM guides.

Indicator $k_{\text{aer.}}$ takes into account the aerodynamic contour allowance and is based on the detail's objective function.

Indicator k_{TK} takes into account the fact that the detail is contacting with a theoretical contour.

Indicator k_N takes into account the details location in relation to conditional plane parts.

By using the method mentioned above, the values of k_i and $k_i\varphi_i$ can be found. These DFM indicators are estimated by experts' evaluation. The sum of the selected coefficients is equal to one.

Therefore the complex DFM value can be calculated by formula (1):

$$p=\frac{0{,}787}{1{,}0}=0{,}787.$$

The resulting DFM value for the aircraft part, namely diaphragm:

- new detail: 0,5;
- calculated by the formula (1): 0,787.

3. Conclusion

Comparing a comprehensive indicator, calculated by the method described above, and the basic indicator for a new part we can conclude that the engineering design is manufacturable and can be accepted for manufacture. This method is use-proven on the design of the aircraft parts stamped from sheet and fully describes the designing and manufacturing process.

Consequently, this method of DFM evaluation can be adapted in practice for the aircraft's parts that are hard to evaluate by the basic indicators due to the lack of information and to the fact that optional indicators can only describe the elements of the construction. Utilization of the method in the authors' DSS "DFM analysis system" at the early design stages for aircraft industry allows easy customization of 3D models. Even when the technological dictionary is changing rapidly, the design dictionary is consistent.

The time rate is changed by the improvement of technological and design engineering methods. 'DFM analysis system' can make it easier for a design engineer when the time rate is changing, but at the same time allows user to utilize these new technological improvements.

Thereby, utilizing "DFM analysis system" can ease the adaptation of the method for a specific manufacturer and accumulate technological knowledge. This provides the method's ability to adapt to the variety manufactures and its peculiarities. The method of feature recognition is suitable for 3D models based on hard-body elements, which are made independently of the model's creation technique.

References

[1] Amirov U D 1990 *Biblioteka konstruktora. M.: Mashinostroyeniye* 768

[2] GOST 14.201-83 1983 Common rules of the design for manufacturability of parts M.: Izdatel'stvo standartov

[3] Kolganov I M 2003 *Ul'yanovsk: UlGTU* 148

[4] Osetrov V G and Amirov Yu D 1980 *M.: Mashinostroyeniye* 205

Maintenance with the use of statistical control charts

E I Gromakov[1], T V Aleksandrova[1] and B P Ivanenko[2]
[1]Tomsk Polytechnic University, 30, Lenina ave., Tomsk, 634050, Russia
[2]Tomsk State University, 36, Lenina ave., Tomsk, 634050, Russia

E-mail: gromakov@tpu.ru

Abstract. The possibility of using statistical process control methods for detection of an abnormal condition of the process equipment at early stages of an emergency is shown in the paper. The authors of the paper has concluded that with the use of Shewhart charts it is possible to monitor the real dynamics of the process equipment condition and make decisions on its maintenance and repair

1. Introduction

The efficient operation of the capital equipment is one of the main factors of technological excellence in production. The lack of timely maintenance entails production stop after a breakdown or an immediate inspection of technical supervision and in the event of accidents - deaths and criminal penalties for official business. Many Russian companies lag behind the western ones in technological maintenance: the equipment can be operated irrationally and can stand unscheduled idle, which leads to its costly repairs.

At present, worldwide the highest priority is given the problems of equipment maintenance and repair because of high losses. These problems are considered in terms of cost control and provision of scheduled industrial work, and from the point of view of minimization of industrial risks and tragic consequences.

The aim of this work is to develop and justify the method of statistical processing and control providing the reliability of measurement interpretation and data on the condition of process installations (PI)arriving at the dispatcher workplace in the automated mode.

There are three basic strategies of maintenance and repair management:

- event-triggered;
- preventive maintenance management;
- repair and maintenance according to assessment of the actual technical state.

2. Event-triggered maintenance

This is a maintenance, during which the repair or replacement of the equipment is done only in case of the emergency fail or ending of its service life. The drawback of the event-triggering maintenance is the unreadiness of the enterprise for the maintenance. Unscheduled shutdowns and downtime of the process equipment due to sudden failures lead to more costly repairs. In addition, in the event of a sudden failure of several different units a simultaneous necessity in repair works may exceed the capabilities of the repair service of the enterprise.

3. Preventivemaintenance management (PMM)

A periodic preventive maintenance underlies the preventive maintenance management (PMM) in accordance with recommendations of the operational and maintenance mechanism documentation, taking into account the existing experience of the equipment operation. Periodically, at certain time intervals, it is necessary to perform the monitoring of equipment condition, repair or replacement of units, parts, lubrication etc. In some cases, these requirements are mandatory, in other cases the decisions on repairs and replacements are made according to the results of equipment examination duringunsealing the mechanism or its components.

The advantage of this method, compared with the event-triggered method,consistsin maintaining of the required reliability of the equipment during operation.

The disadvantages of the preventive maintenance management are as follows:

- it does not take into account the actual condition of the equipment (its technological modes);
- a substantial 'margin of safety' is initially laid in the system;
- unnecessary repairs are often carried out;
- outdated standards of maintenance and repair of equipment are laid in the existing system of preventive maintenance management.

The system of PMM, developed many years ago, does not allow one to determine priorities accurately when selecting objects and volumes of work, which, in conditions of the international market of technological installationsdelivery, entails the necessity of laborious total search for suppliers of components and their subsequent order all over the world.

4. Maintenance based on the actual technical stateor actual state maintenance (ASM)

The essence of this technique lies in the fact that maintenance service is conducted not only depending on how long the mechanism has operated, but also with consideration of its actual current technical state, which is controlled during operation without any disassembly and inspection on the basis of the measurement of relevant parameters of an operating mechanism.

If PMM is based mainly on the operating time of the mechanism, ASM takes into account all the factors that determine the service life of the mechanism. It is the fundamental difference. We should note that ASM takes place in a regular manner due to controlling the condition of the equipment, since whatever factors in whatever combination influence the mechanism there is always a combined reaction to these impacts owing to the change of corresponding indicators of its condition. In this case there is an opportunity not only to monitor the technological installation state, but also to determine the real causes of the changes that take place in each particular situation, and, therefore, to make well-founded decisions for their elimination in future.

The advantage of ASM method is minimization of maintenance and repair works. According to experts evaluations the overhaul period increases from 25 to 40 percent in comparison with PMM method.

The disadvantages of this technology are:

- poor operational budget and technical readiness of maintenance services for the arisen event-triggered problem of the equipment state;
- a relatively high probability of false alarms especially in case of sophisticated equipment;
- increased skill requirements for staff.

5. Combined maintenance

Thus, each of the three strategies mentioned above has drawbacks.

That is why there is a stable trend of companies to combine all basic strategies through the use of so-called combined maintenance aimed at reduction of the total amount of required maintenance and to maximize the service life of the equipment.

The idea behind this service is that at each time horizon of planning of maintenance and repair the proper technologies with a very high planning reliability are used. So the long-term period is used for planning of the milestone indicators of PMM. These indicators are calculated by the direct method. On the operational planning horizons the indicators of maintenance and repair are determined on the basis of statistical processing of data of measuring the condition of the equipment. In case of their absence a financial reserve is formed.

In accordance with the combined maintenance, the consistent refinement of planned performance is carried out as the period of planning horizon reduces. The basis for such refinement is the data of actual equipment state, technological modes of its work, as well as the implementation of plans for equipment maintenance and repair in the prior periods.

The use of combined maintenance on the operational horizon of planning of maintenance and repairs involves a regular evaluation of the technical state of the mechanism without monitoring, in the operating modes and during control using secondary parameters. These requirements can be formulated, for example, as follows:

- controlled parameters should have a single valued quantitative interrelation with the primary parameters of the technical state;
- measurement of the parameters should be provided by the means of the in-use SCADA system as well as, if possible, by simple portable technical devices that do not require special skills from staff;
- during the mechanism operation the range of the monitored parameters from the state of 'good' to the state of 'unacceptable' should be large enough (the parameter can be changed at least 15-20 times) to detect incipient defects early and predict reliably the residual service life of the mechanism;
- The accuracy of the control by the secondary parameters should not be below 80%.

Three methods can be used for statistical processing. The first one is based on Neyman-Pearson criterion and represents a Shewhart control chart [1]. It is historically the very first common tool to control the variable parameters of maintenance processes. The second one is based on the repeated application of Wald sequential analysis results [2]. In practice, it is realised in the form of control charts of cumulative sums. Finally, the third method of detection of the process malfunctions is based on the exponential smoothing of statistical results (EWMA charts) [3]. Hereinafter, the features of the use of the first two methods will be considered.

We believe that in accordance with the recommendations of GOST [1] special statistical processing of regular measurements in the form of Shewhart chart allows solving the problem of identifying the dangerous causes of variations in the equipment state. These algorithms can supplement the algorithms of vibration control technology providing them with additional confidence in the assessment both on the operational and medium-term planning horizons of maintenance and repair.

A control chart, applied to the control of the state, represents a graph of a number of regular samples of measurement of some index, which characterizes technical condition *X*. An outgoing maintenance parameter can be used as *X*, for example, the temperature of the bearing assembly, vibration and others [4]. Then the collected telemetric statistic data can be combined in the form of the diagram of sample son chart g(*X*).

The sample in this diagram represents a sequence of n independent observations $x_1, x_2, \ldots, x_n$of indicator *X*, related to a certain period of time (day, month, year).

The period of measurement samples depending on the planning horizon may be specified in any manner. Moreover, for monitoring of the equipment state one can simultaneously monitor the vibration trends in different spectral ranges and at various intervals of samples.

There are no general rules for selecting the frequency of sampling and sample volumes. Usually 20 - 25 samples with 4 or 5 measurements in each are regarded as acceptable for obtaining preliminary evaluations at each level of planning [5].

Let $f(x,\theta)$ describes a distribution of random variable of observations with some value of the statistical parameter (mean value of samples, scope, etc.) θ.

With the use of the control chart one sets and solves the problem of detecting unnatural variability of the observed parameter, due to which the technical installation have changed. This variability is observed based on the change of the selected for the chart parameter of distribution function $\theta$of consecutive observations x.

A peculiarity of monitoring the technical installation state is its slow changes of variability in the form of trends of operating parameters. Let us assume that the statistical properties of the time series that characterizes these states and properties of the reasons generating its changes remain unchanged or change slowly in the selected time intervals of samples formation. The current control is reduced to monitoring the arrival of the next monitoring of sample and detection of deviation (disorder) of the properties of the observed time series of $\theta$from value θ_0 to θ_1. The necessity in maintenance may occur at unknown point of timet_0. This event will be called a disorder of the operational process.

In a common approach the considered range of applied problems associated with the operational state of the technical installation, is characterized by random sequence $\{x_1, x_2, ..., x_n\} = \{x_1^n\}$, which at moment $t_0$changes its properties that uniquely defined by the vector of parameters θ, $dim\{\theta\}$= r. beginning with$t_0$the parameter vector becomes $\theta = \theta_1$. Moment of disorder $t_0$is detected by the established criterion for assessment of sequence$\{x_1^n\}$, $N \to \infty$, during appearance of the next point x_n.

Each time for a new sample hypothesis $\theta = \theta_0$ is verified, which indicates the absence of the unnatural variability.

The option of the screen form of equipment condition monitoring is shown in Figure 1.

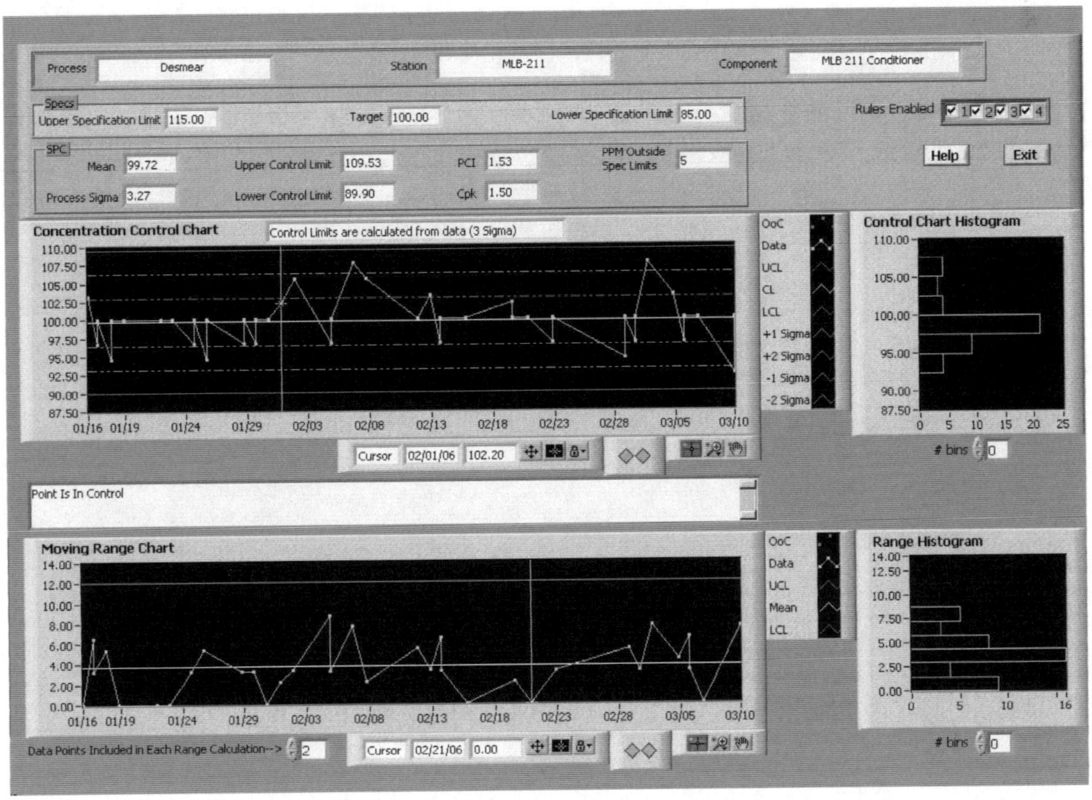

Figure 1. Shewhart control charts displayed on the screen.

To interpret a process flow on the base of Shewhart chart we can use an expert system that is configured to diagnose equipment malfunctions. The recommendations are displayed on the monitor screen. The display form should not be obtrusive except for alarm signals. The choice of a particular decision depends on the operator' will.

IOP Conf. Series: Materials Science and Engineering **124** (2016) 012015 doi:10.1088/1757-899X/124/1/012015

The proposed monitoring method allows implementing the monitoring of the operational state of the technical installation with low error level of decision making at least in three dimensions of real time:

- in the operational real-time daily management of an oil field exploitation;
- in retrospective, archive real-time, with profundity established by the regulatory documents;
- in the forecast time determined by strategic objectives of the enterprise.

6. Conclusion

The temporal analysis of the monitored parameters using Shewhart charts allows tracking the real dynamics of changes and predicting reasonably the timing and content of adjustment and alignment, maintenance and repairs of equipment both of rotary and non-rotary types.

Periodic updating of Shewhart charts on operational planning horizons of maintenance and repairs allows their use in any range of equipment parameterization.

Introduction of operations of statistical condition monitoring of technological equipment according to GOST R 50779 and, if necessary, its adjustment can significantly improve the quality condition of the machinery after undergoing repairs.

References

[1] 1994 *State Standard of Russian Federation* P 50779.42 99
[2] Wald A 2004 Sequential analysis (Mineola, Courier Corporation).
[3] Rabyk L and Schmid W 2015 *Metrika* 1-35.
[4] Aleksandrova T and Gromakov E 2015 *IOP Conf. Ser.: Mater. Sci. Eng.* **93** 012053
[5] Wheeler D and Chambers D 1992 *Understanding Statistical Process Control* (Knoxville, SPC Press)

IOP Conf. Series: Materials Science and Engineering **124** (2016) 012016 doi:10.1088/1757-899X/124/1/012016

A mirror based scheme of a laser projection microscope

F A Gubarev[1,2], Lin Li[1] and M S Klenovskii[1,3]
[1]National Research Tomsk Polytechnic University, 30, Lenin Avenue, Tomsk, 634050, Russia
[2]V. E. Zuev Institute of Atmospheric Optics, Russian Academy of Sciences, Siberian Branch, 1 Academician Zuev square, Tomsk, 634021, Russia
[3]Institute of Electron Physics, National Academy of Sciences of Ukraine, 21, Universiteskaya St., Uzhgorod, 88017, Ukraine

E-mail: gubarevfa@tpu.ru

Abstract. The paper discusses the design of a laser projection microscope with a mirror-based scheme of image formation. It is shown that the laser projection microscope with the mirror-based scheme of image formation is well suited for distant objects monitoring. This scheme allowed obtaining a field of view of more than 3 cm at the distance of 4 m from the brightness amplifier

1. Introduction

The laser projection microscope invented by authors of [1] became the basis for building laser monitors. A laser monitor based on the active media of copper vapor lasers or copper bromide vapor lasers is one of the powerful tools for observing objects and processes shielded by intensive background lighting [2–6]. A number of non-destructive testing tasks require visualization of such processes. These processes occur under interaction of powerful flows of energy (laser radiation, electron beam etc.) with the matter, self-propagating high-temperature syntheses, discharge plasmas etc. The copper vapor lasers and brightness amplifiers, based on them, radiate at two different wavelengths in the visible range of the spectrum, namely 510.6 nm (green) and 578.2 nm (yellow). The temperature of objects under observation by means of the copper bromide laser monitor can reach up to 40000 K and higher [6].

The operating principle of the laser monitor is similar to the operating principle of a laser projection microscope [1] and is as follows. The object under observation is illuminated by amplified spontaneous emission of a brightness amplifier. The radiation reflected from the object is returned then back to the brightness amplifier, amplified by the gain medium, and projected onto a screen or the sensor of a digital camera. The image is conventionally formed by the system of optical lenses.

In the present paper we suggest using the concave mirrors instead of the traditional collecting optical lenses or objectives.

2. Schemes of laser projection systems for distant objects visualization

The conventional laser monitor scheme includes the brightness amplifier, high-speed camera, lenses (or objectives), optical filters, and an object under observation (Figure 1). Physicists pay particular interest to the visualization of distant objects (distances of a few meters or tens of meters) by means of the laser monitor [7]. Figures 1, *b* and *c* shows the modifications of the typical scheme of the laser monitor for visualization of distant objects depending on the relative position of the brightness amplifier, the lens forming image, and an object under observation. The scheme in Figure 1, *b* has the evident disadvantage caused by the need of placing the lens close to the object. Such scheme is not proper for visualization of processes with high temperature or accompanied by the combustion product splashing. Moreover, when the lens is placed at a long distance from the brightness amplifier the area of vision is reduced [8]. The scheme used in paper [8] is depicted in Figure 1, *d*.

The better suited variant of a laser monitor monostatic scheme (based on a single brightness amplifier) for observing distant objects is the one illustrated in Figure 1, *c*. The long-focal-length lens is located far from the object observed. Both schemes for distant objects observation (Figure 1, *b* and *c*) have the principal limitation caused by the gain medium properties. Thus, the duration of existence of population inversion of the laser levels limits the longest possible distance to an object under observation. A typical lasing pulse duration of copper vapor laser is 20-40 ns [9]. In case of the CuBr laser the typical pulse duration of lasing can reach 80 ns [10]. The increase in the pulse duration of lasing is possible by means of varying excitation conditions. The method of the discharge current control [11-13] allowed obtaining the radiation pulse duration of 230 ns for copper vapor laser [12] and 320 ns for CuBr laser [13]. In paper [14] the increase of the lasing pulse duration up to 235 ns was obtained in an unstable discharge mode. In a stable discharge mode the 150-160 ns pulse duration is quite achievable.

Thus, the temporal properties of the active medium of the copper bromide laser can ensure observation of objects located within 25-30 meters. However, the main limiting factor is the optical scheme of the laser monitor. Therefore, a key issue in the design of the laser monitor for the distant objects observation is development of a proper optical configuration.

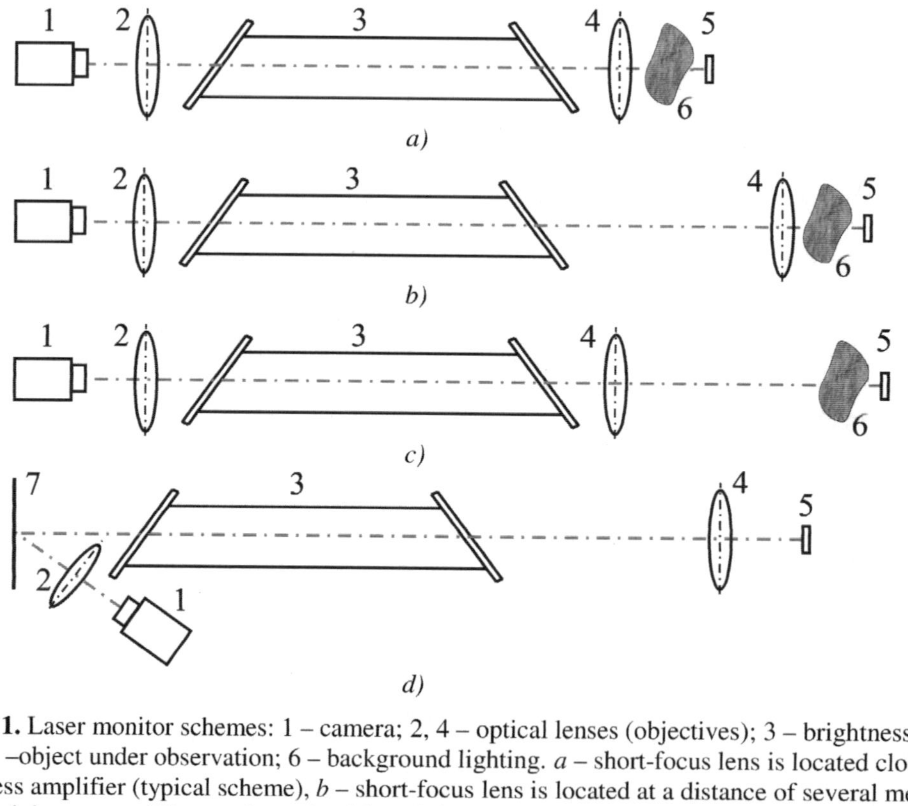

Figure 1. Laser monitor schemes: 1 – camera; 2, 4 – optical lenses (objectives); 3 – brightness amplifier; 5 –object under observation; 6 – background lighting. *a* – short-focus lens is located close to the brightness amplifier (typical scheme), *b* – short-focus lens is located at a distance of several meters from the brightness amplifier, *c* – long-focal-length lens is located close to the brightness amplifier, and *d* – scheme with the image recorded from the screen

3. Mirror based schemes and visualization results

In the present research, we used two CuBr lasers with external heating of the active region of the gas discharge tube (GDT). Such a type of lasers was studied earlier in [15–17]. First, the gas discharge tube (GDT1) has 50 mm in diameter and 90 cm at length and was pumped by means of a circuit of pulsed charge of a storage capacitor and its direct discharge by a thyratron (TGI1-1000/25) [15, 16]. Second, the gas discharge tube (GDT2) has 25 mm in diameter and 40 cm at length and was pumped by the circuit with a solid-state switch [17]. For the image capture the CMOS camera FastCam HiSpec1 was used.

It should be noted that when the observed object is moved away over long distances, it is necessary to increase the signal intensity. Otherwise, the contrast of the object images decreases with the increasing of the distance of visualization. To overcome this problem one needs to use more powerful active elements of copper (copper bromide) vapor lasers. It should be pointed out that a partial issue solution can be obtained by using lenses (4 in Figure 1) with a large aperture. However, the manufacture of large aperture lenses is more difficult. In contrast with lenses, spherical mirrors are easier to manufacture and, in addition, they have no chromatic aberration. The second benefit can provide high contrast of object images obtained simultaneously at both 510.6 nm and 578.2 nm operating wavelengths.

In the present work, the concave mirrors are used instead of the traditional collecting lenses or objectives. Figure 2 shows a variant of the laser monitor scheme for imaging of distant objects with a spherical mirror. We used two types of concave mirrors with radii of curvature (R_c) of 1 m and 3 m, which allowed observing objects moved away over the distance of 1 and 3 m, respectively. The distance between the GDT output window and the mirror was 1 m. The GDT1 is used as a brightness amplifier.

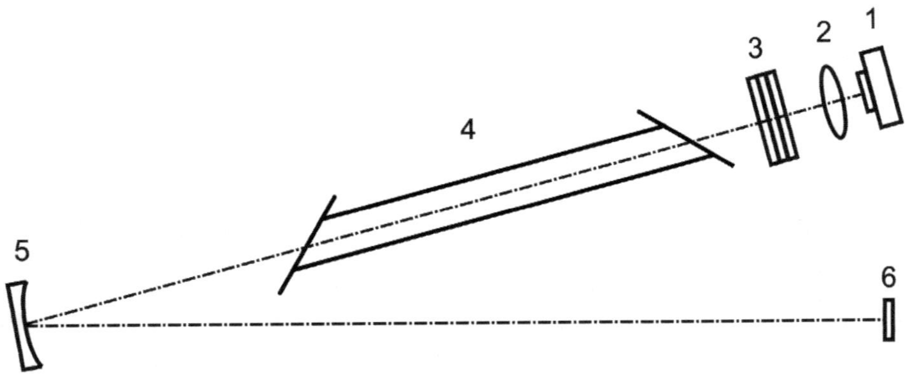

Figure 2. The scheme of a laser projection microscope with one concave mirror: 1 – CMOS camera; 2 – objective; 3 – gray filters; 4 – brightness amplifier; 5 – concave mirror; 6 – object

Figure 3 shows the results of visualization of a steel ruler using this scheme. As one can see the field of view has been obtained to be ~10 mm in case of mirror with R_C=1 m and ~35 mm in case of mirror with R_C=3 m. These values are much greater than those obtained in paper [8], i.e. 4 mm at 0.9 m distance and 3 mm at 1.5 m distance.

We have also tested a variant of the laser monitor scheme with two concave mirrors (Figure 4). This scheme allows us to take away the recording system from the brightness amplifier. In some cases it is necessary to reduce the electromagnetic influence of the discharge on the camera performance. We used mirror 5 (Figure 4) with the radii of curvature of 1 m and 3 m and mirror 3 with the radius of curvature of 3 m. The distance between the GDT output window and the mirror was 0.3 m. We used two types of concave mirrors with radii of curvature (R_c) of 1 m and 3 m, which allowed observing objects moved away over the distance of 1 and 3 m, respectively. GDT2 is used as a brightness amplifier.

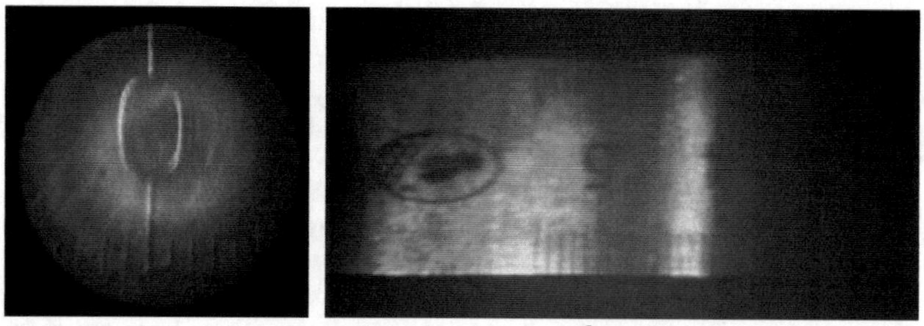

a *b*

Figure 3. Images of a metal ruler obtained with mirrors of different radius of curvature: *a* – 1 m; *b* – 3 m

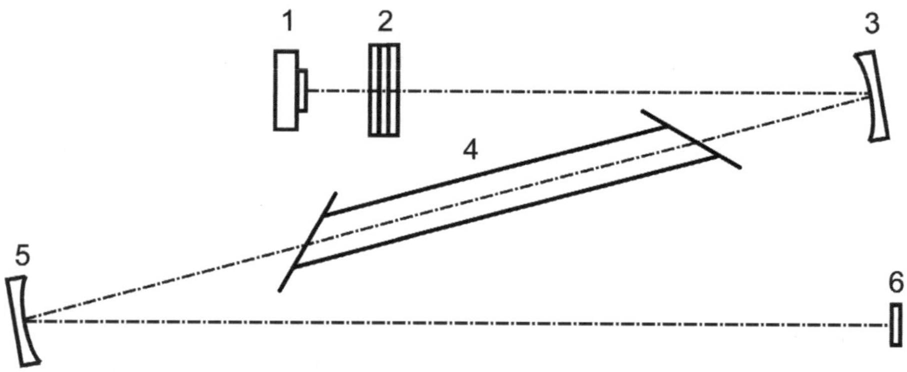

Figure 4. The scheme of a laser projection microscope with two concave mirrors: 1 – CMOS camera; 2 – gray filters; 3 and 5 – concave mirrors; 4 – brightness amplifier; 6 – object

Figures 5 and 6 show the results of visualization of the metal grid (grid size of 0.5 mm) by a two concave mirrors based scheme of the laser projection microscope at different operating wavelengths (510.6 and 578.2 nm).

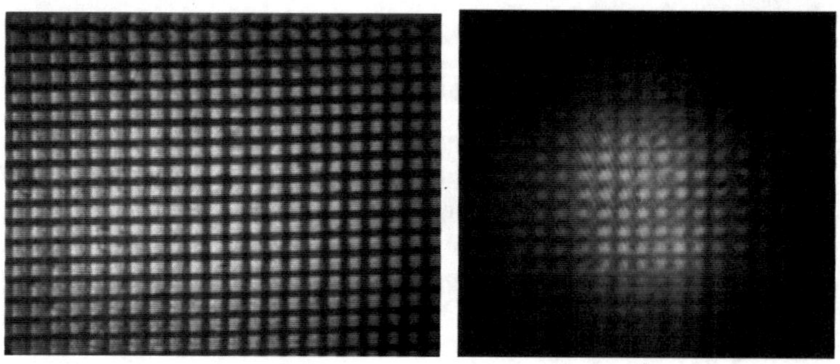

a *b*

Figure 5. Images of the metal grid obtained with mirror 5 (Figure 4) of different radius of curvature: *a* – 1 m; *b* – 3 m. Wavelength of visualization is λ=510.6 nm

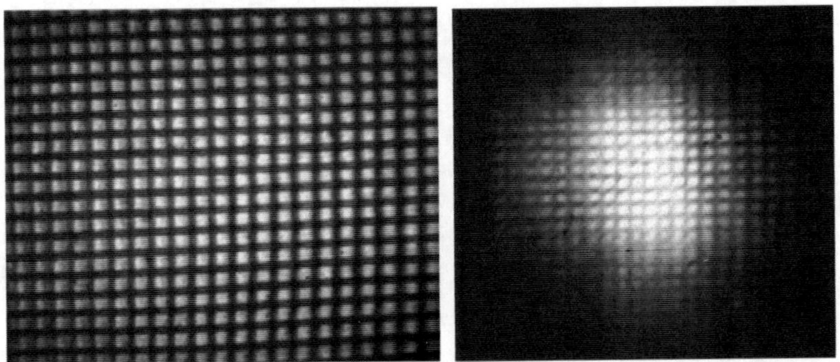

a *b*

Figure 6. Images of the metal grid obtained with mirror 5 (Figure 4) of different radius of curvature: *a* – 1 m; *b* – 3 m. Wavelength of visualization is λ=578.2 nm

4. Conclusions

The presented results have shown the possibility of application of a mirror based scheme of the laser projection microscope (a laser monitor) where concave mirrors are used instead of the traditional collecting lenses or objectives. The scheme can contain two as well as one concave mirrors.

It is shown that the field of view of the projection system decreases with the increase of the radius of curvature of the mirror forming the image (the mirror is between the amplifier and the object). At the same time the discussed scheme provides a greater field of view as compared with the conventional scheme of the laser projection microscope.

The next study will be devoted to improving the image quality of distant objects obtained by a mirror based scheme of the laser projection microscope.

References

[1] Zemskov K I, Isaev A A, Kazaryan M A and Petrash G G 1974 *Sov. J. Quantum electron* **4** 5–6
[2] Batenin V M, Klimovsky I I and Selezneva L A 1988 *Soviet Phys. Dokl.* **33** 949–52
[3] Kuznetsov A P, Buzhinskij R O, Gubskii K L, Savjolov A S, Sarantsev S A and Terekhin A N 2010 *Plasma Phys. Reports* **36** 428–37
[4] Buzhinsky R O, Savransky V V, Zemskov K I, Isaev A A and Buzhinsky O I 2010 *Plasma Phys. Rep.* **36** 1269–71
[5] Gubarev F A, Klenovskii M S and Li Lin 2015 *IOP Conf. Series: Materials Science and Engineering* **81** 012116-1 012116-7
[6] Evtushenko G S, Trigub M V, Gubarev F A, Evtushenko T G, Torgaev S N and Shiyanov D V 2014 *Rev. Sci. Instrum.* **85** 033111-1–5
[7] Buzhinskij O I, Vasiliev N N, Moshkunov A I, Slivitskaya L A and Slivitsky A A 2002 *Fusion Engineering and Design* **60** 141–55
[8] Fedorov K V, Trigub M V and Evtushenko G S 2015 *Int. Siberian Conf. on Control and Communications (SIBCON), IEEE Conference Publications* 1–5
[9] Little C E 1998 *Metal Vapor Lasers: Physics, Engineering and Applications* 620
[10] Yudin N A, Sukhanov V B, Gubarev F A and Evtushenko G S 2008 *Quantum Electron* **38** 23–8
[11] Soldatov A N 1993 *Atmospheric and Oceanic Optics* **6** 650–8
[12] Evtushenko G S 1996 *Dordrecht: Kluwer Academic Publishers* 445–52
[13] Fedorov K V, Fedorov V F, Gubarev F A and Evtushenko G S 2014 *15th Int. Conf. of Young Specialists on Micro/Nanotechnologies and Electron Devices (EDM), IEEE* 333–6
[14] Gubarev F A and Klenovskii M S 2015 *16th Int. Conf. of Young Specialists on Micro/Nanotechnologies and Electron Devices (EDM), IEEE* 292–6
[15] Gubarev F A, Evtushenko G S, Vuchkov N K, Sukhanov V B and Shiyanov D V 2012 *Rev. Sci. Instrum.* **83** 055111–5

[16] Dimaki V A, Sukhanov V B, Troitskii V O and Filonov A G 2012 *Instrum. Exp. Techniq.* **55** 696–700
[17] Evtushenko G S, Kashaev V Yu, Sukhanov V B and Tatur V V 2002 *Proceedings of SPIE* **4747** 198–201

IOP Conf. Series: Materials Science and Engineering **124** (2016) 012017 doi:10.1088/1757-899X/124/1/012017

Description of the traction characteristics of the neodymium compensators of the automatic vibration isolations

E G Gurova, Y V Panchenko and M G Gurov
Novosibirsk State Technical University, 20, K. Marx ave., Novosibirsk, 630073, Russia

E-mail: lena319@mail.ru

Abstract. In this paper the method of calculation of neodymium magnets was presented. The calculation of the neodymium magnets characteristics and stiffness correctors of the vibration isolator according to the requirements for vibration isolation devices with stiffness compensators was performed. This research has been performed with the support of the President scholarship for young scientists, order № 184 of Ministry of education and science of the Russian Federation of the 10th of March 2015.

1. Introduction

In [1] the variant of the device design for decreasing a vibration level is shown. The principle of the action of this device is based on the theory of vibration isolation [2]. Such devices include a resilient element and a compensator of stiffness which are connected in parallel. The design of the proposed compensator is performed with the use of neodymium magnets (supermagnets). These magnets are composed of neodymium, iron and boron. Such magnets have big traction forces and small dimensions. The application of such supermagnets inside the vibration isolation devices simplifies their transport applications for decreasing mechanical oscillations.

2. Calculation and simulations

For designing and calculation of the compensator the vibration and resilient parameters of the whole vibration isolation device should to be taken into account. Therefore, let us calculate the traction characteristic of the stiffness compensator based on supermagnets on the assumption of the fact that the magnitude of vibration is not higher than 1 cm. In [3] the equation of the traction characteristic of the neodymium magnet of a cylindrical shape is presented in (1):

$$F(x) = -\frac{\pi\mu_0}{4}M^2R^4\left(\frac{1}{(2x)^2}+\frac{1}{(2x+2h)^2}-\frac{2}{(2x+h)^2}\right), \tag{1}$$

where $\mu_0=4\pi\cdot 10^{-7}$– magnetic constant, H/m;

M is the residual magnetization of the magnet, A/m;

R is the radius of the magnet, m;

h is the height of the magnet, m;

x is the displacement from the center, m.

Let us assume that there is residual magnetism in the magnet, thus we can find whether force F becomes very big for x close to zero:

$$M = \frac{2B_0}{\mu_0}, \tag{2}$$

where B_0 is a residual magnetic induction of the magnet, T.

In this case the equation of the force dependence on movement of the magnet takes the following view:

$$F(x) = \frac{\pi\mu_0}{4}(\frac{2B_0}{\mu_0})^2 R^4(\frac{1}{x^2} + \frac{1}{(x+2h)^2} - \frac{2}{(x+h)^2}), \quad (3)$$

For calculation of the traction characteristic of the cylinder-shaped neodymium magnet on the basis of equation (3) let us assume the following input data:

- Type of the magnet: N45;
- Residual magnetic induction, B_0: 1320-1380 mT;
- Radius of the magnet, R: 2.25 cm;
- Height of the magnet, h: 2 cm;
- Range of the travel, x: *0* to *10* mm.

The results of the calculation are shown in table 1:

Table 1. Dependence of the force on movement of the neodymium magnet

x, mm	1	3	5	7	9	10
$F(x)$, kN	-267	-27.4	-8.86	-4.02	-2.16	-1.65

According to these data we can formulate the dependence of the repulsion magnet force on movement. Its view is shown in Figure 1.

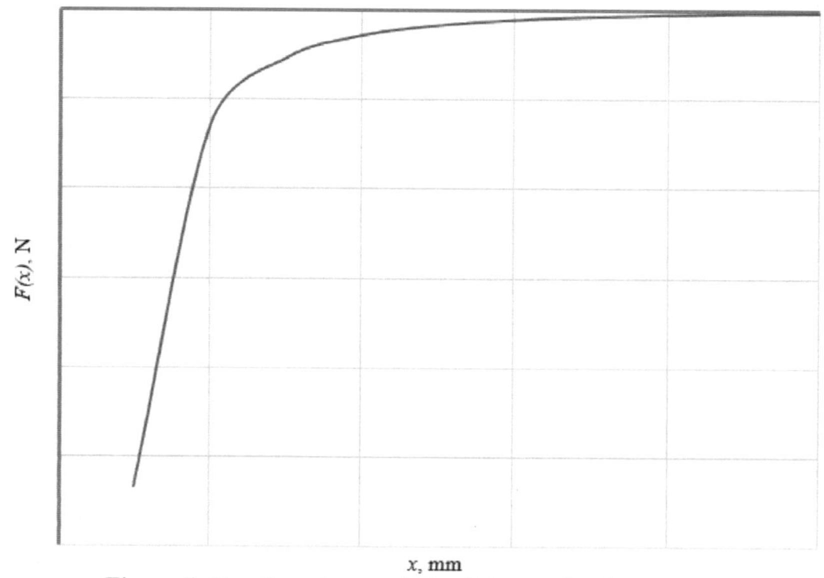

Figure 1. Traction characteristic of the neodymium magnet

Compensators should have an incident force characteristic according to the operating principle of the vibration isolation devices. For this, in the compensator of stiffness based on neodymium magnets it is necessary to use the pairs of oppositely directed magnets. Such position of magnets allows having the incident force characteristic [4].

Now we can calculate the characteristic of two oppositely directed magnets as shown in Figure 2.

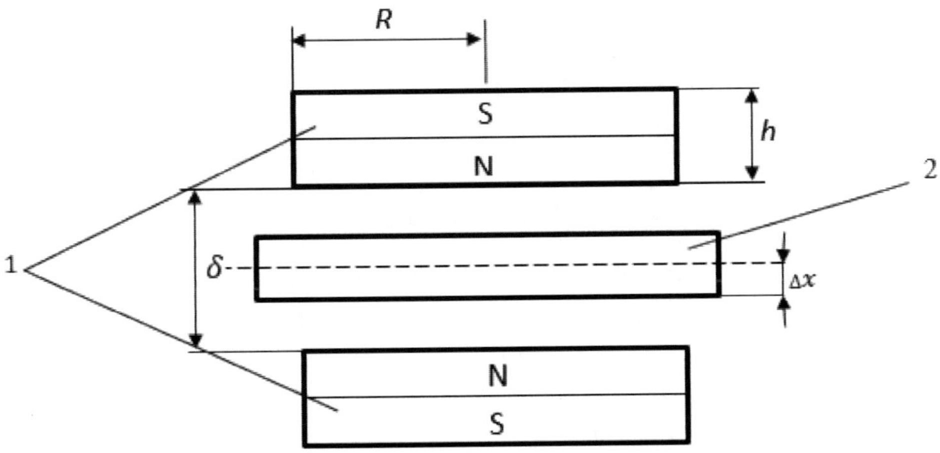

Figure 2. Interaction of two magnets with a flat anchor:
1 – a pair of the oppositely directed magnets; 2 – a flat iron anchor

On the basis of equation (3) for description of the force dependence on movement of two magnets, placed oppositely and relatively each other, that is, the compensator of stiffness itself, we can obtain:

$$F(x) = \frac{\pi\mu_0}{4}\left(\frac{2B_0}{\mu_0}\right)^2 R^4\left(\frac{1}{(2x)^2} + \frac{1}{(2x+2h)^2} - \frac{2}{(2x+h)^2}\right) - \frac{\pi\mu_0}{4}\left(\frac{2B_0}{\mu_0}\right)^2 R^4 \times \\ \times\left(\frac{1}{2(\delta-x)^2} + \frac{1}{(2(\delta-x)+2h)^2} - \frac{2}{(2(\delta-x)+h)^2}\right), \quad (4)$$

where δ is a maximum distance between two magnets, m;

x is a relative movement of the center of the iron anchor with thickness of $2\Delta x$, m.

$$x = x_1 - \Delta x. \quad (5)$$

Considering the fact that the plate should be infinitesimal, value Δx in the formula (5) can be neglected.

On the basis of (4) we can calculate the dependence of the force on movement in the range from 0 to 2 cm:

Table 2. Dependence of the force on movement of the neodymium magnet of the stiffness compensator

x, mm	1	3	5	7	9
$F(x)$, N	2.65E+05	2.34E+04	0	-2.34E+04	-2.65E+05

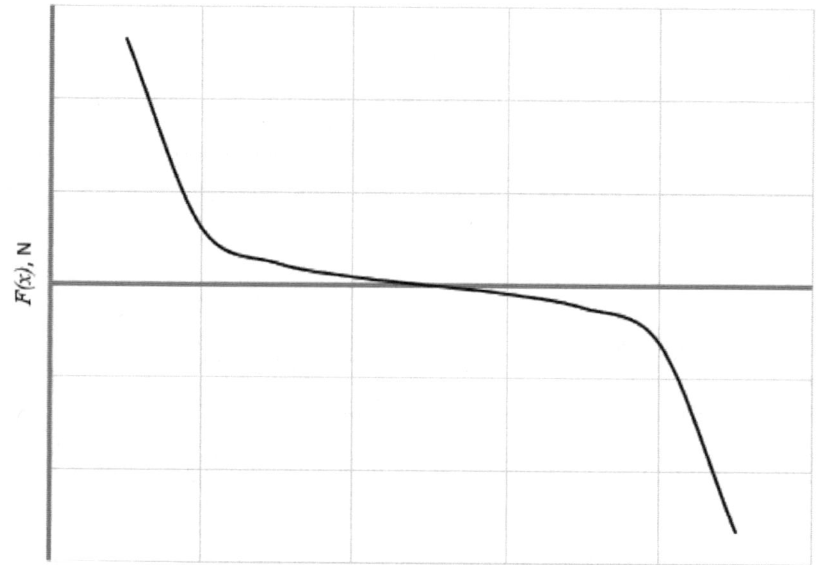

Figure 3. The traction characteristic for two neodymium magnets having a common anchor

The dependence presented in Figure 3 shows that the inclusion of two oppositely directed magnets allows obtaining an incident force characteristic of preset stiffness. The inclusion of the device with a such characteristic corresponds to the operating principles of the vibration isolators.

3. Results and Discussion

However, from (4) we can clearly see that the precise calculation of the characteristic of the compensator is complicated enough and the registration in the control devices by such stiffness compensators is quite difficult either. And it is reasonable to obtain a simplified equation of dependence of the traction characteristic of two oppositely directed neodymium magnets.
For this simplification we should use a polynomial equation of the odd degree:

$$y(x) = A_1x + A_3x^3 + A_5x^5 + \cdots + A_{2n-1}x^{2n-1} \tag{6}$$

where $A_1, A_3, A_5 \ldots A_{2n-1}$ – are constant coefficients.

For the calculation of the constant coefficients we should make up a set of equations of 2n-1.

$$\begin{cases} A_1x_1 + A_3{x_1}^3 + A_5{x_1}^5 + \cdots + A_{2n-1}{x_1}^{2n-1} = y_1, \\ A_1x_2 + A_3{x_2}^3 + A_5{x_2}^5 + \cdots + A_{2n-1}{x_2}^{2n-1} = y_2, \\ \vdots \\ A_1x_{2n-1} + A_3{x_{2n-1}}^3 + A_5{x_{2n-1}}^5 + \cdots + A_{2n-1}{x_{2n-1}}^{2n-1} = y_{2n-1}, \end{cases} \tag{7}$$

Having solved this set of equations let us determine coefficients: $A_1, A_3, A_5 \ldots A_{2n-1}$.

The presented equation (6) allows describing the traction characteristic of two oppositely directed magnets with sufficient precision. It allows a significant simplification of the calculation methods of characteristics and their application. The examples of the accurate (4) and precision (6) equations, presented in this paper, take into account a tuning system in order to obtain the operation of the vibration isolator with the neodymium magnet's corrector of stiffness under different loads.

4. Conclusion

Precision and simplified dependencies for the description of the traction characteristics of stiffness compensators of vibration isolators presented in this paper are a part of SIEDR with the support of the scholarship of the president of the Russian Federation for young scientists (order №184 of the Ministry of science and education of the RF of the 10th of March, 2015), which is directed at the development of the spatial vibration isolation devices based on the theory of the spatial vibration isolation with a zero stiffness phenomenon.

References

[1] Gurova E G and Gurov M G 2014 *Applied Mechanics and Materials* **682** 118-121
[2] Gurova E G, Gross V Y, Kurbatov V S, Makarov S V, Sergeev A A and Shchurov N I 2013 *World Applied Sciences Journal* **22** 44-48
[3] Vokoun D, Beleggia M, Heller L and Sittner P 2009 *Journal of Magnetism and Magnetic Materials* **321** 3758–3763
[4] Gurova E G 2015 *Conf. Series-Materials Science and Engineering* **91** 012091

IOP Conf. Series: Materials Science and Engineering **124** (2016) 012018 doi:10.1088/1757-899X/124/1/012018

Application of mathematical modelling methods for acoustic images reconstruction

I Bolotina, A Kazazaeva, K Kvasnikov and A Kazazaev

Tomsk Polytechnic University, 30, Lenina ave., Tomsk, 634050, Russia

E-mail: bolotina@tpu.ru

Abstract. The article considers the reconstruction of images by Synthetic Aperture Focusing Technique (SAFT). The work compares additive and multiplicative methods for processing signals received from antenna array. We have proven that the multiplicative method gives a better resolution. The study includes the estimation of beam trajectories for antenna arrays using analytical and numerical methods. We have shown that the analytical estimation method allows decreasing the image reconstruction time in case of linear antenna array implementation.

1. Introduction

Currently, ultrasonic non-destructive testing is facing an increasingly relevant problem of considerable improvement of spatial resolution and bringing the dimensions of inspected region closer to the length of probing signal ultrasonic wave λ. This problem can be solved by focusing ultrasonic radiation in the selected point of space. If the inspection depth and ultrasonic transducer dimensions are comparable with wavelength λ, the focusing is practicable only with the use of synthesized aperture having quite large dimensions as compared to the wavelength. Synthetic Aperture Focusing Technique (SAFT) becomes increasingly popular for flaw detection in fine structure materials and medical diagnostics, where it is used for the sake of improved resolution and image quality [1, 2].

SAFT is based on irradiating of a selected point of the inspected region from different directions from the inspection surface, receiving echo-signals from this point also from different directions, and summing the received signals with preliminary addition of time delays to them. This will compensate the differences in distances (time of signal travel) from surface points to the selected point of the inspected semi-space. The generation of a two-dimensional image (tomogram) of the section of the inspected object requires repeating the described procedure for each point of the selected cross-section.

2. Image Reconstruction Using SAFT

Synthetic Aperture Focusing Technique (SAFT) provides for a high rate of the image reconstruction, testing accuracy and an improved signal-to-noise ratio as compared to a conventional phased array.

One of the most significant advantages of SAFT is the capability of increasing a signal-to-noise ratio. Numerous sources of noise determined individually usually form a noise that is characterized by probabilistic parameters. A SAFT processing algorithm implies coherent summation of signals of individual elements comprising the antenna array. In case of normal law of noise probability density distribution, the increase of the signal-to-noise ratio is proportional to the number of elements in the array.

The other benefit of phased antenna arrays is the capability of generating two-dimensional and three-dimensional images of defects. This feature is due to the capability of scanning by controlling an antenna pattern within the required inspection zone. In all probability, the inspection zone can be considered as a linear field; hence, it has the property of additivity. The obtained data comprise a discrete set, which characterizes the state of each separate point of the scanning region. In the method of echo sounding, the regularity of a formed response signal will be primarily based on such characteristic as an antenna pattern. The SAFT-algorithm suggests the realization of the integral value of this parameter in both radiation and reception modes.

Implementation of SAFT is based on computational processing of a large amount of data. All equations describing this method include the embedded summation. In this connection, the application of such equations without computers is practically impossible.

3. Additive and Multiplicative Method for Processing Signals from Antenna Array

SAFT has a large number of modifications that are determined by the method of signal processing. There are such methods as SAFT-C, M-C-SAFT, TS-M-C-SAFT, SPA (Sampling Phased Array) [3, 4], etc.

To estimate acoustic fields, this work has used a multiplication method instead of SPA summation used in the original method. The use of the multiplicative method has the following advantages:

- improved signal-to-noise ratio;
- decreased sidelobe level of the antenna pattern.

The following drawbacks can be mentioned: increased computation time (multiplication is more complex for computers as compared to summation); the necessity of normalized signal received from the transducer stipulated by an overflow caused by multiple multiplications.

Figure 1 demonstrates the antenna patterns of linear antenna arrays received using additive and multiplicative methods of signal treatment. Figure 1 shows that the use of the multiplicative method of signal processing provides a considerably better resolution, while the sidelobe level is lower than that in case of the additive signal processing method.

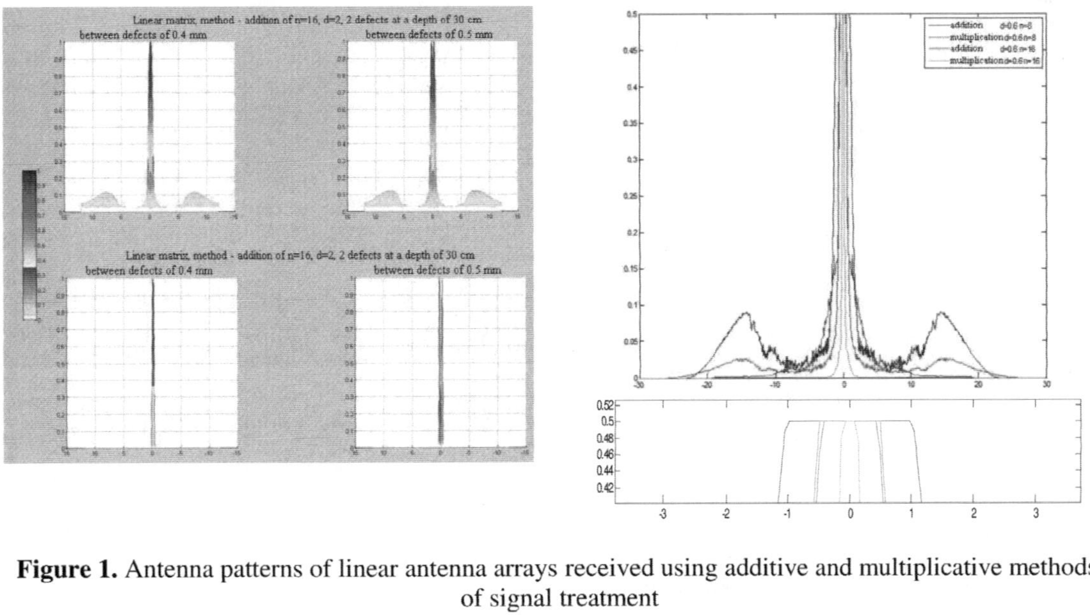

Figure 1. Antenna patterns of linear antenna arrays received using additive and multiplicative methods of signal treatment

4. Real-time Signature Formation

A real-time image reconstruction requires achieving the time of single image generation of at least 1/24 of a second. During the image reconstruction using SAFT, one of the most complex tasks from computational point of view is determination of a trajectory of ultrasound passing from the piezoelectric transducer to the selected inspection point. The necessity of this process is imposed by breaking the inspected object into local regions and considering each local region as a point-like reflecting element. Echo-signals received by each piezo-transducer of the antenna array are shifted back temporarily for the value, which is equal to the time of wave propagation reflected from the inspected local region to a corresponding antenna array piezo-transducer. Then, the time-shifted signals of all antenna array piezo-transducers are summed respectively for each inspected region. The acquired signal amplitude for each local region is mapped by colour and displayed on a monitor [5, 6].

Let us determine the beam trajectory for linear antenna array. It is well known that in the case of the immediate contact of the multi-element antenna array with an inspected object, the width of the intermediate layer of immersion liquid can be neglected (Figure. 2). Then, the calculation of distances from transmitter (point D) to given inspection point (point P) is trivial and comes to the solution of the following quadric equation (1)

$$l_{D-P} = \sqrt{(x_0 - x_d)^2 + {y_0}^2} \quad (1)$$

where x_0,y_0 are coordinates of inspection point P; x_d is the distance from the centre of antenna array to a transducer.

In case when an intermediate agent, for instance, water or a prism, is between the antenna array and the material under inspection, the estimation of paths of ultrasonic waves becomes slightly more complex (Figure. 3).

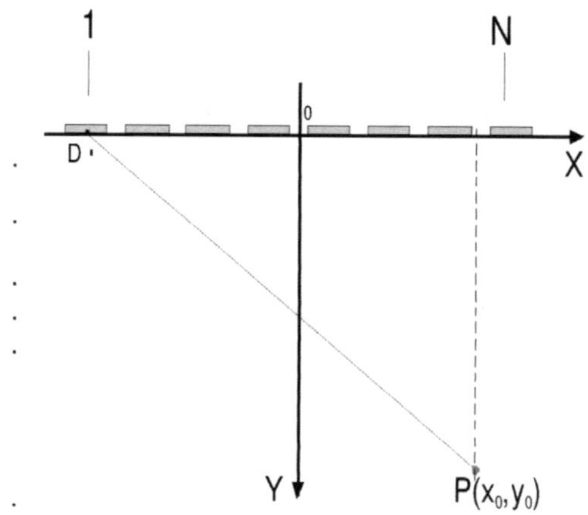

Figure 2. Scheme of ultrasound transmission from the linear antenna array in case of the contact inspection method

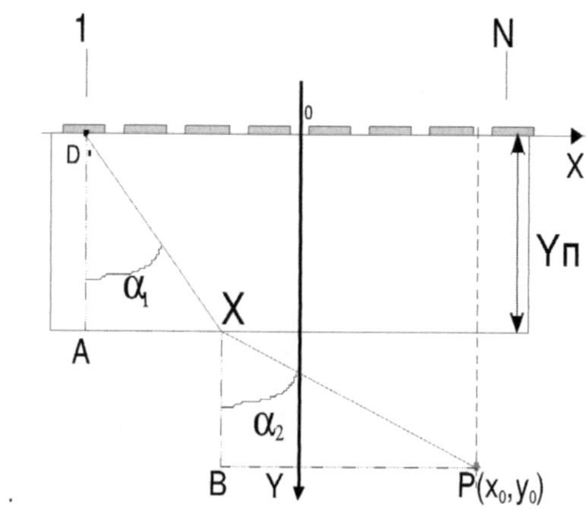

Figure 3. Scheme of ultrasound transmission from the linear antenna array in case of inspection through a prism

Snell's law describes a process of transmission and refraction of waves through an interface by the following equation (2)

$$c_2 \cdot \sin\alpha_1 = c_1 \cdot \sin\alpha_2, \quad (2)$$

where c_1 is sound velocity in the first medium (a prism); c_2 is sound velocity in the second medium (inspection object); α_1 is a wave incidence angle in the first medium (a prism); α_2 is an angle of refracted sound in the second medium (an inspection object);

in order to find point X, we shall express $sin(\alpha_1)$ as a ratio of side *AX* to hypotenuse *DX*, and $sin(\alpha_2)$ as a ratio of side *BP* to hypotenuse *PX* (Figure. 3). Then we shall insert the obtained values into Snell's law and take the square of the right and left parts. In this case,

$$c_2^2 \cdot \frac{(x-x_d)^2}{(x-x_d)^2+y_n^{\,2}} = c_1^2 \cdot \frac{(x_0-x\,)^2}{(x_0-x\,)^2+(y_0-y_n)^2}, \tag{3}$$

where y_n is the height of the prism (an intermediate layer); x is the point of refraction at interface, where angles α_1 and α_2 (Figure.3) shall provide for the compliance with Snell's law.

Having reduced expression (3) to a common denominator and having grouped the coefficients by exponent x, we come to the following expression in the numerator:

$$\begin{aligned}
&x^4 \cdot (c_2^2 - c_1^2) + \\
&2 \cdot x^3 \cdot (-c_2^2 \cdot x_0 - c_2^2 \cdot x_d + c_1^2 \cdot x_0 + c_1^2 \cdot x_d) + \\
&x^2 \cdot (c_2^2 \cdot x_0^2 + c_2^2 \cdot (y_0 - y_n)^2 + c_2^2 \cdot 4x_d \cdot x_0 + c_2^2 \cdot x_d^2 - c_1^2 \cdot x_0^2 - c_1^2 \cdot x_d^2 - c_1^2 \cdot y_n^{\,2} - c_1^2 \cdot 4x_0 \cdot x_d) + \\
&x \cdot (-c_2^2 \cdot 2x_d \cdot x_0^2 - c_2^2 \cdot 2x_d \cdot (y_0 - y_n)^2 - c_2^2 \cdot 2x_d^2 \cdot x_0 + c_1^2 \cdot 2x_0^2 \cdot x_d + c_1^2 \cdot 2x_0 \cdot x_d^2 + c_1^2 \cdot 2x_0 \cdot y_n^2) + \\
&c_2^2 \cdot x_d^2 \cdot x_0^2 + c_2^2 \cdot x_d^2 \cdot (y_0 - y_n)^2 - c_1^2 \cdot x_0^2 \cdot y_n^{\,2} - c_1^2 \cdot x_0^2 \cdot x_d^2 = 0
\end{aligned} \tag{4}$$

Later on, to simplify the expression let us denote the coefficients of the derived equation, such as *a, b, c, d, e*.

Taking into account the non-zero denominator, we obtain $(x-x_d)^2 + y_n^{\,2} \neq 0$ and $(x_0 - x\,)^2 + (y_0 - y_n)^2 \neq 0$

If we adopt

$$p = \frac{8ac-3b^2}{8a^2}; q = \frac{8a^2d + b^3 - 4abc}{8a^3}; r = \frac{b^2c}{16a^3} - \frac{bd}{4a^2} - \frac{3b^4}{256a^4} + \frac{e}{a}, \tag{5}$$

then after solving the acquired quartic equation through Ferrari's solution [7], we get the following roots:

$$x_{1,2,3,4} = \frac{\pm\sqrt{2z_1} \pm \sqrt{2z_1 - 4\left(\dfrac{p}{2} + z_1 + \dfrac{q}{2\sqrt{2z_1}}\right)}}{2} - \frac{b}{4a}, \tag{6}$$

where z_1 is the first real root of the equation.

$$z^3 + pz^2 + \frac{p^2-4r}{4}z - \frac{q^2}{8} = 0.$$

The roots of equations (6) are fairly complex to find. In this connection, one can assume that the numerical method will happen to be less time-consuming. To check this assumption, we will formulate the numerator from equation (3) as

$$f1(x) = c_2^{\,2} \cdot (x-x_d)^2 \cdot ((x_0 - x\,)^2 + (y_0 - y_n)^2) - c_1^{\,2} \cdot (x_0 - x)^2((x-x_d)^2 + y_n^{\,2}) \tag{7}$$

and find the value of x using the numerical method that will correspond to the value of function *f1* being less than 0.1 of the physical dimensions of the inspection point.

If sound velocity in the prism is less than that in the inspected material, then angle α_2 will be larger than angle α_1. Hence, desired point *X* will be between x_d and x_0. Let us take these points as starting ones so that we can search and find the solution through chord method [7].

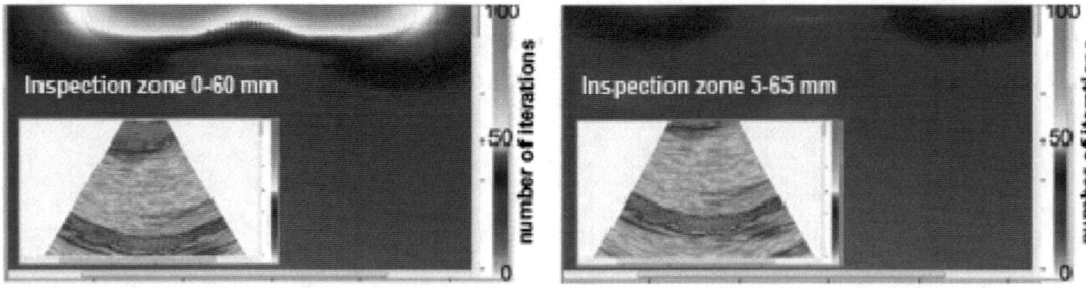

a *b*

Figure. 4. Distribution of the number of iterations for different inspection zones.
a) is an inspection zone from 0 to 60 mm; b) is an inspection zone from 5 to 65 mm.

Figure 4*a* depicts the distribution of the number of iterations necessary for estimating distances from a transducer to the selected point of the inspection zone. The number of iterations is colour-mapped; the principle of colour-mapping is described in the right Figure as a histogram. The insert in Figure 4*a* shows that the reconstruction of the inspection zone is performed on the basis of estimated distances.

We have experimentally established that the largest number of iterations (red zone, more than 100 iterations per point) corresponds to the interfaces. So-called 'dead zone' is also located there. For instance, if we shift the inspection zone by five millimetres into the material, the points with the largest number of iterations will almost disappear (Figure. 4*b*). Moreover, the 'dead zone' is almost completely cut off (an insert in Figure. 4*b*) and the total number of iterations decreases, which leads to shortened estimation time.

5. Conclusions

The use of the multiplicative method of signal processing provides for a considerably better resolution of SAFT, while the side lobe level on the antenna pattern is lower than that in case of the additive signal processing method.

The use of the analytical estimation method allows decreasing the image reconstruction time in case of linear antenna array implementation. The estimation time increases proportionally to the number of displayed points. In case of a large number of displayed points, the real-time image reconstruction is impossible. One of the possible solutions of this problem is shifting to hardware processing on the basis of FPGA logic, which can be used for the analytical estimation method.

Acknowledgments

The study was financially supported by the Ministry of Education and Science of the Russian Federation within the Federal Target Program entitled 'R&D in Priority Areas of Development of Russia's Scientific and Technological Complex for 2014–2020' (unique project ID: RFMEFI57514X0048).

References

[1] Kozlov V N, Samokrutov A A, Yakovlev N N, Kovalev A V and Shevaldykin V G 1989 *Control equipment and systems* **7** 21-23
[2] Thomson R H 1983 *Ultrasonics* **21** 11-18
[3] Badalian V G, Bazulin E G and Vopilkin A Kh 2009 *Moscow: Mashinostroeniye*
[4] Bulavinov A, Kröning M, Pinchuk R and Reddy K M 2009 *Münster: Jahrestagung*
[5] Soldatov A I, Seleznev A I, Soldatov A A, Sorokin P V and Makarov V S 2012 *Russian Journal of Nondestructive Testing* **48** 268-271

[6] Bolotina I O, Kroening H-M, Kvasnikov K G, Sednev D A and Sumtsova O V 2014 *Advanced Materials Research* **1040** 959-964
[7] Korn G and Korn T 2003 *Nauka: Moscow*

Synthesis of a PID-controller of a trim robust control system of an autonomous underwater vehicle

I V Khozhaev and S A Gayvoronskiy
Tomsk Polytechnic University, 30, Lenina ave., Tomsk, 634050, Russia

E-mail: khozhaev.i@gmail.com

Abstract. Autonomous underwater vehicles are often used for performing scientific, emergency or other types of missions under harsh conditions and environments, which can have non-stable, variable parameters. So, the problem of developing autonomous underwater vehicle motion control systems, capable of operating properly in random environments, is highly relevant. The paper is dedicated to the synthesis of a PID-controller of a trim robust control system, capable of keeping an underwater vehicle stable during a translation at different angles of attack. In order to synthesize the PID-controller, two problems were solved: a new method of synthesizing a robust controller was developed and a mathematical model of an underwater vehicle motion process was derived. The newly developed mathematical model structure is simpler than others due to acceptance of some of the system parameters as interval ones. The synthesis method is based on a system poles allocation approach and allows providing the necessary transient process quality in a considered system.

1. Introduction

Modern underwater operations and activities dedicated to scientific, industrial or emergency issues cannot be always performed by a human, especially in harsh environments. This makes a problem of autonomous underwater vehicles (AUVs) development relevant. In order to operate effectively, an AUV must be equipped with a control system, manipulating several parameters: course, roll, trim, translation velocity and depth. In this paper a problem of a trim stabilization control system development is explored.

2. Formulation of a problem

First of all, in order to set the problem properly, a structure of a considered type of AUVs must be stated. AUVs of the considered type are usually cigar-shaped and are being driven by several thrusters: four main thrusters, one horizontal steering thruster and two vertical steering thrusters. The course is manipulated by a horizontal steering thruster, the depth and trim are manipulated by vertical steering thrusters, translation velocity is manipulated by main thrusters. The scheme of thrusters allocation is given below in Figure 1.

The construction of the propulsion and steering system allows considering the trim stabilization problem separately from manipulating the course and roll, so, in this paper it is considered that the course and roll are properly controlled by other subsystems of the AUV.

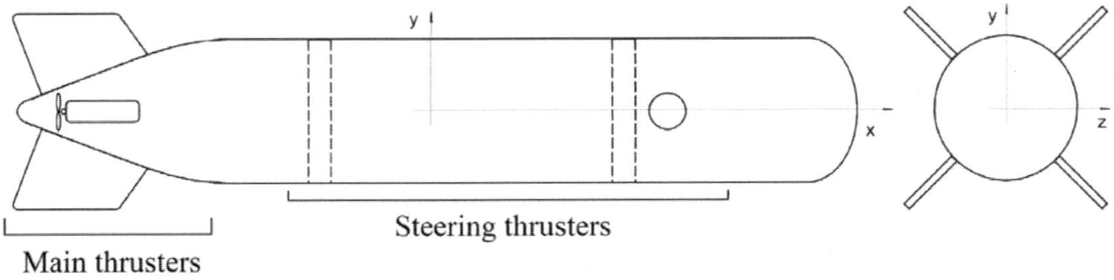

Figure 1. Thrusters allocation

Manipulating the trim of the AUV is considered to be performed by applying the moment of a couple of forces, generated by vertical steering thrusters, to a body of the AUV as it is shown in the Figure 2.

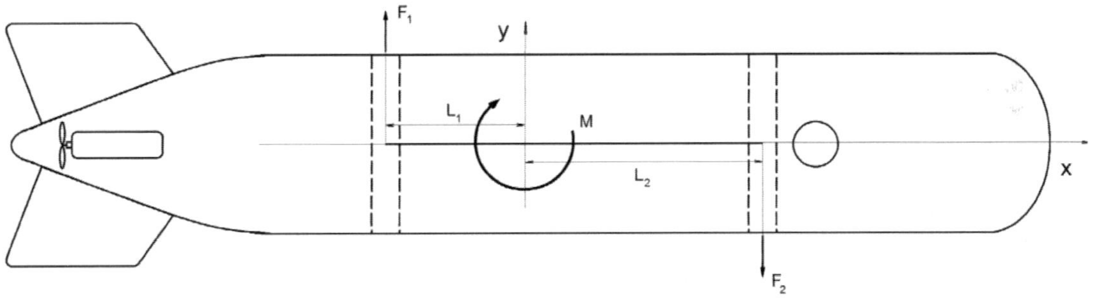

Figure 2. Manipulating principle

A structure of a control loop is shown in Figure 3.

The operating principle of the system shown in figure 3 can be described as follows: a setpoint of the trim is set by a programmable positioner or another control system 1. The error between the actual trim value and the setpoint is calculated by comparator 2 and then arrives to controller 3. The control signal from the output of controller 3 arrives to the propulsion and steering system (PSS) 4. The PSS generates a moment of vertical steering thrusters traction forces, which is applied to the body of AUV 5. After changing the trim value, it is measured by trim sensor 6 and comes back to the comparator. So, the considered system works as a classic closed-loop control system.

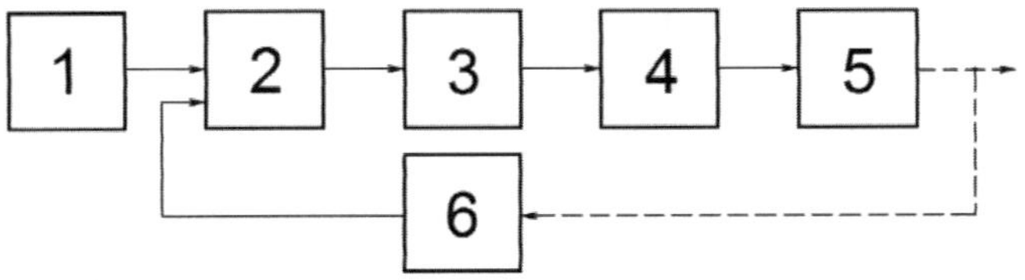

Figure 3. Flowchart of the considered control loop

Finally, the aim of the considered research can be formulated as follows: considering some parameters of the system as interval, it is necessary to synthesize a linear PID-controller for a robust control system shown in figure 3, dedicated to stabilize an AUVs trim during its translation at different angles of attack.

In order to achieve the main aim of the research, following objectives must be accomplished:

- deriving a transfer function of the process of an AUV motion in a longitudinal vertical plane;
- deriving transfer functions of the PSS and a trim sensor;

- developing a controller synthesis method;
- synthesizing a controller, providing a necessary transient process quality.

3. Modelling a process of an AUV motion into a longitudinal vertical plane

The motion of an AUV, having a symmetric body, is fully determined by the following system of differential equations [1]:

$$\left\{\begin{array}{l} (m+\lambda_{11})\cdot\dfrac{d\upsilon}{dt}-(m\cdot y_g-\lambda_{16})\cdot\dfrac{d^2\psi}{dt^2}+(m+\lambda_{11})\cdot\dfrac{d\psi}{dt}+(m+\lambda_{22})\cdot\upsilon\cdot\alpha\cdot\dfrac{d\psi}{dt}-\lambda_{12}\cdot\alpha\cdot\dfrac{d\upsilon}{dt}- \\ -\lambda_{12}\cdot\upsilon\cdot\dfrac{d\alpha}{dt}-\lambda_{12}\cdot\upsilon\cdot\omega_z=T-\dfrac{1}{2}\cdot c_x\cdot\rho\cdot V^{2/3}\cdot\upsilon^2-p\cdot\sin(\psi); \\ -(m+\lambda_{22})\cdot\alpha\cdot\dfrac{d\upsilon}{dt}-(m+\lambda_{22})\cdot\upsilon\cdot\dfrac{d\alpha}{dt}+(m\cdot x_g+\lambda_{26})\cdot\dfrac{d^2\psi}{dt^2}+\lambda_{21}\cdot\dfrac{d\upsilon}{dt}+ \\ +(m+\lambda_{11}-\dfrac{1}{2}\cdot c_y^{\omega z}\cdot\rho\cdot V^{2/3}\cdot L)\cdot\upsilon\cdot\dfrac{d\psi}{dt}=\dfrac{1}{2}\cdot c_y\cdot\rho\cdot V^{2/3}\cdot\upsilon^2-p\cdot\cos(\psi); \\ (J_z+\lambda_{66})\cdot\dfrac{d^2\psi}{dt^2}-(m\cdot x_g+\lambda_{26})\cdot\alpha\cdot\dfrac{d\upsilon}{dt}-(m\cdot x_g+\lambda_{26})\cdot\upsilon\cdot\dfrac{d\alpha}{dt}-(m\cdot y_g-\lambda_{16})\cdot\dfrac{d\upsilon}{dt}+ \\ +\left[m\cdot x_g+\lambda_{26}-\dfrac{1}{2}\cdot m_z^{\omega z}\cdot\rho\cdot V\cdot L-(m\cdot y_g-\lambda_{16})\cdot\alpha\right]\cdot\upsilon\cdot\dfrac{d\psi}{dt}+\lambda_{21}\cdot\upsilon^2=\dfrac{1}{2}\cdot m_z\cdot\rho\cdot V^{2/3}\cdot\upsilon^2- \\ -G\cdot\left[x_g\cdot\cos(\psi)+y_g\cdot\sin(\psi)\right] \end{array}\right.$$

The parameters of this system of differential equations are designated as follows: m – AUVs weight; λ_{ij} – coefficients of added masses of the AUV; α – angle of attack; x_g, y_g – coordinates of the mass center; ρ – water density; V – AUV displacement; c_x, c_y – coefficients of hydrodynamic forces; m_z – coefficient of a hydrodynamic moment; $c_y^{\omega z}$, $m_z^{\omega z}$ – derivative of hydrodynamic forces and moments with respect to angle of attack α; T – traction force of main thrusters; J_z – AUV moment of inertia on Oz axis; p – negative buoyancy; ψ – AUV trim; υ – AUV translation velocity.

The only equation which links rotational characteristics of the AUV motion process with each other is the third one. In this equation the impact of the PSS is not taken into consideration. In order to consider this impact, the moment of a couple of vertical steering thrusters traction forces M was added into the right part of the equation.

Also, this equation describes a multivariable control object. In order to simplify the mathematical model of an object, a translation velocity, an angle of attack and their derivatives were considered as interval parameters. Values of cosines and sinus of the AUV trim are also considered as intervals with minimum and maximum values equal to -1 and 1 accordingly.

Considering all assumptions, the differential equation of a control object can be written as follows:

$$(J_z+\lambda_{66})\cdot\frac{d^2\psi}{dt^2}+\left[m\cdot x_g+\lambda_{26}-\frac{1}{2}\cdot m_z^{\omega z}\cdot\rho\cdot V\cdot L-(m\cdot y_g-\lambda_{16})\cdot[\alpha]\right]\cdot[\upsilon]\cdot\frac{d\psi}{dt}=M+\left(\frac{1}{2}\cdot m_z\cdot\rho\cdot V^{2/3}-\lambda_{21}\right)\cdot[\upsilon]^2-$$
$$-G\cdot\left[x_g+y_g\right]\cdot[-1;1]+(m\cdot x_g+\lambda_{26})\cdot\left([\alpha]\cdot[d\upsilon]+[\upsilon]\cdot[d\alpha]\right)+(m\cdot y_g-\lambda_{16})\cdot[d\upsilon]$$

In this equation some variables are put into square brackets in order to show, that they have interval values. Here [α] – angle of attack, [υ] – AUV translation velocity, [d α] and [d υ] are interval values of their derivatives accordingly. The transfer function of the control object can be written as follows:

$$W_{AUV}(s)=\left\{(J_z+\lambda_{66})\cdot s^2+\left[m\cdot x_g+\lambda_{26}-\frac{1}{2}m_z^{\omega z}\cdot\rho\cdot V\cdot L-(m\cdot y_g-\lambda_{16})\cdot[\alpha]\right]\cdot[\upsilon]\cdot s\right\}^{-1}$$

4. Modelling a trim control loop

Besides the control object, which transfer function was determined previously, a trim control loop includes an actuator – a couple of steering thrusters - and a feedback device – a trim sensor. In order to complete the mathematical model of it, it is necessary to derive mathematical models of thrusters and sensors firstly.

In modern AUVs intellectual thrusters are usually used. An intellectual thruster is equipped with an automatic control system, which manipulates the rotation frequency of a thruster's propeller. So, in fact, the control signal of these thrusters determines a traction force of the thruster but not the rotation frequency of its propeller. The inertness of the thruster should be explored experimentally. The mathematical model of the steering thruster can be written as follows:

$$W_T(s) = K_T \cdot (T_T \cdot s + 1)^{-1}$$

After deriving a single thruster's model, a model of PSS must be derived according to the manipulating principle shown in figure 2. According to this principle, the traction force of each steering thruster should be multiplied by its arm in order to calculate the moment of each thruster's traction force and then, these moments should be summed. So, the mathematical model of the part of PSS, responsible for manipulating the trim, can be written as follows:

$$W_{PSS}(s) = K_T \cdot (L_1 + L_2) \cdot (T_t \cdot s + 1)^{-1}$$

The trim can be described by its transfer coefficient K_{TS} only, because it inertness is negligible. To form a control signal a PID-controller will be used. Its transfer function can be written as follows:

$$W_{PID}(s) = K_P + K_I \cdot s^{-1} + K_D \cdot s$$

After deriving mathematical models of all elements of the system shown in figure 3, a model of the whole control loop can be derived. The transfer function of the considered system can be written as follows:

$$W(s) = \frac{b_2 \cdot s^2 + b_1 \cdot s + b_0}{a_4 \cdot s^4 + a_3 \cdot s^3 + a_2 \cdot s^2 + a_1 \cdot s + a_0}$$

$$a_0 = 2 \cdot K_I \cdot K_{TS} \cdot K_T \cdot (L_1 + L_2); \quad a_1 = 2 \cdot K_P \cdot K_{TS} \cdot K_T \cdot (L_1 + L_2);$$

$$a_2 = 2 \cdot K_D \cdot K_{TS} \cdot K_T \cdot (L_1 + L_2) + [v] \cdot 2 \cdot (\lambda_{26} + m \cdot x_g) + [\alpha] \cdot [\upsilon] \cdot 2 \cdot (\lambda_{16} - m \cdot y_g) - L \cdot V \cdot \rho \cdot \upsilon \cdot m_Z^{\omega_z};$$

$$a_3 = 2 \cdot (J_Z + \lambda_{66}) + [v] \cdot 2 \cdot T_T \cdot (\lambda_{26} + m \cdot x_g) + [\alpha] \cdot [\upsilon] \cdot 2 \cdot T_T \cdot (\lambda_{16} - m \cdot y_g) - L \cdot T_T \cdot V \cdot \rho \cdot \upsilon \cdot m_Z^{\omega_z};$$

$$a_4 = 2 \cdot T_T \cdot (J_Z + \lambda_{66});$$

$$b_0 = 2 \cdot K_I \cdot K_T \cdot (L_1 + L_2); \quad b_1 = 2 \cdot K_P \cdot K_T \cdot (L_1 + L_2); \quad b_2 = 2 \cdot K_D \cdot K_T \cdot (L_1 + L_2);$$

In order to complete the model all parameters of the system must be substituted with their values. For this purpose, the values of interval parameters will be determined. Let us assume that translation velocity υ vary between 0.3 and 3 meters per second; angle of attack α – between -1.5708 and 1.5708 radians. Finally, considering all parameters values, the transfer function of the considered system can be written as follows:

$$W(s, K_P, K_I, K_D) = \frac{K_D \cdot s^2 + K_P \cdot s + K_I}{0.4800 \cdot s^4 + [0.3240; 2.1100] \cdot s^3 + [35.4651 + K_D; 65.0154 + K_D] \cdot s^2 + K_P \cdot s + K_I}$$

5. Synthesizing a robust PID-controller

To synthesize a PID-controller capable of providing the necessary control quality, a newly developed method of synthesis will be used. It is based on a system's poles allocation by applying a robust D-partition method [2].

Before synthesizing a controller, some requirement for the transient process must be determined. It is necessary to provide a 2-3 second long transient process in the considered system. Allocating system's poles allows manipulating the system transient process duration by changing its stability degree. The necessary transient process duration can be provided by achieving a robust stability degree equal to 2 by allocating one system pole at an interval of the real axis of a complex plane between -2.1 and -2; all other poles must be allocated on the left side of the line, which is parallel to the imaginary axis and goes through point (-15;0) of the complex plane.

A necessary poles location can be provided by dividing them into two groups: dominant ones, which location on the complex plane is absolutely fixed and determines a system control quality, and poles with undetermined location, which should be allocated as far from dominant poles as it is possible in order to minimize their influence on the system control quality.

In accordance to aforesaid, an interval characteristic polynomial – denominator of the transfer function – can be divided into three parts as shown:

$$D(s) = A(s, s_0) \cdot B(s, \overline{K}, \overline{q}) + R(\overline{K}, \overline{q}).$$

Here roots of polynomial $A(s, s_0)$ are dominant poles of the system; polynomial $B(s, \overline{K}, \overline{q})$ is polynomial $D(s)$ divided by polynomial $A(s, s_0)$, its roots are poles with undetermined location; $R(\overline{K}, \overline{q})$ is a remainder of this division; s_0 is an interval dominant pole; $\overline{K}$ is a vector of PID-controller parameters; $\overline{q}$ is a vector of system interval parameters. In order to allocate system poles properly with the help of method [2] several steps must be taken:

– the location of the interval dominant pole must be determined;

– $R(\overline{K}, \overline{q})$ equality to zero must be provided;

– coefficients of the polynomial $B(s, \overline{K}, \overline{q})$ must be calculated.

The first step has already been taken, the second step is providing the remainder's equality to zero. To calculate the value of the remainder a following expression will be used:

$$R(\overline{K}, \overline{q}) = \sum_{i=0}^{m} a_i \cdot [s_0]^i = D([s_0])$$

After calculating the interval value of the remainder, in order to provide its equality to zero both minimum and maximum values of the remainder must be equated to zero and put into a system of two linear algebraic equations. For the considered system, these equations can be written as follows:

$$-2.1 \cdot K_P + K_I + 4.41 \cdot K_D + 283.81 = 0$$
$$-2.0 \cdot K_P + K_I + 4.0 \cdot K_D + 125.06 = 0$$

By solving this system, dependencies between controller parameters, which provide remainders equality to zero and a necessary location of the dominant pole, can be derived:

$$\begin{aligned} K_I(K_P) &= 1.0244 \cdot K_P + 1423.70 \\ K_D(K_P) &= 0.2439 \cdot K_P - 387.20 \end{aligned} \qquad (1)$$

After that, in order to place other poles of the system, coefficients of polynomial B must be calculated with the help of the expression written below:

$$[b]_i = [b]_{i+1} + [a]_{i+2} \cdot [s_0]; i \in n-1 \ldots 0.$$

Here $[b]_i$ is a coefficient of $B(s, \overline{K}, \overline{q})$ polynomial; $[a]$ is a coefficient of the original characteristic polynomial of the system. After calculating all coefficients, some substitutions must be made: values of K_I, K_D parameters must be substituted with their functions (1); also substitution $s \to -15 + j\omega$ must be made in order to define the maximal real part of the non-dominant poles. Now, a robust D-partition can be plotted in order to find a stability region on a plane of the K_P parameter. An example of D-partition curve is shown in Figure 4. In Figure 4 the start of the curve is marked with a circle, the end – with a bold point; the chosen value of the controller parameter is marked with a square. After choosing the value of one parameter, it is possible to calculate the rest of the parameters by (1).

Finally, parameters of the PID-controller providing the necessary control quality are $K_P = 1758$, $K_I = 3224.60$, $K_D = 41.58$. In order to estimate the control quality of the considered system a step response in all vertexes of a system coefficients polytope is shown in Figure 5, also a poles location diagram is shown in Figure 6. Figure 5 and figure 6 clearly show that the necessary control quality was successfully provided: the duration of the transient process is less than half a second.

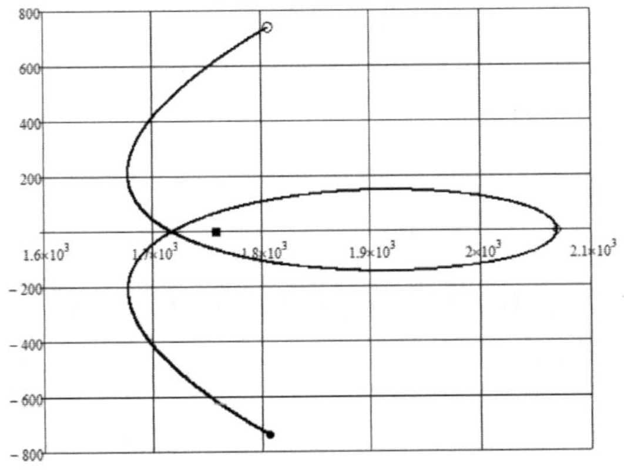

Figure 4. Example of D-partition curve

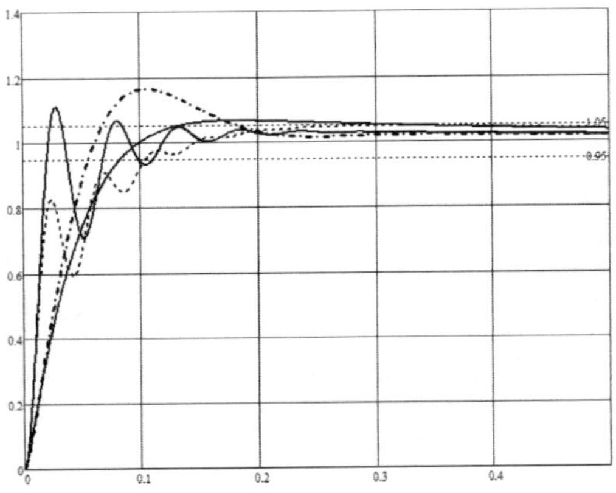

Figure 5. Step responses in all vertexes of coefficients polytope

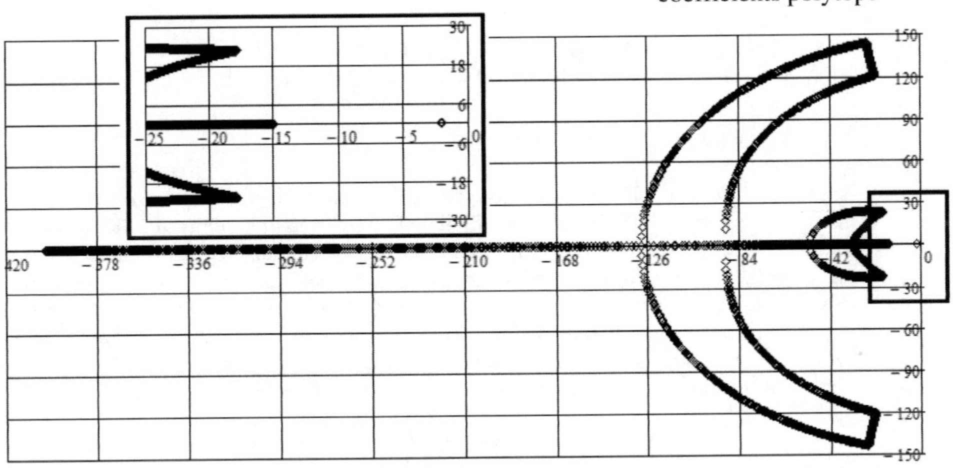

Figure 6. System poles location diagram

6. Conclusion

The main aim of this research has been successfully reached: a robust PID-controller, capable of providing the necessary control quality for the trim control system, has been synthesized. To do this, a new method of synthesis was developed, which allows determining the system robust stability degree and the transient process duration by allocating a single interval dominant pole. This method can only be applied to systems with the simplest interval type of characteristic polynomial coefficients uncertainty. In future papers the method under discussion will be developed to be applicable to systems with an affine or multilinear type of uncertainty. The model of the trim control loop will also be developed later: its interconnection with control loops of the course and roll will be taken into consideration.

References

[1] Pantov E N, Makhin N N and Sheremetov B B 1973 *Saint-Petersburg: Sudostroeniye* 173
[2] Polyak B T and Sherbakov P S 2002 *Moscow: Science* 229

IOP Conf. Series: Materials Science and Engineering **124** (2016) 012020 doi:10.1088/1757-899X/124/1/012020

The course correction implementation of the inertial navigation system based on the information from the aircraft satellite navigation system before take-off

V Markelov[1], A Shukalov[2], I Zharinov[2], M Kostishin[2] and I Kniga[1]
[1] SPb Scientific Design Bureau 'Electroavtomatica' by P A Efimova, 40, Marshala Govorova st., Saint-Petersburg, 198095, Russia
[2] Saint Petersburg National Research University of Information Technologies, Mechanics and Optics, 49, Kronverkskiy ave., Saint-Petersburg, 197101, Russia

E-mail: maksim@kostishin.com

Abstract. The use of the correction course option before aircraft take-off after inertial navigation system (INS) inaccurate alignment based on the platform attitude-and-heading reference system in azimuth is considered in the paper. A course correction is performed based on the track angle defined by the information received from the satellite navigation system (SNS). The course correction incudes a calculated track error definition during ground taxiing along straight sections before take-off with its input in the onboard digital computational system like amendment for using in the current flight. The track error calculation is performed by the statistical evaluation of the track angle comparison defined by the SNS information with the current course measured by INS for a given number of measurements on the realizable time interval. The course correction testing results and recommendation application are given in the paper. The course correction based on the information from SNS can be used for improving accuracy characteristics for determining an aircraft path after making accelerated INS preparation concerning inaccurate initial azimuth alignment.

1. Introduction

In the first place, the aircraft path determining accuracy is specified by the initial exposure accuracy of the inertial navigation system in azimuth. The most widely used azimuth alignment in the INS option is gyrocompass azimuth alignment providing necessary and sufficient ramp heading determination accuracy characteristics. But this option has one essential shortcoming for the INS. It is the process of 10…20 min duration.

If it is necessary to reduce the preparation time for departure, one determines the magnetic heading by means of an inductive sensor and then translates it into the true heading with regard to a given magnetic declination and uses the last path value saved by the on-board navigation system after setting the aircraft into the parking space. In addition, this method does not need additional devices or a fine binding of the aircraft to the parking space.

Whereas there is a clear advantage in preparation time, the data error is about 1°, which is insufficiently satisfactory for execution of some proposed tasks. The course error compensation is usually performed by using an additional course correction on the runway, with moving and running start on centerline of a taxiway strip.

The correction course option based on the information from SNS before take-off is aimed at as an alternative method to specify the course value after reduction of the preparation time for departure. The full capacity deployment of satellite orbital constellation GLONASS allows receiving measured parameters with the required accuracy and enables the use of the course correction based on the information from SNS.

2. Course correction based on the information from SNS before take-off

The course correction based on the information from SNS is founded on the assumption of equality of the track angle and the true angle located on straight sections of aircraft taxiing. It should be kept in mind that this assumption can be observed with some error caused by design features of the aircraft.

The current track angle definition based on the information from SNS and its comparison with the current course measured by INS is provided during the course correction of the taxiing sections.

It is assumed that track speeds and accordingly track angles measurement errors based on the information of SNS during straight aircraft movement along the ground are subjected to the normal distribution law with almost zero expectation (which is acceptable for SNS operating in GLONASS in a GPS mode (with turned-off selective availability).

It is also assumed that course wavering measured by SNS relatively the actual track angle during aircraft movement is subjected to the normal distribution law without noticeable INS azimuth drift.

Thus, the course measured by INS and the track angle defined basing on the information from the SNS mismatch expectation may be taken as the INS course error.

The following conditions are used as a criterion, which allows accepting the course measured by INS and the track angle defined basing on the information from the SNS mismatch expectation as a course error:

- the accordance of mismatch to the normal distribution law;
- the accordance of mismatch dispersion to the sum dispersions measured basing on the information from the SNS track angle and measured by the INS course relatively the actual track angle during aircraft movement;
- the required accuracy is provided by the mismatch measurements number.

3. Implementation and testing course correction based on the information from the SNS mode

The present mode was integrated in the onboard complex software of the aircraft of L-39 type for testing the course correction based on the information from SNS. The mode calculations were implemented by the onboard digital computational system with results registered on the onboard drive and their subsequent processing by the ground workstation.

The course correction based on the information from the SNS mode functioning scheme is represented in Figure 1. It consists of SNS at the evaluation stage of the parking space parameters, the course error according to the taxiing sections calculation and manually through the multifunctional display of the course correction stage.

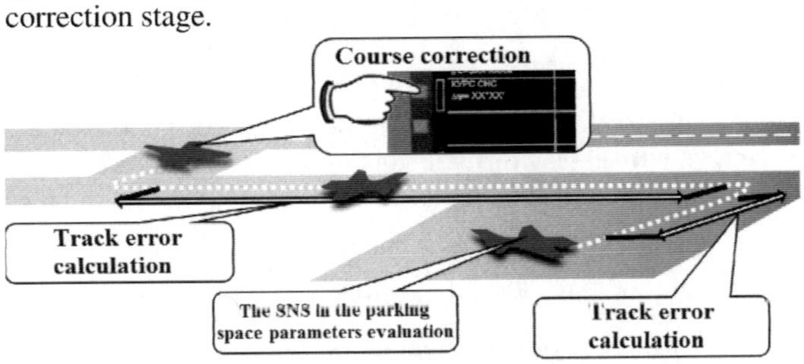

Figure 1. Scheme of the course correction mode.

The SNS information in the GLONASS+GPS mode with the measurement parameters frequency of 10 Hz and geometric factor of less than 2 provided with continuity monitoring and continuity measurement by the SNS data is used as the background information for mode evaluation.

The SNS parameters evaluation of the parking space was implemented in the continuous measurement interval of 10-30 sec with the SNS ground speed values not exceeding 0.2 m/s. Speed components provided by the SNS information, which define a given track angle, are the estimated parameters.

Measurement errors of the northern (WN) and eastern (WE) speed components by the SNS information (σ), which are defined in the measurement interval of 10…30 s with the frequency of 10 Hz, and distribution curves of total speed value expectation (M) in the measurement interval of 10 s and 30 s are represented in Figure 2. The results are obtained from more than 500 implementations, which are spaced in time and place.

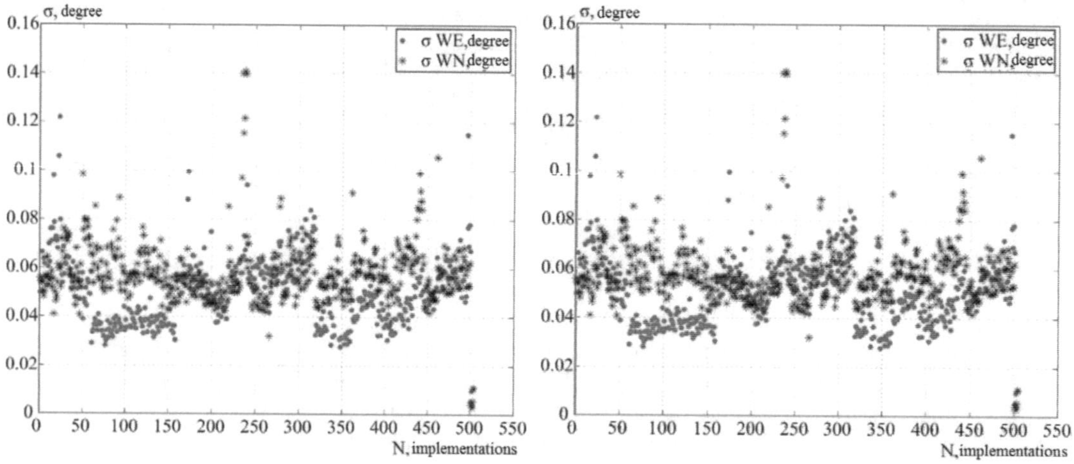

Figure 2. Measurement error of speed components based on the information from SNS and the distribution curves of total SNS speed value expectation.

Speed components measurement errors, which are obtained from the SNS information, are matched to declared characteristics, when SNS operates in the GLONASS mode with the standard accuracy and in the GPS mode with the absence of selective access. Speed measurement errors are up to 0.2 m/s (2σ).

The track error calculation was implemented along the straight section of the airplane movement, which is characterized by the absence of a turn sign and by the track speed obtained from SNS, which exceeded the given threshold value, allowing the obtainment of the sufficient accuracy of the track angle calculation. The INS course changing speed by more than 1 °/sec was accepted as a turn sing. The INS course changing speed value equal to 6 m/s was accepted as a threshold value of speed.

A discrete course error was defined as a difference between the course measured by INS and the track angles calculated by track SNS speed components according to the following formula:

$$\Delta\varphi = \varphi_i^{INS} - \mathrm{arctg}\left(\frac{WE_i}{WN_i}\right)$$

where $\Delta\varphi_i$ – discrete course error of the *i*-th calculation cycle; φ_i^{INS} – true course, measured by INS; WN_i, WE_i – northern and eastern track of SNS speed components.

Track error distribution curves for 11 implementations with more than 400 measurement error cycles in every implementation with total taxi time from 20 s to 120 s are represented in Figure 3. These curves are matched to the normal distribution law with the insignificant deviation, which are determined due to the airplane performance during movement and environmental characteristics. The oscillation amplitude of the course relative track angle measured by INS during airplane movement was less than 1°.

The unknown quantity of the INS error, which is defined according to the information from SNS, is an expectation and is defined along the straight section of the airplane movement according to the following formula:

$$\Delta\Psi_{av_i}=\Delta\Psi_{av\,i-1}+\frac{1}{i}\left(\Delta\Psi_i-\Delta\Psi_{av_{i-1}}\right);$$

where $\Delta\psi av_i$ – track error after the i-th calculation cycle; $\Delta\psi av_{i-1}$ – track error after the $(i-1)$-th calculation cycle.

The number sufficiency of measurements is defined using the track error measurement accuracy calculation (ε):

$$\varepsilon^2=1.96^2\cdot\frac{\sigma_i^2}{(i-1)}$$

where variance (σ_i^2) is defined according to the following formula:

$$\sigma_i^2=\sigma_{i-1}^2\cdot\frac{(i-2)}{(i-1)}+\frac{(\Delta\Psi_i-\Delta\Psi_{av_{i-1}})^2}{i}$$

A sufficient track error measurement accuracy is considered to be 0.05°, which matched more than 400 course measurement error calculation cycles on the average.

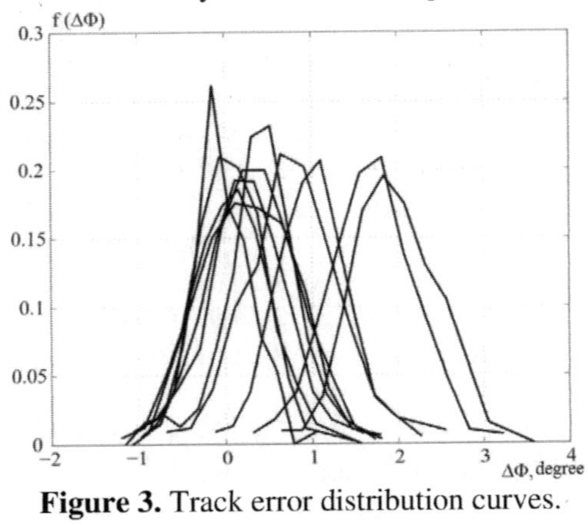

Figure 3. Track error distribution curves.

The mean square deviation of the track error in the measurement of the interval aimed at as substituting criteria of the calculated track error of the normal distribution law and of the absence of bias in its assessment. The criteria value should not exceed 0.3°.

The given mean square deviation is defined according to the following formula:

$$\sigma[\Delta\varphi_{av}]_i^2=\sigma[\Delta\varphi_{av}]_{i-1}^2\cdot\frac{(i\text{-}2)}{(i\text{-}1)}+(\Delta\varphi_{avp_i}\text{-}M[\Delta\varphi_{av}]_{i-1})^2/i$$

where

$$M[\Delta\Psi_{av}]_i=M[\Delta\Psi_{av}]_{i-1}+\frac{1}{i}\left(\Delta\Psi_{av_i}\text{-}M[\Delta\Psi_{av}]_{i-1}\right);$$

$\sigma[\Delta\psi av]_i$ – track error mean square deviation in the measurement interval of the i-th calculation cycle.

The course correction mode evaluation using the information from SNS was implemented by a comparison of the corrected course obtained from the SNS information, leading to the parking space, with a reference value.

The ramp heading value, which is defined through a theodolite, is aimed at as the reference value.

The error value, which was obtained during correction of the information obtained from SNS by all implementations, did not exceed 20′. This fact makes implementation of this correction of the INS inaccurate alignment by azimuth worthwhile.

According to the ground tests result, the course correction mode based on the information obtained from SNS was recommended to use for clarification of the airplane course after the INS accelerated preparing with the initial azimuth defined by the magnetic corrector.

4. Conclusion

The course correction based on the information obtained from SNS before take-off can be used for improving the accuracy characteristics of the definition of the airplane course, if the airplane has an inertial navigation system based on the platform of the attitude-and-heading reference system, after the INS accelerated preparing with the specified initial azimuth determined by a magnetic corrector.

The course error value for correction is defined during airplane taxiing for 20 s with operating SNS in the GLONASS or GPS mode (with turned-off selective availability) and is available using data of integrity monitoring, and with the operating additional parametric control of measured and calculated parameters. The calculation error will not exceed 20′.

References

[1] Gatchin Y A, Zharinov I O, Korobeynikov A G and Zharinov O O 2015 *Journal Modern Applied Science* **9**(5) 197–210

[2] Paramonov P P, Shukalov A V, Raspopov V Ya, Ivanov Yu V and Shvedov A P 2014 *Journal Russian Aeronautics* **57**(3) 319–323

[3] Aleksanin S A, Zharinov I O, Korobeynikov A G, Perezyabov O A and Zharinov O O 2015 *Journal of Engineering and Applied Sciences* 10(17) 7494–7501

[4] Raspopov V Ja, Tovkach S E, Paramonov P P and Sabo J I 2011 *Vertical References for Unmanned Aerial Vehicles* (IEEE Aerospace and Electronic Systems) **26**(3) 42–44

[5] Raspopov V Ja, Tovkach S E, Paramonov P P and Sabo J I 2011 *Journal Gyroscopy and Navigation* **2** 92–98

[6] Raspopov V Ya, Ivanov Yu V, Alaluev R V, Shukalov A V, Pogorelov M G and Shvedov A P 2013 *Journal Automation and remote control* 74(12) 2189–2193

Control system structure design for object positioning system

V Kurgankin, V Zamyatin and S Zamyatin
Tomsk Polytechnic University, 30, Lenina ave., Tomsk, 634050, Russia
E-mail: kurgankin@tpu.ru

Abstract. Object positioning system is intended for load transposition. Its main feature is the control method which is based on operator muscle force. One of the most important object position system development problems is a control system design. The article presents the first step of this problem solution that consists of control system structure design.

1. Introduction

In many industries significant progress and economic efficiency are related to the level of handling mechanical engineering development, which is based on the widespread introduction of production processes automation, the elimination of manual handling and hand-labour.

There is a need for lifting and conveying machines creation and its application because of assembly quality increasing. Cranes are the particular case of those machines. A control of such equipment uses different ways: from crane cabin, using pendant switch, from operator station, etcetera [1, 2]. But sometimes a load positioning quality is increased by an operator muscle force (for example, when there is a need to install a car engine). That is why there is the necessity for an object positioning system (OPS) creation that helps with load transposition [3, 4].

2. The load positioning method

OPS is intended for load transposition. There is a huge amount of its design solutions. In common way developed OPS is a travel crane [3], shown in figure 1. The load hangs on the rope which is reinforced at the movable support and moves in space through actuators help. Units 1 and 2 are X and Y axis slides with actuators. They move load along the horizontal plane using bearing slides 3 and 4. Unit 5 is an elevation. Unit 7 generates control system feedback using information from sensor unit 6. That is why the load moves in space. The structure of sensor unit 6 includes two sensors of rope deflection from the vertical axis and one rope tension sensor.

The system allows one to position the object by one's own direct affecting, as well as to improve the accuracy of work due to the direct operation control.

OPS model in figure 2 describes the method of its control. The load hangs on the rope (figure 2 *a*). Operator defines load track from A to B using one's own muscle force $\overline{F}$, which is directly applied to the load (figure 2 *b*). This is the reason of rope deflection and tension changing (figure 2 *c*). The control system generates the actuator control signals using angle $\overline{\varphi}$ and rope tension $\overline{T}$. That's why the load moves according to the desired trajectory in space (figure 2 *d*).

IOP Conf. Series: Materials Science and Engineering **124** (2016) 012021 doi:10.1088/1757-899X/124/1/012021

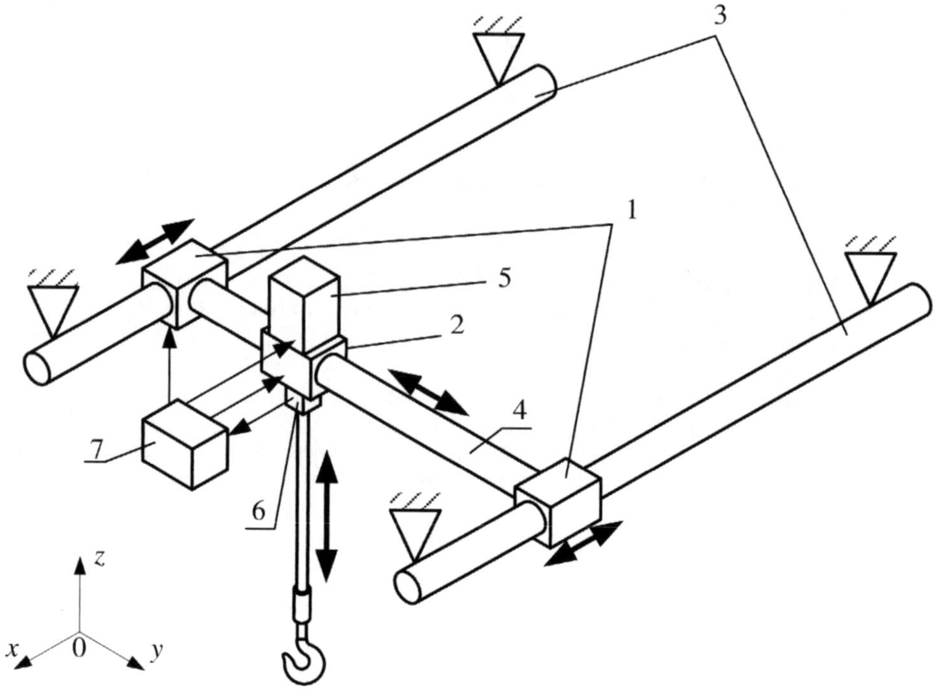

Figure 1. OPS principal scheme

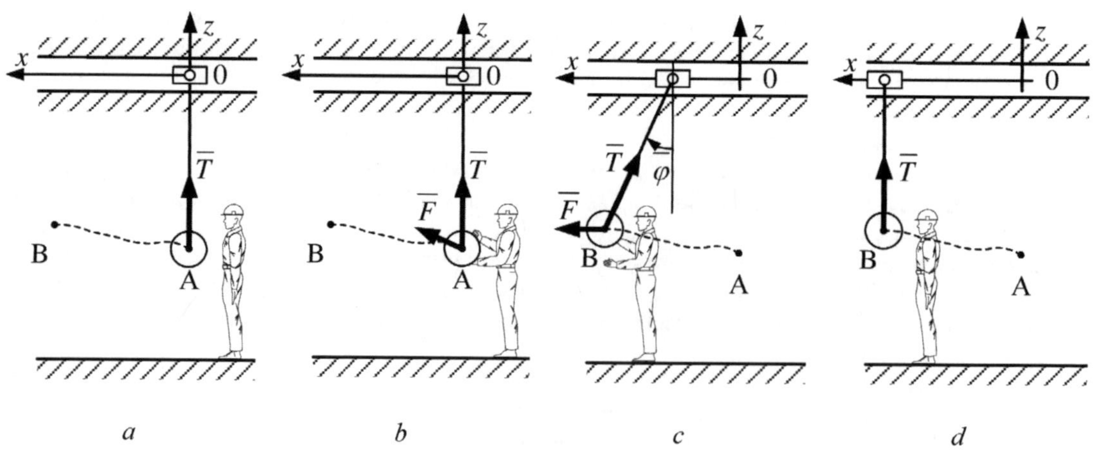

Figure 2.

The control system uses information of three measured OPS parameters. These parameters are rope tension $\overline{T}$ and rope deflection angles $\overline{\varphi}$ and $\overline{\psi}$ (figure 3). These parameters are generic coordinates, which are able to present the desired trajectory.

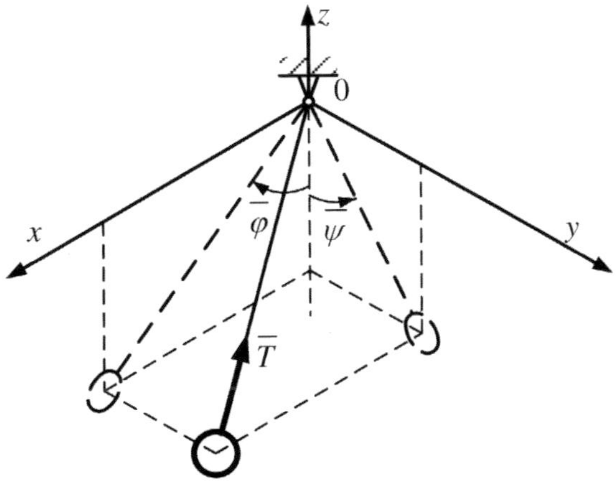

Figure 3. System measured parameters

3. Laboratory bench

The laboratory bench shown in figure 4 is created for OPS processes research. It is able to transpose load, which mass is between 1 and 20 kg. Benches' outline dimensions (LxWxH) are 2500x2500x2500 mm. Operating area (LxWxH) is 2000x2000x2000 mm.

Figure 4. Laboratory bench

4. Control system structure

4.1. Control object block schematic diagram of
OPS (Figure 5) is a multiple-input multiple-output control object, which diagram is shown in figure 5, where $\mathbf{U}(t)$ and $\mathbf{Y}(t)$ are associated matrix input and output.

The specification of the laboratory bench (figure 4) is the following:

$$\mathbf{U}=\left(u_{\dot{x}}(t)\;\; u_{\dot{y}}(t)\;\; u_{\dot{l}}(t)\right)^T \mathbf{Y}=\left(\varphi(t)\;\; \psi(t)\;\; T(t)\right)^T,$$

where l – rope length; $u_{\dot{x}}(t)$ and $u_{\dot{y}}(t)$ – control inputs of actuators that move load along the horizontal plane; $u_{\dot{l}}(t)$ – actuator control input, which changes the height of the load.

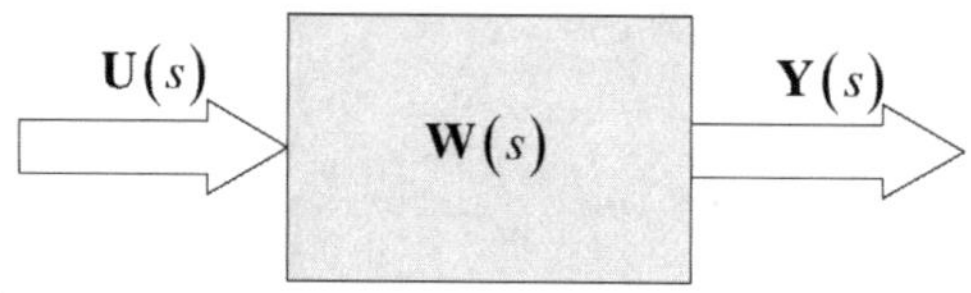

Figure 5. Control object

The following matrix transfer function describes the object math model:

$$\mathbf{W}(s)=\begin{pmatrix} W_{\varphi\dot{x}}(s) & W_{\varphi\dot{y}}(s) & W_{\varphi\dot{l}}(s) \\ W_{\psi\dot{x}}(s) & W_{\psi\dot{y}}(s) & W_{\psi\dot{l}}(s) \\ W_{T\dot{x}}(s) & W_{T\dot{y}}(s) & W_{T\dot{l}}(s) \end{pmatrix}. \tag{1}$$

According to the load positioning method [3, 4] load horizontal speeds $\dot{x}$ and $\dot{y}$ are determined by angles φ and ψ respectively.

Let us assume that $\Delta T = T - T_0$ (T_0 – rest load tension) determines load vertical speed ($\dot{l}$) direction and value. That is why decoupled circuits $\varphi \rightarrow \dot{x}$, $\psi \rightarrow \dot{y}$, $\Delta T \rightarrow \dot{l}$ are obtained. This is the reason that expression (1) transforms to

$$\mathbf{W}(s)=\begin{pmatrix} W_{\varphi\dot{x}}(s) & 0 & 0 \\ 0 & W_{\psi\dot{y}}(s) & 0 \\ 0 & 0 & W_{T\dot{l}}(s) \end{pmatrix} = diag\left(W_{\varphi\dot{x}}(s)\;\; W_{\psi\dot{y}}(s)\;\; W_{T\dot{l}}(s)\right). \tag{2}$$

The math model obtaining problem is to find transfer functions $W_{\varphi\dot{x}}(s)$, $W_{\psi\dot{y}}(s)$ и $W_{T\dot{l}}(s)$ using identification methods of single-input single-output systems (for example, [5]).

4.2. Control system block schematic diagram
An OPS model consists of three transfer functions according to expression (2). That is why the OPS control system has three control circuits and three controllers.

4.2.1. Control system block diagram for load horizontal motion
Load horizontal motion is determined by channels $\varphi \rightarrow \dot{x}$ and $\psi \rightarrow \dot{y}$ that are mathematically equal. That is why the control system block diagram (figure 6) is the same for each channel.

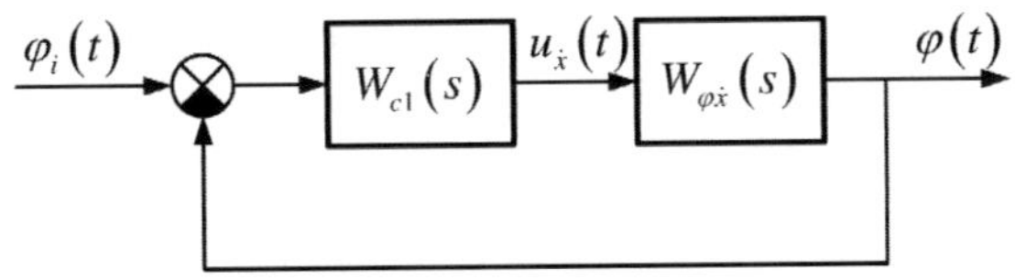

Figure 6. Control system block diagram of channel $\varphi \to \dot{x}$

In figure 6: $W_{\varphi\dot{x}}(s)$ – transfer function of channel $\varphi \to \dot{x}$; $W_{c1}(s)$ – controller transfer function; $\varphi_i(t)$ – control system input.

Such load horizontal motion control systems use angle errors. The control system input is $\varphi_i(t)=0$ when OPS is in an automatic control mode, but the load horizontal speed is determined by the error decay $\varphi_i(t)-\varphi(t)=e_\varphi \to 0$.

4.2.2. Control system block diagram for load vertical motion
According to the way of load positioning [3] the load vertical speed is determined by operator muscle force $\overline{F}$. But there is no possibility to measure force $\overline{F}$ and vertical speed $\overline{V_z}$ in OPS (figure 7).

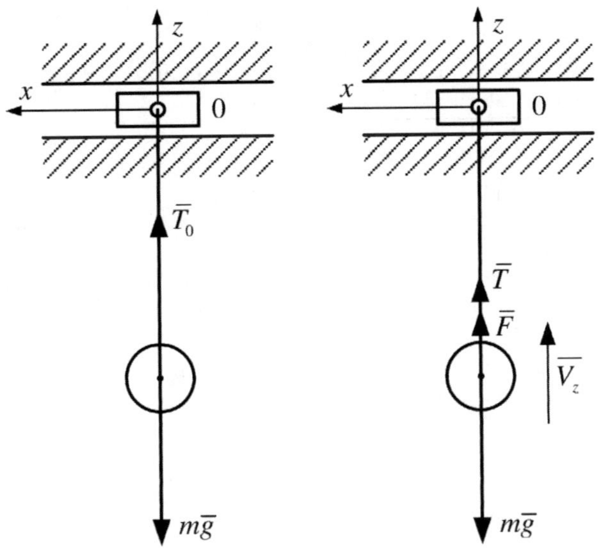

a) Rest state *b)* Operator control

Figure 7. Load acted forces

Let us assume that $|F| \approx |\Delta T|$ and speed is determined by the expression

$$V_z(t) \approx \frac{1}{m}\int \Delta T \cdot dt\,, \qquad (3)$$

where m – load mass.

The expression (3) and our assumption $|F| \approx |\Delta T|$ mean that the designed control system is defined by the dependence $V_z = f(\Delta T)$, which describes a desired load track

The control system block diagram of channel $\Delta T \rightarrow i$ (figure 8) is created on the basis of expression (3).

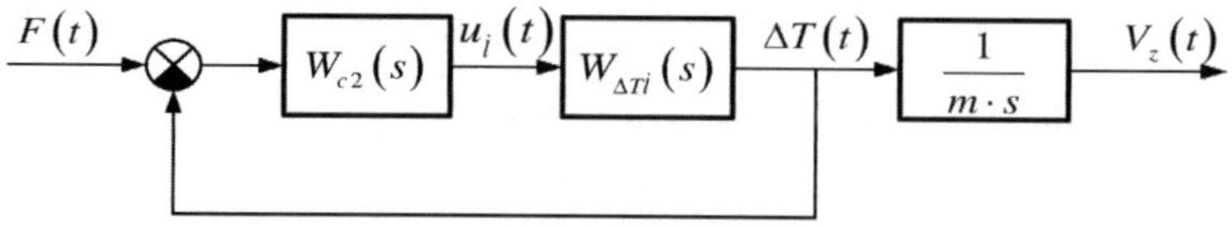

Figure 8. Control system block diagram of channel $\Delta T \rightarrow i$

In figure 8: $W_{\Delta Ti}(s)$ – transfer function of channel $\Delta T \rightarrow i$; $W_{c2}(s)$ – controller transfer function; $F(t)$ – operator muscle force.

5. Conclusion

This paper presents results of a small part of a project which aim is OPS creation and development. The article describes the way of load positioning through OPS and contains an OPS principle scheme. The main result of this paper is block diagrams of the control system, which is the first step of the control system design.

The next steps of OPS development are obtainment of a math model and a controller design on the basis of this paper material.

References

[1] Hicar M and Ritok J 2005 *Journal of ELECTRICAL ENGINEERING* **56** 78–83
[2] Hicar M and Ritok J 2006 *Acta Polytechnica Hungarica* **3**
[3] Zamyatin S V, Kurgankin V V and Zamyatin V M 2011 *RU Patent 2483997*
[4] Kurgankin V V and Zamyatin V M 2015 *Applied Mechanics and Materials: Scientific Journal* **756** 571-574
[5] Zamyatin S V, Kurgankin V V and Rudnicki V A 2014 *Bulletin of The Polish Academy of Sciences: Technical sciences* **62** 341–347

Optimization of Nanostructuring Burnishing Technological Parameters by Taguchi Method

V P Kuznetsov[1,2,3], **A I Dmitriev**[3,4], **G S Anisimova**[2] **and Yu V Semenova**[2]
[1] Ural Federal University, 19, Mira st., Yekaterinburg, 620002, Russia
[2] Kurgan State University, 25, Gogolya st., Kurgan, 640669, Russia
[3] Tomsk Polytechnic University, 30, Lenina ave., Tomsk, 634050, Russia
[4] Tomsk State University, 36, Lenina ave., Tomsk, 634050, Russia

E-mail: wpkuzn@mail.ru

Abstract. On the basis of application of Taguchi optimization method, an approach for researching influence of nanostructuring burnishing technological parameters, considering the surface layer microhardness criterion, is developed. Optimal values of burnishing force, feed and number of tool passes for hardened steel AISI 420 hardening treatment are defined.

1. Introduction

The perspective finishing hardening technology of surface layer processing at multi-purpose treatment of high-precision parts from constructional steel is nanostructuring burnishing. The technology is based on severe plastic deformation development due to repeated frictional loading of elementary volumes of the surface layer material [1, 2]. Nanostructuring burnishing ensures a significant increase of surface layer hardness and a level of compression residual stresses due to grain structure refinement up to a nanoscale state. At this, high plasticity of deformable material as well as dimensional workpiece accuracy remain, and microprofile roughness height decreases several-fold times [1-3].

So far the problem of controlling the nanostructuring burnishing process has not been solved fully because the achievement of the required physical and mechanical properties of the surface layer depends on a significant number of factors: burnishing force and speed, feed and number of tool passes, friction coefficient in the contact area of the indenter and workpiece. The objective of this research aims at optimizing parameters of normal force, feed and number of tool passes at the established magnitudes of the friction coefficient and AISI 420 steel nanostructuring burnishing speed for maximizing surface layer microhardness.

The problem solution was carried out applying Taguchi method, which provides a simple and effective approach to the experiment performance having minimum costs for specimen processing. These particular advantages have caused wide application of Taguchi method for optimizing parameters of various technological processes, including cutting, polishing and burnishing [4-6]. However, till present time, Taguchi method has been applied for optimizing burnishing parameters according to the criterion of improvement of the surface roughness, while nanostructuring burnishing is directed at microhardness increase and structural surface layer modification.

2. Experimental procedure and equipment

Experimental studies were conducted on AISI 420 steel specimens of commercial alloy production containing 0.22 C; 13.06 Cr; 0.17 Ni; 0.28 Mn; 0.30 Si; 0.057 Cu; 0.020 P; 0.008 S, in % wt.; the rest is Fe. The specimens were disk-shaped with the diameter of 75 mm and the thickness of 10 mm. Preliminarily, the specimens underwent quenching and low-tempering at 150°C. The maximum hardness after heat treatment reached 460 HB, the surface layer had a structure of low-tempered lath martensite.

Specimen processing by nanostructuring burnishing was carried out on a multipurpose machining centre Multus B300-W by an indenter with a synthesized diamond in the environment of cooling liquid (Rhenus company, Germany) that provided a friction coefficient in the contact area of the indenter and workpiece surface μ=0.12 [1]. Burnishing speed v_b=25 m/min remained constant in all the experiments.

Surface microhardness after nanostructuring burnishing was measured by means of the ecoHARD XM1270C microhardness tester with load on the indenter of 0.25 N. For each experiment, three values of surface layer microhardness HV were measured and an average value was calculated. The scanning microscopy was carried out with magnification of x2500 by means of a Zeiss CrossBeam AURIGA microscope.

3. Results and discussion

Figure 1 represents the parameters of controlled nanostructuring burnishing. Burnishing force F_b, feed f_b and the number of tool passes n_p were accepted as the variable (input) factors of the process. Surface microhardness *HV* was as a surface layer response to an input action. Burnishing speed v_b and friction coefficient μ in the contact area of the indenter and a workpiece within the frames of this experiment were considered as constant magnitudes.

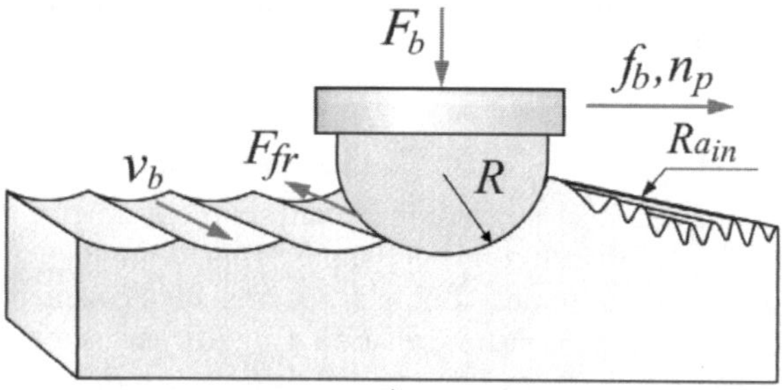

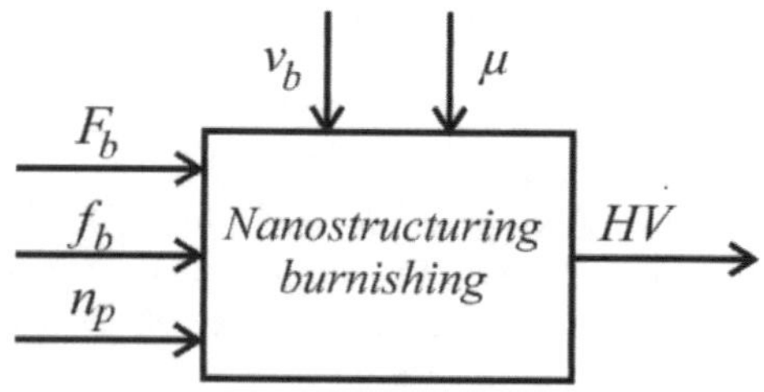

Figure 1. Process diagram (a) and controlling parameters (b) of nanostructuring burnishing: *R* – indenter radius, Ra_{in} – initial surface roughness, F_{fr} – friction force in the contact area of the indenter and a workpiece

The experiment design according to Taguchi method was carried out on the basis of the orthogonal array, the choice of which was defined by the number of assumed variable parameters and assumed levels for each parameter. Thus, for example, for three variable parameters A, B, C, each of which can assume three discrete values (conditionally let us denote them 1, 2, 3), a standard orthogonal array L9 is used, which structure in simple terms is given in Table 1.

Consequently, according to Taguchi method, it is enough to make only 9 experiments to establish the degree of impact of each of three variable parameters on the output characteristic of the process.

The assumed discrete values (levels) of the parameters of controlling nanostructuring burnishing being modified within the frames of the experiment are given in Table 2.

The variation range of nanostructuring burnishing parameters was chosen on the basis of the experimental and theoretical studies having been carried out earlier [1-3]. The minimum force value of burnishing was fixed on condition of full smoothing of the initial roughness, the maximum value was fixed according to the criterion of the material losing shear stability and of essential deterioration of surface layer roughness up to further destruction [3].

Table 1. Standard array L9.

Experiment No.	Parameter values		
	A	B	C
1	1	1	1
2	1	2	2
3	1	3	3
4	2	1	2
5	2	2	3
6	2	3	1
7	3	1	3
8	3	2	1
9	3	3	2

Table 2. Assumed levels of variable parameters

Levels	Variable parameters					
	F_b, N		f_b, mm/rev		n_p	
1	A1	530	B1	0,025	C1	1
2	A2	610	B2	0,04	C2	2
3	A3	690	B3	0,08	C3	3

The resultant orthogonal array considering the assumed values of nanostructuring burnishing variable parameters is given in Table 3.

Table 3. Taguchi orthogonal matrix for carrying out experiments on optimization of nanostructuring burnishing parameters

Experiment No.	Variable parameters		
	A (F_b, N)	B (f_b, mm/rev)	C (n_p)
1	530	0.025	1
2	530	0.040	2
3	530	0.080	3
4	610	0.025	2
5	610	0.040	3
6	610	0.080	1
7	690	0.025	3
8	690	0.040	1
9	690	0.080	2

To define the extent of influence of variable technological parameters on the output characteristic of the process HV, for each experiment, a signal to noise ratio SN_i was calculated. Since by experiment condition the output characteristic of the process has to be maximized, the calculation of a signal to noise ratio was carried out according to the formula:

$$SN_i = -10\log\left[\frac{1}{N_i}\sum_{u=1}^{N_i}\frac{1}{y_u^2}\right], \quad (1)$$

where y is an output characteristic of the process, i is the number of the experiment, u is the number of a measurement in the experiment, N_i is the number of measurements in the i-th experiment.

The calculations were carried out by means of the software for processing statistics, Minitab 15. The results are given in Table 4.

On the basis of the obtained data, the average values of signal to noise ratio for all three levels of each variable parameter of the process are calculated, the difference Δ between maximum and minimum values SN is determined and parameters ranking is carried out. The results are shown in Table 5. The data from Tables 6 are graphically represented in Figure 2. The parameters which are characterized by the greatest value SN are to provide the maximum surface layer microhardness: burnishing force F_b =690 N, feed f_b =0.025 mm/rev, number of tool passes n_p = 3.

Table 4. Calculation results of signal to noise ratio

Experiment No.	Microhardness, HV				Signal to noise ratio, *SNi*
	Mes. 1	Mes. 2	Mes. 3	Mean	
1	897	932	922	917	59.24
2	851	874	862	862.33	58.71
3	788	808	798	798	58.04
4	840	840	862	847.33	58.56
5	862	874	897	877.67	58.86
6	840	862	851	851	58.6
7	1031	1031	1046	1036	60.31
8	909	909	934	917.33	59.25
9	874	897	885	885.33	58.94

The analysis shows that feed f_b and burnishing force F_b have the maximum statistically significant effect of nanostructuring burnishing variable parameters on the surface layer microhardness HV. The number of tool passes has considerably smaller effect on the output characteristic of the process.

Table 5. Parameters ranking

	SN		
Level	F_b	f_b	n_p
1	58,6633	59,37	59,03
2	58,6733	58,94	58,73667
3	59,5	58,526667	59,07
Δ	0,83667	0,8433333	0,333333
rank	**2**	**1**	**3**

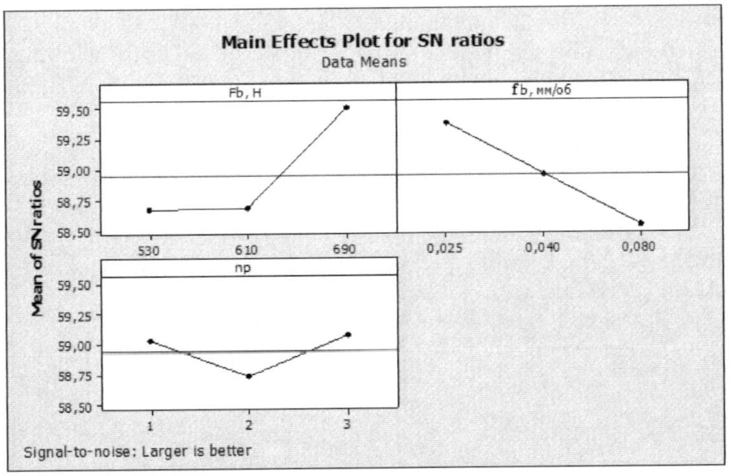

Figure 2. Mean values of signal to noise ratio *SN* for three levels of variable parameters of the process: burnishing force F_b; feed f_b, number of tool passes n_p.

It is established that the surface layer thickness, which underwent shear strain in severe plastic deformation, reaches 10…12 μm (Figure 3).

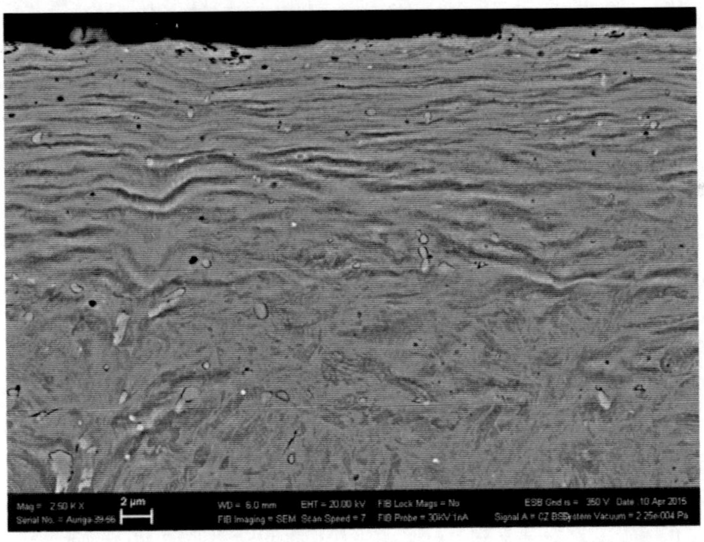

Figure 3. Scanning microscopy of a surface layer of thermostrengthened AISI 420 steel processed by nanostructuring burnishing at optimum parameters F_b =690 N, f_b =0,025 mm/rev and n_p = 3

4. Conclusion

The simplicity and ease of processing outcome results are the advantages of Taguchi method from the point of view of experiment design. The method allows for optimizing nanostructuring burnishing parameters without carrying out an expensive full-factorial experiment. It is established that feed f_b and burnishing force F_b have the greatest effect on surface layer microhardness, and number of tool passes have a lesser influence on hardening. The maximum surface layer microhardness after nanostructuring burnishing with optimum parameter values equals 1036 $HV_{0.025}$.

Acknowledgements
The research is carried out within the frames of the government order for carrying out research and development work No. 01201461774 and backed by the grant RFBR No. 15-08-01511 A.

References

[1] Kuznetsov V P, Makarov A V, Psakhie S G et al 2014 *Phys. Mesomech.* **17(4)** 250-264
[2] Kuznetsov V P, Smolin I Yu, Dmitriev A I et al 2013 *Phys. Mesomech.* **16(1)** 62-72
[3] Kuznetsov V P, Tarasov S Yu and Dmitriev A I 2015 *J. Mater. Process. Technol.* **217** 327-335
[4] Krishankant *et al* 2012 *Int. J. of Engineering and Advanced Technology* **2** 263-274
[5] Shiou FJ and Hsu Ch 2008 *J. Mater. Process. Technol.* **205** 249-258
[6] Pandey A, Garg S, Damle P et *al Int. J. of Computational* **04(12)** 41-45

IOP Conf. Series: Materials Science and Engineering **124** (2016) 012023 doi:10.1088/1757-899X/124/1/012023

Reliability model of fault-tolerant data processing system with primary and backup nodes

P A Rahman[1], E Yu Bobkova[2]
[1] Department of Automated Technological and Informational Systems, Ufa State Petroleum Technological University, Sterlitamak, 453118, Russia
[2] Department of Design and Humanities, Samara Cossack Institute of Food Industry and Business, Samara, 443031, Russia

E-mail: pavelar@yandex.ru

Abstract. This paper deals with the fault-tolerant data processing systems, which are widely used in modern world of information technologies and have acceptable overhead expenses in hardware implementation. A simplified reliability model for duplex systems and the offered by authors advanced model for data processing systems with primary and backup nodes based on a three-state model of recoverable elements, which takes into consideration different failure rates of passive and active nodes and finite time of node activation, are also given. A calculation formula for the availability factor of the dual-node data processing system with primary and backup nodes and calculation examples are also provided.

1. Introduction

In modern world a rapid development of information technologies and their implementation in different spheres of human activity is observed. Almost every day a person has to deal with information. He creates, stores, processes and transmits it using computers and mobile devices. Medium and large-scale enterprises use specialized data storage and processing systems. On their basis a set of information systems operate and assist the business processes of an enterprise.

Data processing systems are widely used in modern enterprises, especially high-availability clusters for database systems, which provide fault-tolerant data processing and storage. A cluster is a set of computers linked by a high-speed communication network and logically united by special software for distributed data processing. In practice the dual-node high-availability clusters with shared storage are used because of the acceptable reliability / cost ratio. To avoid the database access conflicts at any time only one node is active and it processes user requests, the other one is passive in a standby mode. For such systems, it is important to know their reliability for estimation of risks to the business processes. In this situation, the development of reliability models and the analysis of reliability indexes for fault-tolerant data processing systems is quite an important task.

What concerns the reliability models, on the one hand, there are a number of academic books on the reliability theory [1, 2], in which the generalized reliability models of technical systems are discussed, but there are no specific examples related to modern data processing systems, including the high-availability clusters. On the other hand, a number of specialized books [3, 4], dedicated to reliability of computing systems and networks, discuss data processing systems, but the given reliability models for duplex systems are too simplified and provide overestimated values for reliability indexes.

IOP Conf. Series: Materials Science and Engineering **124** (2016) 012023 doi:10.1088/1757-899X/124/1/012023

Within the scope of the scientific research work of authors regarding reliability of data processing and transmission systems [5] a scientific task of development of a specialized reliability model for the dual-node data processing system with primary and backup nodes has been raised. This model is also considered for future application to obtained results in designing of data processing systems for industrial enterprises.

2. A simplified reliability model of the duplex system with independent nodes

In the well-known simplified reliability model of the duplex system [1-4] we consider data processing nodes as the simplest repairable elements with two states: up-state and down-state (Figure 1). From up-state a healthy node can pass to down-state with failure rate λ_A. From down-state a faulty node can pass to up-state with repair rate M_N.

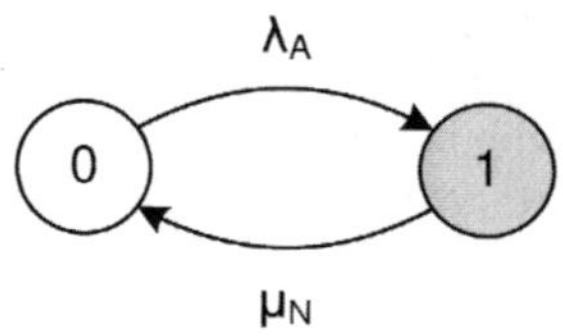

Figure 1. Reliability model of repairable element.

In the duplex system with two independent nodes both of them may be in one of two states independently and the system is considered to be ready for user requests, when at least one node is in up-state. One user request also can be processed on both of healthy nodes simultaneously.

Now let us introduce the following state-space for the duplex system and conditions for transition from one state to another according to a simplified reliability model:

- State 0 (up / up) – both nodes are up, and they are processing user requests. From this state system can pass to state 1 with rate $2\lambda_A$ (failure of any healthy node).
- State 1 (up / down) – one node is up, and it is processing user requests, the other one is down. From this state the system can pass either to state 2 with rate λ_A (failure of the remaining healthy node) or to state 0 with rate μ_N (recovery of the faulty node).
- State 2 (down / down) – both nodes are down, and user requests are not processed. From this state the system can pass to state 1 with rate $2\mu_N$ (recovery of any faulty node).

Figure 2 shows Markov chain, which represents the state-space and transitions conditions according to a simplified reliability model:

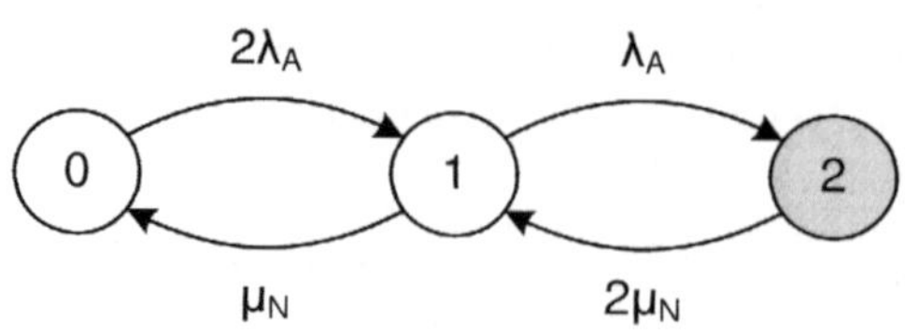

Figure 2. A simplified reliability model of the duplex system with independent nodes.

Accordingly, the differential equations system of Kolmogorov-Chapman for this Markov chain is as follows:

$$\begin{cases} P_0(0)=1; \quad P_1(0)=0; \quad P_2(0)=0; \\ P_0(t)+P_1(t)+P_2(t)=1; \\ \dfrac{dP_0(t)}{dt} = -2\lambda_A P_0(t) + \mu_N P_1(t); \\ \dfrac{dP_1(t)}{dt} = 2\lambda_A P_0(t) - (\lambda_A + \mu_N) P_1(t) + 2\mu_N P_2(t); \\ \dfrac{dP_2(t)}{dt} = \lambda_A P_1(t) - 2\mu_N P_2(t). \end{cases}$$

Considering states 0 and 1 as system healthy states and $t \to \infty$, we obtain the following simple formula for the stationary availability factor of the duplex system with independent nodes:

$$K_{DS} = P_0(\infty) + P_1(\infty) = \frac{\mu_N(\mu_N + 2\lambda_A)}{(\mu_N + \lambda_A)^2}. \tag{1}$$

3. Advanced reliability model of system with primary and backup nodes

In the advanced model of the dual-node system, we consider the finite time of node activation (node initialization and reconfiguration before it will be ready for user requests), and different failure rates for active and passive nodes. We regard nodes as special elements with three states: active, passive and down (Figure 3). Only in the active state the node can process user requests.

From the passive state the node can pass either to the down state with passive failure rate λ_P or to the active state with activation rate γ_N. From the active state the node can pass to the down state with active failure rate λ_A. From the down state the node can pass to the passive state with recovery rate μ_N.

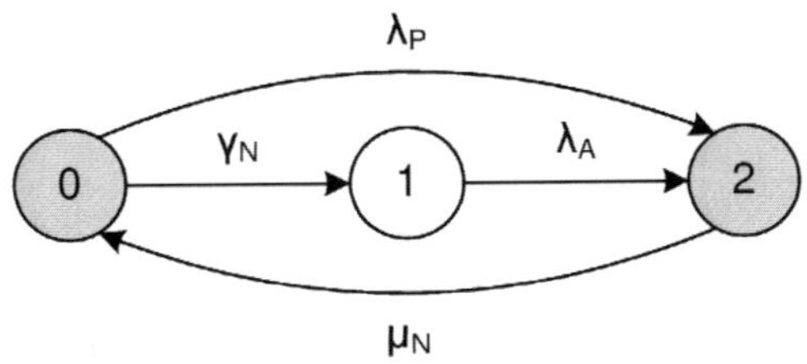

Figure 3. Reliability model of three-state element.

In the advanced model we also suppose that only one of the nodes can be active and the other node must be passive. This is a well-known approach in modern high-availability systems with shared storage, where only one data processing node has access to the system database for processing user requests. So, if both of the nodes are in the passive state and ready to become active, then only one of them (assigned by a system engineer as primary) will pass to the active state, the other node (assigned as backup) will stay in the passive state. As for failure and recovery we assume that both nodes are independent. The advanced model is a further development of reliability models obtained by authors earlier [5].

Now let us introduce the following state-space for the system with primary and backup nodes, and conditions for transition from one state to another according to the advanced reliability model:

- State 0 (passive / passive) – both nodes are passive, user requests are not processed. From this state the system can pass either to state 1 with rate γ_N (activation of the passive node assigned as primary), or to state 2 with rate $2\lambda_P$ (failure of any passive node).

- State 1 (active / passive) – one node is active and the other one is passive, user requests are processed. From this state the system can pass either to state 2 with rate λ_A (failure of the active node), or to state 3 with rate λ_P (failure of the passive node).
- State 2 (passive / down) – one node is passive and the other one is down, user requests are not processed. From this state the system can pass either to state 3 with rate γ_N (activation of the passive node), or to state 4 with rate λ_P (failure of the passive node), or to state 0 with rate μ_N (recovery of the faulty node).
- State 3 (active / down) – one node is active and other one is down, user requests are processed. From this state the system can pass either to state 4 with rate λ_A (failure of the active node), or to state 1 with rate μ_N (recovery of the faulty node).
- State 4 (down / down) – both nodes are down, user requests are not processed. From this state the system can pass to state 2 with rate $2\mu_N$ (recovery of any faulty node).

In Figure 4 Markov chain, which represents the state-space and transitions conditions according to the advanced reliability model, is shown:

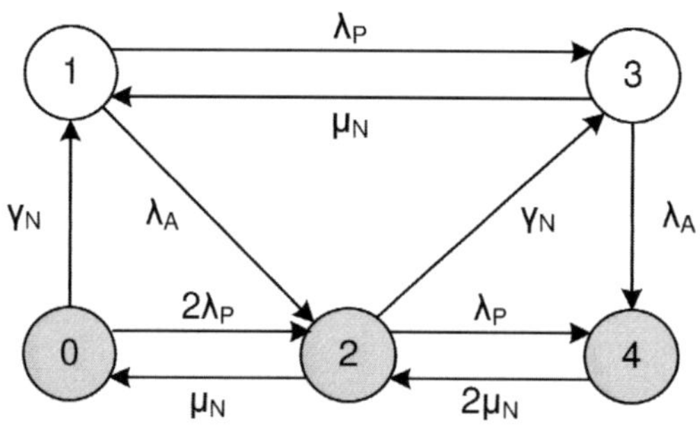

Figure 4. Advanced reliability model of system with primary and backup nodes.

Accordingly, the differential equations system of Kolmogorov-Chapman for this Markov chain is as follows:

$$\begin{cases} P_0(0)=1;\quad P_1(0)=0;\quad P_2(0)=0;\quad P_3(0)=0;\quad P_4(0)=0; \\ P_0(t)+P_1(t)+P_2(t)+P_3(t)+P_4(t)=1; \\ \dfrac{dP_0(t)}{dt}=-(2\lambda_P+\gamma_N)P_0(t)+\mu_N P_2(t); \\ \dfrac{dP_1(t)}{dt}=\gamma_N P_0(t)-(\lambda_A+\lambda_P)P_1(t)+\mu_N P_3(t); \\ \dfrac{dP_2(t)}{dt}=2\lambda_P P_0(t)+\lambda_A P_1(t)-(\mu_N+\lambda_P+\gamma_N)P_2(t)+2\mu_N P_4(t); \\ \dfrac{dP_3(t)}{dt}=\lambda_P P_1(t)+\gamma_N P_2(t)-(\mu_N+\lambda_A)P_3(t); \\ \dfrac{dP_4(t)}{dt}=\lambda_P P_2(t)+\lambda_A P_3(t)-2\mu_N P_4(t). \end{cases}$$

Where,

λ_A is the failure rate of the active node.

λ_P is the failure rate of the passive node.

γ_N is the activation rate of the passive node.

μ_N is the recovery rate of the faulty node.

Considering states 1 and 3 as system healthy states and $t \to \infty$, we obtain the following advanced formula for the stationary availability factor of the system with primary and backup nodes:

$$K_{AP} = \frac{\gamma_N}{(\gamma_N + \lambda_A)\left(1 + \frac{\lambda_A(\gamma_N + \lambda_P)(2\mu_N\lambda_P + (\gamma_N + 2\lambda_P)(\lambda_A + \lambda_P))}{2\mu_N(\gamma_N + \lambda_A)(\mu_N + \gamma_N + 2\lambda_P)(\mu_N + \lambda_A + \lambda_P)}\right)}. \quad (2)$$

Note. In case of infinite activation rate $\Gamma_N \to \infty$ (immediate activation of the passive node), a formula for the availability factor is simplified:

$$K_{AP} \to \frac{2\mu_N(\mu_N + \lambda_A + \lambda_P)}{2\mu_N(\mu_N + \lambda_A + \lambda_P) + \lambda_A(\lambda_A + \lambda_P)}.$$

Moreover, in this case if failure rates for passive and active nodes are also the same $\lambda_A = \lambda_P$, then the formula is simplified to the expression for duplex systems: $K_{AP} \to \frac{\mu_N(\mu_N + 2\lambda_A)}{(\mu_N + \lambda_A)^2}$.

4. Availability factor calculation example

A data processing system with primary and backup nodes is given. The failure rate of the active node is $\lambda_A = 1/8760$ h^{-1} (on average one failure per year). The recovery rate of the node is: $\mu_N = 1/24$ h^{-1} (on average one recovery in 24 h).

For passive node failure and activation rates let us consider three cases:

1) Hot-spare nodes with fast activation: passive node failure rate $\lambda_P = 1/8760$ h^{-1} and activation rate $\gamma_N = 1200$ h^{-1} (on average one activation per 3 s).
2) Warm-spare nodes with intermediate activation: passive node failure rate $\lambda_P = 1/17520$ h^{-1} and activation rate $\gamma_N = 20$ h^{-1} (on average one activation per 3 minutes).
3) Cold-spare nodes with slow activation: passive node failure rate $\lambda_P = 0$ hour^{-1} and activation rate $\gamma_N = 1/3$ h^{-1} (on average one activation per 3 h).

According to the simplified model of the duplex system in all three cases the following value for the availability factor is obtained by using formula 1:

$$K_{DS} \approx 0.9999925349$$

According to the advanced model of the system with primary and backup nodes the following values for the availability factor are obtained by using formula 2:

In case of hot-spare nodes with fast activation:

$$K_{AP} \approx 0.9999924397$$

In case of warm-spare nodes with intermediate activation:

$$K_{AP} \approx 0.9999886897$$

In case of cold-spare nodes with slow activation:

$$K_{AP} \approx 0.9996543268$$

It can be easily noticed that the advanced model (obtained by the authors), which takes into consideration additional reliability parameters of the system with primary and backup nodes, gives more low and realistic estimation of the availability factor rather than the well-known simplified model.

5. Summary

Within the scope of this article a well-known reliability model of duplex system and an advanced model for the data processing system with primary and backup nodes obtained by the authors are presented. The formula for availability factor estimation and calculation examples are also given.

Scientific results obtained by the authors were used for designing of data processing systems of Moscow Power Engineering Institute, Nuclear Power Plant 'Balakovo', Joint Stock Company 'Krasnyi Proletariy' and several other enterprises.

6. Acknowledgements

Authors thank professor A I Kayashev, Sterlitamak branch of Ufa State Petroleum Technological University and professor I I Ladygin, Moscow Power Engineering Institute, for scientific support and theoretical base, also S N Khorkov, a chief system administrator of MPEI corporate network, and R M Gafarov, a chief of information technologies department of Joint Stock Company 'Krasnyi Proletariy', for technical support and experimental base.

7. References

[1] Cherkesov G N 2005 *Reliability of Hardware and Software Systems* (Saint-Petersburg: Piter)

[2] Polovko A M and Gurov S V 2006 *Basis of Reliability Theory* (Saint-Petersburg: BHV-Petersburg)

[3] Shooman M L 2002 *Reliability of computer systems and networks* (New York: John Wiley & Sons, Inc.)

[4] Koren I and Krishna C M 2007 *Fault-Tolerant Systems* (San Francisco: Morgan Kaufmann Publishers)

[5] Rahman P A, Muraveva E A and Sharipov M I 2016 *Key Engineering Materials* **685** 805–810

An Automated Control System for Machinery Parts Machining

D I Petreshin, A V Handozhko and O N Fedonin
Bryansk State Technical University, 7, 50 years of October st., Bryansk, 241035, Russia

E-mail: dipetreshin@yandex.ru

Abstract. The article deals with the problem of creating an automated system for controlling surface layer quality characteristics of machinery parts during machining. An automated system structure, its operation algorithm, mathematical support and work results are provided. The paper proves the necessity of using a self-learning mode in technological systems for providing set values of surface layer quality characteristics.

1. Introduction

Technological support of surface layer quality characteristics of machinery parts during machining consists in determining machining conditions (methods and modes of machining, machinery and tool performance etc.) of work surfaces of machinery parts providing set values with the smallest production cost [1] and their adjustment in workshop environment.

Experience has shown that in most cases during actual processing cutting parameters are usually set up according to standards. However, cutting conditions in actual processing greatly differ from those of standards in which the following aspects are not taken into account: machinery condition, a range of properties of the material being machined and the tool material, fluctuations of machining allowance, etc. It is even more difficult to provide set values of surface layer quality characteristics in case of machining new materials or using new tool materials as data on their cutting conditions are either not available or do not conform to field operating conditions. The latter are usually taken into account by the operator in the workplace according to one's intuition and experience. Besides, the operator is more interested in the increase of manufacturing productivity rather than in working costs or compliance with the drawing. That is why the problem of creating an automated system for controlling surface layer quality characteristics of machinery parts during machining is of vital importance [2-5].

2. Statement of a problem

The task of the Automated Control System is to define cutting parameters directly in the workplace and to update them during machining for providing set values of surface layer quality characteristics.

The operation of the Automated Control System in the mode of adaptive control over surface layer quality characteristics is based on mathematical model [2-5] making a connection between output parameters of the process under control and its input control action. Nowadays there are theoretical and empirical dependences which make it possible to predict surface layer quality characteristics of the machined surface. Theoretical equations are of a general character and have practically no limits, but they do not take into account random factors and have lower accuracy. Empirical dependences have narrow specific application; they predict output parameters of the process in the predetermined experimental conditions rather accurately. However, the analysis has shown that such dependences exist not for all work materials and machining conditions. The situation is even more difficult when machining new materials, using new tool materials, introducing new technological processes, i.e. when reference data are not available or do not conform to field operating conditions. There is uncertainty in providing set quality characteristics of machined surfaces between the existing initial information on the object under control – the cutting process- and the required machined surface finish (initial information includes data on tool geometry, tool and workpiece material, machinery parameters etc.). That is why it becomes necessary to carry out additional experiments which is economically unprofitable for manufacturing enterprises under tough competition when the lead time is substantially reduced. In his paper Professor Tsypkin Y.Z. writes that in case of initial uncertainty the only way out is to eliminate it by means of teaching or self-learning of the system in the process of control and to use the obtained information.

3. Results

The operation of the Automated Control System during its self-learning consists in obtaining mathematical model connecting surface layer quality characteristics with the machining conditions and in using the obtained model for controlling the technological system regarding any quality parameter. Hence, the Automated Control System should incorporate a computing device, a two-way communication channel with the technological system, control sensors of output parameters of the cutting process and also algorithmic software. A functional diagram of the Automated Control System for controlling surface layer quality characteristics of machinery parts is represented in figure 1.

The system incorporates the following elements: a metal-cutting machine; a sensor measuring the main constituent of the cutting force (Sensor Pz); a sensor measuring the current temperature in the cutting zone (Sensor T); a surface finish control sensor (Sensor Ra); sensor interface with a PC and PC interface with a NC unit; a CNC unit controlling the technological system by means of changing feed F and speed S. In most cases CNC units are only designed for controlling the process of workpiece machining. That is why, for building the Automated Control System with the self-learning mode on the basis of the CNC unit it is necessary to use the PC with the corresponding software implementing the system operation algorithm.

The interface incorporated in the system is intended for the data exchange between Sensor Pz and Sensor T and the personal computer and the personal computer with the CNC unit. The computed correction values of feed and speed are transmitted from the PC through the interface controller to the CNC unit.

Programmable logic of the machine is capable of setting up the CNC unit in such a way that it is possible to change cutting feeds and speeds in an automatic mode by means of external programmed control over spindle feed and speed adjustment. Spindle feeds and speeds vary over a range from 0 % to 127 % from initial values (initial values are 100 %) with the minimum programmable movement of 1 %.

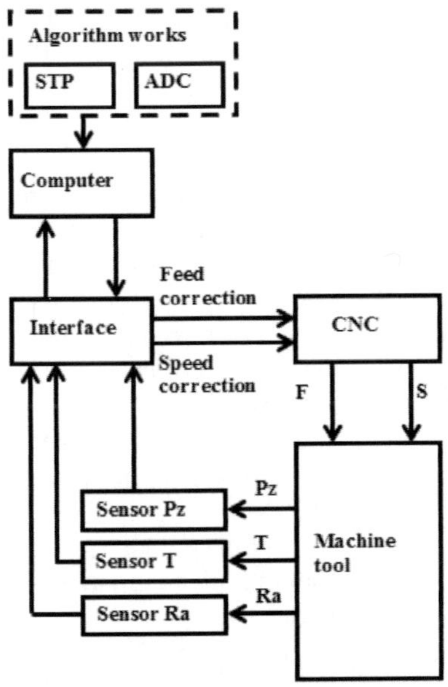

Figure 1. An automated control system for mechanical processing of machine parts.

Nowadays not only hardware but also software and algorithmic support are critical for creating the Systems of Automated Control over technological equipment. That is why the operation algorithm should be considered as an integral part of the self-learning system.

The Automated Control System provides two modes: 'Self-learning' (STP) and 'Adaptive Control' (ADC). A 'Self-learning' mode is intended for parametric identification of mathematical models and storing the results of self-learning in the system database. For parametric identification of mathematical models an active experiment using the Automated Control System is carried out directly in the workplace. The obtained data are handled statistically and parameters for mathematical models are set up and stored in the system database alongside with initial data. Initial data include the following information: the cutting edge geometry, the cutting edge material, the workpiece material and its hardness, required surface layer quality characteristics and machining allowance for the required quality parameter.

During the experiment the Automated Control System automatically determines surface finish value Ra by direct measuring; the values of surface microhardness and surface residual stress are measured indirectly using the values as the main constituent of cutting force Pz and temperature in the cutting zone. The value of complex parameter Cx is calculated on the basis of the obtained data.

An 'Adaptive Control' mode is used to provide the set values of surface layer quality characteristics or complex parameter Cx. Thus, the adaptive control over the specified technological system parameter is carried out. In this mode the measurement data are transmitted from sensors set up near the cutting zone; the obtained information is then studied and on the basis of this study a decision on technological system control is made.

To ensure the operation of the Automated Control System in the mode of adaptive control over surface layer quality characteristics of machinery parts the corresponding process control law is used [6]. In case of changing of the workpiece material and its hardness, the cutting edge material and geometry, self-learning of the technological process should be carried out again if there are no data on the changed parameters in the system database.

The operation algorithm of the Automated Control System includes the following main system operation modes: 1) initial data input and study; 2) knowledge database; 3) learning; 4) operation.

The mode 'Initial Data Input and Study' is intended for inputting and analyzing the data before machining. Initial data include: the workpiece material, tool geometry, cutting modes, technological system rigidity, and the surface layer quality characteristic. It is also necessary to obtain its required value. Preceding from the input data the system identifies the availability of the corresponding mathematical model in the system.

The mode "Knowledge Database" is intended for collecting and storing information on machined materials, cutting conditions and corresponding mathematical models. Thus, knowledge database contains mathematical models for each material being machined and each cutting condition. In case the appropriate model is not available, the system switches on to the mode 'Learning'.

The task of the mode 'Learning' is to carry out an experiment and to obtain a mathematical model. To fulfill this task a test part – prototype – is machined on the machine tool, while the tool, its geometry, the test part material and the cutting depth should correspond to those employed in machining of the next workpiece batch. The prototype is divided into zones and each zone is then automatically machined according to preset cutting parameters. Surface layer quality characteristics are automatically controlled by the system during machining.

The mode "Operation" implements an algorithm of adaptive control over the technological system to ensure the preset values of surface layer quality characteristics. On the basis of the input initial data and required surface layer quality characteristics the system software picks up necessary parameters of mathematical models from the database, chooses corresponding process control law [6], calculates initial feed values F and workpiece rotational speed S for the required surface layer quality characteristics.

Engineering research on self-learning of the Automated Control System of surface layer quality characteristics was carried out. The process task of the system self-learning was a parametric identification of mathematical models specifying: surface finish Ra, the values of surface microhardness Hμ and surface residual stress στ and the value of complex parameter Cx after turning. Several process tasks for different workpiece materials and machining conditions were solved. Initial data and obtained results are given as an example.

Initial data for task №1: workpiece material is steel 20Mn5, workpiece material hardness is HB= 280, cutting edge material is T15K6.Cutting plate geometry is the following: clearance angle - 8°, cutting angle – 7°, major cutting edge angle – 95°, minor cutting edge angle – 15°, tip radius – 0.5 mm, edge radius – 20 μm, cutting parameters:

Longitudinal feed S:

Smin = 0.10 mm/rev;
Smax = 0.4 mm/rev;

Cutting Speed V:

Vmin = 100 m/min;
Vmax = 200 m/min;

Depth of Cut: t=0.4 mm.

The result of parametric identification in the self-learning automatic control system, when processing a workpiece of steel 20Mn5, is mathematical models (1-4)

$$Ra=6.098\cdot S^{0.384}\cdot V^{-0.2017},\ \mu m \tag{1}$$

$$H\mu = 132.27 \cdot S^{-0.147} \cdot V^{0.165}, HV \tag{2}$$

$$\sigma_\tau = 380.5 \cdot S^{-0.486} \cdot V^{-0.222},\ MPa \tag{3}$$

$$Cx = 0.334 \cdot S^{-0.333} \cdot V^{-0.139}. \tag{4}$$

Initial data for task №2: Workpiece material is steel X12CrNiTi 10, workpiece material hardness is HB = 180, cutting edge material is BK8. The cutting plate geometry include: clearance angle – 9°, cutting angle – 6°, major cutting edge angle – 92°, minor cutting edge angle – 10°, tip radius – 0.2 mm, edge radius – 20μm, cutting parameters:

Longitudinal feed S:

Smin = 0.10 mm/rev;
Smax = 0.25 mm/rev;

Cutting Speed V:

Vmin = 77 m/min;
Vmax = 124 m/min;

Depth of Cut: t = 0.4 mm.

The resulting mathematical models of self-learning of the Automated Control System for processing steel workpiece X12CrNiTi 10 are as follows (5-8):

$$Ra = 21.78 \cdot S^{1.355} \cdot V^{-0.073}, \mu\text{m}. \quad (5)$$

$$H\mu = 339.8 \cdot S^{-0.192} \cdot V^{-0.046}, \text{HV} \quad (6)$$

$$\sigma_\tau = 32.09 \cdot S^{0.93} \cdot V^{1.004}, \text{MPa} \quad (7)$$

$$Cx = 0.123 \cdot S^{-0.241} \cdot V^{0.154} \quad (8)$$

In the process of self-learning of the Automated Control System surface finish Ra of the machined surface was measured by a laser optic sensor in an automatic unmanned mode under the monitor program control of the Automated Control System.

On finishing of the self-learning process the obtained parameters of mathematical models are stored in the computer database.

The operation of the Automated Control System was tested in an ADC mode with simultaneous provision of the following parameters of surface layer quality characteristics: Ra = 1.6 μm, Hμ = 420 HV, στ = 450 MPa. This set of parameters corresponds to determined complex parameter Cx = 0.432. So, the task can be formulated as follows: it is necessary to obtain the pre-set value of complex parameter Cx using the adaptive control mode of the Automated Control System.

The initial data for the ADC mode are:

required parameter Cx	0.432
required diameter tolerance	± 15 %
workpiece material steel	X12CrNiTi 10
workpiece material hardness, HB	290
cutting edge material	BK8
clearance angle	9°
cutting angle	6°
major cutting edge angle	92°
minor cutting edge angle	10°
tip radius	0.2 mm
edge radius	20 μm
depth of cut	t = 0.4 mm

While testing the operation of the Automated Control System in the ADC mode five workpieces were machined. After the test parts machining surface finish Ra, and microhardness Hμ of the machined parts were measured. Surface finish Ra was measured with profilometer MarSurf PS1. Microhardness was defined with a microhardness tester. Value στ was calculated. On the basis of measured parameters Ra and Hμ and calculated value στ complex parameter Cx was produced. The average values of the obtained parameters are: Cx = 0.44, Ra = 1.36 μm, Hμ = 421 HV, στ = 432 MPa.

4. Conclusions

The obtained research results show that surface finish value Ra – feeds and speeds are equal – is different for different machining parameters (workpiece hardness, the workpiece material, the cutting edge geometry: the clearance angle, the cutting angle, the major cutting edge angle and the tip radius), that is why it is necessary to carry out the system's self-learning.

It was found that in machining of one and the same material (but with only changing cutting edge geometry, i.e. clearance angle and cutting angle) the value of surface finish does not change much, the maximum error is 2.3 %. This means that for machining of one and the same material usinga tool with different geometry of clearance and cutting angles the technological system self-learning can only be carried out for the tool geometry.

The obtained results prove the efficiency of the Automated Control System in the modes of self-learning and adaptive control over surface layer quality characteristics.

References

[1] Bazrov B M, Suslov A G, Taratynov O V and Shoev A N 2013 *J.Russian Engineering Research.* **33** 651

[2] Daoyuan Y, Zhengcheng D, Jianchun D and Hongwen H 1994 *J. of Materials Processing Technology* **40** 433–42

[3] Durmus Karayel 2009 *J. of Materials Processing Technology* **209** 3125–37

[4] Liu C H, Chen A, Chen C A., Wang Y T 2005 *J. of Materials Processing Technology* **170** 367–73

[5] Galip Ulsoy A and Koren Y1989 Applications of Adaptive Control to Machine Tool Process Control *IEEE Control Systems Magazine* **6** 33–37

[6] Suslov A G, Petreshin D I 2010 *J. Russian Engineering Research* **30** 418–23

Defining a Service Angle for Planar Mechanisms of Manipulators based on the Instantaneous States Analysis

F Pritykin and O Gordeev
Department of Engineering Geometry and CAD, Omsk State Technical University, 11, Mira ave., Omsk, 644050, Russia

E-mail: pritykin@mail.ru

Abstract. The paper provides an algorithm of defining a service angle for planar mechanisms of manipulators having different structures of kinematic chains. The algorithm is based on the motion synthesis in the vector of velocities. The functions defining specified angle determination accuracy by the developed method are obtained.

1. Introduction

The development of universal algorithms that enable us to explore the functionality of manipulators mechanisms is still a crucial task [1-4]. A virtual construction of the manipulator motion implementing the method of instantaneous states allows one to predict fine precision motion of a real mechanism and its interference with known in advance objects of the environment. The method is based on the division of the planned path of the output link into segments and finding a configuration relevant to a motion volume minimization criterion (under motion redundancy). If there are forbidden zones in the service area of a manipulator the use of the configuration relevant to the motion minimization criterion is unlikely to be practical. If that is the case, it is necessary for the intelligent control system to search for different configurations that provide output link shifting to the target point with the required positional accuracy and choose those ones providing avoidance of the forbidden zones, minimum power losses being desirable. A method of searching arrays of configurations defining the service angle based on the instantaneous states analysis of planar mechanisms of manipulators is proposed.

2. The analysis of instantaneous states implementation

The convention is that the value of the generalized velocities vector relevant to the volume minimization criterion is defined as point M^Q. Index Q means pertaining to multidimensional space of generalized velocities Q_n. Figure 1 shows two configurations of a planar seven-link mechanism of the manipulator, links lengths being equal to 120 mm. Figure 1 also represents the results of the instantaneous states implementation under the specified velocity vector of output link $V(V_x,V_y)$ –V_x= 1 mm/s, V_y = 0 and position accuracy $\delta \leq 1$ mm. Parameter r = 2 assigns vector length V, the orientation of the output link not being considered. The sets of configurations relevant to $\delta \leq 1$ mm are obtained by applying AutoCAD and algorithmic programming language AutoLISP. The arrays of configurations are constructed as follows. Point M^Q in the space of generalized velocities Q_n is known to be found with the linear system of equations reflecting dependence of V and Q, where Q is the vector of generalized velocities of length n. In addition to the equations expressing individual components of vector V (output link linear motions) the system (in case $n > r$) uses the equation of complementary mutually perpendicular planes, their amount being equal to the length of p-plane, determining motion redundancy. Normals of the complementary planes reduced to unit vectors Q_{l1}, Q_{l2}, …,Q_{lp} specify an orthonormal basis in coordinate system $O^{Q}{}_{k}$ defining point position N^{Q} in p-plane about point M^{Q}. System $O^{Q}{}_{k}$ coordinates are determined by means of parameters k_1, k_2, … ,k_p defining point N^{Q} motion length towards vectors Q_{l1}, Q_{l2}, …, Q_{lp} accordingly. The vector equation defining point N^{Q} position in space Q_n is as follows:

$$Q_N = Q_M + k_1 Q_{l1} + k_1 Q_{l1} + \ldots + k_p Q_{lp}, \tag{1}$$

Where Q_M and Q_N – vectors defining M^{Q} and N^{Q} points position in space Q_n. Thus, implementation of vector Q_N values with parameter δ allows constructing zone of A^{Q} configurations. Figure 1a, b shows that the angle formed by a gripping device for two different configurations corresponds to zones A^{Q} and equals one and twenty four degrees and it does not detect the service angle.

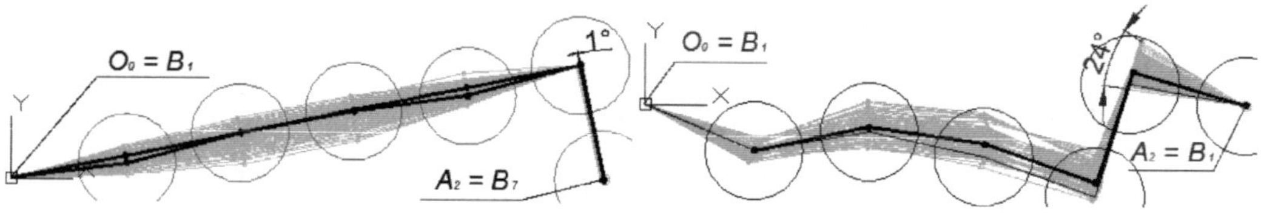

Figure 1. Zone A^{Q} defining manipulator configuration arrays.

3. Defining a service angle implementing instantaneous states

The accuracy of detection of angle U_p, which with certain assumptions detects service angle U_p will be examined with the use of the developed algorithm based on instantaneous states implementation. To configure the mechanism shown in Figure 2a, the service angle specified by the position of segments $\mathbf{B_6^{11}}B_7$ and B_6B_7 for the point of service zone $A_2 = B_7$ is detected by an intersection of two circles. The centers of the circles specified coincide with points O_o and A_2, their radii being equal to$120 \cdot 5 = 600$ mm and 120 mm, respectively. To configure the datum of the gripper unit center position determined by A_2 $(q_1 = 11.5^{\circ}$, $q_{2\text{-}5} = 0$, $q_6 = 90^{\circ})$ the service angle specified equals to 169°, where parameters q_i determine the values of generalized coordinates. Points $B_1 \div B_7$ in Figure 2 define the node points of mechanisms. Thus, to configure the mechanism in Figure 2b the service angle is determined by the position of segments B_6B_7 and $\mathbf{B_6^1}$ B_7. For the mechanism given the limited values of the generalized coordinates prescribing the translational motions of joints are predetermined with inequality 50 mm $\geq \varphi_{1\text{-}2} \geq 0$, values of generalized coordinates being $\varphi_3 = 11.5^{\circ}$, $\varphi_{4\text{-}5} = 0^{\circ}$, $\varphi_6 = 90^{\circ}$, the service angle of the mechanism equals 151°. Point $\mathbf{B_6^1}$ is determined as the tangency point of line l with circle a_o, its radius being defined by segment B_7B_6 and it equals 120 mm. Here $\mathbf{B_3^1} \in l$, $\mathbf{B_3^1} = l_1 \cap a_1$, where $\in$ and $\cap$ denote the relation of the membership and intersection, respectively. In Figure 2b the distance between vertical line l_1 and origin of the fixed coordinate system is 50 mm, a_1 is a circle with the centre at point B_7, its radius being defined by segment B_3B_7.

Let us define the aggregate of angle U_p values implementing instantaneous states for the configurations of the mechanism, whereby point A_i (the gripper centre) is in segment A_1A_3. The length of the segment equals the double length of the link ($A_1A_3 = 240$ mm, wherein$\varphi_{2\text{-}5} = 0^{\circ}$ for the configuration under study).

We shall consider seven different positions of point A_i in segment A_1A_3. Neighboring pointsA_i are at a distance of 30 mm. The output links of two configurations calculated during the start and k iterations (Figure 2a) determine angle U_m^k. The angle between segments B_6B_7 and $B_6^1 B_7$ determines the angle specified, wherein $U_m^k = U_2 - U_1$. We shall construct the aggregate of configurations implementing instantaneous states, the value of angle U_m^k being changedcounterclockwise.This value is more than value U_m^{k-1}, calculated during the previous iteration. The starting position of the mechanism configuration (Figure 2a) for point A_2 corresponds to the values of generalized coordinates $q_1 = 11.5°$, $q_{2\text{-}5} = 0$, $q_6 = 90°$.

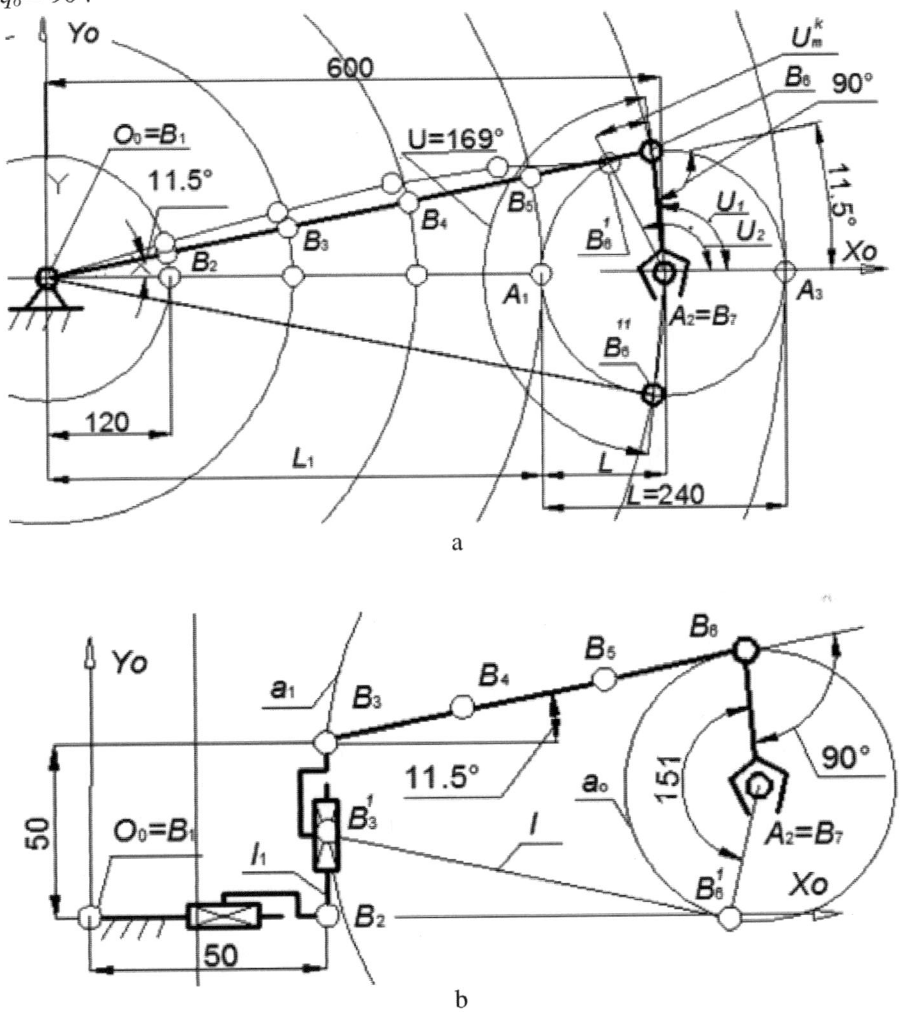

Figure 2. Determination of service angle U using graphic constructions.

Angle $U_m^k = 0$ for the configuration given.An important point is that at each calculation step the maximum value of angle $(U_m^k)^{\max}$ is defined based on values U_m^k and U_m^{k+1}:

$$(U_m^k)^{\max} = \max(U_m^k, U_m^{k+1}).$$

As soon as the configuration satisfying the conditions is defined as

$$U_m^{k+1} > (U_m^k)^{\max}\ , \tag{2}$$

$$\delta < \delta_{min}, \tag{3}$$

the next (k+2) configuration is determined, the list of generalized coordinates $q_{pi} = q_i$ being retained. Parameter $\delta_{\min}$ imposes the prescribed positioning accuracy. Here the (k+1) configuration is counted as the initially prescribed configuration, conditions (1, 2) being satisfied for that configuration. The case considered corresponds to the synthesis of link B_6B_7 motions rotating about point B_7 counterclockwise. If we consider the motion of link B_6B_7 rotating clockwise, condition (1)B_6B_7 is substituted by the following condition:

$$(U_m^k)^{\min} = \min(U_m^k, U_m^{k+1}),$$
$$U_m^{k+1} < (U_m^k)^{\min} \; . \qquad (4)$$

We use variable kk to define the rotation variant: when $kk = 0$, the counterclockwise rotation is considered; when $kk = 1$, clockwise rotation is viewed.

The algorithm of constructing configurations satisfying conditions (2-4) is shown in Figure 3.

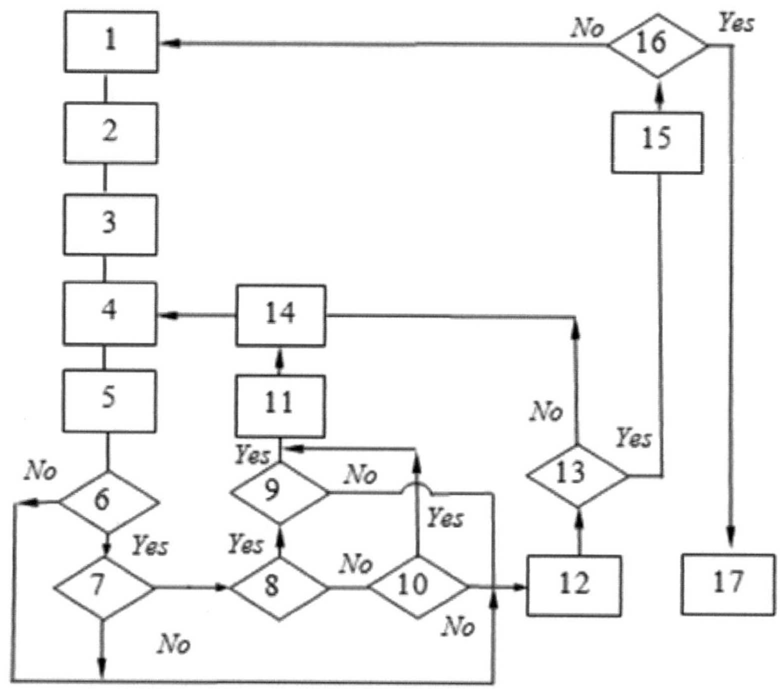

Figure 3. A diagram of constructing the angle U_p algorithm implementing instantaneous states.

The following notations are used in Figure 3: 1 – start, determining the value of vector $\boldsymbol{L}_n$ (q_1, ..., q_6) and parameter kk; 2 – calculation of point A_i of the service zone, the target point and vector $\boldsymbol{V}$; 3 – calculation of the matrix of particular gear ratios $\boldsymbol{J}$ and vector $\boldsymbol{Q}$ with the criterion of minimizing the motion range; 4 – calculation of the configuration $\boldsymbol{L}_n^{k+1}$ position corresponding to vector $\boldsymbol{Q}$, $q_i := q_i + \dot{q}_i$ (assumption $\Delta q_i \approx q_i$is admitted); 5 – calculation of implementations errors δ; 6 – conditional test $\delta < \delta_{\min}$; 7 – conditional tests $q_i^{min} > q_i > q_i^{max}$, where q_i^{min}, q_i^{max}– prescribed limit values of generalized coordinates (calculating test cases q_i^{min}= – 180°, φ_i^{max}= 180°);8 – kk = 0; 9 – $U_m^{k+1} > (U_m^k)^{\max}$; 10 – $U_m^{k+1} < (U_m^k)^{\min}$;11 – configuration of image formation: kk=0 value assignment $\Rightarrow (U_m^k)^{\max} := \max(U_m^k, U_m^{k+1})$,$kk$ = 1 ⇒ value assignment $(U_m^k)^{\min}$:=min(U_m^k, U_m^{k+1}), $q_{pi}=q_i$, where q_{pi}– list of generalized coordinates satisfying conditions (2), (3) or (3), (4), q_i – list of generalized coordinates of the current configuration; 12 – parameters $k_1, k_2, \ldots, k_4$ values change defining coordinates of point N^Q in *p*-plane Γ; 13 – $k_1 > k^{max}$, $k_2 > k^{max}$, ..., $k_4 > k^{max}$(where k^{max}– limit – prescribed largest extremum experimentally determined in research process k^{max} = 20); 14 – calculation of vector $\boldsymbol{Q}$ for $k_i \neq 0 q_i = q_{pi}$; 16 – $U_m^{k+1} = (U_m^k)^{max}$; 17 – end.

In Figure 4 and Figure 5 the synthesis of motions to determine angle U_p using dependences (1-2) for three mechanism configurations studied is shown.

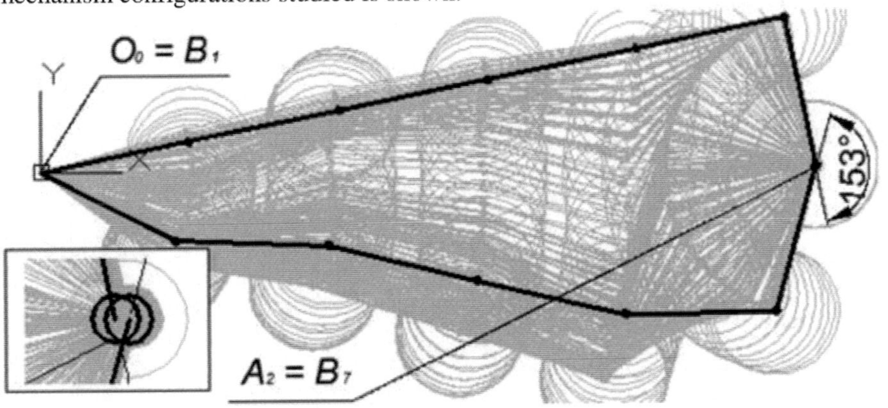

a

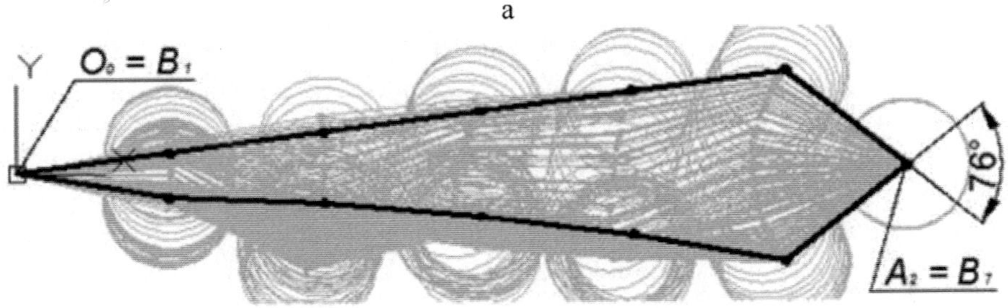

b

Figure 4. Determining angle $U_p$by implementation of instantaneous states: *a*–(q_1 = 11.5°, $q_{2\text{-}5}$ = 0, q_6 = 90°); *b*–(q_1 = 7.4°, $q_{2\text{-}5}$ = 0, q_6 = 45.5°).

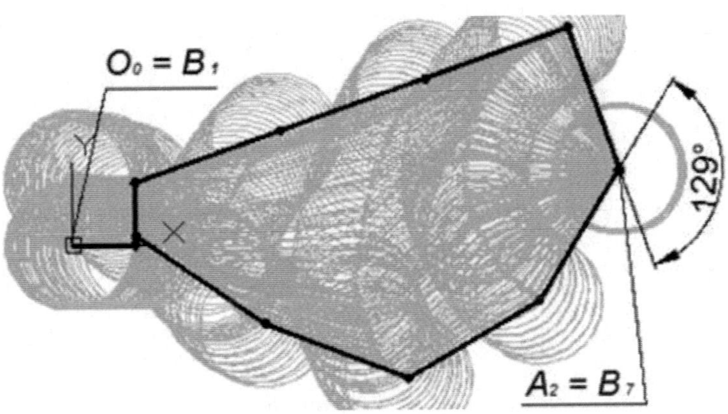

Figure 5. Determining angle U_p by implementation of instantaneous states ($\varphi_{1\text{-}2}$ = 50 mm, φ_3 = 11.5°, $\varphi_{4\text{-}5}$ = 0, φ_6 = 90°).

In Figures 6 and 7 the dependency graphs $U_{p3} = f_1(L)$, $U_{p4} = f_2(L)$, $U_{p5} = f_3(L)$, $U_{p6} = f_4(L)$ and $\mu_3 = f_1(L)$, $\mu_4 = f_2(L)$, $\mu_5 = f_3(L)$, $\mu_6 = f_4(L)$ are shown. These graphs reflect the angle U_p variation and accuracy percentage μ of service angle determination for four-, five-, six- and seven – link planar mechanisms (Figure 2a) having link lengths equal to 120 mm.

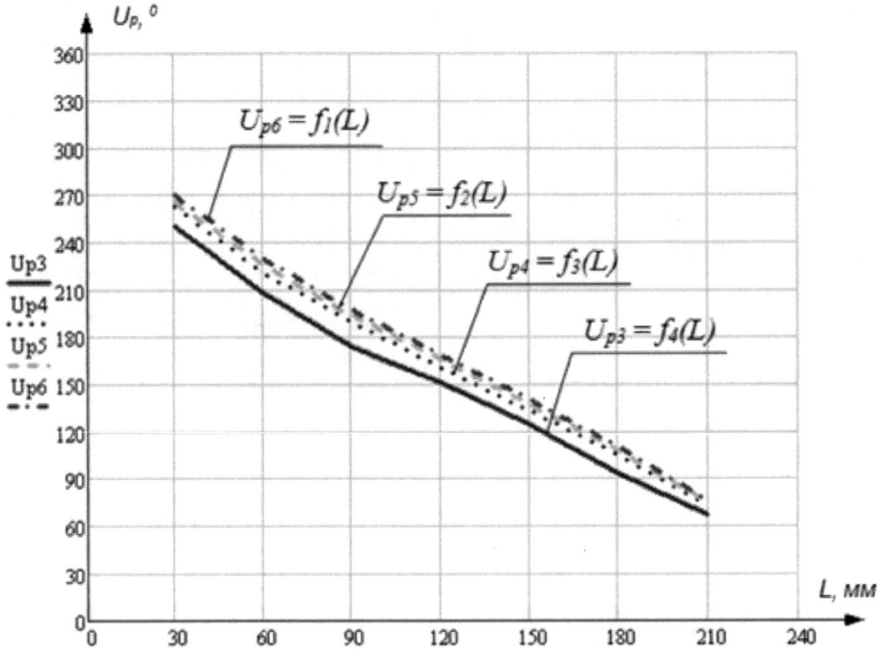

Figure 6. Dependency graphs $U_{pn} = f_1\ (L)$.

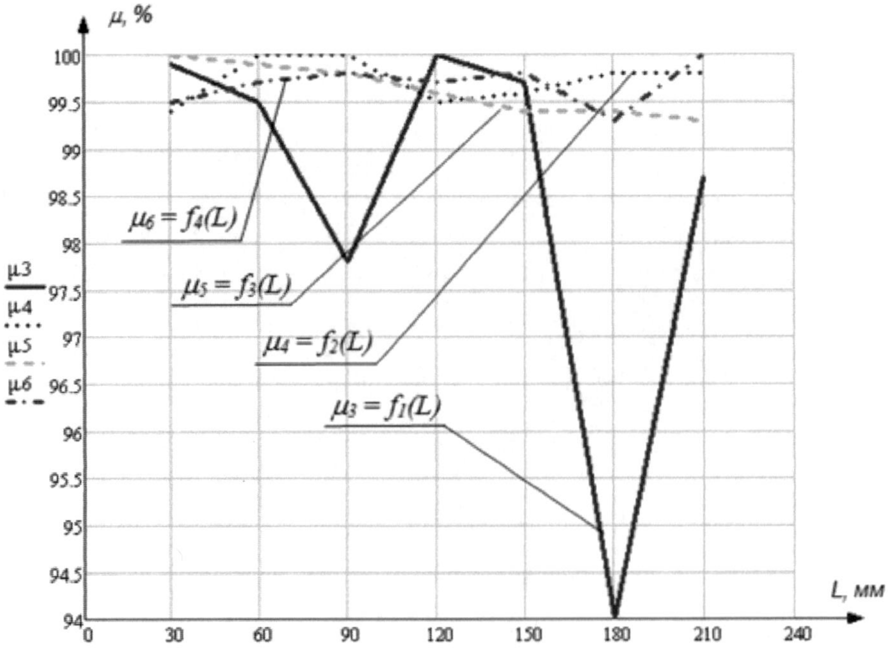

Figure 7. Dependency graphs $\mu_n = f_1\ (L)$.

The accuracy percentage is calculated using dependence $\mu = U_p/U \cdot 100$, where the service angle or parameter U was determined graphically. Parameter L defines deproach of point A_i from point A_l (Figure 2a). Point A_l for the mechanisms specified holds at distance $L_1 = 120 \cdot (n - 2)$ from origin O_o. The analysis of functions graphs shows the accuracy percentage of the service angle defined by the method developed for the mechanism in Figure 2a for $\delta_{min} \leq 1$ mm, where is from 94 to 100 percent depends on the position of the gripper center in segment A_1A_3. The accuracy percentage for $\delta \leq 3$mm is from 80 to 85 percent for the mechanism shown in Figure 2b.

4. Conclusion

The proposed algorithm of determining angle U_p with certain assumption defining the service angle is described as universal and can be used for manipulators having a random structure of planar kinematic chains, a random degree of motion redundancy and prescribed limited values of generalized coordinates. The construction of solid angles projections using the algorithm proposed is possible for spatial mechanisms of robots. The method of the output link reorientation can be applied for intelligent robot control systems to change the output link orientation relative to the prescribed objects of manipulation.

References

[1] Pratt J, Dilworth P and Pratt G 1997 *Robotics and Automation: Proc. IEEE Int. Conf.* **1** 193–8

[2] You B, Zou Y, Xiao W and Wang J 2010 Telerobot control system based on dual-virtual model and virtual force Strategic Technology (IFOST) *Int. Forum* 246 – 50

[3] Hrr J, Pratt J, Chew C, Herr H and Pratt G 1998 *Intelligence and Systems:Proc. IEEE Int. Joint Symp.* 245–51

[4] Tsukamoto H, Takubo T, Ohara K, Mae Y and AraiT 2010 *Information and Automation (ICIA): IEEE Int. Conf.* 729–34

A Single-Loop DC Motor Control System Design with a Desired Aperiodic Degree of Stability

M I Pushkarev, S V Efimov, S A Gayvoronsky and O A Chenkova
Tomsk Polytechnic University, 30, Lenina ave., Tomsk, 634050, Russia

E-mail: pushkarev@tpu.ru

Abstract. The application of the original analytical approach for PI-controller synthesis of a stable second-order plant is considered. This approach allows finding controller parameters without any intensive computing by using the direct expressions. The plant model is obtained on the basis of identification, which is based on the automated real-interpolation method. The results of natural experiments are given.

1. Introduction

The design of automatic control system regulators is one of the most important problems in the control theory. Initial data for the design solution are the mathematical description of the plant and some demands of the designed system functioning. There are a lot of approaches to the automatic control systems design [1, 2]. There is a huge class of automatic control systems, which require aperiodic transient responses, that is, without oscillations. It is also known that a control system of any order and complexity may be described with a tolerable accuracy rate by a system of a reduced order (first or second) [3]. Therefore PI-controller parameters determination for a DC motor control providing the desired behavior of a designed system is considered in this work.

2. Problem Formulation

DC motor 'Dynamo' Sliven PIVT6-25/3 A with a stationary velocity sensor is used as a plant. The motor has the following characteristics: source voltage, 30 V; rotation speed, 3000 rpm; rotating torque, 0.11 Nm; starting current, 4.5 A; electromechanical time constant, 25 msec; electrical time constant, 2 msec; mass moment of inertia, $5.1 \cdot 10^{-5}$ kg m^2; velocity sensor constant, $3\ \mathrm{V} / 1000 \min^{-1}$. A microcircuit chip Pololu High-Power Motor Driver 18v 15 is used as a motor driver.

The view of the control system block diagram is represented in Figure 1.

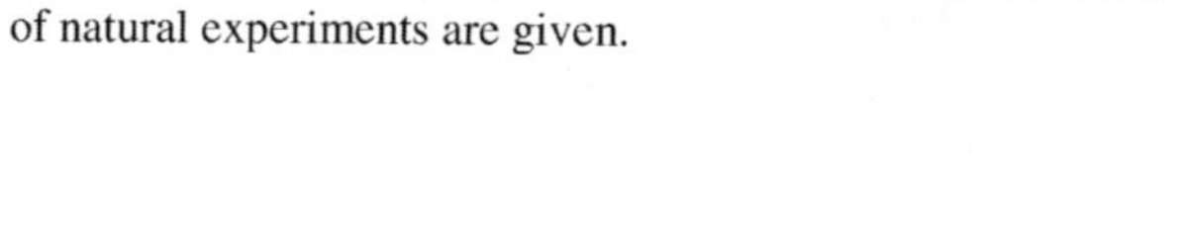

Figure 1. Control system block diagram

PI-controller transfer function is

$$W_C(s) = \frac{k_1 s + k_0}{s} \tag{1}$$

where k_0 is an integral coefficient and k_1 is a proportional coefficients.

The plant transfer function is represented by the following expression:

$$W_P(s)=\frac{B(s)}{A(s)}=\frac{b_m s^m+b_{m-1}s^{m-1}+...+b_1 s+b_0}{a_n s^n+a_{n-1}s^{n-1}+...+a_1 s+a_0},\ n\geq m. \tag{2}$$

It is known that the condition of the aperiodic transient response is such location of closed loop system poles, at which a real dominant pole is the nearest to the imaginary axis, and all other so-called 'free' poles are moved at the sufficient distance $\gamma=[\eta,\alpha]$ to the left and are located in the truncated segment Γ (Figure 2). Therefore, the problem is to place real dominant pole α of the closed loop system in the coordinate $(-1;0)$ and other free poles on the real axis past the boundary $\eta=7$.

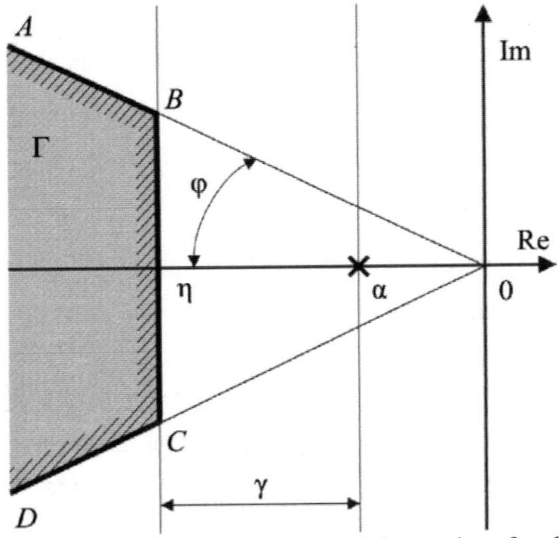

Figure 2. The location of a real dominant pole and free poles of a closed loop system

3. Plant Identification

On the first stage it is necessary to obtain the mathematical description of the plant. It is suggested to make parametric identification, which is mainly used for stable objects with a non-zero static gain.

The plant model is formed on the basis of experimental data applying the approach represented in [4]. The approach is based on the real-interpolation method, which discretizes continuous functions and reconstructs them in a continuous form with the help of interpolation.

As a result of an input test step signal the vector of the plant output data under a load is obtained. In consequence the following model of the plant is obtained:

$$W_{P1}(s)=\frac{4.321\cdot 10^{-3}s+5.126\cdot 10^{-2}}{6.206\cdot 10^{-2}s^2+1.130s+1} \tag{3}$$

It is known that the results of the controller design on the basis of root approaches may have some errors [5]. It is strongly connected with a presence of zeros in the transfer function (2) numerator. This fact may influence in an unforeseen manner the control system behavior and performance measures in the time domain. That is why it is offered to take the transfer function (3) without zeros in the numerator into consideration.

$$W_{P2}(s)=\frac{5.126\cdot 10^{-2}}{6.206\cdot 10^{-2}s^2+1.130s+1} \tag{4}$$

Figure 3 shows step responses of transfer functions (3) and (4), and it is vivid that both plants have similar dynamics.

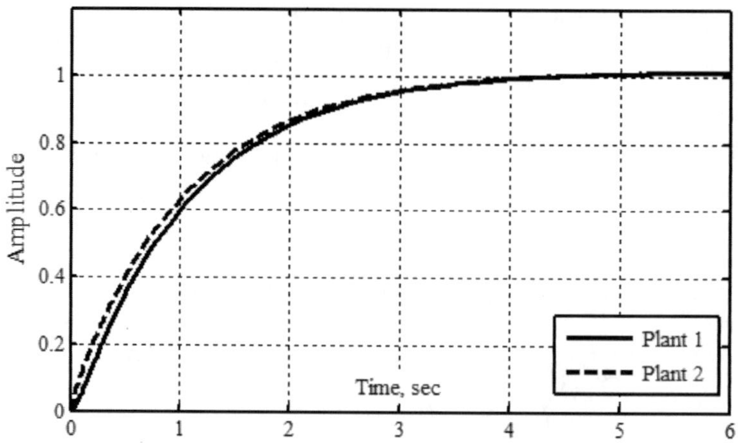

Figure 3. Transfer functions of $W_{P1}(s)$ and $W_{P2}(s)$

4. Controller Design

A controller design method is based on the decomposition of the closed loop system characteristic equation $A(s)=\sum_{i=0}^{n} a_i s^i, a_i>0$ into the dominate polynomial $Q(s)=s+\alpha$, which carries the information about location of the dominate pole and the free polynomial $P(s)=\sum_{i=0}^{n-1} p_i s^i, p_i>0$, which roots are free poles. The result of the decomposition is a residue of division $R=a_0-\alpha p_0$. On the basis of *R*, controller parameter k_0 can be expressed. In general terms the equation may be written as follows: $A(s)=Q(s)\cdot P(s)+R$.

If the characteristic polynomial order of the closed loop system is $n=3$, then the free polynomial order is $n-1=2$. On the basis of the second order equation solution the analytical expressions connecting PI-controller parameters k_0 and k_1 with free polynomial parameters p_i and parameter η the following form can be obtained:

$$\begin{cases} p_1^2-4p_0p_2>0; \\ p_0 \geq \eta p_1-\eta^2 p_2. \end{cases} \quad (5)$$

The verification of design results is conducted through the synthesized closed loop system characteristic equation roots definition.

5. Results of the Design

Let us define the PI-controller parameters for the plant (4). Substituting numeric values in expressions (5), the acceptable region of $k_1 \in (88.353;91.515]$ is obtained. Let us suppose that $k_1=91.515$, then, on the basis of division residue R, $k_0=90.181$, the poles are $s_1=-1$, $s_2=-8.59$, $s_3=-8.63$ and zero is $N_1=-0.985$.

In case of the transfer function (TF) (3) the impact of the zero numerator does not allow placing the closed loop system poles in accordance with requirements: $s_1=-1$, $s_2=-3.73$, $s_3=-19.8$, $N_1=-0.985$, $N_2=-11.9$.

Step responses of the synthesized systems are represented in Figure 4.

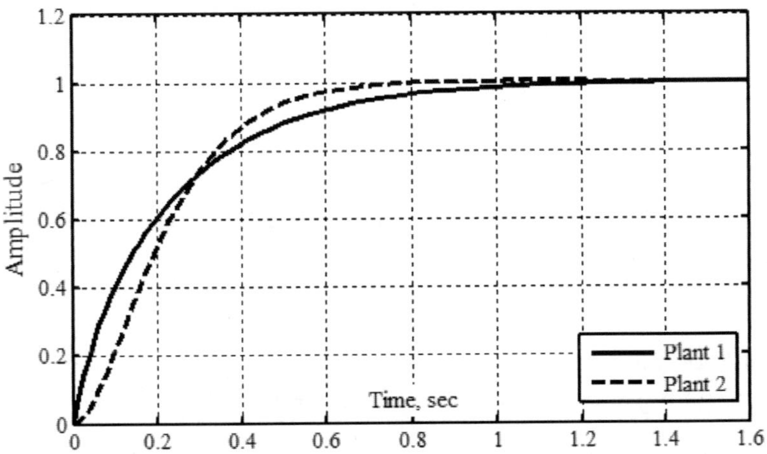

Figure 4. Step responses of the synthesized systems

On the basis of step responses of the synthesized closed loop system (Figure 4) the system performance measures can be estimated. Therefore, for plant 1 there is an overshoot $\sigma = 0.13\%$ and the settling time is $t_s = 0.9$, sec, for plant 2 there is $\sigma = 0.54\%$ and $t_s = 0.6$, sec.

Table 1. Performance measures of identified and synthesized systems.

	DC motor model		Designed Control System	
	$W_{P1}(s)$ TF with zero	$W_{P2}(s)$ Reduced TF	$W_{P1}(s)$ TF with zero	$W_{P2}(s)$ Reduced TF
Overshoot, %	0	0	0.135	0.548
Settling time, sec	4.25	4.17	0.911	0.624

As we can see from the table the overshoot of synthesized system $W_{P2}(s)$ is negligibly higher than $W_{P1}(s)$ overshoot, but at the same time $W_{P1}(s)$ settling time is inconsiderably longer than $W_{P2}(s)$ settling time.

6. Conclusion

As a result of experiments with the DC motor control system it is possible to conclude that both (full and reduced) synthesized systems introduce acceptable dynamics and performance measures. But the introduced method of PI-controller synthesis has some restrictions on its application in the plants with zeros in the transfer function numerator. This fact is inessential in terms of low-order control systems because it does not influence the dynamics of the synthesized system critically.

References

[1] Ogata K 2001 *Prentice Hall, Uppe Saddle River, NJ*
[2] Astrom K J and Murray R M 2010 *Princeton University Press*
[3] Skogestad S 2003 *Journal of Process Control* **13** 291-309
[4] Zamyatin S V, Kurgankin V V and Rudnicki V A 2014 *Bulletin of the Polish Academy of Sciences* **62(2)** 341-347
[5] Efimov S V and Pushkarev M I 2011 *Optoelectronics, Instrumentation and Data Processing* **47(3)** 297-302

IOP Conf. Series: Materials Science and Engineering **124** (2016) 012027 doi:10.1088/1757-899X/124/1/012027

A teleoperation system to control the humanoid robot using an RGB-D sensor

E E Shelomentcev and T V Aleksandrova
Tomsk Polytechnic University, 30, Lenina ave., Tomsk, 634050, Russia

E-mail: see4me@mail.ru

Abstract. This paper presents a concept design of the work algorithm for a teleoperation control system of a humanoid robot. The humanoid robot control system needs to stabilize the robot in a vertical position in order to prevent the robot from falling. The process of design of the control system includes the design of position filter to detect the unstable positions. The application of such a control system enables to control the humanoid robot using motion capture technology.

1. Introduction

Humanoid robots have been actively developed in recent years. There are many different designs of humanoid robot constructions but not all of them have the control system that realizes adequate behavior of the humanoid robot. That is why some researchers propose using teleoperation, [1-4], to control such type of the robots. In this case, the humanoid robot is an avatar controlled by a human operator. Implementation of this technique allows the human operator to manipulate with objects remotely that is helpful, for example, in hazardous environment.

Further, we will consider the algorithm of the teleoperation control system that allows controlling humanoid robots using the RGB-D sensor (Microsoft Kinect). In our research, we use the humanoid robot BIOLOID that have actuators in every joint.

2. Control algorithm

To form the control signals we need to collect the data about the human position. Let us create the operational block that processes the data received from the RGB-D sensor. The objectives of this block are the forming of the packets with the human operator data and the initial data filtering, for example, forming of the packets only with the recognized position of the operator. When the packet is filtered, the system must inform the operator about it.

To control the position of the humanoid robot we need to set the positions of the actuators that are located in the joints of the robot. The positions of the actuators can be mapped from the angles of rotations in the joint of the human operator. It means that we need to receive the control signals that contains the information about angles of rotations.

The human body differs from the robot body because the mass of their limbs is different and it can be a reason of the robot falling that is why it is necessary to ensure the stable vertical position of the robot through the control process to prevent his falling. Thus, we need to filter unstable positions of the robot [5]. Only when all of these operations are mapped and filtered, positions of actuators can be sent to the robot.

Thus, we have an algorithm that can be showed as a row of actions:collecting filtered data, synthesizing actuators positions according to the data, filtering the actuators positions that make robot unstable and sending the actuators positions to the robot (Figure 1).

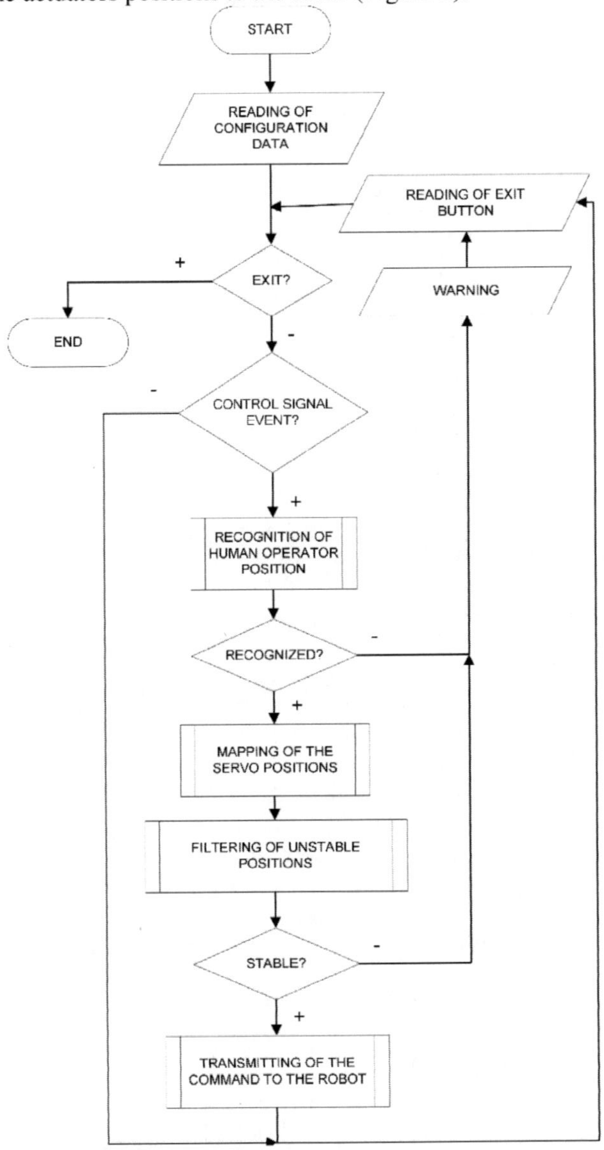

Figure 1. An algorithm of the control system work.

2.1. Collecting data using a RGB-D sensor

To control the position of the humanoid robot we need to set the positions of the actuators that are located in the joints of the robot. The positions of the actuators can be mapped from the angles of rotations in the joint of the robot. It means that we need to receive the control signals that contain the information about angles of rotations. These angles of rotations can be found as angles between the vectors formed by the control points of human operator joints (Figure 2). For example, the angle in the right elbow joint of the robot is an angle between vectors $\bar{a}$ and $\bar{b}$, where $\bar{a}$ is a vector with initial point ELBOW_RIGHT and terminal point WRIST_RIGHT; $\bar{b}$ is a vector with initial point ELBOW_RIGHT and terminal point SHOULDER_RIGHT. Control points of the human operator are received from the RGB-D sensor.The use of special software kits (Kinect SDK, OpenNI) allows one to recognize these control points on human body and to filter unrecognized human positions.

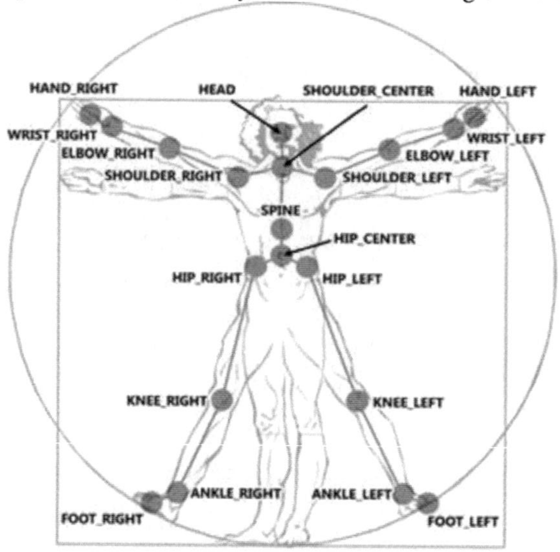

Figure 2. Control points of the human operator

2.2. Actuators control

Now we have the humanoid robot BIOLOID and we have the angles of rotation in every joint.BIOLOID has 18 controllable degrees of freedom that provided by 18 servos. Each servo has an angle constraint (Figure 3) that is it can rotate from 0 to 300 degrees. That is why we need to map our angles of rotations to positions of servos in view of the constraint. After this, we need to input servo positions to model the robot to check the stability of the robot position using position filter [5]. If the position of the robot is unstable then the control system warns the human operator about instability, moreover the control system sends the control signals to robot actuators. Implementation of the filter allows robot to move only from one stable position to another. Thus, the humanoid robot can stay stable even if the control system has time-delay in communication interface.

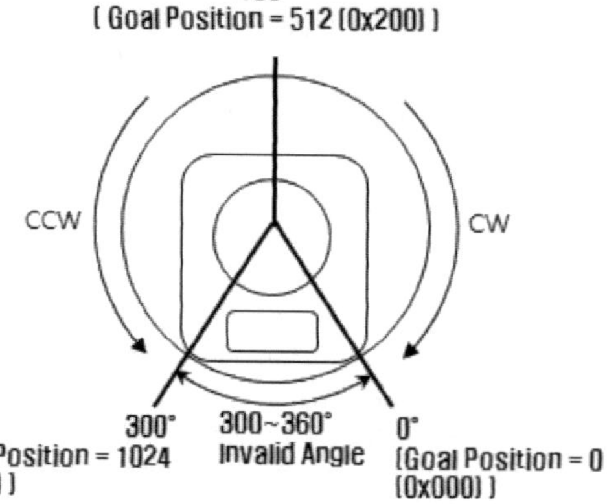

Figure 3. A robot actuator constraint.

2.3. A stability filter
The main idea of the stability filter is the check of a static stability condition [5]. We use the Denavit-Hartenberg convention to model the robot. It allows one to calculate the position of robot CoM (Center of Mass) using the positions of actuators. The CoM position is needed to check the static stability condition.

To check the static stability condition, we use a ZMP (Zero Moment Point) method that allowsoneto determine whether the position of the robot isstable or not considering the effect of all forces [6, 7].

The method comprises calculating the ZMP position,where both inertial and gravity forces do not produce anymoment in a horizontal direction. To evaluate theposition stability of the robot it is necessary to determine that ZMP is located in a supporting polygon (the area of afoot in the single support phase or the area limited by twofeet in the double support phase).

If the influence of the inertial forces is neglected, ZMPis a projection of CoM on a supporting plane. Inthis case, the vector of the gravity force crosses the axes at ZMP, which means that the moment produced by this force is zero. Thus, to assess the position stability of the robot we need to ensure that the projection of CoM is located in the supporting polygon.

2.4. Data transmitting
When we recognize the position of the human operator, generate control commands translating them to the robot model and checkthe robot stability, we need to send correct control commands to the robot. There are two ways of communication with BIOLOID: the use of the ZigBee connection and the use of the cable connection. Both of them apply a UART protocol to transmit the data. That is why we need to realize the UART protocol in our software. Application of the ZigBee connection can cause time-delay in the control system work but it allows one to control the robot more efficiently without movement constraints caused by the cable. Furthermore, time-delay has no critical influence on performance and the robot stability through the control process.

3. Conclusion
In this paper, we have described the concept design of the teleoperation system of the humanoid robot. The designed control system allows the operator to control the robot using the motion capture technology that is the most ergonomic way of the humanoid control [8]. The system includes the stability control that filters all unstable positions of the robot, but it does not prevent the robot falls that are caused by external forces.The whole system allows one to control the humanoid robot effectively through the process of the movement. This paper describes only the concept of the control system and further work is the implementation of this concept in a real control system of the humanoid robot. At this moment, we have realized the basics of the forming of the control signals (Figure 4) and started to realize the stability filter in the control system of the robot BIOLOID.

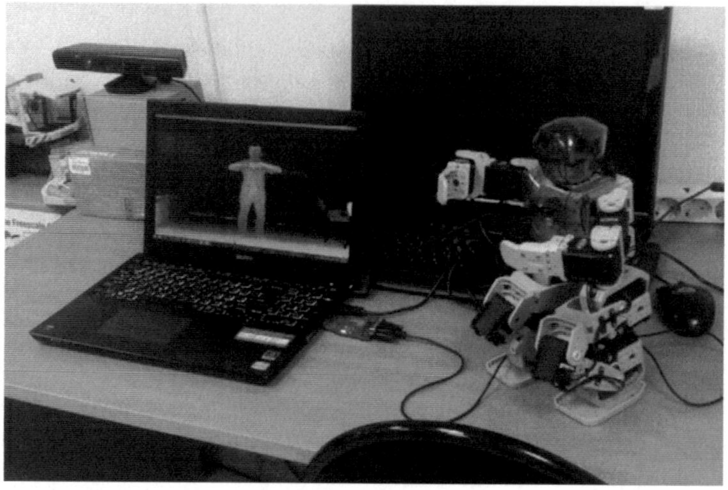

Figure 4. An experiment.

References

[1] Sian N E, Yokoi K, Kajita S, Kanehiro F and Tanie K 2002 *IEEE/RSJ Int. Conf. on Intelligent Robots and Systems* 3

[2] Takubo T, Inoue K, Arai T and Nishii K 2006 *IEEE Int. Conf. on Intelligent Robots and Systems* 4459–4465

[3] Du G, Zhang P, Mai J and Li Z 2012 *International Journal of Advanced Robotic Systems* 1

[4] Du G and Zhang P 2014 *Robot. Comput. Integr. Manuf.* **30(2)** 150–159

[5] Shelomentcev E and Alexandrova T 2014 *IOP Conf. Ser.: Mater. Sci. Eng.* **66** 012027

[6] Vukobratovic M and Stepanenko J 1973 *On the Stability of Anthropomorphic Systems* **15(1–2)** 1–37

[7] Dekker M 2009 Zero-moment point method for stable biped walking (*Internsh. report, Eindhoven*)

[8] Goodrich M A, Crandall J W and Barakova E 2013 *Reviews of Human Factors and Ergonomics* pp 175-226

Adaptive pseudolinear compensators of dynamic characteristics of automatic control systems

M V Skorospeshkin[1], M S Sukhodoev[1], E A Timoshenko[2] and F V Lenskiy[1]
[1]Tomsk Polytechnic University, 30, Lenina ave., Tomsk, 634050, Russia
[2]Tomsk State University, 36, Lenina ave., Tomsk, 634050, Russia

E-mail: smike@tpu.ru

Abstract: Adaptive pseudolinear gain and phase compensators of dynamic characteristics of automatic control systems are suggested. The automatic control system performance with adaptive compensators has been explored. The efficiency of pseudolinear adaptive compensators in the automatic control systems with time-varying parameters has been demonstrated.

1. Introduction

In the systems of automatic control over objects, which characteristics change in the course of time, it is necessary to provide a purposeful change of dynamic characteristics of a controlling device. In most cases, it is realized by parametrization of proportional-integral differential controllers (PID controllers). These approaches are described, for instance, in [1, 2], however their realization concerns either identification or the use of special methods based on calculations according to the curve of the transient process. Both these approaches require a significant amount of time for controller tuning.

The current paper presents the research findings of automatic control system characteristics with PID-controller and cascade adaptive gain and phase pseudolinear compensators of dynamic characteristics. This adaptation method is characterized by the fact that in the process of control system operation the controller parameters do not change and correspond to the tuning, preceding the system start. Thus, only gain or phase shift on a proper compensator changes during automatic control system tuning. This tuning is applied when a process variable has variations connected with the change of the control object properties or disturbances. It allows ensuring a system stability and improving a transient process quality.

Compensators are classified into three groups: linear, nonlinear and pseudolinear. The main disadvantage of the linear compensator is that its parametrization influences both gain and phase-frequency characteristics of the system. In this case, during improvement of the phase-frequency characteristics of the system the gain-frequency characteristics will increase in mid and high frequencies that decreases the system stability. Moreover, when enhancing gain-frequency characteristics of the system, the phase-frequency characteristics of the open-loop system obtain a negative value close to $-\pi$, which significantly decreases system stability.

Investigations [6] show that the linear compensator adaptation procedure is possible only for a limited in range and nature parametrization of the control object even at using modern mathematical tools such as fuzzy logic. The nonlinear compensator usage is problematic due to phase-frequency characteristics dependence on the harmonic oscillations amplitude.

An adaptive pseudolinear compensator allows reaching the required gain- and phase- frequency characteristics. This compensator has two channels (gain and phase) that can be tuned independently. In addition, phase-frequency characteristics of adaptive pseudolinear compensators do not depend on harmonic oscillations amplitude.

Phase-, gain- and phase-gain compensators have become widely spread in comparison with other pseudolinear compensators [5]. The conducted investigations show that gain compensators are the most effective ones for adaptive systems realization. These compensators have no influence on system characteristics in a static mode. However, if variations of the process variable (PV) are present, they reduce the gain without any noticeable change of the phase shift. The application of these compensators provides a required quality of automatic control systems in the wide range of parametrization of the control object and the nature of disturbance influences. This compensator can be considered as a supplementary resource participating in the control action formation and enhancing the control quality.

2. Related work

The structure of the adaptive gain compensator proposed in [7] is shown in Figure 1. This compensator is connected in series with the controller.

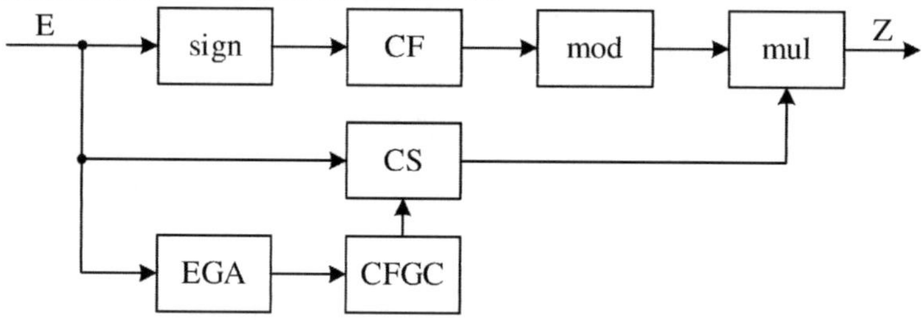

Figure 1. Block diagram of the adaptive gain compensator

Figure 1 has the following symbols: E – compensator input signal; sign – block of sign determination; CF – compensating filter; mod –modulus separator; mul – multiplication block; Z – compensator output signal; CS – compensator static gain; EGA – error gain analysis block; CFGC – correction block of compensating filter coefficient.

This compensator has been made on the base of the pseudolinear filter, which was imparted an additional property of parametric adaptation in addition to gain weakening (of frequency adaptation), consisting in high frequency attenuation.

The parametric adaptation is realized by means of changing the static gain of the compensating filter representing a linear dynamic element. In most cases this is an aperiodic element of the first order. If one uses the element with transfer function

$$W(s) = \frac{K}{Ts+1}, \tag{1}$$

as a compensating filter, where K – static gain of the element, T –time constant of the element, the gain-phase frequency characteristic of the compensator at K=1, obtained as a result of harmonic linearization, will have the following view:

$$W(j\omega) = a + jb, \tag{2}$$

where $a = \frac{8}{\pi^2 \cdot \sqrt{1+T^2\omega^2}} \cdot \left(1 + \frac{1}{3} \cdot \cos(2\theta)\right)$; $b = \frac{8}{\pi^2 \cdot \sqrt{1+T^2\omega^2}} \cdot \left(\frac{1}{3} \cdot \sin(2\theta)\right)$; $\theta = -\mathrm{arctg}\,\omega T$; ω - angular frequency of harmonic oscillations.

Analyzing the analytically presented gain-phase frequency characteristic of (2) it is possible to show that, in case of alternating the frequency, the gain-frequency characteristic of the compensator changes from K to zero. And the change of the phase frequency characteristic at different parameters of the compensating filter occurs no more than by –20°. Moreover, the maximum phase shift of the aperiodic element of the first order (1) can reach –90°. The compensator operates in the following way. At the onset of variation of the output variable of the control object in the automatic control system (ACS) the gain of the compensator reduces automatically. This happens as a result of multiplication of the upper and middle branch signals.

Let us consider the case when K=1. In this case, the gain of the harmonic signal of the upper branch of the compensator will be less than 1. Therefore, the compensator gain overall will be less than 1 as well. In this connection the gain will be reduced with the increase in frequency. In cases when the gain change does not remove oscillations, the middle branch gain of the compensator changes automatically. The correction block of the compensating filter coefficient realizes this process according to the error amplitude value, determined in the block of the error amplitude analysis of ACS. The adaptation mechanism can be different: for example, linear or quadratic. Also, the gain change can be realized on the basis of the fuzzy logic apparatus [6].

Figure 2 shows a block diagram of the automatic control system with the adaptive gain pseudolinear compensator.

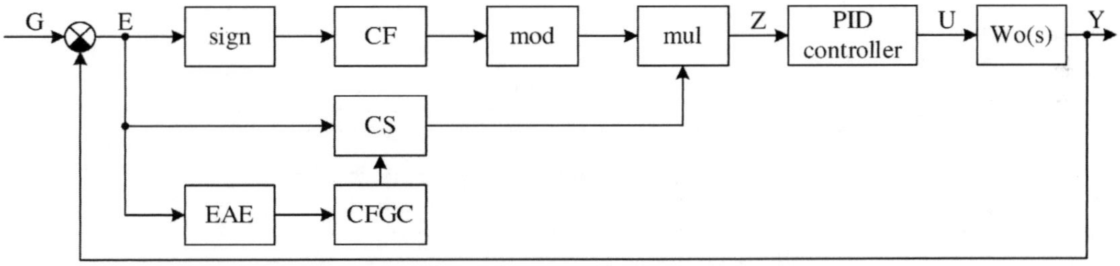

Figure 2. Block diagram of the automatic control system with the adaptive gain pseudolinear compensator

Figure 2 has the following symbols: G – control system setpoint, U – control action, Y – control object output, Wo(s) – transfer function of the control object.

The investigation of the properties of the system with the adaptive compensator was conducted in the MATLAB 6.5 software package. Figure 3 shows acontrol system model with an adaptive gain compensator model performed in MATLAB. The compensator static gain was changed by means of the fuzzy logic apparatus. Defuzzyfication was conducted with the use of the center mass method. The procedures of determination of the error variation amplitude and realization of fuzzy logic have been represented in the form of the separate blocks.

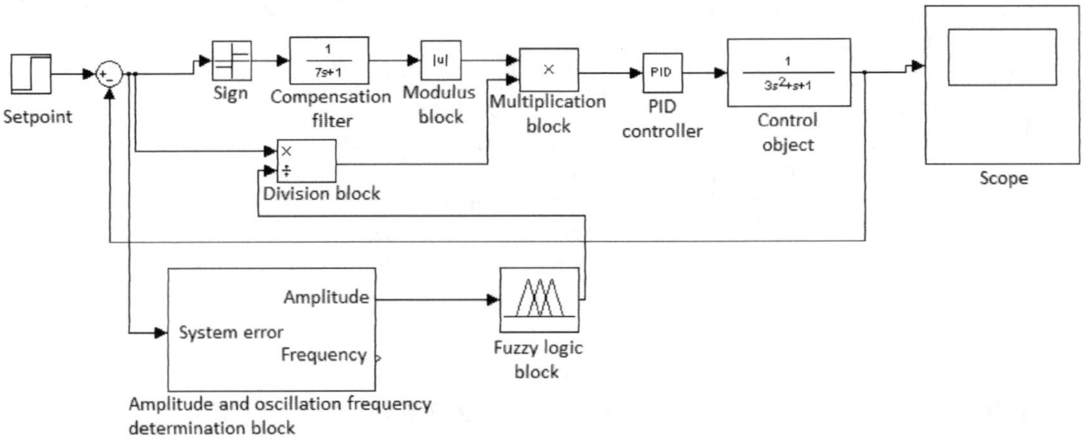

Figure 3. Control system model with the adaptive gain pseudolinear compensator

Figure 4 shows transient processes for automatic systems with the 1st, 2nd and 3rd order objects. These automatic systems consist merely of a PID controller and a PID controller with the pseudolinear gain compensator. These curves represent the adaptation ability of the control system to change the control object parameters.

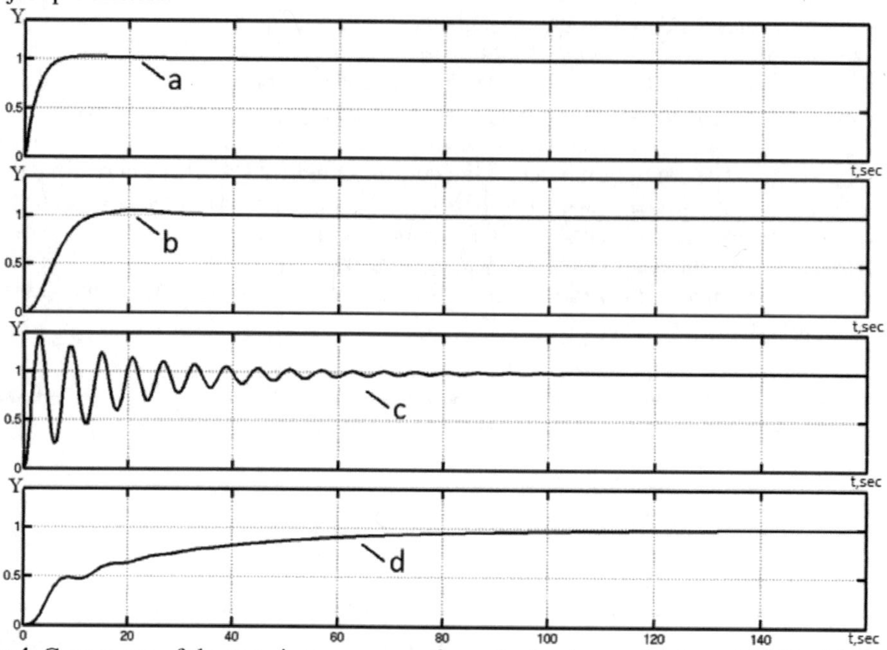

Figure 4. Curvatures of the transient processes for a single step excitation of ACS with the gain compensator

The control object was modelled by the second order element with transfer function

$$Wo(s) = \frac{Ko}{T_1^2 s^2 + T_2 s + 1},$$

where K_O – static gain of the control object, T_1 and T_2 – time constants of the control object.

Figure 4, a shows the nature of the transient process for a single step excitation of ACS with a PID controller without the compensator at the initial tuning of the controller, implemented according to Ziegler-Nichols method.

Figure 4, b shows a transient process for automatic system control with a PID controller and a compensator at the initial tuning of the compensator and controller.

Figures 4, *c* and *d* show transient processes after changing static gain of the control object from 10 to 1 and time constant T_2 from 20 to 1 sec. Value T_1^2 was considered to be equal to 3 sec. In addition, the gain and time constant of integration and differentiation of a PID controller, equal to K_p=2.4249, T_i=5.89, T_d=0.5419 correspondingly, did not change. The type of curve 4 shows that the quality of the automatic control system (ACS) with a compensator (figure 4, d) is much better than that of ACS without a compensator (figure 4, c) at the adjusted parameters of the control object.

Let us consider the adaptive phase compensator with transfer coefficient

$$W(s) = \frac{T_1}{T} \cdot \frac{Ts + 1}{T_1 s + 1},$$

where T_1 and T – time constants of the integral-derivative element.

Figure 5 shows the block scheme of the adaptive pseudolinear phase compensator.

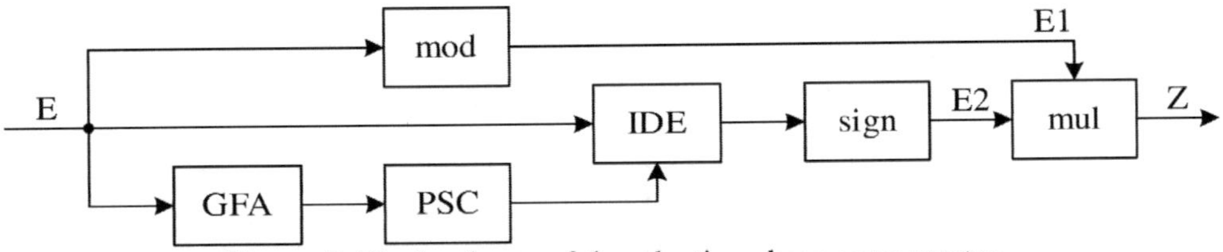

Figure 5. Block scheme of the adaptive phase compensator.

Figure 5 contains the following symbols: GFA – gain and frequency analysis block, PSC – phase shift correction block, IDE – integral-derivative element.

Harmonic linearization coefficients of this compensator have the following view:

$$a = \frac{1}{\pi}(\pi - 2\alpha + \sin 2\alpha) \ ; \ b = \frac{1}{\pi}(1 - \cos 2\alpha), \tag{3}$$

where $\alpha = \mathrm{arctg}\dfrac{\omega T(1-v)}{1+\omega^2 T^2 v}$; $v = \dfrac{T_1}{T}$.

The analysis of the compensator frequency characteristics shows that, when changing frequency and time constants (T_1 from 1 to 0,01 sec and T from 1 to 10 sec.), the phase characteristic undergoes transformations in the range from 0 to 78° and the decibel-log frequency response changes from 0 to –4 dB. Such transformation of the decibel-log frequency response does not influence the stability margins of ACS.

This compensator realizes a phase shift, which magnitude depends on the values of time constants T and T_1. The input signal branches out and goes through three channels. The upper channel provides error modulus extraction. The middle channel consists of an integral-derivative element, a sign determination block and a multiplication block. The lower channel provides determination of the amplitude and frequency of oscillations at the compensator input and changing of the phase shift of the integral-derivative element by correcting time constant T. The output signal (Z) equals the multiplication of the output signals from the upper (E1) and middle (E2) channels.

When the variable variations as well as an error arise in ACS, the compensator increases time constant T automatically, which results in increase of v (3) and this, in its turn, increases phase-frequency characteristics (PFC). This property justifies the PFC analysis of the compensator having the following view:

$$\phi(\omega) = \mathrm{arctg}\frac{\frac{1}{\pi}(1-\cos 2\alpha)}{\frac{1}{\pi}(\pi - 2\alpha + \sin 2\alpha)}.$$

In other words, the variation is suppressed by the rise of phase-frequency characteristic of ACS.

The determination of the time constant providing the required correction of the phase shift occurs in PCS block according to a decision rule, presented in the form of a piecewise linear function and obtained on the basis of the PFC set of an integro-differentiating element, corresponding to different values of time constants *T*. This function is made up on IDE phase-frequency characteristics series that varies with time constant T. The procedure of decision rule formation is based on the piecewise linear function construction that connects time constant T, providing maximal rise of phase-frequency characteristics, and the frequency of harmony oscillations that enter the compensator input [6].

The adaptive pseudolinear phase compensator, as well as the gain one, is connected in series to the control channel. The investigations of the phase corrector properties have been made by the example of ASC of the objects of the first and second order.

Figure 6 represents the curves of the transient process for a single step excitation in the control systems with a PID controller and a pseudolinear phase compensator. Curve 1 describes the transient process in ACS with a PID controller without the compensator and curve 2 describes the transient process in ASC with a PID controller and with the phase compensator. The automatic control system was studied with the same control object as in the case with the gain compensator considered above.

The parameters of the transfer function of the control object conforms to parameters that define the curve of the transient process in Figure 4, c. At that, the PID controller parameters did not change. Thus, the curves of the transient processes, represented in figure 6, characterize the possibility of the adaptive control by means of the pseudolinear phase compensator.

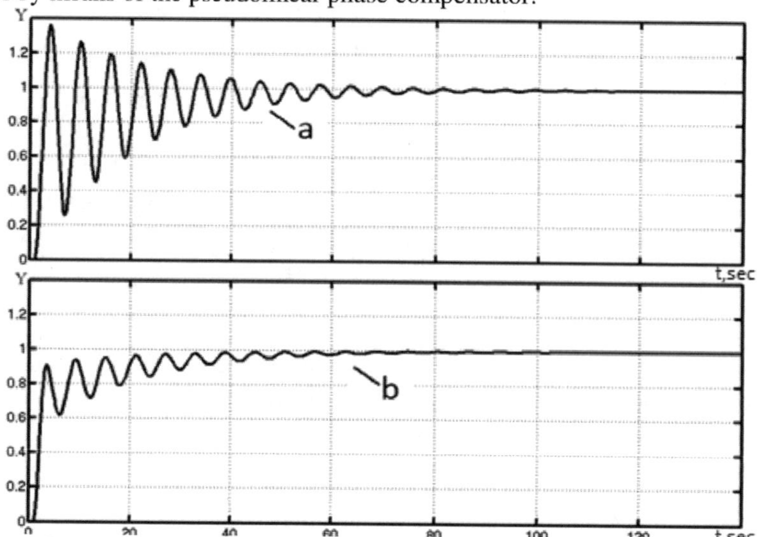

Figure 6. Curves of the transient processes for a single step excitation in the automatic control system with the phase compensator

3. Conclusion

When one applies the above-mentioned compensators practically, the question arises as to when it is better to use the gain compensator and when – the phase one. The investigation results have shown that the phase compensator is more effective in those systems of automatic control that require speed of operation. It is connected with the fact that, in case when the gain compensator is used along with the decrease of the compensator gain, the operation speed of control systems reduces as the reduction of the compensator gain leads to the gain reduction of the open-loop transfer function, and, consequently, to a decreased operation speed. At the same time, the gain compensator allows providing a required ACS quality by approximately 2…2,5 time wider in range of parametrization of the control object.

References

[1] Soldatov V V and Ukharov P E 2004 *Devices and systems. Operation. Control. Diagnostics* **8** 16–20

[2] Shteinberg S E, Zaluckiy I E, Serezhin L P and Varlamov I G 2003 *Industrial ACS and controllers* **10** 43–47

[3] Hlypalo E I 1982 Calculation design of nonlinear compensator devices in automatic systems (in Russian)

[4] Zelchenko V Y, Sharov S N 1981 Nonlinear correction of automatic systems (in Russian)

[5] Topcheev U I 1993 Automated design methods of nonlinear systems (in Russian)

[6] Skorospeshkin M V 2006 *Modern technique and technologies: students, graduate students and young scientists XII international scientific-practical conference* **2** 162–165

[7] Skorospeshkin M V 2005 *Science. Technology. Innovations. Young scientists All-Russian scientific conf.* 35–36

[8] Skorospeshkin M V 2005 *Modern mathematics and computer engineering problems. Young scientists and students IV science Republican conf. 108–110*

IOP Conf. Series: Materials Science and Engineering **124** (2016) 012029 doi:10.1088/1757-899X/124/1/012029

High-frequency trigger generators for CuBr-laser high voltage pumping source

S Torgaev[1,3,4], O Kozhemyak[1], E Yaroslavtsev[1], M Trigub[2,3], I Musorov[1] and D Chertikhina[1]

[1] Department of Industrial and Medical Electronics, National Research Tomsk Polytechnic University, 30, Lenin ave., Tomsk, 634050, Russia

[2] Department of High Voltage Engineering and Electrophysics, National Research Tomsk Polytechnic University, 30, Lenin ave., Tomsk, 634050, Russia

[3] Laboratory of Quantum Electronics, V.E. Zuev Institute of Atmospheric Optics SB RAS, 1, Academician Zuev Square, Tomsk, 634050, Russia

[4] Computer Science Laboratory, National Research Tomsk State University, 36, Lenin ave., Tomsk, 634050, Russia

E-mail: torgaev@tpu.ru

Abstract. In this paper the circuits of high frequency trigger generators of pulses of the nanosecond duration are presented. A detailed study of a generator based on the avalanche transistor with the use of a coaxial cable instead of a capacitor is described. This circuit showed advanced characteristics of the output pulses. A circuit of a generator built on high-speed digital components is also considered. The basic advantages and disadvantages of both generators are presented in this paper.

1. Introduction

There is a multitude of various scientific papers devoted to the study of copper and a copper halides vapor laser [1-5]. The advantages of this laser are: high gain of the laser medium, high average and peak power, a narrow spectral range, etc. This laser radiates two wavelengths: 510 nm (a green line) and 578 nm (a yellow line) and it is widely used in various fields of science and technology [1]. However, using this laser in high speed optical systems [6-8] requires obtaining a high pulse repetition rate (up to 1 MHz and above). Also, a high pulse repetition rate is needed to study the physics of a repetitively pulsed discharge in the laser medium in order to identify the parameters which impair the frequency and energy characteristics of the laser. Today, the maximum frequency of the copper bromide vapor laser is 700 kHz.

In [3, 8] it was shown that for the high pulse repetition rate it is necessary to ensure a low energy input in the mode of a discharge. The low energy input into the discharge can be achieved in various ways: reducing the pump pulse amplitude, an operation of the laser in the low-current mode, reducing the duration of the pump pulse. However, the first and second methods lead to a decrease in pumping efficiency. Therefore, for pumping supply applications the generator of pulses with small duration (about tens of nanoseconds) and high frequency (more than 1MHz) must be designed.

In previous studies [9], we investigated the parameters of the generator circuit built on an avalanche transistor 'GT338B'. This circuit could provide a pulse repetition rate up to 1.3 MHz with pulse duration of 20 ns (FWTM) and of 6-7 ns (FWHM), but these pulses had an oscillation on the falling edge. Besides, during experiments the increase of the fall time of the pulse was observed. This all had a negative effect on the performance of the circuit as a whole. Consequently, it was necessary to study the pulse generator circuit with a better pulse shape. This paper presents results of the experimental study of the pulse generator based on the avalanche transistors, and the other one – with the use of high-speed logic circuits.

2. A generator based on avalanche transistors

Figure 1, *a* shows a typical circuit diagram of the pulse generator based on avalanche transistors [10, 11].

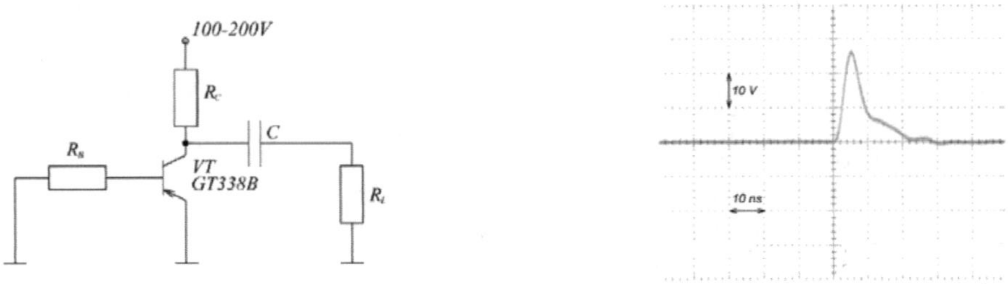

a b

Figure 1. The typical circuit of the pulse generator based on the avalanche transistor (a) and the output pulse of the generator based on the avalanche transistor 'GT338B' (b).

Figure 1, *b* shows the waveform of the voltage across load RL with the following parameters of the circuit: RL=100 Ohm, RB= 470 Ohm, RC=33 kOhm, C=33 pF. Registration of voltage and current pulses, as well as pulses of generation was done with the use of the oscilloscope 'Tektronix 3054C'. Such circuit topology allowed pulse generating with the duration of 6-7 ns (FWHM). However, the output pulse has sufficiently long fall time – more than 10 ns.

To improve the parameters of the output pulses of the generator (to decrease the pulse duration and fall time) various coaxial cables (RG59, RG64, DG113, RG174) were used instead of capacitance C (Figure 2). The parameters of the cables are shown in Table 1.

Table 1. Parameters of coaxial cables.

Parameter	RG59	RG64	DG113	RG174
Impedance, Ohm	75	75	75	50
Capacitance, pF/m	51	67	52	92.1
Copper lead diameter, mm	0.81	0.72	1.13	0.48
Cable diameter, mm	6.02	7.04	6.8	2.55

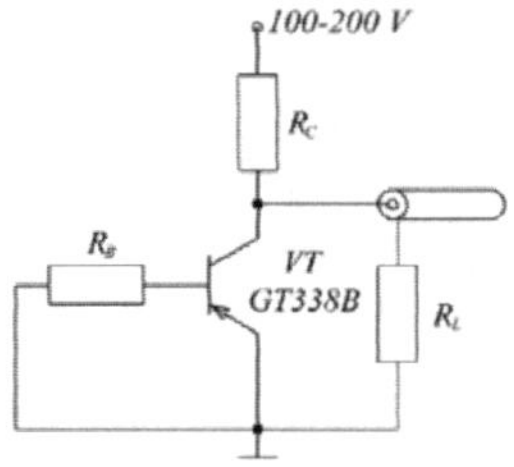

Figure 2. A circuit diagram of the generator based on the avalanche transistor with use of the coaxial cable as a capacitance.

Experimental studies have shown that the use of coaxial cables as a capacitor allows obtaining pulses with a better shape (with shorter fall time) and of short durations. Figure 3 shows the experimental waveforms of output pulses with use of various coaxial cables. The parameters of the circuit elements correspond to those in the experiment presented in Figure 1.

a) b)

Figure 3. The output pulses of the generator based on the avalanche transistor with use of coaxial cables of 10 cm length: a) RG59, b) RG174.

An experimental data show that the use of coaxial cables allows obtaining output pulses of about 2 ns duration (FWHM). At the same time, due to changes in the length of the coaxial cable, it is possible to adjust the frequency range of the generator. Figure 5 shows the waveforms of the output pulses obtained with use of coaxial cable RG174 of 5 cm and 10 cm lengths.

a) b)

Figure 4. The output pulses of the generator based on the avalanche transistor with use of coaxial cable RG174 of various lengths: a) 5 cm, b) 10 cm.

From the diagrams (Figure 4) it is obvious that with increasing length of the coaxial cable the repetition rate of output pulses decreases. Continuous adjustment of the frequency can be implemented by introducing a variable resistor into a collector circuit of the transistor. Figure 5 shows the dependence of the repetition rate of the output pulses on the value of the resistor placed in the collector circuit.

It is seen that an increase in the resistance in the collector circuit decreases the repetition rate of output pulses, and the dependence of the frequency on the resistance is almost linear. As it can be seen from the diagrams shown in figures 3 and 4, the voltage amplitude of the output pulses varies from 22 to 30 V. When using this generator for control of power transistors it is necessary to provide a possibility of adjustment of the amplitude of output pulses. This is due to the fact that the operating voltage on the gate of MOSFETs and IGBTs does not exceed 20 V. The adjustment of the amplitude of the output pulses in this circuit can be implemented by varying load resistance. Figure 6 shows the dependence of the amplitude and frequency of the output pulses on values of the load resistor.

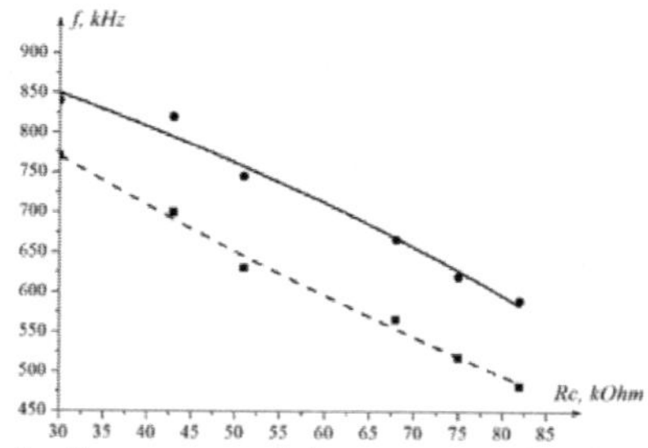

Figure 5. The dependence of the frequency of the output pulses on the value of the resistor in the collector circuit with use of coaxial cable RG64 of various lengths: 5 cm (solid line ●), 10 cm (dashed line ■).

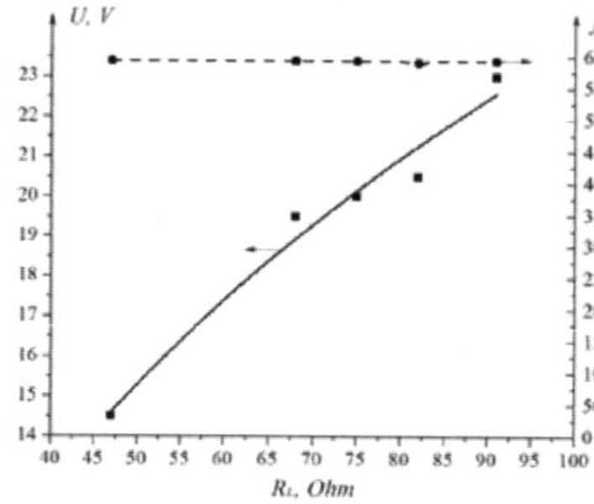

Figure 6. The dependence of the frequency (dashed line, ■) and the amplitude (solid line, ●) of the output pulses on the value of the load resistor with use of coaxial cable DG113.

Presented in Figure 6 the dependence shows the possibility of adjustment of the amplitude of output pulses in a rather wide range. Due to the small value of the load resistor the pulse repetition frequency remains nearly constant, i.e. time constant of the circuit of charge of cable capacitance does not change. However, the change of the load resistance leads to the change of the output pulse duration. One of the ways to eliminate this disadvantage is to build a high-frequency pulse generator on the base of high frequency logic elements.

3. A generator of high frequency pulses based on logic elements

In this part, a high-frequency generator of nanosecond pulses on the base of logic elements was built in the form of the block diagram shown in Figure 7.

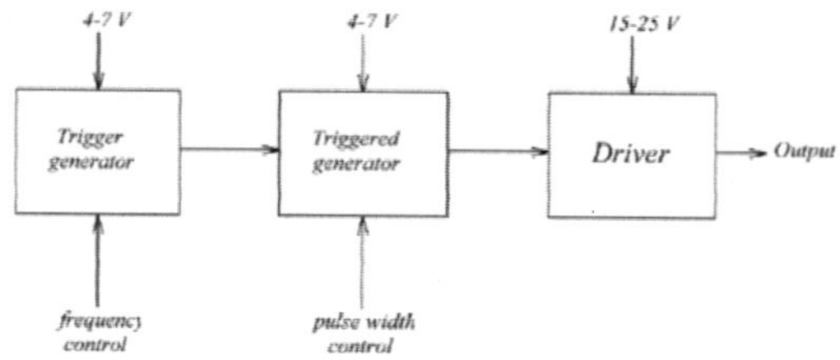

Figure 7. A block diagram of the high frequency trigger generator.

The trigger and triggered generators were built on standard generator circuits with logic elements. The high-frequency logic elements - SN74LVC1G00DBVR – were used in the circuit of the generator. They have a CMOS structure with maximum supply voltage of 5 V. Figure 8 shows the output voltages of the trigger and triggered generators with maximum frequency and the minimum duration of the output pulses.

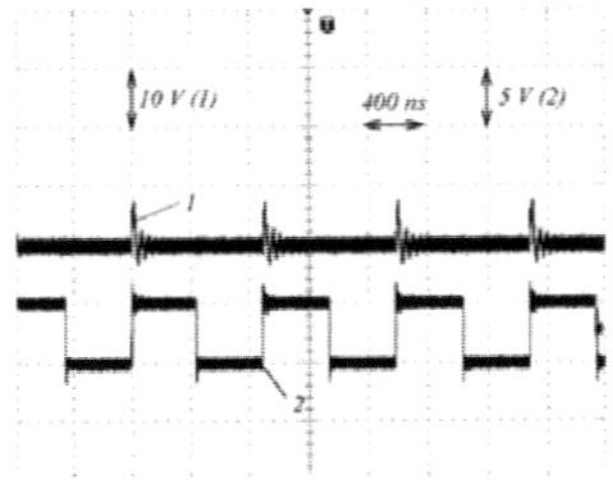

Figure 8. The waveforms of the output pulses of a trigger generator (2) and a triggered generator (1).

Since it is necessary to generate the output pulses with an amplitude of up to 20 V, the generator circuit includes high frequency driver IXDN630CI. This driver allows generating the output pulses with an amplitude of up to 25 V and duration of 10-12 ns (FWHM) (Figure 9).

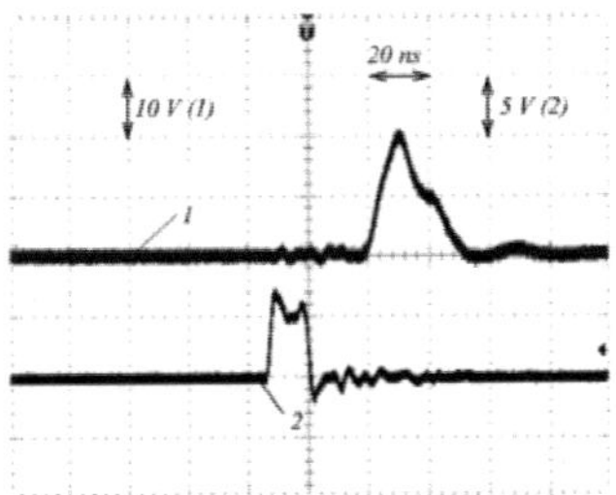

Figure 9. The waveforms of the output pulses of the triggered generator (2) and driver IXDN630CI (1).

Diagrams (Figures 8 and 9) show the possibility of generating the output pulses with frequency greater than 1 MHz and pulse duration less than 20 ns for the circuit of the pulse generator based on logic elements. Such circuit topology allows independent adjustment of the frequency and duration of the output pulses.

4. Conclusion

Generators of high-frequency trigger pulses considered in this paper allow generating control signals for high voltage pumping of the CuBr-laser. Presented generators provide output pulse repetition frequency above 1 MHz with pulse duration of about 4-12 ns (FWHM).

The generator circuit based on the avalanche transistor is the most simple and cheap. The main disadvantage of this circuit is an extremely large variety of transistors parameters (up to 300%). Therefore, the replacement of the transistor will cause the required circuit debugging. The generator circuit based on logic elements eliminates this disadvantage, however, due to the use of high-speed digital circuits and the driver such circuit high frequency generator has substantially greater cost.

Acknowledgment

The work is supported by the Russian Science Foundation, project No 14-19-00175.

References

[1] Little C E 1998 *Metal Vapor Lasers: Physics, Engineering & Applications* (Chichester, UK, John Wiley & Sons Ltd.)

[2] Boychenko A M, Evtushenko G S and Torgaev S N 2011 *Phys. of Wave Phenom.* **19** 189–201

[3] Boychenko A M, Evtushenko G S, Nekhoroshev V O, Shiyanov D V and Torgaev S N 2015 *Phys. of Wave Phenom.* **23** 1–13

[4] Shiyanov D V, Trigub M V, Sukhanov V B, Evtushenko G S and Vlasov V V 2015 *Technical Physics* **60** 571–4

[5] Trigub M V, Shiyanov D V and Evtushenko G S 2013 *Russian Physics Journal* **55** 1152–6
[6] Evtushenko G S, Trigub M V, Gubarev F A, Evtushenko T G, Torgaev S N and Shiyanov D V 2014 *Review of Scientific Instruments* **85** 1–5
[7] Trigub M V, Shiyanov D V and Vlasov V V 2014 Brightness amplifiers with PRF up to 100 kHz Proc. Int. Conf. of Young Specialists on Micro/Nanotechnologies and Electron Devices, EDM 6882534
[8] Nekhoroshev V O, Fedorov V F, Evtushenko G S and Torgaev S N 2012 *Quantum Electronics* **42** 877–9.
[9] Chertikhina D S, Musorov I S, Torgaev S N and Yaroslavtsev E V 2014 *Proc. Int. Conf. of Young Specialists on Micro/Nanotechnologies and Electron Devices, EDM* 23254173
[10] Diyakonov V 2010 *Components and technologies* **8** 49–58
[11] Zbigniew B and Krzysztof C 1996 *Opto-Electronic Review* **4** 58–61

Modification and simulation of the power supply of a metal vapor laser

D N Ogorodnikov[1], M V Trigub[1,2], S N Torgaev[1,2,3], N A Vasnev[1] and T G Evtushenko[3]
[1] National Research Tomsk Polytechnic University, 30, Lenina ave., Tomsk, 634050, Russia
[2] V E Zuev Institute of Atmospheric Optics Siberian Branch of the Russian Academy of Sciences, 1, Academician Zuev sq., Tomsk, 634055, Russia
[3] National Research Tomsk State University, 36, Lenina ave., Tomsk, 634050, Russia

E-mail: ogorodnikov@tpu.ru

Abstract. The modification of a power supply circuit used for pumping metal vapor lasers is analyzed. The results of OrCAD simulation of the processes that occur in the power supply are presented. The effect of the capacitance ratio on the charging process of a storage capacitor is described. The mode which provides more time for the recovery of the thyratron is discussed. The results of the development of the small-size high pulse repetition frequency laser with up to 3 W average output power are presented.

1. Introduction

The range of applications of metal vapor lasers is wide, including a number of medical ones. The main purpose of modification of a laser power supply is optimization of the pumping circuit, and therefore an increase in the laser pulse repetition frequency (PRF) and its efficiency [1-3]. Thyratrons are used most often as the power switches for constructing pumping circuits of metal vapor lasers. A thyratron provides a high rate of the current rise in the gas discharge tube (GDT) to obtain the required parameters of the pump pulse. Short lifetime and the deterioration of switching characteristics over a period of time are the disadvantages of these power switches [4].

Laser pumping circuits based on the direct discharge of the storage capacitor, are described in [4-6]. The disadvantage of these circuits is that a thyratron recovers its electrical resistance when the anode voltage is applied since the storage capacitor begins to charge immediately after its discharge on the GDT. Furthermore, this makes it impossible to increase the PRF of these devices.

The results of research and development of such power supplies at the Laboratory of Quantum Electronics of the Institute of Atmospheric Optics of Siberian Branch of the Russian Academy of Science (IAO SB RAS) are published in [4-5]. The analysis and simulation of a metal vapor laser power supply based on the half-bridge inverter are presented in [6]. The aim of this work is to upgrade the power supply and study its operation modes using OrCAD models. Another aim is to develop a compact high frequency CuBr laser with the pumping circuit based on a TGI1-270/12 thyratron.

2. A power supply with the pulse charge of the storage capacitor

Unlike the circuit described in [6], the upgraded inverter is based on an H-bridge circuit. The functional circuit of the power supply is shown in Figure 1. The storage capacitor charges sinusoidally

IOP Conf. Series: Materials Science and Engineering **124** (2016) 012030 doi:10.1088/1757-899X/124/1/012030

because of inductor L_2 connected in series with capacitor C_2 and primary winding w_p of transformer TV_1.

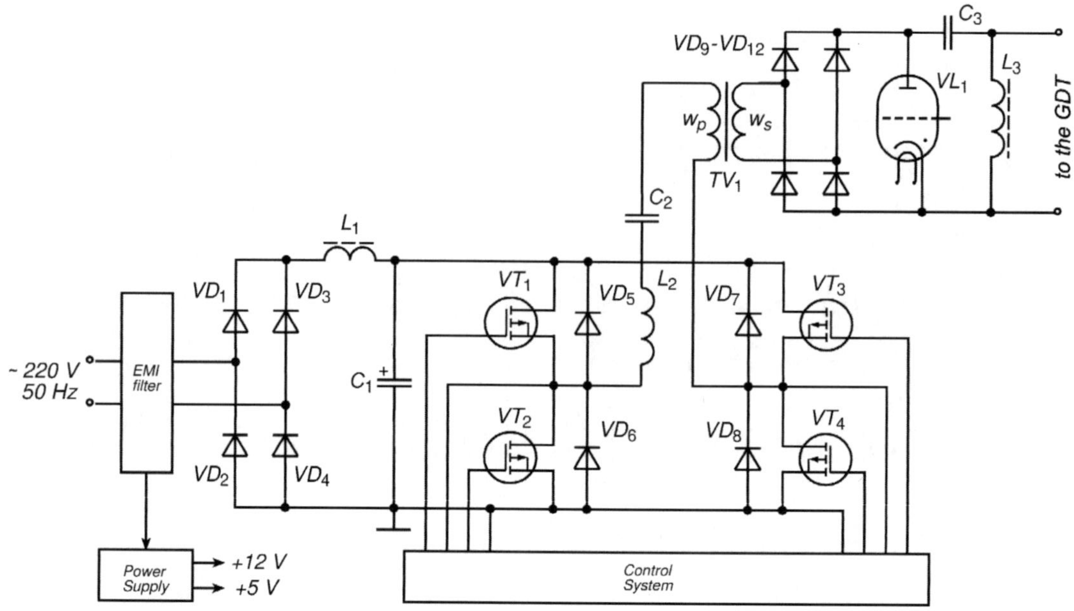

Figure 1. A functional circuit (H-bridge topology) of the power supply.

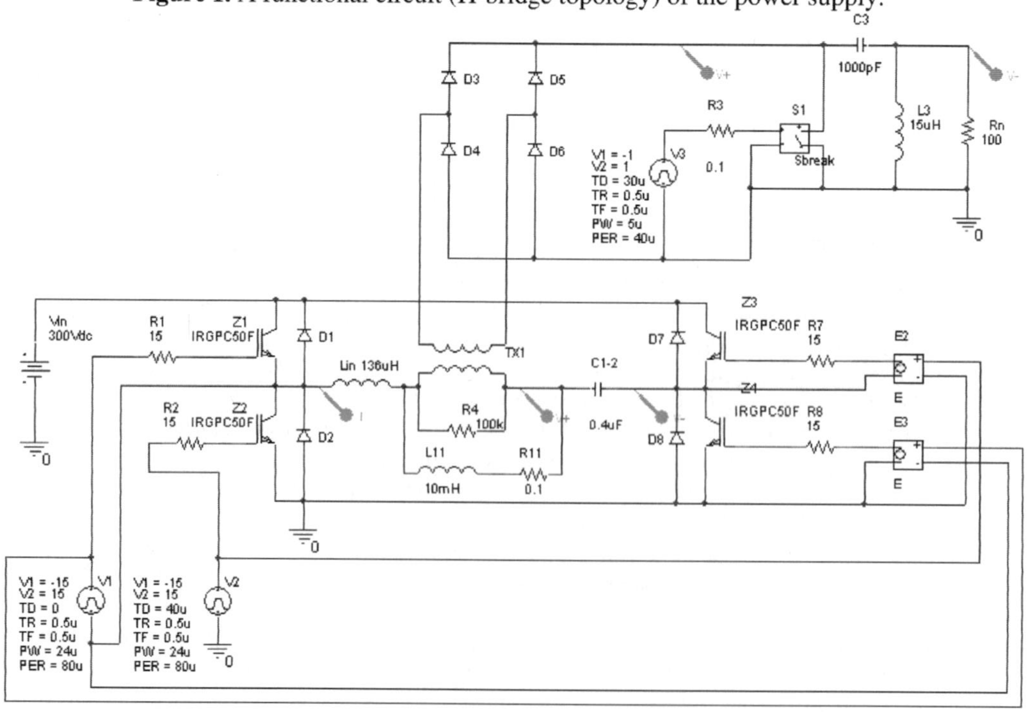

Figure 2. An OrCAD simulation circuit of the power supply (the capacitance ratio $a = 1$).

The control system (CS) provides the algorithms of operation of H-bridge power transistors and thyratron VL_1. Storage capacitor C_3 is charged through a pair of transistors which are switched on. After that the thyratron is triggered by the pulse from the control system. The storage capacitor quickly discharges through the GDT. Then there is a pause during which the thyratron anode voltage is zero, and the device switches off. Such mode ensures strong recovery of the thyratron. The control system forms signals to switch on the transistors of the other pair of the H-bridge. The inductor current and the current of the TV_1 windings start to flow in the opposite direction. The storage capacitor charges again, and the processes repeat periodically.

3. OrCAD simulation of the power supply

The OrCAD simulation circuit is shown in Figure 2. Compared to the half-bridge inverter [6] in this circuit the conversion ratio of TX1 is reduced by half to obtain the same amplitude of the storage capacitor voltage. Consequently, there is a four-time reduction of the value of the storage capacitor for the transformer primary circuit. Hence, when the capacitance ratio remains unchanged there is also a fourfold reduction of the equivalent capacitance of the oscillating circuit. Therefore it is necessary to ensure a fourfold increase of the oscillating circuit inductance to obtain the same resonant frequency.

The simulation results for different capacitance ratios are presented in Figures 3-5. The first waveform is the current of inductor L_{in} of the H-bridge inverter. The voltage across storage capacitor C_3 and the steady state voltage across H-bridge inverter capacitor $C_{1\text{-}2}$ are shown together on the second graph. Let us note that for convenience, the display size of the storage capacitor voltage experiences a 40-time reduction. You can also see that after the end of the storage capacitance charge process (every 40 μs), switch S1 is on, and the capacitance is quickly discharged through the load.

For the case when $a = 1$, we can see that the charge time is minimal, i.e. the resonant frequency is maximal. However, for capacitor $C_{1\text{-}2}$ the voltage swing exceeds 600 V. In this case, the amplitude of the capacitor charging current is maximal. As we can see from Figures 4 and Figures 5, the inductor current amplitude becomes lower when the capacitance ratio is increased, but a capacitor needs more time for charging. Steady state simulation results are shown in Table 1.

Table 1. Simulation Results.

a	C_1 (μF)	U_{C0} (V)	U_{Ctp} (V)	U_{C2m} (kV)	I_{Lm} (A)	tp (μs)
1	0.4	–300	339	11.1	21.2	16.8
3	1.2	–100	88	10.7	16.2	20.5
10	4.0	–26	25	10.1	13.6	22.0

From the simulation results obtained using model parameters we can see that the current amplitude has decreased from 21.2 A to 13.6 A, and the capacitor charge time increased from 16.8 μs to 22 μs. In the modernized circuit the amplitude of the inductor current decreased twice if compared to the half-bridge inverter. Thus, the use of the H-bridge inverter circuit can reduce the size of the pulse transformer, thus reducing the size of the pumping circuit as a whole. This circuit is more preferable for a multistage charging of the storage capacitor.

4. The Power Supply of High Pulse Repetition Frequency

The development of a high pulse repetition frequency of the CuBr laser with PRF up to 50 kHz was the next stage.

The GDT is the self-heated active element with independent heaters for containers with CuBr and for the HBr generator. A laser tube with the active zone diameter of 15 mm and the length of 38 mm is used. The Ne buffer-gas pressure is 30 Torr. The TGI1-270/12 thyratron is used to increase the operating frequency. The storage capacitance of 330 pF is used. For the 1 kW power supply the half-bridge inverter is chosen. The main circuit is similar to that in [5].

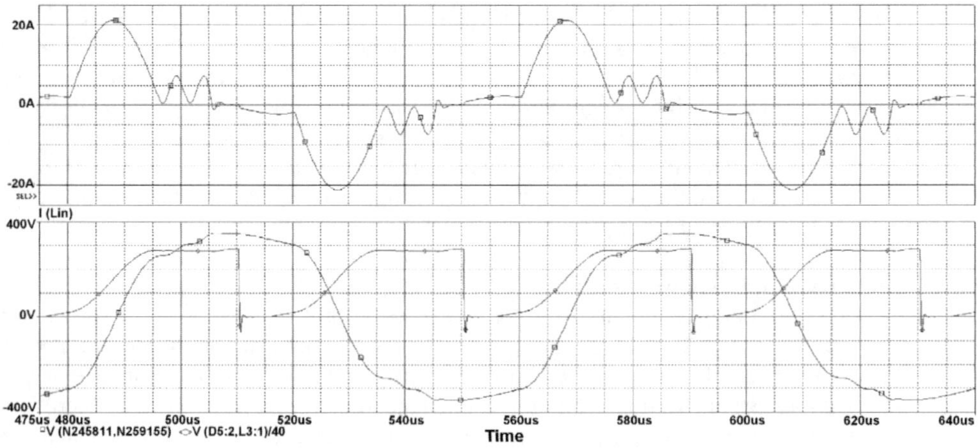

Figure 3. Current and voltage waveforms, capacitance ratio a = 1.

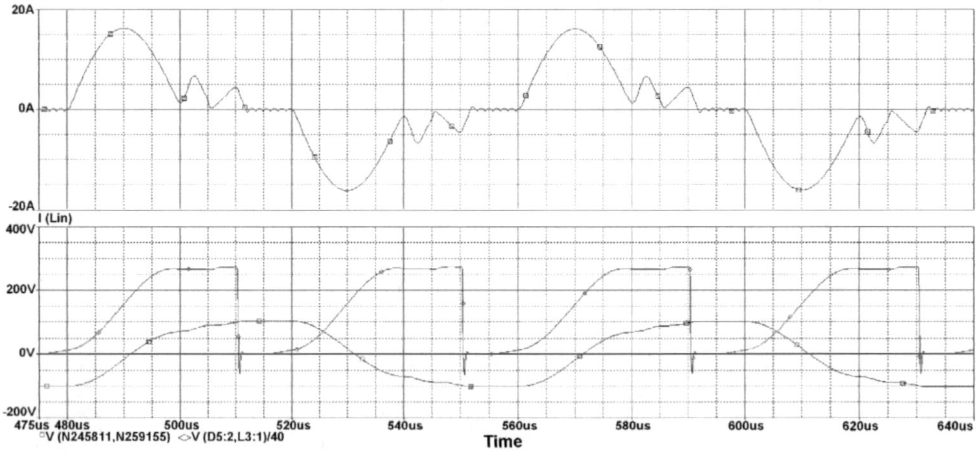

Figure 4. Current and voltage waveforms, capacitance ratio a = 3.

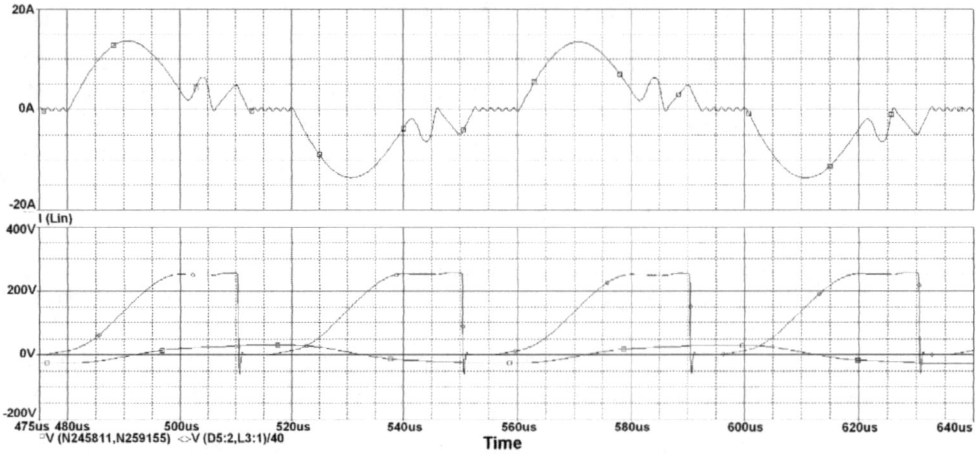

Figure 5. Current and voltage waveforms, capacitance ratio a = 10.

The control circuit is based on an Atmega8 microcontroller; it has the ability to change the width of pulses and pauses between them for driving the power switches of the inverter. The power supply includes a soft start circuit.

The simulation results are experimentally verified. The power supply inverter operates at a frequency of 42 kHz. The results are shown in Figure 6. Waveforms show that the thyratron recovery time is about 8 μs.

A maximal average output power of 2.5 W has been obtained at a maximum pumping power and a frequency of 42 kHz. The practical efficiency is about 0.6 %.

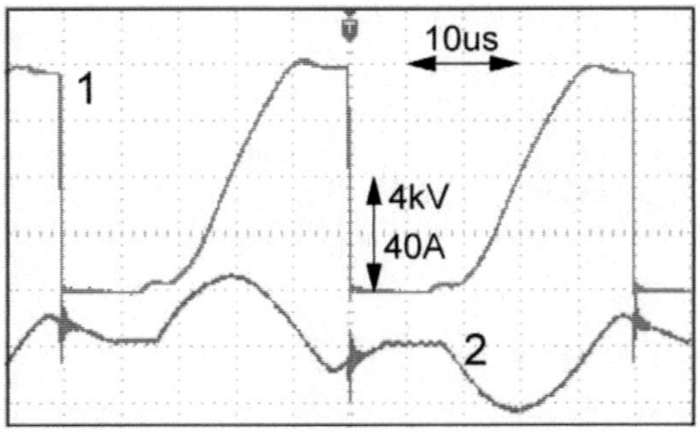

Figure 6. Current and voltage waveforms: 1 – the thyratron anode voltage; 2 – the inductor current (an inverter operates at a frequency of 42 kHz).

5. Conclusion

To sum it up, the results of OrCAD simulation of the processes that occur in the power supply based on an H-bridge inverter are presented. Compared to the half-bridge inverter we can see that the H-bridge topology circuit allows one to reduce the pulse transformer size. The H-bridge circuit is more preferable for CuBr lasers with a multistage charging of the storage capacitor.

The results of development of the high pulse repetition frequency of the CuBr laser are presented. For the half-bridge circuit power supply, an average output power of 2.5 W has been obtained at a frequency of 42 kHz.

Acknowledgment
The work is supported by the Russian Science Foundation, project No 14-19-00175.

References

[1] Batenin V M, Buchanov V V, Ghazaryan M A et al 1998 *Lasers self-contained metal atoms* (Moscow: Science Book)

[2] Trigub M V, Shiyanov D V and Vlasov V V 2014 Brightness amplifiers with PRF up to 100 kHz *Proc. Int. Conf. of Young Specialists on Micro/Nanotechnologies and Electron Devices, EDM* Article number 6882534 301-4 doi: 10.1109/EDM.2014.6882534

[3] Trigub M V, Shiyanov D V, Sukhanov V B and Evtushenko G S 2014 *J. Atmospheric and Oceanic Optics* **27** 321–5

[4] Dimaki V A, Sukhanov V B, Troilskii V O and Filonov A G 2012 *J. Instrum. Experim. Tech.* **55(6)** 696–700

[5] Krasnikov I V, Trigub M V and Evtushenko G S 2012 *Russian Jour. Bulletin of Science in Siberia* **5(6)** 54

[6] Trigub M V, Ogorodnikov D N, Vlasov V V and Evtushenko T G 2014 Design and Simulation of a Metal Vapor Laser Power Supply (*Proc. of 2014 Int. Conf. on Mechanical Engineer., Automat. and Contr. Syst., MEACS 2014)* 6986889

Determination of error measurement by means of the basic magnetization curve

M V Lankin and A M Lankin
Department of information and measuring systems and technologies, Federal State Educational Institution of Higher Professional Education 'M I Platov South Russian State Technical University (NPI)', 132, Prosveshcheniya st., Novocherkassk, 346428, Russia

E-mail: lankinjohn@rambler.ru

Abstract. The article describes the implementation of the methodology for determining the error search by means of the basic magnetization curve of electric cutting machines. The basic magnetization curve of the integrated operation of the electric characteristic allows one to define a fault type. In the process of measurement the definition of error calculation ofthebasic magnetization curve plays a major role as in accuracies of a particular characteristic can have a deleterious effect.

1. Introduction
In the course of the machine tools operation there arises a problem to make these machines available for the self-test drives. For example, to assess the extent of the drive failure we propose using a basic magnetization curve [3-5, 9] as an indication. A method for determination of the basic magnetization curve is based on the solution of the inverse problem of harmonic balance [1] by the natural-model tests [2, 6-8, 10]. An important role in the implementation of the method belongs to the error identification between the model and analyzed characteristics.

2. Materials and methods
All electric machine tools (solenoids, solenoid relays, motors) are movable and fixed parts are represented by the core and at least one working coil. During the operation of electric machine tools the current flows in the work coil, thus generating a working magnetic flux in the stationary part of the magnetic circuit that drives the movable part. The magnetic flux is determined by the design, the mutual arrangement of the working parts of the magnetic core and the coil, as well as the number of turns in it and the magnitude of the flowing current. During the working cycle of the movable part it moves relative to the fixed magnetic circuit, which also leads to a change in the magnetic flux. This suggests that the integral characteristic contains not only information about the operating parameters of the electric machine tools, but it also influences the quality of individual parts of the basic magnetization curve.

For the basic magnetization curve of electric machine tools it is necessary to work on the external changing magnetic field and to measure magnetic flux density occurring in the cross-section of the special sensor. To measure the basic magnetization curve of an electric machine tool we have tested assembled electric machine tools, which made it impossible to use the sensors of magnetic induction.

The method of determining the basic magnetization curve of electromagnets is an inversion harmonic balance method of natural-model tests [2, 6-8, 10], unifying the dimension of the physical

IOP Conf. Series: Materials Science and Engineering **124** (2016) 012031 doi:10.1088/1757-899X/124/1/012031

object and modeling of the object. The solution of the inverse problem of harmonic balance and the method of natural-model tests allow obtaining the basic magnetization curve, in which it is possible not only to determine the state of the electromagnet, but also the type of the fault. The technique is based on the previously conducted studies on obtainment of the basic magnetization curve of a fixed part of the magnetic circuit. The advantage of the proposed method is anopportunity to conduct a diagnosis in the course of an operational procedure and production. The method provides the basic magnetization curve of the operating cycle with the electromagnets error not exceeding 3 %.

Let us search the basic magnetization curve of the approximated electromagnet using the following expression:

$$B = \sum_{m=1}^{n} k_{(2m-1)} H^{2m-1}$$

where B – a value of magnetic induction, $k_{(2m-1)}$ – coefficients of the approximated basic magnetization curve, $m = \overline{(1, n)}$, n – the number of conditions in the approximated curve, i – the current flowing through the coil of the electromagnet, H – magnetic intensity.

In our case, there is an electromagnet with the unknown basic magnetization curve; the known laws of variation of voltage are applied to the electromagnet, as well as to the current flowing through it. It is required to determine the coefficients $k_{(2m-1)}$ expression approximating the basic magnetization curve.

In the experiments with electrical devices without moving parts of the magnetic circuit, which are used as a method to determine the basic magnetization curve of the electrical device, we used an algorithm of the natural-model test, which consists in the following: an electrical unit is supplied with sinusoidal voltage measured by current and voltage sensors, the input (voltage) and the output (current) are measured by an electrical signaling device, then the obtained data are sent to a personal computer, where model inputs are set to obtain the output data model using an optimization algorithm for search, which matches the output of the electrical device and its model. The simplex method of optimization was used as a basis for constructing an optimization algorithm of the control program. The program implements an electromagnet, and an optimization program based on simplex-planning.

The method of finding errors between the basic magnetization curves is explained in Figure 1. On exemplary basic magnetization curve $B_1(H)$ we set points which are arranged perpendicular to the intersection with measured basic magnetization curve $B_2(H)$.

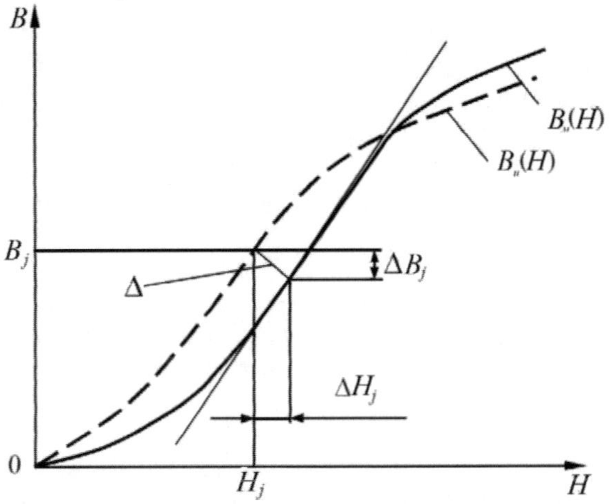

Figure 1. Explanation of the method of finding errors.

The absolute error of magnetic induction B and magnetic intensity H at each test point is determined as the projection pieces made between the model and measured characteristics ΔB and ΔH

on the coordinate axis. The relative error of B and H, as well as complete measurement error characteristics are determined by expressions:

$$\delta_H = \frac{\Delta H}{H}, \quad \delta_B = \frac{\Delta B}{B}, \quad \delta = \sqrt{\delta_H^2 + \delta_B^2}.$$

From the obtained results and the selected maximum we take the basic magnetization curve as a measurement error.

3. Results

The program implements the described technique. In this program a model of the basic magnetization curveis defined by points and measured in the form of a polynomial of an odd degree:

$$B = \sum_1^n k_{2n-1} H^{2n-1}.$$

A diagram of the program is shown in Figure 2.

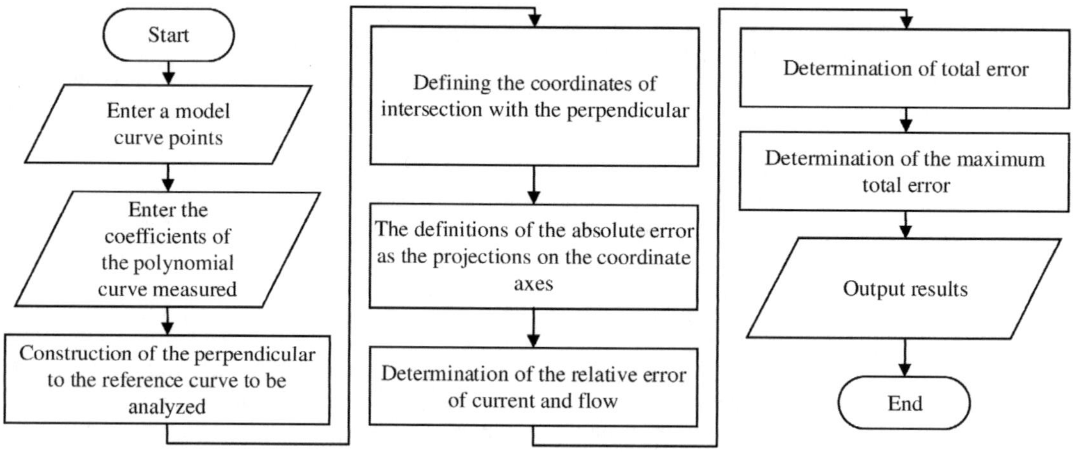

Figure 2. An algorithm for the program.

The algorithm operates as it is shown. The initial data provide an exemplary basic magnetization curve in the form of two column vectors and a measured basic magnetization curve in the form of a polynomial of an odd degree. Let us calculate the equation of the line passing through an i-point, wherein the calculated error and the $(i + 1)$ point inside are

$$y_i = kx_i + b.$$

Then we shall calculate the equation of the perpendicular to this line in the i-point inside

$$y_{i+1} - y_i = -\frac{1}{k}(x_{i+1} - x_i).$$

We seek a point of intersection of the perpendicular and the measured basic magnetization curve given in the form of the polynomial. Knowing the coordinates of the intersection of the perpendicular model and measured characteristics, the determined projection on the coordinate axes (current and flux) is an absolute error:

$$\Delta_{Hi} = x_{oi} - x_{ai};\ .\ \Delta_{Bi} = y_{oi} - y_{ai},$$

where x_{oi} and y_{oi} are points on the exemplary characteristic, and x_{ai} and y_{ai} are points on the measured characteristic. The relative error can be found by dividing the length of the projection on the coordinate axes and coordinate values corresponding to the i-point:

$$\delta_{Hi} = \frac{\Delta_{Hi}}{x_{oi}}, \quad \delta_{Bi} = \frac{\Delta_{Bi}}{y_{oi}}.$$

Further, the total error calculated in the i-point is

$$\delta_i = \sqrt{\delta_{Hi}^{\ 2} + \delta_{Bi}^{\ 2}}\ .$$

After calculation of errors at all points let us find the maximum value, which is an error between the model and the analyzed basic magnetization curve:

$$jj := \begin{bmatrix} j_1 \\ j_2 \\ j_3 \\ j_4 \\ j_5 \\ j_6 \\ j_7 \\ j_8 \\ j_9 \\ j_{10} \end{bmatrix}; ff := \begin{bmatrix} f_1 \\ f_2 \\ f_3 \\ f_4 \\ f_5 \\ f_6 \\ f_7 \\ f_8 \\ f_9 \\ f_{10} \end{bmatrix}$$

$k_1 := k_1; k_3 := k_3; k_5 := k_5; k_7 := k_7; k_9 := k_9$

for i **from** 1 **to** 9 **by** 1 **do** $y0_i := y2_i;\ x0_i := x2_i;\ c_i := solve(\{y1_i = a_i \cdot x1_i + b_i, y2_i = a_i \cdot x2_i + b_i\}, \{a_i, b_i\});\ a_i := solve(c_i[1], a_i);\ b_i := solve(c_i[2], b_i);\ n_i := y_i = solve\left(y_i - y0_i = -\frac{1}{a_i}(x_i - x0_i), y_i\right);\ \mathrm{d}_i := \mathrm{solve}(\{n_i, y_i = k_1 \cdot x_i^1 + k_3 \cdot x_i^3 + k_5 \cdot x_i^5 + k_7 \cdot x_i^7 + k_9 \cdot x_i^9\}, \{x_i, y_i\});\ xx_i := d_i[1];\ xxx_i := solve(xx_i[1], x_i);\ yyy_i := solve(xx_i[2], y_i);\ \mathrm{px}_i := \frac{\mathrm{abs}(xxx_i - x0_i)}{x0_i} \cdot 100;\ \mathrm{py}_i := \frac{\mathrm{abs}(\mathrm{yyy}_i - y0_i)}{y0_i} \cdot 100$ **od**

for i **from** 1 **to** 9 **by** 1 **do** $\mathrm{y1_i := jj[i];\ y2_i := jj[i+1];\ x1_i := \mathit{ff}[i];\ x2_i := \mathit{ff}[i+1]}$ **od**

pxx := *convert*(px, array)

pyy := *convert*(py, array)

Forifrom1 **to** 9 **by 1 do** pz[i]=*sqrt*(pxx[i]^2+pyy[i]^2) **od**

pzmax:=max(pz)

print(pzmax)

Figure 2. Listing of the error search program to analyze the basic magnetization curve.

In addition to this kind of the error, the method must take into account the error of the measurement device. Calculation of the error is reduced to the estimation of the standard deviation of individual links (σ_i) taking into account additional errors from influencing factors and finding the total error of the measurement conversion:

$$\sigma_m = \sqrt{\sum \sigma_i^2}$$

The relative error in the measurement and a processing unit, reduced to the top of the measuring range is δ_{ms} = 1 %, and reduced to the end – δ_{me} = 2 %. The relative error of the shunt is δ_{sh} = 0.5 %. The standard deviation of the error amplification unit consists of three components: a basic error (1 %); pulsation error voltage (0.2 %) and ambient temperature variations (0.66 %). The error switch consists of three components: an error voltage drop of a public key (0.4%),leakage current in each private key channel (0.13 %) and the carrier frequency ripple (0.06 %). ADC at the beginning and the end of a range is δ_{ADCs} = 0.2 % and δ_{ADCe} =0.3 %.

1. We accept a value quintile ratio equal to k = 1.73. Based on a linear scale and formula $\sigma_i = \frac{\delta_i}{k}$ we find σ_{ds} = 0.6 %, and the end of the range, which is σ_{de} = 1.2 %.

2. For shunt δ_{shs} = δ_{she} =0.5 %. Taking the above-mentioned conditions into account, we obtain $\sigma_{shs} = \sigma_{she}$ = 0.29 %.

3. We can obtain the block using $\sigma = \sqrt{\sum \sigma_i^2} = \sqrt{1^2 + 0.2^2} = 1$ %.

Since the developed device has no other errors, the total error is $\sigma_{dev} = \sqrt{\sum \sigma_i^2} = \sqrt{1^2 + 0,66^2} = 1.3$ %.

4. Let us switch, taking into account the conditions of point 1,

$$\sigma_s = \frac{1}{1.73}\sqrt{0.4^2 + 0.13^2 + 0.06^2} = 0.24 \text{ \%}.$$

5. The relative error of the ADC set. Assuming the law of the uniform distribution, we obtain

$$\sigma_{ADCs} = \frac{0,2}{1.73} = 0.13\% \quad \sigma_{ADCe} = \frac{0,3}{1.73} = 0.17\%.$$

6. Finally, the standard deviation for the end of the range will be

$$\sigma_{sume} = \sqrt{1.1^2 + 0.39^2 + 1.3^2 + 0.24^2 + 0.13^2} = 1.76 \text{ \%},$$

and for the beginning of the measuring range it will be

$$\sigma_{sums} = \sqrt{0.6^2 + 0.39^2 + 1.3^2 + 0.24^2 + 0.17^2} = 1.5 \text{ \%}.$$

Conclusion

In the present paper we have developed an algorithm and a program that allows determining the location of the error of the basic magnetization curve of electric cutting machines.

Acknowledgment

The results have been obtained with the support of project № 1.2690.2014 / K 'Methods for solving inverse problems of diagnosis of complex systems (engineering and medicine) on the basis of the natural model experiment' performed in the framework of the project state task using CCU equipment 'Diagnosis and energy-efficient electrical equipment' of South-Russian State Technical University (NPI) after M I Platov.

References

[1] Bakhvalov Y A, Gorbatenko N I, Grechikhin V V 2015 *Measurement Techniques* **58** 336–340
[2] Borovoi V V, Gorbatenko N I, Grechikhin V V 2015 *Measurement Techniques* **56** 614–617
[3] Gorbatenko N I, Lankin M V 2004 *Russian Electrical Engineering* **8** 63–66
[4] Gorbatenko N I, Lankin M V 2004 *Elektrotekhnika* **8** 19–26
[5] Gorbatenko N, Lankin M, Shaykhutdinov D, Gazarov K, Kolomiets A 2011 *Proceedings of the 6th International Forum on Strategic Technology* (Harbin: Heilongjiang) pp 194–196
[6] Gorbatenko N I, Lankin A M, Lankin M V, Shayhutdinov D V 2015 *International Journal of Applied Engineering Research* **3** 6509–6519
[7] Gorbatenko N I, Grechikhin V V, Shaikhutdinov D V 2015 *Measurement Techniques* **56** 609–613
[8] Lankin A M, Lankin M V, Aleksanyan G K, Narakidze N D 2015 *International Journal of Applied Engineering Research* **3** 6489–6498
[9] Shaykhutdinov D V,Gorbatentko N I, Akhmedov Sh V, Shaykhutdinova M V, Shirokov K M 2013 *Life Science Journal* **10** 2698–2702
[10] Shaikhutdinov D V, Gorbatenko N I, Shirokov K M 2015 *Measurement Techniques* **56** 618–620

The study of the current form influence on measurement of the weber-ampere characteristics

A M Lankin and M V Lankin
Department of information and measuring systems and technologies, Federal State Educational Institution of Higher Professional Education 'M I Platov South Russian State Technical University (NPI)', 132, Prosveshcheniya st., Novocherkassk, 346428, Russia

E-mail: lankinjohn@rambler.ru

Abstract. The article describes the study of the effect of measurement error of the current form on determination of weber-ampere characteristics of electric cutting machines. During experimental studies we have developed a method for determining the weber-ampere characteristics of the electric machine tools. However, there is an error in the measurement of the harmonic content of the current, which requires additional studies of the current form effect on weber-ampere characteristics.

1. Introduction

A method for determination of the weber-ampere characteristics of the electric machine tools is based on the methods of harmonic balance and a field-test model [3-8]. The authors have determined weber-ampere characteristics for measuring the form of the supply voltage and current consumption.

2. Materials and methods

Weber-ampere characteristics obtained during the flow of alternating current for operating the electric winding machine tool will have a hysteresis. In the measurement two things must be taken into account: the distortion of the curve, and the phase shift between magnetic flux F and current I in the coil. The relationship between F and I is determined by the shape of the dynamic loop.

The lag phase of a curve flow of the current curve is due to the influence of eddy currents and magnetic viscosity. The angle of lag δ is called the angle of losses.

To consider the effects of the distortion curve shape and the presence of the phase shift dependence, $F = f(I)$, when magnetizing an alternating field, it was suggested that the real dynamic loop is equivalent to ellipse, which satisfies the equation in coordinates F and I:

$$i(t) = I_{max}\sin(\omega t);$$

$$F(t) = F_{max}\sin(\omega t - \delta).$$

If we introduce a component stream, $F_{max1} = F_{max}\cos\delta$, which is in the phase with direction I, component $F_{max2} = F_{max}\sin\delta$ lags by 90° from I, when F_{max1} is associated with reversible processes of energy transformation during magnetization reversal, and F_{max2} – with irreversible magnetization. Expression (2) takes the following form:

$$F(t) = F_{max1}\sin(\omega t) - F_{max2}\cos(\omega t).$$

We propose a method of determining loop $F = f(I) + F_{max2}\cos(\omega t)$, which takes into account its hysteresis. In order to realize this, we will implement the solution of the inverse problem of the harmonic balance [1, 3-5, 9] method of natural-model tests [2, 6-8, 10], the unifying dimension of the physical object and modeling of the object. The method allows determining the shape of loop $F = f(I)+ F_{max2}\cos(\omega t)$, the well-known current flowing through the coil, set in a Fourier series:

$$i(t) = \sum_{m=1}^{n} I_{(2m-1)} \sin((2m-1)\omega t),$$

Where $I_{(2m-1)}$ – amplitude $(2m–1)$ of harmonic current and the known shape, and amplitude U_a of the voltage applied to the coil :

$$u(t)=U_a\sin(\omega t).$$

For the reversible component of the hysteresis of the weber-ampere characteristics of electrical products, we set an approximating expression:

$$F(i)=\sum_{m=1}^{n} k_{(2m-1)}\, i^{2m-1},$$

Where F – value of the magnetic flux, $k_{(2m-1)}$ – coefficients of the approximate expression of the weber-ampere characteristics, $m=\overline{(1,n)}$, n – the number of terms in the approximate expression, i – the current flowing through the coil.

The problem is formulated as follows. There is an electric machine tool with unknown hysteresis, known laws of variation of the voltage applied to the non-linear inductance and current flowing through it. It is required to determine coefficients $k_{(2m-1)}$of expressions approximating the hysteresis of the weber-ampere characteristics and the amplitude of irreversible flow component F_{max2}.

The equation of the circuit of the electrical device is

$$u(t)=Ri+\frac{dF}{dt}.$$

Let us rewrite it taking into account the known laws of current change (4) and voltage (5):

$$U_a\sin\omega t=R\left(\sum_{m=1}^{n} I_{(2m-1)}\sin((2m-1)\omega t)\right)+\frac{d\sum_{m=1}^{n} k_{(2m-1)}\sum_{m=1}^{n}(I_{(2m-1)}\sin((2m-1)t\omega))^{2m-1}}{dt}+F_{max2}\sin(\omega t)$$

The experiments are used as a method for the determination of the hysteresis of the weber-ampere characteristics of the algorithm used in natural-model tests.

Since the measurement range of the current error occurs, we investigated the effect of measurement error on the current harmonics error in determining the weber-ampere characteristics. These studies were conducted by means of the theory of planning a multi-factor experiment. Zero values of the factors are taken as measurement results of the first five odd harmonics of current I_1–I_9. The upper and lower values of each factor are obtained by adding or subtracting a predetermined value of the absolute error of measurement.

The weber-ampere characteristic obtained in the center of the plan, i.e. the measured values of the current harmonics are taken as exemplary characteristics.

Central composite design experiments are built in the shape of an orthogon, one of which is a nucleus from fractional factorial experiment 2^{5-2}. A variation interval for each factor was set so that its values correspond to absolute current harmonics measurement error – 3 %, provided that the size of the stellar arm maximum change was current harmonics ± 4.4 %.

Table 1 shows the scheduling matrix of the experiment. The accuracy was determined by two methods of obtaining the weber-ampere characteristics: natural-model tests (NMT) and a numerical solution of the inverse problem of the harmonic balance (HB).

There is a dependence of additional error methods on the values of the first five odd harmonics of current.

Table 1.The matrix of the experimental design.

№	I_1		I_3		I_5		I_7		I_9		Accuracy	
	Code	Physic A	Code	Physic A	Code	Physic A	Code	Physic A	Code	Physic A	NMT %	HB %
1	-1	4.199	-1	0.928	-1	0.496	1	0.264	1	0.148	0.26	0.78
2	1	4.459	-1	0.928	-1	0.496	-1	0.248	1	0.148	0.44	0.91
3	-1	4.199	1	0.986	-1	0.496	-1	0.248	-1	0.140	0.33	0.83
4	1	4.459	1	0.986	-1	0.496	1	0.264	-1	0.134	0.41	0.57
5	-1	4.199	-1	0.928	1	0.526	1	0.264	-1	0.134	0.31	0.77
6	1	4.459	-1	0.928	1	0.526	-1	0.248	-1	0.140	0.41	0.69
7	-1	4.199	1	0.986	1	0.526	-1	0.248	1	0.148	0.33	0.98
8	1	4.459	1	0.986	1	0.526	1	0.264	1	0.148	0.58	0.81
9	1.47	4.52	0	0.957	0	0.511	0	0.256	0	0.144	0.31	0.53
10	-1.47	4.138	0	0.957	0	0.511	0	0.256	0	0.144	0.41	0.41
11	0	4.329	1.47	0.999	0	0.511	0	0.256	0	0.144	0.18	0.27
12	0	4.329	-1.47	0.915	0	0.511	0	0.256	0	0.144	0.35	0.63
13	0	4.329	0	0.957	1.47	0.534	0	0.256	0	0.144	0.33	0.44
14	0	4.329	0	0.957	-1.47	0.488	0	0.256	0	0.144	0.44	0.21
15	0	4.329	0	0.957	0	0.511	1.47	0.267	0	0.144	0.28	0.68
16	0	4.329	0	0.957	0	0.511	-1.47	0.245	0	0.144	0.07	0.37
17	0	4.329	0	0.957	0	0.511	0	0.256	1.47	0.150	0.31	0.53
18	0	4.329	0	0.957	0	0.511	0	0.256	-1.47	0.138	0.14	0.17
19	0	4.329	0	0.957	0	0.511	0	0.256	0	0.144	0	0

3. Results

For the method based on natural-model tests:
physical values of factors are

$$\delta = 281,52 - 44,37I_1 + 5,16I_1^2 + 52,92I_3^2 + 429,52I_5^2 - 8,09I_7 - 351,59I_9 + 1247,54I_9^2;$$

in the coded values of factors there is

$$Y = 0,148 + 0,038X_1 + 0,087X_1^2 + 0,044X_3^2 + 0,1X_5^2 + 0,029X_7 + 0,033X_9 + 0,023X_9^2.$$

For the method based on the inversion harmonic balance:
physical values of factors are

$$\delta = 774,36 - 82,81I_1 + 9,55I_1^2 - 353,95I_3 +$$

$$+184,21I_3^2 - 406,44I_5 + 400,27I_5^2 + 3161,88I_7^2 - -1608,36I_9 + 5659,55I_9^2;$$

in the coded values of factors there is

$$Y = 0,103 - 0,017X_1 + 0,161X_1^2 - 0,04X_3 + 0,152X_3^2 +$$

$$+0,04X_5 + 0,094X_5^2 + 0,186X_7^2 + 0,093X_9 + 0,106X_9^2.$$

The analysis of regression equations shows that the relationship between variations of harmonic currents and mostly square errors and one of the harmonics are not the dominant influence on the error.

Processing of the results of the experiment was carried out using the software 'Statistica 10' (Figures 1 and 2).

Factor	SS	df	MS	F	p
(1)Var1 (L)	0,052435	1	0,052435	174,2897	0,000000
Var1 (Q)	0,212656	1	0,212656	706,8455	0,000000
(2)Var2 (L)	0,000220	1	0,000220	0,7297	0,398327
Var2 (Q)	0,053454	1	0,053454	177,6740	0,000000
(3)Var3 (L)	0,000071	1	0,000071	0,2365	0,629523
Var3 (Q)	0,286296	1	0,286296	951,6164	0,000000
(4)Var4 (L)	0,031920	1	0,031920	106,0974	0,000000
Var4 (Q)	0,000053	1	0,000053	0,1759	0,677253
(5)Var5 (L)	0,040951	1	0,040951	136,1154	0,000000
Var5 (Q)	0,015230	1	0,015230	50,6246	0,000000
Lack of Fit	0,372211	8	0,046526	154,6489	0,000000
Pure Error	0,011432	38	0,030085		
Total SS	1,077131	56			

Factor	Effect	Std.Err. Pure Err	t(38)	p	-95,% Cnf.Limt	+95,% Cnf.Limt	Coeff.	Std.Err. Coeff.	-95,% Cnf.Limt	+95,% Cnf.Limt
Mean/Interc.	0,147799	0,005273	28,03106	0,000000	0,137125	0,158473	0,147799	0,005273	0,137125	0,158473
(1)Var1 (L)	0,075308	0,005704	13,20188	0,000000	0,063760	0,086856	0,037654	0,002852	0,031880	0,043428
Var1 (Q)	0,173994	0,006544	26,58657	0,000000	0,160745	0,187242	0,086997	0,003272	0,080373	0,093621
(2)Var2 (L)	-0,004873	0,005704	-0,85424	0,398327	-0,016421	0,006675	-0,002436	0,002852	-0,008210	0,003337
Var2 (Q)	0,087233	0,006544	13,32944	0,000000	0,073985	0,100482	0,043617	0,003272	0,036992	0,050241
(3)Var3 (L)	0,002774	0,005704	0,48633	0,629523	-0,008774	0,014322	0,001387	0,002852	-0,004387	0,007161
Var3 (Q)	0,201884	0,006544	30,84828	0,000000	0,188636	0,215132	0,100942	0,003272	0,094318	0,107566
(4)Var4 (L)	0,058757	0,005704	10,30036	0,000000	0,047209	0,070305	0,029378	0,002852	0,023604	0,035152
Var4 (Q)	0,002745	0,006544	0,41945	0,677253	-0,010503	0,015994	0,001373	0,003272	-0,005252	0,007997
(5)Var5 (L)	0,066552	0,005704	11,66685	0,000000	0,055004	0,078099	0,033276	0,002852	0,027502	0,039050
Var5 (Q)	0,046564	0,006544	7,11509	0,000000	0,033316	0,059813	0,023282	0,003272	0,016658	0,029906

Figure 1. Listing of the program 'Statistica 10' (NMT).

Factor	SS	df	MS	F	p
(1)Var1 (L)	0,010076	1	0,010076	29,538	0,000003
Var1 (Q)	0,729024	1	0,729024	2137,154	0,000000
(2)Var2 (L)	0,058325	1	0,058325	170,980	0,000000
Var2 (Q)	0,647760	1	0,647760	1898,926	0,000000
(3)Var3 (L)	0,060433	1	0,060433	177,161	0,000000
Var3 (Q)	0,248627	1	0,248627	728,856	0,000000
(4)Var4 (L)	0,000140	1	0,000140	0,411	0,525522
Var4 (Q)	0,977254	1	0,977254	2864,848	0,000000
(5)Var5 (L)	0,321590	1	0,321590	942,752	0,000000
Var5 (Q)	0,313451	1	0,313451	918,890	0,000000
Lack of Fit	0,561635	8	0,070204	205,806	0,000000
Pure Error	0,129625	38	0,034112		
Total SS	3,942994	56			

Factor	Effect	Std.Err. Pure Err	t(38)	p	-95,% Cnf.Limt	+95,% Cnf.Limt	Coeff.	Std.Err. Coeff.	-95,% Cnf.Limt	+95,% Cnf.Limt
Mean/Interc.	0,103251	0,005614	18,3901	0,000000	0,091885	0,114617	0,103251	0,005614	0,091885	0,114617
(1)Var1 (L)	-0,033012	0,006074	-5,4349	0,000003	-0,045308	-0,020716	-0,016506	0,003037	-0,022654	-0,010358
Var1 (Q)	0,322155	0,006969	46,2294	0,000000	0,308048	0,336263	0,161078	0,003484	0,154024	0,168131
(2)Var2 (L)	-0,079425	0,006074	-13,0759	0,000000	-0,091721	-0,067128	-0,039712	0,003037	-0,045860	-0,033564
Var2 (Q)	0,303670	0,006969	43,5767	0,000000	0,289562	0,317777	0,151835	0,003484	0,144781	0,158888
(3)Var3 (L)	0,080847	0,006074	13,3102	0,000000	0,068551	0,093144	0,040424	0,003037	0,034275	0,046572
Var3 (Q)	0,188134	0,006969	26,9973	0,000000	0,174027	0,202242	0,094067	0,003484	0,087014	0,101121
(4)Var4 (L)	-0,003892	0,006074	-0,6408	0,525522	-0,016188	0,008404	-0,001946	0,003037	-0,008094	0,004202
Var4 (Q)	0,372991	0,006969	53,5243	0,000000	0,358884	0,387098	0,186495	0,003484	0,179442	0,193549
(5)Var5 (L)	0,186501	0,006074	30,7043	0,000000	0,174204	0,198797	0,093250	0,003037	0,087102	0,099398
Var5 (Q)	0,211242	0,006969	30,3132	0,000000	0,197134	0,225349	0,105621	0,003484	0,098567	0,112674

Figure 2. Listing the program 'Statistica 10' (HB).

4. Conclusion

Table 1 and figure 1 show that by varying the values of the harmonic currents in the range of ± 4.4 %, the obtained error of the weber-ampere characteristics does not exceed 1%. The error of the method based on natural-model tests did not exceed 0.6 %.

Acknowledgment

The research has been conducted with the support of project № 1.2690.2014 / K 'Methods for solving inverse problems of diagnosis of complex systems (engineering and medicine) on the basis of the natural-model experiment' performed in the framework of the project state task using the CCU equipment 'Diagnosis and energy-efficient electrical equipment' of South Russian State Technical University (NPI) after MI Platov.

References

[1] Bakhvalov Y A, Gorbatenko N I, Grechikhin V V 2015 *Measurement Techniques* **58** 336–340
[2] Borovoi V V, Gorbatenko N I, Grechikhin V V 2015 *Measurement Techniques* **56** 614–617
[3] Gorbatenko N I, Lankin M V 2004 *Russian Electrical Engineering* **8** 63–66
[4] Gorbatenko N I, Lankin M V 2004 *Elektrotekhnika* **8** 19–26
[5] Gorbatenko N, Lankin M, Shaykhutdinov D, Gazarov K, Kolomiets A 2011 *Proceedings of the 6th International Forum on Strategic Technology* (Harbin: Heilongjiang) pp 194–196
[6] Gorbatenko N I, Lankin A M, Lankin M V, Shayhutdinov D V 2015 *International Journal of Applied Engineering Research* **3** 6509–6519
[7] Gorbatenko N I, Grechikhin V V, Shaikhutdinov D V 2015 *Measurement Techniques* **56** 609–613
[8] Lankin A M, Lankin M V, Aleksanyan G K, Narakidze N D 2015 *International Journal of Applied Engineering Research* **3** 6489–6498
[9] Shaykhutdinov D V,Gorbatentko N I, Akhmedov Sh V, Shaykhutdinova M V, Shirokov K M 2013 *Life Science Journal* **10** 2698–2702
[10] Shaikhutdinov D V, Gorbatenko N I, Shirokov K M 2015 *Measurement Techniques* **56** 618–620

A Batch Feeder for Inhomogeneous Bulk Materials

I S Vislov[1], S N Kladiev[2], S M Slobodyan[2], A M Bogdan[2]
[1] JCS 'SverdNIIhimmash', 32, Griboiedova st., Yekaterinburg, 620010, Russia
[2] National Research Tomsk Polytechnic University, 30, Lenina ave., Tomsk, 634050, Russia

E-mail: kladiev@tpu.ru

Abstract. The work includes the mechanical analysis of mechanical feeders and batchers that find application in various technological processes and industrial fields. Feeders are usually classified according to their design features into two groups: conveyor-type feeders and non-conveyor feeders. Batchers are used to batch solid bulk materials. Less frequently, they are used for liquids. In terms of a batching method, they are divided into volumetric and weighting batchers. Weighting batchers do not provide for sufficient batching accuracy. Automatic weighting batchers include a mass controlling sensor and systems for automatic material feed and automatic mass discharge control. In terms of operating principle, batchers are divided into gravitational batchers and batchers with forced feed of material using conveyors and pumps. Improved consumption of raw materials, decreased loss of materials, ease of use in automatic control systems of industrial facilities allows increasing the quality of technological processes and improve labor conditions. The batch feeder suggested by the authors is a volumetric batcher that has no comparable counterparts among conveyor-type feeders and allows solving the problem of targeted feeding of bulk material batches increasing reliability and hermeticity of the device.

1. Introduction

1.1. Classification of feeders and batchers

Currently, in industry numerous types of feeders and batchers are used in various manufacturing fields. There are many definitions of feeders and batchers; however, we will use the following definitions.

Feeder is a device for regular and controlled feeding of bulk and packaged cargo from bunkers, trays, gutters, infeed magazines and other loading systems to transporting and processing machines (machining units, mills, screeners, etc.) [1].

Batcher is a device for automatic dosing of required mass and volume of liquid and bulk materials. Batchers are used in production of building materials, metallurgy, chemical, foodstuff, pharmaceutical and other industries. They are also used at rail facilities, marine and river transport, in laboratories and trade [2].

These two terms fully describe the principle and intention of feeders and batchers.

Feeders are usually classified according to their design features into two groups [1]:

a) devices similar to certain types of conveyors, having–in contrast to them–small length and enhanced drive motor power. This group includes the following types of feeders:

– belt feeders [3, 4] that are intended for regular feeding of dry materials with the bulk density of less than 2.8 t/m^3 into machines and transporting devices. Such feeders are implemented at mining and metallurgy enterprises, on dry mortar production lines, etc. Belt feeders feed dry materials using a belt that is driven by an electric motor via chain gear or reducing gear. The moving belt that feeds material to transporting of processing machines is located directly under the bunker that pours out the material. The amount of material fed by the feeder is controlled by a special batch gate and by the belt speed;

– plate feeders [5] are intended for transporting and regularly feeding bulk materials at mining and concentration enterprises from one reservoir (bunker, hopper) to other reservoirs, processing machines or storage facilities. The belt of a feeder usually comprises a steel hinge structure forming an integral part of the transporter feeding bulk material at open pits and con-centration plants.

– screw feeders [6, 7] are represented by a tube or a gutter with a working element in it, a screw. A revolving rod placed into a horizontal or inclined chute moves bulk or small-size material along the chute. Throughput is regulated by increasing the rotation rate of the screw. Screw feeders are basically intended for regular and continuous feeding of bulk materials. Such feeders are usually implemented at industrial facilities and processing lines with specified material dosage rate;

– reciprocating feeders [8] are the machines for continuous transportation with a working body represented by an reciprocating chute. They are intended for regular feeding of non-sticky bulk materials from bunkers, hoppers and other reservoirs to processing machines and transporting devices.

– vibrational feeders [9, 10] are intended for regular feeding of non-sticky bulk materials. They are usually installed under bunkers on horizontal sections of material-transporting raceways as loading devices of batchers. The material is transported due to the oscillations of a working element represented by a tube with round cross-section;

b) devices that have no prototype among conveyors. This group of feeders includes:

– drum feeders that suit well for transporting either bulk, granular and small-size materials or large-size materials. Bulk, granular and small-size materials are fed due to smooth internal surface of the drum. Large-size materials are transported owing to ribbed drum surface.

– disk feeders are used for transporting bulk materials. Disk feeders have a loading unit from which the material gets to a pivoting disk. The material is discarded from the disk by a fixed blade. The disk revolution rate is chosen so that the material is not discarded under centrifugal force;

– Ross Chain Feeders [11] are used for large-size materials and have so-called "chain curtain" that spans the outlet of a bunker. The revolution of a drive drum makes the chains to press the layer of material against the chute, which adjusts the material passage rate;

– pneumatic screw feeders [12] are used for feeding bulk dust-forming materials. They differ from conventional screw feeders in the principle of material transportation: at the exit, the material is entrapped and transported by a stream of oxygen.

See Table 1 for a condensed classification of feeders.

Table 1. Classification of feeders.

Feeders	
Conveyor-type	With no prototypes among conveyors
- belt feeders;	- drum feeders;
- plate feeders;	- disk feeders;
- screw feeders;	- Ross Chain feeders;
- reciprocating feeders;	- pneumatic screw feeders.
- vibrational feeders;	

The performance of all the feeders is adjusted by the speed of their working element and the size of outlet slot of a bunker. For vibrational feeders, the performance can also be adjusted by changing the frequency and amplitude of oscillations.

Batchers are usually classified according to batching procedure into two groups [13]:

a) Volumetric batchers [14, 15].

They are used to batch gases, liquids or pastes. Less frequently, they are used to batch solid bulk materials. The dose size varies from fractions of cubic centimeters to hundreds (or thousands, for gases) of cubic meters; the throughput varies from less than cubic centimeters per hour to thousands of cubic meters (dozens of thousands, for gases) per hour. Theses batchers have simple construction and good reliability.

b) Weigh batchers [16, 17].

Such batchers are used to batch solid bulk materials; less frequently, they are used for liquids. Dose size varies from several grams to several kilograms; the throughput varies from hundreds of kg/h to thousands of t/h.

Moreover, depending on the requirement of technological process batchers are classified as single-component batchers for continuous batching of a single material, and multicomponent batchers for continuous batching of several bulk material or liquids. Multicomponent batchers can automatically control the ratio of materials or make corrections according to the specified program. Discrete batchers usually have bunker-like structure, while continuous batchers have a bunker and a belt.

The simplest volumetric batchers do not provide for required accuracy; complex and precise technological process are usually carried out with the use of weigh batchers. Weigh automatic batchers are represented by a complex consisting of weight control sensor, automatic unit for material feeding and system for automatic dosage or mass flow control. In terms of working principle, batchers can be gravitational (hoppers) with or without forced feeding of materials by belt, screw, pan or other conveyors, or plunger, gear or other pumps.

Batchers allow decreasing consumption of initial materials, cutting material losses, expanding flow-line production, eliminating many labor-intensive processes and improving working conditions.

Generally, all feeders work as batchers and vice versa, hence, the design of such equipment shall assume the incorporation of the two functions.

2. Bucket-type batch feeder

Research goes on, and novel processes are discovered that eventually find industrial ap-plication. The development of science imposes new conditions for implementing its results in industry. Correspondingly, the development of science is paralleled with the development of technology.

For instance, working with granular materials having inhomogeneous composition has the problem in the simultaneous presence of particles with the size of several millimeters and particles with the size of several micrometer.

OAO SverdNIIhimmash have developed a bucket-type batch feeder, which found application in industry [18]. Considering the specificity of radiochemical plants, the use of weigh sensors is problematic, which excludes the application of weigh batchers. The said device belongs to weigh batchers having no counterparts among conveyor-type feeders. The scheme of the device of depicted in Figure 1.

The batch feeders includes body 1, working element 2 that contains inlet 3 and bucket 4, bunker 5 for bulk material, loading mouth 6 for material feed from a bunker into the bucket's in-let. The body includes outlet mouth 7 for unloading material from the bucket of working element outside the batch feeder. The working element at the side of bucket is connected with shaft 8 which free end goes into bearing 9 mounted on stand 10.

Figure 1 shows angle α, between shaft axis and vertical axis.

The system also includes reverse turn limiters 11 of working element that are mounted on immovable parts of the batch feeder and limiter 12 mounted on the working element of the batch feeder. These limiters constrain the clockwise and counter-clockwise turn angle of the working element around its middle position.

Figure 2 shows the cross-section of bucket 4 of working element 2 with slot H. The cross-section obviously shows the purpose of elements of the batch feeder bucket that determine its performance considering the angle of repose άrepose of batched bulk material, size of its particles and the length of flat wall section 13.

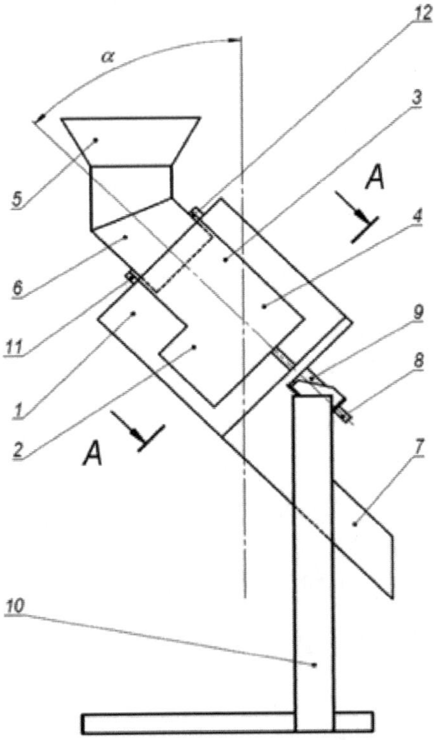

Figure 1. Bucket-type batch feeder.

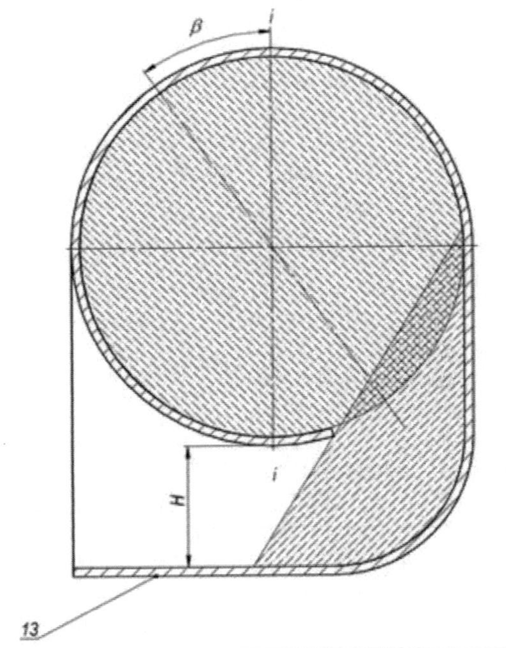

Figure 2. Cross-section of bucket.

Operation of the bucket-type batch feeder includes the following stages:

During the feeding of bulk product into bunker 5, the product fills loading mouth 6, inlet 3 and bucket 4. Nevertheless, it does not pour out from the bucket through a product-discharging slot H, since the product forms an angle of repose $\acute{\alpha}_{repose}$, which generatric line does not go beyond the threads of the second section of flat wall 13 of the bucket.

Let us consider the cross-section of the batch feeder in Figure 2.

The counter-clockwise turn of the bucket for angle β around axis i-i of section A-A with preservation of angle of repose makes the lower edge of slope to shift to the edge of wall 13. This induces the pouring of product into unloading mouth 7.

The amount of material poured out from the bucket depends on turn angle β and length of wall 13.

The maximum dimensions of material's particles are determined by the size of slot H.

After pouring out a batch of the product, the bucket makes a clockwise turn around shaft axis for angle β and gets into initial position.

Further material batching is carried out in a similar way.

The throughput of such device depends on bucket rotation frequency. During testing, the accuracy of batching amounted to + 5 % for the throughput of 5 kilograms and the confidence factor of 0.95.

The application of the suggested device will allow increasing the accuracy of volumetric dosing of bulk material mix containing small and large particles.

The device also enables the targeted feed of specified single batches of bulk material without preliminary unloading of other batches of material to the same destination.

In addition, the design eliminates the jamming of working element inside the body of batch feeder and contamination of the batched material by particles produced by rubbing of body and working element during rotation. This is unachievable in conveyor-type feeders without considerable complication of their design and increase of dimensions. Besides, the body of bucket-type batch feeder is hermetic having simple design, which is not provided in similar feeders that have no prototypes among conveyors.

Thus, the presented device can find application in developing industrial fields, such as radiochemical shops. The suggested device may have the most promising deliverables in processing spent nuclear fuel by high-temperature methods [19], for instance, at the stage of loading milled fragments of fuel units or fuel pellets into chemical reactors for further processing.

References

[1] Spivakovsky A O and Dyachkov V K 1983 *Transporters: Course guide for mechanical-engineering universities. 3rd amended edition* (Moscow: Mashinostroenie) p 487 [in Russian]
[2] Roginskiy G A 2008 *Dosing bulk materials* (Moscow: Chemistry)
[3] Solntsev K E, Ryabtsev A V, Gavrilova M V, Glotov I V and Lebedev Y B 2014 Belt feeder. Patent No. RU 137102
[4] Pirozhenko P P, Bodlo N V, Kobysh O N and Samokhvalov V K 2007 Belt feeder. Patent No. RU 63449
[5] Tarasov Yu D 2014 Heavy-duty plate feeder. Patent No. RU 2518496
[6] Demidenko A A, Romadin V I et al 2013 Feeding bunker with screw unloading for powders susceptible to arching. Patent No. RU 2483999
[7] Romanov V A and Kotelnikov V A 2014 Twin-design screw feeder. Patent No. RU 138263
[8] Tarasov Yu D 2014 Chute feeder. Patent No. RU 2518761
[9] Pleshakov Yu A, Breusenko O V, Sharov S A 2014 Vibrational feeder. Patent No. 137174
[10] Kolyshkin K I, Bashkirov G M, Klokov A N and Pasyuk A V 2013 Vibrational feeder. Patent No. RU 2491212
[11] Tarasov Yu D 2012 Ross Chain feeder. Patent No. RU 2466079
[12] Shinakov V G, Lyalin A V, Travnikov E P and Vinogradov V N 2005 Pneumatic paddle feeder. Patent No. RU 2248703
[13] Sinitsyn B N and Erokhin A S 1982 Continuous batchers, batching automation techniques. Review (Moscow) [in Russian]
[14] Nikitenko G V, Kapustin I V and Zhavoronkov P V 2010 Volumetric batcher of bulk substance. Patent No. RU 100614
[15] Arhangelsky V Yu, Dzhangirian V G and Varenykh N M 1999 Volumetric batcher. Patent No. RU 2133944
[16] Garanin L P, Zamahaev Yu V, Kovtun V E and Teplygin A V 2013 Weigh discrete batcher for powder-lake materials. Patent No. RU 2485450
[17] Streletsky A N and Sannikov A G 2014 Weigh batcher. Patent No. RU 141297
[18] Vislov I S, Lukonin D A, Klenov Yu K and Gamrekeli M N 2014 Batch feeder. Patent No. RU 141559
[19] Ran F, Adamantiades A, Kenton J, Broun Ch, Legasov V A 1989 *Handbook for nuclear electrotechnology* (Moscow: Energoatomizdat) p 723

IOP Conf. Series: Materials Science and Engineering **124** (2016) 012034 doi:10.1088/1757-899X/124/1/012034

Simulation of oxide phases formation under pulsed electron beam

O N Kryukova[1,2] and A L Maslov[1]
[1] Institute of Strength Physics and Materials Science of the Siberian Branch of the Russian Academy of Sciences, 2/4, Akademicheskiy ave., Tomsk, 634021, Russia
[2] Tomsk Polytechnic University, 30, Lenina ave., Tomsk, 634050, Russia

E-mail: okruk@ispms.tsc.ru

Abstract. This paper presents the mathematical model of evolution phase composition in the TiNi+Si system under impulsive electron beam. It was assumed that the initial coating contains small concentration of molecular oxygen. This model is one-demensional, and takes into account the phenomena of diffusion, chemical reactions, and thermal effects of chemical reactions. Results of the numerical modeling had show that the oxide of titanium and triple solution TiNiSi formed in significant amounts, and triple solution preferentially forms in the substrate. Other oxides must be formed in trace amounts.

1. Introduction

Nitinol is well studied and frequently used in practice medicine for a long time. However, its nitinol employment is restricted by use in medicine neat complicated rather the high toxicity of nickel. To avoid the toxic effect of nickel the various protective coatings are used. Special attention has been paid to the adhesive strength of coatings and methods of its increase [1, 2]. The coatings can be formed in by different ways: by chemical, electroplating, electron beam exposure reaction, and etc. One of methods consists in the coating deposition using the ion plasma technique [3] with the subsequent treatment it with the a low-energy high-current electron beam [4-6]. The formation of the protective coating is difficult to observe directly, so hence the important question of the development of a mathematical model of the this process is very important. In this paper introduces a the one-dimensional mathematical model of the dynamics evolution of phase composition in the system under impulse action of a low-energy electron flow. The model takes into account the phenomena of diffusion, chemical reactions, and thermal effects of chemical reactions and allows to study the element's redistribution.

2. Mathematical formulation of the problem

We suppose a that NiTi-sample with the thick silicon coating of thickness h undergoes the pulsed electron beam processing. A small amount of molecular oxygen presents in the surface layer of the coating. Provided that the parameters of the electron beam are identical across the surface of the same sample the problem can be considered a as one-dimensional problem (Figure 1).

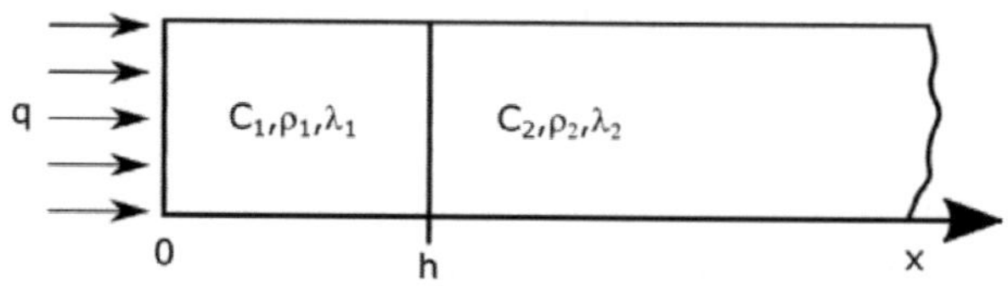

Figure 1. Illustration to the problem formulation.

Chemical reactions during the processing are described to this system of equations:

$$Ti + O_2 \rightarrow TiO_2 \tag{1}$$

$$Ti + Ni + Si \rightarrow TiNiSi \tag{2}$$

$$2Ni + O_2 \rightarrow 2NiO \tag{3}$$

$$Si + O_2 \rightarrow SiO_2 \tag{4}$$

The rates of these reactions are denoted as φ_i where $i = 1\ldots4$.

In such formulation, the dynamics of the specimen temperature in the sample in such a formulation is described by one-dimensional heat thermal conductivity equation:

$$c_k \rho_k \frac{\partial T_k}{\partial t} = \frac{\partial}{\partial x}\left(\lambda_k \frac{\partial T_k}{\partial x}\right) + W_{ch}, \tag{5}$$

where k is the index, $k = 1,2$, c_1, ρ_1 are the density and heat capacity of the coating, c_2, ρ_2 are the density and heat capacity of the substrate, T_1, T_2 are the temperature of the coating and the substrate, respectively, t is the time, x is the spatial coordinate, λ_1, λ_2 are thermal conductivities of the coatings and substrate, W_{ch} is the total heat source associated with the thermal effect of chemical reactions in the specimen, $W_{ch} = \sum_{i=1}^{4} Q_{ch,i}\varphi_i$, $Q_{ch,i}$ is the thermal effect of the i-th reaction.

Concentrations of free compounds elements of (Si, Ti, Ni and O_2) follow from determined by the diffusion equations:

$$\frac{\partial C_j}{\partial t} = \frac{\partial}{\partial x}\left(D_j \frac{\partial C_j}{\partial x}\right) + \sigma_j, \tag{6}$$

where, C_j are the molar concentrations of elements Si, Ti, Ni and O_2, $j = 1,2,3,4$; D_j are their diffusion coefficients, which depend on the temperature corresponding to the Arrhenius law: $D_j = D_{0j}\exp(-E_{dj}/RT)$, D_{0j} are frequency factors, E_{dj} are the activation energies, R is the gas constant, σ_j are the sources or sinks related to chemical reactions.

The concentrations of the reaction products are determined by the kinetic equations:

$$\frac{dC_m}{dt} = \sigma_m, \tag{7}$$

where C_m are the molar concentration of the compounds TiNiSi, TiO_2, NiO, SiO_2; $m = 5,6,7,8$

The system is completed by with boundary and initial conditions:

$$x = 0: \quad -\lambda_1 \frac{\partial T_1}{\partial x} = q(t), \qquad \frac{\partial C_j}{\partial x} = 0, \; j = 1\ldots4$$

$$x = h:\ \lambda_1 \frac{\partial T_1}{\partial x} = \lambda_2 \frac{\partial T_2}{\partial x},\ T_1 = T_2,$$

$$D_j \frac{\partial C_j}{\partial x}\bigg|_{x=h-0} = D_j \frac{\partial C_j}{\partial x}\bigg|_{x=h+0},\ j = 1\ldots4$$

$$x \to \infty:\ \frac{\partial T_2}{\partial x} = 0,\ \frac{\partial C_j}{\partial x} = 0,\ j = 1\ldots4$$

$$t = 0:\ T_1 = T_2 = T_0,\ C_j = C_{j0},\ j = 1\ldots8$$

where $q = E / t_{imp}$ is the external heat flux, t_{imp} – impulse duration.

The reaction rates:

$$\varphi_1 = k_{01}\exp(-E_{a1}/RT)C_2C_4,$$
$$\varphi_2 = k_{02}\exp(-E_{a2}/RT)C_1C_2C_3,$$
$$\varphi_3 = k_{03}\exp(-E_{a3}/RT)C_3^{\,2}C_4,$$
$$\varphi_4 = k_{04}\exp(-E_{a4}/RT)C_1C_4.$$

follow from the equations (1)-(4), the Arrhenius law and the law of mass action. The reaction rate constants, activation energies of chemical reactions () and heats of the reactions was were determined by means of known thermodynamical formulae [10].

According to above we shall obtain:

$$\sigma_1 = -\varphi_2 - \varphi_4,\ \sigma_2 = -\varphi_1 - \varphi_2,\ \sigma_3 = -\varphi_2 - 2\varphi_3,$$

$$\sigma_4 = -\varphi_1 - \varphi_3 - \varphi_4,\ \sigma_5 = \varphi_2,$$

$$\sigma_6 = \varphi_1,\ \sigma_7 = 2\varphi_3,\ \sigma_8 = \varphi_4.$$

In accordance with experiments [2, 5] it was assumed for in the calculations made: the thickness of the substrate is 1 cm, the coating thickness is of 200 nm, the sample is subjected to 5 times the pulse exposure, the pulse duration – is 150 ms, pause time is 0.02, the energy density in the beam was varied, and was about 20 J/cm^2.

Initial concentrations of the reagents are $C_{01} = 0.08296$ mol/cm^3, $C_{02} = 0.0535$ mol/cm ^{3}, $C_{03} = 0.06548$ mol/cm^3, $C_{04} = 0.015$ mol/cm^3 (Figure 2).

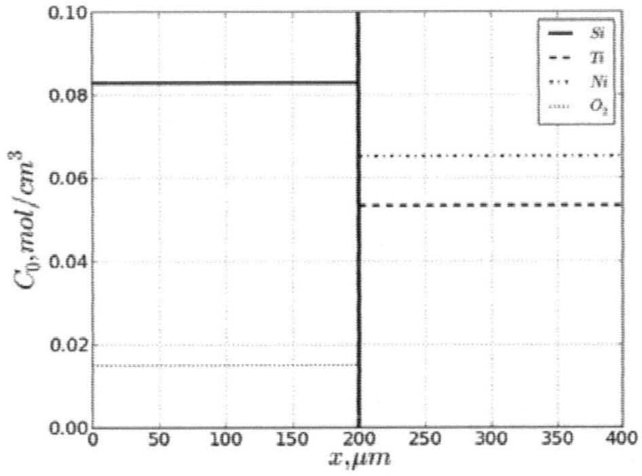

Figure 2. The initial concentrations. The vertical line shows the interface between the coating and substrate.

We assume the following values for thermal parameters: $c_1 = 0.7136 \frac{\text{J}}{mol \cdot \text{K}}$, $\rho_1 = 2.33$ g/cm^3, $c_2 = 0.227 \frac{\text{J}}{mol \cdot \text{K}}$, $\rho_2 = 6.45$ g/cm^3, $\lambda_1 = 1.49 \frac{\text{W}}{cm \cdot \text{K}}$, $\lambda_1 = 0.18 \frac{\text{W}}{cm \cdot \text{K}}$.

The results of the formal evaluation of the formal kinetic parameters of for the reactions are presented in the Table. 1, the values of diffusion parameters values are shown in the Table 2.

Table 1. Formal-kinetic parameters for reaction

Reaction	ΔH, J/mol	E_d, J/mol	$k_0, \frac{1}{\text{s}}$
$Ti + O_2 \rightarrow TiO_2$	-50049	322707	$2.133 \cdot 10^{16}$
$Ti + Ni + Si \rightarrow TiNiSi$	-1430	27650	$0.8 \cdot 10^5$
$2Ni + O_2 \rightarrow 2NiO$	-47818	351900	$1.5 \cdot 10^9$
$Si + O_2 \rightarrow SiO_2$	-40179	316680	$3.188 \cdot 10^9$

Table 2. Diffusion parameters

Substance	$D_{0i}, \frac{\text{cm}}{\text{s}^2}$, $x \le h$	E_a, J/mol, $x \le h$	$D_{0i}, \frac{\text{cm}}{\text{s}^2}$, $x > h$	E_a, J/mol, $x > h$
Si	1800	$11 \cdot 10^4$	$3 \cdot 10^4$	$1 \cdot 10^5$
Ti	6.4	$123 \cdot 10^3$	2	$15 \cdot 10^4$
Ni	190	$185 \cdot 10^3$	$6 \cdot 10^{-4}$	$9 \cdot 10^4$
O_2	0.21	$292 \cdot 10^3$	0.049	$84 \cdot 10^3$

The resulting system of equations was solved numerically. Thermal conductivity heat equations (5) and diffusion equations (6) are solved using an the implicit finite difference schemes and double sweep method, the kinetic equations (7) – was solved by Euler method.

3. Examples

Changing the temperature in the point $x = 0$ is illustrated in Figure 3.

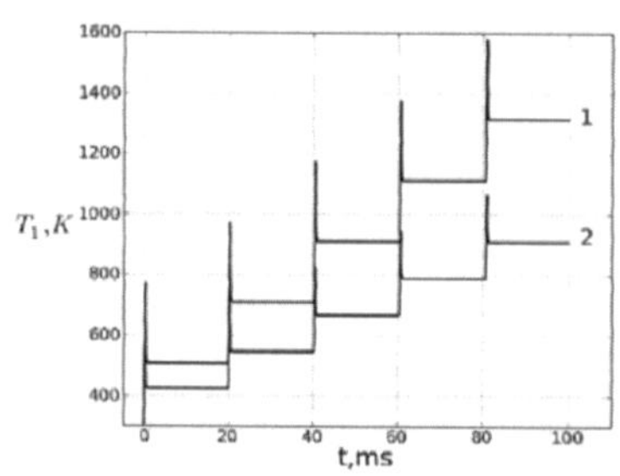

Figure 3. The temperatures at the point $x = 0$.
1 – $E = 25$ J/cm^2, 2 – $E = 15$ J/cm^2.

Quite naturally, the temperature rises sharply in times during of exposures, and rapidly decreases in the pauses between them that is quite naturally. In the during early stages of exposure, diffusion of free elements is insignificant due to the low average temperature of the sample medium of free diffusion of elements is insignificant. Further, the redistribution element concentration near the boundary interface between the coating and the substrate becomes conspicuous, and with further increase in of the temperature accelerates the process accelerates (Figure 4).

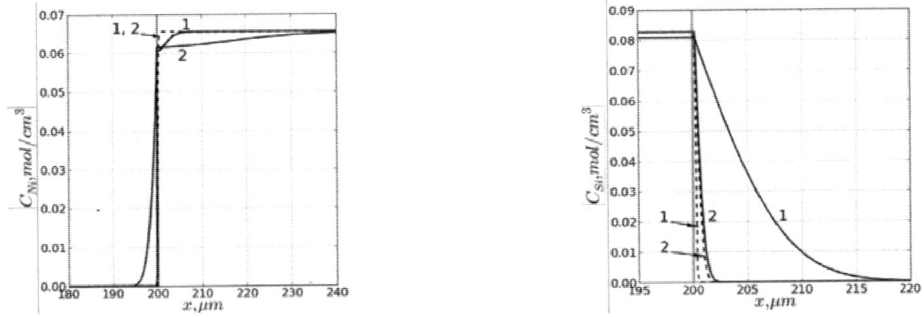

a b

Figure 4. The concentration profiles, a – Ni, b – Si. 1 – $t = 2 \cdot 10^{-2}$ s., 2 – $t = 3 \cdot 10^{-2}$ s The solid line corresponds to the $E = 25$ J/cm^2, dashed – $E = 15$ J/cm^2. The vertical line shows the interface between the coating and substrate.

It was detected that triple solution preferentially forms in the substrate due to the greater mobility of silicon than nickel (Figure 5).

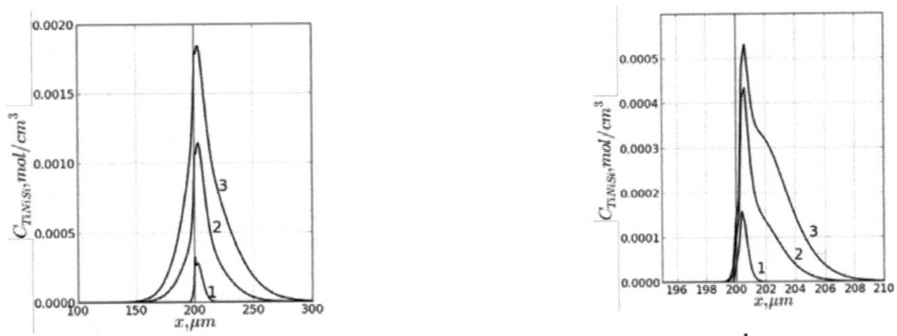

a b

Figure 5. The concentration profiles $TiNiSi$ a – $E = 25$ J/cm^2, b – $E = 15$ J/cm^2. 1 – $t = 3 \cdot 10^{-2}$ s., 2 – $t = 4.5 \cdot 10^{-2}$ s, 3 – $t = 5 \cdot 10^{-2}$ s. The vertical line shows the boundary between the coating and the substrate.

Contained in the coating oxygen reacts with titanium to form titanium oxide (Figure 6). This reaction has the following character: up to a certain point of time titanium oxide practically does not form, then its concentration increases markedly during some microseconds (Figure 6b).

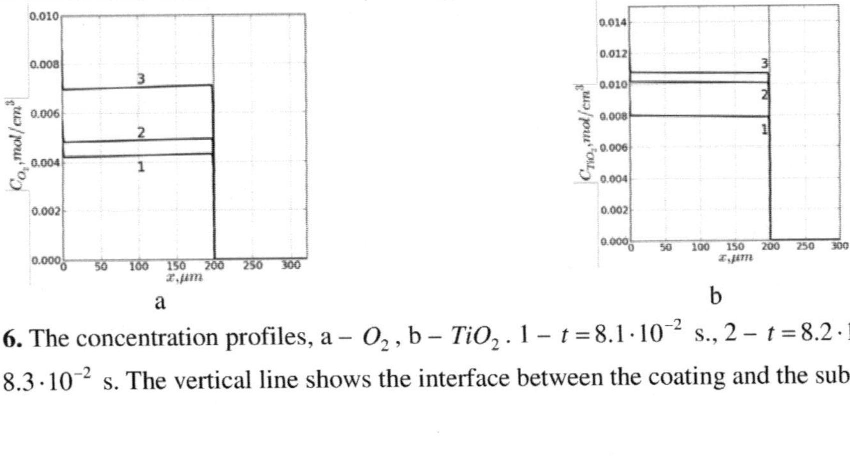

a b

Figure 6. The concentration profiles, a – O_2, b – TiO_2. 1 – $t = 8.1 \cdot 10^{-2}$ s., 2 – $t = 8.2 \cdot 10^{-2}$ s, 3 – $t = 8.3 \cdot 10^{-2}$ s. The vertical line shows the interface between the coating and the substrate.

The oxides of nickel and silicon do not form (in the numerical experiment, their concentrations do not exceed 10^{-9} mol/cm^3).

4. Conclusion

In response a pulsed electron beam on the sample of Nitinol with silicon doped and oxygen near the surface layer of a solid solution and oxide.

Thus, numerical experiments had show that during impulse electron beam action on the sample of Nitinol with silicon coating containing initial Oxygen, solid solution and oxide are generated in the surface layer. Small concentrations of the oxides of nickel and silicon appearing in the heated zone should not affect the properties of the sample.

Modeling suggests that the oxides of nickel and silicon should not affect the properties of the sample, since the assumed values of processing parameters must be formed in trace amounts.

References

[1] Lotkov A I, Psakhie S G, Meisner L L, Korshunov A V, Meisner S N and Diamant V M 2014 *Advanced Biomaterials and Devices in Medicine* **1** 46–52

[2] Psakh'e S G, Lotkov A I, Meisner S N, Meisner L L, Sergeev V P, Sungatulin A R 2013 *Inorganic Materials: Applied Research* **4(5)** 457–463

[3] Psakhie S G, Lotkov A I, Meisner L L, Meisner S N, Matveeva V A 2013 *Russian Physics Journal* **9(55)** 1063–1073

[4] Rotshtein V, Ivanov Yu, Markov A 2006 Surface treatment of materials with low energy, high current electron beams. *Chapter 6 in Book 'Materials surface processing by directed energy techniques'* (*Ed. by Y. Pauleau, Elsevier*) pp. **205–240**

[5] Ostapenko M G, Meisner L L, Lotkov A I, Gudimova E Yu, Zacharova M A 2014 *Bulletin of the Tomsk Polytechnic University* **324(3)** 36–43

[6] Zhang K M, Yang D Z, Zou J X, Grosdidier T, Dong C 2006 *Surface and Coatings Technology* **201(6)** 3096–3102

Scale effects in tribological properties of solid-lubricating composites made of ultra-high molecular weight polyethylene filled with calcium stearate particles

S A Lurie[1,2], **D B Volkov-Bogorodskiy**[1], **A G Knyzeva**[3], **S V Panin**[4] and **L A Kornienko**[4]

[1] Institute of Applied Mechanics of Russian Academy of Sciences, 7, Leningradskiy ave., Moscow, 125040, Russia

[2] Dorodnicyn Computing Centre of Russian Academy of Sciences, 40, Vavilova str., Moscow, 119333, Russia

[3] Tomsk Polytechnic University, 30, Lenina ave., Tomsk, 634050, Russia

[4] Institute of Strength Physics and Materials Science of the Siberian Branch of the Russian Academy of Sciences, 2/4, Akademicheskiy ave., Tomsk, 634021, Russia

E-mail: anna-knyazeva@mail.ru

Abstract. Friction properties being influenced by scale effects are simulated in the paper by the example of polymer composite material made from Ultra High-Molecular Weight Polyethylenes (UHMWPE) filled by calcium stearate (C36H70CaO4). Of interest are the composites whose mechanical properties and tribotechnical characteristics do not depend monotonically on filler (inclusions) weight fraction. In order to describe the influence of scale effects onto frictional properties the model based on Reiss averaging (model of "weak phase") is employed. It is also suggested that when gradient elasticity theory is applicable the formal analogy between effective friction coefficient for surface heterogeneous structures and effective mechanical properties (compliances) for heterogeneous material can take place. Theoretical dependence to describe nonmonotonic change of effective friction coefficient versus filler concentration was obtained for the polymer composites under study. The suggested expressions might be useful for the sake of properties prognosis of antifriction polymeric materilas.

1. Introduction

Composite materials made from Ultra High-Molecular Weight Polyethylene (UHMWPE) obey numerous significant properties including low density, unique ability to absorb shock impacts, chemical and biological inertness, satisfactory strength properties, high wear resistance etc. [1]. Due to their unique properties such composites are of wide demand in mechanical engineering, medicine; they might be as well used at design of engineering structures in oil and gas engineering, mining, power and chemical engineering, etc. High stability of properties within certain temperature range and extremely low friction coefficient [2-6] make these materials very attractive for design of machine parts and equipment to be exploited under extreme conditions of Arctic climate. Particularly, this is related to antifrictional materials.

Numerous papers indicate that in many polymeric composites modified by micro- and nanosized inclusions nonmonotonic dependencies of physical-mechanical properties on filler concentration take place [7,8]. Particularly, according to [8], dependencies of this kind are characteristic feature for strength and tribological properties of modified polymers. Apparently, these nonmonotonic dependencies are related to manifestation of scale effects known for composites of different types.

This paper deals with the development of a model to estimate frictional properties of modified polymers which takes into account scale effects. The ultimate aim of the model is calculation of exploitation properties based on data of filler content that might by supported by the experimental data taken from the literature. It is worth noting that experimentally determined mechanical characteristics such as strength, hardness, etc. as well point at their abnormal nonmonotonic behavior with increase of filler concentration.

2. Modeling of effective friction coefficient

Let us consider that the polymer is modified by inclusions made from calcium stearate with characteristic dimension of r_0 and volume concentration c. The relation between the relative part of inclusion surface and volume concentration is determined by the formulae $S_{slip} = Kc^{2/3}$, where coefficient K is determined by the packing density and shape of the filler particles. For dense packing of spherical particles, the coefficient K can be calculated explicitly and was evaluated to take the value of $K_0 \approx 1.268$. However, the change of shape and surface roughness of the particles is not taken into account. For the case of arbitrary shape of the particles this parameter can be much higher than K_0, but the condition $Kc^{2/3} < 1$ should be true.

For the friction coefficient evaluation it is suggested to use the model being based on the analogy with the "weak link" one where the effective properties of lamellar or nonuniform structures at the serial arrangement of phases are determined (limited) by a weak link. It might be assumed that for the parameters being inverse to the friction coefficient the mixture rule is true that is analogously to the mixture formula for compliances calculation of nonuniform composites in a one dimensional approximation. Though the direct calculations show that above mentioned formula for the effective friction coefficient does not taken into account scale effects and gives monotonic dependence on concentration that does not correspond to experimental data.

At the use of this model the averaged value of friction coefficient does not depend on the modification parameter, i.e. the scale factor (specific size of the inclusions or, that is here the same, on the extent of interphase layer) that naturally appear when mechanical properties are modeled being based on the gradient model. To take into consideration the influence of the scale effect let us continue the above introduced analogy between coefficients of friction and compliance in a heterogeneous composite with micro/nano scale structure. The estimations of mechanical properties for nanocomposites with regard to scale effects were obtained previously in [9,10].

In gradient theory of elasticity, scale factors are determined by the thickness of the interphase zone as well as by the parameter A representing the activity of inclusion surface affecting the crystallization zone of the polymer in the vicinity of calcium stearate particles. As the basic the following formula to evaluate the effective stiffness of lamellar periodic structure will be used [10]

$$\tilde{k} = \frac{k_M H_M L + k_I H_I l}{L+l} = \frac{k_M k_I (k_I \alpha_M + k_M \alpha_I - A_{st} \alpha_I \alpha_M)}{Q}(L+l), \tag{1}$$

where l is the thickness of inclusion layer, $L = l(1-c)/c$,

$$H_M = k_I \frac{k_I \alpha_M + k_M \alpha_I - A_{st} \alpha_I \alpha_M}{Q}(L+l), \quad H_I = k_M \frac{k_I \alpha_M + k_M \alpha_I - A_{st} \alpha_I \alpha_M}{Q}(L+l),$$

$$Q = (k_I L + k_M l + A_{st})(k_I \alpha_M + k_M \alpha_I - A_{st} \alpha_I \alpha_M) - (k_I - k_M - A_{st} \alpha_I)(k_I - k_M + A_{st} \alpha_M),$$

$$\alpha_M = \frac{\kappa_M}{\mathrm{th}(\kappa_M L)}, \quad \alpha_I = \frac{\kappa_I}{\mathrm{th}(\kappa_I l)}, \quad k_{I,M} = 2\mu_{I,M} + \lambda_{I,M}, \quad \kappa_{I,M} = \sqrt{C_{I,M}/k_{I,M}}.$$

Here c is the volume part of inclusions in the representative fragment; $\tilde{l}_{I,M} = (\kappa_{I,M})^{-1}$ is relative thickness of interphase layer in each fragment phase; A_{st} is the surface activity parameter. It is suggested that A_{st} parameter affect the structure of crystallization zone and its stiffness.

Note that relation (10) were obtained in [10] in accordance with the procedure of the asymptotic averaging being generalized for gradient model.

Using the averaging procedure for the values being inverse to friction coefficient and taking into account the formula (1) the following expression for effective friction coefficient can be written:

$$\bar{\tilde{k}} = (1 - S_{slip}) + S_{slip}\left(\bar{k}_{slip}(1 + \tilde{A}_{slip}) - \frac{\left(1 - \bar{k}_{slip}(1 + \tilde{A}_{slip}\alpha_I)\right)\left(1 - \bar{k}_{slip}(1 - \tilde{A}_{slip}\alpha_M)\right)}{\alpha_I \bar{k}_{slip}(1 - \tilde{A}_{slip}\alpha_M) + \alpha_M}\right), \quad \tilde{A}_{slip} = \frac{A_{st}}{r_0}. \quad (2)$$

Thus, in gradient model of tribology the effective friction coefficient is determined by a number of parameters. Among them are, effective coefficient *K* that determines the packaging and geometrical shape of particles to fill the matrix; $\tilde{l}_{I,M}$ - the relative thickness of the interphase zone in a matrix and in an inclusion; $\tilde{A}_{slip}$ - parameter of surface activity for inclusions that accounts for the influence of the excess free energy on the crystalline structure formation on the surface of the phases.

Note that excess free surface energy was included as a thermodynamic parameter (additional generalized thermodynamic force) in the classical thermodynamics [11,12] of surface effects and phase equilibriums as well as in thermodynamics of a media with internal surfaces [13,14] which are described using additional parameter – specific surface area. In doing so, these values are thermodynamic state variables.

After fitting the parameters to provide the best approximation of experimental data [8] there was obtained the following results: $\bar{k}_{slip} = 0.29$, $K = 3$, $\tilde{A}_{slip} = 10$. Parameter of interphase zone thickness in the matrix $\tilde{l}_M = l_M / r_0$ (it is assumed that $\tilde{l}_I = 0$) and its dependence on inclusion concentration are adjusted during the modeling of effective properties of the filled polymer; here r_0 is effective radius of particles with regard to cluster formation. Theoretical estimations for effective mechanical properties (strength) of filled composite were obtained in [9] based on gradient elasticity theory and method of radial factors. It was ascertained that for the composites under study the dependence $\tilde{l}_M = a/(b+c) - d$ can be used when $c \in [0,9]$ and $c \in [9,20]$. The parameters a, b, d were determined through the condition of the best agreement of experimental data [8] and data of the ultimate strength modeling. Following values were found [8]: $a = 3.221$, $b = 0.363$, $d = 6.887$ for $c \in [0,9]$; and $a = 0.00327$, $b = -0.0756$, $d = 0$ for $c \in [9,20]$. In doing so, the tribology model tends to be related to the strength model for the composite.

The graph for theoretical dependence (2) of effective friction coefficient on volume content of inclusions is presented in the fig. 1 (curve 1); solid dots on the graph correspond to experimental data. Curve 2 corresponds to effective friction coefficient being calculated in accordance with the mixture rule $\bar{\tilde{k}} = (1 - S_{slip}) + S_{slip}\bar{k}_{slip}$.

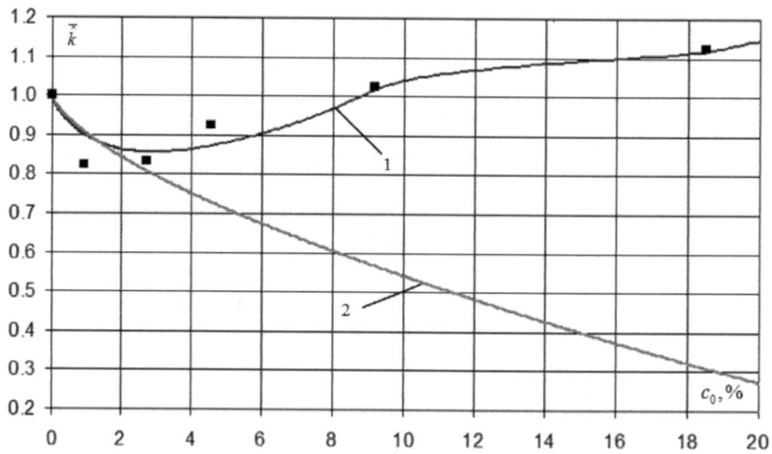

Figure 1. Effective friction coefficient calculated by gradient model of tribology; dots – experimental data; line 1 – corresponds to gradient model; line 2 – corresponds to classical approach.

3. Conclusion

The model for calculating friction properties of modified high molecular weight polymers is offered. It was established that there is a certain analogy between the model of mechanical properties and friction model since scale effects for friction coefficients are taken into account based on the dependence of the interphase layer thickness (the thickness of crystallization zone) on filler concentration that was found during mechanical properties modeling. It was demonstrated that the suggestion on the analogy between friction coefficient and compliance is quite correct. Using this analogy the method of friction properties (friction coefficient) evaluation was suggested for heterogeneous polymeric structures taking into account the nonmonotonic dependence of these characteristics on the concentration of the modifying filler.

References

[1] Briscoe B J, Sinha S K 2002 *Proc. Inst. Mech. Eng. J. Eng. Tribol.* **216** 401
[2] Liu C Z, Wu J Q, Li J Q, Ren L Q, Tong J, Arnell A D 2006 *Wear.* **260** 109
[3] Xue Y, Wu W, Jacobs O, Schdel B 2006 *Polymer Testing.* **25** 221
[4] Kuo Hsien-Chang, Jeng Ming-Chang. 2010 *Wear.* **268** 803
[5] Briscoe B J, Sinha S K 2008 *Tribology of polymeric nanocomposites by K. Friedrich and Alois K. Schlarb (ed. B.J. Briscoe)* 1
[6] Panin S V, Panin V E, Kornienko L A, Puvadin T, Piriyaon S, Shilko S V 2011 *Izvestia vuzov. Chemical and chemical technology* **54(7)** 102
[7] Panin S V, Kornienko L A, Nguen Suan T, Ivanova L R, Korchagin M A, Shil'ko S V, Pleskachevskii Yu M 2015 *Journal of Friction and Wear.* **36(3)** 249
[8] Panin S V, Kornienko L A, Nguen Suan T, Ivanova L R, Poltoranin M A, Shilko S V, Pleskachevskii Yu M 2015 *Materialy. Tehnologii. Instrumenty* **20(1)** 8
[9] Lurie S, Volkov-Bogorodskii D, Tuchkova N 2015 *Acta Mechan.* 10.1007/s00707-015-1422-3
[10] Lurie S A, Tuchkova N P 2009 *Composites and Nanostractures* **2(2)** 25
[11] Gerard A Maugin 1999 *Thermodynamics of nonlinear irreversible behaviors / World Scientific: Singapore* (New Jersey – London, Hong Kong)
[12] Rusanov A I 2005 *Surface Science Reports* **58** 111
[13] Knyazeva A G 2003 *Journal of Applied Mechanics and Technical Physics* 44(3) 373
[14] Knyazeva A G 2002 *Chemistry for sustainable development* **10(4)** 57

Stresses in the material with multilayered coating under impulse thermal loading

S A Shanin
National Research Tomsk Polytechnic University, 30, Lenina ave., Tomsk, 634050, Russia

E-mail: shanin_s@mail.ru

Abstract. The two-dimensional model of mechanical behavior of the specimen with multi layered coating is formulated. The parametric analysis was carried out for various types of boundary conditions. Stress intensity depending on time was studied. The influence of impulse parameters on the temperature and stresses was investigated. It was revealed that radiation heat losses reduces the action of external loading.

1. Introduction

Multilayered materials and coatings are attractive for many applications: in optics and electronics, in semiconductorand metallurgy industries. The combination of layers with different properties allows obtaining unique materials. It relates to the coatings based on TiN and TiC and their solutions [1, 2] and to the coating forming from AlN [3]. Immediate detailed investigation of the structure and properties of multilayered coatings is quite problematically [4-6].

Nowadays, finite-element software are used extensively for engineering calculations. For example, COMSOL, ANSYS, NASTRAN and etc. are widely-distributed. Investigators and developers assert that one can solve any problems and construct predictive models with the help of multiphysical software. In practice it is provided to be changing the objects by some analogies reflecting physical situation quite approximately. We analyze as an example enough simple problem on strain-stressed state calculation of material with multi layered coating in two-dimensional formulation. Using the model presented in [7], we can carry out a series of numerical experiment and obtain the coating with some distributions phases and elements, and residual stresses in the coating. Now it is necessary to investigate the behavior of this coating at the exploitation conditions. We use here COMSOL MULTIPHYSICS [8].

2. Problem formulation

It is assumed that multilayered coating was deposited on iron substrate with the thickness $L = 1$ µm, and the high is $H = 3$ µm. The coating contain the layers from aluminum and titanium nitrides; the thickness of each layer is $h = 0.25$ µm (Figure 1).

The impulse heat flux acts from external sin of the coating. Lateral surfaces of specimen are insulated.

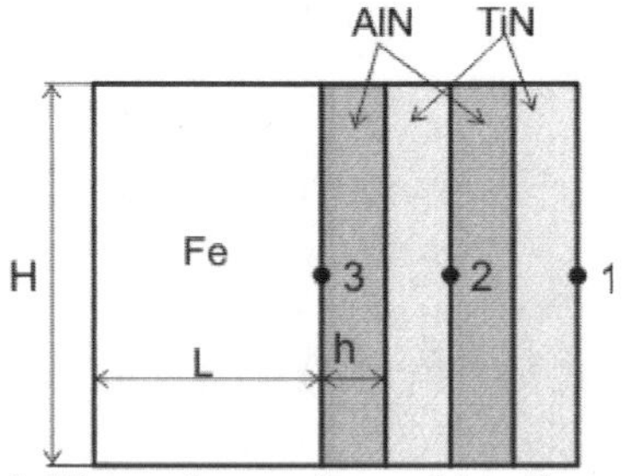

Figure 1. Illustration to the problem formulation.

Mathematical model will include heat and mechanical parts. Temperature field follows from thermal conductivity equation with thermal physical properties different for each layer and for substrate

$$(c\rho)_i \frac{\partial T_i}{\partial t} = \lambda_i \left(\frac{\partial^2 T_i}{\partial x^2} + \frac{\partial^2 T_i}{\partial y^2} \right); \quad i=1\ldots5, \tag{1}$$

and boundary and initial conditions:

$$\left. \frac{\partial T_1}{\partial y} \right|_{x=0} = 0;$$

$$\left. \frac{\partial T_i}{\partial x} \right|_{y=0} = 0; \quad \left. \frac{\partial T_i}{\partial x} \right|_{y=H} = 0; \; i=1\ldots5$$

$$\lambda_j \left. \frac{\partial T_j}{\partial y} \right|_{x=L+i\cdot h} = \lambda_{j+1} \left. \frac{\partial T_{j+1}}{\partial y} \right|_{x=L+(j+1)\cdot h}; \; j=1\ldots4$$

$$T_i = T_{i+1};$$

$$-\lambda_4 \left. \frac{\partial T_4}{\partial y} \right|_{x=L+4\cdot h} = q(t),$$

where

$$q(t) = q_0 f(t) + \sigma\varepsilon(T_0^4 - T^4), \quad f(t) = \begin{cases} 0; & \sin(at) < b; \\ 1; & \sin(at) > b, \end{cases}$$

a, b – are some constants.

Practically, for these conditions, thermal conductivity process will as one-dimensional. Mechanical part of the problem is quasistatic thermal elasticity problem. In this case, only two equilibrium equations are used

$$\frac{\partial \sigma_{xx}}{\partial x} + \frac{\partial \sigma_{xy}}{\partial y} = 0;$$
$$\frac{\partial \sigma_{xy}}{\partial x} + \frac{\partial \sigma_{yy}}{\partial y} = 0, \tag{2}$$

and it is assumed that specimen is very large in the direction z, that is $\varepsilon_{zz}=0$ and $\sigma_{zz}\neq0$.

In thermal elasticity theory,

$$\sigma_{kj} = 2\mu\varepsilon_{kj} + \delta_{kj}\left(\lambda\varepsilon_{ll} - 3K\alpha_T\theta_i\right),$$

where $\theta = T_i T_0$, i=1…5; λ,μ are Lame coefficients; K is bulk module; $K = \lambda + \frac{2}{3}\mu$. All modules and thermal expansion coefficients are different for different materials (indexes are omotted). Modules connect with technical values by relations

$$\lambda = \frac{E\nu}{(1+\nu)(1-2\nu)};\ \mu = \frac{E}{2(1+\nu)}.$$

In the situation under study, we have

$$\varepsilon_{xx} = \frac{\partial u_x}{\partial x};\ \varepsilon_{yy} = \frac{\partial u_y}{\partial y};\ \varepsilon_{xy} = \frac{1}{2}\left(\frac{\partial u_x}{\partial y} + \frac{\partial u_y}{\partial x}\right),$$

$$\varepsilon_{zz} = 0. \tag{3}$$

Hence,

$$\begin{gathered}\sigma_{xx} = 2\mu\varepsilon_{xx} + \lambda(\varepsilon_{xx} + \varepsilon_{yy}) - 3K\alpha_T\theta_i \\ \sigma_{yy} = 2\mu\varepsilon_{yy} + \lambda(\varepsilon_{xx} + \varepsilon_{yy}) - 3K\alpha_T\theta_i \\ \sigma_{zz} = \lambda(\varepsilon_{xx} + \varepsilon_{yy}) - 3K\alpha_T\theta_i \\ \sigma_{xy} = 2\mu\varepsilon_{xy}.\end{gathered} \tag{4}$$

Substituting (4) in (2) and taking into account (3) we come to the equations in displacements

$$\begin{gathered}\mu_i\nabla^2 u_x + (\lambda_i + \mu_i)\frac{\partial}{\partial x}\left(\frac{\partial u_x}{\partial x} + \frac{\partial u_y}{\partial y}\right) - 3K_i\alpha_{Ti}\frac{\partial \theta_i}{\partial x} = 0; \\ \mu_i\nabla^2 u_y + (\lambda_i + \mu_i)\frac{\partial}{\partial y}\left(\frac{\partial u_x}{\partial x} + \frac{\partial u_y}{\partial y}\right) - 3K_i\alpha_{Ti}\frac{\partial \theta_i}{\partial y} = 0,\end{gathered} \tag{5}$$

Note that possible dependencies on spatial coordinate for the properties near the interfaces are not taken into account in (5). Boundary conditions in interfaces between layers are ideal, that is the stresses σ_{xx} for layers and displacements u_x are continuous. Specimen is free on external mechanical loading. Initial stresses and strains are zero. After solution of the problem, stresses and strains are determined using (3) and (4). Physical parameters used in calculations are presented in Table 1 drawn up based on the data from [9].

Table 1. Properties of chemical substances.

Parameter	Fe	AlN	TiN	UM
λ	80.1	36.2	36.2	W/(m·K)
ρ	7.87	4.93	4.93	$\cdot 10^3$, kg/m^3
C_p	25.14	34.23	34.23	J/(mol·K)
m	55.84	61.8	61.8	$\cdot 10^{-3}$, kg/mol
E	190	326	326	GPa
ν	0.28	0.29	0.29	
α_T	4	3.2	3.2	$\cdot 10^{-6}$, K^{-1}

3. Results

Calculations show that temperature distribution and stresses depend basically on duration of external impulse (area of the action) and on its amplitude. Temperature versus time is presented in Figure 2a and Figure 2b in the point number 2 Figure 1 for uniform and for non uniform heat fluxes. With impulse duration decrease and amplitude growth, maximal temperature naturally rises. Maximal temperature growths more quickly for uniform heating.

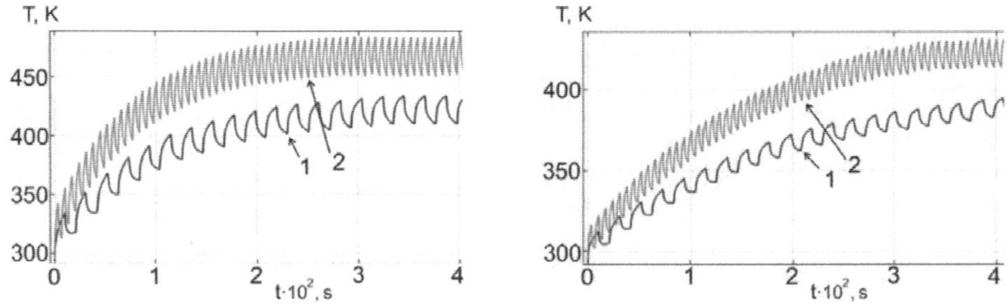

a b

Figure 2. Dependence of the temperature of the point of time for the uniform (a) and nonuniform (b) of boundary conditions 1 – a = 0.3, q_0 = 1·105; 2 – a = 0.9, q_0 = 2·105.

Stresses follow temperature. Impulse duration and its amplitude affect stresses by similarly way. Stress intensity is shown from Figure 3a, b). The values of stresses for presented parameters do not achieve strength limit. Heat losses lead to two-dimensional symmetrical picture of the process development.

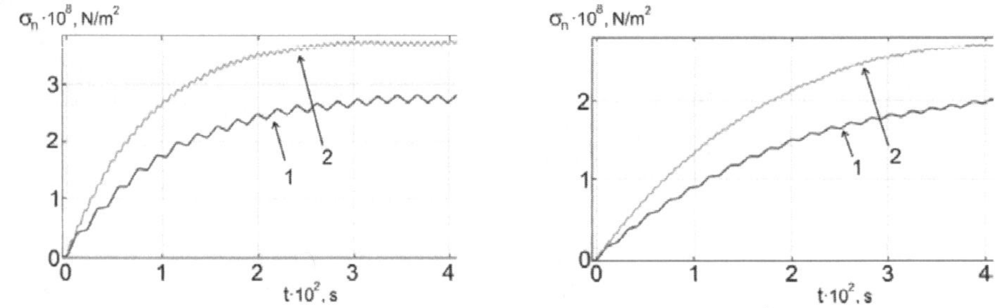

a b

Figure 3. Dependence of the intensity of stresses at the point of time for the uniform (a) and no uniform (b) of boundary conditions 1 – a = 0.3, q_0 = 1·105; 2 – a = 0.9, q_0 = 2·105.

Stress tensor component σ_{xx} growths with time (Figure 4). For uniform heating, difference in maximal and minimal values of stresses during one impulse growths with time, and then it stay constant. For no uniform heating, difference in maximal and minimal values of stresses is small.

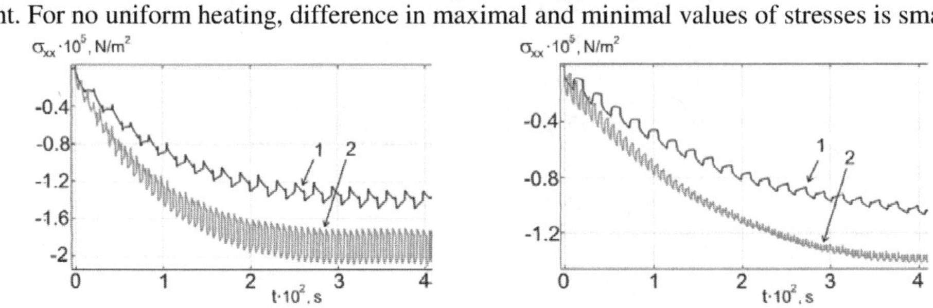

a B

Figure 4. Dependence of the stress σ_{xx} at the point of time for the uniform (a) and no uniform (b) of boundary conditions. 1 – a = 0.3, q_0 = 1·105; 2 – a = 0.9, q_0 = 2·105.

Stress tensor component σ_{yy} changes in a different way. Its value is more in two step then σ_{xx}. (Figure 5a, b). For uniform heating, stationary state (when difference in maximal and minimal values of σ_{yy} does not change) set enough quickly (during 100 s). For no uniform heating, we come to no stationary process.

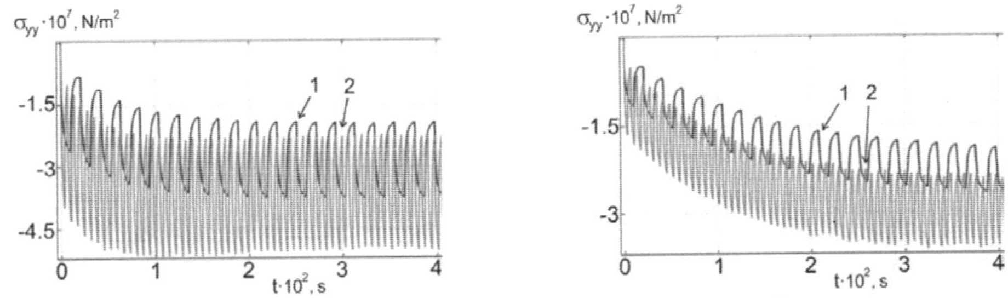

a b

Figure 5. Dependence of the stress σ_{yy} at the point of time for the uniform (a) and nonuniform (b) of boundary conditions. 1 – a = 0.3, $q_0 = 1\cdot10^5$; 2 – a = 0.9, $q_0 = 2\cdot10^5$.

In contrast to stresses σ_{yy} and σ_{xx} , maximal value of tangential stresses is higher for no uniform heating then for uniform one (Figure 6a, b).

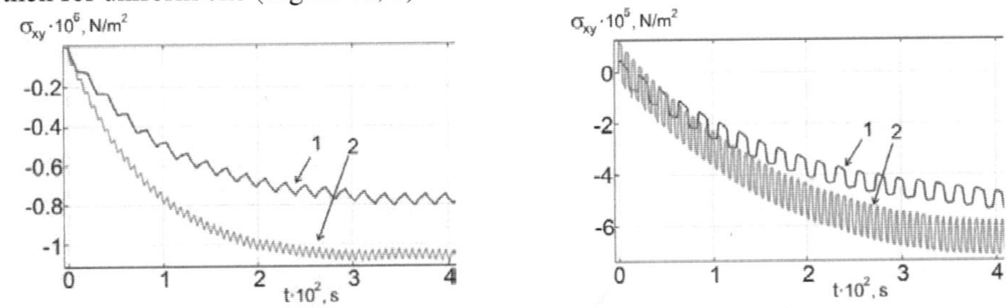

a b

Figure 6. Dependence of the stress σxy at the point of time for the uniform (a) and nonuniform (b) of boundary conditions. 1 – a = 0.3, $q_0 = 1\cdot105$; 2 – a = 0.9, $q_0 = 2\cdot105$.

In (Figure 7a, b) shows the temperature distribution in the sample for some time for the cases of uniform distribution of heat flux along the boundary (a) and for no uniform one (b). For second case, maximal temperature reaches naturally in symmetry axis.

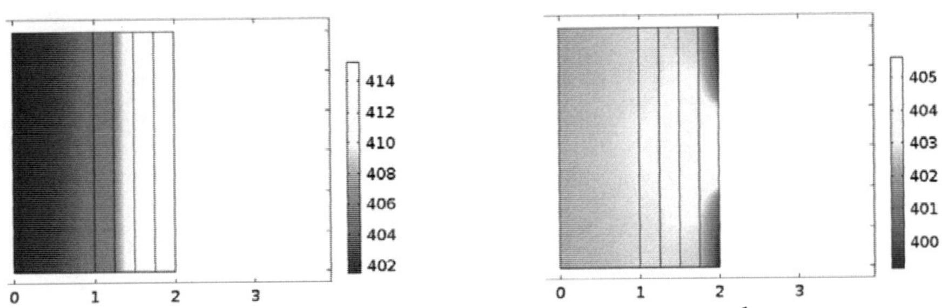

a b

Figure 7. Temperature distribution in the sample at time t = 250 s

4. Conclusion

Thus, in this paper, on the base of two-dimensional thermal elasticity model the investigation of stresses and temperature evolution in the specimen with multilayered coating was investigated for different parameters of the heating. In quasi static formulation, stresses follow temperature: the temperature growth with the heating parameters leads to stresses evaluation. It was marked that strength limit is not achieved for given parameter set. For more realistic picture, the model should take into consideration the no uniform concentration distribution had been obtained in real and numerical experiment with coating deposition.

Acknowledgment
The work was supported by the Ministry of Science and Education of Russia for 2014-2016 No. NIR - 11.815.2014/K.

References

[1] Smolik J, Zdunek K 1999 *Vacuum* **55(2)** 147–151
[2] Restrepo E, Benavides V, Devia A, Olarte S, Arroyave M, and Arango Y C 2004 *Brazilian J. of Physics* **34** 1748–1751
[3] Samani M K, Ding X Z, Khosravian N, Amin-Ahmadi B, Yang Y, Chen G, Neyts E C, Bogaerts A, Tay B K 2015 *Thin Solid Films* **578** 133–138
[4] Conde J P, Chu V, Shen D S, Wagner S 1994 *J. Appl. Phys.* **75(3)** 1638–1955
[5] Schutte C L, Smith P M, Whiteside G M 1991 *J. Mater. Res.* **6(4)** 835–839
[6] McGinnis' A J, Watkins T R, Jagannadham' K 1999 *JCPDS-International Centre for Diffraction Data* **443–454**
[7] Knyazeva A G , Shanin S A 2015 *Acta Mechanica* **1–30**
[8] www.comsol.com
[9] Grigoreva I S, Meilikhov E Z 1991 Physical quantities. Handbook (*Moscow: Energoatomizdat*) p. **1232**

Analysis of some identification problems for the reaction-diffusion-convection equation

G V Alekseev[1,2], D V Mashkov[3], E N Yashenko[3]
[1] Institute of Applied Mathematics FEB RAS, Vladivostok, 690041, Russia
[2] Far Eastern Federal University, 8, Sukhanova st., Vladivostok, 690950, Russia
[3] Far Eastern State Technical Fisheries University, 52B, Lugovaya st., Vladivostok, 690087, Russia

E-mail: Alekseev@iam.dvo.ru

Abstract. Identification problems for a linear stationary reaction–diffusion–convection model, considered in the bounded domain under Dirichlet boundary condition, are studied. Using an optimization method these problems are reduced to respective control problems. The reaction coefficient and the volume density of substance source play the role of controls in this control problem. The solvability of the direct and control problems is proved, the optimality system, which describes the necessary optimality conditions, is derived and the numerical algorithm is developed.

1. Introduction. Statement and solvability of the boundary value problem

Much attention has been given in recent years to the formulation and study of new classes of problems for heat and mass transfer models. The control problems are examples of such problems. There is a number of books [1–3] and papers [4–12] devoted to the theoretical studies of the control problems for models of heat and mass transfer.

Along with control problems, an important role in applications is played by identification problems for heat and mass transfer models. The unknown densities of boundary or distributed sources or the coefficients, involved in the differential equations or the boundary conditions for the heat and mass transfer models under study, are recovered in these problems using the additional information concerning the solution of the original boundary value problem. It is important that the identification problems can be reduced to corresponding extremum problems using a certain choice of the minimized cost functional. Based on this approach, there arise inverse extremum problems. They can be studied using well-known constrained minimization methods. The application of this approach to heat and mass transfer models was described in [13–17]. The papers [18–22], where an analogous approach was used for solving similar inverse problems arising in the study of thermal processes occurring in the Earth's mantle and in technical gas dynamics, should be also mentioned.

In this paper we consider the following boundary value problem for the stationary reaction–diffusion–convection equation:

$$-\operatorname{div}(\lambda\nabla C)+u\cdot\nabla C+kC=f \quad \text{in} \quad \Omega, \quad C|_{\Gamma}=\psi. \qquad (1)$$

Here Ω is a bounded domain in R^d, $d = 2.3$ with boundary Γ, $C(x)$ is the substance concentration, $\lambda = \lambda(x) > 0$ is the diffusion coefficient, $u = u(x)$ is the fluid velocity, function $k = k(x)$ is the reaction coefficient, function f describes the density of volume sources.

Below we shall use Sobolev function spaces $H^s(D), s \in R$. Here, D denotes Ω, its boundary Γ, or certain subdomain $Q \subset \Omega$. By $\|\cdot\|_{s,\Omega}$, $|\cdot|_{s,\Omega}$, and $(\cdot,\cdot)_{s,\Omega}$, we denote the respective norm, seminorm, and scalar product in $H^s(\Omega)$. Expressions $\|\cdot\|_Q$, $\|\cdot\|_{1,Q}$, $(\cdot,\cdot)_Q$ and $(\cdot,\cdot)_{1,Q}$ denote the norms and scalar products in $L^2(Q)$ and $H^1(Q)$, respectively. If $Q = \Omega$ we omit index Ω, setting $\|\cdot\|_\Omega = \|\cdot\|$, $\|\cdot\|_{1,\Omega} = \|\cdot\|_1$, and $(\cdot,\cdot)_\Omega = (\cdot,\cdot)$. Let $X = \{\eta \in H^1(\Omega) : \eta|_\Gamma = 0\}$. The duality relation between X and its dual X^* is denoted by $\langle \cdot,\cdot \rangle$. Let us define $Z = \{u \in H^1(\Omega)^3 : \operatorname{div} u = 0\}$, $L^2_+(\Omega) = \{k \in L^2(\Omega) : k \geq 0 \, \text{in} \, \Omega\}$, $L^\infty_{\lambda_0}(\Omega) = \{\lambda \in L^\infty(\Omega) : \lambda \geq \lambda_0 \, \text{in} \, \Omega\}$, where $\lambda_0 = \text{const} > 0$. We can recall that by Poincare' inequality [3] the following relation holds:

$$|C|^2_{1,\Omega} \geq \delta \|C\|^2_{1,\Omega} \, \forall C \in X$$

where $\delta = \text{const} > 0$.

Let us suppose that the following conditions take place:

(i) $\Gamma \in C^{0,1}$, $\lambda \in L^\infty_{\lambda_0}(\Omega)$, $\lambda_0 = \text{const} > 0$, $u \in Z$;

(ii) $k \in L^2_+(\Omega)$, $f \in L^2(\Omega)$.

We will begin with a brief study of unique solvability of the boundary value problem (1). Let us multiply the equation in (1) by test function $\eta \in X$, integrate over Ω and apply Green's formula. Proceeding as in [17], we obtain

$$a_k(C,\eta) \equiv (\lambda \nabla C, \nabla \eta) + (u \cdot \nabla C, \eta) + (kC, \eta) = (f, \eta) \quad \forall \eta \in X. \tag{2}$$

Here we use notation

$$(\lambda \nabla C, \nabla \eta) = \int_\Omega \lambda \nabla C \cdot \nabla \eta dx, \; (u \cdot \nabla C, \eta) = \int_\Omega (u \cdot \nabla C) \eta dx, \; (kC, \eta) = \int_\Omega kC\eta dx.$$

We shall call solution $C \in H^1(\Omega)$ of the problem (2) a weak solution of the problem (1).

A simple analysis shows that bilinear form $a_k(\cdot,\cdot)$ which is defined in (2) is continuous and coercive on X and the following inequality takes place:

$$a_k(C,C) \geq \delta \lambda_0 \|C\|^2_{1,\Omega} \; \forall C \in X.$$

Based on this fact and using the Lax-Milgram theorem one can prove the following theorem concerning the existence of the unique weak solution of the problem (2).

Theorem 1. Let the conditions (i) take place. Then the following assertions hold:

1) The bilinear form $a_\lambda : X \times X \to R$ given in the left hand of (2) is continuous and coercive on X with coercitivity constant $\lambda_* = \delta_0 \lambda_0$.

2) For any pair (k, f), satisfying conditions (ii), the problem (2) has a unique solution $C \in X$ and this solution satisfies the estimate

$$\| C \|_{1,\Omega} \leq \lambda_0^{-1} \| f \|_\Omega . \tag{3}$$

2. Statement and solvability of control problem. Optimality system

We recall that the boundary value problem (1) contains four variable parameters: diffusion λ, velocity u, reaction coefficient k and source density f. All these functions must be provided for the solution of the problem (1.1) in order to be unique. However there appear situations in practice when some of mentioned functions are unknown and one has to determine them using the additional information about the solution. For example one can use values $C_d(x)$ of concentration C measured in points of some set $Q \in \Omega$.

Get us consider the situation when reaction coefficient k and density f are unknown, and one should determine them together with solution C of (1). This identification problem is studied by applying an optimization method [3]. According to this method the problem is reduced to a corresponding inverse extremum problem (see details in [15-17]). Following this method, we divide the set of input data in the problem (1) into two groups. One consists of fixed data, namely, invariable functions λ and u. The other group consists of two distributed controls, namely, functions k and f. Assume that controls k and f change over sets K_1 and K_2 satisfying the following conditions:

(j) $K_1 \subset Z$, $K_2 \subset L^2(\Omega)$ are nonempty convex closed subsets.

Let us denote $K = K_1 \times K_2$, $u = (k, f)$ and define operator $F = X \times K \to X^*$ acting by

$< F(C,u), \eta >= a_k(C,\eta) - (f,\eta)$.

Let $I : X \to R$ be arbitrary weak lower semicontinuous cost functional. We consider the following constrained minimization problem:

$$(\alpha_0/2)I(C) + (\alpha_1/2)\| k \|_\Omega^2 + (\alpha_2/2)\| f \|_\Omega^2 \to \inf, \; F(C,u) = 0,$$
$$(C,u) \equiv (C,k,f) \in X \times K. \tag{4}$$

As the admissible cost is functional we choose one of the following:

$$I_1(C) = \| C - C_d \|_Q^2 = \int_Q | C - C_d |^2 dx, \quad I_2(C) = \| C - C_d \|_{1,Q}^2.$$

Here $C_d \in L^2(Q)$ (or $C_d \in H^1(Q)$) is a given function in Q. We assume in addition to (j) that the following conditions hold:

(jj) $\alpha_0 > 0$, $\alpha_1 \geq 0$, $\alpha_2 \geq 0$ and K_1, K_2 are bounded sets or $\alpha_1 > 0$, $\alpha_2 > 0$.

The following theorem can be proved.

Theorem 2. Under assumptions (i), (j) and (jj) the control problem (4) for $I = I_m, m = 1,2$ has at least one solution $(\hat{C}, \hat{u}) \equiv (\hat{C}, \hat{k}, \hat{f})$, which is an element of $X \times K_1 \times K_2$.

3. The optimality system. Numerical algorithm

The following stage of studying the control problem is the derivation of the optimality system for the problem (4). The Lagrange principle for extremum problems (see [3, p. 341]) is used for this purpose. Using this principle and techniques of [3,17] we can prove the following theorem.

Theorem 3. Let us suppose that, under conditions of Theorem 2, triple $(\hat{C}, \hat{k}, \hat{f}) \in X \times K_1 \times K_2$ is a solution of the problem (4) and let $I(C)$ continuously be a Frechét differentiable functional with respect to C at point $\hat{C}$. Then there is unique non-zero Lagrange multiplyer $\theta \in X$, so that the Euler–Lagrange equation

$$F_C'(\hat{C},\hat{u})^*\theta + (\alpha_0/2)I_C'(\hat{C}) = 0$$

holds in X^* which is equivalent to the identity

$$\lambda(\nabla\tau,\nabla\theta) + (\hat{u}\cdot\nabla\tau,\theta) + (k\tau,\theta) = -(\alpha_0/2)\langle I_C'(\hat{C}),\tau\rangle \quad \forall\tau\in X \tag{5}$$

and the minimum principle holds which is equivalent to the pair of variational inequalities

$$\alpha_1(\hat{k},k-\hat{k}) + ((k-\hat{k})C,\theta) \geq 0 \quad \forall u\in K_1, \tag{6}$$

$$\mu_2(\hat{f},f-\hat{f}) - (f-\hat{f},\theta) \geq 0 \quad \forall f\in K_2. \tag{7}$$

The direct problem (2), the adjoint problem (5) together with the variational inequalities (6), (7) form the optimality system that describes the necessary optimality conditions. The optimality system plays an important role when studying control problems. Based on the analysis of the optimality system one can apply fundamental properties of solutions to the control problem (see in more details in [13,15,17]). Besides, on the base of the optimality system, the efficient numerical algorithm can be developed for solving the control problem (4). If, in particular, to apply the simple iteration method to the optimality system (2) (5), (6), (7) then one can derive the following iteration algorithm:

$$a_k(C^n,h) \equiv (\lambda\nabla C^n,\nabla h) + (u\cdot\nabla C^n,h) + (k^nC^n,h) = (f^n,h) \quad \forall h\in X, \tag{8}$$

$$\lambda(\nabla\tau,\nabla\theta^n) + (u\cdot\nabla\tau,\theta^n) + (k^nC,\theta^n) = -(\alpha_0/2)\langle I_T(T^n),\tau\rangle \quad \forall\tau\in X, \tag{9}$$

$$\alpha_1(k^{n+1},k-k^n) + ((k-k^{n+1})\cdot C^n,\theta^n) \geq 0 \, \forall k\in K_1, \tag{10}$$

$$\alpha_2(f^{n+1},f-f^n) - (f-f^n,\theta^n) \geq 0 \forall f\in K_2. \tag{11}$$

The proposed algorithm consists of finding k^{n+1} and f^{n+1} for given k^n and f^n by sequentially solving problems (8), (9), (10) and (11). The separate paper of the authors will be devoted to study of the properties of the corresponding algorithm and the analysis of the results of computational experiments on the base of this algorithm.

4. Acknowledgements

This work was supported by the Russian Foundation for Basic Research (project nos. 13-01-00313-a) and by the Department of Education and Sciences (project no. 14.Y26.31.0003).

References

[1] Marchuk G I 1982 *Nauka Moscow* p 319
[2] Samarskii A A and Vabishevich P N 2004 *Editorial URSS*
[3] Alekseev G V and Tereshko D A 2008 *Dal'nauka*
[4] Alekseev G V 1998 *Dokl. Math.* **58** 314–317
[5] Alekseev G V 1998 *Sib. Mat. Zh.* **39** 982–998
[6] Alekseev G V and Tereshko D A 1998 *Journal of Inverse and Ill-Posed Problems* **6** 521-562
[7] Alekseev G V and Tereshko D A 1998 *Sib. J. Indust. Math.* **1** 22–44
[8] Alekseev G V 2002 *Comput. Math. and Math. Phys.* **42** 363–376
[9] Lee H C 2003 *Advan. Comput. Math.* **19** 255–275
[10] Alekseev G V 2006 *Vestn. Novosibirsk. Univ. Ser. Mat.* **6** 7–31
[11] Alekseev G V 2007 *Comput. Math. Math. Phys.* **47** 1007–1028
[12] Lee H C 2010 *Jpn. J. Ind. Appl. Math.* **26** 93–121
[13] Alekseev G V 2007 *Dokl. Math.* **76** 797–800
[14] Alekseev G V Soboleva O V and Tereshko D A 2008 *Prikl. Mekh. Tekh. Fiz.* **49** 24–35
[15] Soboleva O V 2010 *Far. Eastern. Math. J.* **10(2)** 170–184
[16] Alekseev G V Tereshko D A 2011 *Comput. Math. Math. Phys.* **51** 1539–1557

[17] Alekseev G V Vakhitov I S Soboleva O V 2012 *Comput. Math. and Math. Phys.* **52** 1635–1649
[18] Korotkii A I Kovtunov D A 2006 *Trudy Inst. Mat. Mekh.* **12** 88–97
[19] Korotkii A I Kovtunov D A 2010 *Trudy Inst. Mat. Mekh.* **16** 76–101
[20] Fershalov Ju Ya Sazonov T V 2014 *Advan. Mater. Res.* **345** 915–916
[21] Fershalov M Yu Fershalov A Yu Fershalov Yu Ya 2014 *Advan. Mater. Res.* **341** 915–916
[22] Fershalov Yu Ya 2012 *Russian Aeronaut* **55** 424–429

The mathematical simulation of the temperature fields of building envelopes under permanent frozen soil conditions

M V Anisimov[1], M N Babuta[2], U N Kuznetsova[1], E V Safonova[1] and O M Minaeva[3]
[1] Tomsk State University of Architecture and Building, 2, Solyanaya sq., Tomsk, 634003, Russia
[2] National Research Tomsk Polytechnic University, 30, Lenina ave., Tomsk, 634050, Russia
[3] National Research Tomsk State University, 36, Lenina ave., Tomsk, 634050, Russia

E-mail: teploproekt@mail2000.ru

Abstract. The physical-mathematical model of the thermal state of the aired technical underground taking into account the air exchange and design features of construction under permanent frozen soil conditions has been suggested. The computational scheme of the temperature fields prediction of building envelopes of projected buildings and soil under and nearby buildings has been developed. The numerical simulation of the temperature fields of building envelopes changes was conducted during a year. The results of the numerical simulation showed that the heat coming from the technical undergrounds and through the walls does not influence the temperature field of the soil neither under a building nor at a distance from it.

1. Introduction

The prediction of temperature fields in soils and adjoining building constructions is a complex heat engineering problem. It is conditioned by soils heterogeneity, their heat properties and 'change of phase' at their frost penetration. The permanent frozen soils demand a special attention because of the complex thermal processes of 'freezing – defrosting', the formation of the surface soil layer that has a positive temperature during a warm period, thawed soil that has relatively permanent temperatures during a year and a soil layer in a permanent frozen condition.

Very often, when designing and constructing buildings it is necessary to determine the temperature field of the soil. This task is very topical when designing warmth keeping of an underground lead-in water pipeline and a tapping of sewage disposal under conditions of relief with inequalities in the elevation in the site area when soil surface elevations and drops fluctuate. The knowledge of the temperature field in the zone of thermal influence of buildings could allow predicting the water cooling in pipes (including extremely low temperatures in winter) and to recommend the necessary measures in the pipe-line warming.

Underground piping under conditions of extremely low temperatures must be laid taking into account that not only permanent frozen soils influence the medium temperature and may cause pipe frost penetration freezing, but also pipes may cause the temperature rise of these soils. The consequences of natural temperature changes of permanent frozen soils can be unpredictable. The transition from the tough frozen state into a thawed one may entail the loss of lifting properties of piles of building foundations and it will be impossible to operate the construction.

Reviewing the theory of computation of thermal transfer in soils, the contribution of such authors as Bogoslovskiy, A.V. Gindoyan, A.G. Sander, A.A. Ioritis, G.V. Porhayev should be mentioned. Especially the works of Porhayev [1] should be marked as in his research he investigated the thermal interaction of constructions with permanent frozen soils. As G.V. Porhayev's research shows that the soil temperature on the building area differs from the soil temperature on the area free from engineering building.

The contemporary authors such as D.A. Krylov [2-4], P.I. Dyachek [5], U.S. Melnikova [6] AA Sinitsyn [7] and N.I. Sydnyaev [8] should be marked. The foreign experts' works also deserve a special attention. These experts are Lunardini J. Virgil [9], Ebenezer E.K. [10], Ming Zhu and others [11-12].

2. The mathematical simulation of the thermal state of building envelopes deepened into soil

In a general case the problem of the calculation of heat loss through multilayer envelope buildings and the soil of the technical underground is three-dimensional. But, the length of buildings (especially of public buildings) is much greater than their width, that's why when developing the mathematical model of heat exchange in the technical underground a two- dimensional symmetrical problem of thermal conduction in a semi-limited mass with a set of boundary conditions is considered. Along with it, the inner zone of cellar L is separated from the external one by the wall section with thickness δ, m (Figure 1).

For the construction of the mathematical model of heat exchange of the technical underground in non-steady-state conditions the physical model including different media (air, snow, soil) and materials is suggested. Every medium or material having different thermal properties is isolated into a separate material layer: 1 – inner air, 2 – warmth-keeping layer, 3 – external wall, 4 – external layer of the wall envelope, 5 – soil out of a cellar, 6 – snow layer, 7 – air out of a technical underground, 8 – soil inside the technical underground, 9 – permanent snow layer above the blind area.

The schemes of physical and mathematical models with mentioned boundary conditions (BC) are in Figure 1.

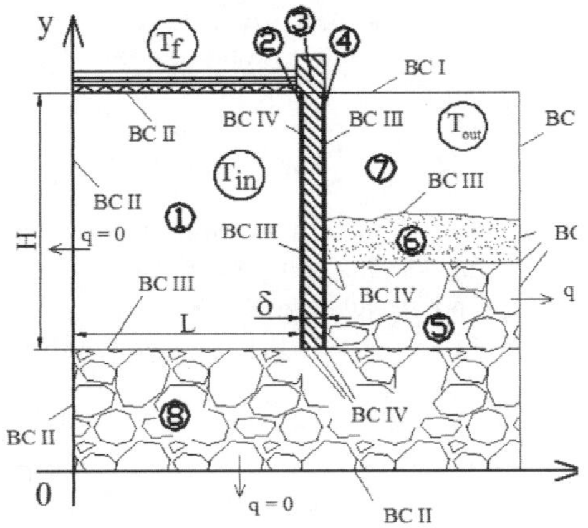

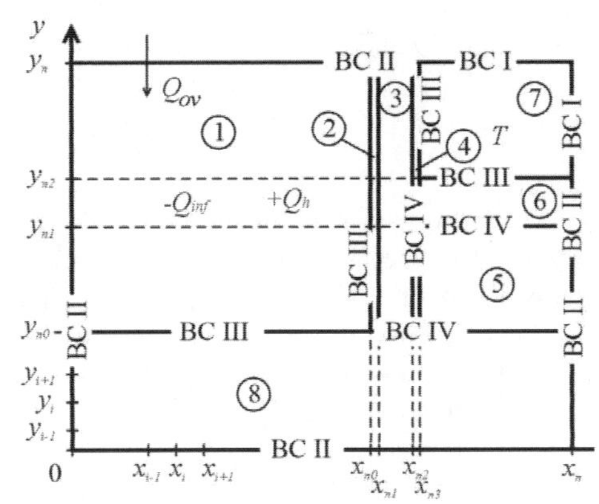

a) b)

Figure 1. The mathematical model design scheme:
a) the physical model of the underground b) the mathematical model of the underground
1 – air inside a cellar; 2 – warmth-keeping; 3 – external wall (reinforced concrete); 4 - external hydroisolation layer; 5, 8 – soil; 6 – snow layer; 7 – air out of the technical underground;
BC – boundary conditions.

It is worth mentioning that such problem statement allows providing enough 'flexibility' in calculations because there is a possibility of calculation of both cellars and technical undergrounds (Figure 1).

At this point it is sufficient to admit that layer 5 has the same thermal properties as layer 6. The heat transfer in multilayer building envelopes of the technical underground and in the solid media such as soil and snow is described with a three-dimensional nonlinear equation of the thermal conduction in the Cartesian coordinate system

$$\rho\, C_p \frac{\partial T}{\partial \tau} = \frac{\partial}{\partial x}\left(\lambda \frac{\partial T}{\partial x}\right) + \frac{\partial}{\partial y}\left(\lambda \frac{\partial T}{\partial y}\right) + \frac{\partial}{\partial z}\left(\lambda \frac{\partial T}{\partial z}\right) + Q(x, y, z), \quad (1)$$

$T = T(\bar{x}, \tau)$ – temperature; C, ρ – thermal capacity and density; λ – heat conduction coefficient; $Q(x, y, z)$ – power of inner heat sources; τ – time.

When constructing the mathematical model the following assumptions have been made:

– the thermal properties of materials (ρ_i, $C_{P,i}$, λ_i), building envelopes, snow and air are considered to be as constant;

– the thermal properties of the soil in layer 8 are assumed to be constant;

– the thermal properties of the soil in layer 5 are taken as constant, based on the results of the conducted research;

– the heat (Q_h) from the utility systems, the heat used for filtered cold air ($-Q_{inf}$) and the heat input through the ground floor (Q_{ov}) is considered equally allocated by the technical underground volume;

– the air temperature inside the ground floor room is equal to critical [13];

– in reference time $\tau = 0$ the temperature in all layers of the mathematical model is equal to the temperature in layer 7 (inner air);

– the coefficient of the heat transfer on the border 'snow layer - inner layer' is calculated in accordance with the recommendations for the definite region's climate conditions.

The equation (1) with proper initial and boundary conditions represents the problem of temperature allocation at each point of the considered area for each moment of time. It is suggested that the initial temperature allocation in the material, the temperature condition on its borders and the power of inner heat sources are known. The problem is symmetrical with respect to the *x*-axis, and in the direction normal to plane *x-y* it is semi-infinite because on these borders the heat streams are insignificant and not taken into account [15]. It allows solving the problem in a two - dimensional statement (figure 1).

Taking into account these assumptions the equation (1) will be the following:

$$\rho C_p \frac{\partial T}{\partial \tau} = \lambda\left(\frac{\partial^2 T}{\partial x^2} + \frac{\partial^2 T}{\partial y^2}\right) + Q, \quad (2)$$

C_p – heat capacity at constant pressure; ρ – material density; λ – heat conduction of the material; T – material temperature; τ – time; Q – inner sources of heat release (this component is taken into account only for layer 1, Figure 1); *x, y* – coordinates in horizontal and vertical directions.

The heat streams allocation on different material layers planes (Figure 1) is revealed from solving the equation (2) with the set of boundary conditions.

The boundary conditions for layers 1 and 2 (the air inside the technical underground, warmth-keeping layer) are:

$$\begin{cases} x = 0;\ y_{n0} \le y \le y_n;\ q = 0; \\ x = x_{n0};\ y_{n0} \le y \le y_n;\ \alpha_{B1}\cdot(T_B - T_{x_{n0}}) = -\lambda \dfrac{\partial T_2}{\partial x}; \\ x_0 \le x \le x_{n0};\ y = y_n;\ q = f(\tau, T_B); \\ x_0 \le x \le x_{n0};\ y = y_{n0};\ \alpha_{B2}\cdot(T_B - T_{y_0}) = -\lambda_8 \dfrac{\partial T_8}{\partial y},\ \lambda_8 = c \end{cases}$$

$$\begin{cases} x = x_{n0};\ y_{n0} \le y \le y_n;\ \alpha_{B1}\cdot(T_B - T_{x_{n0}}) = -\lambda_2 \dfrac{\partial}{\partial} \\ x = x_{n1};\ y_{n0} \le y \le y_n;\ T_2 = T_3;\ \lambda_2 \dfrac{\partial T_2}{\partial x} = \lambda_3 \dfrac{\partial T}{\partial x} \\ x_{n0} \le x \le x_{n1};\ y = y_n;\ q = f(\tau, T_B); \\ x_{n0} \le x \le x_{n1};\ y = y_{n0};\ T_2 = T_8;\ \lambda_2 \dfrac{\partial T_2}{\partial y} = \lambda_8 \dfrac{\partial}{\partial} \end{cases}$$

α_{B1}, α_{B2} – coefficients of the heat transfer beside the basement wall near the soil area, W/(m$^2\cdot$°C); λ_8 – coefficient of the soil heat transfer inside the technical underground, W/(m$^2\cdot$°C).

The boundary conditions for layers 3 and 4 (the building envelope of the technical underground and the finishing layer) are:

$$\begin{cases} x = x_{n1};\ y_{n0} \le y \le y_n;\ T_2 = T_3;\ \lambda_2 \dfrac{\partial T_2}{\partial x} = \lambda_3 \dfrac{\partial T_3}{\partial x}; \\ x_{n1} \le x \le x_{n2};\ y = y_{n0};\ T_3 = T_8;\ \lambda_3 \dfrac{\partial T_3}{\partial y} = -\lambda_8 \dfrac{\partial T_8}{\partial y} \\ x_{n1} \le x \le x_{n2};\ y = y_n;\ q = f(\tau, T_B); \\ x = x_{n2};\ y_{n0} \le y \le y_n;\ T_3 = T_4;\ \lambda_3 \dfrac{\partial T_3}{\partial x} = \lambda_4 \dfrac{\partial T_4}{\partial x}, \end{cases}$$

$$\begin{cases} x = x_{n3};\ y_{n0} \le y \le y_{n2};\ \alpha_{H2} \cdot (T_H - T_{x_{n3}}) = -\lambda_4 \\ x = x_{n3};\ y_{n1} \le y \le y_{n2};\ T_4 = T_6;\ \lambda_4 \dfrac{\partial T_4}{\partial x} = \lambda_6 \dfrac{\partial}{} \\ x = x_{n3};\ y_{n0} \le y \le y_{n1};\ T_4 = T_5;\ \lambda_4 \dfrac{\partial T_4}{\partial x} = \lambda_5 \dfrac{\partial}{} \\ x = x_{n2};\ y_{n0} \le y \le y_n;\ T_3 = T_4;\ \lambda_4 \dfrac{\partial T_4}{\partial x} = \lambda_3 \dfrac{\partial}{} \\ x_{n2} \le x \le x_{n3};\ y = y_n;\ q = f(\tau, T_B); \\ x_{n2} \le x \le x_{n3};\ y = y_{n0};\ T_4 = T_8;\ \lambda_4 \dfrac{\partial T_4}{\partial y} = \lambda_8 \dfrac{\partial}{} \end{cases}$$

λ_2, λ_3, λ_4 – coefficients of the heat conduction of warmth-keeping concrete wall and the plaster; λ_5, λ_6, λ_8 – coefficients of the soil heat conduction out of the technical underground, snow and the soil inside of the technical underground, W/(m^2·°C); α_{H2} – coefficient of the heat conduction on the border 'the inner surface of the basement wall – inner air', W/(m^2·°C).

The boundary conditions for layers 5 and 6 (the soil out of the technical underground and the snow layer):

$$\begin{cases} x = x_{n3};\ y_{n0} \le y \le y_{n1};\ T_4 = T_5;\ \lambda_5 \dfrac{\partial T_5}{\partial x} = \lambda_4 \dfrac{\partial T}{\partial x} \\ x_{n3} \le x \le x_n;\ y = y_{n1};\ T_5 = T_6;\ \lambda_5 \dfrac{\partial T_5}{\partial y} = \lambda_6 \dfrac{\partial T_6}{\partial y} \\ x = x_n;\ y_{n0} \le y \le y_{n1};\ q = 0; \\ x_{n3} \le x \le x_n;\ y = y_{n0};\ T_5 = T_8;\ \lambda_5 \dfrac{\partial T_5}{\partial y} = \lambda_8 \dfrac{\partial T_8}{\partial y} \end{cases}$$

$$\begin{cases} x_{n3} \le x \le x_n;\ y = y_{n2}; \\ \alpha_{H1} \cdot (T_{y_{n2}} - T_H) = -\lambda_6 \dfrac{\partial T_6}{\partial y},\ \alpha_{H1} = f(v, \tau);\ T_H = f \\ x = x_n;\ y_{n1} \le y \le y_{n2};\ q = -\lambda \dfrac{\partial T_6}{\partial x} = 0; \\ x_{n3} \le x \le x_n;\ y = y_{n1};\ T_5 = T_6;\ \lambda_6 \dfrac{\partial T_6}{\partial y} = \lambda_5 \dfrac{\partial T_5}{\partial y}; \\ x = x_{n3};\ y_{n1} \le y \le y_{n2};\ T_4 = T_6;\ \lambda_6 \dfrac{\partial T_6}{\partial x} = \lambda_4 \dfrac{\partial T_4}{\partial x}, \end{cases}$$

λ_5 – coefficient of the soil heat conduction out of the technical underground, W/(m^2·°C); α_{H1} – coefficient of the heat conduction on the border "snow layer – inner air", determined in (10), W/(m^2·°C).

The boundary conditions for layers 7 and 8 (the air out of the technical underground and the soil inside of the technical underground):

$$\begin{cases} x = x_n;\ y_{n2} \le y \le y_n;\ q = 0; \\ x_{n3} \le x \le x_n;\ y = y_n;\ q = 0; \\ x = x_{n3};\ y_{n2} \le y \le y_n;\ \alpha_{H2} \cdot (T_H - T_{x_{n3}}) = -\lambda_4 \dfrac{\partial T_4}{\partial x} \\ x_{n3} \le x \le x_n;\ y = y_{n2};\ \alpha_{H1} \cdot (T_H - T_{y_{n2}}) = -\lambda_6 \dfrac{\partial T_6}{\partial y} \\ \alpha_{H1} = f(v, \tau);\ T_H = f(\tau). \end{cases}$$

$$\begin{cases} x_0 \le x \le x_{n0};\ y = y_{n0};\ \alpha_{B2} \cdot (T_B - T_{y_{n0}}) = -\lambda_8 \dfrac{\partial T_8}{\partial y}; \\ x_{n0} \le x \le x_{n1};\ y = y_{n0};\ T_2 = T_8;\ \lambda_8 \dfrac{\partial T_8}{\partial y} = \lambda_2 \dfrac{\partial T_2}{\partial y}; \\ x_{n1} \le x \le x_{n2};\ y = y_{n0};\ T_3 = T_8;\ \lambda_8 \dfrac{\partial T_8}{\partial y} = \lambda \dfrac{\partial T_3}{\partial y}; \\ x_{n2} \le x \le x_{n3};\ y = y_{n0};\ T_4 = T_8;\ \lambda_8 \dfrac{\partial T_8}{\partial y} = \lambda_4 \dfrac{\partial T_4}{\partial y}; \\ x_{n3} \le x \le x_n;\ y = y_{n0};\ T_5 = T_8;\ \lambda_8 \dfrac{\partial T_8}{\partial y} = \lambda_5 \dfrac{\partial T_5}{\partial y}; \\ x = x_n;\ y_0 \le y \le y_{n0};\ \dfrac{\partial T_8}{\partial x} = 0; \\ x_0 \le x \le x_n;\ y = y_0;\ \dfrac{\partial T_8}{\partial y} = 0; \\ x = x_0;\ y_0 \le y \le y_{n0};\ \dfrac{\partial T_8}{\partial x} = 0. \end{cases}$$

In some authors' [1-3] opinion, at a distance from the construction the heat influence of a building becomes weaker and imperceptible. On these borders the heat exchange equals 0. On some borders of layers 5, 6 and 8 the boundary conditions of the second type (BCII) are taken into account. As additional heat sources Q the following components of heat balance of the technical underground are considered: Q_h – heat of the pipeline, Q_{ov} – heat coming from the underground floor, Q_{inf} – heat loss because of air exchange between the technical underground and the environment, W.

$$Q = Q_h + Q_{ov} - Q_{inf}. \quad (7)$$

Defining the coefficient of the heat transfer on different surfaces of building envelopes the empirical dependence (given in [14]) that describes the heat exchange under the natural convection conditions has been used. The average values of the heat transfer coefficients were assessed by the criteria equation for typical cases.

On account of low temperatures on the surfaces of the building envelopes and, therefore, heat proportion values, we neglect this component of the heat exchange.

Analysing the existing numerical methods of thermal problems solution, the problem of heat exchange of the attic space with the environment is solved using a broadly used sweep method with the implicit difference scheme [16].

The numerical solution of the above-mentioned problem has been implemented by means of the software package developed on the object-based framework Delphi 7.0. When creating the package a modular concept has been used.

A basic software unit realizes the solution of a parabolic type and a general form equation with the boundary conditions of 1 – 4 types in one layer and multilayer areas. It consists of smaller units of calculation of the implicit difference scheme coefficients and solves the difference equation system by the sweep method, tested on the famous analytical solutions.

For testing the module, the analytical solutions with their further comparison with the numerical solution results were obtained.

The results of comparison show that the maximum deviation of the numeric solution from the analytical one does not exceed 0.6 %. It proves the correctness of the numeric scheme choice and its implementation.

3. Conclusion

As a result of the research, the two-dimensional nonstationary symmetrical problem of the heat conduct in a semi-limited mass with a set of boundary conditions has been viewed. The physical-mathematical model of the heat exchange of the technical underground (with the influence of air exchange) and the snow layer out of a construction has been considered.

The results of the numeric simulation showed that the heat coming through the floor and the walls of the technical underground does not influence the temperature field of soils both under the construction and at a distance from it. This deduction fully meets the demands of buildings erection under the climatic conditions with permanent frozen soils [17], which "ban" the heat influence of designed constructions on the natural temperature background of soils.

In accordance with the research the temperatures in soils massive outside of the construction area are equal to their natural values.

The comparison of the results of the numeric solution of the software package (18) with the solution obtained from other thermal computer programs (e.g. Temper-3D) has shown their satisfactory coherence. Nevertheless, the program Temper-3D does not take into account the air exchange in the technical underground, which is very important for the formation of the temperature fields in the building envelopes and soil.

References

[1] Porhaev G V 1970 *Moscow science* **208**
[2] Krilov D A 2010 *Vestnik MGTU* 106–116
[3] Krilov D A, Sidnyev N I, Fedotov A A 2010 *Sarov* **5** 72–76
[4] Krilov D.A. 2012 Available at: http://technomag.edu.ru/doc/354740.html **4** 108–126
[5] Dychek P I 2007 *Energy: International Scientific-Technical and Production Journal* **3** 77–86

[6] Melnikova U S 2012 Available at: http://www.technomag.edu.ru/doc/330390.html
[7] Sinicin A A 2009 *International Scientific and Practical Conference* **12** 99–101
[8] Sidnyaev N I 2010 *Proceed. of the Confer. Mathemat. and Mathemat. Model. Sarov* **6** 86–93
[9] Lunardini J Virgil 1998 *Canada: Seventh Internat. Confer., Collect. Nordicana* **55** 689–695
[10] Ebenezer E K 2006 Predicting *University of Saskatchewan* **183**
[11] Ming Zhu, Radoslaw L Michalowski 2005 *ABAQUS Users' Conference* 1-7
[12] Deru M 2003 A *Colorado: National Renewable Energy Laboratory* **138**
[13] SP 118.13330.2012 *State of rules Minregion Russia* **53**
[14] SNiP 2.06.04-82 1989 *Moscow Gosstroy Russia* **49**
[15] Karaush S A, Anisimov M V 2006 *Vestnik TGASU* **2** 133–141
[16] Samarsky A A *Moscow Nauka* pp 415
[17] Chernadiev V P 1984 *Moscow Nauka* pp 137
[18] Lisak I A, Karaush S A, Anisimov M V 2007 *Certificate of official registration of the computer program.* Register of Computer Programs No. RU 2007611123

A transient free convection study with temperature-dependent viscosity in a square cavity with a local heat source

M S Astanina and M A Sheremet
Tomsk State University, 36, Lenina ave., Tomsk, 634050, Russia

E-mail: astanina.marina@bk.ru

Abstract. Unsteady natural convection inside of a differentially-heated square enclosure filled with a fluid of temperature-dependent viscosity has been numerically studied. A mathematical model formulated in the dimensionless stream function and vorticity has been solved by a finite difference method of the second order accuracy. The effect of dimensionless time and Prandtl number on streamlines and isotherms has been investigated for $Ra = 10^5$. The results clearly demonstrate an evolution of fluid flow and heat transfer in the case of variable viscosity fluid.

1. Introduction

Natural convection in enclosures takes place in many areas of modern life such as cooling of electronic devices, melting of new materials, solar collectors, new heat exchangers and so on [1]. Variable physical properties in these problems (variable viscosity, conductivity) can essentially affect the fluid flow and heat transfer [2-5]. Thus, Astanina et al. [2] have numerically studied natural convection with temperature-dependent viscosity in a porous cavity. They have demonstrated that an increase in the viscosity variation parameter leads to an intensification of the convective flow and heat transfer and a formation of a single-core convective cell while in the case of the clear fluid an increase in the viscosity variation parameter leads to a decrease in the heat transfer rate. Cordoba et al. [3] have analyzed numerically and experimentally laminar natural convection with temperature-dependent viscosity in a differentially-heated cubical cavity. It has been found that numerical and experimental flow patterns illustrate a stable convective flow without any oscillations that is not expected for high Rayleigh numbers. Also the used variable viscosity leads to an essential asymmetry in the flow and heat transfer. Umavathi and Ojjela [4] have investigated the effect of variable viscosity on convective heat transfer in a vertical rectangular duct. They have found that isotherms remain linear for all values of viscosity variation parameter while velocity contours are changed. Kumar and Mahulikar [5] have studied the effects of temperature-dependent viscosity on the flow in a microchannel. They have revealed four different flow regions.

The objective of the present paper is a numerical analysis of the fluid flow and heat transfer evolution in the case of temperature-dependent viscosity and the effect of Prandtl number on heat transfer enhancement.

2. Mathematical model and numerical method

The domain of interest is schematically shown in Figure 1. A square cavity filled with a clear fluid of temperature-dependent viscosity has isothermal vertical walls with temperature T_c, adiabatic horizontal walls and a local heat source of constant temperature T_h ($T_h > T_c$). The heat source is located at the centre of the bottom wall. For the description of this problem the following assumptions have been introduced in the model: the Boussinesq approximation is valid, the fluid is Newtonian and the flow is laminar.

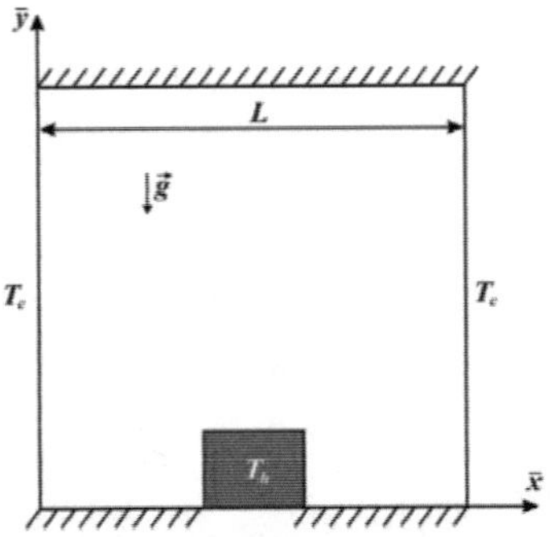

Figure 1. The domain of interest.

A mathematical model has been formulated in dimensionless variables 'stream function – vorticity – temperature' as follows:

$$\frac{\partial^2\psi}{\partial x^2}+\frac{\partial^2\psi}{\partial y^2}=-\omega, \tag{1}$$

$$\begin{gathered}\frac{\partial\omega}{\partial\tau}+\frac{\partial\psi}{\partial y}\frac{\partial\omega}{\partial x}-\frac{\partial\psi}{\partial x}\frac{\partial\omega}{\partial y}=\sqrt{\frac{\Pr}{\mathrm{Ra}}}\left(\frac{\partial^2\,\xi\omega}{\partial x^2}+\frac{\partial^2\,\xi\omega}{\partial y^2}\right)+\frac{\partial\theta}{\partial x}+\\+2\sqrt{\frac{\Pr}{\mathrm{Ra}}}\left[\frac{\partial^2\xi}{\partial x^2}\frac{\partial^2\psi}{\partial y^2}-2\frac{\partial^2\xi}{\partial x\partial y}\frac{\partial^2\psi}{\partial x\partial y}+\frac{\partial^2\xi}{\partial y^2}\frac{\partial^2\psi}{\partial x^2}\right],\end{gathered} \tag{2}$$

$$\frac{\partial\theta}{\partial\tau}+\frac{\partial\psi}{\partial y}\frac{\partial\theta}{\partial x}-\frac{\partial\psi}{\partial x}\frac{\partial\theta}{\partial y}=\frac{1}{\sqrt{\mathrm{Ra}\cdot\Pr}}\left(\frac{\partial^2\theta}{\partial x^2}+\frac{\partial^2\theta}{\partial y^2}\right). \tag{3}$$

Here x, y are the dimensionless Cartesian coordinates; ψ is the dimensionless stream function $u=\partial\psi/\partial y$, $v=-\partial\psi/\partial x$; ω is the dimensionless vorticity $\omega=\partial v/\partial x-\partial u/\partial y$; u and v are the dimensionless velocity components; τ is the dimensionless time; Ra is Rayleigh number; Pr is Prandtl number; ξ is the dimensionless variable viscosity $\xi=\exp\;-C\theta$; θ is the dimensionless temperature.

The initial and boundary conditions for the formulated governing equations (1) – (3) are as follows:

$$\tau=0:\;\;\psi=\omega=\theta=0;$$

$$\tau>0:\;\;\psi=0,\;\omega=-\frac{\partial^2\psi}{\partial x^2},\;\theta=-0.5\;\text{ on } x=0 \text{ and } x=1;$$

$$\psi=0,\;\omega=-\frac{\partial^2\psi}{\partial y^2},\;\frac{\partial\theta}{\partial y}=0\;\text{ on } y=0 \text{ and } y=1;$$

$$\psi=0,\;\omega=-\frac{\partial^2\psi}{\partial \mathbf{n}^2},\;\theta=0.5\;\text{ on heat source surface.}$$

The governing equations (1) – (3) with corresponding initial and boundary conditions have been solved using finite difference method of the second order accuracy. Detailed description of the developed numerical method has been done earlier in [2, 6]. The developed computational code has been verified for several benchmark problems [2, 6].

3. Results and discussion

The numerical analysis has been conducted for the following values of the governing parameters: Ra = 10^5, Pr = 7–700; $0 \le \tau \le 1000$; $C = 2$. Figure 2 shows the evolution of streamlines and isotherms at Pr = 7.

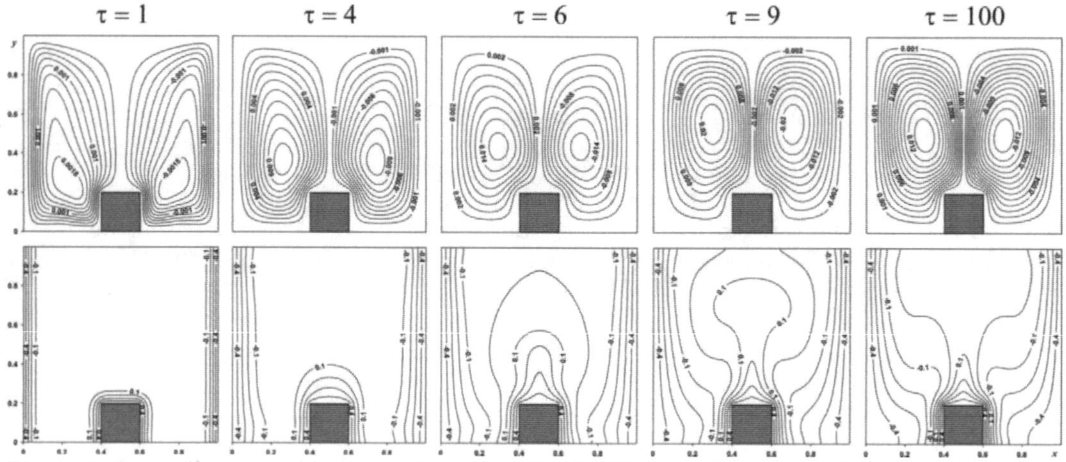

Figure 2. Streamlines ψ and isotherms θ for Pr = 7 and different values of the dimensionless time.

It should be noted that regardless of the dimensionless time values two convective cells form inside of the cavity. An appearance of these vortices is due to the presence of the centered heat source of constant temperature. These two circulations illustrate a formation of the ascending hot flow in the central part of the cavity over the heat source and two descending cold flows close to the vertical walls. At the beginning time level ($\tau = 1$) the dominating heat transfer mechanism is heat conduction. Therefore one can find that isotherms are parallel to the vertical isothermal walls and heat source surfaces. An increase in time leads to an intensification of the convective flow and a displacement of the convective cores along the positive direction of the vertical axis. It is worth noting a formation of thermal plume over the heater for $\tau \ge 4$. Also heating of the cavity leads to a non-monotonic variation of the flow rate taking into account the maximum values of $|\psi|$. Such situation can be explained by more intensive heating of the domain of interest and as a result one can find a decrease in the temperature gradient that illustrates attenuation of the convective flow. The time moment $\tau = 100$ characterizes a steady-state regime because for $\tau > 100$ there are not any variations of streamlines and isotherms.

The effect of the dimensionless time and Prandtl number on the average Nusselt number along the heat source surface is presented in Figure 3. The average Nusselt number has been defined as $Nu_{avg} = \frac{1}{3}\left[\frac{L}{h}\int_0^{h/L}\left(-\frac{\partial\Theta}{\partial X}\right)_{X=L-l/2L} dY + \frac{L}{l}\int_{L-l/2L}^{L+l/2L}\left(-\frac{\partial\Theta}{\partial Y}\right)_{Y=h/L} dX + \frac{L}{h}\int_0^{h/L}\left(-\frac{\partial\Theta}{\partial X}\right)_{X=L+l/2L} dY\right]$, where h is a height of the heat source, l is a length of the heat source.

Taking into account Figure 3 one can find that an increase in Prandtl number leads to both weak intensification of the heat transfer at a steady-state regime and an increase in the time for reaching the steady-state regime.

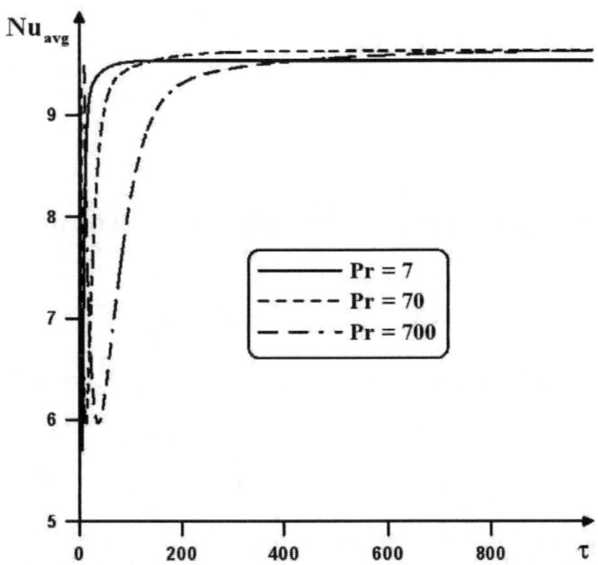

Figure 3. Variation of the average Nusselt number along the heat source surface with dimensionless time and Prandtl number.

4. Conclusion

The numerical analysis of natural convection with temperature-dependent viscosity in a square cavity having isothermal vertical walls and a heater located on the bottom wall has been carried out. The evolution of the streamlines and isotherms has been studied for Ra = 10^5, Pr = 7 and C = 2. It has been found that an increase in dimensionless time leads to an intensification of convective flow and displacement of the convective cores along the vertical axis. This mentioned behavior occurs during the first time level. After that one can find an attenuation of the convective flow due to a decrease in the temperature gradient. In addition, an increase in Prandtl number illustrates the convective heat transfer enhancement and an increase in the time for reaching the steady-state regime.

Acknowledgments

This work was supported by the Grants Council (under the President of the Russian Federation), Grant No. MD-6942.2015.8 and by the Russian Foundation for Basic Research, Grant No. 14-08-31137 mol_a.

References

[1] Jaluria Y 1998 *Design and Optimization of Thermal Systems* (New York: McGraw-Hill)
[2] Astanina M S, Sheremet M A, Umavathi J C 2015 *Transport in Porous Media* **110** 113–126
[3] Cordoba P A, Silin N, Dari E A 2015 *Int. J. Thermal Sciences* **98** 255–265
[4] Umavathi J C and Ojjela O 2015 *Int. J. Heat Mass Transfer* **84** 1–15
[5] Kumar R and Mahulikar S P 2015 *Int. J. Thermal Sciences* **98** 179–191
[6] Martyushev S G and Sheremet M A 2014 *Int. J. Thermal Sciences* **76** 51–67

IOP Conf. Series: Materials Science and Engineering **124** (2016) 012040 doi:10.1088/1757-899X/124/1/012040

Issues of a Computer-Aided Design of Hydraulic Jacks

V I Averchenkov, A V Averchenkov, V V Kolyakinand and O D Orekhov

Bryansk State Technical University, 7, 50years of October st., Bryansk, 241035, Russia

E-mail: aver@tu-bryansk.ru

Abstract. The article deals with the issues of a computer-aided design of hydraulic equipment, namely hydraulic jacks. Design principles of the hydraulic jack CAD system are described. In addition, the possibilities for the system improvement and expansion are considered.

1. Introduction.

Designing hydraulic equipment, namely hydraulic jacks, is a difficult and time-consuming task requiring broad experience of design engineers in the sphere of hydraulic equipment design and production and cooperative effort of a team of engineers, which is extremely impracticable for small enterprises [4].

Although most small enterprises employ CAD systems for jack design, these systems are only used for 3D modeling and preparing design documentation, all the rest is done by engineers without using automated systems [5].

That is why it is essential to develop systems making the whole jack designing process automated.

For automatic implementation of fully or partially formalized tasks it is not at all difficult to create a software aimed at solving not only design preproduction problems but in some cases the problems of manufacturing preparation [2, 6].

2. CAD Principles for Hydraulic Jacks Design

A hydraulic Jacks CAD system simplifies engineer's work. The system provides for calculation of hydraulic jack parameters and automatic construction of 3D model, creation of associative drawings and other required documentation on its basis [2, 4,10].

When developing the hydraulic jacks CAD system a collaborative analysis of requirements and design was carried out:

D=<Zi,Ij,Pk,Hn,Ms> (Figure 1).

Let rs consider an example: *D* = < machine tool building, spring return, 5…10 ton, 50…100 mm, steel>, these requirements are met by several types of jacks, so they are ranked against three criteria: cost, production time and service life.

The developed CAD system is a software product capable of performing a part of the design algorithm independently of other software such as a basic system of geometric modeling used for creating 3D models of parts and hydraulic jack assembly.

Keeping in mind the structure of the hydraulic jack several variants of its assembly units were specified: a connection point of the bearing box and barrel, a piston and a rod, a rod guide ring, a scraper, a spring assembly, a unit for power fluid supply and a portable unit.

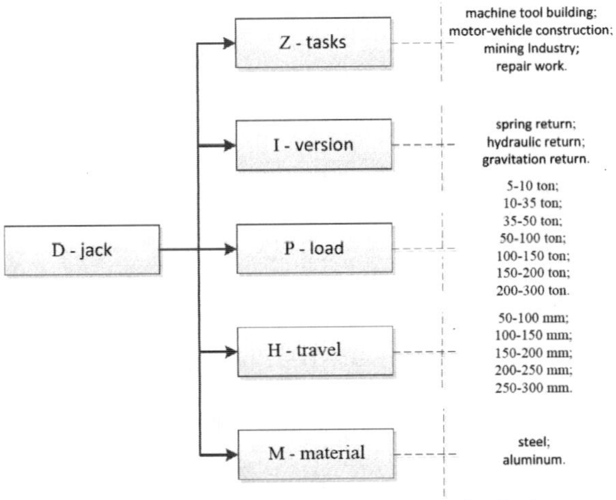

Figure 1. Structure of the hydraulic jack.

The analyzed principal models of jacks being designed can be represented by the system:

D = {Bbg,.Ppsh,.Nnsh,.Ggr,.Hpr,.Lr,.Rtr}, where

Bbg – a set of decisions of the bearing box attached to the rod;
Ppsh–a set of decisions for the piston and rod assembly;
Nnsh – a number of rod way models;
Ggr - a number of scraper models;
Hpr – a set of decisions for the spring assembly:
Lr – a number of variants of the power fluid supply;
Rtr – a number of ways of jack transportation.

From the set of hydraulic jack components containing a set of decisions one assembly variant for each unit is chosen according to the technical specification. This enables the determination of the final arrangement of the hydraulic jack.

Each element in the design scheme has several design solutions, which are stored in the database in the form of parameter-oriented models. A final decision is made on the basis of the construction design.

Having studied the possible solutions in the hydraulic jack CAD system 8 possible modules were developed:

- a module of calculating the basic jack parameters;
- a module of selecting the sealing;
- a module of bearing box calculation;
- a module of piston calculation;
- a module of spring calculation;
- a module of spring locators calculation;
- a module of cylinder tube calculation;
- a module of calculation of standard jack components [1].

In the module of seal selecting, the program gives the engineer a choice of seals suitable for this design; this option gives the engineer a possibility to program the seals which are available (Figure 2).

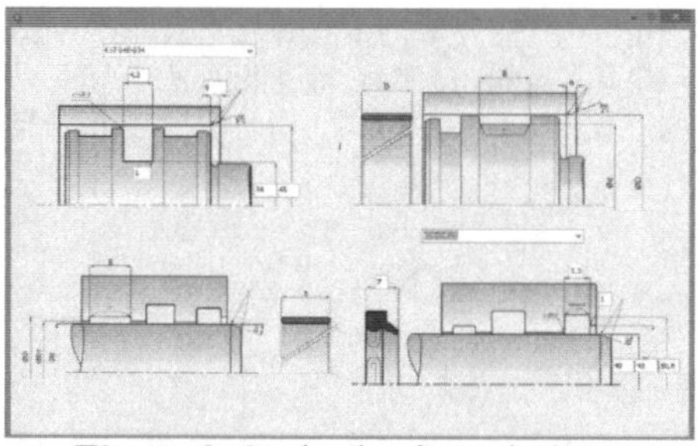

Figure 2. Seal selecting window.

For illustrative purposes when selecting from the available seals the principle parameters are shown. After selecting all suitable seals the engineer proceeds with the calculation of jack principle components. All principle jack components – a bearing box, a piston, a spring, a barrel, a bottom, fixtures and springs – are calculated (Figure 3).

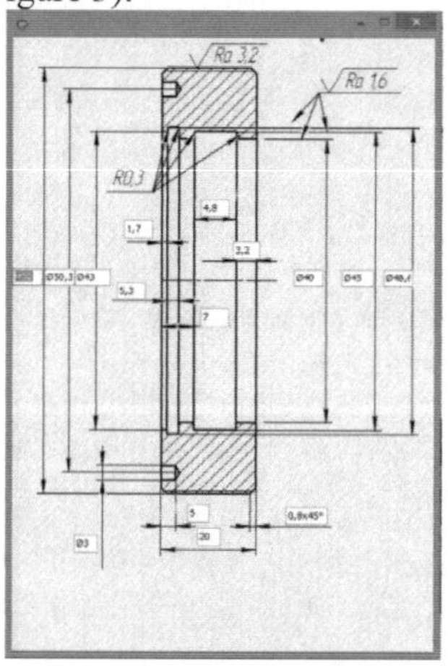

Figure 3. Sketch of the Designed Object.

After calculating all principle jack components the engineer is offered to create a 3D assembly model of the jack and all its components; using the model it is possible to make assembly drawings of the jack and all its components. Also the program helps the engineer to make a report containing all calculation parameters and to prepare a datasheet of the obtained jack.

When developing the CAD system mathematical support for the hydraulic jack design the following calculation algorithms were carried out:

- overall calculation of the hydraulic jack main geometric parameters;
- calculation of the bearing box geometry;
- piston geometry calculation;
- suitable seal selection;
- spring selection and its geometry calculation;
- spring locators geometry calculation;
- barrel geometry calculation and, if possible, selecting a suitable workpiece;
- male and female metric thread parameters calculation;
- selecting parameters of male and female threading relief grooves;
- engineering calculation of thread shear and collapse of screw fillet, etcetera [1].

One of the modules of the hydraulic jack CAD system is a module of calculation of hydraulic jack main geometric parameters.

At the first stage the piston diameters are calculated using the formula (1).

$$D=2\cdot\sqrt{\frac{P}{\pi\cdot\eta m(Pc-\frac{Ph}{\psi})}}, \quad (1)$$

where D – jack piston diameter (mm);

P – rod force (N);

ηm – jack mechanical efficiency, which is in the range of 0.93–0.97 and depends on the jack diameter and the type of sealing;

Pc – rod end pressure (Pa);

Ph– rod end pressure (Pa);

Ψ – intensifier ratio.

The calculated value of piston diameter D is rounded up to the nearest whole number according to GOST 12447-80 (Table 1).

Table 1. Jack piston diameter values regulated by GOST 12447-80.

Main variety, mm	10, 16, 20, 25, 32, 40, 50, 65, 80, 100, 125, 160, 200, 250, 320, 400, 500, 630, 800
Additional variety, mm	14, 18, 22, 28, 36, 45, 50, 70, 90, 110, 140, 180, 220, 360, 450, 560, 710, 930

Jack rod diameter d is calculated using the formula (2):

$$d=D\cdot\sqrt{1-\frac{1}{\psi}} \quad (2)$$

The value of rod diameter d is rounded up to the nearest whole number according to GOST 12447-80. Having the values of the piston and the rod we can calculate the jack wall thickness using formula (3):

$$\delta=\frac{D}{2}\cdot\left[\sqrt{\frac{[\sigma]+Py\cdot(1-2\mu)}{[\sigma]-Py\cdot(1-2\mu)}}-1\right]\cdot n, \quad (3)$$

where – δ– jack wall thickness;

[σ] – yield point;

Py – nominal pressure equal to 1,3*P;

μ – Poisson's ratio;

n – strength factor.

The values of [σ] and μ depend on a cylinder material [4].

One of the most resource-intensive algorithms is the algorithm of selecting the extension spring (Figure 4). This algorithm is based on the extension spring calculation method; the suggested approach reduces the spring assembly design time by several times.

The algorithm starting point is determining a spring type. Then the database is addressed and suitable springs are built into an array.

Then the user selects a suitable spring on the basis of the empiric data and calculatesthe spring working deformation strain.

The next stage is calculation of the number of spring coils. First, the number of active spring coils is calculated and an inspection is carried out. Based on the inspection results either an inability message of calculating active spring coils is produced or a transition to the next stage is made.

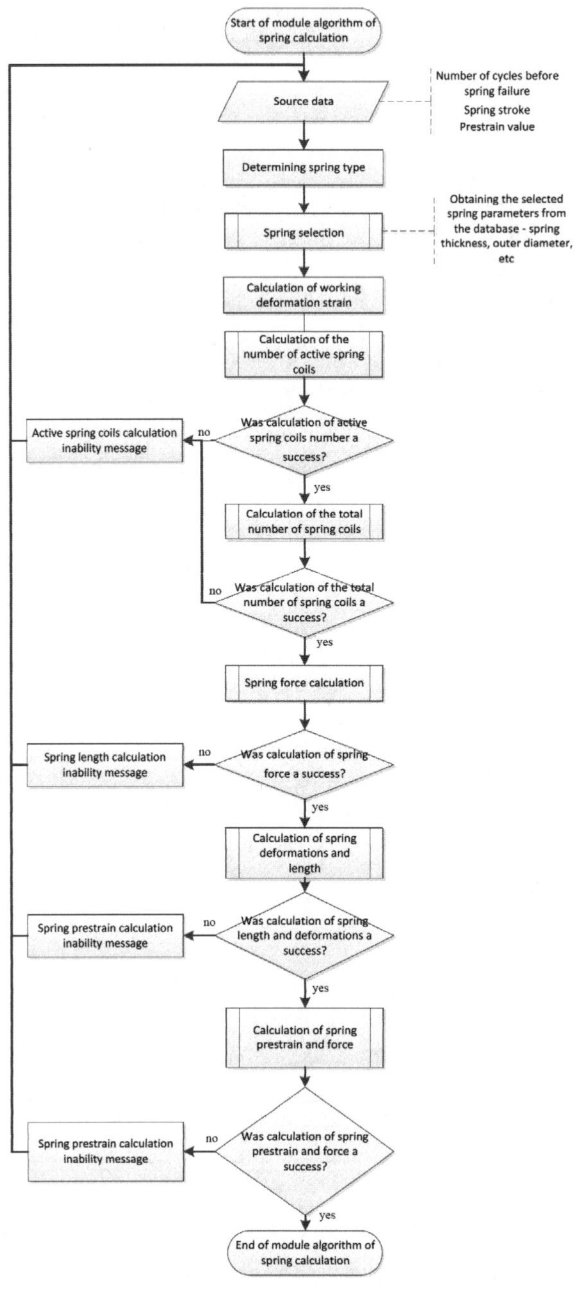

Figure 4. Algorithm of Selecting Extension Spring.

If calculation of the number of active spring coils is a success, the calculation of the total number of spring coils is carried out and its accuracy is then verified. If the total coils number was successfully calculated, a transition to the next stage is made; otherwise calculation inability message is produced.

The next stage is spring force calculation and after that a verification of its results is performed. If spring force was successfully calculated, a transition to the next stage is made; other wise spring force calculation inability message is produced.

After that calculation of spring deformations and length with a subsequent verification of results is carried out. If calculation was not performed successfully, a spring deformation calculation inability message is produced; otherwise a transition to the next stage is made.

The final stage of spring calculation is calculation of spring prestrain and force. If this calculation is not a success, an error message is produced, otherwise the spring is considered to be designed.

3. Changes to the Design Process of Hydraulic Jacks after CAD Implementation.

Before the development and implementation of the hydraulic jacks CAD system design engineers used CAD merely for preparing the design documentation. The algorithm of jack parameters calculation was performed manually using a great number of standards, recommendations and catalogues of standard articles (Figure 5).

The design engineer had to manually calculate a great number of parameters, many of which were to be then selected from the standards.

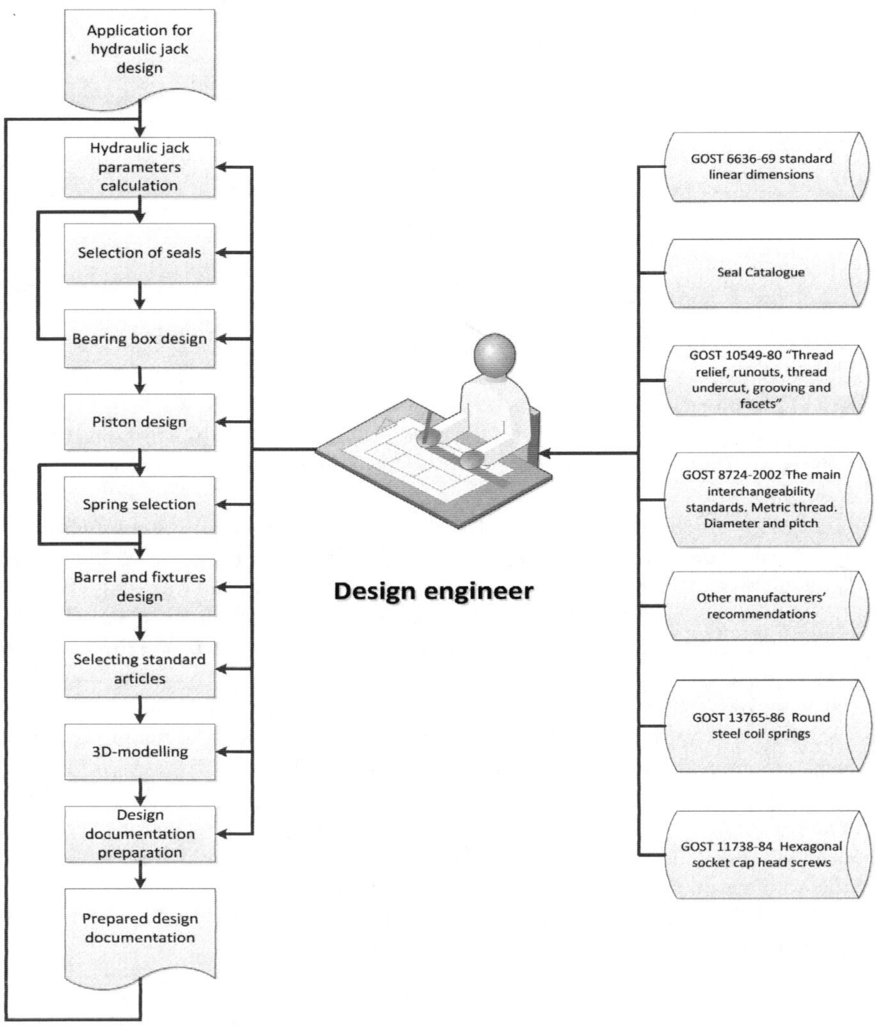

Figure 5. Process of Hydraulic Jack Manual Design.

After implementing the hydraulic jacks CAD system all these calculations and all data handling are performed purely by the CAD system and one design engineer is able to operate it, responding to its questions and finalizing the design documentation (Figure 6).

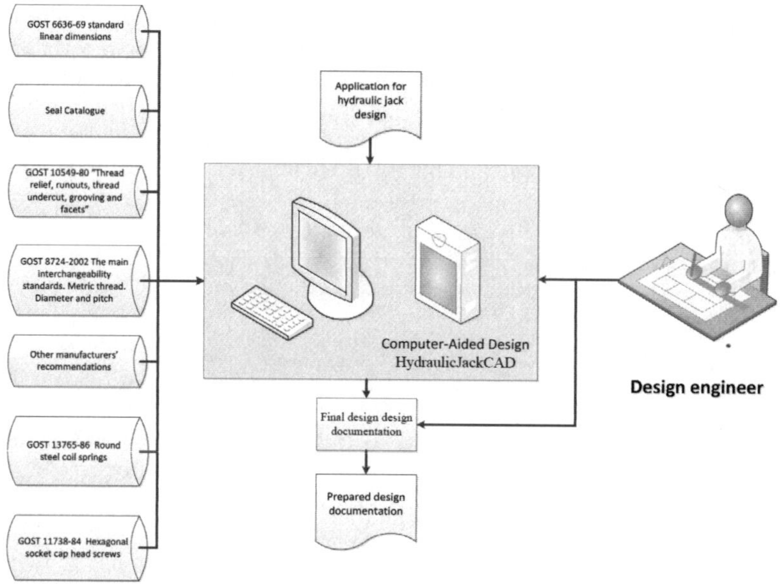

Figure 6. Process of Hydraulic Jack Computer-Aided Design.

4. Results and Further Development

Nowadays the developed CAD system for designing hydraulic jacks with spring return has been tested within the actual projects in the small enterprise and is ready for implementation in other enterprises and further expansion.

The improvement of the hydraulic jack CAD system includes adding new modules of hydraulic jack parameters calculation. This will enable widening of a product range of hydraulic jacks.

Later it is planned to improve the program functionality by adding new types of hydraulic equipment such as a pumped storage plant. At present all calculations necessary for its design have already been structured, and a necessary amount of parameter-oriented models of components of the pumped storage plant have been prepared.

In the process of automation of the pumped storage plant design it is planned to develop the following: a pumped storage plant calculation technique, a library of the main geometric elements of the plant and a selection algorithm of the optimal plant design.

5. Conclusion

When developing the CAD system all its stages were thoroughly elaborated beginning with its technical specification and finishing with its testing and implementation.

The developed CAD system enables the reduction of jacks design time, the period of design documentation preparation as well as other design process costs.

References

[1] Terekhov M V, Kolyakin V V, Leonov J A 2014 *Izvestiya Vyzov. Povolzhsky region. Technical Sciences* **4(32)**

[2] Norenkov I P 2002 *M.: Publishing House MSTU named after Bauman* 336

[3] Kapustin N M 2004 *Moskva: Vysshaya Shkola* 415
[4] Marytov V A 1996 *M.: Machine Building* 191
[5] Eksner H 2003 *Bosch Reksrot A T* 323
[6] Averchenkov A V 2012 *M.* 124
[7] Averchenkov V I 2014 *Btryansk: BSTU* 200
[8] Ustinov A V, Tveritin A L, Bolodurin D V 2015 *Power Technol. and Engineer.* **49(4)** 237–239
[9] Domozhirov L 2015 P*ower Technology and Engineering* **49(3)** 159–164
[10] Ivanov A S, Smirnov N I, Murkin S V, Ermolaev M M 2015 *Rus. Engineer. Resear.* **35(8)** 571–574
[11] Valiev R A, Khairullin A Kh, Shibakov V G 2015 *Rus. Engineer. Resear.* **35(9)** 662–665

Nonlinear dynamics control in single-phase inverter with sinusoidal pulse-width modulation

A I Andriyanov and D Yu Mikhal'tsov
Department of Electronics, Radioelectronic and Electrotechnical Systems, Bryansk State Technical University, 7, 50 years of October st., Bryansk, 241035, Russia

E-mail: mail@ahaos.ru

Abstract. A variant of technical implementation of the control system of a single-phase voltage inverter with sinusoidal pulse-width modulation, based on target-oriented control, which ensures the desired nonlinear dynamic properties of the system, is proposed. The control system under discussion solves the problem of providing a sinusoidal output voltage of the inverter when changing its parameters in a wide range, accompanied by bifurcations. This eliminates unwanted dynamic modes without parametric synthesis of the system. The target-oriented control for managing the dynamics of nonlinear systems with sinusoidal pulse-width modulation is used for the first time and gives a number of advantages compared to other known methods.

1. Introduction

Today, single-phase bridge inverters with sinusoidal pulse-width modulation (SPWM) are widely used in machine engineering that require DC/AC conversion. Such systems include uninterruptible power supplies, passenger car power supply systems, AC welding equipment, variable speed control, induction heats, etc. [1]. The equipment discussed is provided with control loops with voltage feedback that usually have complex dynamics [2, 3]. As a result of bifurcations, when changing settings in such systems, one may experience unwanted dynamic modes, accompanied by a significant harmonic distortion of the output voltage, which causes the spread of noise to the load, current overload of the power switches, or acoustic effects during operation. To prevent such consequences, specific nonlinear dynamics control algorithms are required.

There are many reports on the control of nonlinear dynamics of discrete systems [4-9]. Two key types of control can be mentioned: linearization of the Poincare map [4] and time-delay feedback control methods [5, 6]. There is also a well-known approach based on a so-called resonant perturbation of parameters [7]. The application of above methods in power electronics has been widely reported in the scientific literature. Most of the reports are dedicated to the control of nonlinear dynamics of DC/DC converters, while SPWM converters are mentioned in a fewer number of articles [8-9]. In [8], the possibility of using a resonant perturbation of parameters in the SPWM system in order to stabilize the design dynamic mode is discussed. Here, the control action perturbation was specified as a harmonic signal. In the same report, the possibility of applying a time-delay feedback control was discussed. The application of slope compensation method, which is a version of the resonant perturbation of parameters, was discussed in [9], where the perturbation was specified as a linearly falling signal. The disadvantages of the reports include the lack of analysis of efficiency of the methods discussed when the system parameters are changed over a wide range. For example, the following parameters may vary during an actual voltage converter operation: load impedance, input voltage, or a reference signal, which causes complex dynamic modes, even when the mentioned algorithms are used. When changing system parameters, the control method parameters shall be adapted, that stipulates high demands to microcontroller resources. The linearization of the Poincare map has never been applied to SPWM systems due to the very high demands to computing resources required to calculate the corrective actions.

In [10] the so-called target-oriented control (TOC) is discussed, which allows expanding the possibilities of the design mode stabilization over a wide range of system parameters variation compared to the above mentioned methods. In [11], the possibility of using this method to control the nonlinear dynamics of a DC/DC converter is discussed, where the method has proved its efficiency.

The main objective of this report is creation of the TOC-based control system of a SPWM single-phase voltage inverter, with minimum microcontroller speed requirements, which provides the design dynamic mode over a predetermined range of system parameters variation.

For the purpose of this report, the design dynamic mode means a mode where the process period is the same as the reference sinusoidal signal period. This mode is called 1-cycle [2]. Wherein, the SPWM duty cycle over all cycle periods within the control signal period is: $0<\gamma<1$. Unwanted *m*-cycles are also possible, where the period of processes in the system is *m* times longer than that of the reference sinusoidal signal. Hereinafter, *m* indicator is called a cycle factor.

2. Target-oriented control of nonlinear dynamics

Suppose there is a discrete system described by a function of the stroboscopic map:

$$x_n = f(x_{n-1}), \tag{1}$$

where x_n - variable value in the nth iteration of the map.

According to [10], the unstable fixed point map (1) x^* can be stabilized by adding a stabilizing effect, where the stroboscopic map for the nonlinear dynamic system is as follows:

$$x_n = f(x_{n-1}) + c(x^* - f(x_{n-1})), \tag{2}$$

where c - adjustable factor. According to (2), when the system is in the steady mode, the second term vanishes.

The main problem of this method is the need to precalculate the stabilized fixed point, which, in case of higher order maps, is complicated and involves high computational cost.

3. Single-phase voltage inverter control system

The functional diagram of the SPWM single-phase voltage inverter of the TOC-based control system is shown in Figure 1.

Following designations are used here: R - inductor resistance; L - inductor inductance; C - capacitor capacitance, R_L - load resistance, *VT*1-*VT*4 - power switches; *VD*1-*VD*4 - power diodes; U_{in} - input voltage; α - proportional controller gain; β, β_1, β_2 - feedback loop gains; U_{ref} - reference signal; U_p - power switch control pulses; U_{con} - control signal; U_r - ramp signal; *MCS* - main control system; *ACS* - auxiliary control system; *MC* - master clock; *RG* - ramp generator; *C* – comparator; *SH*, *SH*1, *SH*2 - sample & hold circuits; *NOT* – NOT-gate; *FPC*1 and *FPC*2 - fixed point maps computer; *SB*1 and *SB*2 - subtracters; *LRC* - load resistance computer; K_1, K_2 - gains; U_{err} - *MCS* error; x_{1kref}, x_{2kref} - components of 1-cycle fixed point vector; x_{1clk}, x_{2clk} - components of the feedback vector in terms of state variables in stroboscopic time points; u_{1k}, u_{2k} - components of the control actions vector.

The control system (*CS*) shown in Figure 1 can be divided into two subsystems. The main control subsystem provides stabilization of the average value of the output voltage without regard to nonlinear dynamic properties. The auxiliary control subsystem provides the desired stabilization of the design dynamic mode (1-cycle).

Figure 2 shows signal waveforms when using SPWM of the first kind, applied in microprocessor-based control systems [2]. Designations in Figure 2 correspond to those in Figure 1.

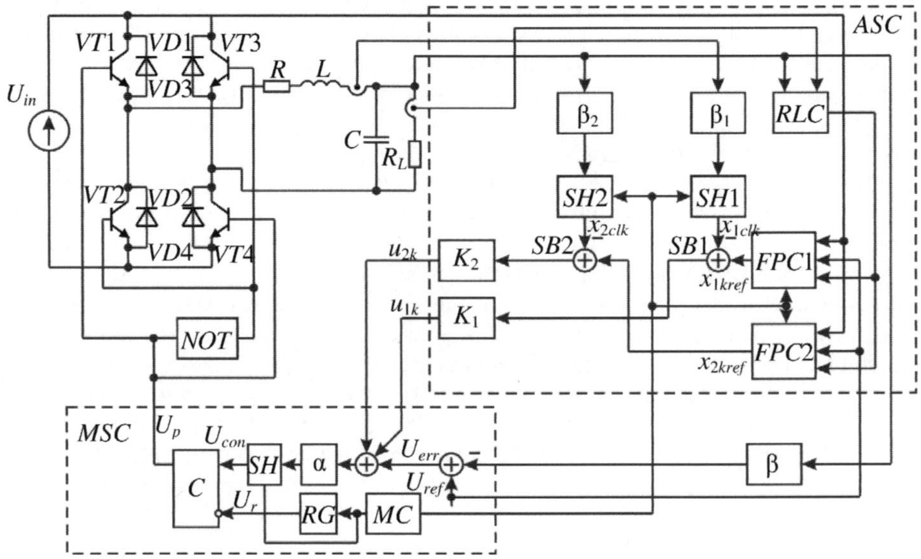

Figure 1. SPWM voltage converter control system

The cycle period is divided into two segments of structure constancy of the power unit (Fig. 2):

1) Segment 1: $0<z<z_k$. In this segment, the power switches, *VT*1 and *VT*4, are closed and positive voltage (U_{fin} in Fig. 2) is supplied to the filter input;

2) Segment 2: $z_k<z<1$. In this segment, the power switches, *VT*2 and *VT*3, are closed and negative voltage is supplied to the filter input.

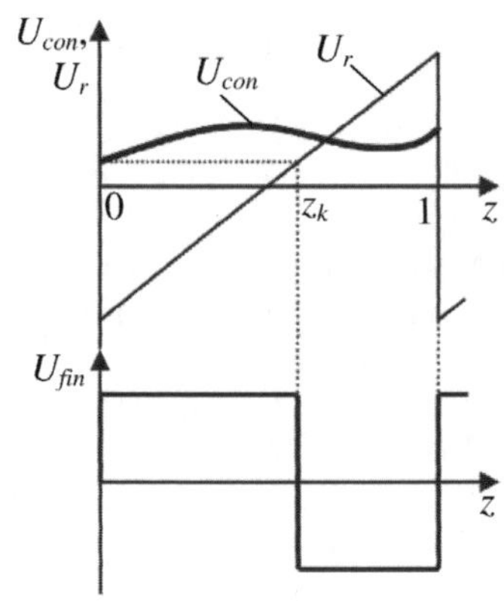

Figure 2. SPWM voltage converter waveforms

The point of switching in the relative time in the kth cycle period is calculated as

$$z_k = \frac{t_k - (k-1)T}{T},$$

where t_k - point of switching in absolute time, T - modulation period, k - number of the cycle period.

At each segment of structure constancy of the power unit, the movements are described by a linear differential equations system that has the following matrix form:

$$\frac{d\mathbf{X}}{dt} = \mathbf{A}_i\mathbf{X} + \mathbf{B}_i, \tag{3}$$

where $\mathbf{A}_i$ - matrix of constant factors in the ith segment, $\mathbf{B}_i$ - forcing vector in the ith segment, i = 1, 2, $\mathbf{A}_1 = \begin{bmatrix} -\frac{R}{L} & -\frac{1}{L} \\ \frac{1}{C} & -\frac{1}{CR_L} \end{bmatrix}$, $\mathbf{B}_1 = \begin{bmatrix} U_{in}/L \\ 0 \end{bmatrix}$, $\mathbf{A}_2 = \mathbf{A}_1 = \mathbf{A}$, $\mathbf{B}_2 = \begin{bmatrix} -U_{in}/L \\ 0 \end{bmatrix}$, $\mathbf{X}=[x_1, x_2]=[i_L, u_C]$ - state variables vector, i_L - inductor current, u_c - capacitor voltage.

The solution of the system (3) in each segment of structure constancy of the power unit is recorded in the analytical form and a segment-to-segment transition is done using the fitting method.

It can be shown that the system behavior within the kth cycle period is described by a stroboscopic map:

$$\mathbf{X}_k = \Psi(\mathbf{X}_{k-1}) = e^{\mathbf{A}a}\mathbf{X}_{k-1} + e^{\mathbf{A}(1-z_k)a}\left(e^{\mathbf{A}z_k a} - \mathbf{E}\right)\mathbf{A}^{-1}\mathbf{B}_1 + \left(e^{\mathbf{A}(1-z_k)a} - \mathbf{E}\right)\mathbf{A}^{-1}\mathbf{B}_2, \tag{4}$$

where $\mathbf{X}_{k-1}$ - state variables vector at the beginning of the kth cycle period.

When using SPWM, the sinusoidal control action period contains q cycle periods, where q is called the quantization factor. Suppose that q is an integer. The stroboscopic map for the SPWM system can then be represented as

$$\mathbf{X}_p = \Psi^{(q)}(\mathbf{X}_{p-1}) \equiv \underbrace{\Psi \circ \Psi \circ \Psi \circ \Psi \circ \ldots \circ \Psi}_{q \text{ times}}, \tag{5}$$

where p - map iteration number.

Point of switching z_k in the kth cycle period can be calculated using the switching manifold equation $\xi(\mathbf{X},z_k)=0$ [2]

$$\xi(\mathbf{X}, z_k) = \alpha\left(U_{ref}(k) - \beta u_c(k)\right) - U_{rgm}(2z_k - 1),$$

where $u_c(k)$ - capacitor voltage at the beginning of the kth cycle interval, U_{rgm} - ramp voltage amplitude, $U_{ref}(k)=U_{rm}\sin(\omega_s(k\text{-}1)T)$, U_{rm} - reference signal amplitude, ω_s – control signal frequency.

When using TOC, we will introduce additional control actions u_{1k} and u_{2k}, which are calculated as follows:

$$u_{1k} = K_1\left(x_{1kref} - x_{1clk}\right);\ u_{2k} = K_2\left(x_{2kref} - x_{2clk}\right).$$

Taking into consideration the above mentioned, the function of the stroboscopic map of the TOC system (3) is as follows:

$$\mathbf{X}_k = \Psi(\mathbf{X}_{k-1}) = e^{\mathbf{A}a}\mathbf{X}_{k-1} + e^{\mathbf{A}(1-(z_k+\Delta z_k))a}\left(e^{\mathbf{A}(z_k+\Delta z_k)a} - \mathbf{E}\right)\mathbf{A}^{-1}\mathbf{B}_1 + \left(e^{\mathbf{A}(1-(z_k+\Delta z_k))a} - \mathbf{E}\right)\mathbf{A}^{-1}\mathbf{B}_2,$$

where Δz_k - duty cycle increment in the kth cycle period.

The mentioned increment can be calculated as (Figure 1)

$$\Delta z_k = \alpha\left(u_{1k} + u_{2k}\right)/U_{pm}.$$

When implementing the discussed control algorithm, the calculation of fixed points arrays of the stroboscopic map (5) is the most important task: $\mathbf{X}_{st1}=[x_{11ref}, x_{12ref}, x_{13ref}, \ldots x_{1qref}]$; $\mathbf{X}_{st2}=[x_{21ref}, x_{22ref}, x_{23ref}, \ldots x_{2qref}]$. The first array is a set of inductor currents (x_{1kref} in Figure 1) over the reference signal period; the second array is a set of capacitor voltages (x_{2kref}). The size of each array is equal to quantization factor q.

To calculate the arrays we need to calculate a fixed map point (5), for which purpose the following equation can be used:

$$\Psi^{(q)}(\mathbf{X})-\mathbf{X}=0, \tag{6}$$

where the components of vector $\mathbf{X}^*=[x_{11ref}, x_{21ref}]$ correspond to the first components in arrays $\mathbf{X}_{st1}$ and $\mathbf{X}_{st2}$. The remaining components of arrays $\mathbf{X}_{st1}$ and $\mathbf{X}_{st2}$ can be calculated using the recurrence formula (4).

The transcendental equation (6) can be solved using Newton's method, which requires significant computing power of the microcontroller. To simplify the task in the microcontroller program, we suggest using two neural networks, each of which calculates its own component of the fixed point of the 1-cycle, X^*. Here, neural networks serve as universal approximators. The calculation of the learning sample and neural network teaching are performed using a support computer, for example, in MatLab. Subsequently, the calculated neuron weight factors and network biases can be used in the microcontroller's program. The values of parameters that change over a wide range during the system operation are supplied to the inputs of neural networks created, that is, U_{ref}, U_{in} and R_L, which are sensor-controlled (Figure 1). *RLC* calculates current load resistance based on measured load voltage and load current. *FPC* units calculate and output the components of arrays $\mathbf{X}_{st1}$ and $\mathbf{X}_{st2}$ depending on cycle period k.

State variables feedback on the stroboscopic time points in the suggested system (Figure 1) is performed using sample & hold circuits (*SH*1 and *SH*2 in Figure 1). As Fig. 1 shows, the output capacitor C voltage and the inductor L current are memorized at the beginning of each cycle period when supplying a strobe pulse from master clock *MC*, which operates in synchrony with ramp generator *RG*, to *SH*1 and *SH*2. Using two subtracters (*SB*1 and *SB*2 in Figure 1), the deviation of the current map point position (4) from the preset value is calculated, followed by respective errors scaling using K_1 and K_2 factors. The calculated control actions, u_{1k} and u_{2k} are added to the *MCS* voltage error (U_{err}), causing duty cycle increase Δz_k (expression (5)), which stabilizes the design mode, in each cycle period. When establishing the design 1-cycle in the system, $u_{1k} = u_{2k} = 0$ and $\Delta z_k = 0$.

The suggested control system structure can be implemented using a relatively large range of modern digital signal microcontrollers or low-cost programmable logic integrated circuits.

4. SPWM voltage converter of nonlinear dynamics study

This segment contains the results of mathematical simulation of a TOC control system with a single-phase SPWM bridge inverter.

The following set of parameters were used for the simulation: L=4 mH; C=3.5 μF; R=1 Ohm; $R_н$=45 Ohm; α=6.9; β=0.075; U_{rgm}=10 V; T=0.0001 s, q=10. The results of the mathematical simulation are shown in Fig. 3 in the form of dynamic mode maps, which show special aspects of division of the system parameter space into stability areas of different modes. Here, $A_{i,j}$ indicates the stability area of a dynamic mode (i - area specific m-cycle; j - area number on the dynamic modes map). In particular, area $A_{1,1}$ is the first stability area of the design mode with frequency $f_{sq}=1/T$ (1-cycle). Areas $A_{X,j}$ correspond to the areas of system parameters, where chaotic inverter modes exist [2, 3].

As Figure 3 (a, b) shows, the area of the design 1-cycle area ($A_{1,1}$) of a system with no nonlinear dynamics control is relatively small.

In case of large amplitude of the reference signal, an unwanted 1-cycle ($A_{1,2}$) is implemented in the system with overmodulation in multiple cycle periods. At high input voltages, the area of quasi-periodic oscillations with periodicity areas can be seen (Figure 3a). Figure 3b is a diagram of the RMS value of parasitic harmonics with numbers other than 1 (U_{pg}). Obviously, at high input voltages, the RMS value of parasitic harmonics is large, which causes significant distortion of output voltage sinusoidality.

Fig. 4a shows the dynamic modes map of the TOC-based automatic control system. In such case, the following *AGS* (as shown in Figure 1) parameters were used: K_1=K_2=-1.5, β_1=β_2=0.05. The area of

IOP Conf. Series: Materials Science and Engineering **124** (2016) 012041 doi:10.1088/1757-899X/124/1/012041

the designed 1-cycle area ($A_{1,1}$) (Figure 4a) is significantly larger compared to the corresponding area in the system with no nonlinear dynamics control (Figure 3a). Also, as Figure 4b shows, the RMS value of parasitic harmonics significantly reduces at large input voltages.

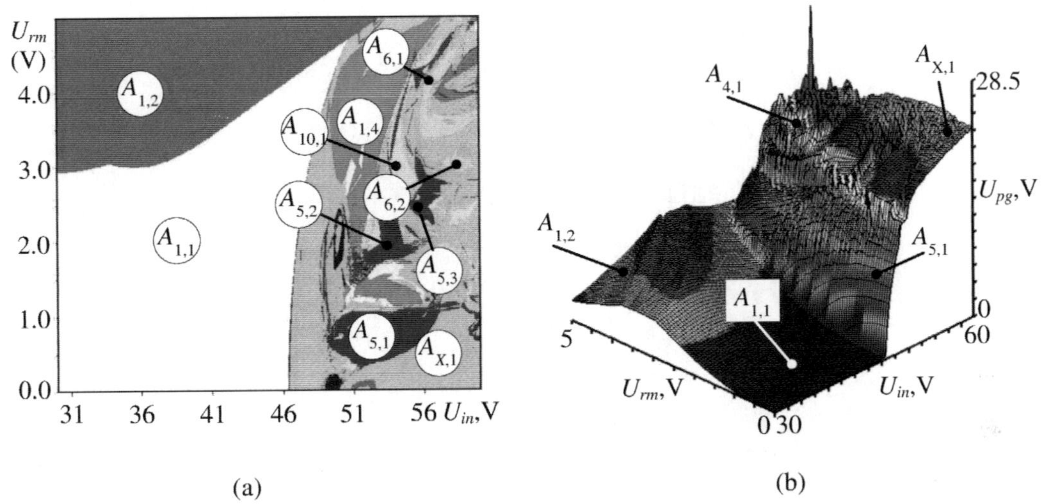

Figure 3. Two-parameter diagrams: (a) dynamic modes map, (b) relative RMS value of parasitic harmonics

As Figure 4a represents, the map still has an area of unwanted 1-cycle $A_{1,2}$ with overmodulation in some cycle periods, making it impossible to eliminate such mode.

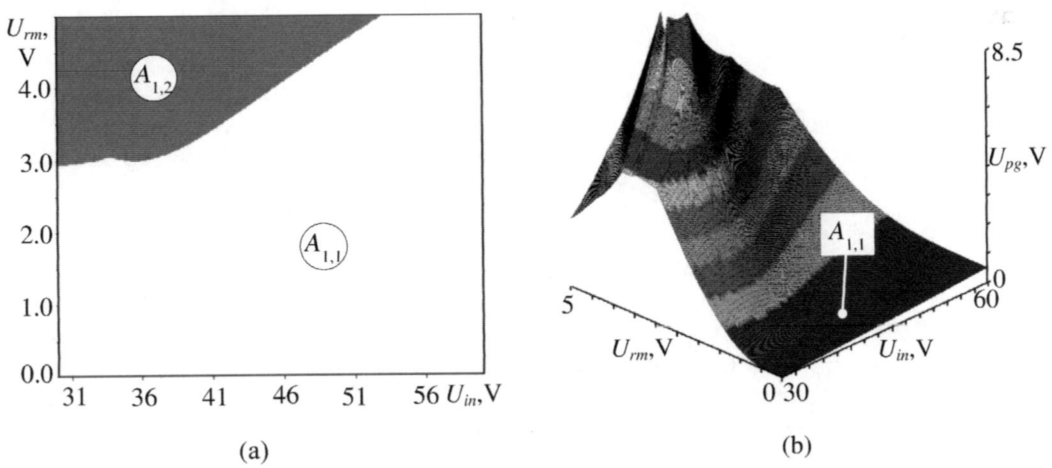

Figure 4. Two-parameter diagrams: (a) dynamic modes map, (b) relative RMS value of parasitic harmonics of the output voltage

Figure 5 is a waveform showing the transition from the unwanted dynamic mode to the design 1-cycle using the discussed control algorithm. The control algorithm starts to operate at t=0.15 s. As the Figure shows, the output voltage oscillation amplitude in the design mode is significantly lower than in the unwanted mode, which indicates the hazard of the latter.

Thus, it can be concluded that the use of TOC has significantly improved the nonlinear dynamic properties of the system, wherein the proportional controller gain remained unchanged, allowing, in this case, keeping a predetermined static error.

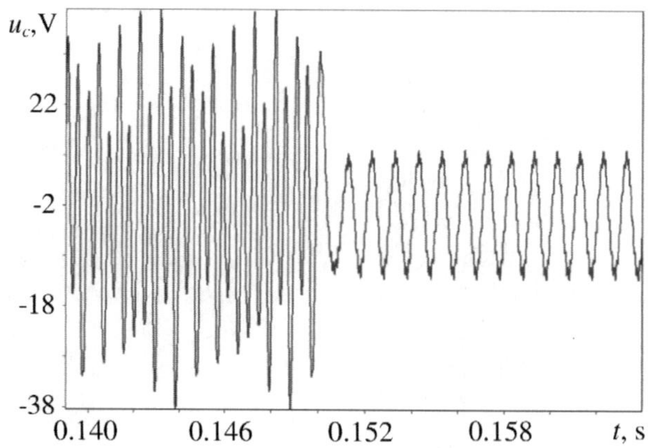

Figure 5. Waveforms at U_{in} = 55 V and U_{rm} = 1 V

5. Conclusion

The target-oriented control of nonlinear dynamics of the SPWM automatic control system has been discussed. The method has been applied to control the SPWM system for the first time. The results show that when using the selected parameters, the TOC system is able to significantly extend the range of the design dynamic mode without having to adapt the method parameters. This control algorithm requires significant computational resources of the microcontroller and can be used for controlling the nonlinear dynamics of inverters with other types of SPWM, with a minor modification stroboscopic map.

References

[1] Ismail B, Mieee S T, Mohd A R, Saad M I and Hadzer C M 2006 *First International Power and Energy. Conference Pecon 2006 November 28-29, 2006, Putrajaya, and Malasia* 437–440
[2] Zhusubaliyev Zh T, Mosekilde E, Andriyanov A I and Shein V V 2014 *Physica D: Nonlinear Phenomena* **268** 14-24
[3] Ming Li,·Dong Dai and Xikui Ma 2008 *Circuits Syst Signal Process* **27** 811–831
[4] Ott E, Grebogi C., Yourke G. 1990 *Phys. Rev. Lett* **64** 1796–1199
[5] Pyragas K 1992 *Phys. Rev. Lett.* A **170** 421–428
[6] Batlle C, Fossas E and Olivar G 1999 *International Journal of Circuit Theory and Applications* **27** 617–631
[7] Zhou Yufei, Tse Chi K, Qiu Shui-Sheng and Lau Francis C M 2003 *International Journal of Bifurcation and Chaos* **13** 3459–3471
[8] Wei Jiang, Fang Yuan and Wen-long Hu 2011 *Applied Mechanics and Materials* **39** 1–6
[9] Naihong Hu, Yufei Zhou and Junning Chen 2012 *Journal of Information & Computational Science* 9 **2** 497–504
[10] Dattani J, Blake J C H and Hilker F M 2011 *Phys. Lett.* A **375** 3986–3992
[11] Andriyanov A I and Krasnov N I 2013 *Izvestiya vysshikh uchebnykh zavedeniy. Priborostroenie* **56** 33–38
[12] Haykin S 2009 *Neural Networks and Learning Machines.* (New Jersey: Prentice Hall)

Activation of mass transfer processes at spark plasma sintering of zirconium dioxide

S A Akarachkin[1], A S Ivashutenko[2] and N V Martyushev[2]
[1] 'SIC 'Polus' JSC, 56B, Kirova ave., Tomsk, 634050, Russia
[2] National Research Tomsk Polytechnic University, 30, Lenina ave., Tomsk, 634050, Russia

E-mail: ivashutenko@tpu.ru

Abstract. The paper presents the results of numerical simulation of thermal and electric fields' distribution in the graphite moulding tool and in the sintered sample of ZrO_2-4%Y_2O_3 in the course of spark plasma sintering (SPS). The reduction of SPS duration is accounted for the largeness of specific thermal flux towards the sample surface, emitted by the graphite moulding tool. The impact of the electric field on the sample structure leads to emergence of the polarizing processes forcing zirconium ions to shift from lattice sites, which is able to reduce the required value of thermal energy necessary for initiation of a diffusion process. The axial pressure at high temperatures of sintering can lead to plastic deformation of the powder particles.

1. Introduction

Engineering ceramics has found its wide application in many branches of industry such as nuclear, chemical, aerospace and in a number of others. The manufacturing procedure of engineering ceramics has been constantly improved towards reduction of expenditures for production unit manufacturing. Of no small importance task is enhancement of service properties of production in order to increase its service life period.

For solution of the set tasks a multiple of technological solutions has been developed. For the purpose of obtaining ceramics with a fine-grained structure the nanopowders with subsequent methods of high-energy influence are applied, for example, impulse pressing [1-3], sintering in glow-discharge plasma [4], sintering in the electric microwave field [5] or in charge-particle beam [6].

One of the traditional methods of activated sintering, which reduces the time of heat-treatment, is hot pressing [7]. The reduction of the sintering time allows decelerating the recrystallization process, which enables obtaining a more refined structure and, consequently, enhanced service properties.

A hot-pressing method has obtained further development. In the literature there are several different names of the new method: sintering assisted by electric field (FAST), sintering activated by electric current (ECAS), sintering by pulse current (PECS) [7-9]. However, the producers of the manufacturing equipment promote a market name of the technology, which is 'Sparkle plasma sintering' (SPS) [7].

Sparkle plasma sintering (SPS) is a method of densification of a porous body with the use of the external pressure as well as thermal and electric fields. There are several theories about the mass transfer mechanism at SPS: plastic deformation, electroplastic deformation, electromigration, local temperature gradient at points of sparkle discharge between particles [8, 10].

Depending on the ratio of electric conductivity of the sintered material and the material of the mould, different of the above-mentioned phenomena can reveal themselves.

The method of hot pressing and SPS are quite similar. In both cases the mould with punches are used. However, SPS and hot pressing have a significant difference in the heating mode. In the installations of hot pressing the array of heating elements is used, which heats punches and matrix (hereafter referred to as 'pressing-tool') as well as the sintered material with heat radiation, convection and/or due to heat conductivity of the latter. In case of SPS the electric charge and Joule heat are directly transferred to the sintered material. As far as electric current, passing through a graphite pressing-tool, can reach high density, the heating rate of the material can reach 10^6 °C /s. The heating rate at hot pressing is much lower and amounts to about 80 °C/min [7].

Many, especially earlier studies, explained the SPS efficiency by creation of discharge arc between the powder particles or plasma formation around particles [7].

In early methods PS plasma could actually form due to applied high voltage to a sample, but in modern installations with conducting graphite matrices and low operating voltage, plasma formation is unlikely [7].

Active studies with application of atomic emission electroscopy, direct observation and high-speed voltage measurement did not reveal plasma presence at SPS. Due to the fact that the heating of the pressing-tool and the sintered material at SPS is a consequence of electric current flow, it becomes quite difficult to differentiate the effects caused by high temperature, current and voltage effects. There were assumptions that the applied electric field increases mobility and concentration of defects in the grain boundary structure and influences diffusion assisted energy. However, it has been shown that electromigration is not a dominating mechanism of diffusion. Theoretical calculations have shown that electromigration can exert significant influence only in the systems with low porosity or in ultrafine powders.

Earlier the authors demonstrated the possibility of sintering of zirconium dioxide powder due to formation of local temperature gradients in the area of sparkle discharges between the particles. SPS methodology assumed the use of high-tension pulse voltage source.

2. Materials and methods

At present the installations for SPS, using conducting graphite matrices and pulsed direct current sources, have gained the widest spread [7]. Therefore, the authors set an urgent task to study the processes occurring at ceramic powders sintering at this kind of installations.

In the experiments on sintering stabilized zirconium dioxide containing 96 mas. % ZrO_2 + 4 mas. % Y_2O_3 the SPS installation of 'HP D 10' brand produced by FCT Systeme Gmbh Company was used. In fig. 1 the general view of the pressing-tool is shown. The powder was poured into a graphite matrix (3) with inner diameter of 20 mm. Inner surface of the matrix and butt ends of the punches (4, 5) contacting the powder (6), were preliminary protected by the graphite foil of 0.2 mm thick. Then the pressing-tool was located in the vacuum water-cooled chamber and fixed by pushers (1, 2) connected to pulse current source. The following sintering process parameters were set: the rate of temperature rise (200 °C/min), soaking time at sintering temperature (10 min.), and cooling rate, in addition the values of the residual pressure in the vacuum chamber (10 Pa) as well as the power of influence of pushers on punches of the graphite matrix (32 MPa) were indicated. After exhaustion of the water-cooled chamber the force was applied to pushers, then the pulse current generator started its operation. The duration of the rectangular current pulse was 25 ms, after which there was a 5 ms pause.

3. Results and discussion

To calculate the parameters of the thermal condition of SPS let us use the basic Fourier conduction law [9]. Since the pressing-tool (Figure 1) has an axially-symmetric shape, the heat-conducting equation can be written as follows:

$$\gamma C_P \frac{\partial T}{\partial t} = \frac{1}{r}\frac{\partial}{\partial r}\left(r\lambda_r \frac{\partial T}{\partial r} \right) + \frac{\partial}{\partial z}\left(\lambda_z \frac{\partial T}{\partial z} \right) + \dot{q}_i \tag{1}$$

where, γ, CP, λr, λz, T, t are density, specific heat, heat conductivity factor along axes r and z, temperature and time correspondingly; $\dot{q}_i$ represents the quantity of heat generated by the inner sources of the system in a unit of volume per a time unit.

Distribution of electric current is described by Kirchhoff equation [9].

$$\frac{1}{r}\frac{\partial (ri_r)}{\partial r} + \frac{\partial (i_z)}{\partial z} = 0 \tag{2}$$

where ir and iz are the current density along axes r and z.

Specific power of heat release $\dot{q}_i$ is connected with current density by Ohm's law [9].

$$\dot{q}_i = i_r \rho_r + i_z \rho_z \tag{3}$$

where ρr and ρz are electric resistance along axes r and z.

When sintering by SPS method Joule heat $\dot{q}_i$ is generated in all components of the system through which the electric current flows. The heat transfer between the parts of the system occurs mainly due to heat conductivity. The heat losses occur because of heat radiation from the graphite pressing-tool as well as heat conductivity from the pusher surfaces coming in contact with water-cooled electrodes.

Stefan-Boltzmann law describes heat transfer at heat radiation.

$$\dot{q}_r = \sigma_S \varepsilon (T_e^4 - T_a^4) \tag{4}$$

where σ_S – Stefan-Boltzmann constant, T_e – temperature of the radiating surface, T_a – temperature of the absorbing surface or environment (300 K [8, 9]), ε – relative emissivity (0.8 [8, 9]), $\dot{q}_r$ - specific power of heat loss by radiation per a time unit from the surface unit.

Heat transfer from the pressing-tool to the water-cooled electrodes is determined as a convective cooling and is described by the following expression:

$$\dot{q}_{conv} = h(T_P - T_W) \tag{5}$$

where T_P, T_W and h – temperatures of the pushers located next to the electrodes, temperature of cooling water (300 K [8, 9]) and a coefficient of the convective heat exchange (880 W/(m·°C) [8, 9]).

Modelling of the heating process was realised along with the replacement of pulse current by direct one. An essential condition of the replacement was the unity of the emitted quantity of heat in the pressing-tool and in the sintered sample per time unit. Thus the connection of the magnitude of pulse current with the equivalent value of the direct current can be described by the following expression:

$$I_C = I_P \sqrt{\tau_И / T_И} \tag{6}$$

where I_C – direct current magnitude, I_P – peak pulse current, τ_H – pulse duration, T_H – pulse repetition period.

The magnitudes of peak pulse voltage U_P and equivalent direct one U_C are related in a similar way.

The solution of the system of equations (1) - (5) in a specified geometric area (see fig. 1) was carried out in COMSOL software package. The operation of this programme is based on numerical solution of a set of differential equations in partial derivatives by the method of finite elements. Before modelling the electrical and thermal characteristics of the following materials were determined: 4Y-ZrO_2, graphite and graphite foil [8, 9].

In a computational model a porous powder body, which represents an initial sample at the beginning of sintering, has been replaced by ceramics of the corresponding composition. A similar approach has shown an excellent accuracy of temperature measurement of all the elements of the pressing-tool and the sintered sample in SPS process [8, 10].

The basic results of modelling are shown in table 1, where I_P and U_P – peak current and peak voltage of power supply, P – total amount of generated energy in volume of the pressing-tool and the sintered sample, P_S – specific heat energy, transferred from the pressing-tool to the sintered sample, T_{Smax} and T_{Smin} – maximum and minimum temperatures in the volume of the sintered sample, Q_S – specific electric charge accumulated by the sintered sample per time of one pulse, ΔU_S – electric voltage drop on the sintered sample.

Table 1 – The results of heat and electric calculations of the experimental model.

I_P, A	U_P, V	P, W	P_S, W/cm^3	T_{Smax}, °C	T_{Smin}, °C	Q_S, C/cm^3	ΔU_S, V
600.00	3.83	1911.60	2.02	734.47	729.52	$8.12 \cdot 10^{-9}$	0.86
648.33	4.11	2219.80	3.28	823.68	819.05	$1.40 \cdot 10^{-8}$	0.93
696.67	4.40	2554.00	4.85	910.78	904.66	$1.55 \cdot 10^{-8}$	1.01
745.00	4.70	2916.30	6.73	995.37	987.08	$1.71 \cdot 10^{-8}$	1.08
793.33	5.01	3309.30	8.90	1077.60	1066.60	$1.66 \cdot 10^{-8}$	1.16
841.67	5.33	3736.10	11.36	1157.80	1143.30	$1.37 \cdot 10^{-8}$	1.24
890.00	5.67	4200.40	14.07	1237.20	1217.70	$1.27 \cdot 10^{-8}$	1.32
938.33	6.02	4706.80	17.04	1315.40	1290.00	$1.36 \cdot 10^{-8}$	1.41
986.67	6.40	5264.40	20.26	1392.70	1360.70	$1.71 \cdot 10^{-8}$	1.50
1035.00	6.81	5871.80	23.61	1469.40	1430.30	$1.93 \cdot 10^{-8}$	1.59

When using SPS method a graphite pressing-tool, through which pulse current flows predominantly, is used for sintering dielectric powders by the primary source of Joule heat. The stabilised zirconium dioxide at temperatures of about 800 °C and higher possesses high ionic conductivity, that is becomes a solid electrolyte [10, 12]. The presence of vacancies in the anionic sublattice of the stabilised zirconium dioxide leads to oxygen transfer according to a vacancy mechanism. Under the influence of pulse electric field there is an oxygen ion drift in the direction of anode in the sample. This is the way the mass transfer process can be activated. Besides the growth of ionic conductivity the stabilised zirconium dioxide acquires abnormal high values of dielectric permittivity ~ $1 \cdot 10^6$ at heating. Electric charge stored in the sample volume can also influence the mobility of ions in the lattice sites.

In paper [13-16] on SPS the composition α -Al_{2-2x} Fe_{2x} O_3 (x = 0.02, 0.05, 0.07, 0.10) was used. Sintering was carried out at temperature of 1350-1450 °C. The samples had a pill-shaped form with the diameter of 8 mm and the height of 2 mm. The voltage on the pressing-tool amounted to 10-20 V.

In the surface layers of samples of 30-40 μm thickness a composite consisting of α-Fe and α-Al_2O_3 particles was found. It is noteworthy that in the cathode fall region the composite was separated from the sample bulk by the layer with the scanty content of Fe particles. At sintering temperature Fe^{3+} ions were transferred towards the cathode under influence of electric field.

On the basis of data from table 1 it is possible to suggest the predominance of thermal diffusion flows over the ion flows conditioned by the electric field influence. Per time unit the amount of specific heat energy obtained by the sample reaches ~ 24 J/cm^3. Specific energy of the electric field accumulated in the sample does not exceed ~ $1 \cdot 10^{-7}$ J/cm^3. Maximum value of electric field intensity in the sample volume is less than 0.3 V/mm. At such low electric field energy it is unlikely that polarized processes, capable of initiating the zirconium ion drift through the existing vacancies in sample lattice structure, will develop.

To evaluate the pattern of electric current distribution between the sample and the pressing-tool let us consider Figure 1. Hereinafter all graphs are presented for the case of usage of amplitude of pulse current I_P = 1035 A, see table 1.

Owing to a significant difference in specific electric conductivity electric current flows preliminary along the graphite pressing-tool. Maximum current density and correspondingly the greatest amount of the generated heat energy fall at graphite punches. Active heat irradiation takes place from the matrix surface by emission, which leads to formation of the temperature gradient between the sample centre and its peripheral area adjacent to the matrix. The magnitude of this gradient increases nonlinearly in accordance with expression (4) at the temperature increase of the pressing-tool, see table 1.

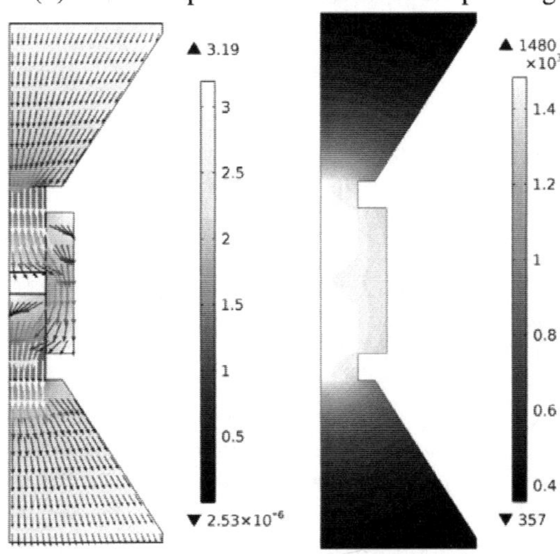

Figure 1. *a)* current density distribution in the experimental model, arrows show the direction of current flow; the arrow length and colour show the magnitude, in A/mm^2; *b)* temperature field in the experimental model, in ° C.

As the installation for SPS uses pulse current source, the process of charging the geometric capacitance at sintering of dielectric materials takes place, which can influence the law of voltage variation on the sintered sample.

For the analysis of this process let us use a simplified calculation model representing a ring segment of the matrix with a sample located inside of it. The height of the matrix segment and that of the sample is equal. The dielectric material of the sample is replaced by the parallel connected capacitor and resistor. The calculation was conducted in MATLAB/Simulink mathematical software package. The values of the numerical values used in the model were specified according to the results of modeling in COMSOL.

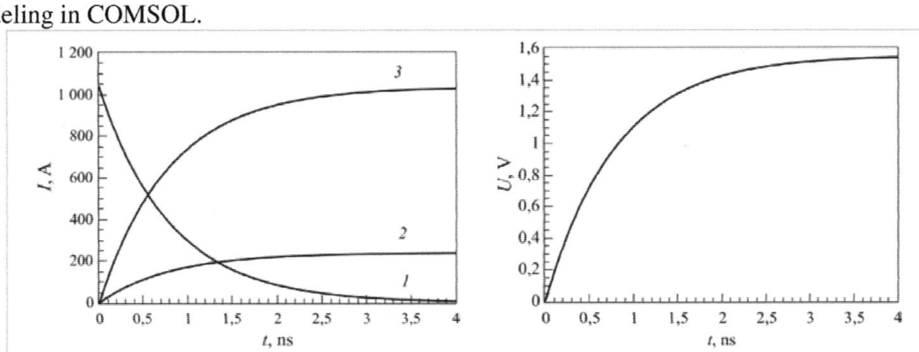

Figure 2. *a) 1* – capacitive current flowing through sample I_{Scap}, A; *2* – conduction current (leakage current) of sample $I_{SR} \cdot 10^5$, A; *3*- current through graphite matrix I_{GR}, A; *b)* voltage U_S, V on the sample.

As Figure 2 shows the conduction current magnitude passing through sample I_{SR} is much less than the magnitude of the current flowing through the graphite matrix I_{GR} (I_{SR} / I_{GR} ~$1{\cdot}10^{-6}$). The heating of the sample due to its own conductivity is not a chief source of thermal energy. At the moment of pulsing capacitive current I_{Scap} can reach peak output current of the pulsed source. In fact, the current curve of pulsed source cannot be instantly converted to maximum value because of the presence of inductivity in the circuit of the feeding transformer of SPS installation [9]. Maximum magnitude I_{Scap} will have less value. The flow of capacitive current is conditioned by the presence of polarized processes in the sample structure. Zirconium ions are displaced relatively the lattice sites. The distance of the resulted displacement is in direct proportion to the applied voltage. Thus the higher the voltage and the pulsed source current, the greater the displacement of zirconium ions. A similar stressed state in the lattice can reduce the amount of activation energy of the thermal diffusion process.

The operating voltage time on sample U_S amounts to about 4 ns. Considering the duration of the unit impulse of current, which is 25 ms, it is possible to state that the mechanical stress in the dioxide zirconium lattice is retained over the period of the whole pulse.

As it was mentioned earlier during analysis of the data from table 1 the energy of the electric field is not enough for formation of independent diffusion flow. The major role in emergence of mass transfer in the sintered sample belongs to thermal energy obtained from the graphite pressing-tool. Surface tension forces at points of contact of powder particles excite the processes of surface and inter-boundary diffusion. Owing to the presence of temperature gradient between the centre and periphery of the sample there is also a diffusion flow directed from the heated areas to cooler ones. This process is graphically demonstrated in fig.3, the arrows show the direction of thermal diffusion flow.

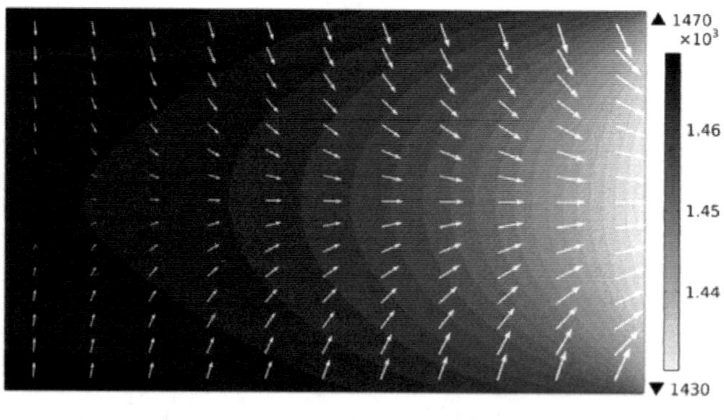

Figure 3. Thermal field in two-dimensional axisymmetric sample model, °C.

Since the axial pressure is applied to the sample in SPS process, plastic deformation [10] contributes to the mass transfer process as well. This phenomenon is the most vividly represented in sintering of malleable materials, for instance, metals.

4. Conclusions

A numerical modelling of thermal and electric field distribution in the graphite pressing-tool and in the sintered sample of stabilised zirconium dioxide in the process of sparkle plasma sintering has been conducted. By the results of calculation the primary mechanism of mass transfer at this sintering method has been determined. The reduction of sintering time at SPS is accounted for the large amount of the specific heat flow in the direction of sample surface arriving from the graphite pressing-tool. According to the data of the developed numerical model the average rate of uncontrolled sample heating made of stabilised zirconium dioxide with the diameter of 20 mm and the height of 6 mm with use of the pressing-tool, shown in Figure 1, amounts to ~ 900 °C/min. For comparison the heating rate of the sintered material when using hot-pressing, which is also based on mutual use of Joule heat and axial pressure, does not exceed 80 °C/min.

The influence of the electric field on the sample structure results in appearance of polarized processes leading to displacement of zirconium ions from the lattice sites. Similar stressed state is capable of reducing the required amount of heat energy to start the diffusion process. In order to identify the signs of electromigration of zirconium ions under the influence of electric field an X-ray phase analysis (by means of 'Shimadzu XRD 6000' diffractometer) of the surface sample layers, sintered at temperature of 1300 °C, has been conducted. The relative density of the resulted ceramics was equal to 95.7 %, at the magnitude of the apparent density amounted to 6.04 g/cm^3. The differences in the structure of the cathode and anode layers in comparison with the structure of the primary array have not been found. The tetragonal modification of zirconium dioxide amounted to 90 %, the rest 10 % belong to monoclinic modification.

The applied axial pressure at high temperatures of sintering can lead to plastic deformation of particles of the sintered powder body, which influences the reduction of time, required for complete removal of pores from ceramics.

5. Acknowledgments
This work was financially supported by the Russian President Grant (SP-1179.2015.1).

References
[1] Annenkov Yu M et al. 2005 *Proceedings of Tomsk Polytechnic University* **7 (308)** 39–43
[2] Ivashutenko A S et al. 2008 *New refractory materials* **10** 35–42
[3] Annenkov Yu M et al. 2006 *Radiation-thermal effects and processes in nonorganic materials: Proceeings of the Vth International scientific conference* vol 1 (Tomsk: TPU) p. 376–381
[4] Matrenin S V, Slosman A I, Myachin Yu V 2005 *Proceed. of Tomsk Polyt. Univer.* **4 (308)** 74–77
[5] Batanov G M 2001 *JTP.* **7(71)** 119–123
[6] Annenkov Yu M, Ivashutenko A S 2011 *Proceedings of universities Physics* **1** 37–39
[7] Grasso S, Sakka Y, Maizza G Electric current activated/assisted sintering (ECAS): a review of patents 1906–2008 *Sci. Technol. Adv. Mater.* **5(10)** Article ID 053001
[8] Molenat G et al. 2010 Temperature Control in Spark Plasma Sintering: An FEM Approach, *Journal of Metallurgy* Article ID 145431
[9] Vanmeensel K et al. 2005 *Acta Materialia* **53** 4379–4388
[10] Kornienko E, Smirnov A, Kuzmin V, 2015 *Applied Mechanics and Materials* **698** 405–410
[11] Skeeba V, Pushnin V, Kornev D 2015 *Applied Mechanics and Materials* **788** 88–94
[12] L I Shevtsova, M A Korchagin, A. Thommes, V I Mali, Anisimov A G, Nagavkin S Yu, 2014 *Advanced Materials Research* **1040** 772–777
[13] Shevtsova L I, Mali V I, Bataev A A, Bataev I A, Terent'ev D S, Lozhkin V S 2013 *The 8 international forum on strategic technologies* **1** 187–189
[14] Lazurenko D V, Mali V I, Anisimov A G, Yartsev P S, Lagereva D I, Shevtsova L I, 2014 *Advanced Materials Research* **1040** 800–804

A method for calculating the productivity of cable communications networks

S N Shulikin[1], M S Shulikina[1], S S Maryin[2], D V Vinogradova[1] and M E Zavgorodnyaya[3]
[1] Tomsk Polytechnic University, 30, Lenina ave., Tomsk, 634050, Russia
[2] Gazprom Transgaz Tomsk, 9, Frunze ave., Tomsk, 634029, Russia
[3] Tomsk State University, 36, Lenina ave., Tomsk, 634050, Russia

E-mail: hupy@tpu.ru

Abstract. A probabilistic-mathematical instrument was used to develop a method for calculating the productivity of a cable line. The effect of deviation of factors from data of recording devices was determined when identifying random stream characteristics. The developed method was used to perform predictive calculation of the productivity of the modern cable communication line.

1. Introduction

One of the tendencies visible in the most recent and modern developments of communication networks is a change in terms of a transmission structure. The traffic has become multimedia-based in large networks. They are characterized as networks with a non-uniform information stream where the emphasis has been shifted towards the use of various network applications. The increased cost of creating and building the system itself necessitates an improvement of the actual quality of the design solutions, particularly the accuracy of the channel throughput determining the attenuation factor, and other productivity characteristics [1–3]. One of the possible approaches to the estimation of these essential design indicators is probabilistic modelling, with the systems being represented as a set of resources used one by one.

The validity of probabilistic modelling of results using the waiting-line theory and other methods is highly dependent on the conformity of the models utilised to the real systems. Network equipment designers and developers need to gather data about the behaviour of networks of various scales, architecture and configuration, as well as about their qualitative characteristics. Therefore, such modelling aids are required to account for all the specific features of the network operation, but also to allow defining initial information and lastly, to obtain reliable network characteristics [4].

2. The principle of communication line productivity modelling

In real cable communication systems the known and time-invariant parameters are those that characterize service facilities. The intensity of incoming/outgoing streams typically varies over time, and these variations are of random nature. However, it is clear that the cable system operation is directly dependent on the intensity of incoming streams or calls: the higher the intensity of incoming streams, the heavier the service mode, which requires, for instance, connection of additional service facilities. In view of this, an important task is to estimate the parameters of event streams at an arbitrary point in time based on observation of this stream.

Based on their randomness, the streams of events with time-dependent intensity can be divided into two types. The first one comprises the streams with intensity $\lambda(t)$, which is a continuous random process. The second type includes only streams having intensity $\lambda(t)$ as a piecewise-constant process with a finite number of states $(\lambda_1, \lambda_2, \ldots, \lambda_n)$. The transition of process $\lambda(t)$ from one state to another occurs at random points in time, with the stream of events behaving as stationary intensity flow λ_i during those time intervals when process $\lambda(t)$ is in state λ_i. Such streams are the most typical for real networks. These piecewise-stationary streams of events can be divided into three categories:

1) *synchronous* twice stochastic streams of events – streams with the intensity for which state transition occurs only at random points in time that are the moments of events occurrence;
2) *asynchronous* twice stochastic streams of events – streams with the intensity for which state transition does not depend on the moments when the events occur;
3) *half-synchronous* twice stochastic streams of events – streams where for one set of states it is 1 that is true, and for the rest of the states it is 2 that is true.

Taking into account all the above-stated requirements for the construction of the model, this paper suggests researching the problem of estimating the parameters of asynchronous streams of events with intensity $\lambda(t)$ being a piecewise-constant Markovian process that takes two values $\{\lambda, 0\}$ of the alternating stream of events, with the events being partially observable. The partial observability is due to the formation of a so-called dead time scheme, implying that an event occurrence in the alternating stream is accompanied by a fixed period of time (a dead time period) during which no other events can be observed. One of the major distorting factors when determining the characteristics of random streams is the dead time of recording devices. Dead time in this respect is the period when a recorded event is being processed, while any other event entered into the system input during this period is lost.

In view of this, the need to estimate the parameters (characteristics) of the real stream of events received at the system input arises. The amount and the character of dead time of certain devices depend on numerous factors. At a first approximation it can be assumed that this period lasts for certain (fixed) time T. All recording devices with a sufficient degree of approximation can be separated into two groups. The first group comprises devices with non-extending dead time, which is independent of the receipt of other events in the range of its action. Non-extending dead time is sometimes called first genus dead time, and corresponding recording devices are called counters, or type I recorders. Devices with extending dead time (type II recorders) belong to the second group. In this case dead time emerges after any event obtained at the system input regardless of it being recorded or not. In terms of data transmission lines, dead time can be represented by disturbances, failures, i.e. any effects on the network operation. Since these effects are also of random nature, recorders of the II type with extending dead time will be used for this study.

Let us consider a stationary stream of events with intensity λ that is observed only at certain time intervals and has two states – 0 and 1. If the control process is in state 1, the stream can be observed. But if it is in state 0, the stream cannot be observed. The lengths of stationary sections within the control process are distributed according to exponential law [5]

$$F_1(t) = 1 - \exp(-\alpha t) \text{ or } F_1(t) = 1 - \exp(-\beta t),$$

where α and β are intensities of the control process transition from state 1 to state 0 and vice versa, respectively.

Thus, the control process induces an alternating stream of events. The alternating stream of events, in its turn, is also partially unobserved. When an event occurs in this stream, there comes a certain time of duration T (dead time) within which other events cannot be observed. Therefore only the alternating stream event that comes after the end of the last period of non-observability will be observed. The occurring situation is shown in Figure 1, where the hatching is the indication of dead time, 0 and 1 are the control process states; t – current time, and $\{t_1, t_2, t_3, t_4, \ldots\}$ – moments of event occurrence in the stream under observation.

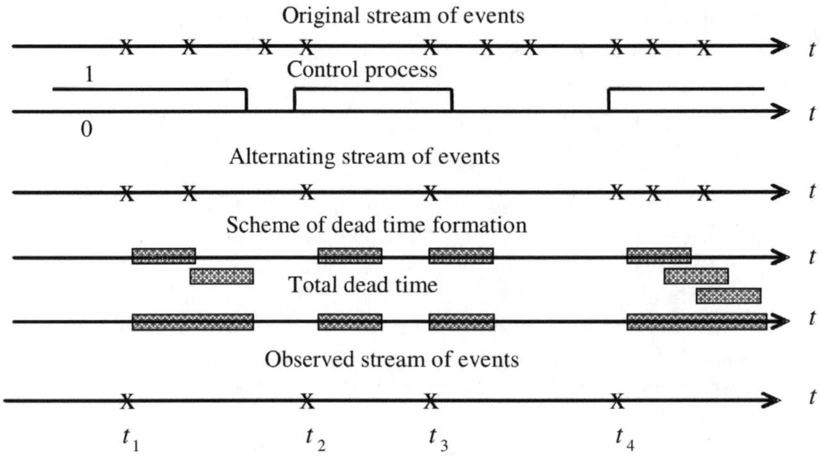

Figure 1. A scheme of the alternating stream.

The observations $\{t_1, t_2, \ldots, t_n\}$ at the final time interval are required to construct the estimation of original stream intensity λ, the estimation of intensities α, β for the control process state transition (i.e. to construct estimates of parameters for the alternating stream of events), as well as the estimation of dead time duration T. Assuming that $\{\tau_1, \tau_2, \ldots, \tau_n\}$ is the time interval between neighbouring events in the stream under observation, four statistical estimates - sampling moments - can be constructed:

$$C_k = \frac{1}{n} \cdot \sum_{i=1}^{n} \tau_i^{\,k}, \quad k = 1,2,3,4.$$

Thereafter the equations for determining parameters estimations $\{\lambda, \alpha, \beta, T\}$ will be formulated as follows:

$$M\left(\tau^k\right) = C_k, \quad k = 1,2,3,4.$$

By solving the probabilistic model, six equations are obtained for determining transmission parameters estimations [6]:

$$M(\tau) = -\frac{\alpha}{\alpha+\beta} \cdot \frac{\left[(z_2-\lambda)e^{-z_1T} + (\lambda - z_z)e^{-z_2T}\right]\cdot\left[2(\alpha+\beta)^2 + \lambda(\alpha+\beta) + \lambda\beta\right]}{2(z_2-z_1)\cdot(\alpha+\beta)^2 e^{(\alpha+\beta)T} + \lambda\cdot(\beta - z_1)\cdot(z_2+\alpha+\beta)e^{-z_1T} + \lambda(z_2-\beta)\cdot(z_1+\alpha+\beta)e^{-z_2T}} -$$

$$- \frac{1}{\alpha+\beta} + \frac{1}{\lambda\beta} \cdot \frac{(z_2-z_1)\cdot(\alpha+\beta) + \lambda\cdot(\beta-z_1)e^{-z_1T} + \lambda\cdot(z_2-\beta)e^{-z_2T}}{(z_2-\lambda)\cdot e^{-z_1T} + (\lambda - z_z)\cdot e^{-z_2T}}, \qquad (1)$$

$$M\left(\tau^2\right) = -\frac{\lambda}{\alpha+\beta} \cdot \frac{2\cdot(z_1+z_2)}{z_1^{\,2} z_2^{\,2}} \cdot \left[\beta\cdot g_\xi(0) + \beta\cdot(\alpha+\beta)\cdot g'_\xi(0) + \alpha\cdot g_\xi\cdot(\alpha+\beta)\right] +$$

$$+ \frac{\lambda}{\alpha+\beta} \cdot \frac{2\cdot\left(z_1^{\,2} + z_1z_2 + z_2^{\,2}\right)}{z_1^{\,3} z_2^{\,3}} \cdot \beta\cdot(\alpha+\beta)\cdot g_\xi(0) +$$

$$+ \frac{\lambda}{\alpha+\beta} \cdot \frac{1}{z_1z_2} \cdot \left[2\beta\cdot g'_\xi(0) + \beta\cdot(\alpha+\beta)\cdot g''_\xi(0) + 2\alpha\cdot g'_\xi\cdot(\alpha+\beta)\right], \qquad (2)$$

$$M\left(\tau^3\right)=\frac{6\lambda}{\alpha+\beta}\cdot\frac{(z_1+z_2)\cdot\left(z_1^2+z_2^2\right)}{z_1^4 z_2^4}\cdot\beta\cdot(\alpha+\beta)\cdot g_\xi(0)+$$
$$+\frac{6\lambda}{\alpha+\beta}\cdot\frac{(z_1+z_2)}{z_1^2 z_2^2}\cdot\left[2\beta\cdot g_\xi'(0)+\beta\cdot(\alpha+\beta)\cdot g_\xi''(0)+2\alpha\cdot g_\xi'\cdot(\alpha+\beta)\right]-$$
$$-\frac{\lambda}{\alpha+\beta}\cdot\frac{1}{z_1 z_2}\cdot\left[3\beta\cdot g_\xi''(0)+\beta\cdot(\alpha+\beta)\cdot g_\xi'''(0)+3\alpha\cdot g_\xi''\cdot(\alpha+\beta)\right], \quad (3)$$

$$M\left(\tau^4\right)=\frac{12\lambda}{\alpha+\beta}\cdot\frac{z_1^4+(z_1+z_2)^2\cdot\left(z_1^2+z_2^2\right)+z_2^4}{z_1^5 z_2^5}\cdot\beta\cdot(\alpha+\beta)\cdot g_\xi(0)-$$
$$-\frac{24\lambda}{\alpha+\beta}\cdot\frac{(z_1+z_2)\cdot\left(z_1^2+z_2^2\right)}{z_1^4 z_2^4}\cdot\left[\beta\cdot g_\xi(0)+\beta\cdot(\alpha+\beta)\cdot g_\xi'(0)+\alpha\cdot g_\xi\cdot(\alpha+\beta)\right]+$$
$$+\frac{12\lambda}{\alpha+\beta}\cdot\frac{z_1^2+z_1 z_2+z_2^2}{z_1^3 z_2^3}\cdot\left[2\beta\cdot g_\xi'(0)+\beta\cdot(\alpha+\beta)\cdot g_\xi''(0)+2\alpha\cdot g_\xi'\cdot(\alpha+\beta)\right]-$$
$$-\frac{4\lambda}{\alpha+\beta}\cdot\frac{z_1+z_2}{z_1^2 z_2^2}\cdot\left[3\beta\cdot g_\xi''(0)+\beta\cdot(\alpha+\beta)\cdot g_\xi'''(0)+3\alpha\cdot g_\xi''\cdot(\alpha+\beta)\right]+$$
$$+\frac{\lambda}{\alpha+\beta}\cdot\frac{1}{z_1 z_2}\cdot\left[4\beta\cdot g_\xi'''(0)+\beta\cdot(\alpha+\beta)\cdot g_\xi^{IV}(0)+4\alpha\cdot g_\xi'''\cdot(\alpha+\beta)\right] \quad (4)$$

$$z_1=\frac{\lambda+\alpha+\beta-\sqrt{(\lambda+\alpha+\beta)^2-4\lambda\beta}}{2},\quad z_2=\frac{\lambda+\alpha+\beta+\sqrt{(\lambda+\alpha+\beta)^2-4\lambda\beta}}{2}. \quad (5)$$

The system of the equations is determined by formulas (1–5), which make it possible to estimate the parameters of system transmission at any point in time. This system contains six equations concerning six unknowns $\{\lambda,\alpha,\beta,T,z_1,z_2\}$. The obtained system has only a numerical solution.

The adequacy of the proposed computational method was verified with a specific data transmission system of JSC "RICT" – "Russian-Italian Company of Telephone Installation" that provides various communication services to the city of Mezhdurechensk, Kemerovo region, such as: city telephony, wire communications, radio-access, a multipurpose duplex pay telephone network, Internet, IP-telephony, additional services based on intellectual networks, etc.

3. Calculation of data transmission network productivity

The original stream for the JSC "RICT" data transmission system was an asynchronous stream of events with intensity $\lambda(t)$, a piecewise-constant process with partially observable events. This study examines carefully a part of the system comprising a data source (a non-uniform stream of events), commutator switches 1 and 2 that were used for measuring intensity $\lambda(t)$, and various sites (residential-accommodation spaces, production plants, etc.). The dependences were measured on a working day, during peak time, from 12 to 2 p.m., when the maximum load in the data transmission network can be observed.

The statistical experiment on estimating the unknown parameters of the observed stream of events was based on simulation modelling of the examined stream of events with parameters $\{\lambda,\alpha,\beta,T\}$ at a fixed interval of observations $(0,T)$, assuming that at some point in time an event occurs in the original stream. The stationarity periods within the control process were generated according to the task definition. If the event occurred at the time point when the control process was in state 1, and the time point fell into the non-observability period, then the event was not recorded as an event of the stream under observation. If the event occurred after the dead time was over, then the event was recorded as an event of the stream under observation. The statistical experiment on estimating unknown parameters $\{\lambda, \alpha, \beta, T\}$ for the given task consisted in constructing the estimations with confidence intervals and looks as follows:

1) the alternating stream model with dead time is implemented and it forms sample $\{\tau_1, \tau_2, \ldots, \tau_n\}$ of observations with random value τ (a time interval between neighbouring events in the stream under observation);
2) statistical estimates C_1, C_2, C_3, C_4 are accumulated;
3) in order to determine the estimations of alternating stream parameters $\{\lambda, \alpha, \beta\}$, a system of equations is formed using formulas (1–5), with this system of equations being solved by a numerical method;
4) estimations $\{\lambda, \alpha, \beta, T\}$ are determined;
5) average values of estimations $\lambda = \frac{1}{N} \cdot \sum_{i=1}^{N} \lambda_i$ and dispersion $D(\lambda) = \frac{1}{N-1} \cdot \sum_{i=1}^{N} (\lambda_i - \lambda)^2$ are formed, which is followed by the calculation of confidence intervals based upon the obtained statistical data [6];
6) network load factor ρ is calculated.

An imitation experiment was conducted to determine parameters $\{\lambda, \alpha, \beta, T\}$ under this procedure. Table 1 summarizes the intensity values obtained through calculating the alternating stream of events in the scheme with extending dead time for commutator switches 1 and 2.

Table 1 – Intensities of data transmission through commutator switches 1 and 2.

Network channels	Commutator switch 1		Commutator switch 2	
	Transmitted to the server, Mbps	Received from the server, Mbps	Transmitted to the server, Mbps	Received from the server, Mbps
Gi0/01	3.28898915	25.53188615	0.67362369	18.54783908
Gi0/02	0.04298392	33.24441975	0.10073978	12.06851600
Gi0/03	0.80975518	35.48801027	63.98897778	15.39330280
Gi0/04	0.12364604	27.12027801	1.33482974	13.78169399
Gi0/05	0.42751007	26.13877361	0.19634786	13.03044206
Gi0/06	0.64386077	25.89963004	0.75870185	10.06545153
Gi0/07	0.88508078	33.38903852	1.07752315	15.33562257
Gi0/08	4.51858631	33.69003122	0.01963478	11.21360999
Gi0/09	0.13307745	26.30807871	1.42522137	16.58389541
Gi0/10	0.93511678	32.63097632	0	0
Gi0/11	15.06527933	23.80922425	0	0
Gi0/12	31.39562755	33.27545442	31.55281399	22.42059607

The developed model adequacy is verified by forming confidence intervals based upon the obtained statistical data. To this end, let us calculate the average of estimations and dispersion. Let us tabulate the obtained numerical dispersion data for commutator switches 1 and 2 in Table 2.

Therefore, the confidence interval of data numerical calculation obtained from commutator switch 1 is equal to 0, while that obtained from commutator switch 2 is $1.236 \cdot 10^{-15}$.

The obtained values of confidence intervals make it possible to confirm the adequacy of the constructed model.

Table 2 – The estimation of dispersion for commutator switches 1 and 2.

Network channels	Commutator switch 1		Commutator switch 2	
	incoming	outgoing	incoming	outgoing
Gi0/01	13.994	8.249	9.725	8.149
Gi0/02	17.24	15.961	10.298	1.670
Gi0/03	16.473	18.205	53.590	4.995
Gi0/04	17.159	9.837	9.064	3.383
Gi0/05	16.856	8.856	10.202	2.632
Gi0/06	16.639	8.616	9.640	0.333
Gi0/07	16.398	16.106	9.321	4.937
Gi0/08	12.765	16.407	10.379	0.815
Gi0/09	17.150	9.025	8.974	6.185
Gi0/10	16.348	15.348	10.399	10.399
Gi0/11	2.218	6.526	10.399	10.399
Gi0/12	14.112	15.992	21.154	12.022
Intensity estimation	17.283		10.398	

The next step was to determine the load factor. Assuming that maximum intensity λ_{max} of the examined line is 100 Mbps, the load factor in our case is the ratio between the intensity, obtained through calculating the alternating stream of events in the scheme with extending dead time, and the maximum load factor:

$$\rho = \frac{\lambda'}{\lambda_{max}},$$

where λ' is the intensity obtained through calculating the alternating stream of events in the scheme with extending dead time; λ_{max} is maximum intensity of information stream transmission within the network. The obtained data are summarized in Table 3.

Table 3 – The load factor for commutator switches 1 and 2.

Network channels	Commutator switch 1		Commutator switch 2	
	incoming	outgoing	incoming	outgoing
Gi0/01	0.0328898915	0.2553188615	0.0067362369	0.1854783908
Gi0/02	0.0004298392	0.3324441975	0.0010073978	0.1206851600
Gi0/03	0.0080975518	0.3548801027	0.6398897778	0.1539330280
Gi0/04	0.0012364604	0.2712027801	0.0133482974	0.1378169399
Gi0/05	0.0042751007	0.2613877361	0.0019634786	0.1303044206
Gi0/06	0.0064386077	0.2589963004	0.0075870185	0.1006545153
Gi0/07	0.0088508078	0.3338903852	0.0107752315	0.1533562257
Gi0/08	0.0451858631	0.3369003122	0.0001963478	0.1121360999
Gi0/09	0.0013307745	0.2630807871	0.0142522137	0.1658389541
Gi0/10	0.0093511678	0.3263097632	0	0
Gi0/11	0.1506527933	0.2380922425	0	0
Gi0/12	0.3139562755	0.3327545442	0.3155281399	0.2242059607

Having calculated the line productivity, a predictive calculation of this communication line productivity was performed. To this end, it was necessary to calculate the maximum data transmission rate within the network so that there would be no overload of servicing facilities, no losses of transmitted and received data calls, as well as to ensure that the service and calls processing time would not lead to server overload. In other words, the formulated task was reduced in order to find a maximum acceptable network load factor, followed by the calculation of optimal intensity for this network.

Let us take a maximum acceptable line load factor as $\rho = 0.99 \ldots 0.95$. Knowing the load factor, the maximum possible transmission rate equal to 1 Gbps and taking into account the number of channels, let us calculate the acceptable transmission rate for each channel. By applying the developed probabilistic-mathematical instrument it was found that, considering the maximum acceptable load factor and ignoring dead time, the maximum rate is within the range of 950 – 999 Mbps. However, taking into account dead time and stated requirements, the data transmission intensity within the network can fall between 193 and 200 Mbps.

4. Conclusion

1. In order to calculate the communication network with a non-uniform stream of information, we have constructed a model that was used as a basis for performing the task of estimating an asynchronous stream with partially observable events of various intensities, which is a piecewise-constant process.
2. It has been established that one of the major distorting factors when determining the characteristics of random streams is the dead time of data recording devices. Upon the event occurrence in the stream there comes a certain time of fixed duration, within which other events cannot be observed, i.e. dead time is the period when a recorded event is being processed, while any other event entered into the system input during this period is lost.
3. The intensity obtained through calculation by a probabilistic-mathematical method and the data transmission intensity in the examined real communication network is two times higher than the existing one, i.e. the communication line is unloaded. This conclusion serves as a recommendation to a possible data transmission rate increase and, hence, the improvement of the communication system productivity as a whole.

References

[1] Shmalko A 2000 *Communication Bulletin* **4** 58–65
[2] Stepanov A 2007 Architecture of computation systems and data-processing networks (Moscow: Stroyizdat) p. 349
[3] Grodnev I and Vernik S 1988 Communication lines *Radio and Communication* (Moscow) p 544
[4] Konnonov A 2008 Study and development of methods for capacity factors of computing machines networks with non-uniform traffic *Thesis for Candidate of Sciences* (Orenburg) p 142.
[5] Ventsel E 1969 The theory of probability (Moscow: Nauka) p. 576.
[6] Zavgorodnyaya M 2002 Estimation of dead time duration and parameters of alternating stream of events *Thesis for Candidate of Sciences* (Tomsk) p. 133

Analytical study of ultrasound influence on the molten metals atomization

A Antonnikova[1], V Arkhipov[1], V Boiko[2], S Basalaev[1], A Konovalenko[1] and N Zolotorev[1]
[1] Tomsk State University, 36, Lenina ave., Tomsk, 634050, Russia
[2] Khristianovich Institute of Theoretical and Applied Mechanics, Siberian Branch of Russian Academy of Sciences, 4/1, Institutskaya st., Novosibirsk, 630090, Russia

E-mail: kai_1@sibmail.com

Abstract. This paper focuses on the study of influence of ultrasound on liquid atomization using ejection nozzles. Two principles of influence of ultrasound on the atomization process such as a change of conditions on gas-liquid boundary during the generation of ultrasound oscillation in the gas and liquid jet (film) disintegration under the action of capillary forces in cases of generation of ultrasound oscillation in the liquid are considered. The optimal values of the ultrasound oscillation frequencies are calculated. Two constructions of the nozzles patented are proposed.

1. Introduction

In metallurgy, the gas spraying technique involving ejection nozzles is widely used to produce the high-dispersive powders from molten light metals [1]. The powder obtained is a poly-dispersive one and the fine fraction yield (the size of particles is less than 10 μm) is equal to 25 %. The problem of enhancement of the fine-dispersive fraction yield is currently important due to stiffening of the obtained atomizer dispersivity requirements. Experience has shown that gas-dynamic molten metals' spraying using hot gas has a limitation of dispersivity of the obtained powders due to stability improving of the small size drops in an airflow. The increase in the obtained powders dispersivity due to increase in temperature or pressure of spraying gas is an expensive approach since the energy consumption for production of powders rises sharply [2, 3].

Work [4] shows an increase in spray dispersivity due to impact of ultrasound oscillation on the atomization process. For this purpose, the wedge connected to a resonant cavity was located in a gas path of the nozzle. During the wedge flow the ultrasound oscillation appeared in gas. This study includes the research of impact of ultrasound oscillation on the liquid spraying process and the search for engineering solutions of its application.

2. Results and discussion

2.1. Excitation of oscillation in a gas flow

During the movement of a liquid jet in gas, the cavitation occurs, known as spraying. The process of liquid spraying (breaking of the liquid surface to form a large number of small droplets) involves an increase of amplitude and the appearance of instability of short waves on the liquid surface in cases of a dynamic impact of the gas flow. The analysis of the problem of the liquid jet disintegration due to a high-speed blowing gas flow showed that the increment of liquid surface oscillation has a maximum at wave number value [5]

$$k_{max} = \frac{2\pi}{\lambda_{max}} = \frac{2}{3}\frac{\rho u_g^2}{\sigma}, \quad (1)$$

where u_g is the relative velocity of gas and a jet at liquid surface; λ_{max} is the wavelength of more instable oscillation.

Using the equation (1), it is possible to obtain the equation for frequency of oscillation of more instable short waves:

$$f_{max} = \frac{u_g}{\lambda_{max}} = \frac{\rho u_g^3}{3\pi\sigma}. \quad (2)$$

The maximum of increment of oscillation is achieved at frequency of oscillation of liquid surface f_{max}:

$$\alpha_{max} = 0.4\frac{u_g^3}{\sigma}\sqrt{\frac{\rho^3}{\rho_l}},$$

where ρ_l is liquid density.

During movement of gas relatively the liquid surface, the turbulent boundary layer appears in it. The wave (roughness) amplitude on liquid surface ζ and the gas velocity in flow core u (that is equal to the nozzle velocity) are linked to the gas velocity on the liquid surface by means of the equation

$$u = u_g \ln\left(\frac{\delta}{\zeta}\right), \quad (3)$$

where δ is a characteristic size of the liquid jet. The amplitude of initial disturbance on the liquid surface does not usually exceed $\zeta=10^{-2}\delta$ [3] so the gas velocity on the liquid surface is equal to $u_g = 0.217\,u$. Substituting this value in equation (1), we obtain the value of oscillation frequency that has a maximum perturbation action on the liquid jet (molten metal)

$$f = f_{max} = 1.1\cdot 10^{-3}\frac{\rho u^3}{\sigma}. \quad (4)$$

Calculating the frequency of the maximum perturbation action on the jet of molten aluminum, we obtain value $f = 87$ kHz. This value of frequency of ultrasound oscillation agrees with the experimental results obtained for the nozzle with generating of oscillation in gas using a resonant cavity [4].

The oscillation of the elastic plate in a blowing gas flow can be used to generate the ultrasound oscillation in gas (in addition to the resonant cavity). During gas movement the plate begins to oscillate with its own frequency determined by its sizes and physical properties of the material [6]:

$$f = \pi^2\sqrt{\frac{D}{\rho_p h}\left\{\frac{G_x^4}{b^4} + \frac{2}{a^2b^2}[\nu H_x H_y + (1-\nu)J_x J_y]\right\}}, \quad (5)$$

where $D = \frac{Eh^3}{12(1-\nu^2)}$ is cylindrical of the plate stiffness; E is the elasticity modulus of the material of a plate; ν is the Poisson number; ρ_p is the density of the material of the plate; a, b, h are the length, width and thickness of the plate; G_x, G_y, H_x, H_y ,J_x, J_y are the coefficients depending on the conditions of plate fixing and the mode of oscillations.

For longitudinal oscillation of the plate that is clamped at one end and the first mode, the equation (5) is simplified as (G_x=0.597, H_x = -0.087, $G_y = H_y = 0$, J_x= 0.471, J_y= $12/\pi^2$) and is given by

$$f = \pi^2\sqrt{\frac{Eh^2}{12(1-\nu^2)\rho_p}\left[\frac{G_x^4}{a^4} + \frac{2}{a^2b^2}(1-\nu)J_x J_y\right]}. \quad (6)$$

For practical calculations the formula (6) can be converted into the following:

$$f = \frac{h}{a^2}\sqrt{\frac{E}{(1-\nu^2)\rho_p}\left[1+9(1-\nu)\frac{a^2}{b^2}\right]}. \qquad (7)$$

Varying the material of the plate (E, ρ_p, ν) and its geometrical sizes (a, b, h), we can state that the frequency range of free oscillation of the plate was in the range of frequency close to the frequency of the maximum perturbation action on the surface of the liquid jet (4). This is the reason of effective liquid jet disintegration (spraying).

The calculated values of sizes of the plate from steel 1X18H9T ensuring its own oscillations with frequency 87 kHz are equal to the following: length $a = 25$ mm, width $b = 12$ mm and thickness $h = 1.92$ mm. Figure 1 shows the construction of the nozzle with six plates of the mentioned sizes [7].

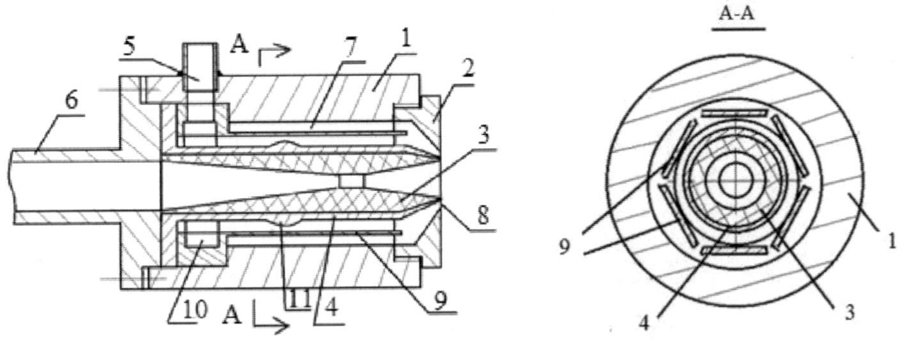

Figure 1. Construction of the nozzle with perturbation of ultrasound oscillations in a gas flow: 1 – nozzle body; 2 – head; 3 – nipple for fuel injection; 4 – metal barrel; 5 – gas port; 6 –metal supply pipe; 7 – gas cavity; 8 – annular nozzle; 9 – plates; 10 – annular cavity; 11 – boss.

2.2. Excitation of oscillation in the liquid film

The surface energy of the liquid jet is not minimal, thus small perturbations acting on the liquid surface are enhanced by capillary forces. It results in disintegration of the jet to the droplets. The analysis of the problem on the liquid jet (film) disintegration under the action of capillary forces has shown [5] that an increment of oscillations of the liquid surface has a maximum at wave number value $k=2\pi/\lambda$ (where λ is the wavelength) that is equal to

$$k = k_{\max} = \left[2\delta^2 + \frac{\mu\delta}{\rho}\sqrt{\frac{\rho\delta}{2\sigma}}\right]^{-\frac{1}{2}}, \qquad (8)$$

where δ is the liquid film thickness; μ is the factor of the dynamic viscosity of the liquid; ρ is the density of the liquid; σ is the surface tension factor of the liquid.

From equation (8), we can find the equation for wavelength:

$$\lambda_{\max} = 2\pi\left[2\delta^2 + \frac{\mu\delta}{\rho}\sqrt{\frac{\rho\delta}{2\sigma}}\right]^{\frac{1}{2}}. \qquad (9)$$

Substituting the ratio among wavelength, oscillation frequency and velocity of the liquid in equation (9)

$$\lambda_{\max} = \frac{u}{f_{\max}},$$

we obtain the equation for oscillation frequency that ensures the best conditions for the film liquid disintegration:

$$f = \frac{u}{2\pi\delta}\left[2+\sqrt{\frac{\mu^2}{2\sigma\rho\delta}}\right]^{-\frac{1}{2}}. \quad (10)$$

The calculations of oscillation frequency that ensures the best conditions for disintegration of the molten aluminum film with the thickness of δ=0.5 mm moving at velocity equal to u=5 m/s give the value f=1.1 kHz. Figure 2 shows the construction of the nozzle acting under the impact of ultrasound on molten metal [8].

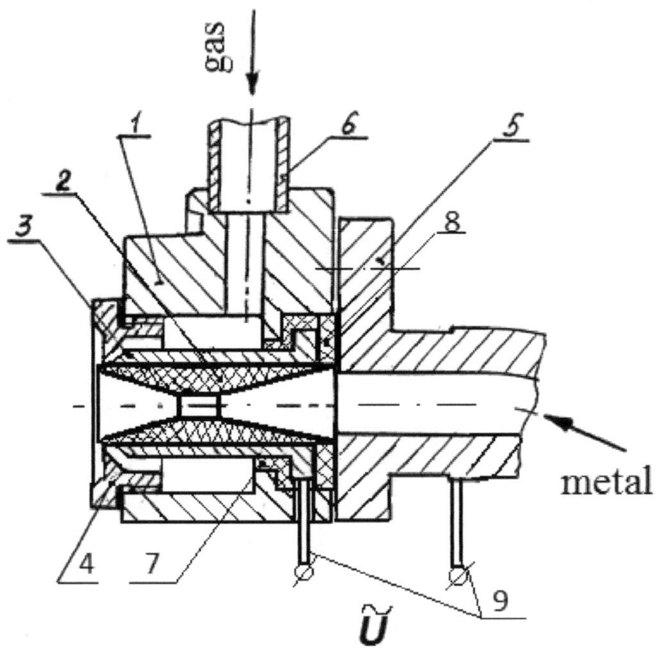

Figure 2. Construction of the nozzle with perturbation of ultrasound oscillations in the jet of molten metal: 1 – nozzle body; 2 – piezoceramic nipple for molten metal injection; 3 – metal barrel; 4 – head; 5 –metal supply pipe; 6 – gas port; 7, 8 – ceramic insulators; 9 – electrodes for alternating voltage supply.

Molten metal moving in the central canal of the spinning nozzle 2 (Figure 2) is mechanically affected by piezoceramic with frequency determined by external variable-voltage generator U. Spreading over the exhaust section of the spinning nozzle, the molten metal forms the film that breaks into droplets under the action of ultrasound oscillations and moves to the pulverization zone.

3. Conclusion

Using the analysis of liquid jet disintegration by the capillary forces action and high-velocity gas flow blowing, the frequencies of ultrasound oscillations influencing significantly the disintegration process are calculated. Based on the research findings, the constructions of nozzles using the ultrasound oscillations for an increase of the molten metal dispersivity are offered.

Acknowledgments
This study was funded by grant from the Russian Science Foundation (Project No. 15-19-10014).

References

[1] Fedorchenko I M and Andrievskiy R A *Kiev: AS USSR*
[2] Arkhipov V A *et al* 2014 *Powder metallurgy and functional coating* **4** 8
[3] Arkhipov V A *et al* 2015 *Powder metallurgy and functional coating* **1** 3
[4] Thomas Duerig, Marcel Escudier, Jakob Keller, Killwangen 1987 *Patent US* 4640806
[5] Levich V G *Moscow: Physmatgiz*
[6] Gontkevich V S *Kiev: Naukova dumka*
[7] Arkhipov V A, Zhukov A S, Zmanovskiy S V, Konovalenko A I, Pavlov A S 2015 Patent No. RU 2559080
[8] Arkhipov V A, Bondarchuk S S, Zhukov A S, Zmanovskiy S V, Konovalenko A I, Litvinov VP 2015 Patent No. RU 2563487

IOP Conf. Series: Materials Science and Engineering **124** (2016) 012045 doi:10.1088/1757-899X/124/1/012045

Mathematical simulation and optimization of cutting modes in turning of titanium alloy workpieces

M Bogoljubova, A Afonasov, V Kozlov and O Sumtsova
Tomsk Polytechnic University, 30, Lenina ave., Tomsk, 634050, Russia

E-mail: mnbogoljubova@tpu.ru

Abstract. A predictive modeling technique of optimal cutting modes in machining of workpieces made of titanium alloys, different from the well-known ones, is offered. The influence of various factors on the process of cutting for the purpose of defining optimal parameters of machining in accordance with certain effectiveness criteria is examined in the present paper. Mathematical models of optimization, algorithms and computer programmes, visual graphical forms presenting dependences of effectiveness criteria on the master schedule parameters have been worked out. Verification of mathematical models in laboratory and working conditions has been carried out.

1. Introduction

Modern machinery production is impossible without data support. Cutting parameters should be optimized for the purpose of increasing effectiveness as well as competitiveness.

One of the ways of increasing effectiveness of manufacturing is the application of optimal cutting modes obtained by simulation modeling. The relevance of these cutting modes should be confirmed by a physical experiment. The investigation of cutting process requires considering a great number of factors, taking into account the impact of cutting modes and geometry of a cutting tool on productivity and net cost of machining. From the position of optimization and manufacture intensification, simulation modeling is more multipurpose and less labour-consuming in comparison with the physical experiment.

Titanium alloys refer to the material groups, which are widely used in engineering. Titanium alloys have a wide application in airspace and chemical industry, shipbuilding, etc. [1]. Titanium alloy BT9 (Ti-6Al-3Mo) is applied in manufacturing of parts working at high temperatures (approximately 500 ^{0}C), on heavy loading, in aggressive environment, etc. It possesses high specific strength and corrosion resistance, low plasticity and heat conductivity, propensity to sticking and strain or work-hardening that causes the increased labour intensity of its machining. Research conducted by many authors [2-6], has presented that cutting modes and geometry of cutting tools in machining of titanium alloys do not always correspond to optimum values.

2. Problem Statement

The research objective is to define the cutting mode parameters and geometry of turning cutters optimizing basic indexes of the production process effectiveness: productivity, net cost, cutting tool life, roughness of the machined surface. Difficult -to-machining titanium alloy BT9 (Ti6Al3Mo) has been selected as a work material. The input data of other work materials can be included into a database for optimization if necessary.

The research method involves mathematical simulation, development of computer software, use of computational methods for solving the optimization task of machining of workpieces made of titanium alloy BT9 as well as the experimental test aiming to the identification of obtained results in working conditions.

3. Development of Mathematical Model

Mathematical models and optimization algorithms reflecting the interrelation of cutting mode parameters with such quality criteria as productivity (P), net cost (C), cutting tool life (T) and roughness of the machined surface (Rz) have been developed.

Dependences of criteria P, T, C, Rz on the parameters of cutting modes are accepted according to formulas (1) and (2) [4] and presented as follows:

$$Rz = \frac{\sqrt[0.4]{\left(8.25 \quad r^{0.15} \big/ t^{0.3} \quad \sin\varphi^{0.4}\right)^2}}{V \quad 8 \quad r} \quad [\mu m], \qquad (1)$$

Where:
r – nose radius at the top of a turning cutter, mm; v – cutting speed, m/s; t – cutting depth, mm; φ – cutting edge angle on the plan, degrees [°].

$$P = \frac{\pi \quad D \quad n \quad t \quad s}{1 + \frac{\tau_{cm}}{T}} \times 10^{-6} \quad [m^3/min], \qquad (2)$$

Where:
D – diameter of a workpiece, mm; n - speed of workpiece rotation (number of revolutions per minute), r/min; s – feed rate of a cutting tool, mm/rev; τ_{ch} - tool change rate, min; T – cutting tool life, min.

$$T = \frac{C_v \quad k_v}{v \quad t^x \quad s^y} \quad [min], \qquad (3)$$

Where:
Cv - coefficient, depending on a group of work and cutting tool materials; k_V – coefficient, depending on strength and chemical composition of work and cutting tool materials, geometry of a cutting tool, use of cutting lubricant; x, y – exponent quantities, depending on the grade of work and cutting tool materials and cutting conditions.

The following function for P is obtained by putting (3) into (2) and taking the change of allowance value *z* at constant *t* for a disturbance effect.

$$P = \frac{\pi \quad D \quad n \quad t \quad s}{1 + \frac{v \quad \tau_{cm} \quad t^x \quad s^y}{C_v \quad k_v}} \quad [m^3/min], \qquad (4)$$

The net cost of manufacturing of a workpiece (a reference area of machining of a workpiece equals 0.1 m^2) is defined using the following formula:

$$C = t_m(\alpha_s + \alpha_m) + \frac{t_{ch}}{Q}(\alpha_s + \alpha_m) + \frac{e}{Q} \quad [\$/piece], \qquad (5)$$

Where:
t_m – cutting (machine-tool) time for machining of a workpiece with a reference area equal to 0.1 m^2 (min,); t_{ch} – tool change time over a period of its tool life (min); a_s - salary of a machine worker per minute, dollar ($); a_m - operating costs of a machine tool (equipment) for a minute of its operation, dollar ($); e - operating

costs of a cutting tool during its tool life, dollar ($); Q – quantity of the machined workpieces during tool life value (pieces). $Q = f(v, s, t, T)$.

The mathematical models under consideration are presented as the set of multiparameter nonlinear functions. These functions are stored on a database and used for the formation of the optimization model as well as the presentation of functions in a required format.

A 'coordinate descent method' for multidimensional functions is used to determine the optimum value of effectiveness criteria. Allowed values of initial parameters are considered as limitations. Optimization is performed due to variations of initial parameters in the given limited ranges.

The optimal solution is represented in view of the adjusted values of parameters when the objective function takes extreme value *F*min (or *F*max) for a specific effectiveness criterion.

Software for the problems to be solved has been developed in Delphi environment [5]. A formalized

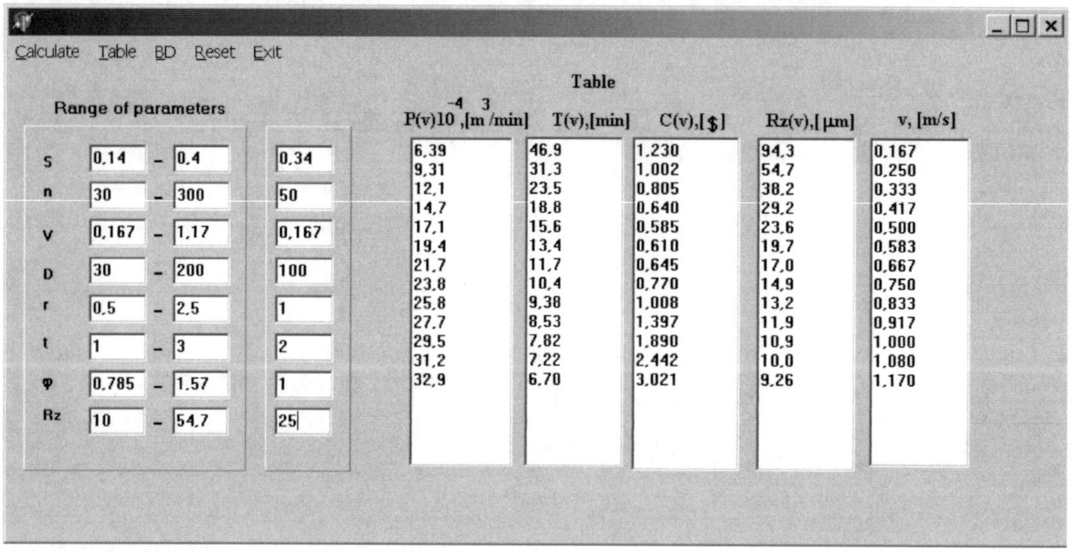

Figure 1. Dynamic pattern of dependences $P = f(v)$, $T = f(v)$, $C = f(v)$, $Rz = f(v,)$ at $s = 0.34$ mm/rev, $t = 1$ mm

presentation of the initial and output information and a range of variation of cutting mode parameters are represented in Figure 1.

The output data include the calculation data of criteria *P, T, C, Rz*, depicted in the form of dependences $P=f(v)$, $T=f(v)$, $C=f(v)$, $Rz=f(v)$. A graphic presentation of the output data is shown in Figure 2.

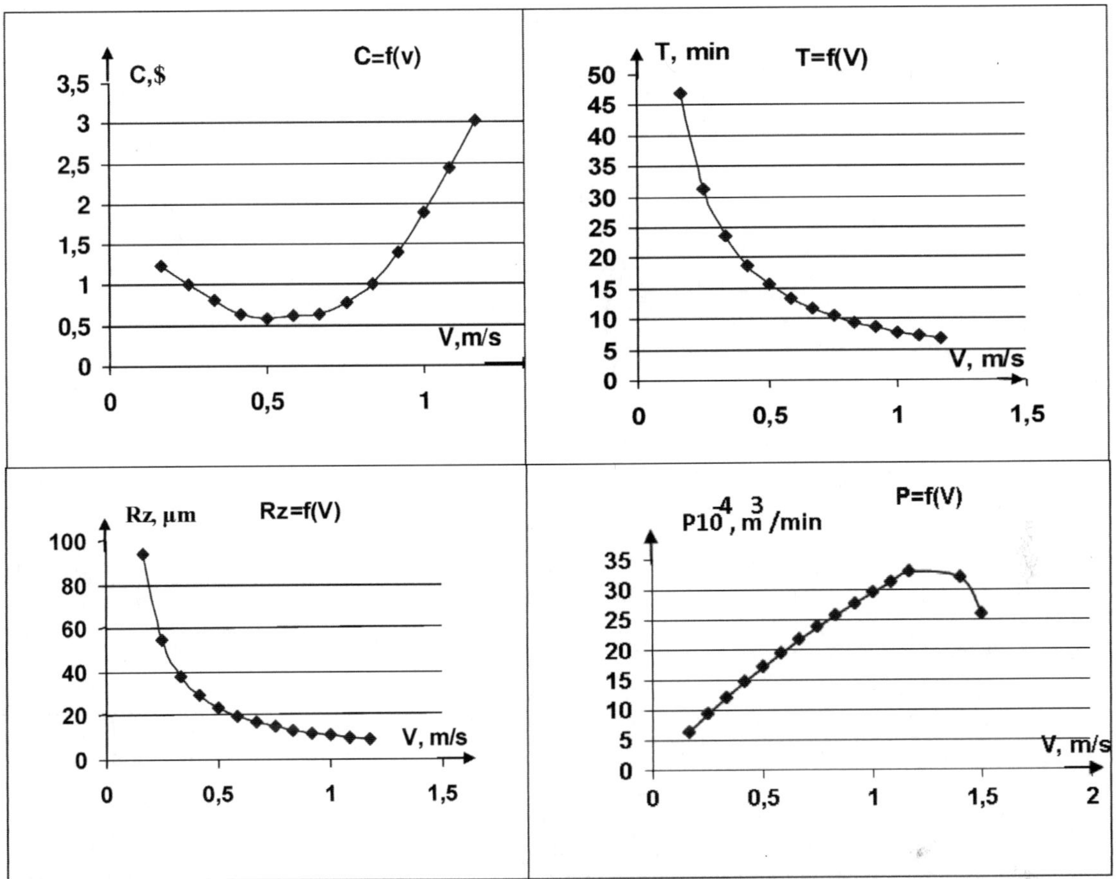

Figure 2. Influence of the cutting speed v on principle characteristics of the cutting process in turning of workpieces made of titanium alloy BT9 at feed rate s = 0.34 mm/rev, depth of cut t = 1 mm.

Calculation data (results) are performed for feed rate s=0.34mm/rev. A unit of measurement P on the graph is presented in m^3. Data for other values s within the given range of feed rate is calculated in a similar way according to the relevant technique. Corresponding to the graphs it is possible to define current values of criteria P, T, C, and Rz when the cutting speed v is changed in limits from 0.17 to 1.5 m/s.

A 'coordinate descent method' is used to determine the optimum value of the effectiveness criterion *P, T, C, Rz*. A calculation of minimum net cost $C = f(s, n, v, D, r, t)$ of machining of the workpiece made of titanium alloy BT9 is demonstrated in Figure 3.

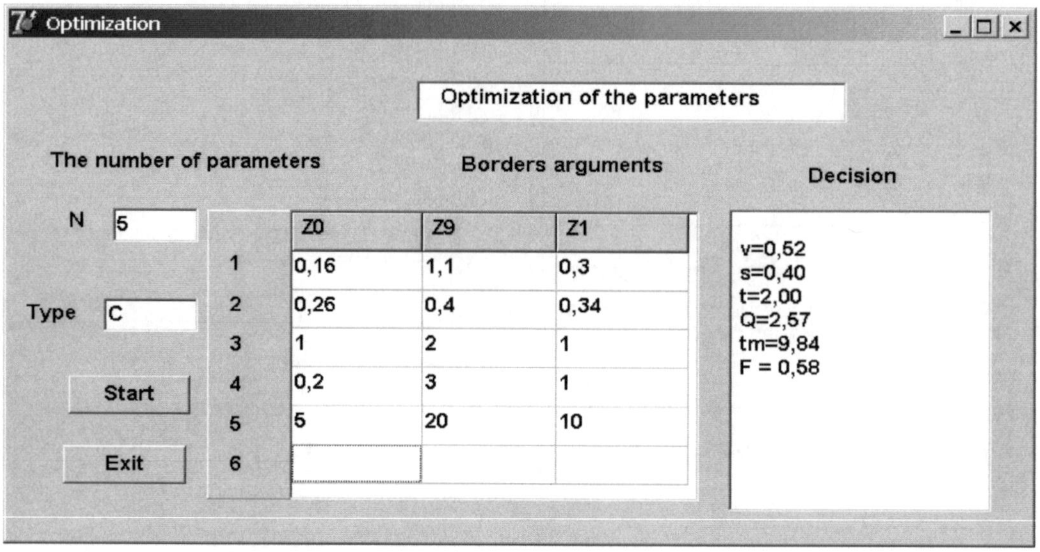

Figure 3. Optimization of cutting mode parameters for net cost $C = f(s, n, v, D, r, t)$.

4. Research results

Mathematical models, numerical methods and graphical interface offer the possibility to determine optimal cutting modes in accordance with the specified effectiveness criteria on the basis of initial parameters and the variation of their values in the given range.

Research is conducted and graphical dependences on cutting speed v and feed rates are plotted for the following criteria: effectiveness, productivity, cutting tool life, net cost, and roughness of the machined surface. It is possible to define optimization parameters for various cutting modes using graphical characteristics. Thus, in machining of the workpiece made of titanium alloy BT9 at depth of cut t=1 mm the minimum net cost $C = \$ 0.585$ will be obtained under the following parameters: productivity $P = 17.1 \cdot 10^{-4}$ m^3/min, cutting tool life $T = 15.6$ min, cutting speed $v = 0.5$ m/s, feed rate $s = 0.34$ mm/rev, depth of cut $t = 1$ mm.

Experimental research was carried out in laboratory and plant conditions on the lathe machine (model 16K20). Cutting speed v was being changed in a range of 0.17…1.17 m/s, feed rate s was being varied in a range of 0.14…0.4 mm/rev.

Straight-turning tools with indexable inserts made of cemented carbide BK8 (WC-8Co) were used as a cutting tool. The conducted experimental research showed the correlation of the theoretical (Figure 1…3) and practical results (Table 1). The criterion of tool deterioration of wear-land h_f was 0.3 mm.

Table 1. Experimental data

№ exp.	Cutting speed V, [m/s]	Cutting tool life T [min]	Productivity $P \times 10^{-4}$, [m^2/min] when t=1 mm	Quantity of machined parts, [pieces]	Machining time, t_m [min]	Net cost, $ [dollar]
1	0.19	40	39.5	2.44	19.75	1.18
2	0.46	18.8	94.5	2.6	9.62	0.6
3	0.66	6.8	134	1.07	7.7	0.628
4	0.9	1.6	183.5	0.284	6.48	1.245
5	1.11	0.8	220	0.139	5.35	2.33

5. Conclusion

The calculation of cutting mode parameters has been carried out by means of mathematical simulation with the subsequent verification in the form of experiment under plant conditions for the purpose of decreasing labour input and reducing high cost of actual test.

The graphical interface of computational modeling for optimization of cutting process parameters for turning of the hard-to-machine titanium alloy has been developed. It allows setting optimal cutting modes based on the range variation of initial parameters.

Guidelines for the cutting process optimization taking into account the restrictions imposed on parameters of the cutting mode in conformity with the specified effectiveness criteria have been worked out on the basis of the present research.

References

[1] Semeniuta V and Krzhenizkii I 2002 *Titanium* **1** 62–65
[2] Vereshhaka A, Anikeev A, Dacheva A 2010 *Mechanical engineering technology* **3** 17–22
[3] Kozlov V and Li X 2015 *Applied Mechanics and Materials: Scientific Journal* **756** 126–131
[4] Makarov A 1976 *Moscow: Mechanical engineering* 278
[5] *Information on Embarcadero Delphi XE2 Lite* http://www.embarcadero.com/
[6] Kirsanov S and Babaev A 2014 *IOP Conference Series: Mater. Science and Engineer.* **661(1)** 1–5

Mathematical and information maintenance of biometric systems

Z Boriev, S Sokolov, A Nyrkov and A Nekrasova
Admiral Makarov State University of Maritime and Island Shipping, 5/7, Dvinskaya st., Saint-Petersburg, 198035, Russia

E-mail: za_mir_b@mail.ru

Abstract. This article describes the different mathematical methods for processing biometric data. A brief overview of methods for personality recognition by means of a signature is conducted. Mathematical solutions of a dynamic authentication method are considered. Recommendations on use of certain mathematical methods, depending on specific tasks, are provided. Based on the conducted analysis of software and the choice made in favor of the wavelet analysis, a brief basis for its use in the course of software development for biometric personal identification is given for the purpose of its practical application.

1. Introduction

Biometric systems are increasingly popular in society. This is due to the problem of accurate personal identification, and, as a result, reliable protection of personal data.

In practical sense, the biometry is considered as a method for identifying a person, based on his/her physiological or behavioral characteristic.

Such biometric parameters as fingerprints or hand geometry are physiological characteristics, which are measured at a specific point of time. It should be noted that there is a variety of non-unique characteristics among physiological characteristics that cannot be used in the personal identification process. For example, such characteristics include weight and growth of a person, because they are subjected to a strong change over time [1-2].

2. Biometric characteristics

Behavioral characteristics are connected with subconscious movements in the course of reproduction or repetition of any common action. Such characteristics include voice, step, signature of a person, and even dynamics of typing [3].

One of the most important characteristics of biometric identification methods is an accuracy, which means an ability of the chosen method to distinguish reliably biometric characteristics, belonging to different people.

More and more biometric parameters, with use of which personal identification and authentication is possible, appear every year. But a lot of time is spent for their study, search of processing methods and algorithms. Therefore, a level of knowledge of biometric parameter directly influences a level of its use.

Having carried out the analysis of personal identification systems, it was revealed that the following biometric parameters are used more often today: fingerprint, iris, hand veins, voice and signature [4-5].

3. Signature identification
An issue of personal identification by means of a signature is considered in this work. Today, there are two main directions of the signature identification:

- a static signature recognition;
- an analysis of dynamic characteristics.

It is possible to get a two- or three-dimensional (if to consider additional characteristics, such as pressing on a pen, pen slope angle, signature rate) signature model by means of a scanner or a media tablet.

Today, there is no single approach to image identification, irrespective of the number of considered signature parameters. A static signature can be transferred to third parties and is not copy-protected: an original picture is copied on a transparent paper, which is then put on a tablet, and this signature is simply outlined. As a result, we have a completely identical copy of the original signature [6][7].

Figure 1.1. Original signature

Figure 1.2. Outlined signature copy

We can see in these figures that the original and copy are almost indistinguishable that means an unreliability of a static signature. Authentication by dynamic characteristics is a solution of this issue of accurate image recognition and protection against a forgery [8].

Biometric systems, based on analysis of dynamic characteristics, can operate according to a general chart:

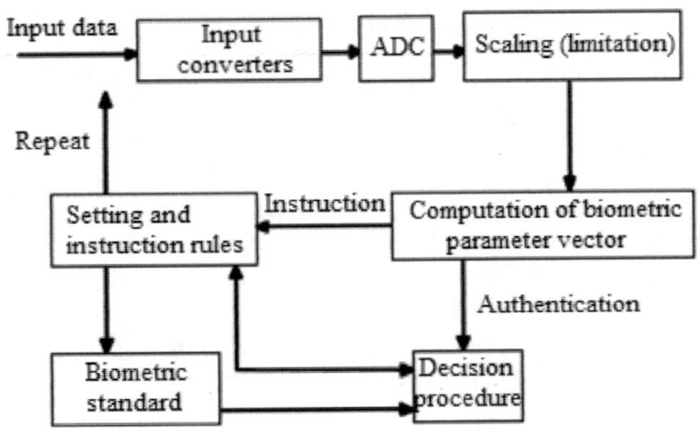

Figure 2. Generalized chart of dynamic biometric systems.

The following general order for dynamic authentication systems can be allocated on the basis of this chart:

- transformation of non-electrical values (coordinates of pen point, noise pressure) to electric signals;
- digitization of input electric signals;
- scaling of input amplitudes, resulting them in a certain reference value;
- adjustment of signals to a single time scale;
- computation of a vector (matrix) for measured biometric parameters;
- the system operation mode (instruction or authentication) determines a set of operations, carried out with a pre-formed parameter vector.

4. Mathematical methods
Mathematical solutions of a dynamic authentication method are more complicated than a method of static signature recognition. It allows reading signature parameters in real-time mode, for example, a

speed of hand movement at different sites, a writing order of particular lines, a line form and a direction, the pressure force and duration of all signature stages. The opportunity of a forgery reduces to zero, because it is almost impossible to repeat the smallest nuances when affixing a signature.

In this case, the input data will be X(t), Y(t), Z(t), where X(t), Y(t) - coordinates of pen tip, Z(t) - an additional parameter (pressure force, slope angle, etc.). All three functions are time-dependent. One-dimensional mathematical models, based on one of parameters, provide an error probability 0,1. The error probability in bi-dimensional mathematical models decreases to 0,01. The error probability for the three-dimensional model respectively will be equal to 0,001.

It is required to compute a biometric parameter vector after receiving an original data and its digitizing. Until recently, the most common method of vector formation was a discrete Fourier transformation. This method is a good enough for recognition of a static signature, as the Fourier coefficient represents a signal behavior for all time of its existence that, in case of dynamic authentication, is unjustified. In this case, the best solution of this issue will be a wavelet transform [8][9].

To refer to wavelets, a class of functions should meet the following requirements:

1. *Tolerance.* Analyzing wavelet ψ (t), also called a mother wavelet, should have a zero-mean value:

$$\int_{-\infty}^{+\infty} \psi(t)dt = 0$$

2. *Similarity.* All class functions are obtained from the analyzing a wavelet by scaling transformation and shift:

$$\psi_{a,b}(t) = \psi(\frac{t-b}{a})$$

The two-parameter class of functions appears as the following: ‘a’ parameter – a function scale (extension), ‘b’ parameter – a function position (shift).

3. *Inversability.* An existence of inverse transform, which definitely restores an initial function, based on its wavelet - transform.

4. Regularity. The ψ(t) function should be well localized both in a physical and in Fourier space.

The wavelets' important feature is a possibility to describe non-stationary signals, consisting of different components, operating in different time intervals. Thus, the X(t), Y(t), Z(t) functions can be arranged, according to its wavelets. As a result, we get a coefficient matrix, which can be used for return to an original function. Besides, the wavelet coefficients tend to zero quicker than the Fourier expansion coefficients, and it is possible to store more information about signature with rather equal data volumes. [10]

A matrix of average coefficients is accepted as a standard in this case. An example of dependence of one of coordinates on time and its approximation using wavelet expansion is shown in the following figure.

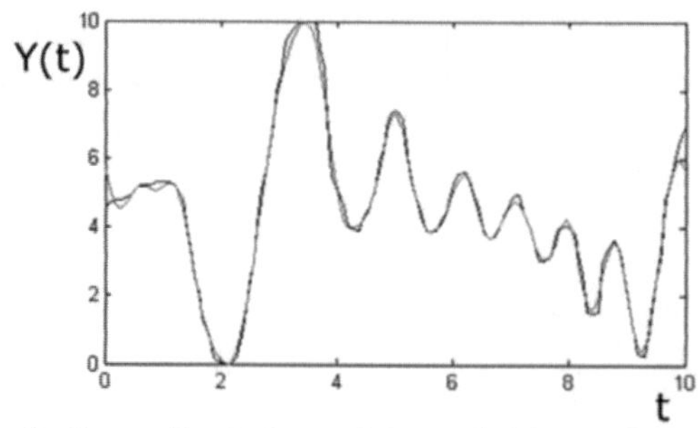

Figure 3. Generalized chart of dynamic biometric systems

It is also possible to use different bases of wavelet expansions, for example, DOG-wavelet, MHAT-wavelet, Gaussian wavelet of the 1st order.

5. Conclusion

The conducted analysis allowed forming a common vision of an issue of mathematical and information maintenance in biometric personal recognition systems. Wavelet transforms proved to be an effective method to solve problems of signature recognition. In the course of work it has been found that processing of biometric signs by means of wavelet transform allows achieving an error reduction. We can say that a computational complexity of algorithms is a lack of use of wavelet transforms. But this problem is easily solved owing to the high development level of mathematical packages. Asoftware, which consists of an instruction and comparison modules, is also developed for the analysis of this method.

References

[1] Boriev Z, Sokolov S 2015 *Materials of the VI International research and practice conference 'Modern tendencies and prospects of the Russian water transport development'* **6** 6–37
[2] Umnov A 2006 *Collection of scientific papers of the 1st International research and practice conference 'World Experience of biometric solutions in complex security systems'* 52–57
[3] Solomatin A 2015 *The St. Petersburg National Research University of Information Technologies, Mechanics and Optics*
[4] Clark R 1994 Personal *Information technologies and people* **3** 6–37
[5] Boriev Z, Sokolov S 2015 *VI International Scientific Practical Conference on Innovative Technologies and Economics in Engineering* **91**
[6] Leus A 2009 *The 'SKUD. Antiterrorism' booklet* 28–29
[7] Chernyi S 2015 *Metallurgical and Mining Industry* **7(3)** 60–65
[8] Nyrkov A, Sokolov S, Chernyi S, Mamunts D 2015 *Metallurg. and Mining Industry* **7(6)** 521–524
[9] Nyrkov A P, Belousov A S, Sokolov S S 2015 *Modern Applied Science* **10(5)** 162–176
[10] Chernyi S, Zhilenkov A, Sokolov S, Titov L 2015 *Metallurgical and Mining Industry* **7(3)** 66–73

Regression analysis for solving diagnosis problem of children's health

Yu A Cherkashina and O M Gerget
Tomsk Polytechnic University, 30, Lenina ave., Tomsk, 634050, Russia

E-mail:cherr999y@mail.ru

Abstract. The paper includes results of scientific researches. These researches are devoted to the application of statistical techniques, namely, regression analysis, to assess the health status of children in the neonatal period based on medical data (hemostatic parameters, parameters of blood tests, the gestational age, vascular-endothelial growth factor) measured at 3-5 days of children's life. In this paper a detailed description of the studied medical data is given. A binary logistic regression procedure is discussed in the paper. Basic results of the research are presented. A classification table of predicted values and factual observed values is shown, the overall percentage of correct recognition is determined. Regression equation coefficients are calculated, the general regression equation is written based on them. Based on the results of logistic regression, ROC analysis was performed, sensitivity and specificity of the model are calculated and ROC curves are constructed. These mathematical techniques allow carrying out diagnostics of health of children providing a high quality of recognition. The results make a significant contribution to the development of evidence-based medicine and have a high practical importance in the professional activity of the author.

1. Introduction

The problem of diagnosing health of children attracts attention of a growing number of researchers. This issue is especially relevant in pediatrics, where the main objective is to identify pathologies and chronic diseases at the early stages of organism development.

It is well known that the human predisposition to various diseases is laid mainly in the first year of life, which is why the prediction of a health status in this period is up-to-date [1].

Therefore, it is necessary to evaluate the state of children's health during the neonatal period to give the necessary recommendations to avoid prepathological and pathological changes at the early stages of organism development.

The aim is to use statistical methods to predict the state of children's health.

2. Characteristics of investigated data

For the research, the data provided by the doctors of Siberian State Medical University (SSMU) and a health center 'Healthy mother – strong baby' were used.

3 hemostatic parameters, 9 parameters of general and biochemical blood tests, the gestational age and a vascular endothelial growth factor, measured at 3-5 days of children's life are available for the diagnosis.

A hemostatic system is a biological system in the organism whose function is to save the liquid state of blood, to stop bleeding in case of damaged vessel walls and to dissolve blood clots which have fulfilled their functions. The investigations of the hemostatic system are easy to perform and have highly informative results. The need to use hemostasis is confirmed by the fact that the hemostatic system disorders are the most dangerous in the neonatal period and reach 10% of the diagnoses of dead newborns.

The gestational age needs to account age-related morphological and functional features of brain vessels of newborns. The vascular endothelial growth factor influences the formation of post-hypoxic structural changes in the brain of the newborn, which are important for central nervous system damages.

A blood test is one of the most important methods of health monitoring. The necessary information for diagnosis of some diseases, disease severity and evaluation of dynamics of treatment can be obtained by the analysis of blood. The composition and concentration of cellular elements in blood vary under different physiological and pathological conditions.

3. Binary logistic regression

A regression analysis is the research of influence of one or more independent variables $X_1, X_2, ..., X_p$ on dependent variable Y [1].

The task of the regression analysis is to build a mathematical model that allows evaluating the dependent variable by the values of independent variables [1].

The main purposes of the regression analysis are

- to predict values of the dependent variable using independent variables;
- to determine the contribution of individual independent variables in the change of the dependent variable.

General regression equation is determined by the equation [4]:

$$Y = F(X_1, X_2, ..., X_p), \tag{1}$$

where F - unknown function to be determined;
Y – dependent variable;
$X_1, X_2, \ldots, X_p$ – set of independent variables;
p – total number of independent variables.

There are two main types of a regression [4]:

- linear;
- nonlinear (hyperbolic, exponential, polynomial, logarithmic, logistics).

Logistic regression was used in the work for the regression analysis.

Logistic regression is a kind of a multiple regression. It is used when the dependent variable can have only two mutually exclusive values, i.e. it is binary (dichotomous). The binary logistic regression method allows studying the dependence of the dichotomous variable on independent variables that can be measured in different scales. In most cases of using dichotomous variables, it is about some events that may or may not happen (0 - Event has not happened, 1 - event happened). Binary logistic regression allows us to calculate the probability of the event, depending on the values of the independent variables. It uses the following regression equation

$$P(Y = 1 \mid X_1, ..., X_p) = \frac{1}{1 + \exp(-(\beta_0 + X_1\beta_1 + ... + X_p\beta_p))} \tag{2}$$

where Y – dependent variable which has values between 0 and 1;
$X_1, X_2, \ldots, X_p$ – set of independent variables;
$\beta_0, \ldots, \beta_p$ – logistic regression coefficients.

Therefore, the equation (2) describes the probability that the event occurs, that is, independent variable Y takes a value of 1, depending on the independent variables of $X_1, ..., X_p$.

If the value of P is less than 0.5, it can be assumed that the event will not occur, otherwise the occurrence of the event can be assumed.

A logistic curve is a dependence linking the likelihood of events and the value of Y. The logistic curve is shown in Figure 1.

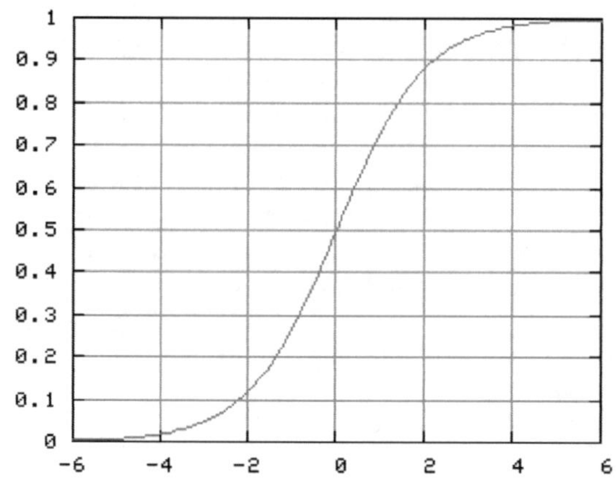

Figure 1. Logistic curve

From formula (2) and figure 1 we can see that regardless of regression coefficients β_0, ..., β_p or variables X_1, ..., X_p, predicted value Y will always be between 0 and 1 in this model.

The maximum likelihood method is used to find the coefficients of logistic regression. The essence of this method is the process of evaluating the regression coefficients coming down to maximization of the occurrence probability of a specific sample.

4. ROC-analysis

The ROC curve is a fundamental tool for the diagnostic test evaluation. ROC curves or Receiver Operating Characteristic curves are taken from the methodology of Signal Detection Analysis. Although the Theory of Signal Detectability (TSD) originally comes from the electronics and electrical engineering, it can also be used in medicine for the analysis of the interaction of sensitivity and specificity of the diagnostic test.

When one considers the results of a particular method in two populations, one population with a disease, the other population without the disease, a perfect separation between the two groups are rarely observed. Indeed, the distribution of the results will overlap, as shown in Figure 2.

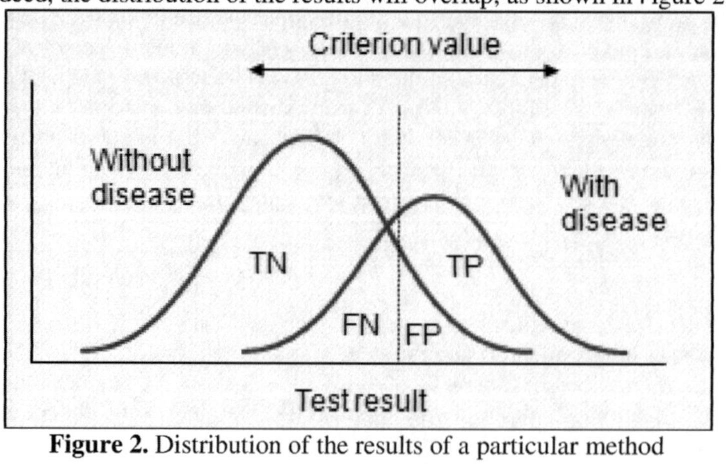

Figure 2. Distribution of the results of a particular method

For every possible cut-off point or criterion value one selects to discriminate between the two populations, there can be some cases with the disease correctly classified as positive (TP = True

Positive fraction), but some cases with the disease will be classified negative (FN = False Negative fraction). On the other hand, some cases without the disease can be correctly classified as negative (TN = True Negative fraction), but some cases without the disease will be classified as positive (FP = False Positive fraction).

Test sensitivity is the proportion of true positive predictions in the total number of patients:

$$Sensitivity = \frac{TP}{TP + FN} \tag{3}$$

This value characterizes the ability of the test to filter out patients with dubious diseases as accurately as possible.

A specificity of the method is the proportion of true negative in healthy patients:

$$Specificity = \frac{TN}{TP + FP} \tag{4}$$

The ability of the method is to detect only the patients with the disease characterized by this index.

Sensitivity and representativeness are reduced to 1 by using the ROC curve. The diagnosed value with a zero degree of prediction is represented by a line inclined at an angle of 45 degrees (diagonal). The more concave the curve of ROC, the more accurate the prediction of test results. The area under the ROC curve is an indicator of this property. The area is equal to 0.5 for the zero-degree prediction, and it is equal to 1 for the maximum degree of prediction.

5. Summary of the results

Experimental investigations were carried out to get a result (whether the child is ill or healthy) and to compare the result with the diagnosis made by a doctor.

The sample is formed by the results of the inspection of 102 children. It is divided into 2 classes (diagnosis): health (31 children) and patients (71 children).

The calculations were performed using the Statistical Package SRSS Statistics [5].

The regression analysis found that only three quantitative variables had an influence on the diagnosis: the period of reaction, gestational age, and total bilirubin. The other variables were not included in the regression equation.

The predicted values of the dependent variable are calculated from the regression equation, compared with the actual values observed in the classification table. The classification table is presented below (Table 1).

Table 1. Classification table

	Predicted		
Observed	health	ill	Percentage correct
health	30	1	96,8
ill	2	69	97,2
Overall percentage	97,1		

If the probability is less than 0.5, it means that the diagnosis is 'healthy' (the value of the variable 'diagnosis' is 0), in opposite case it is 'sick' (the value of the variable 'diagnosis' is 1). Table 1 shows that 97.1% of research subjects predicted the results were correct. This is a good outcome, the percentage of correct recognition of sick children is high.

Taking into account the found coefficients, the regression equation can be written as a formula:

$$P(Y = 1 \mid X_1, X_2, X_3) = \frac{1}{1 + e^{-(65{,}98+0-1{,}27X_1-1{,}59X_2-0{,}02X_3)}}$$

where Y – dependent variable 'diagnosis';

X_1, X_2, X_3 – set of independent variables: reaction period, gestational age and total bilirubin.

To calculate the probability that a child is sick, regression equation variables X_1, X_2, X_3 corresponding to each object (the child) are substituted.

For conducting the ROC analysis it is necessary to calculate the following indicators relying on the logistic regression:

- TP=69;
- FN=2;
- TN=30;
- FP=1.

Let us calculate the sensitivity and specificity of the formulas (3) and (4).

$$Sensitivity = \frac{TP}{TP+FN} = \frac{69}{71} = 97.2$$

$$Specificity = \frac{TN}{TP+FP} = \frac{30}{31} = 96.8$$

The sensitivity characterizes the ability of the test as accurately as possible to filter out patients with questionable presence of the disease. The sensitivity of more than 90% means the ability of the test to have a high ability to recognize sick children.

The specificity characterizes the ability of a test to detect patients with extremely questionable presence of the disease. The value of representativeness is 96%, which means that the test has a high recognition ability of healthy children.

The ROC curve is shown in Figure 3.

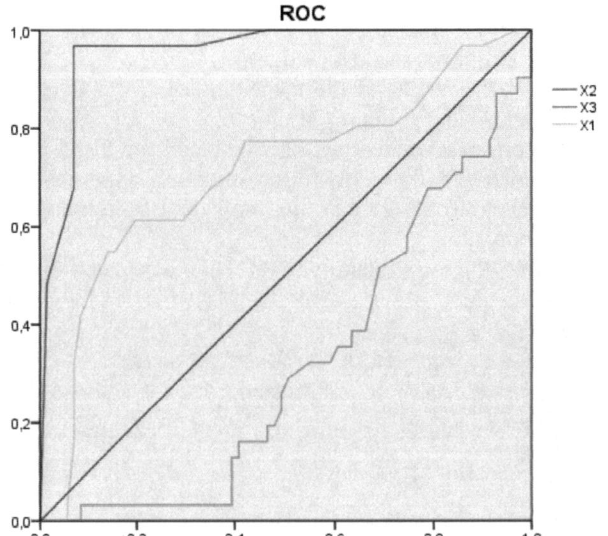

Figure 3. ROC curve

The more curved the curve of ROC, the more accurate the prediction of the test results. An indicator of this property is the area under the ROC, which test for a zero degree of prediction is equal to 0.5, and for the case with a maximum degree of prediction - 1. From Figure 3 it is clear that the largest area is for the index of X2.

6. Conclusion

Logistic regression is of great importance and has a practical application in medicine. The method was tested on the basis of real medical data provided by health professionals. The recognition quality of

97,1% of children was correctly classified to the appropriate class, therefore it can be considered acceptable.

References

[1] Gubler E V 1978 *Medicine* 296
[2] Gerget O M and Kochegurov V A 2002 *Tomsk, Publishing house TPU* 145
[3] Gerget O M and Devyatykh D V 2013 *Computer science and information technologies CSIT'2013: proceedings of the 15th International Workshop, Viena, 15 September-21 October 2013* 126-131
[4] Aguinis H 2004 *The Guilford press* 195
[5] 2015 SPSS software. Predictive analytics software and solutions [electronic resource] Available at: http://www-01.ibm.com/software/analytics/spss/
[6] Gerget, O M, Kochegurov V A, Titarenko E Yu 2015 Dynamic Susceptibility of a System *Applied Mechanics and Materials : Scientific Journal* **756** pp 378-381
[7] Tyurin Y N and Makarov A N 2008 Data analysis on the computer *Forum*

IOP Conf. Series: Materials Science and Engineering **124** (2016) 012048 doi:10.1088/1757-899X/124/1/012048

Techniques for selecting topology and implementing the distributed control system network

S Chernyi
Kerch State Maritime Technological University, 82, Ordzhonikidze st., Kerch, 98309, Ukraine

E-mail: sergiiblack@gmail.com

Abstract. On grounds of reviews devoted to flows analysis methods in the data processing networks within the automated control systems for the technological process and assessment of these methods by the selected set of requirements, one may make conclusion about expediency of using the combination of graph flow algorithms and the queueing theory. The outputs of the research concerning the impact of network dynamics on the drilling platform distributed system control quality prove the fact that the quality of the transient depends upon the frequency of discretization and intensity of flows. With increasing the intensity of flows, the static error of the control enlarges. It was concluded that in order to control the automation objects in the real-time mode it is required to minimize the delays in transmitting packets in the network.

1. Automation in industry

Automation of technological processes made possible to achieve the enhanced quality and increased profits in the various spheres of the industry where it was implemented. Simultaneously, automation of the drilling processes in the oil and gas industry is not a trivial task, the successful solution of which is impeded by a variety of factors, which are negative from the automation viewpoint. Success in this sphere allows accomplishing efficiently the most complicated tasks, namely: to explore most boreholes, drilling and exploitation of which are technically impossible or non-profitable now [1]. It is characteristic that the problems with automation implementation in the drilling process start with the aspects not closely connected with conditions or a site of extraction. Technical supply of the drilling rig comprises a set of components most of which are from different manufacturers. In the context of automation, it complicates a task greatly as cooperation under the control of the unique automation system is a requirement for the controllable equipment. High level of automation inherently assumes the realization of data exchange among all the system components and consequently their complete compatibility in methods, protocols, formats of data transfer and types of control actions as a basis for implementation of automation process.

Now communications between systems of the wellsite monitoring and management are standardized only, the transfer protocol WITSML (wellsite information transfer standard markup language) being a standard for communication. However, it is only a surface, as standard protocol for communications for the drilling rig on its own and for the whole equipment remains under question.

This paper aims at the analysis of the structure of the electric power system of a typical offshore drilling platform and at consideration of methods of selecting topology for the network of distributed system for control of elements of this electric power system at the drilling installation.

2. Description of the structure of the electric power system at an offshore drilling installation

Presence of loads with various power as well as a wide range of possible operations by drilling installations frequently results in necessity of mounting several generators. Synchronous generators produce electric power in autonomous electric power stations of the offshore drilling rigs. Due to their flexibility and versatility, diesel-generator units DGU1, DGU2 (Figure 1) etc., consisting of synchronous generators SG1, SG 2, etc. driven by diesels D1, D2, etc. are used. In general, structure of the electric power system of the offshore drilling installation provides the possibility of the simultaneous operation of all the SG that allows summing up their power on the global bus of the main switchboard (MSB). Thus, under standard conditions for the electric power plant it is possible to choose the optimum structure and to minimize the financial expenses for the power generation. Although with such structure of the electric power plant, availability of the systems for distribution of active/reactive power operating with parallel switch of generators as well as protection system is required.

All the plant generators are supplied with field voltage u_f, and all the diesels are supplied with fuel with discharge rate g (Figure 1) [2,3]. For DGU, which is driven in the group of shunt generators, automated voltage regulating systems and frequency regulating systems are used. For DGUs, which are driven in the group of shunt generators, active and reactive powers are regulated by means of regulators respectively, an automatic regulator of active power and an automatic regulator of reactive power ARAP (an automatic regulator of the active power) and ARRP (an automatic regulator of the reactive power), acting upon the diesel via the fuel system and upon the generator – via the excitation system. Thus, use of electronic control systems allows automatically sharing of active and reactive powers between shunt generators.

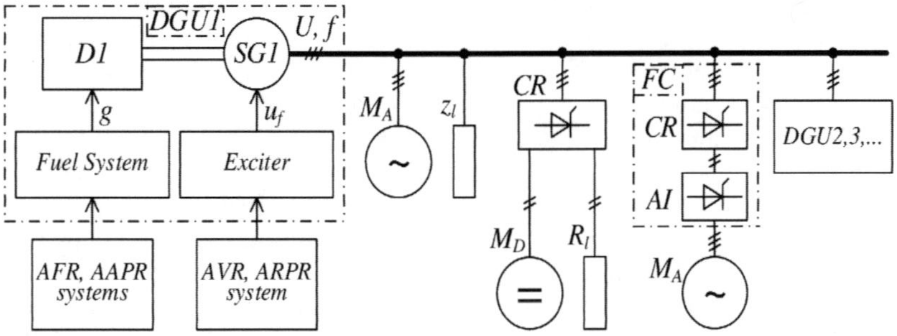

Figure 1. Systems of automated regulation of electric power parameters in the electric power system of the autonomous drilling installation: *DGU1* – diesel-generator unit (diesel *D1* + generator *SG1*); *AFR, AVR* – systems of automated regulation of frequency and voltage in the synchronous generator, DGUn – diesel-generator unit (diesel Dn+ generator SGn, where n – is a number of DGU); AFR, AVR, AAPR and ARPR – systems of automated regulation of frequency, voltage, active and reactive capacity of a synchronous generator; MA and MD – ac and dc motors; zl and Rl – resistance of static load on ac and dc; CR – controlled rectifier; AI – autonomous inverter (of voltage or current); FC – frequency converter.

It should be noted that the autonomous electric power system of the drilling installation is distinguished by a restricted power of the electric plant; commensurability of the converting load power of the variable speed drive and electric plant capacity; distinctive variable character of loads, great change of maximum loads depending on both current performance of the borehole and lithological conditions of drilling; constant changes in configuration of the electric power system of the electric technical complex of the drilling installation connected with the number of operating electric drives due to the conditions of drilling and with changes in the number of shunt synchronous generators due to the drilling conditions. All these require reliable and accurate performance of automated control means for the elements of the electric power system in the drilling installation.

If the energy generated is more than sufficient for switching the powerful load, load of generators is less than certain values, a generator is removed for running in parallel and diesel is stopped.

3. Information support for the drilling rigs distributed systems

The control system for the autonomous electric power network is considered as an integrity of microprocessor systems distributed through the production premises and united by bulk, ring or any other sign. It should provide centralization of the control (collection, processing, submission of the information to the staff at the control points on the technical state of the equipment and quality of technological processes, which should be controlled automatically and remotely). Thus, the system is of a distributed type by arrangement and by topology. The structure of the commutation lines between the computer elements (topology of data network) is determined by considering, as a rule, feasibility of efficient technical implementation. An important role in selecting the network structure is played by analysis of flow intensity data while solving the control problems.

Assessment of the electric power control system in an offshore drilling rig from the hierarchy approach point of view makes possible to determine the topology when creating the control system and to distribute the load for the information channels uniformly and proportionally.

Let us form the control system by the hierarchy principle [4]. In the event of failure of the higher-level elements, the performance of the electric power system is provided by means of the lower level elements [4, 5].

In order to provide the high reliability of the control system, the network should have high connectivity. In case of failure of certain components, the total operability of the control system should be maintained. The choice of network topology is determined, largely, by availability of additional connection between nodes. Thus, linear topology, which is a single-dimension array, combined into a chain, has a number of drawbacks: the time of passing between two nodes depends on the distance between them, and failure of one of the nodes results in failure of a message sending. The use of the ring topology is also expedient as it is not capable for extensibility (addition of a new node requires disassembling of the network), load on the transmit channel between various nodes will be different, backup communications are not available. As the drilling rig electric power control system is formed by the principle of the hierarchy, it is reasonable to use tree-type topology.

If the tree topology is used, local control systems for nodes in every DGU [2] are connected via flow routers for the local automatic means with the systems of the automatic control of every DGU and with flow routers for the local automatic means of the other DGU. These functional complexes are connected in a similar way with each other via individual flow routers for the functional complex control devices and controlled by the complex control system. Local systems, designed, as a rule, on the microprocessor basis, use industrial interfaces *CAN*, RS485; and the complex control system, in its turn, is reasonable to be implemented on the ground of the microprocessor systems of the personal computers enabling to integrate the complex control system under review into the overall ecosystem of the electric power plant of the drilling rig, into existing software-hardware automated control systems for technological processes, etc. Therefore, it is expedient to use Ethernet network at the high level, taking into consideration known specific features and drawbacks: the microprocessor automated control system (ACS) has finite response speed, operation execution time being able to change quickly due to a set of factors; virtual ACS operate with discrete signals, both in time and in amount, therefore replacement of discrete signals for continuous staircase ones is not quire adequate; analogue-digital transformation of signals entails certain distortions in the controlled signal that is connected with the limited band of transmission.

The latter becomes less significant due to the distribution of local control functions and regulation between local microprocessor systems.

4. Methods of network arrangement selection

The main idea of methods for selecting the network arrangement is to analyze the required characteristics of the network, to assess the priority of every network, to choose value judgement for any of the characteristics, to analyze the given battery of parameters and to determine the minimum weight network as the most appropriate for the problem set [6-8].

The methods involve performing the following steps:

- To generate a range Ψ of possible network interfaces;
- To generate a range Ω of analyzed parameters in the network interfaces;
- To generate a range of weights V, where $v_i \in V$, is weight assessment of the characteristics $\omega_i \in \Omega$;
- Reset to zero s_i, for all $s_i \in S$;
- To build up a set of total weights S, where $s_i \in S$, is the total weight of the network interface $\psi_i \in \Psi$;
- To choose the parameter $\omega_i \in \Omega$, $s_i = s_i + F(O(\psi_i \in \Psi, \omega_i \in \Omega)) \cdot v_i$, where $O(\psi_i \in \Psi, \omega_i \in \Omega)$ is a function equal to deviation of the network analyzed parameter $\psi_i \in \Psi$, from the parameter reference value $\omega_i \in \Omega$; $F(x)$ is a serial number in the ascending-sorted set of values for the function $O(\psi_i \in \Psi, \omega_i \in \Omega)$;
- To eliminate ω_i from Ω, if $\Omega \neq \varnothing$, go to paragraph 6;
- To choose $s_{\min} = \min\limits_{s_i \in S} s_i$, that means that the network interface ψ_i is best suited for this problem of automation.

Table 1 is a comparative table of parameters of the most popular local microcontroller networks in respect of the busbar characteristics.

Table 1. Specification comparison of busbars.

Busbar type	*RS-485*	*I²C*	*LonWork*	*CAN*	*MicroLAN*
Max length of wire, m	2000	8	72	40	300
Max number of attachment units	32	14	127	127	2^{56}
Mains supply	-	+	-	-	+
Number of wires in a busbar	2	4	2	2	2
Availability of stand-alone devices	-	-	+	-	+

In analyzing networks as well as their performance and search for optimal data transfer routes, it is helpful to apply the graph theory [6]. The network, may be represented as graph $G(V, U)$, where V is a set of graph vertexes, U is a set of graph arcs. Let set V consist of the network nodes, the network manager as well as points where the direction of the network information flow changes. The weight of the arc in the graph is as follows [7]:

$$u_{ij} = \sqrt{(x_i - x_j)^2 + (y_i - y_j)^2}\,, \tag{1}$$

where x_i, y_i are Cartesian coordinates of vertex v_i; x_j, y_j are Cartesian coordinates of vertex v_j.
In [8] the notion of the vertex separation number is introduced:

$$C(v_i) = \max_{v_j \in V}[d(v_i, v_j)]\,, \tag{2}$$

where $d(v_i, v_j)$ is the shortest distance between vertexes i and j.

$$d(v_i, v_j) = \sum_{k,j \in T_{ij}} u_{kj}\,, \tag{3}$$

where T_{ij} is a graph simple path $G(V, U)$ with the beginning in vertex v_i and the end in vertex v_j [7-9].

Thus, the length of the maximum segment is as follows:

$$C(v_m) = \max_{v_j \in V}[d(v_m, v_j)], \quad (4)$$

where v_m is the vertex where the network manager is located.

The graph radius, being the minimum one of the eccentricities of the connected graph vertexes, is determined with the expression [9]:

$$R_0 = \min[d(v_m, v_j)]. \quad (5)$$

The use of the graph theory makes it possible to search the optimal route for transfer of the information control packets. In searching this route, such options as reliability of the data transfer, network delay, discard probability, channel loading, etc. should be taken into account.

5. Setting the network load simulation problem

The main parameter of the data network is the network traffic. In this system, traffic is a random fuzzy variable. While studying and assessing the performance of the proposed model, one may make load forecasts both for certain network channels and for the network as a whole. This simulation system may be a part of the computer-aided design engineering of data-processing networks.

Simulation time means step-by-step time, i.e. network traffic changes at all the network nodes with every step. Alternators of the network traffic are divided into the following types: expansion-operating generators (generators that go off just once per a finite number of steps of time); and continuously operating ones.

At each regular moment of the simulation time, the unit of the flow corresponding to the port acts at each of the node ports and communication module. The frequency of occurrence of each of the units depends on intensity specified for each of the units as a fuzzy random value. Based on this sequence, integral assessments of the total load of the channel, sub-network as well as the network as a whole may be produced. The methods are given below. The following factors are assessed for every network process:

- The volume of the transmitted data for each class of service;
- Uniformity of the flow of the transmitted data for each class of service;
- Distribution of the operation intensity for 24 hours.

The channel for data transmission is assessed according to:

- Maximum utilization of the channel;
- Reliability of transmission via the channel;
- Dependence of transmission reliability on the traffic volume.

Assessment is produced as a fuzzy random value (FRV) of the following type:

$$T_g = \{A1/P1, A2/P2, A3/P3\}, \quad (6)$$

where A1, A2, A3 are certain fuzzy sets and P1, P2, P3 are a fuzzy membership function.

In engineering systems the description of parameters as a tree-type classification is widely spread. For the sake of convenience while setting these parameters by experts, it is proposed to apply such arrangement in setting the terms for every linguistic variable. As an assessment, the main fuzzy set and values specification with a various range of detailing are used. The sequence of such arrangement is as follows:

- Linguistic variables, required to assess the flow, are identified;
- A number of terms is selected for each linguistic variable and physical values for their marginal conditions are determined;
- For qualitative evaluation a main fuzzy set and qualifying ones with a different refining grade are specified;
- The membership function for the fuzzy set is defined for each term.

The linguistic variable is also defined as a tuple <p,T,X,G,M>, where p is the title or name of the linguistic variable, T is a ground term-set of the linguistic variable or a set of its values (terms), any of them is a title of a certain fuzzy variable; X is a scope of fuzzy variables, which are included into the definition of the linguistic variable p; G is a syntactic procedure which describes the process of new values formation for this linguistic variable from the set T; M is a procedure which makes possible to match a certain substance for every new value of this linguistic variable by forming a relevant fuzzy set.

6. Conclusion

This results in great excess of hardware facilities in the electric power system and its increased cost. Furthermore, the main criterion for choice of topology is its efficiency requirement and cost, the expected upgrading and limits of the existing technologies are of secondary significance. Based on the review of the methods for analyzing the flows in the information processing networks within the automated control systems for the technological process and assessment of the reviewed methods by the selected set of requirements and analysis, one may conclude that it is reasonable to use a combination of graph flow algorithms and queueing techniques.

Moreover, the proposed techniques for choosing the network arrangement may be used which is reasonable to apply in conjunction with modelling of the load for the network under review.

References

[1] Zhilenkov A A and Chernyi S G 2014 *Bulletin of the Samara State University of Posts and Ways* **1(23)** 104-110
[2] Chernii S G and Zhilenkov A A 2014 NTV *SPbGPU. Computer science. Telecommunications. Management* **3(198)** 28-36
[3] Chernyi S and Zhilenkov A 2015 *Transport and Telecommunication* **16** 73-82
[4] Alekseev N A and Baranov A P 2000 *St. Petersburg, Admiral Makarov State Maritime Academy*
[5] Chernyi S 2015 *Metallurgical and Mining Industry*. **7(3)** 60-65
[6] Fatuev V A et al 1998 *Reliability of automated Iformation Systems: manual, Ed., Tula*
[7] Ahuja R K, Magnati T L and Orlin J B 1993 *Theory, Algorithms, and Applications*
[8] Director S and Rocker R 1974 *(Moscow: Mir)*
[9] Gorbatov V A, Smirnov M I and Khlytchev I S 1991 (*Moscow: Energoizdat)*

A behavior model of the viscoelastic system with a melting component under load

Yu A Chumakov[1] **and A G Knyazeva**[1,2]
[1] Institute of Strength Physics and Materials Science of the Siberian Branch of the Russian Academy of Sciences, 2/4, Akademicheskiy ave., Tomsk, 634021, Russia
[2] Tomsk Polytechnic University, 30, Lenina ave., Tomsk, 634050, Russia

E-mail: chya@ispms.tsc.ru, anna@ispms.tsc.ru

Abstract. A numerical investigation of a stressed-strain state powder mixture with a melting component is carried out for the conditions of one-axial loading and nonstationary heating. The state of plane stress was assumed. The influence of the coupling between thermal and mechanical processes on temperature, stress and strains fields was investigated.

1. Introduction

The influence of mechanical stresses on temperature fields and rates of solid-phase chemical reactions has been studied by many authors. Based on experimental data similar to [1-3] theoreticians suggest mathematical models taking into account the features of solid-phase reactions (a heterogeneous character of the reactions, the important role of transport processes and the interrelation between various physical phenomena) [4]. However, the coupled processes of a different physical nature are not included in traditional models. Nevertheless, it was shown in [5-7] that coupling between mechanical, thermal and chemical processes leads to features in the stressed-strain state and reaction front propagation.

This work investigates the stressed-deformed reactive substance state based on melting one of the components, coupling of deformation, temperature and concentration fields.

2. Problem formulation

The fixed plate of the reactive substance under external uniaxial tension is considered. We assume the reactive substance is the compaction of a metallic powder into a plate form. Length of the plate L_x, width L_y, height L_z and conditions $L_z \ll L_x$, $L_z \ll L_y$ are satisfied. The chemical reaction and heating propagates along the X-axis. The external loading direction is perpendicular to that of chemical reaction propagation. We consider the melting one of a component powder mixture. The exothermal reaction is assumed to describe summary schema $A \rightarrow B$.

To estimate a stressed-deformed state of the plate and the influence of coupling field on a solid-phase reaction we consider the uniaxial tension problem (with a chemical reaction), which consists of two parts [7]. These parts are independently solved.

The first part of the problem includes the energy equation for a viscoelastic body [8], the initial and boundary conditions

$$\sigma_{ij}\frac{\partial\varepsilon_{ij}}{\partial t}+c_\varepsilon\rho\frac{\partial T}{\partial t}=\nabla\cdot\lambda_T\nabla T+Q\varphi(\eta,T)-3KT\alpha_T\frac{\partial\varepsilon_{kk}}{\partial t}, \quad (1)$$

$$t{=}0\text{: } T{=}T_0.\ x=0: -\lambda\frac{\partial T}{\partial x}=q(t)\,,\quad x=L_x: \frac{\partial T}{\partial x}=0,\quad y=0: \frac{\partial T}{\partial y}=0,\quad y=L_y: \frac{\partial T}{\partial y}=0\,, \tag{2}$$

where c_ε, ρ and λ_T are the effective heat capacity, density and thermal conductivity, T is the plate temperature, $q(t)=q_0 f(t)$ is an external heat source, q_0 is the maximal density of the heat flux; x and y are space coordinates, $\varepsilon_{kk}=\varepsilon_{xx}+\varepsilon_{yy}+\varepsilon_{zz}$ is the first invariant of the strain tensor, α_T is a thermal expansion coefficient, K is an isothermal bulk modulus, σ_{ij} ε_{ij} are the components of stress and strains tensors. We consider that the heat source varies with the time.

Let us consider the heat capacity, density and thermal conductivity and mechanical properties dependence on temperature and compound, for example

$$c_\varepsilon\rho=c_1\rho_1\chi_1+c_2\rho_2\chi_2+c_p\rho_p\chi_3,$$

where $\chi_{1,2,3}$ are volume fraction of 1,2 component (reactants) and reaction product, respectively.

The melting of one of the components is assumed as

$$c_2\rho_2=(c_2\rho_2)_0+L_m\delta(T-T_m),$$

where index "0" is the solid properties, T_m is the melting temperature, L_m is the heat of phase transition, δ is the Dirac delta function

The present model includes a simple linear viscoelastic model (Maxwell body). The deviatoric stress-strain relation for a Maxwell body is [8].

$$e_{ij}=\frac{1}{2\mu}s_{ij}+\frac{1}{2\eta_v}s_{ij},$$

where e_{ij}, s_{ij} are the dimensional deviatoric strain and stress tensors, respectively, μ is the shear modulus), η_v is the viscosity of the viscoelastic solid.

The second part of the problem describes the mechanical equilibrium of the plate. We consider stressed-deformed state of the plate as a plane stress state. So, $\sigma_{zz}=0$ and two equilibrium equations are [9]

$$\frac{\partial\sigma_{xx}}{\partial x}+\frac{\partial\sigma_{xy}}{\partial y}=0,\quad \frac{\partial\sigma_{xy}}{\partial x}+\frac{\partial\sigma_{yy}}{\partial y}=0\,. \tag{3}$$

The complete stress-strain relation for a Maxwell body may then be stated [8] as

$$\frac{\partial\sigma_{xx}}{\partial t}+\frac{\mu}{\eta}\sigma_{xx}=2\mu\frac{\partial\varepsilon_{xx}}{\partial t}+3\lambda\frac{\partial\varepsilon_{kk}}{\partial t}-3K\left[\frac{\partial\omega}{\partial t}-\frac{\mu}{\eta}(\varepsilon_{kk}-\omega)\right]$$

$$\frac{\partial\sigma_{yy}}{\partial t}+\frac{\mu}{\eta}\sigma_{yy}=2\mu\frac{\partial\varepsilon_{yy}}{\partial t}+3\lambda\frac{\partial\varepsilon_{kk}}{\partial t}-3K\left[\frac{\partial\omega}{\partial t}-\frac{\mu}{\eta}(\varepsilon_{kk}-\omega)\right]$$

$$\frac{\partial\sigma_{xy}}{\partial t}+\frac{\mu}{\eta}\sigma_{xy}=2\mu\frac{\partial\varepsilon_{xy}}{\partial t}$$

$$0=2\mu\frac{\partial\varepsilon_{zz}}{\partial t}+3\lambda\frac{\partial\varepsilon_{kk}}{\partial t}-3K\left[\frac{\partial\omega}{\partial t}-\frac{\mu}{\eta}(\varepsilon_{kk}-\omega)\right]$$

Also, we use the equation of compatibility of strains

$$\frac{\partial^2\varepsilon_{xx}}{\partial x^2}+\frac{\partial^2\varepsilon_{yy}}{\partial y^2}=\frac{\partial^2\varepsilon_{xy}}{\partial x\partial y}$$

We assume that uniform load P along the y axis was applied to the ends of the plate at $x=0$ and $x=L_x$. The boundaries of the plate are free at $y=0$ and $y=L_y$:

$$x=0: x=L_x: \ \sigma_{xx}=0,\ \sigma_{yy}=P,\ \sigma_{xy}=0;$$

$$y=0,\ y=L_y: \ \sigma_{xx}=0,\ \sigma_{yy}=0,\ \sigma_{xy}=0.$$

The strain tensor components follow from (4).
The problem was solved numerically. The differential equations included into system (1)-(3) were approximated by difference equations; the resulting system of linear algebraic equations was solved by a sweep method. In our calculations, we determine the temperature, concentration, stress and strains distributions in the plate at different time instants by varying the physical parameters.

We consider that the reactive plate consists of metal powder Al+Ni. All calculations presented further were carried out for parameters [10]:

1) Al_2O_3: $\lambda_p = 35$ W/(m·K), $\rho_p = 3750$ kg/m^3, $c_p = 858$ J/(kg·K), $L_x = L_y = 0.01$ m, $\alpha_{T,p} = 8\cdot 10^{-6}$ K, $\nu_p = 0.33$, $E_p = 370$ GPa.
2) Ni: $\lambda_m = 91$ W/(m·K), $\rho_m = 8902$ kg/m^3, $c_m = 416$ J/(kg·K), $\alpha_{T,m} = 13.5\cdot 10^{-6}$ K, $\nu_m = 0.3$, $E_m = 200$ GPa.

The effective heat capacity satisfies the mixture rule (this follows from thermodynamics)

$$c\rho = c_M \rho_M (1-\eta) + c_p \rho_p \eta .$$

3. Results and Discussion

Figure 1 shows a special distribution of temperature, the first invariant of the stress tensor at the moment $t = 0.015$ and maximum temperature of the plate. We can observe the impulse character of the heating plate (Figure 2a): at first, the maximum temperature increases smoothly until the end of impulse t_i, then the temperature decreases for some time, that is, until t_p, next impulse leads to further heating (curve 1, Figure 1a) or the heat release due to increased active chemical conversion (curves 2 and 3, Figure 1a). The inhibition of the reaction product layer results in slower increasing of temperature due to the chemical reaction (curve 3, Figure 1a, c).

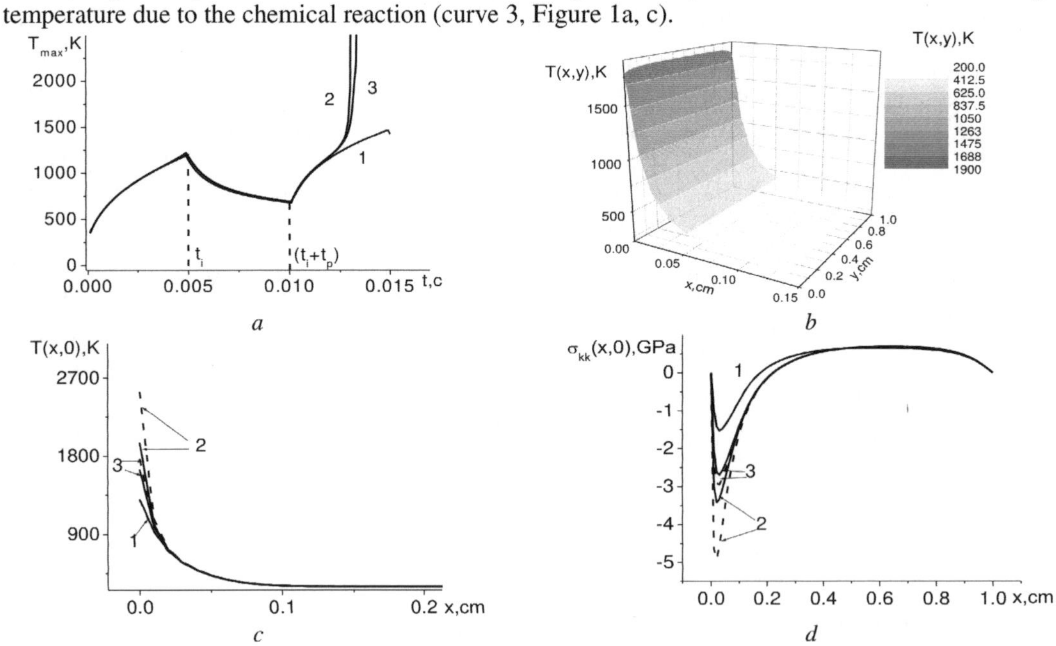

Figure 1. Maximum temperature of the plate (a) and spatial distributions of the plate temperature (b), the temperature (c) and the first invariant of the stress tensor (d) along x-axis at the moment $t = 0.015$ c: $t_i = t_p = 5\cdot 10^{-3}$c, $n = 10$, $q_0 = 10^7$ W/m^2, $P = 1$ GPa. 1. The chemical reactions are absent; 2. The first-order reaction; 3. The inhibition of the reaction products is taken into account in the kinetic function ($m_0 = 2$, $n_0 = 0$).

The assumption of coupled thermal and deformation fields leads to arising of a non-unidimensionality border effect. The temperature profile has a 'bend' due to mechanical dissipation and interaction of thermal and mechanical processes (Figure 1b).

If we consider only thermal stresses, the maximum mechanic stresses reach a value of 1.5 GPa (curves 2, 3, Figure 1d). The concentration expansion of reaction products leads to an increase of stress, that of magnitude becomes comparable to the applied external load (curve 2, Figure 1 d).

4. Conclusion

A model of uniaxial tension plate heating with allowance for interaction of thermal and mechanical processes, as well as the concentration expansion was proposed in the paper. As the example of Ni+Al powder system shows, the thermal and mechanical coupling processes lead to non-unidimensionality border effects and decrease of the maximum temperature. Also it was shown that the internal stresses caused by a chemical reaction can reach values comparable to the applied external load, taking into account the coupling of thermal, concentration and mechanical fields,.

Acknowledgments

This work was supported by state project 'Development of physical principle of metallic and composite material designing with multimodal internal structure' № 01201350716.

References

[1] Butyagin P Yu 1994 *Russ. Chem. Rev.* **63(12)** 965–977
[2] Enikolopyan N S, Voleva V B,Hzardzhyan A A 1987 *Doklady Akademii Nauk USSR* **292(5)** 1165–1169
[3] Enikolopyan N S 1989 *Zhurnal fizicheskoi khimii* **63(9)** 2289–2298
[4] Smolyakov V K 1989 *Combustion, Explosion and Shock* **25(5)** 582–587
[5] Evstigneev N K, Knyazeva A G 2010 *Comb., expl., and shock waves* **46(3)** 307–314
[6] Chumakov Y A, Knyazeva A G 2014 *AIP Conf. Proc.* **1623** 95–98
[7] Chumakov Y A, Knyazeva A G 2015 *Applied Mechanics and Materials* **756** 402–407
[8] Boley B A, and Weiner J H 1995 *Wiley New York*
[9] Sedov L I M.:Nauka
[10] Babichev A P, Babushkina N A, Bratkovskii A M 1991 (*Moscow: Energatomizdat)*

Arbitrary mechanical system description by a symbolic line

O Dmitrochenko[1], A Mikkola[2], A Olshevskiy[1,3]
[1] Bryansk State Technical University, 7, 50 years of October st., Bryansk, 241035, Russia
[2] Lappeenranta University of Technology, 34, Skinnarilankatu, Lappeenranta, 53850, Finland
[3] Konkuk University, 1, Hwanyang-dong, Gwangjin-gu, 143-701, Seoul, Korea

E-mail: dmitroleg@rambler.ru

Abstract. Asingle-line symbolic notation is proposed for description of an arbitrary multibody system. The kinematics is represented by a sequence of elementary transformations, each of those being marked by a reserved alphabetic character. Force and constraint links between the bodies are also defined by reserved characters. The parameters of the system, such as identifiers of degrees of freedom, inertia parameters and others, are assigned default names if not specified. However, user-defined names, parameters and functions can be placed instead if needed. The proposed description in its shortest form is suitable for academic purpose to identify only the essential properties of a multibody system. In an extended form, by explicit mentioning names of variables and parameters and other data like initial conditions, this description can serve as input data for a multibody analysis software. Lots of examples from the academic area and technical applications are given to show the applicability of the description.

Notation

g	applies a uniform gravity force to all bodies (downwards the vertical axis by default)
i, j, k	define a constant or known-as-function-of-time translation along local axes x, y, or z
l, m, n	define a constant or known-as-function-of-time rotation around local axes x, y, or z
o	obtains the orientation from body 0; obtains the orientation from body index 3: $o3$
q	the quaternion; adds four Euler parameters $q_{01},q_{11},q_{21},q_{31}$ as DOFs and a constraint
u, v, w	introduce a translational degree of freedom along local axes x, y, or z, accordingly
x, y, z	introduce a rotational degree of freedom around local axes x, y, or z, accordingly
@, @', '	names for a coordinate and its velocity in a functional definition of a force
>	starts description of the next body relative to the previous one in the kinematical chain
\	starts kinematic description of the next body after body 0; use \5 when starting from body 5
<	starts right-to-left description; sometimes it is convenient to make the description shorter
[]	moments of inertia by default; zero inertia [0]; explicit moments $[I_x, I_y, I_z, I_{xy}, I_{yz}, I_{zx}]$
{}	defines the radius vector of the center of gravity by default; arbitrary translation $\{a, b, c\}$
°	acoordinate in expressions; an initial value for d.o.f. °(x_0); for body °(r_{x0}, r_{y0}, r_{z0})
'	velocity in expressions; initial velocity for d.o.f. '(v_0); for body '(v_{x0}, v_{y0}, v_{z0})
"	an applied force: for d.o.f. "($-c_{11}$@$-d_{11}$@'); for body "(f_x,f_y,f_z, t_x, t_y, t_z), with torque
.	apin/spherical joint: #$body_1$ > $\{a_1, b_1, c_1\}$. $\{a_2, b_2, c_2\}$ < #$body_2$; with joint points {…};
%	rolling without slipping: #$body_1$>{$surface_1$(&₁, &₂)}%{$surface_2$(&₁, &₂)}<#$body_2$
$	$($m_{b7}$) defines identifier m_{b7} to be the mass of the current body
:	a massless link; actuator :($L(t)$); a link with given length and mass :(L, M) or :$ by default
:"	a bipolar force; spring-damper :"($-c_{11}$@$-d_{11}$@'); also, a penalty force in contact

1. Introduction

Denavit–Hartenberg (D–H) method applied in robotics has been developed in [1]. The method allows representing the kinematical structure of a multibody system in a way, which is appropriate for automated generation of kinematical and dynamical equations of the system. The method is based on description of kinematics of a successive body relative to the previous one in the same kinematical chain. In a classic form, the kinematics is represented in the form of a diagram. The diagram representation is not suitable for the use as input data. Better appearance is achieved if Denavit–Hartenberg parameters are arranged in a table. The table already can serve as input data for a multibody software. However, this method is hard to apply for systems with closed kinematical loops; moreover, the table usually contains lots of zero values and rotations by ± 90°, *i.e.* this description is redundant in practice. There have been successful attempts to extend the D–H formulation for mechanisms containing closed loops, for example in [2]. However, this leads to increasing the number of parameters used per kinematic link, and, consequently, to growing the size of the tables describing the mechanism. According to the current research, a short and pictorial text description of the mechanisms can be generated; the description represents kinematics and dynamics of the mechanisms completely and is considerably shorter than any of earlier notations available in literature [3, 4].

2. Symbolic description of single-body systems

Translational motion of a material point is the simplest example of a mechanical system. The point moving along axis x can be described by single character u. A point moving along axis y corresponds to v in the same figure, and, finally, w if along axis z. Note that this single-character notation shows the direction only and leaves other parameters unidentified. If one wants to give all data for this degree of freedom, the description can be extended as follows:

$$u(x_p)^\circ(x_0)\,'(v_0)\,\$(m_p)\,''(-c_1 u + c_3 u^{\wedge 3} - d_1 u' - d_0 \operatorname{sign} u' + a \sin ft) \tag{1}$$

Here, u introduces a degree of freedom along axis x, brackets ( ) rename it to x_p, symbols °( ) define initial value x_0, prime '( ) defines initial velocity v_0, currency sign $( ) gives mass m_p of the body, double prime "( ) imposes the applied force along the degree of freedom. The force is elastic-dissipative with parameters c_i and d_i; it also has dry-friction part $-d_0 \operatorname{sign} u'$ and a time-dependent harmonic part with amplitude a and frequency f.

Two-dimensional point massmovingin plane can be represented by short code uv in Figure 1. Subscript$_g$ indicates that gravity force is to be added by default, see below. **Free point moving in 3D** can be described by uvw_g (Figure 1, right).

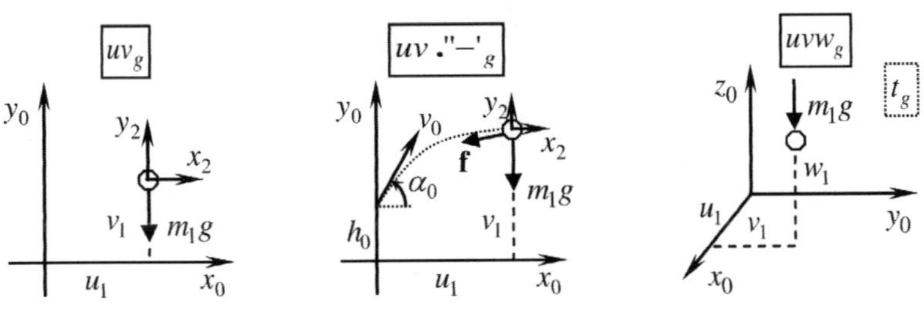

Figure 1. A free point in two and three dimensions.

Figure 1, center, represents a simple 2D ballistic problem taking into account viscous air drag, which is linear in velocities. Its description in terms of the proposed code looks like:

$$u''(-cu')_{\circ 0\,'(v_0 \cos\alpha_0)}\, v''(-cv')_{\circ h_0\,'(v_0 \sin\alpha_0) g} \longrightarrow \begin{cases} m_1 \ddot{u}_1 = -c\dot{u}_1 & u_1(0)=0 \quad \dot{u}_1(0)=v_0\cos\alpha_0 \\ m_1 \ddot{v}_1 = -c\dot{v}_1 - m_1 g & v_1(0)=h_0 \quad \dot{v}_1(0)=v_0 \sin\alpha_0 \end{cases} \tag{2}$$

In equation (2), description $u''(-cu')\,v''(-cv')_g$ contains forces along degrees of freedom separately. Alternatively, the **point force** can be given at once by all its components as $uv''(-cu',-cv')_g$, or, using index notation for degrees of freedom, as $uv''(-c@'_1,-c@'_2)_g$. This can be further shortened to $uv.''-c'_g$, where prime ' refers to the vector of velocity, or, by dropping coefficient c by default, even to shortest notation $uv.''-'_g$ shown in Figure 1, center.

Rotational motion of a simple pendulum is a frequently used benchmark multibody system. In Figure 2, several typical variants are presented. After rotation around axis z by angle ζ_1, translation along axis x_1 by constant value i_1 results in what is usually referred to as an R-link in robotics; in Figure 2 this case is denoted by zi_g. Alternatively, the shift along positive direction of axis y_1 by constant distance j_1 corresponds to inverted pendulum zj_g. If the latter shift occurs along the negative direction of axis y_1, the conventional clock pendulum is meant, denoted by $z\text{-}j_g$ in the figure. Eventually, at least eight pendulum designations are possible including $z\text{-}i_g$, $\text{-}z\text{-}i_g$, $\text{-}zj_g$, $\text{-}z\text{-}j_g$, which are not shown in the picture.

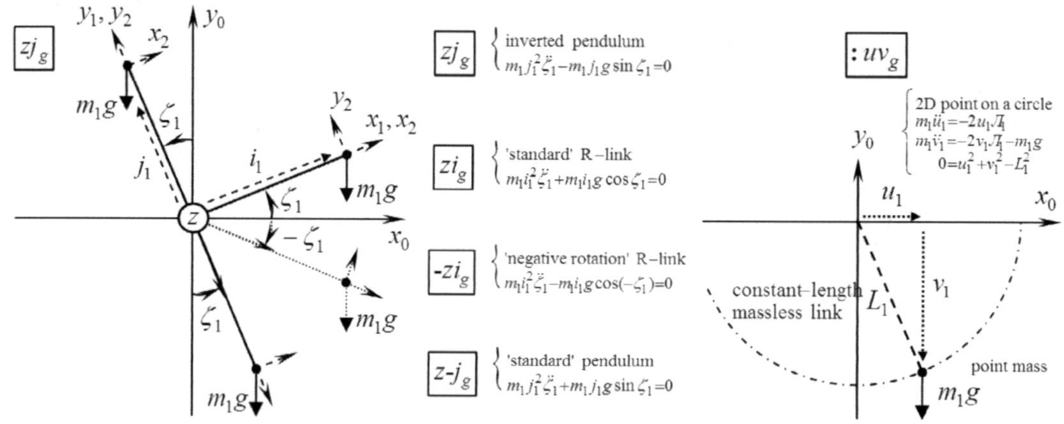

Figure 2. Various descriptions of a simple pendulum.

Moments of inertia are *not included* in the equationsof motion by default *if the last elementary transformation is translational*, as for the case shown in Figure 2. To ultimately define the moments of inertia, square brackets [] are used. The empty brackets introduce three principal moments of inertia and assign default names I_{x1}, I_{y1}, I_{z1} to them. Default names likeI_{z1} can be overridden by any name, let us say C, as follows: $z[C]i_g \rightarrow$ inertia moment C given at the joint point $C\ddot{\zeta}_1 = -m_1 g i_1 \cos\zeta_1$

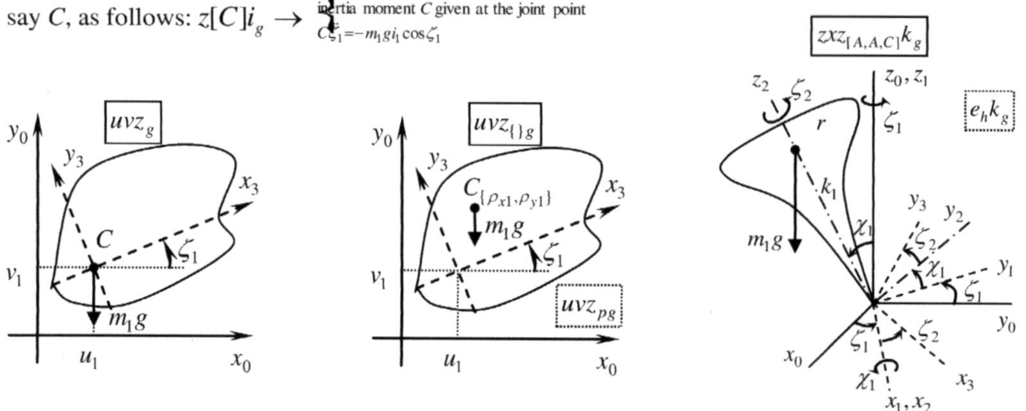

Figure 3. Rigid bodies in two and three dimensions.

Link is a common element of multibody systems. The same simple pendulum can be obtained if, as shown in Figure 2, right, free planar point mass uv_g is constrained to the origin with the help of a massless link of a constant length, denoted by colon **:** so that the description: uv_g (or $uv{:}_g$) holds for the system. In this case, link: contributes constraint $u_1^2+v_1^2-L_1^2=0$ with default length L_1, and Lagrange multiplier Λ_1 into equations of motion, shown in Figure 2, right. The link with given length a has designation **:**(a).

Kinematics of **a rigid body in 2D** is naturally described by combination uvz, showing its three degrees of freedom, Figure 3, left.

The simplest form uvz results in situation when the body centre of mass C is placed in the origin of the rotating reference frame. This leads to decoupled translational and rotational motions. However, a **shifted center of mass** is common in applications. This aspect is designated by curly brackets { }. The empty brackets introduce default values ρ_{x1}, ρ_{y1}, ρ_{z1} for the coordinates of the center of mass in the body reference frame,

Figure 3, middle. **A 3D body** using Cardan angles for orientation can be described by line $uvwxyz_g$ not shown in the figure. Another case is Lagrange's one, shown in

Figure 3, right. As its designation $zxz_{[A,A,C]}k_g$ suggests, a specific symmetry of moments of inertia around the last rotation axis is required, and the center of mass is shifted along this axis.

3. Symbolic description of single-body systems

3.1. Open kinematical chains

The simplest multiple-body system is that consisting of several particles; say three, as in Figure 4, left. Its designation $uv\backslash uv\backslash uv$, or just $(\backslash uv)^{\wedge 3}$, contains degrees of freedom uv for each particle separated by backslash \ meaning that description of a new particle starts from the origin.

This system turns to a general Kepler's three-body problem if so-called bipolar forces are accounted in between the bodies representing Newton's gravitational forces, Figure 4, right.

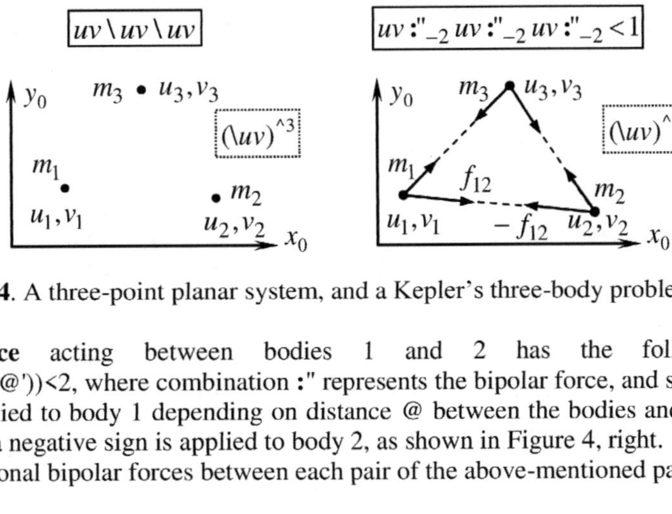

Figure 4. A three-point planar system, and a Kepler's three-body problem.

A bipolar force acting between bodies 1 and 2 has the following general description\1:"(f_{12}(@,@'))<2, where combination :" represents the bipolar force, and scalar function f_{12} defines the force applied to body 1 depending on distance @ between the bodies and its velocity @'. The same force with a negative sign is applied to body 2, as shown in Figure 4, right.

Applying gravitational bipolar forces between each pair of the above-mentioned particles, one can obtain

$$uv\backslash uv\backslash uv\backslash 1{:}"(-m_1m_2G/@^{\wedge 2})<2\backslash 2{:}"(-m_2m_3G/@^{\wedge 2})<3\ \backslash 3\ {:}"(-m_3m_1G/@^{\wedge 2})<1$$

or

$$uv{:}"(-m_1m_2G/@^{\wedge 2})\ uv{:}"(-m_2m_3G/@^{\wedge 2})\ uv{:}"(-m_3m_1G/@^{\wedge 2})\ <1,$$

where the bipolar forces descriptions are inserted between the kinematical descriptions of the bodies. The forces are algebraic of degree –2, which can be shortened to :"$_{-2}$ as in Figure 4, right.

The next simple case is a multiple-pendulum system of two pendulums 1 and 2 that are not interconnected kinematically, depicted in Figure 5, left.

In such case, kinematics of pendulum 2 is described starting from the fixed reference frame, referred to by index 0. The combination of characters \0 or just single backslash \ is used for that.

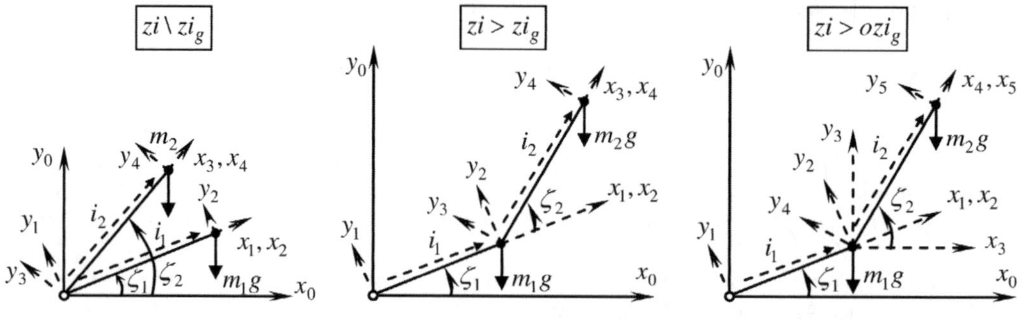

Figure 5. Separate pendulums, relative and mixed connections of two pendulums.

Alternatively, kinematics of pendulum 2 can be described relative to the final coordinate frame of pendulum 1, as shown in the middle of Figure 5. This case of a consecutive connection of bodies is reflected by symbol > and nowthe two pendulums are kinematically interconnected and dynamically influence each other. The third, mixed, situation is possible when next pendulum 2 is attached to previous pendulum 1, but the orientation is described relative to other body, say body 0, as in Figure 5, right. This case is managed by using command $o0$ (value 0 can be omitted).

3.2. Closed-loop mechanisms

Closed-loop systems are usually studied starting with a well-known benchmark, the slider-crank mechanism in Figure 6. The closed loop can be treated in many ways. Starting from the slider body and continuing counter-clockwise along the mechanism's contour, by returning to the origin, reference frame 0, designation $u>zi>izi>iz>0_{g[]}$ is generated, Figure 6, left. Alternatively, going clockwise, description $zi>izi>iz>u>0_{g[]}$ is postulated, Figure 6, middle. Combination >0 corresponds to the loop closure conditions, that, in the planar case, consist of three constraint equations.

A better performance is achieved in Figure 6, right, by cutting the long kinematical chain into two branches, simple pendulum zi, and elliptic pendulum $u>zi$; then, the loop can be closed by means of a **pin joint** denoted by a small dot in the figure. Indeed, dot. is the best suitable symbol to mnemonically denote this type of linkage. Thus, notation $zi>i.i<iz<u_{g[]}$ is suggested here. As one can see, former line $u>zi$ is reversed to be directed towards central dot combination $>i.i<$ representing the pin joint itself and two additional shifts i_2 and i_4 from centers of gravity of each body.

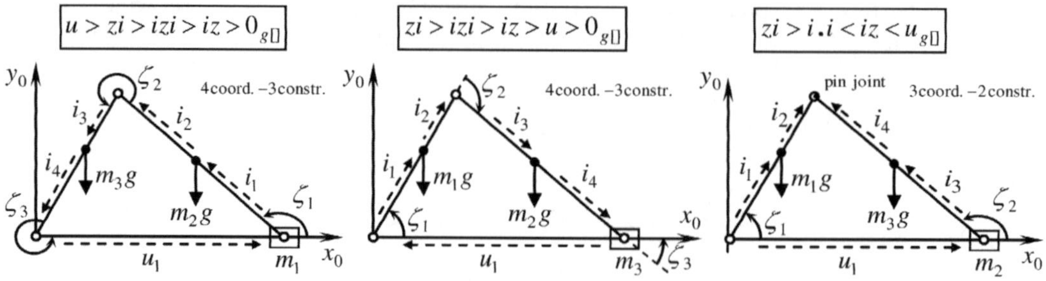

Figure 6. A slider-crank loop mechanism: counter-clockwise, clockwise and mixed (cut by a pin joint).

3.3. Multiple closed loops

Multiple closed loops can also be described in a systematic way. A typical example of such system is a parallel manipulator shown in Figure 7, left. It contains a central body, index 1, which has three degrees of freedom according to description *uvz*, while four kinematic loops are identical, and the description of one of those is given in brackets. Leading backslash \ starts kinematic description from fixed reference frame 0, constant shift *ij* continues with double pendulum *zi* > *izi*; remaining part > *i.ij* < 1 defines the pin joint connecting the last pendulum to body 1, where *i* and *ij* are shifts of a joint point relative to the two bodies.

A three-dimensional parallel mechanism with multiple closed loops is also presented in the same Figure 7, middle, where the famous Gough–Stewart hexapod platform is presented [5]. The position and orientation of the platform (body 1) is given by degrees of freedom *uvwzxy*, while its six legs (cylinders) are modelled by links :" in the brackets; *ij* denote translational shifts w.r.t. to the reference frame of the platform (body \1) and the ground, respectively.

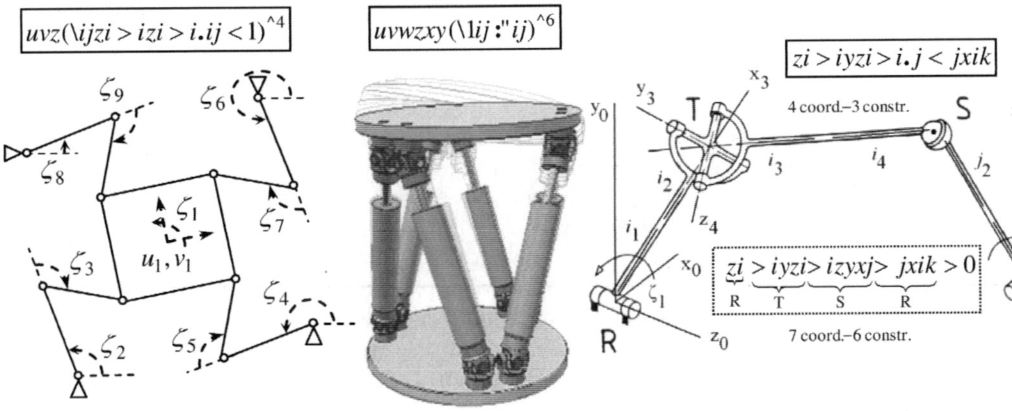

Figure 7. A multiple-loop parallel planar mechanism; Stewart platform in three dimensions.

A spatial four-bar mechanism is shown in Figure 7, right. It is usually referred to as a RTSR mechanism, consisted of abbreviations of its joints: a rotational, a T-joint (or a universal joint), a spherical, and a rotational joint, again. Shortest description *zi* > *iyzi* > *i.j* < *jxik* is obtained, if spherical joint S is modeled by three constraint equations. Almost all elementary transformations are marked in the figure, while the last two, *ik*, are just constant shifts of second rotational joint R w.r.t. the first one in the *xz* plane. Alternative notation *zi* > *iyzi* > *izyxj* > *jxik* > 0 is possible, when spherical joint S is modeled by additional three rotations *zyx*, and the kinematic chain after second rotational joint R around axis *x* and constant shifts *ik* are connected to the origin >0.

```
y#tower4>{-ABx,ABy}z#boom1 >{BGx,BGy}z#jib2>{-GIx+IJx,GIy+IJy}i(JK(t))#arm3>yz-j(RigidLink) #log; g-y[]{}
\#boom1u(BHx)v(BHy)#nodeH:(HF){BGx-FGx,BGy-FGy}<#boom1\#nodeH:(HI){GIx,GIy}<#jib2
\#tower4 {ACx,ACy}:"{BDx,BDy}< #boom1>{BGx-FGx-EFx,BGy-FGy-EFy}:"<#nodeH
; ABx=0.09, ABy=1.115, BGx=2.88, BGy=0.02, GIx=0.095, GIy=0.24, IJx=2.145, IJy=0.008, ACx=0.26, ACy=0.105, BDx=0.3025, BDy=0.104, ...
```

A model of the PATU 655 log crane

Selection of technical examples is concluded here by a model of the PATU 655 log crane considered in reference [6], and represented in Figure 8**Ошибка! Источник ссылки не найден.** together with its main geometrical data.

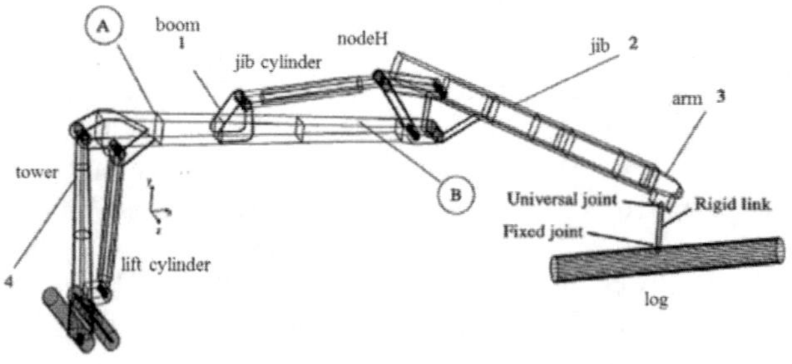

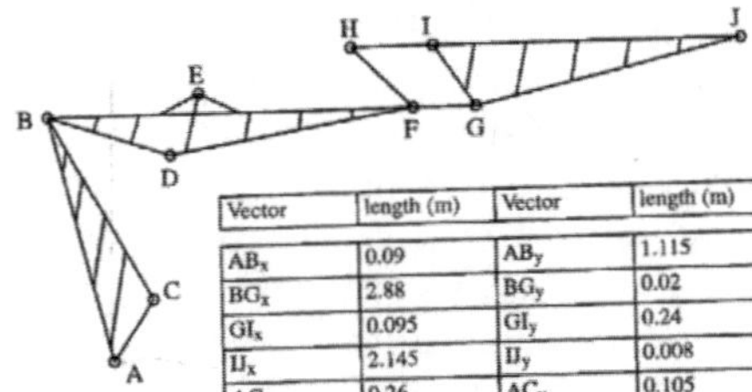

Vector	length (m)	Vector	length (m)
AB_x	0.09	AB_y	1.115
BG_x	2.88	BG_y	0.02
GI_x	0.095	GI_y	0.24
IJ_x	2.145	IJ_y	0.008
AC_x	0.26	AC_y	0.105
BD_x	0.3025	BD_y	1.04
EF_x	1.427	EF_y	0.22
FG_x	0.19	FG_y	0.015
CD_0	0.8	EH_0	1.037

Figure 8. Two models of technical systems.

4. Conclusion

In this paper, a way to describe a multibody system by a short text line, fully representing the kinematics and dynamics of the system, has been proposed. There are two main motivations for this work: for relatively simple mechanical systems this description should be as short as possible to serve as the benchmark designations; and for complex technical systems, it should serve as a language of the input text file for multibody analysis preprocessors.

In the proposed multibody system description, kinematical elementary transformations, various force elements and constraints are represented by reserved alphabetic characters. The parameters of the system, such as identifiers of degrees of freedom, inertia parameters and others, can be assigned default names if not specified. However, user-defined names, parameters and functions can be introduced instead. Examples from the academic area and some technical applications are presented to show the applicability of the proposed text description.

5. Acknowledgement

The research is supported by the Russian Foundation for Basic Research (14-01-00662, 16-51-51022).

References

[1] Hartenberg R S and Denavit J 1955 *J. of Applied Mechanics* **77(2)** 215–221
[2] Jalón J Gd and Bayo E 2005 *Kinematic and dynamic simulation of multibody systems – The real-time challenge*
[3] Dmitrochenko O, Matikainen M and Mikkola A 2014 *Proc. of the 3rd Joint Int. Conf. on Multibody System Dynamics, and the 7th Asian Conf. on Multibody Dynamics* (BEXCO, Busan, Korea)
[4] Dmitrochenko O, Matikainen M and Mikkola A 2014 *Proc. of the ASME 2014 10th Int. Conf. on Multibody Systems, Nonlinear Dynamics, and Control, MSNDC2014* (Buffalo, New York, USA)
[5] Stewart D 1965 *UK Institution of Mechanical Engineers Proceedings* **180(1)** 15
[6] Mikkola A 1997 Studies on fatigue damage in a hydraulically driven boom system using virtual prototype simulations, Ph.D. thesis (Lappeenranta University of Technology, Finland) p 61

Estimation of general and set steps of roof caving in rock mass with excavations at mining. Numerical modelling.

M Eremin and P Makarov
Tomsk State University, 36, Lenina ave., Tomsk, 634050, Russia

E-mail: eremin@ispms.tsc.ru

Abstract. The results of 2D modelling of rock mass elements fracture are shown in the article. The results of modelling are in a good agreement with empirical and theoretical estimations of roof caving steps for the flat-dipping coal seams when the horizons are not so deep (less than 600 m). The estimations of the general and set steps of roof caving are given for the lava conditions at different lengths of the main roof containing sandstone.

1. Introduction

The modern techniques of mining of mineral resources cause a sufficiently non-equilibrium state of rock mass surrounding the underground openings. That is why it is very important to choose the parameters of support machines in order to provide the safe conditions of work for miners, machines and other equipment.

According to different data, a proved coal reserve is enough for approximately 200 years of mining and makes an essential share in energetic complex of many countries which is going to be equal to gas and oil industry by 2030. However, the negative manifestations of rock pressure at mining decrease an economic effectiveness of mines because of different accidents, such as rock bursts, roof collapses and so on. These accidents cause fracture of prepared excavations, coal headings. Practice of rock pressure management demands an understanding of general features of damage accumulation in elements of rock mass at mining of mineral resources, for example, coal. The knowledge of general and set steps of roof caving is necessary for correct choice of lining parameters.

The modern methods of geomechanical modelling provide an opportunity for investigating the stress-strain state evolution of rock mass at different horizons of underlying coal seams if the structural, elastic, non-elastic and strength characteristics are taken into account. Then the steps of general and set steps of roof caving could be estimated. The amount of geotechnical problems is rather wide. The majority of investigations are dedicated to the estimations of a stress-strain state around the underground tunnels or excavations with analytical [1-3] and numerical [4-7] solutions of the boundary problem. The results of these estimations provide good information about disintegration of rocks around the excavations and/or tunnels due to plastic deformations. To our mind, the most interesting papers concern the temporal mechanical behaviour of rock mass with excavations as soon as it supposes the investigation of stability of development workings, rock mass elements (roof and floor) and the influence of rock pressure manifestation during mining [8-11]. The stability and/or the development of inelastic strains/fracture around the underground openings is usually considered as the boundary problem of the media with tunnel of a circle or other cross-section form subjected to multi-axial loading [6, 7, 12]. The minority of the papers is observed concerning the problems of modelling the damage accumulation at coal face moving in rock mass elements [13-15]. This paper is dedicated to numerical modelling of non-elastic deformation and the fracture of rock mass elements at coal face moving.

2. Mathematics

The mathematical model includes the fundamental conservation laws of continuum mechanics: conservation of mass, impulse, energy:

$$\frac{d\rho}{dt}+\rho\, div\bar{\upsilon}=0,\ \rho\frac{d\upsilon_i}{dt}=\frac{\partial\sigma_{ij}}{\partial x^j}+\rho F_i,\ \frac{dE}{dt}=\frac{1}{\rho}\sigma_{ij}\frac{d\varepsilon_{ij}}{dt}$$

The equation of state (EOS) is taken in the form of isotropic Hooke's law.

$$\dot{\sigma}_{ij}=\lambda\left(\dot{\theta}^T-\dot{\theta}^P\right)\delta_{ij}+2\mu\left(\dot{\varepsilon}_{ij}^T-\dot{\varepsilon}_{ij}^P\right)$$

The non-elastic deformation of rock mass elements is descripted within the Drucker-Prager model with a non-associated flow rule.

$$f\left(\sigma_{ij}\right)=\frac{\alpha}{3}I_1+I_2^{1/2}-Y$$

The components of inelastic strain rates tensor are defined according to the Nikolaevskii plastic potential.

$$\dot{\varepsilon}_{ij}^P=(S_{ij}+\frac{2}{3}\Lambda(Y-\frac{\alpha}{3}I_1)\delta_{ij})\dot{\lambda}$$

The fracture of the rock mass elements is considered within the theory of damages accumulation.

$$D=\int_{t_0}^{t}\frac{(\sigma_{\text{тек}}-\sigma_0)^2\text{dt}}{\sigma_*^2 t_*},\ \sigma_*=\sigma_{0*}(1+\mu_\sigma)^n$$

In more details, an applied model is considered in [16].

3. Results and discussion

The geological conditions of coal seams correspond to the flat-dipping formations at different angles. In the investigated case, the depth of the coal horizon is 300 m. In Figure 1, one can see the stratigraphic column. The corresponding physical – mechanical properties of each layer are shown lower.

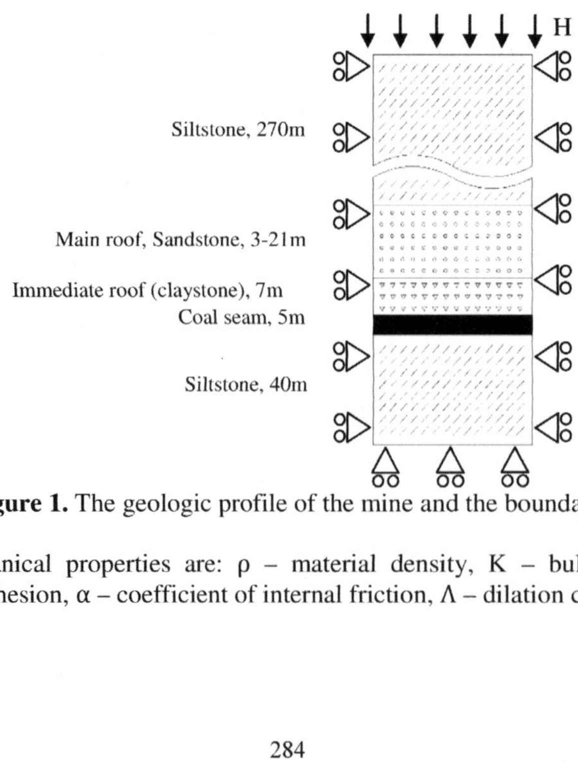

Figure 1. The geologic profile of the mine and the boundary.

The physical-mechanical properties are: ρ – material density, K – bulk modulus, μ – shear modulus, C_0 – initial cohesion, α – coefficient of internal friction, Λ – dilation coefficient.

The parameters of the rock mass layers are as follows:

- Siltstone layer: ρ=2.5g/cm^3, K=9GPa, μ=8.7GPa, C_0=7MPa, α=0.62, Λ=0.22
- Main roof, sandstone layer: ρ=2.2g/cm^3, K=12.8GPa, μ=5.34GPa, C_0=12MPa, α=0.6, Λ=0.12
- Immediate roof, claystone layer: ρ=2.5g/cm^3, K=9GPa, μ=8.7GPa, C_0=6MPa, α=0.62, Λ=0.22
- Coal seam layer: ρ=1.4g/cm^3, K=1.95GPa, μ=0.042GPa, C_0=4MPa, α=0.4, Λ=0.08

Figure 2 shows several patterns of a damage distribution in the main elements of rock mass after the mining process passes the corresponding distances at different heights of main roof. The results of modelling show the correlation between the increase of the main roof length and the increase of the general step of roof caving.

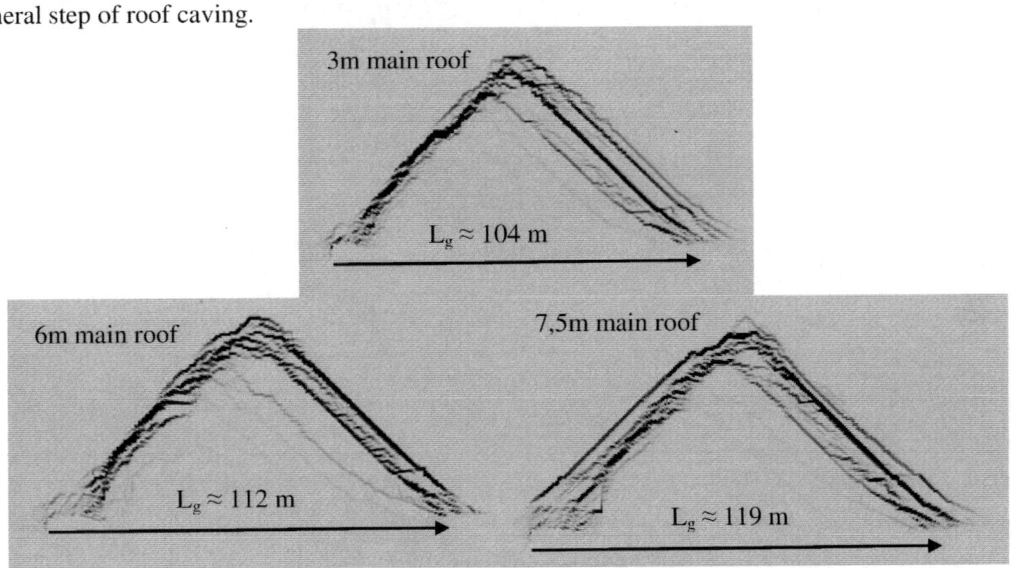

Figure 2. Estimations of the general step lengths for different lengths of the main roof.

In Figure 3 the comparison of the numerical simulation data along with the empirical data and the finite element modelling data is represented. It shows a good agreement.

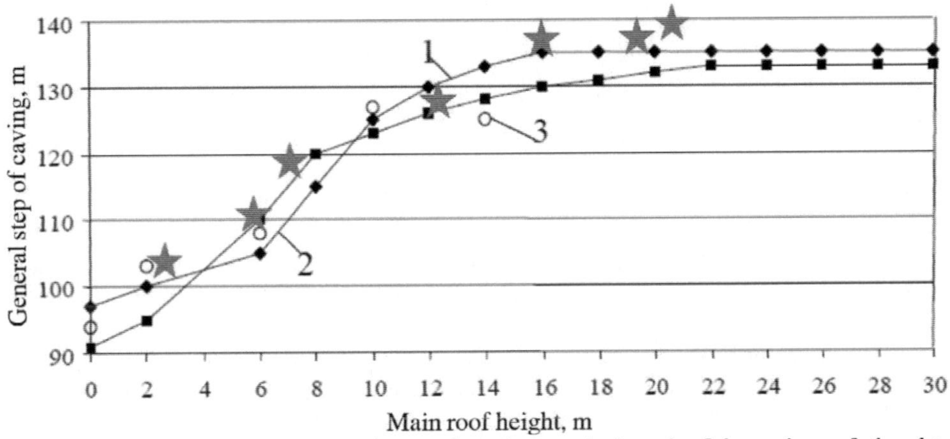

Figure 3. The dependence of the step of general caving on the length of the main roof, the obtained numerical modelling data are represented with red stars, other data are taken from [17].

As an example of testing calculation in 3D, one can see the distribution of inelastic strains accumulated in the roof and the floor of rock mass with excavation (Figure 4). The generalized structural model of rock mass was taken for 3D case. The roof and the floor have the same physical-mechanical properties as those mentioned above. The width of excavation between the airway and a conveyer way is 120m, the horizon of coal seam is 450m, and the direction of excavation is along the Z-axis.

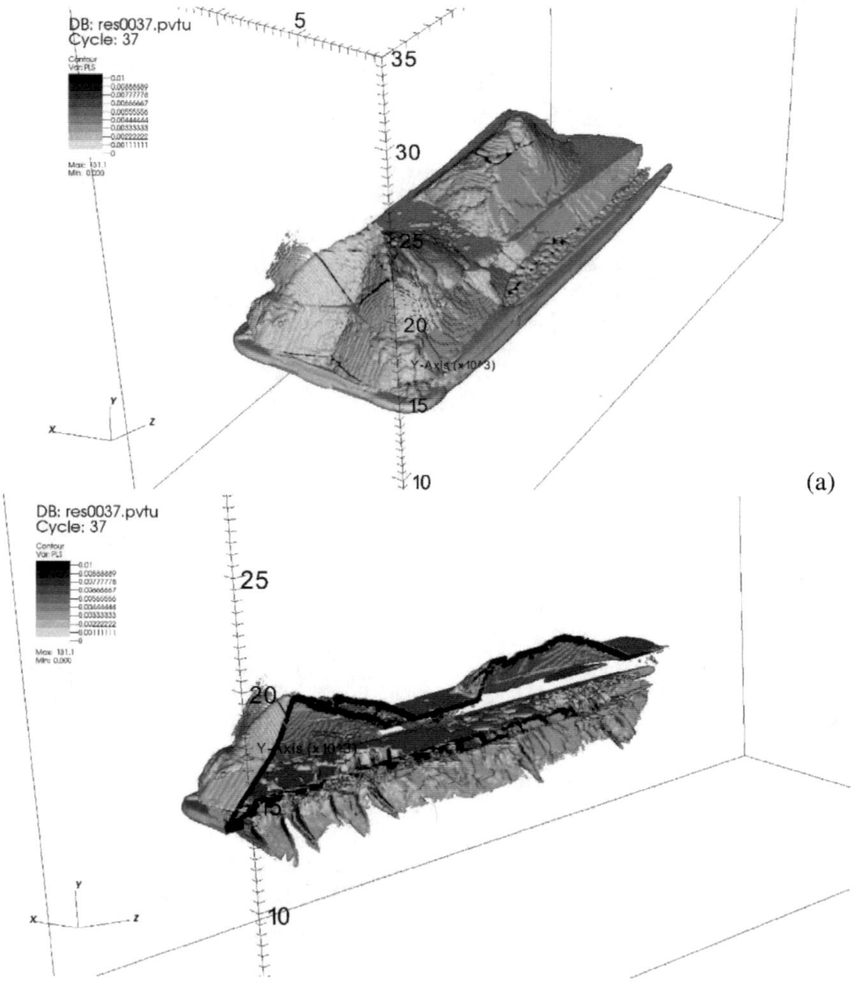

Figure 4. The distribution of accumulated inelastic strains in the rock mass elements in 3D modelling (a), a central cross-section (b).

4. Conclusions

The results of 2D and 3D modelling of damage accumulation in the rock mass elements are represented in the paper. The estimations of the initial (general) and set steps of roof caving are obtained. It is shown that the general and set steps of roof caving obtained in numerical modelling compared to the empirical estimations give the misfit error of about 10%. As the following investigations the developed model of rock mass will be generalized for the 3D case and a different angle of flat-dipping seams will be taken into account in order to estimate the dependency of the general and set steps of roof caving on the angle and depth horizon.

5. Acknowledgements

This work is funded by the Russian Science Foundation (grant No. 14-17-00198).

References

[1] Chanyshev A I and Abdulin I M 2014 *Journal of Mining Science* **50** 18-24
[2] Lu Y and Yang W 2013 *Journal of Central South University* **20** 1397-1404
[3] Demin V F, Nemova N A Demina T V 2015 *Journal of Siberian Federal University. Engineering & Technologies* **1** 74-97
[4] Roatesi S 2010 *Proceedings of the Romanian academy, Series A* **11** 11-18
[5] Kiya O E and Hosseinitoudeshki V 2015 *WALIA Journal* **31** 172-174
[6] Yang H Q, Zeng Y Y, Lan Y F and Zhou X P 2014 *International Journal of Rock Mechanics & Mining Sciences* **69** 59-66
[7] Pardoen B, Collin F, Levasseur S and Charlier R 2012 *ALERT Workshop*
[8] Pavlova L D and Fryanov V N 2004 *Izvestiya Tomskogo Politechnicheskogo Universiteta* **307** 76-79
[9] Shabanimashcool M 2012 Ph. D. thesis, Norwegian University of Science and Technology
[10] Nuric A, Nuric S, Kricak L, Lapandic I and Husagic R 2012 *Engineering and Technology* **6** 316-324
[11] Khoshboresh A R 2013 University LAVAL, Canada 155
[12] Zhu W C, Liu J, Tang C A, Zhao X D and Brady B H 2005 *Tunneling and Underground Technology* **20** 231-247
[13] Verma A K and Deb D 2013 *International Journal of Geomechanics* **13** 181-192
[14] Wang G, Wang H Y, Huang Q M and Su C Q 2014 *Computer Modelling & New Technologies* **18** 26-32
[15] Kwasniewski M 2008 Continuum and Distinct Element Numerical Modelling in Geo-Engineering, Paper: 07-08 1-12
[16] Makarov P V, Eremin M O and Kostandov Yu A 2014 *Physical mesomechanics* **17** 199-203
[17] Hozyaikina N V 2012 *National Rock University, Dnepropetrovsk, Ukraine* 126

Low thrust spacecraft transfers optimization method with the stepwise control structure in the Earth-Moon system in terms of the L1-L2 transfer

M K Fain and O L Starinova
Samara State Aerospace University
34, Moskovskoye shosse, Samara, 443086, Russia

E-mail: fain.maxim@gmail.com

Abstract. The paper outlines the method for determination of the locally optimal stepwise control structure in the problem of the low thrust spacecraft transfer optimization in the Earth-Moon system, including the L1-L2 transfer. The total flight time as an optimization criterion is considered. The optimal control programs were obtained by using the Pontryagin's maximum principle. As a result of optimization, optimal control programs, corresponding trajectories, and minimal total flight times were determined.

1. Introduction

Nowadays the spacefaring nations are developing the missions to achieve the libration points of the Earth-Moon system, especially L1 and L2. The optimal interplanetary trajectories and the trajectories of flights to the Moon pass near libration point L1 of the Earth-Moon system, as shown in works [1-2]. Moreover, the usage of the libration points will help to decrease the fuel expenses for orbit maintaining, to start from the Earth at any moment without choosing the date of start, to monitor the solar wind and to avoid the radiation from the Earth. One of the main problems of such missions is to determine the optimal control structure of the spacecraft transfers.

2. Mathematical Model of the Mission

Let us formulate the general statement of the optimization problem. The following parameters are considered:

$\boldsymbol{x}(t) = (\boldsymbol{r}(t), \boldsymbol{V}(t), m_f(t), \boldsymbol{r}_E(t), \boldsymbol{r}_M(t), \boldsymbol{r}_S(t))^T \in \boldsymbol{X}$ is a system state vector corresponding to boundary conditions, defined by the purpose of the transfer and possible restrictions, where $\boldsymbol{X}$ is a set of admissible state area;

$\boldsymbol{u}(t) = (\delta(t), \boldsymbol{e}(t))^T \in \boldsymbol{U}$ is a vector of control functions, where $\boldsymbol{U}$ is a set of admissible control parameters;

$\boldsymbol{p} = (a_0, j_{sp})^T \in \boldsymbol{P}$ is a vector of optimized design parameters. It is limited by a set of admissible area of design parameters $\boldsymbol{P}$.

Here t is the current time, $\boldsymbol{r}(t)$ is a radius vector of the spacecraft (SC), $\boldsymbol{V}(t)$ is a vector of the SC velocity, $m_f(t)$ is an expended fuel mass, $\boldsymbol{r}_E(t)$, $\boldsymbol{r}_M(t)$, $\boldsymbol{r}_S(t)$ are the radius-vectors of the Earth, the Moon and the Sun, $\delta(t)$ is a function of thrust switching, $\boldsymbol{e}(t)$ is a thrusting direction unit vector, a_0 is a nominal acceleration of the SC in the initial orbit, j_{sp} is a specific impulse of the propulsion system (PS).

The boundary conditions of the flight are shown in Table 1.

Table 1. Boundary conditions of the flights in the Earth-Moon system

	Finishing time	Radius-vector	Velocity vector	Radius-vector of the Earth	Radius-vector of the Moon	Radius-vector of the Sun	Set of admissible state area
L1	(t_i)	$\mathbf{r}_{L1}(t_1)$	$\mathbf{V}_{L1}(t_1)$	$\mathbf{r}_E(t_1)$	$\mathbf{r}_M(t_1)$	$\mathbf{r}_S(t_1)$	$\mathbf{X}_{L1}$
L2		$\mathbf{r}_{L2}(t_2)$	$\mathbf{V}_{L2}(t_2)$	$\mathbf{r}_E(t_2)$	$\mathbf{r}_M(t_2)$	$\mathbf{r}_S(t_2)$	$\mathbf{X}_{L2}$

Optimizing these space transfers with low thrust we need to determine vectors $\boldsymbol{u}_{opt}(t)$ and $\boldsymbol{p}_{opt}$ (vectors of optimal control functions and optimal design parameters correspondingly) that provide the minimum duration of flight T to perform the mission purposes according to Table 1.

$$T = \min_{u \in \mathbf{U}, p \in \mathbf{P}} T \big| m = unfixed, \boldsymbol{x} \in \mathbf{X} \tag{1}$$

The transfers are considered in terms of the barycentric combined inertial (BCI) frame with corresponding motion equations [3]. The following assumptions are made: the eccentricity of the Moon and the Earth orbits around barycenter is neglected; the eccentricity of the gravitational fields of the Earth, the Moon and the Sun are neglected, the SC sometimes moves in the Earth and the Moon shadow.

3. Control laws

In this work we use Fedorenko successful linearization method [1] that accepts the limitation on the composed functions that have Frechet derivatives. The method is based on bringing the variation optimal control problem to the iteration problem of linear programming. The functional for optimization was chosen as a sum of the fuel used during the transfer and the components accounting for the conditions of the spacecraft final orbit insertion:

$$I = m(t_k^{[n]}) + (r^{[n]}(t_k^{[n]}) - r_k)^2 + (\varphi^{[n]}(t_k^{[n]}) - \omega_M t_k^{[n]})^2 + v_r^{[n]}(t_k^{[n]}))^2 + (v_\varphi^{[n]}(t_k^{[n]}) - v_{\varphi k})^2 \rightarrow min, \tag{2}$$

where r_k is a radius of the final orbit, ω_M – Moon rotation rate around the barycenter of the system, $v_{\varphi k}$ – angular rotation rate on the final orbit, n – the number of the final segment of the transfer.

To solve the assigned problem we should find the following variables:

$$\frac{\partial I^*}{\partial T_0^i}, \frac{\partial I^*}{\partial T_k^i}, \frac{\partial I^*}{\partial \lambda_1^i}, \frac{\partial I^*}{\partial a_0}, \frac{\partial I^*}{\partial c_0} \tag{3}$$

The exact solution of the problem was established with the use of the Pontryagin maximum method and the numerical integration. The analysis of that solution shows, that the transfer trajectory has three general segments of work (Figure 1). Each of the general steps of the transfer control program was divided into several segments of work to provide a better accuracy.

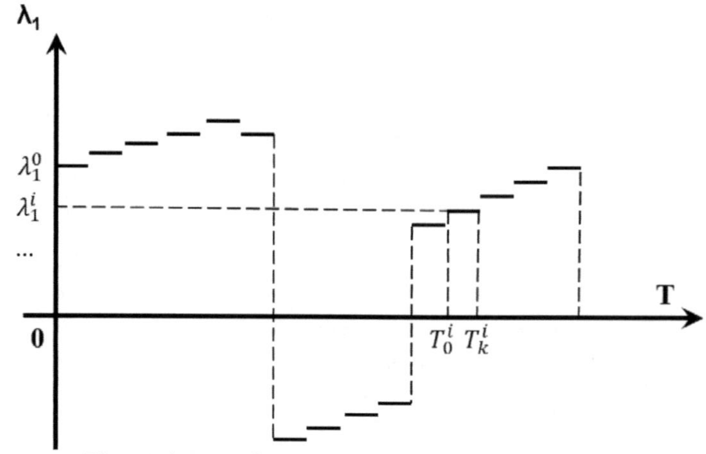

Figure 1. Transfer general stepwise control structure

The thrust is directed at the angle of λ_1^i to the radius-vector of the SC. Thus, ***u*** is a piecewise continuous function which is determined by the following parameters: λ_1^i, $\partial T_0^i, \partial T_k^i$ (each λ_1^i ∂T_0^i and ∂T_k^i is relevant to the corresponding segment of the trajectory).

So, according to notations of the Fedorenko method we have:

$$u^{[i]} = \{\lambda_1^i\},\ p^{[i]} = \{T_0^i\},\ q = \{a_0, c_0\}, \tag{4}$$

where a_0 – spacecraft acceleration, c_0 – exhaust velocity. As one can see from (2) the functional does not include the integral component, it consists only of the terminal component and depends on the following state vector:

$$x^{[i]}(t^{[i]}) = \{r, \varphi, v_r, v_\varphi, m\}^T = \{x_1, x_2, x_3, x_4, x_5\}^T. \tag{5}$$

After obtaining of motion equations [3] with the boundary conditions (Table 1) integrated we get the state vector components relevant to the final transfer segment determined.

Let us derive the Hamiltonian for costate vectors $\psi_r, \psi_\varphi, \psi_{v_r}, \psi_{v_\varphi}, \psi_m$:

$$H = \frac{dr}{dr}\psi_r + \frac{d\varphi}{dt}\psi_\varphi + \frac{dv_r}{dt}\psi_{v_r} + \frac{dv_\varphi}{dt}\psi_{v_\varphi} + \frac{dm}{dt}\psi_m. \tag{6}$$

Then we introduce the right parts of the motion equations into (6):

$$H = v_r\psi_r + \frac{v_\varphi}{r}\psi_\varphi + \left(\frac{v_\varphi^2}{r} - \frac{1}{r^2} + \frac{a_0}{1-m}\delta\ Cos\lambda\right)\psi_{v_r} + \left(-\frac{v_r v_\varphi}{r} + \frac{a_0}{1-m}\delta\ Sin\lambda\right)\psi_{v_\varphi} + \beta\psi_m \tag{7}$$

Here we can estimate the costate functions derivatives by deriving the Hamiltonian with respect to the corresponding state coordinates:

$$\begin{cases} \dfrac{d\psi_r}{dt} = -\dfrac{\partial H}{\partial r} = -\left(\dfrac{v_\varphi}{r^2}\psi_\varphi - \dfrac{v_\varphi^2}{r^2}\psi_{v_r} + \dfrac{2}{r^3}\psi_{v_r} + \dfrac{v_r v_\varphi}{r^2}\psi_{v_\varphi}\right) \\ \dfrac{d\psi_\varphi}{dt} = -\dfrac{\partial H}{\partial \varphi} = 0 \\ \dfrac{d\psi_{v_r}}{dt} = -\dfrac{\partial H}{\partial v_r} = -\left(\psi_r - \dfrac{v_\varphi}{r}\psi_{v_\varphi}\right) \\ \dfrac{d\psi_{v_\varphi}}{dt} = -\dfrac{\partial H}{\partial v_\varphi} = -\left(\dfrac{\psi_\varphi}{r} + 2\dfrac{v_\varphi}{r}\psi_{v_r} - \dfrac{v_r}{r}\psi_{v_\varphi}\right) \\ \dfrac{d\psi_m}{dt} = -\dfrac{\partial H}{\partial m} = -\left(\dfrac{a_0}{(1-m)^2}\delta\ Cos\lambda\ \psi_{v_r} + \dfrac{a_0}{(1-m)^2}\delta\ Sin\lambda\ \psi_{v_\varphi}\right) \end{cases} \tag{8}$$

Scalar function μ determines the finishing time of each segment. In this case each of five μ functions is a null vector, because there are no discontinuous jumps of the state coordinates on the transfer segments boundaries.

To determine the needed derivatives we need to find the following variables:

$$a^{[n]} = -\frac{1}{\mu_x^{[n]} f_k^{[n]} + \mu_t^{[n]}} = -1 \tag{9}$$

$$d = a^{[n]}\left(F_t + F_x f_k^{[n]}\right) = -1\left(\beta + 2(r^{[n]}(t_k^{[n]}) - r_{L2})\right)\Big|_T \cdot v_r\Big|_T + 2\left(\varphi^{[n]}(t_k^{[n]}) - \omega_M t_k^{[n]}\right)\Big|_T + 2\left(v_\varphi^{[n]}(t_k^{[n]}) - v_{\varphi L2}\right)\Big|_T \cdot \left(-\frac{v_r v_\varphi}{r} + \frac{a_0}{1-m}\delta\, Sin\lambda\right)\Big|_T) \tag{10}$$

Then the right parts of motion equations for the state coordinates should be found:

$$\begin{cases} f_t(T_k) = v_r(T_k) \\ f_\varphi(T_k) = \frac{v_\varphi}{r}(T_k) \\ f_{v_r}(T_k) = \left(\frac{v_\varphi^2}{r} - \frac{1}{r^2} + \frac{a_0}{1-m}\delta\, Cos\lambda\right)(T_k). \\ f_{v_\varphi}(T_k) = \left(-\frac{v_r v_\varphi}{r} + \frac{a_0}{1-m}\delta\, Sin\lambda\right)(T_k) \\ f_m(T_k) = \beta \end{cases} \tag{11}$$

Now the costate functions values in the final points could be found:

$$\psi^{[n]}\Big|_T = F_x + d \cdot \mu_x^{[n]}, \tag{12}$$

$$\begin{pmatrix} \psi_r \\ \psi_\varphi \\ \psi_{v_r} \\ \psi_{v_\varphi} \\ \psi_m \end{pmatrix} = \begin{pmatrix} 2(r^{[n]}(t_k^{[n]}) - r_{L2}) \\ 2(\varphi^{[n]}(t_k^{[n]}) - \omega_M t_k^{[n]}) \\ 2\, v_r^{[n]}(t_k^{[n]}) \\ 2(v_\varphi^{[n]}(t_k^{[n]}) - v_{\varphi L2}) \\ 1 \end{pmatrix}. \tag{13}$$

Let us find $\frac{\partial J}{\partial \lambda_1^i}$.

Firstly, let us solve $\omega^i(s)$ (analytical derivatives of the functional with respect to $u^{[i]}$):

$$\omega^i(s) = \psi_i^T(s) f_u^i(s), \quad i = 1, ..., K, \tag{14}$$

where costate functions ψ_i are the solution of the equations:

$$\frac{d\psi_i}{dt^i} = -(f_x^i)^T \psi_i, \quad i = 1, ..., K. \tag{15}$$

So, $\omega^i(s)$ would be as follows:

$$\omega = -\psi_{v_r}\frac{a_0}{1-m}\delta\, Sin\lambda_1^i + \psi_{v_\varphi}\frac{a_0}{1-m}\delta\, Cos\lambda_1^i. \tag{16}$$

$$\omega_{\lambda_1^i} = \int_{t_0^i}^{t_k^i} \omega^i(s)\, ds, \tag{17}$$

Now let us find $\frac{\partial J}{\partial \overline{q}} = \left(\frac{\partial J}{\partial a_0}; \frac{\partial J}{\partial c_0}\right)$.

$$\frac{\partial \overline{f}}{\partial \overline{q}} = \begin{pmatrix} 0 & 0 & 0 & 0 \\ 0 & 0 & 0 & 0 \\ 0 & 0 & \frac{Cos\lambda_1^i \cdot \delta}{1-m} & 0 \\ 0 & 0 & \frac{Sin\lambda_1^i \cdot \delta}{1-m} & 0 \\ 0 & 0 & \frac{\delta}{c_0} & -\frac{a_0 \cdot \delta}{c_0^2} \\ 0 & 0 & \frac{\partial f}{\partial a_0} & \frac{\partial f}{\partial c_0} \end{pmatrix}. \quad (18)$$

$$\frac{\partial J}{\partial \overline{q}} = D = \sum_{i=1}^{n}\left(\int_{t_0^i}^{t_k^i} (\psi^{[i]})^T \cdot \frac{\partial \overline{f}^{[i]}}{\partial \overline{q}} ds \right) + \frac{\partial F}{\partial \overline{q}} + d\frac{\partial \mu^{[i]}}{\partial \overline{q}}; \quad (19)$$

$$\frac{\partial J}{\partial \overline{q}} = \left(\int_{t_0}^{t_k} (\psi_{v_r} \cdot \frac{Cos\lambda \cdot \delta}{1-m} + \psi_{v_\varphi} \cdot \frac{Sin\lambda \cdot \delta}{1-m} + \psi_m \frac{\delta}{c_0})ds; \quad \int_{t_0}^{t_k} (-\psi_m \cdot \frac{a_0 \delta}{c_0^2})ds \right); \quad (20)$$

$$\frac{\partial J}{\partial a_0} = \int_{t_0}^{t_k} (\psi_{v_r} \cdot \frac{Cos \cdot \delta}{1-m} + \psi_{v_\varphi} \cdot \frac{Sin\lambda \cdot \delta}{1-m} + \psi_m \cdot \frac{\delta}{c_0})\, ds; \quad (21)$$

$$\frac{\partial J}{\partial c_0} = \int_{t_0}^{t_k} (-\psi_m \cdot \frac{a_0 \cdot \delta}{c_0^2})\, ds. \quad (22)$$

Now let us find $\frac{\partial J}{\partial \overline{T}} = \left(\frac{\partial J}{\partial T_0^i} \; \frac{\partial J}{\partial T_k^i} \right) \cdot \begin{pmatrix} \frac{\partial T_0^i}{\partial T_k^i} \end{pmatrix}$.

As all the initial boundary conditions are constants, we have:

$$\Pi^0 \partial p^{[0]} = 0. \quad (23)$$

$$\Pi^{[i]} = \left(\psi^{[i+1]}(t_0^{[i+1]})\right)^T \cdot [\frac{\partial \overline{\varphi}^{[i]}}{\partial p^{[i]}} + \overline{c}^{[i]} \frac{\partial \mu^{[i]}}{\partial p^{[i]}} - f^{[i+1]}(t_0^{[i+1]}) \cdot \left(\frac{\partial \tau^{[i]}}{\partial p^{[i]}} + b^{[i]} \frac{\partial \mu^{[i]}}{\partial p^{[i]}} \right)] +$$
$$+ \int_{t_0^{[i]}}^{t_k^{[i]}} (\psi^{[i]}(s))^T \cdot \frac{\partial \overline{f}^{[i]}}{\partial p^{[i]}}(s)\, ds = (\psi^{[i+1]}(t_k^{[i+1]}))^T \cdot [f^{[i]}(t_k^{[i]}) - f^{[i+1]}(t_0^{[i+1]})]; \quad (24)$$

Thus, we have found the analytical expressions for all the functional derivatives (3), therefore, the problem is solved.

4. Results of the optimization

The example of the optimization process is shown in Figure 2. The red points show the terminal position of the spacecraft, the grey points show the terminal position of the Moon and the black line corresponds to the L2 orbit. As one can see from Figure 2, the optimization process makes the spacecraft reach the L2 libration point.

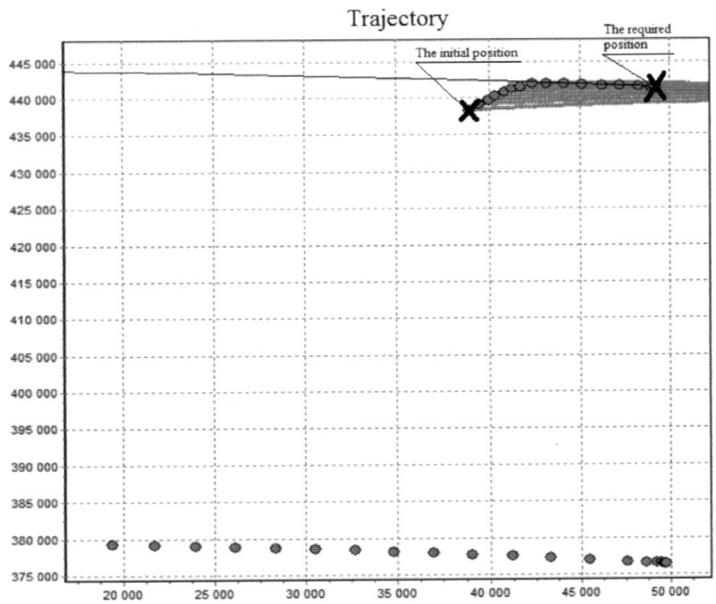

Figure 2. The example of the optimization process

The optimal control program for the L1-L2 transfer and the corresponding trajectory have been obtained. In Figure 3 the red steps correspond to the active segments of the trajectory, and the green lines correspond to the passive segments, where the engine is turned off. In Figure 3 we can see, that the first passive segment of the trajectory is fully degenerated.

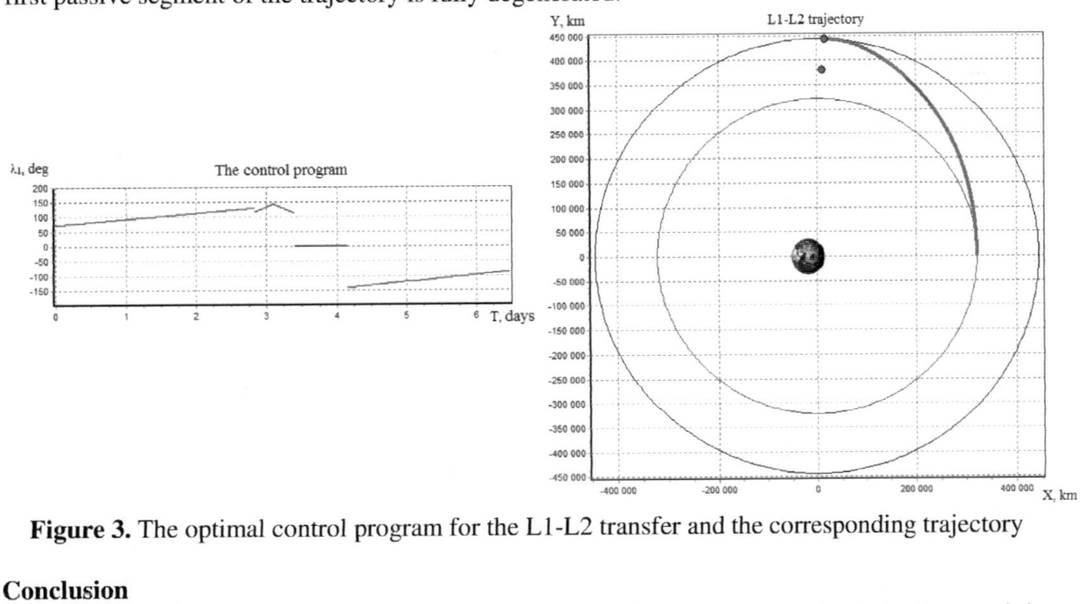

Figure 3. The optimal control program for the L1-L2 transfer and the corresponding trajectory

5. Conclusion

As we can see from the paper the usage of the Fedorenko method to estimate the derivatives and the gradient method to optimize the control laws allows determining the optimal control programs and the corresponding trajectories for the space missions by the example of the L1-L2 mission in the Earth-Moon system. The obtained results are in good agreement with the results obtained by the usage of Pontryagin's maximum principle in the three-body task framework [4]. The accuracy of the calculations was lower than 4%. The applied methods demonstrate their effectiveness for the complex optimization of the SC transfers. The findings may be used to calculate the required design-ballistic parameters of the future space missions.

Acknowledgment
This work was supported by the Ministry of Education and Science of the Russian Federation and Russian Foundation for Basic Research

References

[1] Kazmerchuk P V, Malyshev V V and Usachev V E 2007 *Journal of Computer and Systems Sciences International* **46(1)** 150-161
[2] Starinova O L, Kurochkin D V and Materova I L 2012 *AIP Conf. Proc.* **1493** 964
[3] Starinova O L and Fain M K Ballistic 2015 *Recent Advances in Space Technologies* 95-98
[4] Starinova O L 2014 *AIP Conf. Proc.* **1637** 1056

Determination of parametric reliability of machining technological systems by simulation technique

V P Fyodorov, M N Nagorkin and A V Totai
Bryansk State Technical University, 7, 50 years of October st., Bryansk, 241035, Russia

E-mail: nagorkin@tu-bryansk.ru

Abstract. The article deals with the problem of determining parametric reliability of technological systems for machining parts. The paper describes the methodology of defining the probability of task implementation by quality factors for technological systems by means of simulation technique.

1. Introduction

Current technological systems, aimed at technological processes and operations for machining components, represent a set of interconnected technological equipment, production tools and executives. The fact that technological systems have various interconnections and handling factors which are of probabilistic nature makes it difficult to provide product quality regulated by many factors. Thus, provision of required quality factors and service properties of components mainly depends on reliability of metal working technological systems [1, 2].

The reliability of the technological process (operation) is its property to provide production work-order quantity retaining its quality and permanent parameters of this process.

It is typical of technological systems as any other systems to have different kinds of failures. According to State Standard 27.004-85 technological systems can have parametric failures when they keep their functioning, but run one or more perfection factors of working surfaces away from the limits of the normative document.

So, the product quality depends on the parametric reliability of a technological system, which is characterised by the probability of this system no-failure operation, according to the parameter of the component quality.

Randomness of technological system processes stresses the reasons for applying a simulation technique to determine reliability factors. This paper contains the results of using this technique to estimate the parametric reliability of the technological system for antifrictional treatment of shaft journals.

2. Results and Discussion

In general case, the probability of task implementation according to a product quality factor provided by the technological process (operation) is:

$$P_i(t) = P\{x_{li} \le x_i(t) \le x_{ui}\}, \quad (1)$$

where $x_i(t)$, x_{ui}, x_{li} – are respectively actual, upper and low acceptable values of i-controlled parameter.

The methodology of defining the probability of task implementation for technological systems by means of simulation technique is based on determination of polynomial (2) and multiplicative (3) models of processes for forming component quality factors:

$$Y_i = \beta_0 + \beta_1 X_1 + \ldots + \beta_i X_i + \ldots + \beta_k X_k, \quad (2)$$

$$Y_i = b_0 \cdot X_1^{b_1} \cdot X_2^{b_2} \cdot \ldots \cdot X_k^{b_k}, \quad (3)$$

where Y_i – i-quality factor of the surface after working; X_k – k-working environment factor; b_0, b_1, ..., b_k – regression coefficient values which are allocated in accordance with the normal law of expectation $M\{b_k\}$ and average quadratic deviation $S(b_k)$.

Probability P of task implementation by quality factor Y in a specified interval is taken as a parametric reliability factor:

$$P\left\{Y \in \left(\bar{Y} - \delta\bar{Y}; \bar{Y} + \delta\bar{Y}\right)\right\},$$

where $\bar{Y}$ – an average value of the specified quality factor; δ – a relative acceptable value of its changing ($0 \le \delta \le 1$). Values $\left(\bar{Y} - \delta\bar{Y}\right)$ and $\left(\bar{Y} + \delta\bar{Y}\right)$ correspond to values x_{li}, x_{ui} (1).

To calculate this factor it is necessary to know expectation Y_{i0} and average quadratic deviation σ_{i0} of the component quality factor under study. Simulation techniques of forming parameters Y_i represent so-called "black box models" and are of a statistical nature. It is advisable to use Monte-Carlo method to predict quality factors Y_i and their attributes $M\{Y_i\}$ and $S\{Y_i\}$, which are value estimations Y_{i0} and σ_{i0}.

The use of Monte-Carlo method is reasonable for systems whose models are described by complicated stochastic equations. In such cases when it is impossible to make a model because of the lack of data or the complexity of analytical model, Monte-Carlo method is often the only way to obtain related data.

If the only random variables in simulation models (2) and (3) are coefficients β_0 and β_i, and taking into account the properties of expectation and a random variable we can obtain the following expressions for estimations $M\{Y_{i0}\}$ and $S^2\{Y_{i0}\}$:

– for additive models:

$$M\{Y_{i0}\} = M\{\beta_0\} + \sum_{i=1}^{k} X_i M\{\beta_i\} = b_0 + \sum_{i=1}^{k} X_i b_i, \qquad S^2\{Y_{i0}\} = S^2\{\beta_0\} + \sum_{i=1}^{k} X_i^2 S^2\{\beta_i\};$$

– for multiplicative models:

$$M\{Y_{i0}\} = exp(ln\,M\{\beta_0\} + \sum_{i=1}^{k} ln\,X_i M\{\beta_i\}) = exp(ln\,b_0 + \sum_{i=1}^{k} b_i\,ln\,X_i),$$

$$S^2\{Y_{i0}\} = exp(ln(S^2\{\beta_0\} + \sum_{i=1}^{k} (ln\,X_i)^2 S^2\{\beta_i\}).$$

If models (2) and (3) have one or two working factors X_i as random ones, then it is not possible to find analytical expression for estimations $M\{Y_{i0}\}$ and $S^2\{Y_{i0}\}$. Their determination is connected with data array processing Y_{iN}, obtained as a result of the computer experiment by N executions of simulation models in Monte-Carlo method.

The results of simulation models show that some engineering and technological constraints are used as input data.

There are determined rated values of technological factors X_i, which allow obtaining the desired value of parameter Y_0. At the same time some simulation models use coefficients $b_i = M\{\beta_i\}$. When calculating the machining conditions which provide obtaining value Y in the specified interval $[Y_{min}, Y_{max}]$ we make an obligatory analysis of implementing the requirements for dimensional accuracy and form defects of the working component.

One of the main problems is to choose the number of model executions N during the computer experiment. The number of executions N should satisfy the specified accuracy of value estimates $M\{Y_i\}$ and $S^2\{Y_i\}$ according to the results of computer experiments.

Let the requirements to the expectation estimate be the following:

$$P\{Y_{i0} - d \le M\{Y_i\} \le Y_{i0} + d\} = 1 - \alpha, \tag{4}$$

where $(1 - \alpha)$ is the probability that interval $(Y_{i0} \pm d)$ contains estimate $M\{Y_i\}$.

Assuming the normal allocation of value Y_i obtained as a result of the computer experiment, we assume that the number of executions, which provides dependence (4), can be expressed as follows:

$$N_1 = \frac{(\sigma_{i0} z_{\alpha/2})^2}{d^2}, \tag{5}$$

where $z_{\alpha/2}$ – is two-way standard normal statistics; σ_{i0} – is an average quadratic deviation of quality factor Y_i; d – is allowable difference between the estimate and the true value of the expectation for parameter Y_i.

Thus, the use of correlation (5) is difficult, as it includes unknown value σ_{i0}. Let us express value d as a fraction of σ_{i0}: $d = \sigma_{i0} / a$. Now let us put it into (5) and we obtain:

$$N_1 = \frac{(\sigma_{i0}\, z_{\alpha/2})^2}{\left(\sigma_{i0}/a\right)^2} = a^2 \cdot z_{\alpha/2}^2 .$$

Here for estimate N_1 we have known values and we can use it successfully.

The problem of defining the number of model executions N_2 with the specified accuracy for dispersion estimate can be solved in the following way:

$$P\{(1 - d)\sigma_{i0}^2 \le S^2\{Y_i\} \le (1 + d)\sigma_{i0}^2\} = 1 - \alpha.$$

Here α is the number defining the degree of estimate proximity $S^2\{Y_i\}$ to true dispersion σ_{i0}^2 $(0 \le d \le 1)$.

A transducer of pseudorandom variables is used during the computer experiment to generate random variables X_i, β_{0j}, β_{ij}, ξ_{0i}, ξ_{ij}. At first, there are generated and uniformly distributed values in interval (0, 1), then a normally distributed random variable with parameters $M\{\beta_i\} = b_i$ and $S\{\beta_i\}$ is formed.

For parameter Y_i a simulation model like (3) has the following problem definition for the computer program: "Calculate N values of the function

$$Y_i = f(X_1, \dots, X_j, \dots, X_k;\ b_0, b_1, \dots, b_j, \dots, b_k;\ S\{\beta_0\}, S\{\beta_1\}, \dots, S\{\beta_j\}, \dots, S\{\beta_k\})$$

so that

$$Y_i = b_0 \cdot \prod_{j=1}^{k} X_j^{RNDN_i(b_i, S\{\beta_i\})},$$

where $X_1, X_2, \dots, X_k$ – are values of input parameters of the technological process;
– $b_0, b_1, \dots, b_k$ – are expectations for model coefficients;
– $S\{\beta_0\}, S\{\beta_1\}, \dots, S\{\beta_k\}$ – are their standard deviations;
– $RNDN_i\ (b_i, S\{\beta_i\})$ – is a normally distributed random variable with distribution parameters b_i and $S\{\beta_i\}$.

After calculating N values of parameter Y_i there is a check of statistical hypotheses and estimation of distribution parameters. For this purpose it is necessary to calculate $M\{\tilde{Y}_i\}$ and $S^2\{\tilde{Y}_i\}$:

$$M\{\tilde{Y}_i\} = \frac{\sum_{i=1}^{N} Y_{ij}}{N}, \qquad S^2\{\tilde{Y}_i\} = \frac{1}{N-1}\sum_{j-1}^{N}\left(Y_{ij} - \tilde{Y}_i\right)^2,$$

where Y_{ij} – values of i-parameter obtained in j-execution; $\tilde{Y}_i$ – average value of the quality factor of component Y_i, obtained in calculations during the computer experiment.

According to obtained data $\tilde{Y}_i$ distribution bar charts are made, and there are checked hypotheses on the distribution law by means of criterion χ^2 (Weibull distribution, normal distribution and log-normal distribution are considered). Value χ^2_{calc} is determined by the following expression

$$\chi^2_{calc} = \sum^{\ell} \frac{(\phi_0 - \phi_\ell)^2}{\phi_\ell},$$

where ϕ_0 – is observed frequency for each bar chart interval; ϕ_ℓ – is theoretical frequency for each interval.

If $\chi^2_{calc} \le \chi^2_{table}$ for this level of importance α and a number of freedoms, then there are no reasons to reject null hypothesis, and the distribution law corresponding to this hypothesis is accepted.

After these calculations there is a determined probability of task implementation according to dependences:

$$P\{Y_{i\,min} < Y_i < Y_{i\,max}\} = \Phi\left(\frac{Y_{imax} - M\{\tilde{Y}_i\}}{S\{\tilde{Y}_i\}}\right) - \Phi\left(\frac{Y_{imin} - M\{\tilde{Y}_i\}}{S\{\tilde{Y}_i\}}\right), \quad (6)$$

$$P\{Y_i < Y_{i\,max}\} = 0{,}5 + \Phi\left(\frac{Y_{imax} - M\{\tilde{Y}_i\}}{S\{\tilde{Y}_i\}}\right). \quad (7)$$

Dependence (6) is applied in case of two-way limitation of parameter Y_i, and (7) – in case of one-way limitation.

Taking into account that state parameters of a surface layer are often specified with a symmetrical tolerance variability interval as

$$Y_i = M\{Y_i\} \pm \delta M\{Y_i\},$$

where δ is a relative variability interval width (specified in design documentation), then the estimation of probability of task implementation by the technological system for quality factor Y_i is carried out depending on value δ.

When studying reliability factors of the technological system by means of the simulation technique on statistical models, one should take into account that it is possible to simulate only "past", i.e. data obtained experimentally at the stage of getting simulation models. Thus, to confirm the prediction of the simulation model it is necessary to assume that the basic form of distribution of technological process parameters remains unchanged and its features relevant to a certain time interval will repeat. These assumptions should be made for a normally operating technological process.

Let us take an example of using this method for determination of reliability of tribotechnical characteristics and roughness parameters in the technological system of antifrictional treatment of shaft journals for plain bearings.

The technological process includes finish-turning operations by polycrystalline cbn (cubic boron nitride) cutting tool, layering soft copper-bearing break-in film and diamond burnishing.

Finish turning was made by standard cartridge toolholders of geksanit at cutting speed V_t = 65 – 200 m/min; at feed S_t = 0.05 – 0.15 mm/rotation; at cutting depth t = 0.1 – 0.25 mm; at rigidity j = 2.6 – 16.2 кN/mm.

The soft break-in film was layered by means of the frictional brass coating or chemical copper plating.

Diamond burnishing was carried out by the diamond indenter with tip radius r = 3.5 mm; burnishing power Q_{DB} = 100 – 300 N; processing speed V_{DB} = 65 – 100 m/min; feed S_{DB} = 0.075 – 0.15 mm/rotation.

Tribotechnical characteristics of samples were tested at a special test bench with friction slide bearings working with lubricant at dynamic loads – relative interfacial slip speed V_{run} = 10 – 50 m/min; average linear joint load P_{run} = 30 – 50 N/mm; relative change of the average load at dynamic aging $\Delta p = \pm$ (15 – 25) %.

We studied the following tribotechnical joint characteristics under the conditions of sliding friction: friction coefficient at the beginning f_1 and at the end f_0 of aging; initial wear values h_{01}, h_{02}, aging ways L_{01}, L_{02} and wear rates I_1, I_2 for shafts and bushes, respectively.

For the tested system the corresponding simulation models were made:

$$Y_i = b_0\, K_1\, K_2\, V_t^{b_1}\, S_t^{b_2}\, t^{b_3}\, j^{b_4}\, Q_{DB}^{b_6}\, S_{DB}^{b_7}\, V_{DB}^{b_8}\, V_{run}^{b_{10}}\, P_{run}^{b_{11}}\, \Delta_p^{b_{12}},$$

where Y_i – i-tribotechnical characteristics; K_1, K_2 – coefficients in view of break-in film type M_b and bush material respectively M_{bush}; b_k – regression coefficient.

The reliability of formation of operational parameters was determined by simulation modelling according to the suggested plan.

The results of computer experiments of models showed that tribotechnical characteristics under study can be controlled by changing the factors of the tribotechnological system within wide limits with a great level of reliability (Figure 1).

A statistical data manipulation also showed the possibility to control tribotechnical characteristics (f_1, f_0, (h_{01}+ h_{02}), I_1, I_2) in the following sequence beginning with the most effective factor: $\Delta p \rightarrow V_{run} \rightarrow M_b \rightarrow S_{DB} \rightarrow S_t \rightarrow Q_{DB} \rightarrow j \rightarrow V_t \rightarrow M_{bush} \rightarrow t \rightarrow P_{run} \rightarrow V_{DB}$.

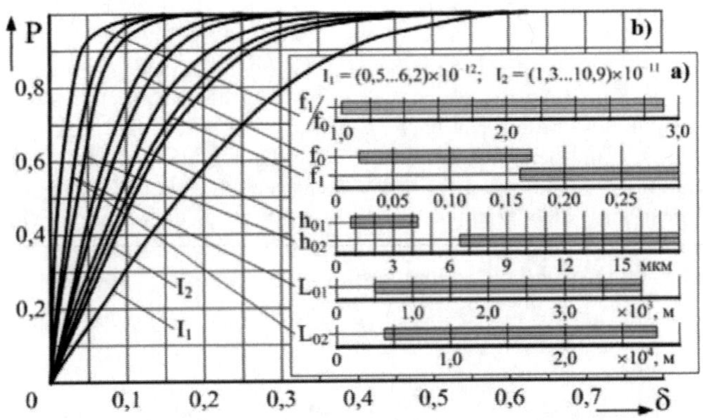

Figure 1. Diagram of changing (a) tribotechnical characteristics and reliability (b) of their providing $P\left(Y_i \in \left(\overline{Y}_i \pm \delta \overline{Y}_i\right)\right)$ depending on value δ in the technological system under study

3. Conclusion

The developed methodology of defining parametric reliability of technological systems by quality factors for working components helps to choose the best optimal technological process of working which has the maximum value of reliability as an optimality criterion that in consequence provides a high quality of working components.

References

[1] Trukhanov V M 2015 *Journal of machinery manufacture and reliability* **44(3)** 254–256

[2] Kuznetsov A P and Kosov M G 2012 *Russian engineering research* **32(5)** 482–490

Modelling of the radial forging process of a hollow billet with the mandrel on the lever radial forging machine

A P Karamyshev, I I Nekrasov, A I Pugin and A A Fedulov
Ural Federal University named after the first President of Russia B.N.Yeltsin, 19, Mira st., Ekaterinburg, 620002, Russia

E-mail: artem.feduloff@ya.ru

Abstract. The finite-element method (FEM) has been used in scientific research of forming technological process modelling. Among the others, the process of the multistage radial forging of hollow billets has been modelled. The model includes both the thermal problem, concerning preliminary heating of the billet taking into account thermal expansion, and the deformation problem, when the billet is forged in a special machine. The latter part of the model describes such features of the process as die calibration, die movement, initial die temperature, friction conditions, etc. The results obtained can be used to define the necessary process parameters and die calibration.

1. Introduction

To obtain the reliable data about the metal-forming processes studied, particularly about the loads on the equipment, the experimental, theoretical and mixed experimental and theoretical research methods are applied. Although experimental research methods allow obtaining the most accurate results, it is sometimes quite difficult to implement them in real working conditions of the studied equipment operation or on rare occasions they are not implemented at all. In that case theoretical study methods of the processes and machines have become more widespread, especially due to the development of mathematical methods of calculation based on generation of finite-element (FE) models [1-3].

At the Department of Metallurgical and Rotary Machines (UrFU) theoretical research of non-stationary metal-forming processes based on the finite-element modelling of the plastic deformation site are carried out. For example, the department staff have conducted a study of several kinds of reduction of solid and hollow billets on lever radial forging machines (LRFM) [4-6].These machines, depending on the type, are intended for different purposes and, as a result, they have a specified design expressed in the disposition and quantity of forging dies, the trajectory of their movement, various modifications of the feeding device and presence or absence of a mandrel for the formation of the required thickness of a hollow billet wall.

2. The study of the radial forging process

One particular task within the research of the radial forging on LRFM was devoted to generating a mathematical model of the multistage hot forging of the hollow billet on the two-die machine with a vertical disposition of levers and with the application of a mandrel. The purpose of the study is to build the mathematical model which would allow defining the stress-strain state, force and velocity parameters of the process, formation of the billet and temperature conditions of the deformation. The development of this model will make it possible to specify rational parameters of the deformation on a particular machine, to select the necessary heating mode and also to lay the foundations of the design of dies calibration.

On the whole, the algorithm of the mathematical model generation could be divided into two main parts. The first part includes a solution to the problem of the billet heat treatment before forging in order to define the boundary conditions of the modelling and the process data analysis. The second part consists of the study of the forging process, which includes formation of the forging dies model with its assigned movement; modelling of the billet feeding process taking into account different velocity modes during the forging process; assessment of temperature fields defined in the first part of the modelling.

2.1. The heat treatment problem

To complete the heat treatment process the boundary conditions on certain surfaces of the billet with a pre-set thermal mode were set. It is assumed that both heating and heat exchange occur on the outer and inner surfaces of the billet and the heating is carried out at a certain distance from the front end of the billet. The billet has the infinite length. The heat treatment for the given length, which corresponds to a volume required to fill the space between dies taking into account the mandrel, is carried out up to the temperature of 1250 °C during 50 seconds to obtain uniform heating.

The temperature distribution along the cylinder surface of the billet is uniform along the full length required (Figure 1).

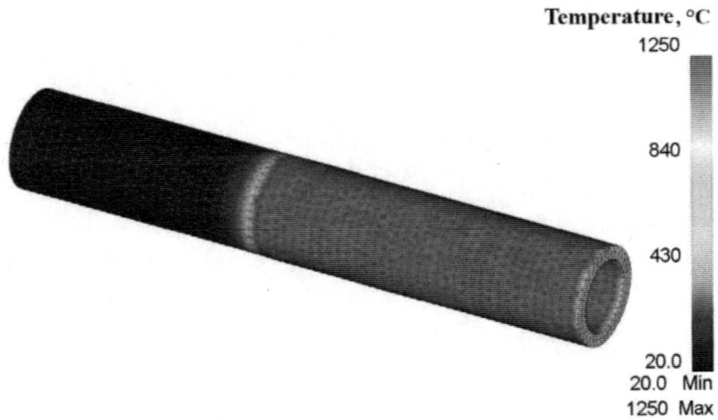

Figure 1. The temperature distribution within the billet

Due to the fact that the billet is considered as anelasto-plastic object it is possible to define temperature-induced deformations of the metal (Figure 2). One of the software features is its capability to interpolate properties from the solution of the previous problem, such as a stress-strain state and temperature distribution, to a new billet.

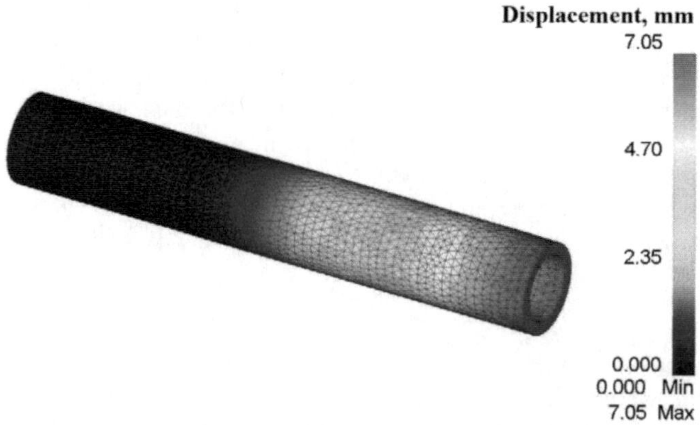

Figure 2. The thermal expansion of the billet

It should also be mentioned that when it exits from the inductor and moves to the axis of the forging machine the cooling process of the billet takes place. Part of the warmth from the heated area of the billet is emitted into the environment. The solution to the problem of the billet cooling starts with the last step of the solution to the heating problem. This step is used to describe the billet stress-strain state, temperature distribution, and volume change after the heating. Boundary conditions are also interpolated with the definition of heat-exchange surfaces.

2.2. The deformation problem

Solutions to the heating and cooling problems used to generate the mathematical model of the forging process which also includes several stages.The forging process is directed to the production of a billet with the required configuration which is achieved by the reduction between two working dies (Figure 3). The striking load of dies is combined with the forward-rotation of the billet produced by the feeding device.

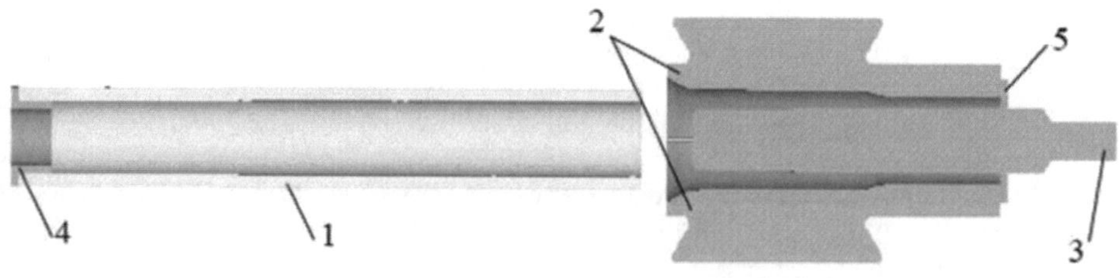

Figure 3. The scheme of the suggested forging process on LRFM: *1* –billet; *2* – forging dies; *3* – mandrel; *4* –object for the feeding device; *5* – thrust collar

Therefore, the scheme of the forging process includes the object from the previous solution step used as a billet, objects for forging dies and the mandrel, heated to working temperature, and an object for the feeding device. The position of the dies and the mandrel in the model and their movement correspond to the real machine kinematics.

To obtain more precise solution results concerning the billet forming during the forging process the local mesh density windows technique is applied in the FE description of the billet object. This allows contracting the areas, most subjected to deformation along with FE, and, at the same time, reducing the whole calculation volume.The choice of the FE quantity is based on the assumption that at least 4 FE should be located in the billet wall thickness at every solution step.Therefore, the part of the billet that does not participate in the forging process has a more rough mesh with a FE size equal to 10 mm, the volume of the billet in the first caliber has a more precise mesh with the size of FE, which is 5 mm, and in the second caliber, where the deformation occurs more actively than in the previous two, it has the size of approximately 3.5 mm.The total number of elements is 195000.

There is also a mechanism of forced re-meshing which depends on the feeding device stroke.After each 5 mm of the feeding device stroke into the deformation site forced re-meshing takes place, which provides the highest accuracy of the results obtained.

The solution to the problem allows getting representation of loads, object velocities, torques and angular velocities correlations depending on the simulation step. The data presented are displayed as discrete points obtained at every step and connected by a smooth curve, which requires further data processing and analysis. The data can also be exported as an array to different software packages. As an example, figure 4 shows the graph of the load placed on the forging die during the forging process.

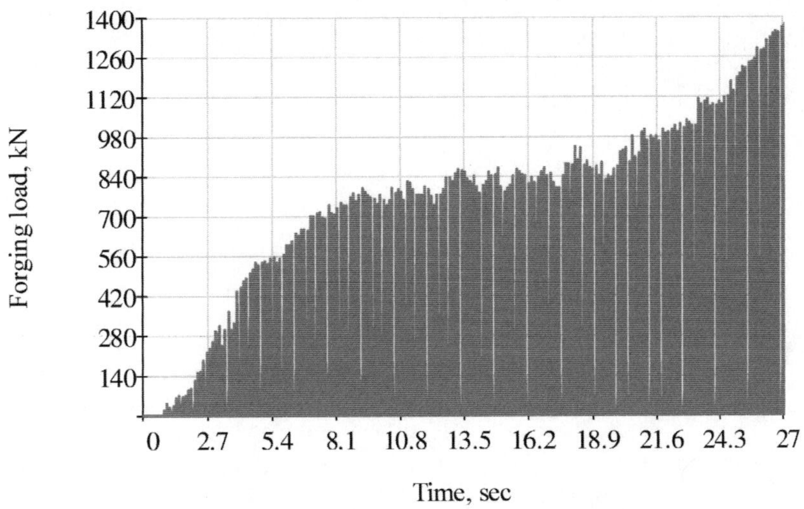

Figure 4. The forging load characteristic graph

The implementation of the mathematical model of the forming process is also carried out as a plain-strain problem which significantly decreases the calculation time. Several examples of forming during the process are shown in Figures 5 and 6.

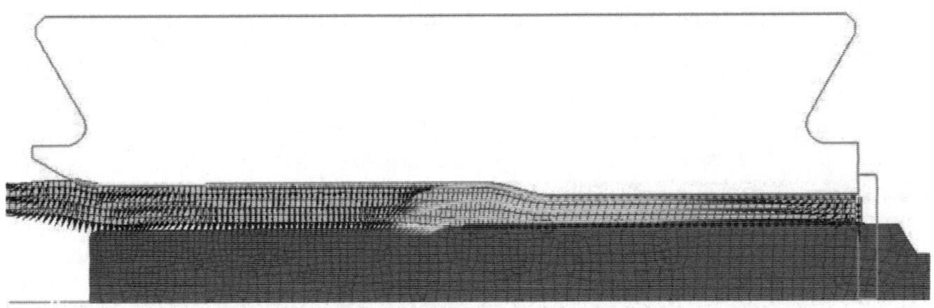

Figure 5. The forming process in the plain-strain problem

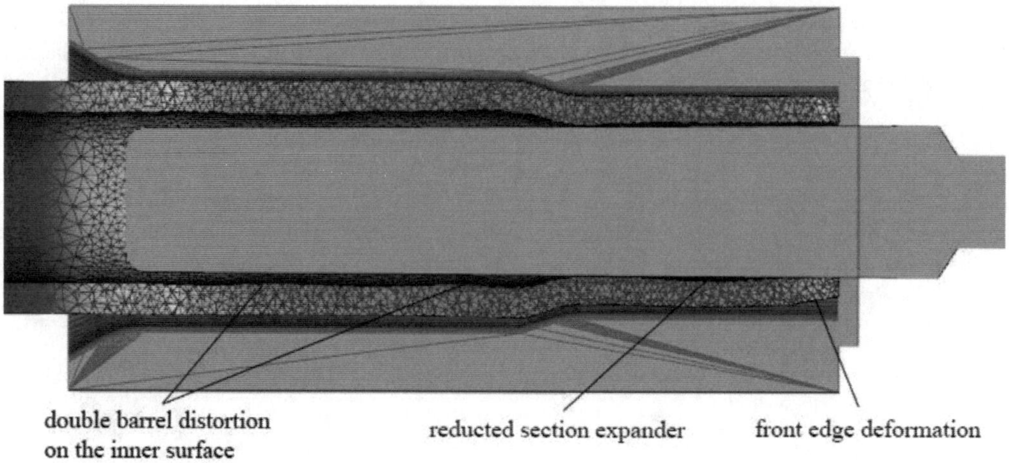

Figure 6. Forming process features

3. Conclusion

3.1. The mathematical model of the definition of the thermal and stress-strain state of the billet has been created. This allows obtaining data about temperature, stress and strain distribution within the billet.
3.2. The mathematical model of the deformation site of the radial forging of the hollow billet located on the mandrel has been created. The model was developed for the lever radial forging machine.
3.3. The created thermal and deformation model of the deformation site on LRFM is used to define the rational process parameters, including reduction and thermal modes, as well as to design dies calibration.

References

[1] Kobayashi S, Oh S and Altan T 1989 *Metal forming and the finite-element method* (OxfordUniversity press, Inc)
[2] Zienkiewicz O and Taylor R 2000 Finite Element Method: fifth edition (Butterworth and Heinemann)
[3] Halici D, Zamani H, Prodinger D, Poletti C, Huber D, Stockinger M and Sommitsch C 2014 *Key Eng. Mat.* **611-612** 99-105
[4] Karamyshev A, Nekrasov I, Parshin V, Pugin A and Fedulov A 2009 *Metallurgist* **53** 623-6
[5] Karamyshev A, Nekrasov I, Parshin V, Pugin A and Fedulov A 2012 *Metallurgist* **56** 116-8
[6] Karamyshev A *et al.* 2014 *WIT Trans. Ecol. Envir.* **190** 169-177

To the theory of mechanisms subfamilies

A Fomin[1,2], L Dvornikov[1], M Paramonov[1] and A Jahr[3]
[1] Siberian State Industrial University, 42, Kirova st., Novokuznetsk, 654007 Russia
[2] IT SB RAS, 1, Ac. Lavrentieva ave., Novosibirsk, 630090, Russia
[3] Düsseldorf University of Applied Sciences, 23 31/32, Gebäude, Düsseldorf, 40225, Germany

E-mail: alexey-nvkz@mail.ru

Abstract. The principles of formation of mechanisms subfamilies based on the usage of different kinds of kinematic pairs within the families of mechanisms are substantiated in the current paper. The division of mechanisms into subfamilies allows defining not only fundamental differences in the structure of mechanisms, but also provides the necessary foundation for the synthesis of new structures. 57 subfamilies of mechanisms have been totally distinguished. Among them, 31 subfamilies – within the zero family, 15 subfamilies– within the first family, 7 subfamilies – within the second family, 3 subfamilies – within the third family and 1 subfamily– within the fourth family. There were separately viewed planar mechanisms of the third family with three general imposed constraints and spatial mechanisms of the second family with two general imposed constraints in terms of their subfamilies. New methods of kinematical and dynamical investigations of mechanisms might be developed according to their analytical equations describing structural organization of different subfamilies of mechanisms.

1. Introduction

The creation of any mechanical system begins with the definition of its structural parameters. This procedure is called a structural synthesis. Despite of sufficient scientific developments in mechanisms and machines science, this part is less investigated.

The synthesis of newly designed structural schemes of mechanisms has been intuitively carried out, using previous experience, by layering the structural groups of zero mobility, called Assur groups [1]. Such solution is not always rational-functional because it does not consider all possible options.

The problem of how to develop the structural schemes with specified number of links and mobility in the system is a multivariate task. This is due to the fact that with the same number of links and kinematic pairs we can find a number of structural schemes differing from each other in structure and in types of kinematic pairs. To solve such problems, it is advisable to consider the structure in terms of the mechanisms imposed constraints and the possible usage of kinematic pairs with different numbers of relative motions.

Professor Artobolevsky I.I. suggested the classification of all the mechanisms into families in 1936 [2]. The number of constraints imposed on mechanism (m) was taken as a general criterion, which enters the universal formula of mobility of kinematic chains described by Professor Dobrovolsky V.V. [3]

$$W_m = (6-m)n - \sum_{k=5}^{k-m=1}(k-m)\,p_k\,, \tag{1}$$

where W_m – number of degrees of freedom of a mechanical system, n – number of movable links, p_k – class of kinematic pairs from 5 to 1. The mechanisms as systems with one degree of freedom (W_m=1), are studied in the current investigation. The introduction of parameter m, accepting positive integer values from zero to four, allowed revealing five families of mechanisms.

2. Classification of mechanisms depending on the number of general imposed constraints

The zero family includes only classically spatial mechanisms without imposed constraints. Links of these mechanisms have motions of a general nature, they are connected by kinematic pairs of all five classes from p_5 to p_1. A structural formula with m=0 introduced in the equation (1) will be written as follows

$$W_0 = 6n - 5p_5 - 4p_4 - 3p_3 - 2p_2 - p_1. \quad (2)$$

and called 'Malishev's formula'. The mechanism of the zero family is shown in Figure 1 [4].

The first family (m=1) includes spatial mechanisms with one general imposed constraint and it is described by the dependence

$$W_1 = 5n - 4p_5 - 3p_4 - 2p_3 - p_2. \quad (3)$$

The mechanisms of this family include all classes of kinematic pairs except five-degree-of-freedom pairs of the first class p_1. Figure 2 provides an angled transmission mechanism which functions within the first family.

Spatial mechanisms of the second family (Figure 3 [5]) with two general imposed constraints (m=2) are constructed by only three classes of pairs: three-, two-, one-degree-of freedom pairs

$$W_2 = 4n - 3p_5 - 2p_4 - p_3 \quad (4)$$

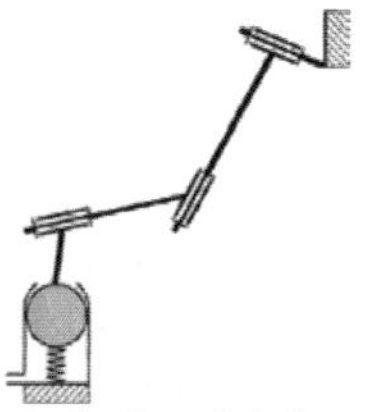

Figure 1. Spatial four-bar jointed mechanism (m=0)

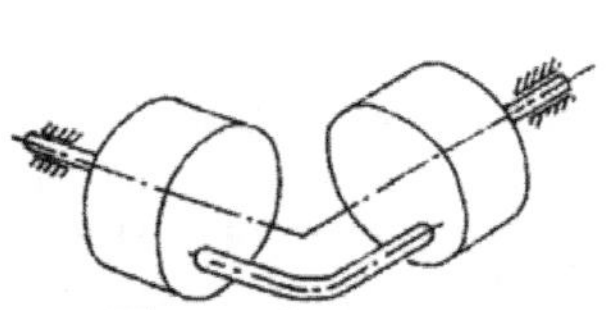

Figure 2. Angled mechanism (m=1)

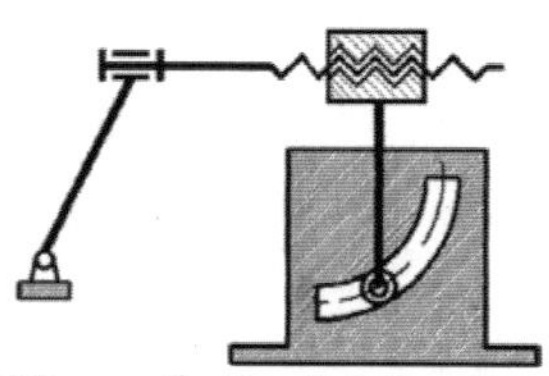

Figure 3. Spatial four-bar mechanism (m=2)

The mechanisms of the third family, planar or spherical, have three general imposed constraints and mobility formula (1) is written with m=3 as following

$$W_3 = 3n - 2p_5 - p_4 \quad (5)$$

.

Figure 4 provides a planar four-bar cam mechanism [6] of the third family, which has three imposed constraints.

The fourth family (m=4), where only wedge and screw mechanisms with four imposed constraints might be synthesized, is organized only by kinematic pairs of the fifth class

$$W_4 = 2n - p_5. \quad (6)$$

There is a mechanism of the fourth family presented in Figure 5 [7].

Thereby the number of general imposed constraints (m) allows the synthesis of mechanisms within five different families in accordance with their structural formulas of mobility (2) - (6), considered numbers of relative motions of links in the Cartesian spaces of coordinates and composition of kinematic pairs.

At the same time one should note the existence of multi-family mechanisms in the technique, which simultaneously includes kinematic chains of different families [8]. Figure 6 shows a kinematic scheme of the multi-family mechanism, which includes the mechanism of the second family (m=2) formed by links 1, 2, 3, 4 and 7, and the mechanism of the fourth family (m=4) made of links 5, 6 and 7.

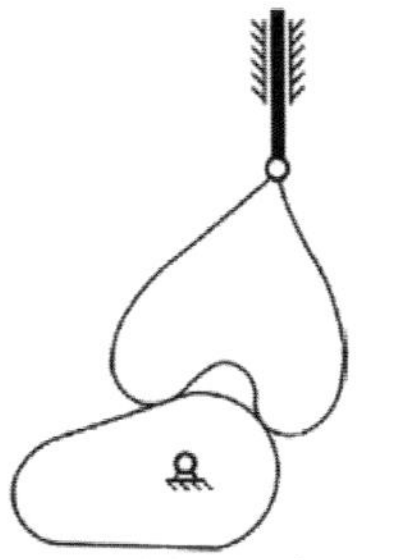

Figure 4. Four-bar cam mechanism (m=3)

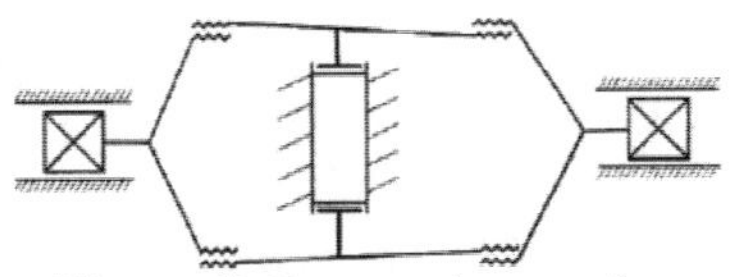

Figure 5. Progressive motion reducer (m=4)

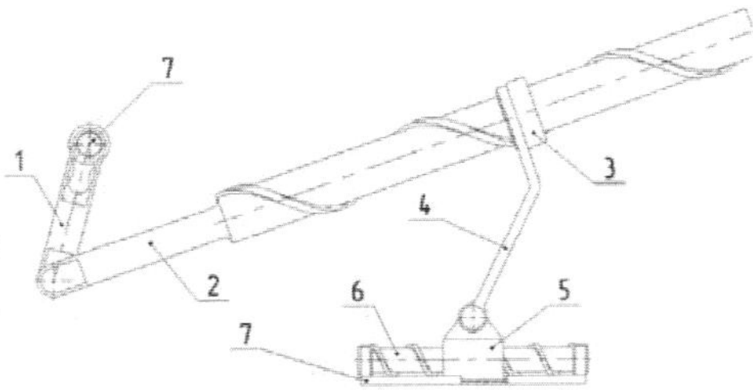

Figure 6. Multi-family mechanism (m=3.5)

The number of degrees of freedom of this mechanism can not be defined with the use of only one formula (2) - (6). It is calculated in accordance with the kinematic chains included in the mechanism's scheme. In this case, for the part of the second family we have mobility

$W_2 = 4 \cdot 4 - 3 \cdot 5 = 1$,

and for the part, belonging to the fourth family, mobility equals

$W_4 = 2 \cdot 2 - 3 = 1$.

To sum up total mobility of this mechanism is calculated as

$W_{2,4} = W_2 + W_4 = 2$,

which means that input motions are set for two links in the mechanism (for links 1 and 6).

3. Mechanisms subfamilies, their structural formulas

The mechanisms have more specific distinctions according to the composition of the applied kinematic pairs of different classes. In this connection it was offered to distinguish different subfamilies of mechanisms. Let us address to the formulas of mechanisms families (2) - (6).

It is possible to distinguish 31 subfamilies with different composition of kinematic pairs within the zero family of mechanisms, which are described by structural formula (1). The most complex subfamily is the first one, describing the equation

$W_{0(1)} = 6n - 5p_5 - 4p_4 - 3p_3 - 2p_2 - p_1$,

when the simplest one is the 31st, which includes only five-degrees-of freedom pairs p_1

$W_{0(31)} = 6n - p_1$.

The mechanisms of the first family include 15 subfamilies. The most complex among them is the first one, which contains pairs p_5, p_4, p_3 and p_2

$W_{1(1)} = 5n - 4p_5 - 3p_4 - 2p_3 - p_2$,

and the simplest one is the fifteenth, which contains only two-degrees-of-freedom pairs p_2

$W_{1(15)} = 5n - 2p_2$.

Only kinematic pairs of three classes – three-, two- and one-degree-of-freedom (p_5, p_4 and p_3) are used in the mechanisms of the second family. So, seven different subfamilies within the second family might be defined. Their structural formulas are provided here

$W_{2(1)} = 4n - 3p_5 - 2p_4 - p_3$, $W_{2(3)} = 4n - 3p_5 - p_3$, $W_{2(5)} = 4n - 2p_4 - p_3$,

$W_{2(2)} = 4n - 3p_5 - 2p_4$, $W_{2(4)} = 4n - 3p_5$, $W_{2(6)} = 4n - 2p_4$,

$W_{2(7)} = 4n - p_3$.

There are three subfamilies within the third family. Pairs p_5 and p_4 are simultaneously used in the first subfamily, while the second and the third subfamilies include either pairs p_5 or p_4 separately
$W_{3(1)} = 3n - 2p_5 - p_4$, $W_{3(2)} = 3n - 2p_5$, $W_{3(3)} = 3n - p_4$.

As the links of mechanisms of the fourth family connect among themselves in pairs of the fourth class, a singular subfamily within the fourth family has been found out. This subfamily is described by the structural formula of the following view
$W_{4(1)} = 2n - p_5$.

It tumbles to a fact that condition W_m=1, which means creation of a mechanism with a singular degree of freedom, cannot be provided in each subfamily. For instance, the sixth subfamily of the second family, which is organized only by kinematic pairs of the fourth class and is described by dependence
$W_{2(6)} = 4n - 2p_4$,
does not allow the synthesis of a mechanism because the coefficients before n and p_4 are even numbers, so condition $W_{2(6)}$=1 cannot be reached. Only mechanical systems with zero or gerade mobility within this subfamily might be constructed.

4. Synthesis of elementary planar mechanisms of the third family within their subfamilies

Let us address to the mechanisms of the third family, which includes three subfamilies. The simultaneous use of one-degree-of-freedom (p_5) and two-degree-of-freedom (p_4) kinematic pairs is the indispensable condition for the first subfamily, which is described by the formula
$W_{3(1)} = 3n - 2p_5 - p_4$
with $W_{3(1)}$=1, we have $n = (2p_5 + p_4 + 1)/3$, where a minimal number of movable links of the mechanism will be equal to two (n=2) with a minimal number of kinematic pairs, which are p_5=2 and p_4=1. According to these parameters the cam mechanism with a higher kinematic pair 'cam-follower' is shown in Figure 7.

Only one-degree-of-freedom pairs of the fifth class might be applied in the second subfamily. Introducing $W_{3(2)}$=1 into the equation of the second subfamily, written as
$W_{3(2)} = 3n - 2p_5$,
and then expressing n, like $n = (2p_5 + 1)/3$, it is possible to find minimal values for n and p_5 which are 3 and 4. Figure 8 provides the simplest mechanism of the third family second subfamily – a four-bar jointed mechanism. The last expression for n has the solution when n=1 and p_5=1, which in fact describes the structure of a driving link – a crank or a slider.

Let us move to the third subfamily. It is formed only by four-degrees-of-freedom pairs p_4. To express parameter n from structural formula $W_{3(2)} = 3n - p_4$ like $n = (p_4 + 1)/3$, we find that the minimal values are n=1, p_4=2. The mechanism synthesized by this solution is shown in Figure 9 [9].

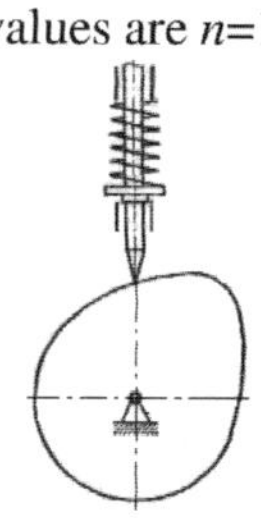

Figure 7. Three-bar cam mechanism

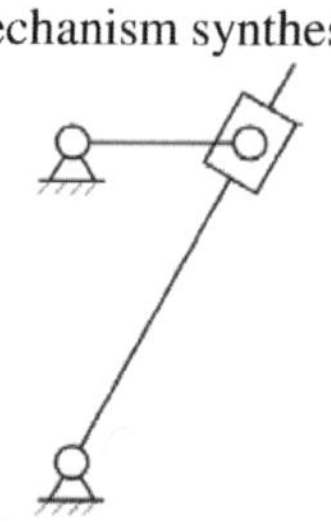

Figure 8. Link gear

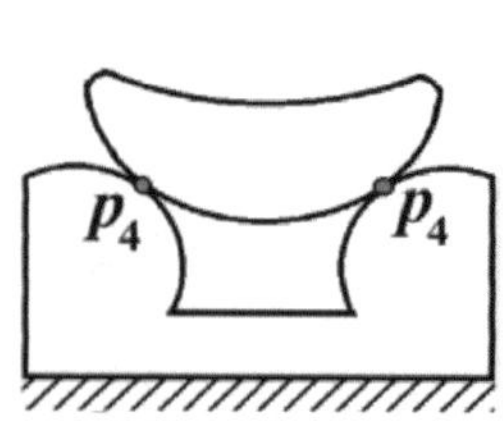

Figure 9. Two-bar mechanism with higher kinematic pairs

Despite the fact that the mechanisms of the third family (Figure 7-9) have the same number of imposed constraints (m=3) and their links move in plane, these mechanisms differ in structure and functioning as they are organized by different types of kinematic pairs.

5. Subfamilies of mechanisms of the second family in the view of their structure

Let us move from planar mechanisms to spatial ones and examine subfamilies of the second family. The first subfamily, described by dependence

$W_{2(1)} = 4n - 3p_5 - 2p_4 - p_3$,

simultaneously includes pairs p_5, p_4 and p_3. Setting $W_{2(1)}=1$, we express a parameter of the number of movable links like $n = (3p_5 + 2p_4 + p_3 + 1)/4$, where a minimum value for n is 2 (when parameters p_5, p_4 and p_3 possess minimum values). In this way sum ($3p_5 + 2p_4 + p_3 + 1$) shall equal to 8. It is possible only when $p_5=1$, $p_4=1$, $p_3=2$. Let us present the scheme of a spatial mechanism of the first subfamily constructed in accordance with the found parameters in Figure 10.

Two kinds of kinematic pairs might be used in the second subfamily, they are p_4 (fourth class) and p_5 (fifth class). From the structural formula of this subfamily

$W_{2(2)} = 4n - 3p_5 - 2p_4$

it follows that the simplest mechanism (with a minimum number of links) is such mechanism which includes two movable links (n=2), one pair of the fifth class (p_5=1) and two pairs of the fourth class (p_4=2). Figure 11 shows a kinematic scheme of this mechanism

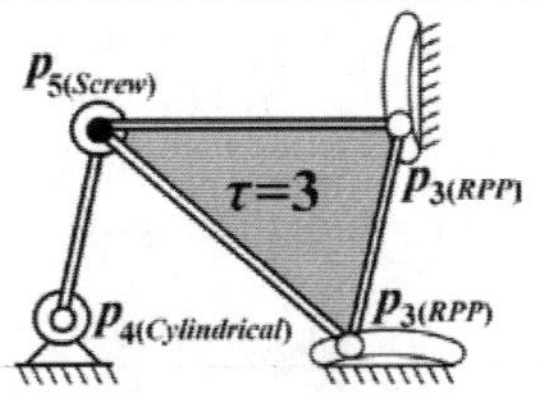

Figure 10. Three-bar linkage

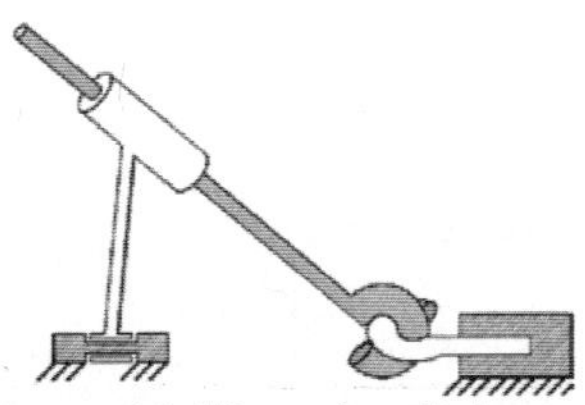
Figure 11. Three-bar linkage

The third subfamily includes mechanisms with three- (p_3) and one-degree-of-freedom kinematic pairs (p_5). From the structural formula of this subfamily

$W_{2(3)} = 4n - 3p_5 - p_3$

It has been found that the minimum number of movable links of mechanism will be n=2, the numbers of kinematic pairs are p_5=2 and p_3=1. This mechanism is shown in Figure 12.

The mechanisms of the fourth subfamily contain kinematic pairs of the fifth class p_5 only. Using their structural formula of mobility

$W_{2(4)} = 4n - 3p_5$

we have found that the simplest mechanism here is the mechanism with four movable links (n=4). Then it is possible to use only five pairs (p_5=5). The mechanism constructed with this solution is shown in Figure 13 [10].

It is impossible to apply one-degree-of-freedom kinematic pairs (p_5) in three residuary subfamilies of mechanisms of the second family. The mechanisms of the fifth subfamily are organized by pairs of the fourth p_4 and third p_3 classes. The formula of mobility of this subfamily

$W_{2(5)} = 4n - 2p_4 - p_3$

with $W_{2(5)}$=1 gives the solution n=1, p_4=1 and p_3=1. The kinematic scheme of the mechanism that corresponds to this solution is shown in Figure 14.

It was mentioned above that it is not possible to create mechanisms in the sixth subfamily as it is impossible to provide one number of degrees of freedom ($W_{2(6)}$=1). So, let us turn to the seventh subfamily, which is constructed by three-degrees-of-freedom kinematic pairs. The first solution of the equation

$W_{2(7)} = 4n - p_3$

with $W_{2(7)}=1$ is $n=1$, $p_3=3$. This describes a two-bar mechanism with one movable link and one fixed link connected by three pairs. Let us present the scheme of the mechanism in Figure 15.

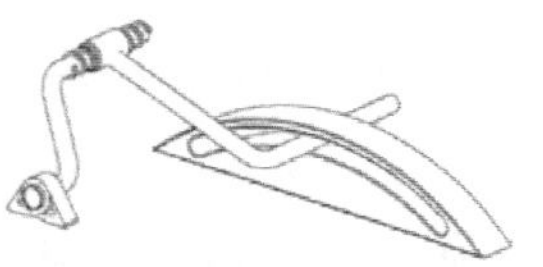

Figure 12. Three-bar linkage

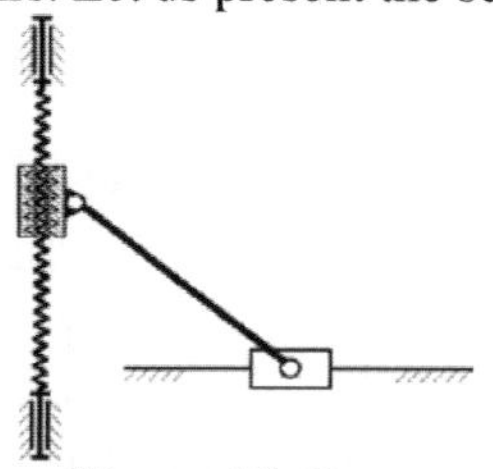

Figure 13. Press mechanism

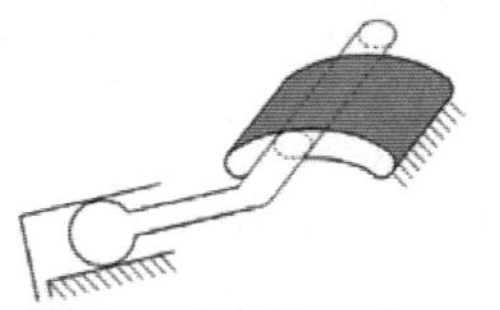

Figure 14. Two-bar linkage

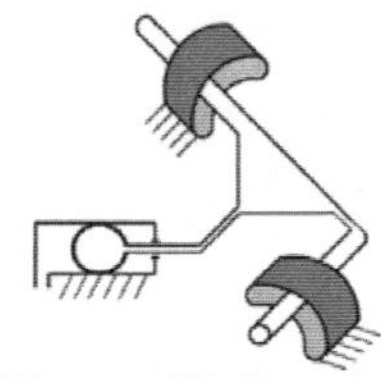

Figure 15. Two-bar linkage

Thereby all diversity of the mechanisms of the second family with two general imposed constraints is created within six different subfamilies. The kinematic schemes of the mechanisms demonstrated in Figure 10-15, have individual design features because of the usage of different kinematic pairs.

6. Conclusions

The division of mechanisms into families, offered in this paper, appreciably simplifies the problem of a mechanisms classification and a structural synthesis task and allows the introduction of more precise parameters of the constructed mechanical system.

The algorithms of a kinematical and dynamical analysis of linkages shall be developed according to their analytic equations.

The theory of mechanisms subfamilies might be developed in the direction of the synthesis of new kinematic chains, in which pairs are located in different sequences with the opportunity to change the complexes of relative motions of kinematic pairs. For example, with the appliance of kinematic pairs of the fifth class it is possible to use screw, rotate or prismatic pairs. This approach allows a structural synthesis of new mechanisms within families and subfamilies of the mechanisms.

Acknowledgments

This study has been supported by the Ministry of Education and Science of the Russian Federation (registration number RFMEF160714X0106).

References

[1] IFToMM Terminology, http://www.iftomm.org
[2] Artobolevski I I 1939 *Prikladnaya matematika i mehanika*, **11(4)**
[3] Dobrovolski V V 1936 *AN USSR*
[4] Dvornikov L T and Obryadin A S 2012 Four-bar jointed mechanism. RU Patent 2446331
[5] Dvornikov L T Fomin A S 2010 Spatial four-bar mechanism with screw pair and roller moving in a slot. RU Patent № 94515
[6] Dvornikov L T Ermolaeva N Yu 2012 *RU Patent* 2456491
[7] Dvornikov L T and Baklushin D S 2007 *RU Patent* 2295080
[8] Dvornikov L T 1997 *Proc. VI scientific and practical conference of the problems of metallurgical and mining machines Novokuznetsk* 15
[9] Adamovich N O 1998 Doctoral thesis Structural synthesis, kinematics and statics of planar non-Assur mechanisms
[10] Ozol O G 1984 *M.: Nauka* 432

IOP Conf. Series: Materials Science and Engineering **124** (2016) 012056 doi:10.1088/1757-899X/124/1/012056

Mathematical simulation of the operating emergency conditions for the purpose of energy efficiency increase of thermal power plants management

O V Gazizova, A V Malafeyev and Y N Kondrashova
Federal State Budgetary Educational Institution of Higher Professional Education, G.I. Nosov Magnitogorsk State Technical University, 38, Lenina ave., Magnitogorsk, 455000, Russia

E-mail: logan_b_7@mail.ru

Abstract. The regulations of the energy saving policy, being implemented at present, require an enhanced reliability of the power supply systems and operations of the motive loading, actuating machines and mechanisms, for the purpose of reduction of power inputs under operating emergency conditions. The development of industrial units of the distributed generation enables putting the generators, equipped with motors of the comparable power, into the operation, which is non-parallel to a power system. To provide static stability under such conditions the complex approach is required since it allows estimating the stability of the generators and the load during parallel and non-parallel operations with the power system and a complex circuit configuration. The mathematical models of units of the distributed generation and motive loading considering the character of the given mechanism are developed. They are implemented in the software package allowing the operative-dispatch personnel of the industrial power stations to forecast similar modes and develop special measures to provide a stable operation of electrical machines.

1. Introduction

The conditions of the operation and control of the power supply system of a large-scale iron and steel enterprise are usually determined by social and economic environment of the world. At present, in conditions of market relations established in the power industry it is more advantageous for industrial enterprises to consume the electric power generated by their own auxiliary power stations being the units of power mechanical engineering. Therefore, one of the modern trends is that large-scale enterprises tend to enhance their own auxiliary generating capacities. Thus, they improve the reliability of electric power supply to consumers and the output quality, which corresponds to the energy saving policy. However, operating conditions of the power-supply system of the industrial enterprise do not eliminate the possibility of the electric power station separation from the power system as a result of triggering the emergency automation. That is why, in order to enhance the reliability of the electric power supply to consumers it is necessary to determine the conditions, when the dynamic stability of the power unit is retained during switching to the non-parallel operation [1-4], and the static stability - during the non-parallel operation immediately. In order to increase reliability of the electric power supply and with due regard to the fact that generators and current-using equipment of the industrial system may have comparable power it is reasonable to apply a complex approach, allowing one to estimate the reliability of the power sources as well as consumers, when analysing the static stability.

Initially, at the time of electrical power engineering evolvement, the policy of power system interconnection was adopted, consequently, scientific researches were aimed at the calculation of the steady-state and transient operation modes of the large interconnected power utilities [5-7]. In this connection, the issues of stability are largely treated relatively the large-scale electric power systems, and restrictedly analyzed concerning the industrial power systems. Consequently, it is urgent to develop simple and convenient methods for calculating the steady-state and transient operating modes as well as stability estimation during off-line operation for the large-scale industrial enterprises, having auxiliary power stations.

As is known, a static stability margin of the industrial power station generators should be provided by the operative-dispatch control of the power system during the operation, occurring in parallel with the power system, and by the operative-dispatch personnel of the enterprise, when switching to the non-parallel operation. A special software allows the forecast of the static stability of the industrial power stations in various modes. The KATRAN software package [8] has been developed at the Department of the Power Supply of Industrial Enterprises of Magnitogorsk State Technical University.

2. Results and Discussion

In calculations of the static stability the main task is to determine the ideal limit of the output electromagnetic power of the generators during parallel operation with the power system and of the actual limit during non-parallel one. This is impossible to implement without a corresponding mathematical apparatus allowing calculating the steady-state modes during the parallel and non-parallel operation with the power system. When calculating the steady-state modes, one of the most convenient calculating methods is the modified method of a sequential network-reduction. It is applicable to calculation of the steady-state modes of the power systems of complex configuration, differing by complex hierarchy and a heterogeneous set of power and load sources. This method forms the basis of the algorithm of the steady-state mode calculation in the KATRAN software.

As is well known, during the non-parallel operation of the power station and the electrical power system, mode parameters may undergo much more significant changes in voltage and frequency. During mathematical simulation of modes, the static parameters of the excitation and speed controllers as well as governing load effect are to be taken into account. When a frequency droop of the automatic generator controllers amounts to about 4–5%, the form of the static load characteristics may very over a wide range depending on the load content. Therefore, during the calculation of these modes parameters it is necessary to determine the content and corresponding governing effect of the complex load, which generally may be positive or negative. In connection with the above-mentioned information, in order to enhance the efficiency of the mode control for the industrial power supply systems, having their own auxiliary power stations, the method of determination of the governing effect of the complex consumer load of the industrial enterprise has been developed, taking into account losses within the distribution networks.

For example, active and reactive power of the asynchronous motor is determined, relatively the nominal unit:

$$P=\frac{k_u^2 r_2^{/} s}{k_f^2 x_k^2 s^2 + r_2^{/2}};$$

$$Q=\frac{P k_f x_k s}{r_2^{/}}+\frac{k_u^2\left(1-0.3+0.3\cdot k_u^2\right)}{k_f x_\mu},$$

where s – slip; $k_f=f_{neworkt}/f_{nom}$; $k_u=U_{network}/U_{nom}$; r'_2 – resistance; x_k – short-circuit reactance; x_μ – magnetizing reactance.

The reactive of the synchronous motor power is determined, relatively the nominal unit:

$$\alpha = \frac{\sqrt{k_{Ff}^2 k_u^2 \left(1 + 2x_d \sin\varphi_{nom} + x_d^2\right) - \left(\beta x_d \cos\varphi_{nom}\right)^2} - \frac{k_u^2}{k_f}}{x_d \sin\varphi_H},$$

where $k_{Ff}=I_f/I_{f\,nom}$; x_d – inductive reactance; β - active power; φ - nominal angle.

This method has been implemented in the KATRAN software package. The study of the static characteristics of the complex load has confirmed the possibility of use of the generalized governing factors during calculation of the steady-state and transient modes in the power supply systems of industrial enterprises having auxiliary power stations. According to the research results, the following limits may be taken for practical calculations in the industrial networks:

- governing effect of the active power according to frequency 1 2.4;
- governing effect of the reactive power according to frequency 0.9 2.1;
- governing effect of the active power according to voltage 0.1 0.3;
- governing effect of the reactive power according to voltage 1.3 2;

As is well known, the presence of the static stability of the synchronous generators may be determined by a small oscillations method and by means of practice benchmarks of the static stability [1], [4]. But these approaches do not allow determining the margin factor of the static stability, which is relevant during the non-parallel operation with the power system. As a rule, the margin factor is calculated using the in-series loading method. During the parallel operation one should use the rotor angle of the synchronous generator as a loading parameter, while loading should be applied during the non-parallel one. The authors of the paper use a direct method of determining the margin of the static stability as it may be conveniently applied in combination with the conventional method of the steady-state mode calculation, that is, the mode parameters are determined at each loading step [9]. During loading the modes are calculated till there is no detected convergence, which justifies the limiting steady-state mode when using the modified sequential equivalent method.

As the steady-state modes are calculated at each loading step considering the governing effect of the load and generators, the obtained values of the mode parameters are the most correct. In this case, the actual limit of the transmitted power is determined as the voltage reduction in the network is taken into account together with the governing load effect. The advantage of this calculation is a possibility to determine the margin of the static stability at any network configuration without formulation of complex differential equation systems. The developed method of calculation of the static stability is shown in Figure 1 and implemented in the KATRAN software package.

As in the modes of non-parallel operation the voltage and frequency may undergo significant changes the load stability is of interest. Generally, the determination of the static stability of the loading is brought to specification of the critical voltage values of the current-using equipment. If these values are reached the equipment lose its stability and is simply shut out by the protection system. Specification of the critical voltage values of the synchronous and asynchronous motors depends on the variation of the characteristics of the electromagnetic powers during corresponding voltage changes as well as the load factor, nature of the drag torque on the shaft and, in case of the synchronous motors, excitation current. Only consideration of the above-mentioned factors gives the idea of the motor stability. The KATRAN software allows determining the factors of the static stability margin of the synchronous and asynchronous motors during the non-parallel and parallel operation of the unit with the power system.

By means of the developed software package the study was conducted for a standard industrial power station interconnected with the power system, for the consumers of the electric power based on the generator voltage and for the generators intended for supplying the auxiliary power consumers of the electric power station with electric current. The power supply diagram of the 3 kV distribution substation supplied by auxiliary generators, connected to the rest of the power system by means of the coupling transformers and supplying auxiliary circuits is shown in Figure 2. Stability has been calculated for a G 1 generator with regard to the action of the automatic excitation control and the field current cavity equal to two. The results of the calculation of marginal steady-state stability when the generators operate in parallel and in non-parallel with power system are shown in Figure 3a, b.

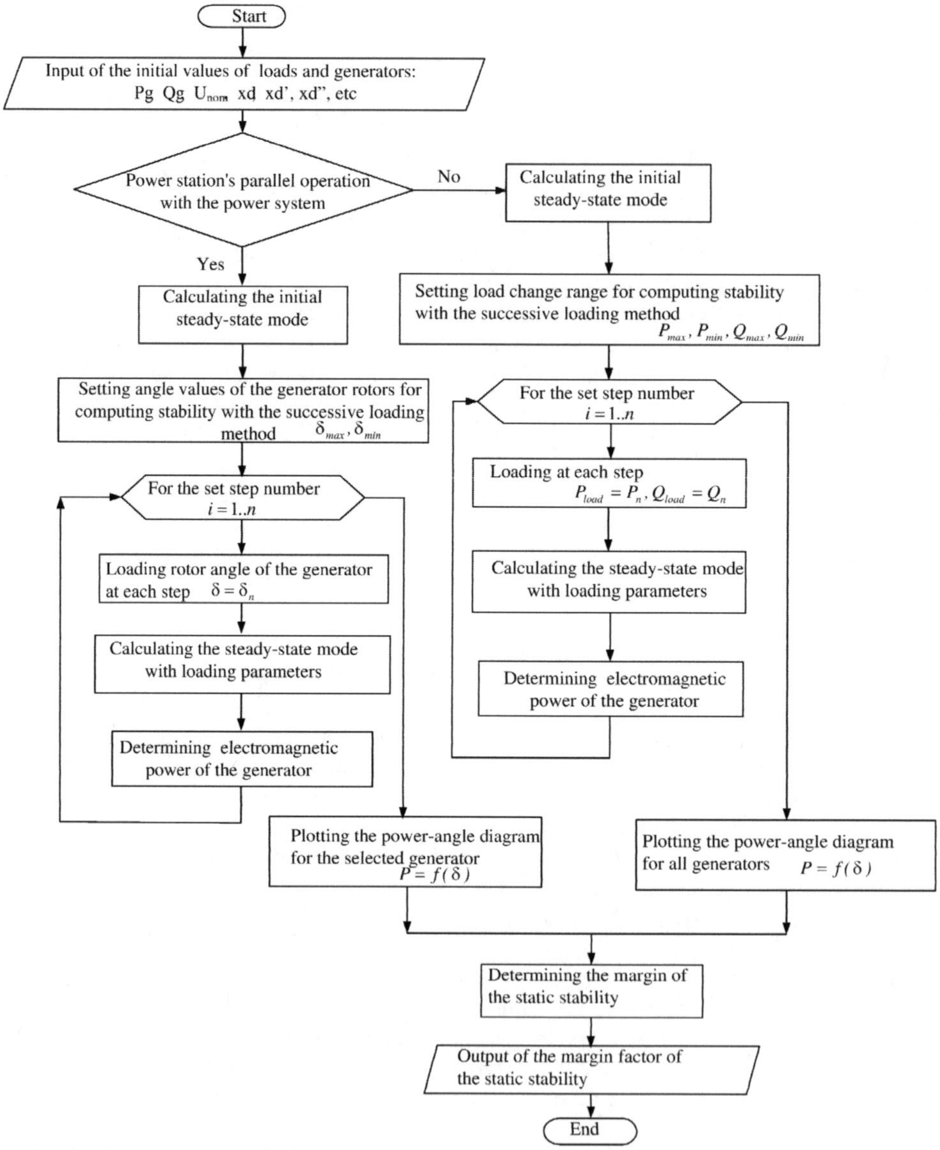

Figure 1. The flow chart of the algorithm for determination of the static stability margin of the synchronous generators according to the active power

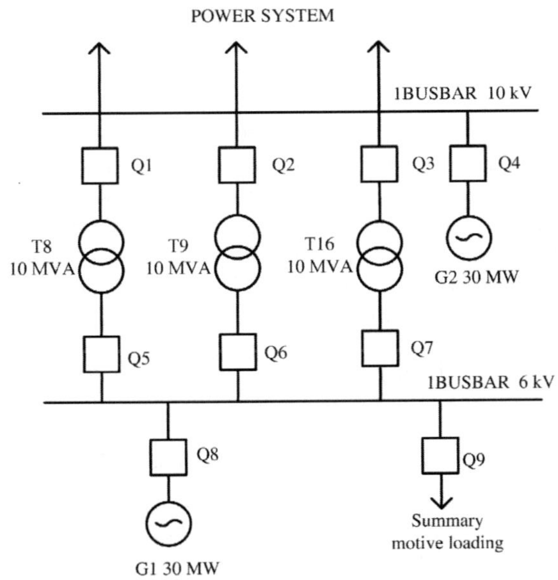

Figure 2. A diagram of the power supply system under consideration

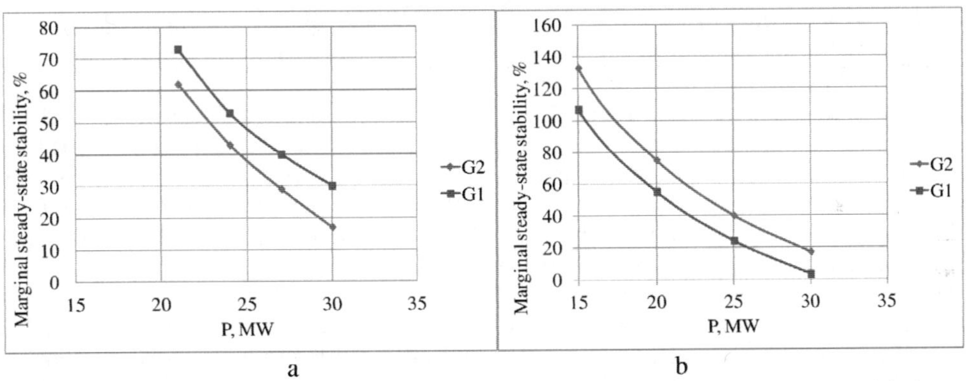

Figure 3. a - marginal steady-state stability during the generators' parallel operation with the power system; b - marginal steady-state stability during the generators' separate operation with the power system.

The generator power-angle diagram during the parallel operation with the power system obtained by means of the developed software is shown in Figure 4. The factor of the static stability margin in this mode is 67%. In case of the network voltage drop or other adverse conditions the separatory emergency automation is actuated and switches this network sector to the non-parallel operation with the power system. During the non-parallel operation in a steady-state mode the power-angle diagram, shown in Figure 5, has been obtained. The obtained results indicate that the factor of the static stability margin in the non-parallel mode is 63 %. This is due to the fact that the generator provides 12 MW of active power during the parallel operation and about 8 MW - during the non-parallel one. Thus, at presence of the high-response automatic excitation control the static stability margin is sufficient for the normal generator operation. The studies of the static stability of the motor load (shown in Figure 2 in a simplified form) have also demonstrated a sufficient value of the static stability during the operation of both generators at the moment of separation from the power system. It should be noted that similar calculations may be performed for the power supply system of a complex configuration with the unlimited number of generators and motors for industrial enterprises of different branches.

3. Conclusion

To provide the operative control of the dispatching department modes a new algorithm for calculation of the steady-state modes of the non-parallel operation of industrial electric power system comprising auxiliary power stations has been developed. It has been developed on the basis of the modified sequential equivalent method. The mathematical models of objects of the distributed generation and motive loading considering the character of the given mechanism are developed. The improved method of determining the governing effect of the complex load has been obtained; it is intended for the analysis of normal, loaded and post-emergency modes under conditions of the power supply system of the industrial enterprise.

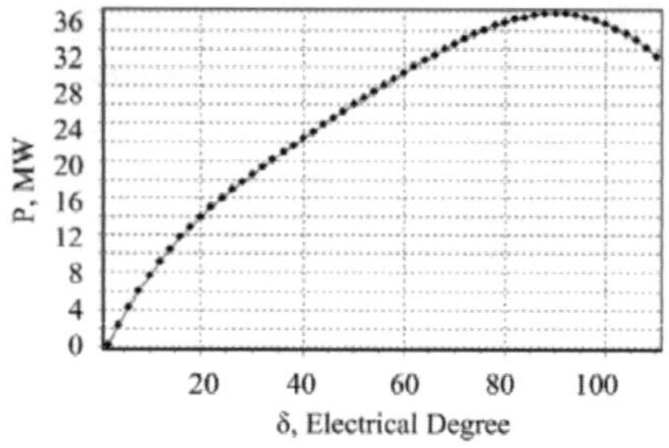

Figure 4. Power-angle diagram of the generator during the parallel operation with power system

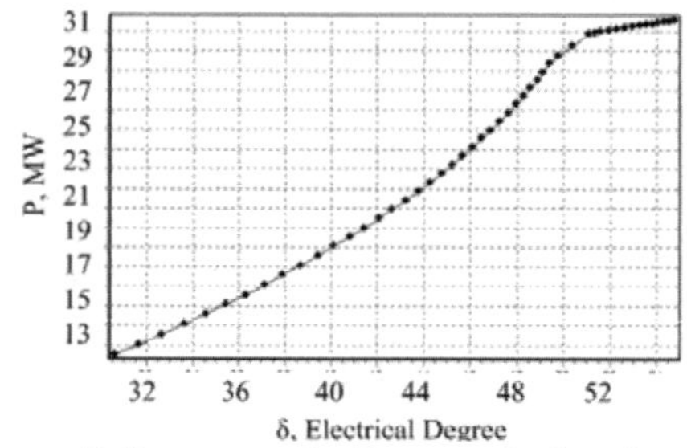

Figure 5. Power-angle diagram during the non-parallel operation

Based on the successive loading method, the procedure of determining the actual limit of the power transmitted during the parallel and non-parallel operations has been developed. The developed original KATRAN software package allows analyzing the stability during scheduling and operative controlling of the modes in the load nodes of large-scale industrial enterprises having auxiliary power stations. Its application enables setting limitations of the active and reactive output power of the generators when detecting an economically feasible operating mode of the power station to provide stability of electrical machines operation. Furthermore, an important condition of the reliable operation for any industrial enterprise is making-out a balance of the power consumption. The developed software allows determination of the critical power imbalances, by which the static stability will retain at the switch to the non-parallel operation.

The operative personnel can use this software package as a convenient tool to forecast the possibility of emergencies arising, to intervene timely into the operation of the power system of the industrial enterprises and, thus, to enhance reliability and provide energy saving.

Acknowledgements
The work has been done with the financial support of the Russian Foundation of Fundamental Research, Grant No. HK 14-07-00200\14.

References

[1] Kothari D P and Nagrath I J 2008 *Hill Publishing Company Limited*
[2] Rotanova J N 2008 *Engineering Science, Magnitogorsk, Russia*
[3] Zaslavets B I, Igumenshev V A, Malapheev A V, Bulanova O V and Rotanova J N 2008 *Bulleting of the South-Ural State University* **1(111)** 3-8
[4] Zhdanov P S 1979 *Issues of the Power System Stability*, Moscow Russia: Energy
[5] Xiufeng Shi and Shiguang Mu 2015 *Applied Mechanics and Materials* **742** 648-652
[6] Hazarika D Oct 2012 *IET Generation, Transmission & Distribution* **6(10)** 977-985

[7] Sujatha Er S, Anitha Dr R, Selvan Dr P and Selvakumar Er S 2014 *Journal of Theoretical and Applied Information Technology* **63(1)** 85-91
[8] Igumenshev V A, Malapheev A V, Gazizova O V, Rotanova J N, Panova E A, Nikolaev N A, Kochkona A V and Zinoviev V V 2012 *Certificate of the Russian Federation* 2012612069
[9] Bulanova O V 2007 Mode Control of Industrial Power Stations at Switching to Non-Parallel Operation. Ph.D. thesis in Engineering Science, Magnitogors, Russia

Effect of the buoyancy force on natural convection in a cubical cavity with a heat source of triangular cross-section

N S Gibanov and M A Sheremet
Tomsk State University, 36, Lenina ave., Tomsk, 634050, Russia

E-mail: fire9n@mail.ru

Abstract. Numerical analysis of laminar natural convection inside a cubical cavity with a local heat source of triangular cross-section has been conducted. The mathematical model formulated in dimensionless variables such as "vector potential functions – vorticity vector" has been solved by the finite difference method of the second order accuracy. The three-dimensional temperature fields, 2D streamlines and isotherms in a wide range of the Rayleigh number from 10^4 to 10^6 have been presented illustrating variations of the fluid flow and heat transfer.

1. Introduction

Natural convection in three-dimensional cavities has received considerable attention in recent years. It is well known that this heat transfer mechanism has many applications in thermal engineering involving cooling of electronic devices, heat transfer in solar collectors and energy storage systems [1]. For example, Martyushev and Sheremet [2] numerically analyzed natural convection combined with thermal surface radiation in a cubical cavity bounded by solid walls of finite thickness and conductivity in the presence of a heat source of constant temperature. They developed an in-house computational code using vector potential functions and a vorticity vector for numerical simulation of natural convective heat transfer. It was found that less intensive cooling of the domain of interest takes place in case of the three-dimensional model in comparison with the 2D-model. Fontana et al. [3] studied numerically mixed and natural convection in a partially open cubical cavity with an internal heat source. They found that the buoyancy forces are dominant at low Reynolds numbers and high Rayleigh numbers. For higher Reynolds numbers recirculation zones in the upper part of the cavity can also appear due to the influence of the horizontal flow. For intermediate values of these dimensionless parameters, an interaction between the buoyancy and inertial forces generates a complex flow and temperature structures with periodic and quasiperiodic regimes. Oosthuizen and Paul [4] have analyzed an unsteady free convective flow in a 3D rectangular enclosure with two heat sources located at the bottom wall. They revealed that the flow, which is steady at low Rayleigh numbers, becomes unsteady at intermediate Rayleigh numbers and then again becomes steady at higher Rayleigh numbers The conditions for unsteady flow were studied.

The purpose of the present study is to numerically investigate the effect of the Rayleigh number on the convective heat transfer inside a cubical cavity with a heat source of the triangular cross-section.

2. Physical and mathematical models

The computational domain is shown in the Figure 1. Two opposite vertical walls ($x = 0$ and $x = L$) are kept at constant low temperature T_c while the heat source located at the bottom wall is kept at constant high temperature T_h. The rest walls are adiabatic. The process of convective heat and mass transfer is described by the system of unsteady three-dimensional Boussinesq equations using the dimensionless variables of 'vector potential functions – vorticity vector – temperature' type [2, 4].

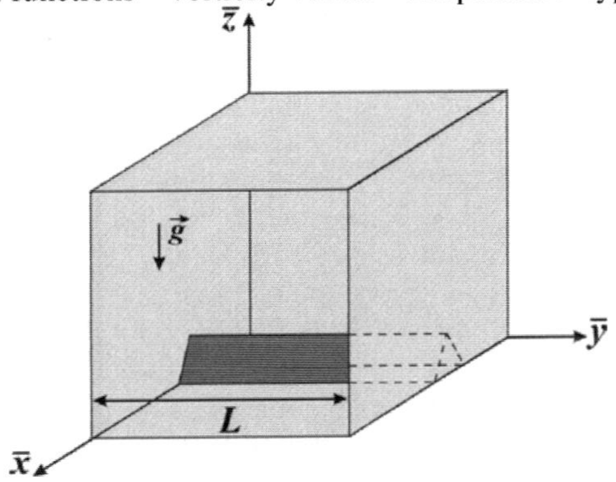

Figure 1. The domain of interest

The mathematical model has been formulated in following form:

$$\frac{\partial\omega_x}{\partial\tau}+u\frac{\partial\omega_x}{\partial x}+v\frac{\partial\omega_x}{\partial y}+w\frac{\partial\omega_x}{\partial z}-\omega_x\frac{\partial u}{\partial x}-\omega_y\frac{\partial u}{\partial y}-\omega_z\frac{\partial u}{\partial z}=\sqrt{\frac{\Pr}{\mathrm{Ra}}}\left(\frac{\partial^2\omega_x}{\partial x^2}+\frac{\partial^2\omega_x}{\partial y^2}+\frac{\partial^2\omega_x}{\partial z^2}\right)+\frac{\partial\theta}{\partial y}, \quad (1)$$

$$\frac{\partial\omega_y}{\partial\tau}+u\frac{\partial\omega_y}{\partial x}+v\frac{\partial\omega_y}{\partial y}+w\frac{\partial\omega_y}{\partial z}-\omega_x\frac{\partial v}{\partial x}-\omega_y\frac{\partial v}{\partial y}-\omega_z\frac{\partial v}{\partial z}=\sqrt{\frac{\Pr}{\mathrm{Ra}}}\left(\frac{\partial^2\omega_y}{\partial x^2}+\frac{\partial^2\omega_y}{\partial y^2}+\frac{\partial^2\omega_y}{\partial z^2}\right)-\frac{\partial\theta}{\partial x}, \quad (2)$$

$$\frac{\partial\omega_z}{\partial\tau}+u\frac{\partial\omega_z}{\partial x}+v\frac{\partial\omega_z}{\partial y}+w\frac{\partial\omega_z}{\partial z}-\omega_x\frac{\partial w}{\partial x}-\omega_y\frac{\partial w}{\partial y}-\omega_z\frac{\partial w}{\partial z}=\sqrt{\frac{\Pr}{\mathrm{Ra}}}\left(\frac{\partial^2\omega_z}{\partial x^2}+\frac{\partial^2\omega_z}{\partial y^2}+\frac{\partial^2\omega_z}{\partial z^2}\right), \quad (3)$$

$$\frac{\partial^2\psi_x}{\partial x^2}+\frac{\partial^2\psi_x}{\partial y^2}+\frac{\partial^2\psi_x}{\partial z^2}=-\omega_x,\quad \frac{\partial^2\psi_y}{\partial x^2}+\frac{\partial^2\psi_y}{\partial y^2}+\frac{\partial^2\psi_y}{\partial z^2}=-\omega_y,\quad \frac{\partial^2\psi_z}{\partial x^2}+\frac{\partial^2\psi_z}{\partial y^2}+\frac{\partial^2\psi_z}{\partial z^2}=-\omega_z, \quad (4)$$

$$\frac{\partial\theta}{\partial\tau}+u\frac{\partial\theta}{\partial x}+v\frac{\partial\theta}{\partial y}+w\frac{\partial\theta}{\partial z}=\frac{1}{\sqrt{\mathrm{Ra}\cdot\Pr}}\left(\frac{\partial^2\theta}{\partial x^2}+\frac{\partial^2\theta}{\partial y^2}+\frac{\partial^2\theta}{\partial z^2}\right). \quad (5)$$

Here x, y, z are the dimensionless Cartesian coordinates; τ is the dimensionless time; u, v, w are the dimensionless velocity components; $\omega_x, \omega_y, \omega_z$ ate the dimensionless vorticity components; ψ_x, ψ_y, ψ_z are the vector potential functions; θ is the dimensionless temperature; Ra is the Rayleigh number; Pr is the Prandtl number.

The initial conditions for formulated governing equations (1) – (5) can be written as follows:

$$\psi_x = \psi_y = \psi_z = \omega_x = \omega_y = \omega_z = 0,\ \theta = 0.5$$

The utilized boundary conditions are:

$$\frac{\partial\psi_x}{\partial x}=\psi_y=\psi_z=0,\ \omega_x=0,\ \omega_y=-\frac{\partial^2\psi_y}{\partial x^2},\ \omega_z=-\frac{\partial^2\psi_z}{\partial x^2},\ \theta=0 \text{ on } x=0 \text{ and } x=1;$$

$$\psi_x=\frac{\partial\psi_y}{\partial y}=\psi_z=0,\ \omega_r=-\frac{\partial^2\psi_x}{\partial y^2},\ \omega_y=0,\ \omega_z=-\frac{\partial^2\psi_z}{\partial y^2},\ \frac{\partial\theta}{\partial y}=0 \text{ on } y=0 \text{ and } y=1;$$

$$\psi_x=\psi_y=\frac{\partial\psi_z}{\partial z}=0,\ \omega_x=-\frac{\partial^2\psi_x}{\partial z^2},\ \omega_y=-\frac{\partial^2\psi_y}{\partial z^2},\ \omega_z=0,\ \frac{\partial\theta}{\partial z}=0 \text{ on } z=0 \text{ and } z=1;$$

$\theta = 1$ on heat source surface.

Governing equations (1) – (5) with corresponding initial and boundary conditions have been solved using the finite difference method. A detailed description of the developed numerical method has been done earlier in [2, 5]. The developed computational code has been verified for several benchmark problems [2, 5].

3. Results and discussion

The numerical analysis has been conducted in a wide range of the Rayleigh number: Ra = 10^4–10^6, Pr = 0.7. Figure 2 shows a three-dimensional velocity and temperature fields for different values of the Rayleigh number. For a small value of the dimensionless buoyancy force (Ra = 10^4) one can find two main convective rolls with clockwise and counterclockwise rotation reflecting the formation of the ascending flow over the heat source and descending flows near the isothermal vertical surfaces. It should be noted that intensive motion can be observed in the central part of the cavity where the effect of the surrounding adiabatic walls ($y = 0$ and $y = 1$) is not essential. Also the temperature field for Ra = 10^4 illustrates development of the heat conduction. An increase in the Rayleigh number leads to an intensification of convective flow and formation of a stable thermal plume.

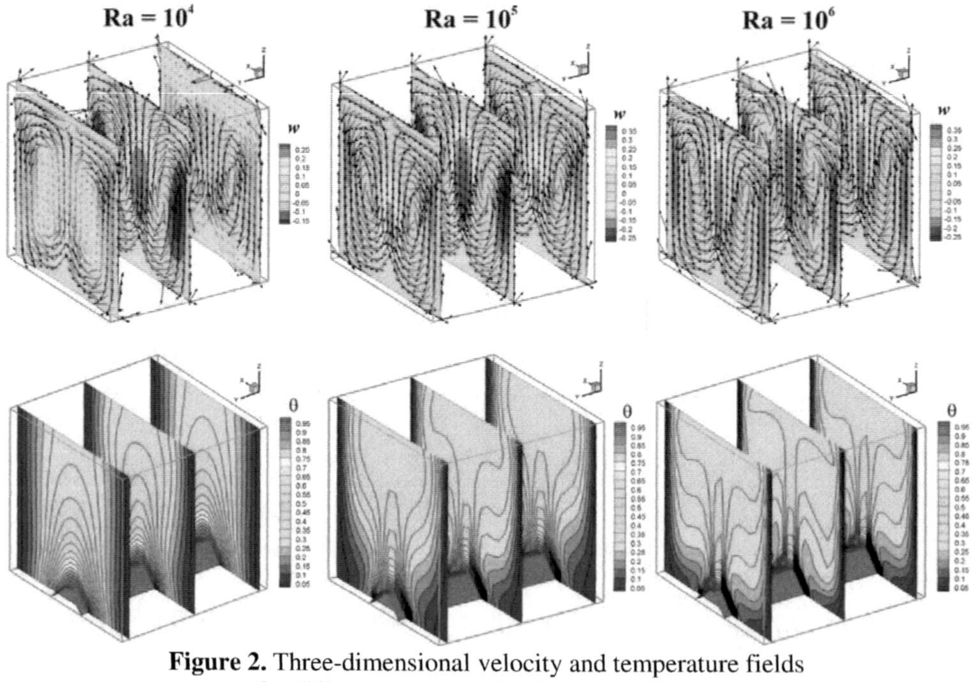

Figure 2. Three-dimensional velocity and temperature fields for different values of the Rayleigh number

Figure 3 demonstrates streamlines (the top row) and isotherms (the bottom row) for 3D (solid lines) and 2D (dashed lines) models for different values of the Rayleigh number. An increase in Ra leads to a displacement of the convective cores to the upper part of the cavity. The thickness of the thermal boundary layers is reduced with Ra. There is an intensification of the convective flow and heat transfer inside the cavity with the dimensionless buoyancy force. Taking into account the results for 2D model it is possible to find some weak differences between streamlines and isotherms for the middle cross-section of the cubical cavity. The main reason for the observed differences is the presence of the third coordinate in 3D case.

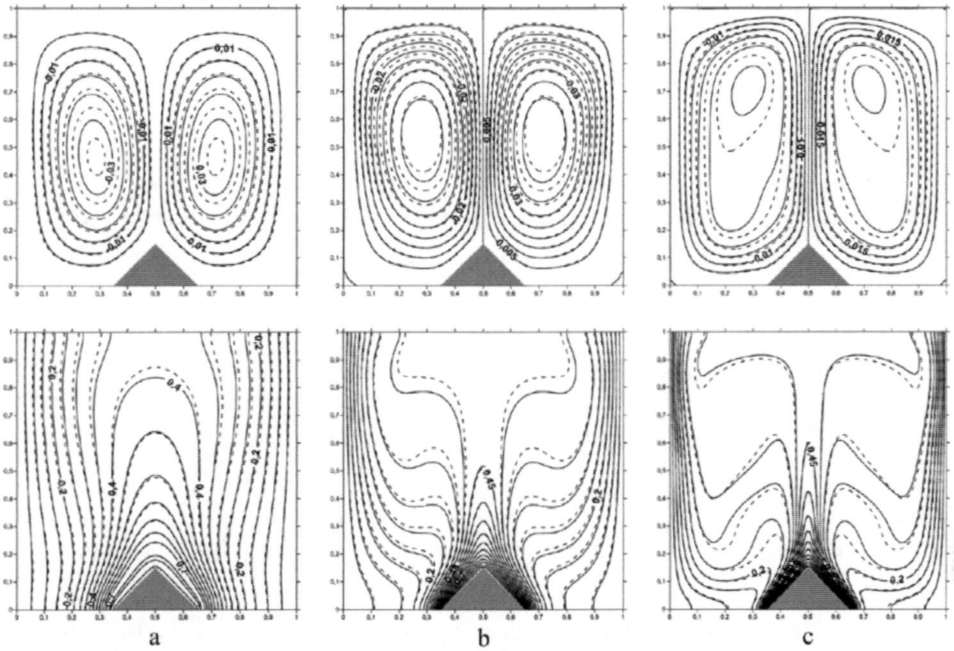

Figure 3. Streamlines (top row) and isotherms (bottom row) for different values of the Rayleigh number: a – Ra = 10^4, b – Ra = 10^5, c – Ra = 10^6

4. Conclusion

A numerical simulation of the natural convection in a cubical cavity with a local heat source of triangular cross-section has been carried out. The effect of the Rayleigh number on three-dimensional velocity and temperature fields and on the streamlines and isotherms in the middle cross-section has been studied for Pr = 0.7. It has been revealed that an increase in Ra leads to an intensification of convective flow and heat transfer. A comparison of streamlines and isotherms with 2D data shows weak differences that can be explained by the presence of the third coordinate in 3D case.

Acknowledgments

This work was supported by the Grants Council (under the President of the Russian Federation), Grant No. MD-6942.2015.8 and by the Russian Foundation for Basic Research, Grant No. 14-08-31137 mol_a.

References

[1] Jaluria Y 1998 *Design and Optimization of Thermal Systems* (New York: McGraw-Hill)
[2] Martyushev S G and Sheremet M A 2014 *Int. J. Heat Mass Transfer* **73** 340–353
[3] Fontana E, Capeletto C A, da Silva A and Mariani V C 2015 *Int. J. Heat Mass Transfer* **81** 829–845
[4] Oosthuizen P H and Paul J T 2005 *Int. J. Heat and Fluid Flow* **26** 587–596
[5] Sheremet M A 2015 *Numerical Heat Transfer, Part A* **68** 243–267

The finite-element behaviour simulation of the rotary-type and frame-type solar sails on the geocentric orbits

I Gorbunova, R Khabibullin, S Chernyakin and O Starinova
Spacecraft department, Samara State Aerospace University, 151, Molodogvardeiskaya st., Samara 443086, Russia

E-mail: gorbunovairina88@gmail.com

Abstract. This paper discusses the research of functioning of different construction types for the spacecraft with a solar sail. Two types of the solar sail are considered, such as frame-type and rotary-type. The research is performed by means of application of the computer-assisted design system. The movement simulation of the spacecraft center mass and the forces acting on the solar sail is described. The finite element models of the two solar sail constructions are developed and compared.

1. Introduction

Nowadays the world leading space nations are actively developing missions focused on the near-Earth space investigation. A promising way of cost reduction for such missions is the use of advanced physical principles of space flight. One of those principles is the development of a solar sailing technology, which allows considerable saving in energy.

The greatest advantage of a solar sailing concept is that it allows the creation of propellant free systems, implying that they can accelerate indefinitely. The absence of propellant means a major saving of mass, together with the fact that there is no engine; the energy to mass ratio of the spacecraft can be very high. Both these factors together mean that the solar sail spacecraft is able to reach high velocities, which is crucial for deep interplanetary travel and especially interstellar travel [1].

2. Solar sail configurations

Solar sails are a form of spacecraft propulsion, which accelerates by means of pressure of stellar radiation on large ultra-thin mirrors.

At present, all solar sails launched into space are carefully folded to fit in the upper stage of the launch vehicle, and deployed when the spacecraft is in space. The sails are generally made of a three-layered film. The layer which faces the Sun is made of highly reflective aluminum, which is fixed to the central layer of a plastic substrate. One commonly used substrate is Kapton. An emissive layer, often chromium, is fixed to the side of the sail directed away from the Sun. Its function is to dissipate the heat produced by the small amount of solar flux which is absorbed by the sail's aluminum front face.

The most reliable and effective solar sail configurations are a frame-type solar sail (Figure 1) and a rotary-type solar sail (Figure 2) [5].

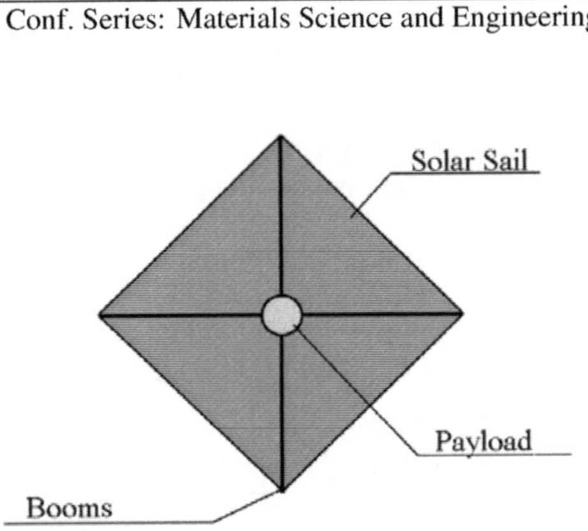

Figure 1. Frame-type solar sail.

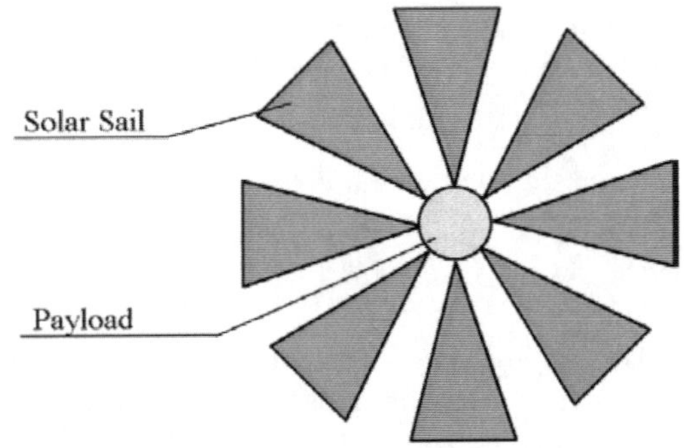

Figure 2. Rotary-type solar sail.

The frame construction of the solar sail is the most simple and reliable structure, that is based on a rigid truss or a frame foundation, to which the parts of the sail are attached. It is a heavy and strong design with a small windage (the ratio of the area of the solar sail to the mass of the spacecraft). The advantage of this construction is reliable sail fixing - they do not curl up and it is easy to control them.

The rotor construction is a more perspective and lightweight structure that provides deployment and support of the sail due to centrifugal forces of inertia. Sails of this type are designed as parts of various shapes attached to the central body of the spacecraft. For some rotating spacecraft small weights are attached to the ends of solar sail vanes or on ropes (which deploys the solar sail), thus increasing the force of inertia. This construction provides a lower mass compared to the frame, and therefore, higher windage.

3. Movement simulation of the centre mass of the spacecraft

The mathematical models and control laws obtained in [4] are used to model the maneuvers to achieve parabolic speed on orbits of different heights Starting circular orbits of different heights are used. Problems of the attitude control were not considered in this paper.

Traditional differential equations in a matrix-vector form are used to simulate the geocentric motion of the spacecraft:

$$\frac{d\mathbf{V}}{dt} = \mathbf{f} + \mathbf{a}\,,\ \mathbf{J}\frac{d\boldsymbol{\omega}}{dt} = \mathbf{M} \tag{1}$$

Here $\mathbf{V}$ is the velocity, $\mathbf{f}$ is the acceleration from gravity, aerodynamics and other disturbing forces, $\mathbf{a}(\lambda_1, \lambda_2)$ is the acceleration from the light pressure forces, $\mathbf{J}$ is the solar sail spacecraft inertia tensor, $\boldsymbol{\omega}$ is the angular velocity, $\mathbf{M}$ is the moment of the external forces applied to the spacecraft.

Sail angle setting (the angle between the normal to the plane of the sail and the direction of the sun) is determined by two control angles: $\lambda_1(t)$ - the angle between the projection of the thrust vector to the instantaneous velocity of the orbit and the radius vector of the gravitational center of the spacecraft, and $\lambda_2(t)$ - the angle between the thrust vector and the normal to the instantaneous plane of the spacecraft path. As the control laws of the solar sail spacecraft are used locally, optimal control laws of setting the sail angle are obtained in [2]. For the first of them, it is supposed that the spacecraft moves in the ecliptic plane and the normal to the plane of the sail also lies in plane ($\lambda_1(t) \equiv 0$), and angle $\lambda_1(t)$ is defined proceeding from maximizing of projection of the accelerate spacecraft on a transversal direction of the object-centred coordinate system. In this case,

$$\lambda_{1opt} = \frac{1}{2}\left[\arcsin\left(\frac{S'_x}{3\left(S'^2_x + S'^2_y\right)^{\frac{1}{2}}}\right) - \operatorname{arctg}\left(\frac{S'_x}{S'_y}\right)\right] \tag{2}$$

Here S'_x, S'_y, and S'_z are components of the vector of the direction to the Sun.
Let us suppose now that $\lambda_l=90°$, that is, a sail rotation axis is directed by the local radius vector. In this case,

$$\lambda_{2opt} = \frac{1}{2}\left[\arcsin\left(\frac{S'_z}{3\left(S'^2_x + S'^2_y\right)^{\frac{1}{2}}}\right) - \operatorname{arctg}\left(\frac{S'_z}{S'_y}\right)\right] \tag{3}$$

This control is convenient to implement travel of the gravity-oriented sail (the spacecraft is a construction extending along axis OX'). Herewith the plane of the orbit must be located as near as possible to terminator ($S'_x= S'_y\approx 0$) [2].

For a vehicle having a solar sail, considered in this project, the mass should be 50 kg, and the area should amount to 200 m^2. Table 1 shows the results of the calculation of the maneuver to achieve parabolic speed for the solar sail spacecraft with these design characteristics for the control laws.

Table 1. Results of the calculation of the maneuver to achieve parabolic speed for control law 1 and control law 2

Control law	**Orbital altitude (km)**	**Required angular velocity (deg/day)**	**(rad/sec)**
Locally optimal law (1)	6000	1135.3	$0.229\cdot 10^{-3}$
	18000	410.4	$0.083\cdot 10^{-3}$
	36000	178.85	$0.036\cdot 10^{-3}$
Locally optimal law (2)	6000	2271.46	$0.459\cdot 10^{-3}$
	18000	821.66	$0.166\cdot 10^{-3}$
	36000	358.56	$0.072\cdot 10^{-3}$

The data in Table 1 show that in order to allow maneuvering the spacecraft at a lower altitude, it is essential that the construction allows sail reversals at higher speeds.

The ability of the solar sail spacecraft to maneuver at low orbits provides additional advantages compared to the spacecraft that can exist and function only at high orbits. First of all, it saves propellant fuel to launch the spacecraft to the start orbit. If the spacecraft is designed to capture the earth's surface, the use of a lower orbit functioning extends the possibility of optics. If the spacecraft is designed for photographing of a part of the earth's surface, it will increase the accuracy of the light direction and the light output [3].

4. Forces acting on the solar sail

Let us introduce a coordinate system associated with the undeformed surface of the solar sail, as shown in Figure 3. Plane xOy coincides with the plane of the undeformed solar sail. The Oz-axis is pointed at the angular velocity of sail rotation ω_l (the sail ensures deployment and maintaining of it in a deployed state). The Ox axis is directed at the angular velocity of programmed reversal of the sail with angular velocity ω_2 (provides program control of the sail).

Let us consider an elementary area at the surface of the sails, which is small enough to neglect the change of the forces and its curvature in moving over its surface. The position of the elementary area is determined by the position of its center $\mathbf{r}_S=\{x_S, y_S, z_S\}$, the position of its plane - by current normal $\mathbf{n}=\{n_x, n_y, n_z\}$, $|\bar{n}|=1$. This area is acted upon by light pressure forces and the force of interaction with the neighboring areas. In case of the rotary-type solar sail the centrifugal force of rotating is considered.

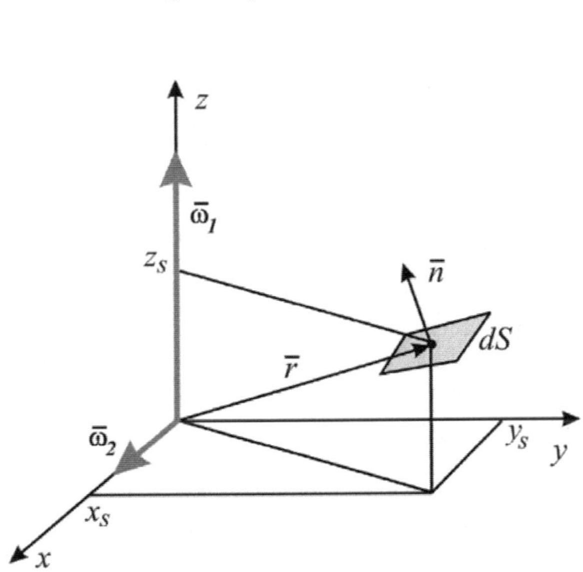

Figure 3. Scheme of the forces acting on the rotating element of the solar sail.

The area is involved in the following movements:

- Motion of the spacecraft as a solid body-controlled movement is assumed to be given, it specifies the amount and direction of light pressure and the desired axis of rotation and the speed controlled reversal;
- Rotation of the solar sail spacecraft relative to the center of mass with angular velocity ω_1 providing the necessary rigidity of the sail surface, it determines the magnitude and direction of the centrifugal forces (in case of the rotary-type solar sail);
- Rotation of the spacecraft relative to the center of mass with angular velocity ω_2 that provides a thrust of a given direction.

If we assume that the Sun is located in the direction of Z, then the value of the light pressure force is defined by the following relationship [5]:

$$dF_C = P_a \cdot n_z^2 \cdot dS \tag{4}$$

The centrifugal force of inertia is determined by vector relation:

$$\Phi_c = m\mathbf{W}_c = -dm \cdot \omega \times (\omega \times r) \tag{5}$$

where $\boldsymbol{\omega}=\boldsymbol{\omega}_1+\boldsymbol{\omega}_2=\{\omega_2, 0, \omega_1\}$ is the total angular velocity vector; $\mathbf{r}=\mathbf{r}_S=\{x_S, y_S, z_S\}$ is the radius vector of the area center relative to the origin of coordinates.

5. Finite Element Models

This paper describes the application of the method of the finite elements analysis applied to frame-type and rotary-type solar sails, with the help of finite elements simulation system.

The description of the frame-type and rotary-type solar sail spacecraft elements is given in Table 2. The finite element model of the frame-type and rotary-type solar sail of spacecrafts is shown in Figure 4 and Figure 5 respectively.

According to (4), (5) and Table 1 the forces influencing the solar sail are determined. Locally optimal law 2 is considered because it has greater angular velocity values. The list of force values acting at distances of 6000 km, 18000 km and 36000 km is presented in Table 3.

Table 2. Frame-type and rotary-type solar sail spacecraft elements description.

	Element of design	Type of element	Material	Properties
Common elements	Spacecraft	element of concentrated mass, absolutely rigid element RBE 2		Mass is 50 kg
	Solar sail	plate element	polyamide film Kapton	Young's Modulus is 2500 MPa thickness is $7.5 \cdot 10^{-3}$ mm
Frame-type only	Booms	beam element	carbon fiber	Young's Modulus is 67000 MPa thickness is 2 mm length is 10000 mm
Rotary-type only	Ropes	beam elements with significantly reduced rigidity	kevlar	thickness is 2 mm
	End weights	elements concentrated mass	carbon fiber	mass is 0.25 kg
	Coupling member between ropes	shell element flexural thick shell	polyamide film	Young's Modulus is 3200 MPa thickness is 2 mm

Figure 4. A finite element model of the frame-type solar sail.

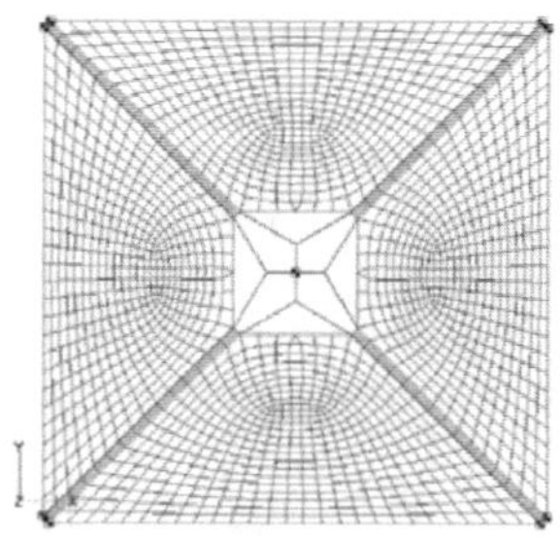

Figure 5. A finite element model of the rotary-type solar sail.

Table 3. Force values acting on the solar sail for locally optimal law 2

Type of the solar sail	Orbital altitude (km)	Max forces of the rotation with ω_1 (N)	Max Sun pressure (N)	Max forces of rotation with ω2 (N)
Frame-type	6000	$F_x=F_y=F_z=0$		$F_x=F_y=73.738 \cdot 10^{-6}$, $F_z=0$
	18000	$F_x=F_y=F_z=0$		$F_x=F_y=9.645 \cdot 10^{-6}$, $F_z=0$
	36000	$F_x=F_y=F_z=0$		$F_x=F_y=1.814 \cdot 10^{-6}$, $F_z=0$
Rotary-type	6000	$F_x=F_y=78.819 \cdot 10^{-3}$, $F_z=0$	$1.856 \cdot 10^{-6}$	$F_x=78.819 \cdot 10^{-3}$, Fy$=77.950 \cdot 10^{-3}$, $F_z=0$
	18000	$F_x=F_y=10{,}354 \cdot 10^{-3}$, $F_z=0$		$F_x=10{,}354 \cdot 10^{-3}$ $F_y=9.546 \cdot 10^{-3}$, $F_z=0$
	36000	$F_x=F_y=1{,}947 \cdot 10^{-3}$, $F_z=0$		$F_x=1.947 \cdot 10^{-3}$ $F_y=1.012 \cdot 10^{-3}$, $F_z=0$

6. Results And Discussion

According to Table 3 the finite element simulation is performed. The strength analysis of the frame-type and rotary-type solar sails at distances of 6000 km, 18000 km and 36000 km is provided. Strain, stress and deformation of the finite element models are calculated according to strain and stress by mean of the computer-assisted design system. The examples of the strength analysis of the frame-type and rotary-type solar sails are shown in Figure 6 and Figure 7 respectively.

Figure 6. Stress distribution [MPa] on the solar sail surface of the frame-type solar sail.

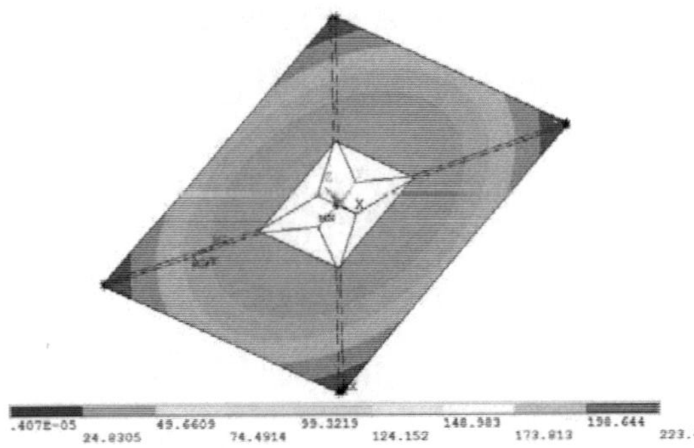

Figure 7. Stress distribution [Pa] of the rotary-type solar sail during deployment.

Having finished the nonlinear static analysis we have obtained the following results:

- the frame-type construction of the solar sail is stronger than the rotary-type construction;
- the frame-type solar sail is capable of flying 36000 km and 6000 km distance;
- the rotary type solar sail can operate with rotation speed of 450 deg/days, consequently the spacecraft should be in 36000 km orbit.

7. Conclusion

In this paper we investigated the interconnection of the behavior of the two different constructions of the solar sail spacecraft during maneuver on the geocentric orbit. The frame-type and rotary type solar sails were compared. The stability analysis of the constructions indicates the necessity of construction or changing the spacecraft control law. Modeling the behavior of the design allows coming to conclusion that the frame-type solar sail is stronger than the rotary-type solar sail, consequently the frame-type construction is capable of flying on lower orbits of 6000 km. In case of the rotary-type construction, the distance of 36000 km is more preferable.

Acknowledgment

This work was supported by the Ministry of Education and Science of the Russian Federation and Russian Foundation for Basic Research.

References

[1] Materials of http://interplanetarytravelandtheuniverse.webs.com
[2] Ishkov S and Starinova O 2005 *Samara: The SSC RAS Publisher* **7(1)** 99–106
[3] Salmin V, Starinova O and Ishkov S 2006 *Samara: The SSC RAS Publisher* 216
[4] Starinova O 2007 *Samara: The SSC RAS Publisher* 196
[5] Starinova O, Khabibullin R and Gorbunova I 2015 *Istanbul: IEEE Publisher* 643-648

Modeling of a single-cycle current generator while forming a quasi-sinusoidal current

V V Grebennikov[1], E V Yaroslavtsev[1], A B Slobodenuk[1] and T G Evtushenko[2]
[1] Tomsk Polytechnic University, 30, Lenina ave., Tomsk, 634050, Russia
[2] Tomsk State University, 36, Lenina ave., Tomsk, 634050, Russia

E-mail: grebennikovvv@tpu.ru

Abstract. The paper presents the results of investigation of the influence of the output voltage magnitude on the operating frequency of the switch in the single-cycle quasi-sinusoidal current generator circuit. Analytical expressions for calculating the time parameters for transients in the circuit under given assumptions have been obtained. The results presented in the paper can be used in the design of converters of this type.

1. Introduction

Sometimes, in order to improve the efficiency of electrochemical processes it is reasonable to use power supplies based on a quasi-sinusoidal asymmetrical current generator [1, 2]. To design a current generator for a rational choice of the element base one must follow the requirements for frequency properties and determine the switching losses of the circuit to achieve the corresponding calculated ratio. The expressions to calculate these parameters have been obtained in [3], but they are transcendental, inconvenient for practical use, since a special mathematical approach is required to obtain a certain result. To simplify the calculation procedure, it is appropriate to obtain more simple approximate analytical expressions that enable the calculations of the switching losses with acceptable errors and set the requirements for their frequency properties.

The quasi-sinusoidal asymmetrical current generator circuit is quite complex [1]; however, to solve the problem it is enough to consider the process of one half-wave current generation, which is realized in a simplified version of the circuit presented in Figure 1a.

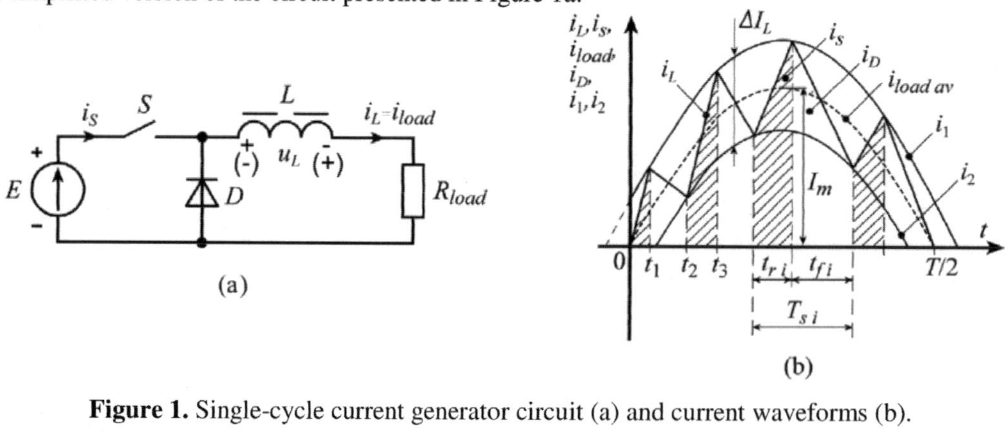

Figure 1. Single-cycle current generator circuit (a) and current waveforms (b).

2. Modeling

The process of generating a quasi-sinusoidal load current is realized by means of regulating turn-on and turn-off states of switch *S* according to the given law of durations and, respectively, of the inductor current *L*. This process is illustrated with the waveforms of currents shown in Figure 1b.

For clarity, the switching frequency is chosen to be relatively low. Further we assume the following: voltage source *E*, diode *D*, inductor *L* and switch *S* are ideal; inductor *L* is a linear element; R_{load} is a constant resistive load; duration of the current *i*-cycle of switch $T_{s\,i}$ is much smaller than the period of the formed sine-wave, i.e. $T_{s\,i} << T$; during the current cycle of the switch the output voltage does not change, i.e. $U_{load\,i} \approx$ const; the load current varies sinusoidally, i.e. its ripples caused by the switch commutations are infinitesimal; in the load there is a half wave of sinusoid with amplitude I_m, angular frequency ω and period *T*, which is the average value of load and inductor currents, are formed as follows:

$$i_{load\,av}(t) = i_{L\,\mathrm{av}}(t) = I_m \sin\omega t;$$

The load current changes between the upper $i_1(t)$ and lower $i_2(t)$ threshold levels, limiting the inductor current ripples with respect to $i_{load\,av}(t)$:

$$\left.\begin{aligned} i_1(t) &= 0.5\Delta I_L + i_{load\,av}(t) = 0.5\Delta I_L + I_m \sin\omega t, \\ i_2(t) &= -0.5\Delta I_L + i_{load\,av}(t) = -0.5\Delta I_L + I_m \sin\omega t \end{aligned}\right\}, \quad (1)$$

with $\Delta I_L = i_1(t) - i_2(t)$ – is the required value of inductor current ripples.

Let us consider the principle of the circuit operation. Let us assume that at time moment $t = 0$ switch *S* is turned on. Voltage *E* is applied to series-connected *L* and R_{load} and diode *D*, providing an off-state of *D*. At this moment, inductor current $i_L(t)$ and the load current are zero, therefore voltage *E* is applied to the inductor with polarity shown in Figure 1a (without brackets). Under given assumptions current $i_L(t)$ begins to increase linearly. The inductance of the inductor is chosen so that the slew rate of inductor current $i_L(t)$ would be a bit higher than the slew rate of load current $i_{load\,av}(t)$. At moment t_1 the inductor current reaches upper threshold level $i_1(t_1)$, and switch *S* is turned off. The polarity of the voltage across inductor *L* is changed to the opposite polarity (shown in brackets in Figure 1a). Flowing through the load and the freewheeling diode, the inductor current decreases linearly. Reaching lower threshold level $i_2(t)$ at moment t_2, switch *S* is turned on again and the inductor current increases once again. Then all the processes, described above, repeat cyclically. Thus, as a result of a large number of operation cycles of the switch, the current is generating in the load. The average (approximated) value of this current corresponds to a half-wave of a sinusoidal signal (in Figure 1b it is shown by the dashed line).

Let us suppose that during time interval $0 - T/2$ *N* cycles are needed for the formation of a half-wave current, and each cycle consists of two transients: rise and fall times of the inductor current, respectively. Let us denote the number of the current cycle as *i*, where $i = 1 \ldots N$ – an integer. Let us assign alphabetic indexes '*r*' or '*f*' indicating the rise or fall time of $i_L(t)$ for current, voltage and time parameters respectively; the numerical index corresponds to the number of the current cycle.

To receive the basic ratios we need to perform the analysis of transients in the circuit. Let us consider an *i*-cycle of the switch operation (Figure 1a), occurring at moment t_i – a fixed moment of time period $0 - T/2$, during which the *i*-cycle of the generator operation takes place.

Taking into account the assumptions that the load current at this point has a definite value we obtain

$$i_{load\,i}(t_i) = I_m \sin\omega t_i .$$

At the stage when the inductor current increases, switch *S* is turned on in the mentioned cycle, the constant voltage with positive polarity is applied to the inductor (polarity is shown in Figure 1a without brackets)

$$U_{Lr\,i} = E - i_{load\,i}(t_i) R_{load} = \mathrm{const}\,. \quad (2)$$

When the switch is turned off during the fall time in the same cycle the current is turned on through freewheeling diode *D*, and the inductor voltage polarity is reversed (voltage polarity is shown in Figure 1a in brackets) and becomes equal to

$$U_{Lf\,i} = -i_{load\,i}(t_i)R_{load} = -U_{m\,load}\sin\omega t_i = \text{const}\,. \tag{3}$$

It is known that the relation between voltage and current of the inductor is as follows

$$u_L(t) = L\frac{\mathrm{d}i_L(t)}{\mathrm{d}t}, \tag{4}$$

from which it follows that if the voltage applied to the inductor is constant, the current of the inductor changes according to the linear law:

$$i_L(t) = I_L(0) + \frac{U_L}{L}t,$$

where $I_L(0)$ – the initial condition for the transient process.

Taking into account this equation and expressions (2) and (3) the inductor current change during the *i*-cycle can be written in the following form (the start of timing is moved to the switching time)

$$i_{Lr\,i}(t) = I_{Lr\,i}(0) + \frac{U_{Lr\,i}}{L}t = I_{Lr\,i}(0) + \frac{E - U_{m\,r}\sin\omega t_i}{L}t,$$

$$i_{Lf\,i}(t) = I_{Lf\,i}(0) + \frac{U_{Lf\,i}}{L}t = I_{Lf\,i}(0) - \frac{U_{m\,load}\sin\omega t_i}{L}t,$$

where $U_{m\,load} = I_m R_{load}$ – the average amplitude of the load voltage.
Initial conditions for rise and fall times, taking into account equation (1), are determined by expressions:

$$I_{Lr\,i}(0) = i_2(t_i) = -\frac{\Delta I_L}{2} + I_{m\,load}\sin\omega t_i,$$

$$I_{Lf\,i}(0) = i_1(t_i) = \frac{\Delta I_L}{2} + I_{m\,load}\sin\omega t_i\,.$$

In formula (4) the differentials for time and current as increments with constant inductor voltage

$$U_L(t) = L\frac{\Delta i_L}{\Delta t} = \text{const}\,; \tag{5}$$

Let us denote $\Delta t = t_{r\,i}$ for the rise time, $\Delta t = t_{f\,i}$ for the fall time, and substituting these parameters into (5) and using expressions (2) and (3), after transformations we obtain:

$$t_{r\,i} = \frac{\Delta I_L L}{E - U_{m\,load}\sin\omega t_i} = \frac{K_{rip}\tau U^*}{1 - U^*\sin\omega t_i},$$

$$t_{f\,i} = \frac{\Delta I_L L}{\left|-U_{m\,н}\right|\sin\omega t_i} = \frac{K_{rip}\tau}{\sin\omega t_i},$$

$$T_{s\,i} = t_{r\,i} + t_{f\,i} = \frac{K_{rip}\tau}{\sin\omega t_i - U^*\sin^2\omega t_i},$$

$$f_{s\,i} = \frac{1}{T_{s\,i}} = \frac{\sin\omega t_i - U^*\sin^2\omega t_i}{K_{rip}\tau},$$

where $U^* = U_{m\,load}/E$ – the normalized amplitude of the output voltage; $\tau = L/R_{load}$ – the time constant of the circuit; $K_{rip} = \Delta I_L/I_m$ – the ripple factor of the load and inductor current.

Let us consider the obtained parameters in a dimensionless form:

$$t_{r\,i}^* = \frac{t_{r\,i}}{T} = \frac{K_{rip}\cdot U^*}{\delta\cdot(1 - U^*\sin\omega t_i)}, \tag{6}$$

$$t_{fi}^* = \frac{t_{fi}}{T} = \frac{K_{rip}}{\delta \cdot \sin \omega t_i}, \tag{7}$$

$$T_{si}^* = \frac{T_{si}}{T} = \frac{K_{rip}}{\delta \cdot \left(\sin \omega t_i - U^* \sin^2 \omega t_i\right)}, \tag{8}$$

$$f_{si}^* = \frac{f_{si}}{f} = \frac{\delta(\sin \omega t_i - U^* \sin^2 \omega t_i)}{K_{rip}}, \tag{9}$$

where $\tau^* = \tau/T$ is the relative value of the time constant; $\delta = 1/\tau^* = T/\tau$ is the inverse value of the relative value of the time constant or the damping ratio of the transient showing how the period of a sine wave exceeds the time constant; $T_{si}^* = 1/f_{si}^* = T_{si}/T$ is the relative value of the cycle duration; $t_{fi}^* = t_{fi}/T$ is the relative value of the fall time of the inductor current; $f_{si}^* = f_{si}/f = 1/T_{si}^*$ is the relative value of the local switching frequency.

3. Discussion

The most important thing for practice is the information about local switching frequency $f_s^*$of the duration of half-period of the generated sine waves, allowing determination of the requirements for the frequency properties of the switches, and estimation of the value of the switching losses in them.

Approximate expression (9) gives a possibility to simply see the tendency of the operating frequency change of the switch during the half-period of the sine wave while changing the normalized amplitude of output voltage U^*. In Figure 2 one can see the ratios obtained with expression (9). It is seen that three local extremums (2 high and 1 low) are observed at values of $U^* > 0.5$; and when $U^* \leq 0.5$ there is only one local extremum and it is the maximum.

Taking into account that the maximum operating frequency of the switch is an important value from the point of view of selecting elements, it is advisable to determine the values of these extremums. Having differentiated equation (9) by the parameter $\omega t = \nu$ and having brought the result to zero, we will find the roots of the resulting equation corresponding to the required extremums:

$$\nu_1 = \pi/2; \tag{10}$$

$$\nu_2 = \arcsin\left(1/2U^*\right); \tag{11}$$

$$\nu_3 = \pi - \arcsin\left(1/2U^*\right). \tag{12}$$

The analysis of expressions (10) – (12) has shown that roots ν_2 and ν_3 exist only for values of $U^* \geq 0.5$, and root ν_1 exists for all values in the range of $0 \leq U^* \leq 0.5$.

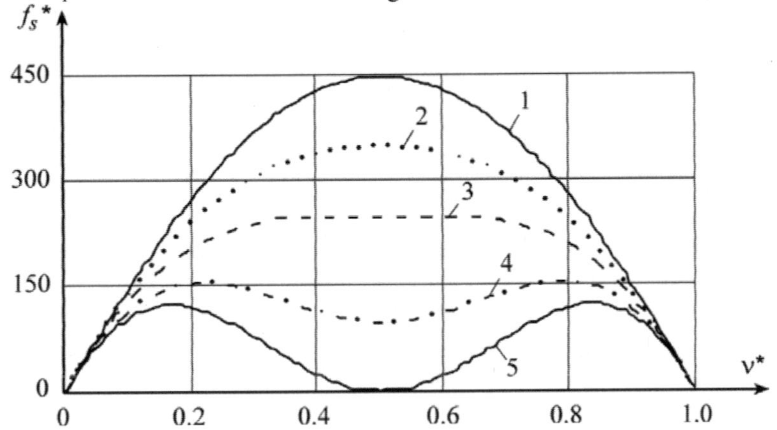

Figure 2. Dependences of the relative value of the local switching frequency on the relative value of the current phase $\nu^*=\omega t/\pi$ when $K_{rip} = 0.2$, $\delta = 100$ and various U^*: 1) $U^* = 0.1$; 2) $U^* = 0.3$; 3) $U^* = 0.5$; 4) $U^* = 0.8$; 5) $U^* = 1$.

We will obtain the expressions for calculating the relative value of the switching frequency at the extreme points. To find out the extremums of the frequency when $U^* \geq 0.5$ we include ν_1, ν_2, ν_3 in equation (9). After transformations for the first and the second maximums of the switching frequency we obtain

$$f_{s\,max1}{}^* = \frac{f_{s\,max1}}{f} = \frac{f_{s\,max2}}{f} = \frac{\delta}{4U^* K_{rip}}, \tag{13}$$

and the expression for the local minimum is

$$f_{s\,min}{}^* = \frac{f_{s\,min}}{f} = \frac{\delta(1-U^*)}{K_{rip}}. \tag{14}$$

Inserting ν_1 to equation (9), we obtain the expression for determining the only maximum of the operating frequency of the switch when $U^* \leq 0.5$:

$$f_{s\,max}{}^* = \frac{f_{s\,max}}{f} = \frac{\delta(1-U^*)}{K_{rip}}. \tag{15}$$

Expressions (14) and (15) are identical. This gives us possibility to claim that when U^* changes from 0 to 1, the maximum goes into the minimum, and the boundary case occurs under $U^* = 0.5$, with $f_{s\,min}{}^* = f_{s\,max}{}^*$.

Equations (6) – (15) are convenient for analysis, though approximated, and may be used for 'quality' assessment of the impact of any parameter on the transient process duration.

A comparison of the results of maximum frequency calculations carried out using approximate formulas with the results obtained by using the exact formulas [1] showed that the error does not exceed 10% if the number of commutate cycles of the switch is $N \geq 12$ when any U^* and $K_{rip} \leq 0.3$.

4. Conclusion

The analysis of the single-cycle current generator allowed deriving approximate calculated ratios specifying the time parameters of transients in the circuit.

These relatively simple expressions made it possible to identify the tendencies and nature of changes of the time parameters of the transients in the circuit.

The investigation results show that under $U^* > 0.5$ the function of the relative value of the local switching frequency versus the relative value of the current phase has three local extremums – two maximus and one minimum; when $U^* < 0.5$ the function has the only extremum (maximum), exceeding the maximum frequency value, which is peculiar when $U^* > 0.5$.

References

[1] Baginskii B A, Grebennikov V V, Nigof B M, Ogorodnikov D N and Yaroslavtsev E V 2001 *Experim. Tech.* **44(2)** 243–5

[2] Baginskiy B A, Grebennikov V V and Obraztsov S V 2000 *Proc. of Int. Scientific and Practical Conference of Students, Postgraduates and Young Scientists 'Modern Techniques and Technology' (MTT' 2000)* 74–76

[3] Grebennikov V V et al. 2015 *Proc. of 2015 Int. Siberian Conf. on Control and Communications, SIBCON 2015* 7147311 1–4

Impact of filtration elements on bioelectrical signal

M Grigoriev and L Babich
Tomsk Polytechnic University, 30, Lenina ave., Tomsk, 634050, Russia

E-mail: Mishatpu@mail.ru

Abstract. Filtering of the wanted signal from the noise caused by power lines, magnetic pickup, interference from body movement, etc. in the field of medical data processing is a weighty problem of cardiography. To date, different types of digital filters have been used to clear the signal from unnecessary frequency bands. This paper presents the results of the filter simulation in MATLAB package.

1. Introduction

Currently, due to the development of microprocessor technology, portable diagnostic devices that can successfully be used in everyday life are invented. The most important tasks are determination of a true diagnosis, detection of pathologies in the early stages of disease and prevention of cardiovascular diseases.

Electrocardiography (ECG) is one of the fundamental methods for the study of the cardiovascular system, which is still the most accessible and popular one for general use.

Recording of high resolution ECG signals with the presence of background interference is the important aspect for studies. The results of these studies will include improved ECG signal. It is necessary to separate some signal components from unnecessary artifacts to facilitate interpretation of the ECG signal.

All recording devices, both analog and digital ones, have properties that make them susceptible to noise. Noise can be random and not coherent, that is not related to the signal itself, or coherent, made by the recording devices and processing algorithms [1].

Noise is random fluctuations of various physical nature notable for complex temporal and spectral structure [2].

Receipt of signal against the background of noise is described as a procedure for signal filtering using the filter; the task in this case is the maximum weakening of noise and interference and minimal distortion of the received signal [3].

Filter (lat. Filtrum 'felt') is notions, devices and mechanisms which extract (or eliminate) some part with preset properties from the initial object [4].

The problem of useful signal extraction against the background of the whole complex of noise and distortion is a major one in the course of modern electrocardiological studies. The presence of artifacts in an electrocardiological signal (ECS) significantly impedes its analysis and identification of diagnostic features. The complexity of this problem solution consists in selection of filtering methods to eliminate certain types of artifacts, as well as the optimization criteria of used algorithms [5].

Standard for cardiographic equipment does not regulate the use of filters for the electrocardiograph, as well as their characteristics in certain models of cardiographs. Many sources contain information on very strong influence of filtering elements on the ECG signal. Several types of filters that are used for different tasks are applied in modern cardiographs, and each of them can be implemented in several variants. This approach creates a large number of possible combinations of filters. For example:

- power-supply noise filter;
- high-pass filter, limiting signal below the cut-off frequency;
- low-pass filter, limiting signal above the cut-off frequency.

Various types of filters are used in various electrocardiographs and, consequently, they will have a different impact on the ECG signal. Exactly various impact of filters on the ECG can explain the difference in the readings of different electrocardiographs. Fairly significant changes may be observed depending on the signal filtering parameters [6, 7].

2. Methods and techniques

The impact of the filter on the signal was studied; in our case the model of the ECG signal was created. The circuit containing the low-pass filter (LPF) was designed in the MatLab package.

Initially the signal was created in the program for work with electronic worksheets Microsoft Excel. (Figure 1).

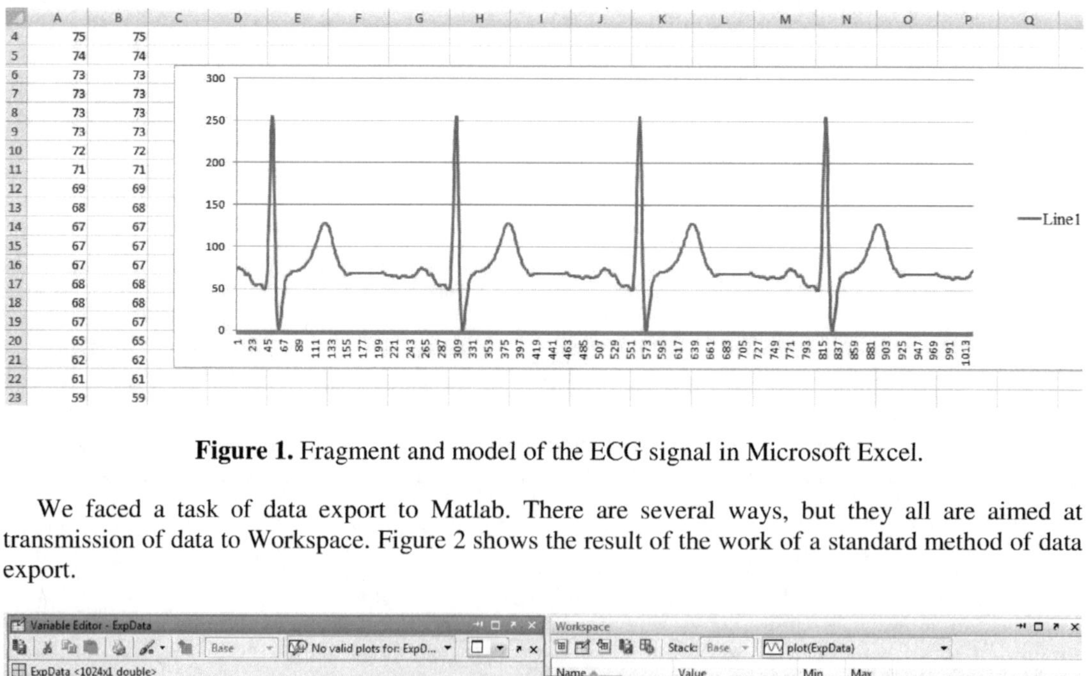

Figure 1. Fragment and model of the ECG signal in Microsoft Excel.

We faced a task of data export to Matlab. There are several ways, but they all are aimed at transmission of data to Workspace. Figure 2 shows the result of the work of a standard method of data export.

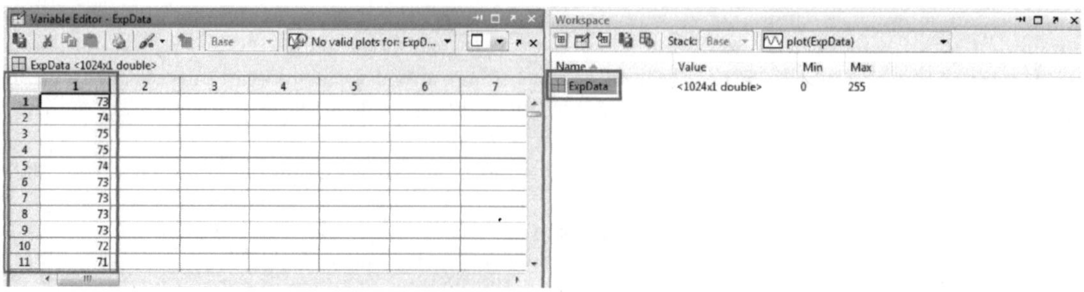

Figure 2. Result of data export using a standard method.

The Figure shows that the data are transmitted as a simple array; this is due to the fact that in many cases it is enough for further operations in the command line. But we need to use the exported data in

Simulink, and a simple array will not be enough, due to the fact that the data transmitted from Matlab in Simulink require special formatting.

We use the method implemented via a simin component to transmit data from Matlab to Simulink (Figure 3).

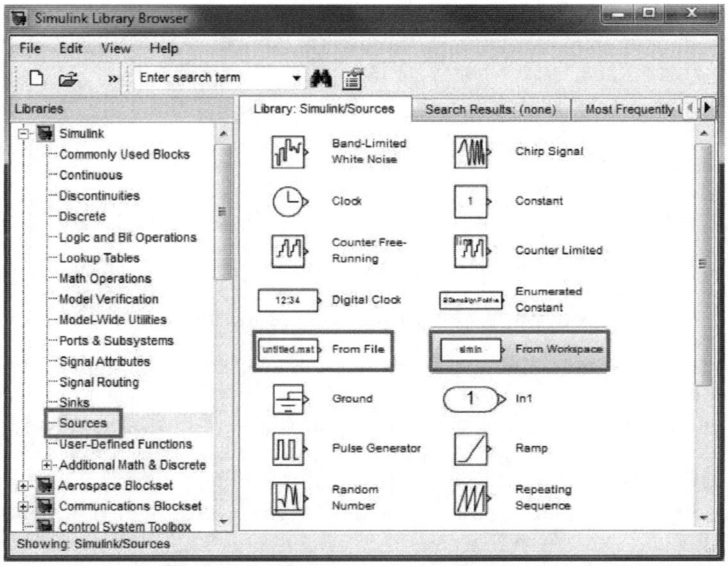

Figure 3. Elements for data transmission.

A simin component (From Workspace) receives data from Workspace and transmits them to Simulink. For correct operation of the component it has to receive a number of parameters, but the most important of them is the name of the variable, from which it is necessary to retrieve data, and correct formatting of data in this variable.

It is necessary to execute a number of commands in the command line in order to format the exported data:

- var.time=[1:length(ExpData)];
- var.signals.values=ExpData;

The result will represent a structure which complies with the requirements of the simin component (From Workspace) (Figure 4).

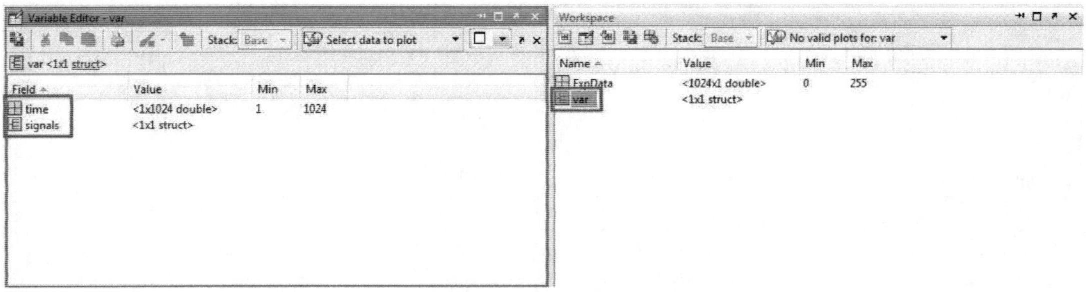

Figure 4. Structure 'var'.

All the necessary work with a signal was made and then it became possible to use it in the Matlab system. A simulated circuit for studying of the impact of filters is shown in Figure 5.

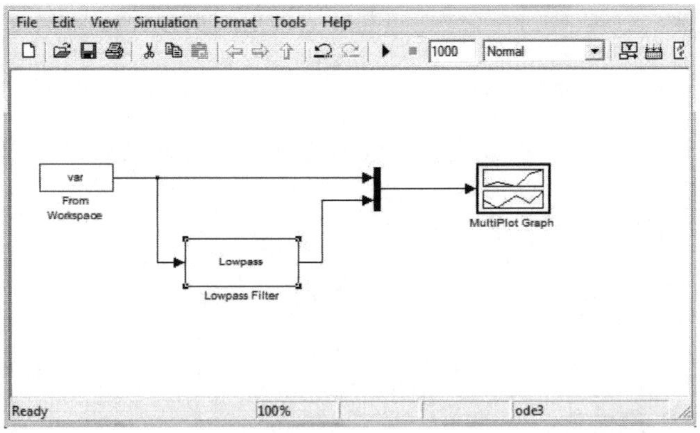

Figure 5. Circuit for the studying of the impact of filters.

This circuit contains: a unit (From Workspace), a low-pass filter (Filter Lowpass), multiplex (Mux), a unit (MultiPlot Graph).

A simulated ECG signal is sent to the circuit input; there is a MultiPlot Graph at the output – this is a unit for graph plotting, which allows for graphical observation of the filter impact on the ECG – signal in the MatLab package [8]. We use multiplex (Mux) in order to combine the graphs of input and output signals for a more convenient analysis. The input signal is represented in Figure 6. The result of the low-pass filter impact is shown in Figure 7–9.

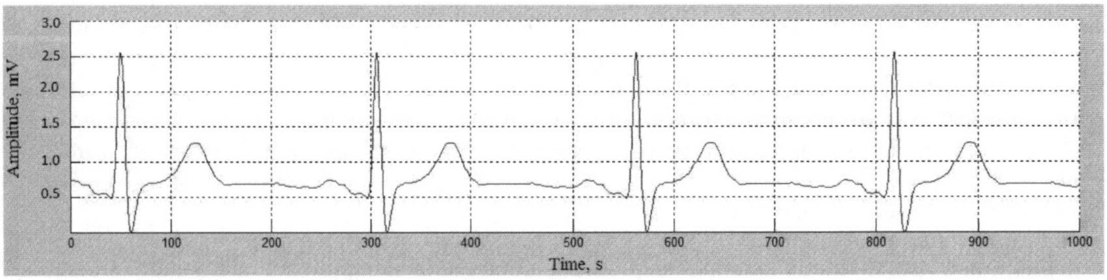

Figure 6. Input signal.

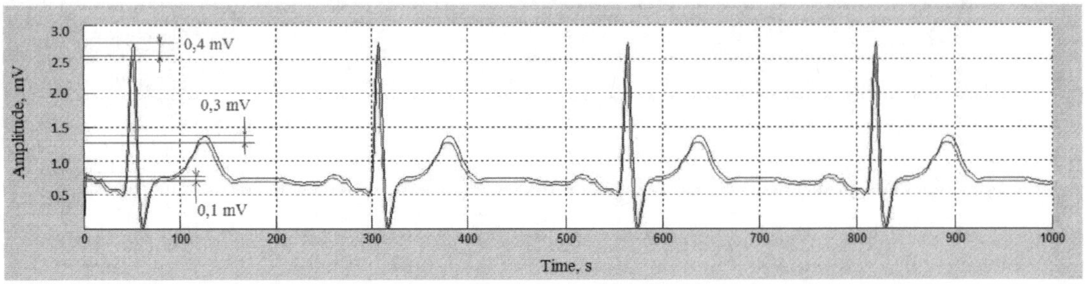

Figure 7. LPF output signal with cut-off frequency of 40 Hz
(Input signal is red, output signal is blue).

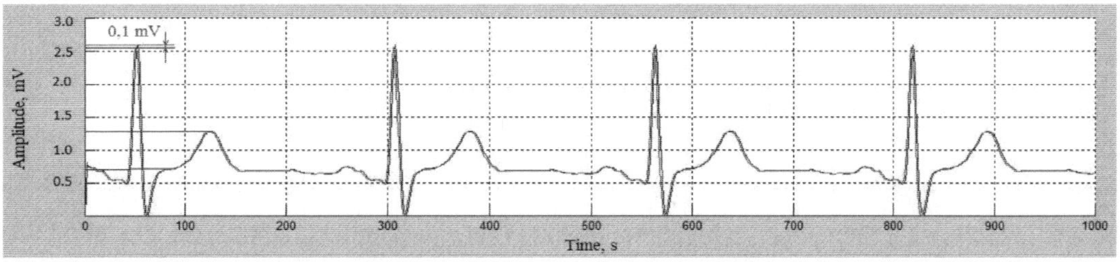

Figure 8. LPF output signal with cut-off frequency of 100 Hz (Input signal is blue, output signal is red).

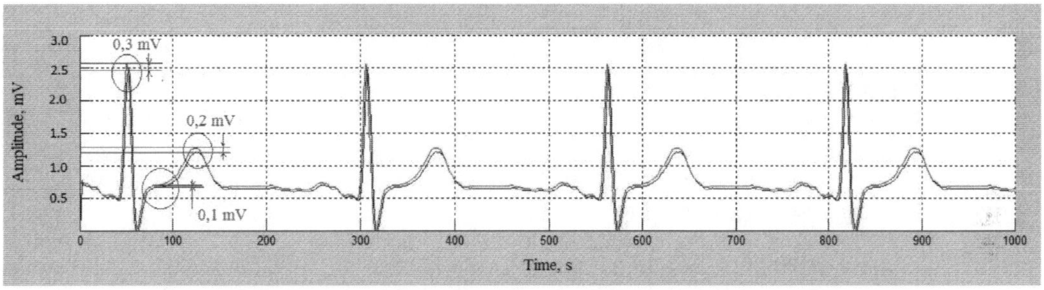

Figure 9. LPF output signal with cut-off frequency of 150 Hz (Input signal is blue, output signal is red).

3. Conclusion

As a result, we can clearly see how the filters affect the signal. Our findings represent a significant deviation when interpreting the electrocardiogram, which will certainly lead to incorrect diagnosis and consequently increase the risk of harm to human health.

In order to register a high resolution signal undistorted by filters in terms of level, it is necessary to standardize the use of filters in electrocardiographs and to develop electrocardiographic equipment of a new generation.

4. Acknowledgement

The research was financially supported by the Federal Targeted Programme 'Research and Development in Priority Fields of S&T Complex of Russia in 2014-2020', Agreement No. 14.578.21.0032 dated 05.06.2014 'Development of the experimental sample of a hardware-software complex for noninvasive recording of heart micropotentials in a wide frequency band without filtering and averaging in real time to early detect the symptoms of a sudden cardiac death', unique identifier of the contract: RFMEF157814X0032.

References

[1] Mäntynen V, Konttila T and Stenroos M 2014 *Physics in Medicine and Biology* **59** 7141–7158
[2] Butcher C, Mareev Y, Markides V, Mason M, Wong T, and Cleland J 2015 *Current Cardiology Reports* **17** 90–103
[3] Mashchenko T and Trubchanova N 2010 *Automatics and professional equipment* **20** 115–123
[4] Fetisova E 2011 *Author's abstract by Candidate of Medicine* 24
[5] Savostin A 2010 *EKSTU Vestnik* **1** 5–10
[6] Grigoriev M and Turushev N 2015 *Applied Mechanics and Materials* **770** 526–530
[7] Grigoriev M and Turushev N 2014 *Advanced Materials Research* **1040** 928–932
[8] Avdeeva D, Turushev N, Ivanov M, Yuzhakova M and Grigoriev M 2014 *Biology and Medicine* **6** 1–6

Quaternion representations of stiffness and momentum of the forces, acting in vibration isolating systems with stiffness compensators

E G Gurova
Novosibirsk State Technical University, 20, K. Marx ave., Novosibirsk, 630073, Russia

E-mail: lena319@mail.ru

Abstract. This research is devoted to development of the spatial vibration isolation devices. The description of the vibration isolation systems has been presented through quaternions of the forces, momentums, and stiffness. The considered method allows taking into account the stochastic vibrations and describes it with the help of the hypercomplex numbers. The theory suggests the development of the vibration isolation devices, which have traction characteristics with zero stiffness area. To obtain such area in traction characteristic, a spatial vibration isolator is presented as a resilient element and the stiffness compensator, which is connected in parallel with it.

1. Introduction

Nowadays one of the most important engineering problems is the control of noise and vibration, which are made by working machines and mechanisms. This problem is acute in all areas of the machining, transport, industry. The bad impact of the vibration consists in destruction of machines and buildings, violation of readings of control devices and measurement tools. Noise and vibration negatively influence the activity and reliability of the devices. But the most detrimental effect is on the human body, causing various diseases.

Today there are many ways to decrease the vibration such as passive vibration isolators (springs, dampers), active vibration isolation systems, dynamic balancing of engines, and etcetera. However neither of these methods fully meets the modern requirements of the spatial vibration isolation, as most of them decrease the vibration only relatively one axis of the space, or the efficiency of the vibration level reduction does not meet sanitary standards.

In connection with the facts mentioned above this research is directed at the development of the spatial vibration isolation devices (SREDI with the support of the scholarship of the Russian Federation president, order №184 of the 10th of March 2015), which will best of all meet the modern requirements of the space vibration isolation, based on the effect of zero stiffness.

In [1, 2] a hypercomplex method of the description of the spatial mechanic oscillations was suggested in order to more precisely register and exclude them with the help of this vibration isolation device. In many foreign and domestic researches the theory of protection against vibration is considered relatively each axis of the space separately. In [2], the suggested technique supposes the spatial vibration isolation with the use of a mathematical apparatus of the quaternions and a notion of the vector space. Hypercomplex numbers allow forming and describing the conditions and requirements of the ideal vibration isolation in space.

2. Calculation and simulations

The ideal vibration isolation in space of a perfectly rigid body against random oscillations is provided (see Figure 1) if at any moment of time the sum of the forces and momentums, influencing the body, will be balanced, that is, the sum vector and sum moment will be equal to zero at any moment of time:

$$\sum F_i = 0, \quad \sum M(F_i) = 0 \tag{1}$$

where - F_i - forces, influencing the body (a protected object). They include: forces, transmitted through a protected device to a vibrating one (F); forces of the elastic interaction of the vibrating and protected objects (F_m); dissipative forces of the objects interaction (F_y); inertia forces of the intermediaries, connecting vibrating and protected objects (F_u). M (Fi) is the moment of the forces affecting the body (a protected object).

Each of the constituents of conditions (1) represents a sum of force projections on random coordinate axes and a sum of their moments with respect to these axes.

This theory assumes the development of the triaxial vibration isolation devices, which have traction characteristics with zero stiffness area. To obtain such area on the traction characteristic the spatial vibration isolator is represented as a resilient element and connected in parallel to the stiffness compensator.

The traction characteristic of the resilient element has a positive coefficient of stiffness relatively all three axes, and a compensator has a negative coefficient of stiffness and equal in modulus of rigidity of the resilient element relatively each spatial axis. To provide the floating point of the zero stiffness at changing of the operation loads the compensator is supplied with the rearranging system, tracking the relative position of the vibrating and protected objects and holding the working elements of the compensator in area of the working travel.

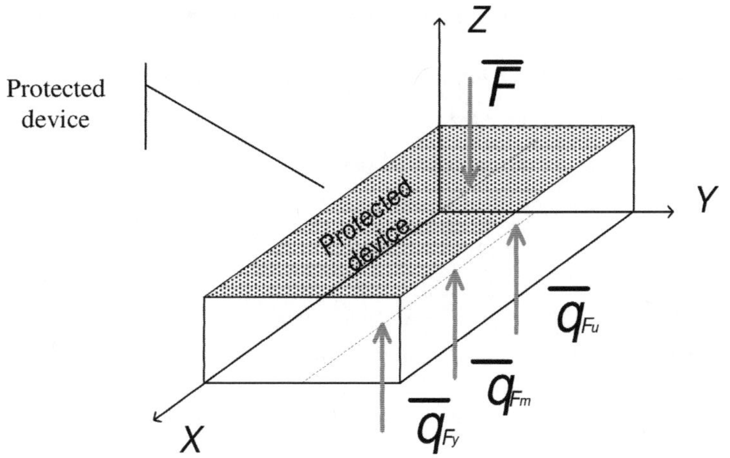

Figure 1. A general case of the forces impact on the protected object for random spatial oscillations.

For description of the vibration isolation theory by means of hypercomplex numbers it is necessary to present mathematical dependencies of stiffness and forces moment in the vibration isolation systems with stiffness compensators.

From theoretical mechanics it is known that stiffness can be described as:

$$C = \frac{F}{l}, \tag{2}$$

where l is the movement.

Also from the theory of mechanics, we know that:

$$M = F \cdot l. \tag{3}$$

Let us represent the force and the movement under random spatial oscillations with the help of quaternions:

$$l = l_o + x \cdot \bar{i} + y \cdot \bar{j} + z \cdot \bar{k}, \tag{4}$$

$$F = F_0 + F_x \cdot \bar{i} + F_y \cdot \bar{j} + F_z \cdot \bar{k}, \tag{5}$$

where $\bar{i}; \bar{j}; \bar{k}$ are unit vectors (orts).

Figure 2 shows the graphical example of one of the forces, acting in the vibration isolation system.

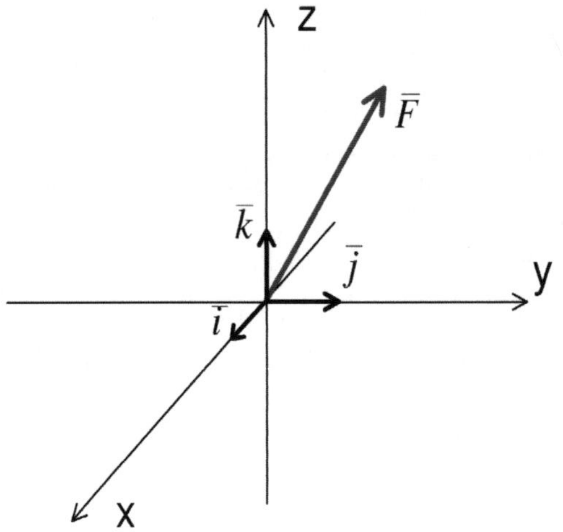

Figure 2. The graphic representation of the force vector in the space.

A hypercomplex description of stiffness and momentum of forces is the following:

$$C = C_0 - C_x \cdot \bar{i} - C_y \cdot \bar{j} - C_z \cdot \bar{k}, \tag{6}$$

$$M = M_0 + M_x \cdot \bar{i} + M_y \cdot \bar{j} + M_z \cdot \bar{k}. \tag{7}$$

Having substituted expressions (4) and (5) into dependence (2) and using algebra properties of quaternions, we will obtain:

$$C = \frac{F_x \cdot x + F_y \cdot y + F_z \cdot z}{x^2 + y^2 + z^2} + \frac{F_z \cdot y - F_y \cdot z}{x^2 + y^2 + z^2} \cdot \bar{i} + \frac{F_x \cdot z - F_z \cdot x}{x^2 + y^2 + z^2} \cdot \bar{j} + \frac{F_y \cdot x - F_x \cdot y}{x^2 + y^2 + z^2} \cdot \bar{k}. \tag{8}$$

Having substituting expressions (4) and (5) into dependence (3) and using algebra properties of hypercomplex numbers, we will obtain:

$$M = -(F_x \cdot x + F_y \cdot y + F_z \cdot z) + (F_y \cdot z - F_z \cdot y) \cdot \bar{i} + (F_z \cdot x - F_x \cdot z) \cdot \bar{j} + (F_x \cdot y - F_y \cdot x) \cdot \bar{k}. \tag{9}$$

The obtained expressions (8) and (9) allow representation of all forces, stiffness and forces momentums of the vibration isolation system with stiffness compensators by means of the quaternions mathematical apparatus relatively all spatial axes simultaneously.

3. Conclusion

The mathematical representation of the forces, stiffness and momentum of the forces in vibration isolation systems with the help of the quaternions sufficiently simplifies the process of description of similar systems in the space. This description will be taken into account when developing the calculations methods and the design of vibration isolation devices with stiffness correctors, for example, such as electromagnetic and supermagnetic stiffness compensators [3]. The development of such devices is supported by the scholarship for young scientists of the Russian Federation president by order №184 of the 10th of March 2015 for three years.

IOP Conf. Series: Materials Science and Engineering **124** (2016) 012061 doi:10.1088/1757-899X/124/1/012061

References

[1] Gurova E G and Gurov M G 2014 *Applied Mechanics and Materials* **682** 118-121
[2] Gurova E G, Gross V Y, Kurbatov V S, Makarov S V, Sergeev A A and Shchurov N I 2013 *World Applied Sciences Journal* **22** 44-48
[3] Gurova E G 2015 *Conf. Series-Materials Science and Engineering* **91** 012091

Polynomial reconstruction of electric charge distribution on the conductive plate caused by external electric field

Yu N I saev[1], V A Kolchanova[1], S S Tarasenko[1] and O V Tikhomirova[2]
[1]Tomsk Polytechnic University, 30, Lenina ave., Tomsk, 634050, Russia
[2]V E Zuev Institute of Atmospheric Optics SB RAS, 1, Academician Zuev square, Tomsk, 634055, Russia

E-mail: ni105@mail.ru

Abstract. The paper proposes an original method of calculating the charge distribution on the surface of the conductive plate introduced into the external electrostatic field. The authors managed to obtain the polynomials which allow to solve the integral equation that establishes the relationship between charge distribution of the conductive plate and the potential distribution of the external field and the potential on the surface of the plate. The proposed algorithms solutions are valid in the presence of axial symmetry of the field and the plate. Examples of calculation of conductor charge distribution in the presence of external field by using a polynomial expansion have been presented. The comparisons of results calculated by the polynomial method and by known analytical solutions have been given

1. Introduction

When a conductive plate is placed into the external electrostatic field, the electric charges commence its induction on the plate surface. A correlation, which connects external field potential $U^0(\mathbf{r}_0)$ and potential on the plate surface U_0 with the distribution of the induced charge, can be described as follows [2, 3]

$$-U^0(\mathbf{r}_0)+U_0=\frac{1}{4\pi\varepsilon_0}\int\limits_D\frac{\sigma(\mathbf{r})}{|\mathbf{r}-\mathbf{r}_0|}d\mathbf{r},\ \ \mathbf{r}=\{x,y\},\ \mathbf{r}_0=\{x_0,y_0\},\ \mathbf{r},\mathbf{r}_0\in D\cdot \tag{1}$$

Thus, with certain charge distribution $\sigma(\mathbf{r})$ on the conductive plate surface we can determine an algebraic sum of potential distribution of the external field $U^0(\mathbf{r}_0)$ and plate surface potential U_0. In natural experiments we usually know the distribution of the external field potential, the potential of the conductor surface can be measured or kept in direct-current voltage. So the real interest is the inverse problem – definition of the charge distribution on the conductor surface. In accordance with expression (1) we need to solve the Fredholm integral equation of the first kind of required relative value $\sigma(\mathbf{r})$. Value $\sigma(\mathbf{r})$ is total charge distribution in the conductor from different sides of the plate, so the next step is to divide the source-faced surface charge and the opposite one. We need to use the normal voltage component of the resulting field to divide the charges:

$$E_n(\mathbf{r})=-(\nabla U(\mathbf{r}),\mathbf{n})=-\frac{\partial}{\partial z}U(\mathbf{r})\cdot \tag{2}$$

Field intensity is related to the correlations charges:

$$2\varepsilon_0 E_n(\mathbf{r})=\sigma_1(\mathbf{r})-\sigma_2(\mathbf{r}),\quad \sigma(\mathbf{r})=\sigma_1(\mathbf{r})+\sigma_2(\mathbf{r})\ , \tag{3}$$

from which we can define:

$$\sigma_1(\mathbf{r}) = \frac{\sigma(\mathbf{r}) + 2\varepsilon_0 E_n(\mathbf{r})}{2}, \qquad \sigma_2(\mathbf{r}) = \frac{\sigma(\mathbf{r}) - 2\varepsilon_0 E_n(\mathbf{r})}{2}. \tag{3}$$

The aim of this work is to replace complicated integral equation (1) with algebraic one with the defined polynomials, in other words, to define analytic correlation, which connects the resulting field potential and the charge distribution in the conductor.

2. Formulation of the problem.

Let us suppose that the external field possesses a symmetry of rotation about the symmetry axis of the disc. Then equation (1) can be described in the following form:

$$U(\rho) = \frac{1}{4\pi\varepsilon_0} \int_0^{2\pi} d\vartheta' \int_o^R \frac{\sigma(\rho')\rho' d\rho'}{\sqrt{\rho^2 + \rho'^2 - 2\rho\rho'\cos(\vartheta - \vartheta')}} \ , \tag{4}$$

where $U(\rho) = -U^0(\rho) + U_0$ – resulting potential.

The latter expression can be rewritten with elliptical integral $K(k) = \int_0^{\pi/2} \frac{d\vartheta'}{\sqrt{1 - k\cos(\vartheta')}}$, then correlation (4) transforms into the following expression:

$$U(\rho) = \frac{1}{4\pi\varepsilon_0} \int_0^R \rho' d\rho' \sigma(\rho') \int_0^{2\pi} \frac{d\vartheta'}{\sqrt{\rho^2 + \rho'^2 - 2\rho\rho'\cos(\vartheta - \vartheta')}} = \frac{1}{\pi\varepsilon_0} \int_0^R \frac{\rho'\sigma(\rho')}{\rho' + \rho} K\left(\frac{2\sqrt{\rho'\rho}}{\rho' + \rho}\right) d\rho'. \tag{5}$$

Taking into account the Landen transformation [2, 3, 4] we can define the following feature of the elliptic integral:

$$\frac{1}{x} K\left(\frac{y}{x}\right) = \int_0^y \frac{ds}{\sqrt{(x^2 - s^2)(y^2 - s^2)}}. \tag{6}$$

This feature allows one to describe an unknown subintegral function with certain potential $U(\rho)$:

$$\sigma(\rho) = -\frac{4\varepsilon_0}{\pi\rho} \frac{d}{d\rho} \int_\rho^R \frac{s ds}{\sqrt{s^2 - \rho^2}} \frac{d}{ds} \int_0^s \frac{U(t) t dt}{\sqrt{s^2 - t^2}}. \tag{7}$$

Let us find the charge distribution, made by potential

$$U(\rho) = \rho^m \tag{8}$$

After substitution of (8) in (7) we obtain the expressions for charge distribution in the form of polynomials $\psi_m(\rho)$ with even and uneven m accordingly:

$$\sigma(\rho) = \begin{cases} \dfrac{M(1,1)}{2} \dfrac{R}{\sqrt{R^2 - \rho^2}} & \text{when } m = 1, \\[2ex] -N(m,1) \displaystyle\sum_{k=0}^{m/2} c\left(\frac{m}{2}, k\right) \frac{\left(R^2 - \rho^2\right)^{k - \frac{1}{2}} \left(\rho^{-2} m R^2 - m - 2\rho^{-2} k R^2 - 1\right) \rho^{-2k+m}}{2k+1} & \text{when } m - \text{an even} \\[2ex] -\dfrac{M(m,1)}{m+1} \left(\dfrac{-R^m}{\sqrt{R^2 - \rho^2}} + \displaystyle\sum_{k=1}^{m-1} \left(\frac{F2\left(\frac{m+1}{2}, k\right)}{2^k F1\left(\frac{m+1}{2}, k\right)} \cdot \left(2k\rho^{2k-2}\sqrt{R^2 - \rho^2} - \frac{\rho^{2k}}{\sqrt{R^2 - \rho^2}} \right) R^{m-2k} \right) \right) + & \\ -\dfrac{F(m) M(m,1)}{2^{\frac{m+1}{2}} \left(\frac{m+1}{2}\right)!} \left[\ln\left(\dfrac{R + \sqrt{R^2 - \rho^2}}{\rho} \right) \rho^{m-1} (m+1) - \left(\dfrac{1}{\sqrt{R^2 - \rho^2}} \cdot \dfrac{\rho^{(m+1)}}{R + \sqrt{R^2 - \rho^2}} + \rho^{(m+1)} \right) \right] & \text{when } m - \text{an odd.} \end{cases}$$

Here

$$F(m)=\prod_{k=0}^{\frac{m-1}{2}}2k+1,\qquad F1(m,k)=\prod_{s=1}^{k}m-s,\qquad F2(m,k)=\prod_{s=1}^{k}2m-2s+1,$$

$$c(m,k)=\frac{m}{(k-m)!k!},$$

$$M(m,s)=\frac{F(m)s^m\pi(m+1)}{2^{\frac{m+3}{2}}\left(\frac{m+1}{2}\right)!},\qquad N(m,s)=(m+1)s^m\sum_{k=0}^{\frac{m}{2}}\frac{(-1)^k}{2k+1}c\left(\frac{m}{2},k\right).$$

Table 1. Description of even and uneven degrees

ρ^m	$\psi_m(\rho)$	ρ^m	$\psi_m(\rho)$
1	$\frac{1}{\sqrt{R^2-\rho^2}}$	ρ	$\frac{1}{2}\pi\left(\frac{R}{\sqrt{R^2-\rho^2}}-\ln\left(\frac{R+\sqrt{R^2-\rho^2}}{\rho}\right)\right)$
ρ^2	$-2\frac{R^2-2\rho^2}{\sqrt{R^2-\rho^2}}$	ρ^3	$-\frac{3}{8}\pi\left(\frac{(R^2-3\rho^2)R}{\sqrt{R^2-\rho^2}}+3\rho^2\ln\left(\frac{R+\sqrt{R^2-\rho^2}}{\rho}\right)\right)$
ρ^4	$-\frac{8}{9}\cdot\frac{4\rho^2R^2-8\rho^4+R^4}{\sqrt{R^2-\rho^2}}$	ρ^5	$-\frac{15}{128}\pi\left(\frac{(2R^4+5R^2\rho^2-15\rho^4)R}{\sqrt{R^2-\rho^2}}+15\rho^4\ln\left(\frac{R+\sqrt{R^2-\rho^2}}{\rho}\right)\right)$

With these tabular correlations we can define the charge distribution (which is of great interest to us) with the certain potential distribution. We need to describe the distribution of the field potentials in the form of polynomials resolution:

$$U(\rho)=\sum_{m=0}^{N}a_m\rho^m. \tag{9}$$

The coefficients of the expansion of series (9) can be easily defined by the method of the least squares. The charge distribution can be defined with the following correlation:

$$\sigma(\rho)=\frac{4\varepsilon_0}{\pi}\sum_{m=0}^{N}a_m\psi_m(\rho). \tag{10}$$

The kind of polynomial of series (8) is defined with the expression from Table 1.

In case of perforated disc in expression (5) the limits of integration are changed:

$$U(\rho)=\frac{1}{\pi\varepsilon_0}\int_r^{\infty}\frac{\rho'\sigma(\rho')}{\rho'+\rho}K\left(\frac{2\sqrt{\rho'\rho}}{\rho'+\rho}\right)d\rho'. \tag{11}$$

In this case it is unnecessary to calculate new polynomials for expansion of the charge distribution. It is enough to realise a necessary transformation of the coordinate system consisting in projecting of the exterior of the circle on its interior. Replacing of the variables we come from integral equation (11) to integral one (5) with the help of conjugate coordinates:

$$\rho'=\frac{r^2}{t'},\ \rho=\frac{r^2}{t}. \tag{12}$$

The result is the integral equation with new limits of integration

$$U1(t)=\frac{1}{\pi\varepsilon_0}\int_0^r\frac{\sigma_1(t')t'}{(t'+t)}K\left(\frac{2\sqrt{t\cdot t'}}{t+t'}\right)dt' \text{ где } \sigma_1(t')=\frac{r^2\sigma(r^2/t')}{t'^3},U_1(t)=U\left(\frac{r^2}{t}\right)\cdot\frac{1}{t} \tag{13}$$

To check the operation of the proposed polynomial algorithm let us consider two examples which have analytical solution [2, 3].

Example 1. Let us define the charge distribution in the grounded disc with radius R = 1m if the disc is in the external field of the point charge, located on the disc axis at a distance of h = 0.3m from the disc surface. Ratio, q/ε_0 had been chosen as 1.

The potential of the point charge in the disc plane and total charge distribution on the disc can be defined with the following expressions accordingly [1]:

$$U(\rho)=\frac{q}{4\pi\varepsilon_0}\frac{1}{\sqrt{\rho^2+h^2}},\quad \sigma(\rho)=-\frac{h}{2\pi^2\left(h^2+\rho^2\right)^{3/2}}\left(arctg\sqrt{\frac{R^2-\rho^2}{h^2+\rho^2}}+\sqrt{\frac{h^2+\rho^2}{R^2-\rho^2}}\right). \tag{15}$$

Let us describe the potential in the form of the series:

$$U(\rho)=\sum_{m=0}^{N}a_m\rho^m, \tag{16}$$

Let us define the coefficients of expansion (12), using the method of the least squares. Let us suppose that the number of the expansion coefficients is N = 10.

$$m=0\,..\,N,\ n=m,\ c_m=\int_0^R U(\rho)\rho d\rho,\ b_{m,n}=\int_0^R \rho^{m+n+1}dr=\frac{R^{m+n+1}}{m+n+1}.$$

Let us define expansion coefficients a_m by calculating the matrix equation:

$$c_m=b_{m,n}a_n \;\rightarrow\; a_n=\mathrm{Re}\left[\left(b_{m,n}+jI_{m,n}\lambda\right)^{-1}\right]c_m,$$

I – unit matrix, $j=\sqrt{-1}$, $\lambda=10^{-19.25}$ – regularisation parameter, decreasing the relative contribution of the noise term in the solution [1, 5, 6].

Table 2. Coefficients of expansion a_m (16) obtained as a result of the solution of the matrix equation

m	*0*	*1*	2	3	4	5	6	7	8	9	*10*
a_m	**0.265**	**0**	**-1.43**	**-1.438**	**29.7**	**-104.7**	**198.4**	**-229.6**	**162.7**	**-65.06**	**11.273**

A desired charge distribution on surface $\sigma(\rho)$ will be defined by the expression with the calculated coefficients: $\sigma(\rho)=\frac{2\varepsilon_0}{\pi}\sum_{m=0}^{N}a_m\psi_m(\rho)$.

The results are given in the form below:

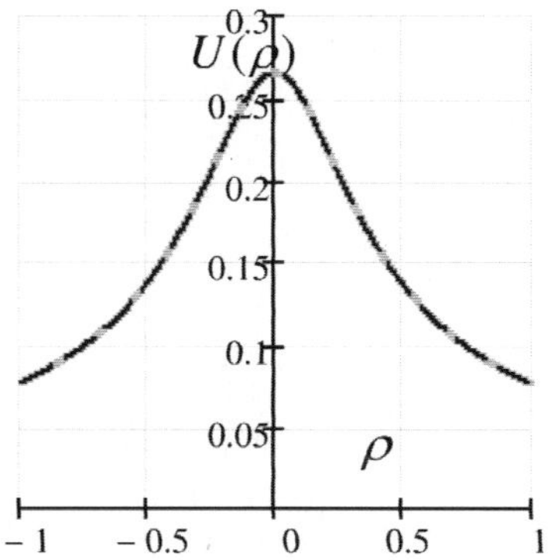

a

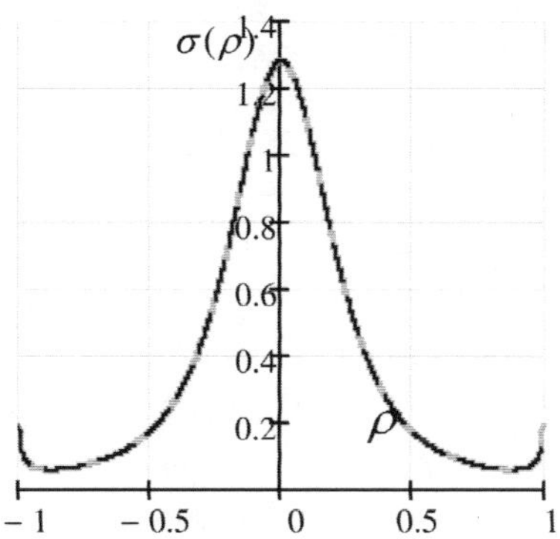

b

Figure 1. a – potential of the external field as an analytical function and the result of its expansion into the series (12); b – charge distribution as an analytic dependence and the result of the return into the series

There is concurrence of the analytic dependences and polynomial method dependences in the forms.

Example 2. Let us define the charge distribution on the surface of the conductive disc with round perforation $r = 0.25$m if the disc is located in the external field of the point charge, located on the disc axis at a distance of $h = 0.4$m from the surface of the disc with radius $R = 1$m.

The potential of the point charge in the disc plane and total charge distribution on the disc can be defined with the following expressions accordingly [2, 3]:

$$U(\rho)=\begin{cases}\dfrac{q}{4\pi\varepsilon_0}\dfrac{1}{\sqrt{\rho^2+h^2}}, & \rho\ge r\\ 0, & \rho<r\end{cases},\quad \sigma(\rho)=-\frac{h}{\pi^2\left(h^2+\rho^2\right)^{3/2}}\left(arctg\frac{h}{r}\sqrt{\frac{\rho^2-r^2}{h^2+\rho^2}}+\frac{r}{h}\sqrt{\frac{h^2+\rho^2}{\rho^2-r^2}}\right). \quad (17)$$

In this case we use equation (5) with the replacement of the variables:

$$\rho'=\frac{r^2}{t'},\ \rho=\frac{r^2}{t}. \quad (18)$$

The result is the integral equation with new limits of integration

$$U\left(\frac{r^2}{t}\right)=\frac{1}{\pi\varepsilon_0}\int_0^r\frac{\frac{r^2}{t'}\sigma\left(\frac{r^2}{t'}\right)}{\frac{r^2}{t'}+\frac{r^2}{t}}K\left(\frac{2\sqrt{\frac{r^2}{t'}\cdot\frac{r^2}{t}}}{\frac{r^2}{t'}+\frac{r^2}{t}}\right)d\frac{r^2}{t'}. \quad (19)$$

Let us simplify equation (19) to obtain the following result:

$$U1(t)=\frac{1}{\pi\varepsilon_0}\int_0^r\frac{\sigma1(t')\,t'}{(t'+t)}K\left(\frac{2\sqrt{t\cdot t'}}{t+t'}\right)dt',\ \text{where}\ \sigma1(t')=\frac{r^2\sigma\left(\frac{r^2}{t'}\right)}{t'^3},\quad U1(t)=U\left(\frac{r^2}{t}\right)\cdot\frac{1}{t}. \quad (20)$$

Let us describe field potential $U1(t)$ in the form of series

$$U1(t)=\frac{q}{4\pi\varepsilon_0}\sum_{m=0}^{N}a_m t^m \quad (21)$$

Using the same method we will define expansion coefficients a_m, and then using the polynomial table and replacement of the variable in form of $t = r^2/\rho$, let us define

$$\sigma1(\rho)=\frac{2\varepsilon_0}{\pi}\sum_{m=0}^{N}a_m\psi_m(\rho),\quad \sigma\left(\frac{r^2}{t}\right)=\sigma1(t)\cdot\frac{t^3}{r^2},\quad \sigma(\rho)=\sigma1\left(\frac{r^2}{\rho}\right)\cdot\frac{r^4}{\rho^3}. \quad (22)$$

Defined charge distribution $\sigma(\rho)$ on the conductor surface is described in form 2, there is also a result of the analytical formula [2, 3].

$$\sigma(\rho)=-\frac{h}{\pi^2\left(h^2+\rho^2\right)^{3/2}}\left(arctg\frac{h}{r}\sqrt{\frac{\rho^2-r^2}{h^2+\rho^2}}+\frac{r}{h}\sqrt{\frac{h^2+\rho^2}{\rho^2-r^2}}\right).$$

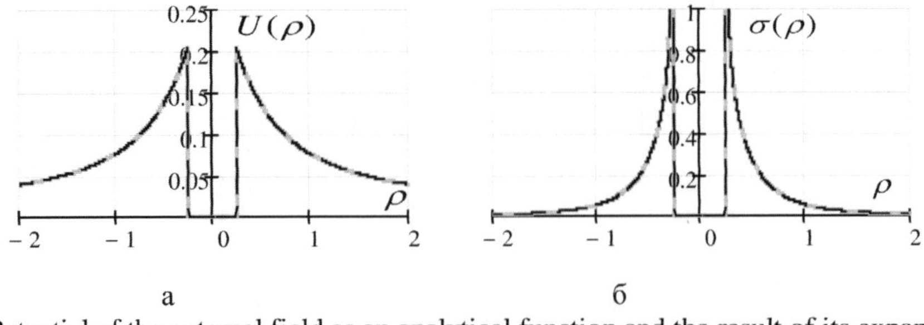

Figure 2. a – Potential of the external field as an analytical function and the result of its expansion into the series (21), b – Charge distribution as an analytic dependence and the result of the return into the series (22), where $\lambda = 10^{-12}$ – a regularization parameter

Conclusion
The authors managed to define the polynomials, leading the integral equation, which connects the charge distribution on the plate with the external static field to a more simplistic algebraic equation. A polynomial algorithm holds with an axial-symmetric external field and an axial-symmetric shape of the plate. In case of using the polynomial method it necessary to do regularization of the expansion coefficients for decreasing the algorithm noises influence.

References
[1] Verlan A F and Sizikov V S 1986 *Kiev: Naukova dumka* 543
[2] Mirolubov N N, Kostenko M V, Levinstein M L and Tikhodeev N N 1963 *M: High school* 414
[3] Pimenov U V 2008 Dolgoprudniy: intelligent 535
[4] Kolchanova V A, Isaev Yu N, Tarasenko S S and Filkov A I 2014 *Proceedings of 2014 International Conference on Mechanical Engineering, Automation and Control Systems, MEACS 2014* **552(1)** 1-4
[5] Shandarova E B and Shwab S A 2014 *Proceedings of 2014 International Conference on Mechanical Engineering, Automation and Control Systems, MEACS 2014* **552(1)** 1-5

Correction for thermal distortions of laser beams with a flexible mirror. Experimental and numerical investigations

F Kanev[1], N Makenova[1], R Nesterov[1] and I Izmailov[2]
[1]Tomsk Polytechnic University, 30, Lenina ave., Tomsk, 634050, Russia
[2]Tomsk State University, 36, Lenina ave., Tomsk, 634050, Russia

E-mail: makenova@tpu.ru

Abstract. A mathematical model of an adaptive optics system was described in the article. The model included two main components: the model of an adaptive mirror and model of beam propagation under conditions of thermal blooming. Results of numerical simulation of adaptive optics systems were compared with data of laboratory experiments. High reliability of the model was shown.

1. Introduction

Adaptive optics systems are extensively used nowadays to correct distortions of laser beams propagating on atmospheric paths [1]. A schematic diagram of such system is shown in Figure 1. As is well known, beam propagating in the atmosphere suffers from random variations of the index of refraction and from thermal distortions induced by heating of the medium. As a result, in the plane of observations the amplitude profile of the beam is severely distorted. In adaptive systems the information about distortions transmitted by the loop to the active element which forms a phase profile of the beam according the algorithm of correction. Effectiveness of compensation for distortions depends on the precision of beam parameter registration, on the control algorithm of the system, on the rate of control, and on the quality of the phase profile formed by the active element. A flexible mirror with continuous reflecting surface is used usually as an active element of adaptive systems [2].

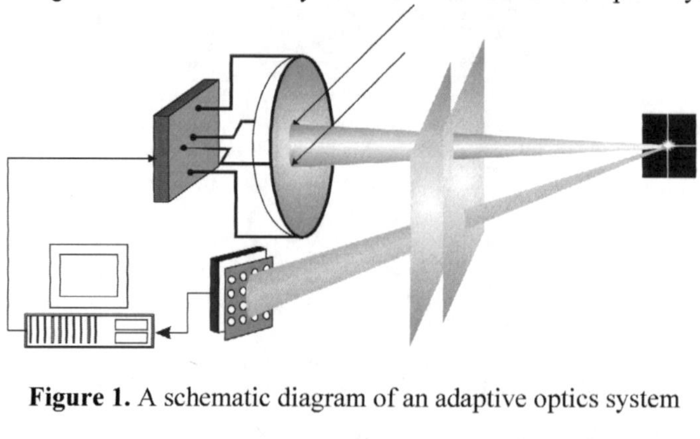

Figure 1. A schematic diagram of an adaptive optics system

In the current paper a model of the flexible mirror is presented along with a model of beam propagation in a medium. The results of numerical simulation are compared with the data of the laboratory experiment, and the possibility of using a developed model in investigations of beam correction effectiveness is assessed.

2. A model of an adaptive optics system

The whole model of an adaptive system, as it follows from the diagram shown above (Figure 1), should include a model of a mirror, a model of beam propagation in a medium, and an algorithm of beam control. The model of the mirror used in our investigations was built on the basis of techniques developed by S. Chesnokov [3, 4]. Theauthorassumedthatan adaptive mirror can be represented as a thin plate controlled by forces applied at discrete points. After that the finite element method was applied to calculate deformations of the plate [4]. Application of the model in problems of adaptive optics was demonstrates in the numeric experiment simulating the multidither system of thermal blooming correction [5].

Following the papers by S. Chesnokov we extended the model. In the first publications on the subject we described the model with an increased number of actuators (from 4 to 20, Figure 2) and assessed its application in the phase conjugation algorithm [6, 7]. Along with the advanced technology of modern computers the limits on calculation grid dimensions were shifted, and control of the mirror surface was realised at 500 points. Also we developed a model of the adaptive corrector controlled by forces and by moments of forces which was used in simulation of the laboratory experiment of correction for thermal blooming [8].

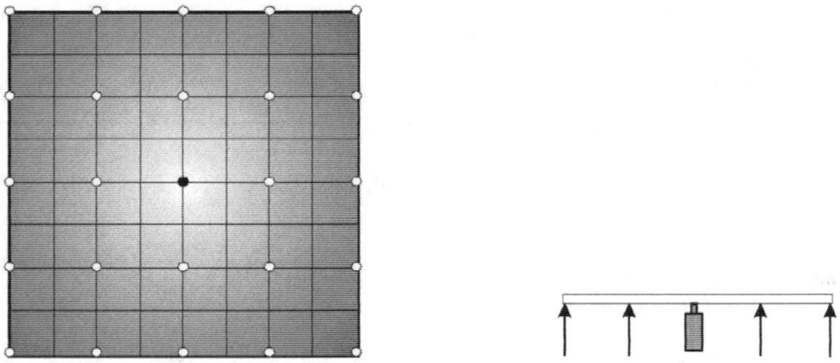

(a) (b)

Figure 2. A model of the mirror controlled by forces applied at discrete points: a) top view, b) side view

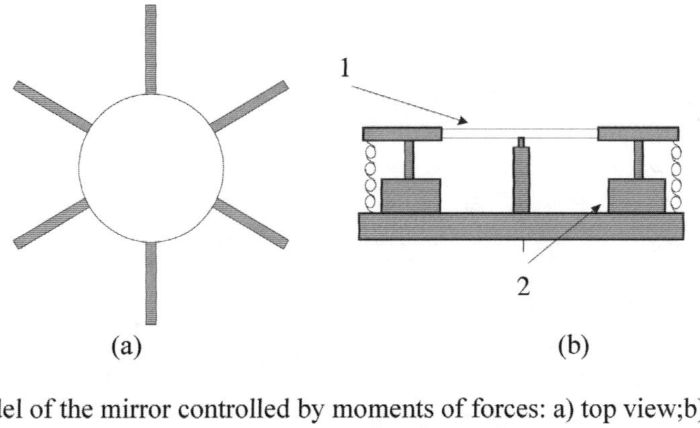

(a) (b)

Figure 3. Model of the mirror controlled by moments of forces: a) top view;b) side view:1-a stepper motor, 2 -a reflective surface

One more model developed for adaptive optics applications wasa model of a dynamic mirror taking into account oscillations of the reflecting surface due to deformation of the plate [9]. In corresponding equations the mass and inertia of the plate were allowed for and transient processes were calculated [10]. Hence it follows that mathematical models of the listed above correctors are presented.

Inapproximationofathin plate static deformations $W(x,y)$ of the corrector can be described by the following equation [10]:

$$D\left(\frac{\partial^4 W}{\partial x^2}+2\frac{\partial^4 W}{\partial x^2 \partial y^2}+\frac{\partial^4 W}{\partial y^2}\right)=f(x,y). \qquad (1)$$

Here$x$and$y$are Cartesian coordinates in the plane of the mirror, $D=Eh^3/(12(1-\sigma^2))$ is cylindrical rigidness, σ is the Poisson coefficient, $E$is Young's modulus, h is thickness of the plate, and $f(x, y)$ are a load applied normally to the plate.

If the plate is controlled by a system of discrete forces P_j, the load can be represented in the form:

$$f(x,y)=\sum_{j=1}^{\nu} P_j \delta(x-x_j, y-y_j) S_j, \qquad (2)$$

x_j and y_j – arecoordinatesofactuators, S_j is its area, and ν is the number of actuators.

Border conditions should be written for all points of plate contour L. Conditions characterizing free edges of the plate can be formulated as:

$$D\left(\frac{\partial^2 W}{\partial n^2}+\sigma\frac{\partial^2 W}{\partial \tau^2}\right)\Bigg|_{Lj}=0, \quad D\frac{\partial}{\partial n}\left(\frac{\partial^2 W}{\partial n^2}+(2-\sigma)\frac{\partial^2 W}{\partial \tau^2}\right)\Bigg|_{Lj}=0.$$

Here $\partial/\partial n$ and $\partial/\partial \tau$ are normal and tangential partial derivatives correspondingly.

If normal force P_j and moment T_j are applied to points L_j on the perimeter of the plate, the border conditions should be changed to the form:

$$D\left(\frac{\partial^2 W}{\partial n^2}+\sigma\frac{\partial^2 W}{\partial \tau^2}\right)\Bigg|_{Lj}=\frac{-T_j\delta\ x-x_j, y-y_j}{l_j},$$

$$D\frac{\partial}{\partial n}\left(\frac{\partial^2 W}{\partial n^2}+(2-\sigma)\frac{\partial^2 W}{\partial \tau^2}\right)\Bigg|_{Lj}=\frac{-P_j\delta\ x-x_j, y-y_j}{l_j},$$

where x_j and y_j are coordinates of the point where the force is applied, and l_j is a distance between points.

Conditions corresponding to the points inside the contour of the plate can be written in the same way, for example, if a point fixed rigidly:

$$W(0,0)=\frac{\partial W(0,0)}{\partial x}=\frac{\partial W(0,0)}{\partial y}=0.$$

Solution to Eq. 1 with the border conditions given above is difficult to obtain analytically, so numerical models of the mirror were developed on the basis of the finite element method [4]. According to the method, all surface of plate $W(x,y)$ is divided into small sections Ω (finite elements), and in each section a local coordinate system with axes Oξ, Oη is set up. So instead of calculation of the whole plate deformation, we determine only deformation of each finite element $W(\xi,\eta)$. Elements should be joined by cinematic and dynamic border conditions, so the total set of finite elements forms the model of the whole plate.

Consecutively, the shape of finite element surface $W(\xi,\eta)$ is approximated by a linear form of base functions $\mathbf{Ш}(\xi,\eta)$:

$$W(\xi,\eta) = \mathbf{Ш}^T(\xi,\eta)\|\mathbf{a}\|\mathbf{q}\,. \tag{3}$$

Here $\mathbf{Ш}^T(\xi,\eta)$ is a vector of base functions, and superscript 'T' signifies the operation of a vector transposing. $\|\mathbf{a}\|$ is a matrix of coordinates transformation and $\mathbf{q}$ is a vector of generalized coordinates. As generalized coordinates shifts W_r and tilts φ_r ,υ_r in nodes of the calculation grid are taken usually as [3, 4], so in each node (the number of nodes is R) the model is characterized by three variables.

The vector of generalized nodal forces

$$\mathbf{Q} = \begin{pmatrix} \mathbf{P} \\ \mathbf{N} \\ \mathbf{T} \end{pmatrix} \tag{4}$$

is conjugated with coordinate vector q. In Eq. 4 **P** is a vector of normal forces and vectors **N** and **T** are vector of force moments relatively axes $o\xi$ and $o\eta$.

The principle of virtual displacements is possible to apply in order to obtain equation relating vectors **q** and **Q** [3]:

$$-\delta U + \delta A_m + \delta A_f + \delta A_Q = 0\,. \tag{5}$$

In Eq. 5 δU is variation of the element potential energy, δA_m is the virtual work of inertia forces, δA_f is virtual work of applied load, and $\delta A_Q = \delta\mathbf{q}^T\mathbf{Q}$ is the work of forces of interactions.

To obtain the equation describing deformation of the whole model we should, firstly, take into account continuity of the generalized coordinates in the nodes of the calculation grid and, secondly, equilibrium of forces in these nodes. As a result,

$$\|\mathbf{M}\|\ddot{\mathbf{W}}(x,y) + \|\mathbf{G}\|\dot{\mathbf{W}}(x,y) + \|\mathbf{K}\|\mathbf{W}(x,y) = \mathbf{Q}_f\,. \tag{6}$$

Matrices $\|\mathbf{M}\|$, $\|\mathbf{G}\|$, and $\|\mathbf{K}\|$ have the same meaning as those described earlier, but now they are written for the whole model. $\|\mathbf{Q}_f\|$ is a vector of external forces applied to the plate. A solution to Eq. 6 is possible to find with application of numeric methods, and in this particular problem Runge-Kutta method was used [11].

Calculation of static deformations allows one to simplify Eq. 6 and obtain the formula:

$$\|\mathbf{K}\|\mathbf{W}(x,y) = \mathbf{Q}_f\,, \tag{7}$$

so the shape of the plate can be found as

$$\mathbf{W}(x,y) = \|\mathbf{L}\|\mathbf{Q}_f\,, \tag{8}$$

here $\|\mathbf{L}\| = \|\mathbf{K}\|^{-1}$ is a matrix inverse relatively to $\|\mathbf{K}\|$.

Another model required in numerical experiments is a model of beam propagation in an atmosphere. Propagation of a beam in the randomly inhomogeneous absorbing medium is characterized by the wave equation written in the approximation of quasioptics:

$$2ik\frac{\partial E}{\partial z} = \frac{\partial^2 E}{\partial x^2} + \frac{\partial^2 E}{\partial y^2} + \frac{k^2}{n_0}\ \tilde{n} + n_{\text{nl}}\ E\,, \tag{9}$$

where $E = E\ x, y, z, t$ is the complex amplitude of the field, $k = 2\pi/\lambda$ is a wave number, λ is a wave length. Variables in the problem were normalized as follows: in the direction of the laser beam propagation (Z-axis) on the diffraction length $Z_d = ka_0^2$ $(Z = Z / Z_d)$, and in the direction perpendicular to the path on the initial radius of the beam a_0 $(x = x / a_0, y = y / a_0)$. In Eq. 9 $\tilde{n}$ is random fluctuations of the refractive index induced by turbulence (not taken into account), and n_{nl} is variations induced by heating of the medium. The influence of heating on the laser beam form is called the effect of thermal blooming. Eq. 8 was solved with the use of the splitting method.

3. Comparison of experimental results with results of numerical simulation

To validate the developed model of an adaptive system the results of numerical experiments were compared with the data found in references. Particularly, variables characterizing the model were put in correspondence to parameters of the mirror built at the Institute of Optics and Electronics, Chinese Academy of Sciences [12]. The parameters of the mirror were the following: the diameter was 20 mm; the maximum shift of the reflecting surface – ±1.0 μm, and the resonance frequency – 30 kHz. The mirror was controlled by 19 actuators attached to the plate at discrete points. Calculation of actuator response functions demonstrated a 10% deviation of numerical results from the published characteristic of the mirror. High precision of the model was assumed.

One more opportunity to compare theoretical and experimental data emerged after publication of the experimental results of correction for thermal blooming [13]. In the optical system employed in the investigations the phase profile of radiation was formed by the flexible mirror controlled by moments of forces. Schematically the mirror is shown in Figure 3. Distortions of the beam were developed as a result of propagation in a cell filled by semitransparent liquid. The flux of the medium was simulated by rotation of the cell. The amplitude distribution of the beam was registered by a video camera, and the main characteristics of the radiation, such as shifts x_C and y_C of the beam centroid in Cartesian coordinate system, radii σ_x and σ_y, as well as maximum intensity I_m are calculated by a specially developed software. The information obtained as a result of calculations was used for generation of mirror control signals. Adaptive correction was realized with the use of the multidither algorithm. The vector characterizing amplitude distribution of the beam was chosen as a goal function of correction. Vector **J** had the following components:

$$J_1 = x_0 - x_C;\ J_2 = y_0 - y_C;\ J_3 = \sigma^2 - \sigma_0^2;$$
$$J_4 = \sigma_x^2 - \sigma_y^2;\ J_5 = \sigma_{xx};\ J_6 = \sigma_x^2 + \sigma_y^2.$$

Here σ_0 is the radius fan undistorted beam.

A qualitative comparison of results was made with the use of photographs of laser beam amplitude profiles taken before and after correction (Figure 4).

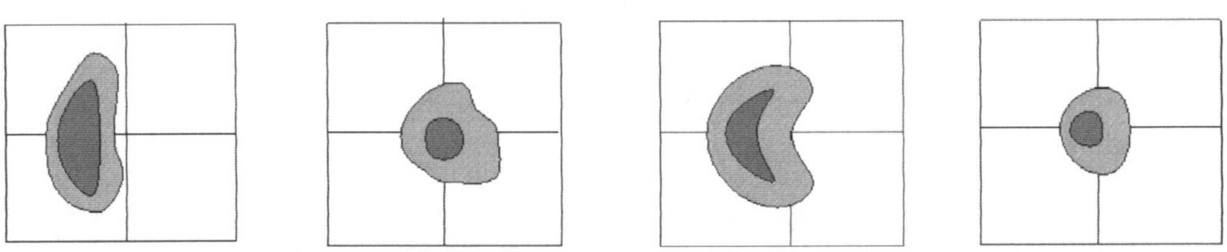

(a) (b) (c) (d)

Figure 4. Comparison of experimental and theoretical results. The beam amplitude profile registered experimentally before (a) and after correction (b). Corresponding results obtained in numerical experiment (c and d)

In the same picture the profiles registered in numerical experiments were presented. As one can see,the beam shape and its changes are practically the same in both cases.

A comparison of results in absolute units was impossible so we introduced a relative criterion characterizing the effectiveness of correction:

$$\eta_J = \left(\frac{J_{opt} - J_0}{J_0} \right) \cdot 100\% ,$$

here J_{opt} is some parameter of the beam (radius, maximum intensity, oracentroid shift) obtained as result of correction and J_0 is the same parameter registered before application of adaptive control. Experimentally obtained criteria $\eta_{\sigma x}$, $\eta_{\sigma y}$, η_{Im} describing changes of radii σ_x, σ_y and maximum intensity of the beam I_m are presented in Table 1. In the same table the results of numerical simulation are shown.

Table 1. Experimental and theoretical results

	$\eta_{\sigma x}$	$\eta_{\sigma y}$	η_{Im}
Laboratory experiment	-24	-55	107
Numerical experiment	-26	-50	344

As one can see the larges discrepancy was registered for maximum intensity of the beam I_m, but this variable is a local characteristic of the beam measured with less precision as compared to its integral characteristics σ_x and σ_y.

4. Conclusions

The experimental data and results of numerical investigations presented here have demonstrated high reliability of the developed model. The model can be used in simulation of real experiments in adaptive optics.

References

[1] Tyson R 1991 *San Diego: Academic Press* 233
[2] Cherezova T *et al* 1999 *Proc. of the 2-rd International Workshop on Adaptive Optics for Industry and Medicine*p 187-192
[3] Vorontsov M and Chesnokov S 1982 *Radiophysics*vol **25** 11
[4] Kandidov V, Chesnokov S and Visloukh V 1976 *Moscow State University Publ. House* 178
[5] Chesnokov S 1983 *Quantum electronics* **10** 6
[6] Kanev F, Lukin V and Lavrinova L 1994 *Atmospheric propagation and remote sensing III Orlando, USA* 57-58
[7] Kanev F, Lukin V and Fortes B 1995 *Internet Geoscience and Remote Sensing Symposium Firenze, Italy* 5
[8] Kanev F and Lukin V 1991 *Atmospheric optics* **4** 12
[9] Kanev F 1990 *Conference Lasers and atmosphere (Obninsk)* 161
[10] Ogibalov P 1958 *Moscow State University Publishing House* 168
[11] Potter D 1975 *Moscow: Mir publishers* 376
[12] Ling N and Guan Ch 1996 *SPIE Proc.* **2828** 472-478
[13] KandidovV, Larionova I and Popov V 1989 *Atmospheric optics* **2** 8

Mathematical modeling of synergetic aspects of machine building enterprise management

O D Kazakov and S V Andriyanov
Bryansk State Technical University, 7, 50 years of October st., Bryansk, 241035, Russia

E-mail: kod8383@mail.ru

Abstract. The multivariate method of determining the optimal values of leading key performance indicators of production divisions of machine-building enterprises in the aspect of synergetics has been worked out.

1. Introduction

Despite the large number of methods and tools of machine-building enterprise management we should note the lack of efficiency of their use.

According to the authors, this inefficiency is largely due to lack of implementation or complete lack of synergetic components in the concept of the investigated tools and approaches. While implementing the tasks of machine-building enterprise management it is necessary to rely not only on organization, but also on self-organization. In this case, there is a non-trivial and important issue of the transition from classical methods of econometrics to modern techniques, implementing the principles of synergetics.

2. Results and discussion

Exploring production divisions of enterprises as a dissipative structure, it is necessary to highlight such components that accompany self-developing systems in synergetics as an attractor, a fluctuation and, in particular, a fractal.

According to the authors' opinion the most relevant and effective method of management of a machine-building enterprise that allows taking into account synergetic effects is the logical-metric modeling. The logical-metric modeling is a systematic set of procedures for the development of metrics, which are interconnected with stochastic and/or functional dependence and allow providing coordinated functioning of production subsystems of a business entity.

In the authors' opinion a fractal as an object that has the property of self-similarity, i.e. consisting of several elements, each of which is similar throughout the whole object. It appears in the logical-metric model of development of a socio-economic system (SES) in the form of a top level balanced scorecard as well as in the procedure of cascading the system to the structural units (Figure 1).

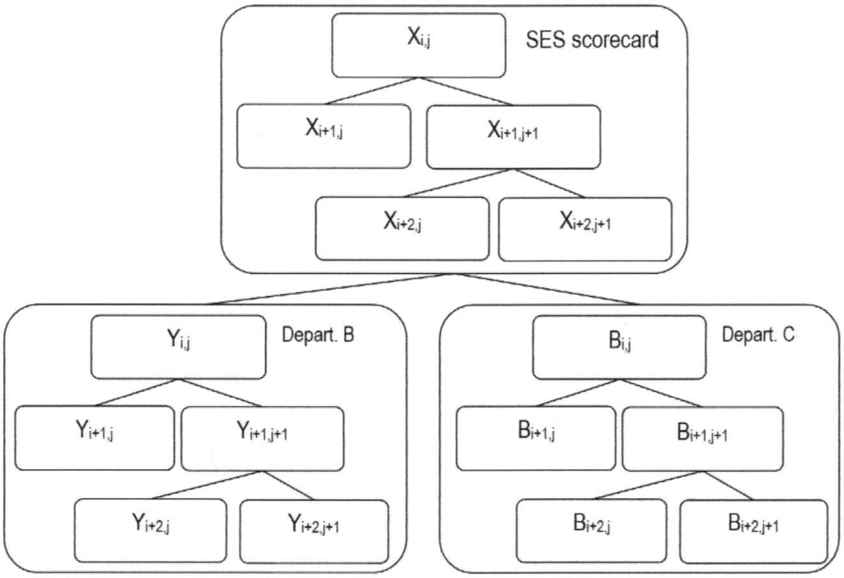

Figure 1. A fractal in the logic-metric model of socio-economic system development.

The authors have investigated the ways of selecting the desired invariant conditions of production systems in accordance with the target as directed self-organization. Therefore, according to the authors, the attractor as the ideal end state of the system development in a logical-metric model of development of a machine-building enterprise manifested in the form of a balanced system of key performance indicators.

In this case, the problem of determining the values of the leading indicators when determining the target value of the indicator at the top level arises. This problem is caused by the presence of stochastic relationships between key performance indicators.

The authors propose the following approach to the determination of target values of key performance indicators of the logical-metric model of the economic entity strategy. At the first stage of the analysis, it is necessary to identify the presence of stochastic relationships using a correlation and regression analysis. If the statistical data necessary for a factor analysis are not available in the development of the logical-metric model of an economic entity evolution, then it is recommended to use expert assessment methods to obtain values of the regression coefficients.

The target value of the resulting financial indicator (y) is set by top managers using methods of forecasting and taking into account the results of a strategic analysis. At this stage, it is necessary to define the planned percentage change in the result indicator (Δy).

The target values of the other factor scores are calculated through their corresponding planned changes needed to achieve target value y:

$$x_k = x_k\left(1+\frac{\Delta x_k}{100}\right), \tag{1}$$

where x_k is the target value of the indicator; Δx_k is the necessary planned changes in the indicator, %.

To determine the required planned change of an indicator the authors propose the following approach.

At first we determine elasticity coefficients which present the percentage change in an average effective index change in y when an x_k factor changes by one percent. These coefficients are calculated by formula

$$e_k = a_k\frac{\bar{x}_k}{\bar{y}}, \tag{2}$$

where e_k is the elasticity coefficient of factor x_k; a_k is a coefficient of multiple linear regression.

Thus, we obtain:

$$\Delta x_1 \times e_1 + \ldots + \Delta x_k \times e_k = \Delta y. \tag{3}$$

From the viewpoint of rationality and economy of resources factor indicators should vary according to the degree of their influence on the resulting indicator. To determine the percentage change of the factor indicator it is supposed to do the following:

A regression equation is transferred to the standardized form:

$$t_0 = \beta_1 \times t_1 + \beta_2 \times t_2 + \ldots + \beta_n \times t_n. \tag{4}$$

Factors and the regression coefficients are related with the respective averages and dispersions by the following relations:

$$t_k = \frac{x_k - \bar{x}}{\sigma_x}$$

$$a_k = \beta_k \frac{\sigma}{\sigma_k}$$

$$t_0 = \frac{y - \bar{y}}{\sigma_y}$$

$$a_0 = \bar{y} - a_1 \bar{x}_1 - \ldots - a_n \bar{x}_n. \tag{5}$$

β-coefficients allow one to give a comparative description of factor significance: the higher the value of a β-coefficient is, the more significant factor is in terms of its influence on the resulting indicator. Thus, the β-coefficients can be used for calculating the degree of change of the factor indicator. In our opinion, the percentage change in factor indicators should relate so as the relationship between their β-coefficients:

$$\frac{\beta_1}{\beta_k} = \frac{\Delta x_1}{\Delta x_2}. \tag{6}$$

It turns out that any change of the factor indicator can be expressed through any change (for example, through the change of the last indicator) and the ratio of the corresponding β-coefficients. Having done this, we can solve equation (3) and calculate the necessary changes of all indicators:

$$\Delta x_k = \frac{\Delta y}{\frac{\beta_1}{\beta_k} \times e_1 + \ldots + \frac{\beta_{k-1}}{\beta_k} \times e_{k-1} + e_k}. \tag{7}$$

In order to take into account the cases when the increase in the result value represents a negative change of the factor, we need to slightly transform equation (7):

$$\Delta x_k = \frac{\Delta y}{\left|\frac{\beta_1}{\beta_k} \times e_1\right| + \ldots + \left|\frac{\beta_{k-1}}{\beta_k} \times e_{k-1}\right| + |e_k|} \times \frac{|e_k|}{e_k}. \tag{8}$$

For enlarging the economic analysis, the authors recommend using the determination coefficient that describes the proportion of variation in dependent variable y, which can be explained by attributes included in the factor model and calculated as follows:

$$D = R^2,$$

where R^2 is the coefficient of multiple correlation.

The coefficient of multiple correlation is calculated by the following formula:

$$R^2 = \beta_1 \times r_{yx1} + \beta_2 \times r_{yx2} + \ldots + \beta_n \times r_{yxn}, \tag{9}$$

where r_{yxk} is the coefficient of correlation between y and x_k

According to the calculation of the determination coefficient, we can conclude that $D \times 100$ % is due to variation included in the model factors and $(1 - D) \times 100$ % is due to the activities aimed at achieving the target, the extent of which is defined as resulting indicator y.

For the purposes of testing the proposed methodology, we determine the optimal values of the KPI of the production hall of the Klintsovskiy truck crane plant.

Let us consider the index of output per shift (y), the target value for 2015, which is equal to 4.5 million rubles. In the developed logical-metric model the factor indicators are:

- an integrated indicator of product quality (x_1);
- an integrated indicator of equipment failure (x_2);
- actual implementation of replacement outfit, % (x_3);
- non-recoverable waste for recycling, % (x_4);

The values of these indicators for 2014 are presented in Table 1.

Table 1. Values of the simulated indicators for 2014.

y	x_1	x_2	x_3	x_4
3400000	3.671	2	93	0.533

The parameters of the regression equation were calculated using the method of least squares:

$$y = 231887.14 \times x_1 + 23972.1 \times x_2 + 63469.72 \times x_3 + 509053.53 \times x_4 - 3644209.78. \quad (10)$$

For determination of the percent change of the factor indicator it is necessary to calculate the β-coefficients for standardized linear regression. The results of the calculation are presented in Table 2.

Table 2. Calculation of the β-coefficients for standardized linear regression.

The dispersion of the factor indicator (σ_k)	β-coefficient		Elasticity coefficient	
	β_k	value	e	value
0.4545	β_1	0.3915	e_1	0.2348
0.4517	β_2	0.0402	e_2	0.0103
2.1665	β_3	0.5107	e_3	1.9021
0.0494	β_4	0.0935	e_4	0.0817

As it was mentioned above, the β-coefficients are used to calculate the change degree of the factor indicators, that is, to define each change of the factor indicator through the change in this index and finding the change of the factor:

$$\Delta x_1 \times \frac{\beta_1}{\beta_4} \Delta x_4 = 4.184 \times \Delta x_4$$

$$\Delta x_2 \times \frac{\beta_2}{\beta_4} \Delta x_4 = 0.4299 \times \Delta x_4$$

$$\Delta x_3 \times \frac{\beta_3}{\beta_4} \Delta x_4 = 5.4585 \times \Delta x_4 . \quad (11)$$

Substituting the values from Table 2, let us find value Δx_4:

$$\Delta x_k = \frac{\Delta y}{\frac{\beta_1}{\beta_4} \times e_1 + \frac{\beta_2}{\beta_4} \times e_2 + \frac{\beta_3}{\beta_4} \times e_3 + e_4} =$$

$$= \frac{32.35}{4.184 \times 0.2348 + 0.4299 \times 0.0103 + 5.4585 \times 1.9021 + 0.0817} = 2.8252 . \quad (12)$$

Using (11) the change of the following factor indicators can be found:

Δx_1=4.184, Δx_4=11.821

Δx_2=0.4299, Δx_4=1.214

$$\Delta x_3=5.4585,\ \Delta x_4=15.421 \tag{13}$$

Thus, the following logical-metric model can be obtained:

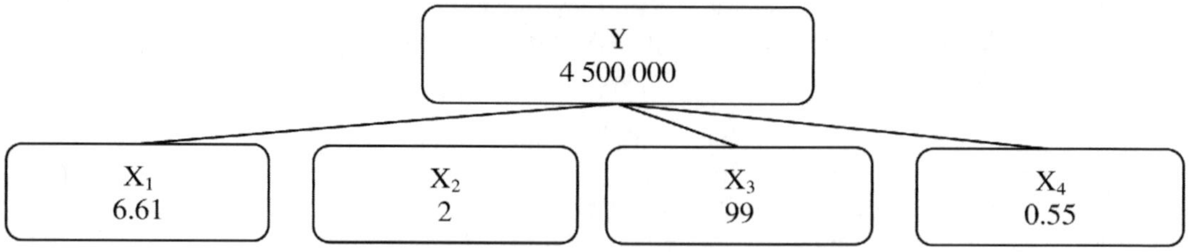

Figure 2. Optimal values of factor indicators of the Klintsovskiy truck crane plant production division

In case of functional dependency between leading indicators and high-level indicators the optimal values of leading indicators are supposed to be calculated using system dynamics methodology.

Mathematically, the basic structure of a formal system dynamics computer simulation model is a system of coupled, nonlinear, first-order differential (or integral) equations:

$$\frac{d}{dt}x(t) = f(x, p), \tag{14}$$

where x is a vector of levels (stocks or state variables), p is a set of parameters, and f is a nonlinear vector-valued function.

Basic elements of a system dynamics model are of four kinds: stocks, flows, variables and links. An application of the system dynamics approach assumes the distribution of the logical-metric model indicators for object classes mentioned above.

System dynamics has ample opportunity for simulating modeling and forecasting based on a stochastic approach that should increase the efficiency of management of the machine-building enterprise.

Further research of the existence of elements of synergetics in the management of production divisions of the machine-building enterprise makes it possible to note that fluctuation as a random deviation from the average values of physical quantities, which characterize the system of a huge amount of particles, can extend (on the results of the correlation analysis) the logical-metric model of indicators for monitoring of operating activities.

3. Conclusion

In summary, it can be noted that the proposed multivariate method of determining target values of leading indicators of the logical-metric models of development of the business entity based on the elements of synergetics should allow one to raise efficiency of management of production divisions of the machine-building enterprise.

References

[1] Malashenko V M, Kozlov D V, Kokunov A V and Putihin Y E 2015 Econometric methods and models *Tutorial* (St. Petersburg: Russia)

[2] Kazakov O D, Andriyanov S V and Kuzovleva I A 2013 Logical-metric modeling of business entity strategy *Monograph* (Bryansk, Russia)

[3] Borisov V N and Pochukaeva O V 2015 Innovative machine engineering as a factor of developing import substitution *Pleiades Publishing* **1** 225-232

[4] Kosarev O I 2014 *Journal of Machinery Manufacture and Reliability* 183-187

[5] Ganiev R F 2015 *Journal of Machinery Manufacture and Reliability* 188-210

Use of anthropometric dummies of mathematical models in the safety and comfortableness analysis of a passenger rolling stock

V Kobishchanov[1], D Antipin[2], S Shorokhov[3] and A Mitrakov[2]
[1] Bryansk State Technical University, 7, 50 years of October st., Bryansk, 241035, Russia
[2] Candidate, Ural State university of railroads, 66, Kolmogorova st., Ekaterinburg, 620034, Russia

E-mail: adya24@rambler.ru

Abstract. Approaches to the safety and comfortableness analysis of a railway passenger rolling stock with anthropometrical dummies of mathematical models usage are offered. There are recommendations about a rolling stock design, based on the analysis of traumatism of passengers and members of train crews, and also based on comfort parameters at various modes of trainmovement.

1. Introduction

Safety and comfort of civil passenger traffic is one of the key problems of railway transport functioning. The present stage of Russian railroads development is characterized by modernization of technical means and implementation of a high-speed rolling stock. That is the result of acceptance of railway transport development strategy till 2030 by the Russian Federation government in 2008.
Implementation of the rolling stock of a new generation into operation could be the reason of increased risks of emergency situation, connected with longitudinal train collisions with obstacles. In this regard, development of high-speed railway traffic requires laboration of actions to enhance safety and comfort conditions of civil passenger traffic.

2. Results and Discussion

Now there are two main approaches in assessment of the passenger rolling stock safety in emergency situations. They are carrying out natural experiments with the use of physical models of anthropometrical dummies and mathematical modeling of emergency situations scenarios [1]. The implementation of the first approach has not been widely adopted because of the high cost of preparation and carrying out passenger rolling stock tests.

With the development of science and equipment, the possibility of creation of hi-tech physical models of human dummies became possible. These dummies have the same form and mass distribution as an alive human being does. The dummy consists of the separate elements joint with hinges capable of accurate reproducing man's body movements in the collision conditions. Anthropometrical dummies like Dummy Hybrid [2], SID, BioSID [3], BioRID [4] are used for vehicle safety assessment in world practice. They allow estimating the degree of a dynamic impact on passenger sand members of train crews with a high degree of adequacy, both during rolling stock movement and emergency situations.

The assessment of person traumatizing levels is based on the following criteria usage: craniocerebral injury, cervical spine traumatizing, chest and hip traumatizing [2].

A craniocerebral injury criterion is determined in accordance with

$$HIC = (t_2 - t_1)\left(\frac{1}{t_2 - t_1}\int_{t_1}^{t_2} a\,dt\right)^{2.5},$$

where t_1, t_2 – time points, measured in seconds. They define an interval between the initial and the final moments of the head contact with aninjuring object, where a HIC value is maximum; a – is complex acceleration, evaluated in G-units.

A cervical spine traumatizing criterion is determined in accordance with

$$N_{ij} = \frac{F}{F_k} + \frac{M}{M_k},$$

where F – axial compression / stretching force; F_k – axial force, used for normalization; M – bending moment; M_k – bending moment, usedfor normalization.

A chest traumatizing criterion is determined in accordance with

$$CTI = \frac{A_{max}}{A_k} + \frac{D_{max}}{D_k},$$

where A_{max} – maximum acceleration; A_k – acceleration, used for normalization; D_{max} – chest pick deformation; D_k – deformation, used for normalization.

A hip traumatizing criterion is determined in accordance with compression pressure that is imparted axially to each dummies hip.

Physical dummies are also used for the specified assessment ofthe passenger comfort level in the course of motion, particularlyfor trains with compulsory inclination of a body in curve tracks.

WithinCEN 12299 the average comfort coefficient is applied to definition of passengers comfort on tangent track:

$$N_{MV} = 6 \cdot \sqrt{\left(a_{XP}^{Wd}\right)^2 + \left(a_{YP}^{Wd}\right)^2 + \left(a_{ZP}^{Wd}\right)^2},$$

where a_{XP}^{Wd} – accelerations, affecting a passenger in the longitudinal direction; a_{YP}^{Wd} – accelerations, affecting a passenger in the broadside direction; a_{ZP}^{Wd} – accelerations, influencing a passenger in the vertical direction.

This dependence is used for specification of the percent of passengers experiencing discomfort:

$$P_{CT} = 100\left\{max\left[A \cdot |\ddot{y}_{ls}|_{max} + B|\dddot{y}_{ls}|_{max} \cdot C; 0\right] + \left(D \cdot |\dot{\varphi}_{ls}|_{max}\right)^E\right\},$$

where $\ddot{y}_{ls}$ – broadside acceleration of the car body, m/s^2; $\dddot{y}_{ls}$ – changes of broadside acceleration of the car body per 1 sec, m/s^3; $\dot{\varphi}_{ls}$ – angular speed, rad/sec; A, B, C, D, E – constant, accepted in accordance with CEN.

It's important to consider theinfluence of motion sickness estimated by a motion sickness dose for a rolling stock, equipped with a system of car body constrained inclination

$$MSDV_z(t) = k_{MSDV}\sqrt{\int_0^t a_{wf}^2(t) \cdot dt},$$

where a_{wf} – frequency-weighted vertical acceleration, m/s^2; $k_{MSDV} = 1/3$ – coefficient for the mixed non-adaptablepublic of the adult women and men, they most probably will be influenced by the motion sicknesseffect.

A discomfort level can be also predicted in accordance with the dose of motion sickness. The value of the 'discomfort level' is graduated on a scale from 0 ('I feel well') to 3 ('I feel absolutely awful') and is determined by the following dependence:

$$IR(t) = \frac{MSDV_z(t)}{50}.$$

Specialists of the 'Railway rolling stock' chair of Bryansk state technical universityhave developedasolid-state computer modelof an anthropometrical dummy of Hybrid III 50th Percentile Male [5] in the area of the program complex of kinematics and dynamics modeling of bodies systems of 'Universal mechanism'. During modeling the dummy was divided into the following elements: a head, a neck, a top part of a trunk, a shoulder, a hip, etc., joint by means of hinges in one model. All elements of the dummy were simulated from rigid bodies with the real weight and geometrical characteristics corresponding to anthropometrical characteristics of the 'average' adult man.

The safety assessment of home car model 61-4440 and electric train ED4M (Figure 1) was carried out using the developed computer model of the anthropometrical dummy.

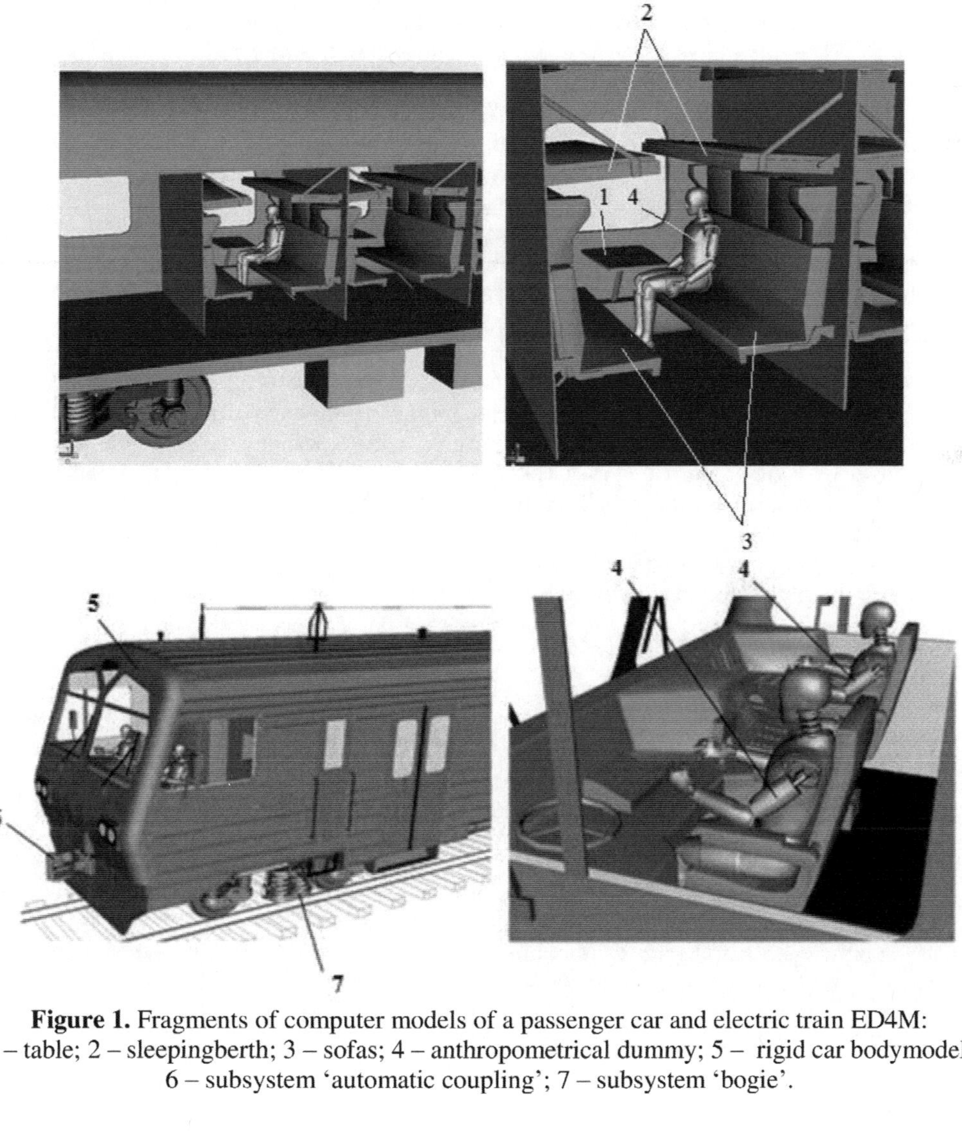

Figure 1. Fragments of computer models of a passenger car and electric train ED4M:
1 – table; 2 – sleepingberth; 3 – sofas; 4 – anthropometrical dummy; 5 – rigid car bodymodel; 6 – subsystem 'automatic coupling'; 7 – subsystem 'bogie'.

For this purpose the computer models of train impacts with the corresponding obstacles [6] were carried out in the area of the program complex 'Universal Mechanism'. Characteristics of impact models corresponded to standard documentation of design of rolling stock mechanical safety systems for civil passenger traffic and were used as test scenarios for its safety assessment.

The comfortableness assessment of a home electric train with systemsof car body constrained inclination in curve tracks was also carried out (Figure 2).

Figure 2. A solid-state computer model of an electric train:
1 – subsystem 'head car'; 2 – subsystem 'trailing car';
3 – subsystem 'railmotorcar'; 4 – solid-state anthropometrical dummy

3. Conclusion

The oscillograms of parameters of dynamic interaction of dummies with interior elements of the car and a control cabinet of an electric train were received as a result of train emergency impact modeling with obstacles. There is traumatizing criteria calculation in accordance with obtained values of dynamic efforts. These values are compared with rated ones, expressing the allowed levels of man traumatizing in emergency situation. The analysis of these results has allowed revealing the most injury-causing elements of rolling stock interior considered units and make the conclusion about their safety.

The parameters of rolling stock comfortableness have been determined in accordance with the given dependences during modelingof electric trainmovementalong various railway tracks. The obtained results have been compared with the rated values defining comfortable conditions of passenger travel of various age categories.The results obtainedfrom the safety and comfortableness analysis of the passenger rolling stockcan be used in developing technical solutions of trains, locomotives. These researches are vital for the enhancement of electric trains mechanical safety in case of emergencyimpacts with obstacles, as well as for the improvement of transportationcomfort with regard to a possible train speedincrease.

References

[1] Caroline VanIngen-Dunn 2003 Commuter Rail Seat Testing and Analysis of Facing Seats. *DOT/FRA/ORD-03/06, U.S. Department of Transportation, Washington, DC* p 195

[2] Kleinberger M, Sun E, Eppinger R, Kuppa S and Saul R 1998 *Development of Improved Injury Criteria for the Assessment of Advanced Automotive Restraint Systems* p 120

[3] Fildes B, LaneJ, Lenard J and Vulcan A 1994 *Passenger cars and Occupant injury: side impact crashes*p 120

[4] Zuby D, Farmer C and Avery M 2003 The influence of crash pulse shape on BioRID response. *Proceedings of the 2003 IRCOBI Conference* pp 327–341

[5] Kobishanov V, Mihal'chenko G, Tihomirov V, Fedyaeva G, Antipin D and Shorohov S 2013 *World Applied Sciences Journal* **24** 86–90

[6] Kobishanov V, Lozbinev V, Sakalo V, Antipin D, Shorohov S and Vysocky A 2013 *World Applied Sciences Journal* **24** 208–212

MEACS2015 IOP Publishing
IOP Conf. Series: Materials Science and Engineering **124** (2016) 012066 doi:10.1088/1757-899X/124/1/012066

Model of user identification of electronic informational-educational environment

N M Kovalnogova, S S Sokolov and A P Nyrkov
Admiral Makarov State University of Maritime and Island Shipping, 5/7, Dvinskaya st., Saint-Petersburg, 198035, Russia

E-mail: SokolovSS@gumrf.ru

Abstract. This article describes the electronic informational-educational environment (EIEE) of a university, components of IEEE, categories of users, a model of user identification.

1. Introduction

Today, E-learning and Distance Learning Technologies (hereinafter DLT) are widely used in world practice. In Russia the background of E-learning and DLT application is regulated by a number of the legal regulations.

According to Federal Law №273, 'The Education Act of the Russian Federation' of 09.12.2012, when implementing educational programs, the university uses different educational technologies including E-learning and DLT. The university should provide suitable conditions for the functioning of the electronic information-educational environment (hereinafter EIEE) while implementing educational programs applying E-learning and DLT. The university EIEE includes electronic information resources, electronic educational resources, a complex of information and telecommunication technologies and technological resources. Besides, it provides a full-scale mastery of educational programs by students irrespectively of the students' location.

Thus the main purpose of the university EIEE establishment is the information support of the educational process in accordance with the requirements for the educational programs implementation.

The information support of the educational process means the following:

- provision of a day-and-night access (via the Internet) to educational materials;
- provision of the opportunities for students to do self-study exercises, perform tasks, prepare for certification;
- provision of opportunities for teachers to render the professional mentoring support for students (through various means of communication);
- organization of the information exchange between lectors and students;
- publication of information for students (about checks and examinations, ways of reporting results);
- storage and structuring of educational material;
- provision of the electronic document flow;
- management of the educational process, etc.

To achieve these objectives, the university EIEE should include many components, which have different software/hardware platforms supporting different data transmission technologies, support different standards for creation and processing of documents. These university EIEE components can be used within the local network, geographically-distributed network, or on the Internet.

The university EIEE includes the following components: web-sites supporting different Content management systems (CMS), Learning Management Systems (LMS), Learning Content Management Systems (LCMS), Electronic Library Systems (ELS), control automation systems of the educational process, different Enterprise Resource Planning (ERP) systems or Product Lifecycle Management (PLM) systems, Customer Relationship Management (CRM) systems, Decision Support Systems (DSS) and etc.

All these EIEE components can be grouped according to architecture, a degree of automation, a data processing nature, a scope of application, an accessibility level of the information they contain.

EIEE users are divided in two categories: internal and external. The internal category includes staff, lectors, students (which in their turn are divided into students of different levels and forms of education). External users are applicants (enrollee), parents, partners, supervisory organizations (government), etc. The same users may have different roles (defining their rights) in the university EIEE components or may not have any roles at all.

2. The model of user identification

To construct a model of user identification the following notations are introduced into EIOS:

1. C_i – *i*-class of EIEE components.
2. S_{ij} – *j*-component in *i*-class.
3. $a_{l_{ij}}$ – *l*-ID attribute of the user definable in a j-component of an *i*-class.
4. $A_{ij} = \{a_{1_{ij}}, a_{2_{ij}}, ..., a_{l_{ij}}\}$ - set of identity attributes of the user definable for a *j*-component of an *i*-class.
5. $A = \{A_{11}, A_{12}, ..., A_r, A_{21}, A_{22}, ..., A_{2r}, ..., A_{q1}, A_{q2}, ..., A_{qr}\}$ – set of all identity attributes for all the EIEE components of all classes.

The user may be authenticated by presenting at least one of the following entities:

- something that the user knows (a password, personal ID, a cryptographic key and etc.);
- something that the user owns (a personal card or other device of a similar purpose);
- something that is a part of the user (voice, fingerprints, etc., that is, one's own biometrics).

For each attribute $a \in A$ domain V_a is defined.

$2^{|V_a|}$ – the total number of possible domain values of attribute *a*.

$V = \bigcup_{a \in A} V_a$

Thus the model of user identification can be represented by couple (A,V).

A set of tuples K_a will be treated as a set being is a direct product of attribute values domains $\{a_1, a_2, ..., a_n\}$:

$$K_A = \prod_{i=1}^{n} V_{a_i} \tag{1}$$

Let $D_m \subseteq A$ be as a set of attributes determining a separate *m*-category of users, and K_{D_m} be a set of tuples of this category.

This way there is a following rule:

$$f : K_{D_m} \to K_A \tag{2}$$

representing a mapping of the set of tuples K_{D_m} on K_A, which gives (associates) a unique element of the set of K_A to each element of the set of K_{D_m}.

Taking into account the necessity and importance of user identification by a specific attribute, which depends on the user's category, the class of EIEE components and the component itself, the domain value that is included into any tuple from set K_{D_m} can have three levels of definition:

1. Explicitly defined.

2. Has value 'null' not defined at the moment.
3. Has value 'x', that is, should not be defined at all.

So, equation

$$K_{D_m} = \left\{ K_{D_m}\left(V_{D_m}\right) | V_{D_m} \in V, D_m \subseteq A \right\} \tag{3}$$

can be viewed as a decomposition of set K_A, which will determine the user category with respect to the components of EIEE and methods of user identification. The attribute values of the model of user identification can change during EIEE development. They can become outdated (can be removed), can be updated (software update of the platform), new components can appear, the types of identification can be changed. In this case, it will be appropriate to update the classification of the components.

In addition to user identification by a particular attribute the identification can happen by a number of alternative signs.

Let a_i and a_{i^a} be pairwise alternative signs, a_j be signs having no alternative analogues, then true identification *True(I)* is possible in case of validity of the following expression:

$$True(I) \equiv \left(a_1 \& a_2 \& \ldots \& a_n\right) \& \left(\left(a_1 \vee a_{1^a}\right) \& \left(\left(a_2 \vee a_{2^a}\right)\right) \& \ldots \& \left(a_m \vee a_{m^a}\right)\right), \tag{4}$$

where n – number of signs not having alternative analogues, m – number of signs having alternative analogues.

Let us make a partition of EIEE into two nonintersecting and independent sets of classes of components (it can be justified including a separation of the information of the open and restricted access) according to the following rules:

1. The first set should consist of pairwise incompatible classes of components.
2. The second set should consist of mutually independent classes of components.

In this case, the total probability of true identification will be determined by expression:

$$P = \sum_{i=1}^{q_1} \left[\sum_{j_1=1}^{n_1} P_{k_{ij1}}(a) + \prod_{j_2=1}^{m_1} P_{k_{ij1}}(a) \right] + \prod_{i=1}^{q_2} \left[\sum_{j_3=1}^{n_2} P_{k_{ij3}}(a) + \prod_{j_4=1}^{m_2} P_{k_{ij4}}(a) \right], \tag{5}$$

where n_1 – the number of pairwise incompatible classes of components in the first set involved in identification, m_1 – the number of independent classes of components in the aggregate in the first set involved in identification, n_2 – the number of pairwise incompatible classes of components in the first set involved in identification, m_2 – the number of independent components in the aggregate in the first set, involved in identification

3. Testing of the model

The model of user identification was tested in one of the Admiral Makarov State Universities of Maritime and Inland Shipping Campuses (in the laboratory of electronic information-educational technologies named after prof. U.M. Kulibanov).

Testing took place in the form of an experiment, within the framework of which three main user groups were chosen: students (three student groups comprising 60 people), lectors (8 people working with these three groups of students in this laboratory) and the staff (5 people including a director of the laboratory and administrators of the EIEE components, who provide a technical support).

As part of the experiment we examined the work of the users with the following EIEE components:

- 54 PC (in the laboratory);
- 1 mobile access device;
- 3 university web-sites (access via the Internet);
- 1 Learning Management System (distant learning system, hereinafter DLS 'FARVATER') (access via the Internet);
- 1 Learning Content Management System ('Educational portal') (access via the Internet);
- 1 Electronic Library System (ELS) (the access to ELS is only within university campus LAN; the access to the ELS web-catalog is via the Internet);

- 1 automation system of the educational process management (access is only within university campus LAN);
- 1 decision support system (the client program working in laboratory LAN);

Due to the fact that the selected categories of users have different access rights to selected EIEE components, they were assigned to the methods of authorization of different complexity using various means of authentication.

The base of the experiment was represented by a matrix of allocation of access rights to EIEE components for different users categories under a predetermined set of attributes taking into account the fact that a specific value from the set is predefined for each user/attribute pair (the attribute value is specified, the attribute value is not specified, the attribute value should not be specified).

Thus temporary guest accounts were created for the students working with PCs, temporary personal accounts – for lectors, permanent staff personal accounts – for staff. Furthermore, in order to confirm the authenticity, the staff have to produce a special key, because a hardware-software module of the trusted boot (or a fingerprint when accessing from a mobile device) was installed on PCs belonging to the staff. A fingerprint attribute for access was selected to increase the number of possible user attributes.

In LMS (DLS 'FARVATER') and LCMS ('Educational portal') special requirements for the password complexity were introduced in order to ensure data security. The complexity includes the use of a minimum number of characters, the use of the letters of different cases, the simultaneous use of letters, figures and special symbols. The level of the password complexity and the frequency of its update depend on the category of users: the highest level is for the staff (administrators of the EIEE components, a tech support team), a lower level is for lectors, and the lowest is for students.

The results of the experiment have shown that the presented model of user identification is adequate and can solve the assigned tasks of:

- forming EIEE users categories;
- partition of EIEE into various locally isolated segments for discrete identification;
- building of a transparent system of access isolation (including a mandatory access control);
- preliminary inspection of compliance (before implementation) of the taken measurements on information security to the claimed requirements (taking into account the necessity of information rating).

4. Conclusion

Improving the quality of education and ensuring equal access to educational resources and services for all citizens are the main aims of educational system today. Development of EIEE and forming mechanisms and prerequisites for the integration of information technology into the everyday educational and scientific practice is a key of integration to an information society.

Information security of EIEE of the educational institution is very important: the number of threats is growing every day, the number of EIEE components is changing, software platforms become outdated.

Data security of EIEE components is provided by EIEE components of the proper software and hardware as well as by organizational and technical measures. The basis for information protection against an unauthorized access is implementation of the principles of access control to resources.

The provision of the correct implementation of the differentiating policy of access to resources is impossible without correct identification of users.

The application of the identification model allows users to control the use of EIEE components.

References

[1] Nyrkov A, Belousov A and Sokolov S 2015 *J Modern Appl. Sci.* **9** 162–76
[2] Chernyi S, Zhilenkov A, Sokolov S and Titov L 2015 *J Metallurg. and Min. Ind.* **7** 66–73
[3] Nyrkov A, Sokolov S, Chernyi S and Mamunts D 2015 *J Metallurg. and Min.* **7** 521–24

[4] Chernyi S 2015 *J Metallurg. and Min.* **7** 60–65
[5] Kovalnogova N and Sokolov S 2015 Electronic informational-educational environment of university complex *Maritime educ.: trad., realities and prosp.* 157–63
[6] Pastuscshak T, Sokolov S, Kovalnogova N and Mankovskaya A 2014 *J Inf. Tech.* **9** 40–46
[7] Boriev Z, Sokolov S and Nyrkov A 2015 *J Materials Science and Engineering* **91**

IOP Conf. Series: Materials Science and Engineering **124** (2016) 012067 doi:10.1088/1757-899X/124/1/012067

Provision of Controlled Motion Accuracy of Industrial Robots and Multiaxis Machines by the Method of Integrated Deviations Correction

O N Krakhmalev, D I Petreshin and O N Fedonin
Bryansk State Technical University, 7, 50 years of October st., Bryansk, 241035, Russia

E-mail: atsys@tu-bryansk.ru

Abstract. There is a developed method of correction of the integrated motion deviations of industrial robots and multiaxis machines, which are caused by the primary geometrical deviations of their segments. This method can be used to develop a control system providing the motion correction for industrial robots and multiaxis machines.

1. Introduction

It is impossible to avoid deviations in the geometrical size and form, known as primary deviations, when producing and assembling parts and units of segments entering into the composition of such complex mechanical systems as industrial robots. These geometrical deviations result in a difference of the motion parameters of a real industrial robot from the motion parameters of its nominal model made in accordance with the design documents (3-D model). To improve the accuracy of controlled motions it is necessary to make corrections of the nominal model of an industrial robot taking into account measuring primary geometrical deviations of its segments [1].

2. A Mathematical Model of Primary Deviations

Primary deviations in the geometrical dimensions of manipulator system segments (Figure 1) can be taken into account by introduction of linear deviations on relevant coordinate axes $\delta_x^{(i)}, \delta_y^{(i)}, \delta_z^{(i)}$, and primary deviations in segment forms and joint cocking – by introduction of the angular deviations between relevant coordinate axes: $\alpha_{xx}^{(i)}, \alpha_{xy}^{(i)}, \alpha_{xz}^{(i)}, \alpha_{yx}^{(i)}, \alpha_{yy}^{(i)}, \alpha_{yz}^{(i)}, \alpha_{zx}^{(i)}, \alpha_{zy}^{(i)}, \alpha_{zz}^{(i)}$ (Figure 2).

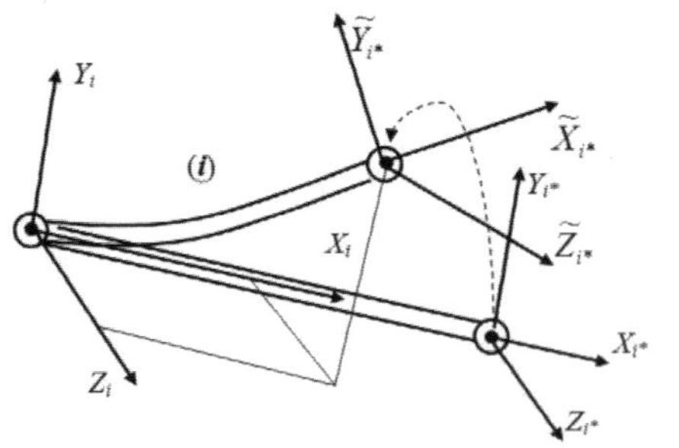

Figure 1. Primary linear deviations.

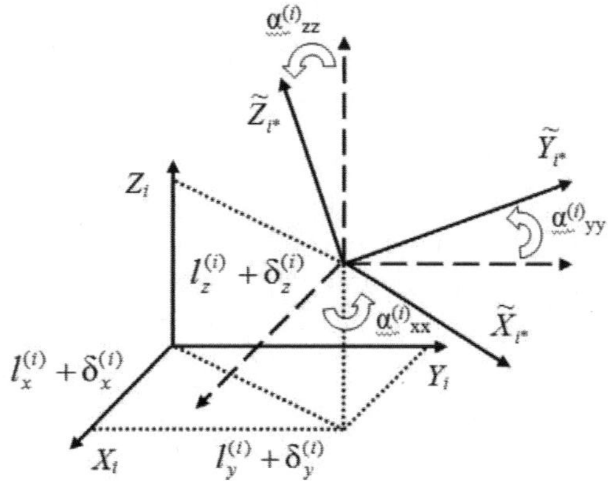

Figure 2. Primary angular deviations.

Matrix Ai,i* that shows a transformation of homogeneous coordinates from coordinate system Si* into coordinate system Si, considering primary linear and angular deviations of the segment, will represent a mathematical model of primary deviations.

$$\tilde{A}_{i,i^*} = \begin{bmatrix} \cos(\alpha_{xx}^{(i)}) & \cos(\alpha_{xy}^{(i)}) & \cos(\alpha_{xz}^{(i)}) & l_x^{(i)} + \delta_x^{(i)} \\ \cos(\alpha_{yx}^{(i)}) & \cos(\alpha_{yy}^{(i)}) & \cos(\alpha_{yz}^{(i)}) & l_y^{(i)} + \delta_x^{(i)} \\ \cos(\varphi\alpha_{zx}^{(i)}) & \cos(\alpha_{zy}^{(i)}) & \cos(\alpha_{zz}^{(i)}) & l_z^{(i)} + \delta_x^{(i)} \\ 0 & 0 & 0 & 1 \end{bmatrix} = \begin{bmatrix} \overset{(3\times3)}{\tilde{M}_i} & \overset{(3\times1)}{\tilde{L}_i} \\ 0 & 1 \end{bmatrix}. \quad (1)$$

Character ~ in expression (1) denotes that the matrix contains primary geometrical deviations.

When drawing up the matrix it is unnecessarily to measure all nine angular deviations. It suffices to measure only three of them. The remaining six angular deviations can be calculated. For this it is necessary to solve six nonlinear equations. These equations represent orthonormal vectors constituting rows and columns of the matrix.

Besides, positioning deviations influence the accuracy of multilink mechanisms. Positioning deviations are associated with deviations of generalized coordinates (Figure 3).

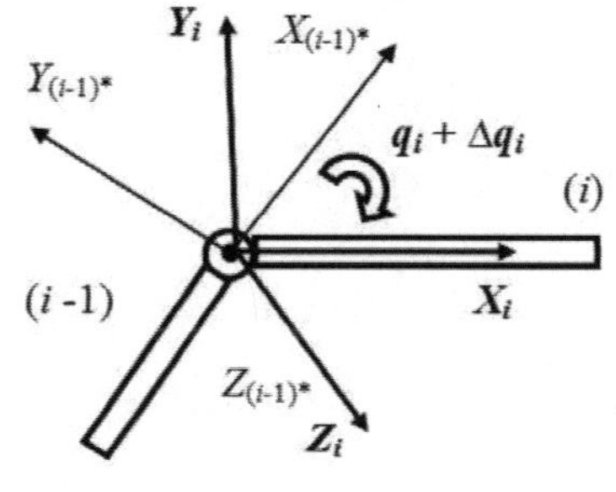

Figure 3. Positioning deviations

The matrix taking into account the positioning deviations has the following view:

$$A_{(i-1)^*,i} = \begin{bmatrix} \cos(\beta_i(q_i+\Delta q_i)) & -\sin(\beta_i(q_i+\Delta q_i)) & 0 & 0 \\ \sin(\beta_i(q_i+\Delta q_i)) & \cos(\beta_i(q_i+\Delta q_i)) & 0 & 0 \\ 0 & 0 & 1 & (1-\beta_i)(q_i+\Delta q_i) \\ 0 & 0 & 0 & 1 \end{bmatrix}, \tag{2}$$

where q_i – generalized coordinate, Δq_i – positioning deviation for the i-th generalized coordinate, β_i – coefficient ($\beta_i = 1$ if q_i – angular; $\beta_i = 0$ if q_i – linear).

A mathematical model represents a matrix multiplication:

$$A_{(i-1),i} = A_{(i-1),(i-1)^*} A_{(i-1)^*,i},$$

$$A_{0,k} = A_{0,1} A_{1,2} \cdot\cdot A_{(i-1),i} A_{i,(i+1)} \cdot\cdot A_{(k-1),k} = \prod_{i=1}^{k} A_{(i-1),i}.$$

3. An Alternative Model

Instead of the mathematical model described by matrix (1), an alternative model can be offered. Let us consider the model in Figure 4.

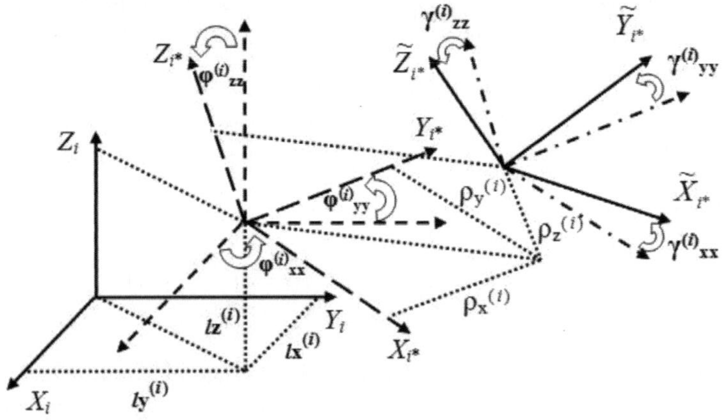

Figure 4. An alternative model.

$$\tilde{A}_{i,i^*} = A_{i,i^*}(\varphi, l)\, \tilde{A}_{i^*,i^*}(\gamma, \rho),$$

$$\tilde{A}_{i,i^*} = \begin{bmatrix} \cos(\varphi^{(i)}_{xx}) & \cos(\varphi^{(i)}_{xy}) & \cos(\varphi^{(i)}_{xz}) & l^{(i)}_x \\ \cos(\varphi^{(i)}_{yx}) & \cos(\varphi^{(i)}_{yy}) & \cos(\varphi^{(i)}_{yz}) & l^{(i)}_y \\ \cos(\varphi^{(i)}_{zx}) & \cos(\varphi^{(i)}_{zy}) & \cos(\varphi^{(i)}_{zz}) & l^{(i)}_z \\ 0 & 0 & 0 & 1 \end{bmatrix} \begin{bmatrix} \cos(\gamma^{(i)}_{xx}) & \cos(\gamma^{(i)}_{xy}) & \cos(\gamma^{(i)}_{xz}) & \rho^{(i)}_x \\ \cos(\gamma^{(i)}_{yx}) & \cos(\gamma^{(i)}_{yy}) & \cos(\gamma^{(i)}_{yz}) & \rho^{(i)}_y \\ \cos(\gamma^{(i)}_{zx}) & \cos(\gamma^{(i)}_{zy}) & \cos(\gamma^{(i)}_{zz}) & \rho^{(i)}_z \\ 0 & 0 & 0 & 1 \end{bmatrix} =$$

$$= \begin{bmatrix} M_i(\varphi) & L_i(l) \\ 0 & 1 \end{bmatrix} \begin{bmatrix} \tilde{M}_i(\gamma) & \tilde{L}_i(\rho) \\ 0 & 1 \end{bmatrix} = \begin{bmatrix} \tilde{M}_i & \tilde{L}_i \\ 0 & 1 \end{bmatrix}. \tag{3}$$

Angles $\varphi(i)$ are set by the initial configuration of the multilink mechanism, and angles $\gamma(i)$ reflect the primary deviation of the links form and distortions in the joints. Linear deviations $\rho(i)$, used in model (3) are associated to previously entered linear deviations $\delta(i)$ by relations

$$\rho^{(i)}_x \cos(\varphi^{(i)}_{xx}) + \rho^{(i)}_y \cos(\varphi^{(i)}_{xy}) + \rho^{(i)}_z \cos(\varphi^{(i)}_{xz}) = \delta^{(i)}_x,$$

$$\rho_x^{(i)}\cos(\varphi_{yx}^{(i)})+\rho_y^{(i)}\cos(\varphi_{yy}^{(i)})+\rho_z^{(i)}\cos(\varphi_{yz}^{(i)})=\delta_y^{(i)},$$
$$\rho_x^{(i)}\cos(\varphi_{zx}^{(i)})+\rho_y^{(i)}\cos(\varphi_{zy}^{(i)})+\rho_z^{(i)}\cos(\varphi_{zz}^{(i)})=\delta_z^{(i)}.$$

As in case of (1) when drawing up matrix (3) is necessary to measure three angular deviations.

4. A Method of Correction of Integrated Deviations

Mathematical models considering primary deviations of segment geometrical parameters in complex mechanical systems with a daisy-chain structure can be used to determine the integrated deviations of segment positions arising because of the error accumulation in the positions of each previous segment of a kinematic chain. An integrated deviation of a complex mechanical system should be defined as the guidepath deviations of segment characteristic points and the segment orientation deviations from the specified orientation caused by primary geometrical deviations (Figure 5).

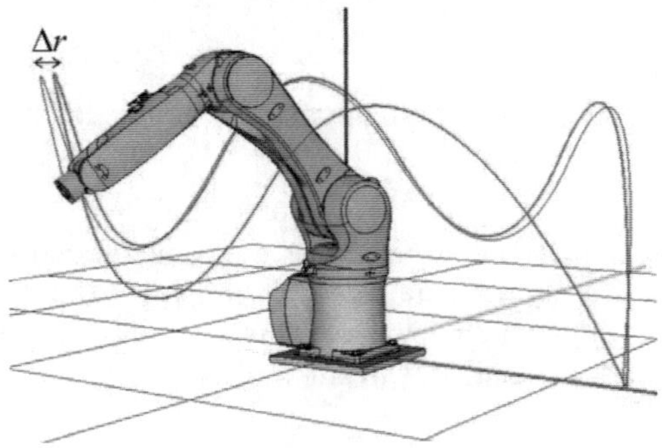

Figure 5. An illustration of integrated deviations

Taking into account primary geometrical deviations, guidepath integrated deviation $\Delta r(q)$ can be defined from the expression which represents a difference between the actual position defined by radius vector $\tilde{r}$, and the program position defined by radius vector r.

$$\Delta r = \tilde{r} - r, \quad \tilde{r} = \tilde{A}_{0,n} r_o,$$

$$\Delta r(q) = \left(\tilde{A}_{0,n} - A_{0,n}\right) r_o = \left[i_0^T \left(\tilde{A}_{0,n} - A_{0,n}\right) r_o \quad j_0^T \left(\tilde{A}_{0,n} - A_{0,n}\right) r_o \quad k_0^T \left(\tilde{A}_{0,n} - A_{0,n}\right) r_o \right]^T,$$

$$i_0^T = [1\,0\,0\,0], \quad j_0^T = [0\,1\,0\,0], \quad k_0^T = [0\,0\,1\,0],$$

where $A_{0,n}$ and $\tilde{A}_{0,n}$ – matrices of nominal and corrected mathematical models, q –generalized coordinates. An integrated deviation, connected with the position defined by vector $\Delta e(q)$ can be determined in a similar way. Integrated position deviations can be determined by complex vector $[\Delta r, \Delta e]T$. the use of the nominal $A_{0,k}$ and corrected $\tilde{A}_{0,k}$ mathematical models allows expressing conditions of the integrated deviation compensation as follows:

$$\begin{Bmatrix} \Delta r(q) \\ \Delta e(q) \end{Bmatrix} = 0 \quad \rightarrow \quad \tilde{q}.$$

The solution of this set of nonlinear equations corresponding to zero value of the complex vector of integrated deviations, provides necessary laws of motion $\tilde{q}(t)$ of a particular actuator of industrial robots or multiaxis machines (Figure 6).

Applying the mathematical support to realize this method of correction in the automatic control systems of industrial robots or multiaxis machines allows improving an enduring accuracy.

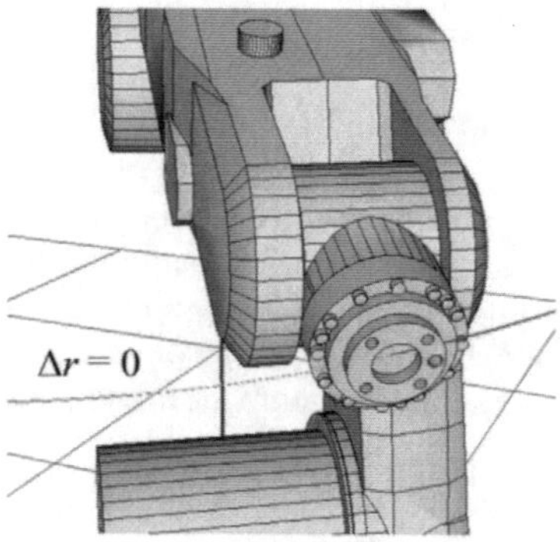

Figure 6. Correction of integrated deviations.

5. Conclusion

Mathematical models (1), (2) and (3) may be used in the parametrization methods of geometric models of multilink mechanisms. Applying mathematical support to realize the suggested method of correction in the systems of automatic control of industrial robots and multiaxis machines allows one to improve enduring accuracy. A detailed realization of the method of integrated deviations correction is given in the papers [2-7].

References

[1] Krakhmalev O N 2015 *Jour New Technologies* **16**(7) 491–496
[2] Krakhmalev O N , Petreshin D I and Fedonin O N 2015 *Joint fund commentaries of electronic resources. Science and Education* **72**(5) 34
[3] Krakhmalev O N , Petreshin D I and Fedonin O N 2015 *Joint fund commentaries of electronic resources. Science and Education* **72**(5) 35
[4] Krakhmalev O N , Petreshin D I and Fedonin O N 2015 *Joint fund commentaries of electronic resources. Science and Education* **72**(5) 36
[5] Krakhmalev O N and Bleyshmidt L I 2014 *J. of Machinery Manufact. and Reliability* **43**(1) 22
[6] Moriwaki T 2008 *CIRP Annals – Manufacturing Technology* **57** 736–749
[7] Schwenke H, Knapp W, Haitjema H, Weckenmann A, Schmitt R and Delbressine F 2008 *CIRP Annals – Manufacturing Technology* **57** 660–675

Dynamic Models of Robots with Elastic Hinges

O N Krakhmalev
Bryansk State Technical University, 7, 50 years of October st., Bryansk, 241035, Russia

E-mail: olegkr64@mail.ru

Abstract. Two dynamic models of robots with elastic hinges are considered. Dynamic models are the implementation of the method based on the Lagrange equation using the transformation matrices of elastic coordinates. Dynamic models make it possible to determine the elastic deviations from programmed motion trajectories caused by elastic deformations in hinges, which are taken into account in directions of change of the corresponding generalized coordinates. One model is the exact implementation of the Lagrange method and makes it possible to determine the total elastic deviation of the robot from the programmed motion trajectory. Another dynamic model is approximated and makes it possible to determine small elastic quasi-static deviations and elastic vibrations. The results of modeling the dynamics by two models are compared to the example of a two-link manipulator system. The considered models can be used when performing investigations of the mathematical accuracy of the robots.

1. Introduction

Let us consider that in robots the elastic elements are conventionally accumulated in the joint nodes of links – hinges. The hinge rigidity can be determined from rigidities of actual elastic elements of the drive established in this hinge by means of its reduction to one of the methods known in the applied theory of elastic vibrations.

Elastic elements are deformed under the effect of static and dynamic loads, which results in an actual motion law that will differ from the programmed one. We denote the magnitude of the deviation of the vector of generalized coordinates from the programmed motion as $\{\Delta q\} = \{\Delta q(t)\}$.

The equation of motion of the manipulator systems of robots allowing for the elastic compliance, which is taken into account in varying the direction of the generalized coordinate corresponding to this hinge, can be derived from [1] by the force of substitution of vector $\{q(t)\}$ by vector $\{q(t) + \Delta q(t)\}$.

$$[M_i(q+\Delta q)]\{\ddot{q}+\Delta\ddot{q}\}+\{\dot{q}+\Delta\dot{q}\}^T[C_i(q+\Delta q)]\{\dot{q}+\Delta\dot{q}\}=Q_{Di}(q+\Delta q)+Q_{Gi}(q+\Delta q)+Q_{Fi}(q+\Delta q),\ i=(1,\ldots,n). \quad (1)$$

In this expression Q_{Di} is the generalized force, which corresponds to the forces developed by an i-drive; Q_{Gi} and Q_{Fi} *are* the generalized forces, which correspond to the force of gravity and external forces.

Equation (1) should be supplemented by the equation associating the forces developed by drives Q_{Di} with elasticity forces, which appear in transmission mechanisms of these drives. If the mass of elastic elements can be neglected and the elastic force linearly depends on Δq, then we have the expression

$$Q_{Di} = w_i \Delta q(q,t),\ i = (1,\ldots,n), \tag{2}$$

where w_i is a stiffness coefficient of an i-hinge.

Equation (2) adds Equation (1) to the set of equations resolvable relatively all generalized coordinates $\{q\}$ and $\{\Delta q\}$. Equations (1) and (2) jointly constitute the mathematical model for modeling the dynamics of manipulator systems of robots with elastic hinges by the composite variables, which are the sum of rigid $\{q\}$ and elastic $\{\Delta q\}$ generalized coordinates.

If rigidities of elastic hinges are sufficiently large, then elastic deformations that appear in them and determine elastic coordinates $\{\Delta q\}$ can be considered as small quantities. In this case, when calculating each composite variable, we are forced to sum the quantities of different orders of magnitude in Equation (1). We can consider that another disadvantage of Equation (1) is the fact that this equation contains slow and rapid variables. The variables, which determine the programmed motion of the manipulator system, belong to a slow group of variables; while the variables, which reflect small elastic vibrations, can be referred to a rapid group. This leads to the fact that when numerically integrating differential Equation (1), we are forced to calculate high-frequency components at long time intervals with a small step.

2. Creating a dynamic model

Based on the assumption of smallness of elastic deformations, we can derive the mathematical model without mentioned disadvantages of the mathematical model. For this purpose, let us expand matrices $[M_i]$ and $[C_i]$ in the vicinity of the programmed motion into the Taylor series at $\Delta q \to 0$ retaining the summands of this series to the first infinitesimal order inclusively:

$$[M_i](q+\Delta q) = [M_i](q) + \sum_{l=1}^{n} \frac{\partial [M_i]}{\partial q_l} \Delta q_l,$$

$$[C_i](q+\Delta q) = [C_i](q) + \sum_{l=1}^{n} \frac{\partial [C_i]}{\partial q_l} \Delta q_l.$$

Matrices $[M_i]$ and $[C_i]$ are determined by the inertial parameters of the manipulator system; therefore, we can consider that they vary weakly with small variations in the vector of generalized coordinates $\{q\}$. Consequently, we can neglect the partial derivatives standing under the summation sign in the obtained expansions and retain only zero approximations of the starting matrices, i.e., we can accept

$$[M_i(q+\Delta q)] = [M_i(q)] + O(\Delta q),$$
$$[C_i(q+\Delta q)] = [C_i(q)] + O(\Delta q).$$

By similar reasoning, we can assume that when $\Delta q \to 0$,

$Q_{Gi}(q+\Delta q) = Q_{Gi}(q) + O(\Delta q)$ and $Q_{Fi}(q+\Delta q) = Q_{Fi}(q) + O(\Delta q)$.

Let us transform Equation (1) rejecting the second infinitesimal order summands and derive

$$[M_i(q)]\{\ddot{q}\} + \{\dot{q}\}^T [C_i(q)]\{\dot{q}\} + [M_i(q)]\{\Delta\ddot{q}\} + 2\{\dot{q}\}^T [C_i]\{\Delta\dot{q}\} = Q_{Di}(q) + Q_{Gi}(q) + Q_{Fi}(q). \tag{3}$$

Total elastic deformation $\{\Delta q\}$, which appears in hinges, can be represented as the following sum:

$$\Delta q_i = \Delta q_i^{ks} + \Delta q_i^{d},\ i = (1,\ldots,n), \tag{4}$$

where Δq_i^{ks} is the quasi-static elastic deformation and Δq_i^{d} is the deformation corresponding to elastic vibrations in an i-hinge.

Using (4), we can decompose Equation (3) into three equations separating slow Δq^{ks} and rapid Δq^{d} variables:

$$\begin{cases} [M_i]\{\ddot{q}\}+\{\dot{q}\}^T[C_i]\{\dot{q}\}=Q_{Di}+Q_{Gi}+Q_{Fi}, i=(1,\ldots,n); \\ [M_i]\{\Delta\ddot{q}^d\}+2\{\dot{q}\}^T[C_i]\{\Delta\dot{q}^d\}=-w_i\Delta q_i^d-r_i\Delta q_i^d; \\ w_i\Delta q_i^{ks}=-Q_{Di}; \end{cases} \quad (5)$$

where r_i is a damping coefficient, which reflects the energy dissipation in an i-hinge.

The set of three Equations (5) is a mathematical model of the manipulator system with elastic hinges at small deformations. The first equation of the system is the equation of motion of a rigid manipulator system. The second equation is the equation of small elastic vibrations $\{\Delta q^d\}$, which appear in the hinges of manipulator systems. The right side of this equation reflects elastic and dissipative properties of hinges. The third equation, according to (2), associates the quasi-statically small elastic deformations $\{\Delta q^{ks}\}$ with forces developed by drives.

3. Test simulation

Let us compare the results of numerical modeling of the dynamics of manipulator systems with elastic hinges obtained using mathematical models (1), (2), and (5). We will perform modeling by the example of the three-link manipulator system (Figure 1).

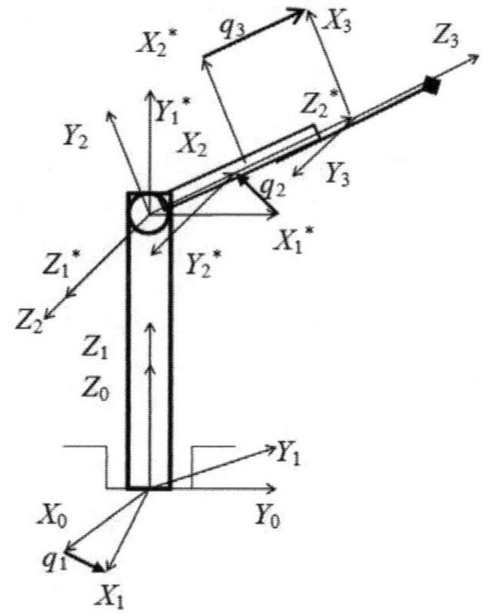

Figure 1. The three-link manipulator system.

The first link of the manipulator system under study is modeled by a thin-wall pipe with length l_1, radius R_1, and mass m_1, which is rotated around the vertical axis coinciding with the central longitudinal axis of the pipe. The second link is pivotally connected with the first one and is modeled by a thin rod with length $2l_2$ and mass m_2, which is rotated relative to the first link in the vertical plane. The third link moves steadily relative to the second link and has length l_3 and mass m_3 concentrated at the link end.

Let us investigate the deviations of motion manipulator system considering the straight-linear uniformly accelerated motion of the point associated with concentrated mass $m3$ of the third link. Let us accept the point with Cartesian coordinates ($x(0)$, $y(0)$, $z(0)$) = (0.6, 0.0, 0.0) as the initial point of the motion trajectory, and the point with coordinates ($x(T)$, $y(T)$, $z(T)$) = (0.0, 0.6, 1.2), where $T = 2$ s is the motion time, as the end point. Solving the inverse kinematic problem, let us find the law of motion manipulator system under study along the specified trajectory (Figure 2).

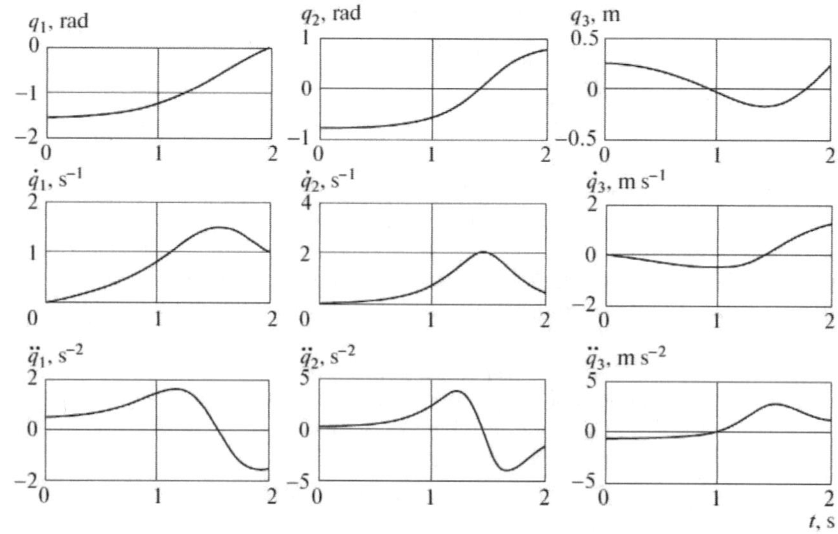

Figure 2. The law of motion manipulator system.

Let us determine the numerical values of geometric, inertial, and elastic parameters of the manipulation system under study as well as the initial values for integration variables. We accept R_1 =0.1 m, l_1 =0.6 m, l_2 =0.3 m, l_3 =0.3 m, m_1=20 kg, m_2 =10 kg, $m3$ =10 kg, w_1 =1.0e4 Nm, w_2 =1.0e4 Nm, w_3 = 1.0e5 Nm, $\Delta q^d_i(0)$ =0.0 (i =1–3), $\Delta\dot q_1^d(0) = \dot q_1$ (0.01) = 0.0075 s^{-1}, $\Delta\dot q_2^d(0)$ = $\dot q_2$ (0.01) = 0.0038 s^{-1}, $\Delta\dot q_3^d(0)$ = $\dot q_3$(0.01) = 0.0095 ms^{-1}, $\Delta q_1(0)$ = $\Delta q_1^{ks}(0.01)$ = –2.201e–4 m, $\Delta q_2(0)$ = $\Delta q_2^{ks}(0.01)$ = 0.0082 m, $\Delta q_3(0)$ = $\Delta q_3^{ks}(0.01)$ = 7.573e–4 m, and $\Delta\dot q_i(0) = \Delta\dot q_i^d(0)$ (i = 1–3).
Differential equations were integrated by the fourth-order Runge–Kutta method. The results of calculations are presented in Figures 3–5. Plot *1* corresponds to the dependence, which reflects the sum of the quasi-static and vibrational components of small elastic deviations $\Delta q_i^{ks}(t)$+$\Delta q_1^d(t)$ (i = 1–3). Plot *2* represents dependence $\Delta q_i^{ks}(t)$ (i = 1–3). Dependencies presented in plots *1* and *2* are calculated on the basis of the set of equations (5). Plot *3* corresponds to dependence $\Delta q_i(t)$ (i = 1–3) calculated on the basis of the set of Equations (1 and 2). Plots *1* and *3* are compared. It should be noted that as the rigidity of hinges wi (i = 1–3) increases, the difference between the results of modeling found by two compared mathematical models decreases.

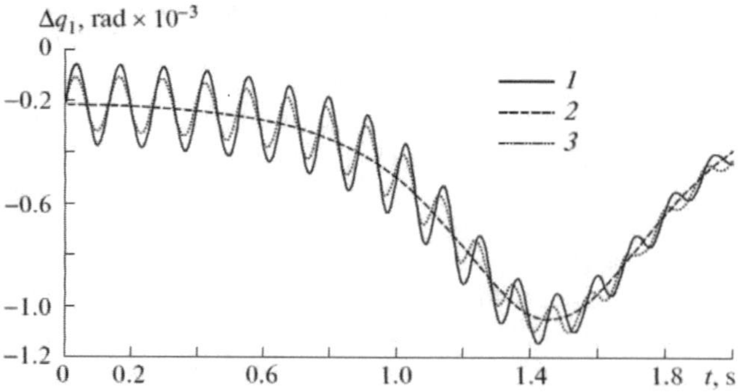

Figure 3. Plot 1. Quasi-static and vibrational components of coordinates q_1.

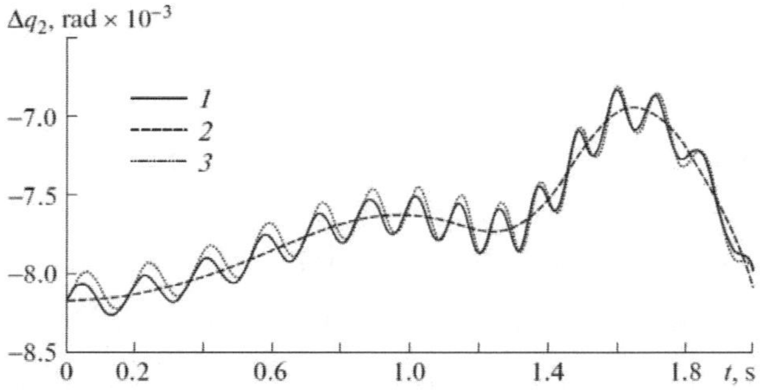

Figure 4. Plot 2. Quasi-static and vibrational components of coordinates q_2.

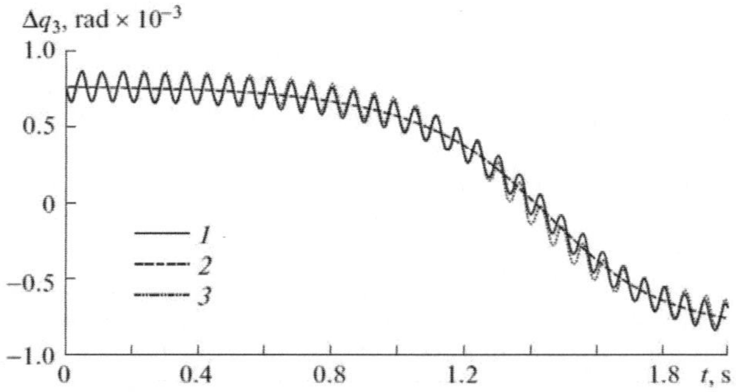

Figure 5. Plot 3. Quasi-static and vibrational components of coordinates q_2.

A detail derivation of the set of equations similar to (5) is presented in [1].

The distinctive feature of the mathematical model of manipulation systems of robots, which is based on the set of Equations (5), is the fact that this mathematical model can be simultaneously used for the synthesis of programmed motions of rigid manipulation systems and for the analysis of deviations from these motions, which appear in actual manipulation systems because of the elastic compliance in hinges.

References

[1] Krakhmalev O N and Bleyshmidt L I 2014 *J. of Machinery Manufact. and Reliability* **43**(1) 22
[2] Chernousko F L, Bolotnik N N and Gradetsky V G 1994 Manipulation Robots. Dynamics, Control, and Optimization (Boca Raton: CRC Press)
[3] Pavlov B I and Ochirov V D 1981 *Computation Experiment in Machines and Mechanisms Dynamics* (Moscow: Nauka)
[4] Kolovskii M Z and Sloushch A V 1988 *Foundations of Industrial Robots Dynamics* (Moscow: Nauka)
[5] Krakhmalev O N 2012 *Mathematical Simulation of Manipulation System Dynamics for Industrial Robots and Articulated Cranes* (Bryansk: Bryansk. Gos. Tekh. Univ.)

IOP Conf. Series: Materials Science and Engineering **124** (2016) 012069 doi:10.1088/1757-899X/124/1/012069

Mathematical simulation application for research of nonuniform distributed-parameter circuit transients

E O Kuleshova[1], A A Plyusnin[1], E B Shandarova[1] and O V Tikhomirova[2]
[1] National Research Tomsk Polytechnic University, 30, Lenina ave., Tomsk, 634050, Russia
[2] V E Zuev Institute of Atmospheric Optics Siberian Branch of the Russian Academy of Sciences, 1, Academician Zuev sq., Tomsk, 634055, Russia

E-mail: kul_2004@mail.ru

Abstract. This paper considers the simulation capability of nonuniform distributed-parameter circuit transients by using MatLab Simulink. This approach is capable of determining currents and voltages of nodes for power networks of any configurations and modes. The paper contains results of nonuniform line simulations in idle, short-circuit and load modes.

1. Introduction

The electric power system reliability during transients is one of the important problems since transient conditions are heavy-duty and result in overload of electrical grid elements and, consequently, power losses and equipment damages [1, 2]. It is necessary to take into account a set of nowadays electrical grid structure features for this problem solution, first of all, nonhomogeneity of power transmission lines on a great bulk of voltage classes.

The accurate information about distributed parameters is required to calculate the transmission line transients. A number of techniques are available to determine the transients in zero-loss circuits [3, 4]. Most of these techniques are based on the state variable approach. They are quite time-consuming in practice. The state variable approach allows obtaining good calculation results of a discrete time model [4-6]. However, it has such disadvantage as a difficulty of sample interval determination.

One of the options is mathematical modeling of the nonuniform distributed parameter circuit in the Matlab Simulink software.

2. A distributed parameters line model

Nonuniform line transients are described by the telegraph equations [5]:

$$\frac{\partial}{\partial x}u(x,t) = -R(x)i(x,t) - L(x)\frac{\partial}{\partial t}i(x,t);$$

$$\frac{\partial}{\partial x}i(x,t) = -G(x)u(x,t) - C(x)\frac{\partial}{\partial t}u(x,t),$$

where $R(x)$, $L(x)$, $G(x)$ and $C(x)$ – parameters of a line length unit; x – coordinate of a selected point, measured from the line beginning.

The Target values will be voltage values $u_1(t)$, $u_2(t)$,…, $u_{N+1}(t)$ incase of N+1 points of the line and current values $i_1(t)$, $i_2(t)$,…, $i_N(t)$ in case of N points of the line. Voltage of the line beginning $u_0(t)$ and current at the line end $i_{N+1}(t)$ are known as boundary conditions.

In the next part of the paper the simulation data of the nonuniform circuit with distributed parameters are described (Figure 1).

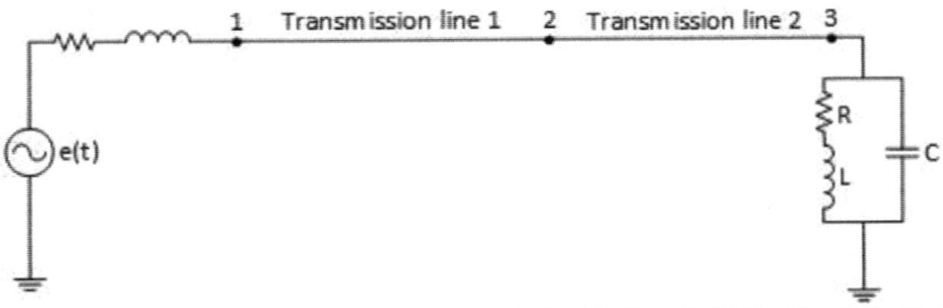

Figure 1. Transmission lines with terminations (L_1 = 1.4 mH, R_1 = 0.42 Ω, C_1 = 8.9 nF, L_2 = 1.33 mH, R_2 = 0.21 Ω, C_2 = 8.5 nF)

3. Nonuniform line simulation

In this part the simulation data of such transmission line modes as idle, short-circuit and load modes are obtained from experiments.

3.1. The idle mode

Here we will consider an open-ended nonuniform line of l = 300 km length, a sending end line breaker is normally opened and a switch is closed when t = 0 s. The voltage and current along the line are zero before the breaker is closed.

A simulation circuit diagram in Matlab Simulink for the idle mode is given in Figure 2. Breaker 1 and Breaker 2 are open (parameter ‘Switching times’ is significant).

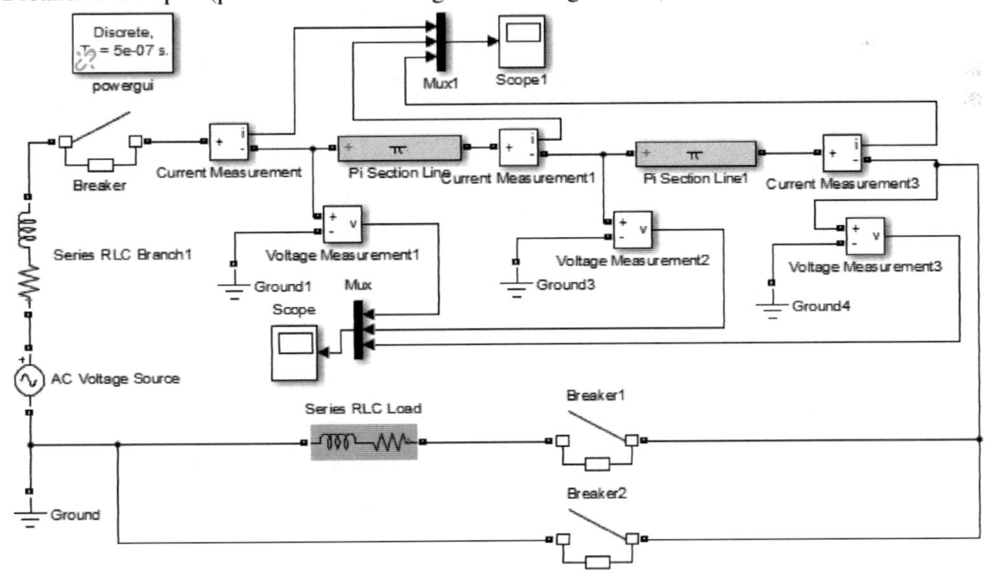

Figure 2. A Matlab Simulink circuit diagram.

The MatLab Simulink software package includes SimPowerSystems – a toolbox of blocks oriented at power systems. The simulation results for circuit diagram nodes 1-3 are shown in Figure 3.

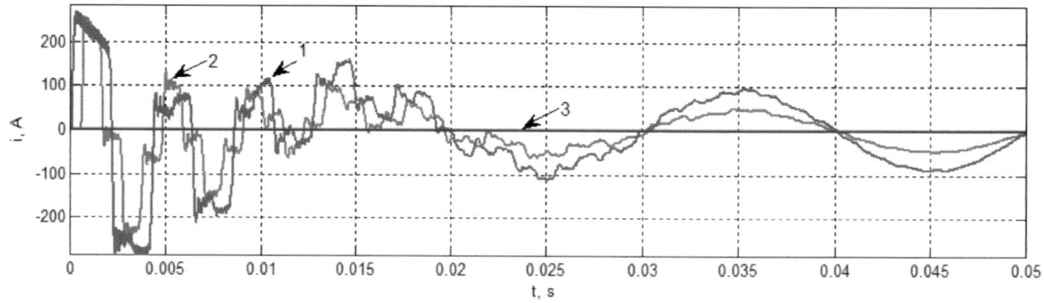

a

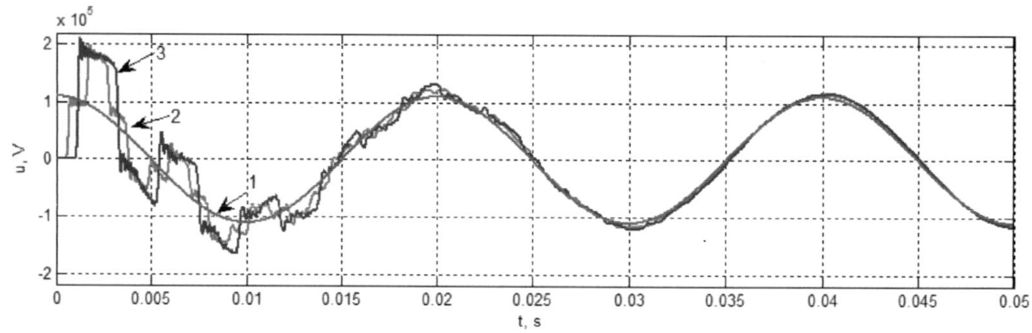

b

Figure 3. The transmission line parameters in the idle mode: *a* – currents of nodes 1-3; *b* – voltages of nodes 1-3

The diagrams show that when the breaker has been closed, the line has been put under alternating voltage. The current incident wave (IW) amplitude is 0.27 kA. When IW reaches node 2 in the time of 0.5 ms and splits from the obstacle into transmitted (TW) and reflected (RW) waves. Since the line characteristic impedance of the second half is less than that of the first, the current TW amplitude is higher and the current RW amplitude is lower than the IW amplitude. When current TW reaches the end, it reflects and changes the sign into the opposite. Therefore, the current of node 3 equals zero.

When voltage TW reaches the end, it is reflected with the same sign. Therefore, the voltage of node 3 is doubled. In a steady-state mode the voltage of node 3 is higher than the voltage of node 1.

3.2. The short-circuit mode

Now we will consider the same line closed at the end.

A simulation of the short-circuit diagram in Matlab Simulink is represented in Figure 2. Breaker 1 is opened, Breaker 2 is closed.

The simulation results for the circuit diagram are shown in Figure 4.

The diagrams make it evident that multiple reflectance overlapping provides multiple current surges. The current peak amplitude reaches almost 0.8 kA and is shaped by these surges. Also, it is obvious that the voltage of node 2 oscillates in between initial and zero values and tends to average out till the end of the transient. So, in a steady-state mode the current of node 2 is two times lower than the current of node 1.

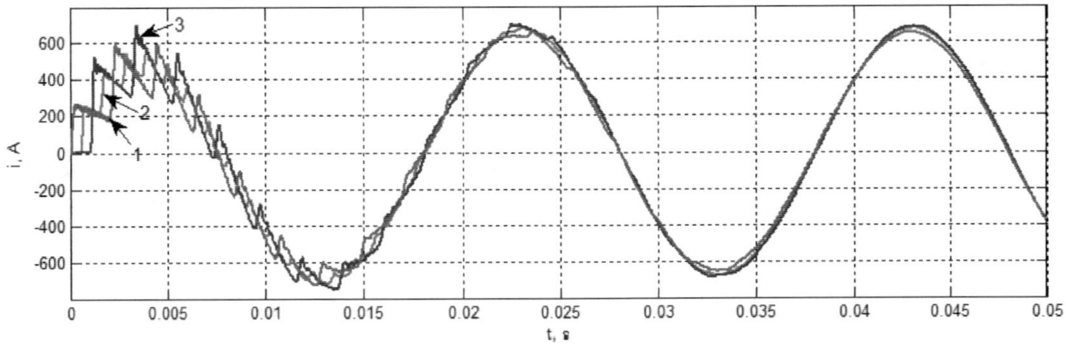

a

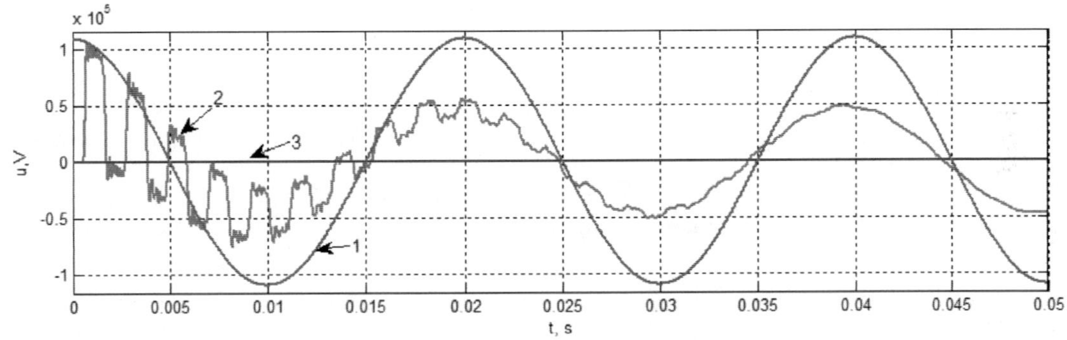

b

Figure 4. The transmission line parameters in the short-circuit mode:
a – currents of nodes 1-3; *b* – voltages of nodes 1-3

3.3. The load mode

Finally, we will consider the same line which is loaded at the end by $\underline{S} = 10 + j8\ MWA$.

A simulation circuit diagram in Matlab Simulink for the load mode is given in Figure 2. Breaker 1 is closed, Breaker 2 is opened.

The simulation results for circuit diagram nodes 1-3 are shown in Figure 5.

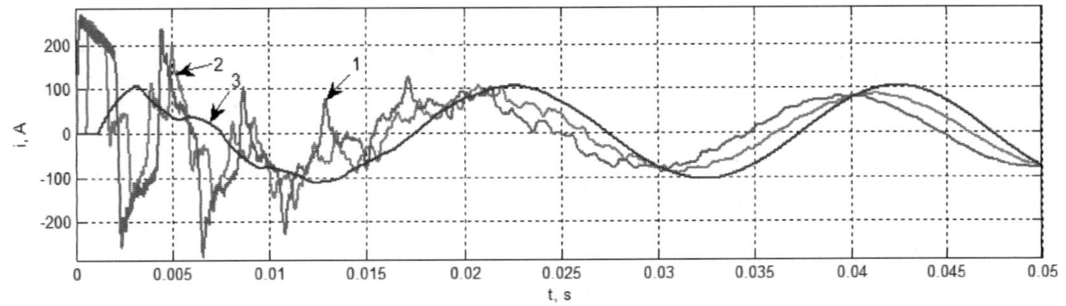

a

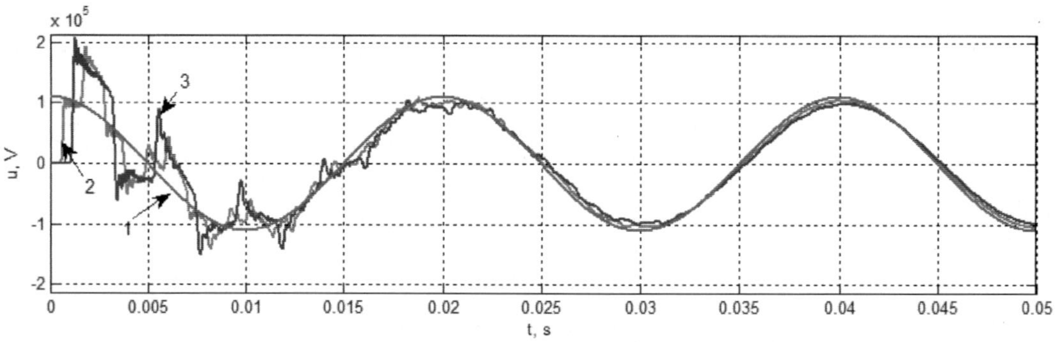

b

Figure 5. The transmission line parameters in the load mode: *a* – currents of nodes 1-3; *b* – voltages of nodes 1-3

It can be seen from the diagrams that the current peak amplitude is 0.27 kA. During the transient current, the TW amplitude (in node 2) is higher than the IW amplitude. But after the transient ending the former is lower than the latter, and the current amplitude in node 3 tends to be a little higher than the amplitude in node 1.

When the TW voltage reaches the end, it is reflected with the same sign. That is why the voltage of node 3 is doubled. In the steady-state mode the voltage of node 3 is a little lower than the voltage of nodes 1 and 2.

4. Conclusions

The mathematical simulation in MatLab Simulink allows calculating the nonuniform distributed-parameter line transients not just in case of the power supply connection, but also during line parameters and load changes. The represented simulation results agree with those obtained in [4]. The capability of reducing the computation complexity for nonuniform distributed-parameter circuit transients by using MatLab Simulink has been shown.

References

[1] Shandarova E B and Shwab S A 2014 Voltage quality of micro-hydroelectric station with rectifier ballast load *Mechanical Engineering, Automation and Control Systems* (*Int. Conf. on 16-18 Oct 1-4* DOI: 10.1109/MEACS.2014.6986872)

[2] Vasileva O V, Budko A A, Lavrinovich A V and Filkov A I 2014 Study of nanosecond pulse switches for low voltage generator for diagnosis of power transformers *Mechanical Engineering, Automation and Control Systems* (*Int. Conf. on 16-18 Oct 1–4* DOI: 10.1109/MEACS.2014.6986867)

[3] Da Costa E C, Kurokawa S, do Prado A J and Pissolato J 2011 *Int. J. Elec. Pow. & Nrg Sys.* **33**

[4] Mamis M S, Kaygusuzi A and Koksal M 2014 *Turk J Elec Eng & Comp Sci* **18**(1)

[5] Kabanova T V, Kuleshova EO, Isaev Y N, Shishka N V and Shishkovskaya Y V 2014 Calculation of Transients in Inhomogeneous Lines by 'The Method of Straight Lines' (*Proc. Int. Conf. on Mechanical Engineering, Automation and Control Systems, MEACS*) 6986903

Operation modes of a hydro-generator as a part of the inverter micro hydropower plant

B V Lukutin[1], E B Shandarova[1], D L Matukhin[1], A F Makarova[1] and I L Fuks[2]
[1] National Research Tomsk Polytechnic University, 30, Lenina ave., Tomsk, 634050, Russia
[2] National Research Tomsk State University, 36, Lenina ave., Tomsk, 634050, Russia

E-mail: shandarovaelena@mail.ru

Abstract. The paper dwells on the selection problem of power equipment for a stand-alone inverter micro hydropower plant, in particular a hydro-generator, and evaluation of its operation modes. Numerical experiments included the modes calculation of hydroelectric units of the same type with various nominal power, supplied to the consumer according to the unchanged electric load curve. The studies developed requirements for a hydro-turbine and a synchronous generator in terms of a speed range and installed capacity, depending on the load curve. The possibility of using general industrial hydroelectric units with nominal power equal to half-maximum capacity of a typical daily load curve in rural areas was shown.

1. Introduction

Diversion micro hydropower plants (micro HPP) with upstream pipeline have become mostly widespread due to their low environmental impact and the possibility of using unregulated turbines combined with semiconductor ballast voltage regulation systems for a hydro-generator, which defines their high performance and low cost of the generated electric energy.

The development of power converter equipment allows designing competitive micro HPPs with a reversible converter and a power storage unit [1, 2]. Such power plants can achieve high-quality output voltage and increase the installed capacity utilization factor of the hydro-generator.

Complex processes of energy conversion of the inverter micro HPP hinder the selection of power equipment, particularly the hydro-generator, and evaluation of its operation modes. Research in this area is very relevant.

The research objective is a diversion micro hydropower plant with a reversible converter and a battery power storage unit. The goal of the research is to study the operation modes of a hydro-generator as part of a stand-alone inverter of micro HPP and to develop the requirements for hydro-turbines and a synchronous generator in terms of a speed range and installed capacity, depending on the electric load curve.

The research methodology rests on the study of power balance in a stand-alone power supply system based on a micro HPP and mathematical simulation of hydro-generator modes by means of the Matlab software package.

2. Operation modes of a hydro-generator

The load curve is a defining factor for operation modes of the stand-alone power supply system. A typical daily domestic load curve of rural households is shown in Figure 1, where P – active load, Q –

reactive load. As shown in the graph, the dark color indicates effective load (electric power) and the white one – ballast load (electric power converted into heat) of the power plant.

Widespread micro HPP with ballast regulation shall be selected due to the condition of providing the consumer with electric power during the maximum load hours, in this graph – evening maximum load is reached when P_m=1 PU. Obviously, the hydro-generator produces the effective load during these hours. The rest of the time its power is distributed between the effective load (according to the load curve) and the ballast load so that gross capacity remains equal to P_m. As follows from the curves shown in Figure 1, the hydro-generator is loaded with active power of the effective load on average at the level of 36-45% compared to its nominal load, depending on the season.

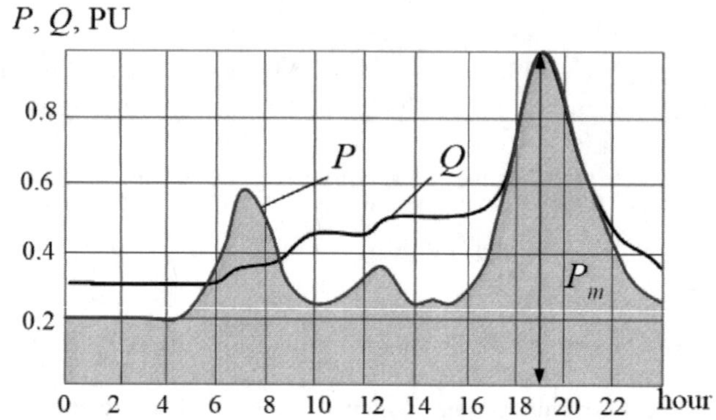

Figure 1. A summer daily domestic load curve.

Thus, using the ballast load of the heating elements, it is possible to use the entire installed capacity of micro HPP, but more than a half of it is used in the form of heat. The load power factor variability prevents the components of gross capacity from stabilizing, which leads to errors in both amplitude and frequency stabilization of the output voltage at micro HPP [3-5].

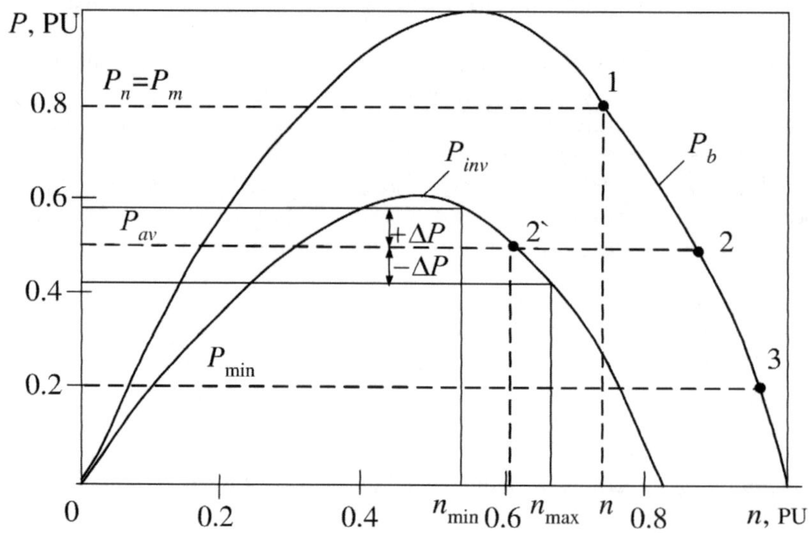

Figure 2. Operation modes of hydro-turbines of ballast and inverter HPP.

In addition to the inverter, the equipment of inverter micro HPP, due to the power storage unit, can be selected for a significantly lower nominal power, as shown in Figure 2, where P_b – characteristic curve of the hydro-turbine with ballast regulation; P_n – characteristic curve of the hydro-turbine of inverter micro HPP; 1 – working point on the curve of P_b=$f(n)$ of ballast micro HPP; 2 – operation mode of the hydro-turbine for the average load of P_{av}; 3 – operation mode of the hydro-turbine for the minimum load of P_{min}.

The figure shows a typical power dependence of the propeller hydro-turbine on speed rate n with constant parameters of the water workflow. The dependence is presented in relative units. The basis values refer to the maximum power of the hydro-turbine with maximum energy of the water workflow and its speed rate in an idling mode. The values of average P_{av} and maximum P_m consumed power correspond to points 1 and 2, respectively. The micro HPP with the ballast stabilization system requires a turbine with the characteristic curve of P_b that can supply consumers with electric energy during the maximum load hours (point 1 in Figure 1).

In accordance with the load curve, the average effective load of hydro-generator P_{av} is characterized by point 2, and the minimum load - by point 3 on the power curve of the turbine.

Assuming that maximum power load is provided by the power storage unit, the hydro-turbine of inverter micro HPP of much less capacity can be selected, which is the average value of daily power load. The curve of the hydro-turbine of inverter micro HPP is indicated as P_{inv} in Figure 2.

It is obvious that the power balance of such plant shall consist of two alternating modes, depending on the ratio of the available capacity of the hydro-generator and the current power load. A charge and a discharge of the power storage unit are characterized by formulas

$$P_{gi} = P_{ni} + P_{chi} \quad \text{and} \quad P_{gi} = P_{ni} - P_{dchi}, \tag{1}$$

where P_{gi} – power of the hydro-generator on the i-th time interval; P_{ni} – power load on the i-th time interval; P_{chi}, P_{dchi} – charge and discharge power of battery power storage units.

The available capacity of the hydro-turbine is limited by the maximum point on its power curve. The power of charge P_{chi} and discharge P_{dchi} of the battery power storage unit provides the power balance and, to a certain extent, stabilizes the operation of the hydro-turbine.

The power storage unit can limit the range of the power load change at a point of $P_{av}\pm\Delta P$. Accordingly, the hydro-generator of lower installed capacity can be selected, with the same power load of a plant. The power curve of the hydro-turbine, which develops less power is shown in Figure 2 and indicated as P_{inv}. The working point, corresponding to the average power load is indicated as 2`. The speed rate of the hydro-generator will vary, to some extent, from n_{min} to n_{max}, depending on the range of the generated power.

The equation for the daily power balance can be shown as

$$W_g = W_n + W_{ch} - W_{dch}. \tag{2}$$

Obviously, at 100% efficiency of battery power storage units W_{ch}=W_{dch} and (2) become equality of W_g=W_n. Taking into account the actual properties of storage units, charge power will always exceed discharge power.

In the joint load operation of the hydro-generator and the power storage unit, the capacity and power contribution of the hydro-generator in maximum load periods shall be limited to their available values. As it was mentioned above, the available capacity of the hydro-generator is limited to the maximum power curve of the turbine, and in case of batteries storage units – the capacitance and the recommended value of its discharge in a cyclic mode.

To control the flows of capacity and power, an intelligent system that monitors the current power load of the stand-alone power consumer and the state of battery storage units is required.

An operation mode of battery storage units, providing acceptable service life in a cyclic mode, shall be limited to the available capacitance to a maximum of 30 % of the total power charge. Thus, the storage units shall provide power for consumers during peak hours, contributing to the available generation level of micro HPP up to the consumption level. The value of this power shall not exceed

30 % of the power of a fully charged battery. The rest of the time, the storage units are charged to the nominal capacitance from the hydro-generator.

A selection of the operation mode is defined by the threshold of power load, denoting the need to include battery energy into the system power balance. The threshold defines the allowable value of the hydro-generator capacity, providing its stable operation under specific conditions [6, 7].

Using daily load curves (Figure 1) and the power curve of the hydro-turbine (Figure 2), the component of the system power balance can be defined.

Daily power, required by a consumer, is defined as

$$W_n = \sum_{i=1}^{48} P_{ni} t_i ,$$

where P_{ni} – power at a half-hour interval; t_i – half-hour interval of time sampling; i –interval number.

Mean consumed power is equal to $P_n = W_n/48$. In view of the charging current of the storage unit, the hydro-generator shall provide the energy system with average power

$$P_g = P_n + P_{ch},$$

where P_{ch} – charge power.

The charge power can be characterized by the recommended value of charging current, which does not numerically exceed 10 % of the nominal capacitance of the battery. Then the charge power of the storage unit can be defined as

$$W_{ch} = \sum_{i=1}^{48} \left(P_{gm} - P_{ni} \right) \cdot t_i ,$$

provided that $P_{gm} > P_{ni}$. Unless the inequality is satisfied, the charge power at the interval is equal to zero.

The amount of the charge power shall exceed the discharge power by the value of battery efficiency – η_{bat}:

$$W_{dch} = W_{ch} \cdot \eta_{bat} .$$

The discharge power is defined as

$$W_{dch} = \sum_{i=1}^{48} \left(P_{gm} - P_{ni} \right) \cdot t_i ,$$

provided that $P_{gm} < P_{ni}$. Accordingly, unless the inequality is satisfied, the battery does not discharge, and the previously described charge of the storage unit occurs.

3. Results and discussion

The study of the described hydro-generator modes has been conducted using a mathematical model, implemented in Matlab by means of the following elements: a synchronous generator with damper winding and an excitation system corresponding to hydroelectric units with the capacity of 85 kVA, 60 kVA and 42.5 kVA, a generic diode bridge was used as a rectifier, an inverter was based on IGBT-transistors shunted by reverse diodes. The maximum power load corresponds to the nominal power of the hydro-generator, which is 85kW. An unregulated turbine was simulated by a transfer function taking into account the water-hammer effect, specific to hydro-turbines [8, 9]:

$$W_t(s) = \frac{1 - a_{dis} T_w s}{1 + 0.5 a_{dis} T_w s} ,$$

where T_w – time constant of the hydro-turbine; a_{dis} – initial opening of the turbine distributor.

Numerical experiments included the modes calculation of hydroelectric units of the same type with various nominal power, supplying the consumer according to the unchanged electric load curve. The calculation results in the form of graphs, interconnecting the hydroelectric unit speed rate and developed power, are shown in Figure 3.

The graphs are given in relative units for three hydroelectric units with different nominal power, such as 1 – 85 kVA; 2 – 60 kVA; 3 – 42.5 kVA. The graphs show that the operation mode of ballast micro HPP is in the vicinity of the nominal mode: $P = 1$ PU, $n = 0.6$ PU. Deviations from the nominal mode are defined by the stabilizing error of gross capacity of ballast and effective load. In view of the operation mode stability of the hydro-generator, the general industrial power equipment, designed for constant speed rate, can be used.

In case of using inverter micro HPP without the power storage unit, the nominal power of the hydroelectric unit remains unchanged compared to the ballast power plant, but the speed range of the hydroelectric unit varies from 0.6 to 0.92 PU in accordance with Figure 3. Consequently, general industrial electric generators cannot be used under the terms of the mechanical strength. Using special generators with the variable speed rate usually leads to micro HPP appreciation.

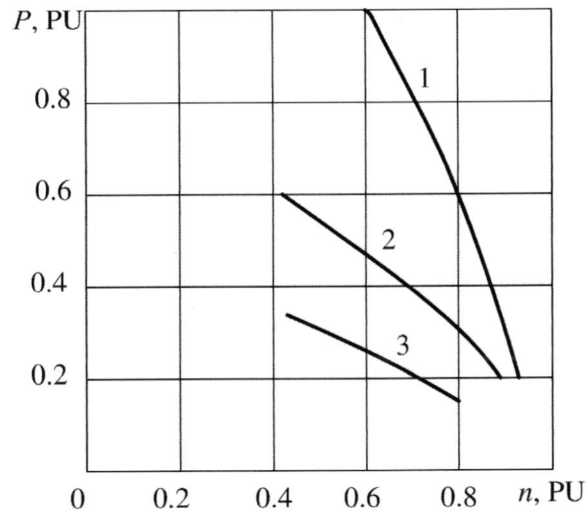

Figure 3. Operation modes of hydro-generators.

The power storage unit as a part of inverter micro HPP allows eliminating the peak load operation of the hydro-generator. Accordingly, its nominal power can be reduced. The curves 2 and 3 in Figure 3 characterize hydro-generator operation modes with nominal power of 0.7 PU and 0.5 PU relative to the basic one.

The current values of power load, with the exception of maximum loads, are within 0.2–0.35 PU (Figure 1). The speed rate of the hydro-generator with the nominal mode of $0.7P_m$ varies from 0.88 PU to 0.75 PU (Figure 3). The hydro-generator with nominal power of 0.5 P_m operates in the frequency range from 0.7 PU to 0.55 PU.

4. Conclusion

Thus, assuming the hydro-generator nominal speed rate of 0.6 PU, it becomes possible to use general industrial electrical equipment, allowing an increase in the speed rate relative to the nominal value under the terms of mechanical strength of 20 %.

Therefore, the studies of hydro-generator operation modes as a part of stand-alone inverter micro HPP, allowed establishing the application of general industrial hydroelectric units with nominal power equal to half-maximum capacity of the typical daily load curve in rural areas.

References

[1] Abbasi T and Abbasi S A 2011 *Renew. Sust. Energ. Rev.* **15** 2134

IOP Conf. Series: Materials Science and Engineering **124** (2016) 012070 doi:10.1088/1757-899X/124/1/012070

[2] Sobczyk T J , Mazgaj W, Szular Z and Wĕgiel T 2011 *Arch. Electr. Eng.* **60** 159
[3] Rajagopal V and Singh B 2011 *Int. Conf. on Power Electronics IICPE* (*New Delhi*) 5728118
[4] Shandarova E B and Shwab S A 2014 *Int. Conf. MEACS* (*Tomsk*) 6986872
[5] Kisseleva O, Balgabaeva L and Kharitonov P 2016 *Asian J. Inf. Tech.* **15** 12
[6] Kabanova T V *et al* 2014 *Int. Conf. MEACS* (*Tomsk*) 6986903
[7] Kolchanova V A *et al* 2014 *Int. Conf. MEACS* (*Tomsk*) 6986868
[8] Bieri M, Schleiss A J and Jordan F 2013 *AMICE Final Conf.* (*Sedan*) p. 61
[9] Babunski D, Tuneski A and Zaev E 2012 *Med. Conf. on Embedded Computing MECO* (*Bar*) 275

An identification algorithm of model kinetic parameters of the interfacial layer growth in fiber composites

V Zubov, S Lurie and Y Solyaev
Computing Centre of Russian Academy of Sciences, 40, Vavilov st., Moscow, 119333, Russia

E-mail: lurieccas.ru

Abstract. This paper considers the identification algorithm of parameters included in a parabolic law that is often used to predict the time dependence of the thickness of the interfacial layers in the structure of composite materials based on a metal matrix. The incubation period of the process and the speed of reaction and pressure are taken into account. The proposed algorithm of identification is based on the introduction of a minimized objective function of a special kind. The problem of identification of unknown parameters in the parabolic law is formulated in a variational form. The authors of the paper have determined the desired parameters, under which the objective function has a minimum value. It is shown that on the basis of four known experimental values of the interfacial layer thickness, corresponding to different values of temperature, pressure and the time of the interfacial layer growth, it is possible to identified four model parameters. They are the activation energy, a pre-exponential parameter, the delay time of the start of the interfacial layer formation, and the parameter determining the pressure effect on the rate of interfacial layer growth. The stability of the proposed identification algorithm is also studied.

1. Introduction

The influence of interfacial zones on physical and mechanical properties of composite materials based on a metal matrix can be significant [1, 2]. A parabolic law is often used to evaluate the interfacial layer thickness at the boundary of fibers and matrix in the metal composite materials [3-9]. This law establishes the relationship that allows determining the variation of the interfacial layer thickness depending on the exposure time of the composite sample at elevated temperature and elevated pressure. In accordance with the parabolic law, thickness of the interfacial layer $h(t)$ is proportional to square root of the process time parameter $\sqrt{t}$, and the proportionality coefficient is rate constant k, which in its turn depends on the temperature and the pressure of the environment. To estimate the rate constant Arrhenius relationship is often used. Thus, the relationship used for estimation of the interfacial layer thickness at the boundary of fibers and the matrix is as follows:

$$h(t) = K_0 \exp\left(-\frac{Q+\kappa P}{2RT}\right)\sqrt{t-t_0}\,. \quad (1)$$

where P is pressure (Pa); Q is the energy of activation of the reaction zone growth process (J / mol); T is process temperature (°K); $R = 8.314$ is a universal gas constant (J / mol / °K); t is exposure time (sec); t_0 is an "incubation" period (sec); h is the interfacial layer thickness (m); K_0 is a pre-exponential parameter (m/s$^{-1/2}$); κ is a parameter determining the effect of pressure on the rate of the interfacial layer growth (m^3/mol).

The aim of this work is to develop an algorithm of identification of parameters, which are included in the parabolic law and characterize the process of the interfacial layer growth around the fibers in the matrix of a metal composite material. It is assumed that there are N data points, each of which is a set of four values: P, T, t, H. The task is to determine unknown parameters of the model K_0, Q, κ, t_0 using N known data points.

2. An identification algorithm of the parabolic model parameters of the interfacial layer growth

Let us suppose that there are experimental data that are a set of N points: $(P,T,t,H)_1$, $(P,T,t,H)_2, \ldots, (P,T,t,H)_N$. The problem of determination of mathematical model parameters is formulated as follows: find such model parameters for which an experimental set of points would be as little different as possible from the theoretical set of points.

Let us pre-convert the equation (1), used to estimate the interfacial layer thickness at the boundary of the fibers and the matrix, to the equivalent form:

$$\ln(h(t)) = \beta - \tilde{Q}\frac{1}{T} - \tilde{\kappa}\frac{P}{T} + \frac{1}{2}\ln(t - t_0), \tag{2}$$

where $\beta = \ln(K_0)$, $\tilde{Q} = Q/(2R)$, $\tilde{\kappa} = \kappa/(2R)$ are new unknown parameters of the model.

We also introduce the functions $f_i(\beta, \tilde{Q}, \tilde{\kappa}, t_0)$, $(i = 1 \ldots N)$:

$$f_i(\beta, \tilde{Q}, \tilde{\kappa}, t_0) = \ln H_i - \ln(h(t_i)) = -\beta + \tilde{Q}\cdot\frac{1}{T_i} + \tilde{\kappa}\cdot\frac{P_i}{T_i} + \ln\left(\frac{H_i}{\sqrt{t_i - t_0}}\right), \tag{3}$$

where i is the number of experimental points, and the value of function $h(t_i)$ is determined by (1) with the use of experimental values P_i, T_i, t_i.

As the objective function we used

$$\Phi(\beta, \tilde{Q}, \tilde{\kappa}, t_0) = \frac{1}{K}\cdot\sum_{i=1}^{N}[\ln H_i - \ln(h(t_i))]^2 = \frac{1}{K}\cdot\sum_{i=1}^{N}\left[f_i(\beta, \tilde{Q}, \tilde{\kappa}, t_0)\right]^2. \tag{4}$$

The model parameters in this approach are determined by minimizing the function $\Phi(\beta, \tilde{Q}, \tilde{\kappa}, t_0)$. In case of the minimum of the objective function its partial derivatives are vanished. This leads to a system of nonlinear algebraic equations in unknown parameters β, $\tilde{Q}$, $\tilde{\kappa}$, t_0. To determine the coefficients of this equation system, we will introduce the following N-dimensional vectors:

$$\bar{V}_1 = \| -1 \;\ldots\; -1 \|^T,$$
$$\bar{V}_2 = \| 1/T_1 \;\ldots\; 1/T_N \|^T,$$
$$\bar{V}_3 = \| P_1/T_1 \;\ldots\; P_N/T_N \|^T,$$
$$\bar{V}_4 = \| 0.5/(t_1 - t_0) \;\ldots\; 0.5/(t_N - t_0) \|^T,$$
$$\bar{V}_5 = \left\| \ln\left(H_1/\sqrt{t_1 - t_0}\right) \;\ldots\; \ln\left(H_N/\sqrt{t_N - t_0}\right) \right\|^T.$$

Now the system of equations can be written as:

$$A_{i1}\cdot\beta + A_{i2}\cdot\tilde{Q} + A_{i3}\cdot\tilde{\kappa} = -D_i, \quad (i = 1..4), \tag{5}$$

where

$$A_{ij} = \sum_{m=1}^{N} (\overline{V}_i)_m \cdot (\overline{V}_j)_m = (\overline{V}_i, \overline{V}_j), \quad D_i = (\overline{V}_i, \overline{V}_5), \quad (i, j = 1..4). \tag{6}$$

All the elements of vector $\overline{D}$, and elements A_{41}, A_{42}, A_{43} of matrix A depend on the unknown quantities, that makes a system of algebraic equations (5) nonlinear. At the same time elements A_{ij} $(i, j = 1..3)$ of matrix A do not depend on unknown variables. Considering the above-mentioned feature of the system (5), we propose the following algorithm to determine the identification parameters:

- Find the minimum value of the experimental observation time, i.e. $t_* = \min\{t_i\}_{i-1}^{N}$.
- Select value $t_0 \in [0, t_*]$.
- Consider a subsystem of the system of algebraic equations (5) with selected value $t_0 \in [0, t_*]$, that consists of the first three equations. This subsystem is a system of linear algebraic equations for β, $\tilde{Q}$, $\tilde{\kappa}$. The solution of this system allows determining these parameters (for any given value of $t_0 \in [0, t_*]$).
- Choosing the value of t_0 in segment $[0, t_*]$, we strive to satisfy the last equation of the system too (5). In addition, for every t_0 parameters β, $\tilde{Q}$, $\tilde{\kappa}$ should be determined again. Since the solution of the last equation of the system (5) can not belong to segment $[0, t_*]$, instead of solving the last equation, it is recommended to minimize objective function $\Phi(t_0) = \Phi[\beta(t_0), \tilde{Q}(t_0), \tilde{\kappa}(t_0), t_0]$ of parameter t_0 on segment $[0, t_*]$. Here we can use any simple algorithm for minimizing a function of one variable.

To solve the problem of identification of the model parameters it is necessary to select the data points in the way that vectors $\overline{V}_1$, $\overline{V}_2$, $\overline{V}_3$, $\overline{V}_4$ are linearly independent, and the number of data points is not less than 4. It should also be noted that the increase of the number of data points helps to ensure more reliable values of the identifiable parameters of the model.

3. Results

With the proposed algorithm we performed the calculations of the model parameters for the experimental data given in [5]. In [5] it was supposed that κ and t_0 are equal to 0, and the data were presented for the two materials: for composite SiC/Ti_2AlNb and for composite SiC/super α_2. These composites reinforced with unidirectional fibers of silicon carbide of the SCS-6 brand were made with the use of two different intermetallic matrices designated as Ti2AlNb and super α2. The investigation of possible chemical reactions, occurring under the composite manufacture during the formation of the interfacial layer at the boundary of the fiber and matrix, was conducted in [5].

Table 1 shows the results of determination of the model parameters for composite SiC / Ti2AlNb (nine experimental points), and Table 2 shows the results for composite SiC / super α_2 (six experimental points).

Fig. 1 and 2 show the dependence of the interphase layer thickness on the exposure time of the sample. Different figures correspond to different temperatures of the process. The points represent the experimental data, the dark line corresponds to the results of [5] and the bright line represents the results obtained using the developed algorithm. It can be seen, that the model parameters obtained by the developed algorithm correspond to the experimental data rather than the parameters specified in [5].

Table 1. Identification of model parameters (1) for composite SiC / Ti2AlNb

$T(^{\circ}C)$	700	700	700	800	800	800	900	900	900
t (h)	1000	1500	2000	1000	1500	2000	50	200	500
h (μm)	0.99	1.09	1.15	1.93	2.27	2.58	1.22	1.95	2.87
The results of the identification:									
	work [5]				developed algorithm				
K_0 (m/s$^{1/2}$)	$14.4 \cdot 10^{-6}$				$6.196 \cdot 10^{-6}$				
Q (J/mol)	175709				154229				

Table 2. Identification of model parameters (1) for composite SiC / super α_2

$T(^{\circ}C)$	700	700	700	800	800	800
t (h)	1000	1500	2000	1000	1500	2000
h (μm)	1.39	1.44	1.52	2.99	3.63	4.04
The results of the identification:						
	work [5]			developed algorithm		
K_0 (m/s$^{1/2}$)	$54377.8 \cdot 10^{-6}$			$3.420 \cdot 10^{-6}$		
Q (J/mol)	317664			139009		

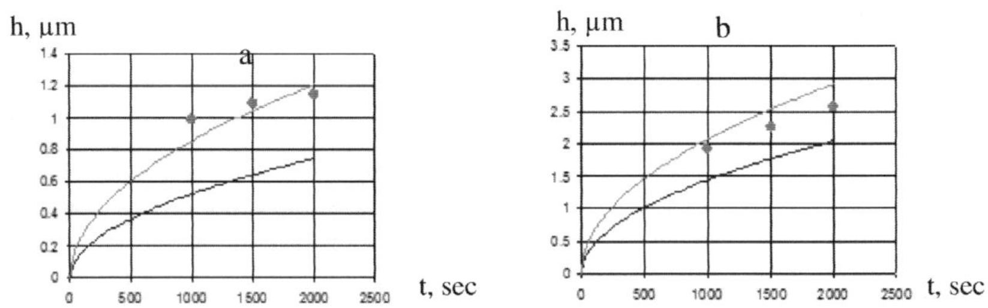

Figure 1. Time dependence of the interfacial layer thickness for the Ti2AlNb composite a: 700°C, b: 800°C.

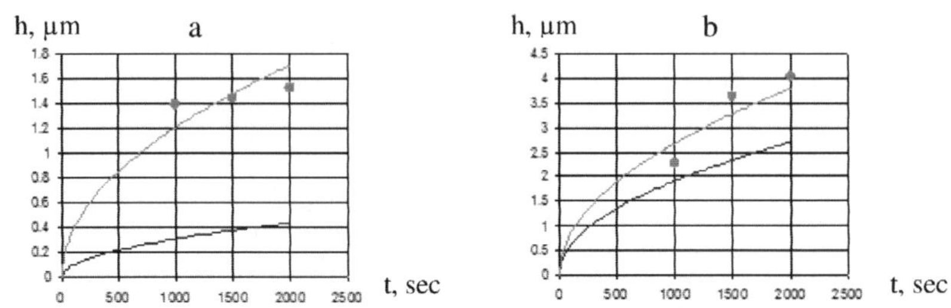

Figure 2. Time dependence of the interfacial layer thickness for the super α2 composite a: 700°C, b: 800°C.

4. Conclusions

The algorithm which allows identifying the kinetic parameters of the parabolic law has been proposed. In order to obtain successful results of the estimations one should use the experimental data consisting of not less than 4 points. More data points will provide better results obtained during identification.

The experimental points should be selected in such a way that the introduced system of N-dimensional vectors would be linearly independent. The appropriate description of the experimental data for the intermettalic titanium-based composites has been shown.

Acknowledgments
The study has been performed with the support of the grant from the Russian Science Foundation (project №14-11-00782).

References
[1] Clyne T W and Winters P J 1992 *An Introduction to Metal Matrix Composites* (Cambridge: Cambridge University Press)
[2] Nalwa H 2001 Handbook of Surfaces and Interfaces of Materials (NY: Academic Press)
[3] Lu X *et al* 2006 *Trans. Nonfer. Met. Soc. Ch.* **16** 77
[4] Yang Y Q et al 1998 *Comp. Part A* **29** 1235
[5] Zhu Y *et al* 2008 *Trans. Nonfer. Met. Soc. Ch.* **18** 733
[6] Guo Z X and Derby B 1994 *Scrip. Metal. Mater.* **30** 89
[7] Martineau A *et al* 1984 *J Mat. Sci.* **19** 2749
[8] McLeod A D and Gabryel C M 1992 *Metallurgical Transactions A* **23** 1279
[9] Knyazeva A *et al* 2014 *AIP Conf. Proc.* **1623** 263

A mathematical model of the vertical dual-mass hydroimpulsive mechanism

F A Simankin, G R Ziyakaev, I V Kuznetsov and R E Lushnikov
Department of Theoretical and Applied Mechanics, National Research Tomsk Polytechnic University, 30, Lenina ave., 634050, Tomsk

E-mail: zgr@rambler.ru

Abstract. In this paper, the model of a hydroimpulsive mechanism for boring machines was presented. Differential equations describing the processes occurring in the mechanism were derived. Keywords: hydroimpulsive mechanism, forced oscillation, natural frequency.

1.Introduction

Papers [3-8] describe the original model of a hydroimpulsive power-actuating mechanism that can be used as a generator of power pulses to intensify drilling of drillholes with a small diameter or to sink piles [1]. One of the features of this mechanism is the mandatory presence of a rigid support, which is replaced by constant compression force F_{COMP} during the modelling. Creation of such support is problematic and pointless for the practical construction of such mechanism. It is much easier to replace it with another additional mass. At the same time, this mass will not remain stationary and will perform periodic oscillations. This raises the question about studying this mechanism in order to determine its main parameters, under which the system operates in a resonant mode and pressure pulses occurring in the hydraulic cylinder will be maximal, and, hence, the stress on the gear will be maximal as well.

2.Results and discussion

Let us consider the vertical dual-mass hydroimpulsive power-actuating mechanism (Figure 1). The liquid pressure pulses are generated in the following way: when the hydropulsator operates, the plunger performs a reciprocating motion that creates the liquid pressure pulses, which are transferred to the hydraulic cylinder. Since the cylinder is preloaded with the elastic force, it oscillates together with inertia masses m_1 and m_2. A periodic conversion of kinetic energy of the masses into potential energy of deformed liquid and a hose takes place, hence the system pressure increases, and vice versa. During the system operation mode close to the resonant one the pressure pulses of a significant value appear. They are transferred to the working surface through the piston and the bar.

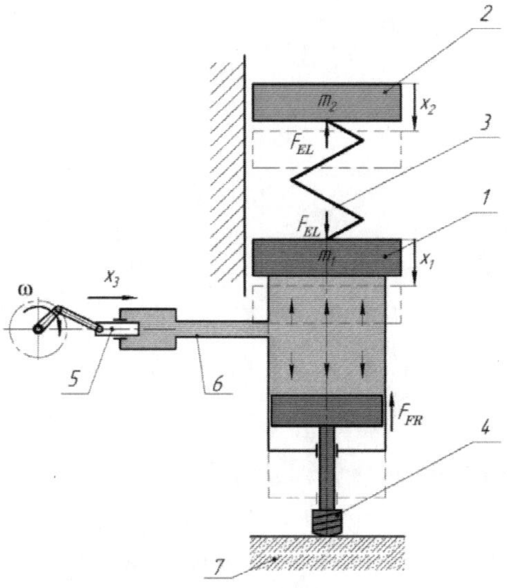

Figure 1. The model of a vertical two-mass hydroimpulsive mechanism: 1 – a hydraulic cylinder with the first active mass; 2 – the second active mass; 3 – a spring; 4 – a boring gear; 5 – a plunger (pulsator); 6 – a high pressure hose (PHP); 7 – rupturing rocks

To derive the equations, we will introduce the following generalized coordinates: x_1 – displacement coordinate of mass 1 with the hydraulic cylinder housing; x_2 – displacement coordinate of mass 2; x_3 – displacement coordinate of the plunger (Figure 2).

The differential equation of the motion of mass 1 with the hydraulic cylinder housing can be represented as:

$$m_1 \frac{d^2 x_1}{dt^2} + k_{FR} \frac{dx_1}{dt} = F_{EL} + m_1 g - p\, S_{HC}, \tag{1}$$

where m_1 – mass 1; k_{FR} – coefficient of friction between the piston and the hydraulic cylinder; p – pressure in the hydraulic cylinder; F_{EL} – elastic force from the spring; S_{HC} – area of the hydraulic cylinder piston.

The differential equation of the motion of mass 2, excluding the viscous drag from the air and the motion block, can be presented as

$$m_2 \frac{d^2 x_2}{dt^2} = -F_{EL} + m_2 g .$$

Spring force is equal to

$$F_{EL} = c\,(x_0 + x_2 - x_1),$$

where c – stiffness of spring; x_0 – the initial spring compression at system equilibrium.

Given that $x_0 = \frac{m_2 g}{c}$, we obtain the following equation after substitution:

$$m_1 \frac{d^2 x_1}{dt^2} + k_{FR} \frac{dx_1}{dt} - c\,(x_2 - x_1) = (m_1 + m_2) g - p\, S_{HC},$$

$$m_2 \frac{d^2 x_2}{dt^2} + c\,(x_2 - x_1) = 0.$$

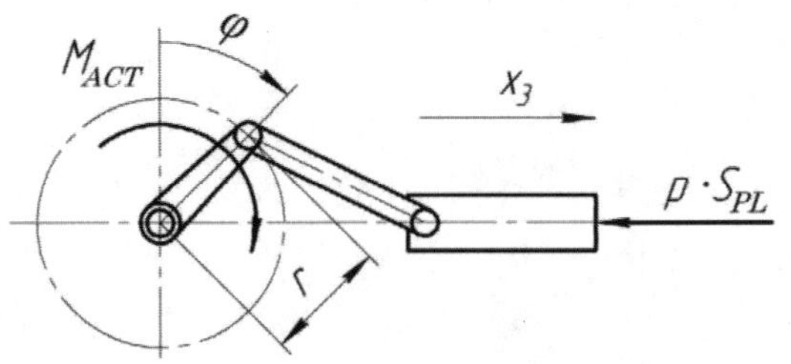

Figure 2. A scheme of the plunger actuator

The equation of the actuating rotor motion is as follows:

$$J_R \frac{d^2\varphi}{dt^2} = M_{ACT} - \chi \frac{d\varphi}{dt} - p\,S_{PL} \cos\varphi\, r\,,$$

where J_R – reduced moment of inertia of the rotating parts of the rotor; M_{ACT} – actuating torque of the motor; χ – damping factor; S_{PL} – area of the plunger, r – radius of the bell crank (Figure. 3).

The equation of the flowrate is

$$S_{HC} \frac{dx_1}{dt} + S_{PL} \frac{dx_3}{dt} = C_V \frac{dp}{dt}, \tag{2}$$

where C_V – coefficient of elasticity of the hydraulic system, which determines the flowrate per the hydraulic system elements deformation and the compressibility of the liquid.

The movement coordinate of the plunger is connected with generalized coordinate φ in the following equation:

$$\frac{dx_3}{dt} = r \cos\varphi \frac{d\varphi}{dt}.$$

After substituting it in equation (2) we obtain:

$$S_{HC} \frac{dx_1}{dt} + S_{PL}\, r \cos\varphi \frac{d\varphi}{dt} = C_V \frac{dp}{dt}.$$

Thus, we obtain the following system of differential equations:

$$\begin{cases} m_1 \frac{d^2 x_1}{dt^2} + k_{FR} \frac{dx_1}{dt} - c\,(x_2 - x_1) = (m_1 + m_2)g - p\,S_{HC}, \\ m_2 \frac{d^2 x_2}{dt^2} + c\,(x_2 - x_1) = 0, \\ J_R \frac{d^2\varphi}{dt^2} = M_{ACT} - \chi \frac{d\varphi}{dt} - p\,S_{PL} \cos\varphi\, r, \\ S_{HC} \frac{dx_1}{dt} + S_{PL}\, r \cos\varphi \frac{d\varphi}{dt} = C_V \frac{dp}{dt}. \end{cases}$$

3. Conclusion

With this system of equations, it is possible to perform modelling of the processes occurring in the hydroimpulsive mechanism. This will make it possible to calculate the required parameters for different operational modes.

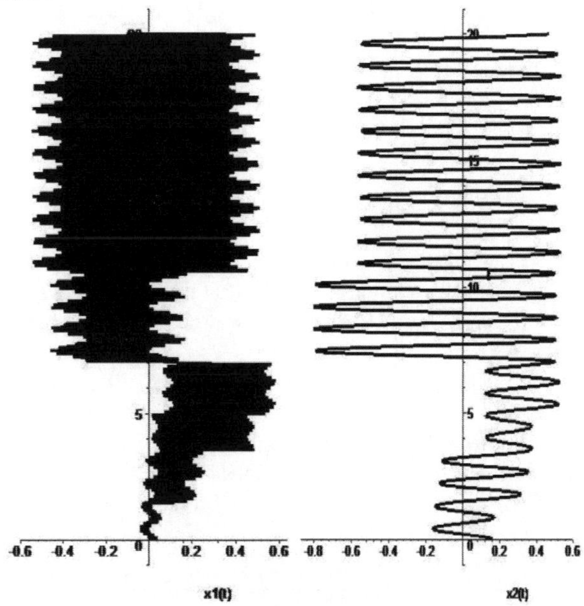

Figure 3. Charts of oscillations of masses 1 and 2 in start-up mode of hydroimpulsive mechanism.

Figure 4. Changes of hydraulic pressure.

Figures 3and Figures 4 show the results of numerical modelling of the resulting system of equations. The following system parameters were taken:

$$M_1\text{=100kg}, M_2\text{=500kg}, k_{FR} = 0.2, c = 100000\ \frac{H}{m}, C_V = 10^{-7}\ \frac{m^3}{Pa}, S_{HC}\text{=0.00785m}^2\ S_{HC}\text{=0.00314 M}^2,$$
$$\chi\text{=0.01}, M\text{=500Hm}, J_R\text{=0.005 kg·m}^2, r\text{ =0.25m}$$

The system is set in motion from the state of rest. Initial conditions are:

$$x_1(0) = 0,\ m_2\ x_2(0) = \frac{m_2 g}{c},\ \varphi(0) = 0,\ p(0) = \frac{(m_1+m_2)g}{S_{HC}},\ \frac{dx_1(0)}{dt} = 0,\ \frac{dx_2(0)}{dt} = 0,\ \frac{d\varphi(0)}{dt} = 0,$$

$$\frac{dp(0)}{dt} = 0.$$

The results of numerical modelling clearly show the effect of moving mass 2 on generated hydraulic pressure pulses - its maximum peaks match up with the chart of weight oscillations. Therefore, by controlling the natural frequency of mass 2, i.e. by varying the value of the mass or value of stiffness of spring, it is possible to receive pressure pulses of predetermined value and frequency. In addition, by controlling speed of the plunger, it is possible to ensure resonant mode of the system. In this case the pressure pulses will be maximized.

Numerical modelling of the resulting system of equations and finding its analytical solutions are a matter of the further study.

References

[1] Butenin N, Luntz J, Merkin D 1979 *Course of theoretical mechanics* **1-2** (Moscow: Nauka)
[2] Lipin A, Tanayno A, Timonin V 2006 *Mining Equipment* (St. Petersburg: Slavutich) pp 116–123
[3] Sidorenko A, Saveliev M, Zhukovsky J 1965 *Mining Electr. Engineer. and Automat.* 3 34–38
[4] Shadrina A, Kazantsev A, Saruev A, Saruev L 2007 Drilling rig for sinking wells in underground conditions. Patent No. RU 69135
[5] Shadrina A, Kazantsev A, Saruev A, Saruev L 2008 Machine for drilling in underground conditions. Patent No. RU 71369
[6] Shadrina A, Saruev L, Saruev A 2009 Dynamic processes in the pipe string in the rotary-percussive drilling small diameter of the underground mine workings (Tomsk: Tomsk Polytechnic University)
[7] Ziyakaev G, Saruev L, Martyushev N 2010 *In the world of scientific discovery* **13** 61–65
[8] Pashkov E, Saruev L, Ziyakaev G 2011 *Mining informational and analytical bulletin* **5** 26–31
[9] Pashkov E, Ziyakaev G, Kuznetsov I 2013 *Volga Scientific Herald* **4** 32–36
[10] Novoseltseva M 2014 *Advanced Materials Research* **1040** 682–685
[11] Novoseltseva M, Pashkov E 2015 *Applied Mechanics and Materials* **756** 47–52
[12] Pashkov E, Ziyakaev G, Novoseltseva M 2013 *Applied Mechanics and Materials* **379** 91–94
[13] Novoseltseva M 2013 *Mining informational and analytical bulletin* **S4** 497–500
[14] Novoseltseva M 2013 *Mining informational and analytical bulletin* **S4** 551–555
[15] Glazov A 2013 *Mining informational and analytical bulletin* **S4** 515–517
[16] Svetashkov A A, Kupriyanov N A, Manabaev K K 2012 An approximate algorithm for solving the problems of linear viscoelasticity (Comp. Cont. Mech.) pp 292–299
[17] Svetashkov A A, Kupriyanov N A, Manabaev K K 2013 *Izvestiya VUZov. Physics* 206–208

IOP Conf. Series: Materials Science and Engineering **124** (2016) 012073 doi:10.1088/1757-899X/124/1/012073

Mathematical and computer modeling of component surface shaping

A Lyashkov
Department of Engineering geometry and CAD, Omsk State Technical University, 11, Mira ave., Omsk, 644050, Russia

E-mail: 3dogibmod@mail.ru

Abstract. The process of shaping technical surfaces is an interaction of a tool (a shape element) and a component (a formable element or a workpiece) in their relative movements. It was established that the main objects of formation are: 1) a discriminant of a surfaces family, formed by the movement of the shape element relatively the workpiece; 2) an enveloping model of the real component surface obtained after machining, including transition curves and undercut lines; 3) The model of cut-off layers obtained in the process of shaping. When modeling shaping objects there are a lot of insufficiently solved or unsolved issues that make up a single scientific problem – a problem of qualitative shaping of the surface of the tool and then the component surface produced by this tool. The improvement of known metal-cutting tools, intensive development of systems of their computer-aided design requires further improvement of the methods of shaping the mating surfaces. In this regard, an important role is played by the study of the processes of shaping of technical surfaces with the use of the positive aspects of analytical and numerical mathematical methods and techniques associated with the use of mathematical and computer modeling. The author of the paper has posed and has solved the problem of development of mathematical, geometric and algorithmic support of computer-aided design of cutting tools based on computer simulation of the shaping process of surfaces.

1. Introduction

The manufacture of products in several branches of mechanical engineering industry is connected with technological processes of shaping of geometrically complex component surfaces. These components include rotors of screw pumps, impellers turbines, compressors and pumps, impeller fans, compressors and other products.

An important role at this is given to the issues of shaping modeling during the process of the cutting tool design. One of the elements of this process is designing its shaping surface. The solution to this problem is considered in many papers. Many of them state that in order to perform the calculation it is necessary to derive the appropriate dependencies as applied to the different initial data. In this case the classical method of differential geometry or the kinematic method is basically used [1]. One of the new approaches was proposed in [2] and to some extent it was used in papers [3, 4]. However the computational dependencies have the form of transcendental equations. Numerical methods are used to solve them, All this complicates the process of the cutting tool profiling.

The definition of the envelope of the family of surfaces is based on the development of mathematical models of the shaping process . This approach does not use differential methods.

However, the effective solution of the considered problem of complex surfaces shaping can be implemented using the methods of geometrical simulation by means of computer graphics and CAD systems [5, 6]. Modern computer technologies, providing the possibility of modeling the shaping process, allow us to determine the influence of various parameters of the tool, as well as its location, on the shape of the component surface. In addition, they provide the opportunity to develop software modeling the movement of the tool and the workpiece in the automated mode. In this case the problem of shaping is performed with necessary accuracy [4, 7].

Such approach provides a solution to a number of problems of technological support of dimensional processing of different surfaces: (1) profiling a shaping element of the cutting tool; (2) assigning the conditions for the relative motion of the tool and the workpiece; (3) forecasting of the possible errors in shaping of the component surface and other tasks.

This paper concerns the study of the shaping processes of technical surfaces by means of using analytical and numerical mathematical methods and techniques associated with the use of geometric and computer modeling. The study, combining the above-mentioned methods, deals with the mappings of the orthogonal projection of two-dimensional surfaces and three-dimensional hypersurfaces, specified by the equation in an implicit form and parametric equations, on the coordinate plane and the hyperplane.

2. Mathematic modeling

Let the original surface be defined by the equation in an implicit form:

$$F(x, y, z) = 0 \tag{1}$$

At points of the criminant of this surface relatively coordinate plane XY the following condition is met:

$$F_z(x, y, z) = 0. \tag{2}$$

Considering equation (2) as an equation of one more surface and having determined its intersection with original surface (1), we obtain surface criminant (1) with respect to coordinate plane XY. The projection of this criminant on coordinate plane XY is surface discriminant (1) and, correspondingly, the envelope of the family of plane congruent curves. Surface (1) is obtained by mapping of these curves in space R^3.

It was found that equation (2), reflecting the condition of tangency of a 'vertical' plane with a specified surface can also be considered as a necessary condition of existence of conditional extrema of some functions. The graphs of these functions are curves obtained at the intersection of surfaces (1) by planes parallel to coordinate planes XZ or YZ (Figure 1).

In order to determine the sufficient conditions for the existence of extrema on the curves under consideration, the second differential of the Lagrangian function is defined:

$$d^2L = -\frac{F_{zz}}{F_y} \cdot dz^2 \tag{3}$$

Then from (3) it follows that if $F_{zz}/F_y > 0$, the point of the studied intersection of the surface by the plane parallel to coordinate plane XZ or YZ, is the point of conditional maximum; if $F_{zz}/F_y < 0$, the corresponding point is the point of a conditional minimum.

Condition $F_{zz} = 0$ requires additional studies.

As a result criminant D of the surface is the combination of a set of extreme points, namely

$$D = \sum_{i=1}^{n} \min f(x, z) \vee \max f(x, z)\Big|_{x=a_i}. \tag{4}$$

The variable is coordinate z in its domain of definition.

The conducted analysis of the total curvature of the surface allowed establishing the following:

- if $F_{zz}=0$ and $F_y \cdot F_{zx} - F_x \cdot F_{zy} \neq 0$ then the Gaussian curvature at points of criminant line of the surface is negative. At the same time the average curvature of the surface in it is directly proportional to the value of the curve curvature obtained at the intersection of the surface by the plane parallel to XY;
- if $F_{zz} \neq 0, |F_x| + |F_y| \neq 0,$ the corresponding point on the surface is hyperbolic. The curve on the surface in a section plane perpendicular to axis z, has a point of inflection;
- if $F_{zz}=0$ and $F_y \cdot F_{zx} - F_x \cdot F_{zy} = 0$, the corresponding point of the surface is parabolic;
- if $F_y^2 \cdot F_{xx} - 2F_x \cdot F_y \cdot F_{xy} + F_x^2 \cdot F_{yy} = 0$ and $F_y \cdot F_{zx} - F_x \cdot F_{zy} = 0$ the corresponding point of the surface is parabolic;
- if $F_x = F_y = 0$, the point is specific on both the original curve and the surface as well as on its discriminant;
- if $2F_x \cdot F_y \cdot F_{xy} - F_{xx} \cdot F_y^2 - F_{yy} \cdot F_x^2 \neq 0,\ F_{zz} \neq 0,\ F_y \cdot F_{zx} - F_x \cdot F_{zy} \neq 0$, the corresponding point of the surface may be either elliptical or hyperbolic.

By analogy with the previous goal we have conducted a study of mapping of the orthogonal projection of the surface specified by the following parametric equations:

$$x = f_1(u,v), y = f_2(u,v), z = f_3(u,v). \tag{5}$$

In this case criminant D of the surface is the combination of a set of extreme points, namely:

$$D = \sum_{i=1}^{n} \min f_1(u,v) \vee \max f_1(u,v)\Big|_{f_2(u,v)=a_i}, z = f_3(u,v). \tag{6}$$

A variable parameter is one of the parameters of the surface in its definition domain.

From (6) it follows that:

$$F(u,v) = f_{1u} \cdot f_{2v} - f_{2u} \cdot f_{1v} = 0. \tag{7}$$

Equation (7) establishes the relationship between parameters u, v. Along with equations (5) they determine the criminant of the surface.

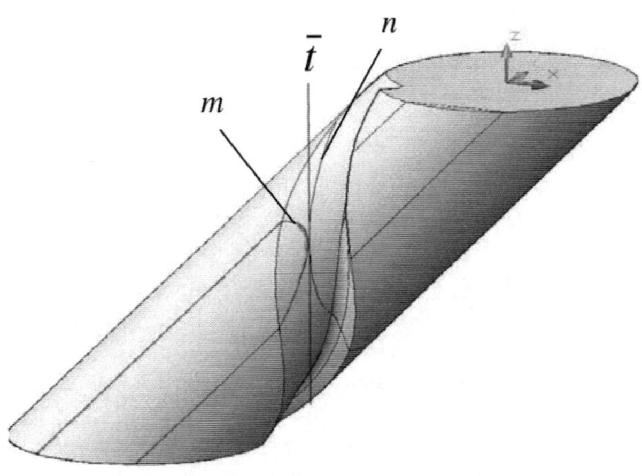

Figure 1. The model of the body and lines of its intersection with the planes parallel to coordinate planes XZ (line m) and YZ (line n); t - tangent to curve m and n

The obtained results are extended to the definition of the envelope of one-parameter family of surfaces. First, let us investigate the mapping of the three-dimensional hypersurface specified in four-dimensional space by the equation in an implicit form:

$$F(x, y, z, t) = 0. \quad (8)$$

For it discriminant D is a combination of the following set of extreme points:

$$D = \sum_{j=1}^{n}\left(\sum_{i=1}^{n} \min f(x, y, t)\Big|_{x=x_i} \vee \max f(x, y, t)\Big|_{x=x_i}\right)_{y=y_j}, \quad (9)$$

where the variable is coordinate t.

Let us now consider the three-dimensional hypersurface in four-dimensional space specified by the parametric equations:

$$x = f_1(u, v, \varphi),\ y = f_2(u, v, \varphi),$$
$$z = f_3(u, v, \varphi), t = f_4(u, v, \varphi). \quad (10)$$

Then criminant D of the surface is a combination of a set of extreme points, namely

$$D = \sum_{j=1}^{n}\left(\sum_{i=1}^{n} \min f_1(u, v, \varphi)\Big|_{y=y_i} \vee \max f_1(u, v, \varphi)\Big|_{y=y_i}\right)_{z=z_j}. \quad (11)$$

A variable parameter is one of the parameters of the hypersurface in its domain of definition.

Using the method of Lagrange multipliers from equation (11) we will obtain the connection of parameters u, v and φ

$$F(u, v, \varphi) = f_{1u}(f_{2v} \cdot f_{3\varphi} - f_{2\varphi} \cdot f_{3v}) - f_{1v}(f_{2u} \cdot f_{3\varphi} - f_{2\varphi} \cdot f_{3u} + f_{1\varphi}(f_{2u} \cdot f_{3v} - f_{2v} \cdot f_{3v}) = 0 \quad (12)$$

This equation along with equations (10) determines the discriminant of the hypersurface. Equation (12) can be regarded as the equation of the two-dimensional surface in cartesian coordinates U, V, Φ. Mapping $\Omega : R^3 \to R^4$ of this surface distinguishes its criminant (a two-dimensional surface) on a specified hypersurface relatively hyperplane XYZ.

Thus, the obtained results allow us to determine the discriminant of the two-dimensional surface and the three-dimensional hypersurface, and, consequently, the envelope of the family of plane curves as well as surfaces from the unified positions. From the analytical position one can realise it by means of conditional extrema of the corresponding functions, based on the method of Lagrange multipliers, and by means of computational methods, that is, without the use of differential surface parameters.

3. Solid-state modeling of components

The next step in the application of computer technology in the problems of surfaces shaping is to develop solid-state models of the components having a geometric shape different from 3D primitives used in the known CAD.

For these purposes one may use in particular the polygonal model of surfaces.

Thus, Figure 2 shows a model of a component with a helical groove. It is derived by cutting off a part of the body with a helical surface from the workpiece. The construction of the solid-state model of the impeller and of the models of removed allowances (Figure 3) is based on the same technology. The very solid-state model of the impeller billet was obtained using standard means of most of CAD systems. For the development of the most effective schemes of allowances removal of the interblade volume it is necessary to analyse the geometric model of the body of this allowance – P. The methods of geometric synthesis and capabilities of modern hardware and software to shape three-dimensional bodies lie in the heart of the proposed approach to the selection of alternatives of removal of the most part of the allowance.

The sequence of selection of rational schemes of removal of the part of interblade volume allowance involves the following tasks:

- analysis of the surface shape bounding the volume under investigation;
- removal of the part of the interblade volume allowance using the operations of geometric modeling;
- selection of a rational technological operation or their combinations of allowance part removal proceeding from the analysis of productivity of each of them;
- selection of the technological process, equipment and tool-set for dimensional preliminary processing of the impeller.

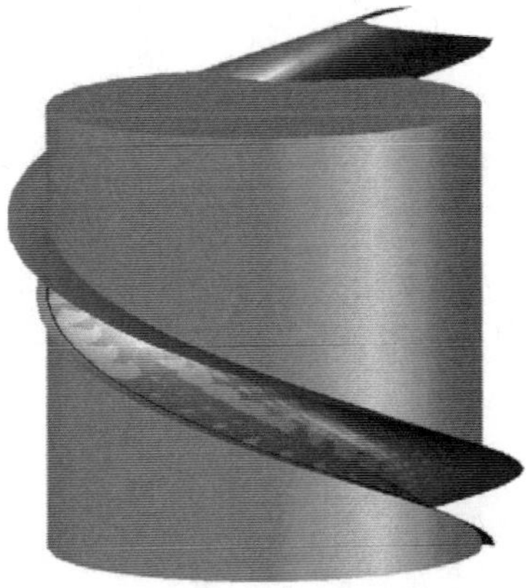

Figure 2. Shaping of the surface of the component model – cutting a segment from the workpiece

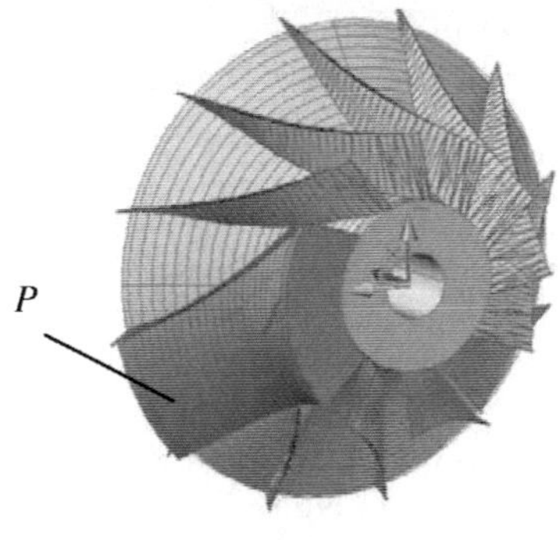

Figure 3. The solid-state model of the impeller; *P* – removable interblade volume

4. Computer Simulation of Shaping

The modeling results obtained at the previous stage are used to develop algorithms and programmes in order to appoint necessary technological conditions of the component shaping using the most rational methods of the dimensional processing. For that one uses the known solid-state modeling operations. So the kinematic operation is used to specify a multitude of tool positions in the process of the component shaping. The choice of setting and the orientation of the tool can be automated using programming languages. This operation allows modeling more complex schemes of formation of surfaces that have a complex profile. Then, using Boolean operations, the required surface is formed.

In the tasks of surfaces shaping of the components processed by the method of centroid or noncentroid rounding one uses effectively the capabilities of modern CAD-systems and the corresponding programming languages. The solution of direct and inverse problems of modeling of surface shaping processed by the method of noncentroid rounding, such as a helicoid surface - rotational surface – helicoid surface, can be one of the examples.

When solving the direct problem, the initial data are a body model with a spiral groove and a workpiece model as a section of the cylinder for the rotary body. The cylinder dimensions and its location are determined by the parameters of setting of the shaping element relatively the workpiece.

Modeling of shaping of the rotary surface on the basis of the previously created model of the product and the installation parameters is performed in accordance with the developed algorithms and programmes (Figure 4).

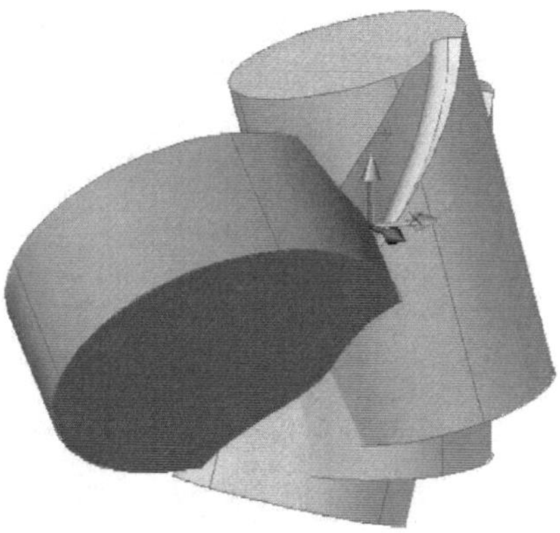

Figure 4. Shaping of the workpiece of a disc tool: an intermediate position of the models

Due to the fact that the rotary body surface is uniquely determined by its axial cross-section, only a part of the workpiece is subjected to shaping simulation. The resulting axial section may be edited both for the purpose of replacing its profile with technological curves and for some other reasons.

This section is used to obtain a solid-state model of the rotary body, on the basis of which the model of a disk cutter and its drawing is created.

The solid-state model of the tool, created on the basis of the axial section, is used to solve the inverse problem of shaping. In this case, the simulation result of shaping is a body with a spiral groove (Figure 5). The same figure shows a model of its end section, which is used for comparative analysis with the original face profile. If the resulting deflection is within the permissible range, the modeling process is completed. Otherwise, it is possible to change the parameters of mutual arrangement of the models of the tool and the product. Along with solution of direct and inverse problems of shaping the developed algorithms and programmes allow obtaining solid-state models of the cut layers in order to select the optimal technological parameters of cutting. Modeling of shaping of the component and cut layers is shown in Figure 6.

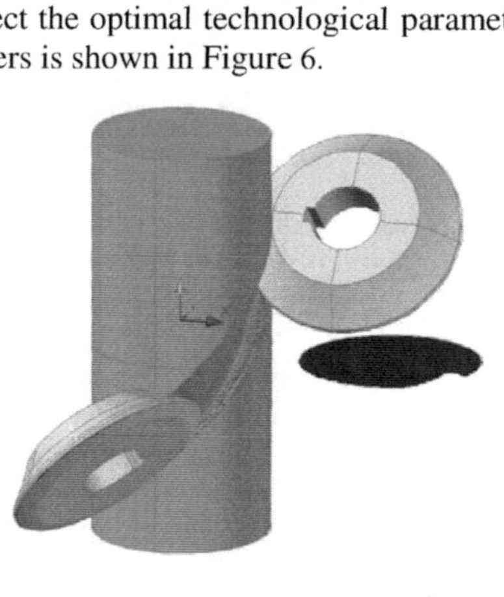

Figure 5. Shaping of the model with a screw surface: a disc tool in the initial and final states and a cross-section of the model

5. Conclusions

On the basis of the above stated information the use of geometric modeling and computer technologies allows performing the following tasks:

- to develop (on the basis of the established laws of geometry) mathematical models of surfaces shaping of the components processed by the method of centroid and decentroid rounding implemented by numerical methods; at this there is no need to obtain dependencies, which establish the connection of the surface parameters and parameters of their family;
- to obtain solid-state models of the component in order to choose the method of dimensional processing corresponding to the removal of the greatest volume of the allowance at the stage of the preliminary treatment,
- to provide the possibility of modeling of the shaping process in an automated mode;
- to assign the conditions of installation, securing and relative movement of the shaping tool and the processed surfaces;
- to predict the conditions preventing technological losses of products during the production process, associated with the failure of the relative position and the relative movement of the workpiece and the tool.

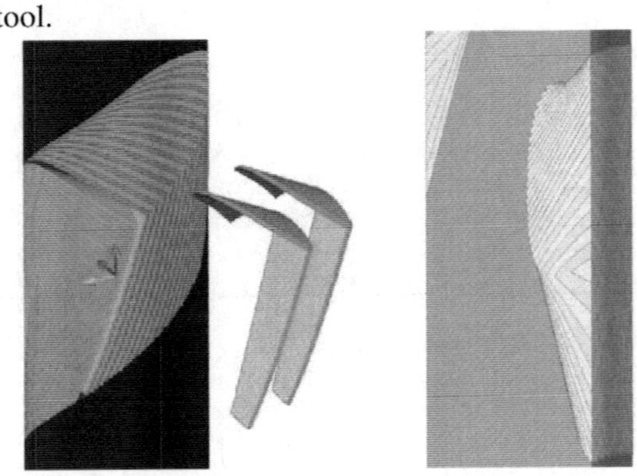

Figure 6. Computer modeling of the component and the cut layers

References

[1] Litvin F L 2004 *Cambridge University Press* 816
[2] Thom R 1962 *J. de math. pur et apple* **41(2)** 177-192
[3] Bruce J W and Giblin P G 1988 *Publisher 'World'* 263
[4] Lyashkov A A 2012 *Russian Engineering Research, Allerton Press, Inc.* **32(4)** 404-406
[5] Nikolaos T 2012 *Journal of Manufacturing Science and Engineering* **134**
[6] Pottmann H and Peternell M 2000 *Comenius University, Bratislava* 3-23
[7] Lyashkov A A and Panchuk K L 2015 *Procedia Engineering* **113** 174-180

IOP Conf. Series: Materials Science and Engineering **124** (2016) 012074 doi:10.1088/1757-899X/124/1/012074

A method of parameter optimization of the equipment configuration with the help of neural net and finite element analysis

A A Sitnikov, N A Makarova and U N Kamyshov
Altai State Technical University, 46, Lenina ave., Barnaul, 656038, Russia

E-mail: nmakarova22@gmail.com

Abstract. The authors have developed the procedure of optimization of the equipment configuration with the help of neural net based on the results of finite element analysis. The dismembrator's characteristics for feeding slop production has been optimized. The results of the virtual experiment are displayed in diagrams of pressure and temperature of the liquid inside the operating devices. The equipment of the optimized configuration has demonstrated a number of advantages compared to the prototype.

1. Introduction

Dismembrators are a variety of flour mills, the operative parts of which are made as a rotor or stator with pegs on their surface. They are widely used in chemical and manufacturing industries and agriculture. In particular, they are used for production of highly efficient feeding slop [1].

There are different approaches to the construction of the operating devices. Numerous scientists (N S Sergeev, U N Kamyshov, S V Zolotarev, V M Chervyakov, U K Sabiev) [2-4] were engaged in objectification of the constructional features of a similar type of equipment, but no suitable procedure of the efficient structural calculations of dismembrators has been developed for now.

Consequently, the aim of the research is to develop a method of optimization of the design parameters of centrifugal-rotor dismembrators for feeding slop production.

2. Theory

An algorithm of optimization of the design parameters of a centrifugal-rotor dismembrator consists of several stages.

Stage 1 represents a computer-based experiment on the operating devices of the equipment. At this stagea procedure of the virtual experiment was developed, enabling us to estimate a module of grain milling, mixture temperature and the efficiency of the equipment.

The system of Navier-Stokes non-stationary equations in their conservative form underlies the mathematical model. Is this case the authors examine the liquid, where the processes of heat transfer and internal friction take place. Therefore, the system includes a heat transfer equation and is concluded by density and specific heat capacity equations as a function of independent variables of pressure and temperature. The system of equations was solved by means of the finite element analysis software tool AnsysCFX.

Stage 2 includes optimization of the configuration with the help of neural net. At this stage, firstly, the rotor and the stator were parameterized. Secondly, the target values of the milling modulus indices, mixture temperature and efficiency of equipment were determined.

It is recommended to mill the grains before feeding. The desired size of the particles in pigs' feed is considered to be 600…700 μm. The amount of particles less than 1 mm must be 75 % of the total mass. Consequently, fine grinding (0.2…1 mm) of feeding stuff is accepted as effective [5].

The appropriate temperature of the mixture is 40 °C, which is slightly higher than the temperature of a pig's body. In case of using unheated feed, it is to be heated up in the pig's stomach before being digested, thereby a part of the nutrients consumed by the pig is wasted. So, the acceptable range of the feed temperature should be 40…60 °C.

To formulate the aim of parameter optimization the authors have specified the following characteristics: the average diameter of the particles after milling (y1, mm); mixture temperature (y2, °C).

The conditions of the operability are the following: $y1 = 0.5 \pm 0.3$; $y2 = 40 \pm 5$.

The controlled characteristics are the following: *L*r – width of the section of pegs of the rotor; *Z* – gap between the section of pegs of the stator and the rotor cavity; *K*r – the amount of rotor sections; A – peg width along the outer side; *B* – cavity width along the outer side: *H*p – height of the peg; *H*d – depth of the cavity; *G* – sharpening angle of the peg; *P* – gap between the stator and the rotor (Figure 1).

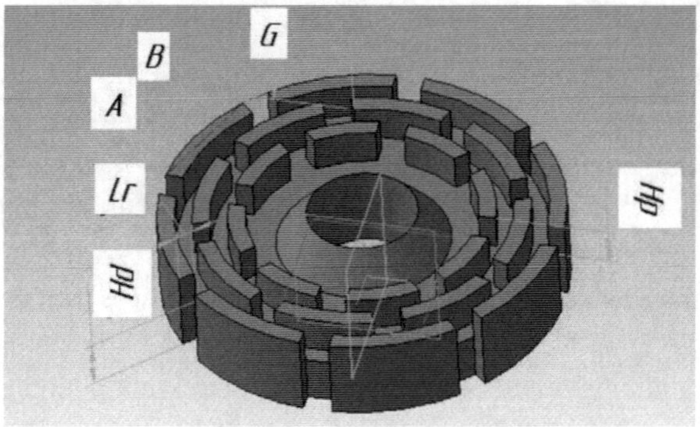

Figure 1. The constructional features of a centrifugal-rotor dismembrator.

The following step is to obtain a training sample. Since it is impossible to conduct a natural experiment by the example of a large number of equipment options, a virtual experiment was preferred.

The authors developed the plan of experimentation, carried out the virtual experiment and obtained the results. Subsequently, on the basis of these results a regression model and the training sample were developed. After application of the neural net, the authors obtained the parameters of the optimized model of operating devices of the dismembrator, allowing one to reach specified target parameters.

3.Results and discussion

The analysis of the modeling results allowed revealing the following regularities:

– the diagrams (Figure 2) reflect the zones of the increased pressure and evacuation. The zones of the increased pressure are formed, when the liquid stream encounters the moving peg of the operating device. At that moment the fluid behaves as a compressible fluid and a pressure jump occurs. The conditions of hydraulic shock origination are created;

– as a result of the liquid compression, at the moment of stream collision with the peg, the temperature of the liquid increases (Figure 3), as the span between the liquid molecules reduces and kinetic energy increases.

After optimization of the design parameters, an optimized configuration of the operating devices of the dismembrator for feeding slop production was developed. Experimental comparison with the prototype revealed the advantages of the optimized configuration.

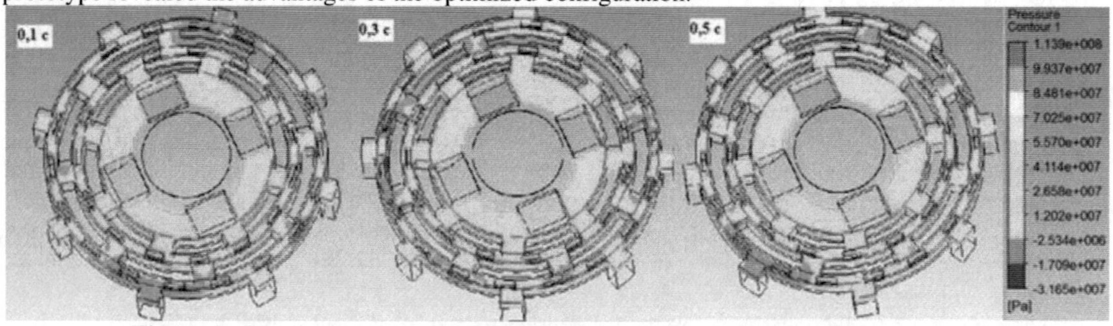

Figure 2. Contour diagrams of the liquid pressure inside the operating devices of the dismembrator.

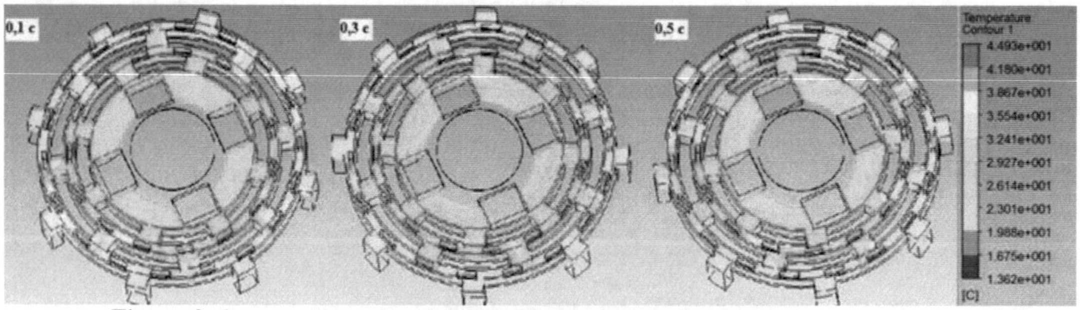

Figure 3. Contour diagrams of the liquid temperature inside the operating devices of the dismembrator.

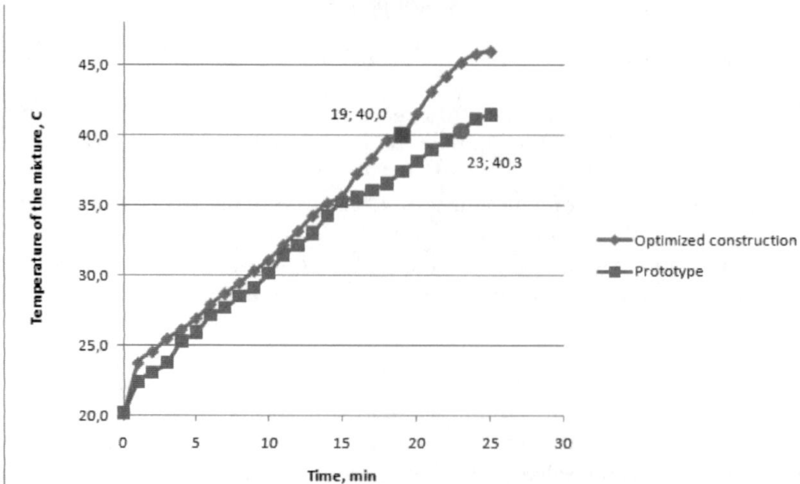

Figure 4. Dependence of the temperature of the mixture on the time of the process behavior.

Figure 4 shows the graphs of the average temperature values depending on time. The graph of dependence displays a sharp rise in heating in case of the optimized configuration. Such result is evidence of a larger size of the pegs compared to the prototype. The mixture is ground by the rotor's and the stator's walls for a longer period of time and, consequently, it is heated more quickly.

One of the most important results of the experiment is that the mixture in the optimized configuration is heated to the specified temperature (40 °C is the recommended temperature of the liquid feed) by 4 min faster (in 19 min), than that in the prototype (in 24 min).

4.Conclusion

Summing up the study mentioned above, it is necessary to note that the authors have succeeded in the development of the optimization method of the dismembrator parameters configuration for feeding slop production. This method has allowed the authors to develop a configuration which is more beneficial than the prototype in terms of the heating rate of the grain mixture.

References

[1] Kamyshov U N 2012 Justification of working bodies' design parameters of dismembrators for production of liquid feed mixture. Abstract of thesis cand. of tech. sci. (Barnaul: Altai State technical University Press.)

[2] Drjuk V A, Saleev F I, Kamyshov U N, Nefjodov K E, Pochter S V, Sitnikov A A, Nefjodov E N 2010 Dismembrator for preparing homogenized products. Patent RF 2466795

[3] Sabiyev U K, Pirozhkov D N, Sabiyev I U 2014 *Bulletin of the Altai State Agricultural University* **12(122)** 132–137

[4] Sergeev S V, Reshetnikov B A, Gordeev E N, Sergeev U S 2007 Method for miling viscous materials. Patent RF 2365469

[5] Sitnikov A A, Kamyshov U N, Makarova N A, Pochter S V 2012 *Bulletin of the Altai State Agricultural University* **3(89)** 85–88

Modelling of the automatic stabilization system of the aircraft course by a fuzzy logic method

T Mamonova, V Syryamkin and T Vasilyeva
National Research Tomsk Polytechnic University, 30, Lenina ave., Tomsk, 634050, Russia

E-mail: mamte@sibmail.com

Abstract. The problem of the present paper concerns the development of a fuzzy model of the system of an aircraft course stabilization. In this work modelling of the aircraft course stabilization system with the application of fuzzy logic is specified. Thus the authors have used the data taken for an ordinary passenger plane. As a result of the study the stabilization system models were realised in the environment of Matlab package Simulink on the basis of the PID-regulator and fuzzy logic. The authors of the paper have shown that the use of the method of artificial intelligence allows reducing the time of regulation to 1, which is 50 times faster than the time when standard receptions of the management theory are used. This fact demonstrates a positive influence of the use of fuzzy regulation.

1. Introduction

A continuous development of the aircraft industry have complicated the range, time and conditions of the flight. An increase in the range of the flight parametric variations, maximum flight altitude and speed have resulted in the new requirements to the modern aircraft equipment. Besides, unattainability of the required flight characteristics at the expense of existing design of the aircrafts, a raise in the standards of the aircraft control have led to the rise of conditions, under which modern and advanced piloted aircrafts are impossible without modern and ideal systems of automatic and semi-automatic control.

Now the development on improvement of the aircraft autopilot is actively conducted. In works [1-3] the main problems mentioned by developers are an automatic control of the azimuth angle, aircraft pitch angle, automatic piloting and development of the devices for aircraft stabilization.

Designers of the automatic control system of the azimuth angle and a restriction of the normal aircraft overload in [6] searched the solution of a problem of ensuring the necessary accuracy of normal overload restriction when the aircraft turns at a specified azimuth angle. They have found a solution having introduced an automatic machine of normal overload restriction by means of the algebraic selector of the minimum signal into the system of azimuth angle automatic control.

The authors of the design consider the device of automatic piloting equipped with the bearing screw, which is proposed as a way of facilitation of the pilot's control of the aircraft equipped with the bearing screw. The bearing screw contains, at least, one pushing propeller. Such way of facilitation allows minimizing the aircraft aerodynamic drag. This device contains the means of starting of the automated piloting mode allowing one to hold an attack angle constant. This means is connected to the processing block, which in its turn controls a collective blade pitch automatically when the automated piloting mode with holding of the attack angle is switched on. In Russia the similar main designs belong to the military aviation. The application of artificial intelligence methods for passenger airplanes, in particular a fuzzy controller, has not been revealed yet. Therefore, the task assigned within this paper is actual.

The system under development has to provide an automatic stabilization of the aircraft course necessary for flight realisation according to the azimuth reference. A fuzzy logic application in the modelled system is substantiated by the following advantages: the lack of necessity in the precise mathematical formulation of the controlled object, an opportunity of relatively easy stabilizer resetting and an opportunity of increasing the input variables number.

2. Development of the automatic stabilization system of the aircraft course by a fuzzy logic method

Flight control of the aircraft consists in retaining of the predetermined motion path of its centre of mass, orientation and stabilization relatively the centre of mass. The solution of these tasks is provided with automatic or semi-automatic control systems, in which the aircraft is rendered as a control object.

The key parameter, defining the aircraft motion direction, is the course – an angle calculated clockwise between the direction of the meridian and a projection of the longitudinal aircraft axis onto a horizon plane. Depending on the type of the meridian (geographical, magnetic, compass) used as a reference line, there are true, magnetic and compass courses [9]. The aircraft automatic control systems include autopilots, gyroscopes, remote control and inertial systems and others.

The basis of the autopilots scheme when operating in the stabilization mode represents a principle of angle control, angular velocity control and angular acceleration control. System “aircraft-autopilot” is a closed loop system of automatic control and management consisting of three automatic control systems with inner cross connections (a channel of the elevators control, a channel of the ailerons control, a channel of the rudder control). All three systems are usually implemented according to identical schemes. The scheme of the rudder control channel is given in figure 1, where δ_H – a deviation angle of the rudder; φ_1 – a preset value of the azimuth angle; φ – an azimuth angle; 1 – a setting mechanism of the azimuth angle; 2, 3 – measuring devices (sensors φ and ω_φ); U_φ, $U_{\varphi 1}$, U_{ω_φ} – stresses proportional to values; 4 – a summator; U_Σ – total stress; 5 – an amplifier; k – an amplifier gain; 6 – a steering engine.

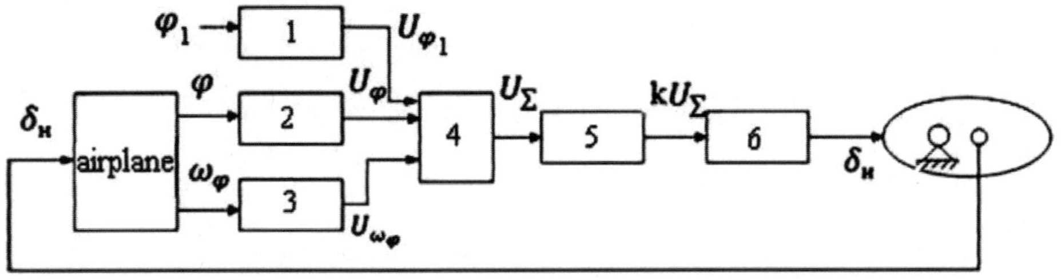

Figure 1. A block diagram of the automatic stability control of an aircraft course.

A channel consists of measuring devices *2* and *3*, summator *4*, amplifying *5* and actuating *6* devices. Measuring devices *2* and *3* of the AP-type autopilot (a banking angle sensor) measure the values of angle deviations and angular speed (azimuth angle φ, azimuth angular speed ω_φ) and convert these deviations to stress values U_φ and U_{ω_φ} After algebraic summation in summator *4*, sum of stresses U_Σ is amplified by amplifier *5* and is fed to the autopilot steering engine, which in its turn deflects the rudder, as a result of which the emerged azimuth deviation angle is eliminated. For setting of a certain value of azimuth angle φ a setting mechanism of the azimuth angle is provided in the autopilot scheme.

For smoother stabilization of the aircraft it is necessary to consider not only angle φ itself, but also the rate of its variation $\frac{d\Delta\varphi}{dt}$. It means that sharper aircraft deviations will correspond to sharper rudders deviations. In this case it is important to take into consideration angular accelerations $\frac{d^2\Delta\varphi}{dt^2}$ as well. In this case the autopilot's equation for one channel of the course has the following view:

$$\delta_H = k_H \Delta\varphi + k'_H \frac{d\varphi}{dt} + k''_H \frac{d^2\varphi}{dt^2},$$

where k_H, k'_H, k''_H – transfer ratios of the autopilot by the angle, angular speed and angular acceleration of the course channel, respectively.

In this case the proportional feedback is unnecessary as the rudder deviation will cause the immediate negative acceleration of the aircraft counteracting the disturbing aircraft deviation. When considering a total time signal, the autopilot stabilizes the aircraft along a trajectory, i.e. returns it to its former trajectory of flight. In this case the autopilot's equation for one channel (of the course) is written as follows:

$$\delta_H = k_H \int_0^t \Delta\varphi dt + k''_H \Delta\varphi + k'_H \frac{d\varphi}{dt},$$

Such control is called astatic.

Nowadays flight gyroscopes, remote indicating magnetic compasses, compass systems are applied as sensitive elements reacting to angular aircraft deviations relatively the center of gravity.

For realization of the autopilot with application of fuzzy logic it is necessary to develop a scheme for aircraft behaviour modelling. An aircraft transfer function, which should be fed with the signal from the fuzzy logic block and the output of which should contain a variable equal to the aircraft rudder angle of rotation, is used as initial data. But any system is affected by external disturbances, for example wind, therefore the output characteristic will consist of both a component having passed through the aircraft transfer function and a component of the external disturbance passing through its transfer function. For provision of management, it is necessary to read off the output characteristic, an aircraft angle of rotation, as well as to measure the rate of the rotation angle change. These characteristics should be transmitted to the fuzzy logic block. As a result we obtain the following block diagram presented in figure 2.

Fuzzy system modelling happens in a fuzzy logic package of Fuzzy Logic Toolbox allowing one to design fuzzy expert and/or controlling systems.

For creation of the stabilization system let us formulate necessary fuzzy rules to control the course rudder using Mamdani inference algorithm.

The input variables are the deviation angle of the control rudder and the rate of the rudder deviation angle change.

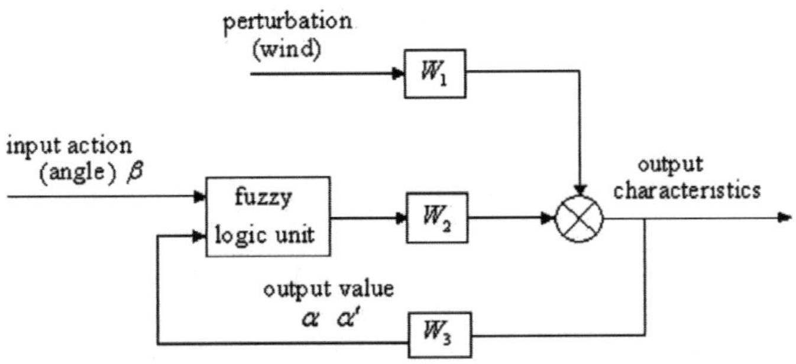

Figure 2. The block diagram of the modelled system.

3. Research of an imitating model of the automatic stabilization system of the aircraft course

For creation of the modelled system it is necessary to assemble an imitating model, being guided by the structural flowchart shown in figure 4. An inspection of the operability of the fuzzy controlling device is carried out in the Simulink environment. The scheme used for operability inspection is presented in figure 3.

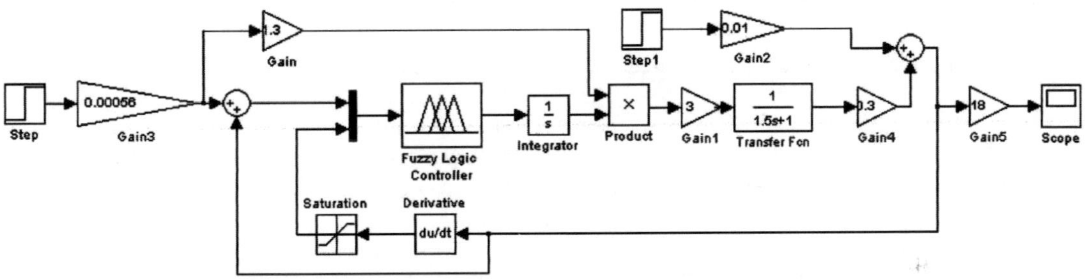

Figure 3. A scheme of the system created in the Simulink environment.

The scheme of the automatic stabilization system of the aircraft course consists of the following blocks [11]:

- Step – a step generator;
- Sum – a block of sum calculation, the sum of the current signal values is calculated;
- Gain – an amplifier, which multiplies an input signal by a constant coefficient;
- Mux – a multiplexer, which unites input signals into a vector;
- Fuzzy Logic Controller – a fuzzy controller;
- Derivative – a block of the derivative calculation;
- Integrator – an integrating block, which integrates the input signal;
- Saturation – a restriction block, which restricts the signal swing/value;
- Product – a multiplication block, which calculates the product of the current signal values;
- Transfer Fcn – a block of the transfer characteristic that sets a transfer function in the form of the polynomial relation;
- Scope – a simulated oscillograph that displays the diagrams of the output signals in the model.

The transfer function of the control object used when modeling the system operation, namely a fuselage, has a view of a lag element of the first order:

$$W(s) = \frac{1}{1.5 + 1}$$

The necessary coefficients used for realisation of the working capacity are:

- a transfer constant of the angle setting mechanism equal to 1/18 = 0.05556 V/grad;
- a coefficient of conversion from the voltage signal into degrees equal to 18 grad / V;
- a transfer constant of the rudder mechanism equal to 3grad /V;
- a transfer constant of a sensitive element being 0.3 V /grad;
- a transfer constant of the perturbation action equal to 0.001 grad/N.
- a signal amplification coefficient equal to 1.3.

A system response to a single step excitation and the wind perturbation action, when excluding the fuzzy regulator from the block diagram, is given in figure 4.

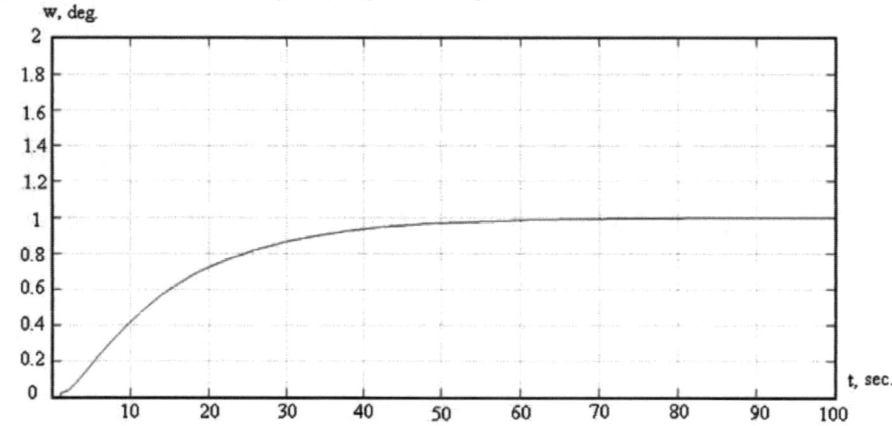

Figure 4. A diagram of the real value of the azimuth angle degree.

As one can see from the diagram, the process is established at the value of 55 s, which is not a suitable value by response speed.

In figure 5 the curve of the transition process in the automatic control system, which includes the adaptive fuzzy correcting device, is presented. The input is fed with a unit stepwise impulse as well, and the perturbation action occurs.

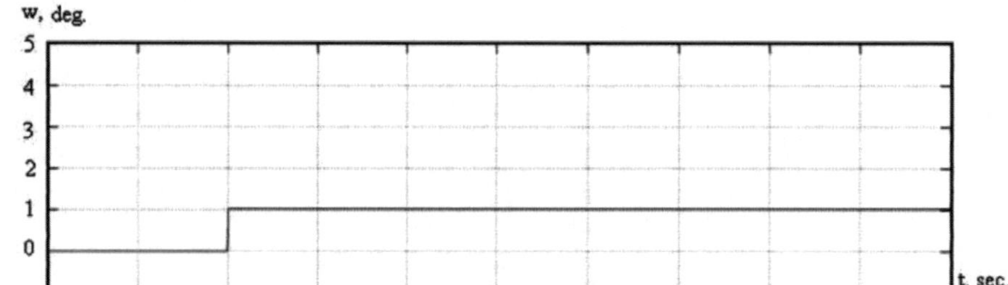

Figure 5. A diagram of the adjustment of the setting value of the azimuth angle degree (in case the perturbation action deviates towards a greater side).

Thus, the results of modelling have showed that the fuzzy regulator provides high response speed of the system and rather a small error in the established mode. Also it should be noted that for the considered system of the automatic stabilization by the aircraft course of the plane having lags, nonlinearity and feedback, the use of the fuzzy regulator is reasonable since the synthesis of the linear digital regulator is quite a complex and labour-consuming task.

The system considered in the paper demands further development and enhancement, namely an increase of the membership functions and corresponding fuzzy rules, so that there is an opportunity to conduct the experiment at the laboratory aircraft stand.

4. Conclusion

The presented work includes modelling of the automatic stabilization system of the passenger plane course with application of fuzzy logic. The advantage of the similar system is that fuzzy logic does not completely replace traditional techniques of control, and on the contrary it is used in combination with traditional techniques and allows one to simplify the creation and to extend the opportunities of traditional methods. The positive moments consist in the following:

- the experience of operators and developers in adjustment of control and management is formalized and united;
- a simple control method for complicated processes is proposed;
- the experience of controlling the processes of this type is considered, regarding all kinds of exceptions and system peculiarities;
- all kinds of initial data are considered and various initial data are combined.

References

[1] Vitzilaios N, Tsourveloudis N 2008 *Submitted to the IEEE International Conference on Robotics and Automation April* (U.S.A.: New Orleans) pp 9975–9981

[2] Tsourveloudis N, Doitsidis L, Valavanis K 2005 *Cutting Edge Robotics* (ARS Press) pp 291-310

[3] Doitsidis L, Valavanis K, Tsourveloudis N, Kontitsis 2004 *Proceedings of the IEEE International conference on robotics and automation* (U.S.A.: New Orleans) 4041–4046

[4] Petrovas A, Rinkevičienė R 2012 *Automatic control theory I, II: a laboratory manual.* (Vilnius: Technika) p 98

[5] Ogata K 2010 *Modern control engineering* 894

[6] Golnaraghi F, Kuo B 2010 *Automatic control systems* (Wiley) p 786

[7] Hassan W, Rahmat M, Mansor S 2011 *Australian Jour. of Basic and Appl. Scienc.* **5(12)** 1065–74

[8] Kisabo A, Agboola F, Osheku C, Adetoro M and Funmilayo A 2012 *Journal of Scientific Reserch & Reports* **1(1)** 1–16

[9] Nurbaiti W, Rahmat M, Jusoff K 2010 *European Jour. of Scientific Res.* **42(2)** 184–194

[10] Lazic Dragan V, Waqar Shahid 2012 *FPGA based longitudinal and lateral controller implementation for a small UAV* World academy of science, engineering and technology p 46

[11] Sivanandam S, Sumathi S, Deepa S 2007 *Introduction to fuzzy logic using MATLAB.* (Berlin Heidelberg: Springer) p 102

Mathematical modelling of the liquid atomization process by cocurrent gas flow

V A Arkhipov[1], V M Boiko[2], V D Goldin[1], E A Maslov[1], S E Orlov[1], S V Poplavskiy[2], A S Usanina[1] and I K Zharova[1]
[1] Tomsk State University, 36, Lenina ave., Tomsk, 634050, Russia
[2] Siberian Branch of Russian Academy of Sciences, Khristianovich Institute of Theoretical and Applied Mechanics, 4/1, Institutskaya st., 630090, Novosibirsk, Russia

E-mail: zharova@niipmm.tsu.ru

Abstract. This paper focuses on the physical-mathematical model of liquid atomization in the spray pattern of an ejection nozzle. A flow field of a gas phase behind the nozzle section is computed using the Ansys Fluent package. Dynamics of molten metal droplets in the gas phase within a trajectory approach is calculated. Using the presented model, numerical calculation results are given.

1. Introduction

In metallurgy, spraying techniques involving ejection nozzles are widely used to produce the high-dispersive powders from molten aluminum [1-3]. The jet of molten aluminum is dispersed by a hot nitrogen flow. It results in formation of poly-dispersed droplets that are crystallized in the form of powder of a termination product in cases of refrigeration in a sedimentation chamber. The morphology of the aluminum particles is significantly affected by time and dynamics of a phase change of the particles.

This study includes the results of mathematical modeling for a two-phase flow in the spray pattern of an ejection nozzle.

2. Results and discussion

2.1. Calculation of a flow field of a gas phase

The problem of gas jet efflux in a sedimentation chamber is considered in a non-viscous formulation using Eulerian equation in a perfect gas [4]. It is assumed that the flow is axisymmetric. Figure 1 shows the scheme of the computational region. The hot gas (nitrogen) is supplied through channel 1 through section AB. In the tapered section of the channel it is accelerated to a sonic velocity that is achieved in the vicinity of cross section CD. Then the hot gas moves in sedimentation chamber 2. In the working chamber the gas velocity in the jet initially increases up to a supersonic one and then it gradually decreases moving from the nozzle.

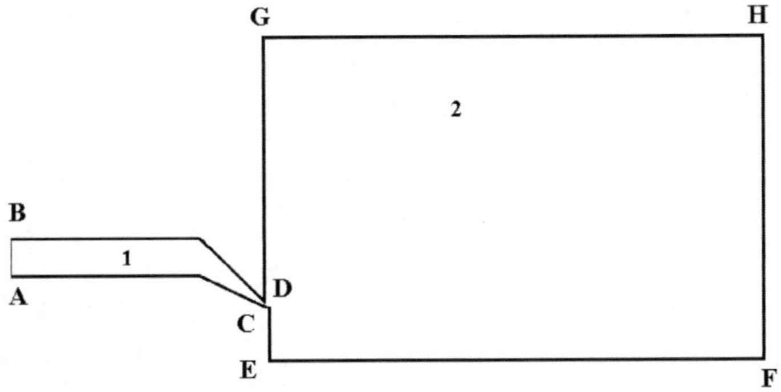

Figure 1. A scheme of the computational region:
1 – gas delivery port; 2 – sedimentation chamber; EF – axis of symmetry.

The flow field is described by Eulerian equations:

$$\frac{\partial \rho}{\partial t}+\frac{\partial}{\partial x}(\rho u)+\frac{1}{r}\frac{\partial}{\partial r}(\rho v r)=0,$$

$$\frac{\partial}{\partial t}(\rho u)+\frac{\partial}{\partial x}(\rho u^2)+\frac{1}{r}\frac{\partial}{\partial r}(\rho u v r)=-\frac{\partial p}{\partial x},$$

$$\frac{\partial}{\partial t}(\rho v)+\frac{\partial}{\partial x}(\rho u v)+\frac{1}{r}\frac{\partial}{\partial r}(\rho v r^2)-\frac{\rho u^2}{r}=-\frac{\partial p}{\partial r},$$

$$\frac{\partial}{\partial t}(\rho E)+\frac{\partial}{\partial x}[u(\rho E+p)]+\frac{1}{r}\frac{\partial}{\partial r}[v(\rho E+p)r]=0.$$

Here x, r are the axial and radial coordinates; t is the time; ρ is the density; u, v are the axial and radial components of the velocity; p is the pressure; $E=h-\frac{p}{\rho}+\frac{1}{2}(u^2+v^2)$ is the total energy per unit mass; $h=C_pT$ is the static specific enthalpy; C_p is the specific isobaric heat capacity; T is the temperature.

The system of Eulerian equations is supplemented by the equation of state:

$$h=\frac{\gamma}{\gamma-1}\frac{p}{\rho},$$

where γ is the adiabatic index.

Eulerian equations are solved with the following boundary conditions.

In cross section AB the values of the stagnation temperature. Mach number and gas pressure were predetermined; on the boundary of regions GH and FH the values of undisturbed pressure were predetermined, $p = 10^5$ Pa; on walls CE, DG the impermeability condition was imposed. The problem was solved using program package ANSYS Fluent 13.0.

A computational grid was composed of about 44000 meshes. Across the channel, 20 meshes were taken; in the working chamber the grid with the size of a mesh less than 1 mm was nonregular.

The problem was solved by explicit Rou method with the first order of both time and spatial variable approximation. Quiescent conditions with undisturbed pressure and temperature values were assigned as an initial field. The Courant number in the calculations performed was equal to 0.8. During solution of the problem, the flow field appearing in cases of ingress of the jet from the initial cross in the channel firstly and then in the working chamber was determined. The calculation was conducted up to physical time $t = 10^{-4}$ s. The calculation time was equal to about 4 hours using PC with CPU frequency of 2 GHz. In this interval of time, the establishment of the flow field was not completely achieved but the picture in the vicinity of the nozzle changed. Figures 2, 3 show the typical results of the numerical calculation.

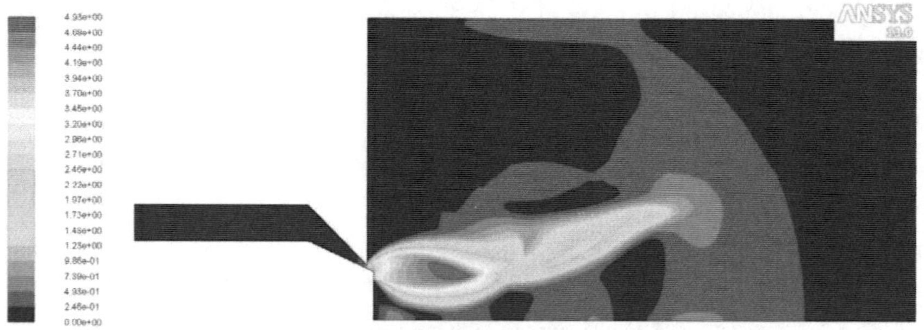

Figure 2. Mach number in the working region.

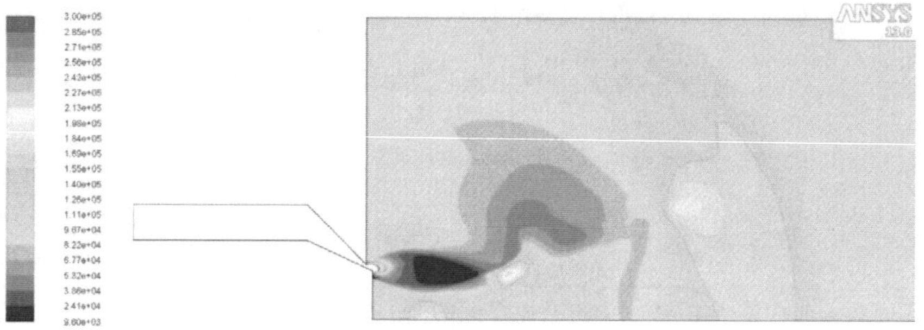

Figure 3. Pressure field in the working chamber.

During the calculations, the geometrical sizes of the region were assigned using the conditions of technical requirements. At the channel input (the cross section AB) the values of stagnation temperature $T = 773$ K, pressure $p = 6.08 \cdot 10^6$ Pa and Mach number $M = 0.055$ were assigned.

Figure 2 shows the distribution of Mach number in the working chamber. In the working chamber the jet comes with a nonuniform velocity profile. Mach number averaged over the cross section is equal to about 1.5. Then in the vicinity of the nozzle the jet is accelerated up to $M \approx 5$ due to nozzle divergence. The velocity decreases along with the increase of the distance from the nozzle. In the vicinity of the symmetry axis the flow always remains subsonic. Figure 2 clearly shows the contours of the jet.

Figure 3 shows the structural contours of the pressure field in the sedimentation chamber. At the outlet from the nozzle in cross section CD the pressure is nonuniform; it changes from 10^6 Pa on the lower edge of the cross section (near point C) to $2.9 \cdot 10^6$ Pa on the upper edge of the cross section (point D). In the sedimentation chamber in the jet core the pressure falls up to $0.96 \cdot 10^4$ Pa, while moving from the jet the pressure increases.

During the quantitative estimation of parameters of the hardening process it is necessary to take into account the change of the particles velocity. The particles that are involved in motion by a gas flow due to drag forces with time tend to equalize their velocity with the velocity of the gas flow. It results in a sufficient decrease in their heat exchange with carrying medium. Both the cooling time of particles (depending on a particles size) and the time of the hardening process essentially grow. As a result, the parametric analysis of the cooling process of the molten metal particles and their hardening taking into account the change of the particle motion velocity at constant values of the gas velocity and temperature has been conducted.

2.2. Calculation of a motion path of a particle

An equation for movement of particles with constant radius r at constant velocity u and density of the blowing gas ρ is as follows:

$$\frac{4}{3}\pi r^3 \rho_p \frac{dv}{dt} = \pi r^2 C_D \frac{\rho(u-v)|u-v|}{2}, \tag{1}$$

$$v|_{t=0} = v_0,$$

where ρ_p is the density of the particles.

The drag coefficient was determined by Klyachko-Mazin's formula:

$$C_D = \begin{cases} \frac{24}{\mathrm{Re}} + \frac{4}{\sqrt[3]{\mathrm{Re}}}, & \mathrm{Re} < 900; \\ 0.44, & \mathrm{Re} \geq 900, \end{cases} \tag{2}$$

where $\mathrm{Re} = \frac{2r\rho|u-v|}{\mu}$ is the Reynolds number; μ is the factor of the dynamic viscosity of the streaming gas.

We integrated equation (1) for the corresponding Reynolds numbers taking into account (2) and obtained dependence Re(t) for the whole period of the particle motion

$$\mathrm{Re}(t) = \begin{cases} 6\sqrt{6}\left[\left(1+\frac{6}{\mathrm{Re}_{t=0}^{2/3}}\right)\exp\left(\frac{3\mu}{r^2\rho_p}t\right)-1\right]^{-3/2}, & \mathrm{Re}(t) < 900; \\ \frac{2r\rho}{\mu}\frac{|u-v_0|}{1+0.165\frac{\rho}{r\rho_p}|u-v_0|t}, & \mathrm{Re}(t) > 900. \end{cases}$$

The comparison of the calculation results on analytical dependence of equation solution (1) and numerical solution showed that they practically coincide at integration step $\Delta t = 10^{-7}$.

Figure 4 shows the typical calculation results for different particles sizes.

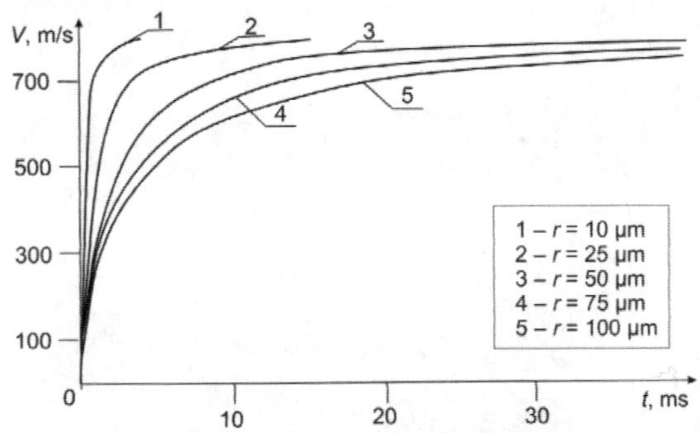

Figure 4. Dependence of the particle movement velocity on time; gas velocity is 800 m/s.

The analysis of the aluminum particles movement taking into account the flow parameters of carrying medium was conducted for specified conditions of the sedimentation chamber. Figure 5 shows the motion pattern that is obtained for the known flow field (without considering the inverse impact of particles on the gas medium) within the trajectory approach. Coordinate of the point of the particles injection in the cross section $x = 0$ is radius $r = 8$ mm that is measured from the axis of the atomizing nozzle.

The evolution of the particle motion pattern is determined by a significant value of the radial velocity of the gas that is directed to the axis of the nozzle due to free jet expansion. Then, the particles movement is determined by the parameters of the jet along with the increase of the distance from the injector exit (Figures 2, 3).

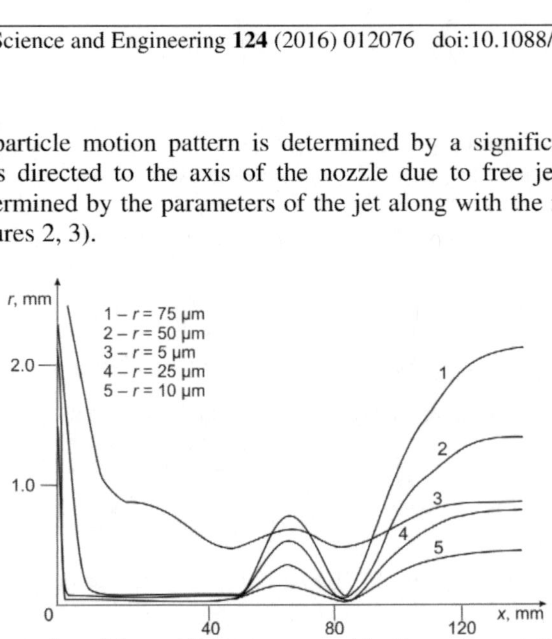

Figure 5. Trajectories of the particles movement in the sedimentation chamber in the vicinity of the nozzle.

2.3. *Experimental study*

The experimental study was carried out to verify the calculation results using the Particle Image Velocimetry method (PIV-method). PIV-method is a widely accepted method for measuring of the velocity field of particles in a fluid and gas flow [5]. Figure 6 shows the structure of the gas jet at different values of the pressures ratio in the supply chamber and ambient medium N_{pr}.

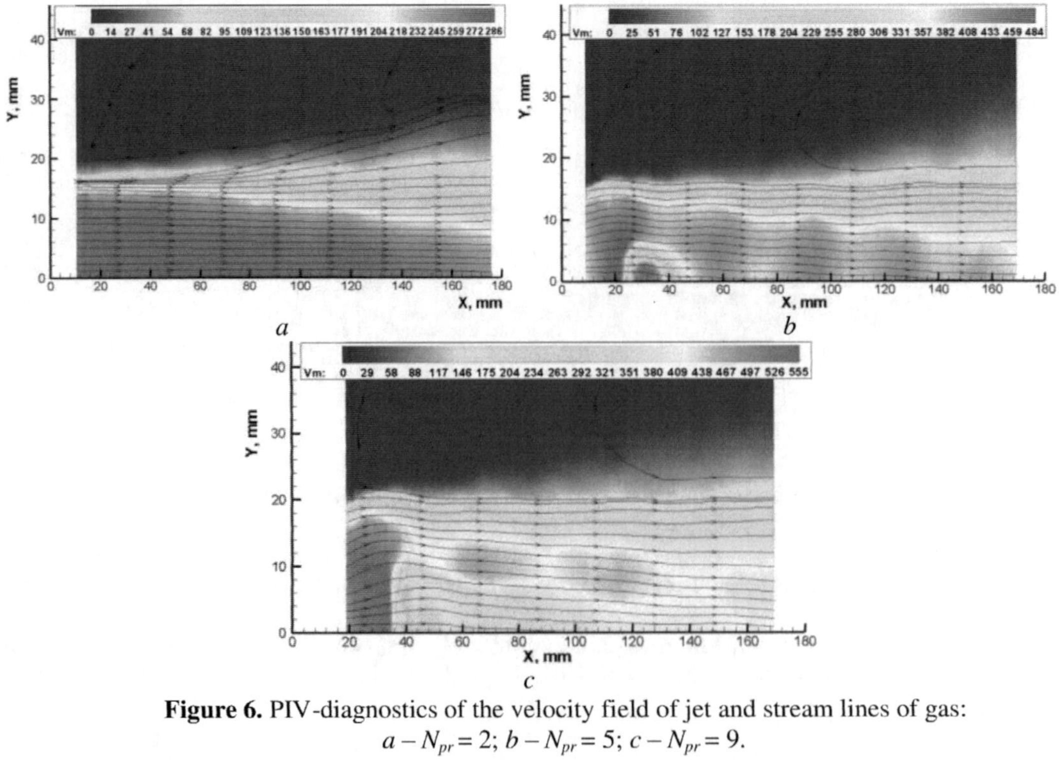

c

Figure 6. PIV-diagnostics of the velocity field of jet and stream lines of gas: $a - N_{pr} = 2$; $b - N_{pr} = 5$; $c - N_{pr} = 9$.

Figure 7 shows a photograph of a gas jet at value $N_{pr} = 9$.

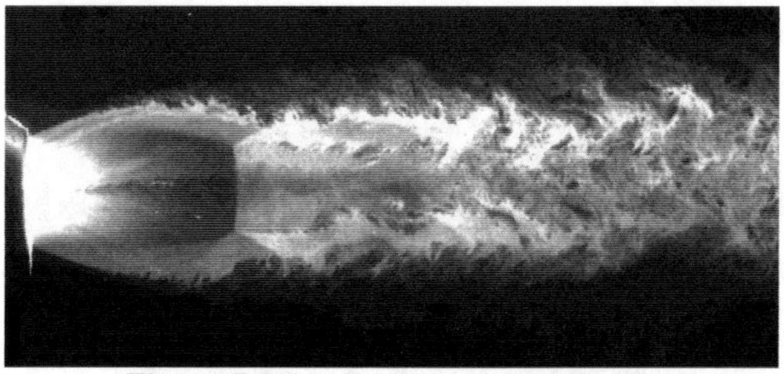

Figure 7. Visualization in a laser knife.

3. Conclusion

The analysis of the numerical results using two approaches to a mathematical description of the crystallization process of the aluminum particles in a spray pattern of the injection nozzle has showed:

– the rate of throughput and stagnation temperature of atomizing gas in the ranges, which are used in the industrial aluminum powder ASD technology, practically does not influence the characteristics of crystallization;

– approximation dependences of the characteristics of the crystallization process (duration of the crystallization process, moment of the time and coordinate along the axis of the jet that corresponds to the beginning of the crystallization, length of the crystallization region) on aluminum particles sizes.

The presented results allow optimizing technological parameters of injection nozzles for obtaining the particles of powder with the specified morphology.

Acknowledgment

This study was funded by the grant from the Russian Science Foundation (Project No. 15-19-10014).

References

[1] Silaev A D and Dishman B D 1983 *Dispersing of liquid metals* (Moscow: Metallurgiya)
[2] Arkhipov V A, Evsevleev M Ya, Zharova I K, Zhukov A S, Zmanovskiy S V, Kozlov E A and Konovalenko A I 2014 *Russian Jour. of Non-Ferr. Metal.* **4** 8
[3] Arkhipov V A, Evsevleev M Ya, Zharova I K, Zhukov A S, Zmanovskiy S V, Kozlov E A and Konovalenko A I 2015 *Russian Jour. of Non-Ferr. Metal.* **1** 3
[4] Loitsyanskii L G 1973 *Fluid mechanics* (Moscow: Nauka)
[5] Boiko V M, Orishich A M, Pavlov A A and Pikalov V V 2009 *Methods of optical diagnostics in an aerophysical experiment* (Novosibirsk: Novosibirsk State University Press)

Mathematical modelling in the computer-aided process planning

S Mitin and P Bochkarev
Yuri Gagarin State Technical University of Saratov, 77, Politechnicheskaya st., Saratov, 410054, Russia

E-mail: ser_gen@inbox.ru

Abstract. This paper presents new approaches to organization of manufacturing preparation and mathematical models related to development of the computer-aided multiproduct process planning (CAMPP) system. CAMPP system has some peculiarities compared to the existing computer-aided process planning (CAPP) systems: fully formalized developing of the machining operations; a capacity to create and to formalize the interrelationships among design, process planning and process implementation; procedures for consideration of the real manufacturing conditions. The paper describes the structure of the CAMPP system and shows the mathematical models and methods to formalize the design procedures.

1. Introduction

The researches related to the computer-aided process planning (CAPP) are relevant throughout the development of mechanical engineering. The reason for that is a regular change of machined parts range, requirements to reduce the period of manufacturing preparation, expansion of technological capabilities of enterprises.

Process planning includes several activities to determine economic and competitive methods for processing workpieces and parts; by means of the methods the parts can be produced at all stages of machine processing [1]. The computer-aided process planning system is an essential part of computer-integrated manufacturing [2]. The aim of CAPP is to automate the development of process plans to a great extent. However, there are no absolutely satisfiable solutions in automated design of machining operations, for instance, formation of the operation structure defining a sequence of manufacturing operations, selection of equipment, tools and fixtures. An approach to decision-making formalization is considered in [3]. There is a description of the decision logic schema of process planning using backward chaining reasoning.

An integration between process planning and production scheduling is also a purpose of CAPP [4]. It is necessary to solve a problem of securing the relationship between the systems of planning and implementation for this purpose achievement. A series of researches considers the possible solutions of this problem. In research [5], the authors used a genetic algorithm optimization methodology during preliminary and detailed process planning. Possible operation sequences are generated by analyzing the constraints and using a genetic algorithm. The architecture of the integrated CAD/CAPP system is presented in [6]. There is a brief discussion of the content and methodology of STEP, and STEP based strategies for CAD/CAPP information integration are pointed out.

Thus, these researches and many other works consider particular tasks for solving the problem of CAPP, but they do not deal with an end-to-end approach to a creation of the integrated CAPP system.

The first part of this paper presents a model of multiproduct machining process planning, and describes its distinctive features and advantages.

The second part considers the subsystem of the machining operation design. Its structure model is given, the models and methods for formalization of the key design procedures are proposed.

The third part describes a database on capabilities and current states of equipment and tools.

In conclusion, the article presents the main findings and proposals to use the results.

2. The model of the process planning system

Currently it is necessary to have a tight link between process planning and implementation. Figure 1 shows the proposed scheme of the computer-aided multiproduct process planning system (CAMPP). The basic approaches to its creation are presented in [7]. The CAMPP system has distinctive peculiarities in comparison with the existing CAPP systems:

1) process plans development at the level of machining operations is fully formalized [8, 9, 10];
2) availability of the capacity to create and to formalize the interrelations among design, process planning and process implementation [11];
3) provision of procedures to consider the real manufacturing conditions [12];
4) availability of the database on capabilities and current states of equipment and machine tools.

The production program is initial data containing the information about the manufactured parts. Every part is represented as a set of features. There are defined dimension interrelations among features and listed accuracy and surface quality requirements. The manufacturing system has necessary information about machine tool resources and equipment capabilities.

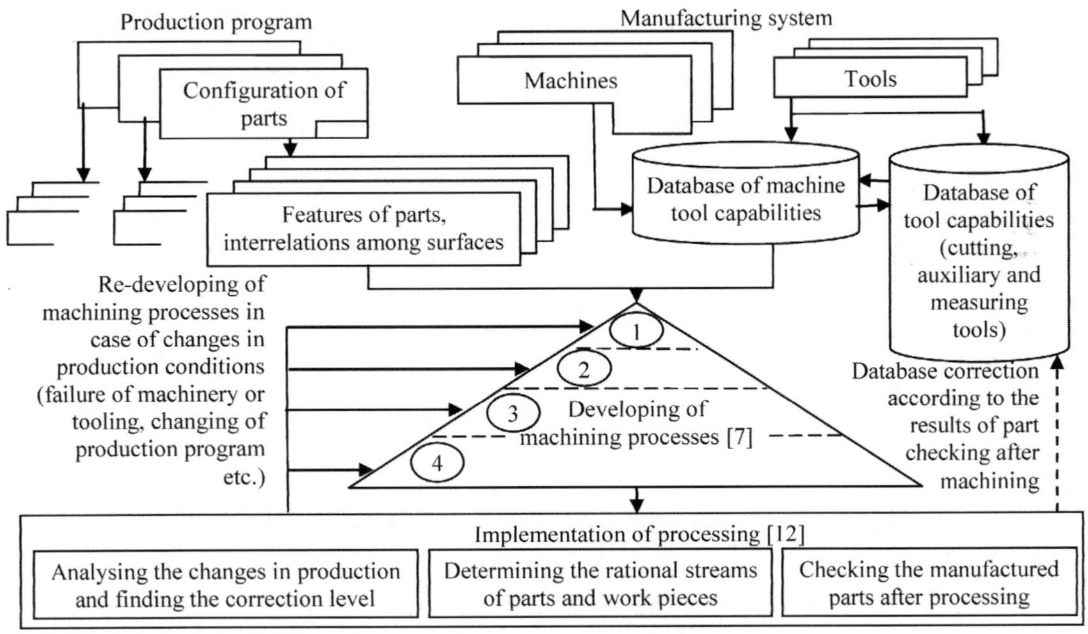

Figure 1. A scheme of multiproduct machining process planning.

The development of process plans has the following order:

1) development of the principal route to parts machining for each feature;
2) formation of the tuples of feasible machining methods;
3) selection of the machine tools;
4) design of machining operations [9, 10].

Each design procedure undergoes automatically three phases: generation of the possible alternatives; elimination of the unpractical alternatives; selection of the efficient alternatives for given manufacturing conditions. This order ensures the reduction of the designing time costs and improvement of the quality of design solutions due to avoidance of subjective factors, i.e. human limited capabilities.

The designed process plans come to the implementation subsystem. Three key procedures secure a regular interrelationships between design and implementation subsystems.

1. Determining the rational streams of parts and workpieces. This secures high loading of machines and low production time.

2. Analyzing the changes in production conditions and finding the correction level. In case of failure of machinery or tooling, changes in production program, the system is able to select among previously generated process plans, that is, to respond to a real manufacturing situation.

3. Checking the manufactured parts after processing. It secures the databases of machines and tools in the actual state. Therefore, the CAMPP system provides the lowest percent of production faults and the highest machining performance.

Each process planning design procedure is performed automatically through the mathematical models briefly described below.

2.1. The model for development of the principal route for machining of parts

The base of this model is in creating the abstract structural scheme of the machining process plan (PP). The defined sequence of process steps φ_1, φ_2 ... φ_n represents this basic scheme. A process step has a set of technological parameters $T_k = (t_{k1}, t_{k2}, \ldots, t_{km})$ carrying hereditary information related to the material and surface quality of the machined part. The set of parameters depends on the selected machining method. The parameters are divided into two groups: the parameters with a constant value from beginning to end of PP and the parameters with a changeable value during PP implementation.

The control parameters are characterized by a finite number of output parameters of the product ξ_i (i = 1, 2, ..., S) fully describing the state of the surface at each process step. Vector of quality $\overline{X}$ defines the surface quality as $\overline{X} = (\xi_1, \xi_2, \ldots, \xi_S) \in K$, where K is a spatial state of the surface quality, which is Cartesian product of tolerances J_i for each ξ_i

Then the task of creating a principal scheme to process a feature is considered as the task of finding possible routes of $\overline{X}$ from some beginning state $X_b \subset K$ to end state $X_e \subset K$ by path $L_j \subset X_N$. In other words, the transition of the surface from X_b to X_e is realized by going through N_j process steps in area X_N.

Fields of feature characteristics obtained after processing are represented as fields of output parameters of process steps in matrix Ω, and fields of initial characteristics as fields of input parameters in matrix Δ. A selection of the machining method consists in finding such process step that has all characteristics $\Omega^j_i(k)$ equaled to or exceeding $\Delta^j_i(k+1)$. This requirement is true for whole PP of the feature. A database on manufacturing capabilities is a base for comparison. As a result, the possible machining routes are formed for each part.

2.2. The model for the formation of the tuples of feasible machining methods

The formation of the tuples of feasible machining methods is a procedure related to the issue of the definition of the process steps that can be jointly accomplished.

A tuple of the feasible machining methods is bundle K = < M, D, C, E >, where M – machining methods; D – variants of datum and fixture; C – characteristics of processed surfaces; E – variants of machines.

Petri net represents a model of the datum surface kit formation. A set of places in designed Petri net $P=\{P_1, P_2, \ldots, P_n\}$ is events to define a possibility of using either machined surface as datum for a given process step and a possibility of including two and more process steps in one tuple. Both execution of transitions $T=\{T_1, T_2, \ldots, T_m\}$ and the net structure consider constraints related to design and engineering requirements of parts and capabilities of machines and tools.

Marking of the net is a code of machining transition, which is assigned to each initial event (P_1, P_2, ...). Due to execution of procedural operations, the net changes marking so that the tokens are moved through the net to form different combinations. The structure of the net supposes the creation of all possible combinations of the given process steps.

The first three phases of movement through Petri net generates a set of technological data. At the next stages, the tuples of process steps are formed and the possibility of merging of several process steps in one operation is considered.

2.3. The model for selection of the processing machines

The graph of queuing system states represents the implementation of the process step tuple on a separate group of machines. This graph considers the characteristics of reliability associated with the emergence and elimination of failures in the manufacturing system. A system of differential equations for the probabilistic states of the system is written in accordance with the labeled graph. The developed model was studied within the limits of the system steady state. Therefore, a system of differential equations was converted into the system of algebraic equations, the solution of which allows defining indicators of machines during parts manufacture in a given range, such as the probability of an idle machines group, the average number of a queued bundle of parts, machine utilization, etc.

The created model defines rational values of the intensities of the input and outputting streams of parts for the considered machine groups and determines the best alternatives of PP.

3. Design of machining operations

A machining operation design is a key element of the process planning system. In this case, the most difficult and responsible decisions are made. Figure 2 shows the generalized structure of machining operation design in CAMPP system. For each design procedure we propose mathematical models described below that can be used to automate the development of machining operations.

3.1. The model for the formation of toolkits

We proposed to solve the problem using Petri nets [13]. A set of features is $S = \{S_1, S_2, \ldots, S_s\}$, a set of equipment items is $E = \{E_1, E_2, \ldots, E_e\}$, a set of cutting tools is $R = \{R_1, R_2, \ldots, R_r\}$ and a set of auxiliary tools is $V = \{V_1, V_2, \ldots, V_v\}$. So, in terms of the Petri nets theory each phase of this design procedure is defined by a set of places P^1, P^2, P^3, P^4, P^5, P^6. Events T^1, T^2, T^3, T^4, T^5, T^6 occur when the decision on the choice of tooling options during each phase is made. During the Petri net transition, one decides whether it is possible to machine with either cutting tool or not. Petri net consists of several levels, where the compliance is verified with one of the parameters: an equipment group; connection parameters of equipment; the hardness of the machined surface; the material of the machined feature; the type of the feature; the accuracy; dimensions of the machined feature.

As a result, a set of feasible tooling is formed to process at the appropriate process step. A criterion of homogeneity is used to eliminate inefficient choices of the formed set of tooling. This criterion ensures the invariance of the operation structure and the high level of interchangeability among machining operations.

3.2. The model for the formation of the machining operational structure

The next stage of process planning is a formation of the rational machining operation structure. A key task is to choose a rational sequence of process steps for each tuple. We used a graph theory to modeling this design procedure.

A set of tuples with process steps $K=\{K_1, K_2, \ldots, K_n\}$ approaches the input of the design subsystem, where n – a number of tuples. A set of machined features is $S=\{S_1, S_2, \ldots, S_p\}$, where p – a number of features. A code of the machined feature is assigned to either vertex of the graph. Each arc between two vertices represents a processing sequence: the outgoing arc defines the feature to be machined before the feature specified in the vertex with the incoming arc. In case of absence of the fundamental difference in the processing sequence of any two features, these two respective vertices are connected by two oppositely directed arcs. The processing sequence of two similar features is shown as a loop in the vertex.

A flock of ordered sets K' is formed after selecting of possible sequences for each tuple of process steps. Each set in the flock of K' represents a possible alternative of the machining operation structure. In case there are several possible sequences for the same tuple, at the next stages of process planning it is necessary to eliminate inefficient alternatives and to choose rational sequences by the criterion of minimizing the processing time due to the combination of process steps and parallel processing.

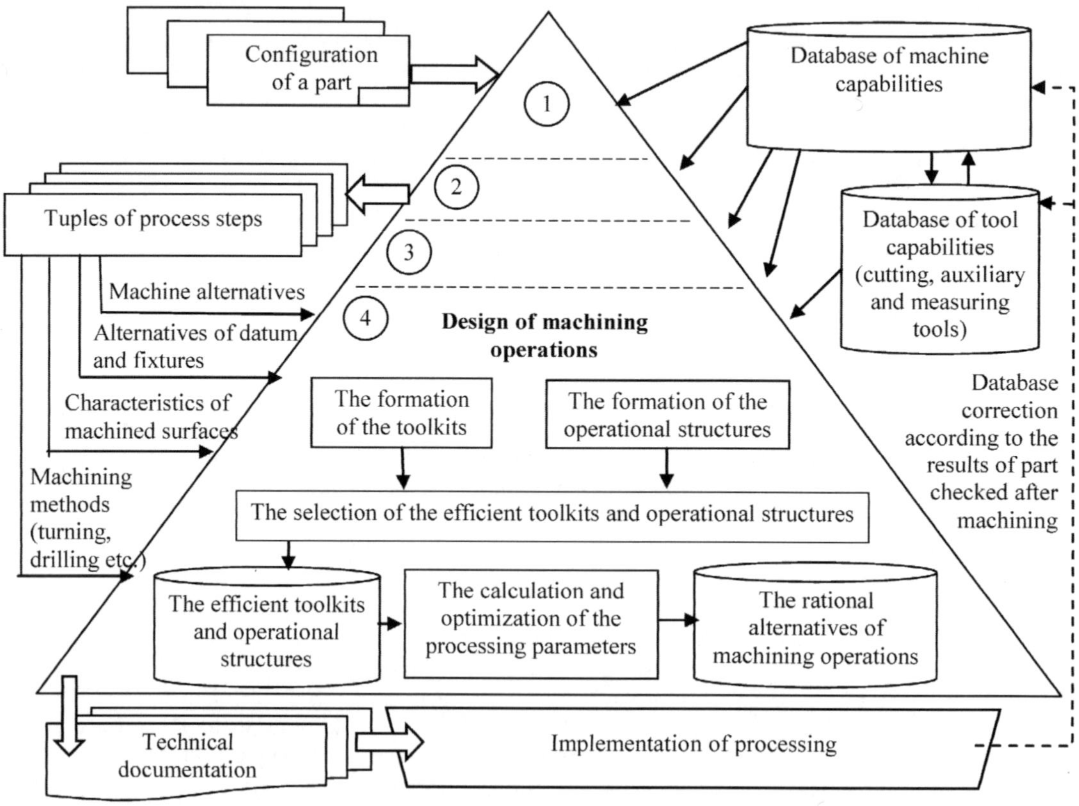

Figure 2. The generalized structure of the subsystem for machining operations design.

3.3. The model for a selection of the rational toolkits and operational structures

The problem of a selection of the rational sets of cutting and auxiliary tools for each type of equipment used in the manufacturing system based on the known methods, is time consuming. It is due to an exhaustive search of all possible alternatives along with the calculation of setup complexity and processing time (figure 6). Therefore, the mathematical apparatus of dynamic programming has been proposed to solve this problem [15]. The advantage of dynamic programming is needlessness of calculating the optimization parameters for each combination of tool alternatives. Furthermore, dynamic programming optimizes the overall system and avoids cases when the optimization of individual elements leads to inoperability of the entire system.

The set of all machining operations is represented as a system with process steps as elements of this system. The set of process steps is $P=\{P_1, P_2, \ldots, P_n\}$, where n – a number of different process steps for all machining operations, and a set of feasible cutting tools is $R=\{R_1, R_2, ..., R_m\}$, where m – a number of different sizes of cutting tools. It is required to choose the option of cutting tools R_j for each process step P_i so that the formation of the kit of cutting tools was the most efficient in current manufacturing conditions.

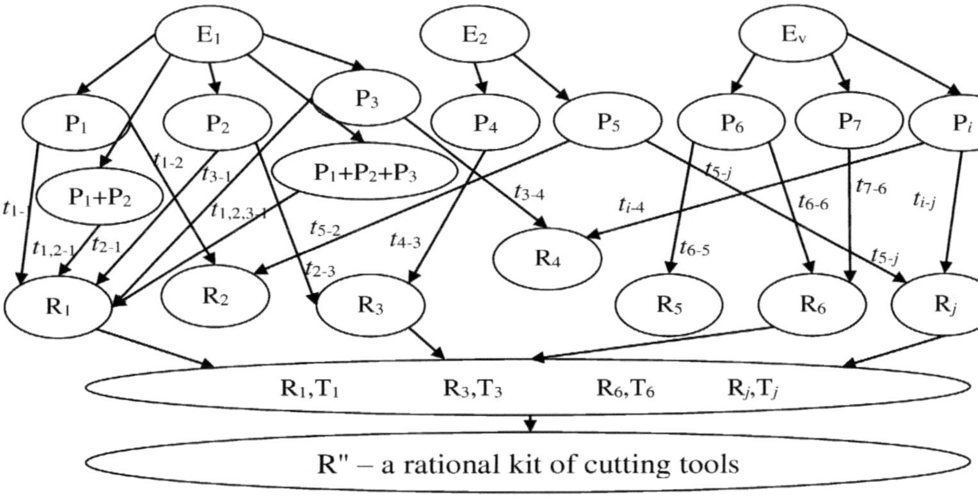

Figure 3. The logical scheme of a selection of the rational toolkits and operational structures.

Step controls $x_1, x_2, ..., x_n$ are decisions on the choice of the cutting tool of any size for processing at the i process stage, where $i=1...n$; n – a number of process steps. In this case, the gain is equal to w_i. The control of the operation is a set of control steps $x=\{x_1, x_2, ..., x_n\}$. It is necessary to find control x^*, in which total gain W becomes maximum. Herewith $x^*=\{x^*_1, x^*_2, ..., x^*_n\}$ is an optimal control consisting of a set of optimal control steps. Optimization parameters are:

1) total machining time $t_{o\Sigma}$ for each considered machining operation that should be minimal;
2) the number of cutting tool changes that should be minimized.

Control $x_i=\{j, t_{ij}, a_{ij}\}$ at the i-stage means that in this process step machining is carried out by cutting tool j with time t_{ij} and number of tool changes a_{ij}. The system state at stage i is characterized by vector $S_i=\{j, \tau_{ij}\}$. It means that before performing process step i, one uses size j of the cutting tool, and it has exhausted resource τ_{ij}.

A solution of the problem begins with the conditional optimization at last stage n, when calculating conditional optimal gain $W_n(S_n) = \max\{f_n(S_n, x_n)\}$ and finding conditional optimal control $x_n(S_n)$ for the possible cutting tools at this stage. Further optimization is performed by determining conditional optimal solutions at previous steps. Finally, unconditional optimization should be performed, following the recommendations received at each step.

4. The database of the manufacturing capabilities

The database of the manufacturing capabilities and capacities of tooling is needed to secure an automatic mode to perform the design procedures. In this regard, a special database structure is developed to reflect the relations among the elements of the manufacturing system and to allow one to keep information about the capabilities of a particular production system up to date.

The database of machine and tool capabilities:

– represents an information basis for the formalization of design procedures to develop and to implement process plans and machining operations;
– supports the compliance with specified requirements to processed parts;
– allows justifying the choice of machines and tools regarding the effectiveness of their use in specific manufacturing conditions;

– allows using the manufacturing capabilities fully;

– provides the capacity to form rational processing tools and auxiliary equipment for maximum efficiency of machining;

– creates conditions for formation of the rational machining operation structure.

All this changes the activity of the process planner: instead of process plan development, one sets up a database on manufacturing capabilities to provide compliance with the specific manufacturing system.

5. Conclusion

In the paper the authors have proposed approaches to creation of an integrated manufacturing preparation system based on fully automated design procedures for process planning and machining operations design. At the same time, the interoperability between the design and implementation processes is ensured, providing an opportunity to reduce the time period required for manufacturing preparation, to reduce manufacturing costs and to increase the competitiveness of organizations.

References

[1] Zhang HC and Alting L 1994 *Computerized manufacturing process planning systems* (USA: Chapman and Hall)

[2] Cay F and Chassapis C 1997 An IT view on perspectives of computer aided process planning research *Comput. Ind.* **34** 307–37

[3] Xu HM, Mi LC and Li DB 2008 *Int. J. Adv. Manuf. Technol.* **38** 1181–91

[4] Kumar M and Rajotia S 2003 *J. Mater. Process. Technol.* **138** 297–300

[5] Salehi M and Tavakkoli-Moghaddam R 2009 *Eng. Appl. Artif. Intell.* **22** 1179–87

[6] Wu Ruirong and Zhang Heming 1998 *Chinese Journal of Mechanical Engineering* **11** 89–95

[7] Bochkaryov P 2002 *Journal of Engineering Technology* **1** 10–14

[8] Mitin S and Bochkaryov P 2013 *Innovation Activity* **4** 36–41

[9] Mitin S and Bochkaryov P 2012 *Science Intensive Technol. in Mechan. Engineer.* **1** 32–39

[10] Bochkaryov P, Shalunov V and Bokova L 2009 *The Bulletin of the Saratov State Technical University* **3** 46–54

[11] Mitin S and Bochkaryov P 2014 *Science Intensive Technol. in Mechan. Engineer.* **11** 41–44

[12] Ivanov A and Bochkaryov P 2011 *The Bulletin of the Saratov State Technical Univers.* **2** 61–69

[13] Razmanova T, Mitin S, Bochkaryov P 2012 *The Bulletin of the Saratov State Technical University* **1** 120–24

[14] Mitin S and Bochkaryov P 2012 *Journal of Engineering Technology* **4** 69–73

[15] Mitin S and Bochkaryov P 2012 *Journal of Machines and Tools* **6** 20–24

Formation of the system indicators analytic dependence during bisubject qualimetric evaluation of arbitrary objects

A Morozova
Bryansk State Technical University, 7, 50 years of October st., Bryansk, 241035, Russia

E-mail: niotiostu@gmail.com

Abstract. The paper presents an analytical model of calculating the values of parametric clusters, a matrix of the parameters compliance and a model of the analyzed parameter values calculation allowing to form an analytical matrix of the indicators system during bisubject qualimetric evaluation of arbitrary objects and to identify the quantitative relationship of the parameters values. The results are useful for solving problems of control over both technical and socio-economic systems for evaluating objects using parameter systems generated by different subjects taking into account their performance and priorities of decision-making.

1. Statement of the research problem

Formalization of information spaces describing an object from the perspective of two different subjects, each of which has its own system of analytical parameters, creates a problem of establishing correspondence between components of these systems and identification of functional dependence between the values of their analytical parameters [1-3]. This task is appropriate to be addressed in several stages.

2. Analytical model of parametric clusters values calculating

Let the system of parameters D_i, $i=\overline{1:n}$, describing a multivariable p th object from the position of the subject Ω be characterized by a set of numerical values reflected in n-dimensional information space N by point $N_p(q_{p1}; q_{p2}; \ldots q_{pr}; \ldots; q_{pn})$.

For each parameter D_i, $i=\overline{1:n}$, we establish a value $Q_i^{\min}$ called a threshold value for this parameter, such one that if parameter D_i describing the multivariable p th object from the position of subject Ω has a numerical value q_{pi} to satisfy condition $Q_i^{\min} \geq q_{pi}$ it is believed that the parameter D_i is not used in the p th object description, i.e. $q_{pi}=0$.

Suppose that the total value of the parametric cluster A_j, $j=\overline{1:m}$, is determined by a subset of parameter values $D_{jg} \subset \{D_i\}, i=\overline{1:n}$, $g=\overline{1:t}$ (Figure 1, position 1–3). Each parameter D_{jg} has a significance level α_{jg}, $g=\overline{1:t}$, which shows what part this parameter brings to the aggregate value of the m-dimensional ($m \rightarrow \infty$) information space M, which is a parametric cluster A_j, $j=\overline{1:m}$.

Thus, the total value of the parametric cluster A_j is formed as a generalized share characteristic of a subset of parameters D_i values.

The level of significance α_{jg}, $g=\overline{1:t}$, for parameter D_{jg}, $g=\overline{1:t}$, included in the subset $D_{jg}\subset\{D_i\}, i=\overline{1:n}$, $g=\overline{1:t}$, in the composition of parametric cluster A_j, $j=\overline{1:m}$, has the following properties: $0\leq\alpha_{jg}\leq 1$ and $\sum_{g=1}^{t}\alpha_{gj}=1$.

The total value of the parametric cluster A_j, $j=\overline{1:m}$, describing multivariable p th object from the position of the subject Ω can be determined in several ways, for example by the maximum, minimum or mean values of all the available parameter values in the set $D_{jg}\subset\{D_i\}, i=\overline{1:n}$, $g=\overline{1:t}$, determining the aggregate value of the parametric cluster A_j.

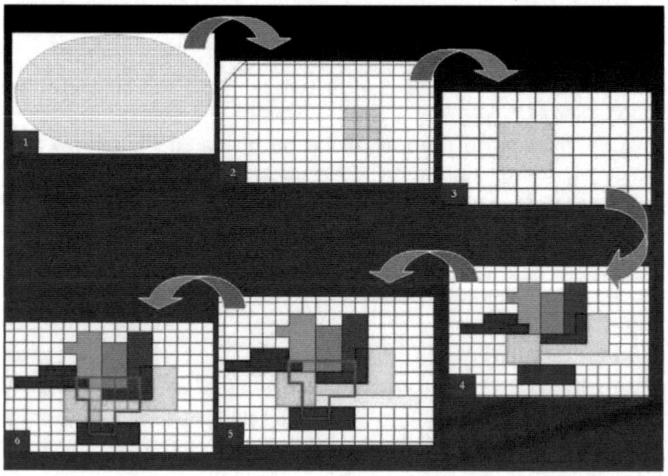

Figure1. Graphic interpretation of bisubject qualimetric evaluation of an arbitrary object.

The level of significance α_{jg} for each parameter D_{jg}, $g=\overline{1:t}$, which is a part of parametric cluster A_j, $j=\overline{1:m}$ is determined in an expert-analytical way.

In this case, expression $s_j=\alpha_{ji}\cdot q_i$ establishes a quantitative relationship between the total value s_j of the parametric cluster A_j, $j=\overline{1:m}$ and the set of coordinates q_i, $i=\overline{1:n}$, of the point N_i of the information space N as a set of values of parameters D_i, $i=\overline{1:n}$, describing the multivariable p th object from the position of the subject Ω. This dependence can be generally represented as: $s_j=f(q_1;q_2;...;q_i;...q_n)$.

3. Formation of the correspondence matrix during bisubject qualimetric evaluation of the arbitrary object

In accordance with the scheme shown in Figure 1, there is a definite relationship between the level of development h_k of the parameter B_k, $k=\overline{1:l}$ system that describes the multivariable p th object from the position of subject Ξ and the values q_i of parameters D_i, $i = \overline{1:n}$, as coordinates of the point N_i of the information space *N* describing the multivariable p th object from the position of the subject Ω.

Then, using a tabular method of function setting one can set the this dependence in a form of a rectangular matrix, thus forming a matrix of relevant parameters D_i, $i = \overline{1:n}$, describing the multivariable p th object from the position of subject Ω and parameters B_k, $k = \overline{1:l}$, that describe the multivariable p th object from the position of the subject Ξ. The presence of '+' sign at the intersection of parameter D_i row and parameter B_k column may indicate that between these parameters there is a certain relationship.

If the resulting matrix is filled with values of equity ratios α_{jg}, characterizing the level of significance α_{jg}, $g = \overline{1:t}$, of parameter D_{jg}, $g = \overline{1:t}$, included in the subset $D_{jg} \subset \{D_i\}, i = \overline{1:n}$, $g = \overline{1:t}$ as a part of parametric cluster A_j, $j = \overline{1:m}$, for each parameter B_k, $k = \overline{1:l}$, we obtain an analytical matrix of correspondence for system parameters D_i, $i = \overline{1:n}$, describing multivariable p th object from the position of subject Ω and parameters B_k, $k = \overline{1:l}$, describing this object from the position of subject Ξ.

In the formation of equity coefficients α_{jg}, characterizing the significance levels of parameters D_{jg}, $g = \overline{1:t}$, as a part of parametric clusters A_j, it is appropriate to apply expert methods, such as the method of hierarchies analysis with application of management decision-making support programs (e.g., Expert Decide).

If the set of parameters D_{jg}, $g = \overline{1:t}$, forms a specific parametric cluster A_j, $j = \overline{1:m}$, each parameter B_k, $k = \overline{1:l}$, is formed as an integrative information subspace determined by the influence of the set of parametric clusters A_j, $j = \overline{1:m}$. Each parametric cluster A_j, $j = \overline{1:m}$ has a different degree of influence on the values formation for the parameter B_k, $k = \overline{1:l}$. This process can be graphically interpreted by the scheme shown in Figure 1: the set of parameters values B_k (positions 5 and 6) describing the object from the position of subject Ξ is generated from the values of parametric clusters A_j (position 4) through integration of the subset parameter values D_i (positions 2 and 3) describing this object from the position of the subject Ω.

4. The model for calculating the values of the analyzed parameter during bisubject qualimetric evaluation of the arbitrary object

Let the matrix of correspondence set that the formation of parameter B_k, $k = \overline{1:l}$, describing the object from the position of subject Ξ, affects a subset of parametric clusters A_j, describing the object from the position of the subject Ω. Of fundamental importance is the fact that every parametric cluster A_j differently affects the formation of parameter B_k value. Thus, we need to introduce a system of coefficients β_{jk} determined using the expert methods and application software, which characterize the degree of the parametric cluster A_j influence on the formation of the parameter B_k values, and it is advisable to use such values β_{jk} that $0 \le \beta_{jk} \le 1$ and $\sum_{k=1}^{l} \beta_{jk} = 1$. Moreover, for each parameter B_k, $k = \overline{1:l}$, there are situations when certain parametric cluster A_j has no influence on its formation. In this case, $\beta_{jk} = 0$.

If throughout the entire column corresponding to the analyzed parameter B_k of the model matrix parameters '+' sign is missing in each line corresponding to the set of parameters D_i in the parametric cluster A_j, it is interpreted as an absence of the parametric cluster influence on the formation of parameter B_k.

Also, for parametric clusters subsystem $A_{j'k}$, $j' = \overline{1:m'}$ and $k = \overline{1:l}$, where $m' \le m$, that influences the formation of the value for parameter B_k, it is advisable to establish such a threshold value $\beta_k^{\min}$ that if $\beta_{j'k} \le \beta_k^{\min}$ it is considered that the parametric cluster $A_{j'}$ describing the object from the position of subject Ω has little influence on the formation of parameter B_k, describing the object from the position of subject Ξ, and therefore we can set $\beta_{j'k} = 0$.

To assess the quality of the analyzed object, it is suitable for parameters B_k, $k = \overline{1:l}$, to set several qualimetric levels: $H_k^{\min}$ – the threshold level; H_k^{v} – advanced level; H_k^{w} – high level. The values of parameters $H_k^{\min}$, H_k^{v} and H_k^{w} are determined by the expert-analytical methods. Furthermore, considering the specific purpose of research in some cases it is advisable to set either only threshold and high level values, or conversely to introduce one more intermediate level such as 'increased' one.

Let parameter B_k, $k = \overline{1:l}$, that characterizes the quality of the p th object from the position of subject Ξ have level h_k. This value is influenced by parametric cluster A_j having value s_{pj} and describing the object from the position of subject Ω. This parametric cluster has an equity level β_{jk} of influence on the formation of parameter B_k. Then, h_k is determined by the relation $h_k = \beta_{jk} \cdot s_{pj}$.

If the formation of the value of parameter B_k that describes the object from the position of subject Ξ involves a subset of parametric clusters $A_{j'}$, $j' = \overline{1:m'}$, describing the object from the position of subject Ω, for the analyzed object the value h_{pk} of the parameter B_k can be characterized in several ways: in terms of the maximum, minimum or mean values of all the available parameters included in the set of parametric clusters $A_{j'} \subset \{A_j\}$, $j = \overline{1:m}$, $j' = \overline{1:m'}$ and determining the total value of the parameter B_k, $k = \overline{1:l}$.

5. Conclusion

The resulting set of values $h_{p1}; h_{p2}; \ldots; h_{pv}; \ldots h_{pl}$ of parameter B_k, $k = \overline{1:l}$, is an analytical matrix in the form of a row vector characterizing the quality of the analyzed object from the position of subject Ξ. It is the result of converting the set of values $q_{p1}; q_{p2}; \ldots q_{pr}; \ldots; q_{pn}$ of parameters D_i, $i = \overline{1:n}$, that describes the quality of this object from the position of subject Ω [2, 5]. So, the received set $h_{p1}; h_{p2}; \ldots; h_{pv}; \ldots h_{pl}$ is an analytical matrix for the system of indicators during bisubject qualimetric evaluation of the arbitrary object. It is the result of establishing the relationship $h_k = g(q_1; q_2; \ldots; q_i; \ldots q_n)$ between the values of parameters D_i, $i = \overline{1:n}$, characterizing the information space N, describing the multivariable object from the position of subject Ω and the values of parameters B_k, $k = \overline{1:l}$, characterizing the information space L and describing the multivariable object from the position of subject Ξ [4, 6, 7].

Acknowledgments
The work was supported by the RFBR grant (research project № 13-01-00147 a).

References

[1] Greeberg B and Busselle R 1996 *J. Quality Assessment of Television* p 169–196
[2] LaRose R 1992 *J. Studies of Broadcasting* **28** 77–119
[3] Morozova A V, Vasilenko Yu V and Tyukhta A V 2015 *Russian J. Bulletin of Bryansk State Technical University* **3(47)** 147–156
[4] Morozova A V, Tarasov D E and Alisov A A 2015 *Multiparametric Social Objects under Conditions of Uncertainty: Neuronet Modeling, Typification and Quality Metering* (Moskau: Publishing House Spectrum) 10.14489/4442-0090-2
[5] Morozova A V , Vasilenko Yu V and Alisov A A 2014 *Russian J. Proceedings of Samara Scientific Center of Russian Academy of Sciences* **16(2)** 449
[6] Rossler P 2004 Qualitat aus Transaktionaler Perspektive. Zur gemeinsamen Model-lierung von 'User Quality' and 'Sender Quality': Kriterien fur Onlinezeitungen (*Germany Munchen: Reinhard Fischer. J. Qualitat in der Online kommunikation)* pp 127–145
[7] Webster J G and Warslag J J 1983 *J.Communication research* **10(4)** 430–446

IOP Conf. Series: Materials Science and Engineering **124** (2016) 012079 doi:10.1088/1757-899X/124/1/012079

Mathematical model of bisubject qualimetric arbitrary objects evaluation

A Morozova
Bryansk State Technical University, 7, 50 years of October st., Bryansk, 241035, Russia

E-mail: niotiostu@gmail.com

Abstract. An analytical basis and the process of formalization of arbitrary objects bisubject qualimetric evaluation mathematical model information spaces are developed. The model is applicable in solving problems of control over both technical and socio-economic systems for objects evaluation using systems of parameters generated by different subjects taking into account their performance and priorities of decision-making.

1. Statement of the research problem

From the standpoint of system analysis, an object of any nature with a certain system of characteristics can be analysed from the perspective of different subjects, each of which generates its own system of parameters by which the analysed object can be described [1, 2, 4]. If such systems of parameters are of numerical nature, there is a problem of inconsistency of the systems parameters that characterize the object from the perspective of various subjects and the lack of evaluation methods and means [6, 7].

In this study, we consider the problem of forming a mathematical model to convert the levels of the object quality measuring in the system of one subject parameters to the levels of its measurement in the system of another subject parameters.

2. Analytical basis of the mathematical model

The mathematical model of arbitrary objects bisubject qualimetric evaluation is based on the following axiomatic basis ensuring the integrity of the subsequent multivariate analysis of data characterizing these objects.

1. There is some n-dimensional $(n \to \infty)$ information space N determining a system of parameters describing a multi-parameter object from the perspective of the subject Ω, m-dimensional $(m \to \infty)$ information space M determining a set of parametric clusters that describe superpositional synergistically generated characteristics of the object from the position of the subject Ω, and l-dimensional $(l \to \infty)$ information space L determining a system of parameters describing a multi-parameter object from the perspective of the subject Ξ.

2. The m-dimensional $(m \to \infty)$ information space M and the l-dimensional $(l \to \infty)$ information space L are subsets of the n-dimensional $(n \to \infty)$ information space N. The elements of information spaces N, M and L are information parameters that describe the studied object. The information parameters are the parameters containing numeric information about the object.

3. There are sets *D*, *A* and *B*, the elements of which are, respectively, parameters D_i, $i=\overline{1:n}$ describing the multivariable object from the position of the subject Ω, parametric clusters A_j, $j=\overline{1:m}$ describing superpositional synergistically generated characteristics of the object from the position of the subject Ω, and parameters B_k, $k=\overline{1:l}$ describing the multivariable object from the position of the subject Ξ.

4. Each *i* th, $i=\overline{1:n}$ axis $O_N N_i$ ($[O_N N_i)$ ray) of the space *N* maps a set of numerical values of the parameter D_i, $i=\overline{1:n}$. On the $[O_N N_i)$ ray, a vector $\overline{e_{Ni}}$ with starting point O_N is marked. If the point is $D_{pi} \in [O_N N_i)$ the vector is $\overline{O_N D_{pi}} = q_{pi} \cdot \overline{e_{Ni}}$, where q_{pi} is a coordinate of the vector $\overline{O_N D_{pi}}$ on the axis $O_N N_i$ interpreted as a numeric value of the parameter D_i, $i=\overline{1:n}$, which characterizes the *p* th object from the position of the subject Ω.

5. Each *j* th, $j=\overline{1:m}$, axis $O_M M_j$ ($[O_M M_j)$ ray) of the space *M* maps a set of numerical values of the parametric cluster A_j, $j=\overline{1:m}$. On the ray $[O_M M_j)$ there is a vector $\overline{e_{Mj}}$ with the starting point O_M. If the point is $A_{pj} \in [O_M M_j)$, the vector is $\overline{O_M A_{pj}} = s_{pj} \cdot \overline{e_{Mj}}$, where s_{pj} is a coordinate of the vector $\overline{O_M A_{pj}}$ on the axis $O_M M_j$, interpreted as a numeric value of the parametric cluster A_j, $j=\overline{1:m}$, which characterizes the *p* th object from the position of the subject Ω.

6. Each *k* th, $k=\overline{1:l}$, axis $O_L L_k$ ($[O_L L_k)$ ray) of the space *L* maps a set of numerical values of the parameter B_k, $k=\overline{1:l}$. On the ray $[O_L L_k)$ there is a vector $\overline{e_{Lk}}$ with the starting point O_L. If the point is $B_{pk} \in [O_L L_k)$, the vector is $\overline{O_L B_{pk}} = h_{pk} \cdot \overline{e_{Lk}}$, where h_{pk} is a coordinate of the vector $\overline{O_L B_{pk}}$ on the axis $O_L L_k$ interpreted as a numeric value of the parameter B_k, $k=\overline{1:l}$, which characterizes the *p* th object from the position of the subject Ξ.

3. The process of the mathematical model information spaces formalizing

Formalization of the arbitrary objects bisubject qualimetric evaluation mathematical model information spaces is performed on its axiomatic basis and is as follows.

1. The set of parameters describing the *p* th object from the position of the subject Ω is displayed in the *n*-dimensional space *N* with a point $N_p(q_{p1}; q_{p2}; \ldots q_{pr}; \ldots; q_{pn})$, where q_{pr} is a projection of the point N_p on the axis $O_N N_r$, $r=\overline{1:n}$.

2. Each *i* th, $i=\overline{1:n}$, axis $O_N N_i$ has a quantitative non-negative scale of values, i.e. each point q_i on the *i* th axis corresponds to a certain numerical value of the parameter D_i that describes the object from the position of the subject Ω. In particular, the value $q_{pr}=0$ indicates that the parameter D_r, $r=\overline{1:n}$, is not used in the description of the *p* th object, and the value $q_{pr} \neq 0$ means that the parameter D_r is used in the description of the *p* th object.

For the further analysis of values q_i, $i=\overline{1:n}$, it is advisable to establish qualimetric levels (numerical values) of the parameter D_i: threshold $Q^{\min}{}_i$, advanced $Q^v{}_i$ and high $Q^w{}_i$. The value q_{pi} at which $q_{pi} \geq Q^w{}_i$, implies that the *p* th object has a high level value of the parameter D_r that describes it from the position of the subject Ω.

A set of characteristics X_p describing the p th object from the position of the subject Ω is functionally dependent on the numerical values of all the coordinates of the point N_i, $i=\overline{1:n}$ of the information space N:

$$X_p = F_1(q_{p1}; q_{p2}; \ldots; q_{pr}; \ldots q_{pn}) \tag{1}$$

In the case where the parameter D_r, $r=\overline{1:n}$, is not used in the description the p th object, the expression (1) becomes:

$$X_p = F_1(q_{p1}; q_{p2}; \ldots; q_{pr-1}; 0; q_{pr+1}; \ldots q_{pn}) \tag{2}$$

3. Each j th, $j=\overline{1:m}$ axis $O_M M_j$ has a quantitative non-negative scale of values, i.e. each point s_j on the j th axis corresponds to a certain numerical value of the parameter A_j that describes the object from the position of the subject Ω. In particular, the value $s_{pw}=0$ indicates that the parametric cluster A_w, $w=\overline{1:m}$, is not used in the description of the p th object, and the value $s_{pw} \neq 0$ means that the parametric cluster A_w is used in the description of the p th object.

Characterization of the p th object by all parametric clusters A_w, $w=\overline{1:m}$, is reflected in the m - dimensional space M with the point $M_p(s_{p1}; s_{p2}; \ldots s_{pw}; \ldots; s_{pm})$, where s_{pw} is a projection of the point M_p on the axis $O_M M_w$, $w=\overline{1:m}$.

For the further analysis of the values s_j, $j=\overline{1:m}$, it is advisable to establish the qualimetric levels (numerical values) of the parametric cluster A_j: threshold $S^{\min}{}_j$, advanced $S^v{}_j$ and high $S^w{}_j$. The value s_{pj}, at which $s_{pj} \geq S^w{}_j$, implies that the p th object has a high level value of the parametric cluster A_j that describes it from the position of the subject Ω.

Superposition characteristics Y_p describing the p th object from the position of the subject Ω functionally depends on the numerical values of all the coordinates of the point M_j, $j=\overline{1:m}$ of the information space M:

$$Y_p = F_2(s_{p1}; s_{p2}; \ldots; s_{pw}; \ldots s_{pm}). \tag{3}$$

In the case where the parametric cluster A_w, $w=\overline{1:m}$, is not used in the description of the p th object, the expression (3) becomes:

$$Y_p = F_2(s_{p1}; s_{p2}; \ldots; s_{pw-1}; 0; s_{pw+1}; \ldots s_{pm}). \tag{4}$$

Consequently, the value s_j of the parametric cluster A_j, $j=\overline{1:m}$, is functionally dependent on the values of all coordinates q_i, $i=\overline{1:n}$, of the point N_i of the information space N corresponding to the numerical values of the parameters D_i, $i=\overline{1:n}$, included in the analyzed parametric cluster A_j in the form:

$$s_j = f(q_1; q_2; \ldots; q_i; \ldots q_n). \tag{5}$$

In the case where the parameter D_r, $r=\overline{1:n}$ is not included in the parametric cluster A_j, $j=\overline{1:m}$ the expression (5) becomes:

$$s_j = f(q_1; q_2; \ldots; q_{r-1}; 0; q_{r+1}; \ldots q_n). \tag{6}$$

In addition, the superposition characteristic Y_p of all parametric clusters describing the p th object from the position of the subject Ω, is also determined by the values of the characteristics set X_p

describing the p th object with a system of parameters D_i, $i=\overline{1:n}$, from the position of this subject, i.e.:

$$Y_p = \Psi(X_p). \tag{7}$$

4. Each k th, $k=\overline{1:l}$ axis $O_L L_k$ has a quantitative non-negative scale of values, i.e. each point h_k on the k th axis corresponds to a certain numerical value of the parameter B_k that describes the object from the position of the subject Ξ. In particular, the value $h_{pv}=0$ indicates that the parameter B_v, $v=\overline{1:l}$, is not used in the description of the p th object, and the value $h_{pv}\neq 0$ means that this parameter is used in the description of the p th object.

Regarding the subsequent analysis of values h_k, $k=\overline{1:l}$, it is advisable to establish the qualimetric levels (numerical values) of the parameter B_k: threshold $H^{\min}{}_k$, advanced $H^{v}{}_k$ and high $H^{w}{}_k$. For example, the value h_{pk}, at which $h_{pk} \geq H^{w}{}_k$, implies that the p th object has a high level value of the parameter B_k that describes this object from the position of the subject Ξ.

The value h_k of the parameter B_k, $k=\overline{1:l}$, is functionally dependent on the values q_i of the parameters system D_i, $i=\overline{1:n}$, as the coordinates of the point N_i of the information space N:

$$h_k = g(q_1;\, q_2;\, \ldots;q_i;\, \ldots q_n). \tag{8}$$

In the case where the parameter D_r, $r=\overline{1:n}$, describing the p th object from the position of the subject Ω is not used as a component of the parameters set B_k, $k=\overline{1:l}$, describing the p th object from the position of the subject Ξ, the expression (8) becomes:

$$h_k = g(q_1;\, q_2;\, \ldots;q_{r-1};\, 0;\, q_{r+1};\, \ldots q_n). \tag{9}$$

5. The system of parameters possessed by the p th object from the position of the subject Ξ is displayed in the l-dimensional space L by the point $L_p(h_{p1};\, h_{p2};\ldots h_{pv};\ldots;h_{pl})$, where h_{pv} is a projection of the point L_p on the axis $O_L L_v$, $v=\overline{1:l}$.

Characteristic Z_p that describes the p th object from the position of the subject Ξ is functionally dependent on the values of all coordinates of the point L_k, $k=\overline{1:l}$, of the information space L:

$$Z_p = F_3(h_{p1};\, h_{p2};\ldots;h_{pv};\ldots h_{pl}). \tag{10}$$

In the case where the parameter B_k, $k=\overline{1:l}$, is not used in the description of the p th object, the expression (10) becomes:

$$Z_p = F_3(h_{p1};\, h_{p2};\ldots;h_{pk-1};\, 0;\, h_{pk+1};\ldots h_{pl}). \tag{11}$$

In addition, the value of the characteristics set Z_p that describes the p th object from the position of the subject Ξ is functionally determined by the value of the characteristics set X_p describing the p th object from the position of the subject Ω by means of the parameters system D_i, $i=\overline{1:n}$, i.e.:

$$Z_p = G(X_p). \tag{12}$$

Consequently, there is a functional relationship between the value of the characteristics set Z_p describing the p th object from the position of the subject Ξ and the total synergetic superposition characteristic Y_p of all parametric clusters describing the p th object from the position of the subject Ω, which can be represented as:

$$Z_p = \Phi(Y_p). \tag{13}$$

4. Conclusion

The developed axiomatic basis of the arbitrary objects bisubject qualimetric evaluation mathematical model is based on the formalization of the two information spaces, one of which is determined by the system of parameters describing the p th object from the position of the subject Ω, and the other one – by the system of parameters describing the same object from the position of the subject Ξ.

In the process of the formed axiomatic basis practical use, subsets of information spaces are determined, which are directly involved in the description of the object from the position of the specific subjects, the system of their parameters descriptions, as well as units of measure. The problem of specific relationship (12) and (13) identification is solved using the actual description of the same object from the positions of different subjects.

This mathematical model is applicable to both technical and social objects [3-5]. For example, the quality of any type of equipment can be described by its technical and technological characteristics, which have a great degree of importance from the standpoint of the manufacturer, and the performance, which is a priority of the consumer. A similar approach can be used when searching for solutions to the following problem: how, using the values of the graduates' educational achievements levels reflected in their high school diploma, to determine the degree of their professional competencies formation, which is essential for employers while filling the vacant posts of technical officers.

Acknowledgments

The work was supported by the RFBR grant (research project № 13-01-00147 a).

References

[1] Greeberg B and Busselle R 1996 *J. Quality Assessment of Television* p 169–196
[2] LaRose R 1992 *J. Studies of Broadcasting* **28** 77–119
[3] Morozova A V, Vasilenko Yu V and Tyukhta A V 2015 *Russian J. Bulletin of Bryansk State Technical University* **3(47)** 147–156
[4] Morozova A V, Tarasov D E and Alisov A A 2015 *Multiparametric Social Objects under Conditions of Uncertainty: Neuronet Modeling, Typification and Quality Metering* (Moskau: Publishing House Spectrum) 10.14489/4442-0090-2
[5] Morozova A V , Vasilenko Yu V and Alisov A A 2014 *Russian J. Proceedings of Samara Scientific Center of Russian Academy of Sciences* **16(2)** 449
[6] Rossler P 2004 Qualitat aus Transaktionaler Perspektive. Zur gemeinsamen Model-lierung von 'User Quality' and 'Sender Quality': Kriterien fur Onlinezeitungen (*Germany Munchen: Reinhard Fischer. J. Qualitat in der Online kommunikation)* pp 127–145
[7] Webster J G and Warslag J J 1983 *J.Communication research* **10(4)** 430–446

IOP Conf. Series: Materials Science and Engineering **124** (2016) 012080 doi:10.1088/1757-899X/124/1/012080

Modeling and optimization of transmission and processing of data in an information computer network

A Nekrasova, Z Boriev, A Nyrkov and S Sokolov
Admiral Makarov State University of Maritime and Island Shipping, 5/7, Dvinskaya st., Saint-Petersburg, 198035, Russia

E-mail: anna14_06@mail.ru

Abstract. The paper presents a comparative analysis of the routing algorithms that allows optimizing the process of transmission and processing of data in information computer networks. A special attention is paid to multipath methods of data transmission coupled with the number of operations necessary for their performance. In additionthe authors have raised the question of a linear programming method for the purpose of the solution of the above-mentioned problem.

1. Introduction

To create an optimal control system of information computer networks (ICN) it is necessary to model the transmission and processing of data. Computer networks can combine a large number of sub-networks with a different structure that is currently a major problem.

There are several key factors that determine the difficulty of modeling:

- Scale. The network integrates multiple computers and servers, which store and transmit data on the communication links;
- A high degree of the elements mismatching includes a variety of computer types, communication equipment, operating systems and applications;
- Information computer networks combine sub-networks by using the assigned lines, satellite, radio, telephone channels, etc.

ICN regulation should change the transmission and processing of information in the network. ICN should also ensure an optimal use of resources, as well as the compatibility of network equipment and forecasting work.

A number of challenges that require solutions in ICN operationhave beenrevealed. These tasks include:

- providing control algorithms adaptability in situations when the available information is not sufficient for decision-making;
- forecasting ICN operation to improve management productivity;
- making decisions that correspond to real life situations;
- providing an operation system with self-management to improve it.

Systems of data transmission often operate in the conditions of sizeable fluctuations of transmission link properties through which data are transferred. Such fluctuations depend on external conditions, as well as on changes in a mutual position of the receiver-transmitter.

When modeling such channels the model of hidden Markov process is often used. A special case of this process is Gilbert model in which the condition of a transmission link is simply described in terms of 'good' – 'bad', and transitions between them are described by a Markov chain.By optimization of the operation of data transmission systems it is usually required to maximize the transinformation content.

Thus coordination of the data transmission rate with the current state of the transmission link gains crucial importance. One of the known developed approaches consists of an identification of transmission link characteristics and a choice of an optimum method of signal coding coordinated with the current frequency response of the transmission link.

Though this method seems to provide results very close to optimum ones, its hardware realization is rather difficult. At the same time management methods of the data transmission rate that allow coordinating it with the current state of the transmission link are already employed in operating communication systems more or less effectively.

For example, it is possible to specify the TCP/IP protocol where a burst packages rate increases under the linear law until there is a loss of a package. After that the burst rate decreases by step in the fixed proportion and then again grows under the linear law. It is necessary to emphasize that this control method is a typical implementation of a stochastic process control on incomplete data. In fact the burst rate of the packages loss is an observed process connected both with a non-observed condition of the transmission link and with the burst packages rate.

The burst packages rate is also under control of the data flow and at the same time - of indirect measurements. In recent years the management mechanisms,in which direct measurement of parameters of a packages losses traffic and the network configured so that to provide the maximum friendliness to the user,have been offered.

The models describing the TCP/IP protocol in terms of the stochastic processes operated by hidden Markov processes have been studied extensively in recent years and proved their solvency when comparing the results of modeling with practice [1].

At the same time a simple transmission of already existing technical solutions by the wireless communication lines will lead to obviously a nonoptimal solution as the level of fluctuations in the wireless communication lines and, most importantly, their temporary characteristicsdiffer significantly from those in fixed fiber-optical carriers and wire communication lines.

On the other hand, in the stochastic theory of management in systems with hidden Markov processes there are results that allow hoping for success during their application in management problems of data transmission systems. The offered approach is based on the following prerequisites:

- thecondition of the channel is described by Markov's walk with a final set of states and known intensity of transitions;
- thedata transmission rate is an operating parameter, and the intensity of packages losses is a known monotonous function from the transmission rate and the channel condition;
- theoperation purpose is a choice of such law of the change of the data transmission rate under which the maximum of an average value of successfully transferred packages is attained [2].

2. Main part

The main problem in the organization of networks is distribution of data flows by the shortest paths. Ways of data transmission, which demand time minimum, or ways with the minimum hindrances belong to such kind of problem. Accordingly, path optimization has to be carried out by any technical and economic criteria, and the chosen paths have to guarantee an effective use of lines and communication vertices. Thus it is necessary to make the comparative analysis of routing algorithms which are used for development and management of information traffics today.

An adaptive routing usage allows one to reduce the average time of the package finding in a network, gives a chance to reduce costs of its delivery to the recipient by networks and to increase the general reliability of the network because of an automatic choice of a route on the basis of network topology data. But all these features hugely load computer centers of vertices. Thereforethe use of adaptive routing is possible in the limited mode.

For this reason there is a problem of software development ofthe path choice optimization for the purpose of load reduction on computer centers, and also receiving the possibility of multipath routing use.

At a choice of the most suitable route for sending packages to the following knot we will use the directive graph. The vertices of the graph are routers, and edges connecting vertices are physical communication lines, to each of which some integral value corresponds. Such value is represented by the 'cost' of a package transfer on it, depending on time wasting during data transmission.

Today there are some algebraic methods, which allow describing a process of receiving results in a form convenient for the subsequent researches. For this purpose it is necessary to analyze known methods for an assessment of the applied importance and restrictions on their use.

2.1 Queueing Theory

Queueing Theory plays an important role in ICN operation modeling. In due time, a theory use has allowed a solution of many applications, for example, modeling of the backbone networks which operate the traffic between the systems.

It should be noted that this theory covers a wide range of models, among which we can mention:

- Jackson networks;
- Markov networks with different classes of requirements and parameters that depend on the state;
- Networks with a random selection of the channel in the unit and the generalized processor sharing;
- Networks of Queuing Systems requirements of different classes with the absolute priority of incoming claims and priority;
- Networks with service discipline 'generalized processor sharing' and different classes of requirements;
- Queuing networks with a variety of channels.

A lot of 'queues' and 'service channels', which receive service demands, are a subject of domain theory of queuing systems (QS). This mathematical algorithm is distributed in the application process for ICN analysis. Predecessors in their works substantiated and derived the general laws of the timing delay request, queue length, etc.., which depend on the QS type.

On the QS basis the formation of new data transmission technologies as well as the emergence of global networks, which in their turn are associated with the information growth and computer security, has led to the need of creating and developing new ICN models [3].

Existing methods generally use the first two moments (mean and variance) of time distribution in the input stream and the service time. But they are quite uncertain, and it is not always possible to obtain reliable results. Therefore, someauthorssuggesttheuseofnewapproaches.

The method of Queuing Theory allows one to determine the sizes of the shortest paths between all vertices as well as the lengths of all possible paths between each set of two vertices of a network. It enables findingthe smallest distances in this network.

2.2 A method of linear programming

To cut it short, these methods are connected with determination of the best program of administrative actions in a case when linear dependence, in which unknowns are in the first degree, acts as an objective function and limitations.

The problem definition of linear programming has an extreme character. It consists of determination of such variable values under which the objective function attains its maximum or a minimum depending on the problem character.

3. Graph algorithms

A special place in mathematical support of traffic choice procedures of information flows transmission is taken by the Dijkstraalgorithm, which is based on the optimality principle.

3.1 Dijkstra algorithm

This method allows finding the shortest path between one node in a graph to all others. The algorithm works only for graphs without edges of negative weight though now there are generalized methods for elimination of this shortcoming (Dijkstra's method with potentials). The essence of the Dijkstraalgorithm consists instep-by-step building of the shortest routes tree from the initial vertex. At the same time it is necessary after addition at each stage of the communication line and a knot the newly formed shortest path was minimal possible on all terminal vertices which are not yet in the tree. There are calculated scales vectors of paths and there are corrected initial vectors of the shortest paths in the process of creation the shortest paths tree. The complexity of the Dijkstra algorithm depends on a way of finding vertex *v*, and also on a way of storage of a set of unvisited vertices and a way of tags updating. In graph *G*, *n* and *m* are respectively the quantity of vertices and the edges for searching the vertex using the shortest path to vertex *v* is looked through the whole set of n.

The operating time of the minimization algorithm is $O(n^2+m)$. For the discharged graphs (for which $m<<n$) when using special algorithms of optimization of the operation speed the operating time can amount to $O(n\log n+m\log n)$ or even $O(n\log n+m)$. Dijkstra method is widely applied in network programming and technologies, for example, it is used in the OSPF protocol (Open Shortest Path First) for elimination of the close paths. The using of the modified Dijkstra algorithm as an effective tool for distribution of input information flows in the main IP networks with the OSFP protocol allows one to improve the network robustness protecting against data overloads. Thus it is possible to use the residual capacity of the link as a criterion of information flows distribution. It is necessary to refer the relative simplicity of practical realization of the method to protocol merits.

3.2 Johnson's algorithm

This method allows finding the shortest ways between all pairs of vertices in a sparse, edge weighted, directed graph. This method operates if the graph contains edges with a positive or negative weight, but there is no cycles with negative weight. Edges are stored in the form of lists of adjacent vertices. If in the Dijkstra algorithm a nondecreasing queue with priorities is realized in the form of Fibonacci heaps, the operating time of Johnson's algorithm is $O(n^2\lg n+nm)$.

3.3 Theory of graphs and network analysis

Graph theory is used in parallel with fuzzy-set theory and probability theory. This compatibility does not allow bringing a large class of tasks to linear programming.

At the present time, graph theory is one of the basic mathematical algorithms, which are used to study ICN behavior at the network level. One of the solutions to the problem of optimized routing and decreasing a flow of traffic in each node of the network is to use graph theory and Petri nets.

A classic Jackson network is studied under the assumption that the duration of a service application has an exponential distribution. The law of the service application time distribution is different from the demonstrated one. There is an urgent problem of analytical apparatus development for the queuing networks study with arbitrary distribution functions of service time, which attracts more and more researchers [4].

There is a method to calculate the stochastic systems characteristics of applications processing based on diffusion models. The functioning process of the application processingnode is regarded as a two-dimensional diffusion process, which consists of two streams: a stream of applications receipt (i = 1) and the exit stream (i = 2).

$$\frac{\partial p(x_i,t)}{\partial t}=\sum_{i=1}^{2}\{a_i\frac{\partial^2 p(x_i,t)}{\partial x_i^2}-b_i\frac{\partial p(x_i,t)}{\partial x_i},$$

where p (x, t) – probability density admission (i = 1), processing (i = 2) requests: a – a diffusion coefficient, b – a drift coefficient. Moreover,

$$a = \frac{1}{\tau}, b = \frac{D}{\tau^{-3}},$$

where τ – average time between the applications receipts (application processing), D – time dispersion between applications receipts (and / or processing of applications).

This approach allows using the number of new algorithms of the stochastic processing applications system.

But the network structure in this case is set apart and the model operates during the average time of receipt and applications maintenance, at the time of arrival variances, at maintenance downtime of applications and a node. Thus, this model is inconvenient [5].

If the dynamics of the applications processing was not analyzed, then it is necessary to compute the stationary distribution of working / non-working hosts of Jackson with the changing number of sites, both for open and for closed networks where nodes may fail or recover.

It has been proved that if a discrete Markov process describing the Jackson network operation with a variable structure is ergodic, its limiting distribution P (n) satisfies the following equation:

$$P(n) = \frac{\prod_{i=1}^{m} \left(\frac{\overline{\lambda_i}}{\mu_i}\right)^n}{\prod_{i=1}^{m} \frac{\mu_i}{\mu_i - \overline{\lambda_i}}}$$

where $\lambda\ i$ and μ_i – the intensities of the input and output stream of applications of the i – type, respectively; n – the number of working units, m – the limit number of different types applications.

4. Results and discussion

1. For the networks with the information traffic operation it is preferably to use Johnson's method as it provides a speed gain in case of the network topology that is described by the big graph.

2. Among one path algorithms of the distance-finding it is possible to note Dijkstra's (OSFP protocol) algorithm. This algorithm is actively used as a part of the OSFP protocol for routing in autonomous systems (the protocol of an internal lock).

3. The matrix algorithm, as well as the other multipath algorithms, has the advantage of one path, consistingin in advance calculated alternative distances that eventually leads to a reliability increase of a network and a possibility of redistribution of loading between communication links.

4. A determination of the shortest path can be carried out by the methods of linear programming that allows one to estimate a temporary network lag effect under minimization of the average time of the message transmission.

5. Conclusion

Optimization of data transmission in information computer networks is connected with difficulties caused by the following factors:

- the necessity of continuous study of new technologies in data transmission, their features and features of the equipment supporting these technologies;
- strict cost and duration requirements of implementation of networks projects, as well asthe users' needs quick change in time that results in the necessity of the organization of teamwork on projects;
- the interactivity of the process of data transmission networks creation connected with continuous emergence of new services, an increase in the clients number of networks and a demand for expansion of the existing users.

Optimized transfer and processing of data in ICN will significantly reduce its work time, as well as the cost of development and support of software products.

The most important thing in the optimization of ICN is a number of information tools that ensure a constant level of reliability.

A reduction of costs on improving server software is accounted for information resources integration into a data processing center (or a few DPCs). In addition, costs for services and data communications are significantly reduced.

Therefore, the optimization of data transmission networks has resulted in the caching and compression of the data transmitted, traffic optimization, a change of transport protocol TSP, which increase the interactivity of the network application performance and reduce the volume of data transmitted.

References

[1] E N Gilbert 1960 *Bell Systems Technical Journal* **39** 1253–1266
[2] E Altman, K Avrachenkov and C Barakat 2000 *Performance evaluation* 42 129–147
[3] Nyrkov A, Sokolov S, Chernyi S and Mamunts D 2015 *Metallurg. and Mining Industry* **7**(6) 521
[4] Nyrkov A P, Belousov A S and Sokolov S S 2015 *Modern Applied Science* **9**(5) 162–176
[5] Chernyi S, Zhilenkov A, Sokolov S and Titov L 2015 *Metallurg. and Mining Industry* **7**(3) 66

Calculation of the surface effect in the ferromagnetic conductor with the harmonic electromagnetic field

G V Nosov[1], E O Kuleshova[1], Yu Z Vassilyeva[1] and A I Elizarov[2]
[1] National Research Tomsk Polytechnic University, 30, Lenina ave., Tomsk, 634050, Russia
[2] National Research Tomsk State University, 36, Lenina ave., Tomsk, 634050, Russia

E-mail: kul_2004@mail.ru

Abstract. The authors of the paper have obtained formulas for analytical calculation of the constants with the harmonic electromagnetic field, which characterize the surface layer (a skin layer) of the ferromagnetic conductors in case of heating and nonlinear magnetic properties, which can be used for practical calculation of the electromagnetic screens, rotors of the electrical machines and inductive heating installations. A nonlinear dependence of the magnetic induction on the magnetic tension of the ferromagnetic conductor is replaced by one or two linear sections. It is considered that the skin layer of the conductor has constant quantities of the specific conductivity and averaged temperature. Linear electrodynamics equations are solved for the conductive half-space. Parameters of the ferromagnetic conductor's surface layer are calculated: magnetic permeability, the thickness of the skin layer and its averaged temperature, exposure time of the electromagnetic field on the conductor with the established maximum temperature on the conductor's surface, pressure of the field on the conductor and its resistance, inductivity of the internal magnetic field in the conductor, the thermal energy capacity. The methods credibility is confirmed with the concurrence of the resistance and inductivity of the ferromagnetic conductor with analogous quantities from other methods.

1. Introduction

It is often necessary to calculate parameters along with the calculation of the electro energetic equipment, which characterize a surface effect in ferromagnetic conductors with the harmonic electromagnetic field (EMF). The calculation are simplified with the parameters, which characterize the conductor's surface layer (a skin layer). These parameters include the thickness of the skin layer and its temperature, EMF pressure on the conductor, and also the conductor resistance and the internal magnetic field inductivity [1-5, 7, 8]. A numerical computation of these parameters, taking into account the heating of the skin layer and the nonlinear magnetic properties of the ferromagnetic conductors, is concerned with huge machine time expenses, and calculations with the use of the known formulas are quite approximate and not enough well-grounded [1-5].

Thus the analytical calculation with harmonic EMF parameters of the skin layer and ferromagnetic conductors in case of its heating and nonlinear magnetic properties is one of the most important problem at present.

The main idea of this paper is the methods development, taking into account heating of the skin layer and its nonlinear magnetic properties, for the practical analytical calculation of the skin layer properties of ferromagnetic conductors in case of harmonic EMF.

To develop the methods the following assumptions are included:

1. The conductor skin layer is characterized by such constant quantities as specific conductivity γ (1/Ω·m), averaged temperature θ (°C), specific thermal capacity C (J/kg·°C), thermal conductivity λ (W/m·°C) and volume density ρ (kg/m^3).
2. The skin layer heating is quite short-term for its processing without heat exchange with the environment [4]. In this case maximum temperature of the ferromagnetic conductor θ_m does not exceed Curie temperature $\theta_K \approx 750$ (°C), thus, the absolute (static) and differential magnetic conductivities do not depend on the temperature [2].
3. The hysteresis losses are insignificant and are not taken into account in comparison with the eddy currents losses [2]. Thus, one-digit magnetization curve $B(H)$ of the ferromagnetic material of the conductor, which is symmetrical to the coordinate origin, is described as a straight line in the first approximation (fig. 1, dependence 1), and in the second approximation – as a two linear section (Figure 1, dependence 2).
4. Conductors sizes and surface curvature radiuses exceed greatly the thickness of the skin layer, thus we will consider the surface effect as penetration of the one-dimensional plane electromagnetic wave into the conductive ferromagnetic half-space [2, 4].
5. Harmonic EMF is specified by magnetic tension on the surface of the conductor, where $z=0$ (the surface of the half-space):
$$H_S(t)=H_m\sin(\omega t), \tag{1}$$
t is time; $\omega=2\pi f$ is angular frequency (1/s).
6. The duration of EMF, to which conductor τ is exposed, greatly exceeds the period of the tension alteration (1), therefore $\tau \gg T = 2\pi/\omega$. Thus the usage of the envelope dependences for the active (root-mean-square) current density, maximum tension and maximum induction is fully well-founded.

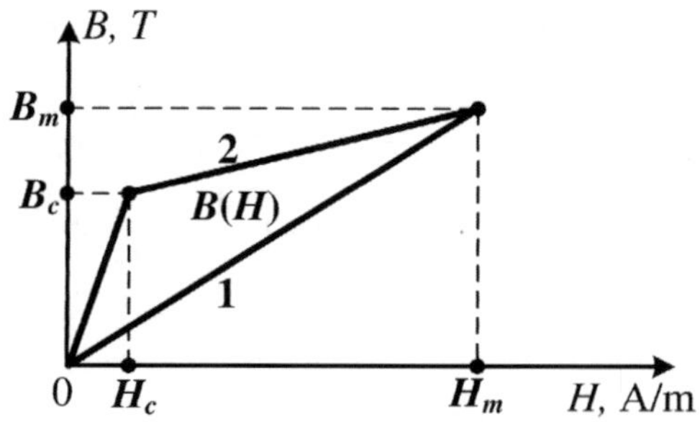

Figure 1. Approximate dependences $B(H)$ of the ferromagnetic conductor: B_m, H_m and B_c, H_c – are maximum and intermediate values of the magnetic induction, magnetic tension.

2. Methods

For the one-dimensional plane electromagnetic wave in the conductive half-space the design equation for magnetic tension $H_{y1,2}(z,t)$ and induction $B_{y1,2}(z,t)$ can be described as [2, 4]

$$\frac{\partial^2 H_{y1,2}(z,t)}{\partial z^2}=\gamma\frac{\partial B_{y1,2}(z,t)}{\partial t}=\mu_{d1,2}(z,t)\gamma\frac{\partial H_{y1,2}(z,t)}{\partial t} . \tag{2}$$

Then the current density will be

$$\delta_{x1,2}(z,t)=-\frac{\partial H_{y1,2}(z,t)}{\partial z} , \tag{3}$$

where the differential magnetic permeability for dependence 1 (Figure 1) is constant and equals static magnetic permeability:

$$\mu_{d1}(z,t)=dB_{y1}(z,t)/dH_{y1}(z,t)=\mu_1(z,t)=B_{y1}(z,t)/H_{y1}(z,t)=\mu_m=B_m/H_m\,. \tag{4}$$

And for dependence 2 (Figure 1) the differential magnetic permeability equals

$$\mu_{d2}(z,t)=\frac{dB_{y2}(z,t)}{dH_{y2}(z,t)}=\begin{cases}\mu_c=B_c/H_c & if\ \ 0<\left|H_{y2}(z,t)\right|<H_c;\\ \mu_d=(B_m-B_c)/(H_m-H_c) & if\ \ H_c<\left|H_{y2}(z,t)\right|<H_m\end{cases} \tag{5}$$

with the static magnetic permeability in accordance with (5)

$$\mu_2(z,t)=\frac{B_{y2}(z,t)}{H_{y2}(z,t)}=\begin{cases}\mu_c & if\ 0<\left|H_{y2}(z,t)\right|<H_c;\\ \mu_d+(\mu_c-\mu_d)H_c/\left|H_{y2}(z,t)\right| & if\ H_c<\left|H_{y2}(z,t)\right|<H_m.\end{cases} \tag{6}$$

With equations (1-4) according to [2, 4] for dependence 1 (fig. 1) let us calculate tension

$$H_{y1}(z,t)=H_m\exp(-z/\Delta_1)\cdot\sin(\omega t-z/\Delta_1) \tag{7}$$

and current density

$$\delta_{x1}(z,t)=\sqrt{2}(H_m/\Delta_1)\exp(-z/\Delta_1)\cdot\sin(\omega t-z/\Delta_1+\pi/4) \tag{8}$$

in case of equivalent depth of the penetration of harmonic EMF into the conductive half-space with magnetic permeability (4)

$$\Delta_1=\sqrt{2/\mu_m\gamma\omega}\,. \tag{9}$$

As a result, on the basis of (7, 8) for straight line 1 (fig. 1) envelope dependences on depth z for the maximum tension equal

$$H_1(z)=H_m\exp(-z/\Delta_1)\,. \tag{10}$$

And for the active (root-mean-square) current density they are equal to

$$\delta_1(z)=(H_m/\Delta_1)\exp(-z/\Delta_1)\,. \tag{11}$$

For dependence 2 (Figure 1) in accordance with (5, 6, 9-11) envelope dependences in the function from z are described as

$$H_2(z)\approx H_m\exp(-z/\Delta_1)\ \text{for the maximum tension,} \tag{12}$$

and for the maximum induction they are

$$B_2(z)=\begin{cases}B_c+\mu_d\left[H_2(z)-H_c\right] & if\ \ 0<z<c;\\ \mu_c H_2(z) & if\ \ c<z<\infty;\end{cases} \tag{13}$$

for the active current density they equal

$$\delta_2(z)\approx(H_m/\Delta_2)\exp(-z/\Delta_2), \tag{14}$$

where the thickness of the skin layer can be found according to the following formula:

$$\Delta_2=\sqrt{2/\mu_p\gamma\omega}\,. \tag{15}$$

In case of design magnetic permeability it is

$$\mu_p\approx\begin{cases}\mu_c & if\ 0.5H_m<H_c;\\ \mu_d+(\mu_c-\mu_d)H_c/0.5H_m & if\ H_c<0.5H_m\end{cases} \tag{16}$$

and the distance from the conductor surface is

$$c=\Delta_1\cdot\ln(H_m/H_c). \tag{17}$$

At that $B_2(0)=B_m;\ B_2(c)=B_c;\ H_2(0)=H_m; H_2(c)=H_c$.

Using (5, 9, 12-17) for dependence 2 (Figure 1) the formulas for the calculation of the maximum energy of the magnetic field inside the conductor [2, 4] are

$$W_m=l_xl_y\int_0^{\infty}\left[\int H_2(z)dB_2(z)\right]dz=\left[1+\left(H_c^2/H_m^2\right)\left(\mu_c/\mu_d-1\right)\right]H_m^2l_xl_y\sqrt{\mu_d^2/8\mu_m\gamma\omega}\,, \tag{18}$$

and the capacity of the thermal energy is

$$P=l_xl_y\int_0^{\infty}\left[\delta_2(z)^2/\gamma\right]dz=H_m^2l_xl_y\sqrt{\mu_p\omega/8\gamma}\,, \tag{19}$$

where *lx*, *ly* – are sizes of the conductor on axis x and y.

In the conductor along axis x there is induced or created by the external source harmonic current, which, in accordance with (1), can be found using Ampere's circuital law [4]:

$$i(t)=l_yH_S(t)=l_yH_m\sin(\omega t)=I_m\sin(\omega t). \tag{20}$$

The function of the linkage of the magnetic flux in the conductor with current (20) can be found using equation (11),

$$K(z)\approx\left(l_y/I_m\right)\int_z^{\infty}\delta_1(z)dz=\exp(-z/\Delta_1), \tag{21}$$

Then, in accordance with (4, 5, 9, 13, 21), the maximum interlinkage for the magnetic field inside the conductor equals [4]

$$\Psi_m=l_x\int_0^{\infty}B_2(z)K(z)dz=0{,}5\mu_dH_m\Delta_1l_x\left[1+\left(H_c/H_m\right)\left(2-H_c/H_m\right)\left(\mu_c/\mu_d-1\right)\right]. \tag{22}$$

For current (20) from the equations described for energy (18), interlinkage (22) and capacity (19) it follows that

$$W_m=0.5I_m^2L_w;\quad \Psi_m=I_mL_\psi;\quad P=0.5I_m^2R.$$

Let us calculate inductions

$$L_\psi=\sqrt{\frac{\mu_d^2}{2\mu_m\gamma\omega}}\left[1+\frac{H_c}{H_m}\left(2-\frac{H_c}{H_m}\right)\left(\frac{\mu_c}{\mu_d}-1\right)\right]\cdot\frac{l_x}{l_y};\quad L_w=\sqrt{\frac{\mu_d^2}{2\mu_m\gamma\omega}}\left[1+\frac{H_c^2}{H_m^2}\left(\frac{\mu_c}{\mu_d}-1\right)\right]\cdot\frac{l_x}{l_y} \tag{23}$$

and resistance of the conductor

$$R=\sqrt{\mu_p\omega/2\gamma}\cdot l_x/l_y\,. \tag{24}$$

At that $L_\psi\geq L_w$ and for linear magnetization curve $B(H)$, which has $\mu_m=\mu_d=\mu_c$, $L_\psi=L_w$ as well.

From the resulting inductance of the internal magnetic field of the conductor let us find the average value of L_ψ and L_w:

$$L=\left(L_\psi+L_w\right)/2=\sqrt{\mu_m/2\gamma\omega}\cdot l_x/l_y\,. \tag{25}$$

Then, in accordance with (24, 25) we can deduce the equation for the conductive half-space with the linear magnetization curve in case of $\mu_m=\mu_p$: $R=\omega L$ [4].

On the basis of equations (11, 13) let us calculate the EMF maximum pressure exerted on the conductor, which is directed along axis z [2]:

$$\sigma_m\approx\int_0^{\infty}\left[\delta_1(z)B_2(z)\right]dz=0.5B_mH_m\left(1+B_c/B_m-H_c/H_m\right). \tag{26}$$

Thus, nonlinearity of dependence $B(H)$ increases pressure σ_m: for rectangular $B(H)$, when $H_c=0;\ B_c=B_m$, we obtain $\sigma_m=B_mH_m$, and for linear $B(H)$ with $H_c/H_m=B_c/B_m$ we have correct value $\sigma_m=0.5B_mH_m$ according to [2].

Without heat removal into the environment the equation of conductor's temperature $\theta_T(z,t)$ and current density (14) will be [5]

$$\rho C \frac{d\theta_T(z,t)}{dt} = \frac{\delta_2(z)^2}{\gamma} + \lambda \frac{\partial^2 \theta_T(z,t)}{\partial z^2}. \quad (27)$$

With temperature $\theta_T(z,0)=\theta_0; \theta_T(\infty,t)=\theta_0$ and (14, 15) the solution of equation (27) is

$$\theta_T(z,t)=\theta_0+\left(H_m^2/4\gamma\lambda\right)\left[\exp\left(4\lambda t/\Delta_2^2\rho C\right)-1\right]\exp\left(-2z/\Delta_2\right). \quad (28)$$

Therefore maximum temperature θ_m of the conductor with nondimensional parameter χ and duration of the EMF exposure τ will be equal to

$$\theta_m=\theta_T(0,\tau)=\theta_0+\frac{H_m^2}{4\gamma\lambda}\left[\exp(\chi)-1\right];\ \chi=\frac{4\lambda\tau}{\Delta_2^2\rho C};\ \tau=\frac{\Delta_2^2\rho C}{4\lambda}\cdot\ln\left[1+\frac{4\gamma\lambda(\theta_m-\theta_0)}{H_m^2}\right], \quad (29)$$

where θ_0 is initial temperature of the conductor.

Let us consider that the thickness of the skin layer equals Δ_2 (15), thus with (28, 29) the equation for the calculation of averaged temperature θ with specified conductivity γ [2, 5] is

$$\theta=\frac{1}{\tau}\int_0^\tau\left[\frac{1}{\Delta_2}\int_0^{\Delta_2}\theta_T(z,t)dz\right]dt=\theta_0+\frac{0.108H_m^2}{\gamma\lambda}\left[\frac{\exp(\chi)-1}{\chi}-1\right];\ \gamma=\frac{\gamma_0}{1+\alpha[\theta-\theta_0]+\beta[\theta-\theta_0]^2}, \quad (30)$$

where γ_0 is the quantity of the specified conductivity with temperature θ_0; α,β are the temperature constant coefficients.

3. Results

The initial data for the ferromagnetic conductor from [2] are shown in table 1. With correlations (1-30) the skin layer parameters of the ferromagnetic conductor were found (Table 2).

Table 1. A ferromagnetic cylinder conductor made of structural steel with diameter D = 0.1 (m) and length 1 (m)

l_x	$l_y=\pi D$	θ_0	ρ	C	λ	μ_0
m	m	°C	kg/m^3	J/kg·°C	W/m·°C	H/m
1	0.314	20	7800	575	42.5	$4\pi\cdot10^{-7}$
γ_0	α	β	ω	θ_m	H_C	B_C	μ_c/μ_0
1/Ω·m	1/°C	1/°C^2	1/s	°C	A/m	T	–
$5\cdot10^6$	0.0055	$9\cdot10^{-6}$	314	700	4000	1.5	298.4

For testing of the developed methods the approximate formulas for the ferromagnetic conductors made of structural steel [2] were used:

$$\mu_S=8130\mu_0\left(H_m/100\right)^{-0.894};\ R=1.32\sqrt{\mu_S\omega/2\gamma}\cdot l_x/l_y;\ L=0.98\sqrt{\mu_S/2\gamma\omega}\cdot l_x/l_y, \quad (31)$$

where μ_S is the static magnetic permeability on the surface of the conductor with $z=0$ and $4000<H_m<400000$ (A/m).

The results of the R and L calculation for (31) with θ and γ according to (30) are shown in Table 2. Resistance R and inductance L were also determined with the help of the computer modulation program 'Elcut' [6], when EMF creates the harmonic current, which runs in the cylinder conductor with diameter $D=0,1$ (m) and length $l_x=1$ (m). Besides, values $\omega,\gamma,\mu_p/\mu_0,\mu_m/\mu_0$ were used from Tables 1 and 2. At that μ_p/μ_0 were used for calculation of R, whereas μ_m/μ_0 were used for calculation of L as the average value of inductances $L_w,L_\psi,L_z=\mathrm{Im}(\underline{Z})/\omega$, which can be found with the help of the Elcut program, where $\underline{Z} = R + j\omega L_z$ is the conductor complex impendence ($j = \sqrt{-1}$). Parameters R and L found with Elcut [6] are shown in table 2.

Resistances R and inductances L calculated with equations (24, 25), [2] and [6] are approximately equal.

Table 2. Design parameters of the ferromagnetic conductor skin layer

H_m		kA/m	6	8	15.9	39.9	79.7	159.4	239.1	318.8
B_m		T	1.550	1.635	1.785	1.985	2.099	2.228	2.338	2.441
μ_m/μ_0		–	205.6	162.6	89.3	39.6	21.0	11.1	7.8	6.1
μ_p/μ_0		–	298.4	298.4	159.6	68.4	35.6	18.5	12.7	9.8
σ_m		kPA	6.05	9.27	22.55	65.56	139.2	292.7	454.2	623.3
τ		s	1.005	0.953	1.548	2.870	4.401	6.143	6.812	6.868
θ		°C	50.84	52.84	58.85	71.10	85.59	106.7	121.7	132.5
γ/γ_0		–	0.849	0.840	0.815	0.767	0.715	0.648	0.605	0.577
Δ_2		mm	2.000	2.010	2.791	4.395	6.309	9.192	11.47	13.38
P		kW	0.666	1.190	3.492	14.84	44.26	134.1	258.7	413.5
R		μΩ	375.0	376.9	279.9	189.0	141.2	106.9	91.7	82.4
L		μH	0.991	0.886	0.667	0.458	0.345	0.264	0.228	0.207
[2]	R	μΩ	414.4	366.3	273.6	187.0	142.1	109.5	94.5	85.1
	L	μH	0.979	0.866	0.647	0.442	0.336	0.259	0.223	0.201
[6]	R	μΩ	434.7	437.0	318.9	208.0	154.6	118.9	104.0	95.2
	L	μH	0.845	0.793	0.546	0.468	0.358	0.272	0.233	0.208

4. Conclusion

The analysis of the forms and calculation results allow formulating the following conclusions:

1. For the harmonic electromagnetic field the methods of analytical calculation of ferromagnetic conductor skin layer parameters in case of the considerable surface effect with the heating and nonlinear magnetic properties were developed. The methods can be applied for the induction heating devices, solid rotors of the electric machines and ferromagnetic screens design.
2. The formulas can be used for calculation of parameters of the nonferromagnetic conductors' skin layer with design magnetic permeability: $\mu_m = \mu_p = \mu_d = \mu_c = \mu_0$.
3. The credibility of the methods can be confirmed with the approximate equation of resistance R and inductance L of the ferromagnetic conductor with analogical quantities found by means of other methods.

References

[1] Brito A I et al. 2016 *Elec. Pow. Sys. Res.* **130**(4421) 132–138
[2] Trip N D et al. 2014 *Conf. Proceedings ISETC* 7010735
[3] Schmülling C, Conradi A 2015 *Lec. Notes in Elec. Eng.* **324** 315–319
[4] Nagsarkar T K 2005 *Basic Electrical Engineering* (New York) pp. 634
[5] D V Richardson, A J 1997 *Caisse. Electric Machinery and Transf. Techn.* (New Jersey) pp. 731
[6] Reutov Yu Ya, Gobov Yu L, Loskutov V E 2002 *Rus. J. of Nondestructive Testing* **6** 425–430
[7] O V Vasileva *et al* 2014 *Int. Conf. MEACS* (Russia: Tomsk) 6986867
[8] Shandarova E B, Shwab S A 2014 *Int. Conf. MEACS* (Russia: Tomsk) 6986872

IOP Conf. Series: Materials Science and Engineering **124** (2016) 012082 doi:10.1088/1757-899X/124/1/012082

Investigation of the input signal frequency effect on the formed pulse of the hydraulic-powered pulse machine

M V Novoseltseva, I A Masson
National Research Tomsk Polytechnic University, 30, Lenina ave., Tomsk, 634050, Russia

E-mail: novmv@tpu.ru

Abstract. Nowadays, a special emphasis is placed on an output signal curve during the analysis of well drilling machines since these machines should have as high energy efficiency as it is possible. This work proposes factors that have an impact of input signal frequency on the formed pulse that are used to find the most efficient frequency for its further applying in the machine. Results of the conducted experiment are obtained by using a mathematical model that is created in Simulink Matlab.

1. Introduction
Nowadays, special emphasis is placed on an output signal curve during the analysis of well drilling machines. Efficiency of existing drilling impact machines is estimated by comparison of impact energy. The value of impact energy is variable and usually dependent on the borehole diameter and the type of a machine. Impact energy is not the only factor that contributes to the overall efficiency of drilling by impact machines; amplitude and the oscillation period should be evaluated too. The current work is dedicated to finding of the best value of amplitude and the oscillation period that eventually can improve key performance indicators of existing impact machines. These indicators include [6]:

- good energy efficiency;
- formation of the applicable pulse curve;
- the highest peak efficiency.

The analysis of impact machines shows several drawbacks of existing impact machines, hence a new type of impact machine is described. This is a hydraulic-powered pulse machine without a pin, which is being developed in Tomsk Polytechnic University [1, 4]. The existing machines have energy losses on friction that occur among the pin, the anvil and the casing. The developed hydraulic-powered pulse machine without the pin is not only less affected by friction, but it also uses a resonance effect in the fluid power cylinder [3].

2. Mathematical modelling of the impact machine
Mathematical models describe the basic properties of an object, process or system, its parameters, internal and external communications with the help of logical-mathematical structures. At the current stage of research, a mathematical model of the developed machine is created to analyze the input signal effect on formed pulse parameters. All experiments are conducted in the MATLAB software. The software offers a variety of methods for findinga solution to Cauchy problem for differential equations: ordinary differential equation (ODE) solvers, special-purpose programs for the model investigation of controlled systems, SIMULINK as an analog simulation tool [2]. The scheme

generated in SIMULINK consists of units that simulate the process, its properties and construction parameters. The flow chart of the impact hydroimpulsive machine that depicts interrelationship of all machine units is shown in Figure 1. The mathematical model covers a hydro-pulsator, a high-pressure hose and impact machine parts.

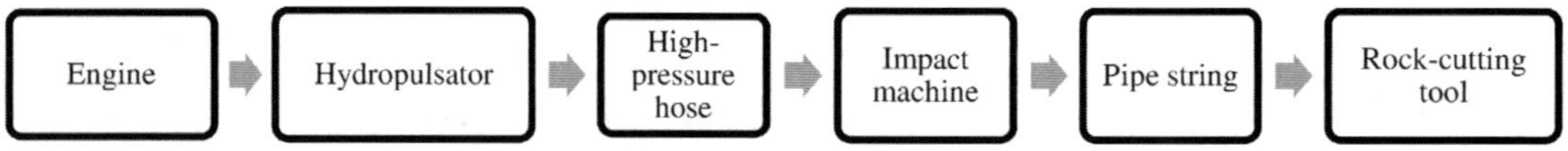

Figure 1. A flow chart of the impact hydroimpulsive machine.

DSP System Toolbox is used for mathematical modeling of the hydraulic-powered pulse machine; it allows predicting the system behavior under different conditions and then simulating it for a more precise system analysis.

3. Impact of input signal frequency on the formed pulse

It is important to determine the value of input pulse frequency, because the system work is based on resonance;therefore, the frequency of its oscillations matches the system's natural frequency of vibration [5]. It is also important to find resonance frequency for different input system parameters.Bode plot, one of the Matlab modules, is used for natural frequency determination. Bode plot itself is a graph of the frequency response of the system. It is usually a combination of Bode magnitude plot and Bode phase plot, both quantities are plotted against the same horizontal axis. Bode magnitude plot expresses the magnitude of the frequency response; ω-axis of the magnitude plot is logarithmic and the magnitude is given in decibels. Bode phase plot is the graph of the phase of the transfer function in degrees plotted logarithmically as a function of ω.

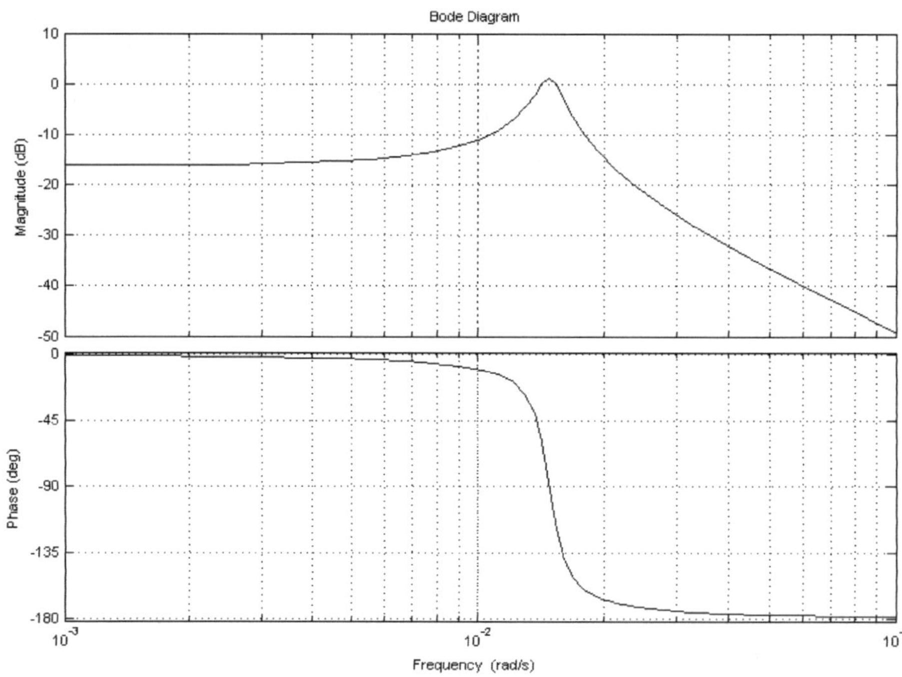

Figure 2. Bode plot of the hydraulic-powered pulse machine without the pin for initial parameters of the system.

The mathematical model is tested for the following system parameters, as depicted in Table 1.

Table 1. Impact machine parameters

	Parameter	Value
c	Spring stiffness	$18639\frac{N}{m}$
m	Active mass	100 kg
Cv	High-pressure hose volume elasticity	$3.441\cdot10^{-9}\frac{m^3}{Pa}$
$\mathbf{k_f}$	Coefficient of friction	0.2

The system base frequency is deduced from bode plot on the magnitude curve as a peak value on X axis. The system base frequency obtainedfrom bode plot (Figure 2) is inserted into the mathematical model input parameters. The input signal curve is sine due to the specified motion law of the plunger [4]. $\boldsymbol{\omega_n}$stands for natural frequency; it depends on spring stiffness, high-pressure hose volume elasticity and active mass. $\boldsymbol{\omega_{in}}$stands for input frequency; it depends on the motion law of the machine drive and can be adjusted.

4. Results and Discussion

For coalescence of frequencies $\omega_n = 2.36$ mHz and $\omega_{in} = 2.36$ mHz, the hydraulic-powered pulse machine operates in a resonance mode as shown in Figure 3. The pulse amplitude reaches the peak value;no instability during the transient condition ordeviationsof the curve form are observed.

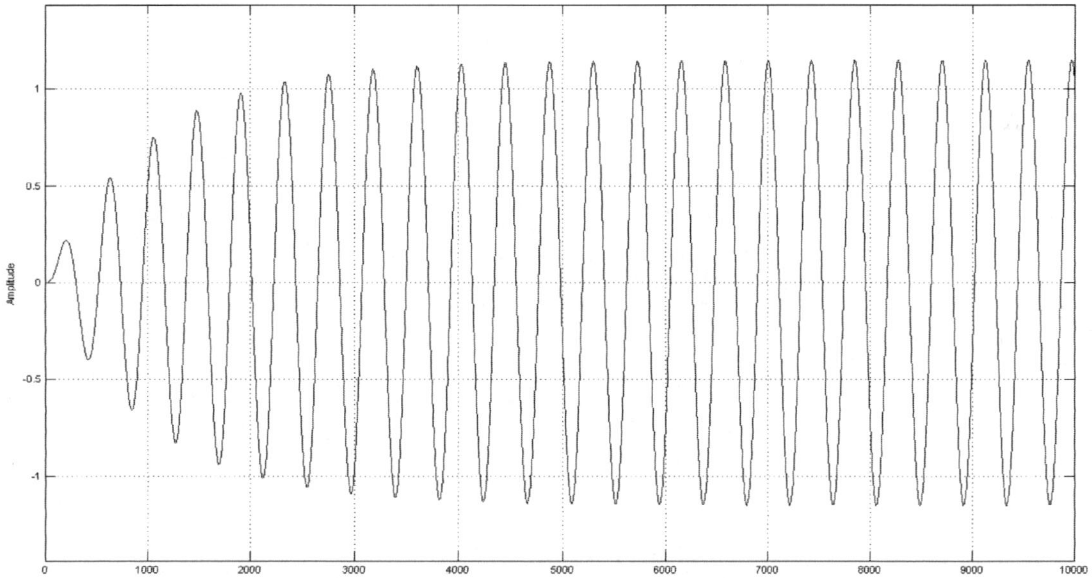

Figure 3. The formed output pulse curve for $\omega_{in} = 2.36$ mHz.

The input frequency value is set below the resonance value of 2.36 mHz. Calculations are made for $\omega_{in} = 1.6$ mHz (Figure 4). The output signal is periodic after some instability in the transient phase; its oscillation period is increased. Though the frequency is slightly changed, the amplitude is almost 4 times lower than thatin the resonance mode.

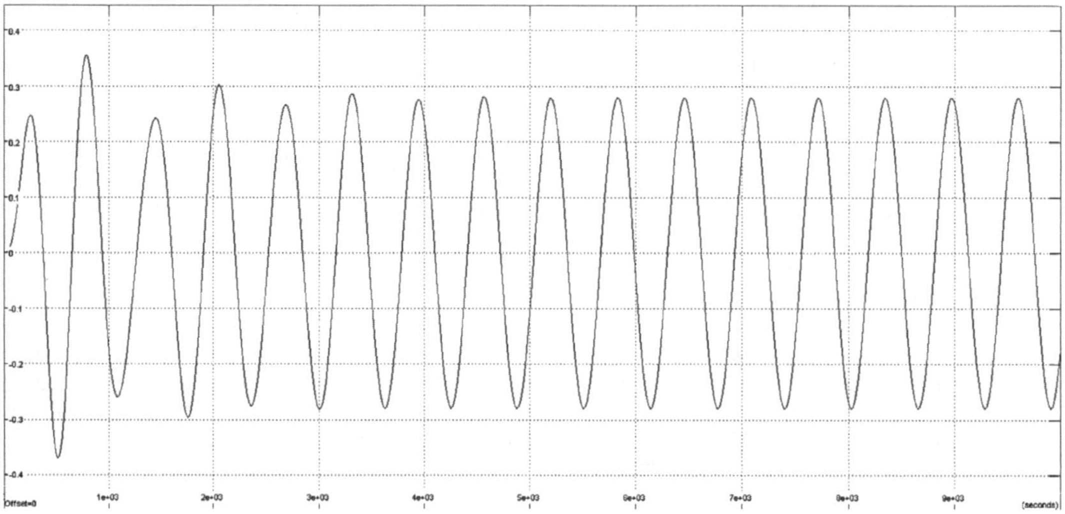

Figure 4. The formed output pulse curve for $\omega_{in} = 1.6$ mHz.

The input frequency value is set above the resonance value of 2.36 mHz. Calculations aremade for $\omega_{in} = 3.2$ mHz (Figure 5). The output signal is periodic after apparent instability in the transient phase; its oscillation period is decreased. Though the frequency is slightly increased, the amplitude is almost 6 times lower than that in the resonance mode.

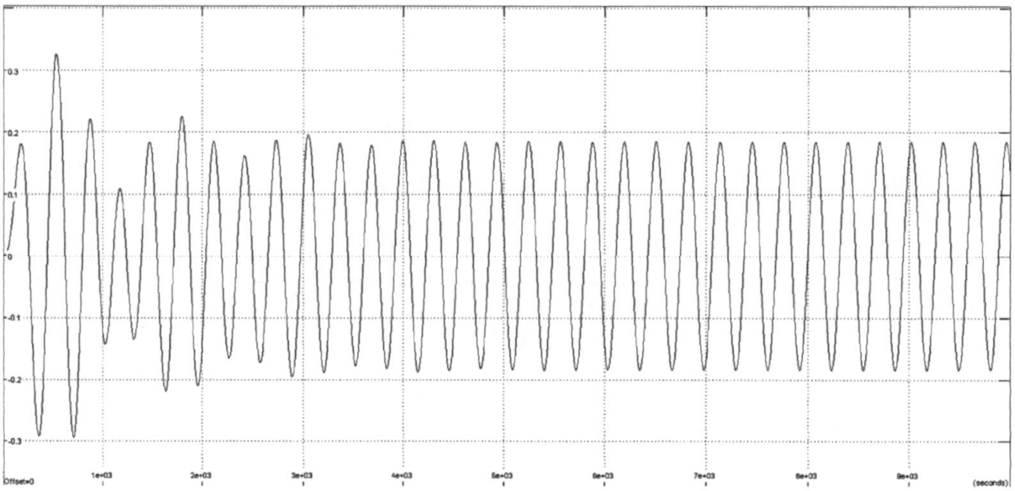

Figure 5. The formed output pulse curve for $\omega_{in} = 3.2$ mHz.

5. Conclusion

The machine mathematical simulation has shown the possibility to impact the pulse parameters in the calculated range by changing frequency as an input parameter.It is of great importance to operate the machine in the resonance mode, because slight changes inthe input frequency strongly reduces the amplitude, i.e. reduces the overall machine efficiency. The conducted experiments show less instability in the mode below the resonance one, whichfollows after the resonance mode;the amplitude is 1.5 times higher during the oscillation period, which is almost 2 times higher than thatfollowing the resonance mode and 1.4 times higher than that in the resonance mode.

References

[1] Pashkov E N, Ziyakaev G R, Novoseltseva M V 2013 *Appl. Mech. Mater.* **379** 91–94
[2] Pashkov E N, Martyushev N V, Masson I A, 2014 *IOP Conf. Series: Mat. Sci. Eng.* **66** 1–5
[3] Angatkina O O, Krauinsh P Y, Deryusheva VN 2015 *Appl. Mech. Mater.* **756** 394–401
[4] Novoseltseva M V 2015 *Adv. Mater. Res.* **1040** 682–685
[5] Glazov A N, Kuznetsov I V 2013 *Mining Informational and Analytical Bulletin* **S4(1)** 515–517
[6] WeiX, Wang C Y, Yuan H L, Xie Z H 2003 *Key Eng. Mater.* **250** 200–208
[7] Yang J D, Wang Y 2011 *Appl. Mech. Mater.* **52** 1930–35
[8] Wang X J, Zhao X X 2014 *Appl. Mech. Mater.* **580** 2096–102

Mathematical modelling of contact of ruled surfaces: theory and practical application

K L Panchuk, A S Niteyskiy
Omsk State Technical University, Department of Engineer's Geometry & CAD, Omsk, 644050, Russia

E- mail: Panchuk_KL@mail.ru

Abstract. In the theory of ruled surfaces there are well known researches of contact of ruled surfaces along their common generator line (Klein image is often used [1]). In this paper we propose a study of contact of nondevelopable ruled surfaces via the dual vector calculus. The advantages of this method have been demonstrated by E. Study, W. Blaschke and D. N. Zeiliger in differential geometry studies of ruled surfaces in space R^3 over the algebra of dual numbers. A practical use of contact is demonstrated by the example modeling of the working surface of the progressive tool for tillage.

1. Fundamentals of theory

The equation of a ruled surface can be represented as a unit dual vector-function with real parameter t: $\boldsymbol{A}_1(t)=\boldsymbol{a}_{01}(t)+\omega\boldsymbol{a}_{11}(t), |\boldsymbol{A}_1(t)|=1, \omega^2=0$, where $\boldsymbol{a}_{01}(t)$ – a unit generator vector of the ruled surface, $\boldsymbol{a}_{11}(t)$ – a momentum vector of the generator with respect to the origin. $\boldsymbol{A}_1(t)$ is $n+1$ times continuously differentiable. It is also known that a trihedron of the ruled surface consists of dual unit vectors:

$$\boldsymbol{A}_1; \boldsymbol{A}_2 = \frac{\boldsymbol{A}_1'}{|\boldsymbol{A}_1'|}; \boldsymbol{A}_3 = \boldsymbol{A}_1 \times \boldsymbol{A}_2,$$

where $\boldsymbol{A}'_1=d\boldsymbol{A}_1/dt$; $H_A=|\boldsymbol{A}'_1|$; $\boldsymbol{A}_2$ – a central normal vector; $\boldsymbol{A}_3$ – a central tangent vector.

And we also used an inner geometric parameter of the ruled surface - an element of a dual arc:

$$ds_A = ds_{A0} + \omega \cdot ds_{A1} = H_A dt, \omega^2 = 0,$$

which means a dual angle between its two infinitesimally close generators.

A distribution parameter of the ruled surface is deduced from the element of the dual arc:

$$p = \frac{ds_{A1}}{ds_{A0}}.$$

Trihedron derivation equations of W. Blaschke [2] are as follows:

$$\boldsymbol{A}_1' = H_A \cdot \boldsymbol{A}_2; \ \boldsymbol{A}_2' = -H_A \cdot \boldsymbol{A}_1 + Q_A \cdot \boldsymbol{A}_3; \ \boldsymbol{A}_3' = -Q_A \cdot \boldsymbol{A}_2,$$

where $H_A = h_{A0}+\omega h_{A1}$, $Q_A = q_{A0}+\omega q_{A1} = (\boldsymbol{A}_1\boldsymbol{A}'_1\boldsymbol{A}''_1)/H_A^2$, $\omega^2 = 0$.

Let us consider the contact of two nondevelopable ruled surfaces parameterized by $\boldsymbol{A}_1(t)$ and $\boldsymbol{B}_1(t)$ with common ruling $\boldsymbol{a}_{01}(t) = \boldsymbol{b}_{01}(t)$, when parameter value $t = t_0$. Initially we will introduce the dual vector of divergence $\boldsymbol{G}(t)$ in their common ruling. Since each unit dual vector function $\boldsymbol{A}_1(t)$ and $\boldsymbol{B}_1(t)$ can be represented as a Taylor series it can be written as:

$$G(t) = A_1(t_0) - B_1(t_0) + [A_1'(t_0) - B_1'(t_0)] \cdot \Delta t + [A_1''(t_0) - B_1''(t_0)] \cdot \frac{\Delta t^2}{2!} + .. \quad (1)$$

We will define the concept of an order of contact of two ruled surfaces in their common generator line. Each of the considered dual vector functions has the following coordinate representation:

$$\begin{aligned} A_1(t) &= i \cdot x_A(t) + j \cdot y_A(t) + k \cdot z_A(t), \\ B_1(t) &= i \cdot x_B(t) + j \cdot y_B(t) + k \cdot z_B(t), \end{aligned} \quad (2)$$

where x_A, y_A, z_A; x_B, y_B, z_B – dual scalar functions. It allows one to write an expression for dual vector $G(t)$:

$$G(t) = A_1(t) - B_1(t) = i \cdot x_G + j \cdot y_G + k \cdot z_G = i(x_A - x_B) + j(y_A - y_B) + k(z_A - z_B). \quad (3)$$

Dual vector $G(t)$ characterizes the closeness of ruled surfaces $A_1(t)$ and $B_1(t)$ in infinitesimal neighborhood for their common generator at parameter $t = t_0$. It is defined by two generators $a'_{01}(t)$ and $b'_{01}(t)$, each of which is shifted in their ruled surface along the same dual arc $ds_A = ds_B$ relatively their common generator $a_{01}(t) = b_{01}(t)$. Let us introduce the notion of the order contact of two ruled surfaces in their common generator. We will estimate an order of infinitesimality of module $|G(t)|$ relatively dual infinitesimal Δs. If n – positive integer then the following condition holds:

$$\lim_{\Delta s \to 0} \frac{|G|}{\Delta s^n} = 0, \quad (4)$$

thus, we will assume that ruled surfaces in their common generator $a_{01}(t_0) = b_{01}(t_0)$ are not lower than the n-th order contact. But if n - maximal then ruled surfaces are exactly in the n-th order contact.

Because of $ds=ds_0 + \omega ds_1=h_0dt + \omega h_1dt$, $\omega^2=0$, where $ds_0 =h_0dt$, $ds_1=h_1dt$, so condition $\Delta s \to 0$ can be replaced by the following two conditions: $\Delta s_0 \to 0$ and $\Delta s_1 \to 0$, that in case of $h_0 \neq 0$ and $h_1 \neq 0$ leads to $\Delta t \to 0$. Indeed ds_0 and dt, ds_1 and dt – pairs of equivalent infinitesimals. As a result we obtain the following condition for estimating the order contact of ruled surfaces in their common generator:

$$\lim_{\Delta t \to 0} \frac{|G(t)|}{\Delta t^n} = 0. \quad (5)$$

Theorem. To fulfill the condition of (5) it is necessary and sufficiently that the following equalities are carried out

$$\begin{aligned} &x_G(t_0)=0,\ x'_G(t_0)=0,\ x''_G(t_0)=0, \ldots, x_G{}^n(t_0)=0; \\ &y_G(t_0)=0,\ y'_G(t_0)=0,\ y''_G(t_0)=0, \ldots, y_G{}^n(t_0)=0; \\ &z_G(t_0)=0,\ z'_G(t_0)=0,\ z''_G(t_0)=0, \ldots, z_G{}^n(t_0)=0; \end{aligned} \quad (6)$$

Sufficiency. Assume that equalities (6) hold. Then from (1) it follows that:

$$x_G(t) = x_A(t) - x_B(t) = x^{n+1}(\delta_1) \cdot \frac{\Delta t^{n+1}}{(n+1)!}; \; y_G(t) = y_A(t) - y_B(t) = y^{n+1}(\delta_2) \cdot \frac{\Delta t^{n+1}}{(n+1)!};$$

$$z_G(t) = z_A(t) - z_B(t) = z^{n+1}(\delta_3) \cdot \frac{\Delta t^{n+1}}{(n+1)!},$$

where δ_1, δ_2, δ_3 – values of parameter t, which satisfy the following conditions: $t_0<\delta_1<t$, $t_0<\delta_2<t$, $t_0<\delta_3<t$.

Thus, the condition of (5) holds.

Necessity. Let us assume that at least one of equalities (6) does not hold, for example $x_G{}^k(t_0) = x_A{}^k(t_0) - x_B{}^k(t_0) \neq 0$; $k<n$.

Thus, $x_G(t) = x_A(t) - x_B(t) = \left[x_A{}^k(t_0) - x_B{}^k(t_0)\right] \cdot \frac{\Delta t^k}{k!} + \ldots,$

hence $\lim_{\Delta t \to 0} \frac{|G(t)|}{\Delta t^n} \neq 0$, $k<n$.

So the condition of (5) does not hold.

Let us consider the primary orders of contact of ruled surfaces.

2. The orders of contact

2.1. Order n=0 of contact of ruled surfaces

$$\boldsymbol{A}_1(t_0)=\boldsymbol{B}_1(t_0);\boldsymbol{A}_1'(t_0)\neq\boldsymbol{B}_1'(t_0). \quad (7)$$

In this case ruled surfaces are intersecting along their common generator.

2.2. Order n=1 of contact of ruled surfaces

$$\boldsymbol{A}_1(t_0)=\boldsymbol{B}_1(t_0);\boldsymbol{A}_1'(t_0)=\boldsymbol{B}_1'(t_0);\boldsymbol{A}_1''(t_0)\neq\boldsymbol{B}_1''(t_0). \quad (8)$$

Ruled surfaces are exactly in the first order contact in their common generator $\boldsymbol{a}_{01}(t_0) = \boldsymbol{b}_{01}(t_0)$.

From equalities of dual arcs $ds_A = ds_B$ it follows that $H_A dt = H_B dt$, i.e. $H_A = H_B$. Since $\mathbf{B}'_1=H_B\cdot\mathbf{B}_2$, where $H_B = |\boldsymbol{B}_1'|$, then from condition $\boldsymbol{A}'_1(t_0)=\boldsymbol{B}'_1(t_0)$ we obtain that in the common generator the following equalities hold:

$$\boldsymbol{A}_1=\boldsymbol{B}_1;\ \boldsymbol{A}_2=\boldsymbol{B}_2;\ \boldsymbol{A}_3=\boldsymbol{B}_3;\ H_A=H_B. \quad (9)$$

For order $n = 1$ of contact ruled surfaces have coincided with trihedrons ($\boldsymbol{a}_{01}$, $\boldsymbol{a}_{02}$, $\boldsymbol{a}_{03}$) and ($\boldsymbol{b}_{01}$, $\boldsymbol{b}_{02}$, $\boldsymbol{b}_{03}$) at the center point of their common generator (see Figure 1, a).

2.3. Order n=2 of contact of ruled surfaces

$$\boldsymbol{A}_1(t_0)=\boldsymbol{B}_1(t_0);\boldsymbol{A}_1'(t_0)=\boldsymbol{B}_1'(t_0);\boldsymbol{A}_1''(t_0)=\boldsymbol{B}_1''(t_0);\boldsymbol{A}_1'''(t_0)\neq\boldsymbol{B}_1'''(t_0). \quad (10)$$

Ruled surfaces are exactly in the second order of contact in their common generator $\boldsymbol{a}_{01}(t_0)= \boldsymbol{b}_{01}(t_0)$ (see Figure 1,b). From relations:

$$\boldsymbol{A}_1''=(H_A\cdot\boldsymbol{A}_2)' = -H_A^2\cdot\boldsymbol{A}_1+H_A'\cdot\boldsymbol{A}_2+H_A\cdot Q_A\cdot\boldsymbol{A}_3;$$
$$\boldsymbol{B}_1''=(H_B\cdot\boldsymbol{B}_2)' = -H_B^2\cdot\boldsymbol{B}_1+H_B'\cdot\boldsymbol{B}_2+H_B\cdot Q_B\cdot\boldsymbol{B}_3,$$

it follows that additionally to the result of contact of order $n = 1$ the following equalities are carried out:

$$H_A' = H_B';Q_A = Q_B. \quad (11)$$

The element of the dual arc of ruled surface $\boldsymbol{A}_3(t)$ on the basis of derivation equation $\boldsymbol{A}'_3 = -Q_A\cdot\boldsymbol{A}_2$ may be expressed as $ds_{A3} = |\boldsymbol{A}'_3|\cdot dt = Q_A\cdot dt$. On the basis of the second equation of (11) it follows that $ds_{A3} = ds_{B3}$. Thus, we obtain the equality of elements of dual arcs of ruled surfaces $\boldsymbol{A}_3(t)$ and $\boldsymbol{B}_3(t)$. And, similarly, on the basis of equations (11) and $\boldsymbol{A}'_2 = -H_A\cdot\boldsymbol{A}_1 + Q_A\cdot\boldsymbol{A}_3$ it follows that $ds_{A2} = ds_{B2}$. We obtain the equality of elements of dual arcs of ruled surfaces $\boldsymbol{A}_2(t)$ and $\boldsymbol{B}_2(t)$.

If order n of contact of initial surfaces is equal 2 then the elements of dual arcs in pairs of surfaces $\boldsymbol{A}_i(t)$ and $\boldsymbol{B}_i(t)$, where i=1, 2, 3, are equal and calculated from common generators in these pairs.

The equality of elements of dual arcs is equivalent to the equality of distribution parameters in the common generators of these pairs of surfaces. In theory of ruled surfaces [3] it is known that $\frac{ds_{A2}}{ds_A}=\frac{1}{\rho_A}$, where ρ_A - a dual curvature radius of ruled surface $\boldsymbol{A}_1(t)$ at centre point M of its ruling. Since order $n = 2$ of contact is characterized by equality $ds_A = ds_B$ and $ds_{A2} = ds_{B2}$ it follows that at this contact ruled surfaces in their common generator $\boldsymbol{a}_{01}(t_0) = \boldsymbol{b}_{01}(t_0)$ have equal dual curvature radius $\rho_A = \rho_B$. Furthermore, since $\rho = \sin R$, where $R = R_0 + \omega R_1$ is a dual angle between the generator of ruled surface $\boldsymbol{A}_1(t)$ and the generator of its first evolute (see Figure 1, b), it follows from equation $\rho_A = \rho_B$ that $R_A = R_B$ and the trihedrons of the first order evolutes coincide.

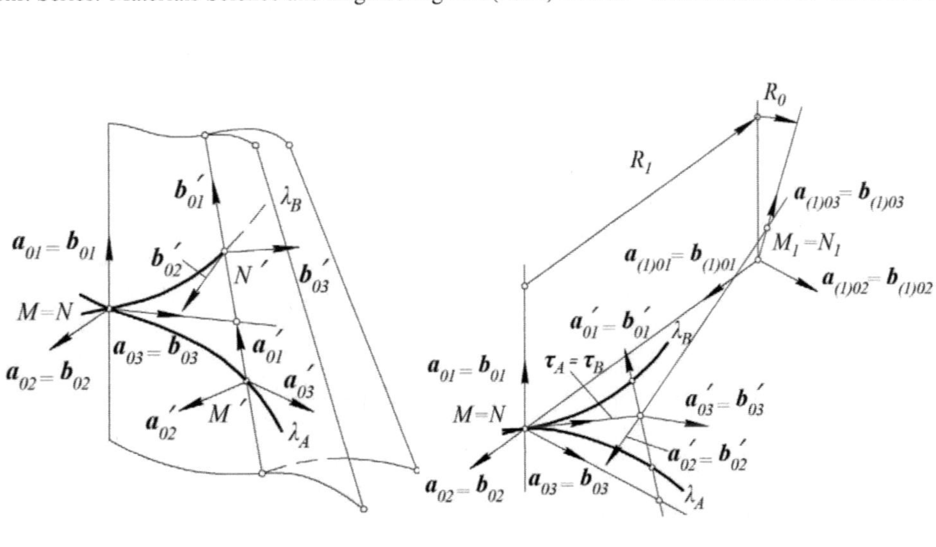

a b

Figure 1. Orders n = 1 (a) and n = 2 (b) of the contact.

2.4. *Order n=3 of contact of ruled surfaces*

$$A_1(t_0)=B_1(t_0);\ A_1'(t_0)=B_1'(t_0);\ A_1''(t_0)=B_1''(t_0);\ A_1'''(t_0)=B_1'''(t_0);\ A_1^{IV}(t_0)\neq B_1^{IV}(t_0) \quad (12)$$

Ruled surfaces are exactly in the third order of contact in their common generator $\boldsymbol{a}_{01}(t_0) = \boldsymbol{b}_{01}(t_0)$. From relations

$$\boldsymbol{A}_1''' = -3H_A \cdot H_A' \cdot \boldsymbol{A}_1 + \left(H_A'' - H_A^{\ 3} - H_A \cdot Q_A^{\ 2}\right)\cdot \boldsymbol{A}_2 + \left(2H_A' \cdot Q_A + H_A \cdot Q_A'\right)\cdot \boldsymbol{A}_3;$$
$$\boldsymbol{B}_1''' = -3H_B \cdot H_B' \cdot \boldsymbol{B}_1 + \left(H_B'' - H_B^{\ 3} - H_B \cdot Q_B^{\ 2}\right)\cdot \boldsymbol{B}_2 + \left(2H_B' \cdot Q_B + H_B \cdot Q_B'\right)\cdot \boldsymbol{B}_3,$$

it follows that additionally to the results of order n = 2 of contact the following equalities are carried out:

$$H_A'' = H_B'';\ Q_A' = Q_B'. \quad (13)$$

Thus in order n=3 of contact the following equations hold:

$$\frac{ds_{A1}}{dt} = H_A = H_B = \frac{ds_{B1}}{dt};\ \frac{d^2 s_{A1}}{dt^2} = H_A' = H_B' = \frac{d^2 s_{B1}}{dt^2};\ \frac{d^3 s_{A1}}{dt^3} = H_A'' = H_B'' = \frac{d^3 s_{B1}}{dt^3};$$
$$\frac{ds_{A3}}{dt} = Q_A = Q_B = \frac{ds_{B3}}{dt};\ \frac{d^2 s_{A3}}{dt^2} = Q_A' = Q_B' = \frac{d^2 s_{B3}}{dt^2}. \quad (14)$$

Central tangent vector $\boldsymbol{A}_3(t)$ has coordinates $\boldsymbol{A}_3(t)=\{x_3(t), y_3(t), z_3(t)\}$.

We define the derivatives of this vector based on the existence of functional relationship $s_{A1}(t)$:

$$\boldsymbol{A}_3' = \frac{d\boldsymbol{A}_3}{ds_{A1}} \cdot \frac{ds_{A1}}{dt} = \left\{ \frac{dx_3}{ds_{A1}} \cdot \frac{ds_{A1}}{dt};\ \frac{dy_3}{ds_{A1}} \cdot \frac{ds_{A1}}{dt};\ \frac{dz_3}{ds_{A1}} \cdot \frac{ds_{A1}}{dt} \right\}.$$

In complex line geometry [3] the following equations are known:

$$\frac{dx_3}{ds_{A1}} = -x_2 \cdot \mathrm{ctg} R;\ \frac{dy_3}{ds_{A1}} = -y_2 \cdot \mathrm{ctg} R;\ \frac{dz_3}{ds_{A1}} = -z_2 \cdot \mathrm{ctg} R, \quad (15)$$

where $\boldsymbol{A}_2(t) = \{x_2(t), y_2(t), z_2(t)\}$. We will consider the obtainment of a second derivative of vector

$\boldsymbol{A}_3(t)$: $$\boldsymbol{A}_3'' = \left\{ \frac{d^2 x_3}{ds^2_{A1}} \cdot \frac{ds_{A1}}{dt} + \frac{dx_3}{ds_{A1}} \cdot \frac{d^2 s_{A1}}{dt^2};\ \ldots;\ \ldots \right\}.$$

Proceeding from equations (15) for vector $\boldsymbol{A}''_3$ we can write:

$$A''_3(t)=\left\{\left(-\frac{x_{(1)2}}{\rho_{A1}}\text{ctg}R+\frac{x_2}{\sin R}\cdot\frac{1}{r_{A1}}\right)\frac{ds_{A1}}{dt}+\left(-x_{(1)2}\,\text{ctg}R\right)\frac{d^2s_{A1}}{dt^2};\ \ldots;\ \ldots\right\} \tag{16}$$

where $\boldsymbol{A}_{(1)2}=\{x_{(1)2}, y_{(1)2}, z_{(1)2}\}$, $\frac{1}{r_{A1}}=\frac{dR}{ds_{A1}}$, ρ_{A1} and r_{A1} – are dual radii of curvature and torsion of ruled surface $\boldsymbol{A}_1(t)$ in the generator of contacting. The analysis of the components of vector $\boldsymbol{A}''_3(t)$ in view of equations (14) and $dR+ds_{A(1)3}=0$, which is peculiar to the evolute of ruled surface $\boldsymbol{A}_1(t)$ [3], allows making certain conclusions. For order n = 3 of contact of the nondevelopable ruled surfaces the elements of dual arcs in pairs of surfaces $\boldsymbol{A}_{(1)i}(t)$ and $\boldsymbol{B}_{(1)i}(t)$, i = 1, 2, 3 are equal. Hence, the evolute of the first order has order n=2 of contact. In addition, the trihedrons of the evolute of the second order coincide and dual radii of curvature $r_{A1}=r_{B1}$ of $\boldsymbol{A}_1(t)$ and $\boldsymbol{B}_1(t)$ with parameter $t = t_0$ are equal.

3. Theoretical examples

As an example we will consider contact of ruled surface $\boldsymbol{A}_1\rightarrow\boldsymbol{R}_A(t,u)$ with elliptical hyperboloid $\boldsymbol{B}_1\rightarrow\boldsymbol{R}_B(t,u)$ (Figure 2):

$$A_1\rightarrow \boldsymbol{R}_A(t,u)=\left\{A(\cos t+t\sin t)+\frac{u\cos t}{m};A(\sin t-t\cos t)+\frac{u\sin t}{m};1+Bt+\frac{u}{m}\right\},$$

$$A=1;B=0.5;m=\sqrt{2};0\le t\le 2\pi;-\infty<u<\infty.$$

$$B_1\rightarrow \boldsymbol{R}_B(t,u)=\left\{c_1\sin t+\frac{u\cos t}{\sqrt{1+c_3^{\,2}}};c_2\cos t+\frac{u\sin t}{\sqrt{1+c_3^{\,2}}};\frac{uc_3}{\sqrt{1+c_3^{\,2}}}\right\},$$

$$0\le t\le 2\pi,-\infty<u<\infty.$$

$$n=1:\quad c_1=\frac{1-2t_0}{2c_3\left(3\cos^2 t_0-4\right)};c_2=\frac{-4\left(1-2t_0\right)}{2c_3\left(3\cos^2 t_0-4\right)};c_3=2;t_0=2.46.$$

$$n=2:\quad c_1=\frac{2\left(1-t_0\right)-\cos^2 t_0\cdot\left(3-4t_0\right)}{2\left(2\cos^2 t_0-1\right)};c_2=-\frac{1-2t_0-\cos^2 t_0\cdot\left(3-4t_0\right)}{2\left(2\cos^2 t_0-1\right)};c_3=1;t_0=4.3.$$

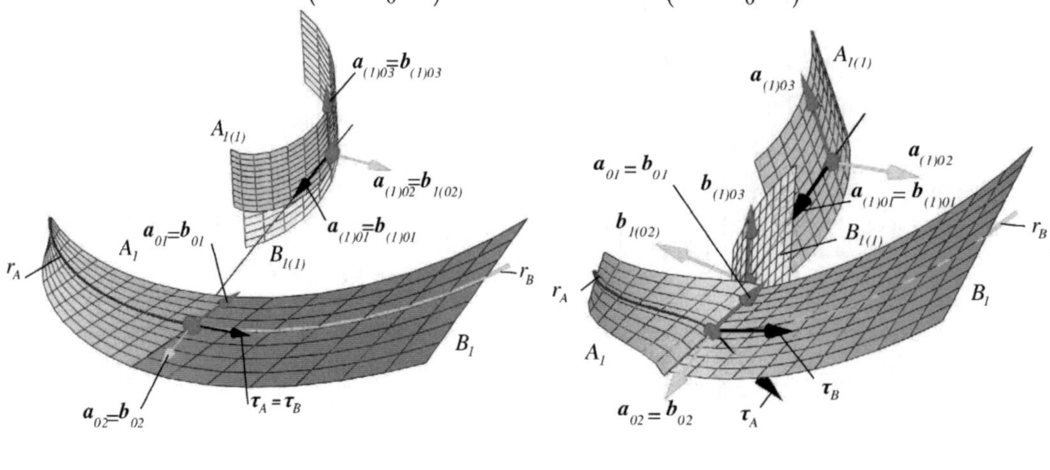

Figure 2. Theoretical examples of orders n = 1 (a) and n = 2 (b) of contact.

4. Practical results

The results of these researches were applied to designing of the working surfaces of the progressive tools for cultivation of the soil. In accordance with the mathematical model of tillage process 'soil-chisel plow' a ruled surface of the chisel plow's rack was designed, which increases the wear resistance and improves the quality of the soil loosening (Figure 3, chisel – 1, rack – 2, knife – 3, working surface – 4).

The developed mathematical theory allows designing the constituent ruled surfaces of technical products with pre-assigned geometric and technical conditions. The mathematical toolkit allows a flexible change of the shape of the surface depending on the type of soil to be treated.

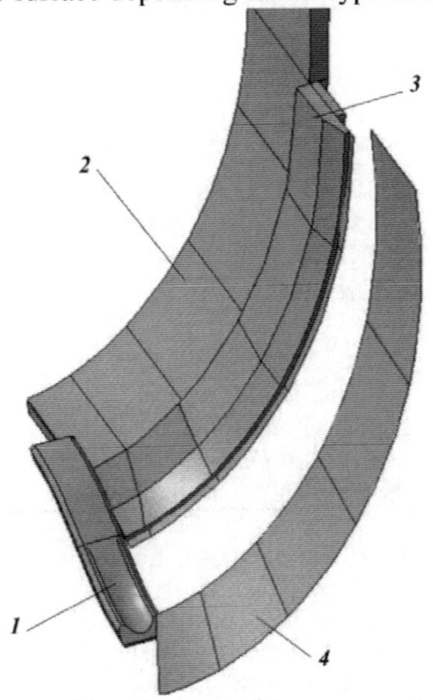

Figure 3. The working surface of the progressive tillage tool with order $n = 2$ of contact.

5. Conclusion

Order n of contact of nondevelopable ruled surfaces is considered. The notion of a dual vector of divergence on the common generator line of two contacting nondevelopable ruled surfaces is introduced via the dual vector calculus, which describes the manifolds of line geometry. To determine this vector we proved necessary and sufficient conditions for order n of contact on the common generator line. Thus, the primary orders of contact have been researched. The results of these researches can be applied in the design of the constituent ruled shells and in CAD systems via approximating issues.

References

[1] Pottmann H and Wallner J 2001 Computational Line Geometry (Springer-Verlag Berlin Heidelberg) p 563
[2] Blaschke W 1921 Vorlesungeon uber Differentialgeometrie. Elementare Differentialgeometrie (Springer Berlin 1. Auflage) vol **1** p 230
[3] Zeiliger D N Complex Line Geometry 1934 (Moscow-Leningrad) p 196
[4] Study E 1903 Geometrie der Dynamen (Teubner Leipzig) p 634
[5] Pottmann H, Peternell M and Ravani B 1999 *Computer-Aided Design* **31** 3–16

IOP Conf. Series: Materials Science and Engineering **124** (2016) 012084 doi:10.1088/1757-899X/124/1/012084

Application of a Lamb waves based technique for structural health monitoring of GFRP undercyclic loading

A Eremin[1, 2], A Byakov[1, 2], S Panin[1, 2], M Burkov[1, 2], P Lyubutin[1, 2] and R Sunder[3]
[1] Institute of Strength Physics and Materials Science of the Siberian Branch of the Russian Academy of Sciences, 2/4, Akademicheskiy ave., Tomsk, 634021, Russia
[2] Tomsk Polytechnic University, 30, Lenina ave., Tomsk, 634050, Russia
[3] Bangalore Integrated System Solutions Ltd, 497E 14th Cross, 4th Phase, Peenya Industrial Area, Bangalore, 560058, India

E-mail: eremin_av@bk.ru

Abstract. A Lamb wave based ultrasonic technique as well as optical image characterization was utilized to estimate a current mechanical state of glass fiber reinforced polymers (GFRP) under cyclic tension. The ultrasonic acoustic method was applied in a 'pitch-catch' mode using piezoelectric transducers adhesively bonded onto a specimen surface. Numerical evaluation of acoustic data was performed by calculating two informative parameters: maximum of amplitude of the received signal and variance of signal envelopes. Optical images were registered and then analysed by calculating Shannon entropy that makes it possible to characterize changing of GFRP specimen translucency. The obtained results were treated in order to find out the relation between the current mechanical state of a specimen and informative parameter values being computed from the acoustic and optical signals.

1. Introduction

Recently advanced composite materials such as glass/carbon fiber reinforced polymers (FRP) have found wide use as structural materials for various engineering applications [1]. The advantages of glass/carbon FRPs over conventional metallic materials results in wide-spread replacement of metals in civil engineering, wind energy, shipbuilding, aerospace and etc. However, due to the lack of ductility the structures made of these composites exhibit brittle (catastrophic) failure and require the application of higher safety factors during design and accurate inspections throughout the entire operating life. Since FRPs tend to find wide spread occurrence the research and development of new methods of their inspection is a relevant task.

The problem of identification of flaws and damages long before failure as well as performing condition-based maintenance requires the information about the structure's mechanical state, which can be provided by Structural Health Monitoring (SHM) systems. Lamb waves techniques are popular for implementation in SHM [2] due to their low attenuation that allows covering large inspection areas [3] and ensures the possibility of detecting damages of various sizes as well as evaluating of the changes of mechanical properties of the material.

However, Lamb waves have a multi-mode dispersive nature, thus the raw captured (acoustic sensing) data are quite complex for interpretation [4]. When a wave is transmitted from an actuator to a receiver a lot of different wave modes appear and each of them propagates with a unique through thickness displacement and stress profile [5]. It might be beneficial to use multi-mode Lamb waves with a different wavelength and, thus, with different sensitivity to defects.

Lamb waves could be generated using piezoelectric actuators or high energy laser impulse and then sensed by various techniques such as optical fiber Bragg cells, non-contact laser vibrometry or piezoelectric sensors. The last one seems to be one of the most versatile technique due to the possibility of simultaneous signals transmitting and receiving. In doing so, large areas could be covered by a network of sensors.

The aim of the present study is development and testing of the Lamb wave based technique combined with the optical method for estimation of the state of glass fiber reinforced polymer specimens under fatigue tests.

2. Methods and materials

Specimens with chamfered tabs (Figure 1) were tested under fatigue conditions with an equal strain ratio but different maximum strains (0.69 %, 0.77 % and 0.85 %). Specimens were made from E-glass fibers with a polyester matrix. A middle layup [0°] was additionally covered by two surface layers oriented at[+45°,-45°].

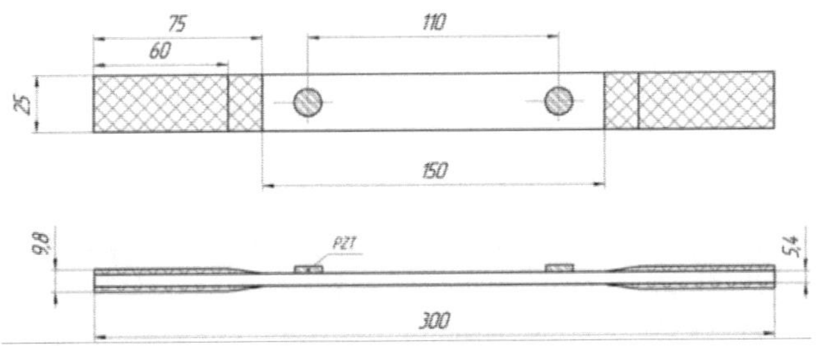

Figure 1. A specimen's design.

During experiments two types of data were acquired: digital optical images of the specimen's surface and acoustic wave signals after their propagation along the specimen area located between two PZT transducers. In doing so one piezo transducer plays a role of an actuator (generates ultrasonic waves) and the other PZT is a sensor (receives acoustic waves). When acoustic disturbance propagates through the bulk specimen it interacts with nearly all large enough heterogeneities and scatters onto those defects which size is larger than or equal to the wave length. These interactions and scatterings give rise to changes of the acoustic response of the material: propagation velocity, amplitude, waveform, spectrum and etc. Detailed analyses of changes taking place in a signal provide the information about various types of defects and make the internal structure reconstruction possible.

Photos were captured by digital camera Canon EOS 750D with resolution of 5184×3456 pixels. Acoustic waves were generated by arbitrary waveform generator AWG-4105 with the amplitude of 10 V. Signal sensing was conducted by USB oscilloscope Handyscope HS4 with the sampling rate of 5 MHz and with the 12 bit resolution. AW1E12G-190EFL1Z (a 9 mm disc with thickness of 0.19 mm) piezoelectric transducers were used for acoustic wave transmitting and receiving. The signal from the generator was recorded by the 1st channel of HS4 to be used as the timing reference to automate the experimental procedure by triggering the data acquisition. The registered signals were processed using a 10…800 kHz band-pass filter.

Figure 2 illustrates the initial signal waveform – 5 cycles of sine modulated by the Hanning window. Such shape of the signal was chosen due to a relatively long duration, bell curve spectral properties and possibilities of recognizing different wave packets in a received signal.

Acoustic and optical data were acquired after every 1000 cycles in the automatic mode. At the same time cyclic loading was not interrupted. The disadvantage of such technique is variation of the mechanical state from one reading to another, since capturing was performed under different values of applied loads. Figure 3 illustrates the fact that cyclic loading causes changes in the acoustic signal wave form and a specimen visual appearance.

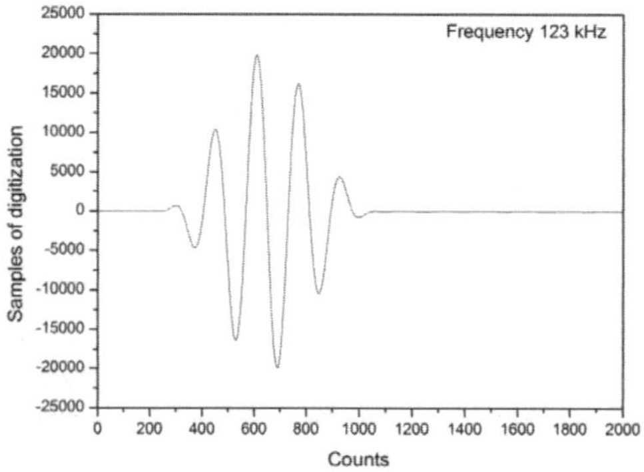

Figure 2. A generated signal waveform (5 cycles of sine modulated by the Hanning window)

Subsurface delamination progresses and it changes specimen transparency, which emerges as white areas being visible on the surface. Damage formation is reflected in the acoustic signal by its amplitude decline; then the signal changes its waveform (which is related to overlapping of different wave packets received, arrival of reflected packets and their interference). Due to the overall phenomena it is possible to evaluate changes in integral properties of the material, predict residual life time or estimate the defect location, size and type.

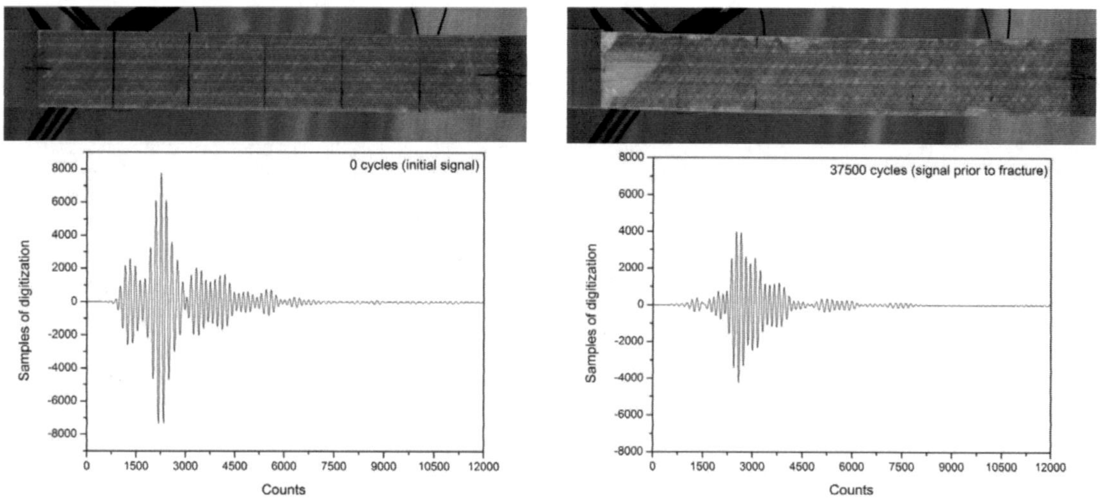

Figure 3. Optical images and acoustic signals waveforms of a GFRP specimen in two states: undeformed (left) and prior to fracture (right)

Recorded data were numerically analyzed and each data acquisition point corresponded to computed values, which characterize the acoustic signal and a surface image. Thus information on the material state becomes available in terms of quantity. One of the calculated parameters used for characteristic of acoustic signals is variance (the 2nd central moment) between signals envelopes. It can be calculated by equation (1).

$$M_2 = D[X] = M[(X - m_x)^2] = \sum_{i=1}^{n} (x_i - m_x)^2 \quad (1)$$

where x_i – difference between two signal envelopes, m_x – expected value of x_i.

A parameter, which characterizes optical images, is Shannon Entropy (or information entropy) which iscalculated usingequation (2).

$$H = -\sum_{i=0}^{255} \frac{HIST_i}{255} \log_2 \frac{HIST_i}{255} \qquad (2)$$

i – pixel intensity value (brightness), $HIST_i$ – sum of pixels of the whole image with an$i$-value of intensity (it is an$i$-value of the image histogram, i. e. brightness distribution over the digital image).

Specimens were tested in slightly distinct initial conditionsdue to experimental procedure scattering induced by the influence of non-vertical alignment of the specimen in grips, some variation in the ultrasonic sensor location, etc. Also, since the distance between PZTs varies from one specimen to another it was necessary to normalize all calculated parameters (relative to a maximum value) as well as the number of cycles. It allows overcoming specimen inequalities and compensating them.

Before the beginning of systematic experimental studies it was necessary to determine the optimal frequency for acoustic wave sounding. For this purpose the frequency range from 10 kHz to 500 kHz was passed through (scanned) and a sensor response was measured in units of spectrum power (Figure 4). The optimum frequencies for this type of piezo-transducers were selected to be equal to 123 kHz and 243 kHz.The corresponding acoustic wave velocities andwavelengthsarepresented in the table 1.

Table 1. Estimated velocity and wavelength of Lamb waves.

Frequency	Frequency-thicknessproduct, MHz·mm	Symmetric - S_0 mode		Antisymmetric - A_0 mode	
		Velocity, m/s	Wavelength, mm	Velocity, m/s	Wavelength, mm
123 kHz	0.66	≈3900	31	≈1800	14
243 kHz	1.31	≈3700	15	≈2400	10

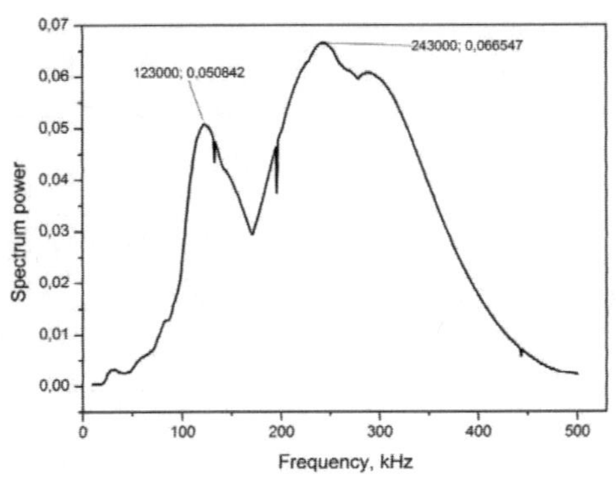

Figure 4. An amplitude-frequency diagram attained in the process of optimum acoustic sounding frequencies finding.

3. Results and discussion

Experiments were performed with the use of 9 specimens (3 specimensof the same type for each strain value). In order to illustrate the most general features and properties, the results for only one specimen from each group are presented in the paper.

Figure 5represents the normalized amplitude of the acoustic signal versus the life time gained during fatigue experiments with three specimen types with various values of maximum strain in a cycle (0.85 %, 0.77 % and 0.69 %).The left plot (a) shows the results when acoustic signal frequency was 123 kHz and the right one (b) – 243 kHz.

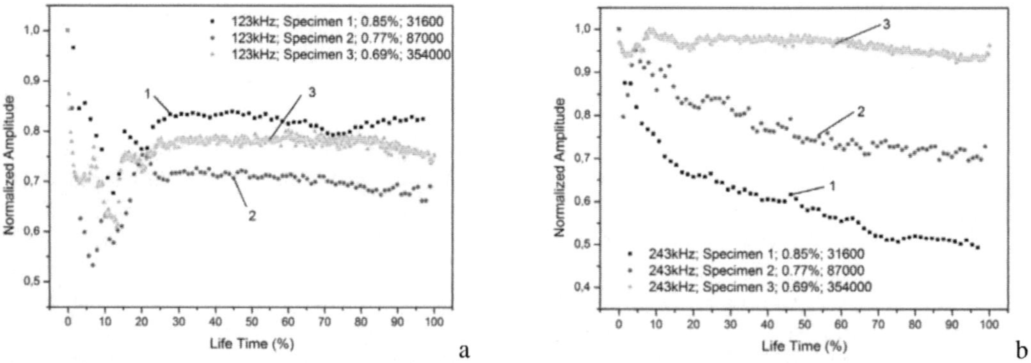

Figure 5. The dependence of normalized signal amplitude versus specimen life time. Specimen 1 – 0.85 %, 31600 cycles; Specimen 2 – 0.77 %, 87000 cycles; Specimen 3 – 0.69 %, 354000 cycles. Frequency of scanning signal: 123 kHz (a) and 243 kHz (b).

As it is seen in Figure 5 (b) the maximum strain value exerts some influence onto the behaviour of acoustic signalsand it depends on the frequency of the acoustic wave. In case of low frequency (123 kHz), which has alarger wavelength and, therefore, lower sensitivity to damages, the dependences do notexhibit any clear correlations between maximum cyclic strain and curve shapes. Whereaswith higher frequency (243 kHz) the wavelength is twice smaller and has higher perceptibility. The latest feature of the acoustic wave stimulatesan increased amplitude response tovariations ofmaximum strains in a cycle. It should be mentioned that the A_0 mode is more informative under signal processing, since the amplitude of the A_0 mode is much higher than that of the S_0 mode. Plots ofthe responses to the high frequency signalshow the downward trendexcluding the first 10 % of life time where some abrupt jumps and drops are registered. The higher is the strain value of appliedload,thesteeperisthe curve of the amplitude-frequency response.

The second parameter of acoustic signals that was computed in order to characterize the specimens' state was variance between two signal envelopes. The dependences of the varianceson the life time are presented in Figure 6.

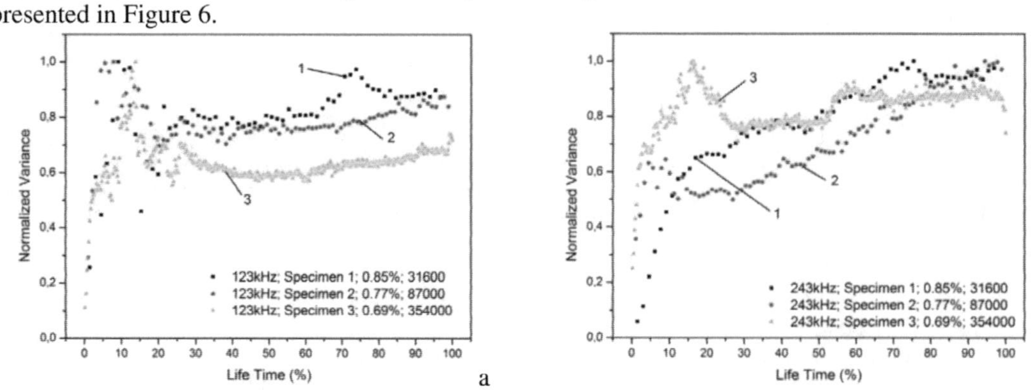

Figure 6. Plots of normalized variance of the acoustic signal versusthe life time for 3 specimens with different strains in a cycle; Specimen 1 – 0.85 %, 31600 cycles; Specimen 2 – 0.77 %, 87000 cycles; Specimen 3 – 0.69 %, 354000 cycles. Frequency of scanning signal: 123 kHz (a) and 243 kHz (b).

Plots characterizing the computed values of the variance for low frequency signalsdepict abrupt rises and drops of the parameter value in presence of the clear upward trend. The latter is a characteristic feature from the beginning of testing and till 20 % of life time. The reason of such behaviour is related to processes of damage formation occurring at the first stage of the fatigue tests. The above mentioned assumption agrees with the optical data, shown in Figure 7. It is seen that after

20 % of life time curves exhibit the smooth trend.The values of variance forthe scanning signal of 243 kHz indicate a sharp increase of the parameter during the first 15 % of life time and then possesses a gradual growth. After 60 % of life time three curves ofthe parameters values tend to intersect.

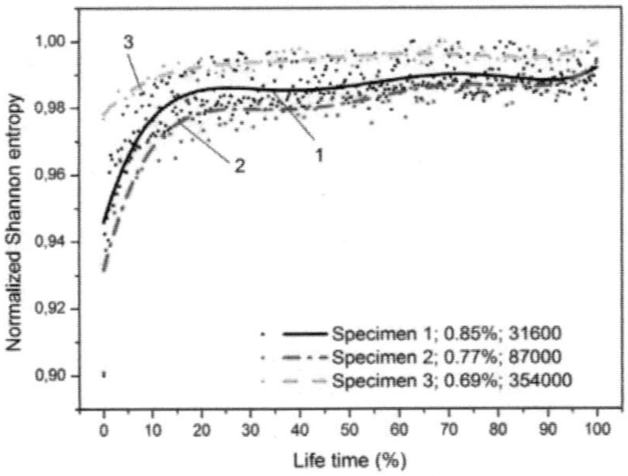

Figure 7. A normalized Shannon entropy (gained from the image processing) dependence on the life time for three specimens with different strains: 0.85 %, 0.77 % and 0.69 %

The results for the optical datacomputationare presented in Figure 7 that indicatesthe dependenceof Shannon entropy for three specimens with a large scattering of the data (smoothed by polynomialfitting and shown by a solid curve) due to imperfection of the image capturing procedure.The analysis of the graphs allows concludingthat the optical transparency suddenly changes after 1000 cycles due to nucleation of micro cracks, loss of adhesion between matrix and fibersas well as delamination. At the last stage of cyclic testing internal changes in the specimens do not substantially influence the optical transparency. Thus, the specimen failure happens without pronounced evidences that might be extracted from the analysis of this parameter.

4. Conclusion

By summarizing the obtained results it is possible to conclude that the combination of the optical and acoustic data analyses provides more comprehensive and robust results for material state characterization for SHM purposes and a condition based maintenance paradigm. An acoustic technique acquiresthe information about the integral mechanical state of a specimensince wave propagation is highly influenced by the specimen stiffness. On the other hand,optical images providethe visualization of cracks and delaminationand allow for location of visible damages. The optical images also provide better sensing of the material structure degradation at first stages of cyclic loading.

A low frequency (123 kHz) ultrasonic inspection does not provide satisfactoryand reliable results fora precise estimation of the material mechanical state, whereas high frequency (243 kHz) seems to be perspectivein theevaluation of residual operating time in future application in the SHM technique. The optical image analysis performed via calculation of theShannon entropy informative parameter clearly indicatesthe structure degradation processes up to 30% of life time,but afterwards itretainsthe value nearly constant up to 90% of life time. Just before fracture the value startsa slow increase.

Prospects of the current investigation arerelated to the following directions of study: a) improvement of data acquisition, processing methodology and techniques, b) development ofthe damage index, which is based on the combination of informative parameters extracted from optical and acoustic signals and c) development of the algorithms for determination of residual operating time and failureprediction.

IOP Conf. Series: Materials Science and Engineering **124** (2016) 012084 doi:10.1088/1757-899X/124/1/012084

References

[1] Gand A, Chan T, Mottram J 2013 Civil and structural engineering applications, recent trends, research and developments on pultruded fiber reinforced polymer closed sections *A review. Frontiers of Structural and Civil Engineering* **7** Issue 3 pp 227-244;

[2] Raghavan A, Cesnik C 2007 Review of guided-wave structural health monitoring *Shock Vib. Dig.* **39** pp 91-114;

[3] Carandente R, Cawley P 2012 A method to estimate the size of corrosion patches with guided waves in pipes *AIP Conf. Proc.* **1430** pp 1929-1936;

[4] Miller C, Hinders M 2014 Classification of flaw severity using pattern recognition for guided wave-based structural health monitoring *Ultrasonics* **54** pp 247-258;

[5] Rose J 1999 Ultrasonic Waves in Solid Media *Cambridge University Press*.

The influence of some model parameters on the impurity distribution implanted into substrate surface

E S Parfenova and A G Knyazeva
Institute of High Technology Physics, Tomsk Polytechnic University, 2a, Lenina ave., Tomsk, 634050, Russia

E-mail: Linsisergg@mail.ru

Abstract. The model for description of the initial stage of ion implantation into the surface layer of the metal is presented. The interdependence of embedded impurity concentration and deformations arising from the impact of particles on the surface is investigated. The model takes into account the particle diffusion, the finite time of mass flux relaxation; the stress appearance due to a composition change of the surface layer and a mass transfer phenomenon under a stress gradient action. It is established that the interaction of mechanical waves and concentration leads to a distribution of concentration not corresponding to a pure diffusion process. The examples of coupled problems solution for different sets of model parameters are presented.

1. Introduction

Researches of the surface treatment process of materials by beams of charged particles are carried out by many authors [1-4]. The composition and structure, and, consequently, the surface properties of materials can be controlled owing to these methods. In spite of extensive experimental and theoretical data, possibilities of the implantation method are not fully exhausted. This is connected with a lack of understanding of physical processes in solids under irradiation [5] for small time moments.

In the process of implantation into the surface substrate, different phenomena occur simultaneously with the diffusion process. For example, mechanical disturbances appear and propagate, structural and phase changes happen, point defects are generated as a result of the particles impact on the surface and appearing mechanical stresses etc. All these phenomena influence each other. Therefore the interrelation between different processes must be considered during modeling [6]. In the literature, the thermoelastic waves caused by exposure to high-energy sources on the surface of materials attract the basic attention. However, it should be noted that there are coupled models of ion implantation which take into account the processes associated with the difference in the properties of the base material and introduced elements; in this case diffusion and physico-chemical phenomena are a rarity. In [7] it was shown that the interrelation between mechanical and diffusion waves leads to a distortion of the deformation (and stress) wave profile, and the concentration distribution does not correspond to a pure diffusion process The purpose of this work consists in the study of the initial stage of the ion implantation.

2. The mathematical formulation

In the present paper, model [7] is used. This model is built with account that temperature is constant. It is assumed that the occurring stresses are elastic and deformations are small. Also this model takes into account the fact that transfer coefficients can be change due to changes in the activation diffusion volume. In the dimensionless variables, the model under deformations takes the form:

$$\tau_r \frac{\partial^2 C}{\partial \tau^2} + \frac{\partial C}{\partial \tau} = \frac{\partial}{\partial \xi}\left[g(C) f_2(\Pi) \frac{\partial C}{\partial \xi}\right] - \omega\gamma \frac{\partial}{\partial \xi}\left[C f_2(\Pi) \frac{\partial e}{\partial \xi}\right], \tag{1}$$

$$\frac{\partial^2 e}{\partial \tau^2} = \frac{\partial^2 e}{\partial \xi^2} - \gamma \frac{\partial^2 C}{\partial \xi^2}, \tag{2}$$

$$S = e - \gamma(C - C_0), \tag{3}$$

$$\boldsymbol{J} = -\left[g(C)\frac{\partial C}{\partial \xi} - C\omega\gamma \frac{\partial e}{\partial \xi}\right] f_2(\Pi) - \tau_r \frac{\partial \boldsymbol{J}}{\partial \tau}, \tag{4}$$

$$\xi = 0: \ \boldsymbol{J} = \beta\varphi(\tau), \ S = S_0\varphi(\tau), \tag{5}$$

$$\xi \to \infty: \ C = 0, \ e = 0, \tag{6}$$

$$\tau = 0: \ C = 0, \ e = 0, \ \frac{\partial C}{\partial \tau} = 0, \ \frac{\partial e}{\partial \tau} = 0, \tag{7}$$

where $S = \frac{\sigma}{\sigma_*} = \frac{\sigma}{3E\alpha_0}$, $\tau = \frac{t}{t_*} = \frac{tE}{\rho D_0}$; $\xi = \frac{x\sqrt{E}}{D_0\sqrt{\rho}}$; $e = \frac{\varepsilon}{3\alpha_0}$ – are dimensionless variables, $\boldsymbol{J}$ – is the mass flux, S, e – are components of the stress and strain tensors in irradiation direction ($O\xi$), τ – is the time, C – is the mass concentration of the implanted component, $g(C) = f_1(C) + \omega\gamma^2 C$ – is the function of concentration, $f_2(\Pi) = -\kappa e S$ – is the function of stress work, the form of function $f(C)$ depends on the "solute" structure formed in the process of implantation, in this paper we used function $f_1(C) = a + bC + dC^2$, ω, γ, τ_r, β – are parameters of the model:

$$\omega = \frac{3mE\alpha_0^2}{RT\rho}, \ \gamma = \frac{(\alpha - \alpha_0)}{\alpha_0}, \ \tau_r = \frac{t_r E}{\rho D_0}, \ \beta = m_0\sqrt{\frac{\rho}{E}}, \ \kappa = \frac{kE\alpha_0}{RT}$$

here ρ – is the density of the base material; α – is the concentration expansion coefficients of the implanted elements, α_0 – is the concentration expansion coefficients of the base material, τ_r – is the relaxation time of the mass flux, D_0 – is the self-diffusion coefficient, R – is the universal gas constant, T – is the temperature, m – is the molar mass, m_0 – is the velocity of the particles in the flux, E – is the elastic module, k – is the constant with the dimension of activation volume.

3. Results and discussions

System (1)-(7) was solved numerically with the help of the explicit difference scheme and a double - sweep method. Below we present examples of calculation.

Values of model parameters are presented in Table 1. Figure 1 shows distributions of impurity concentration (a) and deformations (b) in depth. By the time of $\tau = 1.225$ the concentration wave slightly lags behind the wave of deformations. By the time of $\tau = 3.67$ the deformations wave runs away far forward, and by the time of $\tau = 5.88$ the sloping plot on the wave appears between the extremes, which increases with time. This is a reflection of the weak influence of waves on each other for a long time. During these moments of time the pulse has already no effect on the substrate material. It can be seen from trailing edges of waves.

Table 1 – Values of model parameters.

Parameter	Numerical value
ω	10
τ_r	1.0
β	0.01
γ	-0.09
κ	2.9
$\varphi(\tau)$	$\sin(R\cdot\tau)$, where $R=\dfrac{\pi}{t_{imp}}$
$f_1(C)$	$0.05+0.01C+0.8C^2$

A change of the sign of the strain corresponds to the depth of impurity penetration. This phenomenon is related to the fact that the impurity is introduced at a smaller depth than the mechanical wave. Thus, the maximum range of ions is the distance where compressive deformation converts into tensile (or vice versa, it depends on properties of used materials).

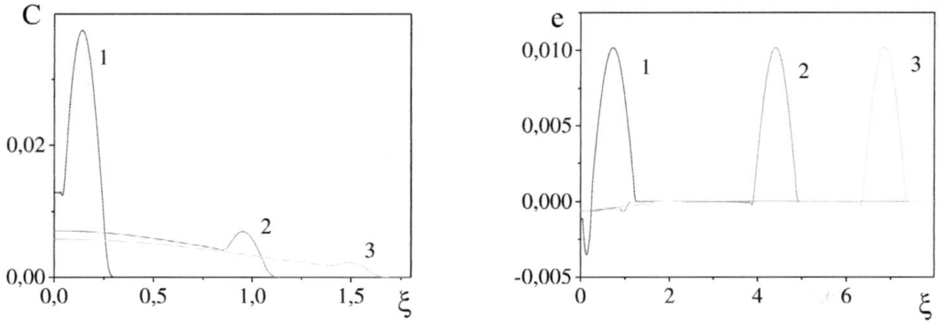

a b

Figure 1. Distributions of impurity concentration and deformations in depth. Pulse time $t_{imp}=1.0$. Times, τ : 1 – 1.225; 2 – 3.675; 3 – 5.88.

Figure 2 shows the influence of model parameters on distributions of impurity concentration. The increase of the relative relaxation time leads to a significant increase in the amplitude of impurity concentration. The reduction of τ_r reduces the penetration depth. The increase of κ leads to decreased values of impurity concentration on the surface (for earlier moments this phenomenon is more significant), the maximum value decreases, the penetration depth of the impurity increases slightly. The change of connectedness coefficient ω results in a significant distortion of wave profiles, in particular the area of waves near extreme. The value of impurity concentration on the sample surface increases. Increase γ leads to augmentation of the maximum and to a decrease of the concentration value on the surface

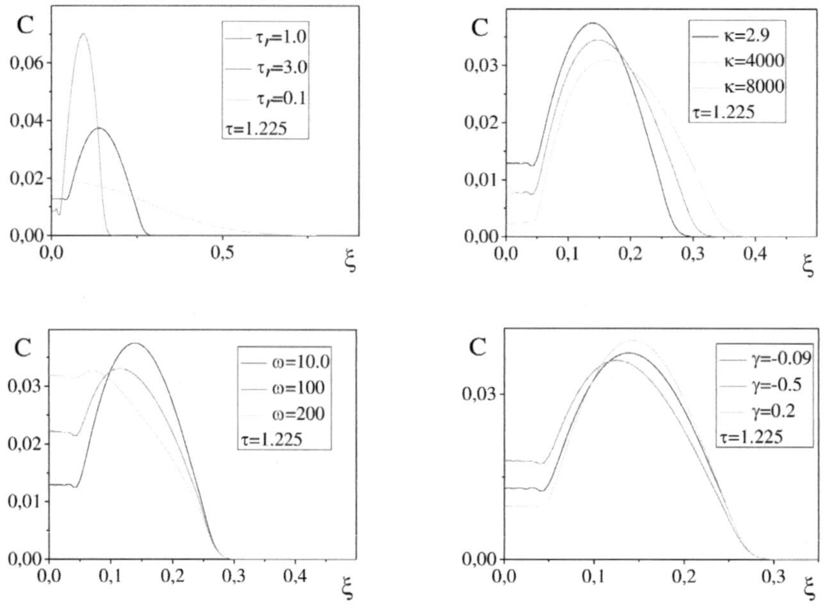

Figure 2. The influence of model parameters on distributions of impurity concentration.

Thus, the variation of model parameters results in qualitative and quantitative changes of the impurity concentration profile. It is necessary to accurately calculate these parameters for obtaining a result corresponding to experimental observations.

4. Conclusion

Thus, in this work the mathematical model of the initial stage of ion implantation is described. Concentration and strain distribution in the waves are strongly dependent on the relations among the parameters of the model.

Acknowledgments

The work was supported by the Ministry of Education and Science of Russia for 2014 - 2016, Research No. 11.815.2014 / K.

References

[1] Townsend P D 1986 *Contempo.Phys.* **3(27)** 241–256
[2] Rosenzweig J B 2002 Fundamentals of Beam Physics (Oxford. University press) p. 300
[3] Popok V N 2012 *Rev.Adv.Mater.Sci.* **30** 1–26
[4] Kurzina I A, Kozlov E V, Sharkeev Yu P 2013 The gradient surface layers based on intermetallic particles: synthesis, structure, properties (Tomsk, Izd-vo NTL) p. 260
[5] Knyazeva A G, 2003 *Journal of Applied Mechanics and Technical Physics* **44(3)** 373–384
[7] Ilina E S, Knayzeva A G 2015 *Applied Mechanics and Materials* **756** 389–393

IOP Conf. Series: Materials Science and Engineering **124** (2016) 012086 doi:10.1088/1757-899X/124/1/012086

Simulation model of processes of servicing refrigerated cargoes in the terminal

A Nyrkov, S Sokolov and L Pavlova
Admiral Makarov State University of Maritime and Island Shipping, 5/7, Dvinskaya st., Saint-Petersburg, 198035, Russia

E-mail: lidiya_pavlova13@mail.ru

Abstract. In the paper the simulation model of servicing processes of refrigerated cargoes in the terminal is described. The mathematical model of the storage process is constructed in view of irregular incoming of cargoes consignments and their casual time of arrival to the refrigerator terminals.

1. Introduction

Historically Russia by right is considered as one of the Great marine Powers due to its geographic location. During the epoch of developing and solidifying of the international connections the foreign trade takes an important place in the national economy. One of the convenient modes of goods shipping at long distances is the marine transport.

According to statistical data the international marine container cargo transportations take the first place in terms of distance and frequency of usage among all other types of cargo delivery. A reason for such great demand is that at present the container carriage is the most unified mode of cargo delivery all over the world.

The international marine container carriage has some indisputable advantages:

- it is the most economic method of cargo delivery. The cost of marine delivery is manifold cheaper than, for example, delivery by air. Such type of delivery will allow loading cargo of various dimensions into the cargo container and delivering it to the destination without problems;
- it is possible to supply cargo practically from any point of the world by means of maritime containers;
- the transit time between ports of loading and discharging is fixed;
- the channel capacity is high;
- cargo safety in a maritime container is assured.

Container terminals for handling refrigerated containers are being developed now. Refrigerated containers can be used during transportations practically by all modes of transport. The use of refrigerated containers allows delivering cargo 'from door to door'.

With the development of the terminal a new organizational structure of control was realized, both methods and means of estimation of job performance standards of the reloading terminal and other sites of multimodal transportations were simultaneously improved [1-3].

Nowadays for exposition of valuation performance of the reloading terminal the determined methods are used. However, such methods, actually, do not reflect all specificity processing of container refrigerated cargoes. The delivery moments of consignment of cargo containers to the terminal present a casual flow of events.

The mathematical model of the storage process is constructed in view of irregular incoming of cargo consignments and their no regular arrival to the refrigerator terminals. In one case, that leads to trailer idle hours, idle and overflow of places in storage depots, in the other – to standing in line.

To analyze and estimate the process of terminals functioning, it is necessary to use both the mathematical models grounded on probabilistic methods and a simulation model, allowing the construction of the models of processes grounded on logical-mathematical description of an object.

The simulation model is a formal description of logic of the researched system functioning during a certain period of time, considering the most significant interactions of system elements and ensuring a possibility of carrying out of statistical experiments.

A suitable method for achieving the assigned tasks in view of the storage process is represented by the method of theory of queues, in the form of a queueing system.

The queueing system is an aggregate of sequentially connected entering streams of requirements (vessels) for servicing, turns of vessels, maintenance channels (conditional-functional sections) and exit streams of requirements after operation.

A simulation model of the queueing systemis the model reflecting system behavior and a modification of its condition status when adjusting streams of requirements that introduce the inputs of the system. Parameters of input streams of requirements are exterior parameters of the queueing system. Exit parameters are the magnitudes describing properties of the system, the quality of its functioning, such as the time of the requirements expectance in queues and maintenance channels etc.

Simulation modeling allows the research of the queueing system with various types of entering streams and a different intensity of inflow of requirements in the system.

The cores stages of simulation for effective utilization of the system are:

- problem statement;
- detection of the basic singularities;
- solution of the problem using analytical methods;
- solution of the problem using an imitative method.

2. Problem statement

Let us suppose that the multi-channel broken queueing system with an unlimited latency period and with the inhomogeneous elementary (stationary Poisson) flow of requirements is set. The elementary flow is characterized by the following singularities:

- inflows of requirements (vessels) in the system for servicing arrive one after another, that is the probability of inflow of two or more requirements (vessels) in one instant is very small, and it can be neglected; □a flow of requirements is ordinary;
- □the probability of inflow of subsequent requirements (vessels) at any moment does not depend on probability of their inflow during the previous moments;□ a flow of requirements is without aftereffects;
- the flow of requirements does not depend on the disposition of a considered interval of time for the time axis;□ a flow of requirements is stationary.

As far as the requirements (requests) of the queueing system have already been analysed, we may considerin coming vessels. As servicing devices belong to a conditional-functional section, the process of cargo-handling is realised there.

Let us analyse the functioning of handling processes of refrigerator containerized cargoes in the reloading terminal, by a mathematical model grounded on probabilistic methods, and a simulation model grounded on logical-mathematical description of a project.

3. Detection of the basic singularities

To solve the research problem and to develop a probability model, it is necessary to research stochastic storage processes of vessels and to make their probability analysis. Functioning of the multi-channel broken queueing system may be described by its all possible condition statuses and by the intensity of passage from one condition status to another.

Main specifications of functioning of the queueing systemare probabilities of a condition status of systems, i.e. each vessel can appear in a conditional-functional sector (event *A*) with the same probability of *p*. And movements of individual consignments may be considered as *f* of independent trials.

If probability *p* is a determination of a shipment in terminal $q = 1 - p$ there is a probability that the given shipment will be in other condition status (in other phases), i.e. no occurrence of event *A*.

We need to find probability (P_n) of shipment in the terminal, which will help to treat *n* shipments, i.e. event *A* in these *f* experiences will appear exactly *n* times as an event.

By using Bayes approach we shall consider event B_n, which shows that event *A* appears equally *n* times in *f* experiences. This event can be carried out by various modes. We shall spread out event B_n as the sum of product of the events consisting in occurrence or nonappearance of event *A* in separate experience. We shall designate A_i as an event corresponding to determination *i* of the consignment in the terminal, and $\bar{A}_i$ – the event corresponding to determination *i* of the consignment outside of the terminal.

Really, each term of sum B_n (occurrence of event) should consist of *n* events A_i and $(f - n)$ events $\bar{A}_j$ with various indexes. Then B_n will be described by the following formula:

$$B_n = A_1 A_2 \ldots A_n \bar{A}_{n+1} \ldots \bar{A}_f + \ldots + A_1 \bar{A}_2 A_3 \ldots \bar{A}_{f-1} A_m + \ldots + \bar{A}_1 \bar{A}_2 \ldots \bar{A}_{f-n} A_{f-n+1} \ldots A_f \quad (1)$$

In the given expression the number of possible modes allows choosing *n* from *f* experiences containing an event. Let us designate this number as C_n^f.

Using the private theorem of recurring experiences, it is possible to discover probability $P_{n_1 n_2 \ldots n_s}$.

$$P_{n_1 n_2 \ldots n_s} = \prod_{i=1}^{s} C_{f_i}^{n_i} p^{n_i} q^{f_i - n_i}, \quad (2)$$

where $C_{f_i}^{n_i}$ – figure of modes , with which from f_i experiences it is possible to choose n_i, in which there was an event (the determination of a shipment of cargo containers in store).

The probability of shipments determination $n_1, n_2, \ldots, n_s$ of stored cargo containers corresponds to the terms of expansion of a binominal formula in a degree equal to a number of shipments of cargo containers.

Therefore, the specified probability distribution is called binomial. By the probability distribution we mean binomial allocating possible occurrences of events *A* (the determination of a shipment of cargo containers stored) at repeated independent trials, in each of some events *A* can be determined with the same probability.

For this purpose we shall present all possible condition statuses of the queueing system in the form of labelled columns of condition statuses (Figure1).

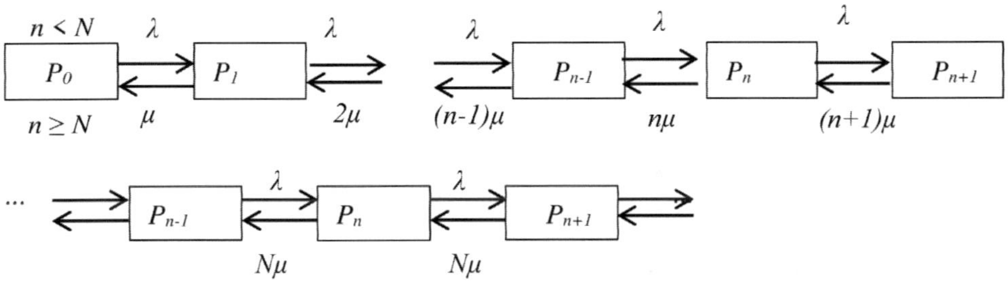

Figure1. Labelled columns of condition statuses of the open multi-channel of the queueing system.

Each rectangle of the column determines one of all possible condition statuses P_n denoting a possibility of presence *n* requirements (vessels) in the system. Arrows on the column specify in what condition status and with what intensity the system can transfer. Thus in the multi-channelof the queueing systemit is necessary to distinguish two cases:

- number of vessels n, arrived in the system is less than the number of servicing channels N, that is, all of them are in service ($0 \leq n < N$);
- number of vessels n, arrived in the system is more or equal to the number of channels of servicing ($N \leq n$), i. e. N vessels are serviced, and remaining r are expected to be in turn ($r = 1, 2, \ldots, n - N$).

Columns with probability P_0 determine a condition status of the system in which all channels of servicing stand idle because of lack of vessels. With intensity μ the system can transfer into condition status P_1 when one requirement appears in it, and also from condition status P_1 into condition status P_0 if the unique requirement that was in the system has been serviced earlier than a new one appeared, etc.

4. The solution of the problem using analytical methods

The important parameter of functioning of the queueing systemis the average latency period in turn for storage $\bar{\tau}_{\text{wait}}$ and an average number of shipments of cargo containers in turn are on deposit $\bar{r}$. The initial parameters describing the queueing system, are:

- number of conditional-functional sections, F;
- intensity of inflow of one requirement for servicing we marked, λ'_i;
- intensity of servicing of a vessel, μ.

Let us consider the system of the queueing systemby the example of the refrigerator container terminal, introducing F for conditional-functional sections with duty cycle φ_i for each type of the vessel on which irregular inhomogeneous traffic flow m_i with resulting intensity λ_i acts. Duty cycle φ_i is directly proportional to the intensity of each incoming consignment λ_i'

Let us designate the magnitude of incoming shipments for all terminal ψ, by:

$$\psi = \frac{\lambda}{\mu}, \tag{3}$$

where λ is an intensity of a flow

$$\lambda = \lambda' m \tag{4}$$

The intensity of vessels servicing is determined as a magnitude in verse to the holding time of one requirement, T_{storage}:

$$\mu = \frac{1}{T_{\text{storage}}} \tag{5}$$

The magnitude of incoming shipments of all terminal (ψ) shall be designated as duty cycle (φ):

$$\psi = \sum_{i=1}^{s} \varphi_i \cdot F \tag{6}$$

The average dwell-time of consignment in terminal $\bar{T}_{\Sigma}$ is an aleatory variable, however expectations of its components, such, as $\bar{T}_{\text{storage}}$ and $\bar{T}_{\text{wait}}$, where $\bar{T}_{\text{wait}}$ – an average latency period of consignment in queueing in the arrangement in the conditional-functional section is known. Thus the first one is considered to be known, and the last one is determined by a method of successive approximations. Expectations of the dwell-time of shipment in terminal $\bar{T}_{\Sigma}$ and expectations of storage time of lot $\bar{T}_{\text{storage}}$ will be identical for all vessels.

$$P_{\Sigma} = \lambda'\left(\bar{T}_{\text{wait}} + \bar{T}_{\text{storage}}\right) = \frac{\lambda}{m}\left(\bar{T}_{\text{wait}} + \bar{T}_{\text{storage}}\right) \tag{7}$$

In subsequent calculations storage time $\bar{T}_{\text{storage}}$ well be accepted equal to 1.Then:

$$P_{\Sigma} = \lambda'(\bar{\tau}_{\text{wait}} + 1), \tag{8}$$

where P_Σ – probability of determination of a lot in the conditional-functional sector (we shall designate it as p).

Let us calculate the parameters of functioning of the queueing system, determination of an average number of cargo containers shipments in turn for storage:

$$\bar{r} = \begin{cases} \sum_{n_\Sigma = F+1}^{m_\Sigma} (n_\Sigma - F) P_{n_1 n_2 .. n_s} & , n_\Sigma > F \\ 0 & , n_\Sigma \le F \end{cases} . \quad (9)$$

The average reduced latency period of the location for storage of cargo containers shipments is

$$\bar{\tau}_{\text{wait}} = \frac{\bar{r}}{\sum_{i=1}^{s} \lambda_i' m_i r_i} . \quad (10)$$

The received probability model allows making the probability analysis of performances of storage of containerized cargoes in container refrigerator terminals.

5. The solution of the problem using an imitative method

To display the basic events of the system we shall represent a process in the form of the graph of functioning of the multi-channel broken queueing system. In Figure 2 the basic processes of the operating multi-channel open for the queueing system are presented [4-6].

Let us describe each event which arises in the modeled system:

- inflow of vessels to the terminal;
- vessels arrival to the roadstead (hoard);
- determining of the conditional-functional section for vessel servicing;
- expectation of the release of one vessel from the conditional-functional sections;
- the exit of the vessel from the accumulator;
- the time of servicing of vessels in the conditional-functional section;
- the release of the conditional-functional section;
- the exit of the vessel from the terminal.

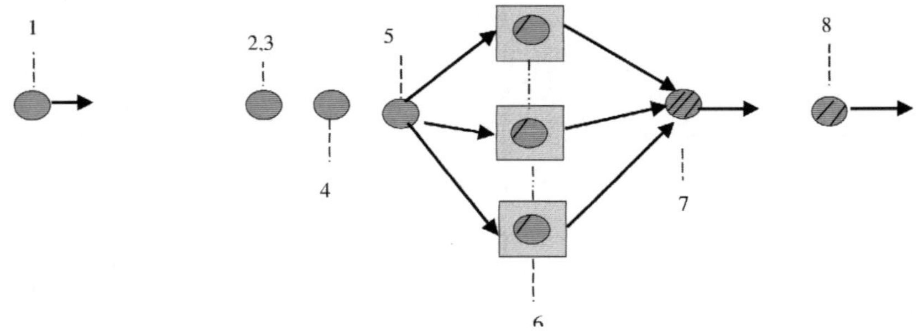

Figure 2. A process of functioning of the open multi-channel of the queueing system.

6. Conclusion

Making the analysis of the results of the work, we have come to a conclusion that the considered simulation model is grounded on logical-mathematical exposition of the plant.

However, the logic part of the system functioning shows the process of modification of condition statuses of work processes of the terminal, during the preset flows of the requirements arriving to inputs of the system.

The mathematical model, is presented by a probability model which allows the analysis of processes of storage of containerized cargoes in refrigerator terminals, considering the specificity of processes functioning of cargo handling. In calculation of the mathematical model, we can discover the following parameters: an average number of cargo containers shipments in turn for storage; the average reduced latency period, the intensity of income of the consignment containers to the terminal and a duty cycle of the conditional-functional sections. When using the ‘classical’ determined method, the following prominent aspects are not considered: an irregular flow of events, the casual time of vessels arrival to the terminal (in the conditional-functional section) and random time of cargo-handling. That leads to inaccuracies in calculations of probability performances of the queueing system.

References

[1] Association of sea trade ports 2015 Association of marine commercial harbors. Available at: http://www.morport.com/rus/news/document1751.shtml
[2] Wiraj-perevozki 2011 Advantages of a container carriage. Available at: http://www.wiraj-perevozki.ru/text_7.html
[3] Tranzito 2014 Virtues of a container carriage and their basic minuses Available at: http://tranzito.com/osobennosti/1094-morskie-perevozki-v-chem-ix-osobennosti.html
[4] Nyrkov A, Sokolov S, Chernyi S and Mamunts D 2015 *Russian Jour. Metallurgical and Mining Industry* **7(6)** 521–524
[5] Nyrkov A, Belousov A and Sokolov S 2015 *Jour. Modern Appl. Science* **10(5)** 162
[6] Chernyi S, Zhilenkov A, Sokolov S and Titov L 2015 *Jour. Metallurgical and Mining* **7(3)** 66

Mathematical modelling of the steam explosion treatment process for pre-impregnated lignocellulosic material

D B Prosvirnikov[1], D F Ziatdinova[1], N F Timerbaev[1], V A Saldaev[1] and K H Gilfanov[2]
[1] Kazan National Research Technological University, 72, Karla Marksa st., Kazan, 420015, Russia
[2] Kazan state power engineering University, 51, Krasnoselskaya str., Kazan, 420066, Russia

Email address: prosvirnikov_dmi@mail.ru

Abstract. The article analyses the physical picture of the process of steam explosion treatment of pre-impregnated lignocellulosic material, on the basis of which a mathematical modelling of the process is done. The mathematical modelling is represented in the form of differential equations with boundary conditions. The obtained mathematical description allows identifying the degree of influence of various factors on the kinetics of the process and producing a rational selection of operating parameters for the considered processes in terms of the set of application tasks.

1. Introduction
The lignocellulosic material, particularly wood, has unique thermal properties, as well as the structure and component composition. One of the ways of solving the applied tasks in the field of complex processing of raw materials to chemical components (cellulose, lignin, hemicellulose) or fibrous semi-finished products is the steam explosion treatment at a high-temperature [1-3], which can also be used as a stage of activation of raw wood prior to chemical processing, for example, sulphate pulping.

For regulation and optimization of process parameters of the steam explosion treatment of the pre-impregnated material in the industrial conditions the physical picture of the process was studied and the mathematical description of the ongoing stages was made.

2. Formalization of the process and the mathematical description
The steam explosion treatment of the pre-impregnated lignocellulosic material includes the following stages: 1) impregnation of the material with water or acid solution; 2) heat treatment of the material with high pressured steam and physical and chemical transformations; 3) decompression of the system and material defibering into separate fibers; 4) product unloading from the reactor. Let us consider these steps in more detail.

2.1. Impregnation stage of the material with water or an acid solution
A lignocellulosic material in the form of crushed particles is pre-treated with water or an acid solution up to complete saturation with moisture. The saturation occurs due to capillary and forced impregnation with symptoms of swelling of the polymer matrix.

Taking into account the assumptions about the absence of hydrodynamic flows and heat generation during impregnation, the pre-impregnation stage is described by the equation of moisture transfer [4]:

$$\frac{\partial U}{\partial \tau} = k_m \cdot \frac{\partial^2 U}{\partial x^2} \quad (1)$$

with initial conditions:

$$U(0,x) = U_0 \quad (2)$$

and boundary conditions:

$$\left.\frac{\partial U}{\partial x}\right|_{x=0} = -\frac{j}{k_m \cdot \rho} \quad (3)$$

The index of mass conductivity k_m (sm^2/sec) of wood in tangential direction may be calculated in the first approximation by the equation:

$$k_{mt} = 0.235 \cdot 10^{-31} T^{10} \gamma^{-3.3} \quad (4)$$

In the radial direction:

$$k_{mr} = k_{mt}(1 + \frac{2P}{100}) \quad (5)$$

where P – is the percentage of the medullary rays.

For pine ratio $\frac{k_{mr}}{k_{mt}}$ is 1.15, for oak - 1.50, for beech - 1.70.

Thus, the lignocellulosic material, pre-impregnated with water or aqueous acid solution with specified initial mass, is subjected to steam explosion treatment by moisture.

2.2. Heat treatment stage of the material with high pressure steam and physical-chemical transformations

Since the treated lignocellulosic material has high humidity after pretreatment, it is necessary to maintain the conditions of thermodynamic equilibrium between the steam and the liquid in the reactor throughout the whole treatment time, which means it is necessary to treat the material with water saturated steam. Let us make the following assumptions for modelling of the heating stage:

- Due to special training of the reactor, we assume that water steam will obey the laws of ideal gases and will not condense on the inner surface of the reactor, we will neglect the pressure and temperature drop at the initial inlet of steam to the reactor.
- During heat treatment of raw materials water steam has constant thermodynamic parameters and is in equilibrium with the liquid on the surface and inside of the material.
- We ignore the location of individual particles of raw materials and believe that all of the particles are evenly and thoroughly washed by the steam flow, the particle having the form of an unlimited plate [5].

The temperature change of an arbitrary point of the particle in the non-stationary heat conduction is determined by the differential Fourier equation:

$$\frac{dt(x,\tau)}{d\tau} = a_x \frac{\partial^2 t(x,\tau)}{\partial x^2} \quad (6)$$

with initial conditions:

$$t(0,x) = t_0 \quad (7)$$

and boundary conditions:

$$r_t j_c = -\lambda \left.\frac{\partial t}{\partial x}\right|_{x=0} \quad (8)$$

where the flow of condensed moisture is determined from the expression:

$$j_c = \frac{dm_c}{\pi d H d\tau} \qquad (9)$$

Thermal diffusivity a_x in (6) is determined by the formula:

$$a_x = \frac{\lambda}{c\rho} \qquad (10)$$

where λ, c, ρ – indexes of heat conduction, heat capacity and density of the material accordingly, determined by empiric formula [4]:

$$\lambda = \sqrt{0.0108 + 0.000773 \cdot t^{0.349} + 1.083 \cdot 10^{-4} \exp\left[t/111.61\right]} \cdot U \cdot \ln(U) \qquad (11)$$

$$c = 1.173\left[U \cdot (1 + \frac{t}{100})\right]^{-0.222-U\cdot 10^{-4}} \qquad (12)$$

$$\rho = m/V \qquad (13)$$

In equations (11) and (12) U is assumed to be equal to the initial humidity of the material.

Under the influence of the saturated steam temperature and due to the heat exchange processes in the wet material the reactions of autohydrolysis of hemicelluloses [6] as well as changes in the structure of lignin with its partial transition into a low molecular weight state [7] take place.

For modelling of the stages of hemicelluloses decomposition and lignin mass loss during steam hydrolysis of raw materials we will make the following assumptions:

- As during hydrolysis water is in excess, the change of its concentration during the reaction is neglected, and the hydrolysis reaction is considered to be as a monomolecular one or the reaction of the first order [8].
- Thermophysical properties of steam, material and liquid contained in it, are believed to be unchanged, despite time-varying treatment of the chemical composition of all three environments.

During treatment the hydrolysis speed of hemicelluloses can be written as [8]:

$$\frac{dC_{xhmc}}{d\tau_{treat}} = k \cdot C_{0hmc} \cdot e^{-k\cdot\tau_{treat}} \qquad (14)$$

In the equation (14) k is calculated as

$$k = \alpha N \delta Q_{10} \qquad (15)$$

where α – relative catalytic activity of the catalyzer (acid); N – normality of the catalyzer, expressed by the number of its gram-equivalents in 1 liter of solution; δ – relative factor of polysaccharide stability relatively hydrolysis that characterizes the strength of glycosidic linkages in polysaccharides; Q_{10} – temperature coefficient.

Changes in the concentration of low molecular lignin will be written in the form of the equation of mass conservation:

$$\frac{dC_{xlig}}{d\tau_{treat}} = -k_2 \cdot C_{0lig} \cdot m_w \qquad (16)$$

where k_2 is an experimentally assignable value.

Since hemicellulose and lignin are binders in the wood substance, they strengthen the bonds. As a result of the destruction of the binder components the intermicellar cohesive forces (and hence the tensile strength of wood), which are directly dependent on the initial composition of the material (wood species), weaken.

The liquid contained in the material from the moment of supply of saturated water steam tends to thermodynamic equilibrium with steam and turns to the overheated (metastable) condition with temperature increase. This is due to the fact that the nascent bubbles in evaporation sites, which are located on the surface of the micelles and fibers, cease to grow, as the pressure above the liquid surface is higher than the steam pressure in the bubbles. Upon reaching the superheated liquid state, it accumulates in itself the same energy in the form of superheating heat.

2.3. Stage of system decompression and material destruction

After soaking the raw material in the reactor at preset parameters the system decompression is performed by opening a decompression valve that leads to external disturbing factor – the difference between the pressure above the liquid and that in the nascent bubbles, resulting in the transition of the liquid, contained in the material, from a metastable state to a stable one [9].

Since the pressure above the liquid and, consequently, above the bubble with steam decreases, the growth of steam bubbles, the overcoming the surface tension of the overheated liquid, the steam release take place. The speed of change of the bubble radius with time equals:

$$\frac{dr}{d\tau} = \left\{ \frac{2}{\rho' r^3} \int_0^{\tau} r^2 \frac{dr}{d\tau} [p'' - p'] dt \right\}^{1/2} \tag{17}$$

The pressure on the surface of the bubble equals:

$$p = p_0 + \frac{\rho_0 R_0^2}{3} \left(\frac{\rho_0}{\rho} \right)^{2/3} \left[\frac{11}{6} (\frac{1}{\rho} \frac{d\rho}{d\tau})^2 - \frac{d^2 \rho}{d\tau^2} \right] \tag{18}$$

where p_0, R_0, ρ_0 – pressure, radius and density at initial time $\tau = 0$.

The density of the steam inside the bubble is:

$$\rho(\tau) = \rho_0 \left[1 - \beta T - (1 - \frac{\rho''}{\rho'}) v'' \right] \tag{19}$$

where β – index of the temperature expansion of the liquid; $v''(\tau)$ – specific volume steam quantity.

The equation of the growth rate of the bubble volume that emerged at time τ is the following:

$$\frac{dv}{d\tau} = kb[p(\tau)](\tau - t)^{k-1} \tag{20}$$

Thus, the greater the vent rate (less than reset time), the greater rate J of boiling centers formation in the superheated liquid inside of the lignocellulosic material.

$$\ln J = 88 - G \tag{21}$$

where $G = \frac{W_*}{RT}$ – a free Gibbs number, and W_* is a work of critical bubble formation:

$$W_* = \frac{16\pi\sigma^3}{3(p_s - p')^2 (1 - v'/v'')} \tag{22}$$

The increase in the pressure drop over the superheated liquid (steam pressure) and in the bubble causes a decrease in the critical work of bubble formation, and therefore Gibbs number reduces and the number of steam formation centers in the superheated liquid increases.

Also, the nucleation intensity increases with a decrease in average waiting time $\overline{\tau}$ of occurrence of the nucleating seed (the time of pressure change) in volume of metastable (superheated) liquid V_{shl}:

$$J = (\overline{\tau} V_{shl})^{-1} \tag{23}$$

That is, from (23) it follows that if the time of pressure drop decreases, the intensity of steam formation, the amount of the steam expanding inside of the material, as well as the energy stored by superheated liquid increase.

All the above-mentioned conclusions give grounds to assert that the greater the pressure drop for the least period of time, the more energy-carrying expanding steam is formed and the larger force of tension arises inside of the material.

Generally, during decompression of the reactor the liquid boils sharply, releasing the accumulated energy. The expanding steam in the pores and cavities provides the mechanical effect (tensile stress) on the weakening as a result of autohydrolysis bonds. The power influence of the expanding steam results in overcoming the tensile strength of these bonds and their rupture. Thus, there is a separation of the material particles (fibers) from each other with the formation of new surfaces as a result of the performed work.

The potential energy accumulated in the system "compressed steam–superheated fluid" prior pressure drop, is spent not only for boiling up the liquid and for the increase in the specific surface of the material (overcoming tensile strength), but also for pushing the material and steam out of the reactor, as well as the overcoming of the hydraulic and pneumatic resistance. The rest unused energy remains in the form of heat.

The emerging force causes tensile stresses along the width of the particle (chip) making efforts to change the width:

$$\Delta A = F_a \Delta a \tag{24}$$

As far as in this case the work is done at the expense of the potential energy store, it equals the loss of potential energy on the segment of axis Δa. Then

$$F_a \Delta a = -\Delta W_p \tag{25}$$

This ratio is valid for any direction in space, so the value of resultant force is:

$$F_p = -gradW_p \tag{26}$$

Turning to the issue of energy balance, it can be written as:

$$W_{shl} = W_1 + W_2 + W_3 + W_4 + W_5 + W_6 + W_7 \tag{27}$$

where W_{shl} – the energy of superheated liquid in the form of heat;

$W_1 = \dfrac{16\pi\sigma^3}{3(p_s - p')^2(1 - v'/v'')^2}$ – energy spent for the formation of critical bubbles;

$W_2 = m_{shl} \cdot r$ – energy spent for steam formation;

$W_3 = pdV$ – energy spent for steam expansion;

$W_2 = \dfrac{\sigma_t a P_{mat} dF_t}{E}$ – energy spent for mechanical deformation of the body, for overcoming the tensile strength across the grain and formation of a new surface;

W_5 – the energy of motion, spent for a filtration flow of steam from inner layers of the material to the outer ones;

W_6 – the energy spent during condensation of the steam at a saturation temperature;

W_7 – energy in the heat form retained in the vapors of boiling.

Let us turn to the problem of modeling of material failure (overcoming the tensile strength of bonds). Boiling liquid, turning into vapor, spends the stored energy for steam formation and the breaking of bonds in the material. And a part of the energy of vapor expansion at constant temperature transforms into work, which is performed at stretching and rupture of the weakened as a result of hydrolysis bonds in the lignocellulosic material, that is, this part of energy is converted into mechanical.

$$dA = -pdV = Vdp \tag{28}$$

where dA – is a change of work of steam expansion.

The work that should be applied to the wood particle for its internal deformation and fracture (that is, the formation of new specific surface area) is equal to [10]:

$$A = \frac{\sigma_t^2 V_p P_{mat}}{2E} \tag{29}$$

where σ_t – is a tensile strength of the material under tension; V_p – the volume of the particle; E - is the Young's modulus. Converting (29) and differentiating it with regard to

$$\sigma_t = \frac{F_t}{bl} \tag{30}$$

where F_t – tensile force; b, l – thickness and length of the particle, respectively, we will obtain:

$$dA = \frac{\sigma_t a P_{mat} dF_t}{E} \tag{31}$$

where a – the width of the particle, P_{mat} – is the porosity of the material.

As at material destruction of the forces of the expanding inner steam affects the individual bonds [11], and the particle has a capillary-porous structure, we can write:

$$V_1 dp = \frac{\sigma_t a P_{mat} dF_t}{E} \tag{32}$$

where V_1 – volume of steam after expansion.

Thus, the change of the tensile force in the particle will appear as:

$$dF_t = \frac{EV_1 dp}{\sigma_t a P_{mat}} \tag{33}$$

And a single tear condition is:

$$\Delta P \geq \frac{\sigma_t^2 V_p P_{mat}}{EV_1} \tag{34}$$

A specific surface of obtained product $\overline{A}_s$ is characteristic of the grinding degree of the fibrous material, obtained as a result of steam explosion. At a first approximation one can imagine, that

$$\overline{A}_s = f(\Delta P, \tau_{dec}) \tag{35}$$

However, in reality this value depends not only on the conditions of decompression pressure drop ΔP and decompression time τ_{dec}), but also on the conditions of the heat treatment of the lignocellulosic material itself, because the change of physico-chemical and mechanical properties of the material contributes to the formation of a new surface. Therefore, this value is not amenable to rigorous mathematical description, and can only be determined on the basis of experimental data.

2.4. Stage of product unloading

For modeling of the process of unloading as a result of decompression, we agree that:

- At the time of pressure relief the possible effects of drying and reduction of the moisture content of the material are not taken into account, since the duration time of these effects is not large, and the processes of condensation begin to dominate right after the reduction of the pressure till the saturation pressure.
- Due to the complex pattern of simultaneously occurring phenomena of liquid boiling, steam expanding, shredding and material unloading for a short time of decompression, we believe that these phenomena occur successively.

During decompression the expanding steam increases in volume. At the exit of the reactor, it creates a rarefaction wave, its energy is partly spent for pushing of the material, the rest of the energy remains in the steam itself in the form of heat.

The amount of kinetic energy obtained when expanding 1 kg of steam can be defined by the equation:

$$W_k = \frac{c^2}{2g} \tag{36}$$

where c – the steam velocity.

On the other hand, when expanding 1 kg of steam from p_0 to p_1 a heat drop will be equal to $i_0 - i_1$. Using the concept of mechanical equivalent of heat [11], we can write that work of 1 kg of steam at its extension will be equal to:

$$W = 427 \cdot (i_0 - i_1) \tag{37}$$

where 427 – is a mechanical equivalent of heat.

Since $W_k = W$, we can determine the average exit velocity of steam:

$$v_{av} = 91.5\sqrt{i_0 - i_1} = 91.5\sqrt{p_0V_0 - p_1V_1} \tag{38}$$

where for steam it is:

$$V_1 = \frac{mRt}{Mp_1} \tag{39}$$

The steam flow passing through the pore channels of the layer, exerts pressure on the individual particles, which is totally expressed in a pressure loss. The increase in the speed of steam flow increases the pressure drop in a fixed layer of the material. The layer retains a fixed state until the weight of the layer per unit of its area becomes equal to a pressure drop. With further increase of speed the layer turns into a movable state. With the increase of velocity the layer expands, that is, the concentration of a solid phase in a unit volume of the layer decreases.

3. Conclusion

The physical picture of the steam explosion treatment of the pre-impregnated lignocellulosic material has allowed developing the mathematical model and the algorithm of process calculation. The results of the calculations can be modeled on a computer and compared with the experimental data that will allow identifying the degree of influence of various factors on the kinetics of the process and to produce a rational selection of operating parameters for the considered processes in terms of the assigned applied problems.

Acknowledgments

The present research results were obtained during the implementation of the grant of the Investment and Venture Fund and the Fund of assistance to development of small forms of enterprises. Research theme cipher is Project No. 07073-SI 'Research of operational parameters of an extraction stage in the process of obtaining the technical cellulose by autohydrolysis'.

References

[1] Kokta B V and Ahmed A 1993 *Wood Sci.Technol.* **27** 271
[2] Garrote G and Parajo J C 2002 *Wood Sci.Technol.* **36** 111
[3] Ziatdinova D F, Safin R G and Prosvirnikov D B 2011 *Bull. Kazan Technol.Univ.* **12** 58
[4] Timerbaev N F, Safin R R, Safin R G and Ziatdinova D F 2014 *Journal of EAAS* **9** 141
[5] Tuntsev D V, Filippova F M, Khismatov R G and Timerbaev N F 2014 *Rus.Jour.of Appl.Chem.* **87** 1367
[6] Abolins J and Gravitis J 2009 *Latv.jour.of phys.and tech.sci.* **5** 66

[7] Ramos L P 2003 *Quim. Nova* **6** 863
[8] Harina M V, Emelyanov V M, Ablaev A R, Mokshina N E, Ibragimova N N and Gorshkova T A 2014 *Chem.of plant raw mat.* **1** 53
[9] Vinogradov V E Study of boiling of superheated liquids and elasticity . Diss . ... Doctor . Sci. Sciences. Ekaterinburg 2006
[10] Prosvirnikov D B, Halitov R A, Lashkov V A 2014 *Bull. Kazan Technol.Univ.* **1** 241

Multifactorial modelling of high-temperature treatment of timber in the saturated water steam medium

D B Prosvirnikov, R G Safin, D F Ziatdinova, N F Timerbaev and V A Lashkov
Kazan National Research Technological University, 72, Karla Marksa st., Kazan, 420015, Russia

E-mail: prosvirnikov_dmi@mail.ru

Abstract. The paper analyses experimental data obtained in studies of high-temperature treatment of softwood and hardwood in an environment of saturated water steam. Data were processed in the Curve Expert software for the purpose of statistical modelling of processes and phenomena occurring during this process. The multifactorial modelling resulted in the empirical dependences, allowing determining the main parameters of this type of hydrothermal treatment with high accuracy.

1. Introduction

Timber is an anisotropic substance, therefore, the physico-chemical processes occurring in it are not always amenable to a precise mathematical description. In this case performance of a natural experiment assists in assessing the influence of different factors of the studied process on the properties and behavior of timber. However, modeling of physical processes with known techniques, such as, for example, planning of multifactorial experiment with the subsequent derivation of the equations of the polynomial nature, have a number of disadvantages. The lack of physical meaning of such equations and impossibility of their usage for similar processes are the major disadvantages.

Statistical modeling [1] uses a biotechnological exponential law as a basis for constructing mathematical models of real physical processes. This law takes into account the impact of various factors and does not require their fixation in the course of the experiment for the purpose of obtaining data that would allow an adequate description of the process.

Implementation of modeling in the Curve Expert software allows choosing the right mathematical relationship for the experimental data and assessing the adequacy of the resulting model with high reliability. Thus, the process of high-temperature treatment of timber in a saturated steam medium [2 – 4] was modeled in a multidimensional way. The main adjustable process parameters are: heating temperature of the material, coefficient of thermal and mass conductivity. The simulation results are given below.

2. The temperature dynamics of the material layers at a high temperature processing with saturated steam

For multidimensional modeling of temperature dynamics t (°C) of material layers at treatment the experiments on the influence of heating time τ (from 0 to 12 minutes, a step is 1 minute) and moisture content of timber W (60, 100, 140 %) in three layers of the processed material were conducted. Initial data of experimental studies are presented in paper [5]. Thus, there are three influencing variables: τ – time of wood heating, min; W – moisture content of the wood at the initial state, %; I – a serial number of the layer. An indicator or dependent factor is the temperature of layer t. Initially single-factor regularities were identified (Figure 1).

The indicator of the dynamics at the single-factor modeling changes in the following way:

$$t = 20.44\exp(0.64\tau^{0.39}) + 3.96\cdot 10^{6}\tau^{5.13}\exp(-13.38\tau^{0.23}) \quad (1)$$

The correlation index in this case is 0.864.

The temperature change depending on the humidity is described by expression (2):

$$t = 124.15\exp(-0.003W) + 0.04W^{1.38}) \quad (2)$$

The correlation index is 0.033, which is much less than the required level of adequacy of 0.3. Therefore, this factor on the multifactorial modeling is not considered. Under the influence of the layer number code rate t changes according to the equation (3):

$$t = 144.06\exp(-0.002I) - 14.14\cdot I^{1.23}) \quad (3)$$

The correlation index in this case is 0.248.

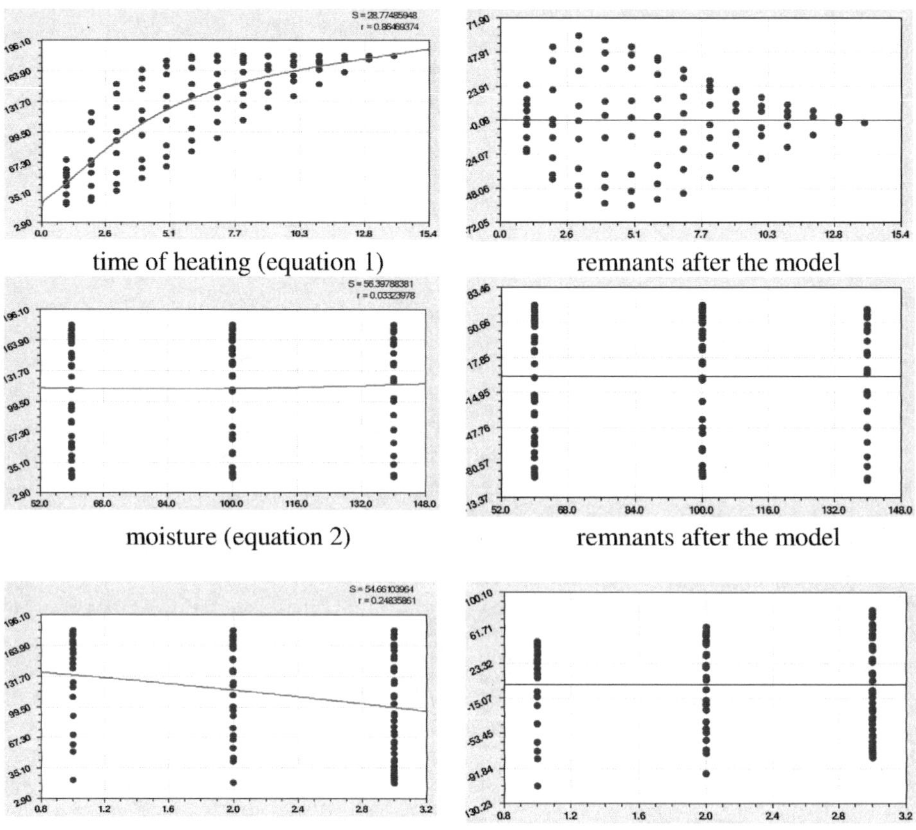

Figure 1. Single-factor influence of parameters on heating temperature

From two remnant factors (τ and I) biotechnical regularities [6] for moisture content W = 60% were obtained (Figure 2).

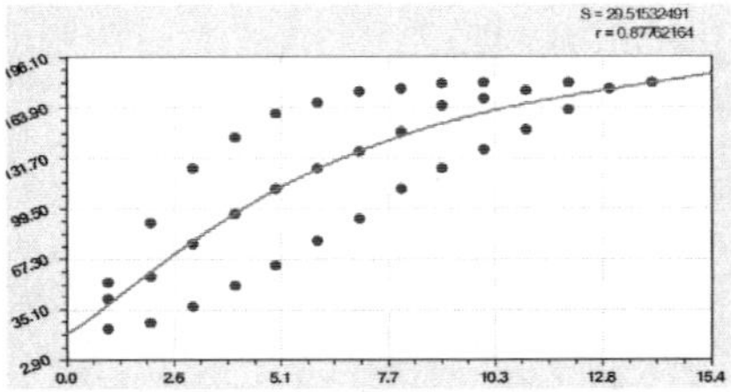

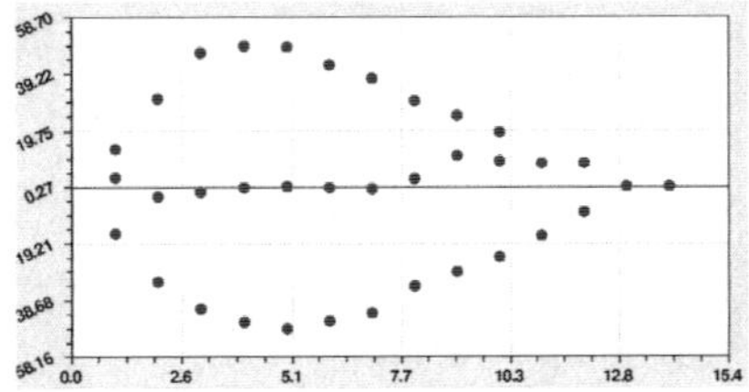

time of heating (equation 4) remnants after the model

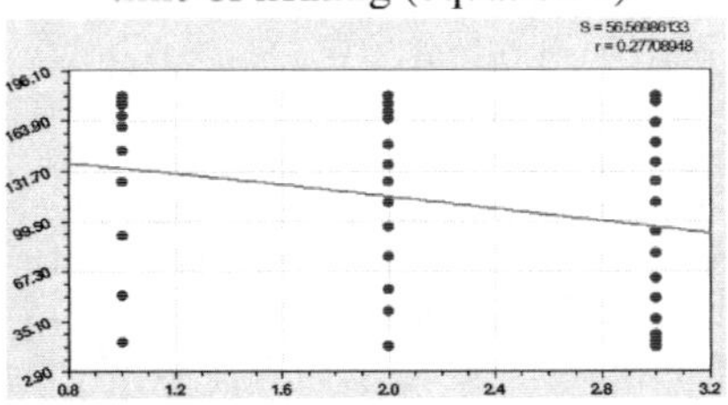

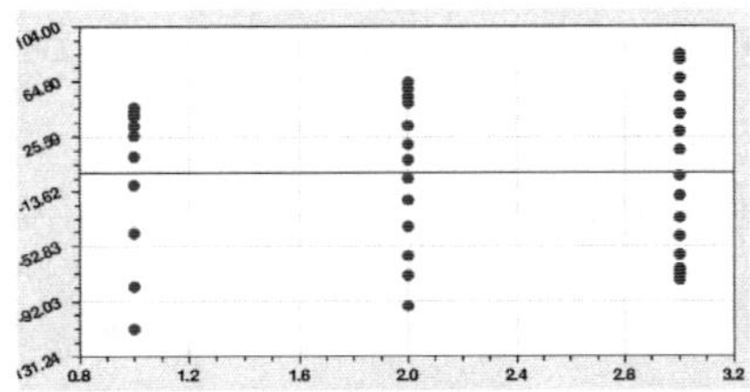

timber layer number (equation 5) remnants after the model

Figure 2. Single-factor influence of parameters on heating temperature at humidity of 60%

For each single-factor modeling the following dependences are composed:

- the influence of heating time on temperature (correlation index is 0.877):

$$t = 20.33\exp(0.0001\tau^{3.02}) + 20.25 \cdot \tau^{1.22} \exp(-0.1\tau^{0.94}) \tag{4}$$

- the influence of layer number on temperature (correlation index equals 0.277):

$$t = 147.73\exp(-0.001I) - 16.45 \cdot I^{1.17}) \tag{5}$$

The exception of a very small impact parameter (the moisture content of the raw material) increased the adequacy of single-factor formulas. To strengthen the validity of the outcome equation of the factors influence on temperature it is necessary to consider remnants from the equations (5) and to place the second explanatory variable (temperature). After identifying [1] we will obtain (Figure 3) the equation:

$$\varepsilon_I = -96.29\exp(-0.21\tau) + 1274.75 \cdot \tau^{2.36} \exp(-5.97\tau^{0.15}) \tag{6}$$

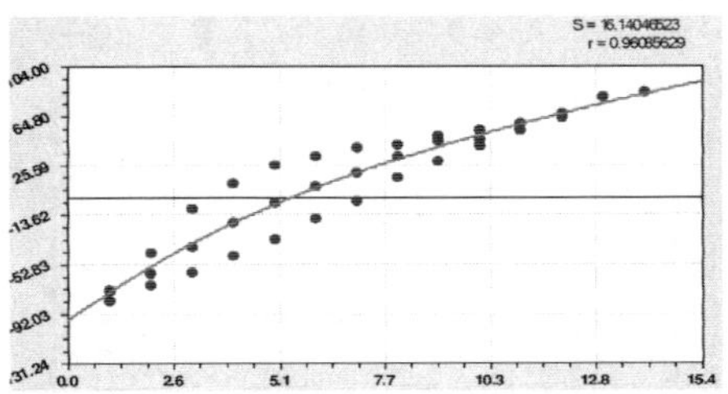

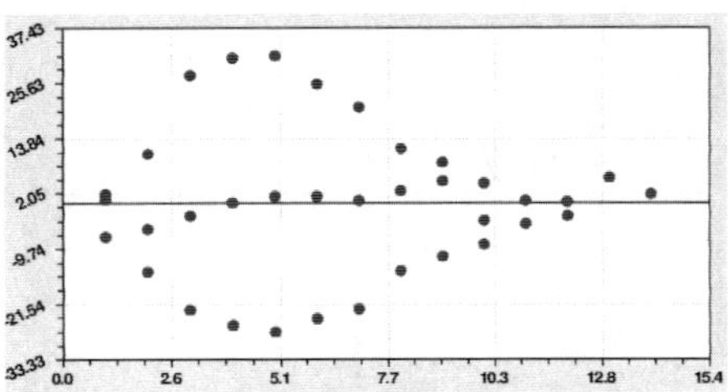

Temperature depending on the remnants after the layer number (equation 6) remnants after the model

Figure 3. Equation diagrams of consistent influence of two parameters on the heating temperature of the material layers at humidity of 60%

The correlation index of two items turned out to be equal to 0.9609. At the ends of curves the points have minimal deviation. Thus, the remnants show the influence of some third (unknown) strong factor. This additional factor may well be an attempt to secure the values of so-called constant parameters by the method of Gauss-Seidel during experimenting.

A general two-factor formula takes the following form:

$$t_{W=60\%} = 147.73\exp(-0.001I) - 16.45 \cdot I^{1.17} - 96.29\exp(-0.21\tau) + 1274.75\tau^{2.36}\exp(-5.97\tau^{0.15}) \quad (7)$$

Similarly biotechnical regularities [6] for material moisture $W = 100\%$ were obtained:

$$t_{W=100\%} = 122.42\exp(-0.22I) + 61.3 \cdot I^{1.57}\exp(-0.73I) - 96.78\exp(-0.28\tau) + 13236.1\tau^{2.47}\exp(-8.17\tau^{0.13}) \quad (8)$$

for material moisture $W = 140\%$:

$$t_{W=140\%} = 167.27\exp(-0.04I) + 28.43 \cdot I^{1.06}\exp(-0.23I) - 98.68\exp(-0.33\tau) + 1120.7\tau^{2.23}\exp(-5.53\tau^{0.16}) \quad (9)$$

When estimating errors of equations (7) - (9) it was revealed that the maximum relative error is shifted to the side of low temperatures in the third layer of the heated material. This fact means that it is necessary to increase the measurement accuracy of layers thickness.

3. Thermal conductivity of timber at high temperature treatment with saturated water steam

For multifactorial modeling of the dynamics of timber thermal conductivity λ (W/m·K) at high temperature treatment of the material layers with saturated water steam, the experiments on the effect of heating temperature t (from 40 to 240 ° C, a step is 20 ° C) and humidity of timber W (70, 100, 130%) for softwood (pine) and hardwood (aspen) were conducted. The initial experimental studies are presented in work [5]. Thus, two influencing variables are known: t – temperature of heating wood, ° C; W – moisture content at the initial state, % for each of the woods. At the same time a wood of timber refers to qualitative factors. An indicator or a dependent factor is the thermal conductivity of timber λ. First, single-factor regularities, which were arranged according to the correlation coefficient ascending, were identified (Figure 4). Then the influencing parameters were added sequentially in the remnants of the previous models.

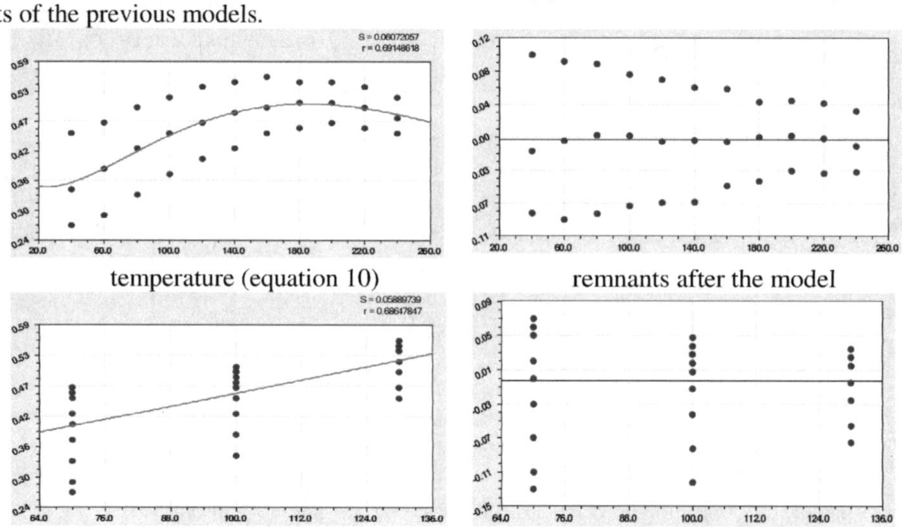

temperature (equation 10) remnants after the model

moisture (equation 11) remnants after the model

Figure 4. Single-factor influence of parameters on the thermal conductivity index

For softwood (pine) of timber the following single-factor regularities were obtained separately (Figure 4):

- effect of thee temperature on the conductivity index

$$\lambda = 0.39\exp(-0.02t^{1.004}) + 0.002t^{1.29}\exp(-0.006t) \quad (10)$$

- influence of timber moisture on thermal conductivity

$$\lambda = 0.25\exp(-0.001t^{1.005}) + 0.002W^{1.02} \tag{11}$$

Comparing correlation coefficients, we notice that for the influence of humidity the coefficient is 0.6865, and for the effect of temperature it is 0.6915. So we take remnants ε_W from function $\lambda = f(W)$ and substitute the values of influencing parameter t. After structural-parametric identification we obtain the equation of the following type (Figure 5):

$$\varepsilon_W = -0.15\exp(-0.004t^{0.99}) + 0.0004t^{1.44}\exp(-0.003t) \tag{12}$$

The correlation coefficient of two summands turned out to be equal to 0.9514.

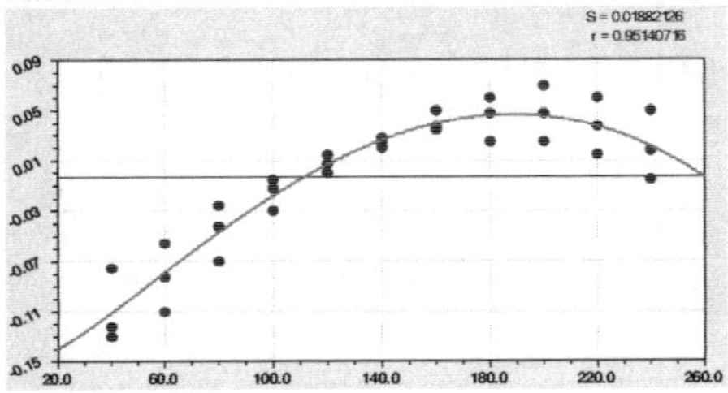

The temperature influence on the moisture remnants (equation 12)

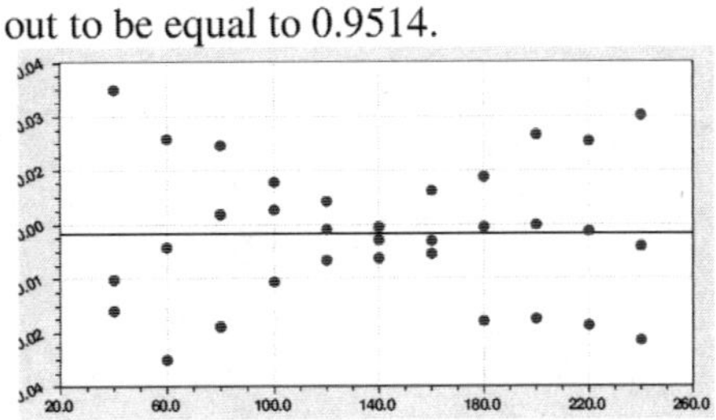

remnants after the model

Figure 5. Equation diagrams of sequential influence of two parameters on the thermal conductivity coefficient of pine

At critical point temperature $t^* = 140$ °C a minimal value of the correlation index results. When increasing or decreasing the temperature from 140 °C remnants increase (Fig. 5). This, apparently, reveals the shortcomings of fixing the non-major factors. Then the overall two-factor equation takes the form:

$$\lambda_{pine} = 0.25\exp(-0.001W^{1.005}) + 0.002W^{1.02} - 0.15\exp(-0.004t^{0.99}) + 0.004t^{1.44}\exp(-0.003t) \tag{13}$$

Similarly, a two-factor dependence of the thermal conductivity of aspen timber on feedstock moisture and temperature of hydrothermal treatment was obtained:

$$\lambda_{aspen} = 0.33\exp(-0.0004W) + 0.002W^{1.02}\exp(-0.001W) - 0.17\exp(-0.004t^{0.99}) + 0.004t^{1.47}\exp(-0.004t) \tag{14}$$

The calculations in the software environment of MS Excel showed that the maximum relative error is equal for pine to 10,76%, for aspen - to 9.11%.

3. The coefficient of mass conductivity of timber at heat treatment with saturated water steam.

The change in the coefficient of mass conductivity k_m (m^2/(sec·%)) of the timber at high-temperature treatment with saturated water steam was modeled. In this example, the number of values of the influencing variable is equal to 18, according to data from the work [5], and that is enough for the obtainment of additional wave components to the trend. The influencing variables are: heating time τ (from 10 to 180 minutes, a step is 10 min), treatment temperature t (60, 140, 220 ° C), medium acidity pH (7.0, 3.0, 2.4). Thus, only the process time has gained a sufficient number of measurements, and the rest two factors were taken into account with the minimum acceptable number of repetitions. As a result we should expect a non-equilibrium statistical model, when the factors will be identified by few measurements according to the elementary regularities. Initially the single-factor regularities were identified (Figure 6).

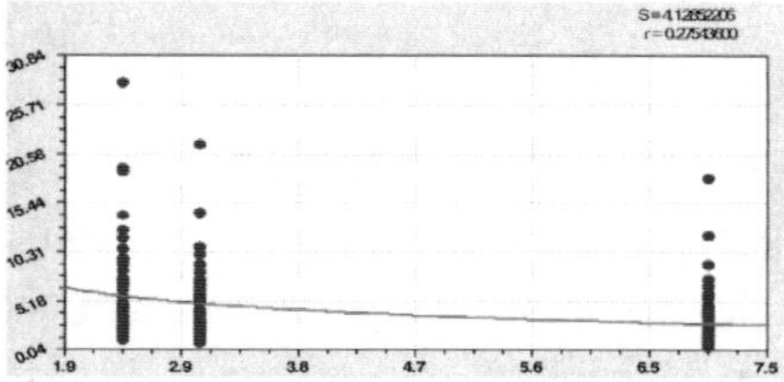

influence of medium acidity (equation 15)

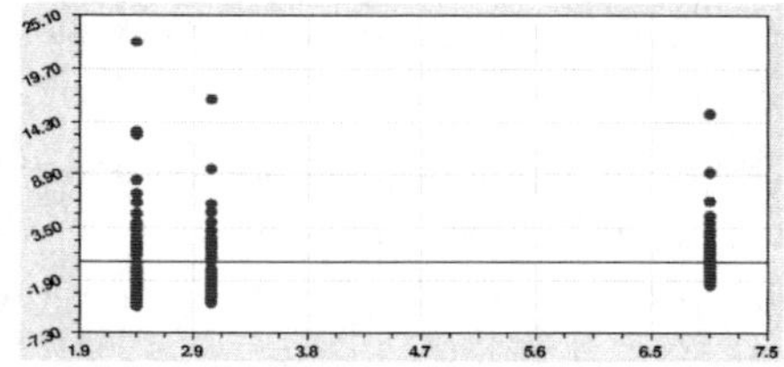

remnants after the model

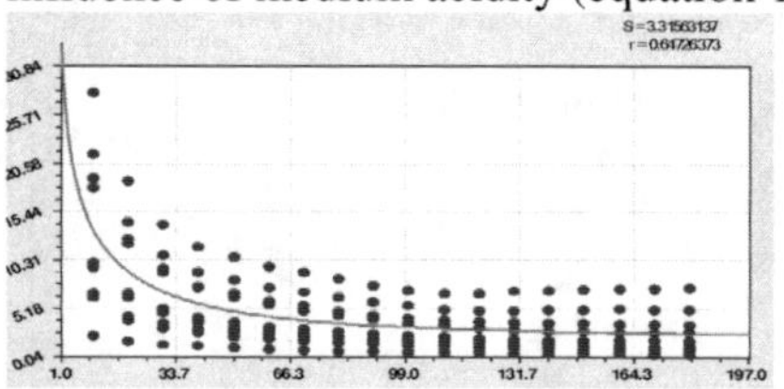

time influence (equation 16)

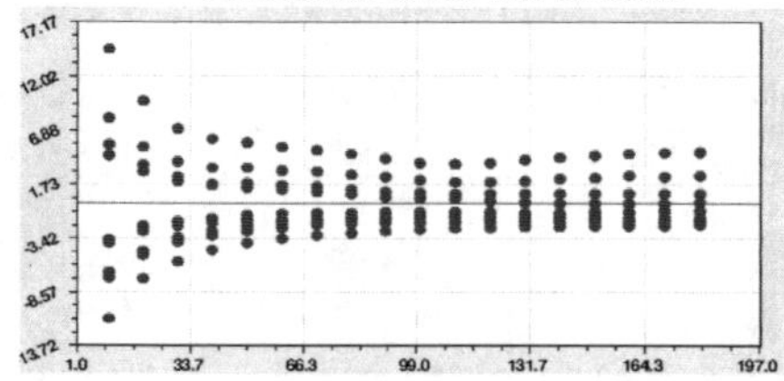

remnants after the model

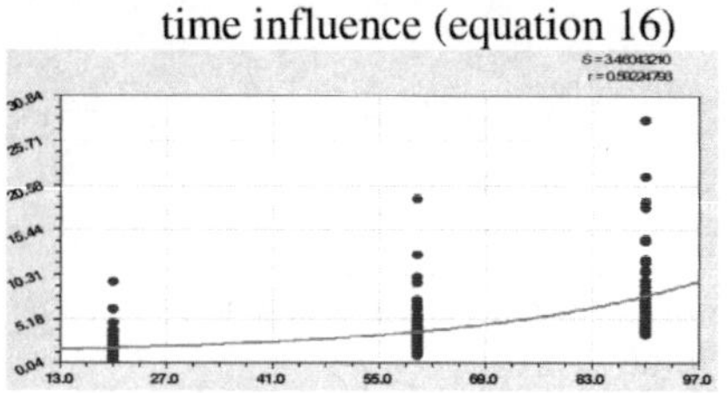

temperature influence (equation 17)

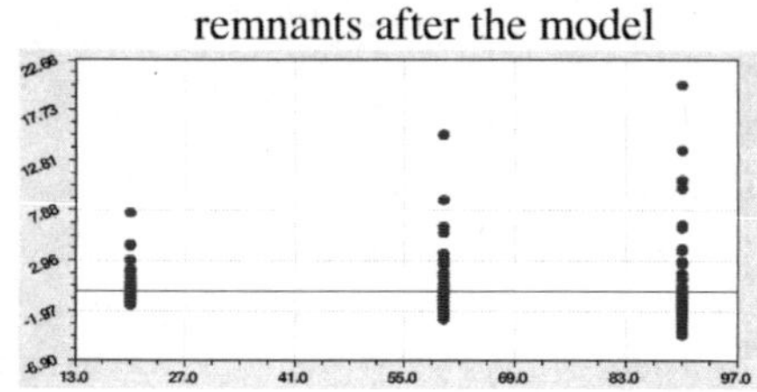

remnants after the model

Figure 6. Equations diagrams of the influence of parameters on the mass conductivity coefficient

The influence of medium acidity is determined by the regularity of the following type (the correlation coefficient is equal to 0.275):

$$k_m = 1.62 \cdot 10^8 \exp(-16.61 pH^{0.03}) \quad (15)$$

The time influence is determined by the equation of the following type (the correlation coefficient is equal to 0.064):

$$k_m = 1555.95 \exp(-0.006\tau^{0.74}) - 1524.13\tau^{0.006} \exp(-0.007\tau^{0.72}) \quad (16)$$

The temperature influence is characterized by the following formula (the correlation coefficient is equal to 0.592):

$$k_m = 1.58 \exp(0.001 t^{1.55}) \quad (17)$$

After that the temperature as an explanatory variable was added to the remnants after the model (15). After identification of the remnants we obtained the following (Figure 7) formula:

$$\varepsilon_{pH} = -2.71 + 2.06 \cdot 10^{-5} t^{2.79}) \quad (18)$$

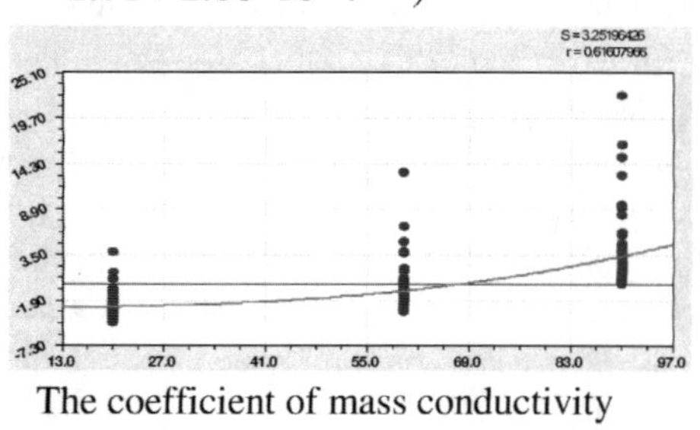

The coefficient of mass conductivity from the remnants after medium acidity

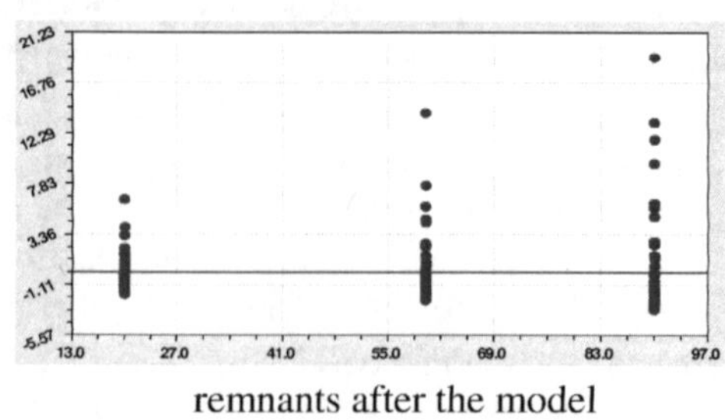

remnants after the model

Figure 7. Equation diagrams of the subsequent influence of two parameters on the mass conductivity coefficient

Again, we take remnants $\varepsilon_{pH,t}$ and place time τ as an influencing variable. Again we identify and obtain the following formula:

$$\varepsilon_{pH,t} = 25.52\exp(-0.18\tau^{0.74}) - 2.34\cdot 10^{-6}\tau^{3.36}\exp(-0.02\tau^{0.97}) \tag{19}$$

The remnants in Figure 8 show that for small time values it is necessary to increase the measurements accuracy sharply. A number of the initial data with variable dispersion of the actual distribution appears with the same accuracy of time measurement.

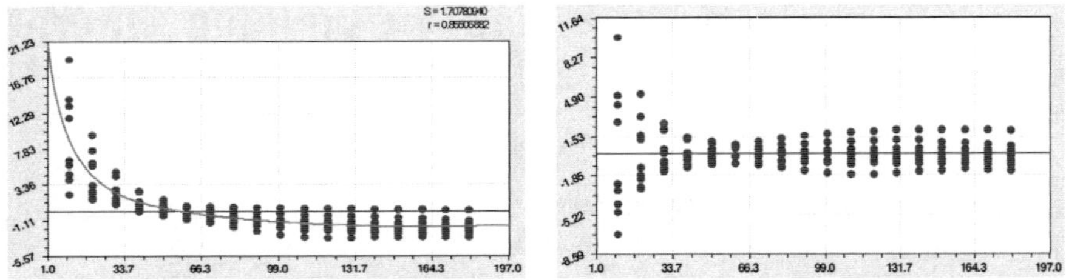

Mass conductivity coefficient from remnants after medium acidity — remnants after the model

Figure 8. Diagrams of the subsequent influence of three parameters on mass conductivity coefficient

A general three-factor model obtains the formula of the following view:

$$\begin{aligned} k_m &= 1.62\cdot 10^{8}\exp(-16.61pH^{0.03}) - 2.71 + 2.06\cdot 10^{-5}t^{2.79} \\ &+ 25.52\exp(-0.18\tau^{0.74}) - 2.34\cdot 10^{-6}\tau^{3.36}\exp(-0.02\tau^{0.97}) \end{aligned} \tag{20}$$

4. Conclusion

For the time being the experiments are carried out according to well-known methods of Gauss-Seidel, when the researchers are trying to modify every single factor at fixed values of other influencing factors. For multifactorial modeling it turns out that secondary factors change very little during the experiment – just in three values. The authors have shown the possibility of multifactorial modeling based on the measurements obtained by the present method of Gauss-Seidel.

Acknowledgments

The presented results were obtained during the implementation of grant of the Investment and venture Fund and the Fund of assistance in development of small forms of enterprises. Research theme cipher is Project No. 07073-SI 'Research of the operating conditions of the extraction stage in the process of technical cellulose production by autohydrolysis'.

References

[1] Mazurkin P M, Safin R G and Prosvirnikov D B 2014 *KNRTU*
[2] Ziatdinova D F, Safin R G and Prosvirnikov D B 2011 *Bull. Kazan Technol.Univ.* **12** 58
[3] Safin R R, Khasanshin R R, Shaikhutdinova A R and Safina A V 2014 *World Appl.Sci.Jour.* **30** 1618
[4] Lashkov V A, Levashko E I and Safin R G 2001 *Jour.of Engin.Phys.* **1** 80
[5] Prosvirnikov D B 2013 Dissertation ... The candidate of technical sciences KNRTU
[6] Mazurkin P M 2009 *The succes.of modern sci.* **9** 152

Models of evaluating efficiency and risks on integration of cloud-base IT-servicesof the machine-building enterprise: a system approach

S Razumnikov and A Kurmanbay
Yurginsky Technological Institute of Tomsk Polytechnic University, 26, Leningradskaya st., Yurga, 652055, Russia

E-mail: demolove7@inbox.ru

Abstract. The present paper suggests a system approach to evaluation of the effectiveness and risks resulted from the integration of cloud-based servicesin a machine-building enterprise. This approach makes it possible to estimate a set of enterprise IT applications and choose the applications to be migrated to the cloud with regard to specific business requirements, a technological strategy and willingness to risk.

1. Introduction

At present cloud computing tends to be developed and integrated rapidly. The most important problems for IT-managers are those of reasonability of transition to a cloud-based platform and an assessment of the economic benefits and risks caused by the integration of cloud computing. The assessment of their economic benefits [1-3] is a compulsory element of feasibility study of an IT-project. Therefore, issues concerning the selection of methods for evaluation of efficiency and risks of IT integration are the ones of particular importance. The suggested methods [4] of the efficiency and risks evaluation of IT and IS do not take into consideration special characteristics of cloud-based technologies.

The paper focuses on design and implementation of a system approach, which involves the assessment of economic efficiency and risks analysis caused by integration of cloud-based IT-services, allowing managers of enterprises to come to a correct decision concerning the integration.

2. A system approach to assessment of efficiency and risks caused by transition to cloud-based IT-services

The problem of the insufficient comprehensive methodological base and tools for supporting decision making, which rely on the processes of the efficiency assessment and risks in conditions of uncertain decision making environment is currently the urgent one in the sphere of cloud-based technologies integration. This issue is relevant for businesses of all branches and levels [5]. To solve a problem of this kind a system approach is to be applied, as well as the method of system analysis. This approach can be reduced to giving a more precise definition of the problem and structuring it in a number of tasks soluble by economic and mathematical methods, revealing criteria for their decision, and making purposes more detailed.

The system analysis is applied to solve hard formalizable and insufficiently structured problems, as a way to turn a complex problem into an interrelated hierarchy of more simplified tasks, which can be solved by formal methods [6, 7].

Figure 1 demonstrates the outline of the system approach to assess efficiency and risks caused by integration of cloud-based IT-services. The first stage ***'Identifications of expenditures and benefits'*** involves calculation of expenditures and benefits from transition to cloud-based services. At the stage of identification of high-level business requirements it is necessary to reveal:

1. Functions of business.
2. Main reasons of the business to integrate the cloud-based services.
3. Cloud-based services, which can support business-processes.
4. Relevant legal standards.
5. It is necessary to determine where the systems providing the services will be located (on the premises of the enterprise, in a particular geographical area) and who will be responsible for provision of services (an external supplier, one's own service, a mixed group, a cloud broker).

The next stage involves identification of the start/basis model of cloud-based service in terms of risk. Here we determine the spheres of risk, which are necessary to take into account; and measures to reduce risk in determined spheres up to the level, acceptable for the enterprise. And then we determine which type of a cloud-based model (SaaS, PaaS, IaaS) is necessary for the enterprise, as well as which model of cloud location (public, private, common, hybrid) is the most suitable for the company.

The measures for risk reduction are the following:

- data encryption by the client necessary to avoid an unauthorized access of the cloud provider personnel;
- development of the strategy of returning to a normal position unless the business of the cloud provider is successful;
- a file backup of data / audit tracking by the client in one's company provided that access to the cloud-based service is lost;
- clearly and perfectly formulated SLA (Service Level Agreement), including a paragraph concerning the rights of auditing;
- compilation and implementation of the internal recovery plan after the emergency.

Preliminary analyzed expenditures include the following items:

1. expenditures on migration from the model available to the cloud-based one (design of applications, data reformatting according to the norms of SaaS provider, adjustment of collective control over identity and access, implementation of cloud controlling).
2. Expenditures on work with a cloud-based model (payments to provider, expenditures on licenses and support, data transfer).
3. Non-recurring and permanent expenditures on the decrease of risks (tools of data encryption, planning and testing of strategies to return to thenormal position, maintenance of the data file backup independently of the provider).

In the course of consideration of other cloud-based models one should take into account the following:

– whether the expenditures will decrease if the model of provision/location of cloud-based services is changed;

– the application of private, public or hybrid cloud will not allow one to take into account any safety measures needed in public cloud;

– the expenditures on a decrease of the risk caused by binding to a definite producer can be reduced due to replacement of SaaS by PaaS or IaaS.

The second stage '**Assessment of expenditures and economic benefits**' concerns the assessment of the model available: one identifies the model, which is currently used to provide services according to functional and legal requirements of business (identified at the first stage). When the risks of the available model of services provision are assessed, the following issues are determined:

1. Spheres, where the risk exceeds its acceptable level for the enterprise and is to be decreased.

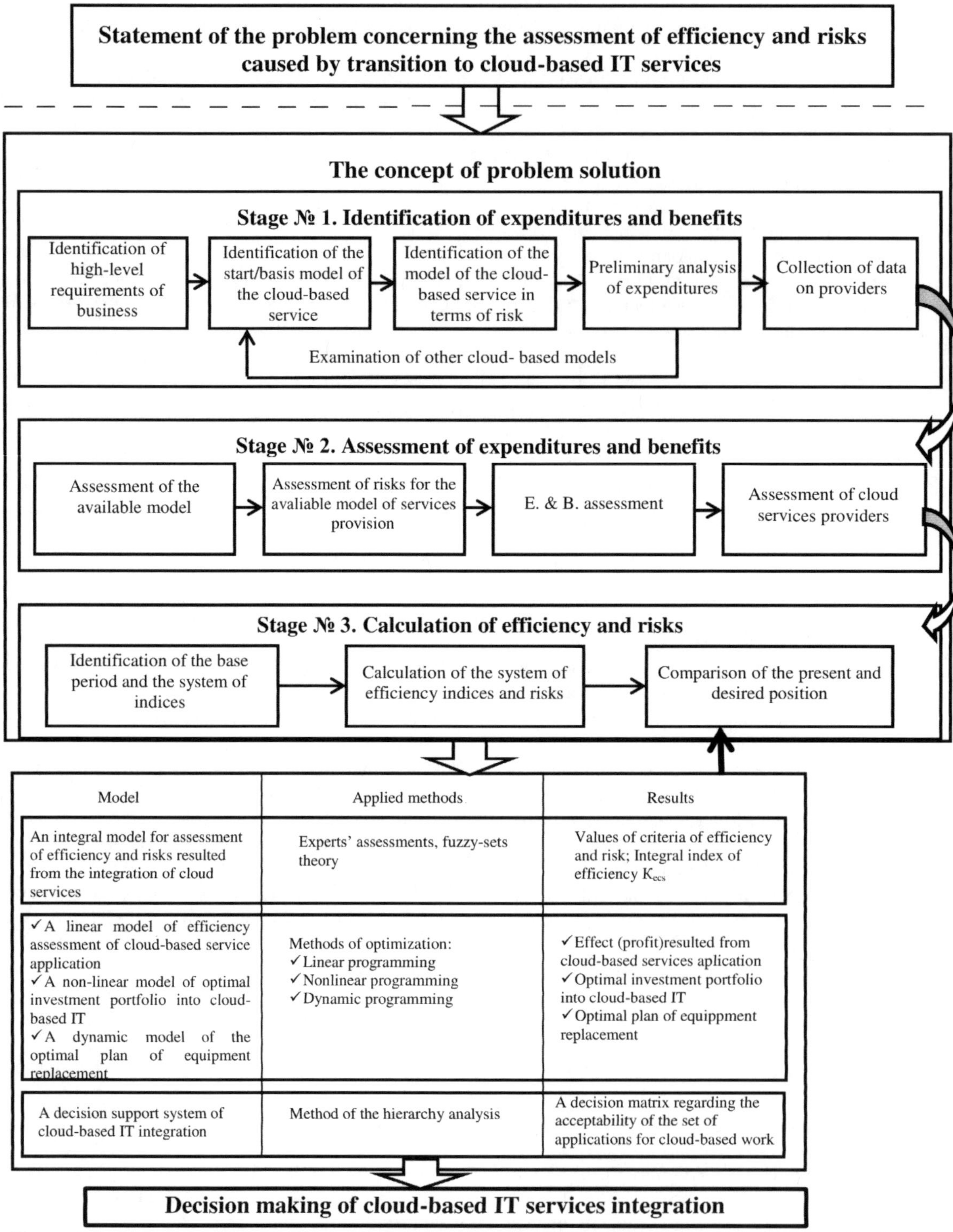

Figure 1. A system approach to the assessment of efficiency and risks caused by transition to cloud-based IT services

2. Measures, which can decrease the risk up to the acceptable level (for example, private cloud can be used in order not to share the platform with other companies; a provider company can be examined, as well as their certificates etc.).

3. Like to like comparison. Spheres of risk typical for the available technology are to be analyzed, and one must be sure that the assessment of both prospective and present positions deals with the same things [5, 8, 9].

The assessment of expenditures and economic benefits includes:

1. Operational expenditures and those for maintenance/service.
2. Latent expenditures including those on the decrease of risks.
3. Financial benefits and intangible effects.

The third stage ***'Efficiency calculation'*** is supposed to determine a base period: a planned period to use cloud-based services (a 5-year period is recommended). Then we calculate criteria and indices of efficiency and risks [9] according to suggested models (Figure 1).

The stage concerning criterion K_{ecs} calculation and risks experts and a financial department are to be engaged in work, as well as corporative standards are to be met.

We compare further the present position and the one that to be achieved:

- Numerical indices for direct and well-estimated benefits are assessed.
- Expenditures resulted from transition to the cloud are calculated.
- Expenditures and economic benefits of the present and desired position are compared.
- Net expenditures and profits for each year are calculated.

The model for calculation of quantity and quality indices for the assessment of corporate applications which can possibly migrate to the cloud-based environment, is considered in paper [1] on the basis of the method of the hierarchy analysis. Paper [7] provides the consideration of linear programming used to assess the efficiency of cloud-based technologies implementation.

Criteria and coefficient 'Efficiency of cloud-based service' are calculated according to the additive formula:

$$K_{ecs} = Is \cdot (a_1 \cdot Eb + a_2 \cdot Fb + a_3 \cdot Tp + a_4 \cdot Dr + a_5 \cdot Pf), \quad (1)$$

whereK_{ecs} – criterion 'Efficiency of cloud-based services';

Eb – the value of criterion 'Efficiency for business';

Fb – the value of criterion 'Financial advantages';

Tp – the value of criterion 'Technical priority';

Is – the value of criterion 'Reliability of work and informational safety';

Dr – the value of criterion 'Measure of risk of cloud-based service';

Pf – the value of criterion 'Psychological factor'.

a_1, a_2, a_3, a_4, a_5 – coefficients of the impact degree.

To assure compliance the criteria have a rank (weighting coefficients). Identifying the coefficients an expert has to pay attention to the range of the criteria scale and average scores of the criterion. The results of research demonstrate that there are differences between the weights determined by experts themselves and those caused by experts' actions. The weights of the most essential criteria can usually be underestimated, whereas those of non-important ones are overestimated. Therefore, when selecting weights a method of a pair-wise comparison is used to reduce subjectivism.

The algorithm for calculation ofefficiency criteria includes:

1. Comparison with required indices and standards regarding the information of the cloud-based IT-service provider. The main principle of comparison is that of compatibility of results in terms of the accepted scale of expert estimates (Table 1).

2. The expert's assessment of the degreeaccording to which cloud computing meets the requirements of safety on the basis of the score assessment and according to the scale. The score of a decimal scale from 0 to 1 is used.

3. The criterion calculation according to Formula 1.

Table 1. A scale of preferred indices.

Index value	Verbal label of an index (criterion) of cloud-based service efficiency
1	The criterion of application efficiency of cloud-based service is very high (it exceeds the standard one in two times and more)
1.00…0.75	The criterion of efficiency is quite high (it goes beyond the standard one by 50-75 %)
0.75…0.5	The criterion of efficiency is probably high (it exceeds the standard one by 75-100 %)
0.5	The mean level of the criterion of efficiency (with respect to the standard)
0.5…0.25	The criterion of the efficiency is probably low (the lag from the standard is 0-25 %)
0.25…0	The criterion of efficiency is quite low (the lag from the standard is 25-50 %)
0	The criterion of efficiency is very low (the lag from the standard is 100 %)

Conclusion

The main feature of decision making regarding the integration of cloud-based technologies is a rather high level of uncertain conditions they are made in, as well as insufficient and imprecise information necessary for the analysis. Unless the initial information is complete and of high quality a decision maker has to change the precise numerical assessments of the quality characteristics. However, the manager tends to obtain the quantity assessments of possible alternatives of development. Consideration of quantitative problems together with the qualitative ones makes the analysis more comprehensible for the analyst, the expert, and the manager.

The paper suggests a system approach to assess the efficiency and risks caused by cloud-based IT-services integration. We have developed an outline of a system approach, which includes 3 stages of the assessment: identification of expenditures and economic benefits, estimation of expenditures and benefits, calculation of efficiency and risks. The models for calculation of the system of efficiency indices and risks includes methods of the fuzzy-sets theory, mathematical programming and the method of the hierarchy analysis. The designed models relying on these methods make it possible to use qualitative expert information alongside with the quantitative information and to determine on their basis the priority of some strategies and measures when the alternatives are evaluated and decisions are made.

Acknowledgments

This work was supported by Russian Foundation for Basic Research as a part of the project 'Development of mathematical models and the software for an assessment of efficiency and risks from transition to cloudy IT services', the project № 15-01-01251.

References

[1] Razumnikov S V, Zakharova A A and Kremneva M S 2014 *Appl. Mech. and Mater.* **682** 600–605
[2] Tceplit A, Grigoreva Aand Osipov Y 2014 *Appl. Mech. and Mater.* **682** 623–630
[3] Zakharova A A2013 *Appl. Mech. and Mater.* **379** 207–213
[4] Yaojun Han, Xuemei Luo 2013 Hierarchical scheduling mechanisms for multilingual information resources in cloud computing, *AASRI Procedia* **5** 268–273
[5] Chernysheva T Yu 2013 *Appl. Mech. and Mater.* **379** 220–223
[6] Maricela-Georgiana Avram (Olaru) 2014 *Procedia Technology* **12** 529–534
[7] Razumnikov S V 2013 *Appl. Mech. and Mater.* **379** 235–239
[8] Kumar Garg S, Buyya R 2012 *Journal of Computer and System Sciences* **78(5)** 1280–1299
[9] Latifa Ben Arfa Rabaia, Mouna Jouinia, Anis Ben Aissab, Ali Mili 2013 *Journal of King Saud University - Computer and Information Sciences* **25(1)** 63–75

IOP Conf. Series: Materials Science and Engineering **124** (2016) 012090 doi:10.1088/1757-899X/124/1/012090

A new pragmatic approach to solve the problems of vector optimization with uncertain parameters

V I Reizlin and A A Nefedova
Department of Informatics and System Design, Tomsk Polytechnic University, 30, Lenina ave., Tomsk, 634050, Russia

E-mail: vir@tpu.ru

Abstract. In this paper, we consider the method for solving problems of multi-criteria optimization with mathematical models that contain a lot of variables. The values of these variables are not regulated by a decision-maker. We introduce a new concept of 'tolerance of the decision variant'. This concept is similar to such concepts as stability, survivability, etc.

1. Introduction

Any decision-making situation is characterized by the following common elements:

a) The set of variables, the values of which are chosen by a decision-maker (hereinafter – DM). Such variables are called **variants of a solution** or easy variants.

b) The set of variables, which are not regulated by DM. We call such variables **the conditions**.

c) The method evaluating quality solutions for each of the conditions. Usually it is one or more functions depending on the options and conditions.

Now let us turn to the mathematical formulation of the problem.

This paper assumes that the solutions are described by n-dimensional vectors $x=<x_1,...,x_n>\in R^n$; conditions are also vectors $p=<p_1,...,p_n>\in R^m$. The components of vectors x are **design parameters**, and the components of vectors p – **external parameters**. A pair of vectors $<x,p>$ is called **a situation**. In other words, the situation is a variant of solution x under condition p.

We also assume that the **parametric restrictions** are as follows:

$$x_i^{\wedge} \le x_i \le x_i^{\vee},\ i=1,...,n,$$

$$p_i^{\wedge} \le p_i \le p_i^{\vee},\ i=1,...,m.$$

These restrictions define rectangular block $B=X\times P$ in the parameter space $R^n \times R^m$, which we call an **original block**. Block X is the block of variants and block P is a block of parameters. In addition to parametric constraint we shall consider functional limitations $0\le g_i(x,p)\le g_i^{\vee},\ 1\le i\le r$.

In the original block they cut some curved region G, which is called a **range of possible situations**. A projection of G to the space of the design parameters is denoted by $G|X$. The set of $G|X$ is a set of alternative solutions allowing one to make a choice. Finally, the quality of situation '**Variant + conditions**' will be evaluated using objective functions (criteria) $f_i(x,p),\ 1\le i\le s$, which are defined in B. In other words, under other equal conditions the situation is better if there are more target functions.

The basic concept for multi-criteria problems is Pareto optimality. Situation $<x, p>$ is Pareto optimal if there is no other situation $<x^*, p^*>$, such that $f_i(x^*, p^*) \geq f_i(x, p)$ for all *i* and at least for one of the objective functions this inequality is strict. We need to find a variant of the solution, which generates an optimal situation for a wide variety of conditions

So, we have the following problem. In area *G*, we need to select 'best' vector *x* that defines a choice of the variant, the quality of which is in any way consistent with a set of objective functions. We have to understand what a 'best' vector is. Let us consider first the simplified version of the problem when external parameters are missing.

2. The existing approaches

The existing methods for solving the multi-criteria optimization can be divided into automatic and interactive.

2.1. Automatic methods

These methods provide the obtaining of a single compromise point (the point of the Pareto set), which is considered to be optimal. To obtain this point in automatic methods of the multi-criteria optimization, the problem is converted into a task with one criterion of efficiency. The most common methods of this type are the methods of the main criterion, the generalized criterion and goal programming [1-4].

2.1.1. A method of the main (primary) criteria. The original multi-criteria optimization problem is reduced to the task of optimizing one of criteria $f_k(x)$, which is considered to be the most important one provided that the values of other criteria should be not less than established values $f_i^{\wedge}$.

2.1.2. Methods of a generalized criterion and goal programming. These methods consist in the convolution of all criteria $f_i(x)$ in a single function, which is called a generalized criterion [5].

In these methods, DM chooses not a particular solution but the generalized criterion. Based on this criterion, the optimal solution will be selected. The obtained solution is considered to be optimal in the sense of the selected generalized criterion. Here DM separates oneself from the immediate values of criteria and can only deal with their normalized values using weighting factors that determine the importance of the criterion. Automatic methods are good because they give one unique solution. However, these methods have significant drawbacks. Consequently for the method of the main criteria it is necessary to:

- identify a set of the most important criteria, which is not possible in all practical tasks;
- set limits for all criteria except for the principal; setting these limitations is difficult for DM if they are objectively unrelated to the problem statement. In addition, when DM chooses large boundary values of the criteria, it often provides an ideal point, i.e. a non-existent one. On the other hand, small values for the constraints resulted from the multi-objective optimization problem lead to a trivial problem of maximization of the criterion, which contradicts to the formulation of the problem of the multi-objective optimization.

2.1.3. An illusion of automatic methods.

The methods of the generalized criterion and goal programming create an illusion of obtaining a single optimal solution, even though DM does not operate directly on the values of functions, but only normalized ones. It is not possible to prove the correctness of the choice of the generalized criterion. The same concerns the justification of the reason why the obtained solution better than any other, apart from the generalized criterion. Automatic methods emerged from disbelief in DM ability to choose the desired compromise. Choosing the main or generalized criterion, generally speaking, is arbitrary. Disadvantages of automatic methods have increasingly convinced us of the need for human involvement in decision-making. Therefore, in recent times interactive methods have been widely adopted.

2.2. Interactive methods
The essence of interactive methods is the same: DM implements a directed search of compromise points to select the point that best meets DM preferences. For compromise points at each step of the procedure automatic methods have been used. The procedure is over when one of the points will satisfy DM requirements. This point is considered to be optimal. However, the analysis of existing interactive methods showed the presence of a paradox associated with the assumptions about DM knowledge: on the one hand, DM decides the best compromising point, i.e. she or he has sufficient knowledge to select it, and on the other hand, it is assumed that DM is not able to clearly indicate what solution she or he wishes to obtain [6].

In [6, 7] there are methods proposed for solving the multi-criteria optimization that require DM to set the desired values of the objective functions. If the point of compromise satisfying specified requirements is absent, the methods require DM to specify setpoints criteria. On the basis of such specification a compromising region narrows until it turns into a point. The proposed procedures are quite flexible, but they demand a very large number of one-criterion extreme tasks for execution of the decision and, therefore, they are only effective for problems with a small number of criteria.

In [8] an interactive method based on sensing of the parameter space was proposed, that is, replacing of the original continuous area of its discrete representation. Moreover, in this method it is assumed that DM has an active dialog with PC. This method has significant advantages, which can be extended to the solution of problems with external parameters.

Let us consider the basic techniques used in the practice of uncertainty removal in case of problems with external parameters. In fact, there are only two.

2.3. A minimax approach
Selection is considered to be optimal if it is optimal for the worst case. In other words, external parameters are excluded from the task by replacing objective function $f_i\left(x,p\right)$ with criterion $f_i^*(x) = \min_p f(x,p)$ and the optimization problem is solved in terms of certainty. The advantage of this approach is that the obtained result is guaranteed, but only in case a solution exists. This approach is very inflexible; it requires a large amount of computation and, besides, it is hardly reasonable to plan the extreme.

2.4. A statistical approach
Selection is optimal if it is optimal average. The basis of this approach rests upon the idea of external parameters as random variables with certain probability characteristics. The uncertainty is removed if we replace the objective functions with their mathematical expectations, calculated using the known distributions of external parameters. However, for the implementation of the statistical approach it is necessary to construct a probability distribution of external parameters, which is often impossible to do.

3. A pragmatic approach
In this paper, we propose a pragmatic approach. Multi-objective optimization tasks are solved not as purely mathematical constructions, but to satisfy very specific needs of design, planning, etc. In order to achieve practical results it is often enough to find an alternative solution, which perhaps not of the best, but of a quite acceptable quality, and keep it within the widest possible set of external factors. Therefore, in each particular situation, the following construction are really meaningful.

For each objective function $f_i(x,p)$ we will assign thresholds – 'worst' quality indicators which are acceptable for DM. In other words, we assign a ***target*** or ***criteria*** constraints. In this case situation $<x,p>$ will be suitable if one all inequalities $f_i(x,p)\geq f_i^{\wedge}$ is performed.

Set $F=\{<x,p>\in G: f_i(x,p)\geq f_i^{\wedge}\}$ will be called an effective region or ***a region of efficiency***.

Let us introduce a new concept, somewhat related to such concepts as durability, stability, insensitivity, etc. Due to the fact that these terms are already engaged and widely used, we will use the word '***tolerance***' using the colors for its values, which mean 'patience', 'ability to accept', 'indifference'.

Let M be a set of non-zero volume, which lies in starting block $B=X\times P$, x is a variant of solution. Let us call set $P(x,M)=\{p\in P: <x,p>\in M\}$ as a ***set of tolerance*** for variant x on set M.

The ratio of the volume of set $P(x,M)$ to the volume of block P $T(x,M)=\dfrac{|P(x,M)|}{|P|}$ we will call ***tolerance*** of variant x on set M. The tolerance characterizes the ability of the variant to keep a situation within set M under control if the conditions will be changed. In other words, it is an indicator of the stability of options for changing of the conditions.

When keeping our problem in mind, we can assume that tolerance $T(x,F)$ is a characteristic of the ability of variant x to retain the values of objective functions inside initial region G within the limits specified by vector $f^{\wedge}=<f_1^{\wedge},\dots,f_s^{\wedge}>$. An option with the greatest tolerance is a pragmatic solution to the optimization problem.

Obviously, the variant of the decision with tolerance of 1 is approximately optimal for the minimax approach. From a statistical point of view, tolerance can be interpreted as a probability of options to keep the situation within the available area. Thus, the concept of tolerance provides the ability to generalize the minimax and stochastic approach to solving optimization problems with uncertain parameters.

Let us note that as set F is a part of the field of G, then tolerance $T(x,F)$ is a function not only of vector x, but also of vectors $f^{\wedge}$ and $g^{\vee}$, which define criteria and functional limitations, respectively.

Let functions $f_i(x,p)$ and $g_k(x,p)$ be continuous in the domain. Then function $T(x,f^{\wedge},g^{\vee})$:

a) is piecewise continuous in variable x,

b) decreases monotonically and is piecewise continuous in variable $f^{\wedge}$,

c) is non-decreasing and piecewise continuous in variable $g^{\vee}$,

d) if all functions $f_i(x,p)$ and $g_k(x,p)$ are not equal to a constant on the set of non-zero volume, $T(x,f^{\wedge},g^{\vee})$ is a continuous function.

Unfortunately, the calculation of tolerance for different variants of solutions using analytical methods is possible only for areas with a simple geometric structure, defined by algebraic equations of the first or second order. But we also need to find an option with the greatest tolerance. From the formal point of view, we have two different problems: the problem of calculating the measure of sets and the task of finding the optimum. We can try to solve both these problems, using good enough evenly distributed sequences [9-11].

We assume that we know how to build the initial portions of uniformly distributed sequences $\{X_k\}$ u $\{P_k\}$ in blocks X and P.

Let A be a perfect (coincides with its closure) subset of block P. Let us suppose also that $P_1, P_2, \ldots, P_N$ are the starting points of sequence $\{P_k\}$, and A_N is a set of those that were in A. Then we have $|A| = \lim_{N\to\infty} \frac{|A_N|}{N}|P|$. For sufficiently large values of N, we have formula $|A| \approx \frac{|A_N|}{N}|P|$. We will use this approximation to calculate tolerance.

4. Schemes for problem solving

There are three fundamentally different cases of the general problem formulation of the multi-objective optimization:

1. Neither the objective functions, no restrictions are independent of the conditions.
2. A part of the objective function depends on the conditions and the other part does not.
3. All objective functions depend on the conditions.

Let us consider each option in detail and offer a scheme of the solution of the corresponding problem.

4.1. Case 1

This is a classical problem of optimization in the conditions of certainty. We offer methods of its solution proposed in [7].

In set G, which coincides with the area of possible options, we select the N probe points $G_1, \ldots, G_N$ uniformly distributed in the X sequence. At each sampling point values for all criteria $f_1, \ldots, f_N$ are calculated. We prepare a table of tests for each criterion. In this table the values are arranged in an ascending order of quality, i.e. the worst values are located before the best values. For large N the edge values of the test table for each piecewise-continuous criterion are close to its minimum and maximum. The table of tests allows estimating not only the extreme values of the criteria, but also makes it possible to judge the frequency of certain values.

Looking through the table of tests, DM assigns threshold criteria, i.e. specifies vector $f^{\wedge}$. Once the thresholds are assigned, there is a selection of effective points – points satisfying criteria constraints. It may happen that for given N there is not a single effective point. This means that if a set of effective options is not empty, then its volume has the order of $|G| / N$. In this situation, we can proceed in two ways: either to change vector $f^{\wedge}$ easing any restrictions, or, if the restrictions do not change, to increase the number of sampling points. If in case of the repeated increase in the number of N the set of efficient options is still empty, then there is reason to believe that the restrictions are not compatible. Of course, one cannot exclude that there is an effective point, but if this is the case, then its neighborhood, in which all the restrictions are applied, has a very small volume. In this case a variant of the solution corresponding to that point will be unstable, i.e. small violations of the allowances will lead to efficiency loss.

Once having found a set of viable options, DM can be pleased with the progress considering that any alternative solution of the problem has been found. However, DM can go further by choosing a built set of points of Pareto optimal.

4.2. Case 2

This is the most general case of multi-objective optimization problems. A vector of objective functions consists of two parts: $f(x,p) = <\varphi(x), \psi(x,p)>$. In blocks X and P we construct a set of sampling points. We form tables of tests for all f and g functions counting them at the points of set $X_N \times P_K$. Let us assign a criteria and functional limitations. For criteria that do not depend on the conditions we find effective options in order to choose Pareto optimal from them, and namely for them we can calculate tolerance. An option with the greatest tolerance will be a solution.

4.3. Case 3
This is another extreme case, in some sense opposite to the first one. Here, all objective functions depend on the conditions.

In blocks X and P sets X_N and P_K of sampling points are constructed. We form tables of tests for all f and g functions calculating them at the points of set $X_N \times P_K$. Then we assign criteria and functional limitations. For each variant of the area of effective options we can calculate tolerance. A variant with the greatest tolerance will be a solution.

5. Conclusion
So, in this paper we have presented a method of solution of the multi-objective optimization with mathematical models that contain many variables, which are not regulated by the decision-maker. This technique is based on the newly introduced concept of ***tolerance of variant solution***. Schemes for three different types of tasks are presented. These procedures can be applied for rather a wide range of problems, but for the problems with a large number of criteria a huge amount of computation may be required. To remove these restrictions, it is possible to apply a technique of parallel computing, which is presented in [12].

References
[1] Podinovskii V V 1981 *The mathematical theory of decision-making in complex situations* (Moscow, Ministry of Defense of the USSR)
[2] Bekey *G A and Masri S F* 1983 *Math. and Computers in Simulation* **25(3)** 210–213
[3] Dubov J A, Travkin S I and Iakimets V N 1986 *Multi-criteria models for the formation and selection of system variants* (Moscow, Nauka)
[4] Peterson M 2009 *An Introduction to Decision Theory* (Cambridge University Press)
[5] Mashunin J K 1986 *Methods and models of vector optimization* (Moscow, Nauka)
[6] Stepanov A V 1991 *Math modeling* **3(5)** 61–73
[7] Sobol I M and Statnikov R B 1981 *Selection of optimum parameters in problems with many criteria* (Moscow, Nauka)
[8] Michalevic V S and Volkovich V L 1982 *Computational methods for the study and design of complex systems* (Moscow, Nauka)
[9] Kuipers L and Niederreiter H 1974 *Uniform distribution of sequences* (New York: John Wiley & Sons)
[10] Orlov V A and Reyzlin V I 2012 *Bulletin of Tomsk Polytechnic University* **320(2)** 24–26
[11] Reizlin V I 2014 *Proc. Int. Conf. on Mechan. Engineer., Autom. and Control Systems* 6986918
[12] Reyzlin V I and Tartakovsky E A 2012 *Proc. of 7th Int. Forum on Strategic Techn.* 6357633

IOP Conf. Series: Materials Science and Engineering **124** (2016) 012091 doi:10.1088/1757-899X/124/1/012091

A sequence of calculation of the modes of dimensional combined processing by an electrode brush

A Yu Ryazantsev[1], O N Kirillov[1], V P Smolentsev[1] and A V Totay[2]

[1] Voronezh State Technical University, 14, Moscow ave., Voronezh, 394026, Russia
[2] Bryansk State Technical University, 7, 50 years of October st., Bryansk, 241035, Russia

E-mail: ryazantsev86@mail.ru

Abstract. In the article the way of calculation of the modes of dimensional processing by an electrode brush is considered. The choice of a liquid working environment is presented. A calculation of tension in electrodes and forces of the technological current realized during processing is given. A choice of a clip of wire bunches in a processing zone, feeding an electrode brush to a non-rigid work piece. The recommended technological indicators of the process of the finishing combined treatment by an electrode brush are presented.

1. Introduction

The complexity of output products continually grows and nomenclature of their production rises. At the same time development, designing and production requirements are getting more rigid. In these circumstances the method of dimensional treatment, by means of an unshaped brush electrode with a working section made from the cantilever fitted wire, which has great elasticity, universality and ability of fast switching to processing of products of a different construction, is advantageous.

As a result of such type of processing a blank part is affected by the combined influence of erosion, chemical and mechanical processes. Their interpenetration allows implementing efficient processing of products made from high-strength and hard materials with high quality and accuracy.

2. Working medium selection

For a combined process of final treatment by means of the brush electrode it is required to choose [1] liquid mediums with low conductivity. They include: manufacturing water, lubricating and cooling fluid, dilute solutions of neutral salts (Table 1).

Experience of the use of working mediums specified in Table 1 has shown that manufacturing water has no stable properties and admixtures (for example, anticorrosion) can change the conductivity. That restricts the application of such composition for final treatment.

Lubricating and cooling mediums can be dielectric (on the basis of oils) and current-conducting and the conductivity depends on the solution concentration. An advantage of lubricating and cooling mediums is the availability of anticorrosion admixtures, for that reason medium preparation comes to concentrated fluid dilution up to required consistency [4].

For experimental works the solutions with the fixed composition and electric conductivity are used more often which allows one to reproduce experimental conditions.

Table 1. Working mediums compositions.

№	Medium composition	Specific conductivity, $Ohm^{-1} m^{-1}$	Admixtures content, %	Toxicity	Fire risk	Availability
1	Manufacturing water with admixtures $NaNO_2$	0.002…0.004	0.5…0.2	No	Nonflammable	Available
2	Lubricant and cooling mediums (Ukrinol 1, Akvol-6)	0.004…0.06	No	Insignificant	Flammable	Expenses required
3	Salt solutions (15-20% Na_2CO_3+oil I-20)	0.008…0.012	10…15	Insignificant	Nonflammable	Available

The temperature of the working medium can influence electric conductivity. As applied to final treatment by means of the brush electrode even during long-term treatment, the temperature changes insignificantly (in the range of 1 ÷ 3 K) and that allows considering it as constant.

3. Electrode voltage

An electrode voltage is the main operating characteristic of the treatment process by means of the brush electrode. In [1, 2] it is shown that the voltage in the mentioned process can be from 2 ÷ 3 V up to 12 ÷ 15 V. Such range of the variation does not allow determining the voltage required for a specific type of treatment of a single blank part.

The voltage according to requirements to the anodic dissolution process, which is performed according to the scheme with a constant inter electrode gap. Although being in the active part of the treatment process, the gap can change from S_k to 0 when the tool approaches the treatment zone up to the range of 0- S_k – at the outlet. With some allowances it is possible to accept gap (S_p) as a half of S_k ($S_p = \frac{1}{2} S_k$). Then the voltage (U) is

$$U = \frac{j_{np} S_k}{2\chi} + \Delta U , \tag{1}$$

where j_{np} – is maximum current density which can be conducted through the wire with diameter d_n. Taking into account cooling of the brass wire by the liquid medium, which is most often used for final treatment [1, 3], we obtain:

$$j_{np} = 8 \div 10 \text{ A/mm}^2$$

If a design value of the voltage meets the recommended range then the obtained value should be rounded (more often up to the value fixed in the generator) to the nearest greater value and calculations of regimes are made in accordance with these values.

4. Manufacturing current strength

Current strength (I), implemented in the process is:

$$I = j_{np} \overline{F}_0, \tag{2}$$

where $\overline{F}_0$ – open area of each beams line, mm^2.

$$\overline{F}_0 = \Pi D_n^{\,2} \cdot N_n / 4 . \tag{3}$$

Then

$$I = j_{np} \cdot \Pi D_n^{\ 2} N_n / 4. \tag{4}$$

Current strength (I_B) is:

$$I_B = \eta_B I, \tag{5}$$

where η_B – coefficient of efficiency of the current source (for electronic sources η_B=0,9÷0,95).

$$I_B = \eta_B \cdot j_{np} \cdot \Pi D_n^{\ 2} N_n / 4. \tag{6}$$

5. Selection of clamping beams (Πp) of the cable for a work piece

A secured contact of the beam end with this place is enough for realization of mechanical cleaning of the working area by the brush electrode. Clamping may be calculated by the formula taking into account tolerances on treatment details (especially non-rigid constructions):

$$\Pi p = [\delta] + \Pi p_0, \tag{7}$$

Where [δ] – the greatest value of tolerance on treatment area; Πp_0 – guaranteed clamping.

From the experience of brush electrode application it is known that cable ends may remain unbent during treatment of non-rigid details, but the cable can amortize in the process of treatment, that is why, during calculation of clamping, it should be considered:

$$\Pi p_0 = \gamma_B \cdot L_z, \tag{8}$$

where γ_B – amortization along the length of the cable relating to the depth of removed material (γ_B= 5 ÷ 15 %); L_z – length of the treatment area of the detail.

If the hardness of the cable is lower than the hardness of the treatment area of the half-finished product, then due to amortization of the cable in beam the reduction of its diameter occurs

$$\Pi p_0 = \gamma_B \cdot \sqrt{\frac{K_{0z} \cdot I_n^2 \cdot \mu_l \cdot l_n^2}{S_0^2 E}}, \tag{9}$$

where I_n – maximum current, which passes through the beam.

Then

$$\Pi p \geq [\delta] + \gamma_B \cdot \sqrt{\frac{K_{0z} \cdot I_n^2 \cdot \mu_l \cdot l_n^2}{S_0^2 E}}, \tag{10}$$

here $I_n = j_{np} \cdot \overline{S}_n$, where $\overline{S}_n$ – beam cross-section.

For circular cross-section $\overline{S}_n = \pi \cdot l_{ep}/4$,

$$I_n = j_{np} \cdot \frac{\pi}{4}\left[\frac{\chi(U - \Delta U)}{j_{\kappa p}} + \left(D_n + \frac{\chi(U - \Delta U)}{j_{\kappa p}}\right)\right], \tag{11}$$

where D_n – diameter of the beam.

6. Brush electrode feed

A speed of feed (W_n) determines all technological parameters of treatment. Reliable calculations and destination of the feed provide the required technological indexes of treatment by the brush electrode (primarily accuracy of the form and sizes and the quality of surface coat). It is necessary to remove the given allowance as a result of each pass Z_1. The removal speed of the allowance is measured as

$$V = V_a + V_{ee} \,, \tag{12}$$

where V_a – speed of anodic dissolution; V_{ee} – speed of electro erosion removal of the allowance.

$$V_a = \frac{\alpha}{\gamma}\eta\chi\frac{2(U-\Delta U)}{S_k} \,. \tag{13}$$

From [2]

$$V_{ee} = K_{ee} A_u \frac{1}{q\cdot\tau_u} \,, \tag{14}$$

where K_{ee} – coefficient defining the effectiveness of input energy of pulse A_u; q – relative pulse duration; τ_u – duration of pulse. When using experimental dimension coefficients ($\overline{K}_0$, m/Vsec) for different materials,

$$V_{ee} = \overline{K}_0 \cdot U \,. \tag{15}$$

Then removal time of the allowance is

$$\tau_n = \frac{Z_1}{V} \,. \tag{16}$$

To provide equal removal of the allowance from the whole surface (L_z) it is necessary to calculate time τ_{obr}.

$$\tau_{obr} = \frac{Z_1}{\frac{\alpha}{\gamma}\eta\chi\frac{2(U-\Delta U)}{S_k} + \overline{K}_0\cdot U} \,. \tag{17}$$

Then the speed feed is

$$W_n = \frac{L_z\left(\frac{\alpha}{\gamma}\chi\frac{2(U-\Delta U)}{S_k} + \overline{K}_0\cdot U\right)}{Z_1} \,. \tag{18}$$

7. Technological indexes of final treatment

The recommended modes of high speed dimensional treatment by the brush electrode are presented in Table 2.

Table 2. Modes of high speed dimensional treatment by the brush electrode.

Parameter name	Value
Voltage, V	4…6
The size of a tool electrode contact and a half-finished product, mm	0…0.2
Radial speed of a tool electrode, m/s	from 35
The diameter of a tool electrode, mm (selected in accordance with equipment technological characteristics)	150…350
The diameter of acable of a tool electrode working part*, mm	0.1…0.4
Operating density of abrush electrode	until 0.5
Speed of the line feed of ahalf-finished product	0.1…5
Movement of a tool electrode and a half-finished product	Opposite, following
Treatment time	Depend on the geometry of the original profile and sizes of the treatment item

* In particular cases during the roughing work of the products with high hardness and nonuniformity of the original profile, the diameter of brush electrode wires may be up to 0.8 mm.

8. Conclusion

In the article a sequence of calculation of modes for final treatment by means of the unshaped brush electrode is shown: electrode voltage, values of wire and blank part beams clamping, brush electrode supply, selection of the working medium [5]. The paper presents the recommended values of calculation that allows one to specify reasonable treatment modes.

References

[1] Kirillov O.N. 2010 *Technology of combined treatment by means of unshaped electrode.* (Voronezh: VSTU)
[2] Smolentsev E.V. 2005 *Designing of electrical and combined methods of treatment* (Moscow: Mechanical engineering)
[3] Kirillov O.N. 2010 *Reporter of Voronezh State Technical University.* **6(9)** 91-94
[4] Baron Y.M. 1986 *Magnetic abrasive and magnetic processing of products and cutting tools.* (Leningrad: Mashinostroenie)
[5] Smolentsev E.V. 2005 *The planning of electrical and combined processing methods.* (Moscow: Mashinostroenie)

The mathematical description of the gasification process of woody biomass in installations with a plasma heat source for producing synthesis gas

A R Sadrtdinov[1], R G Safin[1], M K Gerasimov[1], V I Petrov[1] and K K Gilfanov[2]
[1] Institute of Petroleum and chemical Engineering, Kazan National Research Technological University, 68, K. Marks st., Kazan, 420015, Russia
[2] Department of Automation of Technological Processes and Manufactures, Kazan State Power Engineering University, 51, Krasnoselskaya st., Kazan, 420066, Russia

E-mail: almaz.sadrtdinov@gmail.com

Abstract. The article presents the scheme of processing of plant biomass in the gasification installation with a plasma heat source to produce synthesis gas suitable for chemical industry. The analyzed physical picture of raw materials' recycling process underlies a mathematical description of the process set out in the form of the basic differential equations with boundary conditions. The received mathematical description allows calculating of the main parameters of equipment for biomass recycling and to determine the optimal modes of its operation.

1. Introduction

Researches in the field of plant biomass' recycling to chemical feedstock for further production of useful products of chemical synthesis has become widespread in recent years and it is accelerating with each year [1, 2]. It should be noted that the high growth of development of scientific activity and application of new developments in this industry has the priority point both for the economy and for the environment. Also the development of research on processing of plant raw materials is of great importance for the development of fundamental areas of knowledge, such as theory heat mass exchange processes, chemical kinetics and others.

One of the developing directions of plant raw materials' recycling is the use of gas-producing installations with a plasma heat source. It is known that various gases can serve as a plasma-forming agent: water vapour, hydrogen, air, oxygen, carbon dioxide (CO_2) and other [3]. Such diversity of gasifying agents leads to the fact that each option requires a separate technical solution, both by the design of the plasma device (plasma torch) and by the method of heat supply.

The article discusses and mathematically describes the process of plant raw materials' recycling - wood biomass on gasification installation with a plasma heat source on the basis of air for the purpose of producing synthesis gas suitable for use in the synthesis of various chemical products. The initial data were used by the source, gasifying agent, the type of plasma gas and the characteristics of the biomass such as biomass type, fractional composition, moisture content.

2. Description and formalization of the process

Considering the purpose of the work, based on the known data and on the processes occurring in gas generators, the physical picture of the process, depicted in Figure 1 has been compiled. The figure shows that the synthesis gas is formed by multistage biomass recycling in the gasifier. The resulting synthesis gas is subjected to mechanical cleaning, cooling, compressing, and it passes through the stage of catalytic synthesis of chemical products and other related stages.

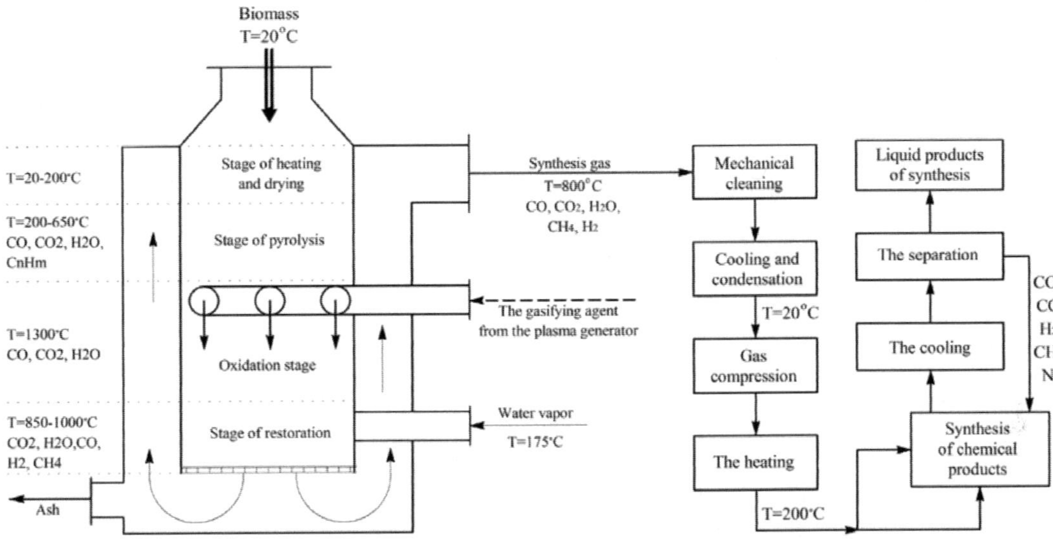

Figure 1. Diagram of the wood biomass' recycling process by the method of gasification and the physical picture of synthesis gas production

Based on the information reviewed and on the process in general, it is possible to conclude that the real picture of the process is complicated and it is imperative to make the following assumptions to receive the constructive mathematical description of the wood biomass recycling process in the gasification installation with a plasma heat source:

- a polydisperse system of particles is conditionally replaced with the equivalent monodisperse system in the form of infinite plates;
- the processes of heating, drying, pyrolysis, combustion, and recovery are discussed by stages, and an overlay of processes in time is not taken into account [4];
- there is a continuous stationary process of heat-mass exchange of the particles with the reactor wall in the area of heating, drying and pyrolysis;
- the change of a particle size during the drying process and pyrolysis is neglected;
- there are no oxidative processes on the heated boundary layer due to lack of oxidizing agents at a temperature equal to the temperature of the end of pyrolysis [5];
- the possible gas-phase chemical reactions between the pyrolysis products are not taken into account [5];
- when considering the stage of pyrolysis it is assumed that the wood consists of two main components - cellulose and lignin in a specific ratio, and the process is taken as a set of processes of thermal decomposition of its main components [4];
- the measures of convective heat transfer in the porous structure of the wood are determined by the mass flow of gaseous products of pyrolysis;
- it should be taken into consideration that there are only homogeneous reactions between gaseous pyrolysis products with an oxidizing agent occuring in the oxidation zone;
- in the reduction zone, the rate of convective transfer in the parallel to the axis reactor direction is much higher than the intensity of molecular diffusion, therefore the latter is neglected;

- the formation of methane (CH_4) in the area of recovery, is not taken into consideration as the rate of formation is significantly lower than that of other reactions [6]. As a result, the reduction zone is defined by the occurrence of the next set of reactions:

$C + O_2 = CO_2$; $2C + O_2 = 2CO$; $C + CO_2 = 2CO$;
$C + H_2O = CO + H_2$; $CO + H_2O = CO + H_2$

- when considering the recovery zone the thermal conductivity and diffusion of gas are negligibly small in comparison with convective heat and mass transfer, so the influence of diffusion and heat conduction is neglected [6];
- the resulting gas-vapor mixture in the reduction zone is subject to the laws of perfect gases due to low pressure and high temperature;
- the particles, a gas-vapor phase and an oxidant at any point in the cross section of the reactor are uniformly distributed;
- the synthesis of chemical compounds on the catalyzer occurs directly from carbon monoxide and hydrogen;
- the change in the chemical composition of the catalyzer under the influence of impurities in the initial mixture is not considered because of its pre-treatment.

Taking into account the accepted assumptions the mathematical description of the process is assuming a classic form from the system of differential equations of thermal and mass conductivity.

3. Description of mathematical model

On the basis of the physical picture (Figure 1), it seems that the stage of heating, drying and pyrolysis all together have an interconnection and in general they can be described by the differential equation of heat transfer [4]

$$c_{layer} \cdot \rho_{layer} \cdot \frac{\partial T_{layer}}{\partial \tau} = \frac{\partial}{\partial x}\left(\lambda_{ef} \cdot \frac{\partial T_{layer}}{\partial x}\right) + q_{layer} \tag{1}$$

where c and ρ –the heat capacity and density, respectively; λ – the coefficient of thermal conductivity; q – heat flow.

For the solution of the differential equation (1) the initial (2) and boundary conditions (3)÷(4) are formulated, shown schematically in Figure 2.

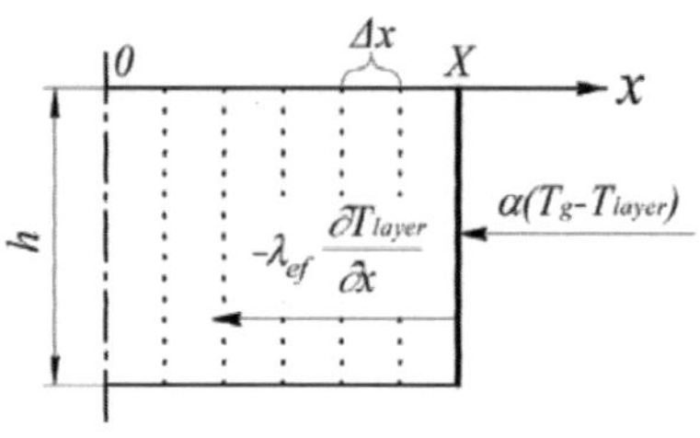

Figure 2. Scheme of boundary conditions

$$T_{layer}(0,x) = T_{layer0} \tag{2}$$

$$-\lambda_{ef}\left.\frac{\partial T_{layer}}{\partial x}\right|_{x=X} = \alpha \cdot (T_g - T_{layer}) \tag{3}$$

$$\left.\frac{\partial T_{layer}}{\partial x}\right|_{x=0} = 0 \tag{4}$$

Taking into account the accepted assumptions and some transformations the stage of drying is described by the system of equations

$$c_{layer} \cdot \rho_{layer} \cdot \frac{\partial T_{layer}}{\partial \tau} = \frac{\partial}{\partial x}\left(\lambda_{ef} \cdot \frac{\partial T_{layer}}{\partial x}\right) + r\varepsilon \frac{\partial U_{layer}}{\partial \tau} \tag{5}$$

$$\frac{\partial U_{layer}}{\partial \tau} = \frac{k_p}{\rho_0}\left(\frac{\partial^2 P_{layer}}{\partial x^2}\right) + a_m \frac{\partial^2 U_{layer}}{\partial x^2} + a_m \delta \frac{\partial^2 T_{layer}}{\partial x^2} \tag{6}$$

where k_p – is the molar transfer coefficient, a_m – is the coefficient of mass conductivity.

To solve the system of equations (5)÷(6) initial and boundary conditions for temperature, humidity and pressure can be written as follows:

$$U_{layer}(0,x)=U_{layer0} \quad (7)$$

$$T_{layer}(0,x)=T_{layer0} \quad (8)$$

$$P_{layer}(0,x)=P_{atmos} \quad (9)$$

$$\left.\frac{\partial U_{layer}}{\partial x}\right|_{x=X}=0 \quad (10)$$

In its turn, the stage of pyrolysis is significantly different from the pre-heating and drying, as it is accompanied by mass decrease and the formation of the gas phase [1]. Based on the above-mentioned, the stage of pyrolysis can be described by a system of equations of energy conservation

$$c_{layer}\cdot m_{wood}\frac{\partial T_{layer}}{\partial \tau}=\frac{\partial}{\partial x}\left(\lambda_{ef}\cdot\frac{\partial T_{layer}}{\partial x}\right)-m_{wood}\cdot c_{layer}\cdot w_{wood}\cdot\frac{\partial T_{layer}}{\partial x}+q_{chem.reac} \quad (11)$$

and mass change, the assumptions may have the following form:

$$\frac{\partial m_{cell}}{\partial \tau}=-k_{cell}m_{cell} \quad (12)$$

$$\frac{\partial m_{lignin}}{\partial \tau}=-k_{lignin}m_{lignin} \quad (13)$$

$$\frac{\partial m_{coal}}{\partial \tau}=\gamma_{lignin}(-k_{lignin}m_{lignin})+\gamma_{cell}(-k_{cell}m_{cell}) \quad (14)$$

$$\frac{\partial(m_{vg}\varepsilon)}{\partial \tau}=-\frac{\partial(w_{vg}m_{vg})}{\partial x}+(1-\gamma_{lignin})k_{lignin}m_{lignin}+(1-\gamma_{cell})k_{cell}m_{cell} \quad (15)$$

where w – the speed; k – the reaction velocity constant; γ – mass fraction of the substance; ε – the porosity of the material, determined from the expression [7]

$$\varepsilon=\left(1-\frac{m_{wood}}{m_{wood0}}\right)\cdot\left(1-\varepsilon_{wood}\right) \quad (16)$$

The boundary conditions for solving the equation (11) are similar to the stages of pre-heating and drying (Figure 2) and are defined by equations (3)÷(4). To solve the system of equations (12)÷(15) the boundary conditions can be written as (17)÷(20), and the initial conditions take the form (21)÷(25).

$$\left.m_{lignin}\right|_{X=0}=m_{lignin0} \quad (17)$$

$$\left.m_{cell}\right|_{X=0}=m_{cell0} \quad (18)$$

$$\left.m_{coal}\right|_{X=0}=m_{coal0} \quad (19)$$

$$\left.m_{vg}\right|_{X=0}=m_{vg0} \quad (20)$$

$$T_{layer}(0,x)=T_{layer0} \quad (21)$$

$$m_{cell}(0,x)=m_{cell0} \quad (22)$$

$$m_{lignin}(0,x)=m_{lignin0} \quad (23)$$

$$m_{coal}(0,x)=0 \quad (24)$$

$$m_{vg}(0,x)=0 \quad (25)$$

Consideration of oxidation stage with the accepted assumptions taken into account is reduced to some simple mathematical expressions, in particular the determination of the theoretically required consumption of oxidant, the temperature of the gas mixture combustion and a quantitative yield of combustion products - CO_2 and H_2O.

A theoretically required amount of oxidant to provide the combustion process at the stage of oxidation can be determined by the formula [4]:

$$L_0 = \frac{1}{0{,}21} \cdot \left[0{,}5 \cdot (CO + H_2) + 2CH_4 + \left(n + \frac{m}{4} \right) \cdot C_n H_m - O_2 \right] \tag{26}$$

And the consumption of the theoretical amount of oxidizing agent is calculated from the equations of material and thermal balance

$$\frac{\partial T_{layer}}{\partial \tau} = \sum q_{chem.reac} \cdot k_i(C_{i_0} - C_i) - w_{vg} \frac{\partial T_{layer}}{\partial x} \tag{27}$$

$$\frac{\partial C_i}{\partial \tau} = w_{vg} \frac{\partial C_i}{\partial x} - k_i(C_{i_0} - C_i), \tag{28}$$

where C_i – is the concentration of the *i*-substance.

Knowing the consumption of the theoretical amount of oxidant and the content of the oxidizing agent in the gasifying agent (plasma gas) it is possible to calculate the total quantity and the required consumption of the gasifying agent supplied to the gasifier. In consequence it is also possible to calculate the amount of heat energy to be carried by a plasma gas for the gasification process and to provide the required operating parameters.

The consumption of the gasifying agent supplied to the gasifier is calculated using the oxidizer (oxygen) provided that the gasifying agent consists of oxygen (O_2), hydrogen (H_2) and water vapor in the percentage ratio by volume of 23÷46÷31 respectively. The heat flow coming together with the gasifying agent is calculated from the law of energy conservation. For clarity, a design scheme is made (Figure 3), according to which the estimated equations have the following form:

$$Q_1 = Q_2 = Q_3 \tag{29}$$

$$Q = c_{wood}.m_{wood}.\Delta t \tag{30}$$

$$Q_1 = -\lambda_{layer}(t_3 - t_4)l_{layer} \tag{31}$$

$$Q_2 = -\lambda_{wall}(t_2 - t_3)l_{wall} \tag{32}$$

$$Q_3 = \alpha(t_1 - t_2) \tag{33}$$

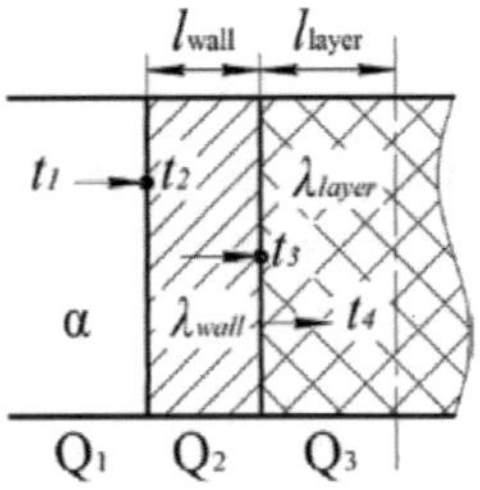

Figure 3. Design scheme of heat flows

where m_{wood} – the mass of the wood; t_1 – the temperature of the gasifying agent, t_2 – the temperature of the outer wall, t_3 – the temperature of the inner wall, t_4 – the temperature of the material.

The recovery phase, taking into account the adopted assumptions, can be written with the system of equations of conservation of matter and energy for a gas stream and carbon residue in the form of

$$w_{stream} \cdot \frac{\partial C_i}{\partial y} = -k_i \cdot C_i \tag{34}$$

$$w_{coal} \cdot \frac{\partial m_{coal}}{\partial y} = -m_{coal} \cdot \sum_{m=1}^{z} k_i \tag{35}$$

$$\rho_{stream} c_{stream} w_{stream} \frac{\partial T_{stream}}{\partial y} = -\alpha_{stream}(T_{coal} - T_{stream})f + \sum_{i=1}^{z} q_i \cdot k_i \cdot(C_{i0} - C_i) \tag{36}$$

$$\rho_{coal} c_{coal} w_{coal} \cdot \frac{\partial T_{coal}}{\partial y} = \alpha_{stream} \cdot(T_{coal} - T_{stream})f - \sum_{i=1}^{z} q_i \cdot k_i \cdot(C_{i0} - C_i) \tag{37}$$

To solve the system of equations (34)÷(37) the boundary conditions on temperature and mass are taken:

$$T_{coal}\big|_{y=0} = T_g\big|_{y=0} \quad (38) \qquad C_i\big|_{y=0} = C_{i0} \quad (39)$$

$$m_{coal}\big|_{y=0} = m_{coal0} \quad (40)$$

When describing the stage of catalytic synthesis of chemical products the equation of consumption of the components of the synthesis gas can be used, it will have the following form:

$$w\frac{\partial C_i}{\partial y} = -k_i \cdot C_i . \quad (41)$$

And the equation of energy conservation will assume the form

$$w \cdot \rho \cdot c \cdot \frac{\partial T_{syngas}}{\partial y} = \sum_{i=1}^{z} (q_i \cdot k_i (C_{i_0} - C_i)) . \quad (42)$$

The time, during which the synthesis gas contacts with the catalyzer (t, sec) at a constant pressure, is calculated according to the formula [8]

$$\tau = \frac{V_k}{V_g} \quad (43)$$

where V_k – the volume of the catalyzer; V_g – the volume of the reaction gas (under normal conditions) passed through the layer of catalyzer per unit time.

4. Conclusions

The mathematical description of the process of wood raw material recycling in the synthesis gas, is presented in the form of the basic differential equations with boundary conditions. Modeling the process according to this mathematical description will allow the following: to calculate the main parameters of the gas generators with plasma heat sources, to determine the optimal modes of their operation and to obtain a synthesis gas suitable for chemical industry.

Acknowledgments
The presented materials are received during the Presidential grant program realization of the Russian Federation under the state support of young Russian scientists. The cypher of the research topic is MK-3434.2015.8 'Development of theoretical foundations, technologies and equipment of integrated thermochemical processing of waste wood and plant biomass into raw materials for chemical synthesis and components of motor fuels'.

References

[1] Tuntsev D V, Filippova F M, Khismatov R G and Timerbaev N F 2014 *Russian Journal of Applied Chemistry* **87** 1367-1370
[2] Safin R G, Timerbaev N F, Sadrtdinov A R and Prosvirnikov D B 2013 *Bulletin of Kazan Technological University Russian Journal* **16** 250-254
[3] Anshakov A S, Urbakh E K, Urbakh A E and Faleev V A 2008 *International Scientific Journal for Alternative Energy and Ecology* **65** 131-135
[4] Timerbaev N F, Safin R R, Safin R G and Ziatdinova D F 2014 *Journal of Engineering and Applied Sciences (EAAS)* **9** 141-146
[5] Safin R R, Khasanshin R R, Shaikhutdinova A R and Safina A V 2014 *World Applied Sciences Journal* **30** 1618-1621
[6] Khuzeev M V, Sattarova Z G and Petrov V I 2014 *Bulletin of Kazan Technological University Russian Journal* **17** 94-96
[7] Safin R R, Khasanshin R R, Timerbaeva A L and Safina A V 2015 *Russian Engineering*

Physics Journal **88** 925-928
[8] Filippova F M, Spiridonova R R, Kochnev A M and Samuilov Y D 2013 *International Polymer Science and Technology* **40** 14-17

Calculation methodology of the heat pump in the process of oscillating vacuum-conductive drying of lumber

R R Safin, R R Khasanshin, A R Shaikhutdinova and I F Khakimzyanov
Kazan National Research Technological University, 72, Karla Marksa st., Kazan, 420015, Russia

E-mail: olambis@rambler.ru

Abstract. The oscillating technologies consisting in alternating of the stage of heating of the material and vacuumization are the most advanced in the process of wood drying. In this regard, the article examines the energy-saving technology of the oscillating vacuum-conductive drying of lumber, during which the thermal energy of the moisture evaporated from the material under vacuum in one chamber by using the heat pump is transferred to the heating of the material in the other chamber. The authors develop the method of calculating the rate of removal of moisture from the heated material at the stage of vacuumization depending on the depth of vacuum, temperature, humidity and thickness of the material, which is the initial condition for calculating the heat pump.

1. Introduction

Drying the materials carried out under reduced pressure allows obtaining a high quality of dried wood and shortening of the process duration. Thus, despite the undeniable advantages, vacuum drying has a problem with the supply of thermal energy to the dried material [1]. The familiar drying technologies existing in various industries, such as microwave, convection, radiation methods do not always allow obtaining a material with the required quality and little energy consumption. Therefore the development of the technologies for accelerated drying with the lowest energy consumption is an important issue [2].

Thus in terms of reducing the energy costs for drying processes most attention is paid to the use of alternative energy sources, particularly waste wood in recent years [3].

At the same time the energy saving techniques well known in other industries, applied to woodworking industry are still not widely used or are of a single character. So, for example, heat pumps used in convective dryers have not received further development and application in other methods of drying, in particular in vacuum technologies [4, 5]. Conducting the process in a sealed condition allows the most effectively catching of heat energy of moisture evaporating from the material. However, the difficulty lies in the fact that the smallest duration of the process of vacuum wood drying is provided by the cyclic modes, consisting of the stages of heating and vacuuming. At the stage of heating, there is the accumulation of heat energy by the material, and at the stage of vacuumization there is the removal of moisture from the material only at the expense of energy previously accumulated by the material, that is, heating the material at this stage is not performed. Such management of the drying process ensures the positive effect of gradients of temperature, pressure and humidity inside of the material, providing accelerated drying without the significant internal stresses. Thus, the energy captured by the heat pump at the stage of vacuumization cannot be directed to the heating of the material at that moment [6].

2. Materials and methods

A programmable vacuum drying oven Memmert 400 was used for experimental studies of the drying rate of lumber heated under vacuum. The input data at the experiment planning were such factors as the residual pressure at the stage of vacuumization, temperature, thickness, average moisture content and basic density of the material. In order to improve the accuracy of calculation output value (the speed of drying), depending on various factors, the process of moisture removal is considered in two intervals: sample drying with moisture, which is above the saturation limit of the cell walls (80-40%) and sample drying with the moisture content, which is below the saturation limit of the cell walls (30-10%).

With the aim of reducing the energy consumption in the processes of vacuum lumber drying the technology of oscillating vacuum-conductive drying with the use of a heat pump was developed. The feature of the proposed technology is the transfer of thermal energy of the moisture evaporated from the material under vacuum in one chamber for heating the material in the other chamber [7, 8].

On the basis of this technology, which allows rational using of energy resources, the technological complex was developed, in which the waste wood, accumulating in large volumes in woodworking industry, can be used as fuel [9]. The process of this complex consists of the following operations: gasification of wood waste to produce synthesis gas, followed by cooling; combustion of synthesis gas in the internal combustion engine to produce mechanical energy to carry out the operation of the compressor of the heat pump [10].

For experimental studies of the drying process according to the developed technology the installation was created (Figure 1).

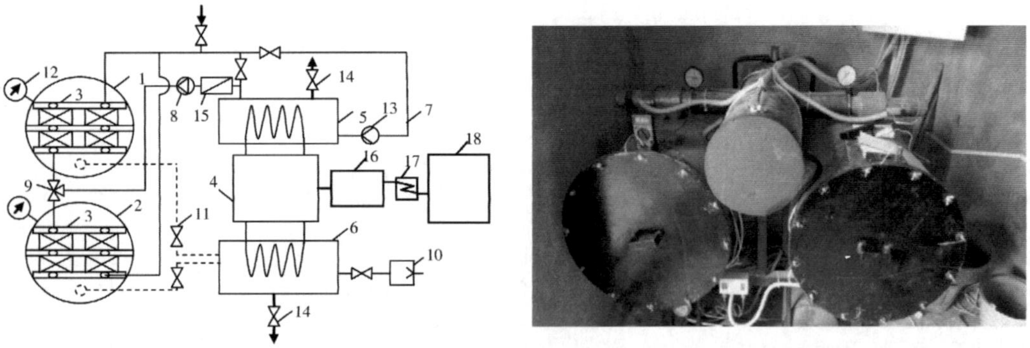

Figure 1. Experimental setup for the study of oscillating vacuum-conductive drying of lumber: 1 - top chamber; 2 - lower chamber; 3 - plate with liquid heating; 4 - heat pump system; 5 - condenser; 6 - evaporator; 7 - pipeline; 8 - circulating pump; 9 - three-way valve; 10 - vacuum pump; 11 - electromechanical valve; 12 - vacuum gauge; 13 - flow meter; 14 - valve; 15 - additional electric heater; 16 - internal combustion engine; 17 - cooling of the synthesis gas; 18 - gasifier

Recently studied cut wood samples are placed in the chambers on the heated plates. Thus the flanks of the wood are painted over with the last quick-drying paint to prevent the intensive removal of moisture from it. The rest of the chambers are sealed with flaps, and after turning on an additional heater and coolant circulation in a small circle, heating of the sample in the first chamber starts. Wood heating continues until the temperature inside of the sample reaches a predetermined value. After heating the first chamber is vacuumized and lumber is kept until it is cooled [11]. Thus the moisture evaporated from the dried material condenses on the evaporator of the heat pump, transfers the thermal energy to the refrigerant, due to what the stage of material heating in the second chamber is realized.

IOP Conf. Series: Materials Science and Engineering **124** (2016) 012093 doi:10.1088/1757-899X/124/1/012093

The duration of the vacuumization stage in one chamber is determined by the duration of material heating in the second one. Furthermore, there is a simultaneous change of stages in the chambers. Drying is carried out until the moisture content of lumber reaches the finite value [12].

3. Results

As the result of the experimental studies of drying rate of lumber heated at vacuum the following regression equations were obtained:

– for the moisture content above the saturation limit of the cell walls:

$$N_1 = -0.06416W - 0.08354T + 0.006635P - 0.0000415TP - 0.000076PW + 0.00142TW + 3.7666, \quad (1)$$

– for the moisture content below the saturation limit of the cell walls:

$$N_2 = -0.001473W - 0.000341\rho - 0.001152S - 0.0000097TP - 0.00001875TW + 0.0000112TS + 0.0000187WS + 0.000018\rho S + 0.000077T^2 - 0.0000282P^2 + 0.00000067\rho^2 - 0.0000368S^2 + 0.303571. \quad (2)$$

where: W – humidity, (%); T - temperature; (° C); P – pressure, (kPa); ρ – reference density, (kg/m^3); S – thickness of lumber, (mm).

Based on the experimental data, kinetic curves of temperature, pressure in the chamber and the humidity of the wood in the process of vacuum-oscillating drying were constructed (Figure 2).

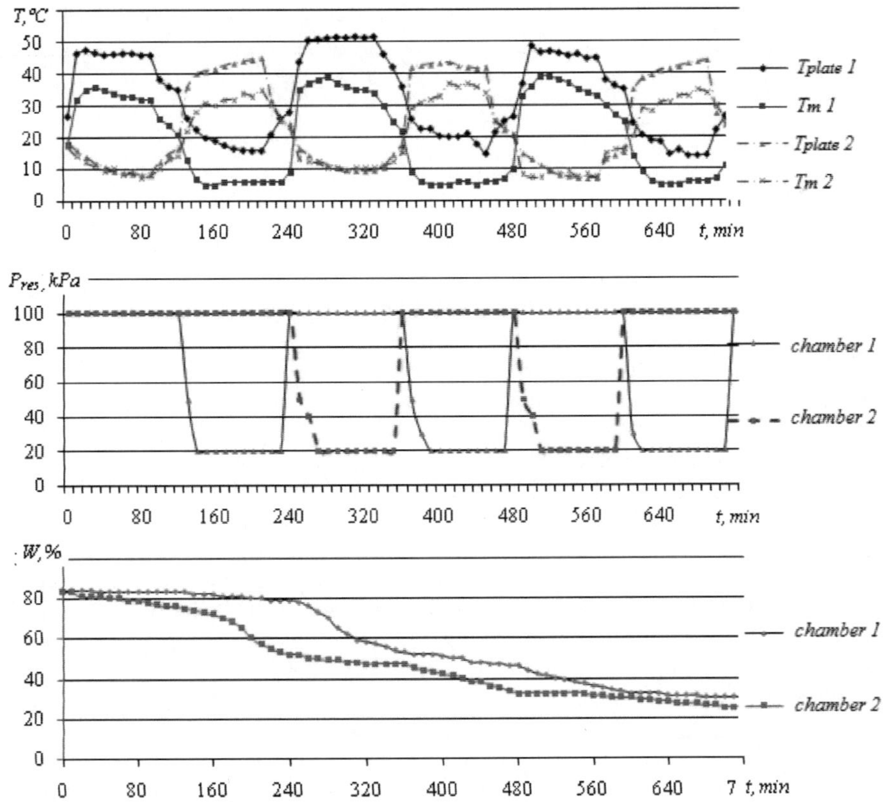

Figure 2. Graphs of the process of oscillating vacuum-conductive drying: a - temperature curve; b – kinematic dependence of the pressure in the chamber; c – drying curve

The studies have shown that the transfer of heat from one chamber to another occurs with partial loss of heat, due to which the optimal parameters of the drying process are retained and the loss of thermal energy must be compensated by the additional heat. The amount of thermal energy for compensating of the heat loss is presented in Figure 3.

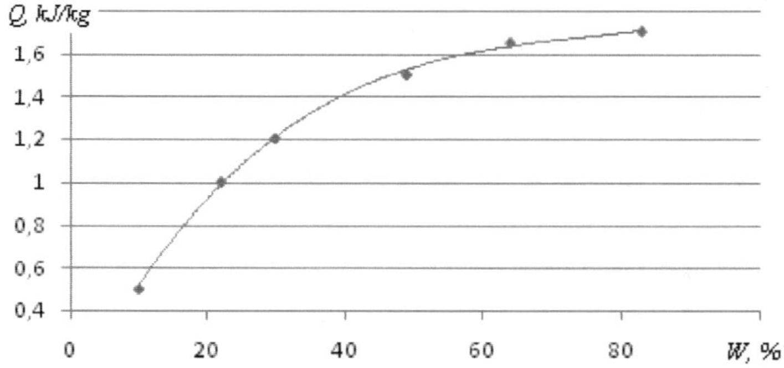

Figure 3. Indicators of additional thermal energy for offsetting heat losses

It was found that a slight reduction of the required additional energy with the decreasing moisture content of the material happens because of the duration reduction of the heating stage due to the increase in the thermal diffusivity of the material.

As a result of the series of experiments it was found that the duration of the vacuumization stage in the drying process increases, while the duration of the heating stage, on the contrary, decreases (Figure 4).

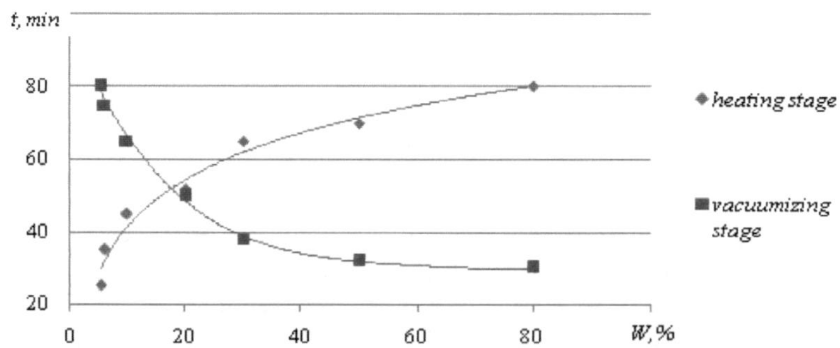

Figure 4. Dependence of duration of heating and vacuumization stages on moisture

The stages change in the chambers should be realized simultaneously and the current moisture content of the material at the vacuumization stage has been studied. Moreover, the potential for further material drying at the vacuumization stage, at the moment the material under heating achieves a predetermined temperature, has been analyzed. It has been established that at this point in time dehumidification of material under vacuum reaches 87 % of the maximum possible value. For this reason it was decided to hold the change of stages when the material reaches the set temperature at the heating stage, which, in turn, reduces the total duration of the drying process of wood material with a slight increase in the number of cycles of 'heating-vacuumization'.

To confirm the conducted research the curves of the dependence of the amount of dehumidification on the duration of the vacuumization stage were built (Figure 5).

With the aim of obtaining data on the energy efficiency of the use of heat pump in the process of vacuum-oscillating drying of wood materials, a comparison of the energy consumption of the established installation for various types of wood with electric and heat pump heating (Figure 6) was conducted. From the graphs we can see that the drying with an electric supply of heat takes place with a significantly higher power consumption (about 3 times). Thus, it has been found that the use of the heat pump in the process of oscillating vacuum-contact drying of wood-based materials is a promising direction of efficient use of heat and electricity.

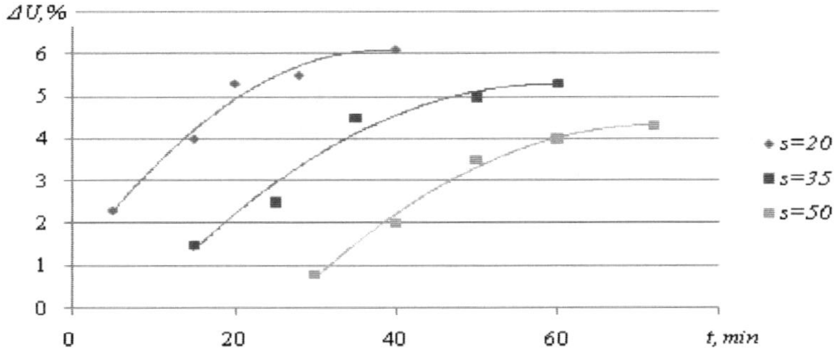

Figure 5. The dependence of the amount of dehumidification on the duration of the vacuumization stage

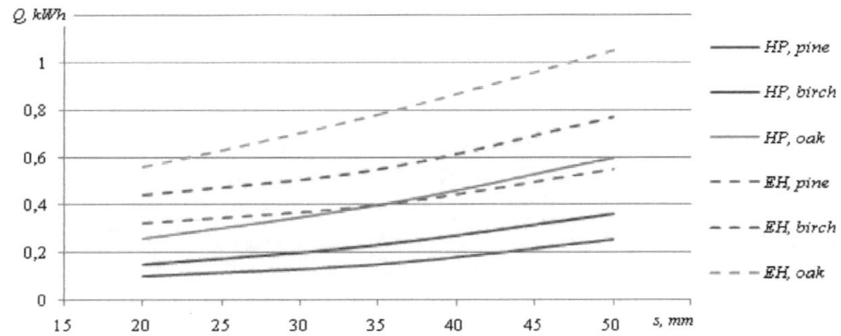

Figure 6. The energy consumption of the installation of the oscillating vacuum-conductive drying for different species of lumber

Taking into account the experimental data of the oscillating vacuum-conductive drying an engineering calculation methodology of the pilot plant using a heat pump was developed.

The data for calculation are as follows:

– heat load Q_{hp} is determined by the rate of lumber drying in the vacuum process:

$$Q_{hp} = N \cdot V_m \cdot \rho_m \cdot r, \tag{3}$$

– low-grade temperature of the coolant at the evaporator inlet is equal to the temperature of the water vapour evaporated from the material, which is located in free space of the drying chamber:

$$T_{l\text{-}g1} = T_v = (B/(A - \lg P)) - C, \tag{4}$$

where: N – speed of drying, (%/min); V_m – volume material of timber, (m^3); r – latent heat of vaporization, (j/kg).A, B, C – coefficients of the Antoine equation.

– the temperature of a high-potential heat carrier (hot water) at the inlet of heat pump $T_{h\text{-}p1}$ and after heat pump $T_{h\text{-}p2}$ is set not less than 15 °C and 5 °C respectively above the temperature of lumber after the heating stage, which was used earlier when calculating the speed of material drying:

$$T_{h\text{-}p2} = T_m + (15 \div 20), \tag{5}$$

$$T_{h\text{-}p1} = T_m + (5 \div 10), \tag{6}$$

The pilot plant with application of the heat pump, the energy efficiency indicators of which are presented in table, is designed according to the developed engineering methodology.

Table 1. The index of the energy efficiency of the estimated heat pump

Heat load density of the heat pump q_{hp}, (kJ/kg)	146
Specific energy consumed by the motor W, (kJ/kg)	27,63
The degree of compression in the compressor ε	7,10
The conversion factor of heat μ	6,95
The conversion factor of electricity $\mu_э$	5,28
Specific consumption of primary energy	0,49

The efficiency of a heat pump with electric drive is determined by the ratio of heat transferred to the hot coolant, to the electric power supplied, to the drive motor of the compressor (the conversion factor of electricity μe) and condition $\mu_e > 1$.

4. Conclusions

The complex, based on the consumption of wood waste by gasification to produce synthesis gas for the operation of the internal combustion engine, bringing heat pump of vacuum drying equipment to action has been developed.

According to recent experimental studies it has been determined that the changes in stages should be carried out at the heating stage when the material achieves the set temperature. The result duration of the drying process reduces with a slight increase in the number of cycles of 'heating-vacuumization'. The proposed technology allows almost three-times reduction in the consumption of energy resources and an increase in the efficiency of the process of lumber drying.

References

[1] Hermawan A, Fujimoto N and Sakagami H 2013 *J. Drying Technology* **31** 587-594
[2] Perre P, Almeida G and Colin J 2011 *Modern Drying Technology* **4** 245-293
[3] Timerbaev N, Safin R, Safin R and Ziatdinova D 2014 *Engineering and Applied Sciences* **16** 141-146
[4] Safin R, Lashkov V, Golubev L and Safin R 2002 *J. Engineering Physics and Thermophysics* **75** 384-389
[5] Vasile M 2015 *J. Drying Technology* **33** 527-540
[6] Chua K, Choua J and Hoa M 2007 *J. Drying Technology* **20** 1579-1610
[7] Vasile M 2012 *J. Drying Technology* **30** 1630-1643
[8] Vasile M 2008 *J. Drying Technology* **26** 1373-1381
[9] Shen B, Mahderekal I and Vineyard E 2013 *HVAC&R Research* **19** 847-856
[10] Vasile M 2013 *J. Drying Technology* **31** 1177-1189
[11] Chua K, Chou S and Yang W 2010 *Applied Energy* **87** 3611-3624
[12] Yang Z, Zongsheng Z and Enlong Z 2013 *J. Drying Technology* **31** 1049-1055

Definition of Locked-up Stresses around a Rectilinear Welding Seam

N L Velikanov, S I Koryagin and O V Sharkov
Immanuel Kant Baltic Federal University, 14 A, Nevskogo st., Kaliningrad, 236041, Russia

E-mail: osharkov@ kantiana.ru

Abstract. In article one of the ways of modelling of the locked-up stresses caused by imposing of a welding seam in a plate is described. These stresses are caused by residual deformations of welding seams. In the settlement scheme the welding seam is presented in the form of an element by the thickness equal to the thickness of a plate, inserted in it with a longitudinal and cross-section tightness. The field of locked-up stresses around the welding seam experiences longitudinal and cross-section shrinkage under cooling and it is defined in work with the use of complex potentials of the theory of elasticity.

1. Introduction

For the definition of locked-up stresses numerical modelling, static methods, hypothesis of flat sections [1-10] is used.

Settlement-experimental methods have found wide application: a variation with use of the common decisions of the theory of elasticity and experimental data of nondestructive tests [3, 7, 8]; a finite element method with use of self-regulation [3, 6, 9], measurements by methods of neutron and x-ray diffraction [10], experimental data about distribution of temperatures and phase transformations during welding [1, 2, 4, 8].

In a number of works the regional problem for not self-counterbalanced locked-up stresses is put, the mechanism of formation of technological residual pressure is considered. An effective analytical method of the definition of locked-up stresses is application of methods of the mathematical theory of elasticity [5, 7]. In particular, reductions to a problem about an insert of bodies with tightness.

2. An analytical model

In the settlement scheme, welding seam 1 (Figures 1 and Figures 2) is presented in the form of an element by the thickness equal to the thickness of a plate, inserted in it with longitudinal $2\Delta_1$ and a cross-section $2\Delta_2$ tightness.

The field of locked-up stresses around the welding seam is caused by its longitudinal and cross-section shrinkage at cooling and is defined in work with the use of potentials of the theory of elasticity:

$$\left.\begin{aligned}\varphi(z) &= \frac{G}{\pi i(\chi+1)}\int_L \frac{g(t)}{t-z}\,\mathrm{d}t\ ,\\ \psi(z) &= -\frac{G}{\pi i(\chi+1)}\int_L \frac{\overline{g(t)}+\bar{t}g'(t)}{t-z}\,\mathrm{d}t\ .\end{aligned}\right\} \qquad (1)$$

where $z = x + iy$ – co-ordinate of a point of a plate; $i = \sqrt{-1}$ – imaginary unit; G – module of shift of a material of a plate; $\chi = (3\mu)/(1+\mu)$ – elastic constant connected with factor of Puassona μ; t – co-ordinate of the point located on the border of a seam; $g(t)$ – jump of moving on the border of welding seam L.

Values of jump of moving on the border of the welding seam $g(t)$, its interfaced values $\overline{g(t)}$ and combinations $t\ g'(t)$ depend on longitudinal and cross-section shrinkage of the welding seam (see table 1).

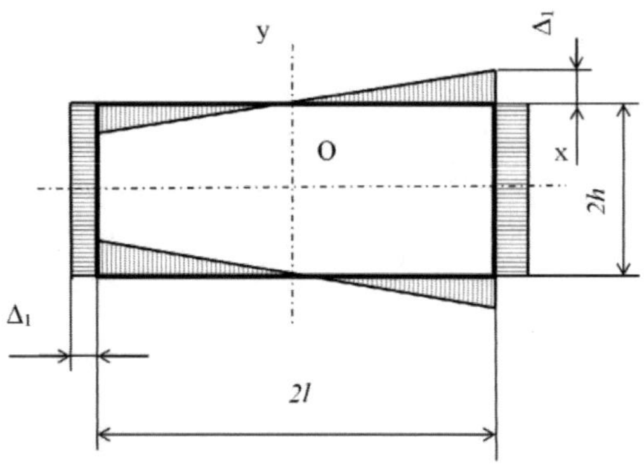

Figure 1. A distribution diagram of longitudinal moving on the seam border.

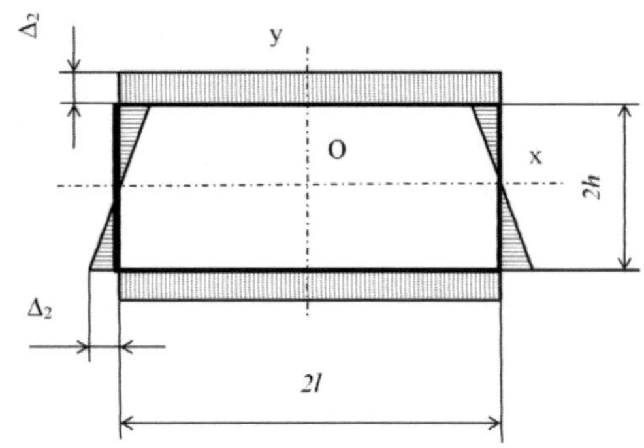

Figure 2. A distribution diagram of cross-section moving on the seam border.

Table 1. A distribution of the jump of moving to the border of the welding seam.

Jump of moving	Line of the welding seam	
	Upper	Lower
t	$x+ih$	$x-ih$
$g(t)$	$(\Delta_1 x/l)+i\Delta_2$	$(\Delta_1 x/l)-i\Delta_2$
$\overline{g(t)}$	$(\Delta_1 x/l)-i\Delta_2$	$(\Delta_1 x/l)+i\Delta_2$
$t\ g'(t)$	$\Delta_1(x-ih)/l$	$\Delta_1(x+ih)/l$

Considering values of the jump of moving, and also accepting $h << 1$, using formulas (1) we will receive the following values of potentials:

$$\left.\begin{aligned} \varphi(z) &= \frac{2G(\Delta_1 h - \Delta_2 l)}{\pi d(\chi+1)} \ln\frac{l-z}{l+z} , \\ \psi(z) &= \frac{-2G}{\pi d(\chi+1)}\left[(3\Delta_1 h + \Delta_2 l)\ln\frac{l-z}{l+z} + \frac{2lz(\Delta_1 h + \Delta_2 l)}{z^2 - l^2}\right] . \end{aligned}\right\} \quad (2)$$

The locked-up stresses expressed through complex potentials (2) using formulas of the theory of elasticity

$$\left.\begin{aligned} \sigma_x &= R_e[2\varphi'(z) - \bar{z}\varphi''(z) - \psi'(z)] , \\ \sigma_y &= R_e[2\varphi'(z) + \bar{z}\varphi''(z) + \psi'(z)] , \\ \tau_{xy} &= I_m[\bar{z}\varphi''(z) + \psi'(z)] . \end{aligned}\right\} \quad (3)$$

where: σ_x, σ_y – normal stresses, τ_{xy} – tangential, Re designates a real part, I_m – imaginary frequency.

3. Numerical results

Initial data for calculation: $2l = 40$ cm – length of the welding seam, $2h = 1$ cm – width of the welding seam, $\Delta_1 = 0.24$ cm – longitudinal shrinkage of the welding seam, $\Delta_2 = 0.24$ cm – cross-section shrinkage of the welding seam, $E = 2\cdot10^5$ MPa – module of elasticity of the material of the plate, $\nu = 0.3$ – factor of Puasson (cross-section deformation).

Results of calculations are given in Figure 3. Normal stresses on line $x = 0$ have turned out to be compressing that corresponds to natural measurements.

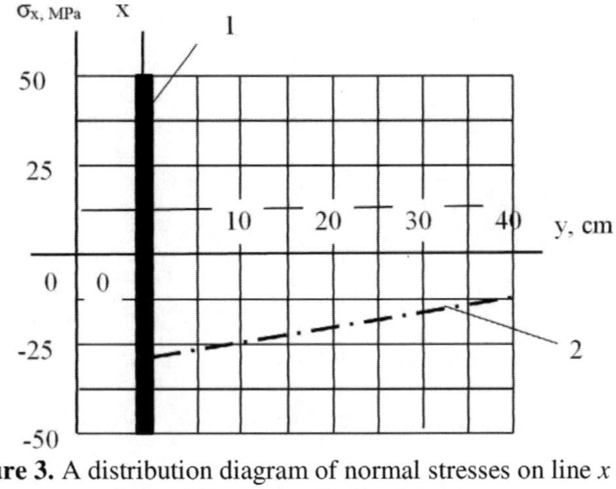

Figure 3. A distribution diagram of normal stresses on line $x = 0$: 1 – welding seam, 2 – load stresses.

4. Conclusions

A scheme of calculation of fields of residual welding stresses is realised. For its further development it is necessary to accumulate experimental data. It will allow one to develop empirical dependences for definition of shrinkages of welding seams.

References

[1] Sule J, Ganguly S, Coules H, and Pirling T 2015 *Jour. of Manufacturing Processes* **18** 141–150
[2] Javadi Y and Ashoori M 2015 *Russian Jour. Materials and Design* **85** 82–90
[3] Iwamatsu F, Miyazaki K and Mochizuki M 2015 *Jour. of Press. Vessel Technol.* **137 (4)** 041402
[4] Maeda A, Jin Y and Kuboki T 2015 *Jour. of Mat. Processing Technol.* **225** 178–184
[5] Chen H, Ma H, Chen X, Jiang S and Wang H 2015 *Jour. of Failure Analysis and Prevention* **15(4)** 563–570
[6] Ikushima K, Itoh S and Shibahara M 2015 Jour. *Welding in the World* **59(4)** 589–595
[7] Kim T-J, Jang B-S and Kang S-W 2015 *Internat. Jour. of Naval Architecture and Ocean Engineer.* **7(1)** 17
[8] Zhu Y, Cheng Y-S and Liu J 2014 *Jour. of Ship Mechanics* **18(10)** 1228–1236
[9] Liu N, Ren H L, Li J Z and Jia L H 2014 *Jour. Advanc. Mater. Res.* **918** 95–100
[10] Vourna P, Ktena A, Tsakiridis P E and Hristoforou E 2015 *NDT and E International* **71** 33–42

IOP Conf. Series: Materials Science and Engineering **124** (2016) 012095 doi:10.1088/1757-899X/124/1/012095

Mathematical simulation of a profile cutter as a surface of revolution

A M Bubenchikov, S M Kazakavitschyus and N R Shcherbakov
National Research Tomsk State University, 36, Lenina ave., Tomsk, 634050, Russia

E-mail: nrs@math.tsu.ru

Abstract. Various types of cutters (spherical, toroidal, etc.) are used in surface processing of parts of a transmission mechanism. The cost of a special profile tool is somewhat higher than that of such cutters. But the increase in the cost of the tool is compensated by a significant reduction in the time of processing parts. The present paper deals with a mathematical model of a profile cutter surface (as a surface of revolution) for processing parts of a cylindrical transmission gear with an eccentrically cycloidal gearing (EC-gearing). A computer program for determining radii of the cutter's circular cross sections for a given set of axial displacements was created.

1. Introduction

In the present paper we will simulate the surface construction of a so-called profile cutter for processing parts of a cylindrical gear with an eccentrically cycloidal gearing (EC-gearing) [1-2]. Profile cutters allow grinding a tooth surface 'in one go' in contrast to spherical and toroidal cutters which perform parts processing by sequential grinding of coordinate lines on the surface of a part [3-5]. The increase in the cost of the profile tool is compensated by a significant reduction in the time of parts processing. The input and output parts of a transmission mechanism are commonly called the 'gear' and the 'wheel', respectively. In the paper we will construct a surface of a profile cutter for processing gear teeth.

2. Constructing the profile cutter surface

Let us write the constants which are included in the equations of parts surfaces:
Aw – the centre-to-centre distance (the distance between parallel axes of parts rotation);
ε – the eccentricity;
z_1 – the number of gear teeth;
z_2 – the number of wheel teeth;
$n = z_2/z_1 + 1$;
ρ – the radius of the gear tooth section circle;
l – the size of a part along the axis of rotation;
r_c – the radius of the cylindrical drum with gear teeth;
K – the number of the cutter's surface sections;
μ_i – the height of the section lift $(i = 0,1...K)$;
r_i – the radii of the cutter's section circle for different values of μ_i;
η – the shift of the cutter's axis.

The surface of a gear tooth in a cylindrical gear with an EC-gearing [1] is formed by circles located in parallel planes, the centres of these circles lying on a helical line which belongs to a cylinder of radius ε (a screw element). Let the axis of the wheel rotation be axis *OZ* and the axis of the gear rotation be directed along the parallel line displaced along the *OX* axis by an amount of *Aw*. Then the parametric equations of the gear tooth surface can be written as a vector function of two arguments:

$$\vec{S}(\upsilon,\alpha)=\begin{pmatrix} Aw-\varepsilon\cos\upsilon+\rho\cos\alpha \\ -\varepsilon\sin\upsilon+\rho\sin\alpha \\ \dfrac{lr\,\upsilon\,z_1}{2\pi} \end{pmatrix}, \qquad (1)$$

where $\alpha = 0,..,2\pi$, $\upsilon = 0,..,2\pi/z_1$.

The surface of the cutter will be built as a family of circles in the planes perpendicular to the axis of the cutter with the centres lying on this axis. The radius of each of these circles will be determined from the condition of its contingence with the flat curve – the section of the tooth by the plane of the circle. The ideal straight line of the cutter's axis intersects the axis of the gear rotation at a right angle and is located at equal distances from the gear teeth. In practice, for the purpose of avoiding a tooth contacting with the adjacent teeth during processing, value η is introduced which is the value of a small parallel shift of the cutter's axis (while maintaining perpendicularity of the gear's axis of rotation). Figure 1 shows the position of the cutter's axis in a flat gear section perpendicular to the axis of rotation for $z_1 = 4$. The four circles are the gear teeth sections having radius ρ, and r_c is the radius of the great circle.

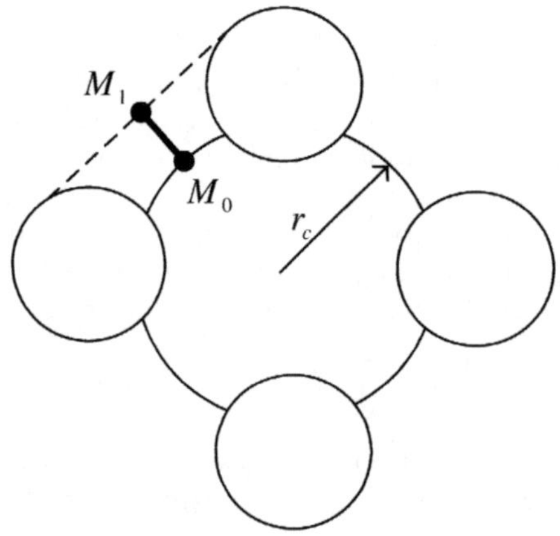

Figure 1. The section of the input part (gear) of a cylindrical transmission gear in an EC-gearing for $z_1 = 4$. The position of the profile cutter's axis and the tangent to the section circle of the first gear tooth perpendicular to the axis of the cutter is shown (dashed line).

With this arrangement of the cutter's axis, its equations in the section plane have the following form:

$$\overrightarrow{Osf}(\mu)=\begin{pmatrix} Aw-r_c\cos\left(\frac{\pi}{z_1}\right)+\eta\sin\left(\frac{\pi}{z_1}\right) \\ r_c\sin\left(\frac{\pi}{z_1}\right)+\eta\cos\left(\frac{\pi}{z_1}\right) \end{pmatrix}+\mu\begin{pmatrix} -\cos\left(\frac{\pi}{z_1}\right) \\ \sin\left(\frac{\pi}{z_1}\right) \end{pmatrix}. \quad (2)$$

When $\mu = 0$, from (2) we obtain the coordinates of point M_0, which is passed through by the surface of the first circle of the cutter's surface section. The position of the last circle surface is defined by intersection point M_1 which is the point of the cutter's axis intersecting the tangent to the section circle of the gear tooth perpendicular to this axis (Figure 1). Let us write the equation of this tangent in the normal form:

$$\cos\left(\frac{\pi}{z_1}\right)\left[x-\left(Aw-\rho\cos\left(\frac{\pi}{z_1}\right)-\varepsilon\right)\right]-\sin\left(\frac{\pi}{z_1}\right)\left(y-\rho\sin\left(\frac{\pi}{z_1}\right)\right)=0. \quad (3)$$

Then distance μ_1 from point M_0 to tangent (3) is obtained by introducing the coordinates of point M_0 in (3):

$$\mu_1=\varepsilon\cos\left(\frac{\pi}{z_1}\right)+\rho-r_c$$

Thus, the planes perpendicular to the axis of the cutter, in which the circles are arranged, forming the surface of the cutter, pass through points M_0 and M_1 of the cutter's axis, i.e. when parameter μ in (2) varies from 0 to μ_1. The equations of these planes family can be written as:

$$\cos\left(\frac{\pi}{z_1}\right)\left[x-\left(Aw-(r_c+\mu)\cos\left(\frac{\pi}{z_1}\right)\right)\right]-\sin\left(\frac{\pi}{z_1}\right)\left(y-(r_c+\mu)\sin\left(\frac{\pi}{z_1}\right)\right)=0, \quad (4)$$

where $\mu = 0,.., \mu_1$.

Now we will get the equation of the family of curves – the sections of the gear tooth surface by the planes of family (4). To do this, we substitute x and y in (4) by the first two coordinates of vector function (1) defining the surface of the gear tooth. As a result, we obtain the relation between parameters υ, α, and μ, from which we can obtain the expression of υ through α and μ:

$$\upsilon(\alpha,\mu)=\frac{-\pi}{z_1}+\arccos\left(\frac{\rho\cos\left(\frac{\pi+\alpha z_1}{z_1}\right)+r_c+\mu}{\varepsilon}\right). \quad (5)$$

Introducing (5) in (1), we obtain the desired equations of the flat curves family – the sections of the gear tooth by the planes of family (4):

$$\overrightarrow{Sem}(\alpha,\mu)=\begin{pmatrix} Aw-\varepsilon\cos(\upsilon(\alpha,\mu))+\rho\cos\alpha \\ -\varepsilon\sin(\upsilon(\alpha,\mu))+\rho\sin\alpha \\ \frac{lr\,\upsilon(\alpha,\mu)\,z_1}{2\pi} \end{pmatrix}. \quad (6)$$

These particular curves must be touched by the circles forming the surface of the profile cutter. These circles are located in the planes of family (4) and their centres lie on the axis of cutter (2) and radii r vary depending on parameter μ of family (4). From this, we can write the equations of such circles family (with arbitrary radius r as yet) in the following form:

$$\overrightarrow{\boldsymbol{Okr}}(\beta,\ \mu)=\begin{pmatrix} Aw-(r_c+\mu)\cos\left(\dfrac{\pi}{z_1}\right)+(r\cos\beta+\eta)\sin\left(\dfrac{\pi}{z_1}\right) \\ (r_c+\mu)\sin\left(\dfrac{\pi}{z_1}\right)+(r\cos\beta+\eta)\cos\left(\dfrac{\pi}{z_1}\right) \\ r\sin\beta \end{pmatrix}. \quad (7)$$

In formulas (5)-(7) $\mu = 0,.., \mu_1$. In each section μ, circle (7) must touch curve (6), i.e. the scalar product of the vectors must be zero:

$$\left(\overrightarrow{\boldsymbol{M}}_0-\overrightarrow{\boldsymbol{Sem}}(\alpha,\ \mu),\ \overrightarrow{\boldsymbol{Sem}}'(\alpha,\ \mu)\right)=0, \quad (8)$$

Here $\overrightarrow{\boldsymbol{M}}_0=\overrightarrow{\boldsymbol{Osf}}(0)$ and $\overrightarrow{\boldsymbol{Sem}}'(\alpha,\ \mu)$ indicates the derivative of vector function $\overrightarrow{\boldsymbol{Sem}}(\alpha,\ \mu)$ with respect to parameter α:

$$\overrightarrow{\boldsymbol{Sem}}'(\alpha,\ \mu)=\begin{pmatrix} \varepsilon\sin(\upsilon(\alpha,\ \mu))\,\upsilon'(\alpha,\ \mu)-\rho\sin\alpha \\ \varepsilon\cos(\upsilon(\alpha,\ \mu))\,\upsilon'(\alpha,\ \mu)-\rho\cos\alpha \\ \dfrac{lr\,\upsilon'(\alpha,\ \mu)\,z_1}{2\pi} \end{pmatrix},$$

$$\upsilon'(\alpha,\mu)=\frac{\rho\sin\left(\dfrac{\pi+\alpha\,z_1}{z_1}\right)}{\left(\varepsilon^2-\left[\rho\cos\left(\dfrac{\pi+\alpha\,z_1}{z_1}\right)+r_c+\mu\right]^2\right)^{\frac{1}{2}}}.$$

For the purpose of solving equation (8) a computer program was created. For each value $\mu_i=\dfrac{\mu_1}{K}i$ (K is a set number of sections of the cutter's surface) this program defines the initial approximation for finding the root of the function in the left-hand part of equation (8) and then, using built-in function *root* of the MathCad pack, defines root α_i. Next, to determine the radius of circle r_i which touches the section of the gear tooth corresponding to μ_i, we require the presence of a common point of these curves, i.e. we equate the coordinates of vector functions (6) and (7). The equality of the second and third coordinates gives a system of two equations:

$$r\cos\beta = \frac{-\varepsilon\sin(\upsilon(\alpha,\,\mu)) + \rho\sin\alpha - (r_c+\mu)\sin\left(\dfrac{\pi}{z_1}\right)}{\cos\left(\dfrac{\pi}{z_1}\right)} - \eta\,,$$

$$r\sin\beta = \frac{lr\,\upsilon(\alpha,\,\mu)\,z_1}{2\pi},$$

from where we find the desired radii of the circles for each value of μ_i:

$$r_i = \left(\left\{\frac{-\varepsilon\sin(\upsilon(\alpha_i,\,\mu_i)) + \rho\sin\alpha_i - (r_c+\mu_i)\sin\left(\dfrac{\pi}{z_1}\right)}{\cos\left(\dfrac{\pi}{z_1}\right)} - \eta\right\}^2 + \left\{\frac{lr\,\upsilon(\alpha_i,\,\mu_i)\,z_1}{2\pi}\right\}^2\right)^{\frac{1}{2}}.$$

Thus, the surface of the profile cutter is obtained as a family of circles of form (7), i.e. as the surface of revolution:

$$\overrightarrow{\boldsymbol{Fr}}(\beta,\,i) = \begin{pmatrix} Aw - (r_c+\mu_i)\cos\left(\dfrac{\pi}{z_1}\right) + (r_i\cos\beta+\eta)\sin\left(\dfrac{\pi}{z_1}\right) \\ (r_c+\mu_i)\sin\left(\dfrac{\pi}{z_1}\right) + (r_i\cos\beta+\eta)\cos\left(\dfrac{\pi}{z_1}\right) \\ r_i\sin\beta \end{pmatrix}.$$

Figure 2 shows a profile cutter, modelled using the above-described scheme. It shows the plane of the cutter's section perpendicular to its axis. In this plane, the section line of the gear tooth which is touched by the circle of the cutter's section is depicted.

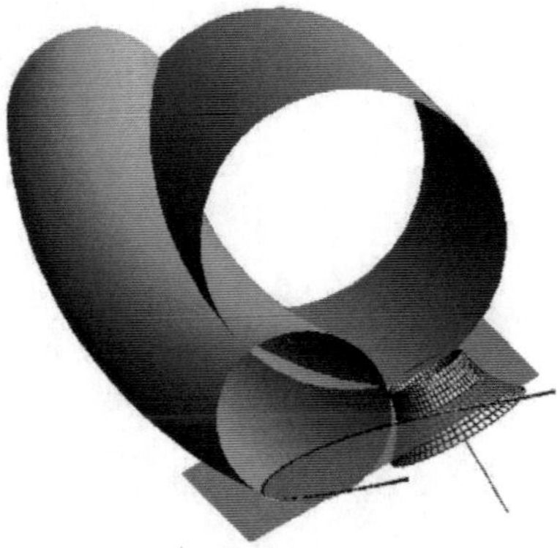

Figure 2. A profile cutter for processing the surface of a gear tooth in a transmission gear with an EC-gearing. The section plane perpendicular to the axis of the cutter and the line of the gear tooth section by this plane are shown.

3. Conclusion

We obtained the parametric equations of the circles family in the planes perpendicular to the cutter's axis of rotation, with the centres lying on this axis. Each circle is found from the condition of touching the flat curve which is the circle plane section of the tooth. The constructed family of circles forms the surface of the profile cutter. The problem was solved for cylindrical gears with an EC-gearing. The work provided substantial assistance in manufacturing EC-engagement parts for gears of various types.

References

[1] Kazakyavichyus S M, Stanovskoy V V, Remneva T A, Kuznetsov W M, Bubentchikov A M and Shcherbakov N R 2011 *Russ. Eng. Research* Vol. **31**(3) 197–199

[2] Stanovskoy V V, Kazakyavichyus S M and Shcherbakov N R 2014 *Int. Symp. 'Theory and practice of gearing – 2014'* Izhevsk Russia pp 220-226

[3] Litvin F L 1989 *Theory of Gearing* (Washington D.C.: NASA RP-1212 AVSCOM 88-CC035)

[4] Litvin F L 1994 *Gear Geometry and Applied Theory* (Prentice Hall, Englewood Cliffs, NJ) p 724

[5] Kamchatniy S A, Skovorodin A V, Stanovskoy A V and Scherbakov N R 2012 *Journal of Mathematics and Mechanics* **4** 15–24

Numerical study of the inverse problem for the diffusion-reaction equation using optimization method

O V Soboleva[2] and R V Brizitskii[1]
[1]Far Eastern Federal University, 8, Sukhanova St., Vladivostok, 690950, Russia
[2]Far Eastern State Technical Fisheries University, 52B, Lugovaya St., Vladivostok, 690087, Russia

E-mail: soboleva22@mail.ru

Abstract. The model of transfer of substance with mixed boundary condition is considered. The inverse extremum problem of identification of the main coefficient in a nonstationary diffusion-reaction equation is formulated. The numerical algorithm based on the Newton-method of nonlinear optimization and finite difference discretization for solving this extremum problem is developed and realized on computer. The results of numerical experiments are discussed.

1. Introduction

The modern machine-building production, based on use of a large variety of materials, technological processes and the large range of products, creates the large quantity of various pollution. There are many methods to control industrial environmental pollution, for example, rational placement of sources of pollution, their localization, cleaning of emissions mathematical modeling, etc. One of the widespread methods is the method of mathematical modeling of the process of transfer pollution [1].

The application of the method of mathematical modeling for studying the distribution process of pollution necessitates the construction and study of the mathematical models describing distribution of pollution in the studied area. The mentioned models contain a number of parameters, which values have to be given for unambiguous finding the solution of boundary value problems. In practice, situations can arise when some of the parameters are unknown and they are to be found along with the solution of the boundary problem. Such problems belong to the class of the inverse problems of identification of unknown density of sources or the coefficients entering used models of transfer pollution. In the latter problems, the unknown densities of boundary or distributed sources or the coefficients involved in the differential equations or the boundary conditions for the model under study are revealed from additional information on the solution of the original boundary value problem. It is important that with a certain choice of the minimized cost the functional identification problems can be reduced to corresponding extremum problems. Following this approach, there arise inverse extremum problems, which can be studied using well known constrained minimization methods. This approach as applied to heat and mass transfer models was described in [1-10] for nonstationary models and in [11-23] for stationary heat and mass transfer models. We should also mention papers [24, 25] where the optimization method is applied for solving the related problems of technical gasdynamics.

In this work we consider the problem of identification of the main coefficient in the parabolic nonlinear diffusion-reaction equation considered at mixed boundary condition, using the additional information on the solution of the initial boundary value problem. The study of this problem can be reduced to the study of the corresponding extremum problem for a certain cost functional [3]. We develop an effective numerical algorithm, based on Newton-method of nonlinear optimization [26] and finite difference discretization of solving this extremum problem, and discuss some results of numerical experiments.

We study the following initial boundary value problem for the nonlinear nonstationary parabolic equation with mixed boundary conditions:

$$\begin{gathered}\frac{\partial \varphi}{\partial t}=\frac{\partial}{\partial x}\left(\lambda \frac{\partial \varphi}{\partial x}\right)-k\varphi+f, \quad (x,t)\in\Omega, \quad x\in(a,b), \\ \varphi(x,0)=\psi(x), \quad \lambda\frac{\partial \varphi}{\partial x}(a,t)=\chi_1(t), \quad \varphi(b,t)=\chi_2(t), \quad t\in[0,T]\end{gathered} \tag{1}$$

which is considered in a bounded domain $\Omega=\{(x,t): a<x<b, \ 0<t<T\}$. Here $\lambda\equiv\lambda(x)>0$ is the variable diffusion coefficient, $k=\text{const}\geq 0$ is a quantity characterizing disintegration of substance by chemical reactions, $f=f(x,t)$ is the volume source density, $\psi(x)$ is a function given in the initial condition, $\chi_1(t)$, $\chi_2(t)$ are functions given at $x=a$ and $x=b$.

2. Statement of Control Problem

The initial boundary value problem (1) contains a number of parameters that must be given to ensure the uniqueness of the solution. In practice, situations can arise when some of the parameters are unknown. For this reason, we need additional information about solution φ of problem (1). For this information we can use, for example, concentration $\varphi_d(x_d,t)$ measured in some set $Q\subset\Omega$. We assume that coefficient λ is an unknown function and we must determine this function together with solution φ of problem (1).

For the study of this identification problem we apply optimization method and reduce solving this problem to the corresponding extremum problem (see [1, 19]). Assuming that $Q=\{(x_d,t), \ 0<t<T\}$ where x_d is fixed we introduce the following cost functional:

$$J(\lambda)=\int_0^T(\varphi(x_d,t)-\varphi_d(t))^2\,dt.$$

Our inverse extremum problem consists of finding two functions φ and λ so that the following conditions take place:

$$\begin{gathered}\frac{\partial \varphi}{\partial t}=\frac{\partial}{\partial x}(\lambda \frac{\partial \varphi}{\partial x})-k\varphi+f, \quad (x,t)\in\Omega, \quad x\in(a,b), \\ \varphi(x,0)=\psi(x), \quad \lambda\frac{\partial \varphi}{\partial x}(a,t)=\chi_1(t), \quad \varphi(b,t)=\chi_2(t), \quad t\in[0,T],\end{gathered} \tag{2}$$

$$J(\lambda)=\int_0^T(\varphi(x_d,t)-\varphi_d(t))^2\,dt\rightarrow\inf, \quad (\varphi,\lambda)\in H^1(\Omega)\times H^s(\Omega). \tag{3}$$

So far there are many works devoted to the theoretical analysis of the inverse extremum problems for the equations of a parabolic type. About solvability of related control problems it is possible to read in works [4, 10], where the existence and uniqueness of solutions of the extremum problems under some restrictions on initial data were proved.

3. Numerical Algorithm

In the work the optimality system was obtained. The optimality system plays an important role in investigating properties of solutions of the control problem. On the basis of the analysis of optimality systems sufficient conditions for the initial data, which provide the uniqueness and stability of solutions to individual extremum problems, can be formulated. Optimality system derived above can be used to design effective numerical algorithms for solving the control problem (2), (3). The simplest numerical algorithm can be obtained by applying a simple iteration method for solving the optimality system. Direct and adjoint problems entering the optimality system can be solved by finite difference discretization method using Scilab [27].

4. Numerical Experiments

Now we should discuss some results of numerical experiments. Concentration of substance φ_d is the analytically given function $\varphi = (x - x^2)(t - t^2)$. We will take functions $f = (1-2t)(x - x^2) + 2(t - t^2) + 0.01(x - x^2)(t - t^2)$, $\psi(x) = 0$, $\chi_1(t) = -(t - t^2)$, $\chi_2(t) = 0$, diffusion coefficient λ=1 and coefficient k=0.01.

The results of the numerical solution of the inverse extremum problem (2), (3) will be presented below. The surface of restored function φ is shown in Figure 1 (right). The received surface coincides with the analytically given function presented in Figure 1 (left). Figure 2 shows restored function λ (dotted line) and required function λ (solid line). Figure 1 and Figure 2 show the case when area Q coincides with Ω.

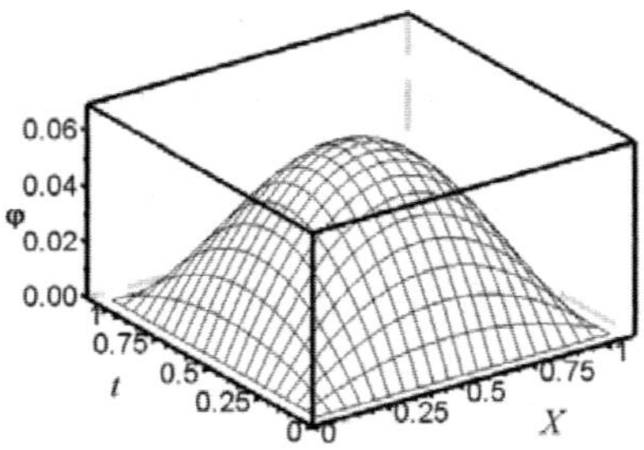

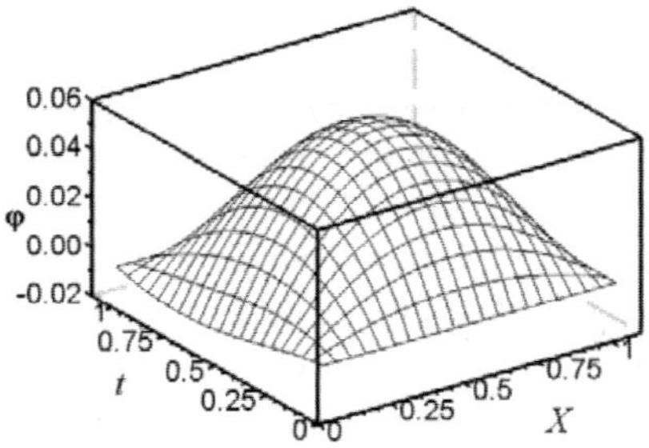

Figure 1. Analytically given function φ (left) and restored function φ (right)

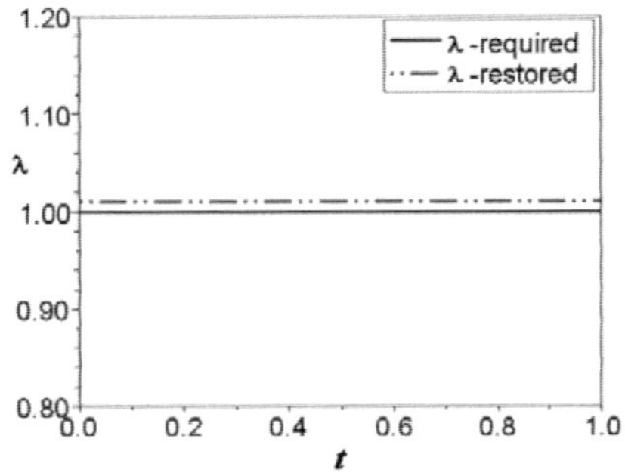

Figure 2. Required (solid line) and restored functions λ (dotted line)

Figure 3 shows restored function φ and required (solid line) and restored functions λ (dotted line) for the case, when observations in the research area are made at one point of the research area in the course of time. The results of calculations were obtained for a case when a source of additional information is the measurements of concentration measured at point x=0.5. The presented graphics show that the developed algorithm of the numerical solution of the inverse extremum problem allows even at a small amount of information a qualitative restoration of the required functions for the non-stationary parabolic equation.

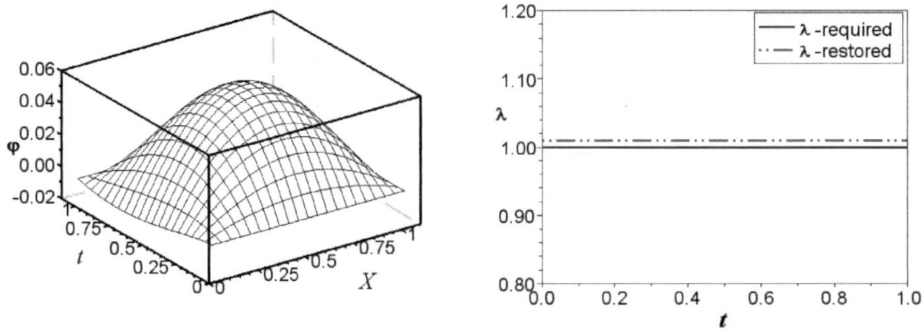

Figure 3. Restored function φ on the left and required (solid line) and restored functions λ (dotted line) on the right

The results of numerical experiments are given above, when the error in measurements of substance concentration is absent. When solving the applied problems of the main coefficient identification of non-stationary diffusion-reaction equation of substance concentration the given measurements are the main source of errors [3, 4]. It is caused by the features of concentration sensors functioning which are inevitable as a consequence of discrete nature of measurements, the accuracy of sensors indication, etcetera. In this regard the problem of identification of unknown coefficient λ with the indignant input data is of interest, that is, there is a need to investigate the resistance of the method for solution of the inverse problems to the errors of measurements.

For these experiments input data were perturbed according to formula $\varphi_d = \varphi + 2\delta(\text{rand}(1,1) - 1/2))$, where $\delta = 0.1$, rand (1,1) – random number generator. As a result of calculations the following function graphs were received (Figure 4). The developed algorithm of the numerical solution of the inverse extremum problem allows a considerable approach of the numerical solution of φ (at the left) and λ (on the right) to the exact solution.

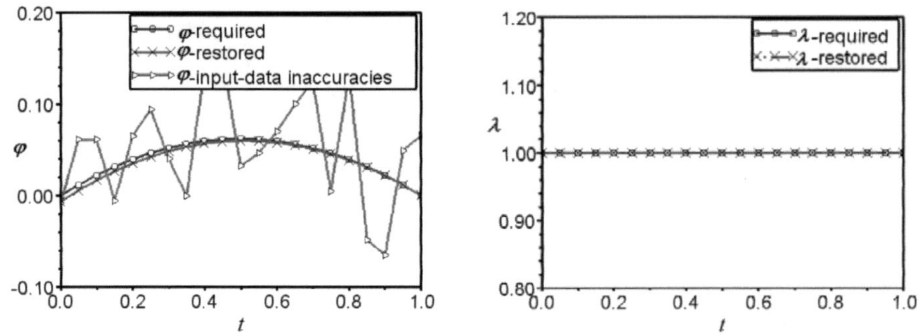

Figure 4. Function φ (left) and functions λ (right) with $\delta = 0.1$

In the next test (see Figure 5) the value $\delta = 0.01$ was choosed in the expression for function φ_d. As it follows from Figure 4 the error between exact and approximate solutions is essentially less than for the case when $\delta = 0.1$.

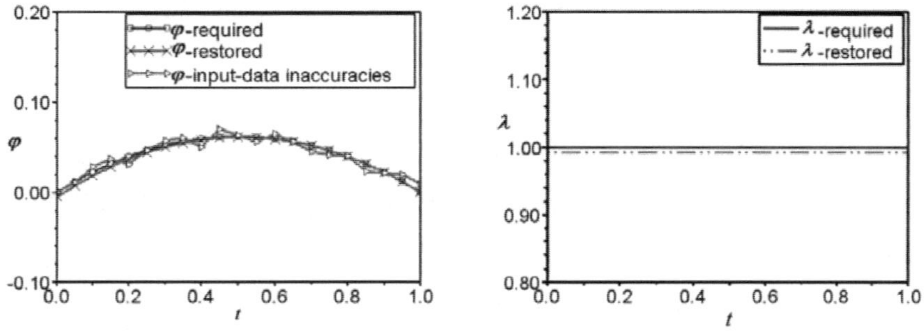

Figure 5. Function φ (left) and functions λ (right) with $\delta = 0.01$

5. Conclusion

In this work the problem of identification of the main coefficient in the nonlinear one-dimensional mass transfer model was formulated and studied. Our model has the form of nonlinear nonstationary diffusion-reaction equation with variable coefficient of the diffusion. This equation is considered to be in the bounded domain under mixed boundary condition. Based on the Newton method of nonlinear optimization we constructed the algorithm and developed a code for numerical solution of the coefficient inverse problem. The results of numerical experiments have shown the effectiveness of the used numerical algorithm and the code of numerical solution of the coefficient inverse problem for the mass transfer model under study.

References

[1] Marchuk G I 1982 *Moscow, Nauka*
[2] Samarskii A A and Vabishevich P N 2004 *Moscow, Editorial URSS*
[3] Borukhov V T, Zayats G M and Tsurko V A 2008 *Informatics* **19** 29–39
[4] Vabishevich P N and Denisenko A Yu 1989 *Mathematical modelling* **1(8)** 116-126
[5] Muzylyov N V 1985 *Inj. phis. J.* **49(6)** 1009-1015
[6] Klibanov M V 1985 *Inj. phis. J.* **49(6)** 1009-1009
[7] Prilepko A I and Solovyov V V 1987 *Theorems of Resolvability and a Method to the Company in the Inverse Probl. for the Equat. of Parabolic Type Different. Equat.* **23(10)** 1791-1799
[8] Frolenkov I V and Romanenko G V 2012 *J. of Siber. Feder. Univ. Mathem. & Phys.* **5(1)** 122-131
[9] Kalinina E A 2004 *Far Eastern Mathematical Journal* **5(1)** 89-99
[10] Savateev E G 1995 *Siberian Mathematical Journal* **36(1)** 177–185
[11] Alekseev G V 1998 *Dokl. Math.* **58** 314–317
[12] Alekseev G V and Tereshko D A 1998 *Sib. J. Indust. Math.* **1** 22–44
[13] Alekseev G V and Tereshko D A 1998 *Journal of Inverse and Ill-Posed Problems* **6** 521-562
[14] Alekseev G V 2000 *Dokl. Math.* **375(3)** 315–319
[15] Alekseev G V and Adomavichus E A 2001 *J. Inv. Ill-Posed Probl.* **9(5)** 435–468
[16] Alekseev G V 2002 *Comput. Math. and Math. Phys.* **42** 363–376
[17] Alekseev G V 2007 *Dokl. Math.* **76** 797–800
[18] Alekseev G V, Soboleva O V, and Tereshko D A 2008 *Prikl. Mekh. Tekh. Fiz.* **49** 24–35
[19] Alekseev G V and Tereshko D A 2008 *Vladivostok, Dal'nauka*
[20] Alekseev G V and Tereshko D A 2010 *J. Appl. Mech. Tecn. Phys.* **4** 72-84
[21] Alekseev G V and Tereshko D A 2011 *Comput. Math. Math. Phys.* **51** 1539–1557
[22] Alekseev G V, Vakhitov I S and Soboleva O V 2012 *Comput. Math. and Math. Phys.* **52** 1635-1649
[23] Alekseev G V and Shepelov M A 2013 *J. of Applied and Industrial Mathematics* **7(1)** 1–14
[24] Fershalov Yu Yu 2012 Russian Aeronaut **55** 424–429
[25] Fershalov M Y., Fershalov A Yu and Fershalov Yu Ya 2014 Advan. Mater. Res. **341** 915–916
[26] Bahvalov N S, Jedkov N P and Kobelkov G M 2003 *Moscov, BINOM*
[27] Home-Scilab WebSite http://www.scilab.org

IOP Conf. Series: Materials Science and Engineering **124** (2016) 012097 doi:10.1088/1757-899X/124/1/012097

The main problems in the mechanical engineering sector and some possible directions of their solution

E Strizhakova
Bryansk State Technical University, 7, 50 years of October st., Bryansk, 241035, Russia

E-mail: kathystr@inbox.ru

Abstract. The article shows the problems of the sector of mechanical engineering in the industrial system in Russia. The author's method of estimating the relative level of risk and the method of determining the de-industrialization degree of the sector based on the aggregated level of adaptability are given. According to them we have analysed the key indicators, such as basic, developed and advanced technologies, and investments in an old or new technology of industrial sectors. The main directions of the impact of industrial policy allowing a change in the current situation in mechanical engineering are given. The results can be applied in practice in formation of directions and actual control actions to improve the overall efficiency of mechanical engineering industry.

1. Modern industrial policy: theoretical aspect

The current situation in the Russian industrial system is characterized by a high degree of dependence on imports of both raw materials and semi-finished products, as well as on means of production, low quality and low profitability of domestic products, the prevalence of resource industries over manufacturing ones.

The reasons for the development of such situation are the decay of existing commercial and technological relations that existed on the territory of the former Soviet Union, a significant decrease in the degree of government influence on the economy, as well as possible errors in the process of adoption and implementation of economic industrial policy. In our view, considerable attention in the analysis of the reasons must be given to the so-called 'paradox of plenty' or 'resource curse'. The so-called 'resource curse theory' was first expressed in the work of Jeffrey Sachs to explain the fact that countries rich in natural resources, especially energy, are lagging behind in their development, compared to others. An explanation of this fact is the statement that the sale of resources brings the country more revenue than production activities, and its redistribution is becoming a priority for economic agents.

However, many researchers give different meanings to the concept of industrial policy. In recent years some new terms have appeared, such as industrialization, and reindustrialization or neoindustrialization that have a significant number of interpretations. Hereafter the national industrial policy and industrial policy (according to the Draft of Federal Law № 98281-5 'On the national industrial policy in the Russian Federation') will be referred to as a system of legal and economic measures and actions of subjects of industrial policy (public authorities, local government bodies, legal entities and individual entrepreneurs, organizations, forming the infrastructure of support for industry, non-profit organizations developing an interest to the industrial activity). These measures originate from the priority to ensure the competitiveness of the national economy, a stable and innovative socio-economic development of the Russian Federation, the Russian Federation subjects and municipalities.

To create a well-functioning industrial policy, in the first place it is necessary to assess the situation in the industry at the moment, to identify the main problems and directions of impact and limitations. It is reasonable to determine the resulting indicators, for which it will be possible to assess the degree of achievement of objectives; the next step is to define the scope, structure and sources of funds for industrial policy. We ought to remember that the same quantity and quality of resources may have different potentials depending on the extent of their use, so we will investigate not only the kinds of resources, but also their ability to produce a beneficial effect.

2. The technology and growth of industries

The rate of growth of the economic system will be higher, the higher is the rate of aggregate investments.

The rate of investment must be greater than the currently generated product and the rate of the change in the share of investment in the product for the growth of any economic system. Thus, we face some problems. On the one hand, investment in fixed assets does not reach the level in the year of 1990 and tends to decrease at the present time (Figure 1). The investments falling to the share of mechanical engineering had decreased from about 15 % to 4 % by 2014 in the total industry volume.

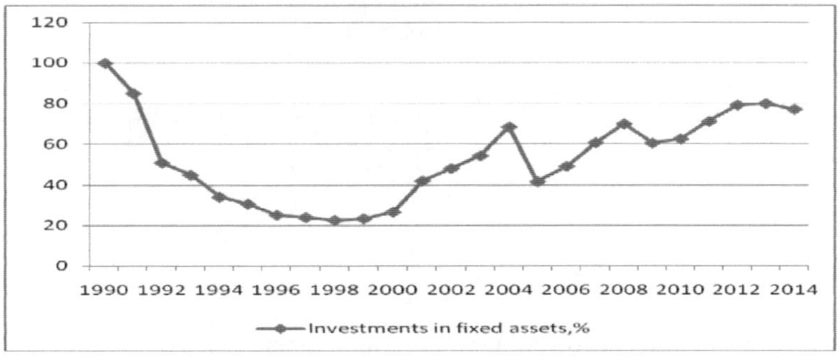

Figure 1. The rate of investments in fixed assets.

On the other hand, the sources of investments must be analyzed. We had discovered that the proportion of borrowed funds in the mechanical engineering is falling from year to year from 40 % in 2009 to 30 % in 2014. This is the result of, first of all, the excessive cost of credit exceeding the level of profitability in the sector, which makes credits unaffordable for most of the enterprises related to mechanical engineering industry. Secondly, this situation arises due to a significant level of risk, which discourages investments [4].

Our method for specifying the relative level of risk includes determination of coefficient β (a general formula is similar to the CAPM method):

$$\beta = \frac{\operatorname{cov}(X_n, Y_n)}{\operatorname{var}(X_n)} \qquad (1),$$

where $\operatorname{cov}(X_n, Y_n)$ – the covariance between X and Y, $\operatorname{var}(X_n)$ – variation and the standard deviation of profitability in the sector, %.

The value of the covariance and variation is as follows:

$$\operatorname{cov}(X_n, Y_n) = \frac{1}{n}\sum_{t=1}^{n}(X_t - \overline{X})(Y_t - \overline{Y}) \qquad (2)$$

$$\text{var}(X_n) = \sqrt{\sum_{t=1}^{n} (X_t - \overline{X})^2} \quad (3)$$

where X_t – the value of profitability in the sector per year, Y_t – the value of return on the average in the industry per year t, $\overline{X}$ – the average value of profitability in the sector, $\overline{Y}$ – average profitability of the system.

We calculated the relative risk level in the mechanical engineering and mining sector of the Russian industry for the period of 2003…2014 using this method.

The risk in the mining depends very much on the price of oil, which is at a very low level nowadays, but in spite of this fact the level of the relative risk is 1.3 times higher in the mechanical engineering than that in mining industry.

Figure 2 also shows the average level of profitability in the industrial system of the Russian Federation in 2003…2014, so we can note that the profitability in mining industry is more than 3 times higher than that in the mechanical engineering, which shows that the excess return in commodity sectors is more than 3 times higher than the yield in manufacturing. As a result of this structural change there is a skew appeal of the sectors to investors.

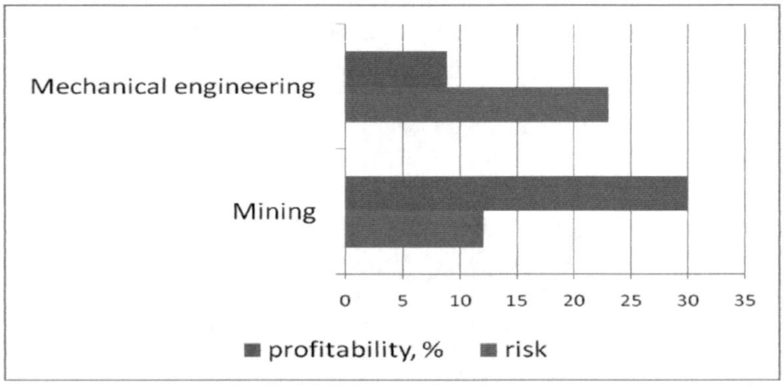

Figure 2. The rate of risk and profitability in mining and mechanical engineering industry.

So, the growth of mechanical engineering and industry as a whole depends on the combination of new and old technologies. However, as it has already been noted, it is unreasonable to conduct the identification borrowed from the obsolete technology (though in terms of borrowing, costs may require fewer resources than creating a new proprietary technology, but this method is not suitable for all countries and this is not always an objective truth).

With increasing speed of diversion of resources and creation of a new resource, the increase of the technological level will accelerate the pace of the economic growth.

We can make the following changes in investments in old and new technologies depending on the parameters of diversion speed of resources from old technologies (α) and creation of a new resource (μ) [2], [5].

An important aspect of analysis of the production system is the evaluation of resources and sources by its growth. According to the theory of industrial advancement developed by Sukharev O S [1], the ratio of the basic and emerging technologies occurs as a change in the structure of the system and the overall rate of its growth. In our research the amount of expenditure on technological innovation was taken as the cost of new technologies, In (t), and the total value of expenses in fixed assets - as all investments, I (t), the difference between the value of total investments in fixed assets and the cost of technology innovation - as the costs of old technology, Is (t), (Table 1).

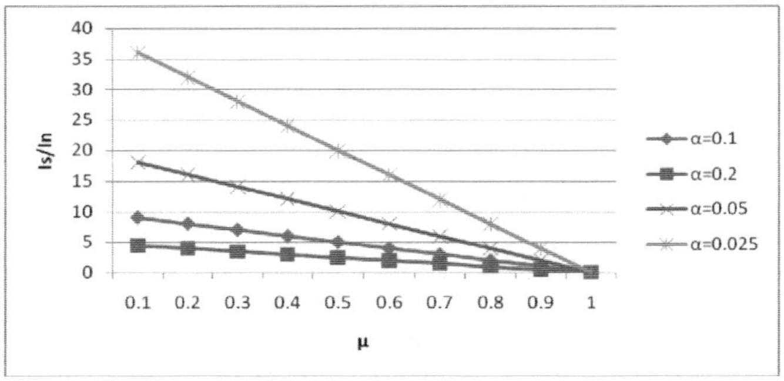

Figure 3. Changes of Is / In depending on α and μ parameters.

Table 1. Values of investments in the sector 'Mechanical engineering'.

Indicators	2009	2010	2011	2012	2013	2014
The total investment in the sector, mln. RUB	183448	201346	204759	276894	349706	392302
The value of investments in technological innovation in the sector, mln. RUB	46672	45411	58445	79634	120050	137305
Percent of the investment in technological innovation in the total investment	25.44	22.55	28.54	28.76	34.33	35.00

As we can see the total amount of investments in the sector is growing, the rate of investments in technological innovation is increasing in absolute terms, their share in the structure increases as well. The analysis of the used technologies and the number of new advanced technologies in the sector indicates the total drop in the number of technologies to 11398 units in the period of 2009...2014 in absolute terms, or by 19.92 %. Thus, despite the overall positive trend of growth in the value of investments, their total amount is not sufficient to improve the technological level.

Structural indicators calculated using the value of fixed assets and the amount of employment in the sector show a general increase in the cost of fixed assets in each of the analyzed sectors and a reduction in the number of employees in the whole industry. The value of investments and sectors attractive to investors are also showing a gradual decline in the manufacturing sector, compared to mining industry, and their movement in the opposite direction, increasing a raw material basis for economic growth and the country's industry. Therefore, the structural policy must include, first, the proportion of the structural changes in the economic system and economy as a whole and in the large-scale industry. A second aspect of the structural policy should cover the distribution of resources and factors of production optimally or approximately. In addition to the obvious misallocation of resources used in the structure there is a problem that the quality of the resources used is of minimum necessity. Unfortunately, the statistical data do not currently contain a residual value and the average age of the equipment used in the production, which makes an objective assessment of value assets impossible. Experts estimate that the average age of used fixed assets is approximately 20 years, reaching a number of economic activities equal to 25, and the value of physical depreciation exceeds 40 %.

The technological level is reduced in all sectors analyzed, the maximum reduction in mechanical engineering is an indicator of the economic activity. Based on empirical data, the situation in the manufacturing sector is the most negative, the situation of the sector is number 5, which describes the classical situation of de-industrialization. Measures to change the structure of the industrial system will be discussed below.

3. Problems of the industrial policy in the region

The process of implementation of industrial policy at the regional and municipal levels is closely related with investment activities, budget and tax policies.

A support for large regional companies, making a significant contribution to the gross regional and domestic products of small and medium-sized enterprises has the advantage of quick adaptation to rapidly changing market conditions. It is the main form of industrial policy at the regional level.

In this case, the main problem the local authorities face is the problem of the budget execution. If funding is insufficient, it is impossible to suggest the performance of programs supporting the industry. As a result of activities involving the modernization of production, additional payments to the budget do not come, and due to implementation of a new investment project, the result of which is to create a new object, the city's budget receives only 0.1 % of its total income. Besides, changes were made in the amount of an interest tax on personal income, fines for violating immigration laws involving incomes to a greater extent in the regional and the federal budget.

The existing and widely used tax incentives granted for a long-term practice are not currently used. The analysis showed that a number of companies receiving tax breaks and subsidized interest rates on loans had not brought additional tax revenues, which generally reduces their value.

Another issue is the aspect of the land owned by the city and the region; the city administration is unable to dispose of the land of the city, which increases transaction costs and time-consuming, violates the principle of 'one window', forcing potential owner to address the issues in the city administration (for example, in the Office of the construction and development of the areas of the Bryansk city in the department of architecture) and the regional administration.

The almost total lack of tax revenue of the city limits resources for industrial policy at the municipal level does not create conditions for the commitment of local authorities to change the current situation. This raises the problem of reforming the existing institutions of local industrial policy, changing the legislation regarding management of tax revenues of the city and the region, as well as adjusting the federal industrial policy, and other similar problems of local governments.

All of the above-mentioned problems can be defined as a dysfunction of the industrial policy in the city of Bryansk. As it is known, the solution is considered to be optimal if the problem solution process is higher than the level of problem origin. Therefore, in order to overcome these contradictions it is necessary to address the developmental level of the whole country. Only in the coordination of federal executive authorities and municipalities, revision of the existing situation becomes possible.

Thus, whereas the necessary technology (related to updating of the physical and human resources) and structural (related to the transformation of the industrial structure) changes are sufficiently clear and unambiguously defined, the institutional problems of industrial policy are not fully identified and, of course, unresolved at the local, regional and federal levels. The main directions of overcoming the dysfunction of industrial policy at the regional level, in our view, are the following.

First, the creation of a single integrated contract system, which is the event of legal, institutional and economic measures, is aimed at ensuring the state and municipal needs through a comprehensive implementation of the interrelated stages of forecasting, planning and procurement (including the execution of contracts), as well as monitoring, surveillance, control and audit. An effective use of this system will allow linking all the activities included in the industrial policy of the state with other state regulation forms of social-economic development of the country and the region. In this case, attention should be focused not only on new industries using innovative advanced manufacturing technologies, but also on the restored base of stereotypical, classic productions ('conservatives') and technology [3].

Second, the promotion of enterprises' associations of different forms of ownership belonging to different types of the economic activity in individual holding companies, will resume economic ties within the region and the country as a whole.

Third, the re-establishment of pre-existing programs of scientific and industrial research, undertaken jointly by companies and universities, included by means of training of diploma students, doctoral and post-graduate practical work, contractual work of the teaching staff in higher education institutions, will intensify efforts in increasing the amount of intellectual capital and facilitate an access to the results of research and development.

Fourth, the adaptation of programs will support the industry at the regional and local levels, taking into account the possible and necessary interaction of industrial systems of different regions.

Fifth, the process of redistribution of resources between regions with regard to their specificity and investment attractiveness, strengths and weaknesses should form a unified state tool of the new industrial policy.

Harmonization and reasonable distribution of functions between the regional and federal institutions, sharing of responsibilities and the resources are to be used to achieve the goal of the efficiency increase of industrial systems at all levels of management, beginning from an enterprise and ending with the country as a whole.

So, the situation in the Russian industrial system and in the engineering sector requires urgent management activities in three basic directions - economic reform, institutional reform and an activity aimed at improving the technological level of the industry.

Economic reforms must be directed at the increase of domestic demand in order to increase the investment in fixed assets. Therefore, the tax system and the credit system must undergo reformation; the cost of loans and taxes should be different for various sectors of the industry.

Institutional reforms must clearly share responsibilities. It is necessary to define the implemention of the industrial policy, consolidating certain institutions to determine their areas of competence, possible management impacts and the amount of available resources.

Actions aimed at improving the technological level of the industry must increase the efficiency of production resources and factors, including renewal of fixed assets, advanced training, development and introduction of new and existing technologies. It is necessary to highlight the fact that there are main production lines, such as machine tools, where Russia has a competitive advantage over other countries. Expanding these sectors allows reconstructing the destroyed production chains.

Thus, there is a hope that the entire set of proposed actions will make it possible to increase the competitiveness of the engineering industry and the entire industrial system as a whole.

References

[1] Sukharev O 2015 *Institutional Theory of Economic Reforms: Basic Imperatives. Ecoforum* **4** 8–25
[2] Sukharev O S and Strizhakova E N 2015 *National interests: priorities and security* **23** 15–37
[3] Sukharev O S 2015 *Institutional economic theory of property and reforms* (Moscow: KRASAND)
[4] Sukharev O S 2013 *Theory of Economic Changes. Problems and Decisions* (Moscow: KRASAND)
[5] Sukharev O S 2014 *Megatrend Revija* **11** 11

Optimal configuration algorithm of a satellite transponder

M S Sukhodoev[1], I I Savenko[1], Y A Martynov[1], N I Savina[2] and V V Asmolovskiy[1]
[1] Tomsk Polytechnic University, 30, Lenina ave., Tomsk, 634050, Russia
[2] Tomsk State University, 36, Lenina ave., Tomsk, 634050, Russia

E-mail: smike@tpu.ru

Abstract. This paper describes the algorithm of determining the optimal transponder configuration of the communication satellite while in service. This method uses a mathematical model of the payload scheme based on the finite-state machine. The repeater scheme is shown as a weighted oriented graph that is represented as plexus in the program view. This paper considers an algorithm example for application with a typical transparent repeater scheme. In addition, the complexity of the current algorithm has been calculated. The main peculiarity of this algorithm is that it takes into account the functionality and state of devices, reserved equipment and input-output ports ranged in accordance with their priority. All described limitations allow a significant decrease in possible payload commutation variants and enable a satellite operator to make reconfiguration solutions operatively.

1. Introduction

The complexity and functionality of the modern satellite payload are constantly increasing. It allows increasing the payload throughput and using satellite positions effectively on the geostationary orbit. Thus, astronautics has an increasing problem concerning the scheme size of the onboard communication equipment (OCE). It is possible to specify the following factors that lead to payload structure complication:

- the number of OCE transponders;
- functional backup;
- the use of scheme with multiple frequency transformations;
- the provision of a straight and reverse communication link.

Modern satellite payload may include up to 100 transponders and each transponder can include up to 20 devices (along with switches). Therefore, payload communication matrix can increase significantly forcing the payload scheme increase up to a huge size. Hence, the manual satellite payload reconfiguration becomes a difficult task for the operator. The current paper suggests determining the algorithm of OCE optimal configuration [5]. The authors offer to consider OCE as an oriented weighted graph. Also this paper describes a mathematical model based on automata theory that allows defining criteria that should be considered during determination of the transponder optimal configuration.

2. Related work

Scientists [2] define 3 tasks arising when determining the optimal satellite payload configuration:

IOP Conf. Series: Materials Science and Engineering **124** (2016) 012098 doi:10.1088/1757-899X/124/1/012098

- Initial satellite payload configuration. This task occurs at the repeater designing stage and consists in defining of initial positions of switches for the optimal initial satellite payload configuration before satellite starts.
- Maintenance configuration. This task requires minimization of the number of switches as well as the number of their revolutions.
- Maintenance configuration with active transponders. This task arises in case of active equipment failure. When solving this task it is necessary to decrease inactive time of active channels and minimize the number of rotating switches.

The authors suggest mathematical models for each of the tasks described above. Also the paper consists of the descriptions of successful algorithms, developed by the authors. Their models allow finding all possible OCE configurations, which are not always necessary for the operator.

In article [1] other scientists suggest using the breadth-first search recurrent algorithm to search all possible ways of channels commutation. The conducted experiments showed the effective results of a small commutation network (seven switches and ten amplifiers) but it is expected that the algorithm will be limited due to a significant time increase for data processing on huge commutation network.

We should note that it is of great importance for the operator to find a single optimal solution for a specific OCE configuration state instead of all possible configurations.

3. The use of automata theory for creating a repeater mathematical model

The application of automata theory allows considering OCE as a mathematical system with input and output channels. This mathematical system can be only in one finite condition and at one discrete instant in time that allows describing the ways of its state transitions. In automata theory the notion of an abstract automaton (AA) is described by a six-component set:

$$A = (\mathrm{S}, \mathrm{X}, \mathrm{Y}, \delta, \lambda, \mathrm{s}_1),$$

where

$S = \{s_1, \ldots, s_z, \ldots, s_Z\}$ – set of states, or AA alphabet;

$X = \{x_1, \ldots, x_z, \ldots, x_Z\}$ – set of input signals, or input AA alphabet;

$Y = \{y_1, \ldots, y_z, \ldots, y_Z\}$ – set of output signals, or output AA alphabet;

δ – transition function of AA, which brings a state–input signal pair (s_k, x_f) into accord with AA state, (s_i), that is, $s_i = \delta(s_k, x_f), s_i \in S$.

λ –transition function of AA, which brings a state-input signal pair (s_k, x_f) into accord with AA state (y_i), that is, $y_i = \lambda(s_k, x_f), y_i \in Y$.

s_1 – initial state of a finite automaton.

The abstract automaton can be represented graphically as a "black box" that converts input signals to output ones [2].

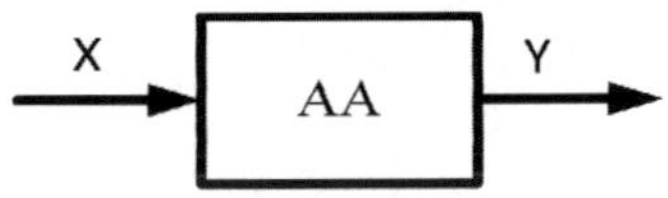

Figure 1. Graphic representation of an abstract automaton

3.1. OCE as a finite Moore machine

In OCE there are two main processes:

- transformation of the input frequency into the output one (receiving device);
- signal power amplification.

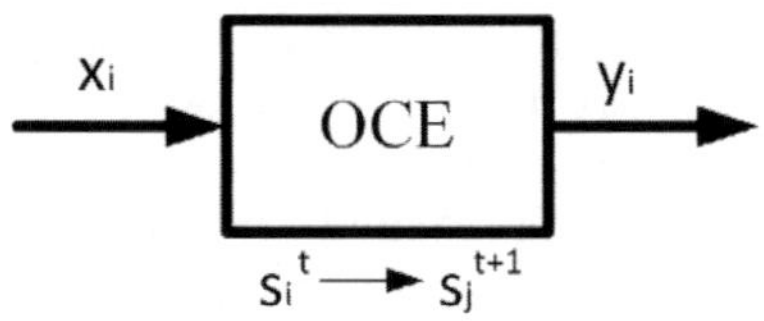

Figure 2. Graphic representation of OCE as a finite automaton

A Mealy automaton has been chosen for description of the mathematical model of the repeater as the repeater state depends on input signal sets. Switches positions and active devices states define OCE state at any moment of time.

$P = \{p_1,\ldots,p_z,\ldots,p_Z\}$ – finite set of switches positions;

$D = \{d_1,\ldots,d_z,\ldots,d_Z\}$ – finite set of device states (on/off);

$T = \{t_1,\ldots,t_z,\ldots,t_Z\}$ – finite set of transponders, which are determined by OCE configuration, switches positions and devices states $t_i = \{P^t, D^t\}$, where P^t – positions of switches belonging to OCE, $P^t \subseteq P$. D^t – states of devices belonging to OCE, $D^t \subseteq D$.

$$S_i = \{t_1,\ldots,t_Z\}$$

Therefore, it may be concluded that the main task of transition function δ is a transponder commutation by calculating the required position of switches– P and devices states – D relatively current initial state – s_k and input frequency – x_f. Automata theory implies that only one symbol from the input set can be fed to the device input at a time. Thus, the main purpose of the transition function is to find the optimal commutation of one transponder. The optimal transponder commutation is a minimal number of switches in the transponder, that is, the power of set P^t tends to minimum

$$|P^t| \to \min$$

where P^t – the number of transponder switches.

The quantity of possible OCE states is finite and differs due to the number of switches of different types, which can take several positions depending on the type and possible states of devices. It is defined as follows:

$$n = \prod_{i=1}^{m}\prod_{j=1}^{n} s(p_j^i)$$

where m – transponders number, n – transponder devices number, p_j^i - transponder switches, s – function determining the total quantity of states of the switch according to its type.

The states transient time is defined by the turn-on time of one transponder, which is dtermined by the characteristics of the devices contained in the transponder.

Input and output sets are the sets of input and output signal frequencies defined by the repeater configuration. The output frequency, which is defined by the frequency transform function, will correspond to input frequency. This function depends on OCE configuration/state s_i. The formula considering a multiple frequency transformation is as follows:

$$x_i - \sum_{k=0}^{d} f(p_k, t_i) + \sum_{m=0}^{u} f(p_m, t_i) = y_i,$$

where d – the number of reducing converters belonging to the transponder, u - the number of increasing converters belonging to the transponder, p_d – frequency decreasing converter $p_d \in P^t$, p_u – frequency increasing converter – $p_u \in P^t$, f – function of calculating the converter frequency.

Modelling task is solved using an OCE design framework [2]. This mathematic model (1) allows the framework to realize the required functions, namely: calculations of the output frequency of the transponder, specifying the criteria of selection of the repeater optimal configuration, which are considered by the shortest-path algorithm described below [7].

4. Description of the shortest-path algorithm

There are many algorithms used for finding as short path as it is possible, such as the algorithms of Dijkstra, Floyd–Warshall, A*, Ant colony [4]. The most suitable algorithm is Dijkstra's algorithm.

The algorithm proposed in this paper is a particular case of the repeater reconfiguration task. The essence of this algorithm is finding the optimal configuration of a transponder taking into account the following limitations:

- use of a minimal number of transponder devices;
- working capacity and the state of onboard equipment (on, off or malfunction);
- back-up equipment
- compliance of input ports with output ports ranged according to their priority.

A program description of the repeater schema is represented as a graph and is stored in the multilinked list. It is connected with the representation of graph nodes by the full-fledged software models of the devices with their preset behaviour and properties. It allows defining the device status, its type or the switch position. The list is formed during the algorithm initialization and has algorithm complexity $O(n^2)$. This list stores the object pointers, as well as the pointers to input and output devices with different degrees of connectivity. These structures are called plexuses.

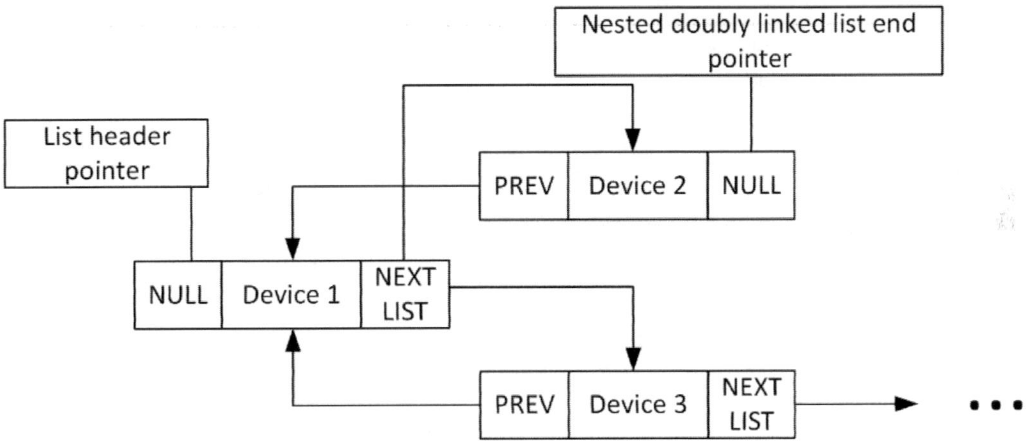

Figure 3. Graph description using plexus

This proposed algorithm uses a priority matrix to optimize the search of the transponder optimal configuration. This matrix consists of input and output ports sorted by priority.

Let us consider a repeater active scheme with an active transponder and malfunction devices (Figure 4). This scheme consists of a receiving device, filters, power amplifiers, an output multiplexer and switches allowing the repeater reconfiguration. The active transponder is marked up by a bold line, the reserved devices - by hatching, and out of service devices are marked up by double hatching.

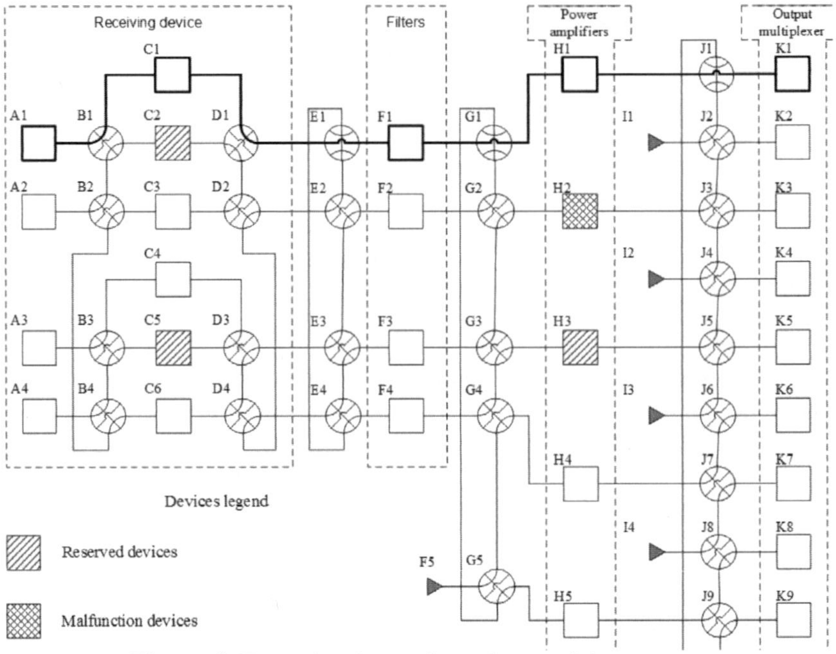

Figure 4. Example of an active scheme of the repeater

The current repeater configuration should be taken into consideration when the repeater scheme transforms into to a weighted directed graph. For this purpose it is necessary to follow the rules given below:

- all devices are marked as graph nodes;
- all possible device links are presented as arcs that connect corresponding graph nodes. At that one considers the directions, at which the signal passes in the transponder;
- the weights (lengths) of the arcs that are included into the reserved devices are set to be equal to a maximal number of switches in one matrix of switches. The rest weights are set to be equal to 1. This weight distribution provides the use of the reserved devices only in case when all possible variants of pass through the basic devices are impossible;
- the links of malfunction devices are not represented in the graph (for example, C_5 and H_2 devices in Figure 5);
- the switches, through which active transponders pass, are represented as two nodes in the graph. Their arcs are connected using the current OCE scheme configuration and state (for example, B1 switch is represented as B1.1 and B1.2 nodes connected respectively).

The weighted oriented graph corresponding to the example of the active scheme of the repeater is represented in Figure 5.

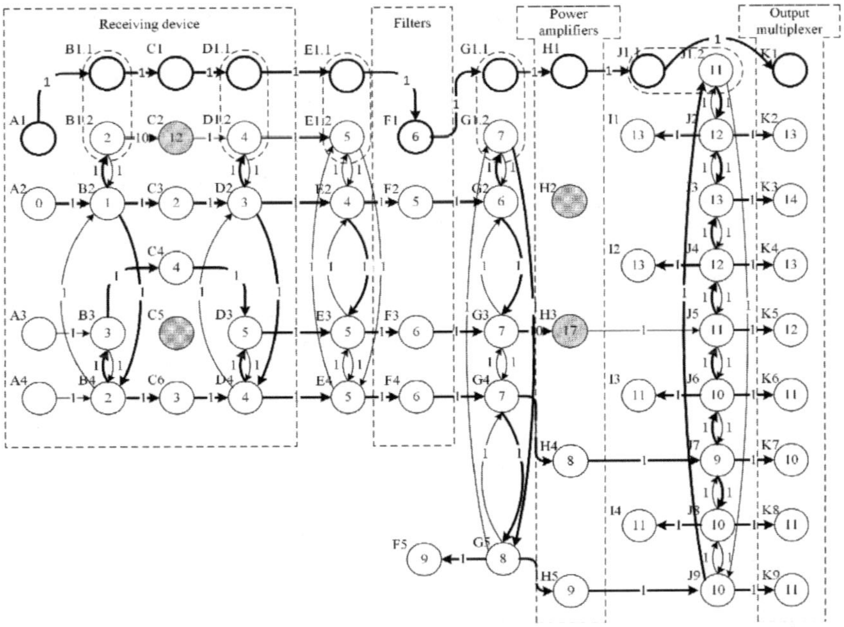

Figure 5. Weighted oriented graph corresponding to the current repeater scheme

Further, it is necessary to identify the shortest route from one node that represents an input port to all possible graph nodes. To solve this task Dijkstra's algorithm is proposed. This algorithm allows finding the shortest routes from one node to all possible graph nodes in the weighted oriented graph for arcs with positive value weight.

Thus, we can identify all possible repeater configurations for the chosen input port. Figure 5 shows the example of this solution for A2 input node. The value of the shortest route to each node is marked up in the center of each node and is graphically represented using bold graph arcs.

Further, the operator is offered to choose one of the possible transponder configurations according to the matrix of the priority output ports. This considers the current repeater configuration and devices states. In case output port K3 has a higher priority for input port A2 the possible shortest routes for the specified transponder are the following:

A2 - B2 - B4 - C6 - D4 - E4 - F4 - G4 - G5 - H5 - J9 - J1 - J2 - J3 - K3
A2 - B2 - C3 - B2 - D4 - E4 - F4 - G4 - G5 - H5 - J9 - J1 - J2 - J3 - K3
A2 - B2 - C3 - D2 - E2 - F2 - G2 - G1.2 - G5 - H5 - J9 - J1 - J2 - J3 - K3
A2 - B2 - B4 - C6 - D4 - E4 - F4 - G4 - H4 - J7 - J6 - J5 - J4 - J3 - K3
A2 - B2 - C3 - B2 - D4 - E4 - F4 - G4 - H4 - J7 - J6 - J5 - J4 - J3 - K3

Thus, the proposed algorithm complexity will be equal to $O(n^2)$ at worst, and at best it will be $\Omega(n)$. It has been experimentally proven that finding the shortest route for OCE, containing about 300 devices and about 200 switches on PC with Intel Core i3 processor, is less than 1 second. This allows the operator to make prompt decisions about OCE reconfigurations or to execute it in an automatic mode.

References

[1] Gulgonul S E, Koklukaya, Erturk I and Tesneli A Y 2012 *Proceedings of the First IEEE Aerospace and Electronic Systems Society (AESS) European Conference on Satellite Telecommunications* 6400090

[2] Stathakis A, Danoy G, Bouvry P, Talbi E and Morelli G 2015 *Engineering Optimization* ***47(12)*** *1709-1734*

[3] Savenko I I, Sukhodoev M S, Tsapko S G and Simonov P K 2015 *International Siberian Conference on Control and Communications, SIBCON 2015* 7147228
[4] Savenko I I, Sukhodoyev M S, Kosmynina N A, Ryzhkov D N and Tsapko S G 2014 *Components of Scientific and Technological Progress* **19(1)** 31-33
[5] Martynova Y A, Martynov Y A, Mustafina D B and Asmolovskiy V V 2014 *Proceedings of 2014 International Conference on Mechanical Engineering, Automation and Control Systems, MEACS 2014* 6986883
[6] Teresa M Braun 2012 *John Wiley & Sons, New York, A.B.D.*
[7] Ozhiganov A A 2013 *Tutorial*

IOP Conf. Series: Materials Science and Engineering **124** (2016) 012099 doi:10.1088/1757-899X/124/1/012099

New effective moduli of isotropic viscoelastic composites. Part I. Theoretical justification

A A Svetashkov, A A Vakurov
National Research Tomsk Polytechnic University, 30, Lenina ave., Tomsk, 634050, Russia

E-mail: svetashkov@tpu.ru

Abstract. According to the approach based on the commonality of problems of determining effective moduli of composites and viscoelastic solids, which properties are time-inhomogeneous, it is assumed that a viscoelastic solid is a two-component composite. One component displays temporal properties defined by a pair of Castiglianian-type effective moduli, and the other is defined by a pair of Lagrangian-type effective moduli. The Voigt and Reuss averaging is performed for the obtained two-composite solid with the introduction of a time function of volume fraction. In order to determine closer estimates, a method of iterative transformation of time effective moduli is applied to the viscoelastic Voigt–Reuss model. The physical justification of the method is provided. As a result, new time effective moduli of the viscoelastic solid are obtained which give a closer estimate of temporal properties as compared to the known models.

1. Introduction

The problems of calculating the stress-strain state of viscoelastic solids are reduced to the solution of a system of integro-differential equations of equilibrium in the region occupied by the solid, with specified boundary loads and displacements. At the same time, these problems can be considered from the viewpoint of composite mechanics because composite properties are inhomogeneous with respect to coordinates and properties of viscoelastic solids, which are time-inhomogeneous. This makes the application of effective modulus theory to viscoelasticity [1-4] possible. Earlier we derived expressions for time effective moduli of the Lagrangian and Castiglianian type [5] and used them to construct transformations for the Voigt–Reuss model [6]. However, this model gives insufficiently close estimates for approximate solutions, as in the case of elastic composites.

The present paper is aimed at obtainment of iteratively transformed time effective characteristics. Earlier we successfully applied this approach for composites [7]. The requirements to iterative transformations are the following:

(i) classical theorems on minimum of strain energy and additional work functionals must be valid, and

(ii) inequalities for effective moduli (Voigt–Reuss bounds) on each iteration step must be satisfied.

The convergence of iterative transformation of time effective characteristics is verified both numerically and by analytical determination of the limit of the iteration sequence.

2. Derivation of expressions for iterated effective moduli

The constitutive equations of a linearly viscoelastic solid is as follows:

$$s_{ij}(t)=\int_0^t R(t-\tau)de_{ij}(\tau)\equiv G^* e_{ij},$$
$$\sigma(t)=\int_0^t K(t-\tau)d\theta(\tau)\equiv K^*\theta. \tag{2.1}$$

Here, s_{ij}, e_{ij} are the deviatoric stress and strain components, and $\sigma(t) = \sigma_{ii}(t)$, $\theta(t) = \varepsilon_{ii}(t)$, $(i = 1, 2, 3)$, $R(t)$, $K(t)$ are the shear and bulk relaxation functions.

The relations inverse to Equations (2.1) have the following form:

$$e_{ij}(t)=\int_0^t \Pi(t-\tau)ds_{ij}(\tau)\equiv G^{*-1}s_{ij},$$
$$\theta(t)=\int_0^t \Pi_1(t-\tau)d\sigma(\tau)\equiv K^{*-1}\theta, \tag{2.2}$$

$\Pi(t)$, $\Pi_1(t)$ are the shear and bulk creep functions, and G^*, K^*, G^{*-1}, K^{*-1} are the shorthand notations for direct and inverse integral operators.

Approximate constitutive equations can be represented as

$$s_{ij}(t)=g_L(t)e_{ij}(t),\quad e_{ij}(t)=g_C(t)s_{ij}(t),$$
$$\sigma(t)=k_L(t)\theta(t),\quad \theta(t)=k_C(t)\sigma(t), \tag{2.3}$$

where the Lagrangian and Castiglianian-type time effective moduli are determined by relations [5]:

$$g_L(t)=G^*h,\quad g_C(t)=\left(G^{*-1}h\right)^{-1},$$
$$k_L(t)=K^*h,\quad k_C(t)=\left(K^{*-1}h\right)^{-1}, \tag{2.4}$$

h is the Heaviside function.

Using Equations (2.4), we constructed the Voigt–Reuss models of time effective characteristics [6]:

$$G_V(t)=\gamma g_C(t)+(1-\gamma)g_L(t),\quad G_R(t)=\left(\frac{\gamma}{g_C(t)}+\frac{1-\gamma}{g_L(t)}\right)^{-1},$$
$$g_C(t)\ge g_L(t),\quad \gamma=1-\alpha\frac{g_C(t)-g_L(t)}{g_C(t)}, \tag{2.5}$$

where α is a numerical parameter [6].

Let us introduce expressions for iteration sequences:

$$G_V^{(n)}(t)=\gamma G_V^{(n-1)}(t)+(1-\gamma)G_R^{(n-1)}(t),$$
$$G_R^{(n)}(t)=\left(\frac{\gamma}{G_V^{(n-1)}(t)}+\frac{1-\gamma}{G_R^{(n-1)}(t)}\right)^{-1},\quad n=1,2,... \tag{2.6}$$

It is possible to show that sequences $G_V^{(n)}(t)$, $G_R^{(n)}(t)$ converge to the same limit at $n\to\infty$:

$$G_V^{(n)}(t)\to G_R^{(n)}(t)\to\sqrt{G_V(t)G_R(t)}. \tag{2.7}$$

Sequences $K_V^{(n)}(t)$, $K_R^{(n)}(t)$ have the same limit at $n\to\infty$:

$$K_V^{(n)}(t)\to K_R^{(n)}(t)\to\sqrt{K_V(t)K_R(t)}. \tag{2.8}$$

3. Physical justification

Let us consider the strain energy and additional work functionals of a homogeneous isotropic elastic solid with time-dependent elastic moduli:

$$\Pi = \int_V W\left(e_{ij}\right)dv - \int_V \rho f_i u_i dV - \int_{S_t} t_i u_i ds,$$
$$\Psi = \int_V \Lambda\left(\sigma_{ij}\right)dv - \int_{S_u} t_i u_s^i ds,$$
$$W\left(\varepsilon_{ij}\right) = \frac{1}{2}\left[\left(K(t) - \frac{2}{3}G(t)\right)\theta^2 + 2G(t)\varepsilon_{ij}\varepsilon_{ij}\right], \tag{3.1}$$
$$\Lambda\left(\sigma_{ij}\right) = \frac{1}{2}\left[\frac{1}{2G(t)}\sigma_{ij}\sigma_{ij} - \frac{3K(t)-2G(t)}{6K(t)+2G(t)}\Sigma^2\right],$$
$$\Sigma = \sigma_{ii,} \quad \theta = \varepsilon_{ii,} \quad (i, j = 1,2,3).$$

Here, $K(t)$, $G(t)$ are the elastic bulk compression and shear moduli, t_i, u_s^i are the stresses and displacements specified on portions of boundary $S = S_t U S_u$, σ_{ij}, ε_{ij}, u_i are the stress tensor, strain tensor and displacement vector components, and f_i, t_i are the bulk and surface force vector components. Functionals Π, Ψ assume minimum values on admissible displacement and stress fields that obey equilibrium and compatibility equations.

Now let us consider Voigt–Reuss bounds

$$G_V \geq G_* \geq G_R, \quad K_V \geq K_* \geq K_R. \tag{3.2}$$

Here, G_*, K_* are the moduli of the homogeneous elastic medium whose specific potential energy is equal to the corresponding potential of a two-component composite with moduli (2.4). Thus, the Voigt and Reuss inequalities give the upper and lower bounds for moduli G^*, K^*. However, using Equations (3.2) it is impossible to calculate G^* and K^* as well as specific strain energy W_* and additional work Λ_*.

Let us now demonstrate that iteration sequences (2.6) can be used to derive expressions for G_*, K_* and W_*, Λ_*.

Since we consider an elastic medium with constitutive equations (2.4), functionals Π, Ψ and specific potentials W, Λ depend on time t. The same is true of sought moduli G_*, K_*. We assume that minimum values $\Pi_* = \Pi_{\min}$, $\Psi_* = \Psi_{\min}$ correspond to moduli G_*, K_* for each particular value of t. We denote the specific potentials corresponding to G_V, $G_V^{(1)}, \dots, G_V^{(n)}$ and G_R, $G_R^{(1)}, \dots, G_R^{(n)}$ by W, W_1, … , W_n and Λ, Λ_1, … , Λ_n. Then, owing to inequalities

$$G_V \geq G_V^{(1)} \geq \dots \geq G_V^{(n)}, \quad G_R \leq G_R^{(1)} \leq \dots \leq G_R^{(n)}, \tag{3.3}$$

the following inequalities are fulfilled:

$$W \geq W_1 \geq \dots \geq W_n, \quad \Lambda \geq \Lambda_1 \geq \dots \geq \Lambda_n. \tag{3.4}$$

Since moduli $G_V^{(n)}$ and $G_R^{(n)}$ for any n satisfy inequality

$$G_V^{(n)} \geq G_* \geq G_R^{(n)}, \tag{3.5}$$

inequalities

$$W_n \geq W_*, \quad \Lambda_n \geq \Lambda_* \tag{3.6}$$

are also satisfied for the specific potentials.
In the limit at $n \to \infty$ we have:

$$W_n \to W_*, \quad \Lambda_n \to \Lambda_*. \tag{3.7}$$

The Voigt–Reuss bounds shrink to a single point in which moduli $G_V^{(n)}$, $G_R^{(n)}$ become equal to the limit (2.7). Thus, the sought limit values of G_*, K_* really correspond to specific potentials $W_*(\varepsilon_{ij})$, $\Lambda_*(\sigma_{ij})$ equal to the specific potential energy and specific additional work of the two-component elastic solid with constitutive equations (2.3).

4. Discussion of results

Expressions for new effective characteristics of viscoelastic solids were derived using the following mathematical procedures:

1. Approximate representation of constitutive equations of viscoelasticity through relations of elastic Hooke's law with time-dependent moduli of Lagrangian and Castiglianian type.
2. Representation of the elastic medium as a two-component one, and formulation of the Voigt–Reuss model for the averaging of properties based on this representation.
3. Construction of iteration sequences for the Voigt–Reuss moduli.
4. Application of minimum variational principles for the functionals of potential strain energy and additional work.

It has been shown that the sequence of iteratively transformed Voigt and Reuss moduli allows the contraction of the Voigt–Reuss bounds. In the limit we obtain effective characteristics of a homogeneous isotropic elastic medium with specific strain and stress potentials equal to the corresponding potentials of a two-component medium. These effective characteristics can be called energetically equivalent.

The reported theoretical findings can be applied to calculate the stress-strain state of structures made of viscoelastic materials, which will be discussed in the second part of this paper [2].

5. Conclusion

Contraction mappings for the Voigt–Reuss bounds were obtained for a two-component homogeneous isotropic elastic medium with time-dependent moduli.

It was shown using theorems on minimum potential strain energy and additional work that the limit of the sequence of iteratively transformed Voigt–Reuss moduli is the effective characteristics of energetically equivalent specific strain and stress potentials of the homogenous medium of comparison. Expressions for these characteristics have been derived.

References

[1] Christensen R M 1979 *Mechanics of Composite Materials* (New York: Academic Press)
[2] Svetashkov A A 2000 *Mechanical of composite materials* **36(1)** 37
[3] Sendeckyj G P 1974 *Composite Materials, Academic Press* 45
[4] Pobedrya B E 1984 *Mechanics of Composites* (Moscow : Izd-vo MGU)
[5] Svetashkov A A 2012 *Applied Problems of the Mechanics of Viscoelastic Materials (*Tomsk : Izd-vo TPU)
[6] Svetashkov A A, Kupriyanov N A and Manabaev K K 2012 *Comp. Cont. Mech.* 292
[7] Svetashkov A A, Kupriyanov N A and Manabaev K K 2013 *Izv. VUZov. Fizika* 206

New effective moduli of isotropic viscoelastic composites. Part II. Comparison of approximate calculation with the analytical solution

N A Kupriyanov, F A Simankin and K K Manabaev
National Research Tomsk Polytechnic University, 30, Lenina ave., Tomsk, 634050, Russia

E-mail: mkk@tpu.ru

Abstract. A new approximate algorithm for calculating a stress-strain state of viscoelastic bodies is used. The algorithm is based on the derivation of the expressions of time-effective modules. These modules are obtained by iterative changes, compressing the fork of Voigt-Reuss. As an example the analytic solution about the action of a concentrated force on the viscoelastic half-space is compared with the approximate solution. Numerical calculations are performed for a wide range of relaxation characteristics of a viscoelastic half-space. Results of the comparison of stresses and displacements with the analytic solution give coincidence within 3...15 %.

1. Introduction

Calculations of the stress-strain state of linear viscoelastic bodies are associated with the solution of the system of integral-differential equations of equilibrium for given stresses or displacements on the border [1]. Approximate approaches are used in connection with the complexity of numerical implementation of such decisions. The most famous of them are: a) the Scheper method [2], b) the method of quasi constant operators [3], an iterative method [4], the approximation method of Il'yushin [1]. An approximate method of the solution based on the concept of time-effective modules [5] is used in the present work.

The purpose of this work is to carry out the calculations of the stress-strain state for a viscoelastic body with new time-effective characteristics [6] by the example of Boussinesq's problem.

2. The action of a concentrated force on the viscoelastic half-space

Exact solutions of the problems of linear viscoelasticity can be obtained on the base of known analytical solutions of elasticity problems. According to the method of Volterra, the elastic constants are replaced by the corresponding integral operators. Then, decoding is performed – some operator function is transformed to an integral equation with the unknown nucleus and the calculation of integral is performed by a given time-function load.

We consider the elastic solution of the Boussinesq's problem [7] about the action of force P on a half-space. We have the following axially symmetric distribution of the components (displacements and stresses) in cylindrical coordinates (r, z):

$$
\begin{aligned}
u(r,z) &= [\frac{rz}{4\pi R^3}\frac{1}{G} - \frac{r}{4\pi R(R+z)} f_1(\nu,G)]P, \\
w(r,z) &= [\frac{1}{2\pi R} f_2(\nu,G) - \frac{z^2}{4\pi R^3}\frac{1}{G}]P, \\
\sigma_r(r,z) &= [\frac{1}{2\pi R(R+z)} f_3(\nu,G) - \frac{3r^2 z}{2\pi R^5}]P, \\
\sigma_\varphi(r,z) &= -(\frac{R}{R+z} - \frac{z}{R})\frac{1}{2\pi R^2} f_3(\nu,G)P.
\end{aligned} \tag{1}
$$

Here r, z are the radial and axial coordinates of an arbitrary point of a half- space, $R^2 = r^2 + z^2$, ν, G are resiliently instant Poisson's ratio and shear modulus.

$$
f_1(\nu,G) = \frac{1-2\nu}{G}, \quad f_2(\nu,G) = \frac{1-\nu}{G}, \quad f_3(\nu,G) = (1-\nu). \tag{2}
$$

Axial σ_z and shearing τ_{rz} stresses do not depend on the elastic constants, therefore, they will not depend on time.

We define material operators of shearing relaxation G^* and creep G^{*-1} as:

$$
\begin{aligned}
G^* x &\equiv G(1 - \lambda \mathcal{Э}^*_{\lambda+\gamma})x, \\
G^{*-1} x &\equiv \frac{1}{G}(1 + \lambda \mathcal{Э}^*_{\gamma})x, \\
\mathcal{Э}^*_{\gamma} x &\equiv \int_0^t e^{-\gamma(t-\tau)} x(\tau) d\tau.
\end{aligned} \tag{3}
$$

Here γ, λ are the function parameters of relaxation and creep.

We need to make substitutions $\nu \to \nu^*, G \to G^*$ in (11) for the construction of an analytical solution of the problem and then we have to decipher the function of operators ν^*, G^*. We assume that the volumetric relaxation is absent, $K^* = K_0 = const$

We introduce the following operators:

$$
\omega^* = \frac{2G^*}{3K_0}, \quad g^*_{1/2} = \frac{1}{1 + \frac{1}{2}\omega^*}.
$$

Here K_0 is the bulk modulus. Then operator ν^* is expressed by the operator of related creep $g^*_{1/2}$ as follows: $\nu^* = \frac{1}{2} g^*_{1/2}(1-\omega^*)$. The product of two integral operators of Volterra type enters the last relation. We use the transformation of the product of two operators into their difference by formula

$$
\frac{1}{2}\omega^* g^*_{1/2} = 1 - g^*_{1/2}. \tag{4}
$$

Then we obtain:

$$
f_1(\nu^*,G^*) = \frac{1}{K_0} g^*_{1/2}, \; f_2(\nu^*,G^*) = \frac{1}{2}(G^{*-1} + \frac{1}{K_0} g^*_{1/2}), \; f_3(\nu^*,G^*) = 2 - \frac{3}{2} g^*_{1/2}. \tag{5}
$$

As one can see, all three operator functions are expressed in terms of the operator of related creep by Il'yushin.

$$g^*_{1/2}x = \frac{2}{2+\omega_0}[1+\lambda\mu Э^*_q]x,$$
$$q = \lambda + \gamma - \lambda\mu, \quad \mu = \frac{\omega_0}{2+\omega_0}, \quad \omega_0 = \frac{2G}{3K_0}. \tag{6}$$

Thus, formulas (5) in conjunction with (3) and (4) give the analytic solution for the problem on loading of the viscoelastic half-space by force $P(t)$.

We obtain approximate solutions on base $G_V(t)$, $G_R(t)$, $G_*(t)$by replacements:

$$f_1^k = \frac{3}{3K_0 + g_k(t)}, \; f_2^k = \frac{1}{2}(\frac{1}{g_k(t)} + \frac{3}{3K_0 + g_k(t)}), \; f_3^k = 2 - \frac{3}{2}\frac{3K}{3K_0 + g_k(t)}, \tag{7}$$

where the index accepts values $k = 1, 2, 3$.

Here $G_V(t) = \gamma g_C(t) + (1-\gamma) g_L(t)$, $G_R(t) = \left(\frac{\gamma}{g_C(t)} + \frac{1-\gamma}{g_L(t)}\right)^{-1}$,

$$g_1(t) = g_L(t), \; g_2(t) = g_C(t), \; g_3(t) = G_*(t) = \sqrt{G_V(t)G_R(t)}, \tag{8}$$

where $g_L(t)$, $g_C(t)$ are known modules of Lagrange and Castigliano.

2. The analysis of numerical results

Numerical calculations of analytical and approximate solutions for Boussinesq's problem were made for three cases, where parameter η that determines the ratio of the resiliently instant module to long-termed module $\eta = G_0 / G_\infty$, takes the values equal to 3, 5, 10.

1) $\lambda = 0.0138$ min^{-1}, $\gamma = 0.0069$ min^{-1}, η =3,
2) $\lambda = 0.0276$ min^{-1}, $\gamma = 0.0069$ min^{-1}, η =5,
3) $\lambda = 0.0621$ min^{-1}, $\gamma = 0.0069$ min^{-1}, η =10.

The boundary load was set as step function $P(t) = P_0\, h(t)$, where $P_0 = 100$ MPa. The values of the elastic constants are $G_0 = 120$ MPa, $K_0 = 360$ MPa. The volume content of one component which properties are determined by time-effective characteristics of Castigliano's type, has magnitude $\gamma = 0.8$. Curves of changes in time-effective modules of Voigt $G_V(t)$, Reuss $G_R(t)$ and $g_k(t)$, $(k = 1, 2, 3)$ are shown in Figure 1.

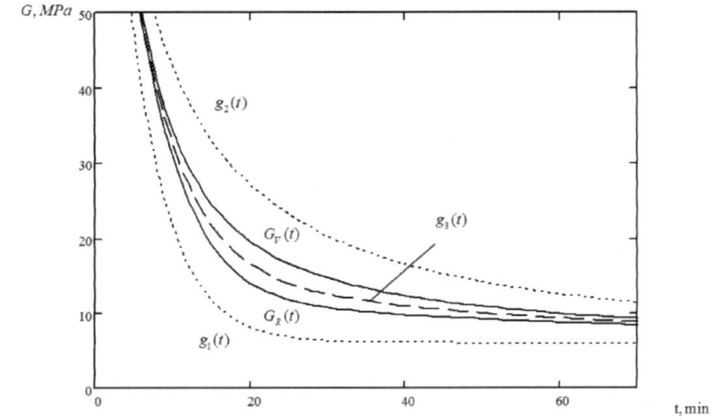

Figure 1. Curves of values for time-effective modules. $G_V(t)$, $G_R(t)$ are evaluations of Voigt and Reis, $g_1(t)$, $g_2(t)$, $g_3(t)$ are modules defined by (8), $\eta = 20$, $\gamma = 0.8$.

The curves of the time variation of the relative deviations of displacements u and stresses σ_r, σ_φ, obtained on the basis of efficient modules $g_n(t), n = 1,2,3$, are shown in Figures 2-4. Figures 2-4 illustrate the temporal character and the deviations of the approximate solutions against analytical solutions for the chosen values $\eta = 3, 5, 10$.

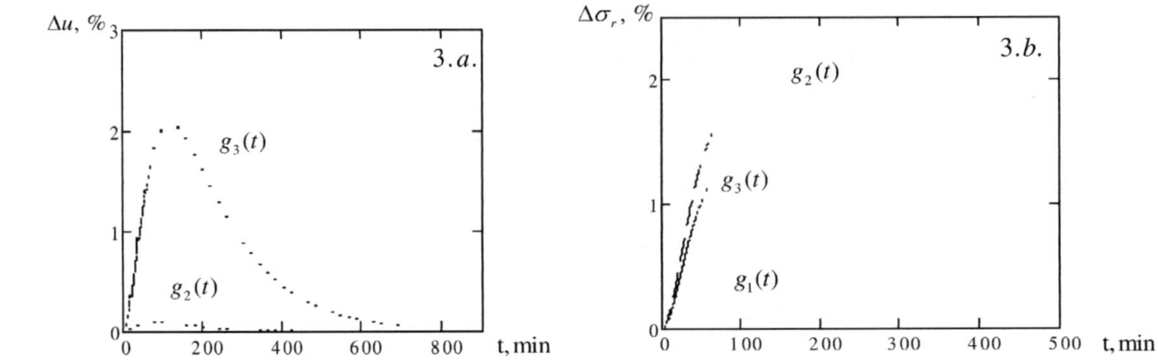

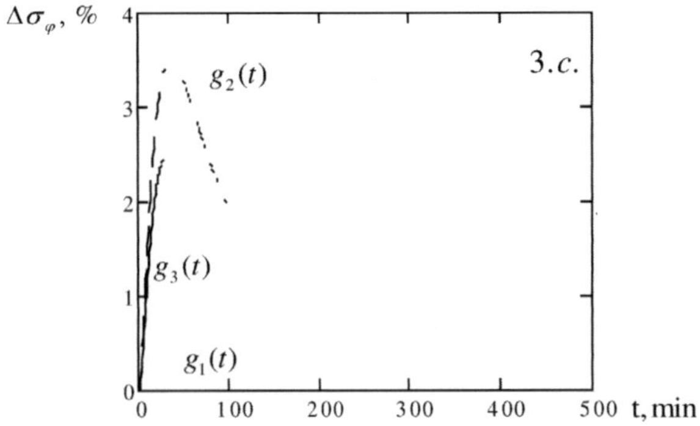

Figure 2. Graphs expressing changes of deviations for displacements u (*3.a*) and stresses σ_r, σ_φ (*3.b-c.*). Calculations for case $\eta = 3$

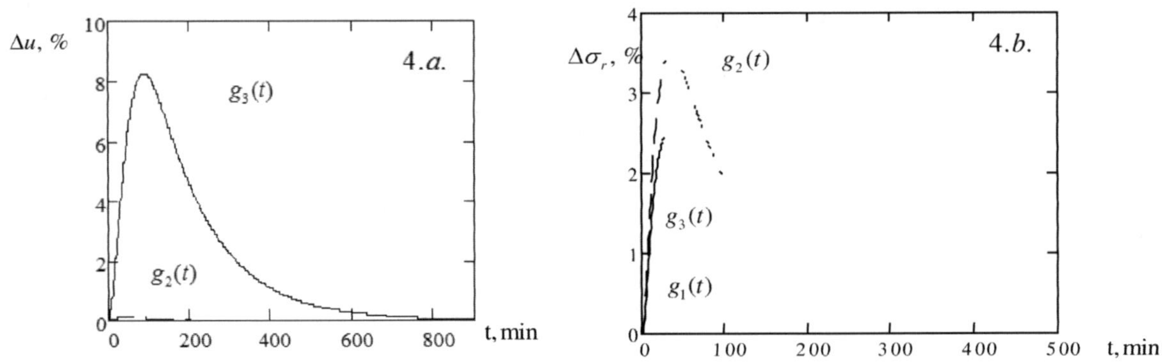

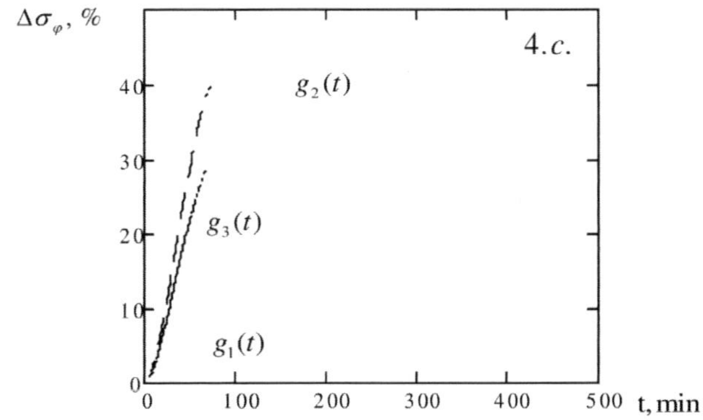

Figure 3. Graphs expressing changes of deviations for displacements u (*4.a*) and stresses σ_r, σ_φ (*4.b-c.*). Calculations for case $\eta = 5$

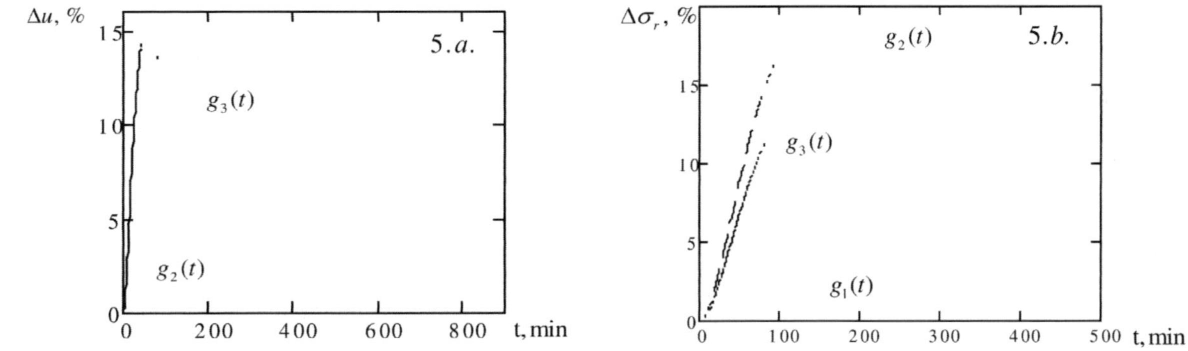

Figure 4. Graphs expressing changes of deviations for displacements u (*5.a*) and stresses σ_r (*5.b.*). Calculations for case $\eta = 10$.

3. Discussing the results

Calculations of the stress-strain state of a viscoelastic half-space (Boussinesq problem) with the help of the approximate method have a feature. It is an error in the calculation of stresses and displacements of different magnitudes. So, for a step-by-step load, the error of the calculation for displacements is practically zero when using time-effective modules of Castigliano's type. An opposite pattern is observed in the calculation of the stress: in this case, the calculations with effective modules of Lagrange's type give the minimum error. In this connection, we can use mathematical techniques applied in the mechanics of composite materials: a selection of the optimal value for the specific volume content of a component from a position of the minimum error for calculations of stresses and displacements; administering an effective time module obtained by compressing the fork of Voigt-Reuss.

4. Conclusions

1. The comparison of analytical and approximate solutions gives the following picture for the distribution of errors. The calculations of stresses and displacements give the minimal error (2…3 %) under $\eta = 3$. The deviations for displacements are less than 8 % under $\eta = 5$. Under $\eta = 10$ deviations for stresses and displacements are less than 15 %.
2. Value $\gamma = 0.8$ corresponds to the minimum magnitude of the error in the calculation of stresses and displacements.

References

[1] Il'yushin A A and Pobedrya B E 1970 *Fundamentals of the mathematical Theory of thermoviscoelasticity* (Moscow)
[2] Scheperi R 1978 *Mekhanika kompozitsionnykh materialov* 102
[3] Malyy V I and Trufanov N A 1989 *Sverdlovsk: UrO RAN* 78
[4] Pavlov S M and Svetashkov A A 1993 Physics Journal 400
[5] Svetashkov A A 2000 *Mechanical of composite materials* **36(1)** 37.
[6] Svetashkov A A, Kupriyanov N A and Manabaev K K 2012 *Comp. Cont. Mech.* 292
[7] Timoshenko S P and Goodier J N 1970 *Theory of elasticity* (McGraw-Hill, N.Y.)

IOP Conf. Series: Materials Science and Engineering **124** (2016) 012101 doi:10.1088/1757-899X/124/1/012101

Computer realization of research into the dynamics of mechanical systems

V Telegin[1] and A Kozlov[2]
Lipetsk State Technical University, 30 , Moskovskaya st., Lipetsk, 398600, Russia

E-mail: vv.telegin@yandex.ru

Abstract. The article considers the application of methods of component modeling and of object-oriented technologies in automated construction of dynamic models of mechanisms in their program realization.

1. Dynamic processes in mechanical systems (mechanisms)

A mechanical system is a set of bodies with elastic and inertial properties making it possible to transform the movements of some bodies into the required movements of other bodies, which is widely used in technological machines for obtaining both the necessary form of the product [1] and the required parameters of surface quality [2]. Dynamic processes are mechanical oscillations in mechanical systems. The method of calculating dynamic processes is known [3, 4]. It is based on the substitution of the real mechanism with a system of concentrated masses, which are connected by inertialess, elastic-dissipative and kinematic connections (a dynamic model). In practice, however, this method is not widely spread. The main reasons are as follows:

- low degree of accuracy and high labor intensity of calculating the model's parameters, primarily its elastic and inertial characteristics;
- high labor intensity of work connected with the development of the dynamic model of the mechanism, its mathematical description (a mathematical model), and the subsequent program realization of calculations on the basis of the mathematical model;
- impossibility to assess the reliability of the obtained results without further experimental research of the ready mechanism.

2. Objective representation of mechanical systems.

Any mechanism can be regarded as a set of several elements connected among themselves by various connections. Through these connections, the elements of the mechanism interact with each other, thus providing its functioning. The term 'element' will designate a separate part of the mechanism, a fragment of a part, or, on the contrary, several details at once. Such definition of the element corresponds to the concept of an object – a basic concept in the theory of component modeling of complex dynamic systems and object-oriented technologies [4].

We will compare each element of a mechanism with an object according to the above-mentioned definition. Any processes in such object can be modeled in any way and, according to the principles of encapsulation and polymorphism of object-oriented technologies, at this stage, their nature and mathematical, program or physical description are disregarded.

Such approach is of interest in solving tasks dealing with development of both dynamic and mathematical models of mechanisms [5, 6] and, which is singularly important, their program realization in further computer processing [1].

Let us imagine a real mechanism as a set of objects connected among themselves by transfer functions. Such objects can be represented by both separate links and sets of links forming elementary mechanisms. Let us connect five sets of characteristics with each object (Figure 1):

- Properties, including masses, moments of inertia and stiffness, dissipation coefficients, parameters of transfer functions and others defining an object as a closed system.
- Responses to disturbances from other objects.
- Power disturbances from other objects.
- Input parameters including kinematic and power.
- Output parameters (kinematic, power, values of criteria of quantitative assessment of dynamic processes in an object).

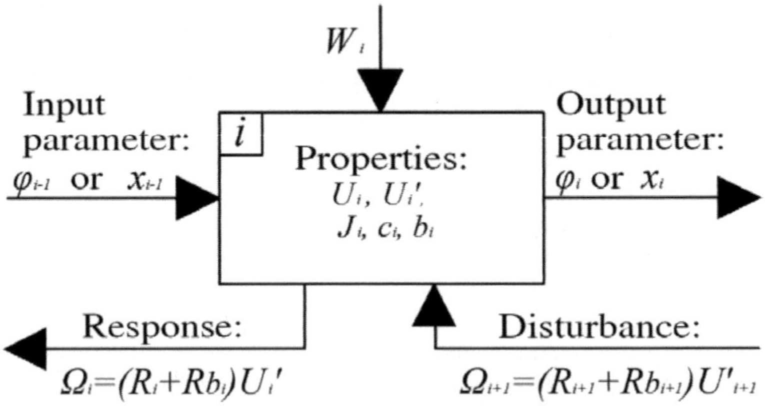

Figure 1. The object's parameters and features

Properties and responses (labels 1 and 2) are values entirely depending on the object. The first constants are independent of external conditions, the second ones depend on the environment in which the object is placed. Power disturbances and input parameters (labels 3 and 4) are the environment of the object. Output parameters (label 5) are a set of the resulting data making it possible to estimate the processes inside the object after it is placed in a certain environment. It should be particularly noted that a set of objects representing a mechanism (object representation) and the mechanism itself completely fit each other. Reality can be substituted with a model only on the level of an object. It means that the accuracy of the results of dynamic research is largely determined by the 'high quality' of the objects which are used in this research. An object and an object model, as is understood from the above, are different concepts. However, in further discussion, the term 'object' will designate its model.

The input parameters of an object are the function of the (angular or linear) relocation of the object preceding it ($\boldsymbol{\varphi_{i-1}}$) and external loads $\boldsymbol{W_i}$ on this object (e.g., forces and moments of friction, constructional and technological ones). The output parameter, respectively, is the function of the relocation of the output link. Properties are a set of constants determining the quantitative characteristics of the model corresponding to this object. Among them there are parameters of position functions and transfer functions (for the entrance link they are $\boldsymbol{U_i}$ and $\boldsymbol{U_i'}$), elastic-inertial and dissipative characteristics. Response, which is in fact an output parameter, is the influence of elastic forces $\boldsymbol{R_i}$ and dissipation forces $\boldsymbol{R_{bi}}$ in the first fragment of the entrance link on the previous object. Disturbance is a response from the following object(s) to the influence of the given object.

3. Computer simulation of dynamic processes in mechanical systems

The system of computer simulation of dynamic processes in mechanisms has been developed in order to decrease the labor intensity of dynamic research of mechanical systems and to increase their reliability. It is based on methods of solid-state modeling [6], methods of component modeling and of object-oriented technologies [4, 5].

The complex of programs combined in a single system (***dam*** – dynamic analysis of mechanisms) includes a number of modules the basic of which is as follows:

1. The creation and opening of the project. A project is a specifically organized set of various files storing information on objects, connections among them and their parameters, as well as the description of the mechanism being researched.
2. The constructor of an object model allows the user in the interactive mode to create an object model of the mechanical system using the database of the existing object models of standard mechanisms and their elements.
3. Modules of input, editing and testing object parameters. The characteristics of these modules (their number is equal to the number of objects) are:

- Determination of the number of masses (complexity) of the model and the output weight. Already at the stage of determining the parameters of the object, its dynamic characteristics, those depending on the complexity of the model, as well as those at different values of its elastic-inertial parameters in the field of their acceptable values can be determined. In fact, such procedure makes it possible to substitute the natural experiment with the numerical one and, thus, to check "experimentally" each of the objects of the object representation of the mechanism for "high quality", which to some extent removes the question of the reliability of the results of dynamic research in the ***dam*** system.
- Input of elastic-inertial and dissipative parameters of the model. The values of these parameters can be determined with a sufficiently high accuracy in designing the mechanism in a system of solid-state modeling [6].
- Availability of the library of standard position functions making it possible to realize the following scheme of the ideal movement of the object's entrance link in various options: rising (forward stroke) – upper dwell – lowering (reverse stroke) – lower dwell.
- Data input on gaps and manufacturing errors.
- Wide opportunities for determining the functions of external loads for each mass of the dynamic model of the object and their division into three groups: friction forces, constructional and technological loads.
- Storing various versions of data separately according to elastic-inertial characteristics, position functions, gaps, manufacturing errors, and external loads for each object.
- A possibility to test the entered data.

4. The dynamic analysis of the mechanism. This module of the application is a set of dynamic models corresponding to the object representation of the mechanism; it transforms systems of differential equations into the mathematical model, solves them with certain results. A reference should be made to wide opportunities in connection with the object data management:

- Managing the productivity of a mechanism.
- Excluding (or changing) different types of loads, dissipative forces, gaps from the simulation model during their calculations.
- Determining the reliability and operability of the mechanism (or its separate object) by a number of criteria and graphic dependences.
- Preserving the results for the purpose of their further comparison in selecting the most rational constructive decision at the stage of designing the mechanism.

IOP Conf. Series: Materials Science and Engineering **124** (2016) 012101 doi:10.1088/1757-899X/124/1/012101

4. Practical application of modeling dynamic processes in the *dam* system

Figures 1 and 2 present a solid-state model of the transfer mechanism of the AB1818 cold upsetting machine and its object model created in the ***dam*** system.

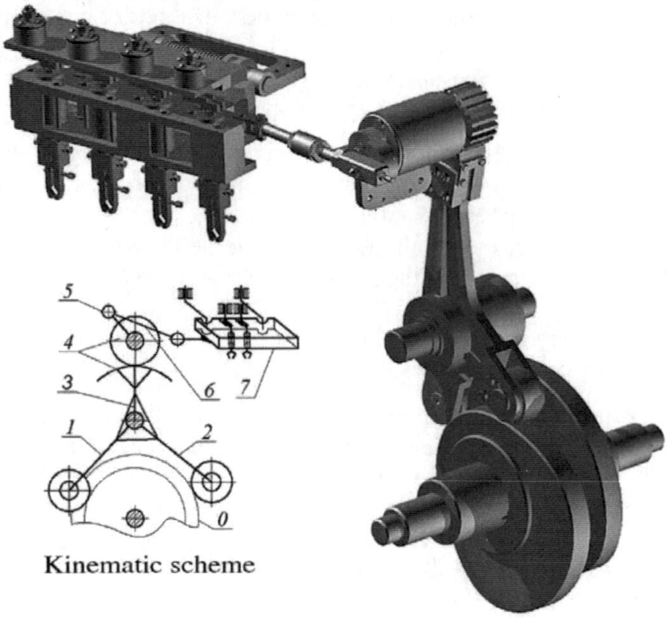

Figure 2. The transfer mechanism of the cold forging machine (a solid state model and a kinematic scheme):
0 – cam unit, 1, 2, 3 – direct and reverse strokes levers, 4 – gear transmission, 5 – crank, 6 – rod, 7 – transfer carriage

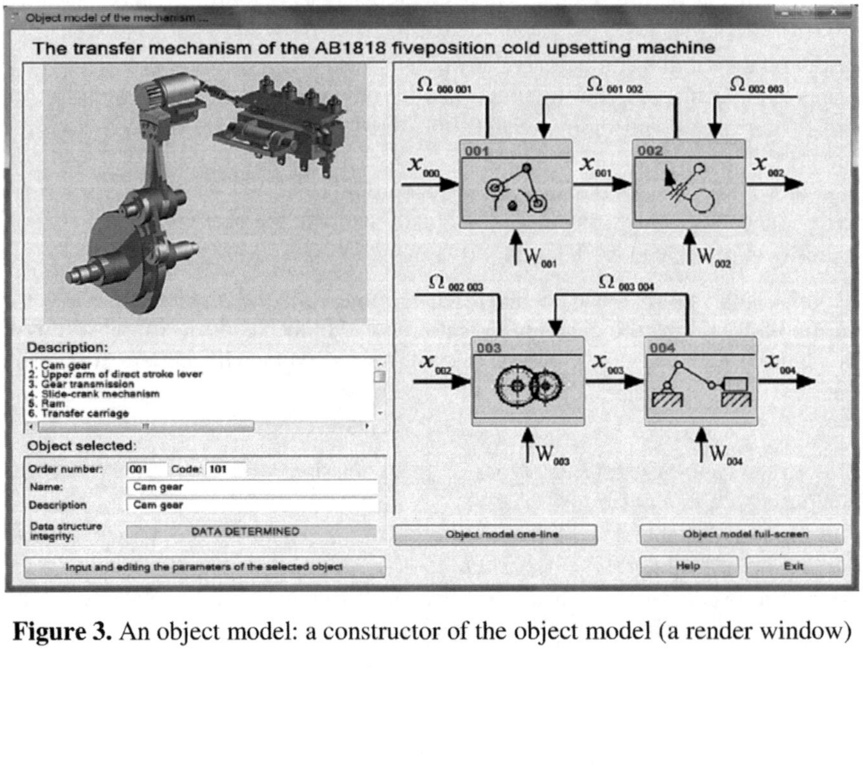

Figure 3. An object model: a constructor of the object model (a render window)

Figure 4 shows a render window for the research of the dynamics of the transfer mechanism of the cold forging machine. What is determined here is the researched object - in Figure 4 it is the slide-crank mechanism of transferring the carriage, the operation speed of the mechanism, its mode of work (single-stroke or automatic), the initial and final positions of the input link. Values of these objects: a number of masses, kinematic, elastic-inertial and dissipative characteristics, loads, parameters of gaps and the manufacturing error are assigned to development of the object model of the mechanism. Some of these data (gaps, manufacturing errors, forces of dissipation and friction, constructional loads and technological efforts) can be excluded from the modeling process.

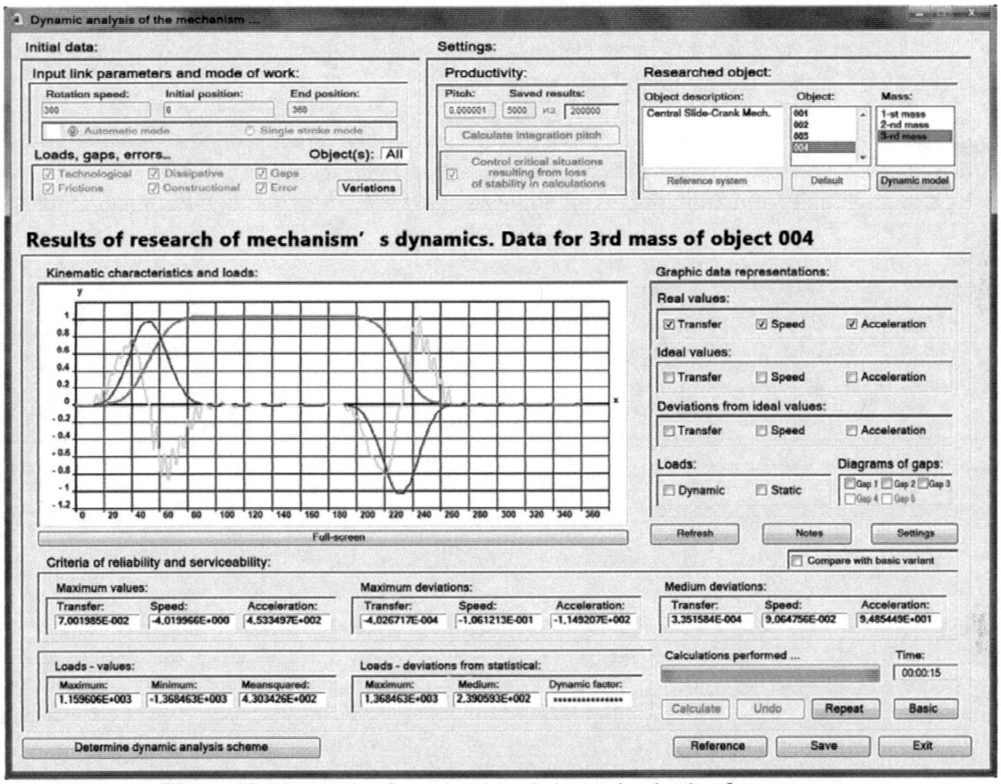

Figure 4. Research of mechanisms' dynamics in the ***dam*** system

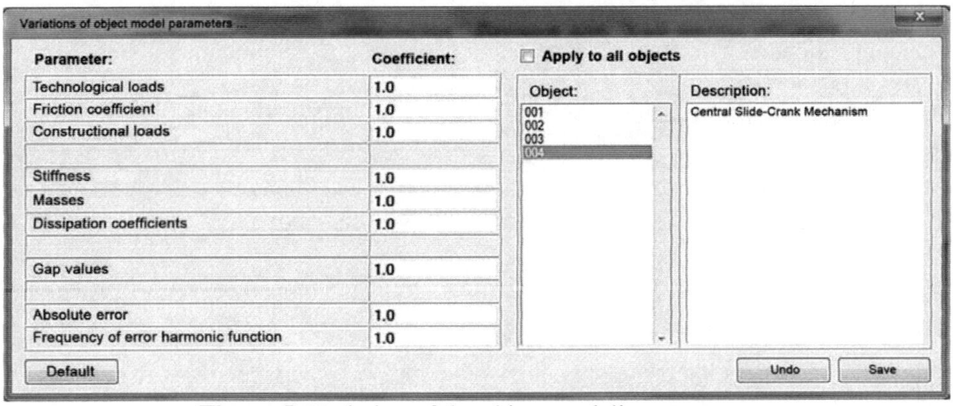

Figure 5. Variation of the object model's parameters

In constructing the strategies of simulation modeling, it is also possible to change the values of most object parameters in the 'Variations of Object Model Parameters' window. Variation possibilities for object model parameters, as is seen from Figure 5, are rather great.

The dynamic analysis of the mechanism is not only the calculation of its kinematic parameters, loads and generalized criteria of quality, but also the tasks connected with forecasting dynamic characteristics depending on the change of external and internal influences and conditions [5]. The realization of similar tasks is possible in the same window ('The dynamic analysis of the mechanism') by means of the button switch 'Determine the dynamic analysis scheme'. The system of the dynamic analysis of the mechanism suggests that the user determine one of 18 types of research each of which is the automation of the dependences construction of the following type: criterion (Y-axis) – changeable parameter (X-axis). After calculations it is possible to determine any of the ten criteria of the quality and to construct corresponding dependences.

References

[1] Kozlov A M and Kozlov A A 2009 *Russian Engineering Research* **29** 743–746
[2] Kozlov A M and Kerusenko E B 2012 *Bulletin of the Ufa state aviation technical University* **4** 117–123
[3] Wolfson I I 1990 *L.: Engineering* 592
[4] Rumbaugh J 1991 *Prentice Hall* 500
[5] Telegin V V 2013 Moscow: The Fundamental Researches **11(5)** 899–904
[6] Telegin V V 2012 *Samara: Publishing house of Samara Science Center of Russian Academy of Science* **14(4)** 1306–1309

Numerical heat fluxcontrol forunsteady viscous fluid flows

D A Tereshko[1,2]

[1] Institute of Applied Mathematics, Far Eastern Branch of the Russian Academy of Sciences, 7, Radio st., Vladivostok, 690041, Russia
[2]School of Engineering, Far Eastern Federal University,8, Suhanova st., Vladivostok, 690950, Russia

E-mail: ter@iam.dvo.ru

Abstract. This paper is devoted to the boundary control problems for nonstationary equations of viscous heat-conducting fluid. In these problems we change the fluid flow by heating or coolingsomeparts of the boundary.Numerical algorithm based on the finite dimensional minimization approach is proposed. Some computational results for vortex suppression in a cavity are presented and discussed.

1. Introduction

Optimization and control problems for viscous heat-conduction fluid flows have a large number of applications in science and mechanical engineering. There are numerous objectives of control, e.g., the flow or temperature field optimization, drag minimization, preventing transition to turbulence, vortex reduction, enhancing or deterring mixing. Usually these problems are formulated as minimization problems for suitable cost functionals and can be analyzed and solved by applying a unified approach based on the constrained optimization theory (see, for example, [1, 2]). The same approach can be applied for inverse and parameter estimation problems. These problems for stationary flow models were studied in [3-7]. An unsteady case is more complicated from theoretical and numerical points of view and requires the development of special numerical algorithms [8, 9].

Numerical optimization for hydrodynamic mathematical models is connected with a number of difficulties. Minimization problems have some constraints because solutions must satisfy the governing equations. Usually a constrained minimization problem is rewritten as an unconstrained minimization problem using the Lagrange multipliers. Then one can formally calculate the derivatives of the Lagrangian and derive the necessary optimality conditions. However, a numerical solution of the optimality system for nonstationary equations is very difficult. Therefore we propose a new numerical algorithm to solve the control problem under consideration. This algorithm is based on the finite dimensional minimization approach where the control is considered as the linear combination of known basis functions with the unknown coefficients. The values of these coefficients are chosen in order to minimize some functional describing the objective of control.

A proposed algorithm is tested on the nonstationary control problem connected with the vortex suppression in the open cavity. We use the heat flux control on the cavity boundaries to obtain the fluid flow close to the potential flow with zero vorticity.

2. A mathematical model

Let Ω be a bounded two-dimensional domain with boundary Γ. For simplicity, we assume that density ρ depends on temperature T as $\rho=\rho^*[1-\beta(T-T^*)]$ according to the Boussines q approximation for the viscous fluid flow. Here ρ^* and T^* are the given reference density and temperature fields, β is the thermal expansion coefficient. As a mathematical model we will consider the following initial boundary value problem:

$$\frac{\partial \mathbf{u}}{\partial t}+(\mathbf{u}\cdot\nabla)\mathbf{u}+\frac{1}{\rho^*}\nabla p-\nu\,\Delta\mathbf{u}=-\beta\,\mathbf{G}(T-T^*),\quad \operatorname{div}\mathbf{u}=0 \text{ in } \Omega,$$

$$\mathbf{u}=\mathbf{u}_b \text{ on } \Gamma_{in},\quad \mathbf{u}=\mathbf{0} \text{ on } \Gamma_0\cup\Gamma_c,\quad \nu\frac{\partial \mathbf{u}}{\partial n}-p\mathbf{n}=\mathbf{0} \text{ on } \Gamma_{out},$$

$$\frac{\partial T}{\partial t}+\mathbf{u}\cdot\nabla T-\lambda\,\Delta T=0 \text{ in } \Omega,$$

$$T=T_b \text{ on } \Gamma_{in},\quad \lambda\frac{\partial T}{\partial n}=\chi \text{ on } \Gamma_c,\quad \frac{\partial T}{\partial n}=0 \text{ on } \Gamma_0\cup\Gamma_{out},$$

$$\mathbf{u}\big|_{t=0}=\mathbf{u}_0,\quad T\big|_{t=0}=T_0 \text{ in } \Omega$$

on the time interval $(0,\ t_{max})$. Here **u**, p and T denote the velocity, pressure and temperature respectively; ν is the kinematic viscosity coefficient, **G** is the gravitational acceleration vector, λ is the thermal diffusivity coefficient. We assume that boundary Γ consists of four parts: inlet Γ_{in}, solid walls Γ_0, Γ_c and outlet Γ_{out} (Figure 1).

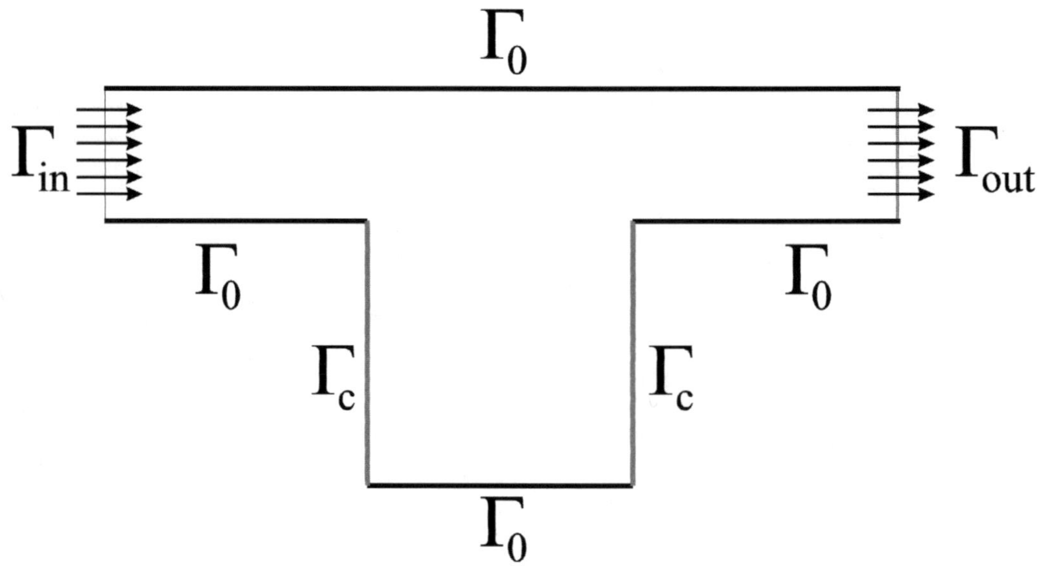

Figure 1. A computational domain.

On the inlet boundary section we prescribe the parabolic inflow profile for the velocity and the temperature of the fluid. We also consider no slip boundary condition $\mathbf{u}=\mathbf{0}$ on the solid walls and 'do nothing' boundary condition on the outlet. The vertical boundary sections Γ_c will be used below for heat flux control.

The open cavity is the common case in many channels and industrial devices. The main problem of these cavities connected with the formation of vortices and stagnant zones. The main flow over the cavity is usually separated from the vortex in the cavity. Small particles can be accumulated in the cavity and in a circular motion would damage the wall surface like sandpaper. So sometimes you need to wash away the contents of the cavity. In many cases it is very difficult to install special injectors in the bottom. Therefore, we need a way to suppress the vortex and change the flow without permeation of the walls. If the fluid density is dependent on temperature, we can do it by means of the buoyancy force with heating and cooling of some boundary sections. In order to find an appropriate heat flux we need to solve the boundary control problem.

3. A numerical algorithm

A proposed algorithm is based on the main idea of paper [10] where an optimal boundary control problem for the stationary Navier-Stokes equations was solved using the finite dimensional minimization approach. Now we will extend this idea on the nonstationary case and a more complicated mathematical model.

At the beginning we split time interval $(0, t_{\max})$ into N parts (t_{n-1}, t_n), $n = 1, \dots, N$. At each time step we shall simultaneously calculate heat flux boundary value χ and the corresponding solution of the initial boundary problem. Now we assume that we already know all values at time $t=t_{n-1}$ and describe the calculation of values at time step t_n. We shall consider heat flux $\chi(t_n)$ as a linear combination of known basis functions χ_i with unknown coefficients k_i, $i = 1, \dots, S$. Here k_i are the new unknowns specific for this time step.

We will use a semidiscrete weak formulation for the initial boundary problem with linear implicit one-step approximation in time and finite element approximation in space. Therefore velocity $\mathbf{u}(t_n)$ can be written as follows:

$$\mathbf{u}^n = w^n + \sum_{j=1}^{S} k_j \mathbf{u}_j$$

Where w^n is the solution for homogeneous boundary condition $\chi = 0$ while $\mathbf{u}_j$, $j = 1, \dots, S$ are the solutions corresponding $\chi = \chi_j$ and homogeneous boundary conditions for velocity and temperature.

If we want velocity$\mathbf{u}(t_n)$ to be close to given field $\mathbf{u}_d$ in some subdomain Ω_d then solution $a_{ij} = (\mathbf{u}_i, \mathbf{u}_j)_{\Omega_d} + \mu(\psi_i, \psi_j)_{\Gamma_C}$of the finite dimensional minimization problem for variables k_i, $i = 1, \dots, S$ can be found by solving the following system of linear algebraic equations:

$$\sum_{j=1}^{S} a_{ij} k_j = b_i, \quad i = 1, \dots, S$$

where

$$a_{ij} = (\mathbf{u}_i, \mathbf{u}_j)_{\Omega_d} + \mu(\chi_i, \chi_j)_{\Gamma_c}, \quad b_j = (\mathbf{u}_d - w^n, \mathbf{u}_j)_{\Omega_d}.$$

Here positive constant μ is a small regularization parameter.

The main steps of the algorithm are the following: find functions w^n, $\mathbf{u}_j$; calculate coefficient a_{ij}, b_j; and solve the linear system for k_j. Then we can find $\chi(t_n)$ and all other values on this time step.

Let us note that this algorithm does not use the first order necessary optimality conditions (see, for example, [1-7]) and is simpler to implement. It can be efficiently parallelized because $S+1$ boundary value problems to find w^n and $\mathbf{u}_j$ are solved independently.

4. Computational results

This numerical algorithm was applied for solving the vortex suppression problem in the open cavity (Figure 1).Fist of all, we find an uncontrolled flow solving the initial boundary value problem with homogeneous boundary condition $\chi = 0$. We solve the non-dimensional equations with Prandtl number $Pr = 7$, Reynolds number $Re = 10$ and Grashof number $Gr = 10000$. At dimensionless time $t=5$ the flow becomes stable. The resulting velocity field in the cavity is shown in Figure 2.

It is clearly seen that most of the cavity is occupied by a large vortex.The main flow takes over only a small part and does not reach the bottom.We want to wash out this stagnant zone from the cavity.

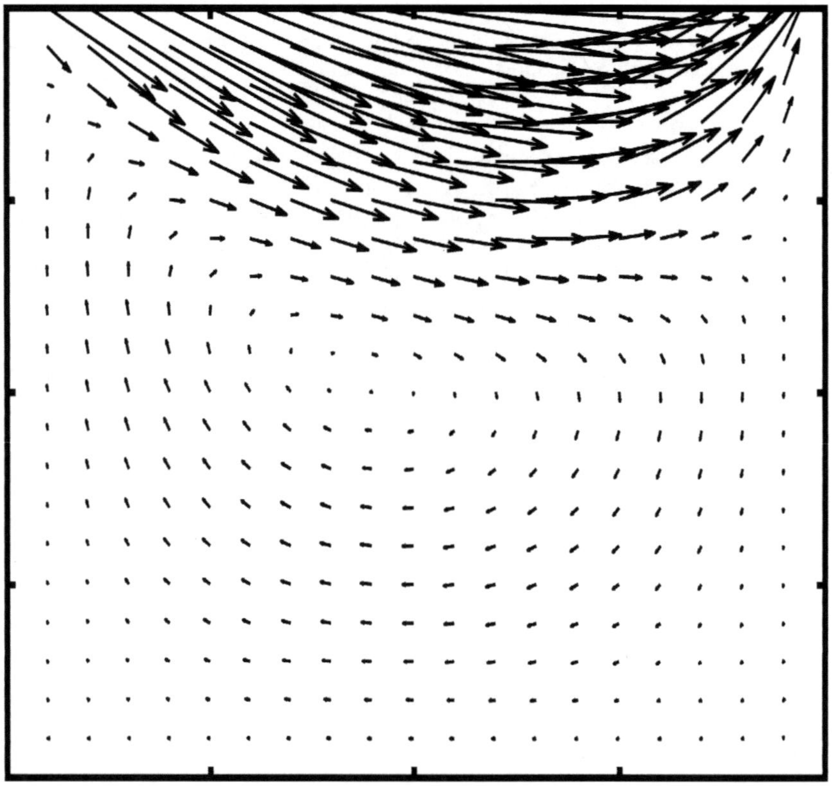

Figure 2. An uncontrolled flow in the cavity.

In order to do this we need to choosethe control, desired velocity field $\mathbf{u}_d$ and solve the control problem. If we want to suppress the vortex then the desired flow is the potential flow with zero vorticity. Heat flux χ on vertical boundary sections Γ_c plays the role of control.

Then we solve the control problem using the algorithm described above.Let us note that this problem is essentially nonstationary.The flow changes gradually at each time step.At dimensionless time $t = 5$ the flow becomes stable. The resulting velocity field in the cavity can be seen in Figure 3. It is easy to notice that in this case we do not have a vortex. The main flow from the channel occupies the entire cavity.This flow can wash out all contents from the bottom.

The analysis of the temperature field shows that we have cooling on the left boundary and heating on the right boundary. Therefore, the cold fluid with a higher density goes down in the left part of the cavity and the heated fluid with a lower density moves upward in the right part of the cavity.

The main goal of the computational experiments was to determine the dependence of the solution accuracy on the choice of the problem parameters. We have considered different values of regularization parameter μ and the different number of basis functions S. The right choice of the control sections on the boundary also plays an important role. Based on the analysis of computational results we can choose optimal values of parameters and develop some recommendations for future applications.

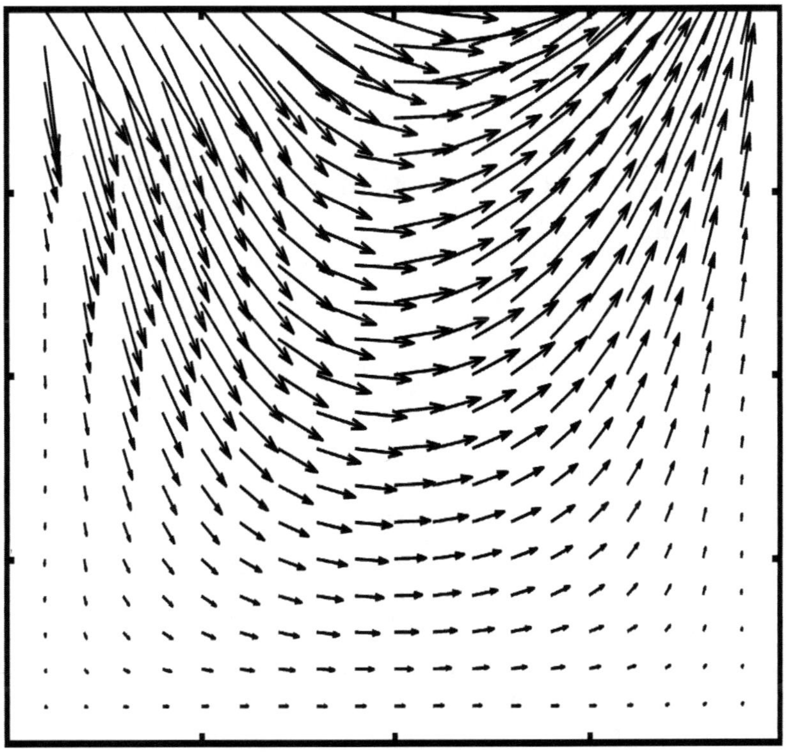

Figure 3. A controlled flow in the cavity.

5. Conclusion

We have considered control problems for nonstationary viscous heat-conducting fluid flows. A new numerical algorithm based on the finite dimensional minimization approach was proposed.This algorithm does not usethe Lagrange multipliersand the first order necessary optimality conditions. It has been successfully appliedto solve the problem of the vortex suppression in the open cavity.

Acknowledgements
This work was supported by the Russian Science Foundation (project no. 14-11-00079).

References

[1] Gunzburger M D 2003 *Perspectives in Flow Control and Optimization* (Philadelphia: SIAM)
[2] Alekseev G V and Tereshko D A 2008 *Analysis and Optimization in Hydrodynamics of Viscous Fluid* (Vladivostok: Dalnauka)
[3] Alekseev G V, Soboleva O V and Tereshko D A 2008 *J. Appl. Mech. Tech. Phys.* **49** 537–547
[4] Alekseev G V and Tereshko D A 2010 *Dokl. Math.* **81** 151–155
[5] Alekseev G V and Tereshko D A 2010 *J. Appl. Mech. Tech. Phys.* **51** 510–520
[6] Alekseev G V and Tereshko D A 2011 *Comput. Math. Math. Phys.* **51** 1539–1557
[7] Tereshko D A 2015 *J. Appl. Indust. Math.* **9** 132–140
[8] Yamazaki D and Kawahara M 2010 *J. Algor. Comp. Tech.* **4** 207–230
[9] Alekseev G V and Tereshko D A 2013 *Appl. Mech. Mater.* **372** 373–376
[10] Alekseev G V and Malikin VV 1994 *Comp. Fluid Dynamics J.* **3** 1–26

Mathematical Modeling of a Solar Arrays Deploying Process at Ground Tests

A Tomilin and I Shpyakin
Tomsk Polytechnic University, 30, Lenina ave., Tomsk, 634050, Russia

E-mail: aktomilin@tpu.ru

Abstract. This paper focuses on the creating of a mathematical model of a solar array deploying process during ground tests. Lagrange equation was used to obtain the math model. The distinctive feature of this mathematical model is the possibility of taking into account the gravity compensation system influence on the construction in the deploying process and the aerodynamic resistance during ground tests.

1. Introduction

An obligatory stage of the spacecraft creation is the testing of its subsystems [1]. The gravity compensation stands are used for ground testing of the solar array deploying process. Mathematical modeling of the solar arrays deploying process during ground tests is carried out to assess the implementation of the stand weightlessness features of the deployable elements and minimize the impact of added masses and other elements of the stand on the solar array's drives [2, 3]. Math model include kinematic and dynamic models of the solar arrays and dynamic models of the stand tracking systems. The requirements to the stands increase with developing of the space constructions and complication of the solar array structure. It leads to the necessity of creation of the more precise mathematical models.

The article considers the method of the mathematical modeling of the solar arrays deploying process, which is directed at the applied task solution of describing the deploying process for more often useable types of the solar array construction. The distinctive feature of this mathematical model is the possibility of taking into account the gravity compensation system influence and the aerodynamic resistance on the construction in the deploying process during ground tests.

2. Selection of a Solution Approach

There are three approaches of the computing modeling of the solar array deploying process.

The first approach consists in drawing up the differential equations of the mechanical system and their analytical decision. In this case a special mathematical software is used [4-6], which allows carrying out all the necessary operations in a symbolic form. The numerical values of parameters are substituted in the resulting symbolic expressions. This approach was applied for describing the deploying process of solar arrays in work [7]. The disadvantages of this approach include the high costs of computer performance.

The second approach is called algorithmic. It consists in numerical integration of the differential equations. It is possible to construct a calculation algorithm of dynamic and kinematic movement system characteristics using the set parameters and initial conditions. This approach is widespread in the applied mechanics and robotics. The algorithms of this approach are based on the physical laws and principles, such as D'Alembert principle, Lagrange equations, and others [8]. An example of the second approach of the solar arrays deploying process modeling is in work [8].

The third approach consists in the block diagram creating [9, 10] or using the mechanical assemblies, which are realised by means of CAD applications [11]. This approach is the most popular and easiest way of computer modeling, as it reduces the possibility of mistakes in the model composition.

Let us apply the second approach of modeling of the solar arrays deploying process during ground tests. We have a holonomic system with n degrees of freedom, which is not relieved from the stationary connections. This system consists of the units with known mass and inertial parameters which are consistently connected with the fifth class kinematic couples. The system dynamic could be described by Lagrange equations [6]:

$$\frac{d}{dt}\left(\frac{\partial W}{\partial \dot{q}_i}\right)-\frac{\partial W}{\partial q_i}=Q_i \qquad i=\overline{1,n}, \tag{1}$$

where W – the kinetic energy of the system, q_i – the generalized coordinates, Q_i – the generalized forces.

The kinetic energy of system W is the sum of kinetic energies W_i of its components:

$$W=\sum_{i=1}^{n} W_i, \qquad i=\overline{1,n}. \tag{2}$$

The number of units is equal to the number of degrees of freedom.

In this article we will consider the active suspensory gravity compensation system [12]. The working principle of the active gravity compensation system is applied to each unit mass center of the weight compensation forces. These forces are generated by the electromechanical actuators. The main requirement for the gravity compensation system is the lack of the additional forces and torques due to interaction of the suspensions and solar arrays. For this purpose, all trajectories and velocities of the mass centers could be accurately calculated for all units of the solar array structure. The calculation and the change of the compensation forces are performed in real time in accordance with the parameters of the motion of the structure.

3. Differential Equations of the Movement

We will form the differential equations of the movement in a horizontal plane for the solar arrays of the 'Express-2000' platform (Figure 1). Angles $\varphi_i, i=\overline{1,4}$ are taken as the generalized coordinates.

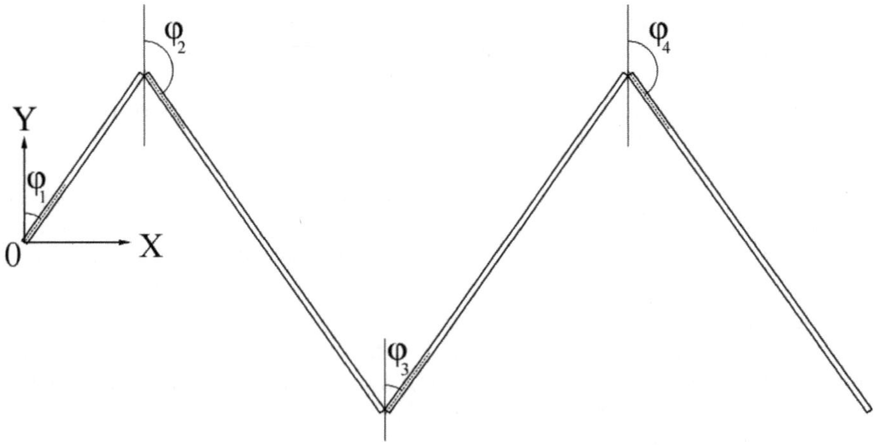

Figure 1. A schematic representation of the solar arrays.

The first unit (rod) makes a rotary motion, the other units make the plane-parallel motions. Let us write the expressions for the kinetic energies of the units:

$$W_1 = J_1 \frac{\dot{\varphi}_1^{\,2}}{2},$$
$$W_i = J_i \frac{\dot{\varphi}_i^{\,2}}{2} + m_i \frac{V_i^{\,2}}{2}, \qquad i = \overline{2,4}, \tag{3}$$

where J_i – the central torques of the inertia, $\dot{\varphi}_i(t)$ – the angles velocities, m_i – the masses, $V_i(t)$ – the linear velocities of the unit mass centers.

The corresponding torques of inertia of the units are:

$$J_1 = \frac{1}{3} m_1 l_1^2, \qquad J_i = \frac{1}{12} m_i l_i^2, \qquad i = \overline{2,4}. \tag{4}$$

We express squares of the linear velocities of the unit mass centers through the generalized coordinates and the generalized velocities:

$$\begin{aligned}
V_2^2 &= [l_1\dot{\varphi}_1 \cos(\varphi_1) + 0.5 l_2 \dot{\varphi}_2 \cos(\varphi_2)]^2 + \\
&\quad [l_1\dot{\varphi}_1 \sin(\varphi_1) + 0.5 l_2 \dot{\varphi}_2 \sin(\varphi_2)]^2, \\
V_3^2 &= [l_1\dot{\varphi}_1 \cos(\varphi_1) + l_2\dot{\varphi}_2 \cos(\varphi_2) + \\
&\quad + 0.5 l_3 \dot{\varphi}_3 \cos(\varphi_3)]^2 + [l_1\dot{\varphi}_1 \sin(\varphi_1) + \\
&\quad + l_2\dot{\varphi}_2 \sin(\varphi_2) + 0.5 l_3 \dot{\varphi}_3 \sin(\varphi_3)]^2, \\
V_4^2 &= [l_1\dot{\varphi}_1 \cos(\varphi_1) + l_2\dot{\varphi}_2 \cos(\varphi_2) + \\
&\quad + l_3\dot{\varphi}_3 \cos(\varphi_3) + 0.5 l_4 \dot{\varphi}_4 \cos(\varphi_4)]^2 + \\
&\quad + [l_1\dot{\varphi}_1 \sin(\varphi_1) + l_2\dot{\varphi}_2 \sin(\varphi_2) + \\
&\quad + l_3\dot{\varphi}_3 \sin(\varphi_3) + 0.5 l_4 \dot{\varphi}_4 \sin(\varphi_4)]^2.
\end{aligned} \tag{5}$$

The rotation angles of the units are accepted as generalized coordinates. Therefore the generalized forces are represented as the torques of the external and internal forces. The spring torques of deploying systems, the aerodynamic torques, the torques of the joint dry friction and the tension torques of the gravity compensation cables impact on the units. As a result the generalized forces take the following form:

$$\begin{gathered} Q_1 = M_1 = M_{S1} + M_{S2} + M_{F1} + M_{C1}, \\ Q_2 = M_2 = M_{S2} + M_{S3} + M_{A2} + M_{F2} + M_{C2}, \\ Q_3 = M_3 = M_{S3} + M_{S4} + M_{A3} + M_{F3} + M_{C3}, \\ Q_4 = M_4 = M_{S4} + M_{A4} + M_{F4} + M_{C4}, \end{gathered} \tag{6}$$

where $M_{Si}, i=\overline{1,4}$ – the spring torques of deploying systems, $M_{Ai}, i=\overline{1,4}$ – the aerodynamic torques, $M_{Fi}, i=\overline{1,4}$ – the torques of a joint dry friction, $M_{Ci}, i=\overline{1,4}$ – the resulted torques, which are created by the respective compensation cables as a result of imperfections in the process of the gravity compensation.

We substitute expressions (2) - (6) in formula (1). The obtained expression can be presented in the short form:

$$\mathbf{\Pi}(\boldsymbol{\varphi},\boldsymbol{\xi})\ddot{\boldsymbol{\varphi}} + \mathbf{b}(\dot{\boldsymbol{\varphi}},\boldsymbol{\varphi},\boldsymbol{\xi}) = \mathbf{M}, \tag{7}$$

where $\mathbf{\Pi}(\boldsymbol{\varphi},\boldsymbol{\xi})$ – the matrix function of a 4×4 dimension, $\mathbf{b}(\dot{\boldsymbol{\varphi}},\boldsymbol{\varphi},\boldsymbol{\xi})$ – the vector function of the fourth dimension, $\mathbf{M}$ – the vector of torques, which are applied to a four-dimensional body, ξ – the vector of solar array parameters.

When the unit achieves a standard position, the corresponding locking mechanism operates. The additional torque (a locking torque) appears in the hinge device. This torque holds the adjacent units in a determined position. We assume that this torque is directly proportional to the angular position of the adjacent units relative to each other.

For the considered model of the solar array deploying the locking torques are:

$$\begin{aligned} M_{L1} &= sign(\dot{\varphi}_1)k_{L1}(\frac{\pi}{2} - \varphi_1), \\ M_{L2} &= sign(\dot{\varphi}_2)k_{L2}(\varphi_1 - \varphi_2), \\ M_{L3} &= sign(\dot{\varphi}_3)k_{L3}(\varphi_2 - \varphi_3), \\ M_{L4} &= sign(\dot{\varphi}_4)k_{L4}(\varphi_3 - \varphi_4), \end{aligned} \tag{8}$$

where $k_{Li}, i=\overline{1,4}$ – the angular stiffness of the appropriate locking mechanism.

4. An Algorithm of Differential Equations Solving

It is necessary to determine the system movement by the set of operating influences and initial conditions. We need to solve a system of nonlinear differential equations of the second order (7) for accomplishing the formulated task.

It can be done by numerical integration of this system. The accelerations are defined and the new values of the generalized coordinates and velocities are extrapolated from set initial values of the generalized coordinates and velocities to the next time point. The solution is a multiple performance of this cycle calculations.

We use the forth order Runge–Kutta method for solving equation system (7). We present equations (7) in the form of Cauchy, i.e. in the form of the first order differential equation system, for using the Runge-Kutta method:

$$\mathbf{AX} = \mathbf{B}, \tag{9}$$

where

$$
\mathbf{A}=\begin{bmatrix}
(\frac{1}{3}m_1+m_2+m_3+m_4)l_1^2, & (\frac{1}{2}m_2+m_3+m_4)\cos(\varphi_1+\varphi_2)l_1l_2, & (\frac{1}{2}m_3+m_4)\cos(\varphi_1-\varphi_3)l_1l_3, & \frac{1}{2}m_4\cos(\varphi_1+\varphi_4)l_1l_4, & 0000\\
(\frac{1}{2}m_2+m_3+m_4)\cos(\varphi_1+\varphi_2)l_1l_2, & (\frac{1}{3}m_2+m_3+m_4)l_2^2, & (\frac{1}{2}m_3+m_4)\cos(\varphi_2+\varphi_3)l_2l_3, & \frac{1}{2}m_4\cos(\varphi_2-\varphi_4)l_2l_4, & 0000\\
(\frac{1}{2}m_3+m_4)\cos(\varphi_1-\varphi_3)l_1l_3, & (\frac{1}{2}m_3+m_4)\cos(\varphi_2+\varphi_3)l_2l_3, & (\frac{1}{3}m_3+m_4)l_3^2, & \frac{1}{2}m_4\cos(\varphi_3+\varphi_4)l_3l_4, & 0000\\
\frac{1}{2}m_4\cos(\varphi_1+\varphi_4)l_1l_4, & \frac{1}{2}m_4\cos(\varphi_2-\varphi_4)l_2l_4, & \frac{1}{2}m_4\cos(\varphi_3+\varphi_4)l_3l_4, & \frac{1}{3}m_4l_4^2, & 0000\\
0 & 0 & 0 & 0 & 1000\\
0 & 0 & 0 & 0 & 0100\\
0 & 0 & 0 & 0 & 0010\\
0 & 0 & 0 & 0 & 0001
\end{bmatrix},
$$

$$
\mathbf{B}=\begin{bmatrix}
M_{S1}+M_{S2}+M_{F1}+M_{C1}+z_2^2l_1l_2(\frac{1}{2}m_2+m_3+m_4)\sin(\varphi_1+\varphi_2)+\\
+z_1z_2l_1l_2(m_2+2m_3+2m_4)\sin(\varphi_1+\varphi_2)-z_3^2l_1l_3(\frac{1}{2}m_3+m_4)\sin(\varphi_1-\varphi_3)+\\
+z_1z_3l_1l_3(m_3+2m_4)\sin(\varphi_1-\varphi_3)+z_1z_4l_1l_4m_4\sin(\varphi_1+\varphi_4),\\
M_{S2}+M_{S3}+M_{A2}+M_{F2}+M_{C2}+z_1^2l_1l_2(\frac{1}{2}m_2+m_3+m_4)\sin(\varphi_1+\varphi_2)+\\
+z_1z_2l_1l_2(m_2+2m_3+2m_4)\sin(\varphi_1+\varphi_2)+z_3^2l_2l_3(\frac{1}{2}m_3+m_4)\sin(\varphi_1+\varphi_3)+\\
+z_2z_3l_2l_3(m_3+2m_4)\sin(\varphi_2+\varphi_3)-\frac{1}{2}z_4^2l_2l_4m_4\sin(\varphi_2-\varphi_4)+z_2z_4l_2l_4m_4\sin(\varphi_2-\varphi_4),\\
M_{S3}+M_{S4}+M_{A3}+M_{F3}+M_{C3}+z_1^2l_1l_3(\frac{1}{2}m_3+m_4)\sin(\varphi_1-\varphi_3)-\\
-z_1z_3l_1l_3(m_3+2m_4)\sin(\varphi_1-\varphi_3)+z_2^2l_2l_3(\frac{1}{2}m_3+m_4)\sin(\varphi_1+\varphi_3)+\\
+z_2z_3l_2l_3(m_3+2m_4)\sin(\varphi_2+\varphi_3)+\frac{1}{2}z_4^2l_3l_4m_4\sin(\varphi_3+\varphi_4)+z_3z_4l_3l_4m_4\sin(\varphi_3+\varphi_4),\\
M_{S4}+M_{A4}+M_{F4}+M_{C4}+\frac{1}{2}z_1^2l_1l_4m_4\sin(\varphi_1+\varphi_4)+\\
+z_1z_4l_1l_4m_4\sin(\varphi_1+\varphi_4)+\frac{1}{2}z_2^2l_2l_4m_4\sin(\varphi_2-\varphi_4)-\\
-z_2z_4l_2l_4m_4\sin(\varphi_2-\varphi_4)+\frac{1}{2}z_3^2l_3l_4m_4\sin(\varphi_3+\varphi_4)+z_3z_4l_3l_4m_4\sin(\varphi_3+\varphi_4),\\
z_1,\\
z_2,\\
z_3,\\
z_4,
\end{bmatrix},\quad
\mathbf{X}=\begin{bmatrix}\dot{z}_1\\ \dot{z}_2\\ \dot{z}_3\\ \dot{z}_4\\ \dot{\varphi}_1\\ \dot{\varphi}_2\\ \dot{\varphi}_3\\ \dot{\varphi}_4\end{bmatrix}.
$$

It is necessary to carry out checking on the performance of the locking operations during the numerical integration process and to apply the additional torques to the model when it is necessary.

The algorithm of numerical integration is presented in Figure 2.

5. Original Assumptions

Modelling always implies certain assumptions which simplify the real system. We use the following assumptions:

- the solar array panels are the absolutely rigid bodies;

- the gravity compensation stand perfectly compensates the gravitational forces during the solar array deploying;
- the hinge devices do not have a backlash and do not create additional friction torques;
- the locking devices do not have a backlash and is performed at the moment of transition to the operating position;
- the aerodynamic torques are linear functions of the angular velocities of the respective panels;
- the deploying process occurs without the synchronizing devices.

We need to take into account the structure and parameters of the locking mechanisms and hinge devices for more accurate modeling. That requires more detailed information about these devices.

The coefficients of the aerodynamic torques depend on the density of environment (air), the square area of panels and the drag coefficients [13]. The drag coefficient can be found by the computer simulation or experimentally.

It is necessary to check the adequacy of the model to the real process of the solar array deploying.

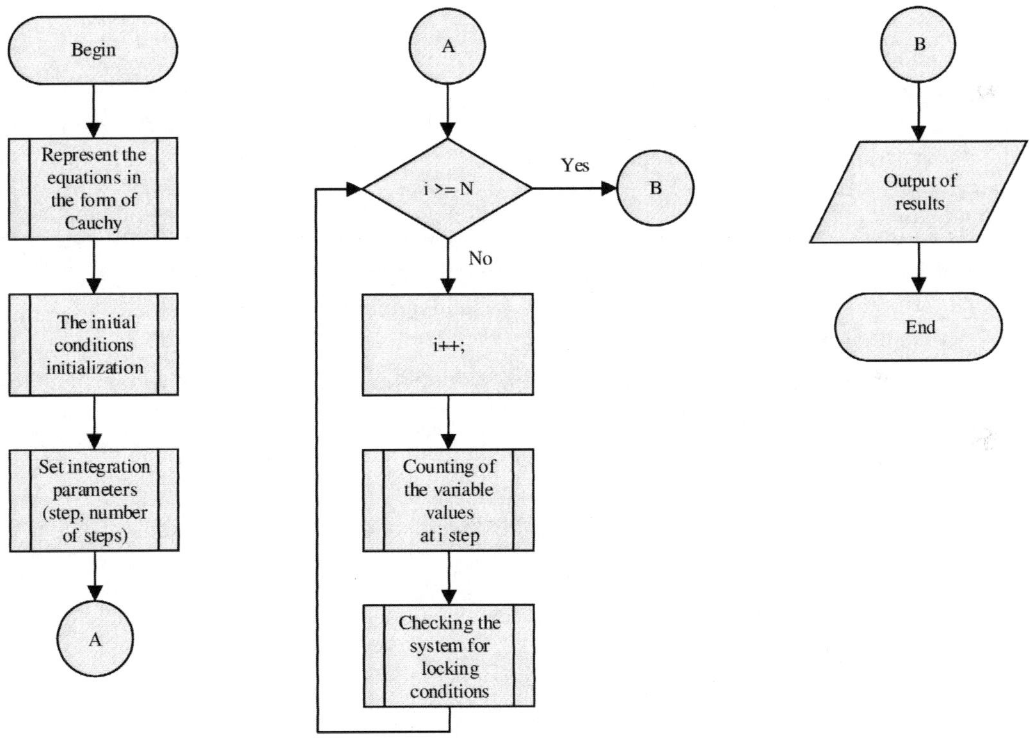

Figure 2. A block diagram of the numerical integration of Runge-Kutta method.

6. An Algorithm Checking on Adequate

We create the model and obtain the numerical solution of the deploying process parameters. The solar array panels of the spacecraft platform 'Express-2000' are used for checking. We will set some numerical values, which are length and weight of the panels:

$$\begin{aligned} &l_1 = 2.5\ m, \quad l_2 = 4\ m, \quad l_3 = 4\ m, \quad l_4 = 4\ m, \\ &m_1 = 10\ kg, \quad m_2 = 30\ kg, \quad m_3 = 65\ kg, \quad m_4 = 27\ kg. \end{aligned} \tag{10}$$

Let us represent the spring torques of the hinge devices in the form of:

$$M_{Si}(\varphi) = 8 - 2\varphi_i, \quad where\ i = \overline{1,4}. \tag{11}$$

The drag coefficient of the panels equals: $\alpha = 0.01$. The function of the aerodynamic torque assumes:

$$M_{Ai}(\dot{\varphi}_i) = -0.01\dot{\varphi}_i, \quad where\ i = \overline{1,4}. \tag{12}$$

We will obtain the following system of equations substituting numerical values (10) - (12) into (9):

$$\mathbf{AX} = \mathbf{B}, \tag{13}$$

where

$$\mathbf{A} = \begin{bmatrix} 2345/3, & 1070\cos(\varphi_1+\varphi_2), & 595\cos(\varphi_1-\varphi_3), & 135\cos(\varphi_1+\varphi_4), & 0\ 0\ 0\ 0 \\ 1070\cos(\varphi_1+\varphi_2), & 1632, & 952\cos(\varphi_2+\varphi_3), & 216\cos(\varphi_2-\varphi_4), & 0\ 0\ 0\ 0 \\ 595\cos(\varphi_1-\varphi_3), & 952\cos(\varphi_2+\varphi_3), & 2336/3, & 216\cos(\varphi_3+\varphi_4), & 0\ 0\ 0\ 0 \\ 135\cos(\varphi_1+\varphi_4), & 216\cos(\varphi_2-\varphi_4), & 216\cos(\varphi_3+\varphi_4), & 144, & 0\ 0\ 0\ 0 \\ 0 & 0 & 0 & 0 & 1\ 0\ 0\ 0 \\ 0 & 0 & 0 & 0 & 0\ 1\ 0\ 0 \\ 0 & 0 & 0 & 0 & 0\ 0\ 1\ 0 \\ 0 & 0 & 0 & 0 & 0\ 0\ 0\ 1 \end{bmatrix},$$

$$\mathbf{B} = \begin{bmatrix} 16-2\cdot\pi-4\cdot\varphi_1+2\cdot\varphi_2+270\cdot z_1\cdot z_4\cdot\sin(\varphi_4+\varphi_1)-595\cdot\sin(\varphi_1-\varphi_3)\cdot z_3^{\ 2}+1190\cdot\sin(\varphi_1-\varphi_3)\cdot z_1\cdot z_3+ \\ +1070\cdot\sin(\varphi_2+\varphi_1)\cdot z_2^{\ 2}+2140\cdot\sin(\varphi_2+\varphi_1)\cdot z_1\cdot z_2, \\ 0.01\cdot z_2-0.01\cdot z_1+16-4\cdot\pi-2\cdot\varphi_1+4\cdot\varphi_2-2\cdot\varphi_3-216\cdot z_4^{\ 2}\cdot\sin(-\varphi_4+\varphi_2)+432\cdot z_2\cdot z_4\cdot\sin(-\varphi_4+\varphi_2)+ \\ +952\cdot\sin(\varphi_3+\varphi_2)\cdot z_3^{\ 2}+1904\cdot\sin(\varphi_3+\varphi_2)\cdot z_2\cdot z_3+2140\cdot\sin(\varphi_2+\varphi_1)\cdot z_1\cdot z_2+1070\cdot\sin(\varphi_2+\varphi_1)\cdot z_1^{\ 2}, \\ -0.01\cdot z_3-0.01\cdot z_2+16-4\cdot\pi-4\cdot\varphi_3+2\cdot\varphi_2+2\cdot\varphi_4+216\cdot\sin(\varphi_3+\varphi_4)\cdot z_4^{\ 2}+432\cdot\sin(\varphi_3+\varphi_4)\cdot z_4\cdot z_3- \\ -(-1904\cdot\sin(\varphi_3+\varphi_2)\cdot z_2+1190\cdot\sin(\varphi_1-\varphi_3)\cdot z_1)\cdot z_3+952\cdot\sin(\varphi_3+\varphi_2)\cdot z_2^{\ 2}+595\cdot\sin(\varphi_1-\varphi_3)\cdot z_1^{\ 2}, \\ 0.01\cdot z_4-0.01\cdot z_3+8-2\cdot\pi-2\cdot\varphi_3+2\cdot\varphi_4-(432\cdot z_2\cdot\sin(-\varphi_4+\varphi_2)-270\cdot z_1\cdot\sin(\varphi_4+\varphi_1)- \\ -432\cdot\sin(\varphi_3+\varphi_4)\cdot z_3)\cdot z_4+135\cdot z_1^{\ 2}\cdot\sin(\varphi_4+\varphi_1)+216\cdot z_3^{\ 2}\cdot\sin(\varphi_3+\varphi_4)+216\cdot z_2^{\ 2}\cdot\sin(-\varphi_4+\varphi_2), \\ z_1, \\ z_2, \\ z_3, \\ z_4, \end{bmatrix},$$

$$\mathbf{X} = \begin{bmatrix} \dot{z}_1 \\ \dot{z}_2 \\ \dot{z}_3 \\ \dot{z}_4 \\ \dot{\varphi}_1 \\ \dot{\varphi}_2 \\ \dot{\varphi}_3 \\ \dot{\varphi}_4 \end{bmatrix}.$$

Write the function of the locking for this type of solar arrays:

$$\begin{aligned} M_{L1} &= 20000\cdot(\frac{\pi}{2}-\varphi_1), \\ M_{L2} &= -8000\cdot(\varphi_1-\varphi_2), \\ M_{L3} &= 6000\cdot(\varphi_2-\varphi_3), \\ M_{L4} &= -2000\cdot(\varphi_3-\varphi_4), \end{aligned} \tag{14}$$

We use a numerical integration algorithm, which is shown in Figure 2. The calculations are made in the Mupad mathematical package of the Matlab software. The results are the sets of graphs. The graphs show the dependence of the generalized angles on time (Figures 3 and Figures 4).

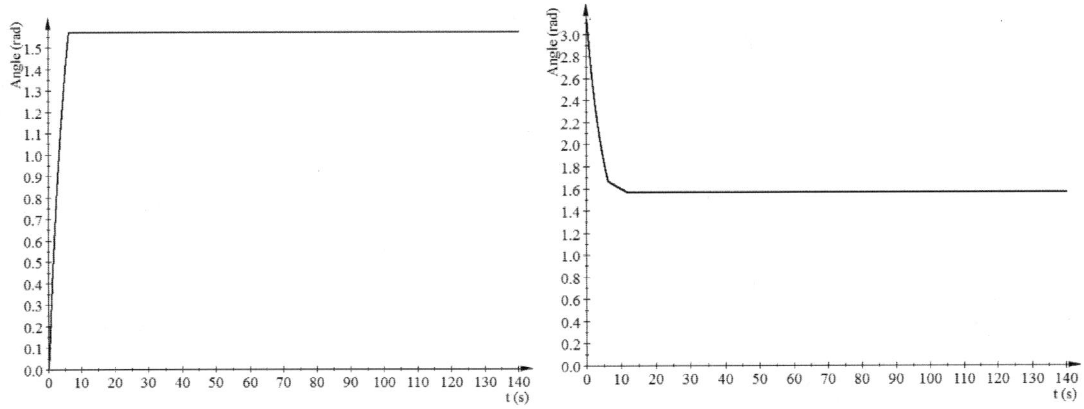

Figure 3. The functions of the first and second hinge angles from time dependence appropriately.

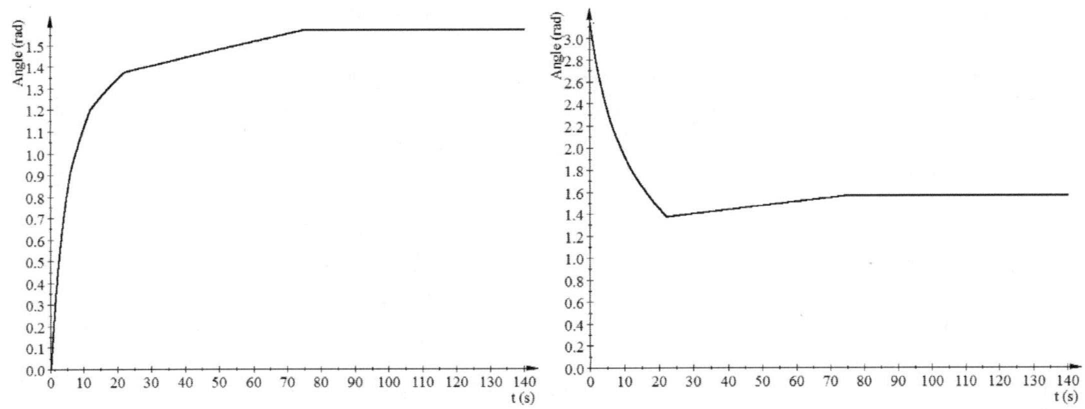

Figure 4. The functions of the fourth and fifth hinge angles from time dependence appropriately.

We can find the kinematic and dynamic characteristics of the motion using these results. In particular, the trajectories of the unit mass centers in the plane are presented in Figure 5.

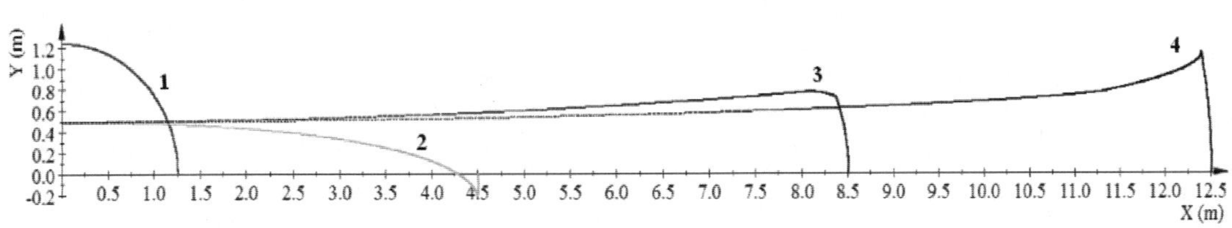

Figure 5. The trajectories of the unit mass center motions in the plane

7. Conclusion

The results allow considering the model as adequate:

- the system is stable and reaches desired conditions;
- the trajectories of the unit mass center motions correspond to reality.

These calculations and figures should be considered in the programming motions of the suspension system carriages for the absence of the additional forces and torques due to the errors of gravity compensation.

This model can be developed and improved, bringing it closer to the real mechanism. In particular, it is possible to take into account the flexibility of the solar array panels and its impact on the properties of the gravity compensation system.

References

[1] 2001 *State Standard 15-203-01. The procedure of development work performing on the creation of products and their components. Main provisions.* (Moscow, Standartinform Publ.)
[2] Bakulin D V *et al* 2004 *Jour. Mathematical modeling* **16(6)** 88–92
[3] Udincev V V *The modeling of deploying processes of the spacecraft multielement constructions.* (Polet) **5** pp 28–33
[4] Doev V S, Doronin F A 2010 *The Theoretical Mechanics Task Collection on Mathcad base* (Rostov-na-Donu: Phoenix)
[5] Kirsanov M N 2010 *The Theoretical Mechanics Tasks with Decisions in Maple 11* (Moscow: Fismatlit)
[6] 2008 *Symbolic Math Toolbox MuPAD Tutorial* (SciFace Software GmbH & Co. KG)
[7] Korendyasev A I, Salamandra B L, Tyves L I 2006 *Theoretical basis for robotic technology. In 2 books.* (Moscow: Nauka)
[8] Vereshchagin A F 1974 *Jour. Engineering Cybernetics* **6** 65–70
[9] Afonin V V, Fedosin S A 2011 *Modeling of systems* (Moscow, Internet-University of Information Technology: BINOM. Laboratory of knowledge)
[10] Alekseev E R, Chesnokova O V, Rudchenko E A 2008 *Scilab: Solution of engineering and mathematical problems* (Moscow)
[11] *Supplemental Adams Tutorial Kit for Design of Machinery Course Curriculum.*URL: http://www.mscsoftware.com/sites/default/files/Book_Adams-Tutorial-ex17-w.pdf
[12] Shpyakin I K, Malyshenko A M 2015 *Comparative Analysis of Gravity Compensation Systems for Ground Tests of Deployable Solar Arrays* (International Siberian Conference on Control and Communications, SIBCON 2015)
[13] Clancy L J 1975 *Aerodynamics* (Pitman Publishing Limited: London)

IOP Conf. Series: Materials Science and Engineering **124** (2016) 012104 doi:10.1088/1757-899X/124/1/012104

Modeling of a data exchange process in the Automatic Process Control System on the base of the universal SCADA-system

D Topolskiy, N Topolskiy, E Solomin and I Topolskaya

Department of Electric Engineering at South Ural State University, 76,Leninaave., Chelyabinsk, 454091, Russia

E-mail:topol69@mail.ru

Abstract. In the present paper the authors discuss some ways of solving energy saving problems in mechanical engineering. In authors' opinion one of the ways of solving this problem is integrated modernization of power engineering objects of mechanical engineering companies, which should be intended for the energy supply control efficiency increase and electric energy commercial accounting improvement. The author have proposed the usage of digital current and voltage transformers for these purposes. To check the compliance of this equipment with the IEC 61850 International Standard, we have built a mathematic model of the data exchange process between measuring transformers and a universal SCADA-system. The results of modeling show that the discussed equipment corresponds to the mentioned Standard requirements and the usage of the universal SCADA-system for these purposes is preferable and economically reasonable. In modeling the authors have used the following software: MasterScada, Master OPC_DI_61850, OPNET.

1. Introduction

Energy saving is one of the modern developing areas of mechanical engineering. It is a global problem which scientists and engineersface continually in all mankind activity fields. The efficient progress of mechanical engineering is also impossible without an integrated energy saving approach taking into account that as a rule the energy part in the cost of a high technological product is relatively high and the actual task is its reduction in fabricating of high competitive products. An energy saving problem solution should be discussed in the wide aspect spectrum. To save energy in mechanical engineering not only the production technology [1], but also all processes of automatic control of the electric energy supply and accounting [2]should be efficient. This energy saving area is characterized by the development of Smart Grids – a new technological type of electric grids with more advanced efficient reliability and controllability parameters.

This approach considers the combined modernizing of electric grids of electric engineering companies [3, 4], automation of power engineering objects, transfer to the adaptive control system, elimination of manual control.

This process includes the usage of the smart electric energy accounting devices, application of digital measuring current and voltage transformers, development of fully automated digital substations design of new solutions, equipment, systems and methods of power transmission lines and transformer substation monitoring. Due to these demands we put forth our efforts in the applied scientific research and experimental works in the development of Adaptive Digital Combined Current and Voltage

Transformer (ADCCVT) being the base of a smart electric energy accounting system of a substation [5].

2. Research Tasks

Substation Automation Systems (SAS) traditionally provide the servicing of the consumption current value; substation buses voltage control; switching apparatus status control; breakers; disconnectors; relay protection device control; telecontrol.

Such function set in the modern conditions is sufficiently limited and does not meet the current requirements in the power engineering object automation area. Since most of the objects recently began using the SCADA technology (Supervisory Control and Data Acquisition), we have obtained the ability to extend the SAS functions. The specific requirements should be applied to the technical software architecture.

IEC 61850 International Standard regulates the presentation of data about the substation as an object of automation, and the protocols of digital data are exchanged between the microprocessor intellectual electronic devices (IED) at the substation. The appearance of the mentioned Standard had become a challenging reason for the development of new generation of substations – a digital substation. A digital substation contains ADCCVT and has a structure as shown in Figure 1.Figure 1 is indicated by: I – Process level: 1 –switching substation, 2 – ADCCVT, 3 – switching devices, 4 – merging unit, 5 – remote terminal unit; II – Bay level: 6 – bay controller, 7 – relay protection and automation, 8 – others IEDs; III –Station level: 9 – server SCADA, 10 – workstation of operator.

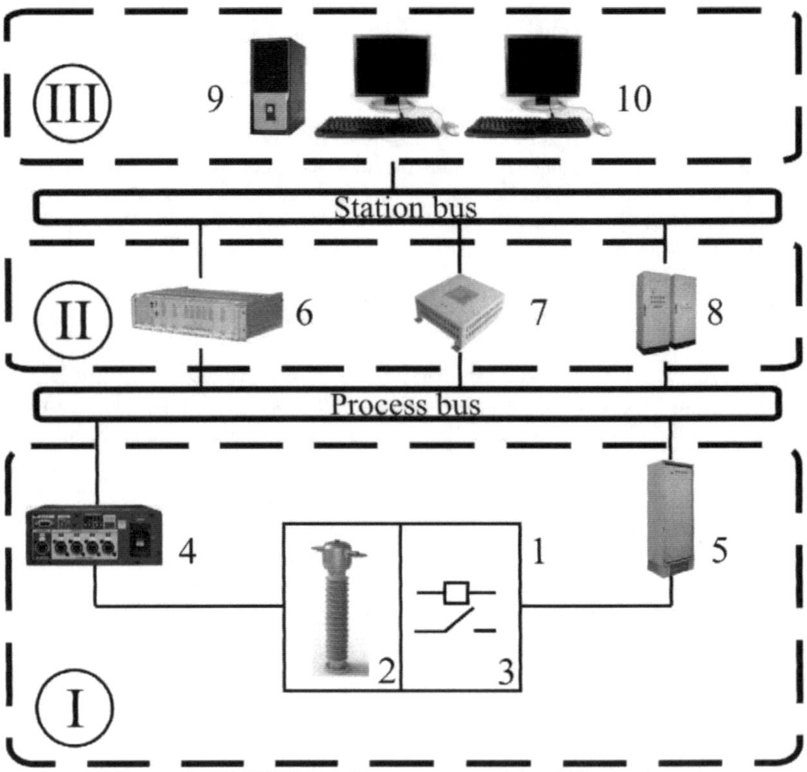

Figure 1.ADCCVT in the structure of the digital substation.

Todayit is high time to improve the whole spectrum of SAS parameters with the help ofmathematic modelling and their deep research: high cost of equipment, long IED development and fabrication time, low universality of equipment. Fabricators of IED often interpret the IEC 61850 Standard requirements in their own or even not in the right way, in the area where the model and the structure of data exchange between different IEDsare being described. In accordance with the differences in Standard understanding, the differences in the IEDs operation also appear. Thus the right modelling of the data exchange process may eliminate the errors or differences at the very basic stage of its development. As a result the ADCCVT – SCADA communication channel should provide the speed performance, synchronizing and noise immunity of metrological information transmission in accordance with the unified platform of IEC 61850 data exchange. The physical topology of Ethernet grid of the digital substation should be taken into account whilst developing its structure.

3. Modeling

SCADA-systems of power engineering objects today collect and archive all incoming data with a minimal available step without losing the information and protecting an unauthorized access, providing the recovery of the system operability after software or apparatus failure. This data set is excessive as a rule, however the information redundancy is directly connected with its quality and processing capabilities. This in turn allows presenting the information about the power object condition on all workstations of operators in the real time mode providing the scalability of systems.

Modeling of the data exchange process allows correcting all time delays, possible data losses, preliminary equipment adjustment. In case of the high accuracy ADCCVT providesa 0.2S accuracy class for current and 0.2 –for voltage, the time delays for the message transmissions are critical, as the measurement frequency is 12800 Hz. Each measurement in the high voltage part of ADCCVT needs 14 bytes. Accordingly the transmission data speed from the high voltage to low voltage part is 1400 Kbit/s. After transmission to the low voltage part, data are being transferred to the merging unit. Stream IEC-61850-9-2LE is being formed in the device – 4 currents and 4 voltages with the step of 80 points per period (for relay protectionand automatic) and 256 points per period (for power quality and measurements). The speeds of transmission are 4 and 9 Mbit/s correspondingly, which require the increase of data processing time and grid operation delays.

We built a full model of the data exchange process in the ADCCVT–SCADA communication channel. In accordance with Figure 1 the structure of data exchange between ADCCVT and SCADA-systems includes a merging unit, process bus, relay protection and station bus. We modelled an OPC-server on the base of Master OPC DI_61850 (Figure 2), which simulated the operation of the mergingunit. These messages of the IEC-61850-9-2 format were received by a universal MasterScada SCADA-system. To model the microprocessorrelay protection operation, MasterScadarecodedthe obtained data into MMS-messages of the IEC-61850-8-1 format.

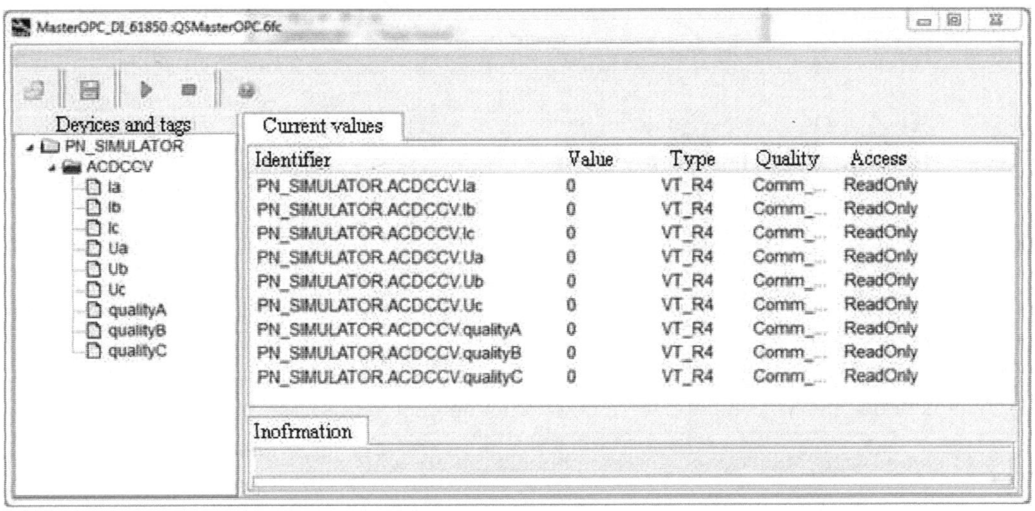

Figure 2. Window of Master OPC_DI_61850 operation.

After recoding the SCADA-system is visualizing the measured current and voltage values. The charts are presented in Figure 3.

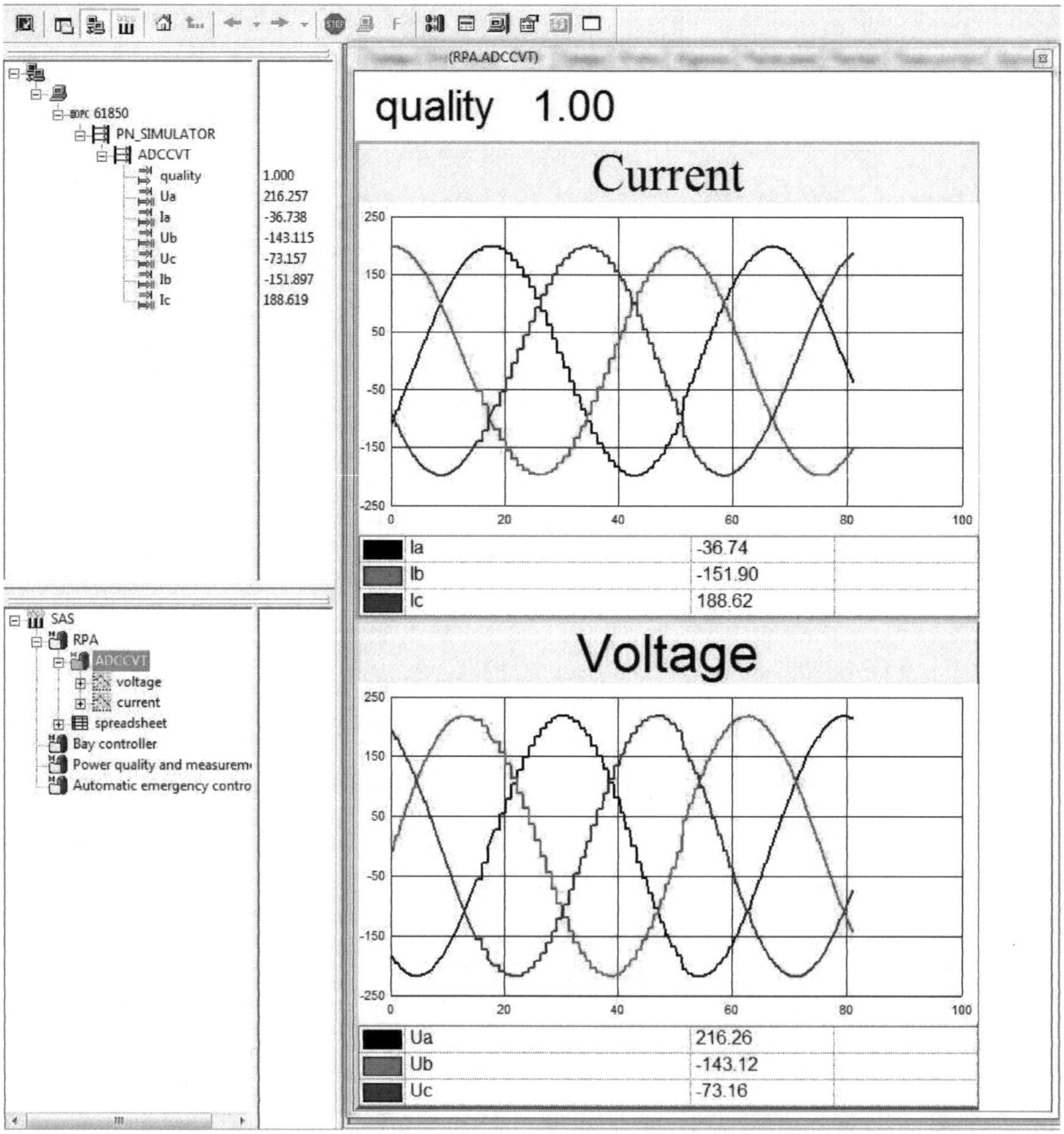

Figure3. Window of SCADA-system operation. The charts of current and voltage.

Modelling also allows the preliminary adjustment of IEDs. To model the IED it is required to generate a SCL file describing in detail the operation of this device. Probabletime delays during the data exchange between IEDs were programmed in the OPNET software. The window of the project is shown in Figure 4.

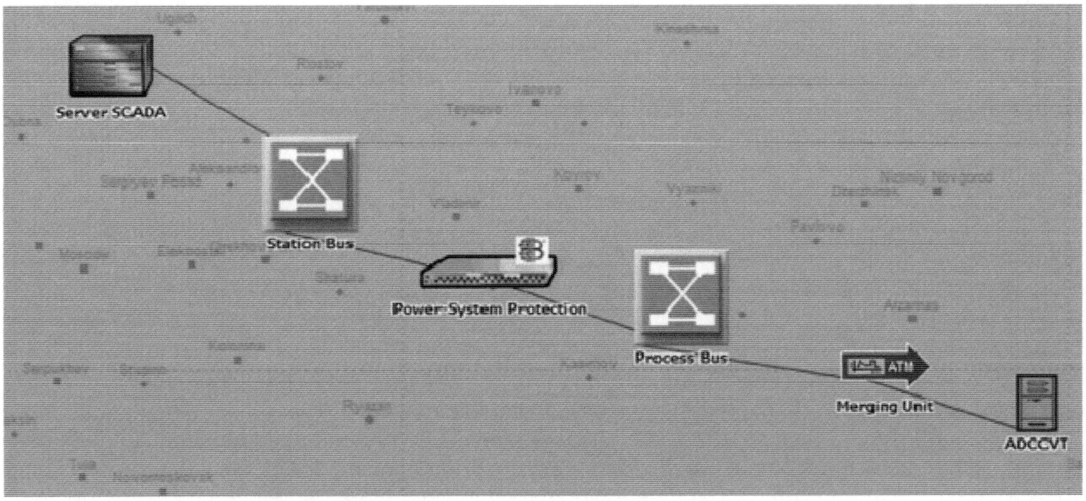

Figure4. Window of the project in the OPNET software.

4. ResultsandDiscussion

After modelling we took into account the following data: power of equipment (both grid and IEDs), minimum and maximum time of delay of the SCADA-system reaction on data received from ADCCVT. The results have shown that all data transmission delays for the used in the model IEDs, meet the requirements of IEC 61850 Standard. The results are shown in Figure 5.

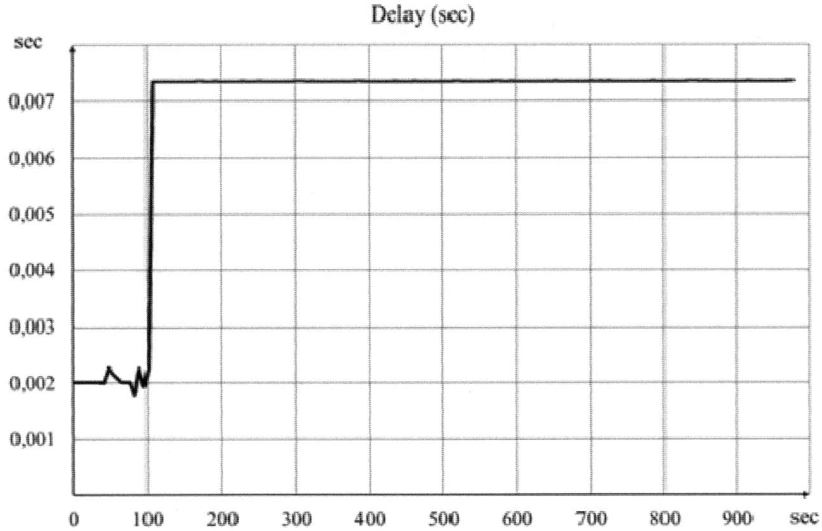

Figure 5. Dependence of delays in the ADCCVT – SCADA system on time whilst increasing the loading.

The specific obtained result is a proved capability of universal SCADA-system usage for digital substation control. This is explained by the fact that although IEC 61850 Standard is oriented at integrated automation of substations and requires the functional compatibility of substations and a software from different developers, the consumer today can purchase mostly the software-hardware complex sets from leading World brands compatible with the mentioned Standard. Notwithstanding the declared compatibility with the Standard the software of the upper level in such complex set 'prefers' as a rule the data exchange with its 'native' devices. Due to the usage of the OPC-server compatible with any equipment meeting the requirements of IEC 61850 Standard, we can provide the capability of usage of universal SCADA-systems from any developers, which as a rule is more functional and less expensive than the software from 'equipment fabricators'.

5. Conclusion

The described research has resulted in the following conclusions:

1. Energy saving is one of the important problems to be solved in mechanical and power engineering.
2. Mechanical engineering has wide perspectives in thesearch of energy saving ways.
3. One of the ways of solving the problem of energy saving in mechanical engineering is the intellectual electric energy accounting.
4. An increase of electric energy commercial accounting automation level could be reached by the development and usage of digital measuring current and voltage transformers.
5. To control the power engineering objects it is preferable to use the universal SCADA-systems which in most cases have more functionality and less price compared with SCADA-systems from equipment developers.

Acknowledgement

This work was supported in part by the R.F. Ministry of Education under Grant 14.576.21.0047 of 22 August, 2014 (ID: RFMEFI57614X0047).

References

[1] Pavlov K S and Khobotov E N 2015 *Journal ofAutomation and Remote Control* **76(2)** 292–303
[2] Furugyan M G 2015 *Journal ofAutomation and Remote Control* **76(3)** 487–492
[3] Genin V S, Koznov V V and Fel'dman S O 2015 *J. ofRussian Electrical Engineer.* **86(2)** 79–82
[4] Berg V R and Gurov A A 2012*J. ofRussian Electrical Engineer.* **83(1)** 35–38
[5] Solomin E V, Topolskiy D V and Topolskiy N D2015Integration of adaptive digital combined current and voltage transformer into digital substation Ethernet grid*Control and Communications (SIBCON), International Siberian Conference* (*ISBN*: 978-1-4799-7102-2)

Principles of E-network modelling of heterogeneous systems

D Tarakanov[1], I Tsapko[2], S Tsapko[2] and R Buldygin[2]
[1] Polytechnic Institute, Surgut State University, 1, Lenina ave., Surgut, 628400, Russia
[2] Tomsk Polytechnic University, 30, Lenina ave., Tomsk, 634050, Russia

E-mail: Tsiv@tpu.ru

Abstract. The present article is concerned with the analytical and simulation modelling of heterogeneous technical systems using E-network mathematical apparatus (the expansion of Petri nets). The distinguishing feature of the given system is the presence of the moduleб which identifies the parameters of the controlled object as well as the external environment.

1. Introduction

When modeling complex heterogeneous systems it is important to choose a scheme (basis) for analytical and simulation modeling [1]. The present paper examines the approach, which allows creating discrete and continuous models for a wide range of processes and systems using uniform methodological positions. It also enables modeling the reliability of particular elements and the whole system.

As the mathematical basis for constructing a computer model, we suggest using E-network apparatus, which expands Petri nets [1]. Due to their properties [2, 3] E-networks meet the initial requirements for being the mathematical basis for constructing complex technical systems, which have structural and parametric configuration changes, numerous elements functioning independently of each other, random processes, discrete and continuous components, etc.

2. Heterogeneous system model architecture

Figure 1 shows the architecture of the modeling system.

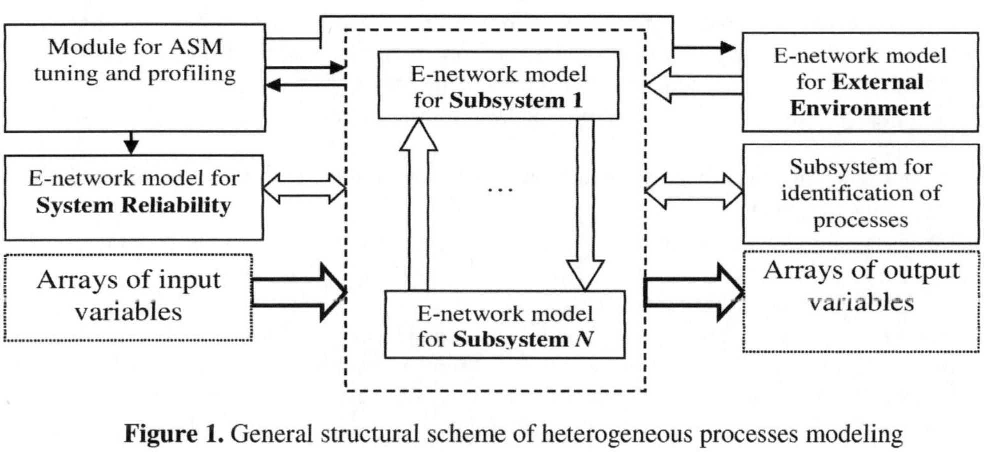

Figure 1. General structural scheme of heterogeneous processes modeling

The module for analytical and simulation model (ASM) tuning and profiling [2] allows configuring the structure and the parameters of the model depending on subsystem working modes, collecting data of modeling results, defining component working time in the model, etc. Every system to be examined is mathematically represented by E-network modeling scheme [1-3], which allows imitating logical and dynamic processes. In order to model continuous processes we propose to use E-network algorithmic modeling schemes [2, 3]. The model is adapted using the subsystem for identification of processes. The given subsystem realizes mathematical apparatus of artificial neural networks, which is widely used in order to identify processes. Besides, in the proposed system it is possible to use ASM for system reliability. The reliability parameters are assessed according to the results of process imitation in other subsystems.

3. Principles of constructing analytical and simulation model

The methodology of constructing analytical and simulation models for complex systems is based on the objective approach to the model organization and consists of the following basic positions:

- Any model is considered as a set of unidirectional objects (hereinafter referred to as components 'Comp') interacting with each other, which allows implementing a causal approach to the model construction;
- Model components are selected according to a minimum number of interconnections between model components and a minimum level of interaction (i.e., intensity of data exchange) between the components;
- Data is exchanged via communication channels;
- Model components are based on the hierarchical principle;
- Several independent processes can be performed in the model, i.e. components can function concurrently and independently of each other;
- Both concrete technical objects of the system and abstract mathematical objects, such as conversion operators, can be associated with model components.
- AIM can be adapted to the alteration in modelling aims, system parameters and structure and the environment.

The existing modelling methods including E-networks are quite effective in problem solving with the help of analytical and simulation modelling [4-6, 8]. Nevertheless, we can experience certain difficulties while creating a unified adaptive model of the system to be studied where all submodels are interrelated and cannot be viewed separately. It makes the interaction of parallel heterogeneous processes necessary. Figure 2 shows the general structure of the analytical and simulation model, which emulates parallel heterogeneous processes.

Figure 2 shows Ma*i* – analytical models of system elements or processes, Mi – analytical models obtained with the help of experimental data.

The given AIM architecture is adaptable according to the changing structure and parameters of the system to be studied as well as the environment. It can be implemented with the help of E-network logical and dynamic scheme of model control. The scheme solves the following tasks:

- Organizing model element interaction including the activation of required levels in the model hierarchy;
- Adapting the model, which includes changing model details of the system elements;
- Verifying the model;
- Teaching neural network elements to identify objects and elements of the system;
- System state backup;
- Calculating quality indicators of the system.

The procedure of determining the analytical dependence of the system elements can be conventionally divided into an analytical approach and an identification method, that is, constructing the model using experimental data. We propose to use both methods in AIM construction concept. After decomposing the system the model developer sets the system elements in one form or another.

One important task is to relate the elements of heterogeneous models to each other. For example, overheating one element of the system results in the change in system reliability and electrical properties. The further use of the object results in altering the quality level of the whole system.

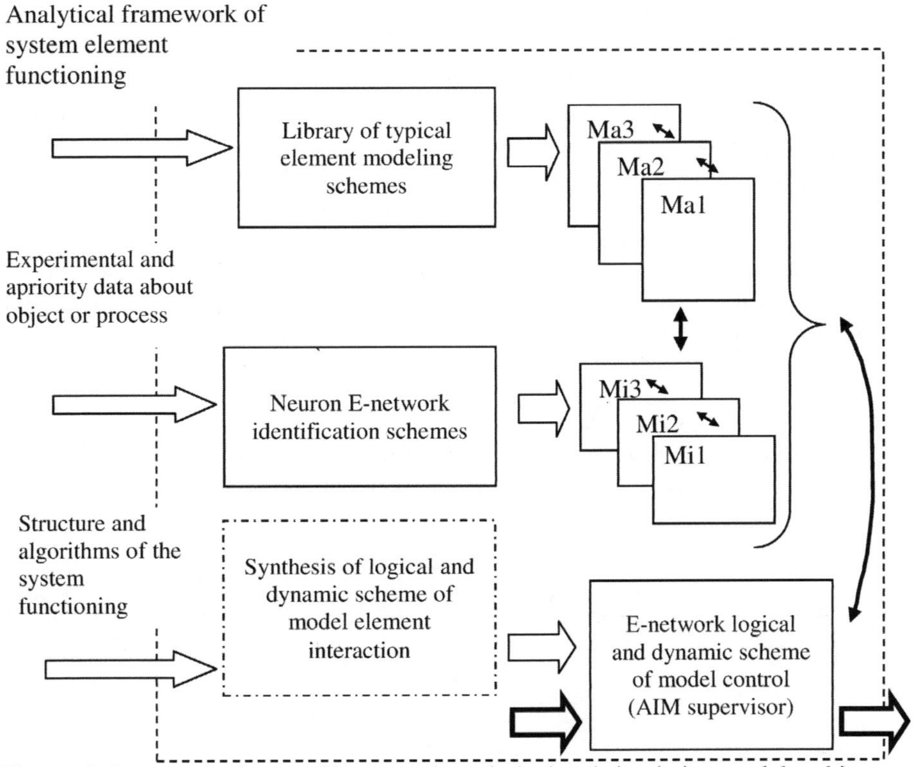

Figure 2. Structural scheme of E-network analytical and simulation model architecture

ASM of the given technical system Σ_0 is presented as the integration of components with certain interrelation. The number of model components is determined by the mode of operation of the multifunctional computer simulator (MCS) and the system complexity. Each component of the technical system model, implemented with E-network simulation apparatus, can be presented by a hierarchical 'input - state - output' structure. The functioning of model components is examined using E-network simulation apparatus at a countable set of times $T = \{t_0, t_1, ..., t_v, ...\}$.

The interaction of model components and the data processing flow from the environment model result in the set of events $E=E^{'}\cup E^{''}$, where $E^{'}$ is a subset of external events $E' = \{E'_1, E'_2, ..., E'_\mu\}$ characterized by input signals $\{X_\mu\}$ and E'' is a subset of internal events $E^{''}=\{E^{''}_1,...,E^{''}_K\}$. The process of sending messages (tokens) to other components, the beginning and the end of the computational process can be considered as internal events. The model operation changes system state vector $\overline{Z(t_v)}$ z∈R. In general, the system state can have a hierarchical structure. The component state is characterized by phase variables $\overline{Z_l(t_v)}, \{z_l, l = \overline{1,L}\}$ at $t_v \in T$. State vector $\overline{Z_l(t_v)}$ is formed by processing the external (input) influences and the internal state of components at t_{v-1}. As stated above, data exchange between components is carried out via communication channels by moving tokens V_i in the network. The value is assigned to input signals $\overline{X(t_\mu)}, \{x_\mu, \mu = \overline{1,M}\}$ only 'outside', that is, while calculating token attributes. The value can be assigned to output signals $\overline{Y(t_\xi)}, \{y_\xi, \xi = \overline{1,\Xi}\}$ only 'inside' the model component unit. Input/output can be represented by control commands (including instruction commands), physical characteristics of components and environment data.

When constructing the analytical and simulation model we propose to define the model as a set of *static* $Stc_i, i = \{1, \dots, N\}$ and *dynamic* Dc_j, $j = \{1, \dots, M\}$ components $Stc \cup Dc \subset M_D$. The subset of static components $Stc = \{Stc_i\}$ of the technical system model includes physical elements, which are *permanently* presented in the technical system and do not depend on the phase of the current technological process. They perform a predetermined function and are in fact a constant resource in the system. Static components are characterized by certain structure *Struc*, functional algorithm *Alg*, conversion operator $H: X \xrightarrow{\Delta t} Y$ and time of the system reaction Δt. Thus static components of the model include the elements of the technological control system (controlling computer, execution mechanisms, sensors, etc.), as well as the elements of the controlled object. Dynamic components are variable resources in the technical system, which are characterized by making qualitative and quantitative changes in physical (information) parameters, depending on the phase of the process. Dynamic components in technical systems include material flows (fuel, electric current, etc.) and information flows with a given set of parameters referring to static components. In E-network apparatus the dynamic components are represented by token V_{DC} with a structured set of attributes: *X*, *Z*, *Y*, *Ind*, where $Ind \in N$ is the identification parameter of the token.

4. Organization of subsystem for identifying processes

In order to realize identification means we propose to use mathematical apparatus of neural networks in the given modeling system. Figure 3 shows the E-network graph, which imitates the work of an artificial neuron.

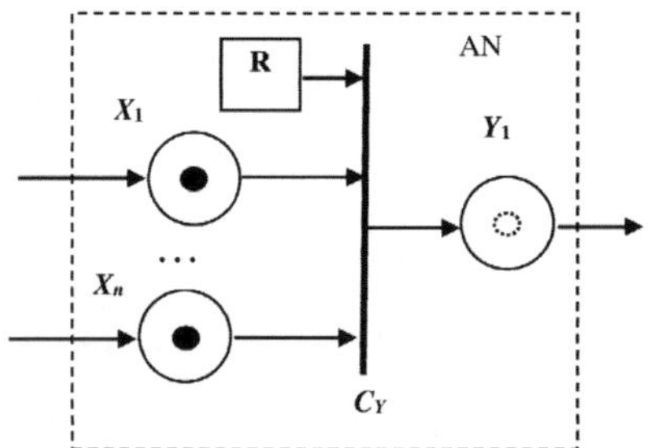

Figure 3. E-network graph of artificial neural network

Any transition of E-network can be described by three parameters [1]:

$$C_j = \left(S, t(C_j), \rho(C_j)\right),$$

where S is the type of transition $S=\{T, X, Y, J, I, F\}$; $t(C_j)$ is the delay function, which determines the interval of the transition phase (i.e. the interval of the event simulated by this transition is calculated); $\rho(C_j)$ is the attribute transformation function.

Firing transition C_j results in:

- changing network layout: layout $M(b^i)$ turns into layout $M'(b^{i+1})$ according to the rule:

$$M'(b^{i+1}) = M(b^i) - I(C_j) + O(C_j);$$

where $I(C_j) = \{b \in B: P_{pre}(b, c) = 1\}$ is the array of input positions of C_j transition, $O(C_j) = \{b \in B_s \cup B_Q: P_{post}(b, c) = 1\}$ is the array of output positions of C_j transition;

- transformation of token attributes.

C_j transition takes the token from the input position and puts the label into the output position. The markup change is indicated as:

M_0 M', C_j

where M_0 is the initial markup.

When modeling systems using E-network, apparatus transitions correspond to events and positions correspond to event conditions. In certain cases dummy activities with zero duration are introduced.

Each position b_i can have a label or can be null (position markup $M(b_i)$ is equal to '1' or '0' correspondingly). The label in the position indicates that the condition described by the given position is fulfilled. Positions with arrows entering the transition are called input positions. Positions with arrows going out of the transition are called output positions. If input positions have labels, the transition is called 'cocked'. If outcome positions are free, the transition is switched to active state after calculating the interval of active state *t*. After the delay the transition moves the label from the input position into the output position and converts attributes of the token.

Y type transitions have the position of condition *R*. The control procedure *S(R)* indicates the conditions which allow moving the token from position *X* to output position Y_i if position X_i is free at a given time, where *i* is the number of the outcome position which is determined by *S(R)*.

$$C_y = B(X_i) \wedge \overline{\boldsymbol{B(Y)}}, i = S(R), i \in \{1,2,\dots,n\}.$$

Let us consider artificial neural networks (ANN) based on E-network principles.

The functioning of artificial neurons can be divided into two stages (Fig.4): teaching and ANN functioning. At the first stage the library of teaching algorithms [4, 5] helps to set the structure, weight coefficients and ANN offsets so that the capacity $J = \sum(P_E - T_E)^2 \rightarrow min,$ where P_E is input experimental data array, T_E is output experimental data array. The given arrays are formed while conducting the experiment.

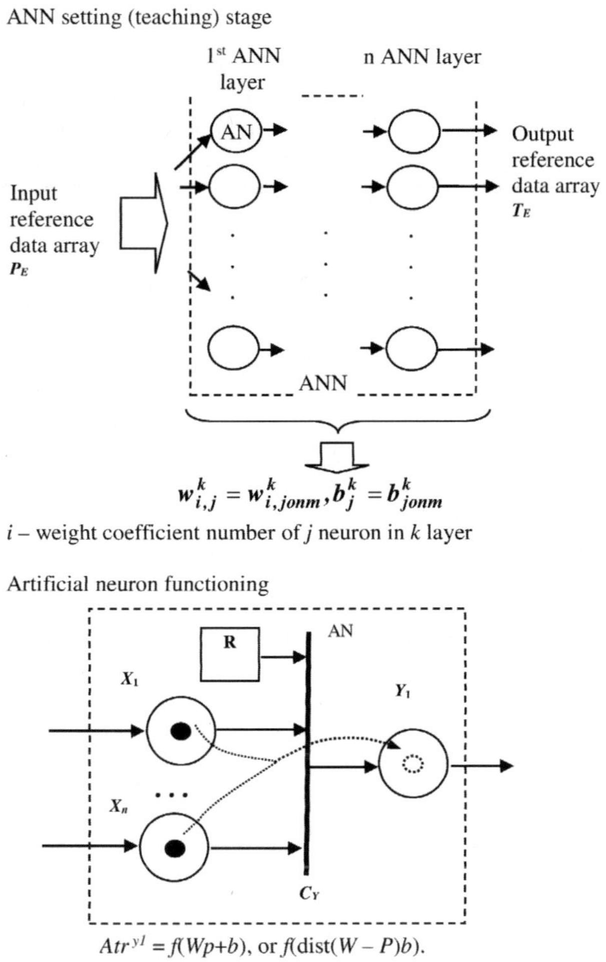

Figure 4. Stages of neural network functioning

The second stage means ANN functioning as the identification system.

When teaching neural networks we propose to use the following tuple of parameters as the attributes of tokens:

$$Atr_{ann} = \{A1, A2, A3, A4, A5\}.$$

The first attribute sets the coordinate of the artificial neuron in the network. The parameter A2 determines the type of activation function *f*, parameter A3 has weight coefficients of neuron $A3 = \{w_1, w_2, \dots, w_n\}$. The fourth attribute determines the value of offset *b*. Position A5 determines the length of the E-network transition delay. Each time the *Y* transition is fired at the stage of ANN teaching, the attributes are recorded in the memory buffer. Subsequently, when the neural network is taught the output activity calculation $a = f(Wp + b)$ will be based on the transition parameters recorded at the last clock period of ANN teaching. In order to teach the neural network the library of typical teaching algorithms is used, for example, the algorithms of reverse error development [2].

At the second stage the input data vector (experimental data or operator activity) is converted by each network neuron. At the stage of ANN functioning the input of the Y transition receives signals (tokens) which have at least one attribute $Atr_{ann} = \{A0\}$. The A0 network parameter has the signal value which goes to the network input. When the *Y* transition is fired the token is moved into the output position with the calculated value of the signal $a = f(Wp+b)$ or $a = f(dist(W - P)b)$.

5. Conclusion

E-network apparatus allows modeling heterogeneous interrelated processes and the functioning of artificial neural networks with different configuration, including dynamic neural networks. The given concept can be used for constructing computer simulators, designing and maintaining technical systems. In a multifunctional computer simulator the use of ANN components in order to identify controlled objects and processes reduces the ASM setting time. The distinguishing feature of the proposed method is that it links ASM and neural networks on the basis of expansion of Petri nets (E-network modeling) using uniform methodological positions.

References

[1] Tsapko G P, Tsapko S G and Tarakanov D V 2005 *Tomsk University Publishing* 228

[2] Best E and Fernandez C 1988 *EATCS Monographs on Theoretical Computer Science* **13**

[3] Pranyavichus G I and Shvanite D Yu 1980 *Mathematics and Mathematical Simulation* 68-72

[4] Sylla A N, Louvel M and Pacull F 2015 *PNSE'15 Petri Nets and Software Engineering* 325-326

[5] Reisig W 1985 *Petri Nets: An Introduction of Monographs in Theoretical Computer Science: An EATCS Series. Springer, Berlin* **4**

[6] Brogi A, Canciani A, Soldani J and Wang PengWei 2015 *PNSE'15 Petri Nets and Software Engineering* 191-205

[7] Jensen K 2013 *Springer Science & Business Media*

[8] Murata T 1989 *Petri nets: Properties, analysis and applications. Proceedings of the IEEE* **77(4)** 541–580

The numerical calculation of the viscous incompressible fluid transfer between contacting surfaces

L G Varepo[1], A V Panichkin[2] and O V Trapeznikova[1]
[1] Omsk State Technical University, 11, Mira ave., Omsk, 644050, Russia
[2] S L Sobolev Omsk branch of Institute of mathematics of the Siberian Branch of the Russian Academy of Sciences, 13, Pevtsova st., Omsk, 644043, Russia

E-mail: larisavarepo@yandex.ru

Abstract. The movement of the thin layer of the viscous incompressible fluid (VIF) between two cylinders is analysed. The numerical calculations results of VIF transfer from the engaged zone of two cylinders to porous substrates are presented. The VIF (ink) is moved along the rubberized top blanket of the first cylinder. The surface of the second cylinder contacts the substrate with some part of the VIF layer transferred from the first cylinder. The fluid is double bounded by the free surface. Images of cylinders boundary deformation and VIF flow areas are shown.

1. Introduction

The process of image printing is VIF (ink) transfer from engaged surfaces of the offset and the printing cylinders.

The known modelling methods of the VIF transfer process onto substrate are characterized by limited functional features: dates are reduced to half splitting of ink with zero absorptive capacity. So, it is conventional that the VIF layer (ink) is divided in a bisected way [1-8].

Therefore, the quantitative evaluation of the VIF transfer ratio onto a substrate in the engaged zone is actual at the moment due to wide and novel assortment of substrates and their different surface parameters.

There are not numerous literature sources about ink dusting evaluation. The information about ink dusting is experimental [9-10].

2. Task definition

As the problem, concerning the part of the VIF (ink) that remains on the ink-carried surface after each cycle of dividing, has not obtained one-valued solution it is necessary to develop the evaluation algorithm of this parameter.

Also we need to formulate the quantitative evaluation algorithm of ink dusting and the ratio of the ink amount immobilized by the printed surface.

3. Experimental methods

A simulated model of ink transfer onto the printed surface in the engaged zone was proposed for evaluation of parameters of VIF (ink) transfer between contacted zones.

The modelling was carried out using a developed algorithm of numerous solving of incompressible Navie-Stokes equations by finite-difference approximation on the two-dimensional grid and the compact template.

The flow is of the laminar nature.

4. Results and discussion

The Navie-Stokes equations for the postulated conditions of printed simulation of coordinates in accordance with system $(x, y)=(R\theta, R\text{-}r)$ and system angular acceleration ε are as follows:

$$\frac{\partial U_r}{\partial t}+U_r\frac{\partial U_r}{\partial r}+\frac{U_\theta R}{r}\frac{\partial U_r}{\partial \theta R}-\frac{(U_\theta+\omega r)^2}{r}=-\frac{1}{\rho}\frac{\partial P}{\partial r}+\nu\left(\nabla^2 U_r-\frac{U_r}{r^2}-\frac{2R}{r^2}\frac{\partial U_\theta}{\partial \theta R}\right),$$

$$\frac{\partial U_\theta}{\partial t}+U_r\frac{\partial U_\theta}{\partial r}+\frac{U_\theta R}{r}\frac{\partial U_\theta}{\partial \theta R}-\frac{U_r U_\theta}{r}+2U_r\omega+\varepsilon r=-\frac{1}{\rho}\frac{R}{r}\frac{\partial P}{\partial \theta R}+\nu\left(\nabla^2 U_\theta-\frac{U_\theta}{r^2}+\frac{2R}{r^2}\frac{\partial U_r}{\partial \theta R}\right), \quad (1)$$

$$\frac{\partial U_r}{\partial r}+\frac{U_r}{r}+\frac{R}{r}\frac{\partial U_\theta}{\partial \theta R}=0,$$

where

$$\nabla^2 V=\frac{1}{r}\frac{\partial U}{\partial r}+\frac{\partial^2 U}{\partial r^2}+\frac{R^2}{r^2}\frac{\partial^2 U}{(\partial \theta R)^2}$$

– Laplasian operator, ν – kinematic viscosity, ρ – fluid density, P – pressure, ω – angular rotating.
The finite-difference methods with input of moving boundary units for the second cylinder boundary and free-boundary fluid, which at the initial moment is placed on the first cylinder without relative movement, are used on the fixed grid.

The computational region is represented as a rectangle with a regular grid and similar steps. For equation (1) in changed coordinate system $(x, y)=(R\theta, R\text{-}r)$ we used the stabilize error-correction scheme with inertial time step τ and grid steps h_x and h_y:

$$\frac{V^{n+1/2}-V^n}{\tau}=\Lambda_1\left(V^{n+1/2}-V^n\right)+\Lambda_0-\frac{Gp^{n+1}}{\rho}+F^n(\varpi,\rho), \quad (2)$$

$$\frac{V^{n+1}-V^{n+1/2}}{\tau}=\Lambda_2\left(V^{n+1}-V^n\right), \quad (3)$$

where

$$\Lambda_1=-u\frac{\Delta_1+\Delta_{-1}}{2h_x}+v_x\frac{\Delta_1\Delta_{-1}}{h_x^2},\ \Lambda_2=-v\frac{\Delta_2+\Delta_{-2}}{2h_y}+v_y\frac{\Delta_2\Delta_{-2}}{h_y^2}, \quad (4)$$

$\Delta_1,\Delta_{-1},\Delta_2,\Delta_{-2}$ – operators of function shift to grid step h_x or h_y up or down to axes x and y; $\Lambda_0=\Lambda_1+\Lambda_2$; $V=(u,v)$ – speed vector with components u, v; Gp –field gradient; $\bar{x}=(x,y)$– point coordinate; ϖ – vector of angular rotation; F– acceleration component from external forces and from the changed coordinate system; n – time iteration number at moment $t=n\tau$.

If approximation of these operators is of the order that does not exceed the second one ($O(h_x^2+h_y^2)$), the pressure gradient approximation (Gp^{n+1} in (2)) will be realized by the fourth order as follows:

$$Gp^{n+1}=\begin{pmatrix}\frac{\Delta_1+\Delta_{-1}}{2h_x}p^{n+1}-\frac{h_x^2}{6}p_{xxx}^{n+1}+O(h_x^4),\\ \\ \frac{\Delta_2+\Delta_{-2}}{2h_y}p^{n+1}-\frac{h_y^2}{6}p_{yyy}^{n+1}+O(h_y^4)\end{pmatrix}, \tag{5}$$

for which approximation of the third-order derivatives with respect to x and y from pressures p_{xxx}^{n+1} and p_{yyy}^{n+1} has been formulated on the compact template as the equation of the second-order after p_{xyy}^{n+1} and p_{yxx}^{n+1} has been changed.

The pressure determination equation has been determined as a divergence-free transformation equation of a low compressive fluid model. After replacing a partial derivative of pressure with respect to time, the equation is:

$$\varepsilon_p\frac{p^{n+1,\,k+1}-p^{n+1,\,k}}{\tau}+\nabla\cdot V^{n+1,\,k}=0, \tag{6}$$

where $\varepsilon_p > 0$ – parameter which was optimally chosen for numerical calculations of the convergence and approximation to the initial equation, k – number of additional iteration for pressure calculation, the speed divergence was presented with the forth-order approximation as:

$$\nabla\cdot V|_{i,j}=\left(\frac{M_1}{12h_x}\right)+\left(\frac{M_2}{12h_y}\right)+O(h_x^4+h_y^4), \tag{7}$$

where

$$M_1=u_{i+1,j+1}+4u_{i+1,j}+u_{i+1,j-1}-u_{i-1,j+1}-4u_{i-1,j}-u_{i-1,j-1},$$

$$M_2=v_{i+1,j+1}+4v_{i,j+1}+v_{i-1,j+1}-v_{i+1,j-1}-4v_{i,j-1}-v_{i-1,j-1}, \tag{8}$$

with main indexes:

$$(i,j), i=1,\dots, N_x-1, j=1,\dots, N_y-1,$$

N_x, N_y – parameters of the computational grid with respect to x and y.

During movement of fluid through the contact zone, some part of fluid is penetrated into the substrate structure due to absorption. The dynamical equation of deformation was defined for some elastic layer (a substrate, a rubberized top blanket) according to arising elastic strains of cylinders.

The parameters of the VIF flow between cylinders were calculated according to:

- the deformation of surface layers of the rubberized top blanket of the first cylinder and the substrate contacting the second cylinder surface due to pressure changing inside the fluid on contact with them;
- the deformational influence on final fluid distribution parameters (ink) along cylinders including absorption into the substrate and dusting;
- the contact width.

The deformation of centres x_c (through the thickness) of the thin layers of rubber and paper were calculated using shortcut evaluation according to a low speed of relative movement of compressed surfaces compared with wave-induced stresses (several hundred times)

$$\rho HS\ddot{x}_c/2=-\frac{E}{H}(\Delta r/2+x_c)\cdot S+(P_1-P_0)\cdot S \tag{9}$$

where Δr – boundary deformation value at some moment t_1, H – initial thickness of the stressed layer, S – small square near a desirable point of layer deformation, E – Young's modulus of rubber or paper elasticity, P_0 – atmosphere pressure, P_1 – pressure on the stressed layer, ρ – fluid density.

Solution (9) with accuracy at current rotating speed for choosing a boundary point with velocity $\dot{x}_{c,1}$ at moment t_1 is:

IOP Conf. Series: Materials Science and Engineering **124** (2016) 012106 doi:10.1088/1757-899X/124/1/012106

$$x_c=\frac{\tau^2}{H\rho}\left[(P_1-P_0)-\frac{E}{H}\frac{\Delta r}{2}\right]/\left(1+\frac{E\tau^2}{H^2\rho}\right)\pm\frac{\sqrt{\frac{\tau^4}{H^2\rho^2}\left[(P_1-P_0)-\frac{E\Delta r}{H\,2}\right]^2+\left[1+\frac{E\tau^2}{H^2\rho}\right]^2\tau^2\cdot\dot{x}_{c,1}^2}}{1+\frac{E\tau^2}{H^2\rho}}. \quad (10)$$

Signs '+' or '–' for (10) are chosen according to the sign of $\left[(P_1-P_0)-\frac{E}{H}\frac{\Delta r}{2}\right]$; it shows the boundary deformation direction with value $\Delta r+2x_c$.

The numerical solutions on 80 × 80 grid are shown in figure 1 calculated for the following conditions: the angular rotation is 10 radians per second at different moments, Young's modulus of rubber or paper elasticity is $E_1=2.9\cdot10^7$ N/m^2 and $E_2=4.8\cdot10^6$ N/m^2 , cylinders radiuses $R_1=R_2=0.30$ m, $P=P_{athm}=10^5$ N/m^2, kinematic viscosity $\nu=0.012$ m^2/sec, density $\rho=10^3$, fluid surface tension $C_n=0.03$ N/m.

The algorithm works with data describing a two-dimensional fluid flow within the required range [11-12]. The presented equation of the ink transfer process was limited by fixed time value $t$during different printing stages (Figure 1).

Images of stress boundaries of contacting surfaces in the VIF transfer areas onto the substrate are presented in the figures below: (1) – ink input in the engaged zone; (2), (3) – ink transfer onto the substrate in the engaged zone; (4), (5), (6) – ink output from the engaged zone. The results of the numerous modeling show that the ink quantity on the offset cylinder is less than 50 % (Table 1).

Table 1. Test results.

Roughness (R_a), μm	Ink amount on offset cylinder,%	Misting ink amount,%	Total ink quantity on print after engagement zone, %
0.427	42.52	2.07	55.31
2.310	35.03	2.24	62.73
3.320	32.13	3.66	64.21

The results of the numerical exam of the ink misting are presented in Table 1. The dependence of the ink quantity on the arithmetical mean deviation value of the profile is R_a (Table 1).

The analysis of the relationship of the ink quantity located on the printed surface in the engaged zone and microgeometry parameters has revealed their high correlation (Figure 2). Results of the dusting evaluation are presented in Table 1.

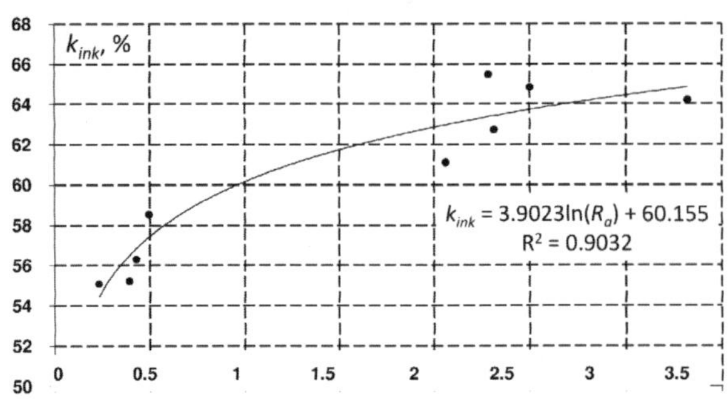

Figure 2. The dependence of the ink quantity transferred onto the printed k_{ink} surface from average profile deviation R_a.

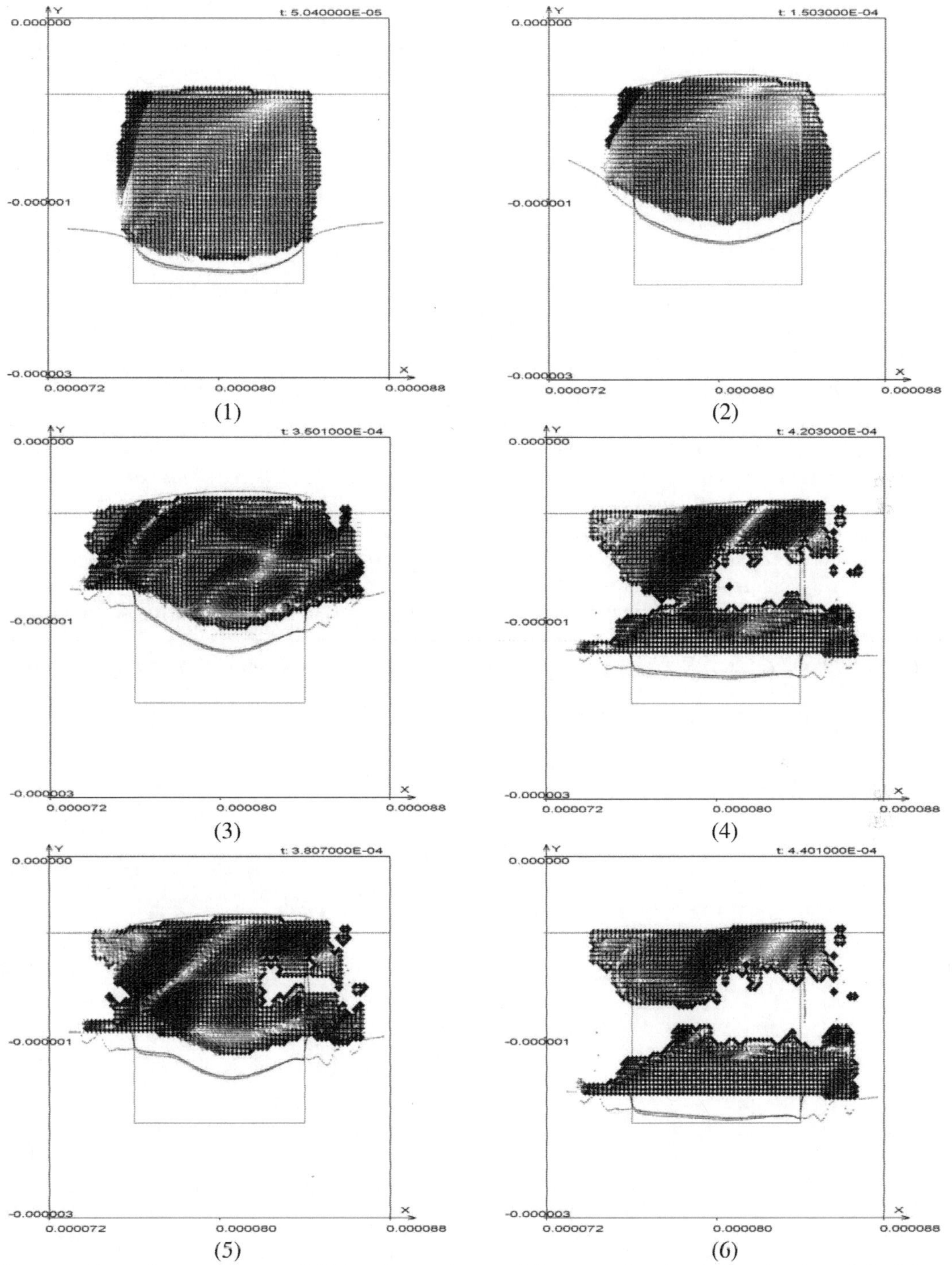

Figure 1. Boundary geometry of VIF and cylinders at different *t*. Graphical visualization of the ink splitting process at the outlet of the printed contact zone.

5. Conclusion

The simulated model of the VIF transfer process between contacting surfaces with different microgeometry according to deformation within the engaged zone has been developed.

The quantitative evaluation of VIF (ink) transfer parameters – the ink ratio on the offset cylinder after the engaged zone; the ink transfer ratio onto the printed material including ink absorbed inside paper pores and the ink quantity immobilized by the printed surface; dusting ratio – has been performed.

Reference:

[1] Zettelmoyer A C, Scarr R F and Schaeffer W D 1958 *International Bulletin for the Printing and Allied Trades* **13** 88–94
[2] Meyers R R, Miller J C and Zettelmoyer A C 1959 *J. Colloid Sci.* 14 **3** 287–299
[3] Fetsko J M, Schaeffer W D and Zettelmoyer A C 1962 *13th TAPPI Testing Conf.* 1–6
[4] Truman A B and Hudson F 1962 *Pulp and Paper Magazine of Canada* **6** 299–306
[5] Mac Phee J 1998 *Fundamentals of Lihographic Printing* (Pittsburgh: GATFPress) p 365
[6] Ozaki Y and Kimura M 2000 *Appita J.* **3** 216–219
[7] Koivula H, Preston J S, Heard P J and Toivakka M 2008 *Colloids and Surfaces A: Physicochemical and Engineering Aspects* 317 **1-3** 557–567
[8] Reis Jr. N C, Griffiths R F and Santos J M 2004 *J. Comput. Phys.* **198** 747–770
[9] Vlachopoulos G, Claypole T and Bould D 2010 *IARIGAI 2010 Proceedings: Advances in Printing and Media Technology* **37** 227–234
[10] Claypole J, Williams P R and Deganello D 2012 *IARIGAI 2012 Proceedings: Advances in Printing and Media Technology* **39** 207–211
[11] Panichkin A and Varepo L 2014 *International Multidisciplinary Microscopy Congress, Springer Proceedings in Physics* **154** 79–83
[12] Varepo L and Panichkin A 2014 *Dynamics of Systems, Mechanisms and Machines* 7005689

IOP Conf. Series: Materials Science and Engineering **124** (2016) 012107 doi:10.1088/1757-899X/124/1/012107

An oscillograms processing algorithm of a high power transformer on the basis of experimental data

O V Vasileva[1], A A Budko[1] and A V Lavrinovich[2]
[1] National Research Tomsk Polytechnic University, 30, Lenin ave., Tomsk, 634050, Russia
[2] Institute of high-current electronics Siberian Branch of the Russian Academy of Sciences Tomsk, 2/3, Academic ave., Tomsk, 634055, Russia

E-mail: vasileva.o.v@mail.ru

Abstract. The paper presents the studies on digital processing of oscillograms of the power transformer operation allowing determining the state of its windings of different types and degrees of damage. The study was carried out according to the authors' own methods using the Fourier analysis and the developed program based on the following application software packages: MathCAD and Lab View. The efficiency of the algorithm was demonstrated by the example of the waveform nondefective and defective transformers on the basis of the method of nanosecond pulses.

1. Introduction
One of the causes of damage is effects of short circuit currents, resulting in the displacement of turns in the windings of the high power transformer [1]. Capacitance and inductance of deformed elements are changed according to the degree of deformations. This leads to a change in the vibration frequency that can manifest itself in the waveform of pulsed currents and voltages [2]. Thus, the development of simulation algorithm processes in the high power transformer for diagnosis is an important problem.

2. Research method
The voltage pulse applied to the matched load was exhibited by an oscilloscope 'Tektronix-TDS2012' using the standard probe of the P2220 type applied to the oscilloscope with a bandwidth of 200 MHz.

A winding turn is presented by an equivalent circuit in Figure 1, where L – leakage inductance coil, considering mutual inductance between the coils, C_1 – capacitance between the turns, C_2 – capacitance between the coil and the grounded part of the transformer, R – resistance of windings.

The flow diagram of the procedure sequence of the high power transformer oscillograms is shown in Figure 2. The efficiency of the developed algorithm was proved by the example of two pulses read under the matched load.

The pulse of the nondefective transformer obtained by connecting the output of the generator with resistance R = 50 Ohm is shown in Figure 3 (curve 1). The pulse obtained by short circuit turns taken in comparison with the pulse previously discussed (curve 2).

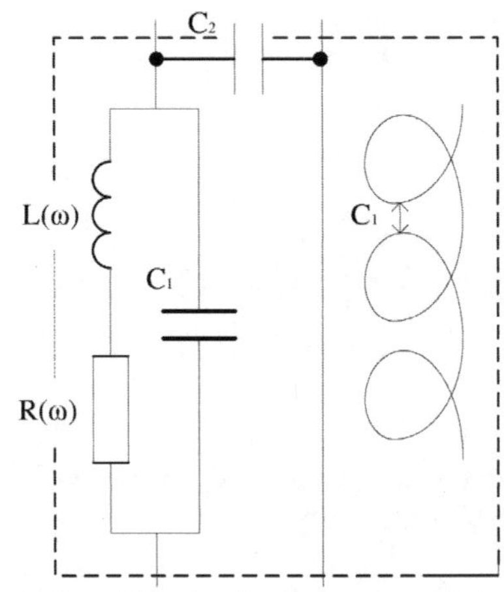

Figure 1. The equivalent circuit of a power transformer winding turn.

The condition of the developed program in the Lab View software program is the *same sample rate condition* of processed pulses [3, 4]. In our case time step was estimated as $h = 4 \cdot 10^{-10}$ ns. Time step can be expressed as:

$$h = t_{11} - t_{12}, \tag{1}$$

where t_{11} and t_{12} – the first column of the first and second rows, accordingly. The original data are similar to other pulses. A coefficient must be brought in the box 'Step, s'.

3. An experimental part

Then the processing of the signals must be tested in order to compare two signals (Figure 4). For example, one waveform might be subtracted from another, provided that sharp rises in both oscillograms are set both at the same moment. The procedure for the above-mentioned example is feasible *after setting simultaneous starting time* of oscillograms having an arbitrary offset relative to each other and receiving pulses difference in a separate box.

The number of entries K must be calculated in order to set simultaneous starting time, which is required to equalize the processed oscillogram relative to the standard sample and enter the coefficient in the box "Number of entries". Number of entries can be expressed as:

$$K = \frac{(t_{11} - t_{21})}{h}. \tag{2}$$

In our case the coefficient was estimated as $K = 110$. Thus, time shift was equal to $t_0 = 44$ ns.

The pulses comparison is given in Figure 4, which justifies that the pulses vary notably. A detailed analysis is required for more thorough comprehension.

After processing (2) the digital data are saved automatically in extension txt.

The study was carried out in the MathCAD software program using the Fourier analysis method for an objective pulses comparison. The method justification is that the discrete Fourier transform allows obtaining the required results quickly and with sufficient accuracy.

The pulse was decomposed into Fourier series F components in the MathCAD software program (Figure 5) [5]:

$$F = FFT(U), \tag{3}$$

where U – instant voltage value on the matched load at the generator exit.

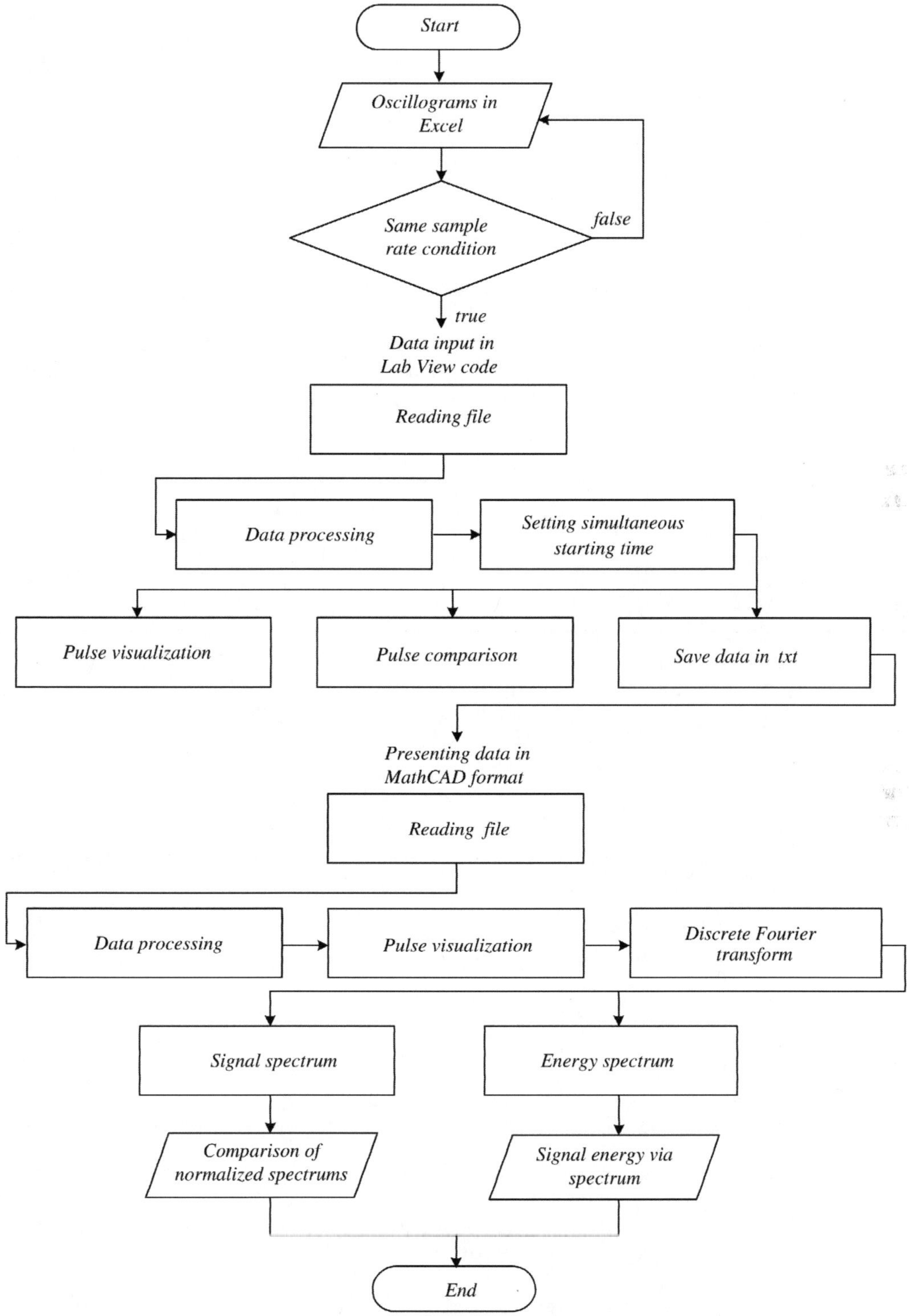

Figure 2. A control-flow chart

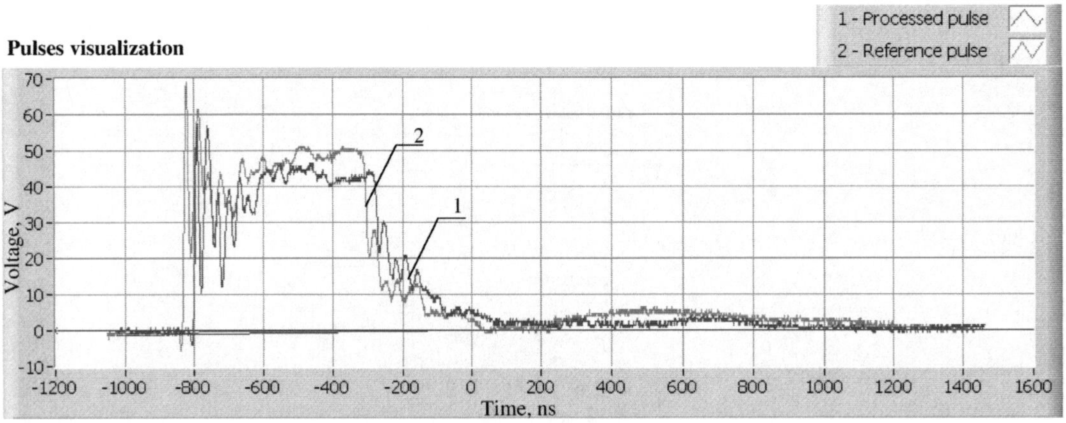

Figure 3. Pulses visualization.

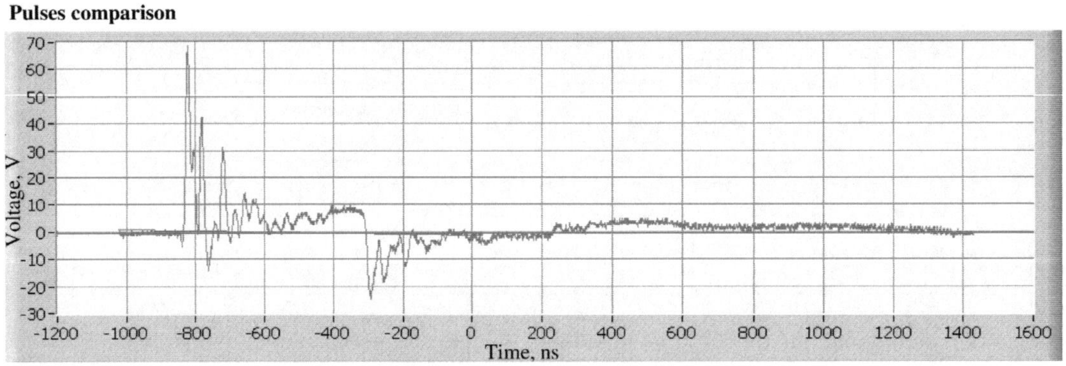

Figure 4. Pulses comparison.

All pulses were normalized by amplitude to unity before Fourier series expansion. The signal of each spectrum was equal to 1 nominal unit. This allows making a comparison of normalized spectrums of different pulses.

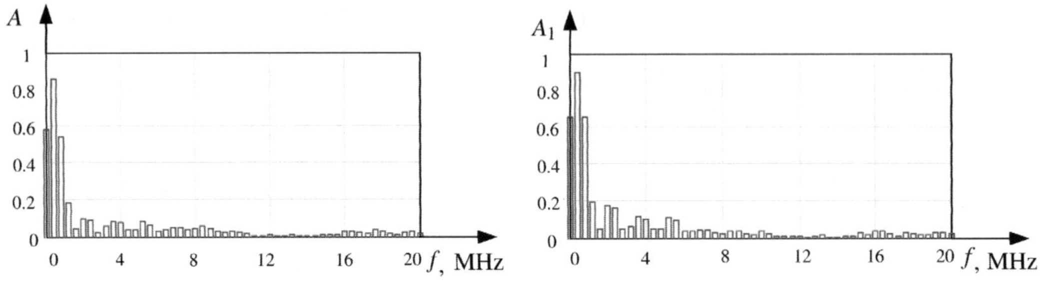

Figure 5. The signal spectrum: *a* – a nondefective transformer; *b* – a defective transformer with short circuit.

Visually the signal spectrums differ from each other a little bit (Figure 5a). Therefore, we suggest comparing the signal spectrums in the form of normalized spectrums of relative divergence (5).

The pulses signal of spectrum A can be expressed as double discrete Fourier function (1) [6, 7]:

$$A_j = |F_j| \cdot 2, \quad (4)$$

where $j = 0 - 50$ – harmonic order, $A_0 = \frac{A_0}{2}$ – steady component amplitude.

A comparison of normalized spectrums ΔA_j can be expressed as:

$$\Delta A_j = \frac{A_j}{A_{1_j}}, \quad (5)$$

where A_j and A_{1j} – the signal spectrum values of the nondefective and defective transformers accordingly, $n = 2^{10} = 1024$ – the number of points.

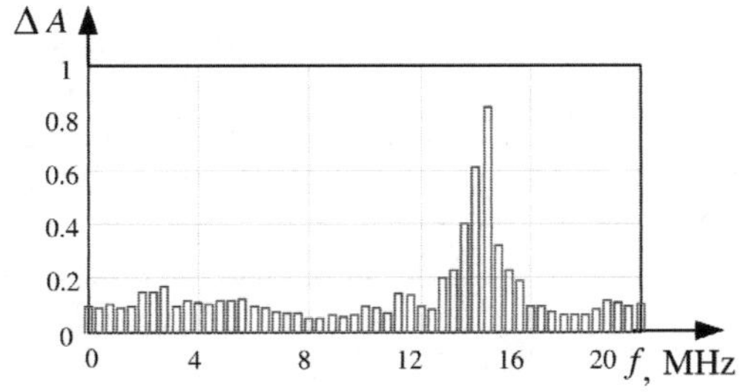

Figure 6. The comparison of normalized spectrums.

The large peak in the latter spectrum demonstrated a lot to the high-frequency components. This indicates obvious damages in the transformer windings, in particular, short circuit.

Analyzing the frequency range it can be concluded that the work of the transformer at frequencies $(12.89 \le f \le 15.71)$ MHz can lead to a breakdown and serious damage of the transformer, period was equal $T = 2.48\ \mu s$, basic frequency was equal $f = 0.4$ MHz.

The total spectrum energy can be expressed as:

$$W = \sum_{j=1}^{n-1} \left|F_j\right| \cdot 2^{\;2} + \frac{\left|F_0 \cdot 2\right|^2}{2}, \quad (6)$$

where $\left|F_j\right| \cdot 2^{\;2}$ – harmonic energy using (3), $F_0 = \frac{F_0}{2}$ – coefficients of the constant component of Fourier series.

Using (6) the total spectrum energy value nondefective transformer was estimated as $W = 898.6$ J and the total spectrum energy value of the defective transformer was estimated as $W_1 = 1121$ J. The large peak in the latter spectrum contributed a lot to the total energy. It gives a physical and numerical concept of the processes.

The direct physical voltage signals were used for test of the algorithm in addition to the experimental data decomposition in Fourier series. The instantaneous experimental values were recorded in the files and then processed and used as an input in the Lab View software program.

This algorithm can be used to simulate the defects of the windings and generate reference signals for various fault operations of the tested equipment.

4. Results

Thus, the proposed algorithm will significantly increase the speed of processing and comparison of the oscillograms while examining transformers by nanosecond pulses. The waveform visualization and a technique, allowing the comparison of pulses by means of Fourier series and the comparison in relative terms of the spectra of initial pulses, have been presented.

References

[1] Lavrinovich V A, Mytnikov A V 2015 *IEEE Transact. on Dielectr. and Electr. Insulat.* 7179164
[2] Vasiljeva O V., Budko A A, Lavrinovich A V and Filkov A I 2014 *Proc. Int. Conf. on Mechanical Engineering, Automation and Control Systems* 6986867
[3] Silva I N, Flauzino R A, Spatti D, Gonzales C G, Carrapato M, Costa P I and Bonfim T 2014 *IEEE Int. Conf. on Instrumentation and Measurement for Sustainable Development* CFP14IMT-USB
[4] Yu Q D, Li Y L and Wang J 2014 *Applied Mechanics and Materials* **562** 5003–5005
[5] Kolchanova V A, Isaev Yu N, Tarasenko S S and Filkov A I 2014 *Proc. Int. Conf. on Mechanical Engineering, Automation and Control Systems* 6986868
[6] Kabanova T V, Kuleshova E O, Isaev Yu N, Shishka N V and Shishkovskaya Y V 2014 *Proc. Int. Conf. on Mechanical Engineering, Automation and Control Systems* 6986903
[7] Shandarova E B and Shvab S A 2014 *Proc. Int. Conf. on Mechanical Engineering, Automation and Control Systems* 6986872

Thermal destruction of vessels with liquid upon heating

V G Zverev[1], V D Goldin[1] and A A Svetashkov[2]
[1] Tomsk State University, 36, Lenina ave., Tomsk, 634050, Russia
[2] Tomsk Polytechnic University, 30, Lenina ave., Tomsk, 634050, Russia

E-mail: zverev@niipmm.tsu.ru

Abstract. A new engineering technique of calculating the heating and thermal destruction of vessels containing liquid under extreme thermal loading conditions is offered. The heating of the shell and the internal vessel volume is described on the basis of the thermodynamic approach. The pressure growth in a vessel is a result of gas heating and liquid evaporation. Stresses within the shell and its destruction conditions are determined, which allows predicting the critical time of destruction upon heating. The calculation and experimental data for pressure growth inside the vessel are in good agreement.

1. Introduction

Modern industry works with potentially hazardous materials that constitute a danger to the lives of workers, infrastructure and environment. The fire and explosion safety of vessels filled with process liquids during their storage, operation and transportation is a very important issue [1, 2].

A typical emergency situation occurs when vessels with liquids are involved in industrial fire. Fire causes heating of vessel walls and contents up to critical temperatures, an internal pressure growth due to gas heating and liquid evaporation, shell rupture and, consequently, leads to destructive effects [1, 2].

Liquefied petroleum gas (LPG) vessels are especially hazardous because they often explode upon heating. The theoretical and experimental aspects of this problem were considered by different researchers [3-9]. In order to reduce the vessel explosion risk and to develop prevention strategies, it is necessary to analyze the heat and mass transfer processes during heating and to provide protection of vessels with liquids subjected to extreme thermal loading of various origin, intensity and duration.

Since full-scale experiments are difficult to perform, a good solution of this task is the mathematical modeling of emergency situations. This would help in determining the critical failure conditions of thermally loaded vessels with regard to the initial vessel state, liquid properties and thermal loading parameters as well as in designing thermal protection of vessels.

The aim of this paper is to develop a new engineering technique of calculating the heating of the vessel with liquid to predict the pressure dynamics, critical shell rupture conditions and the time of vessel resistance to extreme thermal loading.

2. Physical formulation of the problem

Let us consider the local heating of a long cylindrical vessel of diameter D with metal shell thickness d, which is partially filled with liquid. The cross-section of the vessel and its schematic heating are depicted in Figure 1. Shell zones *1*, *2* are subjected to external thermal stress along arc *CBD* defined by central angles β_1, β_2. Zone *1* contacts with gas *6* on the inside, and zone *2* contacts with liquid *5*. The outer shell side along arc *CAD* (zones *3*, *4*) is in heat exchange with the environment.

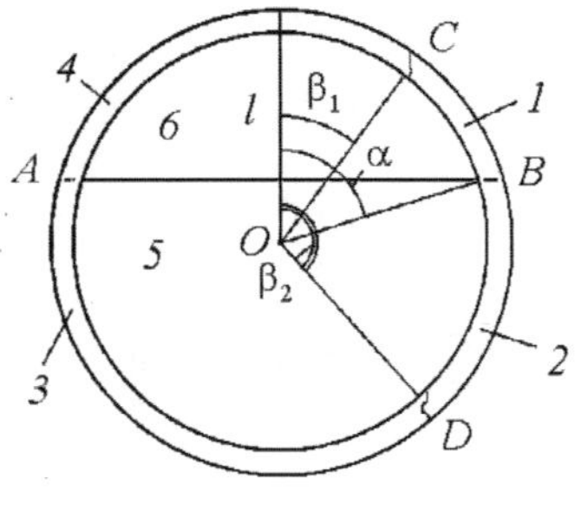

Figure 1. Cross-section geometry and scheme of heating a vessel with liquid under local thermal loading.
Zones *1–4* – vessel shell; *5* – liquid; *6* – vapor-gas mixture; line *AB* – liquid–gas interface; *l* – distance to a liquid level; arc *CBD* – thermally loaded region; arc *CAD* – region contacting with the environment; β_1, β_2, α – central angles of arc *CBD* and liquid level *AB*.

The vessel is partially filled with liquid. The distance from the liquid level (line *AB* in Figure 1 with central angle α) to the upper point of the vessel is equal to l ($l = 0$ corresponds to complete filling). The geometric characteristics of each region (volumes V_i and surface areas S_i, $i = 1$–4) are expressed through the initial parameters of problem D, d, l, β_1, β_2, α, with areas S_i referring to RL and volumes V_i referring to R^2L, where R, L are the radius and length of the vessel.

The physics of the process is the following. The shell, liquid and gas inside the vessel are heated up under thermal loading. This causes partial evaporation of liquid, an increase in temperature and vapor-gas pressure, and the stress growth in the shell. Once the limiting values of these parameters are achieved, the shell ruptures. We need to define the dynamics of the liquid and vapor phase parameters, temperature and stress within the shell, and the critical destruction conditions of the vessel under local thermal loading.

The formulated problem will be solved with regard to the following physical processes: heat exchange between the vessel shell and environment, liquid and vapor-gas mixture; radiation heat exchange between the shells and with the liquid; liquid evaporation. We assume that each region can be characterized by a temperature that depends only on time. Liquid evaporation occurs under equilibrium conditions and is determined by saturated vapor pressure. In order to obtain the upper bound for heating of the regions, we neglect the heat flux over the vessel shell. Its thickness is assumed to be much smaller than the vessel radius.

3. Mathematical formulation of the problem

The problem of determining the temperature of zones *1–6* of the vessel with regard to the made assumptions reduces to the solution of the following system of ordinary differential equations:

$$\rho_s c_s d\frac{dT_{s1}}{dt} = \alpha_e(T_{e0} - T_{s1}) + \varepsilon_{sf}(q_R - \sigma T_{s1}^4) + \alpha_{sg}(T_g - T_{s1}) + \varepsilon_{sl}\sigma(T_l^4 - T_{s1}^4)F_{1-5} + \varepsilon_{ss}\sigma(T_{s4}^4 - T_{s1}^4)F_{1-4}, \quad (1)$$

$$\rho_s c_s d\frac{dT_{s2}}{dt} = \alpha_e(T_{e0} - T_{s2}) + \varepsilon_{sf}(q_R - \sigma T_{s2}^4) + \alpha_{sl}(T_l - T_{s2}), \quad (2)$$

$$\rho_s c_s d\frac{dT_{s3}}{dt} = \alpha_e(T_{e0} - T_{s3}) + \alpha_{sl}(T_l - T_{s3}), \quad (3)$$

$$\rho_s c_s d\frac{dT_{s4}}{dt} = \alpha_e(T_{e0} - T_{s4}) + \alpha_{sg}(T_g - T_{s4}) + \varepsilon_{sl}\sigma(T_l^4 - T_{s4}^4)F_{4-5} + \varepsilon_{ss}\sigma(T_{s1}^4 - T_{s4}^4)F_{4-1}, \quad (4)$$

$$\rho_l c_l R V_5\frac{dT_l}{dt} = \alpha_{sl}(T_{s2} - T_l)S_2 + \alpha_{sl}(T_{s3} - T_l)S_3 + \alpha_{lg}(T_g - T_l)S_5 +$$
$$+ \varepsilon_{sl}\sigma(T_{s1}^4 - T_l^4)F_{5-1}S_5 + \varepsilon_{sl}\sigma(T_{s4}^4 - T_l^4)F_{5-4}S_5 - QR\frac{dm_v}{dt}, \quad (5)$$

$$m_v c_{vv} R\frac{dT_g}{dt} = \alpha_{lg}(T_l - T_g)S_5 + \alpha_{sg}(T_{s1} - T_g)S_1 + \alpha_{sg}(T_{s4} - T_g)S_4, \quad (6)$$

$$m_v = \frac{\mu_v V_6}{R_A T_g} f(T_l), \quad (7)$$

$$p = p_v = f(T_l) = AT_l \exp(-B/T_l) \tag{8}$$

$$t = 0: \quad T_i = T_l = T_g = T_{e0}. \quad i = 1-4. \tag{9}$$

Here the notation is the following: T_{si} is the temperature of the vessel shell zones ($i = 1$–4); T_l, T_g are the liquid and gas temperatures; T_{e0} is the initial temperature of all regions and environment; S, V_5, V_6 are the dimensionless surface areas, liquid and gas volumes; ρ is the density; m_v is the mass of gas in region *6* (referred to R^2L); c, c_v are the specific heat capacity of the material and gas at constant volume; α is the convective heat exchange coefficient; ε is the emissivity factor; σ is the Stefan–Boltzmann constant; F is the angular coefficient of radiation heat exchange between regions; Q is the heat of evaporation of liquefied petroleum gas; μ_v is the molar mass of gas; R_A is the universal gas constant; p, p_v are the gas pressure and saturated vapor pressure in region *6*; $f(T_l)$ is the liquid vapor curve [10]. Subscripts *s*, *l*, *g* are the shell material, liquid and gas; *e* is the environment; *f* is the fire; *sl*, *lg*, *sg*, *ss*, *sf* are respectively the shell–liquid, liquid–gas, shell–gas, shell–shell and shell–fire interfaces.

Equation (1) in differential system (1)–(6) describes heat exchange of zone *1* with the environment and gas-vapor mixture with regard to fire heat flux as well as radiation heat exchange with zone *4* and surface of liquid *5*. The terms in the right-hand side of equation (2) denote external convective heat exchange, thermal loading and internal heat exchange with the liquid. According to equation (3), zone *3* is in heat exchange with the environment and liquid. For zone *4*, equation (4) takes into account the external and internal convective heat exchange, radiation heat exchange with zone *1* and liquid *5*. The latter is in heat exchange with all zones and regions; additionally, equation (5) accounts for liquid evaporation. Vapor-gas mixture heating is expressed in equation (6). The differential system is supplemented with equations (7) and (8) for the determination of vapor mass and pressure in region *6*. Initial conditions (9) correspond to the equality of all temperatures T_{e0}. Angular coefficients F in radiation heat exchange between zones and regions are determined through the geometric parameters of the problem [11].

4. Calculation of stresses within the vessel shell

The vessel destruction conditions will be estimated by considering a problem on the determination of the vessel shell stress state under internal p and external p_{out} pressures uniformly distributed over the lateral surface. In view of axial symmetry, radial σ_r and tangential (circumferential) σ_φ stresses depend only on radius r; tangential stress $\tau_{r\varphi} = 0$. The equilibrium equation for an elementary volume of the vessel shell reads [12]:

$$\sigma_r - \sigma_\varphi + r\frac{\partial\sigma}{\partial r} = 0. \tag{10}$$

The stress and strain are related according to Hooke's law as follows [12]:

$$\sigma_r = \frac{E}{1-\mu^2}(\varepsilon_r + \mu\varepsilon_\varphi), \qquad \sigma_\varphi = \frac{E}{1-\mu^2}(\varepsilon_\varphi + \mu\varepsilon_r), \tag{11}$$

where E, μ are the Young's and Poisson's moduli, and ε_r, ε_φ are the radial and tangential strains. Now let us write expressions that relate strains with displacement u:

$$\varepsilon_r = \frac{\partial u}{\partial r}, \qquad \varepsilon_\varphi = \frac{u}{r}. \tag{12}$$

After substituting equations (11), (12) into equation (10), we have:

$$\frac{d^2u}{dr^2} + \frac{1}{r}\frac{du}{dr} - \frac{u}{r^2} = 0. \tag{13}$$

The general solution of equation (13) is written as

$$u = Ar + \frac{B}{r}. \tag{14}$$

Constants of integration A, B are found from boundary conditions:

$$\sigma_r(r_{in}) = -p, \qquad \sigma_r(r_{out}) = -p_{out}, \tag{15}$$

where r_{in} and r_{out} are the inner and outer radii of the vessel shell of thickness $d = r_{out} - r_{in}$. Hence we come to expressions for stresses σ_r, σ_φ [12]:

$$\sigma_r = \frac{pr_{in}^2 - p_{out}r_{out}^2}{r_{out}^2 - r_{in}^2} - \frac{(p - p_{out})r_{in}^2 r_{out}^2}{(r_{out}^2 - r_{in}^2)r^2}, \qquad r_{in} \le r \le r_{out}, \tag{16}$$

$$\sigma_\varphi = \frac{pr_{in}^2 - p_{out}r_{out}^2}{r_{out}^2 - r_{in}^2} + \frac{(p - p_{out})r_{in}^2 r_{out}^2}{(r_{out}^2 - r_{in}^2)r^2}.$$

Since internal pressure is much higher than external pressure, $p >> p_{out}$, p_{out} can be neglected. Hence equations (16) take on a simpler form:

$$\sigma_r = \frac{pr_{in}^2}{(r_{out}^2 - r_{in}^2)}\left(1 - \frac{r_{out}^2}{r^2}\right), \qquad r_{in} \le r \le r_{out}, \tag{17}$$

$$\sigma_\varphi = \frac{pr_{in}^2}{(r_{out}^2 - r_{in}^2)}\left(1 + \frac{r_{out}^2}{r^2}\right).$$

Maximum stresses are on the inner shell side:

$$r = r_{in}: \qquad (\sigma_\varphi)_{\max} = p\frac{\beta^2 + 1}{\beta^2 - 1}, \qquad \beta = \frac{r_{out}}{r_{in}} \tag{18}$$

$$\sigma_r = -p.$$

Thus, $(\sigma_\varphi)_{\max}$ increases with the growing internal gas pressure and decreasing shell thickness. According to energy theory of forming, the equivalent stress is [12]:

$$\sigma_{\text{eqv}} = \frac{1}{\sqrt{2}}\sqrt{\sigma_\varphi^2 + \sigma_r^2 + (\sigma_r - \sigma_\varphi)^2} \tag{19}$$

Destruction occurs when σ_{eqv} achieves ultimate stress $\sigma_u(T)$ of structural material which depends on temperature. Substituting equation (18) into equation (19) yields:

$$\sigma_{\text{eqv}} = p\sqrt{\xi^2 + \xi + 1} \ge \sigma_u(T), \qquad \xi = \frac{\beta^2 + 1}{\beta^2 - 1}. \tag{20}$$

Calculated vapor pressure p can be used to determine σ_{eqv}. By comparing it with ultimate stress $\sigma_u(T)$ of material (20), we can find the critical temperature and time of vessel explosion under thermal loading.

5. TNT equivalent of thermal destruction of the vessel

The estimation of the mechanical process of vessel destruction requires the knowledge of explosion energy, or TNT equivalent. The TNT equivalent is the measure of energy release expressed in the explosive quantity of trinitrotoluene (TNT) that releases the same amount of energy in detonation.

During destruction of a vessel with combustible gases and liquids, explosion energy E_{expl}, MJ consists not only of chemical energy of the matter, but also of potential energy of pressurized gas [13]:

$$E_{\text{expl}} = Q_{\text{liq}} M_{\text{liq}} + \frac{(p - p_{out})V_g}{(\gamma - 1)}, \tag{21}$$

where Q_{liq}, M_{liq} are the specific heat (MJ/kg) and mass (kg) of combustible liquid, p, V_g are the gas pressure (MPa) and gas volume (m^3) in the vessel at the moment of rupture, and γ is the adiabatic exponent of gas.

If we equate vessel explosion energy E_{expl} to explosion energy $E_{\text{TNT}} = Q_{\text{TNT}} G_{\text{TNT}}$ of TNT of mass G_{TNT}, we can find the TNT equivalent:

$$G_{\text{TNT}} = E_{\text{expl}} / Q_{\text{TNT}}, \text{ kg} \tag{22}$$

where $Q_{\text{TNT}} = 4.52$ MJ/kg.

6. Numerical calculation results and their analysis

The proposed model was validated in modeling calculations for a steel liquefied petroleum gas vessel subjected to fire [8]. The vessel has diameter $D = 0.3$ m, wall thickness $d = 0.004$ m, volume 50 dm^3; it contains 21 kg of liquid propane [8]; the distance from the upper point to the liquid level $l = 0.0651$ m; the vessel is fully involved in fire, $\beta_1 = 0°$, $\beta_2 = 360°$ (zones *3*, *4* in the scheme of heating are absent).

The initial parameters of the problem according to the data of papers [3, 8, 10] are the following: thermophysical properties of the shell material and liquefied petroleum gas – $\rho_s = 7800$ kg/m^3, $c_s = 500$ J/(kg·K), $\rho_l = 500$ kg/m^3, $c_l = 3000$ J/(kg·K); convective heat transfer – $\alpha_e = 25$ W/(m^2·K), $\alpha_{sl} = \alpha_{lg} = \alpha_{sg} = 1200$ W/(m^2·K); radiation properties of materials – $\varepsilon_{sf} = 1.0$, $\varepsilon_{ss} = \varepsilon_{sl} = 0.8$; LPG evaporation characteristics – $A = 6.0 \cdot 10^6$ Pa/K, $B = 2259$ K, $Q = 4.27 \cdot 10^5$ J/kg; initial temperature $T_{e0} = 7$°C; heat flux – $q_R = 60$ kW/m^2.

The system of ordinary differential equations (1)–(9) was solved numerically using an implicit difference scheme. Figure 2 gives the calculation results on heating an unprotected liquid propane vessel. Curves *1*, *2*, *5*, *6* show the temperature of vessel regions, curve *7* is for the vapor pressure in region *6*, and the asterisks denote the experimental data for pressure obtained in full-scale testing [8].

As one can see, the most rapid heating is observed in shell zone *1* subjected to external heat flux and contacting with gas from the inside (curve *1*). The temperature here reaches ~ 200 ^{0}C on the 200th second of the process. A less intensive heating occurs in zone *2* contacting with the liquid because a large part of heat in this zone transfers to the liquid due to intensive heat transfer. The gas mass in region *6* inside the vessel is most heated (curve *6*). However, since gas exchanges heat with the liquid, its temperature is lower than the shell temperature in zone *1*. The large mass and high heat capacity of LPG determine a considerably lower liquid temperature (curve *5*) with an almost linear dependence.

Pressure grows rapidly inside the vessel (curve *7*). Its value reaches ~ 60 atm on the 200th second of heating, which is due to an increase in the partial vapor pressure of liquid propane. The calculated and experimental data for pressure are in good agreement. This means that the mathematical model adequately describes the heat exchange and liquid evaporation because pressure integrally characterizes the thermodynamic state inside the vessel. The calculation results show that the pressure value strongly depends on the method of determining temperature at the gas–liquid interface which enters into equation (8). The heat losses on LPG evaporation influence insignificantly the thermal state of the liquid and gas.

According to equation (20), the pressure growth causes an increase in stress σ_{eqv} within the vessel shell (Figure 3). Its upper bound from the physical standpoint is the ultimate stress of structural material (20). Then explosive vessel destruction occurs. Since ultimate stress $\sigma_u(T)$ of the steel shell decreases with the temperature growth, its lower bound is achieved during heating of shell zone *1* whose temperature is the highest. It is the zone where the vessel rupture is most probable.

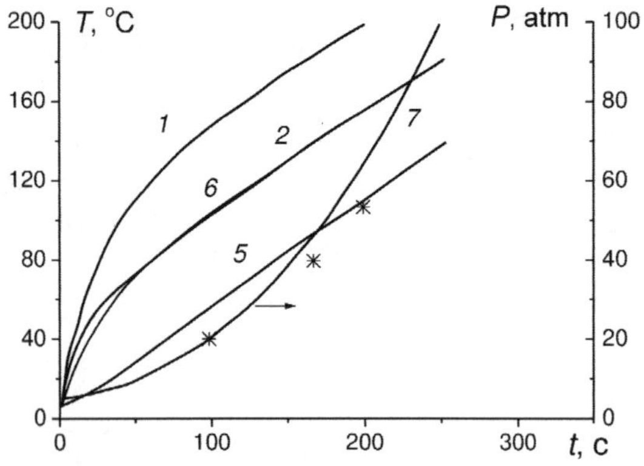

Figure 2. Temperature of shell zones (*1*, *2*), liquid (*5*), gas (*6*), and vapor pressure (*7*) for an unprotected vessel in fire.
∗ – full-scale experiment, vapor pressure [8].

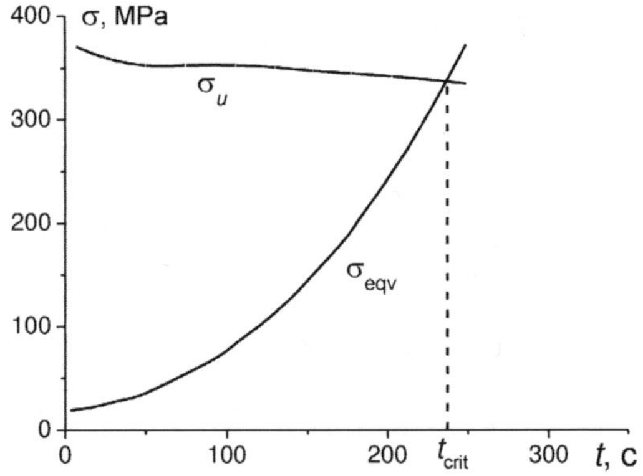

Figure 3. Time-dependence of equivalent stress σ_{eqv} within the shell under heat flux of fire. $\sigma_u(T)$ is the ultimate stress of material. t_{crit} is the critical time when vessel shell rupture begins.

Figure 3 shows curve $\sigma_u(T(t))$ calculated by the temperature of this zone (by curve *1* in Figure 2) for the steel shell. The abscissa of intersection of curves $\sigma_{max}(t)$ and $\sigma_u(T(t))$ gives critical time t_{crit} of shell rupture beginning, and the ordinate gives critical stress σ_{crit}. At the above-given initial data of the problem the critical parameter values are $t_{crit} \approx 240$ s, $\sigma_{crit} \approx 337$ MPa. Using the data of Figure 2 and the obtained value of time t_{crit}, we can find critical values of pressure inside the vessel $p_{crit} \approx 91$ atm and temperature $T_{i,crit}$ of shell zones *1* and *2*, liquid *5*, and gas *6*. Analysis of the obtained data demonstrates that the resistance time of an unprotected LPG vessel in fire with heat flux $q_R = 60$ kW/m^2 is insufficient for fire prevention measures, which points to the necessity of thermal protection.

Of practical interest is to estimate the potential energy of pressurized gas E_{press} at the moment of vessel destruction (the second term in equation (21)). For the initial data of the problem we have pressure $p_{crit} \approx 9.1$ MPa, $p_{out} \approx 0.1$ MPa, volume of the vessel region with gas $V_g = 0.0113$ m^3 at length $L = 1$ m, and adiabatic exponent $\gamma \approx 1.1$ [10]. Calculations yield $E_{press} \approx 1.0$ MJ, which corresponds to $\Delta G_{TNT} \approx 0.22$ kg of trinitrotoluene according to equation (22).

7. Conclusion

1. A new engineering technique of calculating the heating and thermal destruction of vessels with liquid under extreme thermal loading conditions has been proposed.

2. The calculation and experimental data for pressure growth inside the vessel are in good agreement, which indicates that the model is able to adequately describe the heat exchange and liquid evaporation processes.

3. Analysis of the obtained data shows that it is necessary to provide proper thermal protection of LPG vessels to ensure their resistance against industrial fires.

The work was supported by Project No. 9.1024.2014/K of the RF Ministry of Education and Science.

References

[1] Marshall V C 1987 *Major Chemical Hazards* (Chichester: Ellis Horwood)
[2] Abbasi T and Abbasi SA 2007 *J. Hazardous Materials* **141** 489
[3] Shebeko Y N, Filippov V N and Navtsenya V Yu et al 1999 *Fire and Explosion Safety* **89** 33
[4] Davenport J A 1988 *J. of Hazardous Materials* **20** 3
[5] Moodie K 1988 *J. of Hazardous Materials* **20** 149
[6] Kletz T 1977 *Hydrocarbon Processing* **56** 98

[7] Roberts A F and Moodie K 1989 *J. of Oil and Colour Chemical Association* **72** 192
[8] Shebeko Y N, Filippov V N and Gorshkov V I et al. 1998 *Fire and Explosion Safety* **7** 24
[9] ShebekoYu N 1997 *Fire and Explosion Safety* **6** 17
[10] Staskevich N L and Vigdorchik D Ya 1986 *Handbook of liquefied petroleum gases* (Leningrad: Nedra Publ.)
[11] Siegel R and Howell J R 2002 *Thermal Radiation Heat Transfer,4th Ed.* (N. Y.: Taylor&Francis)
[12] Timoshenko S P and Goodier J N 1970 *Theory of Elasticity* (N.Y.: McGraw-Hill)
[13] Baker W E, Cox P A, Westine P S, Kulesz J J and Strehlow R A 1983 *Explosion Hazards and Evaluation* (N.Y.: Elsevier Scientific Publishing Company)

Physical and mechanical properties and thermal protection efficiency of intumescent coatings

V G Zverev[1], V I Zinchenko[1] and A F Tsimbalyuk[2]
[1] Tomsk State University, 36, Lenina ave., Tomsk, 634050, Russia
[2] Tomsk Polytechnic University, 30, Lenina ave., Tomsk, 634050, Russia

E-mail: zverev@niipmm.tsu.ru

Abstract. The new engineering technique for the experimental investigation of physical and mechanical characteristics of thermal protective intumescent coatings is offered. A mathematical model is proposed for predicting the thermal behavior of structures protected by coatings; the model is closed by the studied material characteristics. The heating of a metal plate under standard thermal loading conditions is modeled mathematically. The modeling results are in good agreement with bench test results for metal temperature under the coating. The proposed technique of studying physical and mechanical characteristics can be applied to identify and monitor the state of thermal protective intumescent coatings in the long-term operation.

1. Introduction

The wide application of structural materials requires the improvement of their stability under extreme thermal loading of various origin, intensity and duration, particularly, in industrial fires [1]. Today there is a broad range of materials and compositions for passive thermal protection [2]. Intumescent paint coatings have great potential in this regard.

A 1 ÷ 2 mm thick layer of intumescent coating retains the thermal stability of the substrate for up to 1 ÷ 1.5 h. The protective properties of intumescent coatings are due to a considerable increase of their thickness upon heating. The char produced during intumescence has low porosity, low thermal conductivity and serves as a protective barrier between the substrate and high-enthalpy gaseous medium. This greatly increases thermal resistance, but significantly reduces heat flux to the substrate [3, 4]. Modern intumescent coatings have an expansion ratio of up to 20…40, char porosity amounts to ~ 0.95…0.98, density is ~ 30…50 kg/m^3, and thermal conductivity coefficient is ~ 0.04…0.1 W/(m·K) [1].

Intumescent coatings are of practical interest as they have minimum weight and thickness per unit of protected area, high heat insulating properties of char, and can be deposited by industrial methods. Currently, requirements for the thermal protection of structures have increased. This necessitates the development and application of new, more effective universal materials with high adhesion to any surfaces that are able to resist vibration and weather.

To expand the applications of intumescent coatings, we must investigate their physical and mechanical properties, conduct laboratory and bench-scale studies, develop mathematical models to predict the state of structures and optimize their thermal protection under thermal conditions.

The paper is aimed to develop an engineering technique for studying the physical and mechanical properties of intumescent coatings and to apply a mathematical model for structure stability prediction under extreme thermal conditions.

2. Physical and mechanical properties of intumescent coatings under isothermal heating

Intumescent coatings are usually multicomponent compositions because many practically important factors are needed for their operation, such as adhesion, gas generation, intumescence, char strength and so on. The physical and chemical processes that accompany intumescence as well as formulations of used compositions are very diverse and often unknown. At the same time, intumescent coatings exhibit a common feature associated with thickness (volume) increase during heating. Their study must therefore be based on the characteristics that integrally characterize intumescence and can be adequately determined in a simple laboratory experiment. This would make possible the control of the coating state during the operation. These characteristics are the expansion ratio due to intumescence and the mass of the intumescent coating sample during isothermal heating.

The characteristics were determined using a technique based on the method of thermally stabilized states of the material. The experimental procedure is schematically shown in Figure 1. Samples presented a two-layer coating–substrate system with the initial coating thickness of ~ 2 mm. They were placed into thermally stable glass tubes of ~ 17.5 mm in diameter with a flat bottom. The substrate was used in the experiment in order to model the coating operation in real conditions.

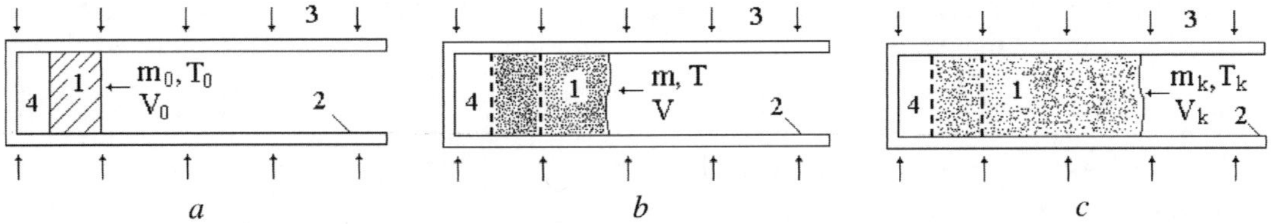

Figure 1. Schematics of the experiment, and various physical and mechanical states of a sample on the substrate during thermostating. *a* – initial (subscript 0), *b* – current, *c* – final (subscript *k*) state (at high temperatures). *m* – mass, *T* – temperature, *V* – sample volume; 1 – sample (an intumescent layer), 2 – thermally stable glass tube, 3 – heat flux, 4 – substrate.

Then the samples were kept for ~ 20 min in a muffle furnace in inert gas (nitrogen) atmosphere at given thermostating temperature T that was maintained automatically with an error of $\pm$ 5 K. The temperature range was from 300 to 900 K with step 25; 50 K. Thermal loading causes changes in the state of the intumescent coating in both its volume and mass. Since the material expands in one direction along the tube, the volume change can be easily determined by the coating sample thickness. To exclude random factors and to define confidence error intervals for measurement results, 3 to 5 samples were kept at the same temperature.

In experiments the initial coating thickness varied from 1 to 6 mm, and the diameter varied from 6 to 40 mm. It was found that relative volume and mass changes in isothermally heated samples do not depend on their initial thickness and diameter, which is in accordance with the notion of elementary material volume and the physical nature of intumescence.

Different coating states were observed depending on thermostating temperature. At T ~ 170 °C the volume changed by a factor of ~ 8 with a dense layer structure. At higher temperatures T > 300 °C the volume changed by a factor of more than 20 with the formation of a loose char structure in the form of elongated intumescent fibers. Thus, each thermostating temperature is characterized by a particular volume, structure and mass of the sample material, which corresponds to quasi-equilibrium processes of intumescence and thermal decomposition.

Upon thermostating we weighed (with an error of ~ 10^{-6} kg) and measured (~ 10^{-4} m) the samples, determined their mass m and length (thickness) h, and calculated relative quantities

$$dm = m_0 - m, \quad dh = h - h_0, \quad \bar{m} = m / m_0, \quad \theta = dh / h_0,$$

where subscript 0 corresponds to the initial state, and θ is the expansion ratio. Density of char ρ and porosity of char φ_g were calculated by formulas

$$\rho = \frac{\rho_0 (m/m_0)}{1+\theta}, \qquad \varphi_g = 1 - \frac{(m/m_0)}{1+\theta}. \tag{1}$$

We ignored initial porosity in the formula for φ_g, assuming that the true char porosity coincides with the initial one. Thus, porosity in equation (1) is fully determined by the loss of mass of the solid material and its intumescence. The set of the discussed physical and mechanical characteristics (θ, m/m_0) individually characterizes the intumescent coating and can be considered as the material response to a thermodynamic state change. This set is necessary for the mathematical modeling of coating heating and for the coating state monitoring during a long-term operation on structures.

Figure 2 displays experimental data for the discussed characteristics of two coatings SGK-2 [5] and SGK-1 [1] depending on thermostating temperature. These coatings differ in composition and a binder-to-filler ratio. Figure 2a demonstrates that intumescence of SGK-2 occurs in two stages: the first stage corresponds to 400 K < T < 600 K, the second stage corresponds to 600 K < T < 800 K. Then the equilibrium curve of the expansion ratio approaches an asymptote with value ~ 24 ÷ 25. Initial coating intumescence is drastic and occurs later than that of SGK-1. The height of the first plateau is θ ~ 16 (~ 10 for SGK-1), which is advantageous of this composition because heat transfer in this case is suppressed more efficiently at low temperatures.

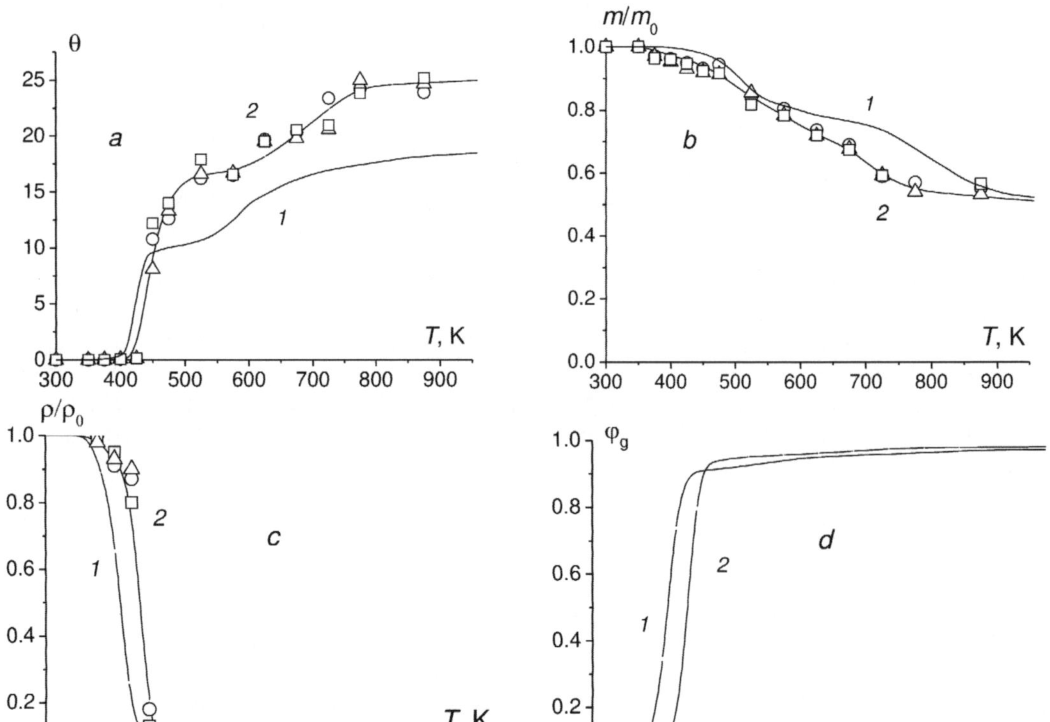

Figure 2. Expansion ratio (*a*), mass (*b*), density (*c*), porosity (*d*) of coating samples vs. temperature. *1* – SGK-1; *2* – SGK-2; Δ, □, ○ – experimental data.

According to Figure 2b, the loss of mass of coating SGK-2 within the temperature range of 400 K < T < 800 K is slightly larger than for SGK-1. The final output of char is nevertheless almost similar, being ~ 0.52. The wave-like shape of curve *2* indicates that thermal destruction of the coating occurs in several stages. Notice that intumescence prevails over the release of gaseous products at the initial stage for both coatings. For example, at T ~ 450 K and 4…5 % of mass loss the layer thickness increases by a factor of ~ 14. This is the main mechanism for the formation of thermal protection properties of the coating.

As one can see from Figure 2c, the SGK-2 coating density changes by a factor of more than 12 at the first intumescence stage and by a factor of only 2 at the second stage. At T ~ 900 K the char density approaches $\rho \sim 20$ kg/m^3. Peeling of density curves *1* and *2* is observed within the range of T ~ 375 ÷ 450 K. At the same mass loss (< 5 %) for these materials this is related to the delay of the SGK-2 intumescence curve by the temperature. Curves *1* and *2* of porosity φ_g increase abruptly up to ~ 0.9 at the first intumescence stage (Figure 2d). The final porosity value for coating SGK-2 is $\varphi_g \sim 0.98$.

The considered characteristics provide information about the state and structure of the intumescent coating layer; they allow us to compare with the same characteristics of other intumescent coatings and to evaluate their thermal protection properties. The expansion ratio increase up to $\theta \sim 25$ is a definite advantage of the SGK-2 composition. The layer thickness strongly affects the thermal protection capacity of the coating during non-stationary heat transfer because the characteristic heating time depends quadratically on this parameter. However, there is no point in further intumescence as it enhances radiation heat transfer in the porous medium and may lead to damage of a char structure due to its strength reduction.

2. Mathematical modeling of intumescent coating heating

Thermal processes in the coating layer will be considered in the framework of a mathematical non-stationary heat transfer model [6, 7]. The model can be applied to a broad range of coatings as it has been proved to be good for practical calculations [8]. The schematic layout of layers and coordinate system are given in Figure 3a. The initial coating thickness is h_0, and that of the substrate is h_b. The intumescent coating will be considered as a porous two-phase, one-temperature medium that contains a solid and gaseous phase. We assume that the material expands in one dimension and does not shrink due to a chemical or mechanical interaction with the heated gaseous medium, which corresponds to the real process.

Without going into details of the intumescence and thermal destruction mechanisms, we determine relative mass and volume changes in an elementary intumescent coating sample through $\bar{m}=m/m_0$ and $\bar{\theta}=V/V_0=1+\theta$. Along with the Euler coordinate system (x), the Lagrangian coordinate system (s) related to the char material will be used.

Using the laws of conservation of mass and energy, the following system of equations is obtained in the Lagrangian coordinate system (t, s) [6]:

$$\rho_0\frac{\partial \bar{m}}{\partial t}+\frac{\partial G_g}{\partial s}=0, \qquad 0\le s\le h_0, \quad x=h_b+\int_0^s\bar{\theta}ds, \tag{2}$$

$$(\rho c_p)_{ef}\,\bar{\theta}\frac{\partial T}{\partial t}+(c_pG)_g\frac{\partial T}{\partial s}=\frac{\partial}{\partial s}\left(\frac{\lambda_{ef}}{\bar{\theta}}\frac{\partial dT}{\partial s}\right)+\rho_0 Q\frac{\partial \bar{m}}{\partial t}, \tag{3}$$

$$\frac{\partial \bar{m}}{\partial t}=-R_s(T,m), \tag{4}$$

$$(\rho c_p)_{ef}=(\rho c_p)_s\varphi_s+(\rho c_p)_g\varphi_g, \qquad \varphi_s+\varphi_g=1, \tag{5}$$

$$\lambda_{ef}=\lambda_s\varphi_s+\lambda_g\varphi_g+\lambda_r\varphi_g, \qquad \lambda_r=4\varepsilon_{ss}\sigma d_p\bar{\theta}^{1/3}T^3.$$

The substrate state is described by the thermal conductivity equation for inert materials:

$$(\rho c_p)_b \frac{\partial T}{\partial t} = \frac{\partial}{\partial x}\left(\lambda_b \frac{\partial T}{\partial x}\right), \qquad 0 < x < h_b . \tag{6}$$

The boundary conditions for equations (2), (3) and (6) read:

$$x = 0: \qquad \lambda_b \frac{\partial T}{\partial x} = 0 . \tag{7}$$

$$x = h_b,\ s = 0: \qquad \lambda_b \frac{\partial T}{\partial x} = \frac{\lambda_{ef}}{\theta}\frac{\partial T}{\partial s}, \quad T|_{x=h_b} = T|_{s=0}, \quad G_g = 0 . \tag{8}$$

$$s = h_0: \qquad \frac{\lambda_{ef}}{\theta}\frac{\partial T}{\partial s} = \alpha_f (T_f(t) - T_w) + \varepsilon_{fw}\sigma(T_f^4(t) - T^4) . \tag{9}$$

$$T_f(t) - T_0 = 345\,\mathrm{lg}(0.133t + 1), \qquad \varepsilon_{fw} = [1 + (\bar{\varepsilon}_w^{1} - 1) + (\bar{\varepsilon}_f^{1} - 1)]^{-1} . \tag{10}$$

The initial conditions for equations (2)–(4) and (7) correspond to thermal equilibrium:

$$t = 0: \qquad T = T_0, \quad \bar{m} = 1, \quad \bar{\theta} = 1, \quad s = x - h_b . \tag{11}$$

Here, t is the time, G_g is the filtration gas flow, ρ, c_p, λ are the density, specific heat capacity and thermal conductivity coefficient, Q, R_s are the thermal effect and rate of thermal coating destruction, φ is the volume fraction of phase, α is the heat transfer coefficient, ε, σ are the emissivity factor and Stefan–Boltzmann constant, and d_p is the initial pore diameter. Subscripts s, g are the solid and gaseous phases of char, w is the coating surface, f is the high-enthalpy gaseous medium (fire), r is the radiation, b is the substrate, 0 are the initial and ef are the effective values.

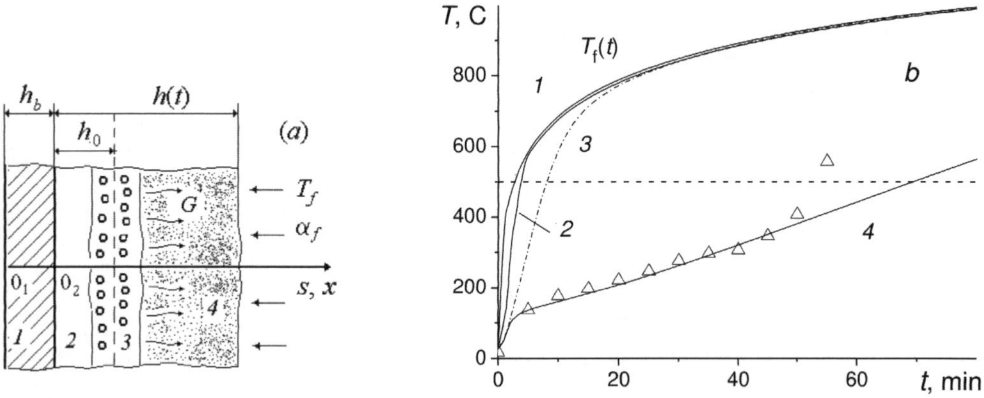

Figure 3. Schematic layout of layers (a) and calculation results on heating (b) of metal substrate with coating SGK-2 in standard fire. h_b = 5 mm, h_0 = 2.2 m. a: *1* – substrate; *2* – initial material; *3* – intumescence zone; *4* – char; h_0, $h(t)$ – initial, current layer thickness: G – filtration gas flow; b: *1* – $T_f(t)$, fire temperature; *2* – $T_w(t)$, coating surface temperature; *3*, *4* – uncoated and coated metal temperature; Δ – experimental data.

Equations (2) and (4) are the law of conservation of mass for the intumescent coating as a whole and for its solid phase; equation (3) is the energy equation with regard to filtration of gaseous decomposition products. Expressions (5) define the thermophysical properties of the porous medium and radiant thermal conductivity in pores [1]. The back side of the substrate is perfectly thermally isolated (equation (7)); the coating–substrate interface is under the conjugate conditions (equation (8)). The coating surface is under the radiative-convective effect of "standard" fire (equation (9)) which temperature changes according to law (10) [9]. The model was closed by defining parameters $\bar{m}$, $\bar{\theta}$ [6].

3. Calculation results and their analysis

Figure 3b shows the calculation results obtained with mathematical model (2)–(11) on the thermal behavior of a 5-mm thick steel plate protected by a 2.2-mm thick layer of coating SGK-2 thermally loaded by standard fire (10) [9]. The coordinate of intersection of curves *3* and *4* with horizontal T = 500° C (critical temperature for steel, a dashed line) gives the fire resistance time of the plate.

It is seen that a relatively thick, but unprotected metal layer cannot resist fire for more than 8 min (curve *3*). The mathematically calculated estimates show that the deposition of a 2.2-mm thick SGK-2 coating increases the fire resistance time up to almost 70 min (curve *4*), which greatly exceeds the practically important fire resistance time limit of one hour. As it follows from the figure, our calculation results and the experimental data of fire tests performed according to fire safety standards show a good agreement up until 50 minutes. The subsequent difference is due to partial spallation of surface char layers in the experiment, which was not taken into account in the mathematical model.

4. Conclusion

1. An original engineering technique is proposed for studying the physical and mechanical characteristics of thermal protective intumescent coatings. It is based on the determination of an expansion ratio and the mass loss of the coating sample under isothermal heating.

2. The technique can be applied to solve problems of identifying and monitoring the state of intumescent coatings during the operation.

3. The obtained coating characteristics are used to mathematically model the heating of a steel plate coated with SGK-2 and to demonstrate a high thermal protection efficiency of the coating. The modeling results agree well with bench test results.

4. The studied physical and mechanical characteristics of coating SGK-2 as well as the proposed mathematical model can be applied to optimize and improve the protection of structures against extreme thermal loading of various intensity and duration.

The work was supported by Project No. 9.1024.2014/K of the RF Ministry of Education and Science.

Reference

[1] Strakhov V L, Krutov A M and Davydkin N F 2000 *Fire Protection of Engineering Structures* (Moscow. Information and Publishing Centre 'TIMR')
[2] Sobur S V 2003 *Fire protection of materials and structures Handbook* (Moscow. Spetstekhnika Publ.)
[3] Griffin G J 2010 *J. of Fire Science* **28** 249
[4] Bartholmai M, Schriever R and Schartel B 2003 *Fire and Materials* **27** 151
[5] TU 7719-171-21366107-02 2002 *Fire-retardant intumescent composition SGK-2* (Moscow. Spetsenergotekhnika Publ.)
[6] Zverev V G, Nesmelov V V, Tsimbalyuk A F and Gol'din V D 1998 *Combustion, Explosion and Shock Waves* **34** 198
[7] Anderson Jr. C E, Wauters D K 1984 *International Journal of Engineering Science* **22** 881
[8] Zverev V G, Nazarenko V A and Tsimbalyuk A F 2008 *High Temperatures* **46** 254
[9] Drysdale D 1985 *An Introduction to Fire Dynamics* (Chichester: John Wiley and Sons)

IOP Conf. Series: Materials Science and Engineering **124** (2016) 012110 doi:10.1088/1757-899X/124/1/012110

Numerical simulation of synthesis gas incineration

A V Kazakov, S A Khaustov, R B Tabakaev and Y A Belousova
Tomsk Polytechnic University, 30, Lenina ave., Tomsk, 634050, Russia

Email: khaustovSA@tpu.ru

Abstract. The authors have analysed the expediency of the suggested low-grade fuels application method. Thermal processing of solid raw materials in the gaseous fuel, called synthesis gas, is investigated. The technical challenges concerning the applicability of the existing gas equipment developed and extensively tested exclusively for natural gas were considered. For this purpose computer simulation of three-dimensional syngas-incinerating flame dynamics was performed by means of the ANSYS Multiphysics engineering software. The subjects of studying were: a three-dimensional aerodynamic flame structure, heat-release and temperature fields, a set of combustion properties: a flare range and the concentration distribution of burnout reagents. The obtained results were presented in the form of a time-averaged pathlines with color indexing. The obtained results can be used for qualitative and quantitative evaluation of complex multicomponent gas incineration singularities.

1. Introduction

One of the main problems of the global energy industry is the problem of low-grade fuels application. Low-grade fuels are characterized by a high moisture values. High fuel humidity causes high costs for its burning associated with the evaporation necessity. It is worth noting that the presence of such moisture amounts significantly facilitates fuel sticking and restricts its production and transportation in winter season. The mentioned disadvantages can be eliminated by heat processing of low-grade fuels into energy products of high quality gaseous fuel, called synthesis gas [1-5]. The aim of this work is the examination of low-grade raw fuel processing to gaseous fuel (synthesis gas) in order to assess the feasibility of this energy application method. The well-known advantages of the gaseous fuels energy use [6–8] (possibility of combustion process automating, ecological compatibility, low abrasive deterioration of heating surfaces and intensive heat transfer due to higher flow velocities available) expand the range of low-grade fuels useful applications.

Despite the fact, that nowadays synthesis gas has a high variety of applications in Power Engineering and Industrial services, there is still no commercially available equipment designed for its utilization. Therefore, the technical challenges concerning applicability of the existing gas equipment, developed and extensively tested exclusively for natural gas, often arise while designing synthesis gas burning power plants.

In this article the turbulent diffusive synthesis gas incineration was investigated using a computer-aided simulation with ANSYS Multiphysics engineering software. The subjects of study were: a three-dimensional aerodynamic flame structure, heat-release and temperature fields, a set of combustion properties: a flare range and the concentration distribution of burnout reagents.

2. Initial data for investigation

The initial data for mathematical modeling are the following: the investigated gaseous fuel composition was simplified to H_2-CH_4-CO-CO_2 compound (Table 1); the oxidant is atmospheric oxygen (the air content is simplified to 0.23 and 0.77 mass fractions of the oxygen and nitrogen respectively); the air excess ratio is 1.02; the fuel-air mixture is supplied with the temperature of 20 °C; the coolant temperature is 368 K.

Table 1. Content (mole fractions) of synthesis gas obtained by thermal decomposition of peat using different processing temperature (T) [9, 10]

T, K	H_2, %	CH_4, %	CO, %	CO_2, %	Q, kJ/m^3
500	30,32	15,9	10,98	42,8	10364,32
600	48,57	27,97	13,65	9,82	17001,7
700	50,1	27,59	7,59	14,72	16264,46
750	50,02	31,57	6,82	11,58	17586,52
800	48,09	37,1	8,17	6,63	19533,08
850	34,52	34,41	30,6	0,47	19938,86

For fuel-air mixing a simplified model of the prevalent automatic gas burner Weishaupt WKGL70 in the embodiment 3LN was analyzed. The investigated burner has 11 MW of thermal power and involves one primary fuel inlet orifice, four secondary fuel inlet orifices, primary and secondary air channels (Figure 1). The virtual cylindrical flame tube with the diameter of 2 m and the length of 10 m was considered as a combustion chamber. The temperature of the screening walls surface was assumed to be equal to the coolant temperature.

A simplified computational geometric model of the investigated burner with the partition into tetragonal finite elements was implemented. The mesh for simulation (Figure 2) was designed by means of the ANSYS Workbench. It consists of 1531011 tetrahedral cells with 196310 nodes. The integral volume of computational cells for combustion is 20 m^3. The boundary zones are:

- 21333 triangular wall faces,
- 5133 triangular pressure-outlet faces,
- 7740 triangular mass-flow-inlet faces.

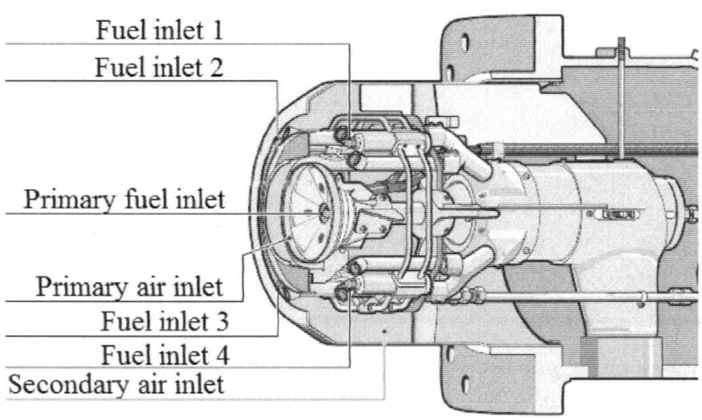

Figure 1. Layout of the investigated burner

Chemical oxidation of fuel components in a turbulent air flow was simulated by a finite-element method using a computer-aided engineering software ANSYS Multiphysics and a widely-used set of mathematical models [7, 11, 12]. Fuel oxidation was supposed to be proceeded by means of these reactions between combustible components and oxygen irreversibly: $CH_4+2O_2=CO_2+2H_2O$; $2CO+O_2=2CO_2$; $2H_2+O_2=2H_2O$. The chemical process in a turbulent flow was simulated with combined "Finite rate/Eddy-Dissipation" model.

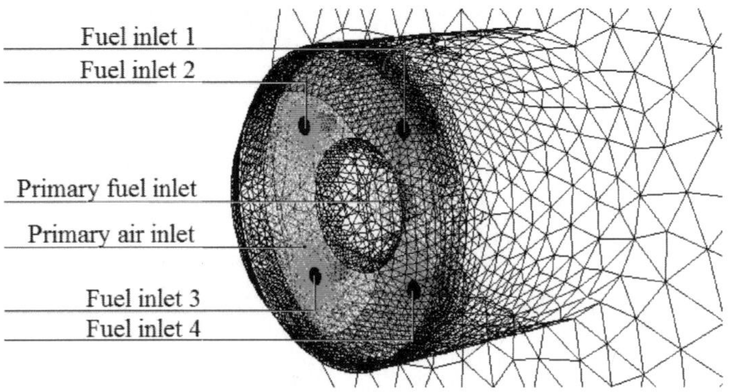

Figure 2. A simplified computational geometric model of the investigated burner with the partition into tetragonal finite elements.

For the mathematical description of the three-dimensional combustion dynamics in the concurrent flows of atmospheric air and synthesis gas the model of nonisothermal incompressible multicomponent gas was adopted. The flow process was considered to be steady, and, therefore, mathematical models of the continuity equation, mass-energy conservation were used in the steady-state formulation in the Euler's approximation.

Heat exchange by radiation was simulated with the spherical harmonics method in the first approximation (P1-model). A standard k-ε model was implemented for calculating pulsational (turbulent) flow characteristics.

The obtained results were presented in the form of a time-averaged pathlines with colored, according to heat release dissipation, temperature and the mass fraction of reactants in the gas mixture.

3. Calculation results

Table 2. Comparative analysis of natural gas and synthesis gas thermal characteristics

Type of fuel	Caloric value, MJ/m^3	The volumetric flow rate of air for the stoichiometric combustion related to 1 cubic meter of gaseous fuel, m^3/m^3	The combustion products volume related to 1 cubic meter of gaseous fuel, m^3/m^3
Natural gas	37,01	9,78	11,29
Synthesis gas	16,2	3,92	4,83

The less, in comparison with natural gas, syngas calorific value (Table 2) at same burner power leads to a higher fuel massflow rate through the nozzle, whereas the amount of air required for stoichiometric combustion is significantly lower because of high ballast content [9, 10], thereby the massflow rate through the air channels reduces. Mathematical modeling is an effective tool to assess the enumerated features impact of the tested fuel on the intensity of mixing processes, flame range and temperature. The approved EBU-model proposed by Spalding [13] was applied for the numerical simulation of chemical reactions behaviour in turbulent flow conditions.

The numerical simulation results display the synthesis gas combustion via the mean pathlines colored according to different thermal characteristics. Despite the existence of massive vortices generated by local flow resistance in a burner slot, the process of combustion takes place in a short section of a straight continuous-flow without large scale eddies (Figure 3).

Figure 3. Synthesis gas burning flame pathlines colored according to temperature (°C).

According to the results of numerical experiments hydrogen burning (Figure 4) occurs in the close area from the burner orifice, due to the low activation energy for its oxidation. As the volume content of hydrogen in synthesis gas is about 50%, then there is a more intensive temperature rise in the region close to the burner slot rather than the combustion of natural gas, where the hydrogen content is much lower. The temperature of the gaseous flow reaches its maximum at the narrow boundaries of the flame wherein there is no intensive burning. The area beyond the boundaries of the flame is filled with combustion products: CO_2 and H_2O.

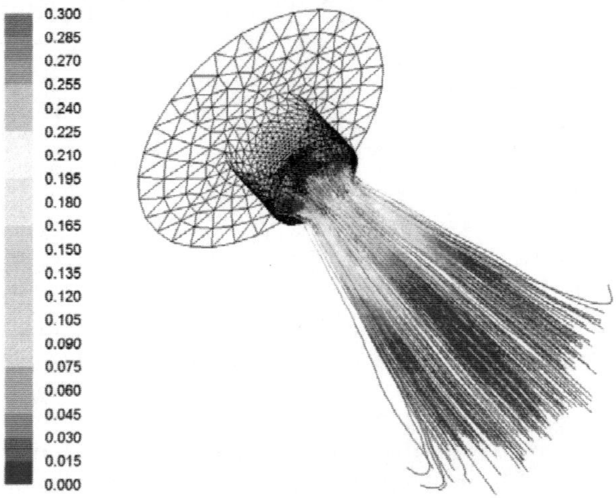

Figure 4. Synthesis gas burning flame pathlines colored according to the reaction rate of $2H_2 + O_2 = 2H_2O$ (kgmol·m^{-3}·s^{-1}).

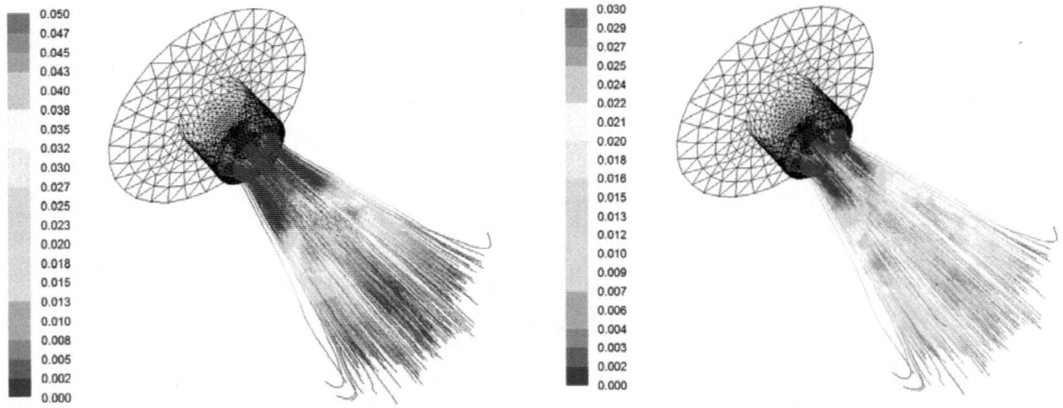

a) $CH_4 + 2O_2 = CO_2 + 2H_2O$ b) $2CO + O_2 = 2CO_2$

Figure 5. Synthesis gas burning flame pathlines colored according to the reaction rate ($kgmol \cdot m^{-3} \cdot s^{-1}$).

The high gas mixture temperature (Figure 3) initiates methane (Figure 5, a) and carbon monoxide (Figure 5, b) ignition in the immediate vicinity of the burner, but the values of the oxidation reaction rate in the combustion chamber is generally lower than those values for the natural gas combustion, due to the burner jet ballasting. A consequence of the described factors is a uniform heat release distribution (Figure 7) for the entire length of the flame, which favorably affects the reliability of the combustion chamber.

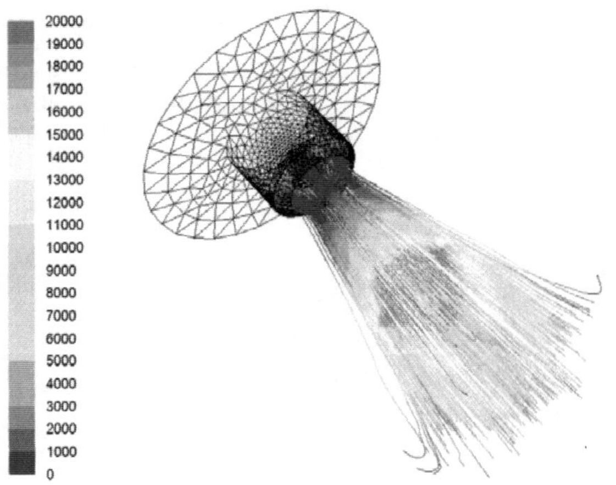

Figure 7. Synthesis gas burning flame pathlines colored according to heat release of three simultaneous reactions in the gas mixture (W).

4. Conclusion

Using the ANSYS Multiphysics engineering software a detailed description of the synthesis gas turbulent combustion is given. The obtained results can be used for qualitative and quantitative evaluation of complex multicomponent gas incineration singularities.

The synthesis gas caloric value is 2.3 times lesser than the natural gas caloric value. For this reason, the volumetric flow rate of synthesis gas through the burner of the same capacity is 130% higher. This increases the velocities in the fuel injectors and reduces its life because of the higher abrasive solids contained in synthesis gas. However, the combustion of synthesis gas requires less air (3.92 m^3/m^3).

For this reason aerodynamic flow resistance and power consumption of burners using VFD fans (with frequency adjustment) is reduced while incinerating synthesis gas.

The amount and composition of the combustion products resulting from the synthesis gas combustion do not significantly change compared to natural gas, so it allows using exhaust fans of the same capacity. In other words, for the synthesis gas incineration in the rarefied gas furnaces (developed for the natural gas combustion with the aerodynamic rarefaction) there is no modernization of draft equipment required.

As a result of the conducted research it is possible to conclude that the low-temperature catalytic conversion, allowing receiving synthesis gas, enriched with hydrogen, from the solid initial raw is a promising way of low-grade fuel recycling. The resulting synthesis gas can be effectively used in Power Engineering to produce heat and electricity without expensive upgrades of gas equipment.

Acknowledgments
The reported study was partially supported by the Ministry of education and science of The Russian Federation (Government Order No. 13.948.2014/K and Government Order No. 2069).

References

[1] Sarker M, Mohammed A S M D, Rashid M M, Schiralli N, Zaman A, Kabir A, Rahman S, Islam K, Molla M 2009 New alternative energy from solid waste plastics. ICDRET: Proc. of 1st Intern. Conf. on the Developments in Renewable Energy Technology, Bangladesh: Dhaka, (5454177) 100–103
[2] Lorincz J, Bule I 2013 Renewable *International Journal of Business Data Communications and Networking* **9(3)** 53–74
[3] Aysu T 2015 *Bioresource Technology* **191** 253–262
[4] Sircar I, Sane A, Wang W, Gore J P 2014 *Fuel* **119** 38–46
[5] Kazakov A V., Tabakaev R B, Novoseltsev P Y and Astafev A V 2014 Low-temperature conversion of low-grade organic raw, part 1 (technical aspects) ed P A Strizhak, A A Kozyreva and R Z Reshetova MATEC Web Conf. 19 01014
[6] Liu G, Chen Z, Li Z, Li G and Zong Q 2015 *Appl. Therm. Eng.* **75** 1034–45
[7] Zavorin A S, Khaustov S A and Zaharushkin R N A 2014 *IOP Conf. Ser. Mater. Sci. Eng.* **66** 012029
[8] Gergelizhiu P S, Khaustov S A, Tabakaev R B, Novoseltsev P U, Kazakov A V and Zavorin A S 2014 *International Conference on Mechanical Engineering, Automation and Control Systems (MEACS) (IEEE)* 1–4
[9] Tabakaev R B, Gergelizhiu P S, Kazakov A V and Zavorin A S 2014 *IOP Conf. Ser. Mater. Sci. Eng.* **66** 012052
[10] Tabakaev R B, Kazakov A V. and Zavorin A S 2015 *Solid Fuel Chem.* **49** 267–73
[11] Zavorin A S, Khaustov S A and Sheikin V A 2014 *International Conference on Mechanical Engineering, Automation and Control Systems (MEACS) (IEEE)* 1–4
[12] Khaustov S A, Zavorin A S, Buvakov K V, Kudryashova L D and Tshelkunova A V 2015 Computer simulation of vortex combustion processes in fire-tube boilers ed G V. Kuznetsov, P A Strizhak, A O Zhdanova and E E Bulba EPJ Web Conf. 82 01041
[13] Spalding D B 1971 *Proc. Combust. Inst.* **13** 649–657

Modeling of a reduction zone of the gasifier installation

V A Lashkov, Z G Sattarova, M A Taymarov, M K Gerasimov and R A Halitov
Kazan National Research Technological University, 72, Karla Marksa st., Kazan, 420015, Russia

E-mail: sattarova@list.ru

Abstract. The development of the mathematical model that describes a reduction zone of the gasifier installation is described in this paper.

1. Introduction

Processes occurring in a reduction zone of the reactor gasification of wood waste, are extremely complex and diverse. They depend on the fractional composition of the processed wood waste and on operating parameters of the recycling process. A recovery zone is the most difficult stage in the gasification process regarding the amount and the quality of substance occurring in chemical reactions. This work considers a stationary layer of material with variable porosity and conventionally constant speed of filtration of a gas phase along the height of the layer.

2. Subject description and method of solution

Both homogeneous (in volume) and heterogeneous reactions occur in the process of gasification (on the surface of phase separation - solid, liquid or gaseous).

Homogeneous reactions are as follows:

$$CO + H_2O = CO_2 + H_2;$$
$$CO + 3H_2 = CH_4 + H_2O.$$

and heterogeneous reactions are:

$$C + CO_2 = 2CO;$$
$$C + H_2O \text{ vapor} = CO + H_2;$$
$$C + 2H_2O = CO_2 + 2H_2;$$
$$C + 2H_2 = CH_4.$$

The gasification process can be described by the following main stages: heating, drying, pyrolysis (release of volatiles), combustion of pyrolysis products and the recovery processes of the producer gas formation [1]. The boundaries between these zones are quite conditional and it is almost impossible to specify the exact location of the end of one process and the beginning of the other.

The stage of heating and the yield of volatile matter are less than 15 % of the total time of the gasification process, unlike the recovery stage, which takes most of the time and is one of the main consumers of energy in the gasifier [2]. The recovery phase is a complex set of interrelated physical and chemical processes of heat exchange and mass transfer of reaction products and chemical reactions, as coal is a porous body, which, as a result of heterogeneous reactions, changes its structure and properties.

The modeling process of the restored area starts with the fact that charcoal enters the reactor and forms a dense layer, which defines the height of the recovery zone. It is known that reduction processes occur at temperatures of 800-1100^{0}C [3]. In this regard, the layer of charcoal is heated to a temperature of 900 ^{0}C through the wall of the reactor, then the reactor is fed with a gas agent (H_2O, CO_2, H_2O+O_2) at a temperature of 900-1000^{0}C (Figure 1).

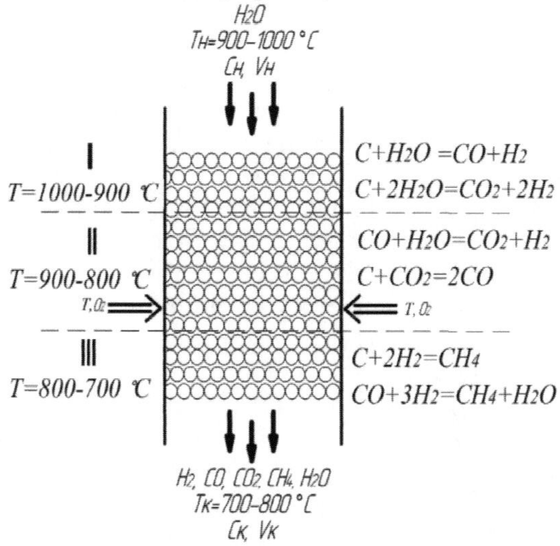

Figure 1. Schematic representation of the recovery zone

Let us divide the recovery area conditionally into 3 phases.

In the first section the superheated water vapor is given to the surface of charcoal, heated to 900÷1000 °C. Upon contact of water vapor with the heated layer of charcoal under high temperature (950 °C and above) the following reaction takes place:

$C + H_2O$ vapor = $CO + H_2$ - 118.8 kJ. (1)

Since the reaction (1) is endothermic, the temperature decreases and at a temperature from 900 to 950 °C the following reaction takes place (2):

$C + 2H_2O = CO_2 + 2H_2$ - 75.2 kJ. (2)

Both reactions are endothermic and lead to a decrease of the temperature.

In the second plot there is a decrease in the speed of passage of water vapor through the charcoal due to the resistance of the layer. Previously formed CO reacts with water vapor to form carbon dioxide and hydrogen in the reaction:

$$CO + H_2O = CO_2 + H_2 +43,6 \text{ kJ}. \quad (3)$$

Also in this plot the reaction of carbon dioxide starts to take place:

$$C + CO_2 = 2CO-162.4 \text{ kJ}. \quad (4)$$

In the third phase the hydrogen formed in the reactions (1) – (3) reacts with carbon, forming methane:

$$C + 2H_2 = CH_4+87,4 \text{ kJ}; \quad (5)$$

$$CO + 3H_2 = CH_4 + H_2O+206,2 \text{ kJ}. \quad (6)$$

The temperature in this zone is about 800-850 °C. The methane content in gas obtained from charcoal is not more than 1%.

Due to the flow of such chemical reactions the charcoal layer decreases.

The reaction (4) goes with the energy consumption due to that fact that the temperature in the reaction zone decreases, and the equilibrium in the reaction shifts towards the formation of carbon dioxide. According to that the heat input either from outside or by internal exothermic reactions is necessary to maintain the temperature level, providing the formation of CO, for example to supply oxygen to the reaction zone

$$2C+O_2=CO+246.4\ \text{kJ}. \tag{7}$$

Also to ensure the conditions for the occurrence of these reactions, it is possible to use catalysts which increase the yield of CO and H by lowering the reaction temperature of 600-700 °C.
The speed of reaction (4) is two times faster than the speed of the reaction (1). The presence of solid carbon limits the reaction zone.
As noted, the stage of heating and the volatile content are less than 15 % of the total time of the gasification process in contrast to the recovery stage, which occupies most of the time and is the primary source and consumer of energy in the gasifier.
The reduction zone is characterized by the occurrence of the following main reactions:

$$C + CO_2 = 2CO, \tag{8}$$
$$C + H_2O = CO + H_2, \tag{9}$$
$$C + 2H_2 = CH_4, \tag{10}$$
$$CO + H_2O = CO_2 + H_2, \tag{11}$$
$$CO + 3H_2 = CH_4 + H_2O, \tag{12}$$
$$C + 2H_2O = CO_2 + 2H_2. \tag{13}$$

The final composition of producer gas is the following: CO_2, CO, CH_4, H_2 and H_2O [4].
The recovery phase is described by the following equations:
-the equation of coal conservation

$$\frac{dm_y}{d\tau} = -m_y \cdot \sum_{m=1}^{z} k_i \tag{14}$$

where k_i - is the rate constant of a chemical reaction, m_y- is the coal mass

- the equation of mass conservation for each component of the gas flow

$$\pm\frac{dC_i}{d\tau} = -k_i \cdot C_i \tag{15}$$

C_i - is the relative concentration of the i-substance.
The "+" sign is for reactions of CO, H_2, CH4 formation.
The sign "-" is for loss reactions of CO_2, H_2O.

- the equations of energy conserving for the gas flow and coal

$$\rho_{g-g} \cdot c_{g-g} \cdot \frac{dT_{g-g}}{d\tau} = -\alpha_{g-g} \cdot (T_c - T_{g-g}) \cdot f + \sum_{i=1}^{z} (q_i \cdot k_i \cdot (C_{i0} - C_i)) \tag{16}$$

$$\rho_c \cdot c_c \cdot \frac{dT_y}{d\tau} = \alpha_{g-g} \cdot (T_c - T_{g-g}) \cdot f - \sum_{i=1}^{z} (q_i \cdot k_i \cdot (C_{i0} - C_i)) \tag{17}$$

where p_c and $p_{g\text{-}g}$ are the specific density of coal and gas stream, respectively; c_y and $c_{g\text{-}g}$ – are heat capacity of coal and gas stream, respectively; $\alpha_{g\text{-}g}$ – is coefficient of convective heat transfer between gas and coal, W/(m^2·K); C_{i0} – is initial concentration of the i-substance; C_i – is the relative concentration of the i-substance; z – is the number of parallel reactions; f – is the specific surface area.
The initial conditions for the system of equations (14) – (17)

$$T_{vapor}=1000\ °C, T_y=900\ °C$$
$$C_{H_2O}=100\%,$$
$$m_{y0}=m_{y\ entr.}$$

To solve the system of equations (14) – (17), the following boundary conditions are accepted:

- according to temperature:

$$T_y\Big|_{\tau=0} = Tentr = 900^oC, \tag{18}$$

- according to weight:

$$C_i\Big|_{\tau=0} = C_{i0}, \tag{19}$$

$$m_y\Big|_{\tau=0} = m_{y0}. \tag{20}$$

The density of the gas flow is determined by the expression

$$\rho_{g-g} = \sum_{i=1}^{z} M_i \cdot C_i \tag{21}$$

The gas permeability of the layer depends on the layer's porosity and is determined experimentally: $K = f(\varepsilon)$.

The calculation of dynamic viscosity of the gas flow is carried out according to the expression

$$\mu_{g-g} = \left(\sum_{i=1}^{z} v_i \mu_i M_i^{0,5}\right) \Big/ \left(\sum_{i=1}^{z} v_i M_i^{0,5}\right) \tag{22}$$

where v_i - is the volume fraction of the i-component in the gas mixture; μi- is the dynamic viscosity of the i-gas, chosen depending on the temperature.

The heat capacity of the gas flow is determined by the expression

$$c_{g-g} = \sum_{i=1}^{z} v_i \cdot c_i \tag{23}$$

where v_i - is the volume fraction of the i-component; c_i - is the heat capacity of the i-component, which is determined depending on the temperature.

The coefficient of convective heat transfer between gas and coal (α_g), is calculated using the Nusselt criterion by the formula corresponding to the flow of gases in the layer of a granular material:

$$\alpha_{g-g} = A \cdot \frac{\lambda_{g-g}}{d_2} \cdot \mathrm{Re}^{0,6} \cdot \mathrm{Pr}^{0,33} \tag{24}$$

where A – is the value depending on the porosity of the layer, $A = f(\varepsilon)$.

The thermal conductivity of the gas flow is calculated by the formula

$$\lambda_{g-g} = \frac{\sum_{i=1}^{z} v_i \lambda_i M_i^{1/3}}{\sum_{i=1}^{z} v_i M_i^{1/3}} \tag{25}$$

where λ_i is the thermal conductivity of the i-component of the gas mixture, which is taken depending on the temperature.

The Reynolds number is defined by the equation

$$\mathrm{Re} = \frac{w_{g-g} \cdot \Delta y \cdot \rho_{g-g}}{\mu_{g-g}} \tag{26}$$

and the Prandtl number according to the equation is

$$\mathrm{Pr} = \left(\sum_{i=1}^{z} c_{pi} C_i\right) \cdot \frac{\mu_{g-g}}{\rho_{g-g} \cdot \lambda_{g-g}} \tag{27}$$

Effective thermal conductivity of the layer depends on the moisture content, porosity and temperature layer: $\lambda_{eff} = f(U, \varepsilon, T)$, which complicates its calculation. Therefore it is determined experimentally.

Let us write the speed of chemical reactions for homogeneous and heterogeneous reactions.

Reactions (11) and (12) are homogeneous.

The speed of the reaction (11) has the form

$$\frac{d[CO]}{dt} = 575RT_g \cdot ([CO] - [CO]_{eq}) \cdot \cdot \left(0{,}5 - p \cdot 10^4 / 253\right)\exp\left(-13971/T_g\right) \tag{28}$$

where T_g - is the temperature of the components of the gas flow, R - is the universal gas constant, equal to 8.314 kJ/(kmolK).

The speed of reaction (12) is written as follows:

$$-\frac{d[CH_4]}{dt} = 312[CH_4]exp\left(-13971/T_g\right) \tag{29}$$

The speeds of heterogeneous reactions are written as

$$R_i^{/} = \frac{dGi}{dt} = A_i'\left(T_c\right)\left(c_i - c_{ieq}\right) \tag{30}$$

where dGi/dt is the quantity of i -gas consumed in a coal wall area of 1 m^2 at the m-reaction; c_{ieq} is the equilibrium concentration of i-gas for this reaction, KMOL/m^3; Am(Tc) is the velocity constant of the reaction, m/s; *Tc* – is the temperature of coal, Ph.

The value of A'i(Ty) is calculated by the formula

$$\frac{1}{A_i'\left(T_c\right)} = \frac{1}{\beta_i} + \frac{1}{k_i} \tag{31}$$

β_i is the coefficient of mass transfer for i-gas. β_i is calculated by the formula

$$\beta_i = Sh \cdot D_i / d_{eff} \tag{32}$$

where Sh is the Sherwood number; D_i is the molecular diffusion coefficient of i-gas, m^2/s .

The Sherwood number is calculated according to the formula

$$Sh = 0{,}023Re^{0,8} \cdot Sc^{1/3} \tag{33}$$

This equation conforms to the turbulent gas flow in the pipe.

Sc in equation (33) is the Schmidt number, which is calculated by the formula

$$Sc = \frac{\nu}{D_i} \tag{34}$$

The speed constant for chemical reactions (8-13) can be written as follows:

- for the reaction (8)

$$k_i(Tc) = k_m \exp(-275000/RTc) \tag{35}$$

$$\log k_m = 0{,}2 \cdot 10^{-4} \cdot 275000 + 2$$

- for the reaction (9)

$$k_i(Tc) = k_m \exp(-200000/RTc) \tag{36}$$

$$\log k_m = 0{,}2 \cdot 10^{-4} \cdot 200000 + 2$$

- for the reaction (10) (37)

$$k_i(Tc) = \frac{k_m \exp(-125000/RTc)}{1500}$$

$$\log k_m = 0{,}2 \cdot 10^{-4} \cdot 125000 + 2$$

- for the reaction (13)

IOP Conf. Series: Materials Science and Engineering **124** (2016) 012111 doi:10.1088/1757-899X/124/1/012111

$$k_i(Tc) = k_m \exp(-200000/RTc) \tag{38}$$
$$\log k_m = 0{,}2 \cdot 10^{-4} \cdot 200000 + 2$$
$$\frac{d[CO]}{dt} = 575RT_g \cdot ([CO] - [CO]_{eq}) \cdot$$
$$(0{,}5 - p \cdot 10^4 / 253)\exp(-13971/T_g)$$

3. Results and discussion:

Due to the fact that in dry charcoal there are cavities and cracks the reaction surface for heterogeneous reactions will differ from the given coal surface. This factor is taken into account in the model by introduction of the enlargement factor of the area, which will vary from 1 to 30 [5].

4. Conclusion

The mathematical model allows determining the influence of operating parameters occurring in the reduction zone according to the parameters of the resulting producer gas.

References

[1] Timerbaev N F, Safin R R, Safin R G and Ziatdinova D F 2014 *Journal of Engineering and Applied Sciences* **9** 141-146
[2] Tuntsev D V, Filippova F M, Khismatov R G and Timerbaev N F 2014 *Russian Journal of Applied Chemistry* **87** 1367-1370
[3] Di Blasi C 1996 *Chemical Engineering Science* **51(7)** 1121-1132
[4] Govind R and Shah J 1984 *AIChE J.* **30(1)** 79–92
[5] Jonson J L 1979 *N.Y.: John Wiley&Sons*

Investigation and Modeling of Recrystallization of Cold Rolled Automotive Steels

P Zhitelev[1], A Vasilyev[1], S Sokolov[1], D Sokolov[1] and R Paligin[2]
[1] Peter the Great St. Petersburg Polytechnic University, 29, Polytekhnicheskaya Str., St. Petersburg, 195251, Russia
[2] Severstal Russian Steel, 3, Mira Str., Cherepovets, 162600, Russia

E-mail: vasilyev_aa@mail.ru

Abstract. Ferrite recrystallization in cold-rolled sheets of automotive steels has been studied using a Geeble 3800 complex. Mathematical models for quantitative description of the process kinetics and prediction of the recrystallized ferrite grain size have been developed. These models enable performing calculations for any arbitrary heating regimes, including those that are used in industrial production practice, and allow taking into account the effects of a fairly wide range variation of the chemical composition of steels.

1. Introduction

Final microstructure and, consequently, mechanical properties of cold-rolled sheets of automotive steels depend strongly on the recrystallization process. In this regard, in recent years considerable attention has been paid to the experimental studies of the process [1, 2], as well as to the development of adequate mathematical models [3, 4].

In this paper we present some results of experimental investigation of recrystallization kinetics for cold-rolled sheets of 10 steel grades produced at the enterprise 'Severstal Russian Steel'. This study was carried out with the help of a Gleeble 3800 complex. Quantitative mathematical models for predicting the process kinetics and the recrystallized ferrite grain size as a function of a cold deformation degree, the parameters of initial steel structure and its chemical composition have been developed. The results of modeling calculations are compared with experimental data.

2. Materials and experimental technique

Investigation of the recrystallization kinetics was performed for cold-rolled sheets of 10 steel grades (DX54D, HX220YD, HX260YD, CR210B2, HX260BD, 08Ю, HX300LAD, 08пс, CR420LA, DP600) (**08Ю, 08пс** are Russian grade marks) with significantly varying contents of carbon, manganese, titanium and niobium ($0.004 \leq C \leq 0.09$; $0.1 \leq Mn \leq 0.72$; $0.002 \leq Ti \leq 0.066$; $0.001 \leq Nb \leq 0.023$ (mass. %)). The degree of deformation during cold rolling varied in the range of 0.5÷0.8 (relative deformation).

Recrystallization kinetics during isothermal annealing at three temperatures in the range of 550÷750°C was investigated using the Gleeble 3800 complex for each of the 10 steel grades. The samples were rapidly heated (heating rate is 50°C/s) to the annealing temperature, and held for a definite period (≤ 3000 s). Softening of the studied samples due to recrystallization was estimated by measuring their hardness after cooling to room temperature at a sufficiently high rate (10°C/s). The recrystallized fraction was estimated basing on the hardness data and on metallographic analysis.

3. Mathematical models describing the recrystallization kinetics and a recrystallized ferrite grain size

When constructing the models, the following assumptions based on experimental data were used:

- potential nuclei of the recrystallized grains ('sustainable' subgrains) are formed at the original ferrite grain boundaries and have a high angle boundary from the initial stage of structure evolution;
- all subgrains have the same initial size determined by the value of strain hardening during cold rolling;
- subgrains differ in dislocation density and their distribution in dislocation density is taken to be normal;
- recrystallized ferrite grains are formed when growing subgrains reach critical size;
- the form of grains and subgrains is supposed to be spherical;
- the final size of a recrystallized ferrite grain is determined by its original structure (ferrite grain size after hot rolling) and by the degree of deformation during cold rolling.

In accordance with the above assumptions, subgrains nucleated after cold deformation have different dislocation density [4] and their distribution in this density is considered to be normal. At the first stage of modeling a set of subgrain classes characterized by different dislocation density is created. The number of the classes is equal to N_{class}. Further modeling of the recrystallization kinetics reduces to description of the subgrains growth in all classes with deferent dislocation density, which determines the corresponding different driving pressure.

Upon reaching the critical size by a growing subgrain it is assumed that recrystallized grains are formed in the corresponding class and continue to grow with higher interface mobility than the previous subgrains.

The so-called extended recrystallized metal volume fraction, $X_{ext}(t)$, is calculated as follows:

$$X_{ext}(t) = \frac{4\pi}{3}\sum_{i}^{N_{class}} N_g^i \left(\bar{R}_g^i(t)\right)^3, \tag{1}$$

where $N_g^i = N_{sg}^i$ is a number of recrystallized grains in the i-th class with an average radius equal to $\bar{R}_g^i$. The actual value of recrystallized volume fraction is calculated as:

$$X(t) = 1 - \exp\left(-X_{ext}(t)\right). \tag{2}$$

3.1. Calculation of the driving pressure of recrystallization

At the stage of the subgrains growth the driving pressure of recrystallization, $G_i(t)$, corresponding to the i-th class of their distribution in dislocation density, is calculated as:

$$G_i(t) = \frac{1}{2}\rho_d^i(t)\mu b^2, \tag{3}$$

where μ is shear modulus of ferrite; b is Burgers vector modulus of dislocations.

To determine the initial value of the average dislocation density as a function of strain hardening the following well known formula is used:

$$\bar{\rho}_d = \left(\frac{\Delta\sigma}{\alpha_\rho M_T \mu b}\right)^2, \tag{4}$$

where $\Delta\sigma$ is the strain hardening after cold rolling; M_T is the Tailor factor assumed to be 2.7 for ferrite; α_ρ □ 0.33.

A change of the dislocation density due to recovery in the subgrains belonging to the i-th class is determined on the basis of calculated values of strain hardening, $\Delta\sigma_i$, decreasing with time using the equation similar to (4). Calculation of the strain hardening change with time is performed using kinetic equation [5]:

$$\frac{d\Delta\sigma_i}{dt} = -\frac{64\Delta\sigma_i^2 v_d}{9M_T^3\alpha_\rho^2 E}\exp\left(-\frac{U_a}{RT}\right)\sinh\left(\frac{\Delta\sigma_i V_a}{RT}\right). \tag{5}$$

where U_a and V_a are, respectively, the activation energy and volume of recovery that are assumed to be independent of temperature and chemical composition; v_d is Debye frequency; E is Young modulus; R and T have their usual meaning.

3.2. Calculation of the subgrains volume density and a recrystallized grain size

As noted above, the model assumes that nucleation of recrystallized subgrains occurs at the boundaries of the original ferrite grains. Accordingly, for the volume density of subgrains, N_{sg}, we can write:

$$N_{sg} = \frac{\alpha_{sg} S_{GB}(D_\alpha^0,\varepsilon)}{S_{sg}^0}, \tag{6}$$

where $S_{GB}(D_\alpha^0,\varepsilon)$ is an area of deformed ferrite grain boundaries per unit of its volume; D_α^0 is an initial ferrite grain size; ε is a degree of deformation; S_{sg}^0 is an initial area occupied by the subgrain at ferrite grain boundaries.

The dependence of a specific area of deformed ferrite grain boundaries on a degree of deformation is taken into account as follows [6]:

$$S_{GB}(D_\alpha^0,\varepsilon) = \frac{24}{\pi D_\alpha^0}\left(0.491e^{\varepsilon} + 0.155e^{-\varepsilon} + 0.143e^{-3\varepsilon}\right). \tag{7}$$

This formula takes into account that in the process of rolling deformation initially spherical (equiaxed) ferrite grains change their shape into ellipsoidal.

It was shown that the final size of recrystallized ferrite grains weakly dependends on the annealing temperature when it changes in the range of 600÷750°C, i.e. according to the kinetics of the process [4]. Accordingly, this size should be determined by the initial structure of ferrite and the degree of deformation during cold rolling. The analysis of the experimental data obtained in the present study confirms this statement. On this basis, assuming that area $S_{sg}^0 \square \bar{G}^{-2}(t=0)$ for the subgrain volume density under discussion we obtain:

$$N_{sg} = \alpha_{sg} S_{GB}\left(D_\alpha^0,\varepsilon\right)\left(\bar{G}(t=0)\right)^2, \tag{8}$$

where $\bar{G}(t=0)$ is the driving pressure of recrystallization at the initial moment, calculated using the average density of dislocations.

An important parameter of the recrystallized structure is the final grain size of ferrite, D_α^{rex}. This size may be calculated from the volume density of nuclei as follows:

$$D_\alpha^{rex} = N_{sg}^{1/3}. \tag{9}$$

3.3. Calculation of the growth rate

In the present model calculations of grain and subgrain growth rates are performed using the following equation:

$$\upsilon(t) = M_{GB}^{sg;g} G(t), \tag{10}$$

where $\upsilon(t)$ is the growth rate; M_{GB}^{sg} and M_{GB}^{g} are effective mobility values for subgrain and grain boundaries, respectively; $G(t)$ is the driving pressure. The radius of the subgrains in the i-th class under time moment t is:

$$R_{sg}^{i}(t)=R_{sg}^{0}+\int_{0}^{t}M_{GB}^{sg}G_i(t)dt, \tag{11}$$

where R_{sg}^{0} is the initial sub grain size; $G_i(t)$ is the driving pressure for sub grain of the i-th class.

The size of the grain, which is nucleated by time τ, at time moment t, is:

$$R_{g}^{i}(t)=R_{c}^{i}+\int_{\tau}^{t}M_{GB}^{g}\bar{G}(t-\tau)dt, \tag{12}$$

where R_c^i is the initial grain size for the i-th class that is equal to a corresponding critical subgrain size; $\bar{G}(t)$ is an average value of driving pressure.

The recrystallized grain boundary mobility is calculated as:

$$M_{GB}^{g}(T;Y_{AE})=M_0^g\exp\left(\frac{S_{GG}(Y_{AE})}{R}\right)\exp\left(-\frac{Q_{GG}(Y_{AE})}{RT}\right), \tag{13}$$

where $Q_{GG}(Y_{AE})$ and $S_{GG}(Y_{AE})$ are, respectively, the activation energy and entropy of the diffusion lattice reconstruction, that controls the movement of recrystallized grain boundaries; $Y_{AE}=\{y_C^*;y_{Mn};y_{Si};y_{Ni};y_{Cr};y_{Mo};y_{Nb};y_{Ti};y_V\}$ is the set of average alloying elements concentrations in a solid solution; y_C^* is an effective value of carbon concentration in the moving boundary, estimated taking into account cementite particles dissolution upon heating; M_0^g is a constant parameter. The entropy of grain growth activation is calculated as, $S_{GG}(Y_{AE})=\beta_{GG}Q_{GG}(Y_{AE})$, where β_{GG} is an empirical model parameter.

The effective activation energy of the process is assumed to be equal to the activation energy of grain boundary self-diffusion, which is proportional to the volume self-diffusion activation energy (SDAE) $Q_{SD}(Y_{AE})$: $Q_{GG}(Y_{AE})=\alpha_{GG}\,Q_{SD}(Y_{AE})$, where α_{GG} is an empirical parameter. SDAE depends on the chemical composition of the solid solution as follows [7]:

$$\begin{aligned}Q_{SD}(Y_{AE})&=311691-278242\left(1-\exp\left(-0.394y_C^*\right)\right)+88752y_{Mn}^{0.31}+22801y_{Si}-6490y_{Cr}+\\&84864y_{Mo}^{0.65}-38575y_{Ni}^{0.3}-7298y_V+132594y_{Nb}^{0.263}+82128y_{Ti}^{0.401}\ (J/mol)\end{aligned} \tag{14}$$

Subgrain boundaries mobility is calculated using equation:

$$M_{GB}^{sg}(T;Y_{AE})=\alpha_M^{sg}M_{GB}^{g}(T;Y_{AE}), \tag{15}$$

where $\alpha_M^{sg}<1$ is parameter of the model.

An optimal set of empirical values for model parameters was obtained using an experimental data base for the steels investigated in the present study.

3.4. A brief description of the numerical simulation algorithm

In the numerical model under discussion continuous time evolution of subgrain and grain systems is modeled as a series of changes that occur over short time intervals (steps) $\delta t^{(k)}$. At each step an increment of subgrain/grain sizes is calculated for all classes. In the process of calculations the subgrains in the i-th class are transformed into recrystallized grains with the same class index at the time moment when their size reaches critical value R_C^i, i.e. under condition: $R_{sg}^i\geq R_C^i=2\gamma/\bar{G}(t)$, where γ is an effective value of specific energy of the recrystallized grain boundaries.

5. Comparison of simulation results with experimental data

The results of modeling calculations of the recrystallization kinetics for some investigated steels are compared with experimental data in Figures 1, 2.

One can see that the developed model allows one to achieve good agreement with the experiment.

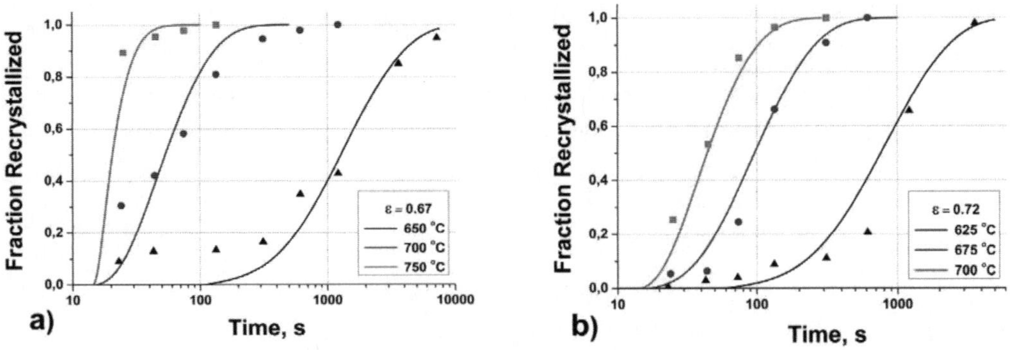

Figure 1. Comparison of the results of recrystallization kinetics calculation with the experimental data for IF-steels: a) HX220YD; b) HX260BD.

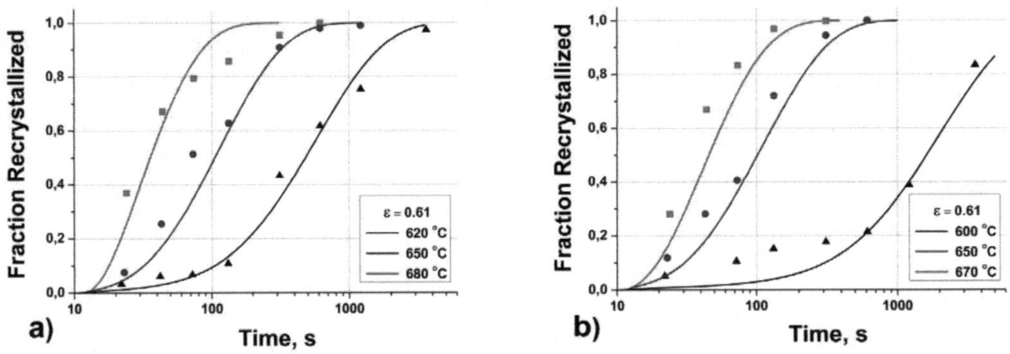

Figure 2. Comparison of the results of recrystallization kinetics calculation with the experimental data for investigated steels: a) CR420LA; b) DP600.

The results presented in Figure 3 demonstrate that the complementary model provides good prediction accuracy for a ferrite recrystallized grain size comparable with the error of experimental determination of this parameter.

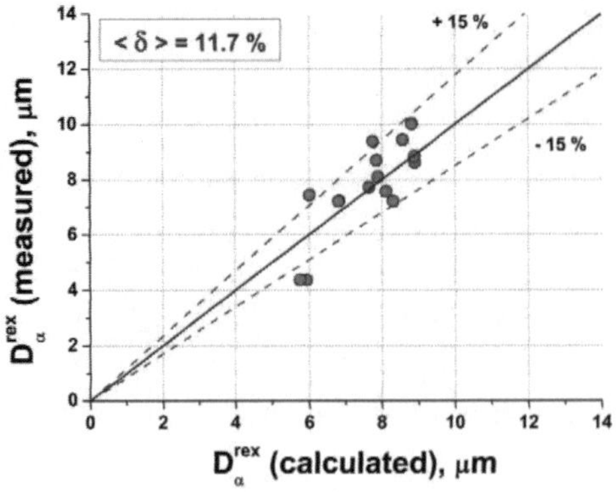

Figure 3. Calculation results for a recrystallized ferrite grain size in comparison with the experimental data.

Conclusions

Thus, basing on the presented results, the following conclusions can be made:

- mathematical models of recrystallization processes in cold-rolled sheets of investigated steels are developed and described with satisfactory accuracy both the kinetics of the process under isothermal annealing and a recrystallized ferrite grain size;
- the developed recrystallization model predicts with reasonable accuracy the crystallization kinetics in conditions of heating at an arbitrary rate and subsequent isothermal holding, which are employed when annealing cold-rolled sheets in industrial production lines;
- the developed recrystallization model adequately takes into account both the effect of content of carbon and the main substitution alloying elements (Mn, Si, Nb, Ti) on the kinetics of the process, which distinguishes it from the currently available analogs in the published literature.

References

[1] Ferry M et al 2001 *ISIJ Int.* **41** 1053
[2] Kulakov M et al 2013 *Metall. Mater. Trans. A.* **44A** 3564
[3] Senuma T 2012 *ISIJ Int.* **52** 679
[4] Zhu B et al 2012 *Mater. Sci. Eng.* **20** 1
[5] Verdier M et al 1999 *Acta mater.* **47** 127
[6] Lefevre-Schlick F et al 2009 *Mater. Sci. Eng. A.* **502** 70
[7] Vasilyev A et al 2012 *Physics of the Solid State* **54** 1669

Fabrication of the Ni_3Al-based alloy formed by spark plasma sintering of VKNA powders

L I Shevtsova[1], A S Ivashutenko[2], N V Martyushev[2] and R I Kuzmin[1]
[1] Novosibirsk State Technical University, 20, Karla Marksa ave., Novosibirsk, 630073, Russia
[2] National Research Tomsk Polytechnic University, 30, Lenina ave., Tomsk, 634050, Russia

E-mail: edeliya2010@mail.ru

Abstract. The material based on Ni_3Al intermetallic has been obtained from the industrial powder of a VKNA type by the method of spark plasma sintering. Materials sintering was conducted at the temperature of 1100°C, compacting pressure of 20 MPa, and during soaking time equal to 5 minutes. The heating rate of samples amounted to 50 and 200 °C/min. It has been established that the material obtained by sintering at the rate of 50 °C/min possesses a maximum value of density (5.93 g/cm^3) and a maximum level of bending strength (~ 400 MPa).

1. Introduction

A Ni_3Al compound is a basic strengthening phase of heat-resisting nickel alloys used in large quantities when fabricating high-duty parts in chemical industry, power engineering as well as in aircraft industry and rocket production. The content of the Ni_3Al intermetallic in such materials reaches 70 %. The presence of this phase allows using nickel alloys when manufacturing parts operated at elevated temperatures [1]. This is accounted for the combinations of such Ni_3Al properties as heat resistance, oxidation resistance, corrosion resistance at elevated temperatures as well as the high melting temperature (1395 °C) and relatively low density (7.5 g/cm^3) [2-4].

However wide application of the Ni_3Al intermetallic is impeded by a number of peculiarities of nickel aluminides including a low level of plasticity and crack resistance. This peculiarity accounts for intractability of materials by cutting. At room temperature the samples of polycrystalline Ni_3Al intermetallic are brittle [4].

In the past decades the alloys based on nickel aluminide of a VKNA type (Ni_3Al phase content is up to 90 %) developed in the All-Russian Institute of aviation materials (AIAM) have been extensively used. In the technical literature the structure of alloys of the VKNA-type in the as-cast state is described in detail and some information on their mechanical properties are given [2]. However there are not enough data on the structure and properties of such materials obtained by the powder metallurgical technique.

It has been experimentally established that an effective way of obtaining powder compacted materials based on intermetallics is spark plasma sintering (SPS technology) [3]. Owing to a short-term high temperature impact on the powder mixture, an optimal ratio of heating and material deformation modes as well as peculiarities of electric current flow through the sample, SPS technology allows obtaining high strength items with low values of porosity and residual mechanical stresses. The application of this technology enables preserving a fine-grained structure of the sintered material.

The purpose of this paper is a formation of the material on the basis of nickel aluminide during spark plasma sintering of the powder of the VKNA type at different rates of heating.

2. Materials and methods

The powder of nickel aluminide of the PN75Yu23V (VKNA) grade was used as initial material. The average size of the powder particles was 20 μm. The basic phase of these materials was Ni_3Al intermetallic. The chemical composition of the applied powder is presented in table 1. The material of the VKNA type is applied in industrial production for obtainment of heat and wear resistant coatings, applied to the parts of the machines operating in the hostile environments.

Table 1 – The chemical composition of the nickel aluminide powder of the VKNA grade.

Powder designation	Mass fraction of the elements, % by mass									
	Ni	*Al*	*Cr*	*Co*	*W*	*Ti*	*Mo*	*Ca*	*Fe*	*C*
PN75Yu23V	71.7 -74	19- 22.3	2.93	0.8	1.16	0.32	0.79	0.08	0.15	0.07

The powder was sintered in the spark plasma sintering system 'SPS10-4' produced by Advanced Technology Company (USA). The powder sintering process was implemented as follows: the powder mixture was poured into a current-conducting graphite press mould with the inner diameter of 30 or 40 mm, which was subsequently placed in the vacuum chamber of the installation (10^{-2} Pa).

Sintering was conducted according to the modes presented in table 2. The range of the sintering modes was selected on the basis of literary data and personal studies of the authors of the project.

Table 2 – Modes of spark plasma sintering of the powder of the VKNA type.

Mode number	Powder mixture	Sintering temperature, °C	Compacting pressure, MPa	Soaking time, min.	Heating rate, °C/min	Vacuum, Pa
1	VKNA	1100	50	5	50	10^{-2}
2					100	
3					200	
4			30		100	
5			40			
6			60			

3. Results and discussion

Analysis of the literary data justifies the fact that the process of spark plasma sintering, the structure and properties of the obtained materials is influenced by the following basic parameters of sintering: heating rate, sintering medium, electric current magnitude, compacting pressure, temperature and duration of isothermal soaking [4-7]. Since the most significant difference of SPS technology from hot pressing is the high rate of heating; the problem of its influence on the resulting density of the sintered materials is of great interest. High heating rates allow reducing the sintering process duration and restrict the powder particle growth during the consolidation.

M.S. Boldin's paper [8-10] contains a qualitative analysis of studies on evaluation of the heating rate impact on the structure and properties of the sintered ceramic materials, which specifies the fact that a heating rate increase up to ~ 300 °C/min enables an increase in the density of the finished workpiece. During a subsequent increase of the heating rate, the areas with closed porosity are formed in the material structure, which is accompanied by the density reduction. In this case the density decrease effect is intensified while increasing heating temperature [11]. When studying the nickel powder sintered at different heating rates (from 90 to 1100 °C/min), the authors of the paper note that high heating rates during sintering lead to heterogeneous compacting of samples and crack formation. On the basis of the analysed literary data VKNA powder heating was realised at the rates of 50, 100 and 200 °C/min. The compacting pressure was varied from 30 to 60 MPa.

Based on the data of studies conducted by the project implementator, the most reasonable sintering temperature of nickel aluminide of the PN85Yu15 grade (an alloy on the Ni_3Al basis) was determined; this temperature allows one to obtain qualitative workpieces possessing the best set of mechanical properties by SPS technology. In this work the sintering temperature remained stable and amounted to 1100 °C.

It is possible to use shorter soaking time compared to customary sintering one for obtaining high-density compacts during SPS. The reduction of soaking time enables reduction of microstructural evolutions as a result of diffusion control of phase transformations. The soaking time for all series of experiments amounted to 5 minutes.

The density of sintered materials was calculated by the method of measurement of sample linear dimensions and by the method of hydrostatic weighing [12]. The obtained averaged values of density of the resulted samples are shown in Table 3.

Table 3 – The influence of compacting pressure and heating rate on the density of the sintered materials.

Mode number	Powder mixture	Compacting pressure, MPa	Heating rate, °C/min	Density, g/cm^3
1	VKNA	50	50	5.93
2			100	5.71
3			200	5.86
4		30	100	5.26
5		40		5.92
6		60		6.23

It is seen from the table that maximum density (6.23 g/cm^3) is reached for the material obtained by VKNA powder sintering at compacting pressure equal to 60 MPa. The reduction of heating rate influences favourably the density of the finished material.

Structural studies of the VKNA sintered materials at different scale levels (with the use of optical metallography, scanning electron microscopy and diffractometry of the X-ray radiation).

Structural studies were conducted under optical and scanning electron microscopes. The scanning microscope of the Carl Zeiss Axio Observer A1m type was used for metallographic studies in the mode of the bright field. For the purpose of structural study under higher magnifications and for X-ray microanalysis a scanning electron microscope of the Carl Zeiss EVO 50 XVP type was used.

The microstructure of the compact sintered by mode №4 is homogeneous throughout the whole volume. Such defects as large pores or cracks were not revealed in the material structure. Formation of the large quantity of micropores in the structure of this material can be explained by the presence of different refractory elements. This testifies the fact that such sintering modes do not provide the formation of high-quality compacts from VKNA powder. The average particle size after sintering by this mode amounts to 30-50 μm. As a result of sintering of the PN85Yu15 type powder under specified modes, a compact with the density of 5.9 g/cm^3, which amounts to 92 % of the Ni_3Al intermetallic density, was obtained.

Figure 1 shows a material structure obtained using the scanning electron microscopy methods. The results of the energy dispersive X-ray microanalysis of the local regions in the sintered material is evidence of the refractory elements allocation (figure 1b).

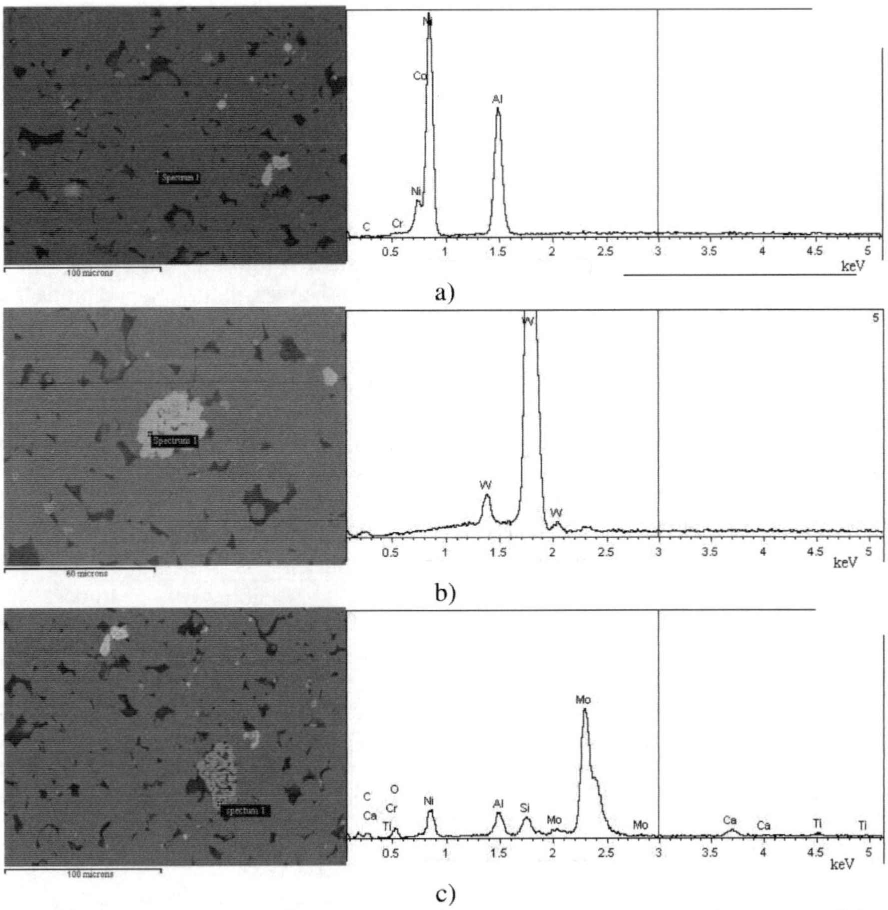

Figure 1. Results of the X-ray microanalysis of the local regions in the material obtained by spark plasma sintering of VKNA powder by mode № 4.

The X-ray phase analysis was conducted using an ARL X'TRA diffractometer. Diffractograms were shot by means of CuKα–radiation with the scanning pitch of 0.02 degree. The accumulation time amounted to 1 s per one point. The X-ray diffraction data analysis allowed determining that the material under study consists of the Ni3Al phase by 95 % with the average size of the coherent-scattering regions equal to 64.5 nm and with microstresses of $1.1 \cdot 10^{-4}$. The remaining 5 % part is an X-ray amorphous phase.

Mechanical tests of the obtained materials comprise determination of the strength properties (durometric studies, strength tests of the materials under conditions of uniaxial tension and three-point bending) and impact toughness.

The hardness indices of most of materials correlate closely with such important mechanical property as the magnitude of ultimate strength. In this case the advantage of durometric studies is simplicity of measurements, applicability of small-size samples, immediacy of experimental data acquisition.

The microhardness of the materials obtained by means of VKNA type powder under different modes was evaluated by means of the microhardness tester, Wolpert Group 402 MVD, intended for Vickers hardness testing. The indentor of the microhardness tester is a tetrahedral diamond pyramid. The indentor pressed on the pyramid with the force of 19.6 N. The measurements were conducted on the crosscut metallographic sections.

Three-point bending tests were conducted to evaluate strength properties. Tests for strength determination under uniaxial tension failed to succeed. This failure is due to increased brittleness of the sintered samples. Bending tests were conducted at room temperature at the installation 'Instron 3369'. The rate of travers travel was 0.5 mm/min. In order to conduct mechanical tests, flat samples with dimensions of 3×4×30 mm were cut out at the electroerosion cutting installation 'Sodick AG 400L'. Testing in the conditions of uniaxial tension of the obtained materials failed to succeed due to increased brittleness of samples.

Impact strength of the sintered samples characterises the energy used for destruction of the workpieces during their dynamic loading. Tests were conducted at room temperature using an impact pendulum-type testing machine 'CAST 9050' (Instron), with the maximum value of impact energy of 25 J. Samples of a rectangular shape with dimensions of 3×4×25 mm (without a notch) were applied for testing. Samples were cut out of the sintered workpieces at the electroerosion cutting installation 'Sodick AG 400L'. In the process of testing the values of fracture energy were fixed.

The results of compacting pressure and heating rate influence on the strength properties are reflected in tables 4 and 5, respectively. The maximum value of the impact strength level is registered on the basis of the material obtained by sintering of VKNA powder under 30 MPa.

Table 4 – The compacting pressure influence on strength properties of sintered VKNA powder (at heating rate of 100 °C/min).

Powder mixture	Compacting pressure, MPa	Microhardness, HV1	Bending strength, MPa	Impact strength, J/cm^2
VKNA	30	290	220	3.14
	40	480	330	0.8
	50	370	225	0.86
	60	520	490	0.8

Table 5 – The heating rate influence on strength properties of sintered VKNA powder (under compacting pressure of 50 MPa).

Powder mixture	Heating rate, °C/min	Microhardness, HV1	Bending strength, MPa	Impact strength, J/cm^2
VKNA	50	490	390	1
	100	370	225	0.86
	200	415	310	1.05

The results of the fractography studies of the VKNA powder sintered by different modes after bending impact tests indicate a domination of the intercrystalline nature of the fracture in all sintered samples (Figure 2).

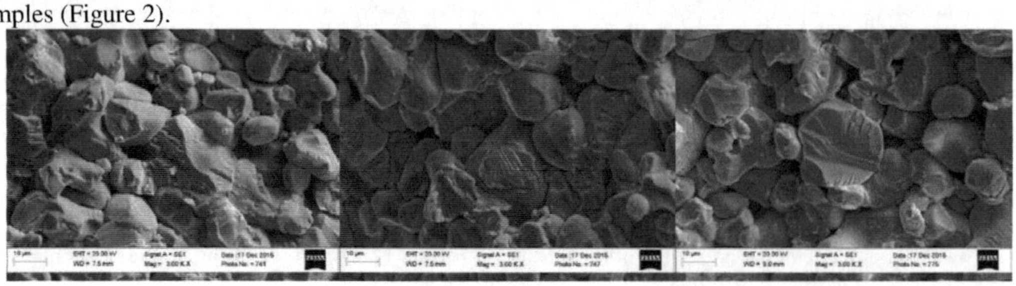

a b c

Figure 2. Fractures of samples sintered at different heating rates after bending impact tests: a – 50 °C/min; b – 100 °C/min; c – 200 °C/min.

4. Conclusion

Thus it has been experimentally established that spark plasma sintering is an effective method of obtaining low defective compacts from the materials based on nickel aluminide with the enhanced set of mechanical properties. This technology allows preserving a fine-grained structure of the sintered material. Optimization of the spark plasma sintering modes was conducted using literary data and experimental research. It has been revealed that such sintering parameters as compacting pressure and heating rate play an important part in formation of the structure and properties of the sintered material. It has been established that an increase in the compacting pressure enables an increase in the relative density of the material and strength properties of sintered VKNA powder. The maximum value of bending strength (490 MPa) is registered in the material obtained by sintering at the pressure of 60 MPa. The reduction of the heating rate also influences favourably the density and strength properties of the sintered material.

Based on the obtained experimental data it has been established that it is expedient to conduct VKNA powder sintering in the following mode: the heating temperature of the sample should be 1100 °C; the average heating rate is 100 °C/min; compacting pressure and soaking time under pressure at sintering temperature should be equal to 60 MPa and 5 minutes, respectively.

5. Acknowledgements

The work has been performed with financial support of the Russian Foundation for Basic Research, grant № 15-33-50845 (mol_nr) and Russian President Grant (SP-1179.2015.1).

References

[1] Deevi S C, Sikka V K 1996 *Intermetallics.* **4** 357–375
[2] Lazurenko D V, Mali V I, Shevtsova K E, 2014 *Applied Mechanics and Materials* **682** 132–137
[3] Shevtsova L I, Bataev I A, Mali V I et al. 2003 *Metal Work.: Technol., Equipm., Tools* **4 (61)** 35–42
[4] Tokita M 1993 *J. Soc. Powd. Tech. Jap.* **11 (30)** 790–804
[5] Munir Z A 2000 *J. Mater. Syn. Proc.* **8** 189–196
[6] Groza J R, Zavaliangos A 2003 *Rev. Adv. Mater. Sci.* **1 (5)** 24–33
[7] Kim J S, Choi H S, Dudina D, Lee J K, Kwon Y S 2007 *Solid State Phenomena.* **119** 35–38
[8] Shevtsova L I, Korchagin M A, Thommes A, Mali V I, Anisimov A G, Nagavkin S Yu, 2014 *Advanced Materials Research* **1040** 772–777
[9] Terentyev D, Bataev A, Bataev I, Burov V, Nikulina A, Bannov A 2013 *Advanced Science Letters* **19** 3695–3696
[10] Ivashutenko A S, Martyushev N V, Vidayev I G 2014 *Advanc. Mater. Res.* **1040** 819–823
[11] Shevtsova L I, Sameyshcheva T S, Munkueva D D, 2014 *Appl. Mech. and Mater.* **682** 188–191
[12] Guo Y, Bataev I, Georgarakis K, Jorge A M Jr, Nogueira R P, Pons M, Yavari A R, 2015 *Intermetallics* **63** 86–96

IOP Conf. Series: Materials Science and Engineering **124** (2016) 012114 doi:10.1088/1757-899X/124/1/012114

The influence of modification by superdispersed powders on the lead-tin-base bronze structure

I V Semenkov[1], N V Martyushev[1], A I Popelyukh[2], A T Alpeisov[3], Yu Yu Drozdov[1] and A P Zykova[1]
[1] Tomsk Polytechnic University, 30, Lenina ave., Tomsk, 634050, Russia
[2] Novosibirsk State Technical University, 20, Karka Marksa ave., Novosibirsk, 630073, Russia
[3] Kazakh National research Technical University after K I Satpayev, 22a, Satpaev str., Almaty, 050013, Republic of Kazakhstan,

E-mail: martjushev@tpu.ru

Abstract. The paper presents data on the influence of additives of the pre-treated aluminium oxide powder on the structure of cast lead-tin-based bronzes. Different quantities of the modifier, based on the superdispersed aluminum oxide powder, were added to the bronze melt. The studies have shown that addition of a small amount of aluminum oxide powder (0.07…0.25 %) allows modifying the microstructure of the obtained castings. This modification includes grain refinement, reduction of the matrix dendrites size of tin solid solution in copper, as well as formation of spherical inclusions of the low-melting phase – lead. In this case, the addition of such modifier influences weakly the morphology and the quantity of solid eutectoid inclusions based on electron compound $Cu_{31}Sn_8$.

1. Introduction

In the modern mechanical engineering a significant number of parts are fabricated from lead-tin-base bronzes. They include a series of parts which should possess sufficiently advanced strength characteristics (sealings and piston rings, oil-seal and expander rings). In order to enhance tribotechnical characteristics, lead is introduced into these materials. Lead reduces friction coefficient, enhances tribotechnical characteristics, however, it reduces strength significantly.

At present one of the promising trends of enhancing a set of service properties of such bronzes is alloying them with superdispersed powders (SDP). Introduction of their small amount into the melt before a crystallization process allows increasing strength propeties of castings [1, 2]. But a mechanism of interaction with lead-tin-based bronzes, as well as the process regularities of such modification, is not studied profoundly. However, such modification of copper alloys is promising from several points of view. First, the nanopowder particles introduced into into the melt will serve as centres of crystallization. As a result, the castings microstructure will be more fine-grained. Such fine-grained structure will possess more advanced strength properties compared to an ordinary microstructure. Second, introduction of additional particles of the nanopowder – grain nuclei – allows narrowing the size of the temperature interval of alloy crystallization. Thereby a reduction of the shrinkage porosity of castings is achieved. The casting becomes more dense with less quantity of defects. This results in the growth of strength properties. Third, the nanopowder particles of oxides and nitrides of metals possess high melting temperature and increased hardness. They will not dissolve in the casting's volume and remain in the form of nanodimensional inclusions. In the process of plastic deformation such inclusions will impede a dislocation glide. Thus, homogeneously distributed particles will create dispersion hardening of the casting. Besides, the growth of matrix strength can provide enhancement of antifriction properties (reduction of friction coeffcient and wear decrease).

This paper presents an investigation of the influence of different content of additives of the pre-treated aluminium oxide powder on the structure of lead-tin-base bronze under formation.

2. Materials and methods

The lead-tin-base bronze of the BrO10S10 grade (Russian grade abbreviation) was used as a material for investigation. This bronze contains 10 % wt. of lead, 10 % wt. of tin and 80 % wt. of copper. The multicomponent bronzes under study were melted in the induction high-frequency furnace in crucibles. The crucibles' material is silicicated graphite. Melting was conducted using the components of technical grade. Cathode copper of the Mlk grade (GOST 859-78), sheet lead of the C-2 grade (GOST 3778-77); rod tin of the O1 grade (GOST 860-75) were used as a charge mixture. The phosphorous-copper alloy of the MF1B grade (GOST 4515-93) was used as a deoxidizer. The preliminarily placed in the copper foil powder-modifier was introduced into the melt.

The aluminum oxide powder, being wrapped into the copper foil, was introduced into the bronze melt after its treatment in the copper powder mixture in the ball planetary-type mill. The method of powder treatment is described in [3, 4]. The pouring was realised into the graphite moulds at room temperature. The content of the powder was 0.07; 0.15; 0.25; 0.5; 0.75 and 1.5 % wt.

Metallographic analysis of the structure was realised using optical microscopes 'MIM-8M', 'ZEISS AXIO Observer.A1m'. An 'Observer.A1m' microscope is equipped with a built-in photocamera. The obtained images were analysed, using the programme of digital image autoprocessing 'System KOI', developed in NR TPU (National Research Tomsk Polytechnic University) [5]. This programme allows performing calculations of the inclusion voume fraction, an average size of the particles by the microstructure photos. The operation of the 'System KOI' programme is described in detail in [6-8].

3. The study of the structure of the modified lead-tin-base bronze

Reasoning from theoretical premises, the main consequence of nanopowder introduction into the melt should be refinement of the macro- and microstructure, as the powder particles must serve as nuclei of new grains. Figures 1 and 2 show photos of the microstructure of cast samples from the bronze of the lead-tin bronze grade, both modified by SDP of aluminium oxide and without modifier addition. The phase composition of the bronze under study represented in photos of the microstructure is a solid solution of tin in copper, lead inclusions and eutectoid inclusions based on electron compound $Cu_{31}Sn_8$.

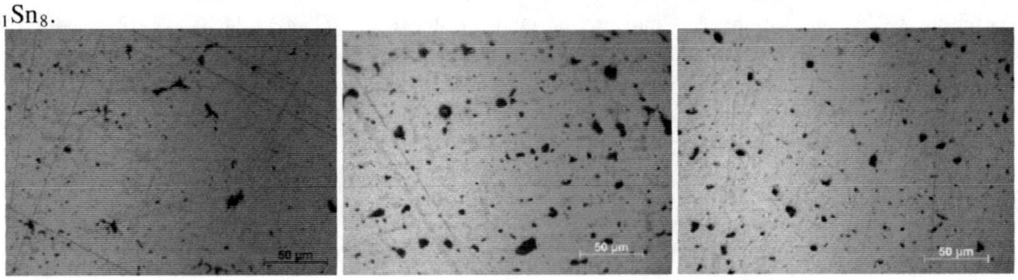

a b c

Figure 1. A microstructure of BrO10S10 bronze castings with different content of the alloying additive of the Al_2O_3 powder: a – without the powder; b – 0.07 % of the Al_2O_3 powder; c – 0.75 % of the Al_2O_3 powder.

As it is seen from Figures 1 and 2, as a result of modification of crystallization conditions, a different morphology of inclusions of the low-melting phase and eutectic inclusions in the castings is formed. In addition, from the studies conducted earlier [3], it is known that the cooling rate influences considerably both the shape and the quantity of eutectoid under formation. The higher is the cooling rate, the less amount of eutectoid is formed. But the data on the impact of modifying additives on the change of the content and the shape of eutectoid inclusions are virtually absent. By means of the obtained photos of the microstructure, a quantitative calculation of a sphericity coefficient and an average size of particles, both for lead inclusions and for eutectoid inclusions, was performed. The eutectoid content was determined in a simiar way.

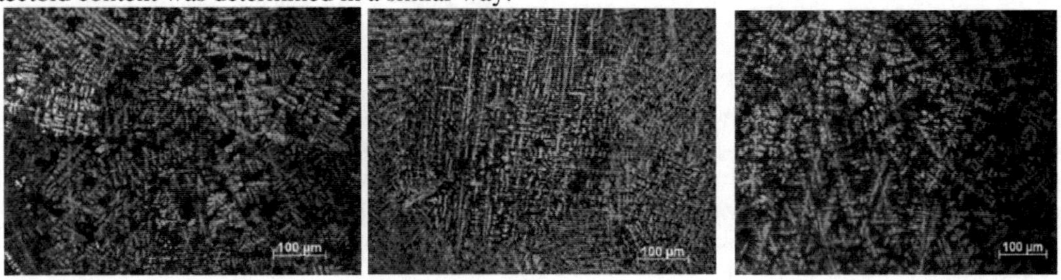

a b c

Figure 2. The etched microstructure of BrO10S10 bronze castings with different content of the alloying additive of the Al_2O_3 powder: a – without the powder; b – 0.07 % of the Al_2O_3 powder; c – 0.75 % of the Al_2O_3 powder.

Figures 1 and 2 show that when introducing a small amount (0.25 %) of the aluminium oxide powder, there is a 2.5 time decrease of the distance between the dendrites axes of the second order, and a ~ 1.5 time decrease of the grain size. This implies that a considerable amount of powder particles served as effective crystallization centres. When increasing the content of SDP of Al_2O_3, the structure started to coarsen relatively that which was obtained, using low concentrations of the powder. With its 1.5% perscentage, the differences of its structure from the unmodified one are insignificant. This is connected with coagulation of the powder particles in case of the sufficiently high (>0.5 wt. %) powder concentrations. It is necessary to note that the microstructure refinement was homogeneous throughout the casting volume, which indirectly implies that the modifer distribution was also homogeneous enough and it had not floated to the casting surface.

Table 1. Dependence of the sphericity coefficient of lead inclusions and average sizes of bronze lead inclusions on the modifier concentration

Al_2O_3 powder content, %	Sphericity coefficient of lead inclusions	Average size of lead inclusions, μm	Eutectoid content, %	Sphericity coefficient of (α+δ) eutectoid inclusions	Average size of (α+δ) eutectoid inclusions, μm
0	10.2	6.9	15	-	-
0.07	2.2	5.1	13	8.2	20.1
0.15	2.8	7.3	12	9.1	21.8
0.25	2.6	9.3	14	11.3	16.7
0.5	3.5	9.5	13	12.7	18.6
0.75	4.1	15.8	10	-	-
1.5	6.9	18.2	13	-	-

The qualitative analysis of the cast samples' structures has shown that we have not succeeded in detecting a significant difference in the eutectoid amount under various concentration of the modifier. By means of the metallographic studies it has been found that the change in the eutectoid content is not so significant as it was during the influence of different cooling rates on the same bronze. The maximal difference (a decrease from 15 % to 10 %) was detected in case of the sample, which contained 0.75 % of the modifier. At that, the samples with 0.75 and 1.5 % addition of the modifier possessed such vastly branching morphology of eutectoid that it did not allow determining a spericity coefficient and an average size for them.

In case of the lead inclusions the following tendency was traced. With the decrease of the modifier content, lead inclusions became more sperical (Table 1), the sphericity coefficient increased approximately threefold. As to their sizes, it has been found that in case of low modifier concentrations, lead inclusions were 1.5 time coarser than those in the unmodified sample; in case of average concentrations, lead inclusions were similar to those in the unmodified sample, and in case of high concentrations – three-time coarser.

4. Conclusion

After conducted studies of the lead-tin-base bronze structure of the BrO10S10 grade, it is possible to conclude that the modification with superdispersed powders influences considerably its structure. It is particularly typical for low concentrations of the modifier (up to 0.25 %).

When introducing a small amount of the aluminium oxide powder, the distance between the axes of the second-order dendrites and the average grain size reduces. This implies that a considerable part of the powder particles is effective crystallization centres. When increasing the nanopowder content, the structure starts coarsening relatively that, which was obtained using low powder concentrations. Introduction of the large quantity of the modifier leads to its coagulation and reduction of its influence on the structure.

The eutectoid content does not undergo such significant changes. The maximal difference (a reduction from 15 % to 10 %) has been detected in the sample containing 0.75 % of the modifier. At this, the samples containing 0.75 % and 1.5 % of the modifier possessed such vastly branching morphology of eutectoid that it was impossible to determine a sphericity coefficient and an average size for them.

The modifier, based on aluminium oxide, influences maximally on the inclusions of the low-melting phase – lead. The most impact is exerted by small quantities of the powder (0.07…0.25 %). In this case lead inclusions become more spherical and the sphericity coefficient decreases threefold.

5. Acknowledgments

The reported study was funded by the Russian Humanitarian Science Foundation (RHSF 16-16-70006) and RFBR, according to the research project No. 16-38-60146 mol_a_dk.

References

[1] Martyushev N V, Petrenko Y N 2014 *Applied Mechanics and Materials* **682** 231–235
[2] Terentyev D, Bataev A, Bataev I, Burov V, Nikulina A, Bannov A 2013 *Advanced Science Letters* **19** 3695–3696
[3] Martyushev N V, Semenkov I V, Petrenko Y N 2014 *Advanc. Mater. Res.* 872 112–117
[4] Stepanova N V, Razumakov A A 2013 *The 8 international forum on strategic technologies (IFOST 2013)* **1** 240–242
[5] Martyushev N V, Egorov Y P 2003 *Proceedings of the 9th International Scientific and Practical Conference of Students, Post-graduates and Young Scientists - Modern Techniques and Technologies, MTT'* **1** 1438190
[6] Martyushev N V, Semenkov I V 2014 *Advanc. Mater. Res.* **880** 272–275
[7] Bataev I A, Bataev A A, Mali V I, Pavliukova D V 2012 *Materials and Design* **35** 225–234
[8] Martyushev N V 2012 *Proceedings 2012 7th International Forum on Strategic Technology, IFOST-2012* **1** 6357597

Investigation of the structure and microhardness of 'Mo-Fe-C' coatings obtained by the electron beam injected in the atmosphere

D O Mul[1], S N Sorokova[2], N V Martyushev[2], N V Belousova[1], V G Burov[1] and M C Koroleva[1]
[1] Novosibirsk State Technical University, 20, Karla Marksa ave., Novosibirsk, 630073, Russia
[2] Tomsk Polytechnic University, 30, Lenina ave., Tomsk, 634050, Russia

E-mail: ddariol@yandex.ru

Abstract. In this work 'Mo-Fe-C' coatings fabricated on medium carbon steel by non-vacuum electron beam cladding were investigated. The structure of coatings and transition zones were studied by scanning electron microscopy (SEM). It was shown that an increase of Fe percentage in the cladding mixture led to a decrease of the eutectic volume fraction in the coating and was accompanied by the formation of the gradient structure between the coating and a substrate material. Measurements of microhardness in the cross section of samples revealed that the cladding of a 'Mo-C' powder mixture contributed to a 4.5-fold increase of microhardness.

1. Introduction

One of the main tasks of mechanical engineering is enhancing wear-resistance of the parts functioning in abrasive wear conditions. The surface layers of parts contacting an abrasive material are subjected to wear. Therefore a rational solution of the tasks is hardening of the product surface.

Formation of high-resistant particles surrounded by a viscous matrix in the surface layers of the work-piece provides high resistance to wear. When cladding wear-resistant coatings, the alloys of a carbide family or carbide-based powders are used. The coatings obtained by cladding with expensive tungsten carbide, (wolfram carbide) WC [1-4], have the highest resistance indices at abrasive wear. However, the carbides of the transition metals, belonging to 4 – 6th groups of the chemical elements of the periodic system, possess high hardness and yield to other elements in wear-resistant indices insignificantly. The use of refractory carbides, such as TiC, CrC, TaC, VC, MoC, NbC, as surfacing materials seems a more rational solution [5-12].

Nowadays there is a great number of ways of coating formation using a cladding technique. The best prospective methods refer to cladding with the use of high-energy sources of heating: plasma, laser and electron beam [13-18]. A combination of high productivity of the process of non-vacuum electron-beam cladding along with the possibility of processing large-sized products distinguishes this method from others. Per one beam pass the coatings with thickness of several millimetres are formed. Besides, during inhibition of relativistic electrons a large amount of heat, sufficient for melting of virtually any materials, is emitted, which is why, at further cooling, the synthesis of carbides in the molten pool becomes possible [9].

The aim of the present paper is to study structural-phase transformations, occurring at formation of molybdenum-containing coatings realised by the method of non-vacuum electron-beam cladding.

2. Materials and methods

Molybdenum carbides are distinguished by their increased hardness (14 GPa) and melting temperature (2700 °C). At high energy impact as a result of non-vacuum electron-beam cladding a synthesis of molybdenum carbides is possible. Besides, molybdenum, present in steel, enhances steel viscosity, which decreases the embrittlement influence of the cladded coatings. Therefore the powders of molybdenum and graphite have been chosen as main alloying elements for the cladding process.

Samples made of 40X steel with dimensions of 50×100×12 mm have been chosen as a basis. The powder mixture, consisting of molybdenum, graphite, iron, and flux – in one case, and molybdenum, graphite, iron, and flux – in the other, was applied to a preliminary smoothed surface of the sample. The pouring density of the powder mixtures amounted to 0.33 g/sm^2. The ratio of the molybdenum content to the carbon content in the mixture corresponded to 1:1 (at.). The iron powder was introduced into the cladding mixture in order to increase moistening of molybdenum particles in the molten pool in the amount of 30 % wt. The CaF_2 flux was used for protection against oxidation in the process of cladding in the amount of 40 % wt. The chemical composition of the original materials is shown in Table 1.

Non-vacuum electron-beam cladding was conducted at the Institute of Nuclear Physics of the Siberian Branch of Russian Academy of Sciences (SB RAS) named after G I Budker (Novosibirsk) using the electron accelerator, ELV-6. The electron-beam processing of the powder mixtures was realised in a scanning mode. The electron beam energy amounted to 1.4 MeV and the beam current was 28 mA. The sample moved relatively the beam in the direction, which is perpendicular to the plane of beam scanning, at the rate of 1 sm/s.

Table 1 – Chemical composition of original materials, wt. %.

	C	Mn	Si	P	S	Ni	Cr	Cu	Fe	Mo	O
Steel 40X	0.41	0.86	0.32	0.03	0.02	0.17	1.11	0.16	96.92	-	-
Powder Mo	-	-	-	-	-	-	-	-	-	90.40	9.60
Powder C	100	-	-	-	-	-	-	-	-	-	-
Powder Fe	-	-	-	-	-	-	-	-	98.61	-	1.39

A structural analysis of materials was conducted using the scanning electron microscope, Carl Zeis EVO 50 XVP. The structure was revealed using a three-percent solution of nitric acid in ethyl alcohol. The element composition of the coatings was determined by the energy-dispersive analyser, INCAX-ACT (OxfordInstruments). For determination of the phase composition of the cladded coatings the ARL X`TRA diffractometer was used. The diffraction patterns were registered in increments of 0.05 ° at accumulation time of 3 s.

Durometric properties of coatings, interface regions and the base metal were determined with the use of the microhardness tester of Wolpert Group 402MVD type. The indentation load amounted to 0.98 N. The microhardness measurements were conducted on the basis of the crosscut sections in the direction from the coating surface to the base metal.

3. Results and discussion

The conducted metallurgical studies have not revealed any cracks, pores in the coatings. The modified layers have the thickness of ~ 2.3 mm at cladding of 'Mo-C' and 'Mo-C-Fe' powder mixtures. Delamination signs along interfacial area 'coating-base metal' were not registered (Figure 1). Additional introduction of the iron into the cladding mixture has reduced the concentration of the alloying elements (molybdenum and carbon) in the alloyed layers. As a result the interfacial area has become more blurred (Figure 1B).

The structure of the cladded layer represents an iron matrix, which grains have a dendrite shape. The eutectic, consisting of alpha-iron carbides, excreted along the dendrite boundaries (Figure 2). The volume fraction of eutectic excretions for the coating, obtained by cladding with the molybdenum powder mixed with graphite, amounted to 27.7 %. Hereupon, the reduction in the concentration of the alloying elements in the coating has affected the decrease of the volume fraction of eutectic up to 12.3 %.

Using the method of the X-ray phase analysis one has fixed four phases: α-Fe, γ-Fe, Mo3Fe3C and Mo6Fe11C5 in both coatings. It is necessary to note that the peaks corresponding to γ-Fe phase in the layer, clad by the 'Mo-C-Fe' powder mixture, are lower in comparison with the second coating. In the 'Ti-C' coating high-alloyed martensite crystals, surrounded by the residual austenite, have grown across the whole volume of dendrites (Figure 2 A).

In the volumes of eutectic in addition to iron and complex carbides, containing iron and molybdenum, there is also oxygen, detected by the X-ray spectral microanalysis method (Figure 2, Table 2). It should be mentioned that the oxygen was detected during the elemental analysis of the initial molybdenum powder, where the oxygen was found probably in the form of a molybdenum oxide (Table 1). During the process of dendrite formation the oxygen was forced out on the grain boundary.

An introduction of the iron into the cladding mixture has resulted in reduction of molybdenum in eutectic. At the same time the quantitative analysis has shown that the amount of carbon has increased (Table 2). These results indicate that in this type of coating the volume fraction of the $Mo_6Fe_{11}C_5$ phase has increased.

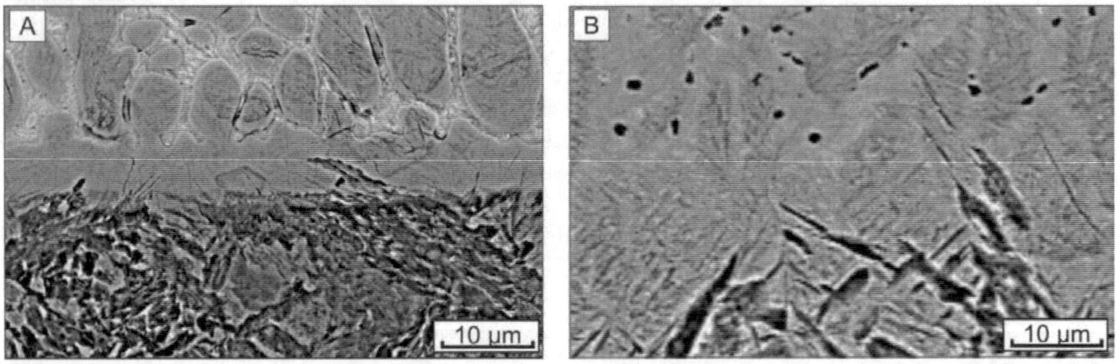

Figure 1. Interfacial area of the coating – a base metal for the materials obtained by cladding of the powder mixtures of 'Mo-C' (A) and 'Mo-C-Fe' (B).

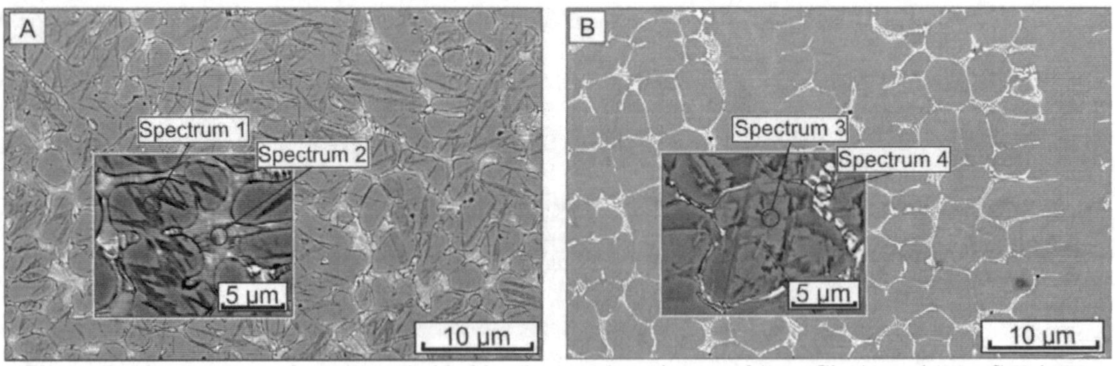

Figure 2. The structure of coatings cladded by the powder mixture of 'Mo-C' (A) and 'Mo-C-Fe' (B)

Table 2 – Elemental composition of local zones of molybdenum coatings.

Element / Zone	Fe, at. %	Mo, at. %	C, at. %	Cr, at. %	O, at. %
Spectrum 1	78.68	6.53	13.76	1.02	-
Spectrum 2	42.12	22.83	20.78	1.00	13.27
Spectrum 3	83.74	2.59	12.84	0.83	-
Spectrum 4	48.19	15.16	22.02	1.81	12.82

The coating formed on steel by means of electron-beam cladding of the 'Mo-C' powders possesses a microhardness of ~ 9 GPa (Figure 3). The decrease of the volume fraction of eutectic in the molybdenum-containing coating from 27.7 to 12.3 % has influenced the reduction of the microhardness of up to 4.5 GPa. However, in the second case the hardness of the cladded layer is more than two times higher than the hardness of the base material.

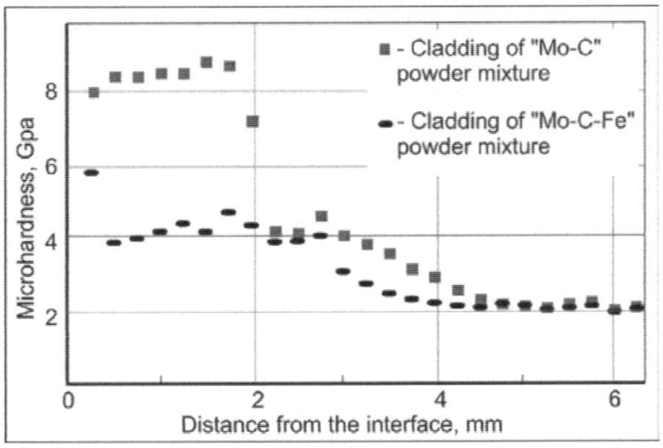

Figure 3. Microhardness of the materials obtained by means of electron-beam cladding with molybdenum-containing powders.

4. Conclusions

The method of non-vacuum electron-beam cladding of molybdenum-containing layers on steels shows the effectiveness of obtaining of the coatings with the thickness of ~ 2.3 mm with enhanced microhardness. These coatings are remarkable for absence of macroscopic defects in the form of pores and cracks. Structural studies of the modified layers have shown that the introduction of iron into the cladding mixture is accompanied by blurring of the interfacial area – 'coating-base metal', but, at the same time, it affects the decrease of the eutectic volume fraction. Electron-beam cladding of the 'MO-C' powder mixtures ensures enhanced microhardness of the material up to 9 GPa.

Acknowledgements

This work has been supported by the Russian Foundation for Basic Research (Contract No. 15-33-50993).

Reference

[1] Gao Y, Jing Y, Zhang Z, Chen G and Xue Q 2002 *Wear* **253** 563–71
[2] Gnyusov S F and Tarasov S Y 2014 *Appl Surf Sci* **293** 318–25
[3] He J and Schoenung J M 2002 *Surface and Coatings Technology* **157** 72–9
[4] Zhang G S, Xing J D and Gao Y M 2006 *Wear* **260** 728–34
[5] Geng Z D, Wu H B, Zhao A M, Wang X P, Zhang Y and Cheng J Y 2015 *Gongcheng Kexue Xuebao* **37** 905–12
[6] Kartavykh A V, Gorshenkov M V, Danilov V D, Tcherdyntsev V V and Andreev N V 2015 *Tribology International* **90** 270–7
[7] Ma H, Zhou P and Du J 2015 *Jinshu Rechuli* **40** 159–63
[8] Qin S, Liao B, Xiao F and Mao L 2015 *Mater Manuf Process* **30** 116–21
[9] An G and Liu G 2011 *Rare Metals* **30** 262–6
[10] Hu J, Liu G, Yan W and An G 2009 *Xiyou Jinshu* **33** 196–9
[11] Zhang N, Wang S C, Jiang X, Qu S L and Chen H 2014 *3rd International Conference on Energy Materials and Material Application, EMMA 2014* 7–10
[12] Lee J, Euh K, Lee S and Kim N J 2001 *Curr. Appl. Phys.* **1** 467–71
[13] Ding Q, Wang L and Hu L 2015 *Laser Surface Engineering,* ed J L G Waugh: Woodhead Publishing) pp 405–22
[14] Dobrzański L A, Tański T, Dobrzańska-Danikiewicz A D, Jonda E, Bonek M and Drygała A 2015 *Laser Surface Engineering,* ed J L G Waugh: Woodhead Publishing) pp 3–32

[15] Fadeev S N, Golkovski M G, Korchagin A I, Kuksanov N K, Lavruhin A V, Petrov S E, Salimov R A and Vaisman A F 2000 *Radiation Physics and Chemistry* **57** 653–5
[16] Abboud J H and West D R F 1992 *Journal of Materials Science Letters* **11** 1479–82
[17] Bataev I A, Bataev A A, Golkovsky M G, Teplykh A Y, Burov V G and Veselov S V 2012 *Surface and Coatings Technology* **207** 245–53
[18] Bourithis L and Papadimitriou G 2003 *Materials Letters* **57** 1835–9
[19] Lee J, Euh K, Oh J C and Lee S 2002 *Materials Science and Engineering A* **323** 251–9

The use of titanium alloys for details of downhole hammers

A I Popelyukh[1], A A Repin[2], S E Alekseev[2], N V Martyushev[3] and Yu Yu Drozdov[3]
[1] Novosibirsk State Technical University, 20, Karla Marksa ave., Novosibirsk, 630073, Russia
[2] Chinakal Institute of Mining, Siberian Branch, Russian Academy of Sciences
[3] Tomsk Polytechnic University, 30, Lenina ave., Tomsk, 634050, Russia

E-mail: aip13@mail.ru

Abstract. The influence of cementation technology of titanium alloy Ti-Al-Mn on its wear resistance is studied. It is established that after lubrication a friction pair with mineral oil the wear resistance of the cemented titanium alloy is comparable to wear resistance of the tempered steel 12HN3A, and in water medium surpasses it by 1.5 times. Decrease in the tendency to seizure with steel is the main reason for increase of wear resistance of titanium alloy. Industrial tests of the ASH43 hammer have shown that the use of titanium alloys for the manufacture of hammer strikers allows to increase impact capacity by 1.5 times and to increase drilling rate by 30 % compared to hammers with steel strikers.

1. Introduction

At present the most efficient, cost-effective and productive drilling methods are based on the use of downhole hammers. The hole hammers must be as compact as possible to accommodate the limited space, and herewith to effectively fracture the rock, they must have a considerable impact capacity. One of the most promising materials for vital parts of impact mechanisms are high-strength titanium alloys [1, 2]. The cost of titanium strikers is higher than of the steel strikers, however, they have lower specific gravity and high corrosion resistance [3]. This allows to increase the speed and the frequency of impact loading, as well as to use water as an energy resource medium. Titanium alloys have a high attenuation of acoustic waves and effectively extinguish reflected impact pulses [4].

The major disadvantage of titanium alloys is low wear resistance in the friction pair with steel. It is possible to increase the wear resistance of titanium alloys by rational choice of the conjugated component material, for example to produce it from brass, or by modifying the surface of a titanium alloy. The most rational method of modifying the surface of the impact machine parts is a thermochemical treatment. Unlike coating processes it ensures high adhesive strength and reliability of components under dynamic load. Nitriding, cementation and oxidation are the most rational technologies of titanium alloy surface modification used in industry [5]. The main disadvantages of nitriding and oxidation of titanium alloys are small thickness of the hardened layer (10…30 μm) and a significant reduction in the resistance to fatigue failure. A hardened diffusion layer with the thickness of 80…100 μm can be obtained by cementation, so this process is best suited to improve the wear resistance of titanium parts, working with the steel in friction pair. The purpose of the work was to evaluate the effectiveness of the cementation process to improve wear resistance of parts made of titanium alloys and to identify the possibility of using titanium alloys for the production of impact mechanism strikers.

2. Materials Methods and Results of Research

A high-strength titanium alloy OT4-1, which is the most manufacturable deformable titanium alloy, was used as a material for studies. It has high fracture toughness (K_{IC} = 106 MPa · √ m) and exceeds the majority of high-strength titanium alloys by this measure [6]. The alloy's chemical composition is shown in the Table 1. For increase of wear resistance indicators the surface of a titanium alloy was subjected to cementations by the following modes: heating of samples in hard carburizer up to the temperature of 900 °C with soaking at this temperature for 3 hours and the subsequent cooling together with the furnace. Charcoal with addition of 20…25 % $BaCO_3$ and $CaCO_3$ was used as carburizer. Samples of steel 12XN3A were used for comparison of mechanical and tribotechnical properties of the modified titanium alloy; such steel is widely applied for production of details of impact mechanisms. Samples of steel 12XN3A were subjected to cementation and hardened with tempering to HRC 40 hardness.

Static tension tests were carried out on flat proportional samples with use of Instron 3369 complex. Impact strength was defined on impact pendulum-type testing machine MetroCom on samples of 10x10x55 mm in size with the concentrator of U-type at the maximum energy of impact of 300 J. The wear resistance assessment of friction couples was carried out by the method of crashing indenter in conditions of sliding friction according to the 'disk – plane' scheme. The essence of the method is to wear a sample of the material by rotating disk (indenter) with subsequent determination of the volume of worn-out material. The friction distance was 706 m. Rectangular samples of 35×10×4 mm were studied. 3 samples were tested in each series. Relative wear resistance was calculated as the ratio of mean value of wear volume from reference sample to mean value of wear volume amount on the surface of test sample. Discs with a diameter of 50 mm and a width of 10 mm made of hardened steel 45 (HRC 50) or brass L63 were used as indentors. Rotating speed was 300 revolutions per minute. The load on the friction pair was equal to 100 N. Tests were carried out in water and mineral oil M8B. For metallographic research light microscope Carl Zeiss Axio Observer A1m operating in the magnification range from 25 to 1000 was used.

Table 1 – Chemical composition of used materials [wt. %]

Material	Al	C	Cr	Cu	Fe	Mn	Mo	Ni	Si	Zn	Ti
OT-4	1.822	0.007	0.015	0.003	0.079	1.981	0.003	0.014	0.011	0.0007	Rest
Steel 12XN3A	0.02	0.13	0.78	0.07	Rest	0.42	0.03	3.06	0.27	-	-
Steel 45	0.02	0.45	0.01	0.03	Rest	0.44	0.01	0.03	0.21	-	-
Brass L63	-	-	-	Rest	0.13	-	-	-	-	28.38	-

Metallographic studies showed that after cementation a hardened layer with the depth of 350...400 μm is formed on the surface of the alloy OT4-1 (Figure 1). Near the surface it consists of equiaxed grains with sizes 50…70 μm. In the bulk of a sample the structure is of lamellar type representing the plates of alpha phase of titan separated by thin layers of beta phase. The maximum value of microhardness of the cemented layer near the surface reaches 700…720 HV, and decreases to 325 ….350 HV with moving away from the surface of titanium alloy. Mechanical testing of OT4-1 alloy showed that its strength is 2 times lower in comparison with steel 12XN3A (Table 2). However, titanium alloy has higher ductility and impact toughness.

Table 2 – Mechanical properties of steel 12XN3A (HRC 40) and alloy OT4-1

№	Property	Steel 12XN3A	Alloy OT4-1
1	Ultimate tensile strength, MPa	1180	543
2	Yield stress, MPa	1062	508
3	Relative elongation, %	14	17
4	Impact toughness, J/sm^2	59	81

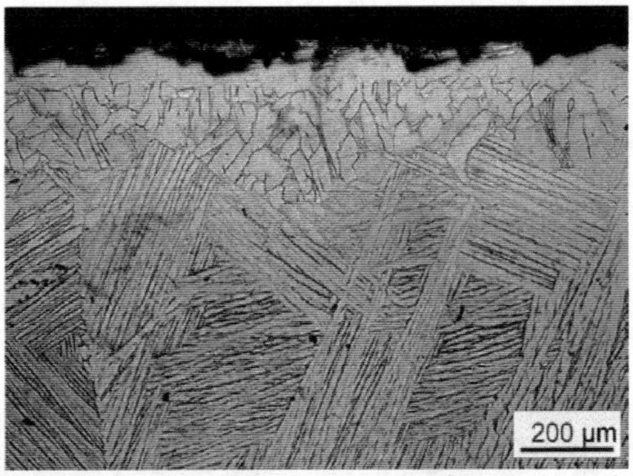

Figure 1. Structure of surface layer after cementation.

Wear resistance is one of the most important indicators of the structural strength of impact machine parts. Surface wear of friction pair "striker – body" leads to pressure reduction in the working chamber of impact machine, reduction of the speed and frequency of loading and, consequently, to decrease of impact capacity and efficiency of impact mechanism. Tribological testing showed that friction pair of 'titan – steel' has low properties, both in the conditions of friction in water, and when using mineral oil as the lubricating environment. In a friction pair with steel 45 the wear resistance of alloy OT4-1 is 50…100 times lower comparing to steel 12HN3A. Cementation of titanium alloy surface can significantly increase the resistance to wear caused by friction with steel. Being lubricated with mineral oil the wear resistance of cemented titanium alloy is comparable with the wear resistance of hardened steel 12HN3A, and it is 1.5 times higher in water medium. A friction pair of 'cemented titanium alloy – brass' has the highest thibological properties. In this case, the wear resistance of titanium alloy is 4 times higher than the wear resistance of steel 12HN3A operating in the friction pair with steel 45.

Table 3 – Tribological properties of different friction pairs

Sample material	Counterbody material	Medium	Friction ratio	Volume of wear crater, mm^3	Relative wear resistance
Steel 12XN3A	Steel 45	Water	0.39	0.119	1.00
Titanium alloy OT4-1	Steel 45	Water	0.37	6.197	0.02
Titanium alloy OT4 (cemented)	Steel 45	Water	0.15	0.074	1.61
Titanium alloy OT4	Brass L63	Water	0.17	0.080	1.50
Titanium alloy OT4 (cemented)	Brass L63	Water	0.12	0.068	1.75
Steel 12XN3A	Steel 45	Oil	0.16	0.044	1.00
Titanium alloy OT4-1	Steel 45	Oil	0.20	7.600	0.01
Titanium alloy OT4 (cemented)	Steel 45	Oil	0.13	0.040	1.10
Titanium alloy OT4-1	Brass L63	Oil	0.13	0.018	2.41
Titanium alloy OT4 (cemented)	Brass L63	Oil	0.14	0.011	4.02

Research of wear crater topography of titanium alloy OT4-1 without cementation showed that a developed relief with scratches of 7 μm in depth and local regions of seizure with counterbody material (Figure 2a) is formed on the surface. The average value of the surface roughness of wear craters calculated by measuring ten points is Ra 2.2 μm. On the surface of counterbody material (steel disk) a significant surface damage associated with the contact sticking and adhesion of weld-in particles of titanium alloy is found. Wear crater surface of titanium alloy after cementation has smoothed form with average roughness Ra 0.26 μm (Figure 2b). There were no significant traces of adhesive wear on the steel disk surface. Thus, the reducing of tendency to seizure is assumed to be the main reason for increasing of titanium alloy wear resistance.

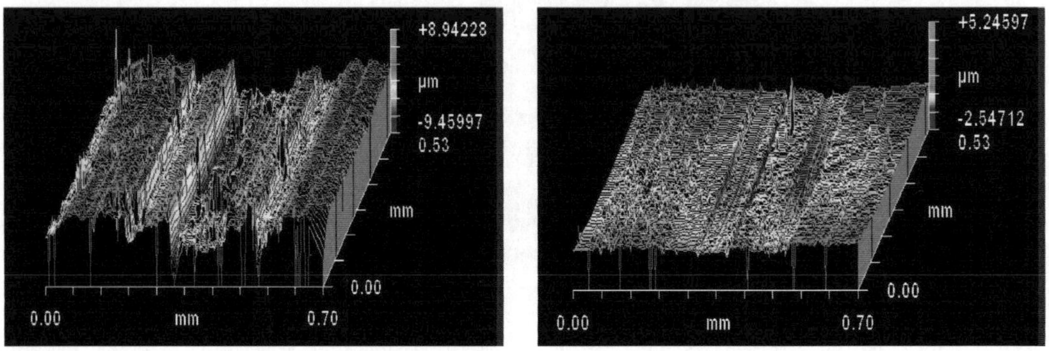

a b

Figure 2. 3D-image of wear crater surface of titanium alloy OT4-1 Without cementation (a); after cementation (b).

Two samples of small-sized hammer ASH43 were manufactured to evaluate the effectiveness of application the cementation process for details made of titanium alloys and the possibility of their usage for the production of the strikers of impact machines (Figure 3). It allows drilling holes with diameter of 46 mm. In one hammer the striker was made from steel 12XN3A and had the weight of 365 g. The striker of another hammer was made of titanium alloy OT4-1 with subsequent cementation. Its weight was 205 grams. Industrial tests showed that at equal energy of a single impact the power of hammer with steel striker was 299 W. In this case the speed of drilling of granite was 70 mm/min. The power of hammer with striker made from titanium alloy was 444 W, and drilling speed was 90 mm/min. The conducted researches showed efficiency of power increase of pneumatic impact hammers due to decrease of striker weight.

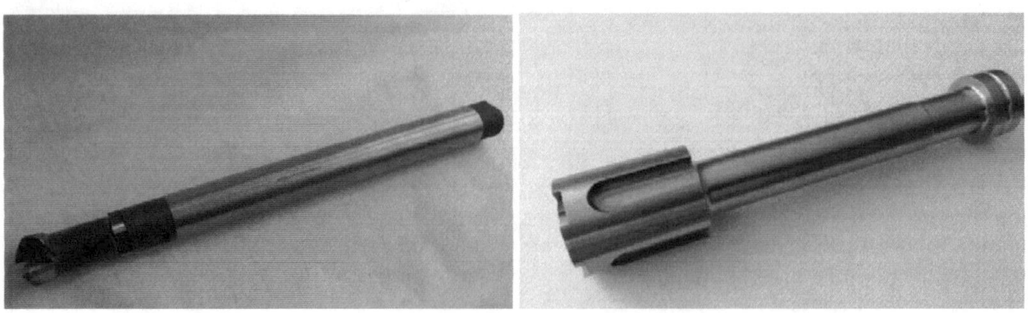

a b

Figure 3. Pneumatic impact hammer ASH43 (a) and striker made of OT4-1 alloy (b).

4. Conclusions

1. As a result of cementation a hardened layer having thickness of 350...400 μm with the highest microhardness near the surface 700...720 HV is formed at the surface of the titanium alloy.
2. Cementation of the titanium alloy OT4-1 is an effective way of improving the wear resistance in the friction pair with steel. Being lubricated with mineral oil the wear resistance of cemented titanium alloy is comparable with the wear resistance of hardened steel 12HN3A, and it is 1.5 times higher in water medium. The reducing of tendency to seizure is assumed to be the main reason for increasing of titanium alloy wear resistance in the friction pair with steel.
3. Results of industrial tests of the hammer ASH43 have shown that the use of titanium alloys for the manufacture of hammer strikers allows increasing impact capacity for 1.5 times and increasing of drilling rate for 30 % in comparison to hammers with steel strikers.

References

[1] Oparin V N, Freidin A M, Tapsiev A P, and Filippov P A 2013 *Journal of Mining Science* **49(4)** 670–676
[2] Repin A A, Alekseev, S E, Popelyukh A I. 2012 *Journal of Mining Science* **48 (4)** 669–674
[3] Leyens C, Peters M 2003 Titanium and Titanium Alloys. Fundamentals and Applications. Wiley-VCH Verlag GmbH & Co. p. 513
[4] Moiseyev V N 2006 Titanium Alloys- Russian Aircraft and Aerospace Applications. Taylor and Francis, LLC. p. 196.
[5] Matthew Jr, Donachie. 2000 Titanium. Technical Guide. ASM International. p. 366.
[6] Ilyin A, Kolachev B A, Pol'kin I S, 2009 Titanium alloys. The composition, structure, properties. Guide. M.: VILS-MATI. p. 520

The influence of non-vacuum electron-beam facing on the structure of Ti-Ta layers formed on the surface of VT1-0 alloy

V V Samoylenko[1], O G Lenivtseva[1], I A Polyakov[1], I S Laptev[1] and N V Martyushev[2]
[1] Novosibirsk State Technical University, 20, Karla Marksa ave., Novosibirsk, 630073, Russia
[2] Tomsk Polytechnic University, 30, Lenina ave., Tomsk, 634050, Russia

E-mail: samoylenko.vitaliy@mail.ru

Abstract. The influence of electron-beam facing modes on the structure of Ti-Ta layers formed on the surface of commercially pure titanium VT1-0 has been studied in the paper. The mode of the electron-beam treatment of alloying powder mixture, by which there were no defects in the pad, has been identified. The methods of optical microscopy and scanning electron microscopy have shown that in pads there is dendritic segregation typical for the process of initial crystallisation. At greater magnifications it is possible to observe a structure of the laminar type. The X-ray phase analysis of titanium-tantalum layers justifies the presence of two phases: a hexagonal α'-phase and a cubic β-phase of titanium.

1. Introduction

At present titanium and its alloys are widely used in different branches of industry [1-3]. An excessive interest in this material is conditioned by its enhanced set of mechanical characteristics in combination with a low rate of corrosive attack in most of the corrosive solutions. However titanium alloys are subjected to significant corrosion in boiling solutions of strong acids which does not allow its use in the manufacture of machine components and high-duty structural elements operating in similar conditions [4, 5]. In order to increase the corrosion resistance of titanium to the influence of boiling acid solutions it is alloyed with such elements as tantalum, niobium and molybdenum [5, 7]. Titanium-molybdenum alloys possess high corrosion resistance in boiling solutions of sulphuric and hydrochloric acids [8]. Nevertheless these alloys are subjected to significant corrosion in boiling solutions of nitric acid, the rate of which increases with the rise of molybdenum concentration [5]. The introduction of niobium and tantalum in titanium enables the increase of corrosion resistance of the alloy in boiling nitric acid [7, 9]. At that the greatest effect is achieved if titanium is alloyed with tantalum [5]. It is known that introduction of a small amount of tantalum into the titanic alloy facilitates a significant decrease of its corrosion rate in boiling acids [6]. When introducing tantalum in the amount of 40 % wt and more the corrosion resistance of the alloy in 80 % sulphuric acid, heated up to the temperature of 75 °C, becomes comparable to commercially pure tantalum [10]. An essential drawback of tantalum is its high cost and high density [11]. In order to save costly tantalum and to increase corrosion resistance of titanium, surface alloying of titanium alloys with tantalum is a reasonable. It should be noted that the obtainment of tantalum-containing layers on the surface of titanium workpieces does not lead to a significant increase in the weight in comparison with three-dimensionally alloyed workpieces.

The studies undertaken earlier indicate high efficiency of the method of non-vacuum electron-beam treatment during formation of surface-alloyed layers on titanium workpieces [12-14]. For the realisation of this technology the industrial electron accelerators providing extraction of electron beam into the air atmosphere can be used. Such accelerators include, for example, the electron accelerator 'ELV-6' developed by the Institute of Nuclear Physics after G I Budkera, Siberian Branch of the Russian Academy of Science. In papers [14-17] the facing of tantalum-containing layers was applied to the workpiece 50 mm wide by means of the electron beam. The increase in the amplitude of scanning of the electron beam was achieved due to installation of the electro-magnetic scanner on the exhaust diaphragm of the accelerator. However Ti-Ta facing on the surface of titanic workpieces without using the electro-magnetic scanner (in a track mode) has not been applied earlier. In this connection the purpose of this paper is to study the influence of the modes of electron-beam alloying of commercially pure titanium VT1-0 with tantalum at treatment in a track mode.

2. Materials and research techniques

Tantalum-containing layers were faced on the laminae of commercially pure titanium VT1-0 with dimensions of 100×25×12 mm. The mixture of tantalum, titanium and flux powders (75 % CaF_2 + 25 % LiF) in the amount of 40 % wt of Ta, 24 % wt of Ti, 36 % wt of flux was used as an alloying material. Prior electron-beam treatment the powder mixture was applied to the surface of the titanium workpiece in the amount of 0.45 g/sm^2. Then the workpiece with the formed powder layer was mounted on the movable table, where it moved forward relatively the outlet with the rate of 10 mm/s. The distance from the outlet to the workpiece surface equaled to 123 mm. The initial energy of the electrons in the beam amounted to 1.4 MeV. For the purpose of studying the influence of the facing modes on the structure and properties of the surface layer different values of beam current, which are 7, 8, 9 and 10 mA, were used.

The structural peculiarities of the pads were studied by means of the optical microscope 'Carl Ziess Axio Observer A1m' and the scanning electron microscope 'Carl Zeiss EVO 50 XVP'. The concentration of alloying elements in the surface-alloyed layer was determined by means of the energy-dispersive electron probe microanalyser 'X-ACT' (Oxford Instruments).

The phase composition of the pads was analysed by means of the X-ray diffractometer 'ARL X'TRA'. The diffraction patterns were recorded using $CuK\alpha_1/\alpha_2$ radiation. The accumulation time for one point amounted to t = 3 s, a scanning pitch was $\Delta 2\theta = 0.05$ °.

3. The results and discussions

By means of the methods of optical metallography it has been established that the minimum thickness of the pad corresponds to the sample obtained at 7 mA current, which amounts to 0.6 mm (Figure 1a). The increase of current of the electron-beam treatment enables thickness growth of the surface-alloyed layer. Thus for samples faced at current of 8 and 9 mA the layer thickness equals 1.2 and 2.0 mm correspondingly (Figure 1b, c).The maximum thickness of the pad is fixed for samples obtained at the electron beam current of 10 mA and is equal to 2.5 mm (Figure 1d).

In the structure of the surface-alloyed layer faced at 7 mA current there are pores as well as unfluxed tantalum particles. The increase in the current of electron-beam treatment up to 8 mA leads to formation of a structure without any pores. However there is a small amount of unfluxed tantalum particles. The structure of the layers formed by the electron beam at currents of 9 and 10 mA has a homogeneous and defect-free structure. The difference between the layers obtained by facing at these points consists in the chemical composition. The titanium-tantalum alloy formed on the titanium surface at 10 mA current is less alloyed with tantalum in comparison with the sample obtained at 9 mA current as a result of a greater depth of melting of the base material and dilution of the pad with titanium. The results obtained by the method of the electron probe microanalysis justify this fact (Table 1).

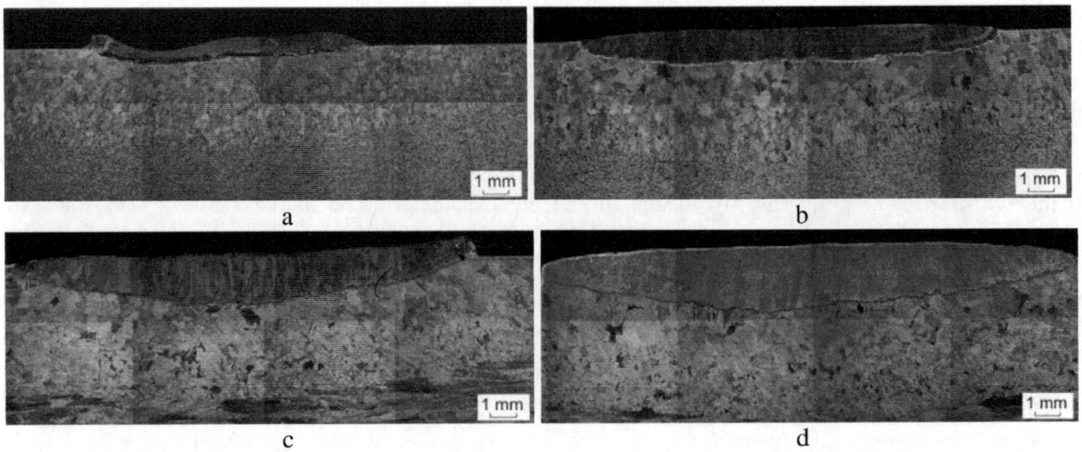

Figure 1. Micrographs of the layers obtained by means of non-vacuum electron-beam facing of Ti-Ta powder mixture using currents of 7 (a), 8 (b), 9 (c) and 10 mA (d).

The tantalum concentration in the layer formed at 9 mA current amounts to 23.2 % wt. The increase of the current of the electron beam up to 10 mA contributes to reduction of tantalum content in the pad up to 15.5 % wt. Thus in order to obtain defect-free coatings with maximum tantalum concentration we should use treatment with the beam current of 9 mA.

Table 1 – Chemical composition of the surface-alloyed layers faced by the electron beam at currents of 9 and 10 mA

Current intensity	Content, % wt.	
	Ta	Ti
9 mA	23.2	76.8
10 mA	15.5	84.5

The structure of the pad formed by the electron beam at 9 mA current is represented in figure 2. In the pad there is a significant dendritic segregation typical for the materials obtained in the process of the primary crystallisation (Figure 2a). Dendritic crystals grow from the boundary containing unfluxed base material in the direction of the surface. During etching the former grain boundaries of a β-phase emerge against the dendritic background. The grains have an elongated shape and are comparable in size to dendritic crystals. The study of the material at greater magnifications has shown that the material has a structure of a lamellar type (Figure 2b). A similar structure results at cooling from β-area of titanium alloys alloyed with beta-isomorphic stabilizers. At this the cooling rate must be insufficiently high for fixing a pure α' or α'' phase.

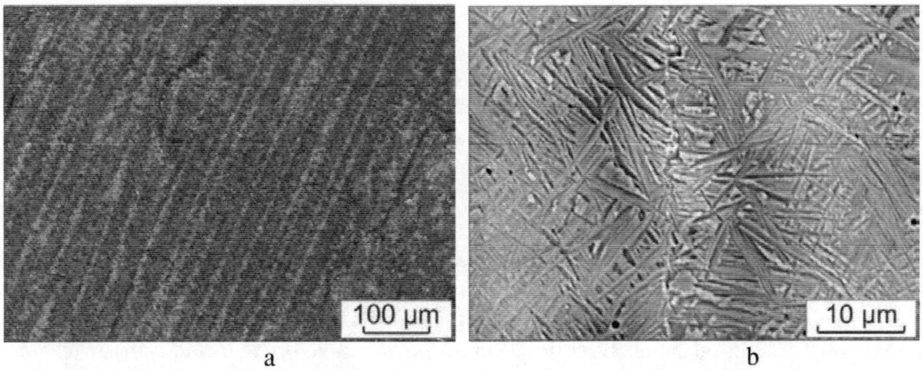

Figure 2. The structure of a Ti-Ta layer obtained by the electron-beam facing of the alloyed powder mixture at 9 mA current.

In order to determine the phase composition of the surface-alloyed layer obtained at the beam current of 9 mA, an X-ray diffraction studies have been conducted. The X-ray phase analysis has shown the presence of peaks of a hexagonal α (or α')-phase and a cubic β-phase of titanium (Figure 3). The analysis of the X-ray picture of the sample is evidence of the difference in intensity of experimental peaks from intensity of the peaks typical for the materials with a random distribution of orientation. Redistribution of peaks intensity is probably connected with texture formation typical for the process of initial crystallisation.

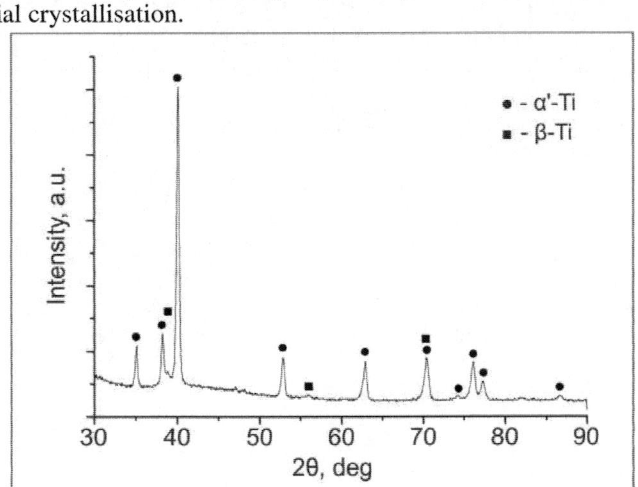

Figure 3. The X-ray picture of Ti-Ta alloy formed by the electron-beam facing at the beam current of 9 mA.

Conclusions

In the course of the work the mode of the electron-beam alloying of titanium with tantalum, providing the formation of the defect-free layer with high concentration of tantalum, has been selected. The surface-alloyed layers obtained by means of this mode are characterised by the absence of such defects as pores and undissolved tantalum particles. During the study of the structure of titanium-tantalum layers a significant dendritic segregation has been identified. It has been stated that a faced alloy has a structure of a lamellar type. The analysis of the x-ray pictures justifies the presence of two phases: a hexagonal α (or α')-phase and a cubic β-phase of titanium.

Acknowledgements
This work has been supported by the Russian Foundation for Basic Research (Contract No. 15-33-50955).

References

[1] Matthea J and Donachie Jr 2000 *Titanium: A Technical Guide. ASM International, Materials Park*, Ohio **381**
[2] Lütjering G and Williams J C 2007 *Titanium (Engineering Materials and Processes). Springer* **442**
[3] Leyens C and Peters M 2003 *Titanium and Titanium Alloys Fundamentals and Applications.* WILEY-VCH Verlag GmbH & Co. KGaA, Weinheim **532**
[4] Raj B, Mudali U K, Prog. Nucl. Energy. 2006 *Materials development and corrosion problems in nuclear fuel reprocessing plants* **48** 283–313
[5] Tomashov N D 1985 *Metallurgy* **80**
[6] Guo D, Yang Y, Wu J, Zhao B, Zhao H, Su H, Lu Y 2013 *J. Electron. Spectrosc. Relat. Phenom* **189** 122–126
[7] Ravi Shankar A, Dayal R K, Balasubramaniam R, Raju V R, Mythili R, Saroja S, Vijayalakshmi M, Raghunathan V S 2008 *J. Nucl. Mater.* **372** 277–284
[8] Shreir L L, Jarman R A and Burstein G T 1976 *Corrosion Metal/Environment Reactions.* Newnes-Butterworths, London.
[9] Kapoor K, Vivekanand Kain, Gopalkrishna T, Sanyal T, De P K 2003 *J. Nucl. Mater.* **322** 36–44
[10] De Souza K A, Robin A 2007 *Mater. Chem. Phys.* **103** 351–360
[11] De Souza K A, Robin A 2003 *Mater. Lett.* **57** 3010–3016
[12] Lenivtseva O., Golovin E., Samoylenko V., Mul D., Golovin D. 2014 *Adv. Mater. Res.* **1040** 784–789
[13] Lenivtseva O G, Bataev I A, Golkovskii M G, Bataev A A, Samoilenko V V, Plotnikova N V 2015 *Appl. Surf. Sci.* **355** 320–326
[14] Golkovski M G, Bataev I A, Bataev A A, Ruktuev A A, Zhuravina T V, Kuksanov N K, Salimov R A, Bataev V A 2013 *Mater. Sci. Eng. A.* **578** 310–317
[15] Ruktuev A, Golkovski M, Samoylenko V, Komarov P, Bataev I, Bataev A 2014 *Adv. Mater. Res.* **1040** 759–763
[16] Ruktuev A A, Samoylenko V V, Golkovski M G 2014 *Adv. Mater. Res.* **682** 100–103
[17] Samoylenko V V, Lazuenko D V, Lenivtsiva O G, Lozhkin V S 2015 *Appl. Mech. Mater.* **698** 424–429

The influence of granulated powder temporary organics composition on ceramic structure and properties

S Veselov, N Belousova, A Zamyatina, I Sokolov and A Felofyanova
Novosibirsk State Technical University, 20, K. Marks ave., Novosibirsk, 630073, Russia

E-mail: belousova_ns@mail.ru

Abstract. The influence of granulated powder binder/plasticizer ratio on epy microstructure and properties of sintered ceramic specimens is shown using submicron alumina. The fracture mode, microstructure and properties analyses of a green part and sintered samples obtained with different organic ratio contained in granulated powder are presented. It was stated that the usage of PVA (grade 16/1) and PEG 400 at an equal ratio results in a uniform distribution of granulated powder in a die and high green density. It is also shown that an addition of 2 wt. % of PVA and 2 wt. % of PEG400 into granules of ceramic press powder allows obtaining Al_2O_3 ceramic with apparent density of about 3,91 g/cm^3 and bending strength of up to 410 MPa.

1. Introduction

A ceramic production technological procedure that contains a dry pressing technique includes temporary organic additives (binders and plasticizers) usage. These additives act as a glue for ceramic particles in granules and can change physical-chemical properties of press powder. Organics exert main influence on the granulated powder behavior during the green part formation. Green part defects related with incorrect organics selection are partially saved in sintered ceramic and significantly affects mechanical properties. However the question related with a proper organic composition selection has not been paid enough attention in the literature and the existing information is of general nature that is applicable to a wide range of organics. The temporary additives composition is determined in relation with raw ceramic powder characteristics and almost each technological stage of ceramic production. Regarding this the revealing optimal composition of organics for widely spread advanced fine-grained alumina ceramic obtained using a dry pressing technique is a very important task for industry.

2. Materials and experimental procedure

Alumina powder brand CT 3000 SG (Almatis) was used as a raw material for ceramic production. According to the preliminary investigation [1] typical particle size characteristics of the powder are d50 < 0.6 μm, d90 < 2.0 μm. Ceramic specimens were produced by using a widely spread manufacturing technology including the following steps: ceramic powder water suspension dispersing, granulation with a spray drying technique, dry pressing and free sintering. Detailed technological parameters of samples preparation and their proofs were published in [2, 3].

At the granulation stage the suspension was mixed with 10 wt. % water solution of PVA brand 16/1 and PEG 400, which were used as a binder and a plasticizer correspondingly. Several series of samples with different binder (PVA16/1)/plasticizer(PEG400) ratios (0:4 for the 1st series, 1:3 for the 2nd series,

2:2 for the 3[d] series, 3:1 for the 4th series, 4:0 for the 5th series) were prepared and analyzed in the paper. Each press powder contained 4 wt. % of the total organic amount.

The granulation process was performed with a Mobile Minor 0.8 (GEA Niro) spray dryer which allows obtaining press powder with a particles size in the range from 10 to 150 μm. The granulated powder behavior was studied on the green cylinders with the height of 10 mm and about 40 mm diameter. The forming uniaxial pressure was 100 MPa. To find out the strength of sintered ceramics the green bars of 5×5×55 mm size were also made. Sample sintering was performed in the LHT 02/17 (Nabertherm) laboratory furnace under 1600 °C temperature during 3 hours of soaking.

The green sample fracture surface and the sintered ceramic microstructure were analyzed with the Carl Zeiss EVO 50 scanning electron microscope. The green density was estimated with geometry and weight measurements. Apparent density measurements were performed in accordance with GOST 2409-95. To reveal mechanical properties sintered ceramic was subjected to bending strength measurements. Mechanical tests were done according to GOST 24409-80 under a 3-point loading condition on an Instron 3369 testing machine.

3. Results and discussion

A preliminary investigation of the organic additives (binder/plasticizer) ratio influence on a ceramic producing process was carried out using the data of the green samples fracture analysis. Figure 1 shows fracture surfaces of the green samples with 1 wt. % and without PVA in the granules. A distinctive feature of these samples is almost total absence of granule boundaries and junctions (Figure 1a). This effect is related with low granules strength that maintains their shape only with capillary forces. Applying any pressure to the granulated powder results in their complete destruction. At the same time low press powder strength results in the occurrence of strong pressure gradients throughout the green sample cross – section and unpressed granules in local areas. It was stated that at

Figure 1. Fracture surface of green ceramics obtained with a small amount of binder in the press-powder: a, b, c – 0 wt. % PVA 16/1 + 4 wt. % PEG 400, d – 1 wt. % PVA + 3 wt. % PEG 400.

the initial stage of forming the granules cannot be redistributed in the die and provide a uniform pressure application throughout the cross – section. As a confirmation of this hypothesis the presence of a small amount of initial granules in the local areas (Figure 1b, c) and macro cracks caused by local over pressing (Figure 1d) were revealed.

The additional confirmation of this effect was obtained with a compaction curve analysis that was made for the first and the second series of samples (Figure 2). According to the theory [4] at the initial forming stage, it is important to provide press powder redistribution in the die, which is accomplished with a horizontal area at the compaction curve. If the quantity of the organic binder in the press powder is not enough its granules start to collapse immediately under any load, and this effect is accomplished with a gradual density increase on the plot.

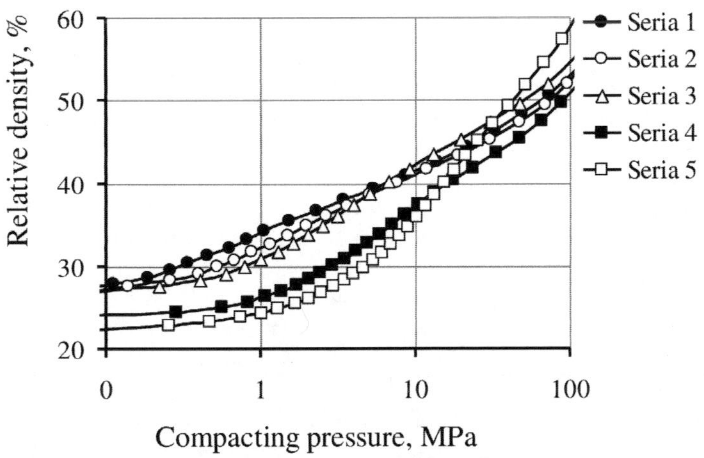

Figure 2. Die-fill density and compaction response for granulated powder with different organic ratios.

The green sample gradient density and macro cracks exert influence on the microstructure and properties of sintered ceramic. Figure 3 shows the structure of local areas for the samples of the 1…2 series. According to the obtained results, the sintered samples contain such defects as boundaries and whole granules in the structure.

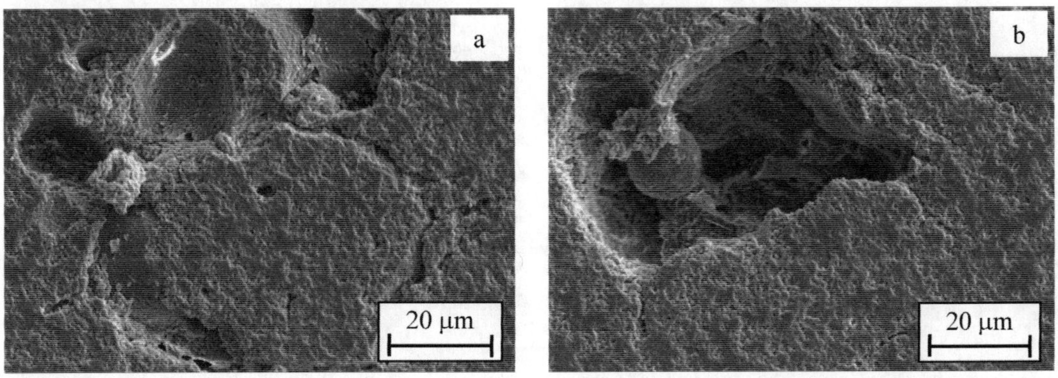

Figure 3. Local areas microstructure of sintered ceramic that does not contain PVA in the press.

The absence of the granule rearrangement process also influence the ceramic properties distribution. If the powder occasionally filled the die uniformly the ceramic bending strength can reach

330 MPa with the density of about 3.93 g/cm^3 (Table 1, series 2). However the probability of a such state is very low and most of the time samples strength will be no more than 220 MPa with apparent density of about 3.84 g/cm^3 (Table 1, series 1).

Table 1. Green and sintered ceramic properties.

Series N	PVA, wt.%	PEG 400, wt.	Total amount of organic, wt. %	Green density, g/cm^3	Apparent density, g/cm^3	Bending strength, MPa
1	0	4		2.18	3.84	220
2	1	3		2.17	3.93	330
3	2	2	4	2.18	3.91	410
4	3	1		2.12	3.81	280
5	4	0		1.96	3.61	260

The analysis of compaction curves for samples with more than 2 wt. % PVA in the powder indicates the occurrence of the granules rearrangement process at the initial forming stage. The greater the fracture of the binder in the granulated powders, the more evidently the horizontal area is seen on the compaction curves (Figure 2, serial 3...5). If the powder contains 4 wt. % of PVA than rearrangement lasts up to 2 MPa formation pressure. When applying a higher load to the samples of the fifth series, a linear growth of green density is shown. According to the literature [4] this region characterizes granule deformation in the green body.

However, the fracture surface analysis of the 3…5 series samples shows that the more fraction of the binder in the press powder, the greater is the volume of undestroyed granules in the green part (Figure 4a, b). The presence of initial granules negatively influences green and sintered sample properties. It was stated that at an unloading stage of the forming procedure the springback effect in green samples appears due to elastic deformation. The measured density of green samples is lower

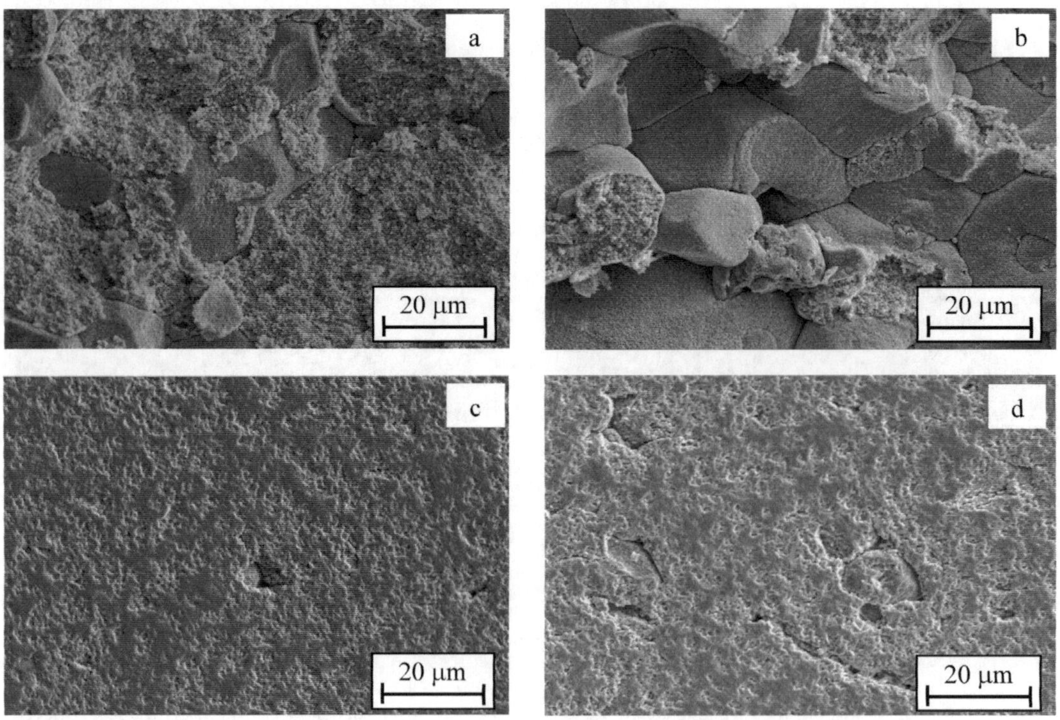

Figure 4. Green part fracture surface and sintered ceramic microstructure of the 3rd (a, b), 4th (c, d) samples series.

than the calculated one according to the compaction curve (5 series, Table 1, Figure 2). Moreover, a microstructure analysis shows that the more PVA in the initial powder, the greater the volume fraction of defects in sintered ceramic (Figure 4 c, d).

The microstructure analysis results concerning 3 – 5 series of samples were confirmed by apparent density and bending strength examination. It was stated that the more PVA in granulated powder, the less density of green and sintered ceramic (Table 1). An increase of the micro cracks volume fraction in sintered ceramic related with granule boundaries results in strength decreasing. The best bending strength obtained in ceramic of series 3 was about 410 MPa.

4. Conclusion

Thus the development of the binder/plasticizer temporary organic additives ratio should be done in connection with providing the complex behaviour of granulated powder at the shaping stage of ceramic producing. On the one hand, granules should possess enough strength to provide their rearrangement at the initial pressing stage. On the other hand, the quantity of the binder should be less than the defined limit that determines granules destruction at the maximum applied formation pressure.

If one uses polyvinyl alcohol 16/1 as a binder and polyethylene glycol 400 as a plasticizer with 4 wt. % of the total amount of organic in granulated powder, the optimal ratio is 2 wt. % PVA + 2 wt. % PEG 400. At this ratio of organic additives in the press powder, the microstructure of sintered ceramic is characterized as one having a minimum quantity of saved granule boundary defects. Moreover, the flexural strength of this ceramic is about 410 MPa, which is the highest magnitude for the analyzed series.

Acknowledgments

The work was performed at the financial support of the Ministry of education and science of the Russian Federation in the framework of implementation of the complex project devoted to the development of high-tech manufacturing at Novosibirsk State Technical University (contract № 02.G25.31.0060 of 12.02.2013).

References

[1] Belousova N, Goryaynova O and Melnikova E 2014 *Applied Mechan. and Mater.* **682** 32–34
[2] Belousova N, Veselov S, Kuzmin R and Melnikova E 2015 *Appl. Mechan. and Mater.* **698** 513–518
[3] Veselov S, Belousova N, Anufrienko D and Goryainova O 2015 *Appl. Mechan. and Mater.* **698** 472–477
[4] Glass S and Ewsuk K 1997 *BULL* **22(12)** 24–28

Microstructure and mechanical properties of eutectic nickel alloy coatings

V P Bezborodov[1,2] and Yu N Saraev [2]
[1]Tomsk Polytechnic University, 30, Lenina ave., Tomsk, 634050, Russia
[2]Institute of Physics and Strength of Metals, Siberian Division of the Russian Academy of Sciences, 2/4, Acadimicheskiy ave., Tomsk, 634021, Russia

E-mail: val@ispms.tsc.ru

Abstract. The paper discusses the peculiarities of a structure and a coating composition after reflow. It was established that the structure of coatings from nickel alloy is a solid solution based on nickel, the eutectic of γ-Ni+Ni3B composition and dispersed reinforcing particles. The content of alloying elements in the initial powder material determines the type of the coating structure and the formation of hypoeutectic or hypereutectic structures. The influence of formation conditions on the structure and physical-mechanical properties of the coatings is considered in this paper.

Key words: alloy, property, coating strength, hardness, structure.

1. Introduction

At present the plants both in Russia and abroad use alloys for the production of wear- and corrosion-resistant coatings by surfacing or spraying with subsequent melting. These alloys are characterised by: a relatively low melting point (950-1350 °C); efficient wetting of the surfaces of steels and other materials; dissolution of oxide films; the ability to form strong bonds with the substrate metal; high physico-mechanical and service properties, such as strength, hardness, wear resistance in the conditions of abrasive friction and friction with lubrication; chemical resistance in many media and atmospheres; antifriction properties; creep strength and etcetera.

This unique combination of properties of nickel alloys often makes them irreplaceable in protection and handling of components and equipment. Therefore, irrespective of the current large increase in the price of materials produced from nickel alloys, the application of these materials in reconditioning and hardening of expensive components is highly promising. Thus the problem of production of high-quality coatings requires urgent solution. The solution of this problem with special reference to specific components and materials has been treated in a large number of scientific and technical investigations, published in the last couple of decades. However, because of the multifactorial nature of the processes of spraying and surfacing with melting, the existence of a large variety of complicated compositions of materials of nickel alloys the coating formation has been studied insufficiently. In the present work, investigations were carried out into the effect of the main parameters of the process of melting materials produced from nickel alloys of the Ni- Cr-B-Si system (heating temperature and holding time) on the formation of the structure of coatings. In order to select the optimum melting conditions of coatings of different compositions, the experiments were carried out to determine the effect of temperature, time and rate parameters of heating on the structure and strength properties of the coatings and of the 'coating-steel base' composite. The significance of these investigations is conditioned by the difficulties in controlling the influence on the process parameters in the actual conditions of melting coatings on components.

The heating rate and temperature are very important in surfacing and in carrying out operations of melting of the coating. The process of melting by external heat sources (plasma and gas flame torches, laser and electron beams) is characterised by the supply of heat from the side of the surface of the coating. In order to avoid cracking, splashing, dripping and other negative phenomena, formed at high power, the rate and temperature of heating causing delamination and failure of the coatings, it is important to select the melting conditions for any type of treatment.

The holding period of the coatings at the melting temperature has a significant effect on the porosity of the coatings. In case of short period holding (several seconds), the residual porosity may exceed 10 %, especially in case of thick coatings (thicker than 1 mm). This is associated with high viscosity of the melt and inefficient degassing [1]. With an increase of the holding period in melting, the porosity decreases to 1-3 %. In addition to this, in case of rapid growth of the grains of the y-solid solution (up to 60 mm) and in long period holding (3-4 min), the hardness and strength greatly decrease (by 20-30 %). The short holding time makes it possible to fix the maximum dispersion of all structural components of the coatings. However, this is accompanied by extensive heterogeneity of the coating, inherited from the initial powder and the sprayed coating. Most importantly, by regulating the energy effect parameters of the heat source for a specific thickness and composition of the coating, it is necessary to ensure the uniform favourable distribution of temperature in the cross- section. In this case, it is possible to ensure suitable conditions for efficient penetration to the depth (without melting the hardening phases), maximum degassing and prevention of grain growth in the structure or, on the other hand, decrease in the dimensions and number of carbide and boride particles as a result of superheating when exceeding the optimum temperature in heating or the holding period, causing a decrease in the coating hardness.

2. Results and Discussion

In the initial condition, the powder materials of nickel alloys have a structure consisting of disperse grains of the solid solution based on nickel, a fine branched eutectic of the solid solution based on nickel Ni_3B, fine primary carbides of chromium and boron, and borides of nickel and chromium [2-7]. The results of investigations carried out in many studies show that after melting, the structure of the thermally sprayed coatings of nickel, like the structure of the materials of the powders, is represented mainly by the same four groups of the structural components. In many cases, the contradicting nature of the investigation results obtained by various authors in determination of the relationship of the main structural components in the coatings is associated with the factor that the phase and structural composition of the coating of a specific chemical composition depends on the degree of deviation of the coating from the equilibrium condition and the melting conditions (temperature, cooling rate and the cooling medium). The indeterminacy of the structural and phase composition results contradicts an understanding of the formation processes of the thermally sprayed coatings and the production of high-quality coating- substrate compositions. In this work, the investigations were carried out in order to determine the dependence of the volume fraction of the hardening phases on the melting temperature and the holding period at this temperature, and also on the content of alloying elements in the coating alloy. The conditions of coatings formation for producing a specific structure were varied by changing the melting conditions in a furnace, gas-flame, plasma, laser and electron beam methods. The results show that an increase in the content of alloying elements in the alloy increases the content of the eutectic and hardening phases and also the hardness of the gamma solid solution based on nickel as well as the hardness of the entire coating (Figure 1, Table 1). An increase in the content of the eutectic component in the coating is accompanied by a large increase in the hardness of the coating, with a simultaneous small decrease in strength (Figure 2). The volume content of the hardening phases and colonies of the eutectic in the alloys greatly increases after heating to the temperatures in the range of the solidus line of the Ni-Ni_3B system (Figure 3 and 4). At the same time, with an increase in the holding period at the melting point of the coatings, both in the furnace and in plasma heating and also in gas flame heating, the colonies of the eutectic rapidly grow with a relatively small growth of the particles of hardening phases (Figure 5 b)

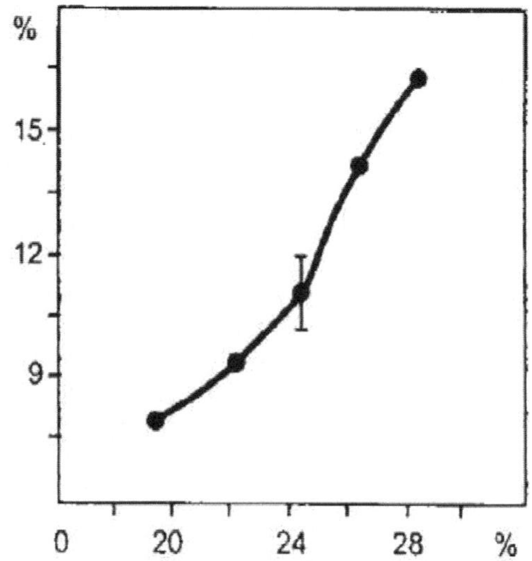

Figure 1. The dependens of the volume fraction of the hardening phases on the content of alloying elements in the coatings

In this work, the results of measurements of the structural parameters of coatings in melting are analysed.

The rate of growth of eutectic colonies and crystals of the carbide and borides is determined by the transformation temperature. In furnace heating, the melting of the coatings starts at the eutectic temperature and takes place in the equilibrium conditions, whereas in rapid heating with an increased heating rate, starting with gas flame, plasma heating and ending with laser and electron beam heating, it is necessary to heat the material to temperatures greatly higher than the eutectic temperature (100-300 °C).

It is well-known[5] that the main role in the formation and growth of the eutectic colonies and the growth of hardening phases is played by the diffusion processes taking place in the liquid phase between the sections of the melt, adjacent to the solid crystals, in our case - Ni_3B and the y-solid solution based on nickel.

Table 1. The content of alloying elements and structural components in the coatings

Coating	Content of allouing elements, %	Content of solid solution, %	Content of eutectic, %	Content of hardening phases, %	Hardness, HRC
PG-12N-01	12.5-23.6	50	43	7	35-40
PG-SR3	17.4-28.5	44	46	10	45
PG-12N-02	18.4-30.8	38	48	14	45-50
PG-10N-01	22.4-32.9	22	51	17	56-62
PG-10N-04	3.8-5.3	82	18	-	18-20

Therefore, the probability of formation of these excess phases and the extent of the eutectic reaction are determined by the degree of deviation of the heating temperature from the equilibrium conditions, the heating and cooling rate, and by supercooling of the alloy as well.

In most cases, melting of the majority of the alloys of the Ni-Cr-B-Si system during heating in the furnace is accompanied by the structure formation of the hypoeutectic of alloy type with the presence of primary phases of the gamma solid solution based on nickel, the presence of eutectic based on this solution and Ni_3B and precipitates of independent particles of Ni_3B and chromium carbides $Cr_{23}C_6$ and Cr_7C_3 (Figure 6 a, b).

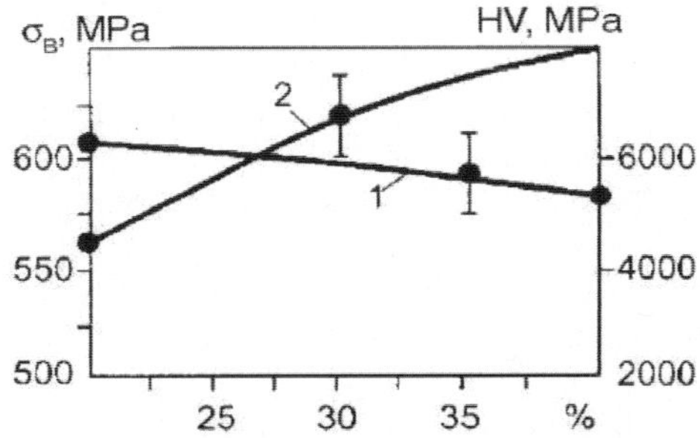

Figure 2. The dependens of the ultimate strength of (1) PG-10N-01-coating steel composite and (2) the hardness of the coating

In this case, depending on the heating temperature and time, as well as the cooling conditions, the resultant alloys may be characterised by the preferential content of a specific structural component of the coating. The formation of the eutectic takes place when the rate of its formation is higher than the rate of formation of the individual phases of the gamma solid solution and Ni_3B. Otherwise, the rapid process of formation of these separate phases does not make it possible for the eutectic colonies to grow.Increasing the heating temperature in the range of the solidus line of the alloys and the holding period at this temperature, the size of the eutectic colonies, formed during cooling, increases and the structure of these colonies may be characterised by a single developed formation penetrating through the entire volume of the coating and forming a unique frame (Figure 6c). This coating is characterised by high hardness and reduced plasticity. This is associated with the fact that the content of the Ni_3B hardening phase in the eutectic is very high and this phase controls the physical-mechanical and service properties of the coatings because the hardness of the eutectic is almost twice the hardness of the y-solid solution based on nickel. Therefore, the properties of the coatings strongly depend on the ratio of the structural components.

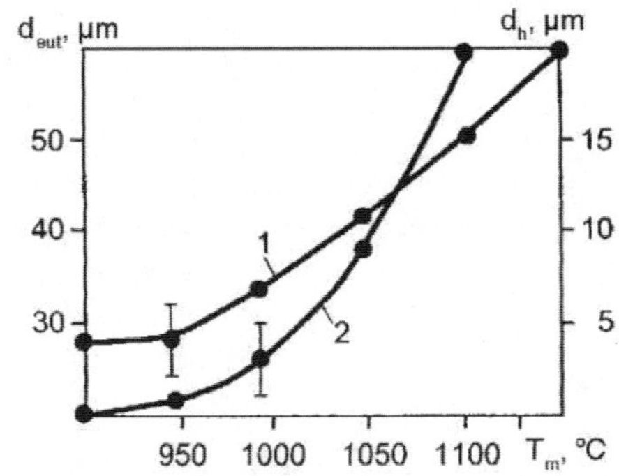

Figure 3. The effect temperature on the variation of the size of eutectic colonies (1) and hardening phases (2)

In the case of low degrees of supercooling, the examination showed the preferential fixing of the gamma solid solution based on nickel with the carbide and boride hardening phases distributed in the volume of the solid solution. In many cases, the alloys are characterised by the formation of a dendrite-like structure (y-solid solution based on nickel), penetrating through the entire coating (Figure 6d). In this case, the disperse eutectic (y-Ni + Ni_3B) fills the space between the dendrites. The material of this coating is characterised by relatively high strength and reduced hardness.

The process of crystallisation of the alloy of a specific composition with a high degree of supercooling is accompanied by almost complete suppression of the processes of formation of excess phases (y-solid solution based on nickel and Ni_3B), with the fixation of the eutectic (y-Ni + Ni_3B). The rate of formation of the eutectic for the melt of the given composition was higher. The large increase in the cooling rate of the coating-substrate composition using compressed air and a spraying device in a liquid results in the refining of the structure of the alloy of the coating but also causes an increase in the level of internal stresses and the probability of formation of microcracks. With the efficient application of powerful heating sources it is usually possible to produce structures with smaller grains in comparison to heating in the furnace, although the relationships of formation of the structural components remain unchanged. In this case, the structure of these alloys usually contains only a small amount of eutectic, with the formation of a protective structure with columnar grains of the gamma solid solution based on nickel. In the case of high cooling rates of compositions, a small content of the hardening phases is retained in the coatings.

The experimental results show that hardening of the coatings as a result of structure refining and efficient formation of a large amount of the eutectic and hardening phases, increases the wear existence of the alloys (Tables 2 and 3).

For this purpose, production tests were carried out on blades for cutting the wires of nonferrous metals, hardened with coatings on the cutting edge, and on punches for stamping foil. The durability of the blades and punches was determined by the operating time between two sharpening cycles (grinding). The experimental results obtained for the components with coatings produced by different methods and having different structures, were compared to the durability of the components produced from heat-treated Kh12F1 steel.

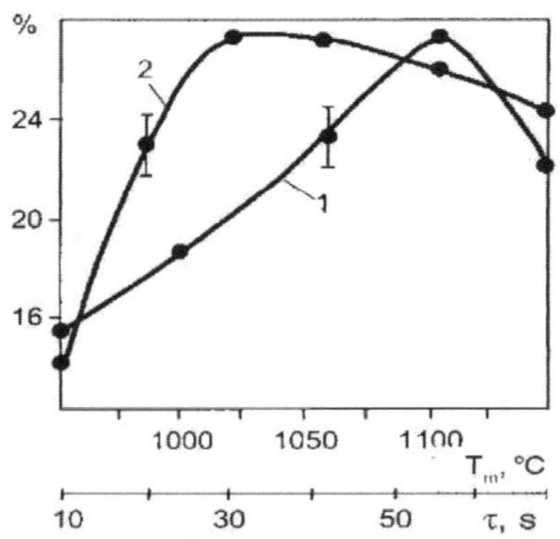

Figure 4. The dependence of the volume content of hardness in the PG-10N-01 coating on temperature (1) and the duration (2) of gas flame melting

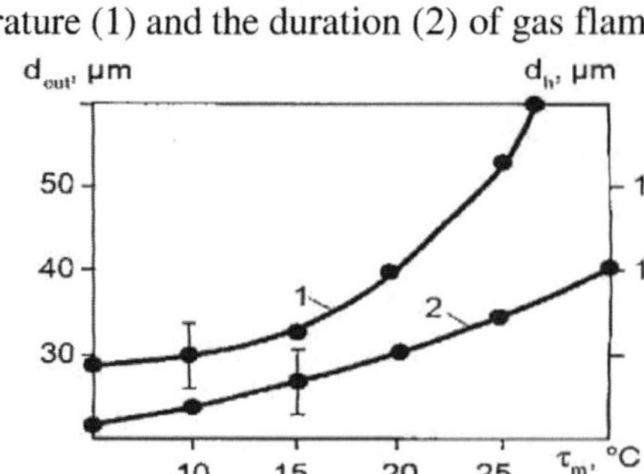

Figure 5. The dependens of the dimensions of eutectic colonies (1) and hardening phases (2) on holding time in gas flame melting

In the process of wire and foil cutting, the coating operates in the conditions of impact and abrasive loading, and the edges of the blades and punches are characterised by the formation of a nonuniform stress field, the nucleation and propagation of fatigue cracks and subsequent brittle failure and chipping, which are the main features of operation of these components. Therefore, the special features of service of these components determine the following most important technological properties: high hardness, strength, toughness and wear resistance. In addition, when selecting the material of the coating, one should take into account that the given coatings of the nickel alloys are characterised by their susceptibility to seizure, the capacity to polish during dry friction of the metal and by a low friction coefficient.

The analysis of the worn surfaces of the cutting edges shows that the wear of the coatings takes place due to damaging of soft components (initially the gamma solid solution based on nickel and, subsequently, the eutectic y-Ni + Ni_3B), exposure and subsequent chipping of hard carbide and borides phases (Figure 7a).

Therefore, the amount of hardening phases, the eutectic and the degree of hardening of the solid solution together determine the failure resistance of the coating material. The worn surfaces of the coating are characterised by the development of striations. During the wear of the coating with a more disperse structure and containing a large amount of eutectic and hardening particles (carbide and borides phases), the size and number of the striations on the surface decrease and the total wear rapidly decreases (Figure 7b).

Thus, the results of the tests carried out on the blades and punches show that the intensity and nature of wear of the coatings greatly depend on the type of the structure. Hardening of the coatings, obtained by means of refining the structure and purposeful formation of a large amount of the eutectic and hardening phases, increases the level of wear resistance.

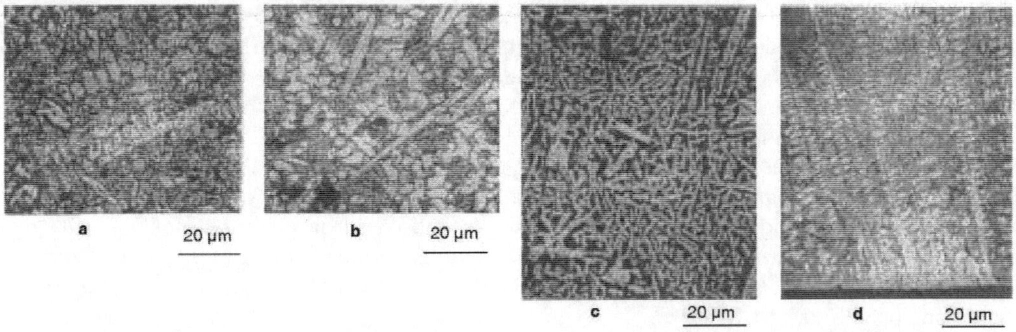

Figure 6. The structure of the coatings Ni-Cr-B-Si alloys after melting: a) preferential of the gamma Solid solution based on nickel and the γ-Ni+ Ni_3B eutectic; b) the high content of the carbide and boride hardening phases; c) eutectic frame; d) continuous dendritic structure

Table 2. Thermal processing of materials, their structure and properties

Materials of base and coating	Treatment	Structure	Hardness, HRC	Number of Shifts to regrinding	Hardness, HRC
Steel	Quenching and tempering	Martensit	54	12	83
PG-10N-01	Furnace	Eutectic	56	16	62
PG-10N-01	Gas-flame	Eutectic	58	18	56
PG-10N-01	Electron beam	Eutectic	60	23	43
PS-12N-VK	Gas-flame	Eutectic and carbides	62	25	40

Table 3. Thermal processing of materials, the structure and properties of the working edge

Materials of base and coating	Treatment	Hardness, HRC	Structure	Wear, μm	Number of shifts to regrinding
Steel	Quenching and tempering	54	Martensit	1.5	66
PG-10N-01	Furnace	56	Eutectic	1.4	71
PG-10N-01	Gas-flame	58	Eutectic	1.3	77
PG-10N-01	Plasma	54	Eutectic	1.7	58

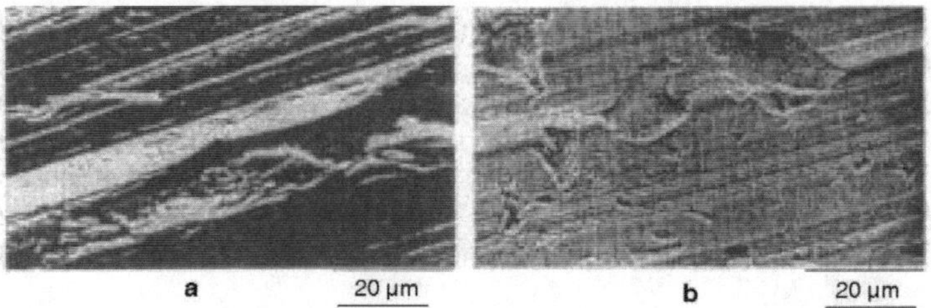

Figure 7. The worn surfaces of the coatings Ni-Cr-B-Si alloys PG-10N-01: a) coarse structure; b) disperse structure.

3. Conclusions

The increase in the content of the alloying elements in the alloys from 12.5 to 32.9 % after melting in the furnace is accompanied by an increase in the volume fraction of the carbide and boride hardening phases from 7 to 17 %. The volume fraction of the Y-Ni+Ni_3B eutectic also increases from 43 to 51 %. An increase in the hardness of the coatings from 35 to 62 HRC accompanies this process as well.

The increase in the volume content of the Y-Ni + Ni_3B eutectic in the PG-10N-201 coating from 20 to 40 % increases its hardness from 563 to 650 MPa.

The increase in the temperature of gas flame melting by 50-200 °C above the solidus of the alloys and the increase in the holding period from 10 to 30 seconds result in the growth of both the amount of colonies of the Y-Ni+Ni_3B eutectic by 1.5-2 times and the volume content of the carbide and boride hardening phases. When heating coatings to higher temperatures and increasing the holding period, the hardening phases start to dissolve.

The wear resistance of the PG-10N-201 coatings increases by 1.3-1.9 times in case of the blades for wires cutting along with hardness increase of the content of the hardening phases and eutectic, whereas the wear resistance of the punches for foil cutting increases by 8-17 %.

The melting of the nickel alloy coatings is highly efficient in reconditioning and hardening of the components operating in the conditions of friction and impact-abrasive wear due to a two-time increase of wear resistance of the surfaces in comparison to the steel.

Acknowledgments

The work was performed in accordance with the program of basic scientific research of the State Academies of Sciences of the Russian Federation for the period of 2013-2020 and the program of the Presidium of RAS 'A basic research of the development of the Arctic zone of the Russian Federation', code: 'Arctic-2014'.

References

[1] Shmakov A M and Ivshina N N 1992 *Fiz. Khim. Obrab. Mater.* **5** 90-100
[2] Bezborodov V P et al 2000 *Perspektivny Materialy* **4** 64-68
[3] Saraev Y N, Bezborodov V P and Selivanov Y V 2009 *Welding International b* **23(12)** 916-920

[4] Saraev Y N, Bezborodov V P and Selivanov Y V 2009 *Welding production* **4** 20-25
[5] Borisov Yu S et al 1985 *Poroshk Metall* **9** 22-26
[6] Spiridonov N V et al 1988 *Poroshk Metall* **1** 10-14
[7] Elliot R 1987 *Metallurgiya Moscow*

Mechanical activation influence on the morphological properties of La_2O_3-TiO_2-B

O Dolmatov, V Zakusilov, M Kuznetsov, N Pimenov and S Chursin
Tomsk Polytechnic University, 30, Lenina ave., Tomsk, 634050, Russia

E-mail: kms@tpu.ru

Abstract. The influence of mechanical activation of the powder mixture used to obtain the high-perfomance cathode for accelerating engineering with the SHS-method has been explored. The mechanically processed mixtures have been morphologically analyzed. The optimal modes of mechanical activation have been determined for the mixture.

1. Introduction

Accelerating engineering requires creation of new and more powerful plants, which generate flows of charged particles. This fact has a significant influence on the industrial, machine-building, energetic development and other fields of science and engineering. A great attention is paid not only to a plant modernization as a whole, but to its individual parts. The most important element of the devices generating flows of charged particles is a cathode. The use of hexaborides of rare earth metals as a cathode secures the maximal effectiveness of the cathode assembly by some parameters. The main advantages of the cathode made of lanthanum hexaboride are: the minimal work function (~2.7 eV); the high density of obtained emission current ($\approx 2\cdot 10^3$ A/cm^2, at T=2000 K); the high melting point (2800 K); the ability to retain emission properties in the conditions of low vacuum and intensive ion bombardment; the absence of air pollution; high exploitation characteristics [1].

The methods of powder metallurgy used for obtaining lanthanum hexaboride have some disadvantages: poisoning of the finished product as a result of a fusion process; technically complicated equipment; great energy expenditures for a long-lasting synthesis and annealing of products. One of the alternative methods is a self-propagating high-temperature synthesis permitting to obtain materials with specified properties [1].

A self-propagating high-temperature synthesis (SHS) is an exothermic reaction initiated locally as a result of specific chemical transformations, which occur in the mixture in the mode of a combustion wave (self-propagation) with the formation of solid products.

SHS has a controllable character and is realized in three stages: preparation of burden components, realization of SHS and cooling of finished products. A preparation stage is regarded as the most technological one, at which burden parameters can be changed. So, the possibility to control terminal properties of synthesized materials has appeared. One of such parameters is a mechanical activation [2].

The mechanical activation is a formation of a substance with a higher chemical activity owing to preliminary mechanical processing. The energy acquired during the activation time is assimilated by solid in the form of point and line defects. The accumulated dump energy is released during the synthesis. In addition, during mechanical activation a particle size is decreased, what increases the contact area. The process of mechanical activation can be observed in devices capable of grinding in impact, impact-rubbing or rubbing modes (the VSI, planetary and jet mills and other apparatus, where high values of frequency and force of mechanical impact are combined). Compared with other apparatus planetary mills have a higher efficiency. As a result of its use the maximum accumulation of structure defects is observed, the curvature value of surface is increased [3].

As a result of mechanical activation of reagents the decrease of the initial particle size and the increase of reactivity of solid reagents are observed; the temperature of reaction initiation is reduced; the structure becomes more uniform; the mechanical properties of materials are improved (the porosity is reduced, the strength is risen, the plastic properties are improved) [4-6].

This article represents a study of property changing of the mixture (La_2O_3+TiO_2+B) under the action of mechanical activation with the purpose of obtaining of finished product based on lanthanum hexaboride and titanium diboride and meeting the properties of a high-performance emitter.

2. Experiment

During the experiment burden parameters were studied according to the rotational frequency of the mill activating. In order to study the influence of mechanical activation parameters a special mixture was used: a mixture of chemically pure (99.9 mass %) fine-dispersed powders of lanthanum oxide, titanium oxide and boron (La_2O_3+TiO_2+B) mixed in a stoichiometric proportion for reaction propagation:

$$La_2O_3+3TiO_2+24B \rightarrow 2LaB_6+3TiB_2+3B_2O_3. \quad (1)$$

The initial reagents were being thoroughly intermingled in a pin cubical mixer with the help of the universal drive ERWEKA AR 403S during 30 minutes. The powder mixture had been dried before being placed in a muffle furnace during 5 hours at 100 °C and at atmospheric pressure. The mechanical activation was carried out in the planetary ball mill AGO-2S. The rotational frequency was varied within the limits of 10 to 40 Hz while the mechanical activation time was from 5 to 60 min. The grinding objects were metal balls with a diameter of 4 mm. The ratio of the ball mass to the mass of the processing substance was 10/1.

3. Results and discussion

The analysis of initial reagents was carried out on the BET-analyzer of specific surface META SORBI-M. Table 1 presents the results of the BET-analysis.

Table 1. The measurement of the specific particles surface area of the mixture (La_2O_3+TiO_2+B) (the activation time – 10 min).

Rotational frequency of mill (Hz)	Measured value of A_{sp} (m^2/g)
0	1.19 ± 0.02
10	2.00 ± 0.01
20	3.16 ± 0.04
30	5.14 ± 0.05
40	5.03 ± 0.05

In the pictures of powders before mechanical activation, taken by means of the raster electronic microscope Philips SEM 515, the large-sized particles are distinctly observed (Figure 1).

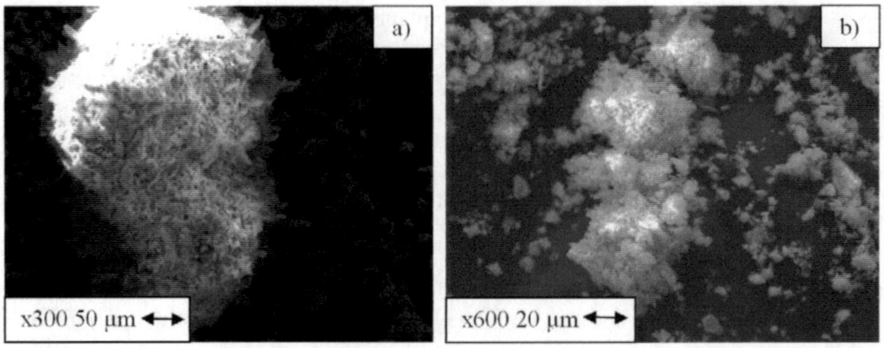

Figure 1. Photographs of the powder mixture ($La_2O_3 + TiO_2 + B$) before mechanical activation at magnification a) of 300 times; b) of 600 times.

After mechanical activation (Figure 2) the particle size has noticeably diminished as a result of particle rubbing and collision in the planetary mill.

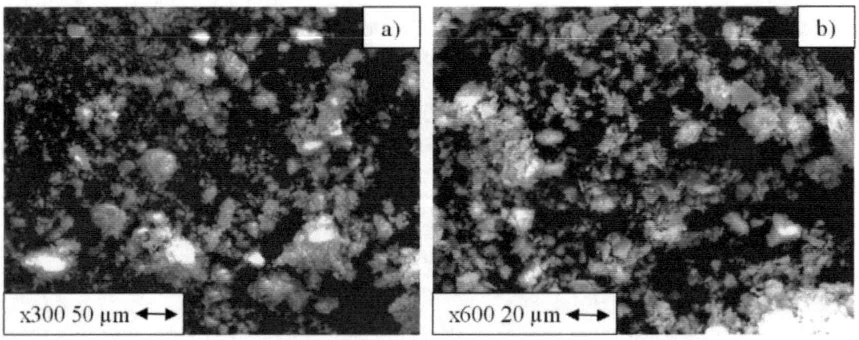

Figure 2. Photographs of the powder mixture (La_2O_3+TiO_2+B) after mechanical activation at magnification a) of 300 times; b) of 600 times.

In order to estimate the derived information it is necessary to plot the dependence of the specific surface area of processing burden on the rotational frequency of the activating mill. The graph is presented in Figure 3.

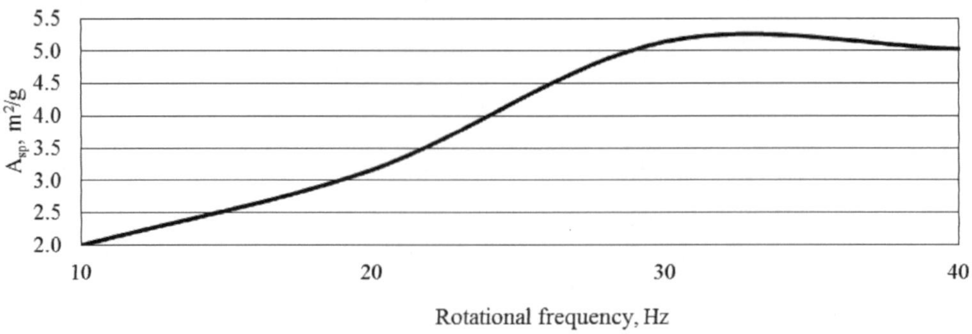

Figure 3. Dependence of the specific surface area of processing burden on the rotational frequency of the activating mill.

It can be seen from the plot that for the mixture (La_2O_3+TiO_2+B) the maximum value of the specific surface area is observed at the frequency of 30 Hz. This value decreases if the rotational frequency increases. In order to understand this process it was necessary to carry out the granulometric analysis of burden composition with the use of the laser analyzer of a particle size SALD-7101.

Figure 4 presents the particle size before grinding in the planetary ball mill.

Figure 5 presents the particle distribution according to sizes after mechanical activation with the frequency of 30 Hz during 10 min.

Figure 6 presents the particle distribution according to sizes after mechanical activation with the frequency of 40 Hz during 10 min.

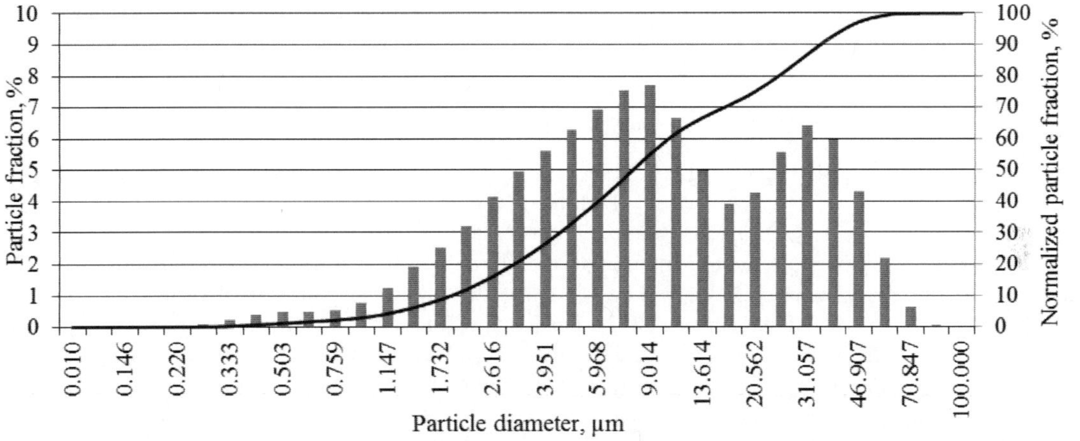

Figure 4. Granulometric composition of burden without mechanical activation.

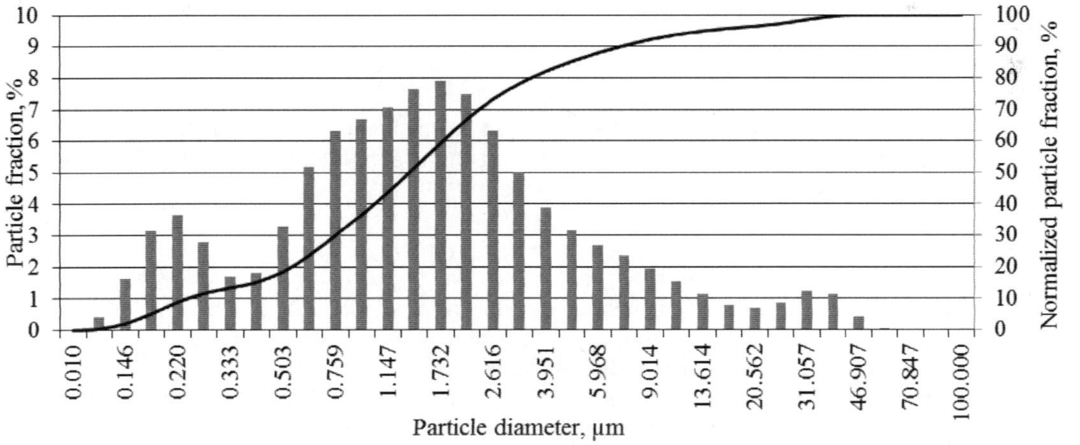

Figure 5. Granulometric composition of burden after mechanical activation with the frequency of 30 Hz during 10 min.

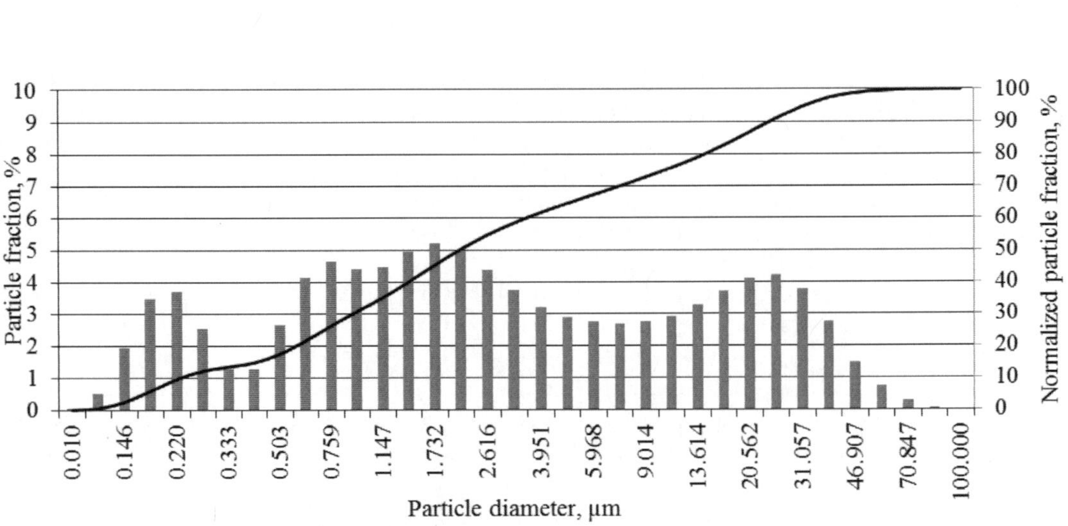

Figure 6. Granulometric composition of burden after mechanical activation with the frequency of 40 Hz during 10 min.

The comparison of Figure 4 and Figures 5, 6 clearly demonstrates the decrease of a particle size as a result of mechanical activation. It can be seen that the maximum value of the particle fraction is displaced to the less side. The decrease of the particle size increases the surface contact area reducing a temperature gradient, which has a negative influence on obtaining of finished products. However, when comparing the particle sizes obtained at the rotational frequency of 30 Hz (Figure 5) and the rotational frequency of 40 Hz (Figure 6) one can notice that the maximum particle size is increased, so the particle fraction with a larger size is raised. This process is also observed in Figure 3 in the frequency interval from 30 to 40 Hz. A similar negative influence of mechanical activation during the increase of rotational frequency is explained in the following way: the particle size stops diminishing under the action of impact-rubbing forces; particles start to collide with each other forming agglomerates (the particles of a larger size).

The integral curve in Figures 4–6 allows defining the particle fraction among all mixture compositions by a specified maximum size. In addition, it is possible to define the average particle size of the mixture, which serves as a more descriptive parameter. An average particle size is a universal characteristic for all modes of mechanical activation what allows using derived results in all devices, where mechanical activation is possible to realize.

4. Conclusions

During the research it has been discovered that as a result of mechanical activation of reagents the average particle size is decreased. The small particle size increases the surface contact area of particles thereby increases a reaction rate and improves the uniformity of combustion wave passing, thus promoting the obtainment of uniform and high-quality samples.

As a result of the research the dependence of the average particle size on the rotational frequency of the planetary mill has been found experimentally. The optimal value of rotational frequency of activating mill equaling to 30 Hz has been derived for the system (La_2O_3+TiO_2+B). It has been ascertained that the unregulated increase of rotational frequency may lead to the so-called «saturation threshold». In other words a mechanical activation will have a negative influence on SHS starting from certain values.

Mechanical activation allows controlling SHS what gives the opportunity to obtain material with specified properties: hardness, strength, porosity, density. In order to obtain a high-quality product with the help of mechanical activation it is necessary to calculate a frequency value individually for every reagent composition.

References

[1] Kresanov V S et al. 1987 *M., 'Energoatomizdat'* 152
[2] Merzhanov A and Mukasyan A 2007 *M.: Torus Press* 336
[3] Avvakumov E G 2009 *Novosibirsk: Izd. Sib. Otd. RAS*
[4] Rogachev A S, Shkodich N F, Vadchenko S G, Baras F, Chassagnon R, Sachkova N V, and Boyarchenko O D 2013 *Int. J. Self-Propag. High-Temp. Synth.* **22(4)** 210–216
[5] Korchagin M A, Grigorieva T F, Barinova A P and Lyakhov N Z 2000 *Int. J. Self-Propag. High-Temp. Synth.* **9(3)** 307–320
[6] Bernard F and Gaffet E 2001 *Int. J. Self-Propag. High-Temp. Synth.* **10(2)** 109–132

IOP Conf. Series: Materials Science and Engineering **124** (2016) 012121 doi:10.1088/1757-899X/124/1/012121

A new type of antifriction and wear resistant malleable cast iron

S V Davidov and A O Gorlenko
Bryansk State Technical University, 7, 50 years of October st., Bryansk, 241035, Russia

E-mail: davidov69@tu-bryansk.ru

Abstract. There is developed a technology of malleable cast iron modification on the basis of complex chemical compound of surface-active elements and their solid solutions with other elements. Silicon high content in malleable cast iron helped to develop a power efficient technology of graphitizing annealing which has considerably lower annealing temperature and complete renunciation of the second graphitizing annealing stage at the expense of its change by controlled cooling up to ferrite structure or by air cooling for perlite structure.

1. Introduction

Malleable cast iron (MCI) is a fine structural material with a wide range of mechanical, technological and service properties. Malleable cast iron wear resistance is various. Perlite MCI has good wear resistance in the conditions of lubricated friction at pressure up to 20 MPa and is worn quickly at dry friction. Perlite-ferrite MCIs have comparatively bad antifriction properties in the conditions of lubricated friction and rather good ones at dry friction. However, there is a reduction of MCI cast products and their replacement with high-strength iron in recent years. The reason for MCI replacement in cast structural and antifriction materials is high power inputs fore cast graphitizing annealing. Depending on MCI chemical composition, its production technology and furnace for heat treatment, graphitizing annealing lasts: for the first stage – 3…15 h, for the second stage – 6…20 h, so the total MCI annealing lasts from 9 to 35 h.

2. Problem statement

The development of centers of crystallization that occur in super cooled melt and a modified, and, consequently, the final structure of the alloy, it is possible to control both the thermodynamic and thermo kinetic factors. Thermodynamic factors affecting the speed and density of the diffusion flow to the growing nuclei of the crystallizing substance through its thermodynamic activity. Thermo-kinetic factors affect the critical size and the rate of formation of nuclei, their number and form. As a rule, in real conditions the crystallization of the melt and the thermodynamic and thermo kinetic factors are mixed and their individual contribution to the character and type of the emerging structure of the alloy is not easy to establish.

The first task of modification is solved by introducing stabilizing (antigraphitizing) modifiers into hot cast iron, the second – by one-time or divided element introduction – graphitizers. The most interesting are the effects produced by elements of Va and VIa periodic table groups which are surface-active elements (SAE). These effects are of thermodynamic, kinetic and mixed character. Bi, Te, S, N_2, Ce, Mo, Y, B are used as stabilizing modifiers for white cast iron. Their effectiveness reveals in extremely small amounts, especially of bismuth and tellurium: from 0.0003 % to 0.01 %. It should be stressed that Bi and Te have an advantage over other antigraphitizing elements as they do not influence annealing time if the number of nucleidoes not change. At the same time, N_2, Cr, Mn or Al surplus determine the time of the annealing process themselves independently from nuclei amount.

3. Research methodology

To reduce the time of MI annealing there was developed two-stage technology of its modification on the basis of the modifier which is represented by a complex chemical compound of surface-active elements and their solid solutions with other elements. The modifier consists mainly of surface-active elements of Va and VIa periodic table groups and also aluminum, antimony, tin, copper and other elements which are added to the modifier in accordance with technological tasks. The modifier is added to iron melt in the amount of 0.005…0.01 % of hot iron mass. This amount is enough to crystallize iron according to metastable system. This so-called counter modification counteracts fading of inoculation effect during keeping iron in the ladle before filling casting molds and is based on adding small amounts of surface-active elements to typical inoculating graphitizing modifiers. Surface-active elements adsorb on the surface of graphite nuclei and prevent their premature solution in considerably overheated melt during its homogenization.

There is blocking of cast iron graphitization centers at the first stage as surface-active elements are tensioactive in hot iron relatively to the phase of nonmetallic inclusions including such clusterings as carbide-like atomic segregations, fullerenes in liquid solution of cast iron. In this method of complex melt modification antigraphitizers are elements of Va and VIa periodic table groups – Sb, Bi, S, Se, Te (their surface–active properties cause cast iron chilling); and graphitizers are C, Si, A1, Cu.

If to use classical approach, then counter modification represents combined modification of I type (formation of surface film on stimulated nuclei) and modification of II type (formation of additional stimulated nuclei).

4. Research results

Using counter modification gave the opportunity of considerable increase of carbon and silicon content in MCI (Table 1) that in its turn sharply reduced graphitizing annealing duration. There were developed two types of cast iron with average carbon content 2.9…3.3 % (in standard MCI 2.4…2.8 % C): type I, average silicon content – 1.8…3.0 % Si and type II, high silicon content – 3.0…4.0 % Si (in standard MCI 1.0…1.6 % Si). Table 1 gives some MCI compositions with reduced annealing mode and their mechanical properties.

High content of silicon in MCI according to table 1 allowed to develop a power efficient technology of graphitizing annealing. Moreover, it became possible to decrease annealing temperature by 100 °C and completely renunciation the second graphitizing annealing stage changing it with controlled cooling for obtaining ferrite structure (Figure 1) or with air cooling for perlite structure (Figure 2).

Besides, controlling annealing temperature and cooling rate gives the opportunity to control the annealing graphite morphology, its distribution in matrix and quantity. High degree of graphite compactness provided rather high level of mechanical properties (Table 1). Table 2 represents time modes of MCI annealing. Researching graphitization process in MCI of a new type we found out the effect of cementite selective solution and graphitizing activity of sulfide-manganese phase (Figure 3).

Each annealing graphite inclusion has nonmetallic inclusion (Figure 3a, b). Absorption of iron-carbide fullerene complexes activating nonmetallic inclusion surfaces is represented at Figure 3c and Figure 3d, where sulfide co crystallizes with cementite and is graphitized only where perlite is. Sulfide surface inside cementite is not activated and there is no decay of complexes. The effect discovered [2] was used to increase the number of kinds of MCI of a new type with compact graphite form and to develop stable-mottled cast irons (Figure 4).

Table1. MCI Chemical composition and mechanical properties.

Iron type	Chemical composition, % to mass					Mechanical properties		
	C	Si	Mn	S	P	σ, MPa	δ, %	HB, MPa
I	2.92	2.54	0.41	0.28	0.08	710	2.0	2300
	3.21	1.74	0.60	0.28	0.07	735	1.8	2600
	3.05	2.95	0.60	0.24	0.06	580	3.0	2200
	3.28	2.76	0.44	0.35	0.10	550	2.6	2100
II	2.90	3.55	0.40	0.33	0.09	470	5.5	2000
	3.12	3.97	0.37	0.40	0.08	490	6.5	1800
	2.92	3.70	0.33	0.37	0.09	610	2.2	2100
	3.27	3.00	0.44	0.47	0.12	510	2.0	2000

Table 2. Modes of MCI heat treatment.

Iron type	Optimum time of graphitizing annealing, hour			
	For perlite*		For ferrite	
	850 °C	950 °C	850 °C	950 °C
I	2.0	0.5	2.0/120**	0.5/120
II	0.5	0.3	0.5/240	0.33/240

*Air cooling.
**The first stage of graphitization as numerator, these condstage of graphitization as denominator, replaced by cooling (V_{cool}, °C/h).

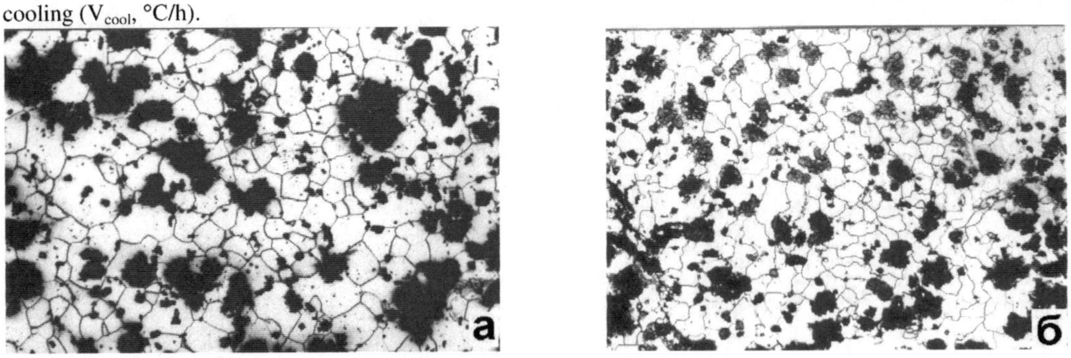

Figure 1. Ferrite MC Is type I, after annealing (*a* – 950 °C, x100; *b* – 850 °C, x100) within 30 min. and cooling rate 120 °C/h (nital etching)

During graphitization only outer cementite shell decays because of destabilization by endohedrals on the basis of bismuth, tellurium and copper. For this purpose we added to MCI with higher content of sulphur (0.2…0.4 %) extra chromium (up to 2.0%) to obtain stable cementite and extra copper (0.3…5.0 %) for complete perlite matrix of cast iron and further modification according to the technology of counter modification. The choice of alloying elements was based on the following:

1. Cast iron is alloyed simultaneously by graphitizing elements (Si, Cu) and antigraphitizing elements (Cr, S).
2. Cast iron alloying with copper is carried out for several reasons:

Firstly, it is known that copper in white cast iron increases its heat conduction and most of its tribotechnical properties.

Secondly, copper in contrast to silicon destabilizes cementite because it partly enters into its composition replacing iron atoms and forming compound $(Fe_{2n}Cu_n)C$.

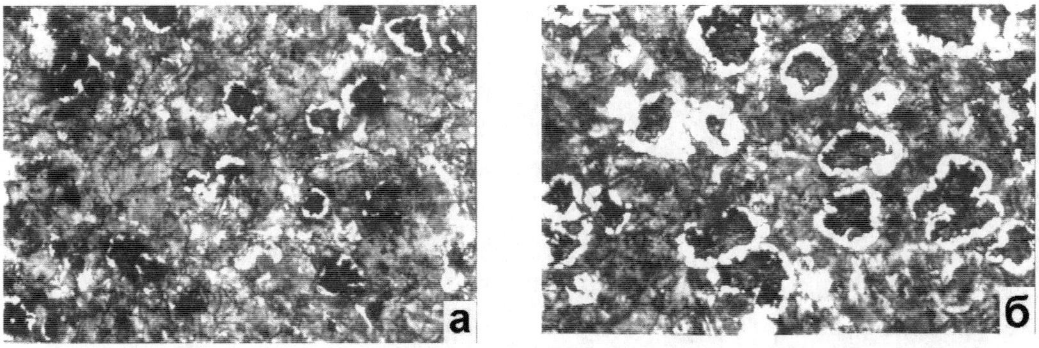

Figure 2. Perlite-ferrite MCI softype II with compact graphite form after (*a* – 850 °C, × 250; *b* – 950 °C, ×250) within 30 min. and cooling rate 120 °C/h (nital etching)

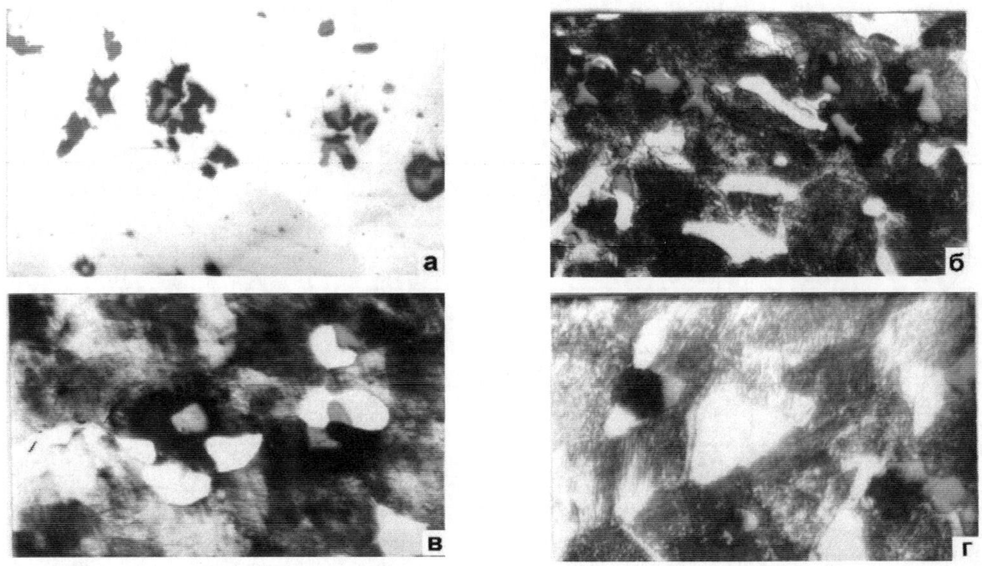

Figure 3. Sulfide-manganese phase influence on MCI graphitization, × 630 (nital etching)

Cementite destabilizing is a direct formation of copper with fullerenes of cuprous endohedrals and bismuth and tellurium. It means that the first portions of crystallizing cementite are enriched with chromium (inner nuclei of cementite inclusions) and the last portions – with copper, which are also enriched with silicon, though with a very small amount because of its bad solubility in cementite. There is polarization of carbide phase – one of its parts becomes highly graphitized and the other (inclusion nuclei) – sharply lower.

In such cases cast iron is additionally modified by strong chilling additives of bismuth and tellurium bismuth and tellurium. Adding chromium into cast iron (about 2 %) sharply stabilized cementite structure of cast iron (Figure 4a).Total perlite structure was obtained only after 3 h at heating up to 1000 °C, and 8 h were needed for decay of ledeburite frames on isolated cementite inclusions (Figure 4a, b).

The experimental results also showed that there was intensive release of annealing graphite on sulphide phase (Figure 4c). In general, the structure obtained is a composite one: evenly distributed isolated cementite inclusions in perlite matrix. Tribotechnical tests were carried out according to the standard technique on friction machine SMC-2 using counter face made of hard steel 40X at block pressure to the roller 120MPa in distilled water simulating the conditions of half-floating condition. Table 3 represents data on the wear of stable-mottled cast irons.

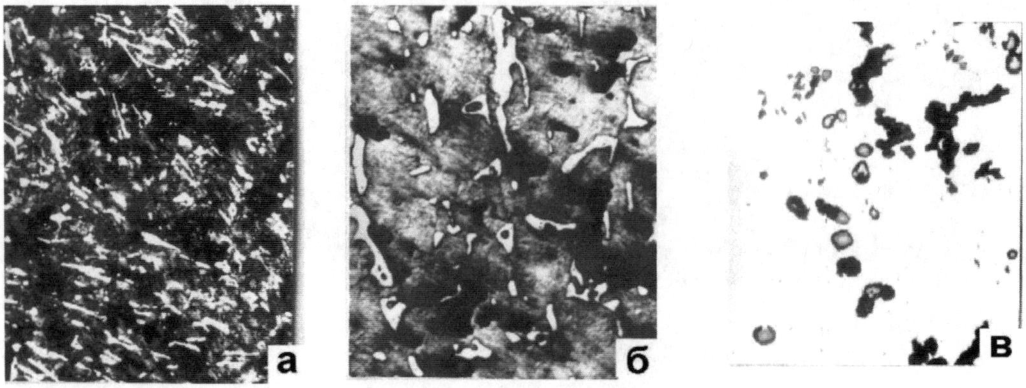

Figure 4. Microstructure of stable-mottled castiron subjected to graphitizing annealing at 1000 °C during 8 h: *a* – × 100; *b*, *c* – × 400 (*a*, *b* – natal etching, *c* – without etching).

Table3. Chemical Composition, Tribotechnical and Mechanical of Stable-mottled Cast Irons.

Chemical composition, % to mass				Wear,	Friction	σ_B,	HB,
C	Si	S	Cu	mg/(cm^{2}km)	coefficient	MPa	MPa
2.80	4.40	0.92	Сл.	7.0	0.10	280	2010
3.23	3.40	1.11	Сл.	5.8	0.09	292	2210
3.60	2.40	1.39	Сл.	4.4	0.07	271	2100
1.80	4.40	0.92	4.99	6.7	0.09	490	2400
2.22	3.41	1.11	3.02	5.4	0.07	382	2260
3.56	2.41	1.37	0.30	4.3	0.07	245	2210
2.82	4.36	0.92	0.33	8.2	0.11	492	2100
2.17	3.37	1.12	3.22	6.1	0.10	390	2010
1.85	2.41	1.40	4.96	6.6	0.08	578	1970
Note: Cr up to 2.0%; Mn up to 0.4%; P up to 0.3%							

5. Conclusion

On the basis of cast iron counter modification there was developed a new group of constructional cast irons having unique technological, functional, mechanical and service properties, first of all antifriction and wear resistant. This group includes cast irons with reduced annealing mode, their structure being controlled by annealing temperature or cooling rate during graphitization; high-strength cast irons with compact graphite form and stable-mottled cast irons with controlled amount of graphite and carbide phase.

6.References

[1]. Zhukov A A and Davydov S V 1983 Malleable cast iron with short mode graphitizing annealing *Steel in Translation* **6** pp 100-103

[2]. Zhukov A A and Davydov S V 1983 On the graphitizing effect of sulphide inclusions in stable-half the iron *Met Sci Heat Treat.* **4** pp 36-37

Friction Surface Modification by Nanodiamonds of Denotational Synthesis

S V Davidov and A O Gorlenko
Bryansk State Technical University, 7, 50 years of October st., Bryansk, 241035, Russia

E-mail: davidov69@tu-bryansk.ru

Abstract. The paper deals with the technology and equipment for making a hardened antifriction surface layer of machine elements on the basis of using carbon nanomaterials in order to improve wear resistance. The paper presents the results of tribotechnical tests of pins according to standard techniques using an automated system for research.

1. Introduction

The perspective directions of solving the problem of increasing wear resistance of friction surfaces of machine parts include the use of nanodiamonds as a reinforcement (reinforcing) of the surface layer of particles of a small size.

The diversity of principally new properties of nonmaterials allows using them for new applications in various industries including friction pairs of machine elements. In technologies of obtaining and applying nonmaterials of 1…500 nm carbon clusters are becoming more and more popular, the most promising of them are carbon nonmaterials, super disperse nanodiamonds of denotation synthesis in particular.

Synthesized in highly non-equilibrium conditions, nanodiamonds do not have a clear cut crystal. The rounded shape of nanodiamonds along with modification of their surface during detonation synthesis provides an effective application to improve mechanical characteristics of materials, and, in some cases, it provides unique tribological properties due to the formation of spatial grid of physical connections at the interface of the material structure with nanoparticles with high adsorption properties.

Making surface layers with implanted nanodiamonds as reinforcing nanoparticles of the layer increases its density, strength and provides a nanostate, which gives improved wear resistance during friction.

The micro- and nanostate of friction surface is formed by electromechanical working (EMW) for making surface layers with implanted highly consistent disperse particles consisting of coagulated nanodiamonds. The distribution pattern and morphology of the disperse strengthening phase (a strengthening composition effect as a result of decay of supersaturated solid solutions of the material structure) have the main influence on wear resistance of the worked surface. Electromechanical working is characterized by wide technological possibilities to control micro- and nanostate and tribotechnical parameters of friction surfaces of machine elements and also by reduction of cost, labor and energy intensity by 3…5 times in comparison with other more widespread technological methods, in particular laser and plasma working. At the same time plant-performance Figures of components rise by 1.5…3 times.

The technology of electromechanical working is based on the combination of thermal and power influence on the surface of the work piece, that leads to changing physical-mechanical micro geometric parameters of the surface layer (strength increase, reduction of roughness parameter, etcetera) and, as consequence, improvement of plant-performance Figures of components, in particular wear resistance, contact stiffness and strength, endurance limit, heat resistance, fretting resistance.

A combined electromechanical working (CEMW) technology was developed to get wear-resistant surface layers on friction surfaces of machine elements with implanted nanodiamonds of denotational synthesis. This technology includes implanting nanodiamonds of denotational synthesis into a friction surface and subsequent reinforcing the worked surface. This technology is referred to additive technologies [1].

If we consider any tribosystem functioning under the conditions of adhesive, fatigue, oxidative and abrasive wear, than nonmetallic hard materials – carbides, nitrides, borides, composite and nanocomposite materials, cermets and diamond are the most perspective [2].

2. Research technique

The technology additionally includes a special set representing the technological complex: a universal machine (used for mechanical treatment of billets) with appropriate tools and devices for fixing workpieces and supplying an electric current of great strength and low voltage; a power conversion unit for electric current; the control unit processing modes; and switching means for supplying lubricating and cooling technological environment; the interface unit with the PC. Implanting nanodiamonds into the surface layer during EMW is done under certain conditions in a coagulated state (powder fractions of (200…250) nm).

They are coated to the surface before working, having been mixed with consistent graphite grease in a certain proportion. Most of coagulants penetrate into the formed surface layer reinforcing it. Then EMW is performed on the same surface to harden it.

Complex comparative tests of samples were performed at an automatic plant based on the friction machine MI-1M and intended for tribotechnical tests of pins made of metals and alloys. The tests give the opportunity to determine surface tribotechnical characteristics in the conditions of sliding friction and with boundary lubrication.

The tests were carried out under the following conditions: sliding velocity υ = 1 m/s; normal loading force N = 100 ± 0.5 %, H (corresponds to pressure in Hz, about 150 MPa) ;initial contact – concentrated plastic contact; boundary lubrication; lubrication type – dipping; key wear type - fatigue wear; lubricant – industrial oil I – 20A ГОСТ 20799 – 88; indenter material - hard alloy VK8.

The complex comparative tests of wear resistance were performed on outside cylinder surfaces of samples made of medium-carbon steel 45 and in accordance with the following technology:

- combined electromechanical working (CEMW) – forming a surface layer with implanted nanodiamonds and subsequent reinforcing the worked surface, microhardness is 950 HV.
- bulk hardening, backing, hardness is 55 HRC;
- electromechanical reinforcing of the worked surface, microhardness is 750 HV.
- nanocoating of system Si-O-C-N, performed by using the technology of plasma finishing in Ltd "NPF "Plazmatsentr", St. Petersburg, coating nanohardness is 23GPa.

According to the results of parameters analysis we determined the following values of tribotechnical properties:

- aging time t_0, in hours, defined as the time from the test start till the moment when wear curve comes to the piece of normal wear;
- aging wear h_0 , in microns, an approach value, defined at the moment of aging finishing t_0;
- average friction coefficient during normal wear f;
- f_0 / f – ratio of a maximum friction coefficient during aging f_0 to its average value during normal wear f;
- average value of wear intensity in the period of normal wear:

- $I_h = (h - h_0) / (L - L_0)$, where h, micron, - total value of sample wear at the test time; L, micron, - friction distance of the sample surface at the test time; $L_0 = 3.6 \cdot 10^9 \cdot t_0 \cdot \upsilon$, micron, - friction distance of the sample surface at the aging time.

3. Research results

There is austenization of the steel surface layer in the contact zone in the process of high temperature plastic deformation and under the influence of high temperature and pressure. Carbon of the coating consisting of graphite and nanodiamonds is diffused during a solid phase saturation process into the surface layers increasing the carbon content in the austenite.

With the increase of the carbon content and a higher cooling rate the steel matrix structure acquires a higher share of pearlite with intensive decrease of ferrite fringe thickness around pearlite grains. Intensive cooling forms a steel structure, which consists of alternate layers of various structure arrangement – normalized ferrite-pearlite and sorbate-troostite (hardening structure) zones (Figure1).

Thus, in the hardening process by CEMW steel is simultaneously subjected to normalizing with faster cooling and formation of the ferrite-pearlite structure and hardening with formation of the ferrite-sorbate-troostite structure. As a result of high temperature plastic deformation there is a formed sintered graphite layer with implanted nanodiamonds on the steel surface (Figure 1, e).

This layer consists mainly of sintered mass of graphite and nanodiamonds (Figure 2, a). A thin layer of over strained metal is formed on the outer surface of this layer, which is removed by diamond smoothing (Figure 2, b). The structure of the nanodiamond graphite layer (Figure 2, a) consists of the base, sintered graphite, nanodiamonds of various size groups and different carbide inclusions, cementite for example. This layer has good tribotechnical characteristics as its base (sintered graphite) is a high-performance antifriction material, reinforced by implanted nanodiamonds and disperse carbide particles of various types, for example, cementite. Figure 3 shows the structure of a sintered layer.

4. Discussion of results

The underlying hardened ferrite-sorbite-troostite layer (Figure 1) has a wear-resistant fine-grained textured structure. A significant component of this structure is milled grain ferrite obtained by recrystallization of heavily deformed austenite during Electromechanical hardening. The grain size of the ferrite compared to the base material is 15-fold reduced (Table 1).

Table 1. The geometrical parameters of the hardened layer

№ p/p	Layer	Size, mcm	Positioning in Figure
	The thickness of the layers		
1	Ferrito-pearlitic	350...500	1,б
2	Ferrite-sorbite-troostite	300...450	1,в
3	Transition	30...70	1,г
4	Graphite with nanodiamonds	200...250	2,a
5	Fragments overstrained steel	100...120	2,б
	The total thickness of the hardened layer	850...1200	
	A ferritic grain size of steel (medium)		
6	Matrix steel main	30	1,a
7	Ferrito-pearlitic layer	20	1,б
8	Ferrite-sorbite-troostite layer	2	1,в

The reinforced lower ferrite-sorbate-troostite layer (Figure 1, c) has a fine grain-oriented wear resistant structure. The significant component of this structure is a ground ferrite grain, obtained as a result of recrystallization of greatly strained austenite by electromechanical reinforcement. The ferrite grain size in comparison with the base material decreases by 15 times.

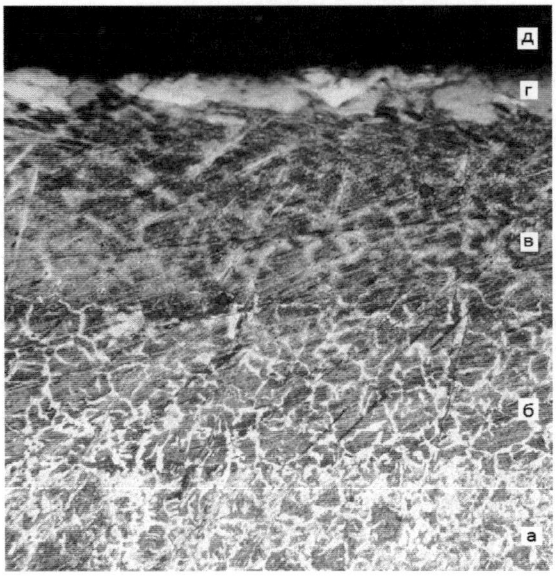

Figure 1. Multilayer structure of the surface content of steel 45 after forming the layer with implanted nanodiamonds and subsequent electromechanical reinforcing (×100): a - lower layer – material base, standard initial steel structure; b - 1st following layer – normalized structure formed by high cooling velocity when heating; c - 2nd following layer – a zone with the higher carbon content consisting of hardened sorbate-troostite structures; d - transition layer or 'white' layer; e - upper layer – sintered graphite layer implanted with nanodiamonds

Table 2. Results of tribotechnical tests of samples made of steel 45 after different working techniques

Tribotechnical property	Index	Values for different working techniques			
		Backing	EMW	Plasma finishing	CEMW
	t_0, h	2.75	2.85	1.94	2.13
Running ability	h_0, mcm	11.5	6.0	3.5	2.5
	f_0 / f	1.58	2.94	3.28	2.60
Antifrictionality	f	0.33	0.17	0.035	0.098
Wear resistance	$I_h \cdot 10^{-10}$	2.41	0.8	0.96	0.75

Microhardness of the surface layer (by medium-carbon steel working) reaches 1000 HV on the surface. The depth of the general reinforced layer is up to 1.2 mm. Besides, there is blending of reinforced layer hardness from the surface with the non-reinforced core of the element.

So, it does not cause delaminating at dynamic loads because of the integrated phase condition of the layer metal matrix, different in its structure. The obtained microgeometrical parameters of the worked elements are: $Ra = 0.2 \ldots 3.2$ micron; $Sm = 0.025 \ldots 0.36$ mm; $tm = 50 \ldots 70$ %; $Wz = 0.4 \ldots 8.0$ micron, $max = 6 \ldots 20$ micron.

The results of tribotechnical tests of samples made of steel 45 after different working techniques are given in Table 2. The results of tribotechnical tests showed that the wear resistance of samples with implanted nanodiamonds with subsequent electromechanical reinforcing in comparison with heat-treated samples rises by 3.2 times, in comparison with reinforced by EMW – by 1.1times, in comparison with plasma finishing – by 1.3 times.

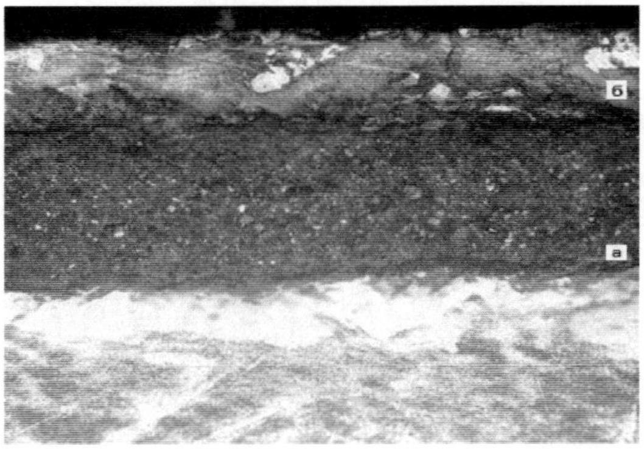

Figure 2. Sintered graphite layer with implanted nanodiamonds on the friction surface of steel 45(a), covered with layer fragments of overstrained steel (b) during CEMW, (x500)

5. Findings

1. Modification of the steel friction surface by means of the forming sintered graphite layer, implanted and reinforced by nanodiamonds and carbides of various nature along with the formation of the reinforced ferrite-sorbate-troostiteand ferrite-pearlite fine-grained structure of lower layers by CEMW allows improving wear resistance of friction surfaces, which is proved by tribotechnical tests.

2. A multilayer structure with the general metal matrix of the base material of a homogeneous phase provides monolithic cohesiveness of reinforced layers without continuity violation during wear at heavy loads, which is also proved by tribotechnical tests.

3. It is economically sound to use relatively cheap medium-carbon steels for critical friction pair instead of expensive high-carbon and high-alloy steels. It is possible to get the same microhardness and higher wear resistance of medium-carbon steels, which is not possible by using other heat processing and surface reinforcing methods. This is the main advantage of CEMW. The coefficient of reinforcement reaches $k = 3.5 \ldots 4$.

6. Conclusion

This research urgency and scientific significance consist in a system approach to the study of the formation of the surface layer quality, its phase, micro- and nanostructural states, tribotechnical characteristics of machine element surfaces. This approach is a methodological foundation of the research. It allows applying scientific ways in order to choose, standardize and optimize the modes of combined electromechanical working taking into account functionality of the worked elements.

References

[1] Grigoriev N and Tarasova T 2015 *Met Sci Heat Treat.* **10(724)** 5-10
[2] Stanski D et al 2015 *Met Sci Heat Treat.* **7(721)** 77-83

Basalt fiber insulating material with a mineral binding agent for industrial use

T Drozdyuk, A Aizenshtadt, A Tutygin and M Frolova
Lomonosov Northern (Arctic) Federal University, 17, Severnaya Dvina Emb., Arkhangelsk, 163002, Russia

E-mail: t.drozdyuk@narfu.ru

Abstract. The paper considers a possibility of using mining industry waste as a binding agent for heat insulating material on the basis of basalt fiber. The main objective of the research is to produce a heat-insulating material to be applied in machine building in high-temperature environments. After synthetic binder having been replaced by a mineral one, an environmentally sound thermal insulating material having desirable heat-protecting ability and not failing when exposed to high temperatures was obtained.

1. Introduction

Basalt fiber insulation is commonly used [1]. In mechanical engineering such kind of thermal covering is applied for heat and acoustic isolation of heat-treatment machinery, heating and hardening furnaces, heat pipelines, and etcetera. In addition, more than 55% of insulating materials perform under temperatures below 200 °C, about 25 % – at 180…400 °C, 5 % – in the range of 401…600 °C, and only 0.1 % of them is fitted on sites where temperature values exceed 600 °C.

The most part of its drawbacks is caused by the application of such synthetic organic binding agent as phenol-formaldehyde resin having a narrow temperature operational range (max 250 °C). In this case, such toxic compounds as phenol, formaldehyde, methyl isocyanate, etc. are emitted to the environment due to the binding agent oxidative decomposition, so the mineral wool item loses its structure rigidity [2].

In this regard, producing light noncombustible environmentally sound and cost effective heat insulating materials operating at temperatures above 600 °C is very important.

High-plasticity bentonite clay should be used as a binding agent to eliminate the disadvantages of phenol-formaldehyde mineral wool insulation. This method of clay application is implemented into manufacturing at the ‘Zavod izolatsii’ (‘Insulation plant’, Irpen, Ukraine).

The saponite containing material (SCM) is a typical example of bentonite clays. SCM is a clay mineral, layer silicate (montmorillonite group), its solid mineral particles in suspension including 63 % of saponite, 10 % of quartz, 10 % of dolomite, other minerals (chlorite, hematite, calcite, apatite, etc.) not exceeding 2…3 % [3]. The SCM chemical composition determination by X-ray fluorescence spectroscopy [4] revealed harmful impurities being absent. However, the fact of such chemical compounds (expressed as oxides) as SiO_2 (52 %), MgO (19 %), Al_2O_3 (10 %), CaO (4 %) present in the pilot samples implies that hydrosilicates out of mechanically pre-activated saponite, containing raw material, are to be formed while hydrating, after SCM having been mechanically dispersed in a ball mill to an ultrafine state [5-6]. Binding properties of SCM were evaluated by calorimetric method of hydration heating value measuring [6].

The aim of the proposed research is to evaluate the ability of SCM to be applied as a binding agent in mineral wool industrial insulating.

2. Method

The initial SCM sample was pre-dried to constant weight at 105 °C, and then grinded with a planetary ball mill (PM 100, RetschGmbH) using a dry mechanical dispersion method. The grinding machine operating parameters are the following: there were 25 tungsten carbide grinding bodies (ø = 20 mm), rotation velocity was 420 rpm, grinding periods were 15, 30, 45, and 60 min.

The granulometric composition of superfine SCM samples was determined with Delsa Nano (a sub-micron particle size analyzer) by photon correlation spectroscopy based on the dynamic light diffusion principle.

The average basalt fiber diameter was estimated by Lasentec D600 (a laser particle size analyzer) provided with PVM V819 (a laser and video microscopical investigation system) using a technology based on the focused beam reflection method (FBRM).

To measure hydration enthalpy changes in the SCM samples the Expert-001K-2 installation was used, calorimetric method being applied. While pretesting, the calorimetric system constant (k) was estimated by introducing a specified portion (2.5040 g) of potassium chloride salt (k = 580 J/K). To determine the sample hydration specific heat, SCM dry weight (4…5 g) measured by counterbalance was used, data being read up to hundredth values. The distilled water volume in the calorimeter inner chamber was 100 ml. Calorimetric measuring (final stage) was being carried out for two hours, the system temperature premeasuring periods vary between 10 and 15 minutes, tested mixture temperature changes being recorded up to 0.01 °.

Test samples of insulating material were procured by layer-by-layer deposition of SCM aqueous suspension (10 %) sprayed onto mineral wool. Then, the sample was processed in the drying oven (200 °C) until moisture was removed completely. Drying oven soaking temperature was selected on the basis of an ordinary production cycle of thermal insulation material manufacturing, the synthetic binding agent polymerization chamber maintaining max 240 °C.

To specify density, thermal conductivity, combustibility, and destruction temperature characteristics of the binding agent, sample testing was carried out on the basis of standard procedures.

3. Result and discussion

After having analyzed fiber ultra-photos (Figure 1) the fiber diameter range was found to be 3…7 μm.

The dimensional characteristics of the SCM samples obtained after different grinding periods are shown in Table 1.

Table 1. Characteristics of saponite containing material fraction after dry planetary ball mill grinding

Grinding time, t, min	Fraction size, nm / Content, %							Average size, nm
	№ 1	№ 2	№ 3	№ 4	№ 5	№ 6	№ 7	
15	1270/16	1375/22	1489/21	1613/15	53774/2	58235/2	63066 /1	5259
30	1111/8	2809/35	4134/6	36018/8	57274/4	66850/1	78028/1	8898
45	900/9	968/13	1041/15	1119/14	1245/22	1390/8	1502/9	1269
60	964 /20	1061/14	1142/14	1228/13	1321/11	1421/8	1528/6	1230

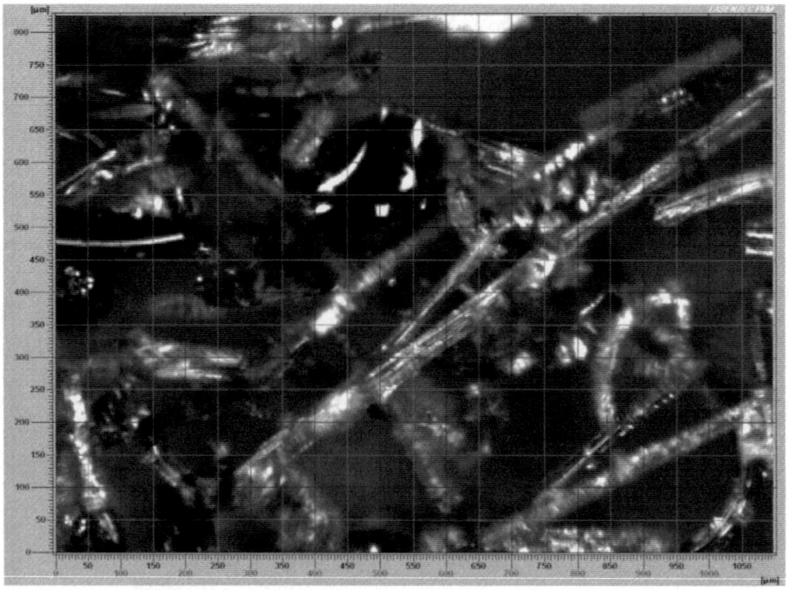

Figure 1. Fiber insulating material (ultra-photo).

The above-mentioned data (Table 1) proved that the SCM samples milled for 15 and 30 min to include coarse fraction, which can possibly cause uneven distribution of the binding agent along insulation fibers, thus increasing SCM hydration period. SCM samples milled for 45 and 60 min demonstrated more even degree of dispersion proportionate to the fiber insulation average diameter.

To specify an optimal SCM dispersion mode the experiments determining the hydration enthalpy values (ΔH) for highly dispersed saponite, containing material samples milled for 45 (SCM1) and 60 min (SCM2), were carried out. The results of the experiment are included in Table 2.

Table 2. SCM hydration specific enthalpy calorimetric measuring results.

Sample	m, g	k, kJ	Temperature, °C		Δt, °	$-\Delta H$, kJ/kg
			Input	2 h after experiment start		
SCM1	5.06	0.58	23.50	25.50	2.00	230
SCM2	5.02	0.58	23.50	25.50	2.00	230

Specific hydration heating (for 2 h) for SCM samples is 230 kJ/kg and is proportionate to hydration heat values of the main clinker mineral (belit $\beta - 2CaO \cdot SiO_2$) equal to 260 kJ/kg, thus proving the saponite containing material binding properties.

The experiments, measuring the SCM samples specific hydration enthalpy, proved the optimal grinding period for a planetary ball mill to be 45 min while dry grinding.

Figure 2 demonstrates a prototype of the thermal insulating material on the basis of the mineral binder.

The main quality characteristics of the specimen, produced under laboratory conditions, and those of the mineral wool mat (P35) are illustrated in Table 3.

The data of Table 3 demonstrate that the thermal conductivity value of the thermal mineral binder insulating material sample is 0.0379 W/(m·K). It is 4 % higher than that of the similar synthetic binder insulating material, but it does not exceed the standard values for this insulation grade.

Table 3. Test results

Sample	Density, kg/m^3	Humidity, %	Thermal conductivity, W/(m·K)	Combustibility group	Binding agent destruction temperature, °C
Mineral binder thermal insulating material sample	36.80 ± 0.20	1.2 ± 0.1	0.0379 ± 0.0020	noncombustible	over 1000
P35, mineral wool mat	34.49 ± 0.20	1.0 ± 0.1	0.0365 ± 0.0020	noncombustible	300...350

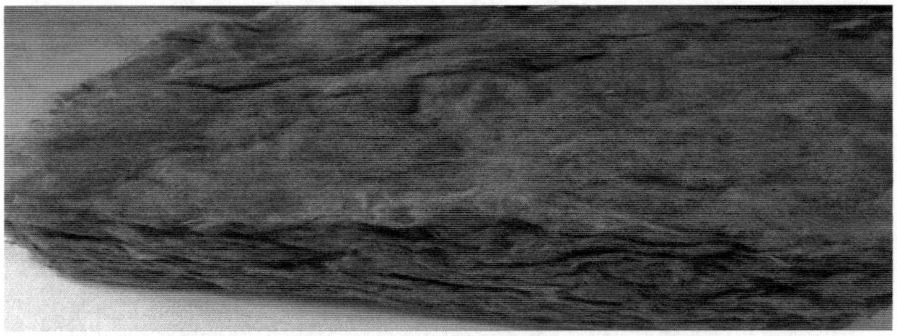

Figure 2. Mineral binder insulating material.

Moreover, the mineral binding agent can withstand minimum 1000 °C making it possible to operate with mineral binder insulating materials in high-temperature environments without emitting carcinogenic substances.

4. Conclusion

Summarizing the above-mentioned data and results it is possible to state the following: while using SCM as a binding agent it is possible to produce thermal insulating material in industrial scale without changing a mineral wool insulation production mode. In addition, such type of thermal insulation should be completely ecologically sound and should not fail by heating.

Acknowledgments

The investigation has been carried out with the financial support of the Ministry of Education and Science of the Russian Federation in the framework of the base section of the state task in research.

References

[1] Papadopoulos A 2005 *Energy and Buildings* **37(1)** 77–86
[2] Gunschera J *et al* 2013 *Building and Environment* **64** 138–145
[3] Tutygin A S, Aisenstadt M A, Aisenstadt A M, Makhova T A 2012 *Geoekologiya* **5** 379–383
[4] Tutygin A *et al* 2012 *J. of International Scientific Publications: Ecology & Safety* **6(1)** 45–54
[5] Glaser A M 2002 *Journal of Russian Chemical Society* **5** 57–63
[6] Veshnyakova L A, Aisenstadt A M, Frolova M A 2015 *Physics and Chemistry of Mater. Processing* **2** 68–72

High-Temperature TEM investigation of the phase composition and structure of the Zr-Y-O / Si-Al-N multilayer coatings

M Fedorischeva[1, 2], M Kalashnikov[1, 2], V Sergeev[1, 2] and I Bozhko[1, 2]
[1] Institute of Strength Physics and Materials Science of the Siberian Branch of the Russian Academy of Sciences, 2/4, Akademicheskiy ave., Tomsk, 634021, Russia
[2] Tomsk Polytechnic University, 30, Lenina ave., Tomsk, 634050, Russia

E-mail: fed_mv@mail.ru

Abstract. Deposition of nanostructured multi-layered coatings on the basis of Zr-Y-O was implemented by the pulse magnetron methods. Structural-phase statesand morphology of the nanostructured coatings were investigated by SEM and the high-temperature TEM method. Thehigh-temperature TEM revealed the presence of reversible phase transition of the tetragonal phase to the monoclinic phase, which can ensure stress relaxation and closure of surface cracks.

1. Introduction

Ceramics on the basis of Zr holds the leading place among fire-resistant constructional materials as it retains high mechanical properties up to 0.8…0.9 of the melting temperature of 3173 K. That is why ZrO_2coatingsare mainly used as thermal barrier coatings in hot turbine sections and other engine units.

It is known that ZrO_2 has three stable crystal structures depending on temperature: the monoclinicphase from the temperature of 1170 °C, the tetragonal phase within 1170 °C…2370 °C and the cubic phase from the temperature of 2370 °C. Mechanical properties of zirconium dioxide ceramics are known to be the function of phase composition and structure. Tetragonal zirconium dioxide possesses high strength and toughness. Various methods and technologies were improved for production and stabilization of the tetragonal phase of the zirconium dioxide material [1-2]. The most common method is the production of tetragonal zirconia by adding to it some stabilizing compounds. Analysis of published data shows that the yttrium oxide Y_2O_3 is commonly used as a stabilizing agent, which contributes to transformation of the tetragonal phase to monoclinic one at cooling.

A special attention is paid to reversible martensitic transformations in metal alloys (the so-called transformational conversion) due to the potential of their practical use in many fields of science, technology, medicine and industry. These alloys belong to the group of the so-called ‘smart’ functional materials, as they allow controllingtheir behavior [3-4]. These are transformation-hardening materials which are widely used in engineering practice as structural materials. The majority of such ceramic materials are developed on the basis of zirconium dioxide that is partially stabilized in the tetragonal phase [5-7]. The tetragonal phase is capable of a monoclinic martensitic phase transition. The phase transition is accompanied by development of the shear and volume strain, stress relaxation, and closure of surface cracks. The hardening effect can result in ceramic materials with the strength properties (fracture toughness and strength) comparable to that of structure materials.

Experiments of the direct observation mechanisms of martensitic transformations are rare. In [8] 'in-situ' by TEM was observed martensitic transition from the tetragonal ZrO_2 (*t*) to a monoclinic (*m*) phases. In this experiment, defect-free tetragonal ZrO_2 single crystals 0.2 μmin size were investigated. Phase transition of the tetragonal phase to monoclinic was initiated by some heterogeneity in individual particles, when the amplitude of the stress exceeds a critical value. Then the motion of interphase boundaries observed between the tetragonal and martensitic phases 'in-situ' [9].

The paper deals with investigation of change in grain structure, grain and interphase boundaries, structure and phase composition during heating mode 'in-situ' in the microscope column and followed by thermos elastic phase transition in the multi-layer coating based on Si-Al-N / Zr-Y-O in ZrO_2 layer.

2. Experimental

Deposition of the coatings was carried out in the KVANT-03MI [10] unit equipped with a magnetron with a mosaic zirconium-yttrium target. The magnetron was powered from a pulse source at the frequency of 50 kHz. The samples were placed in the chamber on the rotating table. The sample temperature during the deposition was 573 K.

The structural-phase state of the surface layers of the copper substrate was investigated by transmission electron microscopy (TEM) using the JEM-2100 microscope (Jeol Ltd., Japan). Foils were prepared by the 'cross-section' method using the ION SLISER-EM-09100IS installation (Jeol Ltd., Japan). The elemental composition of the coatings was determined by energy dispersive X-ray (MRSA) using the INCA-Energy microanalyzer (EDX) (Oxford Instruments) with the built-in SEM LEO EVO-50XVP.

3. Results

Figure 1 shows a cross-section of the Zr-Y-O / Si-Al-N coating obtained by TEM. SEM investigation using the mapping method of cross section of multilayer coatings shown that the layers of the multilayer coating actually consist of elements Zr, Y, O, Si, Al, N (Figure 1b, the red rectangle shows research area).

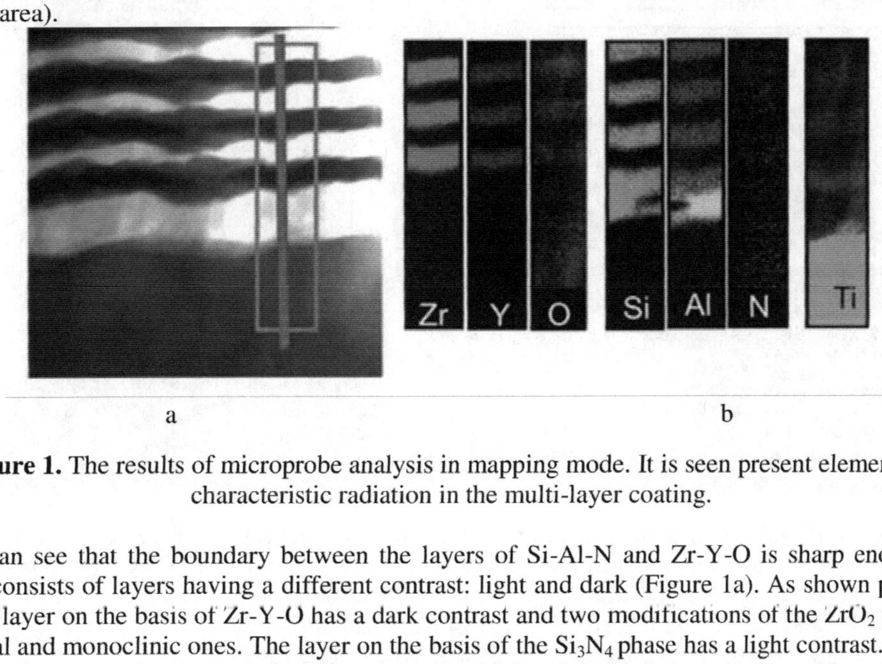

a b

Figure 1. The results of microprobe analysis in mapping mode. It is seen present elements in characteristic radiation in the multi-layer coating.

We can see that the boundary between the layers of Si-Al-N and Zr-Y-O is sharp enough. The coating consists of layers having a different contrast: light and dark (Figure 1a). As shown previously [11], the layer on the basis of Zr-Y-O has a dark contrast and two modifications of the ZrO_2 phase: the tetragonal and monoclinic ones. The layer on the basis of the Si_3N_4 phase has a light contrast.

The grain structure of the coating layer on the basis of Zr-Y-O is columnar (Figure 2a). The longitudinal size of the grains of the ZrO_2 phase coincides with the axis of growth of the coating. The height of the column is determined by the layer thickness, and the average cross dimension of the grains is less than about 80 nm. The height of the columns depends on the deposition time of the layer and may be adjusted depending on the application. In this particular case, the column height of the coating layer is about 1000 nm (Figure 1a).

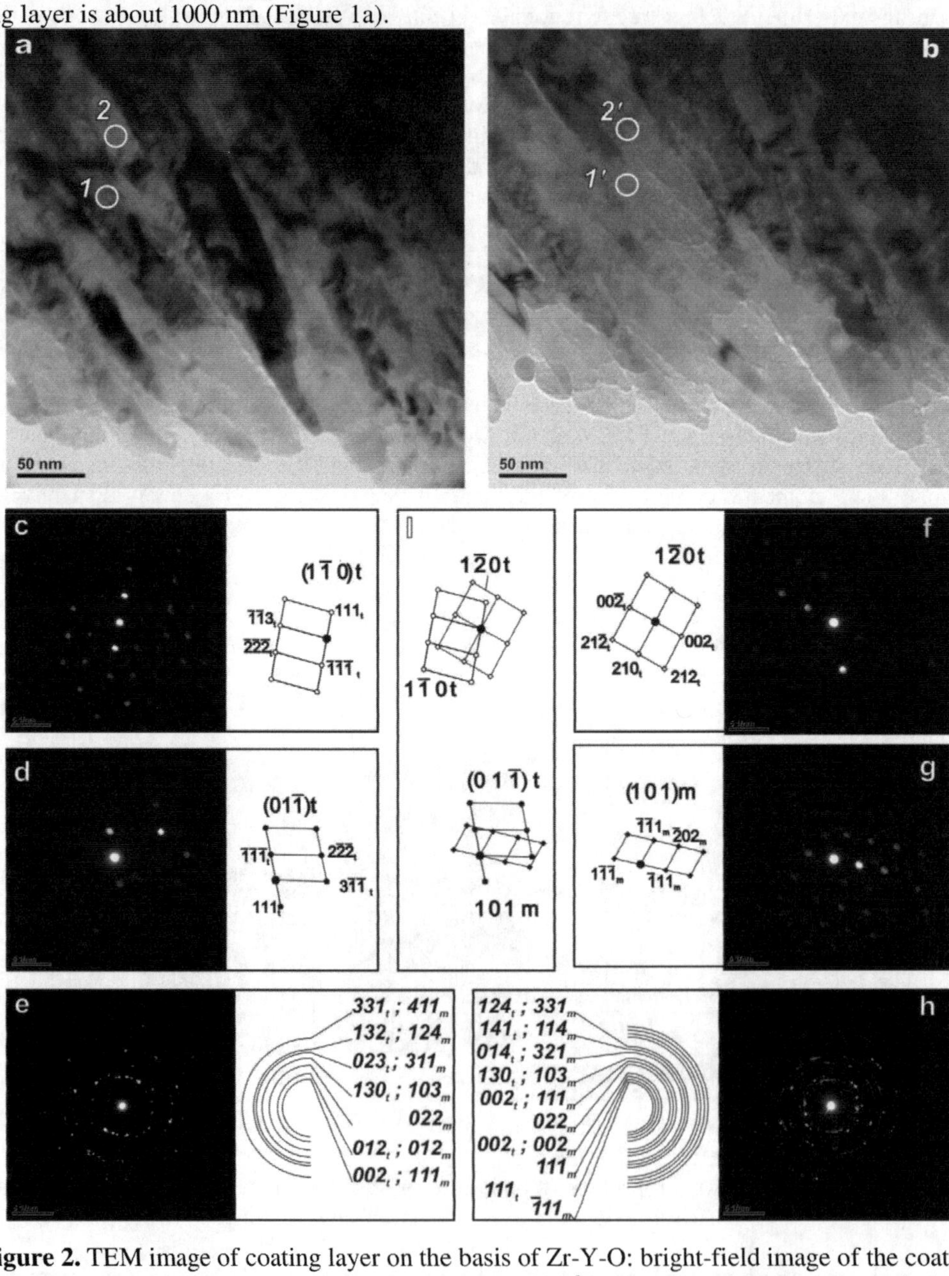

Figure 2. TEM image of coating layer on the basis of Zr-Y-O: bright-field image of the coating structure in the initial state (a) and after heating up to 900 °C (b); micro diffraction patterns for the initial state (c, d, e) and their indexing schemes; microdiffraction patterns and their indexing schemes for coatings, heated to 900 °C (f, g, h); schemes of misorientations of crystalline lattices of phases in the initial state and at $T = 900$ °C at sites 1 and 1' and 2 and 2'.

Figure 2a and Figure 2b show the bright field image and microdiffraction patterns obtained in the nanodiffraction mode from sites 1 and 1' (sites marked by circles in bright field image) and their indexing scheme for the initial state and at a temperature 900 °C (Figure 2c, f). In the initial state at RT we observes reflexes belonging to the plane of the reciprocal lattice ($1\bar{1}0$) of the ZrO_2tetragonal phase, then at 900 °C we see plane ($1\bar{2}0$) the same modification.

In this case, the heating led to a reorientation of the crystal lattice of the columnar grains of tetragonal ZrO_2. The angle of reorientation of grains is 18 degrees (Figure 2I). Figure 2 (f, g, h) shows diffraction patterns and indexing scheme obtained from the second site, marked on the bright-field image. It is evident that in the initial state at the micro diffraction pattern there are reflexes belonging to the plane ($01\bar{1}$) tetragonal phase ZrO_2. At heating foil up to 900 °C a tetragonal phase transfer to monoclinic one under the thermal stress. This phase transition is a martensitic.

Figure 2g shows micro electron diffraction with reflexes belonging to the plane 101 of the reciprocal lattice of the ZrO_2 monoclinic phase. These data are confirmed by the data obtained at interpretation of ring diffraction patterns. Figure 2e and Figure 2h shows microdiffraction patterns obtained in initial state and at a temperature of 900 °C. They can be identified as belonging to the tetragonal and the monoclinic phases ZrO_2. As shown in [12], microscopic mechanism of formation of a martensitic phase consists in the movement of dislocations of transformation along interphase boundary causing local reorganization of a lattice from high-temperature in a low-temperature phase. At the reverse martensitic transformation dislocations move in the opposite direction.

The reversible movement of phase boundaries between the low-temperature and high-temperature phases $m\text{-}ZrO_2 \leftrightarrow t\text{-}ZrO_2$ was found at research of a coating on the basis of Zr-Y-O is in the basis of the mechanism of transformation hardening [12].

Locally occurring martensitic transformation $t \rightarrow m$ in ZrO_2 contribute to closing the edges of the cracks formed during thermal cycling or other loads, thanks to what intensity of concentrators of tension near defects decreases. As a result the existing or again formed microcracks become stable at preservation of level of external loading [9]. It is promoted by martensitic phase transformations which are followed not only development of shifts, but also essential change of volume of material [11]. The transition from the tetragonal to the monoclinic phase is accompanied by an increase in phase by 4 % [12].

4. Conclusion

By SEM and TEM it has been established that coatings on the basis of Zr-Y-O produced by the magnetron sputtering methods have a nanograin column structure where the columns are spread through the entire coating thickness.

In the initial state layers on the basis of Zr-Y-O are two-phase and consist generally of the tetragonal phase ZrO_2 with a small amount of monoclinic one.

At heating layer in a column of TEM in the 'in-situ' mode we can observe: 1) turns of grains of the main phase together with change in the angle of disorientation crystallographic planes, 2) martensitic transition of the tetragonal phase to the monoclinic 3) modification of grain boundaries – their total length increases, the form of grains changes, in initial column grains there are cross boundaries, i.e. there is a process of fragmentation of grains.

Acknowledgments

The results were obtained using equipment of Center of collective use 'NANOTECH' of ISPMS RAS (CCU TSC SB RAS)

The work was supported by The Program of Basic Scientific Research of the State Academies of Sciences for 2013–2020. This work was partially financed from the grant of the President of the Russian Federation NH-2817.2014.1 awarded to The School of Academician V E Panin.

References

[1] Hannink R H J 2000 *J. Am. Ceram. Soc.* **83** 461
[2] Evans A G and Cannon R M 1986 *Acta Metall.* **34(5)** 761

IOP Conf. Series: Materials Science and Engineering **124** (2016) 012124 doi:10.1088/1757-899X/124/1/012124

[3] TamarinY 2002 *Protective Coatings for Turbine Blades* (USA ASM International)
[4] Hannink R, Kelly P and Muddle B 2000 *J. Am. Ceram. Soc.* **83** 461
[5] Langjahr A, Oberacker R and Michael J 2001 *J. Am. Ceram. Soc.* **84** 1301
[6] Huang X, Zakurdaev A and Wang D 2008 *J. Mater. Sci.* **43** 2631
[7] JiZ HaynesJ, and VoelklE 2001 *J. Am. Ceram. Soc.* **84** 929
[8] Chen I and Ciao Y 1985 *Acta Metallurgica* **33(9)** 1827
[9] Ciao Yand Chen-Wei I *Acta Met. Mater.*1990 **38(6)** 1163
[10] Sergeev V, Yanovsky V, Paraev Yu,. Kozlov S and Zhuravlyov S 2004 *Phys. Mesomech.* 7 333
[11] Fedorischeva M, Sergeev V, Kalashnikov M and Voronov A 2014 *Physical Mesomechanics of Multilevel Systems* 2014, AIP Conference Proceedings 1623, edited by V. E. Panin, et al. (American Institute of Physics, Melville, NY) pp 147–150.
[12] Daniel R, Martinschitz K, Keckes J and Mitterer B 2010 *Acta Mater.* **58** 2621

Combined treatment of steel, including electrospark doping and subsequent irradiation with a high-intensity electron beam

V D Klopotov[1], Yu A Denisova[2], A D Teresov[2,3], E A Petrikova[2,3], V V Shugurov[2], M A Seksenalina[1], Yu F Ivanov[1,2,3] and A A Klopotov[3,4]
[1] National Research Tomsk Polytechnic University, 30, Lenina ave., Tomsk, 634050, Russia
[2] Institute of High Current Electronics of the Siberian Branch of the Russian Academy of Sciences, 2/3, Akademicheskiy ave., Tomsk, 634055, Russia
[3] National Research Tomsk State University, 36, Lenina ave., Tomsk, 634050, Russia
[4] Tomsk State University of Architecture and Building, 2, Solyanaya sq., Tomsk, 634003, Russia

E-mail: yufi55@mail.ru

Abstract. A thermodynamic analysis of phase transformations taking place during doping of steel with tungsten and titanium has been performed. The studies on the surface layer of steel modified using the combined method (electrospark doping and the subsequent electron-beam treatment) have been carried out. Formation in the surface layer of a multi-phase submicrocrystalline structure with high strength properties has been revealed.

1. Introduction

The method of electrospark doping (ESD), based on the phenomenon of electrical erosion and polar transfer of the anode material to a cathode (a work piece) during a pulsed electric discharge between them in a gaseous medium, differs from other surface treatment methods in its simplicity, reliability, and cost-effectiveness [1]. Depending on the anode material, an extended surface layer with high strength, tribological, and other properties is formed on the work piece. The disadvantages of ESD are as follows: high surface roughness of the treated surface of the work piece, the presence of cracks, discontinuities, and micropores. To reduce the roughness introduced by electrospark doping, the methods based on the mechanical impact on the modified surface are used (surface-plastic strain, running the ball, nonabrasive ultrasonic treatment, etc.) [2, 3], as well as treatment with concentrated energy flows (plasma flow, electron and ion beams, laser beams) [4, 5].

The paper is devoted to the analysis of the structure and properties of the surface layer of carbon steel subjected to electrospark doping and subsequent treatment with a high-intensity pulsed electron beam.

2. Material and methods of study

The material of the study was carbon steel 65G. Samples with dimensions of 25x25x50 mm were hardened at the temperature from 820 °C in oil and tempered at 350 °C. The electrospark treatment was carried out in air using installation 'Elitron 22A' at the discharge current of 2 A and the voltage on the doping electrode of 33 V. Technically pure tungsten was used as an electrode. The electron-beam treatment was performed on the basis of installation 'SOLO' [5]. The studies on the structure of the modified layer were carried out using methods of optical (NEOFOT-32) and scanning electron (SEM-515 Philips) microscopy, the X-ray diffraction analysis (diffractometer XRD 6000). The microhardness (PMT-3, the load on the indenter is 0.1 N) and the durability (Tribometer CSEM, Switzerland) were also measured.

3. Research results and their discussion

Doping of steel with tungsten allows a significantly improvement of its durability. This is due to the fact that doping of wolfram (W) increases the volume share of tungsten carbides during quenching at normal temperature. It is important to note that an increase in the volume fraction of tungsten carbides leads to formation of a finer grain and, therefore, to the improvement of mechanical properties, in particular higher viscosity.

In system Fe-W-C, the atoms of Fe and W belong to transition metals of VI*A* group, and carbon is a nonmetal [9]. In system W-C, the formation of three compounds has been established: $W_2C(\beta)$ (prototype Fe2N, space group (sp. gr.) *P63/mmc*), γWC_{1-X} (prototype NaCl, sp. gr. $Fm\bar{3}m$) and WC(δ) (prototype WC, sp. gr. *P6m2*) (Figure 1).

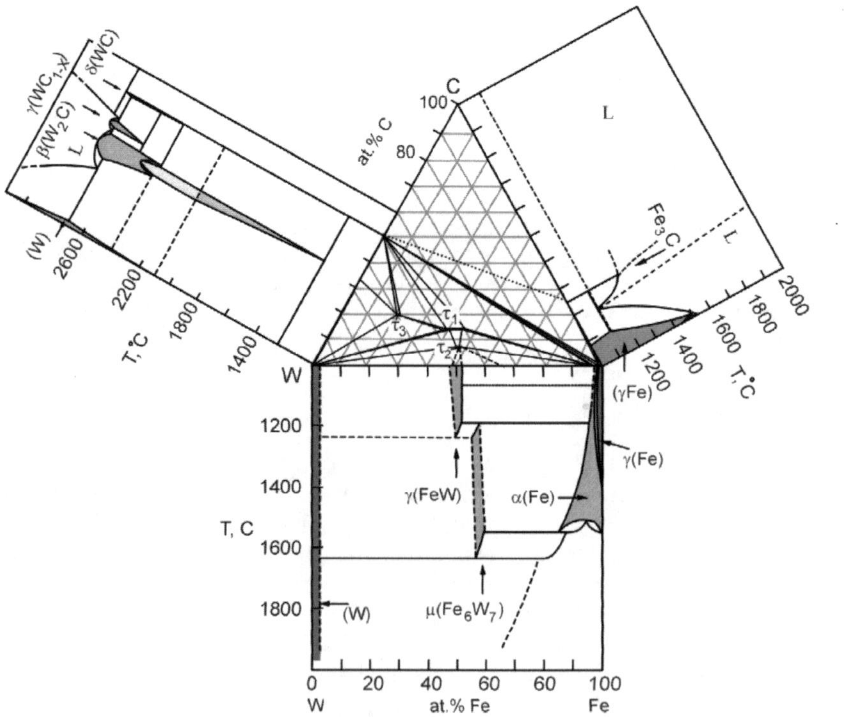

Figure 1. Binary diagrams of systems Fe-W, Fe-C, WC [7] and the isothermal section of ternary system Fe-W-C at 1000 °C [8].

In system Fe-C the following has been revealed: the formation of stable compound Fe_3C (prototype Fe_3C, sp. gr. *Pnma*) and solid solutions (α-Fe) and (γ-Fe), and, at increased pressures, the stabilization of compound Fe_7C_3 (prototype Cr_7C_3 , sp. gr. *Pnma*,) take place. Carbon is a nonmetal and belongs to group IV*B* of the periodic table. Carbon atoms have four electrons on outer shells *s*- and *p*- ($2s^22p^2$). As compared to the main alloy-forming metals, carbon atoms are much smaller and when alloyed with metals they form interstitial phases ($R_C = 0.0916$ nm, $\delta_R = (R_W - R_C)/R_W = 0.38$ and $\delta_R = (R_W - R_C)/R_{Fe} = 0.28$). Since treatment of the material with an electron beam may lead to formation of new metastable compounds, we shall briefly mention metastable iron-based carbides. Thus, in [7, 9] it has been found that higher carbides Fe_7C_3 and Fe_2C stabilize at increased pressures. Carbide Fe_7C_3 has hexagonal structure $D10_1$ (prototype Cr_7C_3, sp.gr. *Pnma*). Compound Fe_2C is metastable with a hexagonal structure (ε-carbide, sp. gr. $P6_322$) and is released during tempering of steel [10]. Moreover, [9] represents the data on the release of metastable carbides ηFe_2C (sp. gr. *Pnnm*) and χFe_5C_2 (prototype B_2Pd_5, sp. gr. C2/c), which are found during tempering of steel.

In system Fe-W the following has been discovered: three intermediate phases $\lambda(Fe_2W)$ ($P6_3/mmc$, prototype $MgZn_2$)), $\mu(Fe_7W_6)$ (prototype Fe_7W_6, sp. gr. $R\bar{3}m$), γ(FeW) (prototype MoNi, sp. gr. $P2_12_12_1$), and solid solutions (W), (α-Fe) and (γ-Fe). In this system, alloy-forming elements have size factor $\delta_R = 0.139$ ($R_W = 0.148$ nm, $R_{Fe} = 0.1274$ nm), which is significantly less than 0.2; however, during alloying metals Fe and W form solid solutions with a very limited solubility. The solubility in W (α-Fe) with a decrease in the temperature reduces from 14.3 % (at.) at 1548 °C up to 4.6 % (at.) at 1190 °C [7]. The solubility of Fe in (W) < 2.6 % (atm.) is at a temperature of 1677 °C [7].

In ternary system Fe-W-C a whole range of intermediate stable and metastable compounds with complex crystalline structures that exist in different temperature and concentration ranges has been discovered (Figure 1).

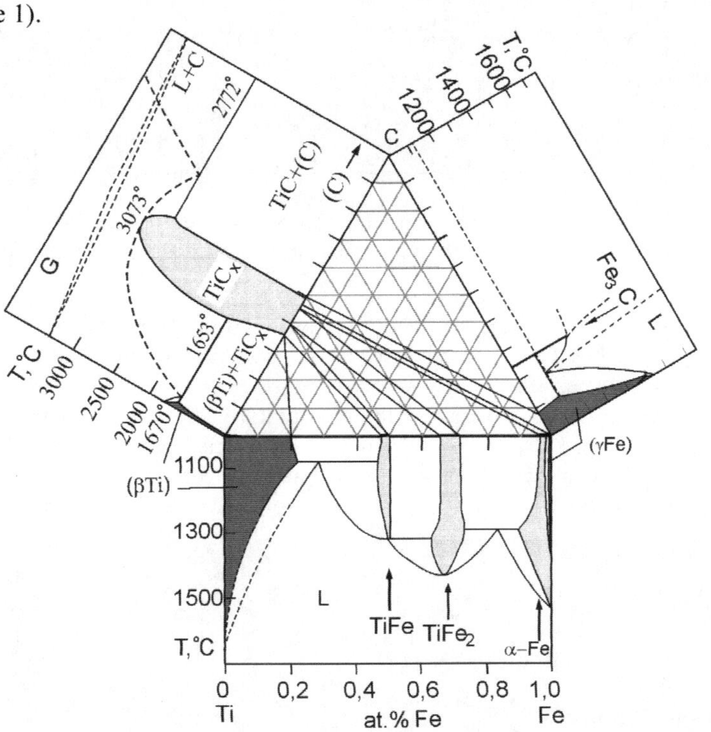

Figure 2. Binary diagrams of systems Fe-Ti, Fe-C, Ti-C [7] and the isothermal section of ternary system Fe-Ti-C at 1000 °C [12, 13].

Two intermediate phases, $TiFe_2$ (structure C14, prototype $MgZn_2$, space group $P6_3/mmc$) and TiFe (structure B2, prototype CsCl, sp. gr. $Pm\bar{3}m$), are formed in system Fe-Ti (Figure 2). In regard to initial components, there are solid solutions based on Fe and Ti. Solid solutions based on Fe and Ti exhibit transformations which are caused by polymorphic transitions in pure metals Fe and Ti [7].

In system Ti-C there is one compound of carbide TiC(δ) (structure B1, prototype NaCl, sp. gr. *Fm*3*m*) with a wide range of homogeneity. Compound TiC(δ) melts congruently at 3073 ± 25 °C. The homogeneity range of phase TiC near the solidus phase occupies from 32 up to 50 % (at.) [7]. From a structural point of view, compound TiC(δ) is a solid solution of non-metal atoms in octa-interstitial sites of the FCC-lattice of metal [11]. Consequently, in the literature carbide TiC (δ) is often referred to as TiC_x.

Within the isothermal section of system Fe-Ti-C there are no ternary compounds formed based on binary compounds of corresponding binary systems. It is important to note that in system Fe-Ti-C there are no intermediate ternary compounds, as it occurs in ternary system Fe-W-C.

Electrospark doping of steel, as mentioned above, is accompanied by formation of a highly-defected surface layer containing microcraters and microcracks, as well as buildups of metal. Our case has also shown formation of a similar structure of the doping surface (Figure 3a and Figure 4a).

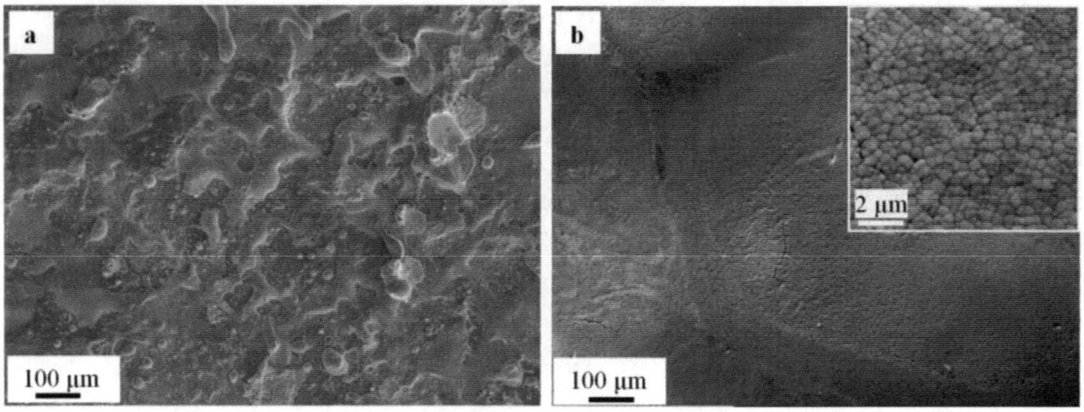

Figure 3. The structure of the steel surface subjected to electrospark doping with tungsten (a) and the subsequent irradiation with a high-intensity pulsed electron beam (b).

The subsequent irradiation of the electroexplosive doping surface with a high-intensity electron beam leads to melting of the surface layer with a thickness of up to 30 microns. The doping surface smooths microcraters and buildups of metal are not registered (Figure 3b and Figure 4b). In some cases, microcracks are detected on doping surfaces. High-rate crystallization of the molten layer is accompanied by formation of a submicrocrystalline structure, the characteristic image of which is shown in Figure 3b and Figure 4b. It should be noted that average sizes of crystallites formed in the surface layer of steel are as follows: doped with tungsten – ≈ 300 nm; doped with titanium alloy – ≈ 900 nm (Figure 3 and Figure 4).

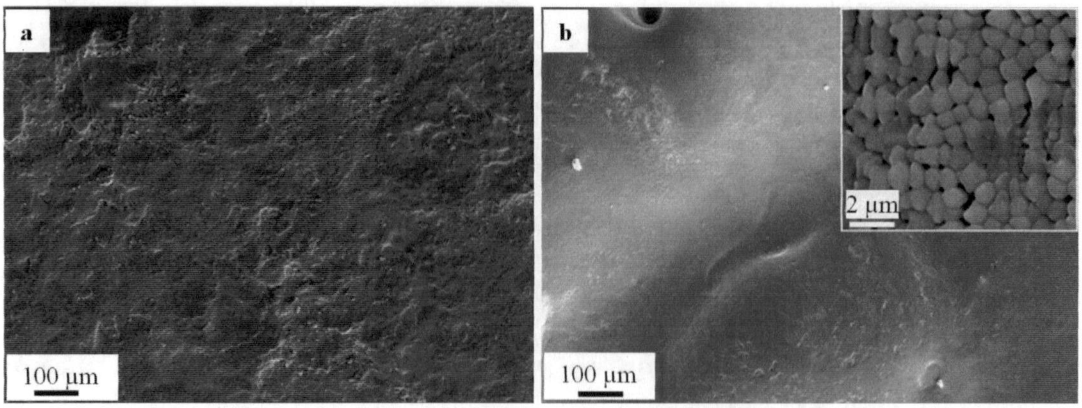

Figure 4. The structure of the steel surface subjected to electrospark doping with titanium (a) and the subsequent irradiation with a high-intensity pulsed electron beam (b).

The elemental composition of the doped layer has been studied using the electron microprobe analysis method. The studies have shown that after electrospark doping with tungsten, the concentration of the alloying element in the steel surface layer was ≈ 7 al. %; after doping with titanium – (30…40) at. %. Additional irradiation of the modified layer with a high-intensity electron beam led to remelting of the surface layer. Consequently, the concentration of the doping element in the surface layer of steel decreased and in system Fe-W-C it amounted to (3…5) at. %; in system Fe-Ti-C it was equal to (10…18) at. %

Mechanical and tribological tests of the modified steel have been carried out. It has been established that formation in the surface layer of a multiphase nano- and submicrocrystalline structure leads to a substantial increase in mechanical and tribological properties of steel.

4. Conclusion

A thermodynamic analysis of phase transformations taking place during doping of steel with tungsten and titanium has been performed. It has been shown that doping of steel with carbide-forming elements results in formation of solid solutions, inclusions of carbide phases, and metal compounds. The study of the surface layer of steel subjected to electrospark doping has been carried out. Formation of a high-relief modified layer containing microcraters and microcracks has been revealed. It has been shown that irradiation of the doped layer with a high-intensity electron beam leads to smoothing of the modification surface and to formation of a submicron and nanocrystalline structure. It has been established that formation in the surface layer of a multiphase submicrocrystalline structure leads to a substantial increase in mechanical and tribological properties of steel.

Acknowledgments

The study is carried out with the financial support of the grant of the Russian Foundation for Basic Research (project number 14-08-31706) supported by 'The Tomsk State University Academic D.I. Mendeleev Fund Program' based on grant (№8.1.42.2015).

References

[1] Lazarenko N I 1976 *Electrospark doping of metal surfaces* (Moscow: Mashinostroenie)
[2] Tarelnik V B *et al* 2011 *Compressor and power engineering* **1** 50
[3] Ledkov E A *et al* 2010 *Fundamental and applied problems of engineering and technology* **5** 72
[4] Radek N and Bartkowiak K 2011 *Physics Procedia* **12** 499
[5] Laskovnev A P *et al* 2013 *Modification of the structure and properties of eutectic silumin using the electron-ion-plasma treatment* (Minsk: Nauka)
[6] Kornilov I I 1964 *Metalloids and the interaction between them* (Moscow: Nauka)
[7] Lyakishev N P 1996-2000 *State diagrams of binary metallic systems* **1-3** (Moscow: Mashinostroenie)
[8] Pollock C B and Stadelmaier HH 1970 *Metall. Trans* **1** 767
[9] ViUars P and Calvert L D 1985 *Pearson's Handbook of Crystallography Data for Intermetallic Phases* **1** (Ohio: Metals Park)
[10] Jack K H 1951 *J. Iron Steel Inst.* (L.) **169** 26
[11] Rempel A A 1992 *Effects of ordering in nonstoichiometric interstitial compounds* (Yekaterinburg: Nauka)
[12] Raghavan V 2003 *J. Phase Equilib* **24** 62
[13] Jonsson S 1998 *Metall. Mater. Trans. B* **29B** 371

The character of fracture of iron based thermal coating during fretting

Zh G Kovalevskaya[1,2], **E A Kovalevskiy**[1,3] **and M A Khimich**[1,3]
[1] Institute of Strength Physics and Materials Science of the Siberian Branch of the Russian Academy of Sciences, 2/4, Akademicheskiy ave., Tomsk, 634021, Russia
[2] National Research Tomsk Polytechnic University, 30, Lenina ave., Tomsk, 634050, Russia
[3] National Research Tomsk State University, 36, Lenin ave., Tomsk, 634050, Russia

E-mail: zhanna_kovalevskaya@mail.ru

Abstract. The character of destruction of thermal coatings during fretting has been investigated. An iron based plasma coating has been tested with oscillation amplitude from 30 to 200 microns. The tests were conducted in air. It has been determined that the main factor influencing the rate of the wear of the coating during fretting corrosion is the size of the coating area involved into the wear process. The coating exhibits high wear resistance when the amplitude of the oscillation is commensurate with the size of the sprayed particles. During destruction of the coating the leading role belongs to fatigue-oxidation processes. The wear of the coating acquires a catastrophic character when coating macro defects – pores and interlayer boundaries – are involved into the wear process.

1. Introduction

The stress-strain state of the system under friction representing a composition with a coating is defined by two interfaces: 'counterface – coating' and 'coating – substrate'. During wear of the composition with a thermal coating it is necessary to consider the interfaces presence in the coating which are formed by layering monolayers of a sprayed material on each other [1, 2]. The main bulk of solid oxide inclusions and pore space is situated at interlaminar boundaries. It leads to arising of stress concentrators on internal interfaces. The structural features of the thermal coating appear during arising of all kinds of wear including fretting [3-5]. Fretting is a kind of corrosion-mechanical wear of contacting bodies at small amplitudes of relative vibratory motions of rubbing surfaces. Staging is typical for the development of destruction processes during fretting. Wear is a result of the process of setting and fatigue failure in the first stage of fretting. In the second stage the surface destruction is due to the fatigue-oxidation processes. The processes of fatigue and corrosion failure are solidary in the third stage. Each stage of the process feature is defined by the nature of contacting materials, their structure and physical and mechanical properties. The fracture mode of iron based plasma coatings and a role of structural components of the coating in this process are considered during coating wear in conditions of fretting.

2. Materials and methods

Thermal coatings obtained by the method of plasma spraying were investigated [2]. The coating of the iron-based alloy doped with 2.0 mas. % C, 8.2 mas. % Si, 5.1 mas. % Al was sprayed on the substrate of the preliminary sandblasted carbon steel. The coating thickness was about 2 mm, the coating had a heterogeneous structure consisting of particles with non-identical internal structure form and properties [6]. The average value of coating microhardness was 5800 MPa.

Tribological tests of investigated coatings for fretting resistance were carried out in conditions of air vibratory displacement. A "plane-plane" test circuit with the motion amplitude of 30 μm was used [7]. The alloy steel was a material of the counterface. The material was pre-quenched martensite tempered to sorbite. The counterface microhardness was 4440 Pa. To evaluate the role of macrostructure elements of the coating involved in the process of wear during fretting, the tests in the vibratory displacement range of 30, 60, 100 μm and the load equal to 25 MPa, as well as in the ranges of 100, 150, 200 μm and the load of 50 MPa were carried out. The test duration was $6{\cdot}10^5$ cycles. Total linear wear was being continuously recorded during tests. A friction coefficient was calculated in the area of the steady friction torque. Then, in order to determine the final amount of linear wear of the coating, a friction track of the specimen with the coating was registered by means of a strip chart recorder. Thus, the main criteria for evaluation of tribological properties were linear wear of a tribopair and coating. The wear character was evaluated by the metallographic analysis of friction tracks fragments of the coating and counterface with the characteristic tribo-failure.

3. Results and discussion

When the material has a heterogeneous structure consisting of individual particles, the main factor affecting the intensity of the coating wear under fretting is the size of the coating area involved in the wear process. The minimum of oscillation amplitude established in the experiment is a small part of the longitudinal size of sprayed particles [6]. Hence, the interaction and the contact of individual claws of a counterface micro pimple arise to the elements of the coating microstructure. The structure with the large assurance coefficient is formed inside the sprayed particles during the process of plasma spraying. The main phase of solid solution hardening and the secondary phases of hardening by fine particles promote high deformation and fracture resistance of the coating particles [6]. Macrostructural elements of the coating can serve as a main source of failure during the oscillation amplitude increase. These elements represent pores and boundaries between particles and interlaminar interfaces [6].

The analysis of obtained results showed that an increase in vibratory displacement amplitude enhances the wear of the coating and the tribopair. The growth of vibratory displacement amplitude from 30 to 100 μm under the load of 25 MPa leads to twice as much of wear increase of both coating and tribopair (Figure 1).

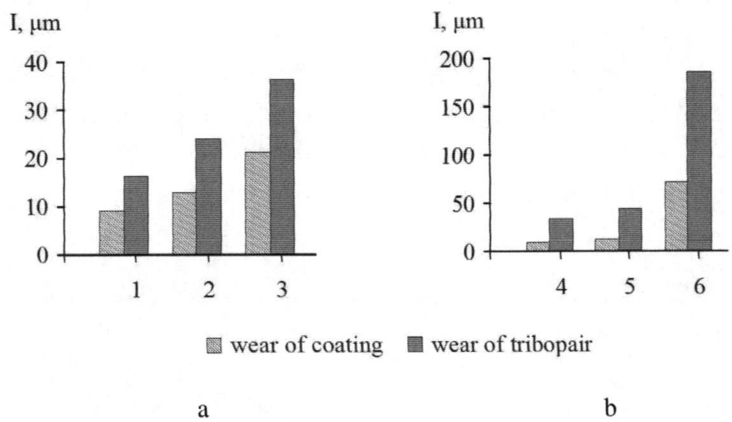

Figure 1. Wear values (I) of tribopairs and coatings under the test load of 25 MPa (a) and 50 MPa (b) and vibratory displacement amplitude: 1 – 30 μm; 2 – 60 μm; 3, 4 – 100 μm; 5 – 150 μm; 6 – 200 μm.

The load increase from 25 MPa to 50 MPa with vibratory displacement amplitude of 100 μm has practically not affected the wear. The wear of coating and tribopair increases during vibratory displacement amplitude increase up to 150 μm (Figure 1). The wear increase is due to the intensification of processes of setting and abrasion of wear particles on the friction surface. A wear pattern which is typical for fatigue-oxidation processes is observed during tests with vibratory displacements amplitude of 100 μm. Caverns and small amount of adhered wear particles are observed in the area of the friction track. (Figure 2a, b). A fatigue-oxidation wear mechanism is confirmed by optical metallography of the coating cross section after wear. As a result of fatigue-oxidation processes during fretting, the thickness of oxidized layers in the coating can be tens of micrometers. The other important outcome of the worn coatings structure analysis is the that roughening of coating fragments is of a wave nature. The size of these fragments is comparable to the motion amplitude [7].

Along with the fatigue-oxidation damages, the traces of abrasive damages with the increase in vibratory displacement amplitude up to 150 μm are clearly visible (Figure 2c, d). The friction coefficient increases with increasing vibratory displacement amplitude. Friction coefficient has values 0.35, 0.4, 0.5 and 0.55 when vibratory displacement amplitude equals 30, 60, 100 and 150 μm, respectively. The friction coefficient increase is due to the fact that wear yield from the friction zone is facilitated under large amplitude values. Thus the conditions for the direct contact of friction surfaces in the large area are created. In turn, the contact area increase leads to the increase of friction force and, hence, the friction coefficient.

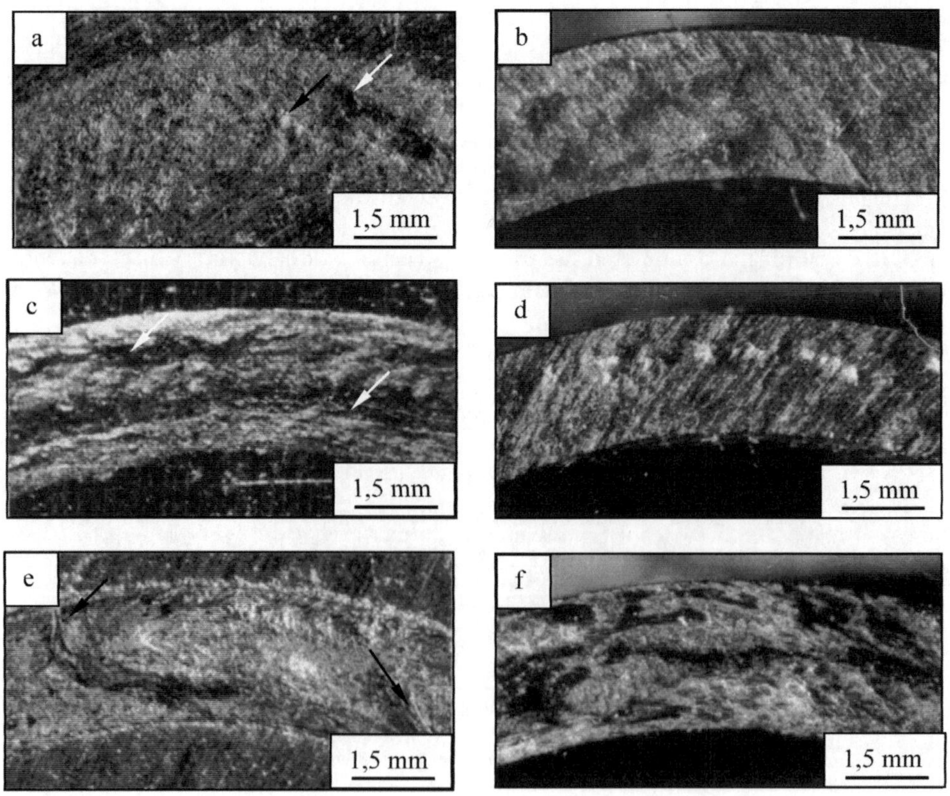

Figure 2. Fragments of coated friction tracks (a, c, e) and counterface (b, d, f). An experiment with amplitude vibratory displacement: a, b – 100 μm; c, d – 150 μm; e, f – 200 μm. Arrow marks on the photo show the following: white – cavity; black – adhered wear particles (a) should be abrasive damage (c), burrs (d).

The coating wear increases up to 70 μm and the tribopair wear increases up to 180 μm with increasing vibratory displacement amplitude up to 200 μm. A friction coefficient becomes 0.95. In this case the coating fragments are involved in the setting process. These fragments exceed the size of individual sprayed particles. The process of interaction of the coating material with the counter body proceeds to macro. Catastrophic wear of the coating illustrates the nature of the tribo damage of the friction surface (Figure 2e, f). Multiple teasers, cut away places and mutual material transfers are observed on the surface of the coating and its associated counterface.

The catastrophic wear of coating can be described as follows. Stress oscillations, enhanced in the places of contact of the counterface with the coating surface, occur at the first stage of fretting. As a result, crack extension occurs in pores at the interfaces between the particles. Formed fragments are fragile crumbles from the coating surface, and large vibratory displacement amplitude intensifies wear yield of the friction zone. The defect structure accumulation in the coating barrier layer occurs at the second stage of fretting. Then the formation and the rotational movement of large fragments and their separation in the form of wear particles occur. In this case few monolayers of the coating are involved in the wear process. Such intensive wear leads to the total coating wear.

It is not recommended to work with vibratory displacement amplitude of more than 150 μm. Therefore, these parameters exceeding significantly the fragments of the coating macrostructure are involved in the wear process. The size of these fragments is comparable with the size of several sprayed particles and interparticle defects. That leads to catastrophic wear.

4. Conclusions

The main role in the iron-based coating destruction during fretting has fatigue-oxidation processes. The main factor influencing the coating wear intensity is the size of the coating area involved in the wear process. The coating shows high wear resistance which is comparable to and exceeds the wear resistance of the counterface material when vibratory displacement amplitude is less than 150 μm, which is comparable to the sprayed particles size. Coating macrodefects (pores and interlayer boundaries) are involved in the process of wear when vibratory displacement amplitude increases up to more than 150 μm. In this case the coating wear becomes catastrophic.

References

[1] Pawlowski L 2008 Science and Engineering of Thermal Spray Coatings (Wiley-Blackwell)
[2] Hamidreza S J 2012 Advanced Plasma Spray Applications (InTech)
[3] Chaudhry V, Satish V Kailas 2013 *Jour. Wear of Materials* **301** 524–539.
[4] Attia H, Meshreki M, Korashy A, Thomson V and Chung V 2011 *J. Tribol. Internat.* **44** 1407
[5] Carrasquero E J, Lesage J, Puchi-Cabrera E S and Staia M H *J. Surface and Coatings Technology* **202** 4544–51
[6] Klimenov V A, Ivanov Yu F, Perevalova O B and Senchilo Z G 1997 *J. Materials and Manufacturing* 12 849–861
[7] Kovalevskaya Zh G *The Bulletin of the Tomsk Polytechnic University* **315** 128–133

The analysis of speed increase perspectives of nanostructuring burnishing with heat removal from the tool

V P Kuznetsov[1,2], A S Skorobogatov[1], V G Gorgots[3] and A S Yurovskikh[1]
[1] Ural Federal University, 19, Mira st., Yekaterinburg, 620002 Russia
[2] Tomsk Polytechnic University, 30, Lenina ave., Tomsk, 634050, Russia
[3] Kurgan State University, 25, Gogolya st., Kurgan, 640669, Russia

E-mail: wpkuzn@mail.ru

Abstract. The work deals with investigation of opportunities of speed increase of nanostructuring burnishing due to heat removal from the contact area of severe plastic deformation using friction via the tool. The work has analyzed changes of the structure, thickness and quality of the modified layer, while increasing treatment speed of the tool without heat removal and with heat-cooling system. It is established that the cooling system of the tool indenter with a cooling capacity of 10 W allows for the two-times increase of the critical speed of burnishing, and when exceeding this speed, the deformable material does not turn into a nanostructuring state. It is shown that heat removal provides for a stable maintenance of the indenter temperature, an increase in thickness of the nanostructured layer up to 5 μm and roughness up to Ra = 150 nanometers at processing speed up to 0.17 m/s.

1. Introduction

The linear speed of the nanostructuring burnishing and the temperature in the contact area of the tool have an essential effect on stability of structure dispersion and the surface layer quality. The work has investigated the influence of thermostrengthened steel processing speed on roughness and shear elasticity of the surface layer material [1]. The critical speed of nanostructuring burnishing was determined, at the increase of which an essential deterioration of processed surface layer roughness, loss of shear stability and material destruction occur. In works [2-4], it is shown that, at severe plastic deformation of constructional materials, the temperature increase in the area of deformation leads to a decrease in microhardness and in extent of structure dispersion.

The problem of nanostructuring burnishing speed increase at finishing processing of high-precision parts is connected with ensuring an economically reasonable processing rate and tool life

The objective of the research is aimed at identification of a possibility to increase speed of nanostructuring burnishing of quenched and tempered cemented 20Cr4 steel by means of heat removal from the tool indenter. The work solves the problem of experimental determination of dependence of structure refinement, thickness and quality parameters of the modified layer on the linear processing speed and indenter temperature in the conditions of dry friction and application of the tool with the heat-cooling system.

2. Methodology and experiment equipment

The experimental investigation is conducted when processing a disk-type specimen made from 20Cr4 steel (Fe rest; C 0.21; Cr 0.84; Ni 0.08; Mn 0.57; Si 0.27; Cu 0.18; P 0.017; S 0.019 (wt.%)) with a diameter of 80 mm and 12 mm thick. During preliminary preparation, the specimen underwent cementation, quenching, low tempering and turning. Quenching was carried out at a temperature of 850 °C, low tempering – at 120 °C. After chemical heat treatment, the material had a structure of lath martensite with the size of laths equal to 0.3…0.5 μm and the hardness of 64…65 HRC. The turning of specimen flat surfaces was carried out by a cemented metal carbide tips of Sandvik WNANOMETERSG 080408-SM at a speed of 80 m/min and feed of 0.06 mm/rev. After turning, the surface roughness amounted to Ra = 0.25 μm, and microhardness – 900…950 $HV_{0.5}$.

Nanostructuring burnishing in the conditions of dry friction was carried out on Takisawa EX-310 CNC turning center (Figure 1b) at normal force F_b on indenter 200 N and tool feed f_b=0.04 mm/rev. At this, spindle rotational speed n was maintained at 150 rpm. In proportion to the tool feed, linear processing speed v_b was changed within the range of 0.06…0.6 m/s. The indenters were used with semispherical tips from ultrafine grained cubic dense boron nitride (DBN) with a 2 mm radius and with working surface roughness equal to Ra=50 nm. Friction coefficient μ of the indenter in contact with the processed surface amounted to 0.3…0.34 and friction force F_{fr} equaled 60…68 N. One plane of the specimen was processed by the burnishing tool without the heatsink, another by the tool with a heat-cooling system.

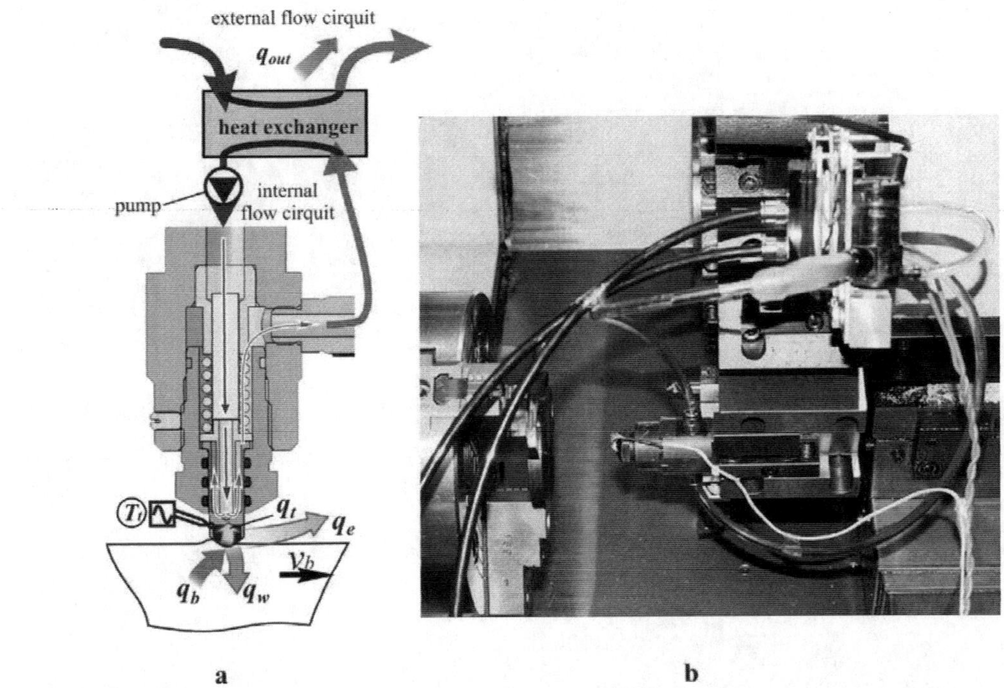

Figure 1. The heat flow diagram at nanostructuring burnishing by the tool with a two-circuit liquid cooling system (a) and the snapshot of the experimental set-up (b).

In the course of the nanostructuring burnishing of 20Cr4 steel with friction force of 60…68 N and speed up to 0.6 m/s, maximum heat capacity q_b amounted to 36…41 W.

A burnishing tool with a two-circuit liquid cooling system (Figure 1a) was used for heat removal q_t passing through the indenter [5]. In the first circuit due to coolant circulation by means of a pump, heat transfer occurs from the indenter into the heat exchanger. In the heat exchanger Peltier thermoelectric modules, which cool the coolant, are used. The cooling of Peltier thermoelectric modules is carried out by the coolant of the second circuit (Russian Federation Patent No. 150111). The cooling capacity of the cooling system reaches 10 W.

The measurement of the indenter tip temperature was carried out by the thermocouple of a K-type. The research of the surface layer structure after nanostructuring burnishing was conducted by means of microscopes JEOL JEM 200CX, JEOL JEM 2100 and Zeiss CrossBeam AURIGA by TEM and SEM methods. The measurement of surface microhardness was carried out by Vickers method using a microhardness tester Leica at loading of 0.5 N. The roughness was determined by optical 3D – profiler Veeco WYCO NT1100 by PSI method.

3. Results

The transmission electronic microscopy of the surface layer foils after processing by the tool with the heat-cooling system has shown that, at burnishing speed from 0.06 to 0.12 m/s, the nanocrystalline structure with grains of 40…100 nm in size (Figure 2a) is formed.

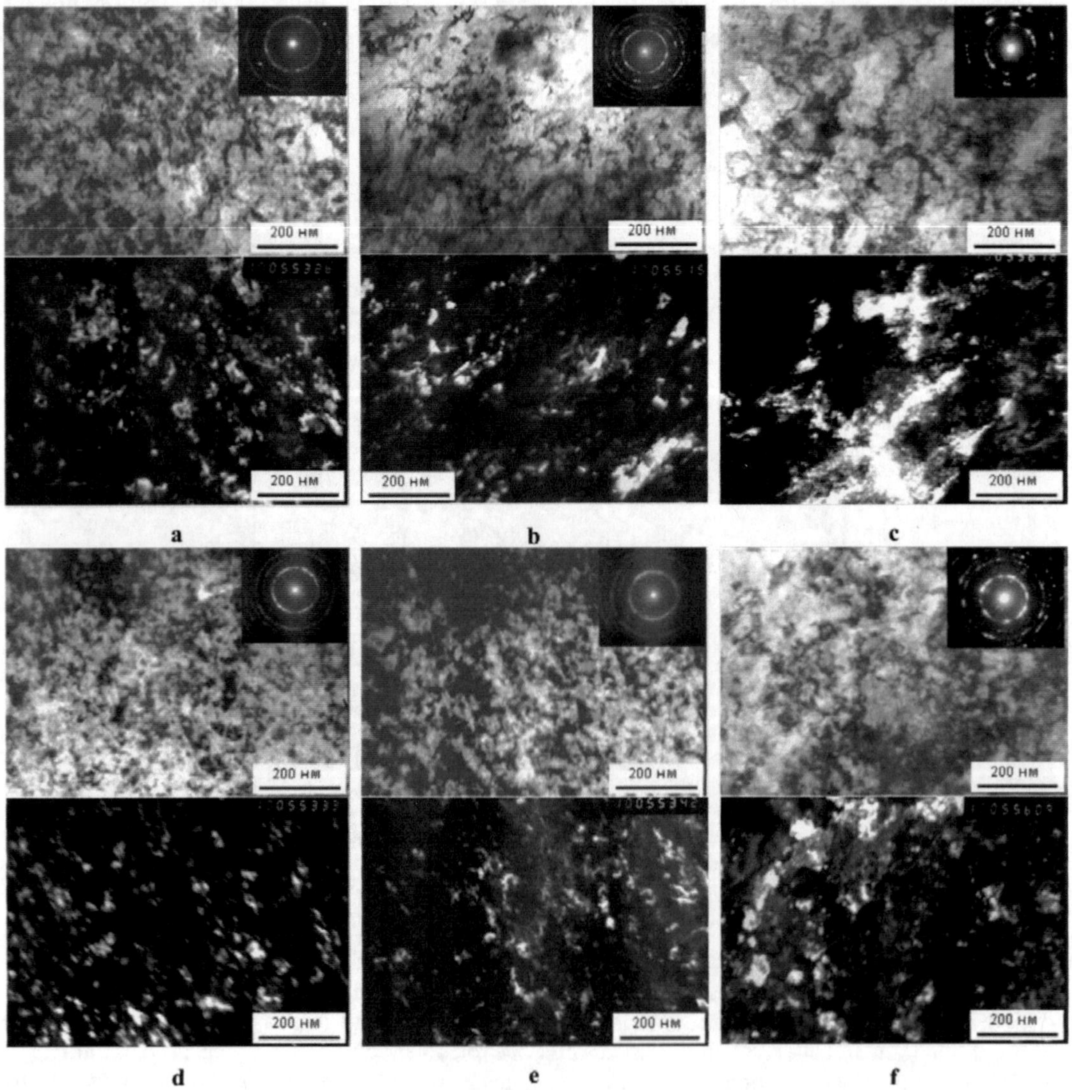

Figure 2. Light-field, dark-field images and diffractions (110α) of the specimen surface layer when processing by the tool without the heatsink at a speed of 0.1 m/s (a), 0.17 m/s (b), 0.2 m/s (c) and by the tool with the heatsink system at a speed of 0.1 m/s (d), 0.22 m/s (e) and 0.38 m/s (f).

At a speed increase up to 0.17 m/s, the grain sizes amounted to 50…300 nm (Figure 2b). The speed increase up to 0.22 m/s and higher leads to formation of a grain structure of 200…600 nm in size (Figure 2c). In the course of tool processing with the heatsink system at a speed from 0.06 to 0.22 m/s, formation of the nanocrystalline structure of 30...70 nm in size occurs (Figure 2d, e). At a speed increase up to 0.38 m/s, the sizes of grains reach 50...300 nm (Figure 2f). Further increase of speed leads to an increase in the grain size up to 200…600 nm and, practically, to the absence of the nanostructural state of the deformed material.

The scanning electron microscopy of the surface layer after nanostructuring burnishing by the tool without the heatsink system shows that, at a speed up to 0.12 m/s, the thickness of the modified layer equals 2...3 μm (Figure 3a). A speed increase up to 0.17 m/s leads to a decrease in thickness of the layer up to 0.8…1 μm (Figure 3b). At further speed increase, the modified layer is not observed (Figure 3c). While processing by the tool with the heatsink system at a speed up to 0.16 m/s, the thickness of the modified layer amounts to 4…5 μm (Figure 3d). The thickness of the layer decreases to 2 microns (Figure 3e) at a speed increase up to 0.27 m/s. The increase in speed up to 0.41 m/s leads to layer thickness reduction up to 0.7…1 μm (Figure 3f). At further speed increase the modified layer is not observed.

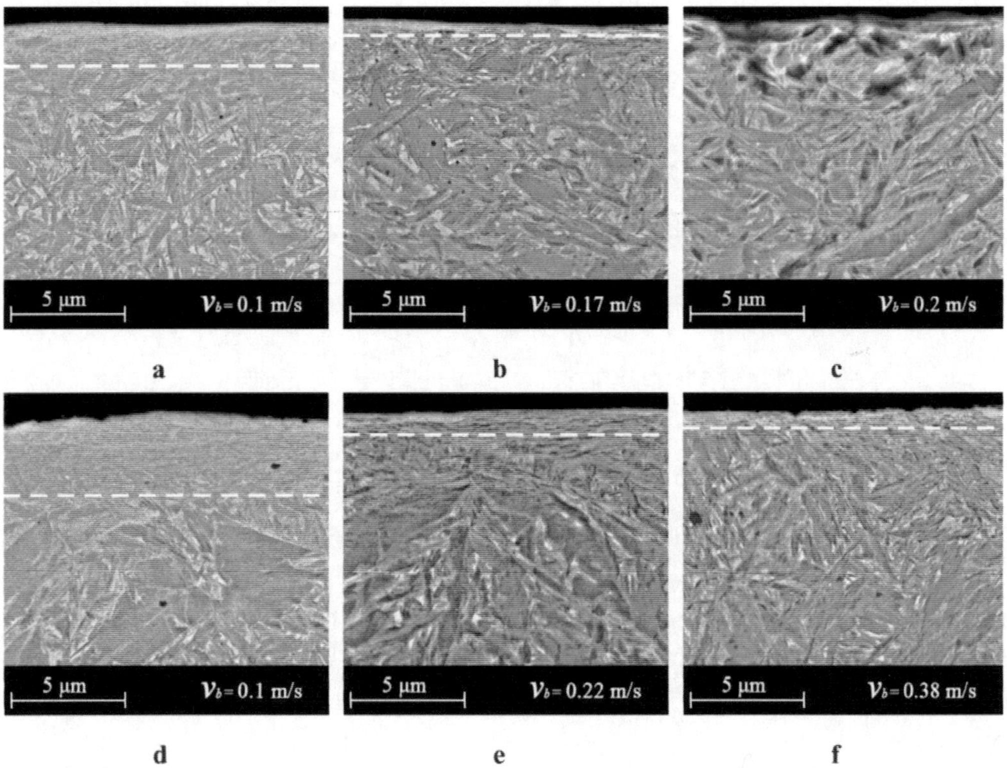

Figure 3. Scanning electron microscopy of the surface layer after nanostructuring burnishing: by the tool without the heatsink (a, b, c); by the tool with the heatsink system (d, e, f)

After nanostructuring burnishing without the heatsink, the microhardness of the processed surface amounts to 1120...1250 $HV_{0.5}$ (Figure 4a). When processing by the tool with the heatsink system, the microhardness equals 1160…1220 $HV_{0.5}$, which, taking into account the measurement accuracy, makes it possible to consider that it has an identical result. Besides, there is a decrease in the surface microhardness by 50...60 $HV_{0.5}$ at a speed increase up to 0.22 m/s when being processed by the tool without the heatsink.

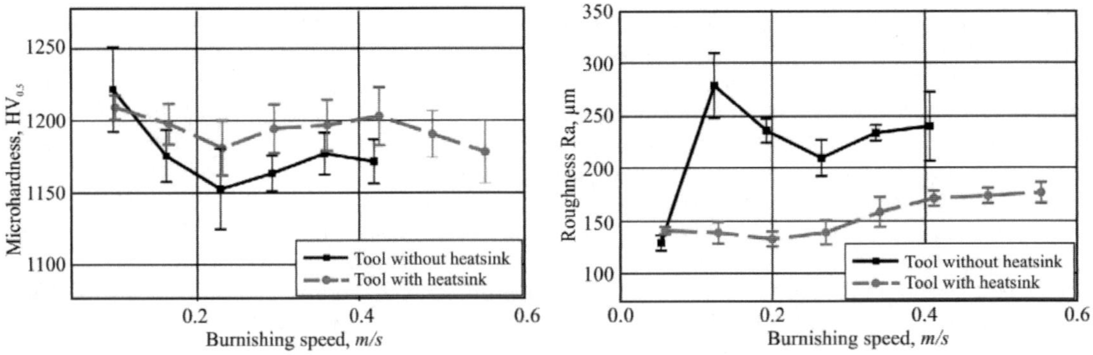

Figure 4. A change of the microhardness (a) and roughness (b) of the surface layer at an increase of nanostructuring burnishing speed.

When processing by the tool without the heatsink at a speed over 0.1 m/s, the roughness worsens from Ra = 100…150 nm to the level of Ra = 250... 300 nm, that is more than in the initial state of the specimen after turning (Figure 4b). At the same time, the surface roughness after treatment by the tool with the heatsink system amounts to Ra = 140…170 nanometers within all the range of the investigated speeds.

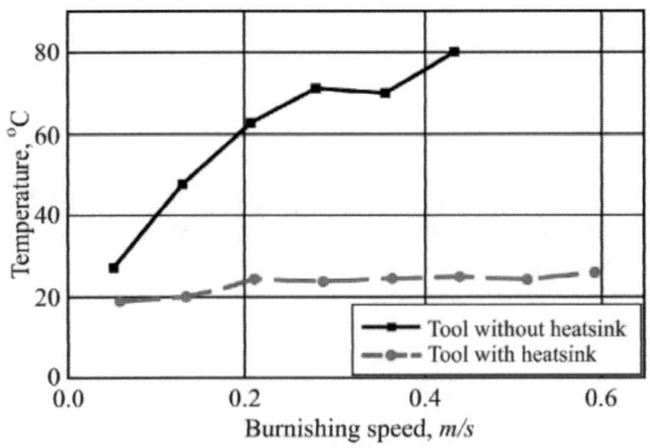

Figure 5. A change of the indenter tip temperature at increasing speed of nanostructuring burnishing.

The indenter tip temperature (Figure 5), when processing by the tool without the heatsink and when changing speed from 0.06 to 0.4 m/s, increases from 28 to 80 °C. The application of the tool with the heatsink system provides stable maintenance of the temperature at the level of 20…26 °C within all the range of the investigated speeds.

4. Discussion

According to the concept of a fast-moving source of thermal generation [6], an increase of nanostructuring burnishing speed leads to an increase in thermal flow resistance q_w in the specimen and to the reduction of the heated layer thickness in the contact area. The essential heating-up of the deformed material can lead to changes of the friction force in contact and of the rotational and shear nanostructuring mechanism.

The given research showed that at nanostructuring burnishing of 20Cr4 steel, the tool without the heatsink and with an indenter tip with the radius of 2 mm, the formation of the layer with a nanodimensional crystal structure 2…3 µm thick is possible only at heat generation capacity in the deformation center up to 8.2 W. This capacity corresponds to the burnishing speed of 0.12 m/s at friction force in contact of 68 N. The application of the tool with the two-circuit heat-cooling system allowed for increasing thickness of a nanostructured layer up to 4…5 μm at the abovementioned capacity of heat generation. The formation of a nanostructural state of the material in the surface layer 2…3 μm thick applying the tool with the heat-cooling system is provided at heat generation capacity up to 18.4 W that corresponds to the burnishing speed increase up to 0.27 m/s.

To form a nanocrystalline grain structure of the surface layer of parts from constructional materials at a burnishing speed of more than 0.3 m/s, it is necessary to increase, first of all, the efficiency of heat removal via the tool. To implement this, two ways are possible: 1) Coolant temperature decreases in the first circuit due to an increase in cooling capacity of the cooling system when using more powerful Peltier thermoelectric modules; 2) Contact thermal resistance decreases due to an increase of the contact area by increasing the radius of the indenter tip.

5. Conclusion

Heat removal from the contact area via the tool allows for ensuring stability of the nanocrystalline grain structure formation in the surface layer of constructional steels at the increase in a linear processing speed in the conditions of dry friction.

The application of the tool cooling system with a cooling capacity of 10 W allows for increasing the critical speed of nanostructuring burnishing of 20Cr4 steel from 0.12 to 0.27 m/s.

Acknowledgements

The research has been carried out within the frames of the government order for carrying out research and development work No. 01201461774, backed by grant RFFI No. 15-08-01511 A and the grant of the Department of Education and Science of the Russian Federation No. 2015-218-06-055.

References

[1] Kuznetsov V P, Tarasov S Yu and Dmitriev A I 2015 *J. of Mater. Process. Techn.* **217** 327–335
[2] Vorhauer A, Kleber S and Pippan R 2005 *Mater. Science and Engineer. A* **410-411** 281–284
[3] Kamikawa N and Tsuji N 2012 *Materials Transactions* **53(1)** 30–37
[4] Dedyulina O K et al 2014 *IOP Conf. Series: Materials Science and Engineering* **63(1)** 012054
[5] Kuznetsov V P, Skorobogatov A S and Gorgots V G 2015 *Appl. Mech. and Mater.* **770** 449–455
[6] Laraqi N, Bairi A and Segui L 2004 *Applied Thermal Engineering* **24(17–18)** 2567–2581

Influence of Electron Beam Treatment Regimes on the Structure and Properties of Intermetallic Clads Obtained on Titanium Substrates

D V Lazurenko, I A Bataev, O G Lenivtseva, Iu N Maliutina, O E Mats, V V Samoylenko and I A Polyakov
Novosibirsk State Technical University, 20, K. Marks ave., Novosibirsk, 630073, Russia

E-mail: pavlyukova_87@mail.ru

Abstract. Intermetallic clads were obtained on the basis of the titanium workpieces. The cladding was carried out using a non-vacuum electron beam treatment technique. A powder mixture which contained 16.97 Al – 28.27 Ti – 41.07 CaF_2 – 13.69 LiF (wt. %) was used for cladding. Two regimes with the beam current of 16 mA and 18 mA were used. The beam current influenced significantly the structure and phase composition of clads. In case of the electron beam treatment with a beam current of 16 mA a lamellar structure consisted of a mixture of $AlTi_3$ and AlTi was fabricated. An increase of the beam current to 18 mA led to the formation of an $AlTi_3$ acicular structure. Microhardness and wear resistance of the intermetallic coating was significantly higher in comparison with cp-Ti. The maximum microhardness of clads was 480 HV. A wear rate of clads was 10 times lower than that of cp-Ti.

1. Introduction

Currently titanium and its alloys are widely used in many areas of industry such as airspace engineering, automotive industry and chemical engineering due to its enhanced corrosion resistance, high specific strength and plasticity [1, 2]. However, low wear resistance and high-temperature oxidation resistance of titanium alloys restrict significantly the areas of their application. They are almost unsuitable for fabrication of the elements of friction pairs and details, which operate at high temperatures (higher than 500 °C). Surface modification of titanium alloys by elements, entering into the reaction with titanium which results in the formation of intermetallic compounds, can be the reliable solution of the aforementioned problems. Particularly, coatings consisted of titanium aluminides are considered to be the most efficient due to a combination of such properties as high-temperature strength, oxidation resistance, low density and high hardness [3, 4]. The above-mentioned statement suggests that titanium aluminides can possess the enhanced wear resistance level in comparison with titanium alloys.

Nowadays many technologies can be applied to produce Ti-Al intermetallic coating on the surface of the titanium workpiece. Among them there are mechanical alloying [5], aluminizing [6, 7], different methods of deposition [8-10] and others [11-13]. However, technologies based on the application of highly concentrated energy flows such as laser [14-16] or electron beam cladding are supposed to be the most prospective. Application of these technologies allows fabricating thick layers (up to 3 mm per one pass) with a good adhesion to metallic substrates [17]. However, application of non-vacuum electron beam cladding for formation of Ti-Al intermetallic coatings has not been yet described in the literature. Thus, the aim of this study was the investigation of the possibility of using the electron beam treatment for producing the titanium aluminide coatings. The influence of electron beam cladding regimes on the structure and properties of the cladded samples is discussed.

2. Materials and methods

Ti (98.69 wt. %) and Al (97.28 wt. %) powders and cp-Ti substrates were chosen as starting materials for cladding. Intermetallic coatings were fabricated by non-vacuum electron beam cladding using an ELV-6 industrial electron accelerator produced by Budker Institute of Nuclear Physics. The powder mixtures consisted of Al, Ti, CaF_2 and LiF were uniformly poured on cp-Ti substrates with a thickness of 12 mm. After that surfaces of the workpieces were heated by an electron beam. A velocity of a sample movement with respect to the electron beam was 10 mm/s. Transversal scanning with a frequency of 50 Hz and an amplitude of 25 mm was used to increase the area of the treated surface. A distance from the sample to the outlet was 90 mm. Electron beam cladding regimes are shown in Table 1.

Table 1. Electron beam cladding regimes.

Powder composition, wt. %	Flux composition, wt. %	Beam current, mA	Initial energy of the electrons, MeV	Velocity of a sample movement, mm/s	Distance to the outlet, mm	Scanning frequency, Hz
16.97 Al – 28.27 Ti	41.07 CaF_2 – 13.69 LiF	16	1.4	10	90	50
		18				

Structural investigations of obtained materials were carried out by means of scanning electron microscopy (SEM) using a Carl Zeiss EVO 50 XVP microscope in a back-scattered electron regime. Preparation of samples was carried out using the following approach. Materials molded into the resin were grinded by abrasion papers of different grit (from 100 to 2500) using a Struers LaboPol-5 machine and polished by alumina with a grain size of about 3 μm. Final polishing was carried out using a colloidal silica solution.

The elemental composition of coatings was estimated by an INCA X-ACT (Oxford Instruments) energy dispersive X-ray (EDX) analyzer. The investigation of the phase composition was carried out using an ARL X'TRA X-ray diffractometer (XRD). Diffraction patterns were obtained in a step mode with a step size of 0.05 ° and a dwell time of 6 s using Cu $K\alpha_{1,2}$ radiation.

A WolpertGroup 402 MVD semiautomatic tester measured the microhardness level. A load on a diamond indenter was 100 g. Tribological tests in the conditions of dry sliding friction were carried out by an II 5018 friction machine according to GOST 23.204-78 using a "disc-on-plane" scheme. Dimensions of samples were equal to 20x10x10 mm. A load on a sample was 100 N and a disc rotation speed was 100 rpm. Preliminary preparation of samples was carried out by a surface grinder. Surface roughness of samples before tests was equal to 0.32 μm. The quenched carbon steel disk (HRC 55) was used as a counterbody. An outer disk diameter was 50 mm; its width was 10 mm. The dimensions of the wear crater were measured to estimate the wear rate.

3. Results and discussion

Structural investigations showed that thicknesses of the layers obtained by cladding at a beam current of 16 mA and 18 mA was 900 and 1200 μm, respectively. The increase of the layer thickness was obviously caused by the increase of the energy input in the material. The beam current can play a critical role in formation of a structure and a phase composition of the clads. The increase of the beam current causes the increase of the depth of the molten substrate. This in turn leads to a significant decrease of the concentration of alloying elements due to mixing with the substrate material. This statement could be easily confirmed using an EDX analysis. An average Al and Ti content in the coating obtained at 16 and 18 mA was: 48 Al – 52 Ti and 35 Al – 65 Ti (at. %), respectively.

The analysis of the Ti-Al phase diagram indicates that the aforementioned concentrations corresponded to different phases. In case of 48 at. % Al an AlTi compound with a wide homogeneity range (48…63 at. % Al) forms. In the range between 23 and 35 at. % Al the formation of an $AlTi_3$ intermetallic compound should appear. In case of 35…48 at. % Al a mechanical mixture of two aforementioned compounds can be observed. Thereunder, we can suppose that the coating obtained at 16 mA contained γ-AlTi, while the electron beam treatment at 18 mA led to formation of the coating consisted of α_2-$AlTi_3$. For a more sophisticated and nuanced consideration of the structural and phase compositions of the coatings, the SEM and XRD analyses were applied.

The coating fabricated by the electron beam with a current of 16 mA possessed mainly a dendritic structure (Figure 2a). A difference in the contrast between dendritic arms and the interdendritic space can indicate the variation of the composition in these areas. It was mentioned in [19] that in the alloys containing 45…48 at. % Al the bright dendritic arms were enriched with Ti, while the interdendritic space consisted mainly of γ-grains. Microphotographs obtained at higher magnifications give the evidence that the coating had a lamellar structure. According to [19], such structure can be represented by the mixture of α_2- and γ-aluminides, as confirmed by the XRD analysis data (Figure 1). This coating contained 48 at. % Al, which corresponded to the line between a γ-phase and the mechanical mixture of γ- and α_2-phases on the phase diagram. Since the XRD analysis revealed the presence of two phases, then it can be concluded that formation of AlTi close to equiatomic composition occurred and excessive Ti was used to form $AlTi_3$.

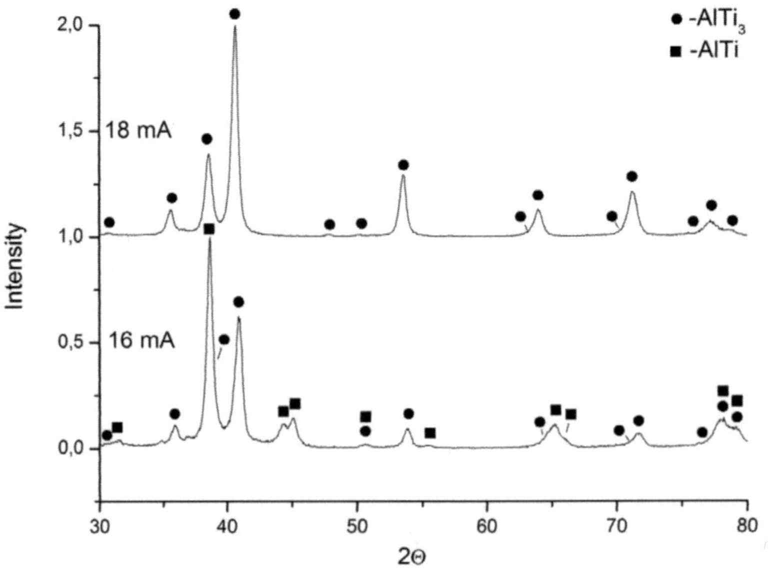

Figure 1. Results of the XRD analysis of coatings fabricated on cp-Ti substrates by the electron beam cladding.

The EDX analysis revealed segregation of P, Cu and Si (Figure 2a, Table 2) in the interdentritic areas. Their appearance was probably caused by the presence of the aforementioned elements in the initial powders.

The increase of the beam current led to a deeper melting of the material and consequently to stronger dilution of the alloying material in a substrate material. Thereby, a single phase coating consisted only of the α_2-phase was formed (Figure 1). A quantitative analysis showed that $AlTi_3$, which composed the coating, was enriched with aluminum. An acicular structure of the coating is shown in Figure 2b.

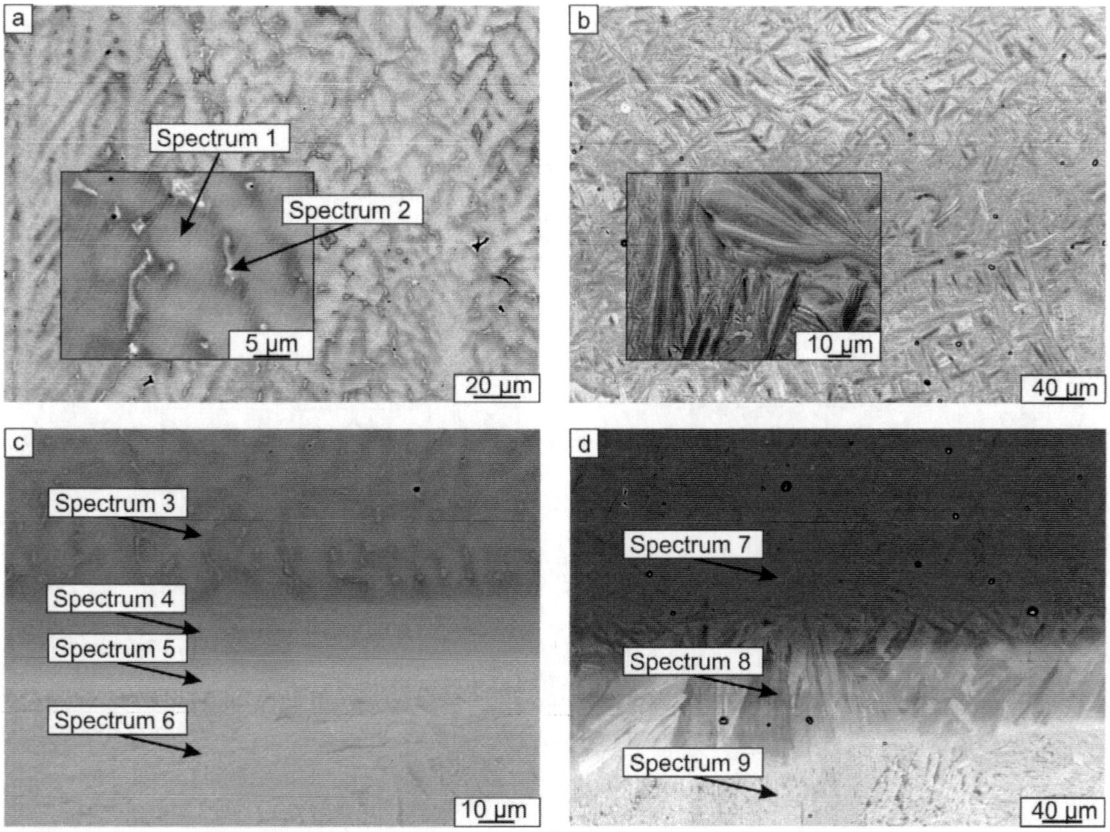

Figure 2. A structure of coatings cladded at *a* – 16 mA, *b* – 18 mA. Transition layers between the substrate and a coating obtained at *c* – 16 mA, *d* – 18 mA.

Table 2. The elemental composition of the local areas of intermetallic layers.

Zone \ Element	Al, at. %	Ti, at. %	P, at. %	Si, at. %	Cu, at. %
Spectrum 1	46.15	53.85	–	–	–
Spectrum 2	47.56	45.45	4.97	0.35	1.68
Spectrum 3	47.58	52.42	–	–	–
Spectrum 4	33.58	66.42	–	–	–
Spectrum 5	23.72	76.28	–	–	–
Spectrum 6	3.48	96.52	–	–	–
Spectrum 7	36.42	63.58	–	–	–
Spectrum 8	26.97	73.03	–	–	–
Spectrum 9	–	100	–	–	–

The contrast of transition zones of both samples in SEM microphotographs was different from the substrate and a cladded material (Figure 2c, d). The transition layer between the base material and the coating obtained at 16 mA consisted of two interlayers which corresponded to $AlTi_3$ enriched and depleted with aluminum (Figure 2c, Table 2). The transition area between titanium and intermetallic obtained at 18 mA corresponded to $AlTi_3$ with almost a stoichiometric composition (Figure 2d, Table 2).

The microhardness and the wear resistance of the clads were measured to estimate the properties of the sample (Figure 3). It was revealed that the microhardness of the intermetallic layers cladded according to the different regimes was 3.7…3.9 times higher compared to cp-Ti (Figure 3a). Besides, the microhardness of the coating obtained at 18 mA was slightly higher than the microhardness of the other coating. However, the wear resistance of the coating obtained at 18 mA was significantly higher, than that of the coating, obtained at 16 mA.

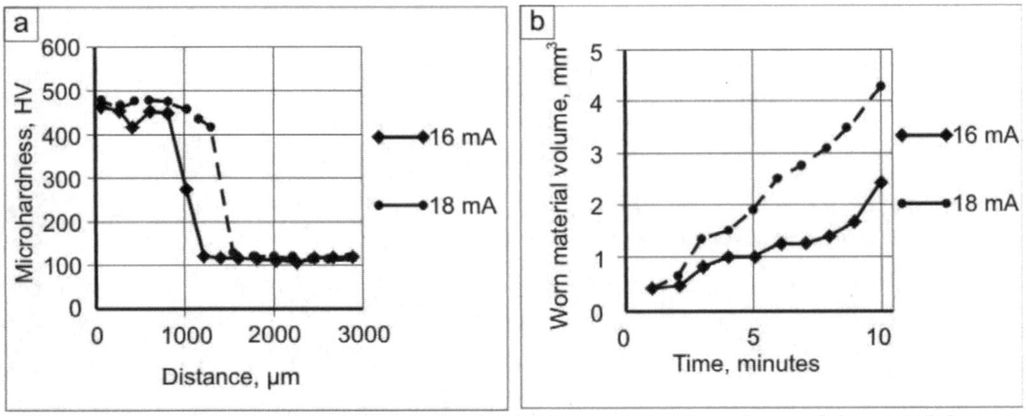

Figure 3. Microhardness of materials with cladded layers (a) and a volume of the material worn during the sliding friction tests (b).

Wear resistance of cp-Ti in the conditions of sliding friction was significantly lower compared to that of both coatings. The volume of the worn material equaled to 3 mm^3 was reached already after one minute of testing. The similar volume of the coating material (2.4…4.2 mm^3) was worn only after 10 minutes of friction.

4. Conclusions

Structural investigations of coatings obtained on cp-Ti substrates by non-vacuum electron beam cladding of a 16.97 Al – 28.27 Ti (wt. %) powder mixture according to different regimes revealed that a beam current value influenced significantly structure and phase compositions of cladded layers. Treatment by the electron beam with a current of 16 mA led to the formation of a two-phase lamellar coating consisted of AlTi and $AlTi_3$. An increase of a beam current to 18 mA led to the formation of a single-phase layer, which contained only $AlTi_3$. Formation of the intermetallic layers on the surface of cp-Ti led to a significant increase of its hardness (up to 480 HV) and wear resistance. The volume of the worn material in samples obtained at the beam current of 16 and 18 mA after 10 minutes of testing was 2.4 and 4.2 mm^3, respectively. Similar values of the worn material were reached after 1 minute of titanium testing.

Acknowledgements

This study was supported by RFBR, research project No. 15-38-20776 mol_a_ved.

References

[1] Leyens C and Peters M 2003 *Titanium and Titanium Alloys: Fundamentals and Applications* (New York: Wiley-VCH Verlag GmbH & Co. KGaA)

[2] Lütjering G and Williams J C 2007 *Titanium (Engineering Materials and Processes)* (Berlin: Springer)

[3] Loria E A 2000 *Intermetallics* **8** 1339-1345

[4] Franzén S F, Karlsson J, Dehoff R et al 2011 *The Minerals, Metals & Materials Society* **3** 455

[5] Romankov S, Kaloshkin S D, Hayasaka Y et al 2009 *J. of Alloys and Compoun.* **483** 386–388

[6] Zhang Z G, Peng Y P, Mao Y L et al 2012 *High Temperat. Mater. and Process.* **30** 519–525

[7] Goral M, Moskal G and Swadzba L 2009 *Intermetallics* **17** 669–671
[8] Leyens C, Peters M and Kaysser W A 1997 *Surf. Coat. Technol.* **94-95** 34–40
[9] Liu Y, Wang D, Deng C et al 2015 *J. Alloys Compd.* **628** 208–212
[10] Yumoto A, Hiroki F, Shiota I et al 2003 *Surf. Coat. Technol.* **169-170** 499–503
[11] Itoh Y, Itoh A, Azuma H et al 1999 *Surf. Coat. Technol.* **111** 172–176
[12] Borgioli F, Galvanetto E, Iozzelli F et al 2005 *Mater. Lett.* **59** 2159–62
[13] Li Y, Zhou T, Luo P et al 2015 *Internat. J. of Minerals, Metallurgy and Mater.* **22** 868–875
[14] Guo B, Zhou J, Zhang S et al 2007 *Appl. Surf. Sci.* **253** 9301–10
[15] Cárcel B, Serrano A, Zambrano J et al 2014 *Physics Procedia* **56** 284–293
[16] Maliutina I, Si-Mohand H, Piolet R et al 2015 *Metall. Mater. Trans. A* 1-10 *(in Press)*
[17] Lenivtseva O G, Bataev I A, Golkovskii M G et al 2015 *Appl. Surf. Sci.* **355** 320–326
[18] Bataev I A, Bataev A A, Golkovski M G, et al 2013 *Appl. Surf. Sci.* **284** 472– 81
[19] Appel F, Paul J D H and Oehring M 2011 *Gamma Titanium Aluminide Alloys: Science and Technology (*New York: Wiley-VCH Verlag GmbH & Co. KGaA)

Formation of deformation substructures in FCC crystals under the influence of point defect fluxes

M V Matveev[1], O V Selivanikova[1] and D N Cherepanov[2]
[1] Department of Nuclear Power Plants, Tomsk Polytechnic University, 30, Lenina ave., Tomsk, 634050, Russia
[2] Department of Higher Mathematics, Tomsk State University of Architecture and Building, 2, Solyanaya sq., Tomsk, 634003, Russia

E-mail: ohgres@yahoo.com

Abstract. The article deals with substructural transformations in FCC crystals under the influence of point defect fluxes. Different relationships between accumulation of point defects in crystal and substructure transformations, in particular during the process of fragmented dislocation structure formation have been received.

1. Introduction

The most productive approach to the study of staging of metal and alloys plastic deformation is to identify substructural transformations and related hardening mechanisms [1]. In mathematical modeling of plastic deformation the occurrence of individual elements of every substructure at all stages of deformation is expected, wherein the substructural transformation means element accumulation of a new substructure to such significant quantities that these elements become predominant among all others.

2. Overall dislocation density

Figure 1 shows calculations of dependencies of different types of dislocation densities on the deformation degree in a model with dynamic generation of point defects. For comparison, there are experimental data of dislocation densities, obtained by the method of electron microscopy by different authors. Triangles correspond to dislocation density, calculated according to formula $\rho = 7.8 \cdot 10^6 \tau^2$ (E. Gottler, 1973, [2]) based on the hardening curve of a copper single crystal oriented in the [100] direction, deformable at room temperature and at constant rate of shear deformation $a = 10^4$ s^{-1}. Snowflakes correspond to dislocation density, calculated using formula $\rho = 7.8 \cdot 10^6 \tau^2$, based on hardening curve of a copper single crystal oriented in the [001] direction, deformable at $T = 293$ K and at constant speed of shear deformation $a = 10^4$ s^{-1}.

Other dataare taken from [3], which gives and overview of the experimental data for dislocation density obtained by different methods. Bright circles are obtained from the data on energy, accumulated in a copper single crystal (D. Roennpagel, Ch. Schwtnk, 1978). Dark circles are the data on electron microscopy of the copper single crystal (E. Gottler 1973). Bright squares are the data on the X-ray analysis of the copper single crystal (T. Ungár, H. Mughrabi, D. Roennpagel, M. Wilkens, 1984). Bright diamonds are the data on the X-ray analysis of the copper polycrystal, obtained by the authors of [3].

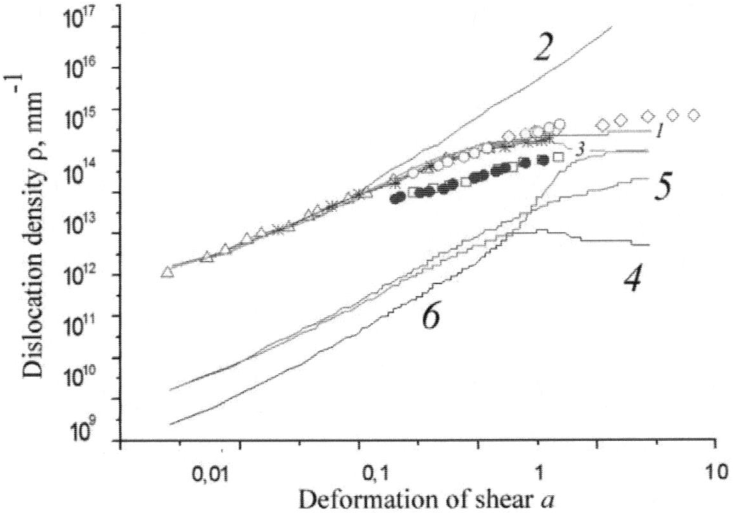

Figure1. Dependence of dislocation density on the deformation degree.

Curve 1 is in good agreement with the experimental data. There is satisfactory quantitative and qualitative agreement with the experimentally observed character of changes in total dislocation density $\rho = \rho_m + \rho_d + \rho_{\pm}$ with the order of shear deformation.

Curve 2 in Figure1 is a dependence of total dislocation density on the deformation degree, calculated on the assumption that the process of dislocation annihilation and transformation is absent. This condition in case of a small deformation degree practically has no effect on calculated curve ρ(*a*), which is evident from a comparison of curves 1 and 2, in case of high deformation degrees leads to a significant discrepancy in the experimental data: there is an unlimited increase in dislocation density with the further increase in the deformation degree. This suggests that both mechanisms play a significant role in the process of dislocation annihilation: a mechanism of screw dislocation annihilation, due to cross-slip, and a mechanism of edge dislocation annihilation due to climbing. The absence of one of these mechanisms leads to an infinite increase in the dislocation density.

3. Modeling results for components of overall dislocation density

Calculated curves 3, 4, 5, b in Figure 1 reflect the nature of the change in the dislocation density deformation: shearforming ρ_m (curve 3), in dynamic dipole configurations $\rho_d = \rho_d^v + \rho_d^i$ vacancy ρ_d^v and interstitial ρ_d^i types (curve 4 and 5, respectively), in dislocation walls $\rho_W = N_W \cdot d_W^1$ (curve b). Figure 1 shows that the main contribution to the overall dislocation density (curve 1) cause shearforming dislocations (curve 3), which density is comparable in order of power with overall dislocation density (comparing curves 1 and 3). Dislocation density in dynamic dipole configurations of vacancy (curve 4) and interstitial (curve 5) types as well as density in walls (curve 6) contribute more in case of a higher deformation degree.

Figure 2 shows the calculated dependenciesof dislocation densities on the deformation degree: in a dipole configuration of vacancy and interstitial types (Figure 2a, b), during shearforming (Figure 2c), in dislocation walls (Figure 2d).

The presented results suggest that in the initial stages of deformation a non-disoriented cellular substructure is formed, which elements are dislocation dipoles formed by dislocation in the dynamic dipole configuration, dislocation scraps, and shearforming dislocations. Available individual dislocation walls and dislocating clusters are also rearranged to form the dipole configuration with the closest dislocation of the opposite sign.

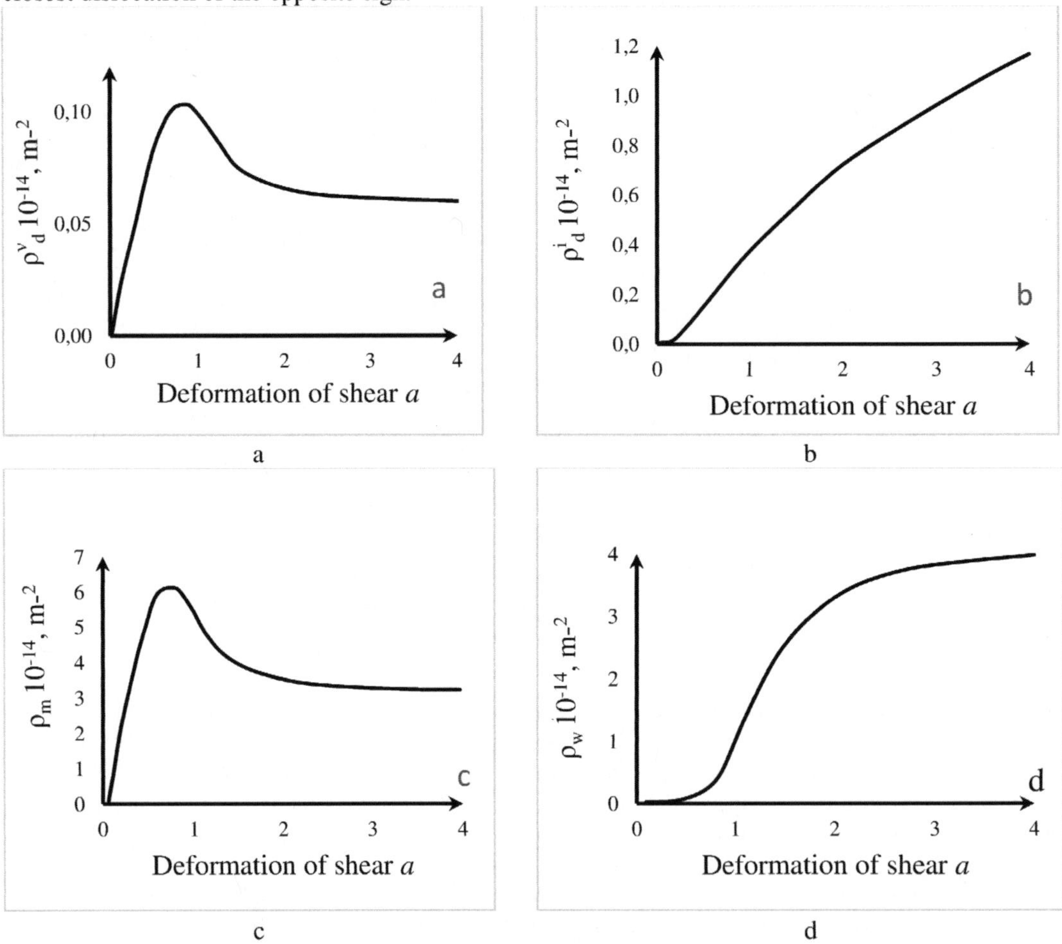

Figure 2. Dependencies of dislocation density in a dipole configuration of the vacancy type (a), dislocation density in a dipole configuration of the interstitial type (b), shearforming dislocation density (c), dislocation density in walls (d) depending on the deformation degree.

The greater the deformation is, the greater number of dislocation becomes integrated into dislocation walls, and dislocations together with dislocation clusters form excess dislocation density on the boundaries of shear zones. The boundaries of the fragments formed from dislocation walls, which are sprouting through the volume of a crystal as a result of absorption of shearforming dislocation, meet and form bodies, which shape is close to the convex polyhedron thatconnects in the junctures.

Figure 3a provides calculated and experimental data (Δ) [4] for the excess dislocation density. The excess growth of the dislocation density is directly related to the increase in the angle of disorientation, because it is inherent in the formula for the continuous disoriented angle.

Figure 3b provides calculated data for generated (curve 1), destroyed (curve 2) and accumulated (curve 3) densities of dislocation walls. At the initial stage the density of accumulated walls (curve 3) is greater than the density of the destroyed ones (curve 2), but in the late deformation stages the density of the destroyed walls is significantly higher than the density of the accumulated ones. The density of generated dislocation walls ten times greater than current density (curve 3), whichindicates high intensity process of the dislocation walls destruction.

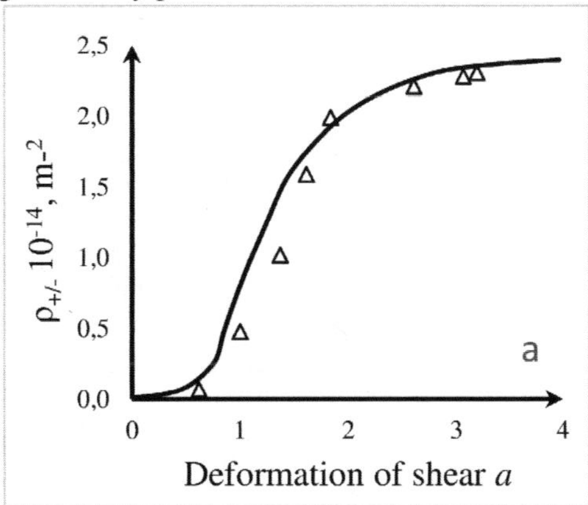

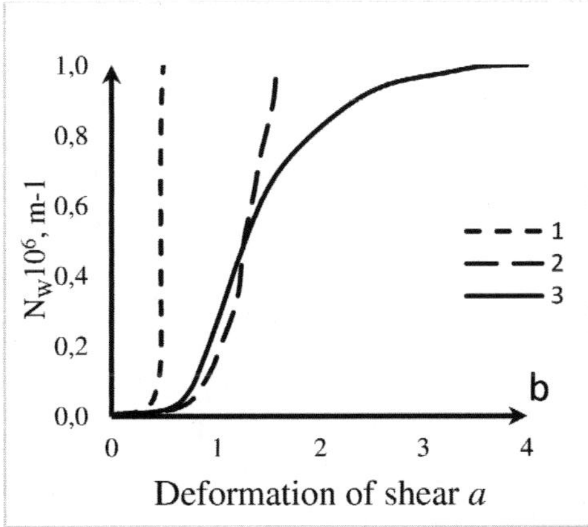

a b

Figure 3. Dependence of excess dislocation density (a), generated densities (1) and destroyed dislocation walls (2), and current density (3) (b) on the deformation degree.

The intensity of dislocation absorption by walls is significantly higher than the annihilation intensity of dislocations due to the deposition of point defects (Figure 3b). Both of them contributein intensity of decreasing of shearforming dislocation density, that the density at the deep deformations becomes less than the dislocation density in walls.

Figure 4 shows the dependences of generated density (curve 1), annihilated (curve 2) and accumulated (curve 3) shearforming dislocations on the deformation degree.

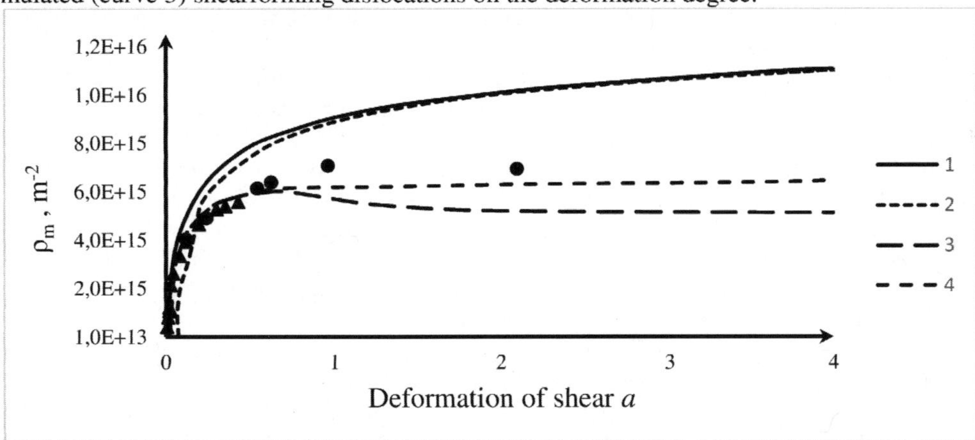

Figure 4. Dependences of generated (1), annihilated (2) and accumulated (3) shearforming dislocation densities, and total dislocation density (4) on the deformation degree.

For comparison, there is a dependence of total dislocation density on the deformation degree (curve 4). Curve 3 is calculated considering the annihilation of both screw and edge dislocations, as well as the mechanism of the dislocation walls relaxation growth. Points show the experimental data for single (Δ) and polycrystalline copper (○) according to [1, 2].

Figure 5, a shows the dependence of the density of generated dislocation wallson concentrations that gone to point defect sinks. Monovacancies – curve 1, divacancies – curve 2, interstitial atoms – curve 3.

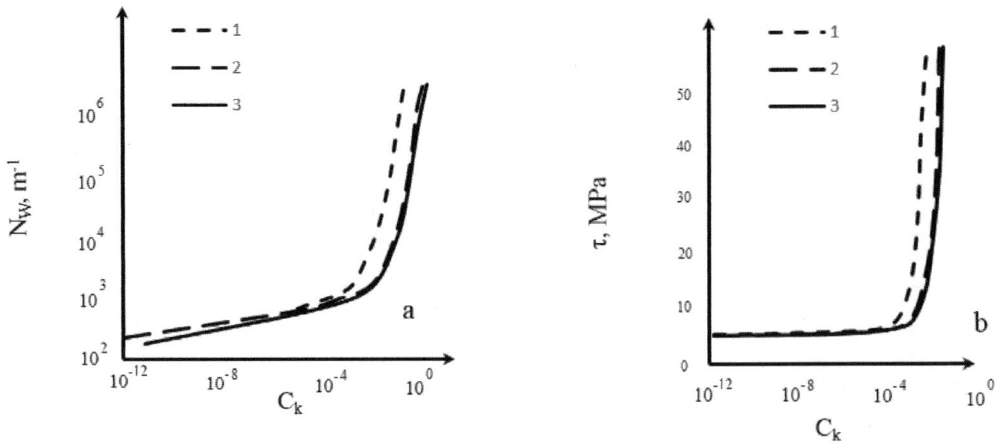

a b

Figure 5. Dependences of the density of generated dislocation walls (a) and creep stress (b) on the concentrations that gone to point defect sinks. Monovacancies – curve 1, divacancies – curve 2, interstitial atoms – curve 3.

There is also a direct link between the formation of the fragmented structure and the departure of point defects on sinks. This direct relationship is revealed only in the simulation, and has not been explicitly incorporated in the model. It is evident that the more intensive deposition of point defects on dislocations, gives the more intensivefragmentation.

Conclusion

In conclusion, it should be mentioned that modelling allows discovering the relationships between accumulation of point defects in crystal and substructure transformations, particular during the process of fragmented dislocation structure formation. The overwhelming majority of the generated defects involved in relaxation processes, was associated with the substructure transformation.

The process of the disoriented structure formation is a significant factor in the balance of deformation defects. A significant part of interstitial atoms is involved in the process of dynamic formation of dislocation wall germs. Annihilation processes play the main role in accumulation of interstitial atoms.

High values of the point defects concentration in the absence of their annihilation mechanisms allows suggesting that formation of amorphous deformation structures is possible in suppressed diffusion (in case of plastic effects at low temperatures of high-speed deformations).

Therefore, on the one hand, point defects absorption by dislocations prevents anamorphization process, and on the other hand, a major role of impact of point defect fluxes of high intensity on dislocations in dislocation substructure evolution becomes clear. Considering the connections between structure elements, that are carried out by point defects, is particularly important in the analysis of the nanostructure formation processes under the influence of deep deformations.

References

[1] Koneva N A and Kozlov E V 2003 *Russian J. Tambov Univer. Messenger, Series: Natural and Engineer. Scien.* **8(4)** 514
[2] Göttler E 1973 *Phys. Stat. Sol.* **28** 1057–76
[3] Ungár E, Schafler E, Hanák P and Bernstorff S 2007 *J. Mater. Scien. and Engineer. A* **426** 398–401
[4] Starenchenko V A 1991 *Experimental research and mathematical modeling of deformational and thermal hardening of FCC single crystals in pure metals and alloys with a superstructure* (Tomsk: dissertation, Phys. Mat. PhD) p. 797

Additional heat treatment of non-porous coatings obtained on medium carbon steel substrates by electron beam cladding of a 'Ti-Mo-C' powder composition

D O Mul, E A Drobyaz, T A Zimoglyadova, V A Bataev, D V Lazurenko and L I Shevtsova
Novosibirsk State Technical University, 20, Karla Marksa st., Novosibirsk, 630092, Russia

E-mail: ddariol@yandex.ru

Abstract. The structure and microhardness of surface layers, obtained by non-vacuum electron beam cladding of Ti-Mo-C powder mixture on a steel substrate after different types of heat treatment, were investigated. After cladding samples were heat treated in a furnace at 200…500 °C, as well as quenched at 860 ° C and then underwent high-temperature tempering. Heat treatment of cladded coatings induced tempering of martensite and precipitation of cementite particles (Fe_3C). Transmission electron microscopy of the samples after heating and holding at 300 ° C revealed precipitation of nanosized cubical TiC particles. The formation of hard nanosized particles led to the surface layer microhardness growth. The highest level of microhardness (which was 1.2…1.5-fold higher in comparison with coating microhardness after heat treatment) was achieved after heating of the claded material at 300 °C and 400 °C Additional quenching of samples at 860 °C did not increase the microhardness level.

1. Introduction

Recently the surface-modified metallic materials fabricated by using high-energy laser, electron or ion beams have been widely investigated [1-6]. Irradiation of metallic surfaces covered with powders (carbides, borides or carbide forming metals) by a high-energy beam leads to the formation of a melt, where added or synthesized ceramic compounds are distributed. After crystallization of such melt the structures, which provide enhanced microhardness and wear resistance of cladded materials compared to a substrate, can be formed. Subsequent heat treatment contributes to improvement of the structure and properties of cladded layers obtained on steel surfaces by cladding. Residual internal stresses can be relieved by temper of materials obtained by the methods of cladding [7, 8]. Moreover, quenching and temper of surface-hardened workpieces can be applied to improve the structure and properties.

It is well-known that temper of quenched steels alloyed with tungsten, vanadium and molybdenum at 500…600 °C is accompanied by precipitation of fine specific carbides. The decrease of the alloying degree of retained austenite provides a more complete martensitic transformation. Secondary hardening of martensite contributes to increase of hardness and thermal resistance of materials used at the temperatures up to 500 °C.

The current investigation is aimed at revealing the mechanisms of improvement of surface properties of mild-carbon steels treated by electron beam followed by heating in the temperature range between 200 and 500 °C, as well as post-quenching at 860 °C and subsequent temper at 500 °C.

2. Materials and methods

At the preliminary stage the workpieces with a size of 16×100×50 mm were cut out of the mild-carbon steel (0.41 % C, 1.11 % Cr, 0.86 % Mn, 0.32 % Si, 0.02 % S, 0.03 % P). Further these workpieces were covered by a powder mixture consisted of titanium, molybdenum, graphite and a flux (CaF_2 + LiF). The weight ratio of the aforementioned components was as follows: 41.8:4.4:13.8:33.3:6.7. Surface density of a powder mixture was 0.33 g/cm^2. Electron beam cladding was carried out using the electron accelerator ELV-6 (Novosibirsk). A beam energy value was 1.4 MeV. The distance from the outlet to a workpiece was 90 mm. The diameter of the electron beam correspondng to this distance was 12 mm. A workpiece was moved relatively the outlet with a rate of 10 mm/s. An increase of a treated square was reached by applying a scanning regime. The electron beam scanned a surface in the range of the workpiece width (50 mm) with a frequency of 50 Hz. A current beam was 28 mA.

Heat treatment of samples with coatings was carried out in a SNOL7.2/1300 muffle furnace. Heat addition was carried out at the temperatures of 200 °C, 300 °C, 400 °C and 500 °C during three hours, and oil quenching was conducted at 860 °C with subsequent drawing-back at 500 °C.

The structure of obtained materials was studied by the methods of optical microscopy, scanning electron microscopy (SEM) and X-ray diffraction (XRD) analysis. Metallographic investigations were carried out using an Axio Observer Z1m optical microscope. The study of the structure of the cladded and heat treated layers at higher magnifications and estimation of the elemental composition of local microvolumes of materials was carried out using a Carl Zeiss EVO 50 XVP scanning electron microscope equipped with a EDX X-Act microanalyzer. Transmission electron microscopy (TEM) was carried out using a FEI Tecnai G2 20 TWIN microscope. The phase composition of coatings was investigated by an ARL X'TRA diffractometer.

3. Results and discussion

3.1. The structure of cladded materials after heat treatment

Cladding of titanium, molybdenum and graphite on a steel workpiece led to the formation of a non-porous modified layer with a thickness of 2.4 mm (Figure 1).

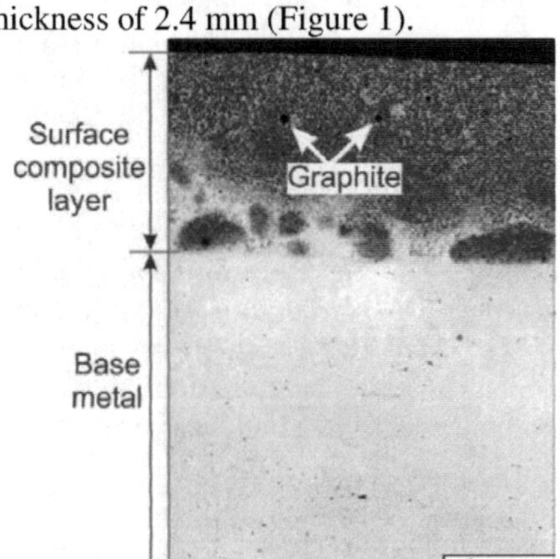

Figure 1. Cross section of samples after non-vacuum electron-beam cladding.

During the cooling of the cladded material the segregations of carbides were formed in the microvolumes enriched by titanium and carbon. These conglomerates appear as black spots on the bright background in the microphotograph. Moreover, in the cross section of the coating a small number of microvolumes with a size of 50 μm (or less) corresponded to undissolved graphite was revealed. The more detailed investigations showed that the structure of the coating was represented by a martensitic matrix, precipitated carbide particles (TiC) and retained austenite (Figure 2a). The volume fraction of carbides in the coating was equal to 30 %. The formation of such a matrix was caused by the chemical composition of the material and crystallization conditions. A cooling velocity of the material was about 200…1000 °C per second. Carbon content in the coating was higher than 2 % (taking into account the losses occurring during the electron beam treatment). Such carbon percentage in combination with alloying elements caused formation of the enhanced quantity of retained austenite.

Quenching at 860 °C with subsequent high tempering induced the formation of a fine martensitic structure with retained austenite and a lot of carbide particles in the coating (Figure 2b).

Heating a cladded material in the temperature range between 200 and 500 °C during three hours was accompanied by martensite temper (Figure 2c-f). Wherein the structure of retained austenite was also transformed and a carbide phase was modified partially. White crystals were precipitated in the coatings. Many of them had a lamellar shape (in Figure 2 these particles are shown by arrows). The size of these precipitations equaled to 10…15 μm. The temperature rise induced the increase of the thickness of lamellar particles.

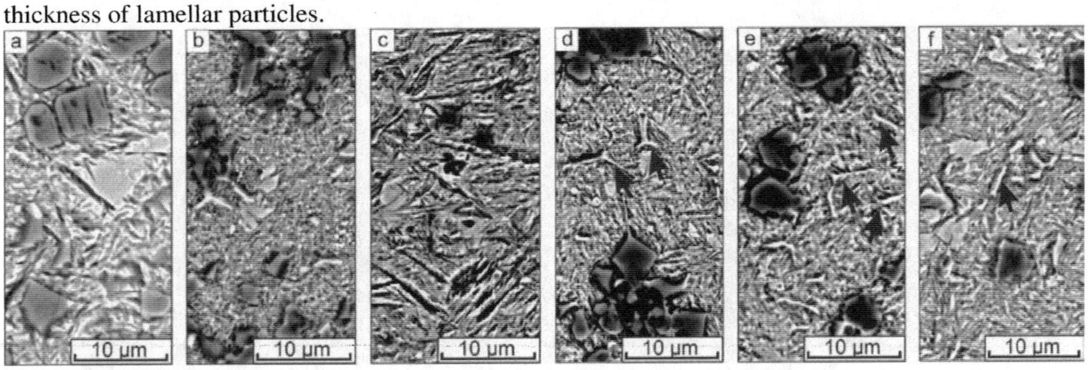

Figure 2. Scanning electron microscopy of the surface layer after cladding (a), after cladding and quenching at 860 °C with subsequent temper at 500 °C (b), after cladding and heating at 200 °C (c), 300 °C (d), 400 °C (e), 500 °C (f).

XRD analysis of heat treated coating revealed the presence of the Fe_3C phase (Figure 3). The austenite amount in the coating after additional heating was significantly reduced which was evidenced by decrease of the corresponding diffraction maximums. The same conclusion can be made based on the results of the metallographic analysis (Figure 2). Cementite precipitation is associated with the processes of austenite and martensite decomposition.

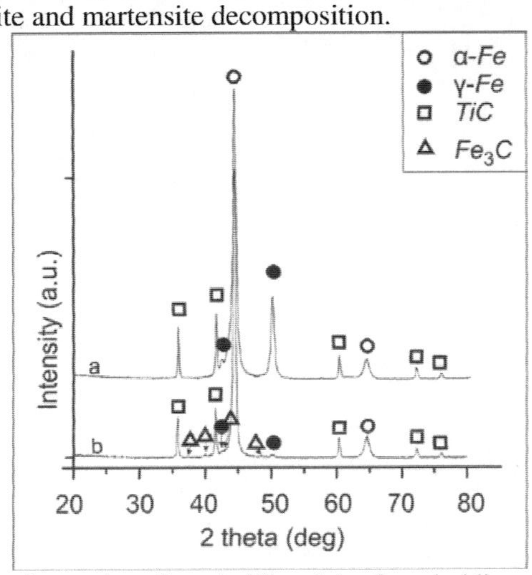

Figure 3. XRD analysis of samples after cladding (a), after cladding and heating at 300 °C (b)

A more sophisticated and nuanced view of the structure and a phase composition of the cladded layer before and after heat treatment at 300 °C was obtained by the electron diffraction method (Figure 4). The diffraction analysis showed that the matrix structure of the coating obtained by the electron beam treatment was represented by the mixture of martensite and retained austenite (Figure 4a, b). Titanium carbides precipitated during crystallization acted as a strengthening phase.

Additional heating at 300 °C during three hours led to the structural transformations of austenite accompanied by precipitation of the carbide particles of two types. Except the lamellar iron carbides, cubic-shaped carbide particles were revealed by TEM (Figure 4c, d). These particles distributed in α-Fe were characterized by the size of 5...10 nm and the phase composition of TiC.

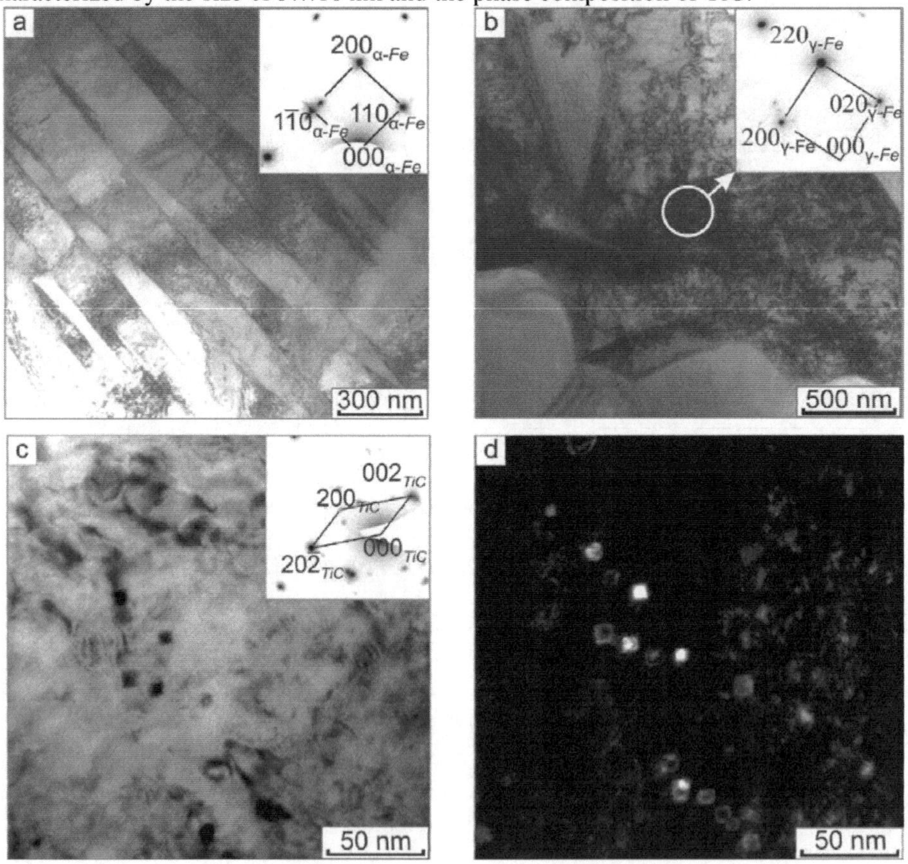

Figure 4. The fine structure of surface layers, obtained by cladding of titanium, molybdenum and graphite on steel substrate (a, b) after cladding followed by heating at 300 ° C (c, d), the dark-field image of carbide particles (d) obtained from electron-diffraction pattern is shown in Figure 5 (c) in 002 reflex.

3.2. Influence of heat treatment on microhardness of surface-hardened layer.

Microhardness of the cladded layer, heat affected zone and a base material of samples with cladded layers were measured after additional heat treatment of materials at 200 °C, 300 °C, 400 °C and 500 °C during three hours. The analysis of the results obtained revealed that the additional heating of materials at the temperatures of 300 °C and 400 °C induced 1.2...1.5-fold increase of microhardness in comparison with untreated coatings (Figure 5). A heat affected zone was characterized by the martensite temper. Increase of the temperature caused decrease of microhardness of these areas.

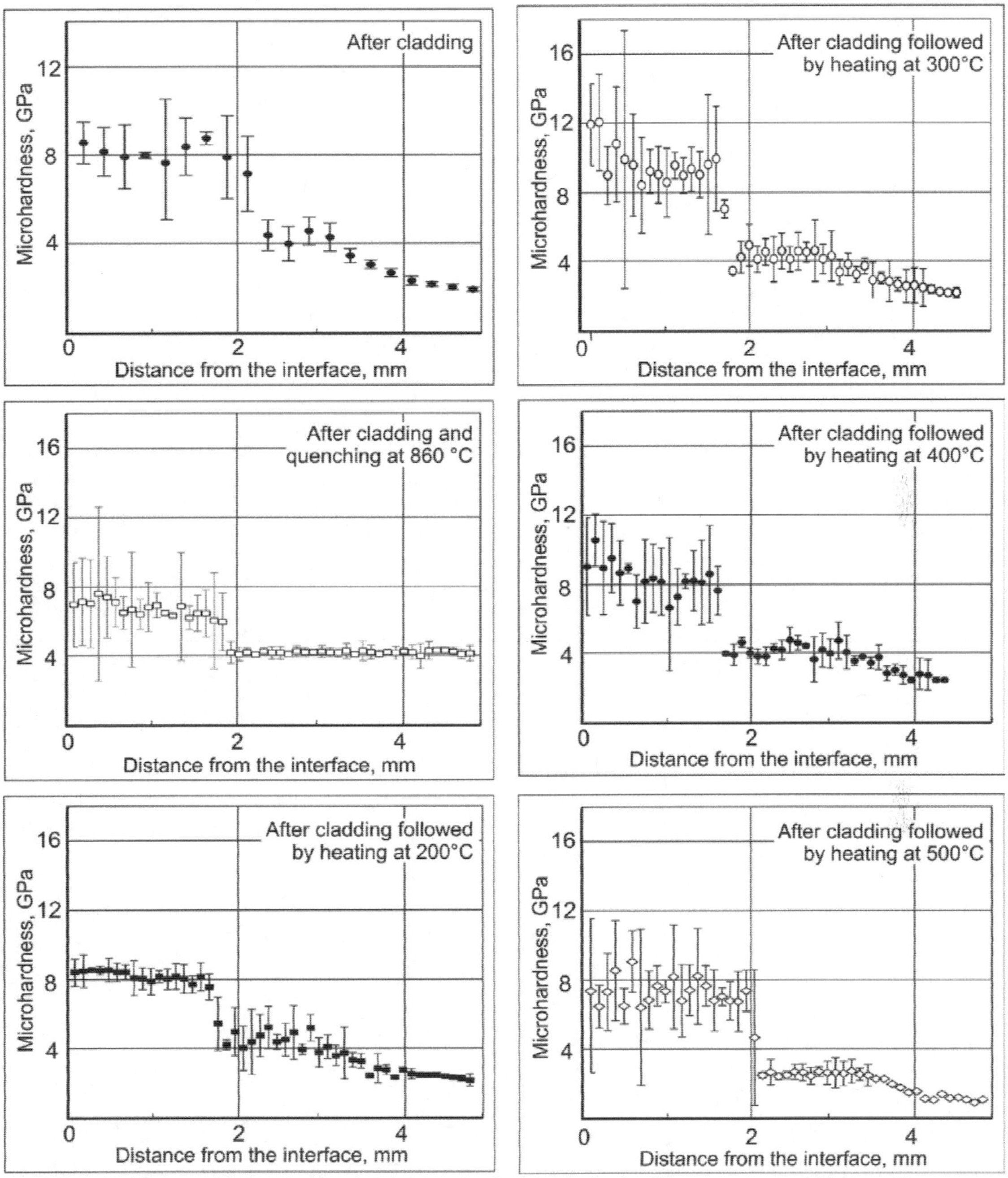

Figure 5. Microhardness distribution in depth of a layer, obtained by non-vacuum electron beam cladding of Ti-Mo-C powder mixture, after cladding and additional heat treatments.

The multiplicity formation of fine nanosized hard TiC particles possessed high hardness contributed in an increase of the cladded layer microhardness. The maximum microhardness level was reached after heating materials at 300…400 °C (Figure 5). The presence of hard nanosized TiC particles in the matrix can promote stabilization of a grain growth at higher temperatures.

The enhanced heating temperature (500 °C) initiated growth of the secondary carbides Fe_3C; the degree of the structure dispersity decreased and consequently microhardness of the coating dropped (Figure 5).

It should be emphasized that subsequent quenching of the cladded material from 860 °C did not induce increase of its microhardness (Figure 5). It can be explained by the fact that the temperature optimum for quenching of medium carbon steel (860 °C) does not allow forming homogeneous austenite in the cladded layer. The aforementioned temperature is corresponded to an intercritical temperature range for the cladded material.

4. Conclusions

The peculiarities of structural transformations of the surface alloyed layer obtained by non-vacuum electron beam cladding of the powder mixture consisted of titanium, molybdenum and graphite on the mild-carbon steel and additional heat treatments were investigated. The following characteristics of the coating heated at the temperature range between 200 and 500 °C were revealed:

Heating of the surface-hardened material at the temperatures from 200 to 500 °C was accompanied by precipitation of carbide particles of two types in the cladded layers. The first-type carbides had a lamellar shape; their length was 10…15 μm. The shape of the second-type carbides was cubic; the size was about 5…10 μm. XRD analysis and TEM investigations revealed that the phase composition of cubic particles corresponded to TiC and the lamellar particles were represented by cementite. Decomposition of martensite and retained austenite was the main factor contributed in the formation of carbides in the cladded layers.

Decomposition of the oversaturated solid solution occurring during the additional heat treatment of surface modified layer was accompanied by an increase of microhardness of the cladded material. The maximum microhardness growth (1.2…1.5-fold) was observed after heating the material at 300…400 °C during 3 h.

The presence of nanosized titanium carbide particles homogeneously distributed in the iron matrix can be considered as a factor, which induced stabilization of the grain growth in materials working at high temperatures.

Quenching of the obtained materials from the temperature of 860 °C and subsequent temper at 500 °C was accompanied by a decrease of the retained austenite volume fraction and precipitation of a large number of carbide particles in the cladded layer. It led to a decrease of the microhardness value from 8.5 GPa to 7 GPa.

Acknowledgements

This study was supported by the Russian Science Foundation (project No. 15-19-00230).

Reference

[1] Oh J C, Yun E, Golkovski M G and Lee S 2003 *Mater. Sci. Eng. A* **351** 98-108
[2] Yilbas B S, Patel F and Karatas C 2015 *Laser Surface Engineering Waugh J.L.G. Woodhead Publishing* 97-105
[3] Leyens C and Beyer E 2015 *Laser Surface Engineering Waugh J.L.G. Woodhead Publishing* 181-192
[4] Kusinski J et al 2012 *Bulletin of the Polish Academy of Sciences: Technical Sciences* **60** 711-728
[5] Nam D, Do J and Lee S 2009 *Scripta Mater.* **60** 695-698
[6] Golkovsky M G et al 2011 *Guangzhou* **314-316** 23-27
[7] Euh K et al 2003 *Mater. Sci. Eng. A* **346** 228-236
[8] Lee D G, Lee K and Lee S 2006 *Surf. Coat. Technol.* **201** 1296-1301

Thermal treatment of dissimilar steels' welded joints

A A Nikulina, A S Denisova, I N Gradusov, P A Ryabinkina and M V Rushkovets
Novosibirsk State Technical University, 20, K. Marks ave., Novosibirsk, 630073, Russia

E-mail: a.nikulina@corp.nstu.ru

Abstract. In this paper combinations of chrome-nickel steel and high-carbon steel, produced by flash butt welding after heat treatment, are investigated. Light and electron microscopic studies show that the welded joints after heat treatment have a complex structure consisting of several phases as initial welded joints. A martensite structure in welded joints after thermal treatment at 300…800 °C has been found.

1. Introduction

The aim of heat treatment of welded joints is to change the structure, the phase composition and the stress state of the material. It may include various types of thermal influence, and the main features of the process will be determined by the structure of steels before welding, the weld metal composition, the purpose of the structure and the conditions of its operation [1]. When similar steels are welded, thermal treatment allows changing the structure of the heat affected zone and the weld, resulting in increased reliability. In the case of dissimilar steels welding the possibility of using heat treatment is limited, as it can lead to an unequal or even opposite effect in welded steels [1–4].

One of the most common combinations in the composite structure is a combination of austenitic and pearlitic steels. An example is a railway frog, which is welded with the rail by means of a layer of austenitic chromium-nickel steel allowing to combine the welding conditions of austenitic steel, included into the structure of the crossbar, and the pearlite rail steel [5]. A construction consisting of dissimilar steels is performed by flash butt contact welding. Although stainless steel allows forming the weld seam between chromium-nickel and carbon steels, its fracture toughness will be low [5, 6].

As thermal treatment is one of the ways to improve the reliability of welded joints, the aim of this work was to study the structural transformations occurring during heat treatment of the welds between the pearlite and high-chromium-nickel austenitic steels obtained by flash butt welding.

2. Materials and methods

The chemical composition of base steels is given in table 1. Samples of the welded joints were cut perpendicularly to the plane of joining. Thermal treatment of welded joints was carried out in furnace SNOL 7.2/1300 at the temperatures of 300…1000 °C. After heating to the temperatures of 300…800 °C samples were cooled in the air, and after heating to 1000 °C – in the furnace. Structural investigations were carried out using the Carl Zeiss Axio Observer Z1m light microscope and scanning electron microscope Carl Zeiss EVO 50XVP. Microhardness measurements were taken by the Wolpert Group 402MVD tester.

Table 1. Chemical composition of steels (wt %).

Steel	C	Mn	Si	Cr	Ni	Ti	P	S
High-carbon steel	0.72	0.78	0.14	–	–	–	0.020	0.012
Chrome-nickel steel	0.11	0.58	0.53	17.1	9.7	0.38	0.021	0.012

3. Result and discussion

The investigated steels are dissimilar and, in addition to the differences in the chemical composition, they also have different initial structures. The welds of these steels obtained by flash butt welding are characterized by considerable heterogeneity and the presence of high-strength areas. Their formation is a result of mechanical agitation of the welded materials and diffusion processes. This leads to the formation of local areas with a low content of alloying elements compared with the initial chrome-nickel steel, which is confirmed by the EDS analysis (Figure 1a). When the weld cools, the low alloyed areas gain an austenitic-martensitic structure (Figure 1b). The microhardness of these areas is on average 600 HV, whereas the level of the initial microhardness of steels does not exceed 250…300 HV. The width of such weld is approximately 1000 μm. The martensitic areas reduce significantly fracture toughness of welded compositions 'high carbon steel – chrome-nickel steel', because either they or their boundaries are preferred places for crack propagation [6].

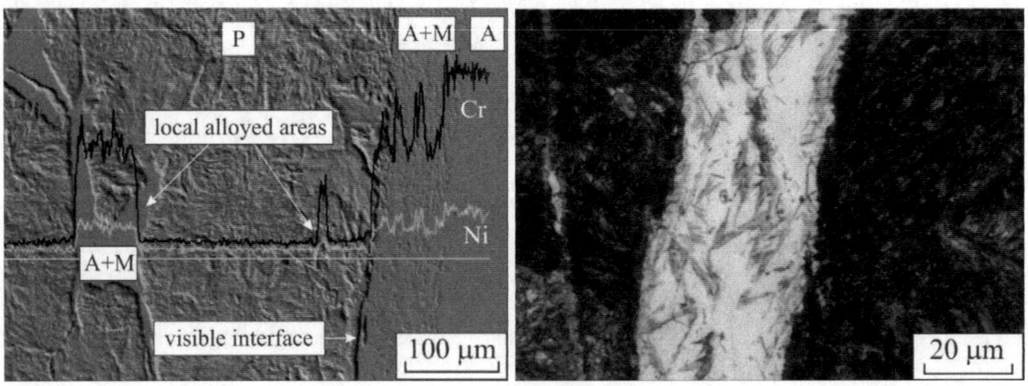

Figure 1. A general view of the weld in the initial state with the distribution of elements (*a*) and local alloyed area (*b*). P – pearlite, A – austenite, M – martensite.

After tempering at temperatures of 300…500 °C major structural changes in the welds are not observed: areas having austenitic-martensitic structure are preserved (Figure 2), and their microhardness corresponds to the initial state. The width of the weld does not change either.

Heating of welds to a temperature of 800 °C results in substantial changes in the structure of samples under investigation. The width of the weld increases up to 1200…1800 μm, and ferrite and ferrite-pearlite areas appear as well (Figure 3a). This is due to the diffusion process, which actively takes place at high temperatures. However, the austenitic-martensitic zones are also present (Figure 3b). The presence of areas with the reduced content of alloying elements compared with the initial chrome-nickel steel was confirmed by the EDS-analysis of welded joints (Figure 4). However, this temperature was not enough to significantly reduce the concentration of alloying elements in these areas and cooling leads to the formation of the martensitic structure with maintaining a substantial fraction of the residual austenite at the same time. Microhardness of the ferrite layer is equal to 150 HV, and the level of microhardness of the adjacent layer which has a ferrite-pearlite structure is 260 HV. The hardness of the martensitic-austenitic regions varies from 350 to 700 HV. Such deviations in the microhardness level of martensite are caused by the difference in the chemical composition of the austenite microvolumes from which it was formed.

After heating at 1000 °C samples were cooled by the furnace. Such heat treatment leads to a significant change in the structure of welded joints (Figure 5a). Active diffusion causes increasing of the weld width, and finally it can reach 5000 μm. The structure of the weld from the carbon steel part is characterized by a successive ferrite and then ferrite-pearlite structure with localized austenitic areas. On the other side of the visible boundary austenite grains were observed with dense carbide mesh (Figure 5b).

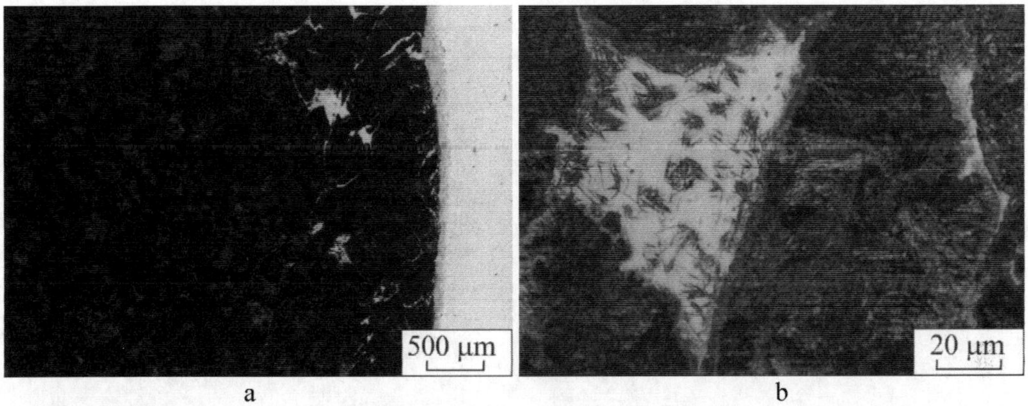

Figure 2. A general view of the welded joint (*a*) and the local austenitic-martensitic area (*b*) after tempering at 400 °C.

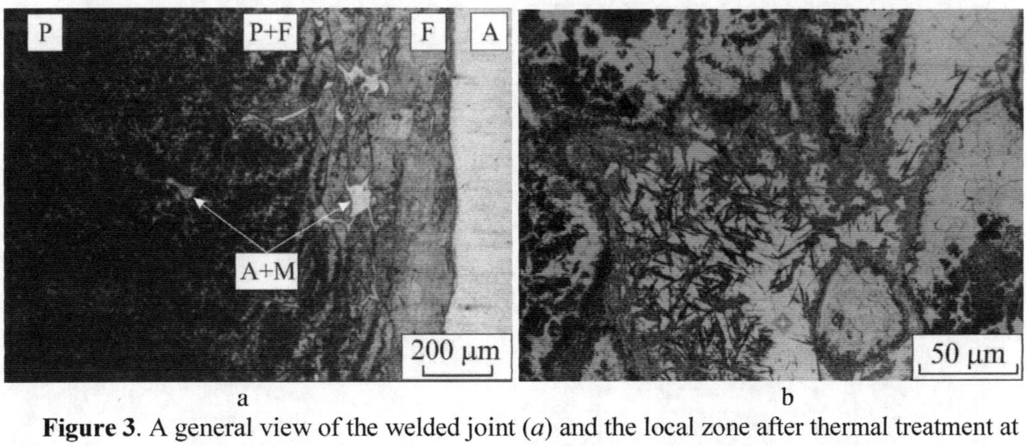

Figure 3. A general view of the welded joint (*a*) and the local zone after thermal treatment at 800 °C (*b*). P – pearlite, A – austenite, M – martensite, F – ferrite.

Metallographic studies showed the absence of martensite in the weld after its annealing. The level of microhardness of the weld is characterized by significant heterogeneity, and the range of its changing even increases: microhardness of the ferrite layer is 150 HV and microhardness of carbides formed (Figure 6) reaches 800 HV (Figure 7). The presence of the ferrite layer with low hardness and strength in the weld is undesirable because it leads to decreasing in weld strength and reduction of its service life due to fatigue spalling of the ferrite layers [7]. A carbide network in the weld is also an unfavorable factor and may lead to premature failure of welded constructions [1]. Using of low-carbon barrier layers can be an alternative to heat treatment in order to decrease martensite hardness formed in the welding process [8].

4. Conclusions

Structural studies and microhardness evaluation results indicate that heat treatment at temperatures of 300…500 °C does not lead to significant changes in the structure of the welds. Areas with the austenitic-martensitic structure are present, and their microhardness corresponds to the original state before heat treatment. A martensitic structure is also formed in the weld zone and after heat treatment at 800 °C. In general, the thickness of the weld increases from 1000 to 1800 μm. Microhardness of austenitic-martensitic areas depends on the initial chemical composition of austenite and varies from 350 to 700 HV.

Active diffusion during heat treatment at 1000 °C leads to an increase of the width of the weld up to 5000 μm.

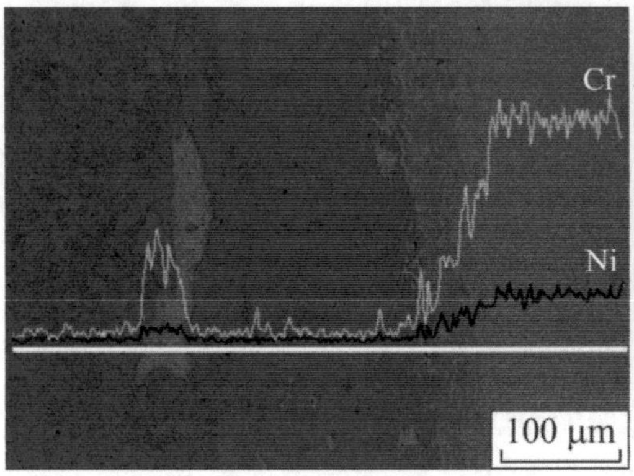

a

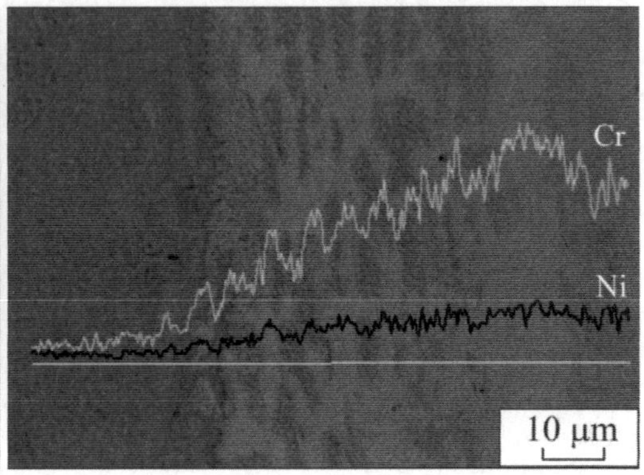

b

Figure 4. Line distribution of chemical elements perpendicularly to the welded joint:
a – a general view; *b* – at the visible border of steels.

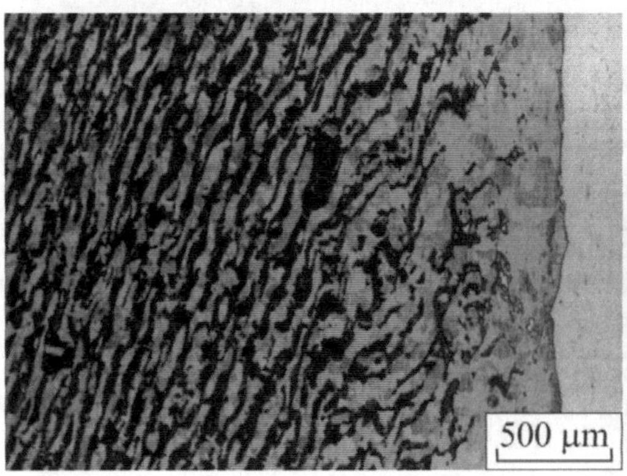

a

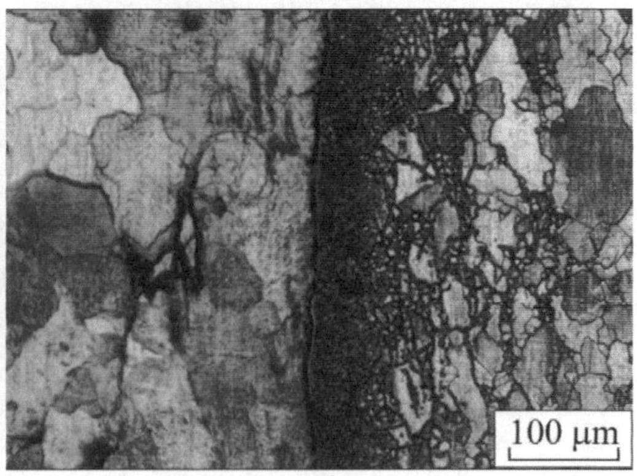

b

Figure 5. A general view (*a*) and a visible border of steels (*b*) after thermal treatment at 1000 °C.

The range of the microhardness level becomes even wider as microhardness of carbides formed reaches 800 HV, and the adjacent ferrite layer is only 150 HV. Thus, heat treatment of welded joints of austenitic chromium-nickel steel and pearlite carbon is not suitable.

Acknowledgement

The authors gratefully acknowledge the financial support from the Russian Ministry of Education and Science (research task No.2014/138, project 257).

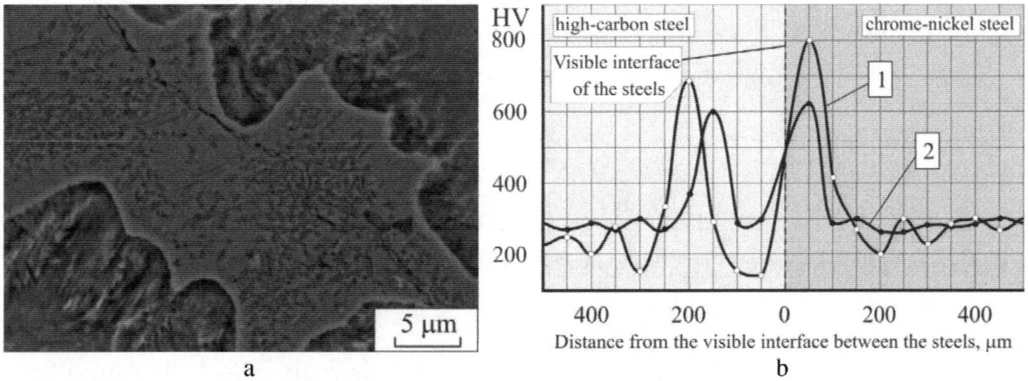

Figure 6. Carbides in localized alloyed areas after heat treatment at 1000 °C (*a*) and microhardness of the welded joint (*b*): *1* – after heat treatment at 1000 °C, *2* – initial state.

References

[1] Olson D L, Siewert T A, Liu S, Edwards G R 1993 *Welding, Brazing, and Soldering ASM Handbook vol 6* (Asm International) p 873

[2] Robert W and Messler Jr 2004 *Joining of Materials and Structures* (Burlington: Elsevier) p 815

[3] Dawson K E, Tatlock G J, Chi K and Barnard P 2013 Changes in precipitate distributions and the microstructural evolution of P24/P91 dissimilar metal welds during PWHT *Metallurgical and materials transactions A* **44A** 5065–80

[4] Falat L, Ciripová L, Kepic J, Buršík J and Podstranská I 2014 Correlation between microstructure and creep performance of martensitic/austenitic transition weldment in dependence of its post-weld heat treatment *Engineering Failure Analysis* **40** 141

[5] Zhang F, Lv B, Hu B and Li. Y 2007 Flash butt welding of high manganese steel crossing and carbon steel rail *Mater. Sci. Eng. A* **454–455** 288

[6] Nikulina A A, Bataev A A, Smirnov A I, Popelyukh A I, Burov V G and Veselov S V 2015 Microstructure and fracture behaviour of flash butt welds between dissimilar steels *Science and Technology of Welding and Joining* **20** 138

[7] Ed Lewis R, Olofsson U 2009 *Wheel/rail interface handbook* (Cambridge: Woodhead publ Ltd) p 864

[8] Chevakinskaya A A, Nikulina A A and Plotnikova N V 2015 Reliability increase of dissimilar steel welded joints *Applied Mechanics and Materials* **698** 378

Formation of intermetallics at the interface of explosively welded Ni-Al multilayered composites during annealing

T S Ogneva[1], D V Lazurenko[1], I A Bataev[1], V I Mali[2], M A Esikov[2], A A Bataev[1]
[1] Department of Materials Science, Novosibirsk State Technical University, 20, Karla Marksa ave., Novosibirsk, 630073, Russia
[2] Lavrentyev Institute of Hydrodynamics, 15, Lavrentieva ave., Novosibirsk, 630090, Russia

E-mail: ogneva@corp.nstu.ru

Abstract. The Ni-Al multilayer composite was fabricated using explosive welding. The zones of mixing of Ni and Al are observed at the composite interfaces after the welding. The composition of these zones is inhomogeneous. Continuous homogeneous intermetallic layers are formed at the interface after heat treatment at 620 °C during 5 h. These intermetallic layers consist of $NiAl_3$ and Ni_2Al_3 phases. The presence of mixed zones significantly accelerates the growth rate of intermetallic phases at the initial stages of heating.

1. Introduction

Development of multilayered composites with a high complex of mechanical and performance properties is one of the most significant problems of the advanced materials science. There are many special requirements such as high specific strength, stiffness and low density that are imposed on the materials used in aircraft, rocket building and power engineering [1-2]. Some types of intermetallic compounds meet these requirements and may become a promising basis for manufacturing of the details applied in critical conditions at high load and temperature. A special attention is given to chemical compounds such as aluminides. These materials have unique properties that allow their use in the advanced aerospace industry. At the same time, their extensive use is limited by increased brittleness, especially at low temperatures [2]. Reliable utilization of the products manufactured of these materials is only possible under such conditions that enhance ductility, for example, heating [3].

One way to solve this problem is to form laminated composites, which consist of alternated plates of dissimilar alloys that are inclined to form intermetallics. The growth of intermetallic layers occurs during the heat treatment of such composites at the boundaries of dissimilar metals [4]. High mechanical properties of the material are provided by a combination of hard and stiff intermetallic layers and ductile metallic layers [5, 6].

Processes frequently used for production of the composite materials with intermetallic layers are reaction sintering of metal sheets [7, 8] and roll bonding of sheets with subsequent heat treatment of obtained multilayered samples [9, 10]. Furthermore, a lot of researchers experimentally showed that explosive welding is an effective method of connecting similar and dissimilar metals [11-13]. Features that are distinctive to the explosive welding process, in particular, local heating of workpieces to high temperatures (up to the melting point), high pressure during the interaction between welded plates, promote intensification of the chemical compounds formation [13]. It should be noted that the published data on the problem of intermetallic compounds formation during explosive welding are fragmentary. Many features of this process are still not investigated. The subject of this research is to study the processes of intermetallic compounds formation during explosive welding of metal plates and subsequent heat treatment of welded multilayer samples. Ni – Al systems were chosen as objects of the study.

2. Materials and methods

Ni-Al composites with intermetallic layers were produced by explosive welding and subsequent heat treatment. Explosive welding of alternatively stacked 4 plates of commercially pure (cp-) nickel (100 × 50 ×1 mm) and 3 plates of cp-aluminum (100 × 50 × 0.5 mm) was carried out according to the scheme with parallel orientation of workpieces. The distance between the plates during explosive welding was 1 mm. Ammonite 6GV was used as an explosive material. The chemical composition of initial metals is shown in Table 1.

Table 1. Chemical composition of initial materials, % wt.

	Fe	Cr	Ni	V	Si	Cu	Al
Cp-aluminum (alloy A5)	0.32	0.002	–	0.006	0.14	0.002	rest
Cp-nickel (alloy NP2)	0.029	–	rest	–	0.03	0.142	0.029

The intermetallic interlayers were fabricated by annealing of explosively welded Ni–Al composites at 620 °C. The duration of annealing varied from 5 min to 5 h.

The samples for characterization of the structure were prepared by the standard procedure including mechanical grinding and polishing. Metallographic studies were carried out using light microscopy (LM) and scanning electron microscopy (SEM). The elemental composition of the local zones was studied using energy dispersive X-Ray spectrometry (EDX). The phase analysis of metallic – intermetallic composites was performed using X-ray diffractometer ARL X'TRA by means of CuKα radiation in a step mode with $\Delta 2\theta = 0.02°$. The dwell time was 5 s.

3. Results and discussion

3.1. The structure of explosively welded samples

The cross section of the explosively welded composite is shown in Figure 1a. Continuous mixing zones are observed at the interface between dissimilar plates. The formation of these zones during explosive welding is known and described in many researches [18-20]. The thickness of the interlayer depends on the interface location in the composite. Two types of the interfaces can be distinguished in the composite, namely 'Ni/Al' type and 'Al/Ni' type (Figure 1b). If during collision the nickel plate is above aluminum plate the 'Ni/Al' interface is formed, otherwise, when the aluminum plate is above the nickel plate the 'Al/Ni' interface is formed (Figure 1a). The average thickness of mixing zones at 'Al/Ni' interfaces is in the range of 20…70 μm. At the 'Ni/Al' interfaces this value is in the range of 7…25 μm. A typical peculiarity of explosively welded materials is formation of the wavy interface. However, no waves are observed at interfaces of both types. 'Ni/Al' interfaces are nearly flat, while at the 'Al/Ni' interfaces a periodic variation of the mixing interlayer thickness is observed.

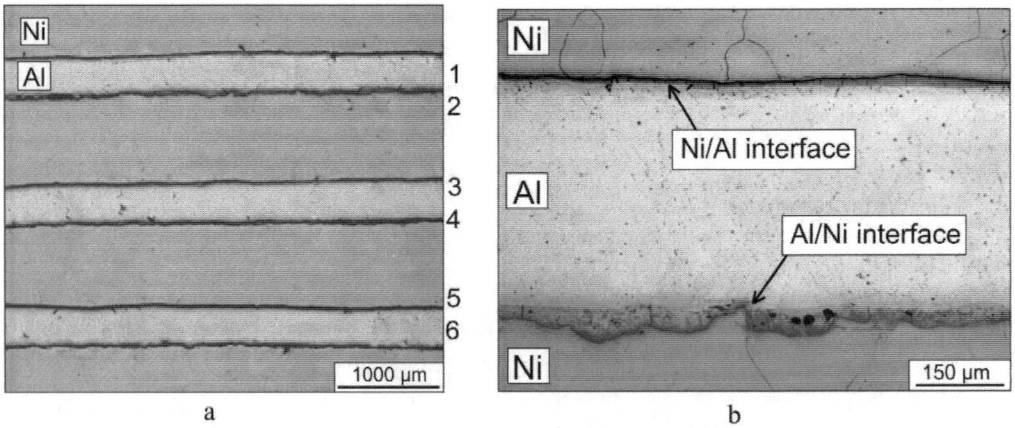

Figure 1. A structure of explosively welded multilayered Ni-Al composites.

Characterization of Ni–Al interfaces, performed by means of LM and SEM revealed that mixing zones possess a complicated non-uniform structure. Multiple dendrites with dimensions of 0.5…3 micrometers are observed (Figure 2a), which indicates that the mixing zones appear from the liquid state. Some particles have a complex fragmental shape with the size exceeding 10 μm. Results of the EDX analysis of the mixing zone, corresponding to Figure 2b are presented in Table 2. Different chemical composition within the intermixing regions evidences that the solidification rate of the material in the mixing zones is very high since homogenization does not have time to occur. A detailed TEM research revealed that such mixing zones contain metastable intermetallic compounds based on nickel and aluminum [13].

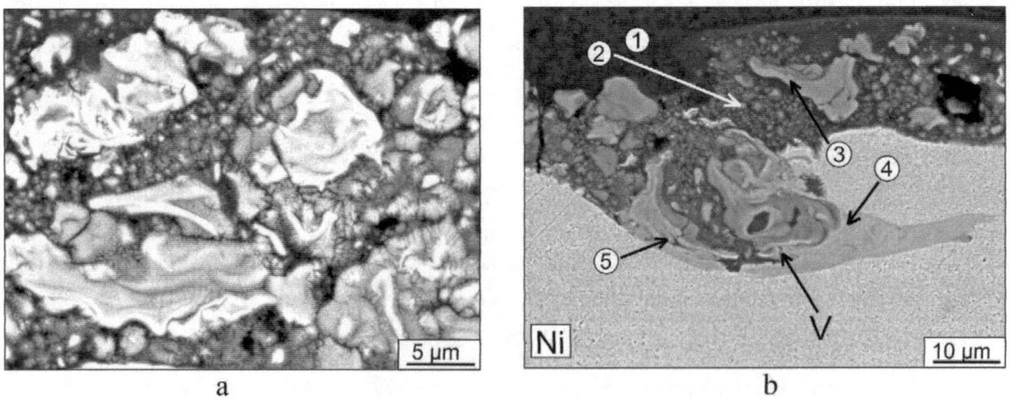

Figure 2. Mixing zones at the Ni–Al interfaces.

Table 2. The content of the chemical elements at points marked in Figure 2b

Zone #	Al, % at.	Ni, % at.
1	93.5	6.5
2	75.8	24.2
3	62.1	37.9
4	43.2	56.8
5	30.5	69.5

3.2. Structural transformation occurring at the interfaces during annealing

During heat treatment the nonequilibrium interlayers at the interfaces are gradually replaced by continuous intermetallic layers. This process occurs due to a reaction between Ni and Al and is greatly intensified at elevated temperature. The intermetallic $NiAl_3$ layer grows near aluminum. The Ni_2Al_3 intermetallic compound layer adjoins nickel. After 5 hours of heating only two intermetallic layers $NiAl_3$ and Ni_2Al_3 are observed along the interface. This is evidenced by the EDX analysis (Figure 3a, Table 3) and X-ray diffraction results (Figure 3b).

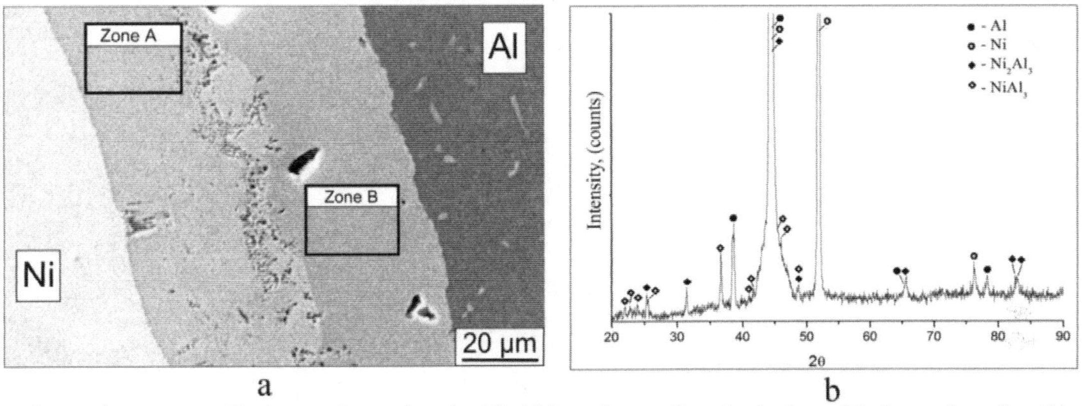

Figure 3. Intermetallic layers formed at the Ni-Al interfaces of explosively welded samples after 5 h of annealing (a) structure (b) X-ray diffraction pattern.

Table 3. The content of the chemical elements in areas A and B, indicated in Figure 4.

one #	Al, % at.	Ni, % at.
Zone A	60.4	39.6
Zone B	75.2	24.8

The difference in thickness and morphology of mixing zones having interfacial types 'Al/Ni' and 'Ni/Al', which is described above, leads to different rates of intermetallic layers growth. That is seen in the structure of samples annealed during 3 hours (Figure 4). The dependence of intermetallic layers growth at 'Al/Ni' and 'Ni/Al' interfaces on time is demonstrated by the graph in Figure 5a.

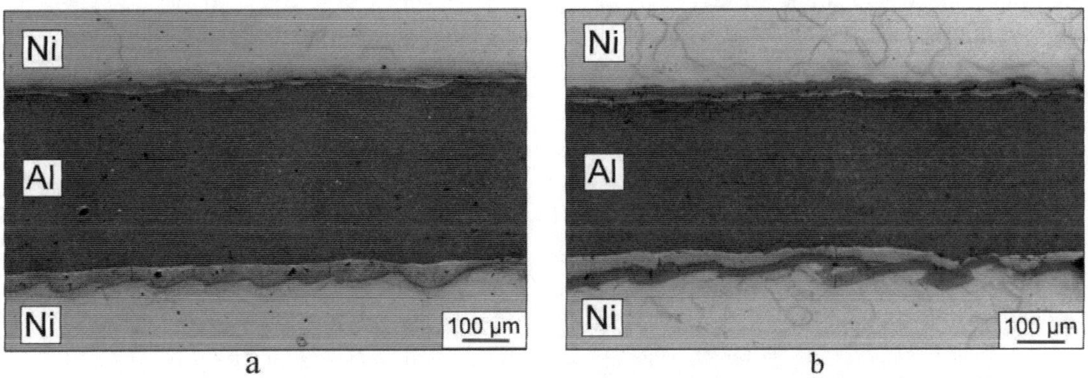

Figure 4. Intermetallic layers formed at the interfaces of Ni-Al composites after heating during 5 min (a) and during 3 h (b).

The intermetallic layer growth rate in explosively welded composites was compared with that in casted composites, which were fabricated as it is described in [14]. The results of structural studies suggest that processes, which take place during explosive welding, promote active nucleation and growth of intermetallic compounds during subsequent annealing. The acceleration of intermetallic formation is caused by the presence of the above-mentioned mixing zones. Another important factor that facilitates fast growing of intermetallic layers is removing of oxides and impurities from the surfaces during explosive welding. Comprehensive studies of the annealed samples confirmed that the explosively welded structure enables faster intermetallic growth than that in the composites produced by means of the casting technique (Figure 5b).

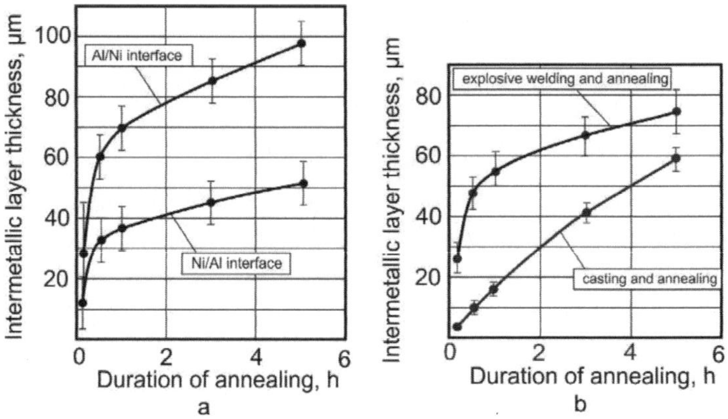

Figure 5. *a* – intermetallic layers growth in explosively welded composites at interfaces Al/Ni and Ni/Al, *b* – comparison of intermetallic growth during annealing at 620 °C in the samples produced by different techniques (explosive welding and casting).

4. Conclusion

The Ni-Al multilayered composite was successfully welded using the explosive welding technique. Zones of materials mixing were observed at the interfaces between Al and Ni. The mixing zones had a non-uniform chemical composition due to the nonequilibrium processes of explosive welding.

The $NiAl_3$ and Ni_2Al_3 intermetallic layers were synthesized at the interface after heat treatment of the composite at 620 °C. The intermetallic growth rate in the explosively welded composite was faster in comparison with that in the casted composite. The factors that accelerated the formation of intermetallic layers in explosively welded Ni–Al multilayered composites during annealing were purification of the surface of the plates from contaminations and oxides by a cumulative jet and formation of zones with intensively mixed Ni-Al.

Acknowledgements

This work has been supported by the Russian Ministry of Education and Science (No 11.1892.2014/K, project code 1892).

References

[1] Zhang F, Lu L, Lai M Froes F 2003 *J. of Mater.Sci.* **38**(3) 613–619
[2] López G.A, Sommadossi S, Gust W, Mittemeijer E J, Zieba P 2002 *Interface Sci.* **10** 13–19
[3] Fan, Q; Chai, H; Jin, Z 2001 *Intermetallics* **9** 609–619
[4] Konieczny M , Mola R, Thomas P, Kopcial M 2011 *Arch. of Metallurg. and Mater.* **56(3)** 695–702
[5] Rawers J, Perry K, 1996 *J. of Mater. Sci* **31** 1901–1906
[6] Kim H Y, Chung D S, Enoki M, Hong S H, 2006 *J. Mater. Res.* Vol. **21**(5) 1141–1149

[7] Zhu P, Li J C M, Liu CT 1997 *Mat. Sci.and Eng.* **A239–240** 532–539
[8] Kim H Y, Chung D S, Hong H 2006 *Scripta Materialia* **54** 1715–1719
[9] Chih-C H, Ming-Shou S, and Weite W 2012 *Met. Mater. Int.* **18**(1) 1–6
[10] Srivastava V C, Singh T, Kumar S, Choudhary S G and Jindal V *2012 J. of Mat. Eng. and Perf.* **21(9)** 1912
[11] Deribas A 1980 Physics of Explosive Hardening and Welding (Nauka: Novosibirsk)
[12] Bataev I, Hokamoto K, Keno H, Bataev A, Balaganskii I, Vinogradov A 2015*Met. Mater. Int.* **21** 713–718
[13] Bataev I A, Ogneva T S, Bataev A A, Mali V I, Esikov M A, Lazurenko D V, Guo Y, Jorge Junior A M 2015 *Mater. & Design* **88** 1082–1087
[14] Sameyshcheva T S, Bataev I A, Bataev A A, Yartsev P S, Polyakov I A 2012 *Proc. of IFOST* **1** 275–278

A method for determination of heat storage capacity of the mold materials using a differential thermal analysis

E Ol'khovik
Admiral Makarov State University of Maritime and Inland Shipping, 5/7, Dvinskaya st., Saint-Petersburg, 198035, Russia

E-mail: olhovick@gmail.com

Abstract. The article proposes a method for determining of the heat storage capacity of the mould materials. Modern materials for mouldsare made using a variety of technologies, and the manufacturers of binders and additives ensure thermal properties of certain materials only when using a certain recipe.In practice, for management of the casting solidification process (creation of the volume or directed mode) it is favorable to apply various technological methods, including modification of one of the important properties of the casting mould, which is heat storage capacity. A rather simple technique based on the application of the differential thermal analysis was developed for its experimental definition. The obtained data showed a possibility of industrial application of the method.

1. Introduction

Improving the quality of steel castings and their obtaining with good mechanical properties is an important scientific and technical problem of foundry production. The successful solution of this problem is possible through the use of efficient technologies.

Detailed information about the course of the physical processes happening during solidification and subsequent cooling of the metal of casting in the mold is necessary in order to elaborate the technological solutions. The thermal processes happening both in the metal and on borders of metal-mold and metal-core define the formation of the majority of main defects. The ability to managethermal processes and to predict the possible types of their interactions allows the process engineer to develop a package of measures [1], providing the obtaining of high quality cast details.

The most important property of the mold materials is their heat physical properties: thermal capacity, a coefficient of heat conductivity and heat storage capacity which in turn depends on the technology of mixture preparation (the structure of initial components, existence of additives, etc.) and conditions of the molds production (extent of consolidation, hold time after molding, drying, etc.).

2. Experimental work. A method of measurement

In this work, due to development of special techniques of the differential thermal analysis, the research of heat physical properties of themoulding materials, including a specified analysis of heat storage capacity of a sandmold, is proposed.

The differential thermal analysis is based on simultaneous measurement of temperature in several points of an alloy and a mold, thereby, it is possible to receive a fuller picture of metal solidification of in the mold.Thus there is an opportunity for analytical calculation of such magnitudes as thermal capacity, heat conductivity and heat storage capacity of the materials participating in cast molding.

The differential thermal analysis allows obtaining important data about kinetics of phase transformations from the results of measurement of the difficult temperature field. The technique developed in this work allows making the process of measurements continuous. Besides, it allows usto study even the fast-proceeding phase transformations during metalsolidification.

The intensity of heat exchange between the casting and the mold is various in the different periods of solidification. At the beginning of contact of the liquid metal with the cold mold the intensity of heat removal from the surface of the casting is high, and then, in the process of warming up of the mold, the intensity of heat exchange decreases. The nature of the intensity change of heat removal from the casting entirely depends on heat physical properties of the mold.

Obviously, the use of molding materials with various heat physical properties is one of the means of regulation of the castings properties. According to work results, the corresponding choice of heat physical properties of the molding mixture allows adozens of time change of the speed of the casting solidification.

The basic heat physical constants are: heat conductivity coefficient – λ, specific heat capacity - C, and density – ρ , coefficient of heat storage capacity – $b_{\Phi} = \sqrt{\lambda \cdot c \cdot \rho}$ and thermal diffusivity - $a = \dfrac{\lambda}{c \cdot \rho}$.

A direct measurement of heat capacity of the disperse materials in general and molding [2] mixtures in particular is connected with special difficulties. The matter is that the existing methods of direct determination of the coefficient of heat capacity belong to the methods of usual calorimetry. Because of small heat conductivity of the disperse materials the temperature balance in the calorimeter is established extremely slowly, and the losses of heat by the calorimeter in environment are considerable and it is difficult to keep records of them. In this regard for definition of heat capacity of the molding mixtures we have used the indirect methods that allowcalculatingof the coefficient of heat capacity according to experimentally found other heat physical characteristics [3]. But, despite the possibility of λ calculation of other heat physical coefficients, some researchers prefer a direct determination of the coefficient of heat conductivity of the molding mixtures [4].

The developed scheme for carrying out of the thermal analysis of alloys is represented in Figure 1. For implementation of the thermal analysis, the mold (Figure 2) made of quartz sand (96 %) with addition of binding was prepared in the following proportions: a polyvinyl acetate emulsion – 1 %, divinyl-styrol latex – 1 %, water – 2 %, in which one of the thermocouples previously painted with special non-stick paint was located.The control thermocouples were also located in the moldat a pre-determined distance from the experimental casting [5].

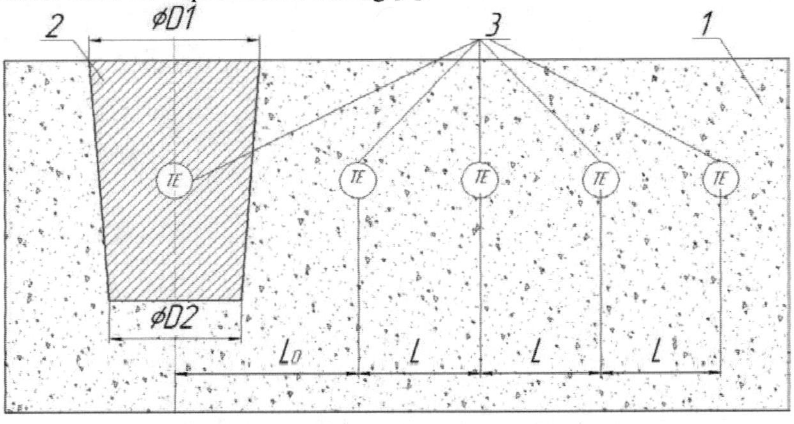

Figure 1. A principal scheme for determination of heat storage capacity of the mold material with the use of differential thermal analysis: 1 – casting mold, 2 – experimental casting, 3 – set of thermocouples (TE), *L* – distance between the thermocouples.

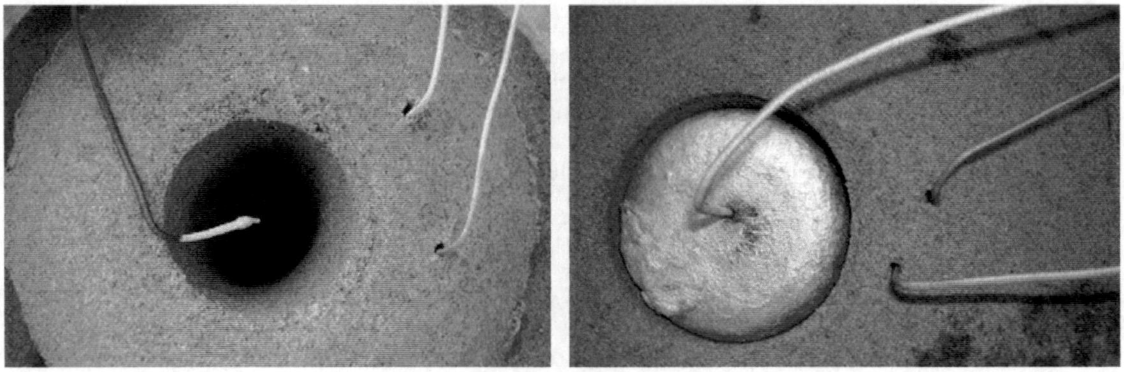

a b

Figure 2. The casting mold with defined thermocouples (a) and the experimental casting (b) for determination of the heat storage capacity of the mold.

The volume of the casting was previously calculated and included into the volume of metal solidifying within 3 minutes. In the electric furnace the aluminum alloy made of the Aluminum 6063-T6 (ISO)grade was prepared and filled into the mold (Figure 2b). Installed thermocouples are measuring the change of temperature in the casting sample and in the mold.With the use of the multichannel analog-digital converter (ADC) the data from each thermocouple were fixed in the personal computer with the frequency of one measurement per second.

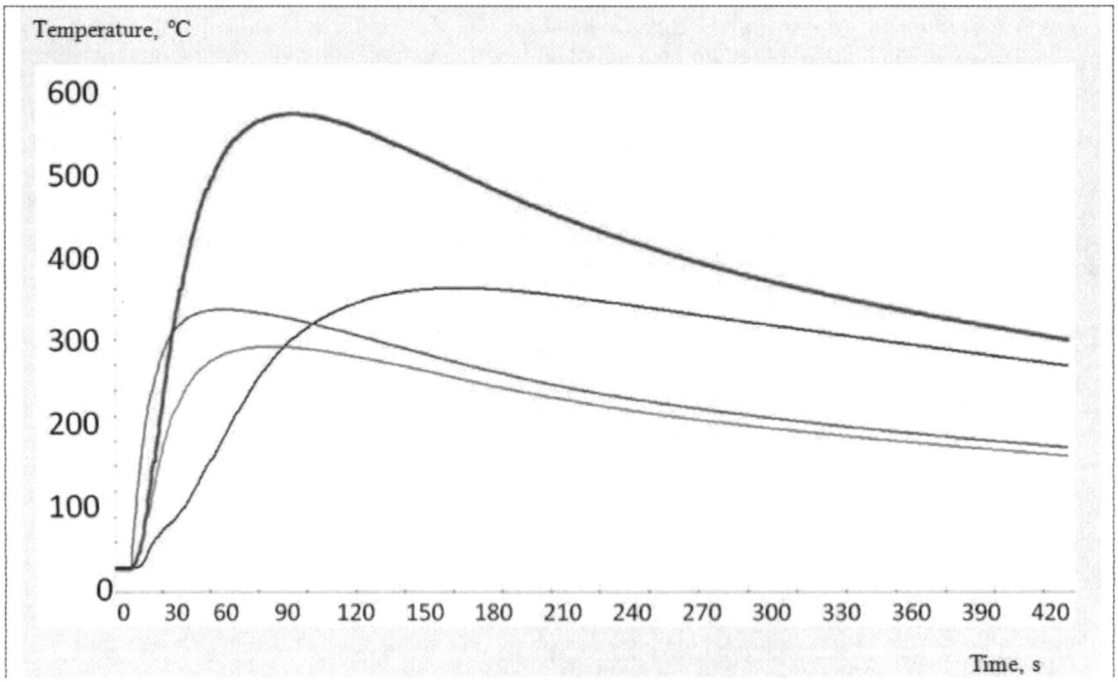

Figure 3. Experimental data of the differential thermal analysis
——Temperature in the center of the experimental casting sample (Al-alloy)
—, —,—Temperature in the mold (alpha-set)

3. Results

For experimental determination of heat physical properties of the moldingmaterials it is necessary to possess a qualitative mathematical apparatus which is capable of analyzing the data obtained by results of the differential thermal analysis (Figure 3).

For this purpose the computational model of the process of cooling of the casting in the moldon the basis of the differential equation of heat conductivity has been developed. In the finite-difference calculations of the heat transfer between the casting and the mold the method of elementary balances has been applied, when the equation of thermal balance for an *i*-element of splitting of the wall acquires the following view:

$$\sum_{j=1}^{m} \cdot k_{i,j}^{n}(T_i^n - T_j^n) F_{i,j} \Delta\tau = c_i G_i (T_i^n - T_i^{n+1}) \tag{1}$$

where k – the heat transfer coefficient between elements i and j, m – the number of non adiabatic element faces; F – the contact area between elements i and j; G – the specific heat of the material element and its mass; $\Delta\tau$ – a time step; n – a time index.

When the i element contacts a j element the coefficient of heat transfer will be determined by formula

$$k_{i,j} = 1/(\frac{L_i}{\lambda_i} + \frac{L_j}{\lambda_j}) \ , \tag{2}$$

for an extreme element of the conditionally allocated part of the mold,

$$k_{i,\alpha} = 1/(\frac{L_i}{\lambda_i} + \frac{1}{\alpha_i}) \ , \tag{3}$$

where L – distance from the center of gravity of the *i*-element (with the coefficient of heat conductivity of 1) to the next element (or to the border of the conditionally allocated part of the mold) in the direction of the movement of the thermal stream.

The thickness of the mold is allocated conditionally, for the purpose of reduction of the computational area, and it can be of any value.Heat diffusionon the boundary of mold and in the allocated part is expediently presented as an equality of thermal streams at variable coefficient of heat transfer α_m. During warming up of the whole conditionally allocated thickness it is necessary to consider heat transfer conditions on the conditional border of the mold. The results of calculation of heat transfer are carried out in the mold of temperature fields at various time points until full solidification of the casting.

4. Conclusion

Thus, at each temporary step of the process of casting solidification there is information about the temperature for each of the thermocouples located both in the experimental casting and in various sites of the mold. In this case, it is possible to make a system of the equations (1…3) where one can take measured temperatures as unknown values, heat physical characteristics of the mold(heat conductivity coefficient – λ and specific heat capacity – C)as constants. We have considered that the density of metal and the density of the material of the moldare initially known with the sufficient accuracy. Jointly solving not less than 3 equations, it is possible to execute calculation λ, C that is sufficient information for calculation of heat storage capacity of the mold.

The conducted experiments have showed a good convergence of results. For the material of the mold, the described above adequate results in comparison with the literary data have been received.

The advantage of the developed method is a possibility of creation of dependence of the value of heat storage capacity of the moldmaterial, both from the point of the metal temperature and the mold temperature. The offered method can be used for the increase of the efficiency of foundry technologies, including the creation of the mode of the volume or directed solidification [6, 7] and for definition of heat physical characteristics of new mold materials.

References

[1] Adedayo A V 2010 *Jour. of the Brazilian Society of Mechan. Sci. and Engineer.* **32** **2** 171–175
[2] Bulankin V P andVozzhin Yu V 1975 *Russ Cast Prod* . **6** 241–250
[3] Svidró J T, Diószegi A and Tóth J 2014 *Jour. of Thermal Analys. and Calorim.* **1151** 331–338
[4] Desnitskaya L V, Ol'khovik E O, Pirainen V Y, Glebov S M, Semyonova Y M and Matveyev Y A 2010 *Foundry. Technologies and Equipment* **8** 25–28
[5] Ol'khovik E O, Desnitskii V V and Molchanyuk R A 2007 *Steel in Translation* **375** 422–424
[6] Ol'khovik E O 2015 *Applied Mechanics and Materials* **770** 419–423
[7] Ol'khovik E O 2015 *IOP Conference Series: Materials Science and Engineering* **91** 012–022

Prospects for improvement of mining machines' cutting picks

S Prokopenko[1,2] , I Kurzina[3] and Yu Lesin[4]
[1] Yurginsky Technological Institute of Tomsk Polytechnic University, 26, Leningradskaya st., Yurga, 652055, Russia
[2] JSC 'NC VostNII', 3, Institutskaya st., Kemerovo, 650003, Russia
[3] Tomsk State University, 36, Lenina ave., Tomsk, 634050, Russia
[4] Kuzbass State Technical University, 28, Vesennyaya st., Kemerovo, 650026, Russia

[5]E-mail: sibgp@mail.ru

Abstract. Advantages and disadvantages of tangential rotary picks widely used in mining machines are described. A prospective line of picks design improvement has been found based on a change in the principle of destruction. A new design of blade-type cutting picks with a reinforcing tip is proposed having extended service life, requiring less energy for rock breaking, reducing yield of fine dust and increasing the grade of the mined coal.

1. Introduction

Mining machines currently used in mines of Russia, China, the USA, Australia and other countries are equipped mainly with tangential rotary picks (TRP) for rock breaking. Such picks are fixed on screw-, bar-, and crown cutting tools of shearers and road headers. In terms of design a pick presents a steel holder made up of a cylinder shaped tail and a tapered head (Figure 1).

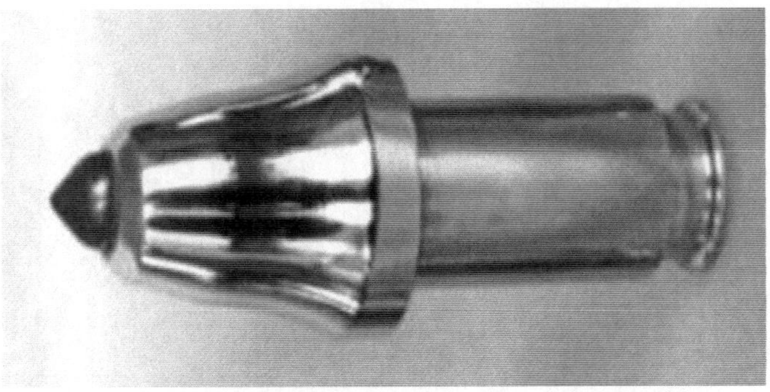

Figure 1. A tangential rotary pick of a mining machine.

The holder is reinforced with a hard tip of the high-tensile tungsten and cobalt alloy. The reinforcing tip is made in a cylinder- or mushroom shape. The protruding end of the reinforcing is pointed (Figure 2).

IOP Conf. Series: Materials Science and Engineering **124** (2016) 012134 doi:10.1088/1757-899X/124/1/012134

Figure 2. Hard-alloy tips of picks: *a* – cylinder shape; *b* –mushroom shape.

The following advantages of the picks have ensured their widespread use:
– ease of installation and replacement;
– even wear due to rock hardness while breaking and self-sharpening;
– small effect of the bending moment on the reinforcing due to the pick inclined to the cut surface;
– relatively long service life.

Despite the named advantages the tool has significant disadvantages. One of which is that such picks are not suitable for re-use. As a result, after the tool is 20…30 % worn out, the remaining 70…80 % go to waste, which is rather improvident from the view point of resource efficiency requirements to the modern production. Besides, a lot of dust and small particles are generated during coal mining, which decreases the grade of coal products.

2. Methodology of simulation

The research methodology is based on the study of the picks wear pattern and degree in terms of underground mining in Kuzbass. The research was conducted in eight mines at more than thirty roadheaders and shearers of models GPKS, KP-21, P-110, KSP-35, SL-500, Sandvik, Joy, EL-3000. More than five hundred picks were examined. For measurements the laboratory equipment of Tomsk Polytechnic University was used. The research base was replenished by the research results and analytical materials of other researchers dealing with improvement of rock breaking tools. References were made to industrial catalogues of cutting tools manufactured by the leading Russian and foreign manufacturers, such as Kopeysky Machinery Works, Kuznetskiy Machinery Works, Gorniy Instrument, Kennametal, Sandvik, BETEK [1-5].

3. Results and discussion

The research indicates that when a TRP impacts the mass coal (or rock) contacting the tip, at first it is compacted, then it collapses and breaks. Fine dust and relatively large (∅ 10…50 mm and over) shapeless particles are generated. The mechanism of such pick breaking the rock is presented in Figure 3 [6].

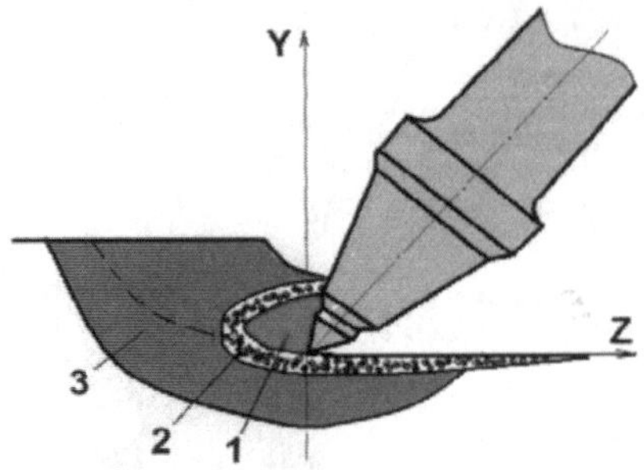

Figure 3. A mechanism of rock breaking by a tangential rotary pick:
1 – compacted core (fines) in triaxial compression; 2 – zone of rock collapsing; 3 – zone of elastic deformation of surrounding rock sectors.

For breaking and chipping of the rock by the tool the high energy consumption and considerable efforts are required to penetrate into the rock; it causes great resistance which results in accelerated destruction of the pick tip and friction sparkling.

High dust formation and friction sparkling are major causes of ignition of explosive dust and methane mixtures in road headings. According to the scientists of the State Technical University of Kuzbass the scope of use of TRPs due to their design features is restricted to low abrasive rock (less than 15 mg) with hardness of up to 60 MPa [7, 8].

Currently works on picks improvement are in progress. They include an increase in surface strength of the tool, a change in the form of the head, protection of the highest wear degree zone [9-14]. However, for considerable extension of service life and an increase in efficiency the picks should be re-designed.

Understanding of the concept 'cutting pick' shows that such a tool, by definition, should not tear, but rather cut the rock dividing it into pieces by even cuts. This will reduce the yield of fines and coal and rock dust, as well as sparking in the mine. Large (but properly sized) and smooth parts are more convenient for transportation and storage operations. The cutting process requires less energy costs for the rock breaking than crushing and separation of parts.

According to the above mentioned information TRPs shall be further improved with regard to the change in the operating principle: from rock breaking to rock cutting. The principle may be changed if the pointed tip is replaced with a cutting blade. This blade can be made of a hard metal plate placed in the axial slot of the pick head. In this respect a part of the head with the cutting blade should be able to rotate on the holder about its axis. This will ensure that the movement direction of the cutting blade will be aligned with that of the pick in the rock. The penetrating power of such reinforcing of the pick fixed on the cutting tool of the mining machine will be higher as compared to the currently used pointed tips. Rock will be cut with less efforts and energy costs. Parts of the cut rock will have a more regular shape, which will increase the grade of the mined coal. In addition, the yield of fines and dust will be significantly reduced. The proposed improvement of the tool ensures correspondence with the meaning of the word 'cutting pick'.

The design of the pick with a hard-alloy reinforcing plate having one cutting edge is shown in Figure 4a. The plate protruding above the head is designed as a blade with a cutting angle of 50...70 degrees and rounded edges. The blade is fixed in the head by brazing or by screws, which makes it possible to replace the plate once the blade is blunted. The working part of the head equipped with a hard-alloy plate is made rotatable about the axis of the pick. In case of excessive wear of the plate and damage to the working part of the head the junction allows for replacement of the latter, and the pick may be used further on. A general view of a pick with a replaceable rotating head is shown in Figure 4b.

The cutting blade is much longer than the pointed tip of the currently used hard-alloy inserts. It ensures a longer service life of the insert. Furthermore, after the rotating part of the head is worn out it can be replaced with a new one, which significantly extends the service life of the pick. Industrial tests showed that the pick can be used for up to 8...10 cycles provided the worn rotating heads are replaced [15].

Two rotating assemblies available at the proposed picks (a pick in the pick holder and a replaceable head in the head holder) provide for easy rotation of cutting blades and their alignment with the direction of the pick moving along the rock surface. Due to the friction forces the cutting edge of the plate will turn along the line of the least resistance and the pick will cut the rock, instead of crushing and chipping it out as the currently used picks do.

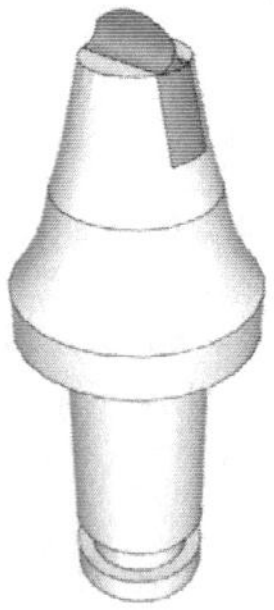

a

b

Figure 4. Design of a pick with a cutting blade (a) and a pick for 1-GPKS roadheader with a replaceable rotating cutting part of the head (b).

The cutting power of the pick maintained for a long time (owing to the great length of the cutting edge) and a small area of contact with the rock reduce friction sparkling, heating of the pick, the risk of explosions in the coal and methane atmosphere, the energy costs of the rock breaking are reduced as well. Changing of the principle of destruction from deformation and chipping out to cutting decreases the yield of fines and rock dust, and increases the grade of the coal mined.

4. Conclusion

The conditions and mechanism of rock breaking by a pick in mines have been analyzed. The advantages and disadvantages of the current pick design have been identified. Reusable blade-type tangential rotary picks having good prospects have been designed. The proposed picks are more cost efficient, have more extended service life, increase the grade of the coal mined and improve safety of underground mining operations.

References

[1] Catalogue of tools 2014 *Gorniy Instrument* 19
[2] Catalog of BETEK tools Available at: http://www.betek.de/ru/ productprogramme/ Mining-tunneling.html (Last accessed date 17.03.2014)
[3] Chain and wheel trenching tools 2012 *Catalogue of Kennametal products* 59
[4] Modern Sandvik Equipment for Underground Mines 2011 *Mining Industry* **2**
[5] Picks for Mining and Road Equipment 'Sandvik Mining and Sandvik Construction'. Available at: http://www.mining.sandvik.com/ (Last accessed date 12.01.2014)
[6] Kolesnichenko E A, Lyubomischenko E I and Demura V N 2012 *Coal* **6** 39–42
[7] Khoreshok A A, Mametyev L E, Borisov A Yu and Vorobyev A V 2015 *IOP Confer. Series: Mater. Sci. and Engin.* **91** 012084
[8] Khoreshok A A, Mametev L E, Borisov A Yu and Vorobev A V 2015 *Applied Mechanics and Materials* **770** 434–438
[9] Cutting tool for rocks 2014 Mining. RC 100 *Sandvik product catalogue* 52

[10] Aksenov V V, Lavrikov S V and Revuzhenko A F 2014 *Applied Mechanics and Materials* 682
[11] Prokopenko S, Sushko A and Kurzina I 2015 *IOP Confer. Series: Materials Science and Engineering* **91** 012058
[12] Prokopenko S A and Ludzish V S 2014 *Gornyi zhurnal/ Mining Journal* **1** 47–49
[13] Chinakhov D A 2011 *Applied Mechanics and Materials* **52–54** 442–447
[14] Bolobov V I, Bobrov V L, Talerov M P and Bochkov V S 2012 *Notes of Mining Institute* **195** 238
[15] Prokopenko S A, Ludzish V S, Kurzina I A and Sushko A B 2015 *Gornyi zhurnal/ Mining Journal* **5** 67–71

IOP Conf. Series: Materials Science and Engineering **124** (2016) 012135 doi:10.1088/1757-899X/124/1/012135

A study of the composition and microstructure of aluminum matrix composites reinforced with alumina fibers

D Zolotova[1], V Serpova[2], M Prokofiev[1], L Rabinskiy[1] and A Shavnev[2]
[1] Moscow Aviation Institute, 4, Volokolamskoe shosse, Moscow, 125993, Russia
[2] All-Russian Scientific Research Institute of Aviation Materials, 17, Radio st., Moscow, 105005, Russia

E-mail: f9_dec@mai.ru

Abstract. This article presents the results of a study of the microstructure and the composition of aluminum-based metal matrix composites (MMC) reinforced with continuous alumina fibers. An Al-Mg-Cu alloy similar to that of AA 2024 was used. X-ray diffraction and X-ray fluorescence analyses were used for investigation of a probable volume fraction of a spinel phase in MMC. Scanning electron microscopy and an X-ray microanalysis were used to study a change of the elemental composition of the composites microstructure on the polished cross sections. The constant mass fractions of magnesium (0.65 wt. %) and copper (1.25 wt. %) were found in the interphase area within radius of 1 μm around fibers.

1. Introduction

Enhancement of metallic materials strength nowadays is a topical issue of materials science due to a sharp rise of requirements in the leading-edge aerospace engineering. One of the effective ways of solution of this problem is design of composite materials based on a metallic matrix and, in particular, MMC based on an aluminum matrix reinforced with continuous Al_2O_3 fibers.

The main method of production of Al/A_2O_3 fiber composites is vacuum impregnation. The main problem of this method is to ensure a reliable contact between fibers and the matrix [1]. The interface properties, first of all adhesion between fibers and the matrix, determine the level of macroscopic properties of composites and preserve it throughout the lifecycle. Local stresses in a composite reach maximum values exactly near or immediately at an interface where material destruction usually starts. Therefore, to reach high mechanic performance, it is necessary to provide effective load transfer from the matrix to the reinforcing agent in the composite material. An adhesive bond at the interface should not be destroyed under the impact of external loading and in case of thermal and shrinkage stress arising due to the difference of matrix and fibers coefficients of thermal expansions. High strength of MMC is achieved in case the reinforcing component is properly moistened with the matrix alloy. However, as a rule, the interphase interaction causes the formation of intermetallic compounds that account for brittle fracture of MMC [1].

In case of production of $Al-Al_2O_3$ system of MMC, interphase layers consisting of $MgAl_2O_4$ and/or $CuAl_2O_4$ (depending on the matrix composition), spinel phases may be formed around fibers in the matrix [2-7]. There is an assumption that if magnesium is added to an aluminum alloy, it is segregated at the inter-phase border thus improving moistening and adhesion in Al/Al2O3 MMC [7-9].

The $MgAl_2O_4$ spinel may be formed by two identical reactions: in case of interaction of magnesium with aluminum and magnesium with Al_2O_3 fibers or α-Al_2O_3 polycrystals are formed in case of aluminum oxidation [2]:

$$Mg + 2Al + 2O_2 \rightarrow MgAl_2O_4$$
$$3Mg + 4Al_2O_3 \rightarrow 3MgAl_2O_4 + 2Al.$$

The purpose of this work is to study the detailed elemental composition and thickness of the interfacial zone in the structure of MMC aluminum based composites armed with alumina fibers.

2. Experimental procedure

Experimental MMC samples were manufactured using a liquid-phase vacuum impregnation technology. Reinforcement was performed using unidirectional continuous Al_2O_3 fibers. Russian D16 Al-Mg-Cu system aluminum, which is an analog of 2024 aluminum alloys, was used as a matrix. The volumetric content of fibers in MMC was 55 % wt.

The process operations used to produce MMC samples were: fibrous workpiece formation of the continuous Al_2O_3 fiber using the winding method, vacuum-compressor impregnation of a fibrous workpiece with a melt of a matrix alloy, machining.

The X-ray analysis was performed using the ALR X'TRA (TermoFisher) device of Cu-K_α radiation with the θ-θ scanning method using the PDF-2 data base and the Crystallographica Search-Match Version 3.1.0.0 software. The analysis was performed with the samples of longitudinal and transversal arrangement of fibers towards the radiation source. The quantitative X-ray phase analysis was performed using the Siroquant v.3 program. Aluminum oxide fibers have been also studied separately. This test was performed for estimation of sensitivity and precision of the performed X-ray phase analysis of a composite material.

The chemical X-ray fluorescent analysis was performed using ARL-Optim-X (TermoFisher) along with the UQuant program. At the same time, substances with a content of aluminum less than 0.2 % were not taken into consideration. The analysis was performed with allowance for backward scattering. The equipment used does not enable revealing a volume fraction of light elements, hydrogen and fluorine. In this case, the volumetric content of oxygen is not explicitly taken into consideration. The analyzing program only enables estimation of the volumetric content of elements or oxides thereof.

The composite microstructure was studied by the example of the cross-sections of samples using the EVO 40 Karl-Zeiss microscope. The X-ray microanalysis of the composition was performed using electronic scanning microscope Versa 3D LoVac DualBeam.

3. Results

For unidirectional samples oriented longitudinally and transversely to the radiation source X-ray diagrams were obtained (Figure 1a, b). Data given in table 1 has been acquired as a result of the analysis of X-ray diagrams. The results of intensity measurements for two types of samples are slightly different. It was established that a sample is mostly two-phase and consists of (-Al_2O_3 and a solid solution of copper in Al.

The results of the chemical X-ray fluorescent analysis are given in Table 2. It contains the weight fractions of components that are a part of the material in terms of the metallic phase and in terms of oxides. The analysis has revealed that a fraction of copper and magnesium (or oxides thereof) in a metal composite is approximately half as high as compared with the original fraction in the matrix alloy.

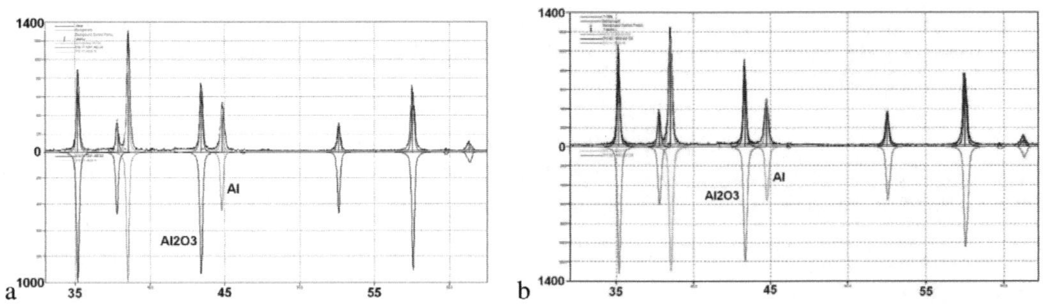

Figure 1. X-ray analysis results in case of longitudinal orientation of fibers (a), transversal orientation of fibers (b). In the upper part of the figures (blue lines) – X-ray diagrams obtained from experiments; given below – from the data base (green and black lines).

Table 1. Results of the X-ray analysis of composites.

	Aluminum Oxide	Aluminum
Longitudinal Arrangement of Fibers		
Formula	α-Al_2O_3	Al
PDF number	010-71-1241	010-71-4625
Concentration, % wt.	58	42
Total Peaks	25	5
Transversal Arrangement of Fibers		
Formula	α-Al_2O_3	Al
PDF number	010-82-1468	010-71-4624
Concentration, % wt.	59%	41%
Total Peaks	25	5

Table 2. Results of the X-ray fluorescent analysis in terms of oxides and in terms of metallic phase.

In terms of Metallic Phase		In Terms of Oxides	
Element	Content	Oxide	Content
Al	94.14	Al_2O_3	96.30
Cu	2.03	CuO	1.01
Si	1.15	MgO	0.941
Mg	0.908	SiO_2	0.752
Fe	0.667	Fe_2O_3	0.395

We could assume that all founded magnesium and copper oxides may be the parts of spinel phases. In this case, based on the known values of density of compounds and weight content of elements that are part thereof, we could establish that the maximum volume fractions of spinel in a composite shall not exceed 7 %. The probable areas of location of spinel particles are interfacial zones near fibers [2-7]. The average probable thickness of spinel layers on fibers were calculated considering the known average diameter of one fiber (11 μm) and the value of the volumetric content of the spinel phase was found. This value equals, approximately, 0.74 μm. For these areas, we further perform an X-ray microanalysis.

Micrographs of composite cross-sections and individual fibers are given in Figure 2 and 3. Light areas in Figure 2b are the areas of high concentration of copper and iron and the black areas among fibers are the areas of high concentration of carbon admixtures. Sets of points at different distances from the fiber surface were chosen for the X-ray microanalysis (Figure 3a). The elemental composition was defined at each point. Totally, 12 such sets have been studied for several fibers around which a sufficiently thick layer of matrix alloy is located without the visible content of admixtures. Each set contained 3 points: located near the fiber surface; located at a distance of 0.5 μm and those 1 μm distant. Figure 3b shows the established change of the volumetric content of elements in the matrix distant from the fiber surface. The given results are averaged for 12 sets of point. The variation of the concentration values has been found at points located at the same distance from the fiber surface amounted to a maximum of 50 %. The aluminum concentration is not shown on this graph as it amounts to the remaining value of 100 % of the volumetric content.

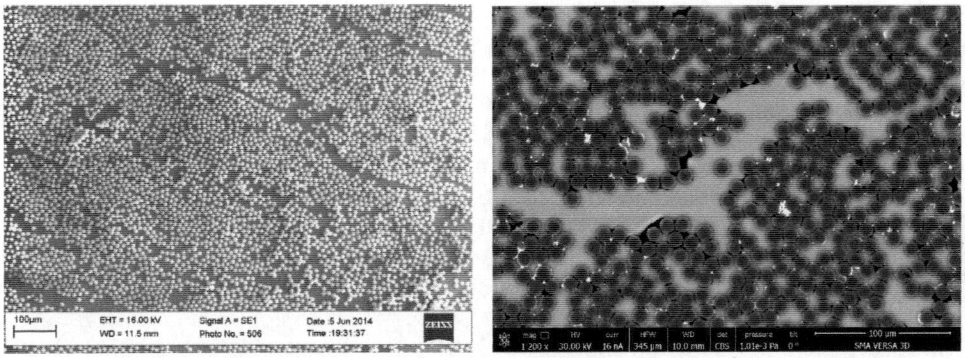

a b

Figure 2. The overall view of the microstructure of the composites cross-sections (a) and the presence of admixtures in the matrix (b).

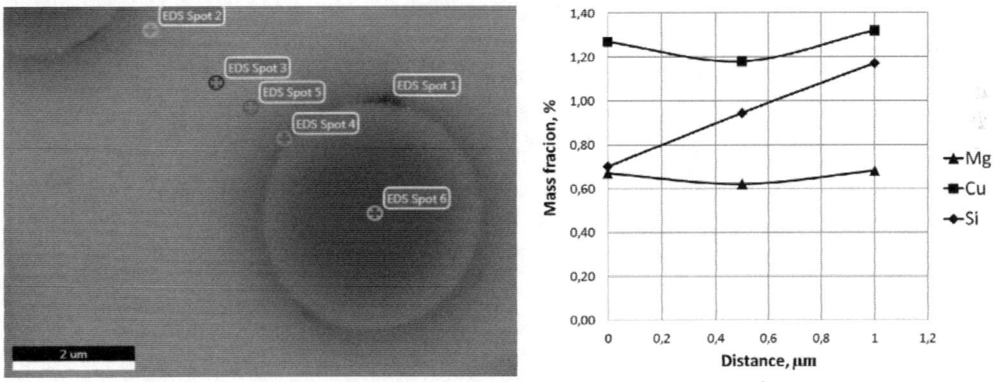

a b

Figure 3. Points of the X-ray microanalysis in the interphase area between the fiber and the matrix (b) and their results (b).

4. Conclusions

We presented the results of studies of the composites based on Russian alloy D16 (similar to AA in 2024) reinforced with alumina continuous fibers. According to the data of the X-ray analysis, only Al and Al_2O_3 phases have been found in MMC samples. $MgAl_2O_4$ and $CuAl_2O_4$ spinel phases have not been found, which may be a consequence of the method error that equals 5 % wt. Based on the acquired results of the analysis of the sample chemical composition, an estimation of the possible maximum volumetric content of interphase areas in the MMC structure has been performed. The estimation has shown that the phase of spinel in a composite could equal up to 7 vol. %. Therefore, the average thickness of interphase areas could equal up to 740 nm. As a result of the X-ray microanalysis, it has been established that near the fibers almost no change of the weight content of magnesium and copper elements occurs. The mass fractions are 0.65 % and 1.25 %, respectively. This may mean that in the studied composite samples there were no spinel phases formed near the surface of the fibers. Otherwise, the spinel crystals could be formed locally not causing formation of a solid interphase layer with constant thickness. The X-ray microanalysis has also revealed the fact that near the fibers the mass fraction of silicon is 0.75% and the farther it is from the fiber, the higher it gets. The composition analysis has revealed the fact that the alloy structure contains a large amount of carbon admixtures.

Acknowledgments
The authors acknowledge the funding support from FP 'Research and development in priority areas of scientific-technological complex of Russia, 2014 – 2020' event 1.3, agreement № 14.577.21.0171.

References

[1] Chawla N and Chawla K 2006 *Metal Matrix Composites* (NY: Springer)
[2] Munitz A *et al* 1979 Metall. Mat. Trans. A **10** 1491
[3] Zhong W *et al* 1995 Metall. Mat. Trans. A **26** 2625
[4] Moser B *et al* 2004 Acta Materialia **52** 573
[5] Sreekumar V *et al* 2008. Metall. Mat. Trans. A **39** 919
[6] Shen P *et al* 2004 J. Am. Ceram. Soc.. **87** 2151
[7] Levi C *et al* 1978 Metall. Trans. A. **9** 697
[8] Jonas T *et al* 1995 Metall. Trans. A. **26** 1491.
[9] Hallstedt B *et al* 1993 Mater. Sci. Eng. A. **169** 149

The Structure and Properties of Cast Iron Alloyed with Copper

A A Razumakov, N V Stepanova, I A Bataev, O G Lenivtseva, Iu Iu Riapolova and K I Emurlaev
Materials Science in Mechanical Engineering Department, Novosibirsk State Technical University, 20, Karla Marks ave., Novosibirsk, 630073, Russia

E-mail: stepanova@adm.nstu.ru

Abstract. Cast iron with 3 wt. % Cu was prepared by induction melting and casting in sand molds. The structure of the samples was studied using light microscopy (LM), scanning electron microscopy (SEM) and transmission electron microscopy (TEM). The addition of Cu promoted formation of pearlite and slightly decreased the volume fraction of graphite. No Cu inclusions were found by LM and SEM. The nanoprecipitations of ε-Cu in lamellar pearlite were observed by TEM. The properties of the Cu-alloyed cast iron were compared with the properties of cast iron not alloyed with Cu. The hardness of cast iron after alloying with Cu increased and the friction coefficient decreased in comparison with the reference sample.

1. Introduction

Generally, copper is not considered as one of the major alloying elements in steels and cast irons. However, the effect of minor additions of copper on mechanical and functional properties of ferrous alloys was reported in multiple studies (see review in reference [1]). Particularly, minor additions of copper (typically bellow 1 wt. %) significantly improve atmospheric corrosion of low carbon steels [2]. Copper is an important alloying element in high-strength low-alloy steels, which are widely applied in shipbuilding due to their high toughness and perfect weldability [1]. Additions of more than 0.6 wt. % of Cu to steel provides the effect of precipitation hardening during aging. Typical concentrations of Cu in HSLA steel is in the range of 1…2 %. Minor amounts of Cu (typically up to 1 %) are often added to ductile iron to increase the volume fraction of graphite [3]. And the content of Cu slightly above 1 % (e.g. 1.13 wt. %) in ductile cast iron induces formation of the predominantly perlite matrix with a large volume fraction of graphite particles and consequently improves hardness, and ultimate tensile strength [4]. From the above mentionedit follows that an influence of minor concentrations of copper on properties of ferrous alloys is studied relatively well. However, the effect of significant additions of copper on properties of iron is poorly reported in the literature. Due to this reason, this study describes the structure and the properties of cast iron with 3 wt. % addition of copper.

2. Materials and methods

The samples of cast iron were received by induction melting. At the first stage, low carbon steel and graphite were melted and liquid cast iron was produced. After that, commercially pure Cu and Al were added to the molten alloy. The melt was casted in sand molds at 1350 °C.

The elemental composition of samples was measured using an ARL 3460 optical emission spectrometer. Six measurements in different parts of the sample were taken to find the average values of concentrations. The structure of the samples was studies by light microscopy (LM) using a Carl Zeiss Axio Observer Z1m microscope, scanning electron microscopy (SEM) using a Carl Zeiss EVO 50 XVP microscope equipped with an energy dispersive X-ray (EDX) analyzer (Oxford Instruments X-Act), and transmission electron microscopy (TEM) using a Tecnai G2 20 microscope. The samples for LM and SEM were grinded using SiC grinding papers and polished using a diamond suspension. Polished samples were etched with 5 % nital solution. The samples for TEM were grinded using SiC grinding papers before the thickness of 100 μm. After that the dimple was formed in the sample by a Gatan Dimple Gridner apparatus using fine alumina suspension. The final stage of TEM sample preparation was ion milling using a Gatan Precision Ion Polishing System.

The averaged hardness of the samples was estimated using a Brinell hardness tester at a load of 3000 kg. A 10 mm diameter quenched steel ball was used as an indenter. The microhardness of local zones was measured using a Vickers Woulpert Group 402 MVD hardness tester with a diamond four-side indenter at a load of 100 g.

The coefficient of friction was measured according to a disk-on-liner scheme using an II 5018 apparatus. The cast iron sample was machined in the form of a ring segment (liner). The steel disk with hardness of 50 HRC and the diameter of 50 mm was used as a counterbody. Mineral oil LUKOIL 10W40 was used for lubrication. The test was carried out at 50 rpm. A load on the friction pair was gradually increased and the friction coefficient was registered at different loads. The convenient cast iron with low copper content (less than 0.1 wt. %) was used as a reference sample.

3. Results and discussion

The elemental composition of the sample, measured using the optical emission spectrometer is shown in Table 1. The cast iron sample alloyed with Cu and the reference samples had similar concentration of carbon and minor alloying elements, such as Al, Mn, Si, P, S, etc.

Table 1. The elemental composition of the samples, wt. %.

	C	Mn	Si	P	S	Ni	Cr	Cu	Al
Referencesample	3.13	0.53	0.26	0.017	0.016	0.04	0.04	0.09	1.05
Cast iron with copper	3.11	0.53	0.25	0.017	0.018	0.04	0.03	3.35	1.09

Figure 1 shows the images of unetched samples. It can be noticed, that the volume fraction of graphite after addition of approximately 3 wt. % Cu has slightly decreased. The results of the computer analysis of multiple images showed, that the average volume fraction of graphite in a sample, which was alloyed with Cu, was 7.5 %, while in the reference sample it was equal to 7.9 %. These data contradict to the results of Hsu and Lin [3], who showed that the minor addition of Cu promotes graphite formation in ductile iron. However, in the current research the concentration of Cu in cast iron was considerably higher in comparison with the experiment of Hsu and Lin.

Unalloyed cast iron had a ferrite-pearlite matrix (Figure 2a). The matrix of the alloyed sample was solely pearlitic (Figure 2b), which is consistent with common opinion, that Cu is a good pearlite former element in cast irons [5]. It is known, that Cu is an austenite stabilizer and increases the time of austenite decomposition [6]. During the continuous cooling at the time when pearlite transformation starts, the temperature of the copper-alloyed sample was lower in comparison with the reference sample. The diffusion of carbon at low temperatures is slower, which, according to conventional wisdom in the eutectoid transformation theory, leads to a decrease in the interlamellar distance of pearlite. Some decrease in the interlamellar distance in pearlite can be noticed in SEM images (Figure 3).

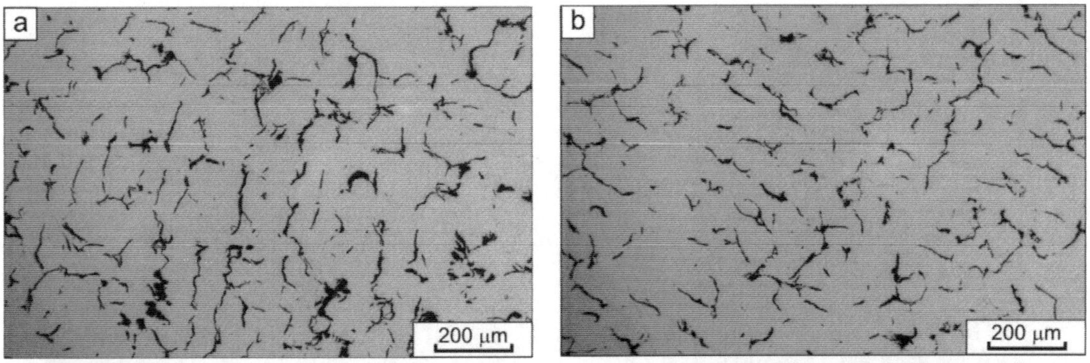

Figure 1. The structure of a reference sample (a) and a copper-alloyed sample (b) (unetched).

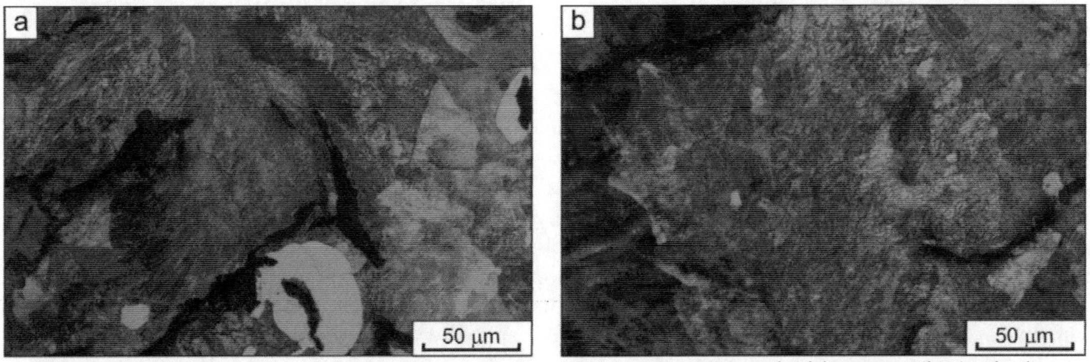

Figure 2. The structure of the cast iron sample without copper (a) and with copper (b) (etched).

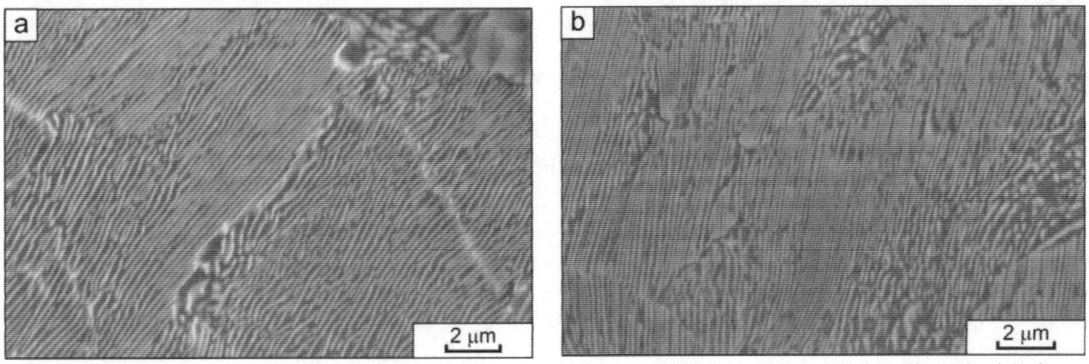

Figure 3. SEM images of pearlite colonies in a reference sample (a) and a sample, alloyed with Cu.

It is interesting to note, that according to the 'Fe-Cu' phase diagram the solubility of Cu in Fe at room temperature is below 1 %. However, no Cu-rich inclusions were found using SEM or LM. TEM investigations revealed that copper concentrated in the pearlite fine structure in the form of nanoscale precipitations ofa ε-phase (Figure 4). The ε-phase is a solid solution of Fe in Cu with FCC structure.

The decrease of interlamellar spacing and precipitation of nanoscale copper inclusions led to a noticeable increase of cast iron hardness. The Brinell hardness increased from 217 HB to 248 HB. The hardness of the pearlite increased from 330 HV to 368 HV.

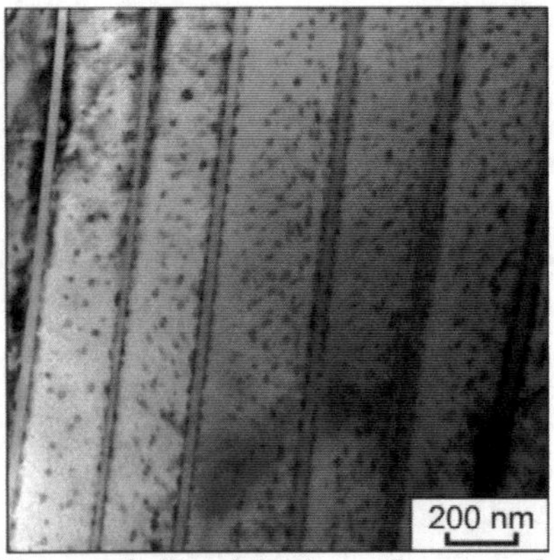

Figure 4. Precipitation of nanoscale Cu particles in lamellar pearlite in cast iron, alloyed with 3 % of Cu.

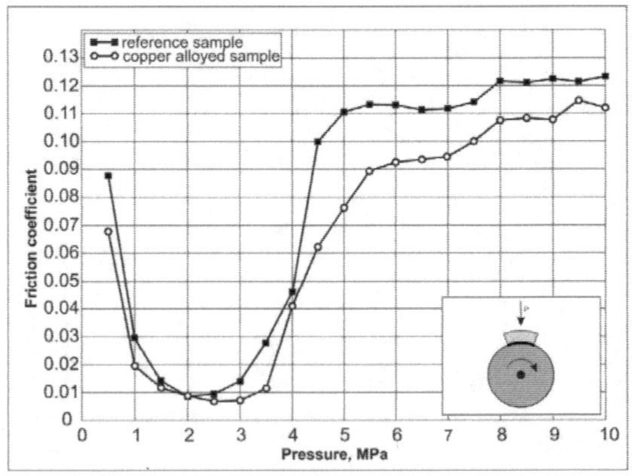

Figure 5. Variation of the friction coefficient with the pressure.

The dependence of the friction coefficient on pressure is shown in Figure 5. According to the existed theory of lubricated friction, the curves, shown in Figure 5, can be dividedinto 3 parts corresponding to different stages of friction. At the first stage, when the pressure is not high (in this case below 3 MPa), the coefficient of friction rapidly drops down to ~0.01 and stabilizes at this level. At this stage the hydrodynamic lubrication occurs and the surfaces of the sample and the counterbody are separated with a layer of lubricant. At the second stage, the friction coefficient rapidly increases due to local discontinuity of the lubricant film. At the third stage, which is called the stage of boundary lubrication, the coefficient of friction stabilizes at a relatively high level and the continuous layer of lubricant does not exist anymore. At this stage, the friction coefficient depends not only on the properties of the lubricant but also on tribological properties of friction surfaces. The boundary lubrication is typical for heavily loaded friction elements, for instance, in sliding bearings. It follows from Figure 5 that a friction coefficient of Cu-alloyed cast iron is lower in comparison with the reference sample especially in the boundary lubrication mode.

4. Conclusions

1. Addition of 3 % of Cu in cast iron increases the hardness of the alloy and decreases the friction coefficient in the boundary lubrication mode.
2. The Cu in cast iron predominantly precipitates as a nanoscale ε-phase and as a consequence it isinvisible for LM and low resolution SEM.
3. In the Cu-alloyed sample, the volume fraction of graphite is lower in comparison with the reference sample. The Cu alloying changes the cast iron matrix from ferrite-pearlite to purely pearlite.

Acknowledgements

This study was supported by Russian Science Foundation (project No. 15-19-00230).

References

[1] May I Le and Schetky L McD 1982 *Copper in Iron and Steel* (New York: Wiley Interscience)
[2] Buck D M 1913 *Ind. Eng. Chem.* **5** 447–452
[3] Hsu H and Lin K T 2011 *Materials Science and Engineering* **A528** 5706–5712
[4] Dasgupta R K et al 2012 *J. Mater. Eng. Perform.* **21** 1728–1736
[5] Totten G E, Xie L and Funatan K 2003 *Handbook of Mechanical Alloy Design* (New York: Marcel Dekker, Inc.)
[6] Popov A A and Popova L E 1965 *Spravochnik termista izotermicheskie i termokineticheskie diagrammy raspada pereohlazhdennogo austenita* (Moskva: Metallurgiya)

Optimization of aluminumand its alloys doping by ionic-beam-plasma coating

M Rygina[1], O Krisina[2], Yu Ivanov[1,2], E Petrikova[2], A Teresov[2]

[1]Tomsk Polytechnic University, 30, Lenina ave., Tomsk, 634050, Russia
[2]Institute of High Current Electronics, Siberian Branch of the Russian Academy of Sciences, 2/4, Akademicheskiyave., Tomsk, 634055, Russia

E-mail: L-7755me@mail.ru

Abstract. The surface morphology, chemical composition, microstructure, nanohardness, and tribological properties of systems were investigated. The paper considers the methodology offilmpplicationusingionic-beam irradiationby means of the installation'Solo' with different exposure modes. Irradiation modes which allow an increase in the microhardness of the material and a decrease in its wear rate are defined. Physical substantiation of this phenomenon is given.

1. Introduction

Alloying is the best way to improve characteristics of metals. However, the alloying process demands a large amount of dopant, but the application of plasma methodsallows reducing this amount.In this method heating and cooling rates are sufficiently high for a rapid growth ofa crystal. In addition to it, there is a method of ion implantation allowing one to introducealloying atoms directly into the surface layers of solids [1].Silicon and titanium are the main alloying elements. Silicon is one of the main additives in creation of cast alloys (silumin). Silumins usually contain from 5 to 14% of Si, i.e. a few percent higher or lower than the eutectic concentration Sr and Sb [2], V and Mo [3], W [4] are normally used as additives. The addition of these elements influencesbeneficially on the structure and mechanical properties of alloys.

We have revealed the presenceof intermetallic compounds of aluminum and titanium (TiAl, Ti_3Al, Al_3Ti) in the alloys. The main disadvantage of these compounds is low plasticity and impact strength. This drawbackcan be reduced owing to a decrease of the grain sizes and a creationof a columnar structure.

The aim of our study is to choose an optimal mode of sample influence, to analyze the structures and strength properties of commercially pure aluminum of the A7 grade and siluminof the AK12 grade alloyed with silumin (25 wt. %)and titanium.

2. Materials

Samples of commercially pure aluminum of the A7 grade, silumin of the AK12 grade, and silumin (25 wt. % Si) were used as amaterialfor study. Samples were subjected to electronic-ionic-plasma treatment on the'SOLO'installation. At the first stage samples were evaporated witha thin silumin film (h = 0.5 μm and h = 1.0 μm) by the method of sputtering of the silumin AK25 sample. After it, samples were irradiated by a high-intensity pulsed electron beam at different irradiation modes.

Another system is represented by theTi film /Al substrate system,the thickness of the Ti film being 0.5 μm. Thin (0.5 μm)Ti films were synthesized on a specialvacuum installation 'TRIO' in the low-pressure arc plasma [4].The film/substrate system was modified by the high-intense pulsed electron beam on the SOLO installation [5].The electron beam energy density was 10 J/cm^2 and 15 J/cm^2;the pulse repetition frequency was 0.3 Hz anda number of radiation pulses were varied in the range of 3...30.

The samples were characterized by microhardness (a PMT-3 device, the indenter load is 0.1N). The study of the wear rate of the film/substrate system was conducted in the disc-pin geometry by means of a

CSEM tribometer, Switzerland, at room temperature and air humidity. The wear volume of the materialwas estimatedafter conducting a profilometry of the formed track using a laser range optical profilometer 'MicroMeasure 3D Station' (Stil, France). The elemental and phase compositions of the modified surface were analyzed by means of an optical microscope (NEOFOT-32) and a scanning electron microscope (Philips SEM-515).

3. Research Results and Discussion

Let us analyze the results for the film (Al-25wt.% Si) / substrate (Al-12wt.% Si)system presented in Figure 1.The silicon inclusions in the thin near-surface layer have a globular shape, if the film thickness is 0.5 μm (Figure 1a, b), and an island shape, if the film thickness is 1.0 μm (Figure 1c, d).

Table 1. Dependence of the hardnesson the mode exposure.

Processing mode	**Hardness**
Silumin (12 wt.%)	125 kg/mm^2
E_s =15J/cm^2, h=0.5μm	84 kg/mm^2
E_s =20J/cm^2, h=0.5μm	104 kg/mm^2
E_s =15J/cm^2, h=1μm	88.7 kg/mm^2
E_s =20J/cm^2, h=1μm	165 kg/mm^2

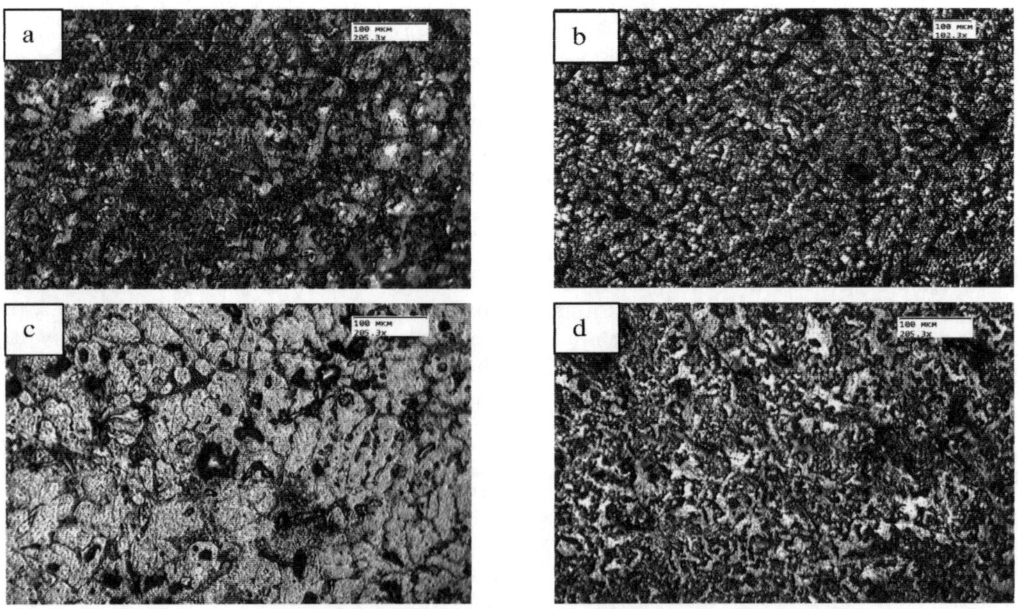

Figure 1. The surface structure of samples of the film (Al-25wt.% Si) / substrate (Al-12wt.% Si) systemirradiated by an electron beam pulse; *a*) E_S = 15 J/cm^2, h = 0.5 μm; b) E_S = 20 J/cm^2, h = 0.5 μm; c) E_S = 15 J/cm^2, h = 1 μm; d) E_S = 20 J/cm^2, h = 1 μm

Table 2. The wear resistance and friction coefficient of the surface layer of silumin (12 wt. %) modified by sputtered silumin and subsequent processing by the electron beam.

Mode	**V, 10^{-6}**	**V_0/V**	**<μ>**	**<μ$_0$>/<μ>**
Silumin (12wt.%)	24322 mm^3/(H·m)		0.612	
15 J/cm^2, 0.5 μm, Al-Si	18630 mm^3/(H·m)	1.31	0.648	0.94
15 J/cm^2, 1 μm, Al-Si	22228 mm^3/(H·m)	1.10	0.632	0.97
20 J/cm^2, 0.5 μm, Al-Si	19416 mm^3/(H·m)	1.25	0.633	0.97

20 J/cm^2, 1 μm, Al-Si	21124 mm^3/(H·m)	1.15	0.628	0.97

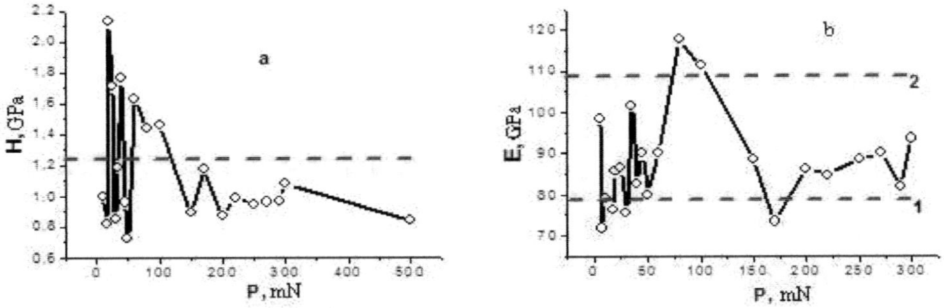

Figure 2. Dependence of hardness H and Young's modulus E on the indenter load (a nanoindentation method) of the surface layer of the AK12 grade silumin modified by silumin sputtering (the layer thickness is 0.5 μm) and subsequent treatment with an electron beam (in the mode of 15 J/cm^2, 150 ms, 5 imp., 0.3 Hz). Dash-dotted lines indicate the characteristics of the original AK12 grade silumin (12wt.%). Infigure b: 1 –siluminYoung's modulus 2 –silicon Young's modulus

The data on microhardness for another film(25 wt.% of silumin)/substrate (Al) system are represented in Table 3.

Table 3. Dependence of hardnesson the mode exposure.

Processing mode	**Hardness**
Mode of processing	250 MPa
Pure Al	425 MPa
E_S =15 J/cm^2, h=0.5μm	370 MPa
E_S =20 J/cm^2, h=0.5μm	548 MPa
E_S =15 J/cm^2, h=1μm	470 MPa

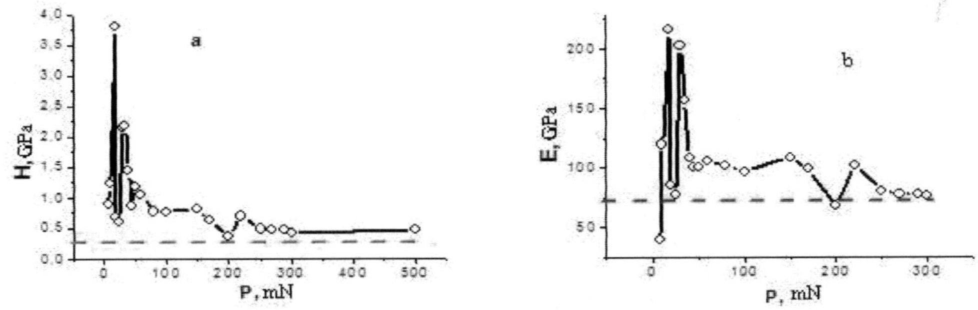

Figure 3. Dependence of hardness H and Young's modulus E on the indenterload (a nanoindentation method) of the surface layer of aluminum modified by the method of silumin sputtering (layer thickness is 1 μm) and subsequent treatment with an electron beam in the mode of 20 J/cm^2, 150 ms, 5 imp., 0.3 Hz. Dashed-dotted lines indicate the characteristics of the original A7 grade aluminum.

The wear resistance and friction coefficients of the surface of samples are given in Table 4. It has been established that high wear resistance belongs to samplestreated in the mode of20 J/cm^2 and the film thickness of 0.5 μm. Thelowestfriction coefficient is acquired by the materialin the mode of 20 J/cm^2 and the film thickness of1 μm.

Table 4. The wear resistance and friction coefficients of the surface layer of commercially pure A7 grade aluminum modified by siluminsputtering and subsequent processing by the electron beam (150 ms, 5 imp., 0.3 Hz.)

Mode	V, 10^{-6}, $mm^3/(H \cdot m)$	V_0/V	<μ>
Pure Al	7592	-	0.483
15 J/cm^2, 0.5 μm	6080	1.25	0.385
15 J/cm^2, 1 μm	6566	1.16	0.4
20 J/cm^2, 0.5 μm	5466	1.39	0.353
20 J/cm^2, 1 μm	4790	1,59	0.311

The research results of the hardness of the thirdTi film/Al substrate system have shown that hardnessincreases more than three timesduringthe electron beam irradiation underthe mode of 15 J/cm^2, 30 pulses, 50 μs, and 0.3 Hz (Figure4, curve 2).

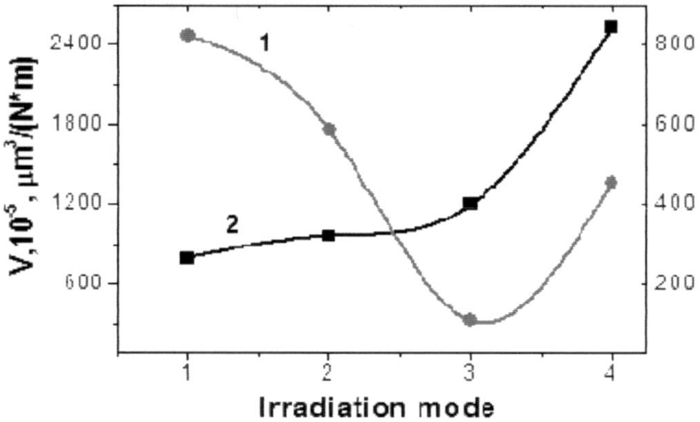

Figure 4. The wear rate (curve *1*) and the surface microhardness (curve *2*) of the film (Ti)/substrate (Al) system after different modes of treatment: 1 – substrate (commercially pure A7 aluminum); 2...4 the film/substrate systems irradiated with the electron beam with parameters of 10 J/cm^2, 50 μs, 10 pulses (lower index *2*);with parameters of 15 J/cm^2, 50 μs, 3 pulses (lower index *3*); and with parameters of 15 J/cm^2, 50 μs, 30 pulses (lower index *4*)

The best resulting properties are attainedduring the electron beam irradiation of the film/substrate system with parameters of 15 J/cm^2, 3 pulses, 50 μs, and 0.3 Hz (Figure 4, curve *1*); the wear rate of the modified layer is more than 7.5 times lower compared to the substrate.

The surface alloy, while displaying high hardness, reveals a decline in wear resistance, which is apparently due to structural features of the formed film/substrate system. This is due to the appearance of intermetallic compounds in the structure.

Intermetallic compounds of aluminum and titanium (TiAl, Ti_3Al, Al_3Ti) feature a unique combination of properties: high values of specific strength, heat resistance, thermal stability, elastic modulus, creep resistance, and anomalous temperature dependence on the mechanical properties [1].

Typical images of the surface structure of the film / substratesystem irradiated with the intense electron beam are shown in Figure5. When irradiating with the electron beam with the energy density of E_S = 10 J/cm^2 (10 imp.) and 15 J/cm^2 (3 imp.), the Ti film deposited on the aluminum surface is preserved. However, at the same time, the Ti film is fragmented withmicro-cracks (Figure5a, b).The sizes of fragments underE_S = 15 J/cm^{2}are 3...4 times less than those in the sample irradiated underE_S = 10 J/cm^2. A microcracks volume with cross-sectional dimensionsof ~10 microns is filledwith the melt of aluminum.

The crystal lattice parameter of aluminum, when E_S = 10 J/cm^2,is a= (0.4041±0.0001) nm, which corresponds to the lattice parameter of pure aluminum. In addition to the diffraction maximum of aluminum and titanium, the X-ray patterns show the diffraction maximums of the Al_3Ti phase.Thevolume fraction of the intermetallic is ~20 %, titanium – 28 %. Irradiation underE_S = 15 J/cm^2 (3 imp.) leads to an increase in the lattice parameter of aluminum up to a = (0.4050±0.0001) nm, which indicates the formation of alloy Al-Ti.

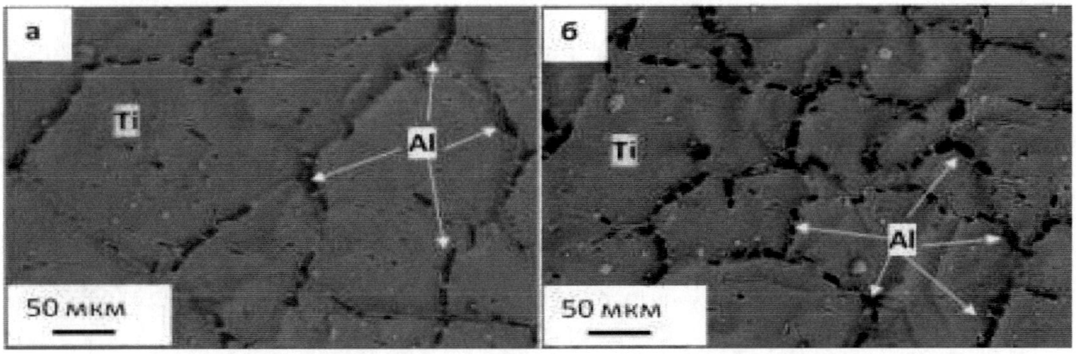

Figure 5. The surface structure of the film/substrate system irradiated with an electron beam under: a– 10 J/cm^2, 10 imp., b– 15 J/cm^2, 3 imp.

The volume fraction of the Al3Tiintermetallic compound reduces to ~3 %,the volume fraction of titanium remainsvirtually constant (~30 %).

4. Conclusion

The maximum microhardness is achieved by means of influence on the surface of the silumin (25 wt. %) film/substrate (Al) system and the energy density of the electron beam ofE_S = 15 J/cm^2. An increase of the energy density up to 20 J/cm^2 (regardless of the deposited film thickness) leads to a decrease of the microhardness of the irradiated surface.A percentage of silicon in the surface changes in a similar way. The hardness of the modified layer depends on the percentage of silicon in it.A better wear resistance was shown by a sample irradiated with an electron beam underE_S = 20 J/cm^2, h = 0.5 μm. The minimum friction coefficient was detected in the sample under E_S = 20 J / cm^2, h = 1μm.

The maximum microhardness is identified in the silumin (25 wt. %) film / silumin (12 wt. %) substrate system irradiated by the electron beam with the energy density of E_S = 20 J / cm^2. When using a substrate, wear resistance reduces irrespective of the irradiation modes and the layer thickness for 12 % silumin.

The Ti film/Al substrate system subjected to high-energy impact under 15 J/cm^2, 3 pulses, 50 μs, 0.3 Hz demonstrates a 7.5 time increase in the wear rate. The irradiation under 15 J/cm^2, 30 pulses, 50 μs, 0.3 Hz provides a more than 3 time increase of the system microhardness of the material compared to the substrate.

Acknowledgments

The work has been performedwith the support of the RNF grant (project № 14-29-00091).

References

[1] Mashkov Yu, PoleschenkoKN,Povoroznyuk C H and Orlov PV 2000*Friction and modification of materials tribosystems* (M: Nauka) p 280
[2] Szymczak T,Gumienny G and Pacyniak T 2015*Archives of Foundry Engineering***15**(1) 105–108
[3] Szymczak T,Gumienny G and Pacyniak T 2015*ArchivesofFoundry Engineering***15**(4) 81–86
[4] Szymczak T, Gumienny G, Pacyniak T and Walas K 2015 *Archives of Foundry Engineering***15**(3) 61–66
[5] GrinbergBA and IvanovMA2002 *Intermetallic Ni_3Al and TiAl compounds: microstructure and deformation behavior*(UrB RAS: Ekaterinburg)

Modification of the sample's surface of hypereutectic silumin by pulsed electron beam

M E Rygina[2], Yu F Ivanov[1,2], A P Lasconev[3], A D Teresov[1,2], N N Cherenda[4], V V Uglov[4], E A Petricova[1,2], M V Astashinskay[4]

[1] Tomsk Polytechnic University, 30, Lenina ave., Tomsk, 634050, Russia

[2] Institute of High Current Electronics, Siberian Branch of the Russian Academy of Sciences, 2/4, Akademicheskiy ave., Tomsk, 634055, Russia

[3] Physical-Technical Institute, Belarus National Academy of Science Minsk, 66, Independence ave., Minsk, 220072, Republic of Belarus

[4] Belarusian State University, 4, Nezavisimosti ave., Minsk, 220030, Republic of Belarus

E-mail: L-7755me@mail.ru

Abstract. The article presents the results of the analysis of the elemental and phase composition, defect substructures. It demonstrates strength and tribological characteristics of the aluminium-silicon alloy of the hypereutectic composition in the cast state and after irradiation with a high-intensity pulsed electron beam of a submillisecond exposure duration (a Solo installation, Institute of High Current Electrons of the Siberian Branch of the Russian Academy of Sciences). The research has been conducted using optical and scanning electron microscopy, and the X-ray phase analysis. Mechanical properties have been characterized by microhardness, tribological properties – by wear resistance and the friction coefficient value. Irradiation of silumin with the high-intensity pulsed electron beam has led to the modification of the surface layer up to 1000 microns thick. The surface layer with the thickness of up to 100 microns is characterized by melting of all phases present in the alloy; subsequent high-speed crystallization leads to the formation of a submicro- and nanocrystalline structure in this layer. The hardness of the modified layer decreases with the increasing distance from the surface exposure. The hardness of the surface layer is more than twice the hardness of cast silumin. Durability of silumin treated with a high intensity electron beam is ≈1,2 times as much as the wear resistance of the cast material.

1. Introduction

The application of hypereutectic silumin is limited due to its structure instability conditioned by, in particular, its uneven distribution in the castings of relatively large crystals of primary silicon and intermetallic inclusions. Segregation of primary silicon crystals in hypereutectic silumin leads to worsening of castings cutting, anisotropy of the alloy properties in the cross section of castings, a relatively low level of mechanical properties and, first of all, ductility, wear resistance reduction of products. In order to suppress primary crystallization of a silicon phase it is proposed:

- to overheat the melt above the dome of collapse of the metastable colloid thus obtaining quasi-eutectic [1].
- to use vibration [2], melt processing by a periodic (cyclic) pulse unipolar electric current [3] and other energetic effects on the melt [4].

In [5] it is offered to provide a combination of the high crystallization rate of the alloy, a modification with phosphorus, and additional heat treatment, in case of obtainment of castings for subsequent plastic deformation, to produce the high quality castings of hypereutectic silumin, containing fine, less than 15 microns, crystals, which are uniformly distributed in the volume of the primary silicon.

The purpose of this paper is to analyze the results obtained in the study of structure and properties of the surface layer of the composition of hypereutectic silumin irradiated with a high-intensity pulsed electron beam of a submillisecond exposure duration.

2. A Material and a Research Technique

A material for the study was samples of the aluminum-silicon alloy (silumin) of the hypereutectic composition. The average concentration of silicon amounts to 28.5 wt. % (the concentration of silicon in the range of 22.1…35.5 wt. %). The concentration of silicon was measured by the electron microprobe analysis of cast samples (averaging according to the results of the elemental analysis of five areas with sizes of 1,2x0,95 $mm^{2)}$.

The samples surface was modified by the high-intensity pulsed electron beam on the 'Solo' installation (Institute of High Current Electronics, the Siberian Branch of the Russian Academy of Sciences) [6]; the energy density of the electron beam is 40 J/cm^2, pulse repetition frequency is 0.3 Hz, the number of pulses were 20; the electron beam pulse duration is 200 μs, and the energy of accelerated electrons is 18 keV. The study of the elemental and phase composition, a defect structure of the modification surface and transverse sections was performed by the methods of optical and scanning electron microscopy, and the X-ray analysis. Mechanical properties of the material were characterized by microhardness. The study of the silumin wear resistance was conducted in the geometry of disc-pin at room temperature and humidity (a CSEM pin-on-disk tribometer, Switzerland). The volume of the material wear-and-tear was estimated after conducting a profilometry of the formed track (a MicroMeasure 3D Station, Stil, France).

3. Research Results and Discussion

Silumin in its original state is a very nonhomogeneous material. The analysis of the structure of the section, realized by the methods of scanning electron microscopy, has allowed one to reveal rough inclusions of silicon and intermetallic compounds. Their average sizes may reach 50 microns (Figure 1). The inclusions have different shapes; they are distributed in the volume of the material unevenly. Melting of samples was accompanied by the formation of many micropores in the volume of material in a chaotic manner (Figure 1).

The electron microprobe analysis of the original sample (averaging according to the results of the elemental analysis from five areas with the sizes of 1.2×0.95 mm^2) has revealed a silicon concentration of 28.5 wt. % (when changing the silicon concentration in the range of 22.1…35.5 wt. %). Investigations of the phase composition of the original alloy have revealed the presence of the two main phases – a solid solution based on alumina and a solid solution based on silicon – in the material. The research was carried out with the methods of X-ray diffraction. The X-ray pattern of the original sample has shown weak lines obviously belonging to intermetallic compounds. The X-ray pattern analysis showed that these intermetallic compounds are the phases of Al_9FeSi_3 or $FeAl_3Si_2$ compositions. The results of X-ray phase analysis of the main phases structure of the studied alloy have showed that initially the parameters of aluminum and silicon lattices are close to the tabulated values; the dimensions of the areas of coherent aluminum scattering are significantly greater than those of silicon.

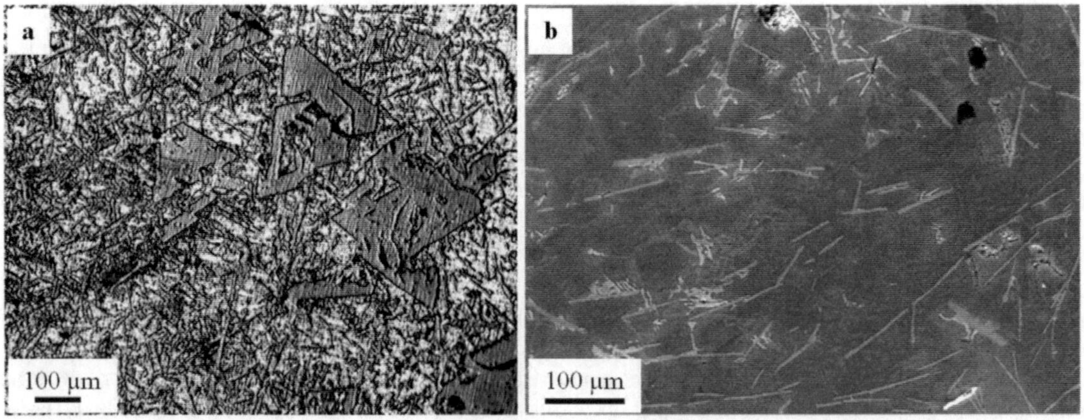

Figure 1. A structure of the cast silumin sample prior to irradiation with the high-intensity pulsed electron beam: *a* – optical microscopy; *b* – scanning electron microscopy.

The samples of the aluminum-silicon alloy were irradiated, as previously performed calculations of the temperature field showed [7-9], in the melting mode of all present in the material phases. Figure 2a shows micrographs demonstrating the sample surface structure being treated with an electron beam. It is clearly seen that a homogeneous layer forms along with the structure of the high-speed crystallization. The crystallite sizes vary in the ranges of 0.4...0.5 microns. The microprobe analysis indicates that the silicon concentration in the surface layer varies in the ranges of 10...12 at. %. It corresponds to the composition of the aluminum-silicon alloy eutectic [10].

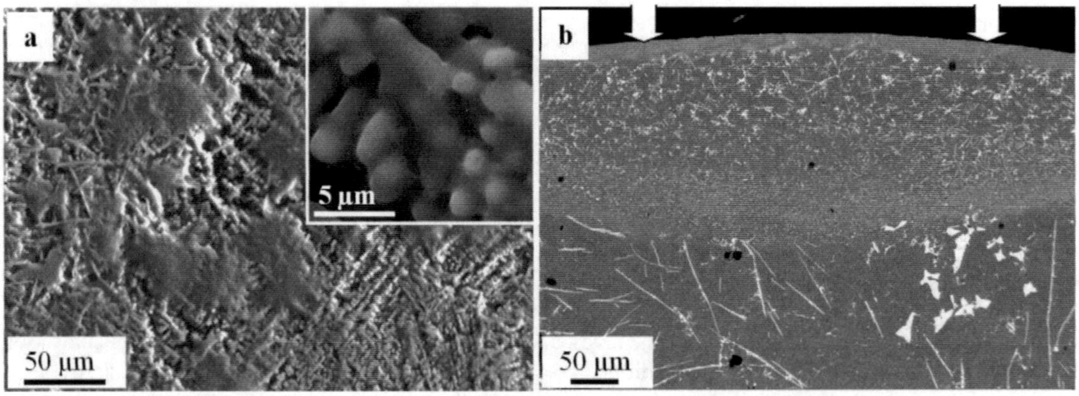

Figure 2. An electron microscope image of the surface structure (a) and a transverse section (b) of silumin, irradiated with the high-intensity pulsed electron beam. In (b) the arrows indicate the surface of radiation exposure.

The X-ray phase analysis of the modified sample has showed that the influence of the electron beam leads to a significant reduction in the sizes of coherent scattering regions (OCD) of both aluminum and silicon. The parameter of the crystal lattice of aluminum and silicon increases. The increase of the parameters of the aluminum and silicon lattice is probably due to dissolution of impurity atoms (Mg, Fe, Cu) in them. This assumption is confirmed by the absence of diffraction lines of the corresponding intermetallic phases on the X-ray patterns of the irradiated samples.

The analysis of the structure of the transverse section irradiated by the electron beam has revealed the formation of a multi-layer structure (Figure 2b), such as:

- a surface layer (Figure 3a) formed as a result of melting and speed crystallization of all phases of the material (aluminum, silicon and intermetallic precipitates);

- an intermediate layer (Figure 3b) containing not completely dissolved, particularly large inclusions of silicon and intermetallic compounds;
- a transition layer (Figure. 3c) containing not completely dissolved inclusions of silicon and intermetallic compounds.

The total thickness of the modified layer is up to 1000 microns. A silumin layer modified by the electron beam becomes dense and comprises a minimum number of micropores and microcracks. This result can serve as a recommendation to irradiate silumin with the high-intensity pulsed electron beam as a method, which enables one to remove micropores and microcracks formed in the process of material melting from a relatively thick surface layer.

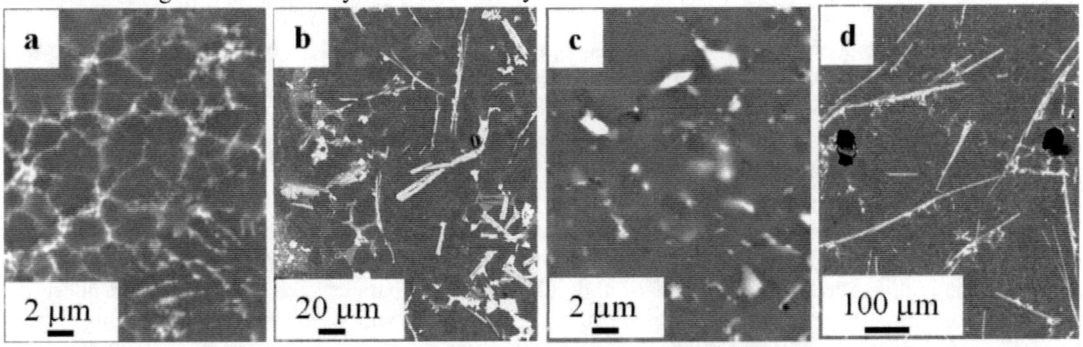

Figure 3. A silumin volume structure modified by the electron beam: *a* – a surface layer; *b* – an intermediate layer; *c* – a transition layer; *d* – the bulk of the material. Scanning electron microscopy of the cross etched polished section

The structure of the surface layer, which has been formed as a result of high-speed crystallization, is represented by crystallites (obviously, a solid solution based on aluminum) surrounded by relatively thin interlayers of the second phase (obviously, silicon and, possibly, intermetallics) (Figure 3a). The crystallite sizes vary within the limits of 2 microns to 5 microns. The transverse dimensions of interlayers amount to a few tenths of a micrometer. The thickness of the surface layer reaches 100 microns. The intermediate layer is characterized by the presence of a large number of intermetallic inclusions, which number exceeds the similar characteristic of the original material per unit of the section surface (Figure 3b). This fact allows one to suggest that the intermediate layer has been enriched with the intermetallic phase previously located in the surface layer and transferred to the intermediate layer under the action of gravity forces, when melting the material by the electron beam. The thickness of the intermediate layer reaches 400 microns. The transition layer is the longest one, its thickness reaches 500 microns. This layer is characterized by the presence of relatively fine silicon inclusions, predominantly of a globular shape, and intermetallic compounds, which sizes are 2…5 microns. These sizes are significantly less than the sizes of the inclusions disposed in the intermediate layer and in the bulk of the material (Figure 3c).

The mechanical properties of silumin were characterized by the microhardness value. The accomplished studies have shown that irradiation of silumin with the high-intensity pulsed electron beam in the mode of melting of the surface layer leads to the formation of the multilayer structure. Its hardness naturally decreases when it is distant from the surface of treatment. After treatment with the electron beam the microhardness of the surface of irradiation is ≈1.9 GPa; the hardness of the surface layer of the silumin modified by the electron beam is ≈1.60 GPa, the hardness of the intermediate layer is ≈1.08 GPa, the hardness of the transition layer – ≈0.97 GPa. The microhardness value of the original material is ≈0.8 GPa. The microhardness of the surface of irradiation exceeds the microhardness of the original material more than 2 times.

It has been found that the wear resistance of silumin treated by the high-intensity electron beam is ≈1.2 times greater than that of the cast material; the friction coefficient of irradiated silumin is ≈1.5 times less than that of the cast material.

4. Conclusion

The results of the analysis of the elemental and phase composition, the defect substructure, strength and tribological characteristics of the hypereutectic silumin composition in the cast state and after irradiation with the high-intensity pulsed electron beam of a submillisecond exposure duration (a SOLO installation, Institute of High Current Electronics, the Siberian Branch of the Russian Academy of Sciences) have been represented.

The studies have been performed by using optical and scanning electron microscopy and X-ray diffraction. Mechanical properties were characterized by microhardness, tribological properties – by resistance to wear and friction coefficient. A characteristic feature of the alloy in the cast state is the presence of a big number of large (up to 50 micrometers in diameter) particles of excess silicon, intermetallic inclusions (Al_9FeSi_3 or $FeAl_3Si_2$) and micropores. Irradiation of silumin by the electron beam has resulted in the modification of the surface layer up to 1000 microns. The modified volume has a multilayer structure devoid of micropores and microcracks. The surface layer with the thickness of up to 100 microns is characterized by melting of all phases present in the alloy. In this layer high-speed crystallization leads to formation of the submicro- and nanocrystalline structure. Underlying sublayers contain both insoluble inclusions of initial phases and compounds formed during high-speed crystallization and cooling of the material. The hardness of the surface layer is more than two times greater than the hardness of cast silumin. When moving away from the irradiated surface, the hardness of the modified layer naturally reduces. The wear resistance of silumin, treated with the high intensity electron beam, is ≈1,2 greater than the wear resistance of the cast material in 2 times. The friction coefficient of irradiated silumin is 1,5 times less than that of the cast material.

Acknowledgments

The work has been performed with the support of grant RNF (project № 14-29-00091).

References

[1] Korchagin O A, Brodova I G, Viscosity and Nikitin V I 1991 *Melt.* **1** 10–17

[2] Ivanov AA and GG Krushenko 1991 *Foundry* **3** 7–8

[3] Petrov S S, Prigunova A G, Prigunov S V and Klyuchnik D N 2015 *Metal Physics and Advanced Technologies* **37(3)** 367–78

[4] Belov M V, Ten E.B. and Batyshev K A 2007 *Proceedings of the universities. Ferrous metallurgy* **5** 42–4

[5] Brodova I G, Popel P S and Barbin N M 2005 *Melts as the basis for the formation of structure and properties of aluminum alloys (*Ekaterinburg, Ural Branch of Russian Academy of Sciences) pp. 369

[6] Koval N, Ivanov Yu 2008 *Proceedings of the universities Physics* **5** 60–70

[7] Ivanov Yu F, Koval N N, Vlasov V I, Teresov A D, Petrikova E A, Shugurov V V, Ivanova O V, Ikonnikova I A, Klopotov A A 2013 *High Temperature Material Processes.* **17(4)** 241–56

[8] Ivanov Yu F, Petrikova E A, Ivanova O V, Ikonnikova I A, Tereso A.D., Shugurov V V, Krysina O V. 2015 *Russian Physics Journal* **58**(3) 73–9

[9] Ivanov Yu, Petrikova E A, Ivanova O V, Ikonnikova I A, Tkachenko A V 2005 *Proceedings of the universities. Physics* **58(4)** 46–51

[10] Laskovnev 2013 A P *Modification of the structure and properties of eutectic silumin electron-ion-plasma treatment* (Minsk: Belarus. Navuka) p 287

Boron-modified Ni_3Al intermetallic compound formed by spark plasma sintering of mechanically activated Ni and Al powders

L I Shevtsova[1], T S Ogneva[1], D O Mul[1], M A Esikov[2], A Yu Larichkin[2] and V N Malikov[3]
[1] Department of Materials Science, Novosibirsk State Technical University, 20, Karla Marksa ave., Novosibirsk, 630073, Russia
[2] Lavrentyev Institute of Hydrodynamics, 15, Lavrentyeva ave., Novosibirsk, 630090, Russia
[3] Altay State University, 61, Lenina ave., Barnaul, 656049, Russia

E-mail: edeliya2010@mail.ru

Abstract. A Ni_3Al intermetallic compound was obtained by spark plasma sintering of mechanically activated Ni and Al powders in atomic ratio 3:1 respectively. Samples with boron addition of 0.1 and 0.2 % (wt.) and samples without boron were obtained. The maximum value of the relative density (~ 99 %) has been obtained for the material by sintering of mechanically activated mixture powders modified with 0.1 % of boron. No differences have been found between the structure of boron-modified Ni_3Al and Ni_3Al without boron addition. The maximum level of bending strength (2200 MPa) has been achieved for Ni_3Al with 0.1 % (wt.) of boron. This value is almost 3 times the bending strength of the sample of Ni_3Al sintered without boron addition.

1. Introduction

Development of the materials with enhanced physical, chemical and mechanical properties is one of the topical issues of advanced materials science. Particular attention is paid to the investigation of nickel aluminides due to their high mechanical properties, high corrosion resistance and oxidization resistance at high temperature [1-5]. Among all aluminides Ni_3Al has anomalous temperature dependence of the yield strength in a certain temperature range. This phenomenon is advantageous from the point of application of the intermetallic compounds as a structural material in aircraft and rocket industry, chemical engineering and power engineering [1-3].

However, the widespread industrial application of nickel aluminide is constrained by a number of features including the low level of ductility and fracture toughness. This fact explains the poor machinability of materials. Polycrystalline Ni_3Al is brittle at room temperature [4].

In recent years, heightened interest has been aroused in the obtaining of materials based on intermetallic compounds produced by mechanical activation of powder mixtures and subsequent spark plasma sintering [3-9]. Mechanocomposites with significant dispersion of reagents are formed as a result of plastic deformation during mechanical activation of powder mixture in a planetary ball mill. Mechanocomposites are characterized by an increasing contact area, and a large number of nonequilibrium defects and internal stresses [8]. Due to short duration of high temperature exposure, an optimal ratio of heating and deformation modes, and features of current flow through the powder mixture, SPS-technology allows producing high-strength materials with the low level of porosity and low residual stresses. Application of this technology promotes the formation of a fine grain structure. However, intermetallic compounds obtained using the technology described above still has low ductility [6, 7, 9-12].

It is well known, that one way to improve the ductility of the intermetallic Ni_3Al while increasing its mechanical properties is microalloying. The analysis of published data shows that the most effective alloying element for Ni_3Al is boron. The positive effect of boron on the plasticity of the intermetallic is related to its segregation to the grain boundaries. However, this method works only as long as the content of aluminum in the alloy does not exceed 25 % (at.). In addition, the grain size should be sufficiently fine [13-15].

In paper [15] the authors investigated the dependence of boron concentration on the mechanical properties. Samples with 0.2 % (wt.) of boron demonstrated the highest level of mechanical properties. Further increasing of boron concentration led to the embrittlement of Ni_3Al because of the borides formation. Thus, in this paper the effect of boron microadditions on the structure and mechanical properties of intermetallic compound Ni_3Al formed by SPS-sintering of mechanically activated powders of nickel and aluminum were investigated.

2. Materials and methods

Carbonyl nickel powder PNK UT3 (99.85 % Ni) and aluminum powder PA-4 (98 % Al) with the average particle size of 10 μm and 60 μm, respectively, were used as initial materials. The ratio of nickel and aluminum powders was 3:1, which provided the formation of the $Ni_{75}Al_{25}$ % (at.) compound. Amorphous boron was chosen as a modifying element. The amount of added boron was 0.1 and 0.2 % (wt.).

Preliminary activation of the reaction mixtures was carried out in two steel drums of the planetary ball mill 'AGO-2'. Each drum was filled with 10 g of the powder mixture. The weight ratio of the balls and the powder mixture was 20:1. In order to prevent oxidation of the materials the drums were filled with argon, the pressure of the gas was 0.3 MPa. Centrifugal acceleration of balls reached 400 $m \cdot s^{-2}$.

Experimental results and literature data revealed the optimum time (3 min) for mechanical activation which ensures the formation of single-phase intermetallic compound Ni_3Al during the subsequent sintering of this mixture [8, 10, 11]. Mechanocomposites consisting of flake shaped nickel and aluminum particles were prepared after three minutes of mechanical activation of powder mixtures (Figure 1).

a

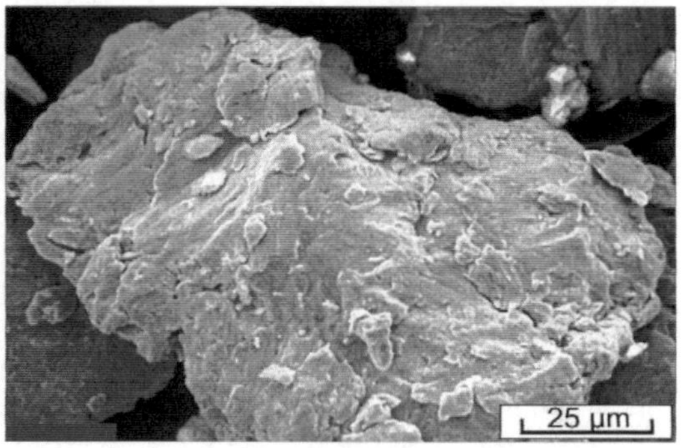

b

Figure 1. Mechanocomposites based on nickel and aluminum formed after 3 minutes of mechanical activation.

SPS-sintering of mechanocomposites was carried out using the SPS Labox-1575 equipment in a graphite mold with the inside diameter of 30 mm. Heating speed of the powder mixture was 100 °C/min, pressing pressure was 40 MPa. The samples were sintered at 1100 °C during 5 min. The density of the obtained materials was calculated by determining their mass and linear dimensions.

Structural studies of sintered materials were performed using scanning electron microscopy. The phase composition of the samples was determined using an X-ray diffractometer 'ARL X`TRA' with the copper tube X-ray source. Diffraction patterns were recorded with a step of Δ2θ and the dwell time of 5 sec.

The value of bending strength was used as the main indicator of the strength properties. Samples for the bending test (a rectangular shape with dimensions of 3×4×28 mm) were cut using an electrical discharge machine 'Sodick'. Mechanical three-point bending tests were performed at room temperature using the 'Instron 3369' machine. Crosshead speed was 0.5 mm/min.

3. Results and discussion

Radiographs of materials obtained by sintering of mechanically activated Ni and Al powder mixtures without boron addition and with small amount of boron are shown in Figure 2. The X-ray analysis of sintered materials free from boron indicates the formation of only a Ni_3Al phase. The results confirm that mechanical activation and subsequent sintering according to the present modes provide the formation of single-phase nickel aluminide. Radiographs of sintered materials modified with boron also showed only a peak of nickel aluminide. Reflexes corresponding to boron compounds were not detected. This is due to the amorphous structure of boron and its small amount (up to 0.2 % (wt.)).

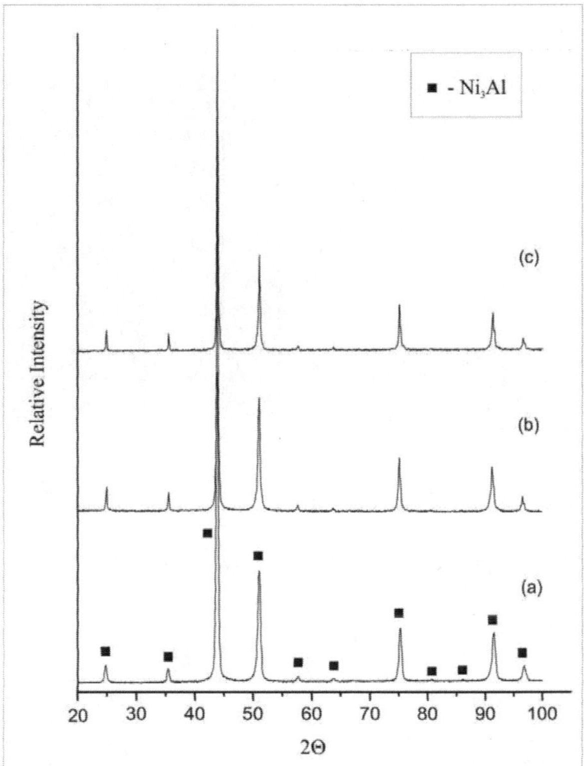

Figure 2. The XRD patterns of samples obtained by sintering of the mechanically activated powder mixture of nickel and aluminum with no boron addition (a); with boron addition of 0.1 % (wt.) with boron addition of 0.2 % (wt.)

The effect of boron addition on density of the sintered compounds is represented in Table 1.

Table 1. The effect of boron addition on density and relative density of Ni_3Al obtained using an SPS-technology.

Boron content, % (wt.)	Density of sintered samples, g/cm^3	Relative density, %
–	7.29	97.2
0.1	7.41	98.8
0.2	7.39	98.6

The results in the table indicate that boron has a positive effect on the density of sintered intermetallic Ni_3Al. It should be noted that increasing the boron concentration from 0.1 to 0.2 % (wt.) did not significantly affect the density of the sintered sample. The maximum value of relative density (98.8 %) was observed in the sample obtained by SPS-sintering of mechanically activated samples with 0.1 % (wt.) of boron addition. This fact allows expecting that the resulting material will have an increased complex of mechanical properties as compared to Ni_3Al intermetallic compound sintered without modifying elements.

The structure of the boron-modified sintered material was revealed using a scanning electron microscope. No significantly structural differences from the boron-free materials were found. As an example, the structure of the material obtained by the sintering powder mixture with 0.1% addition of boron is illustrated in Figure 3.

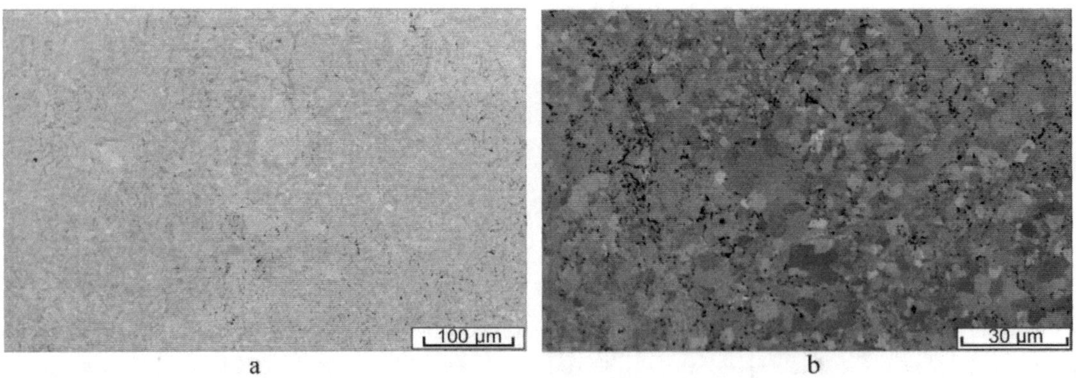

Figure 3. The structure of the samples obtained by SPS-sintering of the mechanically activated powder mixture of nickel and aluminum with 0.2 % (wt.) addition of boron.

The influence of boron addition on strength characteristics of aluminide nickel produced by the SPS-sintering method is shown in Table 2.

Table 2. The influence of boron on the bending strength of sintered Ni_3Al

Composition of the material	Bending strength, MPa
Ni_3Al	790
Ni_3Al + 0.1 B % (wt.)	2200
Ni_3Al + 0.2 B % (wt.)	1720

The value of bending strength of the Ni_3Al intermetallic compound obtained by the spark plasma sintering technology of nickel and aluminum mechanically activated powders was ~ 800 MPa. A typical fracture behavior of the material is shown in Figure 4. The fracture pattern revealed brittle fracture of the sample.

As it has been noted above, the modification of the nickel and aluminum powder mixture with 0.1 ... 0.2 % (wt.) of boron improves the density of sintered Ni_3Al intermetallic. The maximum level of strength properties was obtained when testing the material containing 0.1 % (wt.) of boron. The bending strength (2200 MPa) of this sample is almost 3 times the bending strength of nickel aluminide sintered without boron addition. The fracture surface of the sample sintered with 0.1 % (wt.) of boron contains pits that are typical of ductile fracture. Areas with signs of brittle fracture were considerably reduced (Figure 4b).

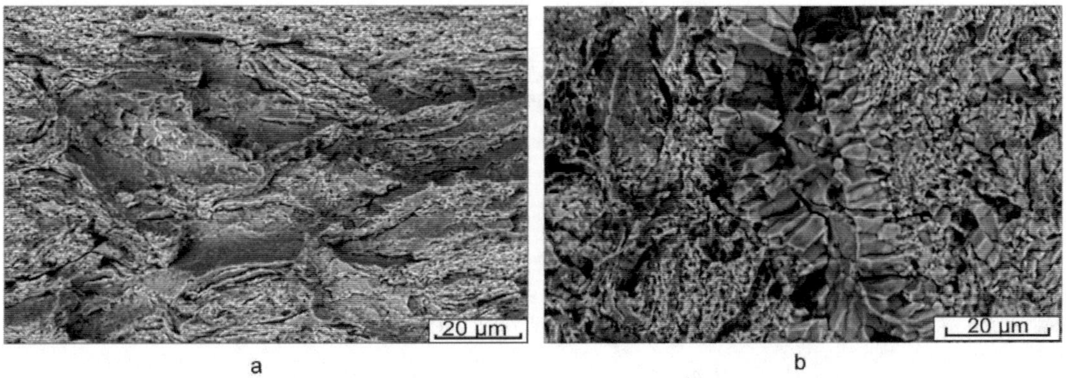

Figure 4. Fracture surface of the samples of Ni_3Al, obtained by SPS-sintering of the mechanically activated powder mixture without boron addition (a), and with addition of 0.1 % (wt.) of boron (b)

4. Conclusion

Applying the processes of mechanical activation of powder mixture free from boron, and with 0.1 and 0.2 % (wt.) of boron addition and subsequent spark plasma sintering allows obtaining a Ni_3Al intermetallic compound.

Addition of small amount of boron to the initial mixture of nickel and aluminum improves the density of the sintered Ni_3Al intermetallic compound. The maximum value of the relative density (~ 99 %) belongs to the material obtained by sintering of the mechanically activated powder mixture with addition of 0.1 % of boron. The only peaks corresponding to Ni_3Al were revealed in XRD patterns of materials sintered with boron addition. Reflexes corresponding to boron compounds were not found due to their small amount and amorphous structure.

The analysis of the obtained data indicates that addition of small amount of boron to the powder mixture of nickel and aluminum had a positive effect on the mechanical properties of the intermetallic compound obtained by the SPS-method. The samples containing 0.1 % (wt.) of boron showed the highest level of bending strength (2200 MPa). This value is almost 3 times the bending strength of nickel aluminide sintered without boron addition. The fracture surface contains pits that are typical of ductile fracture. Areas with signs of brittle fracture were considerably reduced.

5. Acknowledgements

This work has been supported by the Russian Ministry of Education and Science (No 11.1892.2014/K, the project code 1892).

References

[1] Deevi S C, Sikka V K 1996 *Intermetallics* **4** 357–375
[2] Stoloff N S, Liu C T, Deevi S C 2000 *Intermetallics* **8** 1313–1320
[3] Norbega B N, Ristow W, Machado R 2008 *Powder Metal* **51** 107–110

[4] Kim J S, Choi H S, Dudina D, Lee J K, Kwon Y S 2007 *Solid State Phenomena* **119** 35–38
[5] Udhayabanu V, Ravi K R, Murty B S 2011 *J. of Alloys and Compounds* **509** S223–S228
[6] Meng J, Jia C, He Q 2008 *Powder Metallurgy* **51** 227–230
[7] Cao G, Geng L, Zheng Z, Naka M 2007 *Intermetallics* **15** 1672–1677
[8] Filimonov V Yu, Korchagin M A, Smirnov E V, A A, Yakovlev V I, Lyakhov N Z 2011 *Intermetallics* **19(7)** 833–840
[9] Bokhonov B B, Dudina D V, Ukhina A V, Korchagin M A, Bulina N V, Mali V I, Anisimov A G 2015 *J. Phys. and Chemistry of Solids* **76** 192–202
[10] Shevtsova L I, Sameyshcheva T S, Munkueva D D 2014 *Applied Mech. and Mater.* **682** 188–191
[11] Shevtsova L I, Korchagin M A, Thömmes A, Mali V I, Anisimov A G, Nagavkin S Yu 2014 *Adv. Mater. Res.* **1040** 772–777
[12] Munir Z A 2000 *J. of Mater. Synthesis and Processing* **8(3–4)** 189–196
[13] Aoki K, Izumi O 1979 *J. Japan Institute of Metals* **43** 1190–1196
[14] Ball J, Gottstein G 1994 *Intermetallics* **2(3)** 205–219
[15] Liu C T, White C L, Horton J A 1985 Acta Metallurgica **33** 213–229

The equipment for controlling the structure and functional properties of nanostructured composite films

V G Burov, N V Plotnikova, E V Prokhorenko and A I Smirnov
Novosibirsk State Technical University, 20, K. Marks ave., Novosibirsk, 630073, Russia

E-mail: v.burov@corp.nstu.ru

Abstract. The article is devoted to the creation of an instrumental system allowing evaluating the functional properties and current-voltage characteristics of nanostructured composite films at different temperatures and other environmental parameters. The system is based on an assessment of current-voltage characteristics of a nanostructured film material. The main components of the system are a chamber and a unit for current-voltage characteristics measuring. The stage with the test material and the contact system are provided with a heating element and a cooling system thus allowing warming to 150 °C and fast cooling to negative temperatures by liquid nitrogen circulating. The chamber body leak proofness against the external environment allows forming a composition of the atmosphere at a predetermined humidity level, which is essential for the measurement of current-voltage characteristics of polymer materials. The article describes the design features of the instrumental system and results of its application used for determining the properties of polymer nanostructured composite films.

1. Introduction

Laboratory evaluation of the structural state of the thin film made of composite materials provides sample preparation for X-ray diffraction and electron microscopic studies. In contrast to the research laboratories conditions preparation of samples at factories and plants the evaluation of the material's structural state is incorporated into the technological process chain only in the form of sampling. Counting the fact that the formation of nanostructured thin film materials is the so-called 'thin technologies', there is the need to have equipment that enables controlling the process without preparation of special samples. For example, to evaluate the structural state of the surface layer of engineering products, the measurement of hardness or the damping parameters of ultrasonic vibrations, which are sensitive to the structural state of the material, are widely used. Under the conditions of technology compliance with these characteristics it is possible to reliably assess the compliance of the material structure with initial parameters and to judge about accordance of its surface layer properties with specified ones in the manufacturing documentation. Benefits of using structure-sensitive characteristics are the possibility of taking measurements at various stages of the manufacturing process of each product without disrupting its integrity and availability.

Hardness testers and ultrasonic devices are of little use for assessing the structural state of the nanostructured thin film materials. In this case, measurement of electrical resistance which is a structure-sensitive characteristic would be the most convenient way to assess the state of the material [1-6]. The use of the current-voltage dependence allows one to evaluate the structural condition of the nanostructured thin film material, if one knows the relationship between them. Since the electrical conductivity and current-voltage characteristics of all types of nanostructured films are affected by the temperature at which measurements are taken, the equipment for the control of current-voltage dependencies must ensure its variation and stabilization during measurement. For films with polymeric components, a chemical composition and environment humidity are essential, and for the light-sensitive semiconductor films the measurements at certain level of light should be provided. Thus, we have solved the task of equipment creation which allows: simultaneous measurement of conductivity (current-voltage characteristics) and X-ray analysis for film structure assessment; measurement of conductivity (current-voltage characteristics) under different external conditions. Equipment was originally designed for evaluating the photovoltaic characteristics of the solar cell semiconductor polymer materials, as well as for studies of the structure and properties of the active material from which these elements are made. The created equipment is applicable to assess the structural state of nanostructured thin films of various compositions.

As thermal treatment is one of the ways to improve the reliability of welded joints, the aim of this work was to study the structural transformations occurring during heat treatment of the welds between the pearlite and high-chromium-nickel austenitic steels, obtained by flash butt welding.

2. Description of the developed equipment

The main devices for monitoring the current-voltage characteristics of nanostructured thin film materials are an in-situ chamber and the software-hardware system. The in-situ chamber (Figure 1) is a leak-proof prismatic box (position 1), bounded with X-ray transparent plastic film windows from four sides (position 2). All windows are removable and are attached to the chamber through sealing. The front window is connected to the stage (position 3), which is supplied with a ceramic heating element and the thermocouple.

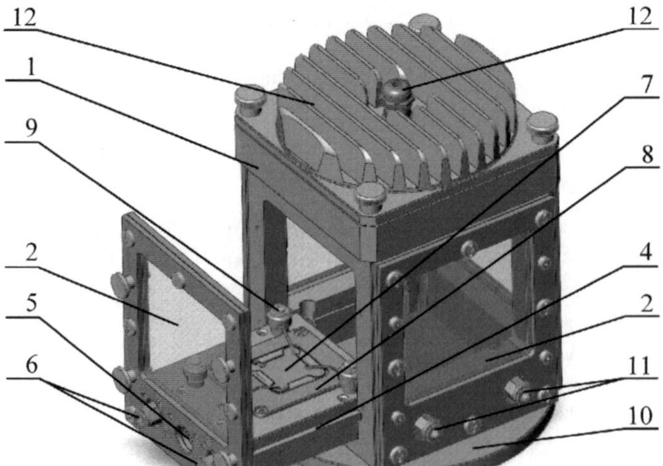

Figure 1. An in-situ chamber for control of structure and current-voltage characteristics of thin composite film materials.

The heating element is located under the top plateau (position 4) and thermocouple (position 5) is fixed in the center of the stage by a threaded connection. The stage design provides its cooling by liquid nitrogen pumping through the channel located in the lower part under the heater. Pumping of nitrogen is carried out through unions with collet fastening (position 6), which are connected by heat-resistant polymeric tubes with a liquid nitrogen tank on the one hand and with the pump – on another one. All materials used in the chamber should work trouble-freely in the temperature range from –40 to +150 °C.

The substrate up to 20 × 20 mm in size with a controlled film is placed on an insulator (position 7), which is on the top plateau (position 8) of the stage. Limiting the size of the controlled film and, accordingly, the size of the chamber is caused by its use in the step of finding the relationship between conductivity or current-voltage characteristics of the films with their structure. Such studies are carried out using an X-ray diffractometer with limited space to place the chamber. The insulator is made of PTFA and has a cave with conductive coating of gold. Four electrical contacts in the form of leaf springs are brought to the surface of the controlled film or to the conductive coating through the isolator (position 9). The chamber is installed in the X-ray diffractometer with high intensity radiation with positioning at the base (position 10). The leak proofness of the chamber allows making necessary environmental conditions by pumping gas composition through four fittings (position 11) mounted on the sides of the chamber. In the study and control of film materials, which functional properties related to photovoltaics, an installation of a sunlight simulator is provided in the chamber which ensures adjustable intensity of the film surface radiation up to 1500 W/m^2. The printed-circuit board (PCB) of the sunlight simulator was developed by the innovative enterprise 'Ekoenergotekh' at Novosibirsk State Technical University and executed on an aluminum substrate with an oxide insulating layer. The PCB incorporates RGB + UV + IR light-emitting diodes providing a range with a wave band from 0.3 to 1.3 μm which is close to sunlight in intensity of the spectral structure. The need to use a simulator of original development was caused by the requirement to ensure sufficient chamber leak proofness. Power supply of the simulator is carried out through the leak proof cable input of PG7 (position 12) from an adjustable source of direct current for the purpose of illumination pulsation exception. The PCB simulator is protected from the chamber environment by PET glass which passes all range of sunlight.

The designed and manufactured chamber is used together with a hardware-software system which consists of the block of current-voltage characteristics measurement and the microprocessor control block; the former is guided by the latter. The hardware-software complex functions firstly in the mode of voltage generation in the range from 0 to 10 V with an accuracy of ± 1 mV and current measurements in the range of 1 pA to 1 mA with an accuracy of ± 1 pA in the range of 1-1000 nA measurement and ± 10 % within measuring of 0.001…1 mA and, secondly, in the mode of measurement of current-voltage characteristics of photo cells at various levels of electric load.

3. A sequence of equipment use

The current-voltage characteristic of the material and its structure is in complex functional relationship, which has to be established at the stage of film formation technology development.

This step is a research and it requires the use of analytical tools and, above all, the X-ray analysis while studying the nature of changes in conductivity or current-voltage characteristics under various conditions. The second stage is designed for the production conditions and consists in monitoring of the film material state at various steps of the process. It will be enough to measure the current-voltage dependence to assess the structural state of the nanostructured film material at this stage.

3.1. The mode of the studying of the relationship of current-voltage characteristics and the film structure

An example of using the developed equipment was the research of technological modes for creation of nanostructured films on the basis of polymeric mixtures P3HT/PCBM which are applied for organic solar elements creation.

Figure 2 shows the current-voltage characteristics of nanostructured films with simultaneous recording of X-ray, allowing judging the structural changes taking place in the film material. GIWAXS study of the film revealed that there is a two-stage increase in the current strength corresponding to the two-stage dependence of the crystallinity degree. The determination of these characteristics for films P3HT: ICBA showed that the intensity of peak (100) and the current increase linearly with the film deposition before achieving the maximum in the solid state. The highest conductivity values are observed in the solid state. This is due to better compatibility of P3HT and ICBA, whereby ICBA generates smaller domains which inhibit crystallization of the polymer during the solvent evaporation to a lesser degree. The structure of polymer films of a different chemical composition in the initial state (after casting) has a low degree of crystallinity, and its value does not exceed few percent, and the sizes of the crystalline domains do not exceed an average of several tens of nanometers. So, the studies have revealed the optimal mode of preparation of the active material for solar cells to increase their efficiency.

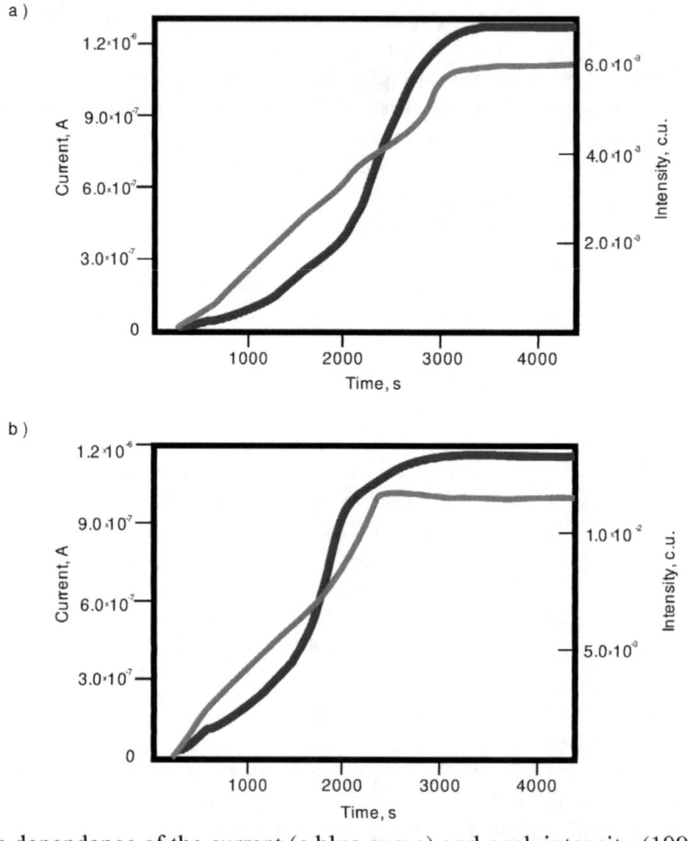

Figure 2. The time dependence of the current (a blue curve) and peak intensity (100) (a red line) in the films of P3HT: PCBM (a) and P3HT: ICBA (b) during crystallization.

3.2. The mode of assessment of the structural state of the film material by the current-voltage characteristic

During production the designed equipment may be used for controlling the properties of the final product – the nanostructured composite film, and in the mode of measuring of the current-voltage characteristics of the film material, made at various stages of the process. Deviation of current-voltage dependence obtained at this step from the one obtained at the step of researching the relationship between structure and current-voltage characteristics indicates deviation from the required structure. The level of current values may thus differ insignificantly in case of controlling the films, sensitive to X-rays, for example, in assessing the current-voltage characteristics of the semiconductor films.

IOP Conf. Series: Materials Science and Engineering **124** (2016) 012140 doi:10.1088/1757-899X/124/1/012140

4. Conclusions

The developed equipment is designed to assess the structural state of nanostructured thin films of conductive and semiconductor materials with minimization of the sample preparation operations. The use of the equipment in production provides the ability to embed the process of nondestructive testing of film material into it.

Acknowledgement

The authors gratefully acknowledge the financial support from the Russian Ministry of Education and Science (the program 'Research and Development in Priority Areas of Development of the Russian Scientific and Technological Complex for 2014–2020', grant agreement No. 14.574.21.0097, unique identifier of the agreement RFMEFI57414X0097).

References

[1] Hemalatha K, Somashekarappa H and Somashekar R 2015 *Advanc. in Mater. Phys. and Chem.* **5** 408-418

[2] Meenakshi V, Sayeed A, Subramanyam S V *Materials Science Forum* **223-224** 307–310

[3] Sharkeev Yu P, Legostaeva E V, Eroshenko A Yu et al 2009 *Composite Interfaces* **16** 535–546

[4] Garsia-Belmonte G, Bisquert J and Miranzo P 2002 *Solid State Ionics* **146** 367–376

[5] Bataev V A, Burov V G, Grigorian S, Ivanov D A, Plotnikova N V and Smirnov A I 2015 *Applied Mechanics and Materials* **788** 301–305

[6] Kvamen K, Grigoryan S, Anokhin D V, Bataev V A, Smirnov A I and Ivanov D A 2015 *Nanotechnologies in Russia* **10** 600–605

IOP Conf. Series: Materials Science and Engineering **124** (2016) 012141 doi:10.1088/1757-899X/124/1/012141

Magnetic pulse cleaning of products

V P Smolentsev[1], S V Safonov[1], E V Smolentsev[1] and O N Fedonin[2]
[1] Voronezh State Technical University, 14, Moskovskiy ave., Voronezh, 394026, Russia
[2] Bryansk State Technical University, 7, 50 years of October st., Bryansk, 241035, Russia

E-mail: vsmolen@inbox.ru

Abstract. The article deals with the application of a magnetic impact for inventing new equipment and methods of cleaning cast precision blanks from fragile or granular thickened surface coatings, which are difficult to remove and highly resistant to further mechanical processing. The issues relating to a rational use of the new method for typical products and auxiliary operations have been studied. The calculation and design methods have been elaborated for load-carrying elements of the equipment created. It has been shown, that the application of the magnetic pulse method, combined with a low-frequency vibration process is perspective at enterprises of general and special machine construction, for cleaning lightweight blanks and containers, used for transporting bulk goods.

1. Introduction

The magnetic methods of processing, whose foundations were laid in [1], are being developed in various fields of mechanical engineering. One of the perspective trends in the development of these methods is the application of their theoretical points for combined cleaning of products from polluting coatings. In our case, the magnetic method is implemented as a source of a low-frequency impact upon the striker, which causes the mechanical wall vibration in products having fragile or granular layers intended for removal. In mechanical engineering, the application of magnetic pulse processing makes it possible to replace the labor-intensive operations, which are intended for removal of the remnants of ceramic molds from blanks, obtained by precision molding (investment casting, shell-mold casting, etc.).

Many kinds of bulk goods (e.g. mixtures formation, construction materials, etc.) are delivered in sheet metal containers, where thick layers of transported products remain on walls after unloading. The cleaning of such containers, performed by vibration test systems, is quite cumbersome and not always feasible, since it can damage the products. The magnetic pulse method is characterised by versatility, and it is an effective means of mechanical cleaning of non-rigid blanks, steel sheet containers (up to 3 mm thick) and non-metallic vessels (with walls of up to 5…8 mm thick).

2. The scheme of magnetic pulse cleaning

The scheme, presented in Figure 1, shows that polluting coatings are removed by vibrations, caused by the striker, which produces the impact on the wall and is actuated by an electromagnetic force.

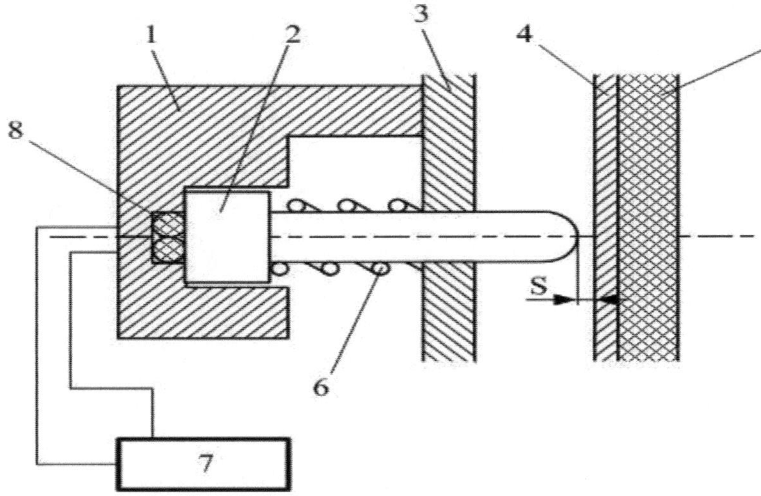

Figure 1. The scheme of a magnetic pulse installation for removing of pollutants from surfaces of products: 1 – chamber casing; 2 – striker; 3 – machine casing; 4 – sheet metal blank (shell); 5 – coating (pollutant) to be removed; 6 – resilient element (spring); 7 – magnetic pulse generator; 8 – inductor; S – initial gap between the striker and the shell.

The scheme presented in Figure 1, contains the novelty aspects and is protected by patent. Polluting coating 5 (Figure 1) is removed due to a difference in movement speed between shell 4 and coating 5, at the time of the pulse impact caused to shell 4 by striker 2. Inductor 8 causes the abrupt movement of striker 2, due to the action of high current flow (up to 300 KA) supplied by pulse generator 7. RC – circuits (R – adjustable resistance, C – adjustable capacitance). When striker 2 hits shell 4, its reset is induced by spring 6. Before the process starts, striker 2 is installed so as to leave a gap varying in the range from 0.2 to 1.0 mm.

3. An optimal degree of product cleaning

While cleaning the workpiece walls, it should be noted that burnt particles of forming mixtures or ceramic forms can be found on workpiece segments, inaccessible for the striker action, or in places where stiffness of segments does not allow generating the workpiece vibration. If a coating is brittle, it can flake off even by action of weak shock waves. However, some kinds of pollution can only be removed by flushing, including the one performed in aggressive liquids. The impact of the continuous magnetic pulse can completely remove fragile and compressed granular pollutants. However, this method is not economically justified, since it does not require the removal of all foreign particles for subsequent mechanical processing, and the final part of the operation is the most labor-intensive.

The container design usually includes the stiffening elements (stiffening ribs, wall couplings, etc.) which restrict the transposition of shell components, and such containers require multiple pulses for cleaning their walls. Figure 2 demonstrates the economically attainable degree of steel container cleaning (wall thickness – 0.5 mm, volume – 2 m^3) depending on the number of pulses (single-impulse energy – 0.5 kJ). Similar results can be obtained for workpieces of the same rigidity.

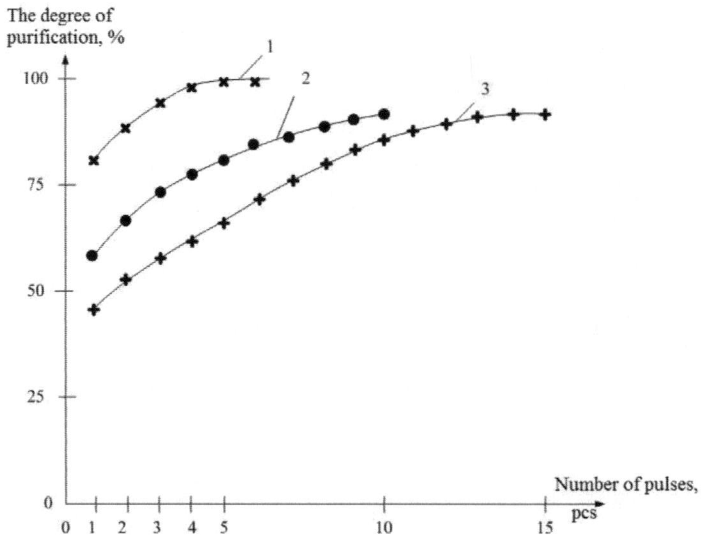

Figure 2. Cleaning of workpiece segments of various rigidity (as exemplified by containers with walls covered with the forming mixture: 1 – the segment having no stiffening elements; 2 – the segment close to container wall couplings; 3 – the segment close to stiffening ribs (45×45 mm angle bars).

The degree of cleaning (Figure2) is the ratio between the areas of a clean blank or container walls, and the total area of a workpiece wall or a segment to be cleaned.

To carry out the cleaning process presented in Figure 2, the pulse impact was induced in the resonant mode. As it is shown in Figure 2, the removal of pollutants from segments having no stiffening elements (curve 1) occurs after 3…4 pulses, i.e. within several seconds. In this case, the area of the clean segment is 0.15…0.25 m^2.

In places where container rigidity varies, the cleaning intensity decreases sharply (2, 3 in Figure 2). For instance, the complete cleaning of places located near wall couplings, is 2…2.5 times slower compared with case 1 (Figure 2). Meanwhile, even 15 pulses are not sufficient to perform full cleaning of places located near the metal angle-bar frame, since the frame almost entirely damps the wall vibrations.

Figure 3 shows the results of using various methods of container cleaning, under conditions laid down in Figure 3.

The resonant mode (A, as presented in Figure 3) permits threefold acceleration of the cleaning process, especially for segments adjoining the stiffening elements (2, 3, as presented in Figure 3).

Consequently, it is necessary to constantly vary the pulse repetition rate in order to maintain the resonant mode for all container segments subjected to cleaning.

It is not always practicable to achieve complete removal of pollutants by the pulse impact. For this purpose, flushing or mechanical cleaning of containers is required, e.g. the one performed with a chisel and a hammer, which can often cause permanent damage to the product. Therefore, in the present case, the adequate cleaning implies the removal of most coatings (primarily, thick ones) from container walls. Usually, a supplier and a customer agree on removing of 95...98% of the total mass of bulk materials, after which a container is recognized suitable for reuse. If this condition is fulfilled, the container cleaning time for certain transported products is 4...5 times less. In this case, the cost saving can exceed the material losses by several times, due to the lost value of cargo particles remaining on container walls after their cleaning.

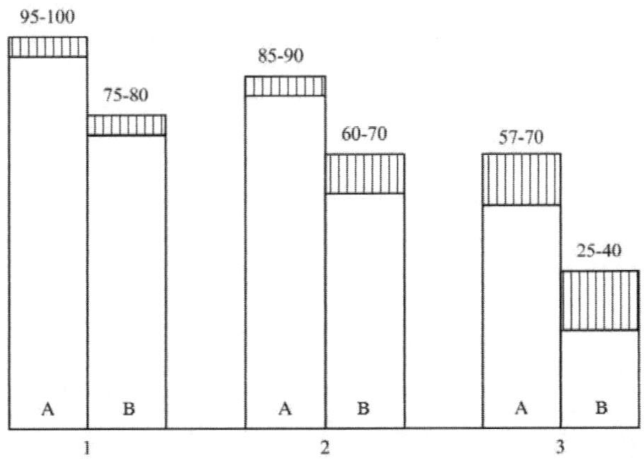

Figure 3. The degree of container cleaning from pollutants (forming mixture): *A* – the cleaning performed in the resonant mode; *B* – the cleaning performed in the non-resonant mode (with the constant frequency pulse impact); 1 – the segment having no stiffening elements; 2 – the segment close to container wall couplings; 3 – the cleaning in places close to stiffening ribs (45×45 mm angle bars).

4. The feasibility study of rational application of techniques and equipment for removing thick coatings

Studies have shown, that the most efficient method of cleaning of containers and openwork products is the contact magnetic pulse vibration method, involving the impact produced by the tool (i.e. the striker). The pulse energy of the tool is regulated within the limits to ensure elastic deformation of products to be cleaned, avoiding residual shell deformation.

If the efficiency ratio of magnetic pulse processing is low (2…20 %), it is advisable to use satisfactorily light strikers. In a replaceable striker, the construction includes all moving parts affecting the workpiece walls (i.e. the shell).

Figure 4 presents the curve demonstrating the dependence of magnetic pulse processing parameters on the mass of the striker construction. From this figure, it can be seen that the increasing mass of the striker system causes the rise of pulse energy, while reducing the frequency of the striker impact on the shell. In such case, the total input energy per time unit remains practically unchanged.

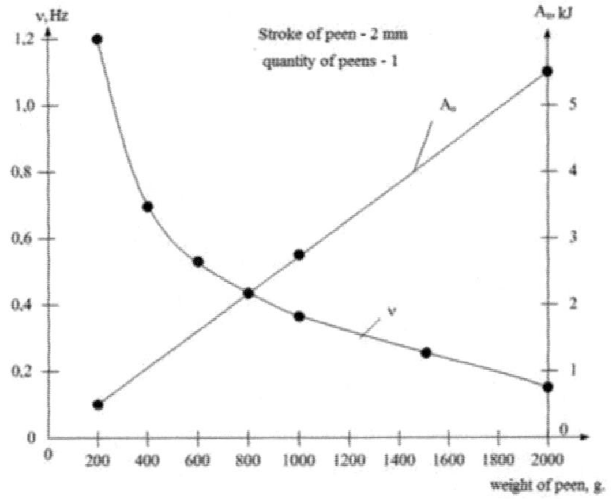

Figure 4. The dependence of the striker construction mass (m) on operating pulse frequency (υ) and pulse energy (Au).

The conducted measurements have revealed (Figure 5) the optimal number of strikers, depending on frequency and efficiency of input energy utilization in an impulse.

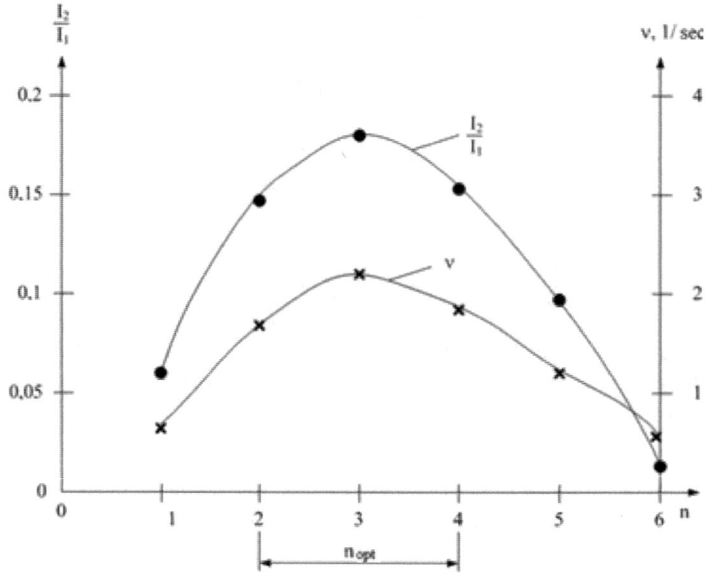

Figure 5. The efficiency of utilizing multiple inductors.

Dependence I_2/I_1 demonstrates the efficiency of input energy utilization, i.e. the nature of efficiency change, which varies from 0.02 to 0.2 (according to [2-4]). From Figure 5, it can be seen that the optimal number of inductors (n_{opt}), supplied with strikers, is between 2 and 4. The further increase of 'n' does not produce any gain in the use of input energy, though it somewhat improves the operational performance. However, it should be taken into account that the growing number of inductors, powered by a single generator, disrupts the process stability, which can lead to product damage, resulting from the striker impact, along with the destruction of the mechanical part of the magnetic pulse vibration processing system.

5. Conclusion

The analysis has been conducted to study the contemporary magnetic energy utilization in systems, producing magnetic pulses for cleaning non-rigid workpiece segments and container walls, made of sheet materials. The conducted analysis has shown that the method of vibration is effective for cleaning the cast sections from the remnants of forming mixes (burned layers, ceramic shell fragments, etc.). In addition, the openwork elements of blanks are not damaged improving the safety. The mechanization of the large-scale casting cleanup makes it possible to resolve both technical and social problems, related to blank production (e.g. the improvement of the working conditions, the possibility of casting cleanup process automation, the reduction of environmental pollution, etc.)

The newly invented constructions of magnetic pulse systems have been proposed. The study has demonstrated the advantages of the cleaning method under review, and the efficiency of applying the newly created installations for removing the remnants of transported bulk goods from container walls made of sheet materials.

The methodological basis for calculating the parameters of the load-carrying element (striker) for magnetic pulse installations has been worked out.

The new technological processes of cleaning openwork blanks have been examined, which consider the geometric features of cast products, including the blanks for non-rigid precision components of perspective rocket and space equipment.

References

[1] Baron Y M 1986 Magnetic abrasive and magnetic processing of products and cutting tools (Leningrad: Mashinostroenie) p 176

[2] Smolentsev V P 1983 *Electrophysical and electrochemical methods of material processing* (Moscow: Vyschaya Shkola) **2** 208

[3] Smolentsev E V 2005 *The planning of electrical and combined processing methods* (Moscow: Mashinostroenie) p 511

[4] Smolentsev V P, Grebenschikov A V, Nikolaenko I B and Kaluzhin S L 2008 The installation for container cleaning from pollution. Patent No, RU 69787

[5] The Ironworker's Guide 1976 (Moscow: Mashinostroenie) **1** p 768

Microstructure and mechanical properties of silicon carbide ceramics reinforced with multi-walled carbon nanotubes

A Barmin[1], V Bortnikova[1], A Ivanov[1], V Kornev[2], S Lurie[2] and Y Solyaev[2]
[1] Keldysh Research Center, 8, Onezhskaya st., Moscow, 125438, Russia
[2] Institute of Applied Mechanics Russian Academy of Sciences, 7, Leningradsky ave., Moscow, 125040, Russia

E-mail: yos@iam.ras.ru

Abstract. A microstructure, a composition and mechanical properties of multi-walled carbon of nanotube-reinforced silicon carbide ceramics were examined. The amount of carbon nanotubes was up to 1% wt. Samples was prepared by spark plasma sintering. It has been found that the optimal sintering temperature is 2000°C with an exposure duration of 5 minutes and a pressure of 50 MPa. The effect of the CNT mass fraction on the Young modulus of silicon carbide ceramics composites was investigated for different temperatures and processing conditions of samples using ultrasonic techniques. It has been established that Young's modulus of ceramics decreases due to addition of CNT. Elastic properties of the composites cross section were characterized using nano-indentation. It has been revealed that the stiffness of the ceramics intergranular phase decreases due to addition of CNT.

1. Introduction

Composite ceramics reinforced with carbon nanoparticles is a new class of advanced materials for spacecraft engine elements [1, 2]. In this work, we will study the problem of production of composite ceramics based on silicon carbide and reinforced with multi-walled carbon nanotubes. The spark plasma sintering method is used to prepare samples. The purpose of the study is to improve mechanical properties of ceramics within a temperature range of up to 1200 °C.

It is known, that addition of carbon nanotubes may provide enhancement in strength and toughness of different types of ceramics [1-8]. A considerable number of SiC-CNT ceramics studies [3-8] have examined various sintering technologies including spark plasma sintering, precursor infiltration and pyrolysis, a hot-press method, etc. Bending strength of such composites is usually 200…400 MPa and fracture toughness is 2…7 MPa $m^{0.5}$.

This work contains a study of the microstructure and the composition of SiC-CNT small samples. The relationship between the CNT volume fraction and Young modulus of silicon carbide ceramics composites was studied. It is noteworthy, that the elastic modulus value is directly included in the formula for calculation of fracture toughness by the indentation tests [9]. It should also be noted that the elastic modulus value considerably affects the results of the notched samples tests [10]. Therefore, the reliable determination of composite ceramics Young's modulus could provide an update of measured values of fracture toughness. Besides, the values of ceramics effective modulus and elasticity modulus of grains (measured by nano-indentation) enables the identification of important micromechanical parameters required for modeling of effective properties of ceramics composites [11].

2. Experimental procedure

Green silicon carbide powder (SiC) with an average size of particles of 5 μm and amorphous boron were used as basic material. Multi-walled carbon nanotubes with the following specification were used for reinforcement: the outside diameter is 50…80 nm, the inside diameter is 5…15 nm, the length is 10…25 μm, the specific surface area is 90…160 m^2/g, density is 2.1 g/cm^3. The composition is C – 97.32 %, Cl – 0.21 %, Fe – 0.56 %, Ni – 1.88 %, S – 0.03 %.

For disaggregation and formation of a monodisperse system, multi-layer carbon nanotubes were dispersed using an ultrasonic unit during one hour. A polyvinyl alcohol 0.5 % solution was used as dispersion medium; the ratio of the number of nanotubes versus the solution is estimated at 1:5. The SiC ceramic powder with sinter-activating additives (non-crystalline boron, 1% wt.) was mixed in a rotary drum mixer during 5 hours in distilled water. Further, the suspension was dried at a temperature of 70…90 °C allowing it to reach a moisture level of 3…5 %; subsequently, the powder was rubbed through No 0063 sieve. The dry mix powder with additives, the polyvinyl alcohol solution and the carbon nanotubes suspension (1:5) were mixed using a planetary mill in corundum drums with corundum balls with a mass ratio of material, milling agents and medium of 1:1:2 respectively for optimal homogenization. The homogeneous suspension was dried at a temperature of 80…90 °C to so that it reaches a moisture level of 5 %; the powder was subsequently rubbed through No 01 sieve.

The initial powder was compacted using the spark plasma sintering method in the FCT HP D5 machine. Consolidation was performed in a cylindrical graphite matrix with an internal diameter of 20 mm in a dynamic vacuum medium with a residual pressure of 3 Pa. In the course of the sintering process, the matrix was continuously compacted using a pressure of 30 MPa while heating up to 600 °C and a pressure of 50 MPa at higher temperatures. A 5-minute long degassing break took place at a temperature of 600 °C. Before sintering temperatures (1900…2000 °C) were reached, heating was carried on at a constant rate of 100 K/min. These temperatures cover the optimal temperature range of sintering of silicon carbide doped with non-crystalline boron. The exposure to the sintering temperature was 5 minutes.

Three series of samples were made with the varying content of reinforcing additives. The dry mix for the first series of samples was prepared without adding multi-wall carbon nanotubes; the second series contained 0.5 % wt. of MWCNT, the mass fraction of nanotubes in the third series of samples was 1 % wt. There was a variation in the sintering temperature within one series. Each of the series contained three samples sintered at different temperatures: 1900, 1950 and 2000 °C. The parameters of the samples of ceramics are given in Table 1. It has been found that the total porosity in the samples synthesized at a temperature of 1900 °C amounts to 18 %, at 1950 °C – 16 %, at 2000 °C – 8 %.

Table 1. Properties and Parameters of Ceramics Samples Consolidation.

Sample No	Composition	T_{sinter}, °C	Picnometric density, g/cm^3
1	SiC+1%B	1900	3.11
2	SiC+1%B	1950	3.07
3	SiC+1%B	2000	3.01
4	SiC+1%B+0.5% MWCNT	1900	3.10
5	SiC+1%B+0.5% MWCNT	1950	3.08
6	SiC+1%B+0.5% MWCNT	2000	3.10
7	SiC+1%B+1% MWCNT	1900	3.12
8	SiC+1%B+1% MWCNT	1950	3.09
9	SiC+1%B+1% MWCNT	2000	3.11

Investigation of microstructure and X-ray microanalysis of the composition were performed on samples using a scanning electronic microscope 'FEI Versa 3D LoVac DualBeam'. Studies of elastic behavior of samples were performed using the ultrasonic spectroscopic method and the nano-indentation method. Samples for acoustic measurements were shaped as thin round-section discs with a diameter of 20 mm and a height of 2.4 mm. Young's modulus was measured in a temperature range of 20 to 1200 °C. All the measurements were performed in high vacuum, which was provided by a diffusion pump. The heating rate was 12 °C/min. The temperature was measured using tungsten-rhenium thermocouples. We suggested that the value of Poisson's ratio equals to the typical value for this type of ceramics (0.19).

Nano-indentation was performed on the cross-section of sample with 1 % wt. of CNT. We measured the elasticity modulus of grains in the structure of the composite. NanoTest 600 with a Berkovich's diamond indentor was used. The results of measurements were not affected by the difference of samples and the indentor temperature due to preliminary holding of samples inside the chamber for 12 hours. Both before and after the indentation tests, the sample surface was scanned in order to recognize its structure, locations of pores and to establish the indentation points. The tests were performed with a force of 50 mN and 100 mN in closely located non-overlapping points. Two values of the force had to be made to verify that the material volumetric properties were measured, and not the surface properties. Value of Young's modulus E_s was calculated by a well-known approximate formula for reduced modulus E_r:

$$\frac{1}{E_r} = \frac{1-\nu_s^2}{E_s} + \frac{1-\nu_i^2}{E_i}, \tag{1}$$

here, E_i = 800 GPa is the indentor material elastic modulus found by the calibration on quartz samples, ν_i = 0.07 is the diamond indentor Poisson's ratio picked up from the table data; ν_i is silicon carbide grains Poisson's ratio assumed to be amounting to 0.125.

3. Results

Micrographs of the composites structure are given in Figure 1 and Figure 2. It is clear that CNTs are positioned as large agglomerates and fill up the intergranular space and voids (Figure 2). The Micro X-ray phase analysis of the chemical composition of the samples surfaces confirmed the fact of a high volumetric content of carbon in the intergranular space area: 88% wt. as compared with 30 % wt. in the ceramics grain area (Figure 3). It was also established that in the CNT location area, there is a high oxygen concentration (4.7 % wt.) as compared with 1 % wt. in the grains area. This may indicate nanotubes oxidation in the process of cleaning or processing or in the process of powders sintering.

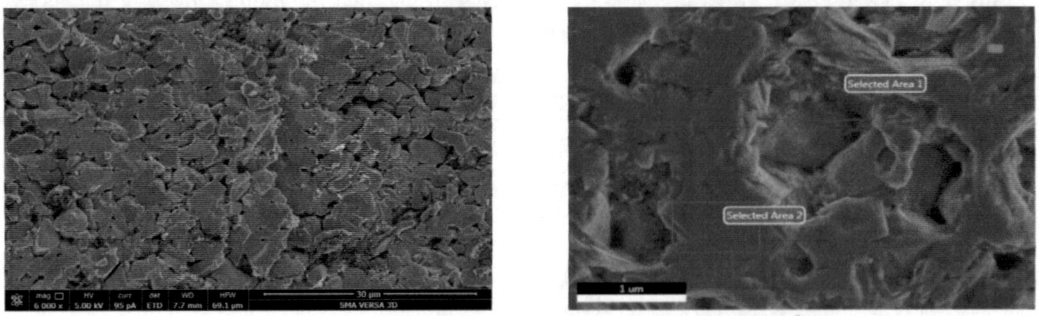

a b

Figure 1. A microstructure of SiC + CNT composite ceramics: *a* – an overall view, *b* – areas of the X-ray microanalysis.

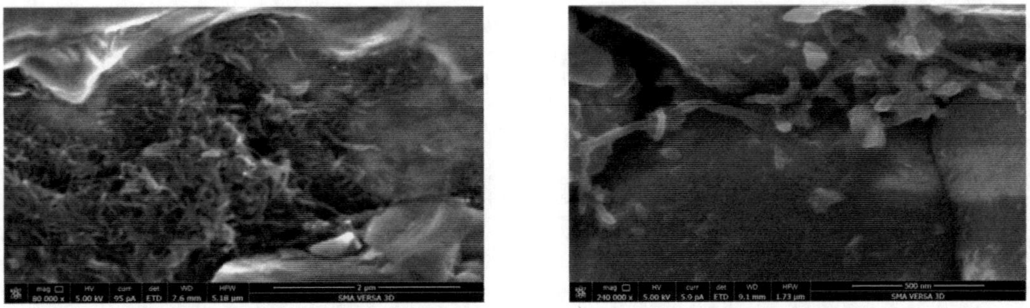

Figure 2. Agglomerates of CNT in the intergranular space of ceramics.

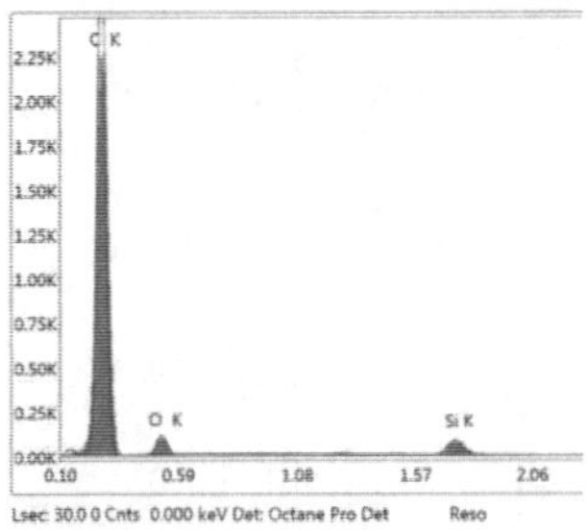

Element	Weight %	Atomic %	Net Int.	Error %
C K	88.04	93.01	893.71	5.39
O K	4.65	3.69	37.17	11.96
SiK	7.3	3.3	44.11	10.59

a

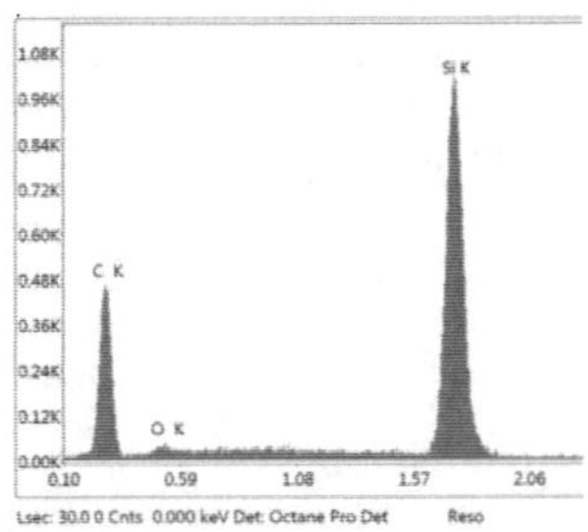

Element	Weight %	Atomic %	Net Int.	Error %
C K	30.38	50.24	188.38	11.1
O K	0.99	1.23	11.79	22.61
SiK	68.63	48.53	564.87	6.23

b

Figure 3. Results of X-ray microanalysis of samples surface: a – the results of the analysis in the area of the pore containing CNT; b – the results of the analysis in the SiC ceramics grain area.

The measurement results of Young's modulus of nanocomposite samples reinforced by CNT at room temperature and the temperature of 1200 °C are given in Figure 4.

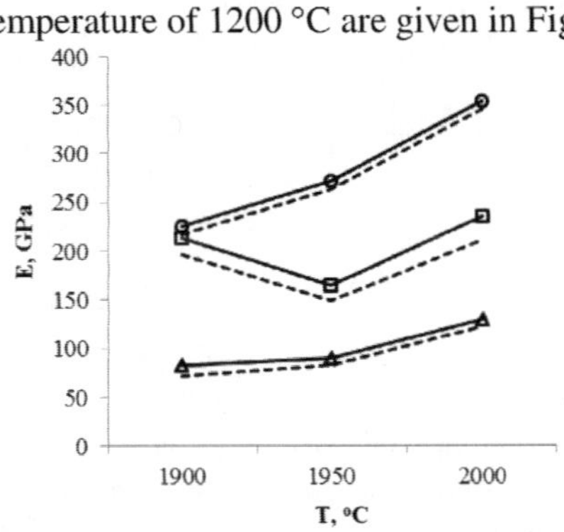

Figure 4. Correlation of Young's modulus with the sintering temperature and a CNT mass fraction at room temperature (a continuous line) and at high (a dashed line) temperature.

An estimated measurement error is ± 3 %. It was found that the increase in the sintering temperature of ceramic composites results in an increase of the Young's modulus. The maximum value of the modulus takes place for the composite sintered at a maximum temperature in the absence of nanotubes. Such behavior of Young's modulus may be explained by the impact of residual porosity at lower sintering temperatures and by the degradation effect due to CNT agglomerates.

The nano-indentation results are given in Figure 5 and Table 2. Figure 4b shows the indents on the sample surface after experiments. The dependence of indentor penetration depths on the applied force was revealed. Based on these data the NanoTest 600 system software enabled automatic estimation of a reduced elastic modulus value and a hardness value in the indentation point (Table 4). The resultant average value of the ceramics grains reduced elastic modulus amounted to 344 GPa. The spread of values for six points reached 10…14 %. The grains Young's modulus found by formula (1) amounted to 591 GPa.

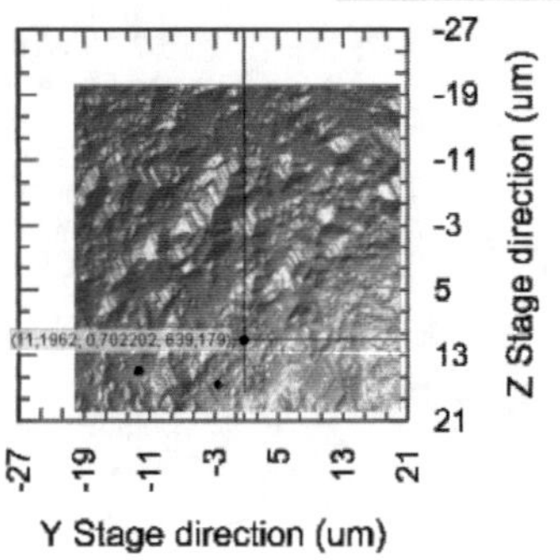

a

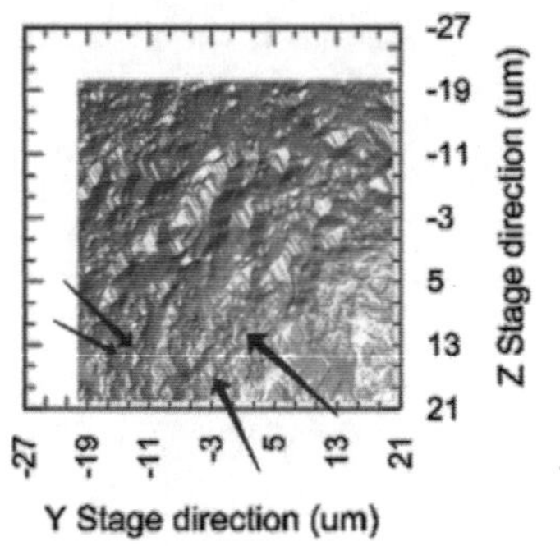

b

Figure 5. The ceramics cross section surface structure scanned with NanoTest 600: *a* – indentation points; *b* – indents on the sample surface.

Table 4. A reduced elastic modulus and hardness of ceramics grains.

Indent No	Maximum Depth, nm	Force, mN	Depth Of Trace nm	Hardness, GPa	Reduced Modulus, GPa
1	322.2	50	248.4	29.9	336.9
2	361.3	50	290.2	22.3	301.9
3	341.4	50	268.0	25.9	315.4
3	442.6	100	340.0	33.1	360.3
4	401.9	100	308.1	39.9	432.7
5	469.5	100	359.1	29.8	317.9
Mean				**30.2**	**344**
Mean deviation:				4.2 (14%)	34 (10%)

4. Conclusions

The study of SiC-CNT nanocomposite ceramics samples by means of spark plasma sintering has been conducted. It has been established that carbon nanotubes in the structure of sintered composites are located as agglomerates in the intergranular space of ceramics. An addition of CNT causes the composite ceramics elastic modulus to decrease. As it has been revealed from nanoindentation measurements, the silicon carbide grains stiffness characteristics turn out to be equal, approximately, to the theoretical value of ideal SiC crystals stiffness. It has been found that the best option of the ceramics sintering modes is exposure to 2000 °C. In this case, composites have the smallest volumetric fraction of porosity and the highest values of elastic modulus. The addition of CNT causes the ceramics Young's modulus to reduce. Therefore, the nanotubes may act as defects in the structure of composite and intergranular phase stiffness characteristics decrease due to the presence of multi-walled CNT agglomerates.

References

[1] Ahmad I, Unwin M, Cao H, Chen H, Zhao H, Kennedy A 2015 *Nanomaterials* **5** 90
[2] Arsecularatne J A and Zhang L C 2007 *Rec. Pat. Nanotech.* **1** 176
[3] Ma R Z, Wu J, Wei B Q, Liang J and Wu D H 1998 *J Mat.Sci* **33** 5243
[4] Wang Y, Voronin G A, Zerda T W and Winiarski A 2006 *J. Phys: Cond. Mat.* **18** 275
[5] Yu H, Zhou X, Zhang W, Peng H, Zhang C and Sun K 2011 *Mat. & Des.* **32** 3516
[6] Yadhukulakrishnan G B, Rahman A, Karumuri S, Stackpoole M M, Kalkan A K, Singh R P and Harimkar S P 2012 *Mat. Sci. Eng.* **552**, 125
[7] Datye A, Wu K, Gomes G, Kumari L, Li W and Lin H 2010 *Ceram. Envir. & Energy. Appl.: Ceram. Trans.* **217** 117
[8] Carlson T A, Marsh C P, Kriven W M, Stynoski P B and Welch C R 2013 *Composite Materials and Joining Technologies for Composites* (NY: Springer)
[9] Anstis G R, Chantikul P, Lawn B R and Marshall D B 1981 *J. Am. Ceram. Soc.* **64** 533
[10] ASTM C1421–15
[11] Lurie S A, Kasimovskii A A, Solyaev Y O, Ivanova D D 2012 *Nanmech. Sci Tech.* **3** 1

IOP Conf. Series: Materials Science and Engineering **124** (2016) 012143 doi:10.1088/1757-899X/124/1/012143

The investigation of the matrix structure of ceramic brick made from carbonaceous mudstone tailings

A Stolboushkin[1,2], **O Fomina**[1] **and A Fomin**[1,2]
[1] Siberian State Industrial University, 42, Kirova st., Novokuznetsk, 654007, Russia
[2] Kutateladze Institute of Thermophysics of the Siberian Branch of the Russian Academy of Sciences, 1, Ac. Lavrentieva ave., Novosibirsk, 630090, Russia

E-mail: alexey-nvkz@mail.ru

Abstract. The study of the matrix structure of ceramic brick made from carbonaceous mudstone tailings of Korkinsky coal opened pit mine is presented in the current paper. This study includes a thin sections analysis by the polarizing microscope, X-ray, SEM and infrared spectra investigations. It has been discovered that processes of solid- and liquid-phase sintering with the formation of new mineral phases occur inside and on the surfaces of granules during firing. It is shown that a liquid phase is formed in the matrix. It fills inter-grain gaps and connects mineral particles between themselves. It has been found that the advanced physical and mechanical properties of ceramic bricks obtained by creation of the matrix ceramic crock structure, intensive forming of a glass phase on the boundary of the section medium of ceramic composite and temperature reduction of the processes of solid-phase sintering.

1. Introduction

The waste after coal extraction and cleaning can be a valuable raw material for production of construction materials such as wall ceramics. Up to the present moment, however, it has not been widely utilized and bulk anthropogenic wastes worsen the environment in coal-mining districts [1, 2]. Utilization of such waste is especially crucial for Chelyabinsk region where the biggest in Europe Korkinskiy brown coal open pit mine is located. During the period of its operation since 1934, hundreds of millions ща tons of waste, that containы argillites, siltstones, coaly argillites, sandstones and shales, have been accumulated in the disposal areas [3].

The constraint factors of use of waste from Korkinskiy open pit coal mine in the wall ceramics technology are unstable, the content of carbon and non-plasticity of anthropogenic raw material are often increased. This problem could be solved via reducing the coal content in the rock [4, 5], the application of semi-dry pressing technology and directed formation of the spatially-organized structure of a ceramic crock for non-plastic raw materials [6].

The formulated technological idea was verified in industrial conditions of a brick plant LLC 'Krasniy Kirpich' of semi-dry pressing (Krasnoyarsk Region) [7]. Physical-mechanical tests of the ceramic brick lot produced from coaly argillites of Korkinskiy open pit coal mine showed that the new method of burden preparation, which includes waste granulation with subsequent powdering of granules by clay, makes it possible to obtain high-quality ceramic products according to the Russian State Standards.

The purpose of this work is to carry out the petrographic study of the formation mechanism of a strong structure of a ceramic brick pilot lot produced from coal mine waste with the coal content below 11 %.

2. Raw materials selection

The tailings of the brown coal (carbonaceous mudstones) from Korkinsky coal opened pit mine have been investigated as the initial raw for ceramic bricks production. In this mine the coal content is from 10 up to 18 %, depending on the size of the wastes. Coal mine wastes are not soaked in water, the stone-like clay material is made as a result of compression, hydration and cementation of clays deposited in the coal thickness and is not ductile in the natural conditions.

The chemical composition (see the Table) of the waste belongs to the semiacid group of a raw material (Al_2O_3 more than 16 %), with a high content of coloring oxides ($Fe_2O_3 + TiO_2$ more than 10 %). The study of the phase composition according to the data of the X-ray powder diffractometry showed that the waste consists of quartz, kaolinite, siderite, muscovite, there is also dolomite, hematite, and field spar.

Table. A chemical composition of the waste from Korkinskiy coal open pit mine.

Mass fraction of the components in % per absolutely dry weight								
SiO_2	Al_2O_3	Fe_2O_3	MnO	TiO_2	MgO	CaO	R_2O	*LWF**
40.8-42.8	16.2-17.9	9.97-12.6	0.2-0.26	0.8-1.4	2.5-2.8	2.1-4.4	1.9-2.2	17.8-19.2

**LWF* – loss when firing.

3. Experimental procedure

The production of the ceramic bricks with matrix structure from raw material has been partially described in paper [8]. Dispersed particles are aggregated into granules, clayey material covers their surface, and samples from these granules are pressed and fired. The spatial cellular carcass is formed as a result of such process, it serves as disperse medium after bricks pressing. Disperse medium produces a liquid-phase during firing. This phase is introduced into a peripheral zone of disperse medium (a boundary layer) and forms a strong matrix structure after crystallization.

The granulated molding powder for ceramic brick production was prepared by granulation of fine-dispersed coal waste and subsequent powdering of granules by the mixture of loamy clay and crushed glass. The burden composition (mass %) was the following: coal mining waste – 65; Novokuznetsk loamy clay – 25; crushed glass – 10.

4. The investigation of the ceramic bricks matrix structure

The macrostructure study of a ceramic product burnt at 900…1000 °C (Figure 1) revealed the presence of spatial surfaces of phase boundaries in the common body of the material distinguished by a more intensive brown color and forming a regular continuous carcass with the directed orientation of cells, allows to refer it to the class of ceramic matrix composites [9]. The matrix of the composite material is formed from the clay component of the burden, which plays a role of the disperse medium after pressing and burning of products, and the disperse phase is formed from the coal mine waste with addition of crushed glass.

The disperse phase has a concentric-zonal structure, the granule body is represented by mono- and poly-mineral fragments of minerals (Figure 2). Mono-mineral fragments of quartz preserve their isometric angulated shape; some of them are of a ditrigonal form and remain almost unchangeable during sintering. Small and fine fragments on the surface of large crystals transit into crystobalite or tridimite. Large polygonal fragments of field spar decay and re-crystallize during sintering. Predominantly, solid-phase reactions result in the formation of anorthite-bytownite plagioclases and potassium-sodium field spars.

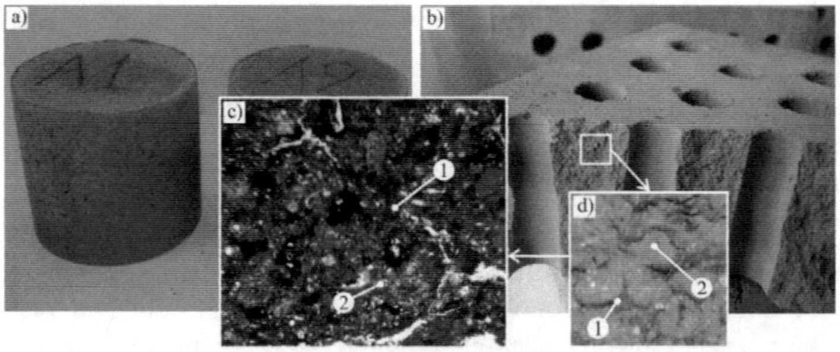

Figure 1. A macrostructure of ceramic bricks produced from coal wastes
a – experimental samples; *b* – fracture of brick; *c* – thin section, transmitting light, × 20; *d* – polished section, reflecting light, × 5; 1 – boundary layer between granules; 2 – granule body

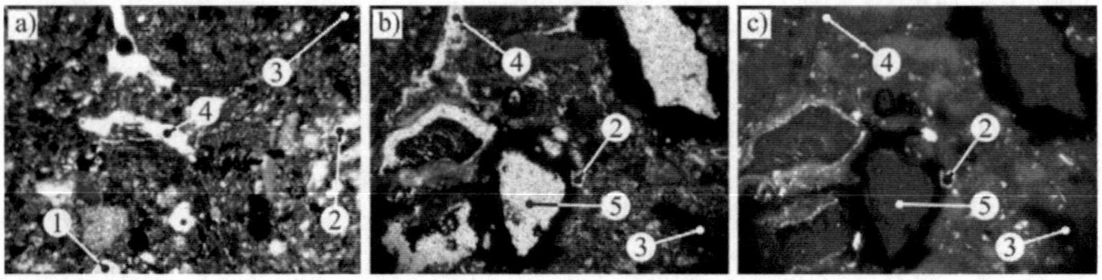

Figure 2. Micrographs of the matrix structure of ceramic bricks produced from coal wastes
a – nicols II, × 25; *b* – nicols II, × 100; *c* – nicols X, × 100; 1 – fritted quartz particles; 2 – field spar; 3 – ore minerals; 4 – pores; 5 – complex pyroxene with inclusions of cryptocrystalline minerals covered with the brownish coat

During the analysis of thin sections by a polarizing microscope, ore minerals of a spherical, oval, or sometimes polygonal form can be seen due to dehydration, dissociation, and re-crystallization of initial components during sintering, that is especially typical for sulfide minerals. Ore minerals are partially fused, which is proved by the evident isometric shape of limonite crystals (Figure 3b, e). In the periphery large fragments are covered by a brownish coat. The gaps between large fragments are filled with the cryptocrystalline and amorphous-glassy substance forming the cementing mass, a crystal phase of which is represented by pyroxenes of a variable structure (Figure 3e), quartz and field spars, sometimes by olivines and decomposed garnets. Thus, the granule body has a fine-brecciated texture and consists of relict minerals and newly formed crystal phases bound into a comprehensive whole by the glass phase.

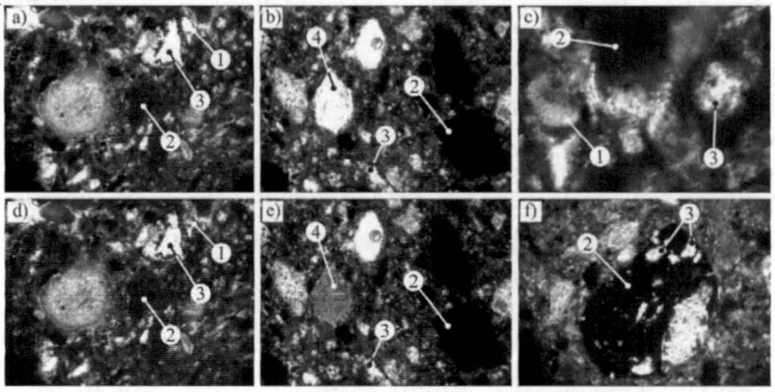

Figure 3. Micrographs of individual parts of the disperse phase in the composite material: a – thin section, transmitting light, nicols II, × 50; b – nicols II, × 70; c – nicols X, × 200; d – nicols X, × 50; e – nicols X, × 70; f – nicols II, × 100; 1 – augite; 2 – ore minerals; 3 – field spar; 4 – rhombic pyroxene

The X-ray (Figure 4a) and microscopic investigations of the disperse phase vindicate the solid-phase mechanism of sintering with the formation of new mineral phases. Decoding of diffraction patterns enabled identification of quartz, hematite, field spars, mullite-like phase, and augite in the granules. In the plane-polarized light, the pyroxenes are mainly represented by augite of the variable composition (Figure 3c, d), sometimes by diopside and newly formed pseudo-wollastonite. From the X-ray point of view, the powdering layer after sintering is represented by quartz, hematite, mullite, field spars, and spinel.

Infrared spectra of absorption of the ceramic matrix composite based on coal waste prove the presence of the found mineral phases (Figure 4b). In the low- and intermediate-frequency areas of spectrum (up to 1.800 cm^{-1}) the composite material has the absorption maximums of 470, 545, 1090, 1170 cm^{-1} typical for hematite. The presence of quartz is confirmed by the characteristic doublet of 770, 790 cm^{-1}. The absorption maximums (605, 1090 cm^{-1}) correspond to mullite. Spinel in the IR-spectrum of the disperse medium is registered at the peaks of 575, 710, and 1735 cm^{-1}, augite in the disperse medium – at peaks of 560, 670, and 1490 cm^{-1}.

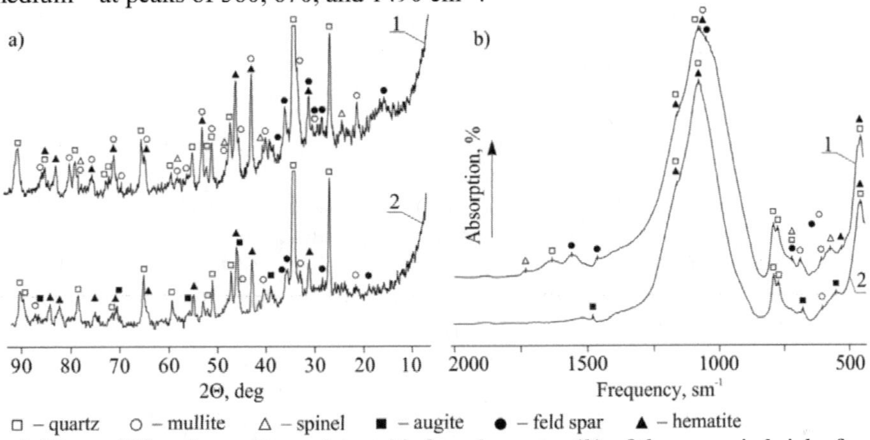

Figure 4. X-ray diffraction patterns (*a*) and infrared spectra (*b*) of the ceramic bricks from coal wastes: 1 – disperse medium; 2 –disperse phase

The pores with the sizes from 0.1 to 400 mμ are evenly distributed over the ceramic crock cross section. Normally, the pores are of a spherical form and have a rough uneven surface (Figure 5), and a significant quantity of them are of a cell shape (Figures 1, 2).

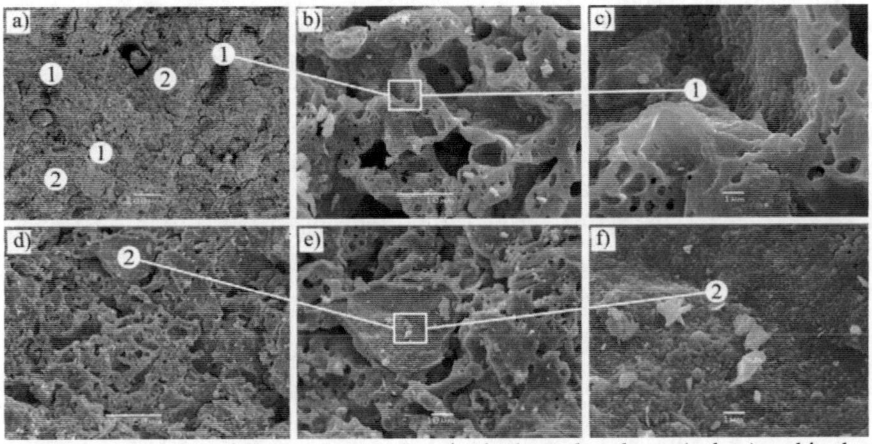

Figure 5. SEM-image of the microporous structure in the boundary layer (a, b, c) and in the body of a granule (d, e, f) of ceramic brick from coal wastes 1 – disperse medium; 2 –disperse phase

5. Results and discussion

It was experimentally established that the granule body has a fine-brecciated texture and consists of relict minerals and newly formed crystal phases bound into a comprehensive whole by the glass phase. The disperse medium (a granules boundary layer) of the matrix composite, formed by argillites during sintering, represents the cryptocrystalline aggregate continuously transiting from one granule to another. In the disperse medium, large fragments are not present. The medium consists mainly of microscopic crystals and fragments of transparent yellowish-white minerals evenly distributed in the glass phase; there are also fine cryptocrystalline aggregates, brownish-red in most cases (Figure 2c).

Expansion of the powdering layer results from the interaction with the boundary layer of the granules during sintering. Under oxygen deficiency, needed for burn-out of carbonic residual from the waste, the iron oxides recover into the protoxidic form and react with the amorphous silica, which are generated in the process of decomposition of clay minerals in the boundary layer. Thus, the glass phase is formed intensively on the granules boundary and it stimulates solid phase reactions and intensifies the cementing action of the cellular carcass.

The vividly expressed X-ray amorphous halo vindicates a significant quantity of glass in the disperse medium (up to 15…20 %) which agrees with the observations in the thin sections. Particles of remnant quartz and field spar, along with the new mineral formations, are the reinforcing carcass of the matrix, having a fine hematite nature.

The increased amount of pores in the granule body is explained by the burn-out in the process of dispersed coal mass burning. In the boundary layer the porosity increases to some extent, which is caused by the air compression as the granulated mass is compacted, and also by water squeezing-out into the powdering layer during adobe brick molding. It should be noted that pore walls in the boundary layer are formed by the amorphous vitreous mass veined with cryptocrystalline minerals (Figure 5c). Evidently such internal fusion of pore space walls increases the fraction of closed porosity and decreases the water absorption ability of the ceramic crock.

6. Conclusions

The petrographic analysis of the ceramic brick test lot produced from the wastes of Korkinskiy coal open pit mine shows that the advanced physical and mechanical properties result from:

– the matrix structure of the ceramic crock in which the disperse medium i.e. a re-crystallized bond of the amorphous and mineral phases fills inter-grain gaps and binds mineral particles to each other;

– intensive formation of the glass phase on the boundary of the disperse medium and the disperse phase which promotes solid-phase reactions and intensifies the cementing action of the matrix;

– reduction of the temperature of solid-phase sintering processes followed by the formation of anorthite-bytownite plagioclases, hematite, augite, spinel and other mineral phases.

Acknowledgement

This work was supported by the Ministry of Education and Science of the Russian Federation (registration number RFMEF160714X0106).

References

[1] Shpirt M J, Ruban V A and Itkin Yu V 1990 Rational use of waste from mining and coal preparation (Moscow: Nedra)
[2] Storozhenko G I, Stolboushkin A Yu and Mishin M P 2013 *J. Construction Materials* **4** 57
[3] Nikolayenko M A, Khodykin E I and Lyutenko A O 2008 *Collection of Papers for the General Meeting of the Rus. Ac. of Arch. and Constr. Sc.* **2** 377–381
[4] Stolboushkin A Yu 2011 *J. Construction Materials* **4** 43
[5] Kotlyar V D, Ustinov A V and Kovalev V Y 2013 *J. Construction Materials* **4** 44
[6] Stolboushkin A Yu 2011 *J. Construction materials* **2** 10

[7] Stolboushkin AYu, Stolboushkina O A and Ivanov A I 2012 *Collection of the Reports of the Fourth International Scientific and Practical Conference* (Novokuznetsk: Russian Federation) pp 176–181
[8] Stolboushkin A Yu, Storozhenko G I, Ivanov A I, Berdov G I and Stolboushkina O A 2013 Raw mix to manufacture wall ceramics and method of its production. RU Patent 2500647
[9] Mecholsky J J 1986 *American society-bulletin* **2** 315

A temperature change in elastic and acoustic properties of auricupride

E P Tesleva[1] and B P Sorokin[2,3]
[1] Yurga Institute of Technology of National Research Tomsk Polytechnic University, 26, Leningrad st., Yurga, 652050, Russia
[2] Moscow Institute of Physics and Technology, 9, Institute st., Dolgoprudny, 141700, Russia
[3] Technological Institute for Superhard and Novel Carbon Materials, 7a, Central st., Moscow, Troitsk, 142190, Russia

E-mail: tesleva@mail.ru

Abstract. Changes in acoustic and elastic properties of auricupride approach within the high temperature interval and close to an order-disorder phase transition have been discussed. Calculations of the phase velocities of purely transverse and longitudinal elastic waves, elastic moduli and Poisson's ratios based on elastic constants $c_{ij}(T)$ of the crystal have been executed. Temperature changes of Grüneisen parameter along crystallographic directions [100], [110] and [111] have been defined.

1. Introduction

Elastic properties of crystals are the most important properties of solids. The elastic properties of crystals are characterized by different modules and elastic constants adequately revealing the nature of interatomic bonds, which is one of the main problems of the solid state physics. The elastic properties anisotropy allows estimating on the interatomic bond strength in different crystalline planes.

Elastic properties and anharmonic effects in metallic, covalent, ionic, ion-covalent and molecular cubic single crystals with lattices of various types were studied in [1-3]. Anisotropy of elastic properties and anharmonicity of atomic interaction in the auricupride – binary alloy Cu_3Au with the positional order-disorder in the low-temperature region were considered in [4]. In the low temperatures area the given alloy has a cubic face-centered lattice with a $Pm\bar{3}m$ space group where face centers are occupied by copper atoms and cube corners by aurum atoms. A high-temperature disordered phase of Cu_3Au alloy is observed at the temperatures over $T_c = 661$ K. It has a face-centered cubic lattice where Cu and Au atoms are distributed in all positions without a long-range order. Elastic properties of a single crystalline alloy were studied by the method of a compound piezoelectric vibrator within the temperature range of 293…723 K (especially thoroughly near T_c) in [5].

2. Calculated ratios

Use the data on $c_{ij}(T)$ [5] of the auricupride crystal we have executed calculations of phase velocities of purely longitudinal and transverse bulk acoustic waves (BAW) in three crystallographic directions, elastic anisotropy factor (A), Cauchy relation (Δ) both for the ordered and disordered forms within a wide temperature range. The formulas for calculating phase velocities along directions [100], [110] and [111] are given in [4].

The stiffness constants were measured with the following inaccuracies: c_{11} – 0.8 %, c_{12} – 1.13 %, c_{44} – 0.23 %. According to these inaccuracies the maximal error of phase velocity determination in any direction does not exceed 0.5 %.

An elastic anisotropy factor determining the crystal isotropy measure was found according to ratio $A = 2c_{44}/c_{11} - c_{12}$ (for elastically isotropic body $A = 1$). Couchy relation which defines a measure of the central interatomic forces (under $\Delta = 1$ all interaction forces in the crystal must be central) equals to $\Delta = c_{12}/c_{44}$. The elastic properties of isotropic solids are characterized by elastic moduli B (bulk modulus of elasticity), E (Young modulus), G (shear modulus) and were found by the Voigt-Reuss-Hill (VRH) approximation (a cubic system) [4]. Elastic compliances s_{ij} were used for calculating the elastic moduli of the Cu_3Au alloy in the various crystallographic directions. The formulas for calculating the Poisson's ratio along three specific directions $\sigma_{\langle hkl \rangle}$ were presented in paper [6].

3. Results and discussion

Temperature changes of BAW phase velocities, elastic constants, etc. of the auricupride alloy within the interval of 300…725 K are linear both in the ordered and disordered phases with the exception of a narrow temperature interval near $T_c \approx 661$ K (Figures 1-3). Stiffness constants, phase velocities and elastic moduli do not decrease much with the temperature growth but Poisson's ratios demonstrate the tendency to an increase.

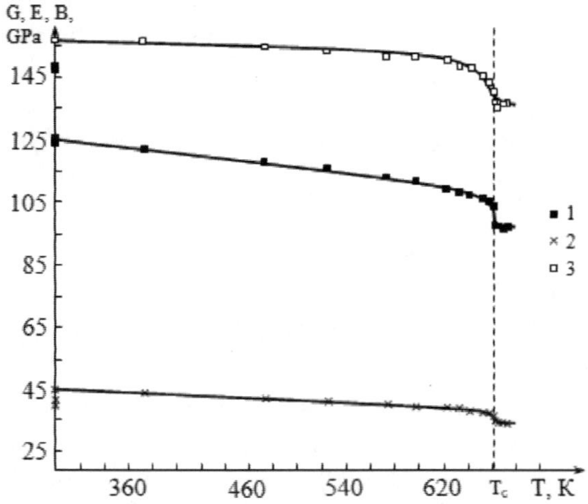

Figure 1. Change of elastic moduli of the auricupride polycrystal conditioned by temperature alteration: 1 – Young modulus E, 2 – shear modulus G, 3 – bulk modulus B.

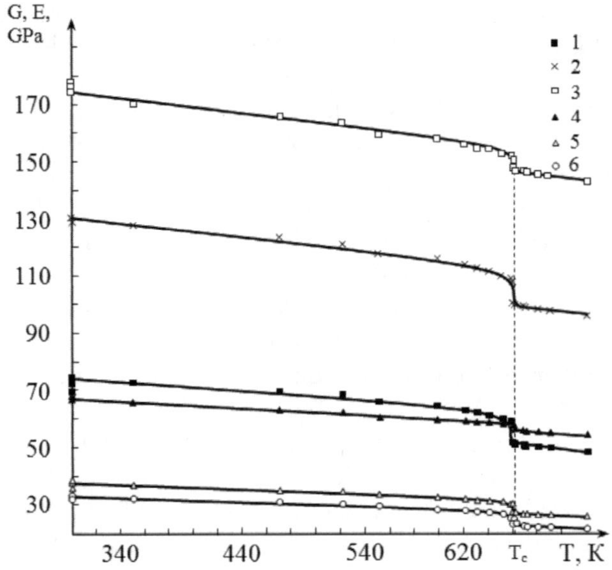

Figure 2. Change of elastic moduli of the auricupride monocrystal conditioned by temperature alteration in various crystallographic directions:
$1 - E_{\langle 100\rangle}$; $2 - E_{\langle 110\rangle}$; $3 - E_{\langle 111\rangle}$; $4 - G_{\langle 100\rangle}$; $5 - G_{\langle 110\rangle}$; $6 - G_{\langle 111\rangle}$.

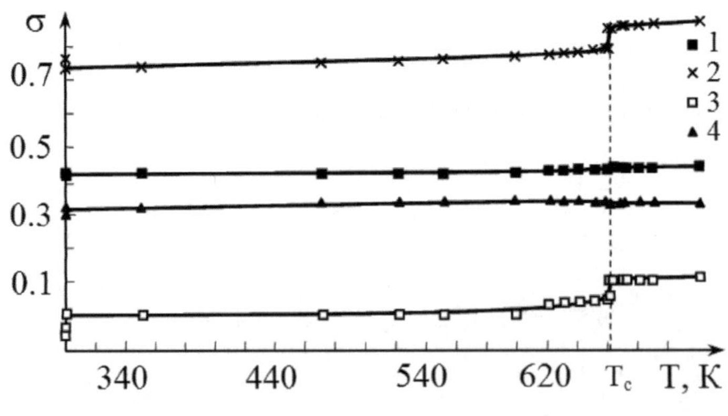

Figure 3. Change of Poisson's ratios monocrystal conditioned by temperature alteration in various crystallographic directions: $1 - \langle 100\rangle$; $2 - \langle 110,001\rangle$; $3 - \langle 110,1\bar{1}0\rangle$; $4 - \langle 111\rangle$.

From Figure 1 it is possible to see that all three effective elastic moduli (*G*, *E*, *B*) of auricupride near T_c demonstrate anomalous behavior ('softening' when a 'step' decreases by almost 10 %) which is associated with the phase transition into the disordered state.

Anisotropy of elastic properties of the auricupride alloy influences the relation between Young's and shear moduli in various crystallographic directions (Figure 2). Inequations between elasticity moduli for a given alloy are similar to the known dependencies between magnitudes $E_{\langle hkl\rangle}$ and $G_{\langle hkl\rangle}$ for the set of lithium halogenides under $A > 1$: $E_{\langle 111\rangle} > E_{\langle 110\rangle} > E_{\langle 100\rangle}$ and $G_{\langle 100\rangle} > G_{\langle 110\rangle} > G_{\langle 111\rangle}$. Temperature changes of Poisson's ratios in the auricupride alloy for various crystallographic directions are presented in Figure 3. In the temperature range from 300 to 725 K the following correlation is fulfilled: $\sigma_{\langle 110,001\rangle} > \sigma_{\langle 100\rangle} > \sigma_{\langle 111\rangle} > \sigma_{\langle 110,1\bar{1}0\rangle}$. Near T_c, Poisson's ratios increase abruptly (approximately by 10 %) under auricupride crystal deformation in two directions ($<100>$ and $<110,1\bar{1}0>$). The described peculiarities of acoustic and elastic properties near T_c demonstrate that the phase transition in auricupride is partly isothermal (a first-order phase transition) and partly smooth (a second-order phase transition).

4. Conclusion

We have applied the date on stiffness constants to calculated the BAW velocities along three special crystallographic directions of single crystalline alloy Cu_3Au. It has been established that among the longitudinal wave velocities $\upsilon_{L[111]}$ have the maximal value and the velocity of transverse waves is maximal in the [100] direction for all the temperature ranges. Temperature-related elastic moduli (Young, shear, bulk) changes are linear (the moduli decrease with the temperature growth with the exception of the temperature region near T_c). Thermal dependences of Poisson's ratios are also linear, although σ slightly increases with the temperature growth.

References

[1] Lebedev Y N, Kozlov E V and Starostenkov M D 1975 *Soviet Physics Journal* **10** 1419
[2] Belomestnykh V N, Tesleva E P, Soboleva E G 2008 *Technical Physics Letters* **34** 867
[3] Belomestnykh V N, Soboleva E G 2014 *Applied Mechanics and Materials* **682** 170
[4] Tesleva E, Belkova T 2014 *Applied Mechanics and Materials* **682** 519
[5] Siegel S 1940 *Phys. Rev.* **57** 537
[6] Belomestnykh V N, Soboleva EG 2012 *7th International Forum on Strategic Technology (IFOST – 2012)* **1** 499
[7] Leibfried G, Ludwig W 1963 *Theory of anharmonic effects in crystals* (Moscow: Foreign literature publishers)
[8] Belomestnykh V N 2004 *Technical Physics Letters* **30** 91

The development of a modified composition of ceramic mass for the production of bricks

V F Torosyan, E S Torosyan, V A Yakutova and V K Antyufeev
Yurga Institute of Technology, Tomsk Polytechnic University, 26, Leningradskaya st., Yurga, 652055, Russia

E-mail: torosjaneno@mail.ru

Abstract. The need to improve the technical level of production of construction materials, their product range, to improve product quality and reduce its cost requires the expansion of the raw material base, the use of resource and energy saving technology and design solutions. To implement all these it is necessary to conduct a more detailed study of the properties of ceramic materials and to investigate the behavior-modifying components of their formulations. This paper presents the development of the composition of ceramic mass for the production of bricks, a modified silicon-waste production of ferrosilicon.

1. Introduction

Foreign and Russian authors have significant experience in manufacturing multi-purpose construction materials with the use of micro-silica. The chemical composition of the latter, peculiarities of its structure, composition and amount can influence colloid-chemical, structural and mechanical processes, a character and kinetics of hydration of different materials in conditions of thermal impact [1, 2].

Characteristics of ceramic materials are determined by properties of raw materials and a technology of their manufacturing. Clay materials are the basic ones for ceramic brick manufacturing. This issue viewed both as fundamental and applied one have attracted attention of many researchers (A I Avgustinik, A S Berkman, F D Ovcharenko, M I Rogovoy et al).

Sweden scientists have developed a method of manufacturing of high-strength cellular ceramic materials from clay and gas-treated dust, dust produced by electrofilters, utilized in manufacturing of ferrosilicium and silicochromium. A charge with this chemical composition can be used for manufacturing of construction elements, whereas a raw product is reasonably to be covered into foil keeping gas-reducers [3]. There are papers where in the authors apply micro-silica as a binding component in masses for manufacturing of ash-ceramic products [4]. A combination of micro-silica with titanium dioxide when sintering causes formation of a heat-resistant material due to the reaction of amorphous silica with titanium dioxide $TiO_2 + SiO_2 = TiSiO_3$ [5].

Refractoriness of micro-silica determines its application in manufacturing of fire-proof and heat-resistant materials which are highly resistant to thermal shock and do not shrink much [6, 7].
A mineralizing effect of micro-silica while crystal and amorphous phases are formed in heat-resistant and fire-proof materials makes it possible to control a special chemical composition in products of their synthesis [8].

Microsilica as a modifying component in fired ceramics has an effect on physical and chemical processes of structural-and-phase construction. A lot of scientists carried out fundamental and applied research into issues of phase transformations of clay raw materials used in manufacturing of ceramics (A I Avgustinik, A S Berkman, F D Ovcharenko, M I Rogovoy, L N Tazki et al). The description of the main characteristics of these processes isgiven below.

While finish drying at temperature of 80-130 °C physical and mechanical bound water vaporizes and a product is uniformly warmed up. During the first phase of firing, source minerals are destroyed because of dehydration, dissociation and amorphization of components.

Within the temperaturelimit of 200…600 °C organic admixtures and combustible additives are burnt out at temperature 500…600 °C physical and chemical bound water escapes from clay and other minerals; therefore, clay loses its plasticity, mass is reduced, a crystal lattice of the mineral is damaged, mechanical strength fails and products shrink. Dehydration of kaolin proceeds with formation of metakaolinite $Al_2O_3 \cdot 2SiO_2 \cdot 2H_2O \rightarrow Al_2O_3 \cdot 2SiO_2 + 2H_2O$.

After shrinkage, the clay becomes more concentrated because low-melt compounds fill the pores among high-heat particles and due to further compaction – firing.

Within the temperature limit of 500…700 °C, minerals containing ferrum, sulphides and sulphates are dissociated and at temperatures of 700…900 °C – those containing carbonates. Burning out of organic admixtures and dissociation of minerals, emitting volatile gases (CO_2, SO_2 et al), are to be finished before a crock is fired in order to prevent its cracking and swelling. At temperature of 700 °C and above, alkalis in the clay interact with other components constituting a melt, where (as A. I. Avgustinik stated) hard particles are attracted, minerals dissolve and new thermo-dynamically stable crystal phases form. Further heating of the material above 700 °C results in decomposition of metakaolinite into free oxides Al_2O_3 and $2SiO_2$, sillimanite formation ($Al_2O_3 \cdot SiO_2$) and SiO_2 and metakaolinite ($Al_2O_3 \cdot 2SiO_2$), and at temperature of 950…1150 °C and above – mullite ($3Al_2O_3 \cdot 2SiO_2$). Mullite is the most stable compound, strengthening products, impacting them and making them thermally resistant.

Clays and loams are often sensitive to drying, after firing they have low compression and flexural strengths and frost resistance. All these disadvantages make it impossible to use the materials without improving additives. Moreover, as many researchers emphasized [3-5], it is rather difficult to find an additive which can eliminate complex technological problems. Therefore, charge materials necessitate correction by introducing several additives or mineralizers into ceramic masses, consequently the price of finished products rises and materials consumption of production increases. In addition, the growing number of components that ceramic mass consists of causes further difficulties, which can have a negative effect on the quality of baked products.

This paper is aimed at the analysis of micro-silica behavior viewed as a modifying component in fired ceramic materials based on low-grade clays. The use of the latter for manufacturing ceramic bricks can improve their physical, mechanical and operational characteristics.

2. An experimental part

Stiff clay of Bolotnoye deposit and micro-silica produced by the public limited company 'Yurga ferroalloys" were applied for research. Loam of Bolotnoye deposit is a low-melt clay raw material with high concentration of Fe_2O_3 = 5.23 % and a semi-acid loam ($Al_2O_{3\ calc}$= 14.22 %.). Its technological characteristics are listed in table 1.

Table 1. Technological characteristics of loam.

Loam	Fraction 0.5 mm, %	Fraction maximum 0.5	Chemical composition
Bolotnoye deposit, explored by Rosgeolnerudrazvedka in 1969	6.5…14.4	Smaller than 0.005 mm – 17.3…20.6%; 0.005…0.05 mm – 61.1…63.8 %; 0.05…1 mm – 15.5…21.6 %	SiO_2– 61.8...64.2; Al_2O_3– 12.0...14.66; Fe_2O_3 – 5.23...5.84; CaO– 3.09...5.21; MgO– 1.31...2.24; P_2O_5– 17.27...20.5; others – 6.91...9.57

Microsilica is condensed silicate dust, which is a sideline product of crystal silicon or silicon alloys manufacturing. Some silicon monoxide SiO becomes gaseous and, being exposed to oxidation and condensation, it forms an extraordinary small product shaped like ball-shaped particles containing a lot of amorpous silica. Microsilica as opposed to conventional high-silica raw materials, for example, quarzt sand, is a light fine-dispersed material containing small spheres of amorphous silica with 0.01 to 0.7 micron particles.

Despite differences in chemical composition, colour and carbon concentration, fine dust particles, being the wastes of silicon and ferrosilicium manufacturing, have some commoncharacteristics:

- They are condensates of silicon vapors (silicon monoxide).
- They mainly consist of globules with the average size of 0.1…0.2 microns (100 times smaller than cement particles).
- They are amorphous.
- Concentration of SiO_2 is high (84…98 %).
- They are difficult to store and transport.

The dust colour varies from light-grey to wholly black and depends generally on carbon concentration and less on ferrum. The density is 2200…2300 kg/m^3. This coefficient can exceed the above-mentioned value in production of other silicon containing alloys, for example, ferrochrome silicon or calcium-silicon.

Melting temperature of various silicon containing alloys is different. In production of crystal silicon the temperature is higher than in melting of ferrosilicium with silicon concentration of 40…50%. Hence, the chemical and granulometric composition is different. It is significant that particles of the stopped dust are not separate micro-globules but conglomerates of finest particles of 0.01…2 microns in diameter.

Exceptionally high dispersibility of micro-silica and its structure are the reasons of its low apparent density (150…250 kg/m^3). These characteristics cause the difficulties in its application and transportation. Table 2 provides the information on the chemical composition and physical-mechanical characteristics of micro-silica, produced by the public limited company "Yurga ferroalloys".

Table 2. Physical and mechanical characteristics of micro-silica, produced by the public limited company 'Yurga ferroalloys'.

Raw material	Chemical composition	Concentration in mass %	Density, kg/m^3	Apparentdensity,kg/m^3		Porosity of material layer, %		Specific surface,s m^2/g	Natural moisture, mass %
Micro-silica	SiO_2	94.2	2180	150	300	82	79	25000…30000	2…3
	Al_2O_3	0.47							
	Fe_2O_3	0.6							
	CaO	0.36							
	MgO	0.3							
	K_2O	0.69							
	Δm	3.43							

Experimental chemical compositions of ceramic mass were prepared in laboratory conditions. Raw material preparation included the following phases of treatment:

- drying of raw materials, grinding in a dry-grinding mill, sieving;
- samples in the form of a cube of 30x30x30 were made by the method of soft forming;
- when the soft forming mixture was moistened up to the necessary forming moisture, then soured during 24 hours till adsorbed hydrate films formed completely;

- samples were preliminary dried in natural conditions during 7 days at ambient temperature of 23…25 °C;
- baked in a muffle furnace at temperature of 1000-1050°C for an hour.

Low-melt loams with high concentration of Fe_2O_3 = 5.23 % are used as a raw material at Yurga brick factory for brick manufacturing. These loams are acidorsemi-acidones (Al_2O_{3calc}= 14.22 %.). Their technological characteristics are similar to those of Bolotnoye deposit. A raw-brick formed at Yurga brick factory was taken for comparison with finished products divided into cubes of 30x30x30 and burnt with experimental samples in the muffle furnace at temperature of 1000-1050 °C for an hour. Sulfate soap was used to increase cohesion between particles of masses. All samples were subjected to physical and mechanical tests. Table 3 shows the chemical composition of ceramics based on a low-melt clay raw material, containing micro-silica.

Table 3. Ceramics chemical composition.

Components	Amount, mass %					
	1	2	3	4	5	6
Clay	89	87	84	80	75	70
Micro-silica	10	12	15	19	24	29
Sulfate soap	1	1	1	1	1	1

3. Results discussion

Water absorption is one of negative characteristics of construction materials. It reduces the strength of bricks, they become more heat-conducting, and a mould can develop on their surface. A moistened brick is subjected to destruction in winter because of constantly repeated cycle freezing – defrosting. Micro-silica in ceramic masses promotes water extraction when heating and an amorphized component of micro-silica crystallizes at high temperatures. In general, water absorption of bricks is reduced and their strength increases. Table 4 provides the information on physical and mechanical properties of samples made on the basis of materials, listed in Table 3.

Table 4. Physical-mechanical properties of samples.

Property	Chemical composition					
	1	2	3	4	5	6
Compression strength, MPa	33.5	32.7	30.2	28.3	23.5	18.7
Mean density, kg/m^3	1710	1660	1620	1580	1510	1480
Frost resistance, cycles	min60	min 60	min 60	min 40	min 40	min 35
Water adsorption, mass %	18.5	19.2	20.0	22.4	26.5	28.3
Temperature of burning, °C	1050	1050	1050	1050	1050	1050

4. Conclusions

1. Micro-silica in ceramic masses is10…15 %: water adsorption is 18.5…20.0%; 19…29% of water adsorption changes from 22.4% to 28.3%.
2. The number of frost resistance cycles of samples is increased, starting with 35 cycles under conditions of 29% of micro-silica up to more than 60 cycles under conditions of 10% of micro-silica.
3. Fracture strength of samples improves when samples are compressed up to 33.5 MPa.
4. Volumetric coloring of bricks containing micro-silica as a modifying component can be used in manufacturing of high quality facing ceramic bricks. Micro-silica has an advantage in this

process because there is no need to grind it as opposed to additives utilized in traditional methods of brick coloring.

References

[1] Kaprielov S S, Sheinfeld A V, Gazizulin V M and Voronov Yu N 1992 *J. Steel* **5** pp 83–85
[2] Milovanova R G, Semyonova V E and Chupina A I Building mortar *Certificate of authorship* № 637359 C04B13/06
[3] Mironyuk I F, Chuyko A A, Khoma M I, Kisly P S, Khaber N V and Ostafiychuk V K Ceramic material *Certificate of authorship* № 962262 C04B35/14 35/80
[4] Fredrik Wilhelm Anton, Dr. Lidingo Method for producing fired expanded clay *Products 19.03.71* (Sweden) ДЕ2213231 C3 C04 B21/04 Short
[5] Pak N V, Artyomov L N and Makarov L M 1990 *J. Energy construction* **3** p 38
[6] Kurz F W and Rudmark H 1978 Method of manufacturing porous ceramic products by reacting flue gas dust and filter dust with clays of the like, such as expanded clay *Patent* 4071369
[7] Shpirko N V and Chumak L I 1989 Heat-insulating materials on the basis of pearlite with the use of industrial wastes *Proceedings of higher educational institutions Building and architecture* (Tambov: Tambov State Technical University) pp. 62–66
[8] Onasenko Yu A and Peschanskaya V V 2013 *J. Technol. Auditand Product. Reserv.* ½ 25–28
[9] Lokhova N A, Makarova I A and Patramanskaya S V 2002 Fired materials on the basis of micro-silica (Bratsk State Technical University) p 163
[10] Torosyan V F and Torosyan E S 2014 *J. App. Mechanics and Materials* **682** 480–484

IOP Conf. Series: Materials Science and Engineering **124** (2016) 012146 doi:10.1088/1757-899X/124/1/012146

Control of the electrode metal transfer by means of the welding current pulse generator

A Knyaz'kov[1], O Pustovykh[1], A Verevkin[1], V Terekhin[2], A Shachek[1], S Knyaz'kov[1] and A Tyasto[1]
[1] National Research Tomsk Polytechnic University, 30, Lenina st., Tomsk, 634050, Russia
[2] Seversk Technological Institute - Branch of State Autonomous Educational Institution of Higher Professional Education 'National Research Nuclear University 'MEPhI', 65, Communistic ave., Seversk, 636036, Russia

E-mail: bos1983@tpu.ru

Abstract. The paper presents a generator of welding current pulses to transfer an electrode metal into the molten pool. A homogeneous artificial line is used to produce near rectangular pulses. The homogeneous artificial line provides the minimum heat input with in the pulse to transfer the electrode metal, and it significantly decreases the impact of disturbances affecting this transfer. The pulse frequency does not exceed 300 Hz, and the duration is 0.6 ÷ 0.9 ms.

1. Introduction

A number of power supplies have been developed for pulsed arc welding with a consumable electrode. They are made as separate pulse generators, which operate in aggregate with standard power supplies or in combination with a normal power supply [1, 2].

The easiest way to form welding current pulses is by means of direct or transformed discharging current or a capacitor bank of charging current through the arc space.

Figure 1 shows an equivalent circuit to form welding current pulses in this type of devices.

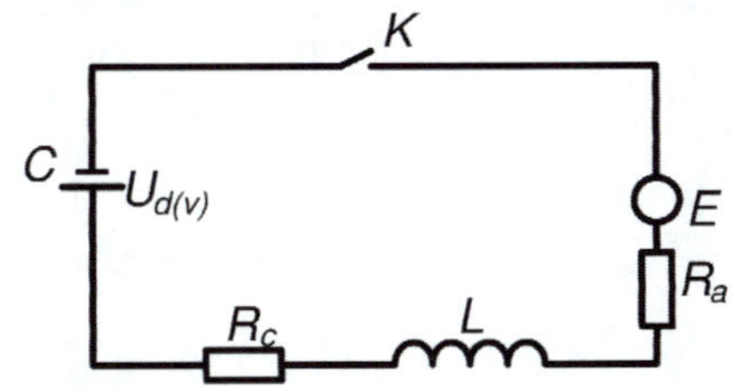

Figure 1. An equivalent welding current pulse-forming circuit, where L is the total inductance of the circuit; C is a capacitor bank; K is a half-controlled switching device; $U_{d\,(v)}$ is capacitor charge voltage; R_c is circuit resistance, R_a is arc resistance.

When welding with a consumable electrode in shielding gas environment, the arc static characteristic is of a restoring character. In the working area, it may be represented in the form of emf and arc resistance R_a connected in series.

The process occurs in a decaying oscillatory or aperiodic manner, which depends on the circuit parameters. The pulse amplitude in these devices is adjusted by changing the capacitor charge voltage, and the pulse length is adjusted by changing the capacitance value or inductance that simultaneously affects the current pulse amplitude [3].

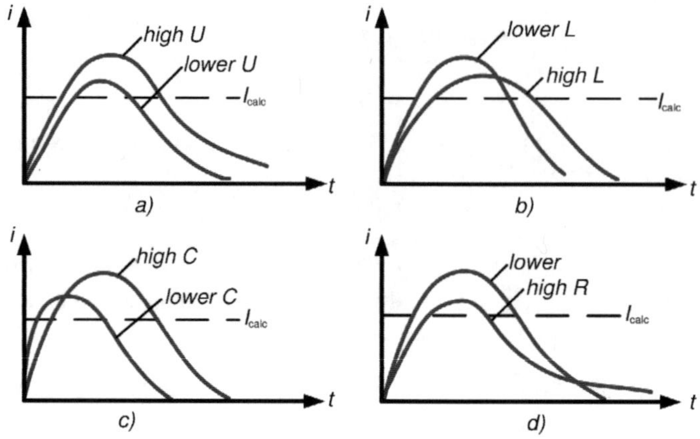

Figure 2. Dependence of the aperiodic discharge current waveform on adjustable parameters; a) capacitor voltage, b) inductance, c) capacitance, d) real resistance.

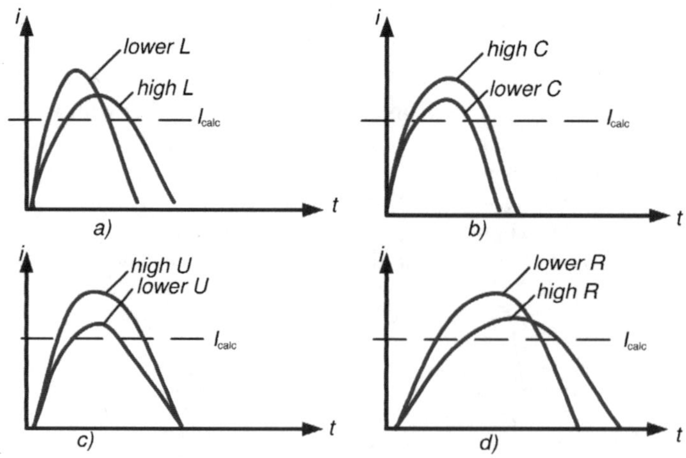

Figure 3. Dependence of the current waveform on adjustable parameters under decaying oscillatory discharge; a) inductance, b) capacitance, c) capacitor voltage, d) real resistance.

Figures 2 and 3 show the curves of the current change over time and the impact of the change in one of the variables on the current curve for aperiodic and oscillatory processes, respectively.

The required pulse amplitude of the welding current, which detaches a droplet from the electrode, is determined by equation

$$I \geq 2\pi \cdot \sqrt{\frac{4.2(T_{cr} - T_e) \cdot R_e}{\left(\frac{M}{\gamma}\right)^{\frac{2}{3}} \mu_0 \cdot \ln\left(\frac{R_c}{R_e}\right)}}$$

where $T_{c\gamma}$ is critical temperature of the metal electrode wire, °K; T_e is a melting point of the metal electrode wire, °K; R_e is an electrode wire radius, m; M is molecular weight of the metal electrode; γ is specific weight g/cm^3; R_c is an arc column radius, m.

To enhance the stability of the electrode material transfer, the current pulse amplitude value should have a 1.2 … 1.3 time increase (1), i.e.

$$I = (1.2 \div 1.3) I_{calc.}$$

In case of large disturbances in the arc voltage, the pulse amplitude is to be increased.

Figures 2 and 3 show a conventionally plotted horizontal line corresponding to the calculated value of the desired pulse amplitude value. The time during which the current pulse has an effective impact on the molten metal at the end of the electrode is determined as the time during which the pulse current exceeds the calculated current value.

Figures 2 and 3 demonstrate that the total pulse duration measured at the base does not provide data on the time of the effective impact on a droplet. This may explain a wide spread of the values of pulse duration required for droplet detachment in the published data, as it is determined as the total duration of the pulse at the base.

As can be seen in Figures 2 and 3, a change of a single parameter causes a change in both amplitude and duration of the welding current pulses, therefore, the amplitude and duration of the current pulses cannot be controlled independently.

To obtain a uniform criterion when assessing the current pulse duration required for detachment of a droplet, it is recommended to use the concept of the active current pulse duration for any pulse shape. It is to be formed for the current value equal to the calculated current value.

According to the suggested criterion, the active duration of a sinusoidal current pulse is 0.378 … 0.445 I of the total duration. Due to the pulse, undesirable heat input into the product occurs within the rest of the time, which deteriorates the quality and energy data of welding.

A major shortcoming of sine and aperiodic current pulses is insignificant front and edge steepness. Under the impact of external disturbances, such kind of steepness disturbs the transfer stability as the active duration changes significantly (this pulse shape is not optimal).

Figure 4 shows the shapes of the welding current pulse, which can be formed depending on the principles and techniques used.

A horizontal line plotted in Figure 4 corresponds to the calculated value of the current pulse amplitude.

The analysis of the presented pulse shape shows that virtually all of the pulses introduce the energy in the arc space, which exceeds the energy required to detach a droplet and accelerate its transfer to the molten pool. In Figure 4, it is shown in shaded areas. Trapezoidal pulses are the most preferable to this effect. The active duration of these pulses t_{ap} under the calculated pulse amplitude of the welding current pulse significantly differs from total pulse duration t_{tp}. Hence, it can be concluded that for energy reasons near rectangular or at least near trapezoidal pulses are most preferable.

The energy of the pulse of this shape is minimal if compared to other pulse shapes, which is advantageous in terms of both quality and energy. The stability of the electrode metal transfer is considerably higher, since the active pulse duration during welding under arc voltage disturbance changes insignificantly.

A near trapezoidal current pulse shape can be formed using the artificial forming line [4, 5]. Figure 5 shows an electrical schematic of the pulse generator power unit based on a homogeneous artificial line.

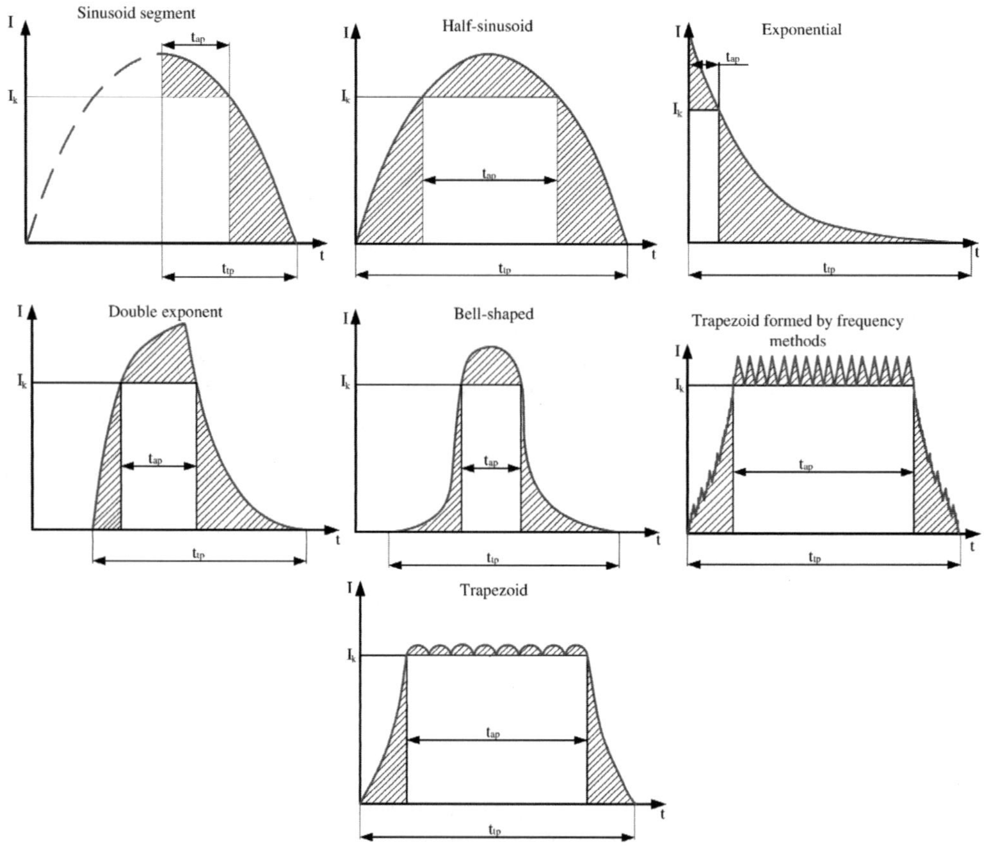

Figure 4. Shapes of the welding current pulse.

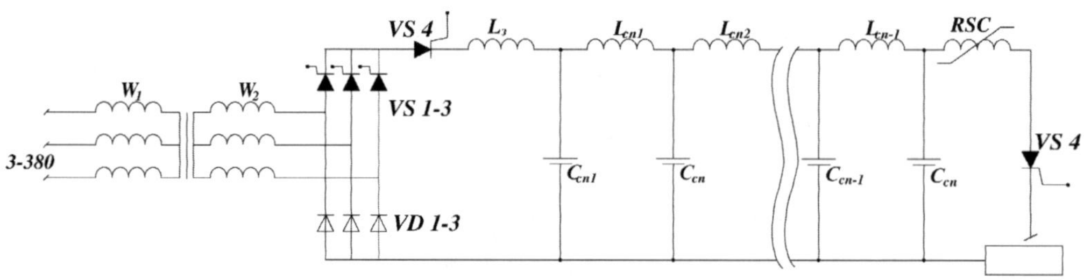

Figure 5. An electrical schematic of the generator power unit.

Artificial line is a three-phase step-down transformer, primary windings, secondary windings, a discharge thyristor, a charging choke coil, cell choke coils, cell capacitors, a rapidly saturated choke coil.

The artificial forming line consists of homogeneous elements with lumped parameters. Each of the elements is a cell consisting of choke coil L_c and capacitor C_c connected in series.

Rectangular-shaped pulses are generated due to the discharge of the pre-charged line to the welding arc. A rectifier with stepped regulation of the idling mode for varying the line charge voltage is used to charge the forming line, and, hence, the amplitude of current pulses. The inductance value of charging choke coil L_3 is chosen such that the minimum duration of the pulse repetition could exceed the half-period of the charging circuit natural oscillations, since small value $L_э$ has no significant effect on the process of capacitor charging.

The discharge of the forming line to the arc space is carried out when thyristor VS4 is open. The shape of the current pulse is near trapezoidal with the amount of ripples at the top equal to the number of cells n. At first, the direct wave of the forming line discharge travels causing the discharge of cell capacitors C_c by one half. Then, a backward wave travels causing the complete discharge of the capacitors and the overcharge of the capacitors with an opposite polarity up to the voltage sufficient to turn off thyristor VS4 by reverse voltage.

For a larger number of cells ($n > 8$), the depth of ripples at the top of the pulse is insignificant, but this causes a high rate of the current rise both at the beginning of the pulse front and during the current edge decrease. This can cause failure in the arc zone shielding, electrode metal spattering and weld deterioration.

Theoretical and experimental studies [6, 7] proved that a rapidly saturated choke coil used as the last cell choke coil with a rectangular hysteresis loop results in a bell-shaped pulse. The inductance of this type of the choke coil is comparable with the inductance of cell L_c in a saturated state. The rapidly saturated choke coil at the beginning of the pulse front is unsaturated and its inductance is greater than that of cell L_c. This ensures a smooth current change at the beginning of the front and at the end of the pulse edge.

The combination of the parameters of the device, such as a forming line and a rapidly saturated choke coil, which is the choke coil of the last cell of the forming line, results in a bell-shaped pulse with a cut and almost flat top. This fact provides the optimum pulse shape, minimum energy input during a pulse, an increase in the electrode metal transfer stability, and improvement of the welded joint quality.

The experimental test has shown that in Al-based alloy welding using a 1.2 mm diameter wire, the developed pulse shapes provide the total duration of 0.6…0.9 ms, and the duration of sinusoidal-shaped pulses is 1.5…2.3 ms. In addition, the stability of the electrode metal transfer into the molten pool is provided.

The parameters of L_c–C_c cells should be fulfilled according to the following relation:

$$\sqrt{\frac{L_c}{C_c}} = \rho > R_c ,$$

where L_c is cell choke inductance; C_c is cell capacitance; ρ is cell wave resistance; R_c is welding circuit resistance.

The current pulse amplitude is determined by

$$I = \frac{U_c - U_a}{2\rho} ,$$

where U_c is capacitor charge voltage in the artificial forming line; U_a is arc space voltage.

The pulse duration is determined by

$$t_p = 2n\sqrt{L_c - C_c} .$$

The maximum current pulse frequency is 300 Hz.

It can be smoothly varied within the range of 0…300 Hz.

The advantages of the device are as follows:

– the possibility to smoothly adjust the current pulse frequency;
– input of the minimum energy required to transfer a droplet within the pulse duration;
– independence of the active pulse duration of the welding current on the arc space voltage disturbances.

The experimental studies have shown that application of the considered pulse shape formed by means of an artificial forming line can be extended. In particular, the considered pulse shape can be used to control electrode metal transfer for gas mixtures and self-shielded flux-cored wires in a straight polarity manner. The pulse actual shape is shown in Figure 6.

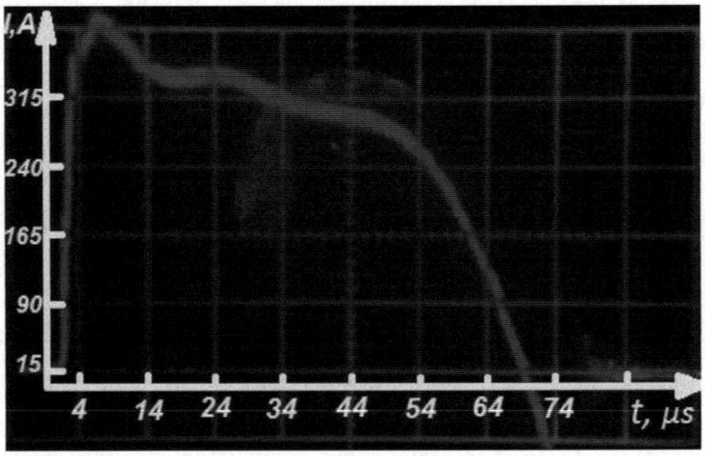

Figure 6. The actual shape of the pulse

The discharge of the artificial forming line to the arc space is a near trapezoidal pulse with a number of pulses at the top equal to the number of cells *n*.

Conclusion

A homogeneous artificial line is the most prospective for the development of a welding current pulse generator with energy storage.

According to the pulse shape estimated by the criterial parameter, the trapezoidal shape is optimal, and the edge duration is about 3 times greater than the pulse front duration.

Presence of the rapidly saturated choke coil in the welding circuit allows reducing both the rate of the pulse current rise at the initial stage and the rate of the decrease in the pulse edge current at the final stage. This fact has a beneficial effect on the welding process.

Active duration of the current pulse does not change significantly under the arc voltage disturbances. It is possible to adjust the welding current pulse frequency smoothly. The input of the minimum pulse energy required to transfer a droplet is provided.

References

[1] T. Cunha, A. Voigt, C. Bohorquez 2016 *Journal of Materials Processing Tech* **231** 449–455
[2] Brumm S, Burkner G *Conference MEFORM 2015*
[3] Zaitsev A, Knyaz'kov A 1969
[4] Knyaz'kov A, Knyaz'kov S, Lolyu Ya, Pronyaev A Patent No. RU 2294269
[5] Itskhohi Ya, Ovchinnikov N 1972 *Pulsed and digital devices* (Moscow:IzdSoviet radio)
[6] Knyaz'kov A, Knyaz'kov V 2010 *Welding International* **24(12)** 955–957
[7] Knyaz'kov A, Dementsev K, Knyaz'kov V 2013 *Welding International* **27(2)** 147–149

Technology for melting amber chips to produce a solid block

A S Vikhareva, A G Melnikov and O M Utyev
National Research Tomsk Polytechnic University, 30, Lenina ave., Tomsk, 634050, Russia

E-mail: any94_94@mail.ru

Abstract. This research is relevant, because the bulk of the mined amber comes in amber chips. Therefore, we have decided to review the current ways of melting amber chips to develop the most technologically efficient algorithm and to use it further for producing decorative items. The purpose of the work is to perfect the technology of obtaining whole-piece amber from amber chips and to explore the usability of the obtained material in decorative items and jewelry.

1. Introduction

The Roman author, Pliny the Elder, was the first scientist to prove that fossilized tree resin is amber. He noticed the resinous smell and smoky flame produced when amber was burning. Moreover, transparent amber often contained particles of plants and fossil insects. Later on, German scientists tried to challenge this conclusion. In the XVI century, for instance, Georgius Agricola claimed that amber consisted of liquid bituminous substance; it was formed in the interior of the Earth and hardened when out on its surface. There was also a hypothesis at the beginning of the XVIII century that amber originated from oil combined with mineral acids. Ancient Russians knew amber well and called it alatyr or latyr-stone, attributing miraculous properties to it. Mikhail Lomonosov gave more reasons in favor of the organic origin of amber in his works ‘First Principles of Mining Science’ (1742) and ‘A word on the formation of metals from earth tremors’ (1757).

Figure 1. Amber chips.

Today we generally accept that amber (Figure 1) is a mineral of organic origin. It is fossilized resin of conifer trees that grew 30 to 60 million years ago. The first stage of amber formation was an abundant discharge of conifer resin, which scientists associate with dramatic climate warming at that time. Resin is formed in the cells and it usually fills round cavities and elongated channels inside a plant. The second stage of amber formation involved resin deposition in forest soils. In dry, well-aerated soil, resin hardened and became more chemically stable when exposed to oxygen. The third stage of amber formation included the washout, transfer and deposition of fossilized tree resin into water basins. The transformation of resin into amber is facilitated by oxygen-containing potassium-enriched alkaline silt water. The water reacts with the resin, which results in the formation of succinic acid and its ethers. The final stages of this process involve the formation of not only amber but also glauconite – a mineral that always accompanies amber deposits. Most of the mined amber is small rocks of up to 3 cm in cross section; very occasionally, 4…7 kg pieces can occur [1].

There are several patented techniques for processing amber chips. The first patented technique [2] for creating decorative items with amber filler (up to 20 mm thick and produced from amber or amber scrap) involves the preparation of a molding compound with the following composition: filler (10…15 %), caustic magnesite (20…25 %), sodium chloride solution (20…25 %), quartz sand (the remaining percentage). The obtained mixture is poured into a mold of a required size and is placed on a shake table for 0.5…1.0 min. Then the mold is dried in a chamber at 13…30 °C for 1…3 days, depending on the thickness of the item. After hardening the compound is removed from the mold and polished with diamond abrasives and polishing materials. Another technique suggested in the patent [2] involves creating an amber filler for decorative items by filling the casting mold with a fluid amber compound with a polymer binder, accelerator and hardening agent. Low-quality amber of various grades as well as amber scrap serves as a filler without grinding. The filler is grouped by color and size during preparation. Then each group is treated with a gel coat, and the amber to polymer binder mass proportion is calculated for each group. After that, the molding compound laid into designated molds is prepared, and the compound with the following mass percentage is obtained: filler (40…70 %), gel coat (3…4 %), unsaturated polyester resins (55…25 %), hardening agent (1…0.5 %), and accelerator (1…0.5 %). The product is formed via manual or vibration treatment for 20…40 min. After hardening, the product is removed from the mold and dried in a drying chamber for 3…4 h at 50…70 °C. This technique has a drawback that limits its practical use – it is very labor intensive, including the retrieval of the product from the mold. It is hard to retrieve due to the high adhesive power of the compound. Moreover, the surface of the cast product has relatively low cleanliness. In order to facilitate the removing of the cast product from the mold, its inner surface is covered with a protective gel coat or other materials that will reduce the adhesion. The used pigmented resins (gel coats) are classified as harmful in both the production process and subsequent product use. The patent [3] describes the production of large shapes from fine grains of natural amber. This technique involves loading ~10 mm amber grains into airtight molds, heating them up to 140 °C with subsequent gradual increase in pressure from 0 to 100 atm, while maintaining the press force in the process. This technique allows obtaining products that preserve the physical, chemical and aesthetic properties of natural monolithic amber. In this case grains of natural amber of varying sizes are used as a raw material: large pieces of amber, amber pebbles and chips, flawed crafted items as well as previously pressed amber. The raw materials can vary in terms of transparency and sizes. The raw material heated to 110…130 °C is loaded into the mold block pre-heated to 160…230 °C. The temperature of the amber and that of the mold determine the color of the final product which can be anywhere between that of the natural amber and dark cognac color. After reaching the operating temperature, the mold is placed into the hydraulic press. The punch die is forced down to the stop by the press in 30…40 s. After sealing the mold and breaking down of the raw material, the volume of the shape-forming mold block is maintained by locking the die punch with a retention mechanism. The pressure in the heated amber breaks intermolecular bonds, turning it into a heterogeneous mass. When the heated amber reaches the temperature of the mold, it expands and fills the entire volume of the part-forming mold cavity. The baking process occurs due to the inner reaction of the raw material confined within the part-forming cavity. Natural cooling is preferable to refrigeration, because the latter may cause the product embrittlement.

2. Results and Discussion

The goal of this work is to use the patent data as well as strengths and weaknesses of each method to perfect the technology of obtaining whole-piece amber from fine grains, to research the properties and prospective use of the material obtained for decorative items. Two types of amber chips – polished and unpolished – of various grain sizes are used as the raw material. This technique involves loading 1.7 g of amber chips into the mold. Then the mold is heated to 140…180 °C for 1 to 2 h. After that, the mold is exposed to the pressure of 31.8 MPa or 63.7 MPa, depending on the experience, using a mechanical screw press, while controlling the press force with a dynamometer. Then the product cools under natural conditions. After the cooling completion, the samples are retrieved from the mold to measure their diameter, mass and height. Then the cross-section area, volume and material density are calculated. Normally, the density of natural amber is between 1.05…1.09 g/cm^3. Three-point bending tests are conducted on the GP 30 kN DLC test machine. See the Table 1 below for the research data:

Table 1. The dependence of density on temperature at 31.8 MPa.

	Dist. temp.	Pressure MPa	Heatin g time	Bending strength, MPa	Density, g/cm^3
Polished	140	31.8	1 hour	5.90	1.03
			2 hours	4.32	1.02
	160	31.8	1 hour	5.16	1.08
			2 hours	5.08	1.05
	180	31.8	1 hour	9.99	1.08
			2 hours	9.19	1.06
Unpolished	140	31.8	1 hour	2.11	1.01
			2 hours	2.47	1.03
	160	31.8	1 hour	3.26	1.07
			2 hours	2.17	1.04
	180	31.8	1 hour	6.73	1.08
			2 hours	6.12	1.05

Table 2. The dependence of density on temperature at 63.7 MPa

	Dist. temp.	Pressure MPa	Heating time	Bending strength, MPa	Density, g/cm³
Polished	140	63.7	1 hour	3.72	1.04
			2 hours	3.93	1.03
	160	63.7	1 hour	4.71	1.04
			2 hours	6.09	1.06
	180	63.7	1 hour	7.73	1.06
			2 hours	6.82	1.06
Unpolished	140	63.7	1 hour	0.85	1.00
			2 hours	1.32	0.99
	160	63.7	1 hour	3.05	1.04
			2 hours	3.23	1.06
	180	63.7	1 hour	4.76	1.05
			2 hours	5.08	1.06

Figure 2 shows dependence of density on heating temperature of the samples exposed to deformation stress of 63.7 MPa for 1 h and 2 h.

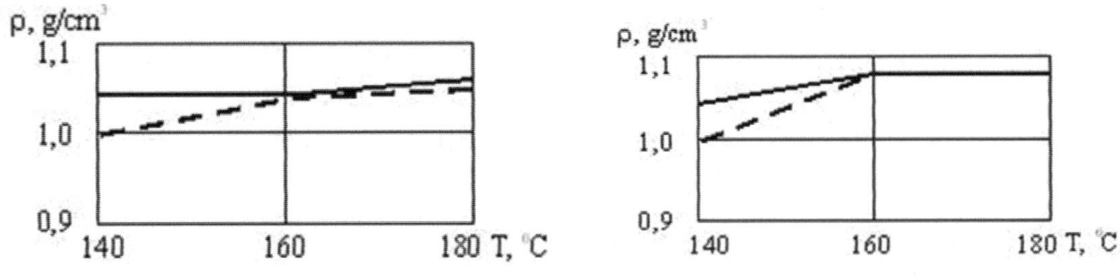

Figure 2. Dependence of amber density on the distortion temperature: heating time – 1 h (a), 2 h (b)

It follows from the diagrams that the density of samples increases with rising temperature. The heat distortion temperature markedly affects the color of amber. With rising heating temperature, the color of the samples changed from natural to dark cognac (Figure 3a, b).

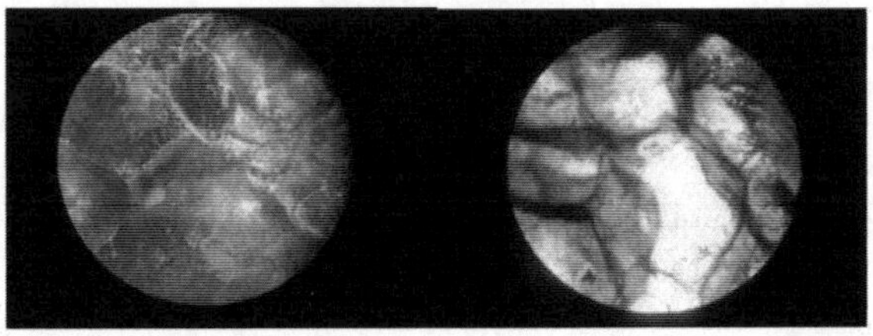

a b

Figure 3. A visual appearance of samples produced at the temperatures of 140 °C (a) and 180 °C (b) from unpolished amber.

The variety of jewelry made of amber is very diverse. Most often, amber serves as a filler in jewelry. For a decorative item, we chose a composition of a cast brass mount with an inserted piece of amber. The brass substrate was produced by means of investment casting [4]. The item was designed in the CorelDRAW graphics editor (Figure 4), and then the model was laser cut from a 4 mm acrylic sheet by means of the Speedy 100 fiber Laser Engraver. We then obtained wax models using a rubber die mold, connecting them with a gating system, formed them, poured the metal and obtained casts, which were then mechanically processed for marketable appearance. We chose unpolished amber for the insert, which makes it possible to obtain a wider variety of colors than with the polished one (Figure 5).

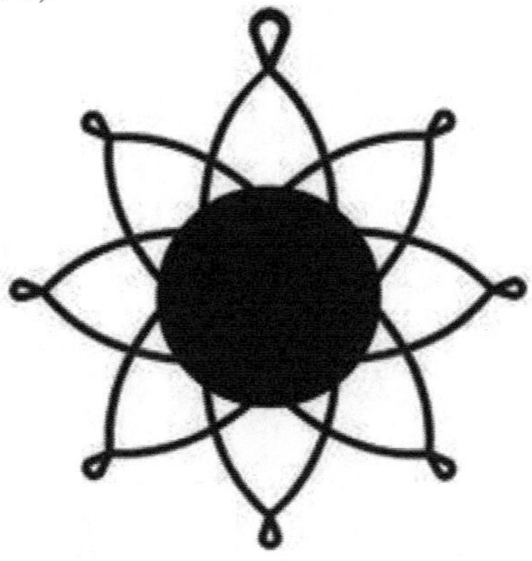

Figure 4. A brass mount design.

Figure 5. A decorative item

3. Conclusion

We have developed a technology for melting amber chip to produce a solid block with the same density as that of natural amber. Experiments with various pressure have shown that the density of amber increases with the rising heat distortion temperature, but the color of the sample becomes darker. We have also produced a decorative item with an amber insert.

References

[1] Nilova D 2010 Remedial resins: Shilajit, propolis, amber (Moscow: Krylov Publishing Company) p 6

[2] Shubnikova A A et al 2012 Method to make decorative items with filler from amber and/or wastes of its processing. Patent No. RU 2452715
[3] Chernohatsky O N et al 2002 Method to produce amber items. Patent No. RU 2191704
[4] Tuzovskaya A V 2013 *Proceedings of the XIX International Scientific and Practical Conference of Students, Post-Graduates and Young Researchers* **3** 363–364

Radiation-induced deposition of transparent conductive tin oxide coatings

S Umnov, O Asainov and V Temenkov

National Research Tomsk Polytechnic University, 30, Lenin ave., Tomsk, 634050, Russia

E-mail: usp@tpu.ru

Abstract. The study of tin oxide films is stimulated by the search for an alternative replacement of indium-tin oxide (ITO) films used as transparent conductors, oxidation catalysts, material gas sensors, etc. This work was aimed at studying the influence of argon ions irradiation on optical and electrical characteristics of tin oxide films. Thin films of tin oxide (without dopants) were deposited on glass substrates at room temperature using reactive magnetron sputtering. After deposition, the films were irradiated with an argon ion beam. The current density of the beam was (were) 2.5 mA / cm^2, and the particles energy was 300-400 eV. The change of the optical and electrical properties of the films depending on the irradiation time was studied. Films optical properties were investigated by photometry in the range of 300-1100 nm. Films structural properties were studied using X-ray diffraction. The diffractometric research showed that the films, deposited on a substrate, had a crystal structure, and after argon ions irradiation they became quasi-crystalline (amorphous). It has been found that the transmission increases proportionally with the irradiation time, however the sheet resistance increases disproportionally. Tin oxide films (thickness ~ 30 nm) with ~ 100% transmittance and sheet resistance of ~ 100 kOhm/sq. were obtained. The study has proved to be prospective in the use of ion beams to improve the properties of transparent conducting oxides.

1. Introduction

The transparent conductive oxide (TCO) films are widely used in various industries. These optically transparent electrodes are intended for displays, solar panels, photoelectric devices, touch panels, etc. Currently, one of the most widely used TCO substances is indium tin oxide (ITO) because it has high electrical conductivity and optical transparency.

Recently, because of the high cost of indium,the research has been conducted to find an alternative replacement for ITO. In particular, this is a compound based on tin oxide (dopant-Sb, F, As, Nb, Ta), zinc (dopant- Sn, Ge, Mo, F, Ti, Zr, Hf, Nb, W, Te), and other elements. Although, SnO_2 and ZnO impurities increase their conductivity, but it does not reach the values obtained for the ITO films. However, it is not necessary in some practical applications (very large conductivity). So, in paper [1] pure oxide films (SnO_2) prepared by electron beam evaporation had transmittance of 90% and a surface resistance of 200 kOhm /sq. (substrate temperature is 100°C). Originally, pure SnO_2 films are the conductors if they are deposited in a shortage of oxygen, so they are non-stoichiometric.

For deposition of thin TCO films various methods are applied, including thermal, electron- beam, laser vaporization, cathode, a high-frequency or magnetron sputtering and others. Wesuggest focusing on the film deposition by magnetron sputtering with additional useof the ion beam.The use of ion beams in the film coatings deposition leads to almost all films properties change [2-3].

Papers [4-8] present data showing that the use of an assisting ion leads to the changes in the optical and electrical properties of the deposited ITO thin films. It means that ion beams allow achieving the optimum relationship of the metal and the oxygen components in the film. So, the transmission of the films treated in that wayis about 90%, and the surface resistance of the best samples is 20 Ohms/sq. According to the authors, this is due to such factors as changes in the mobility of ad atoms and crystal lattice damage.

In their works, the authors used different types of ion sources that had been used under different deposition conditions (the ion beam parameters, the temperature of the substrate). In the papers mentioned above,assisting ion beams were used in the deposition of the doped TCO oxides.

Whereas, the energy formation of oxygen vacancies and tin interstitials in SnO_2 is very low and, thus, these defects form readily, explaining often observed high conductivity of pure, but non-stoichiometric, SnO_2 [9], so the ion beams could be a good tool to stimulate this process.

In this paper, we have studied the influence of the irradiation of the pure tin oxide (without dopant) films argon ions after deposition ontheir the optical and electrical properties.

2. Experimental details

Tin oxide films with thickness of about 30 nm were deposited by reactive magnetron sputtering of the target made of tin (99.5 wt.%) on the glass substrates at room temperature. Before the film deposition the glass substrates were purified by a chalk suspension in isopropyl alcohol.

The substrates were placed on a carrier that was able to carry out a reciprocating motion under the magnetron and the ion source in a vacuum chamber. The speed of the carrier movement was 3.62 cm / second. The working gas was a mixture of oxygen and argon (partial pressure ratio is O_2 – 60%, Ar – 40%). The pressure in the vacuum chamber during sputtering was 0.3 Pa. The gas flow was controlled by an electronic meter and gas pressure was measured by an ionization and thermocouple gauge. After the deposition, tin oxide films were irradiated with argon ions.

The ions were generated by the ion source with a closed electron drift of 2800 V voltage and a 0.28A discharge current. The greatest number of argon ions in the beam hasthe energy of 400-600 eV. The chamber pressure in such case was 0.166 Pa. For measuring surface electrical resistance Rsq. the samples were deposited with aluminum strips, which provided a reliable contact with the film. The thickness of the deposited films was measured using a profilometer'Micro Measure 3D Station'. To measure the transmission spectrum in the range of 300-1100 nm the spectrophotometer 'SF-256'was used. The structure and morphology of the films werestudied using the X-ray diffractometer'XRD6000' and an atomic-force microscope 'NT-MDT'.

3. Results and discussion

All freshly prepared tin oxide films were transparent dielectrics. This is due to deposition by reactive magnetron sputtering that was performed in an atmosphere with the excess oxygen content. Under these conditions, there was enough oxygen to have all the metal oxidised to a state of complete oxide, so all the oxygen vacancies, creating conductivity, are busy and a SnO_2 semiconductor film is deposited as a dielectric. After the deposition of the tin oxide films, the samples were irradiated with argon ions. Figure 1 shows the results of measurement of the films transmittance of both non-irradiated and irradiated samples. As it is seen from the figure the transmission is strongly dependent on exposure time.

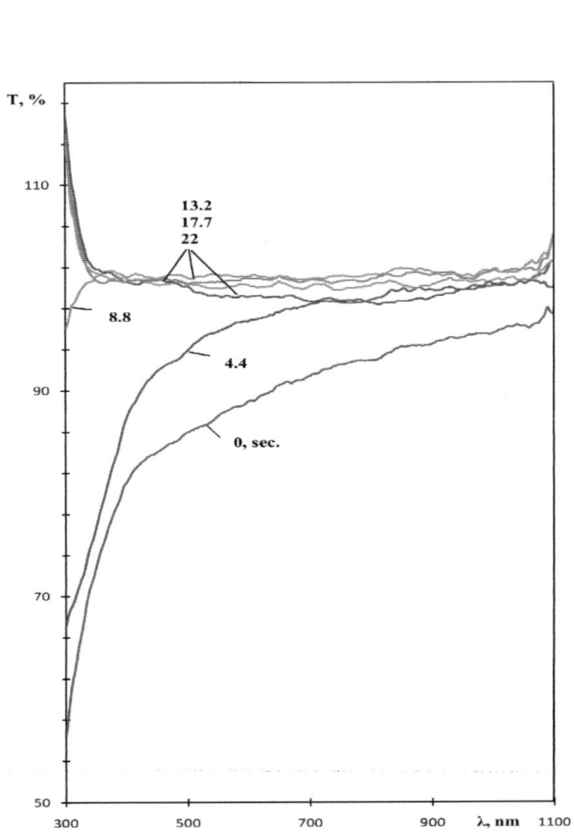

Figure 1.Dependence of the transmissionon the irradiation time

So, when the exposure time is less than ~ 4.5 sec. the transmission slightly improves. For longer times (≥ 8,8 sec.) the transmission is increased to 100% in almost the entire range of the wavelengths. It should be mentioned that when the irradiation time is 13 - 22 seconds, there is an increase in transmittance under wavelengths less than 360 nm and a slight increase in the infrared region. The increase in transmission, especially in the short-wavelengths region, of the irradiated films is related to the change of the plasma frequency [10-12]. This change is a consequence of increasing the electrons concentration.

However, the nature of the transmission is defined by the frequency of the incident light wave and the plasma frequency.

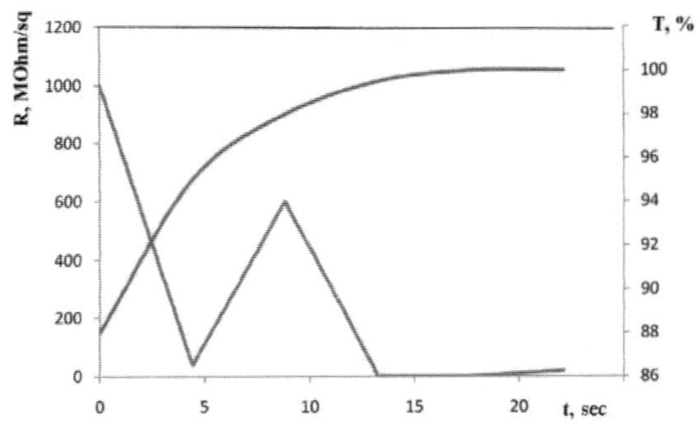

Figure 2. Dependence of the transmission on the irradiation time

Figure 2 shows the results of the surface resistance measurements, an average transmittance value at various irradiation times. The figure shows that the film surface resistance is disproportionately

IOP Conf. Series: Materials Science and Engineering **124** (2016) 012148 doi:10.1088/1757-899X/124/1/012148

dependent on the exposure time. With the exposure time increase, the resistance initially decreases and then increases again. The film transmittance almost uniformly increases and quickly reaches 100%.

As the deposition was carried out under the excess oxygen content, the films were deposited in the SnO_2 stoichiometric form which is a good insulator.

During the Ar ions irradiation there is their collision with the film surface and the power transfer in the local area to the neighbouring atoms of the crystal lattice, imbedded and adsorbed on the grains boundary. As a result, there are physical and chemical processes that lead to break or reduction of chemical bonds, causing desorption or chemical reaction that alters the structure of the coating.

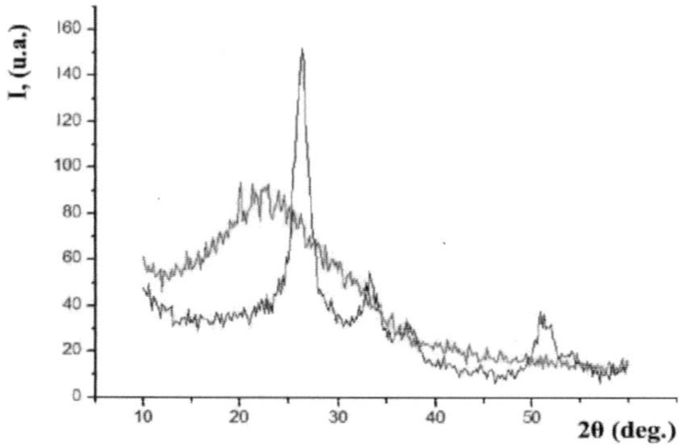

Figure 3.XRD diagram samples of SnO_2

The concentration of the increasing number of electrons is connected with both the break of the Sn-O chemical bonds and desorption of oxygen, which leads to the formation of non-stoichiometric tin oxide. Thus, the film exposure leads to an increase in the number of oxygen vacancies, which, in its turn, leads to a reduction in the films electric resistance. However, a further irradiation time increase leads to the film resistance increase. This is, possibly, connected with the formation of Sn self-point defects [11-12]. Whereas, the point defects interact with all conductive electrons, so the electrons, located near the defects, are 'bound' and do not participate in conductivity. As a result, the film resistance increases.

Figure 3 shows the results of the study of the deposited films structure. The figures show that the deposited tin oxide films have a crystal structure (tetragonal) with the grain size bottom of ~ 9.8 nm. After argon ions irradiation the structure becomes quasi-crystalline (amorphous - a red curve). This is due to the opening of the crystal lattice chemical bonds and forming new ones, which lead to formation of a quasi-crystalline structure.

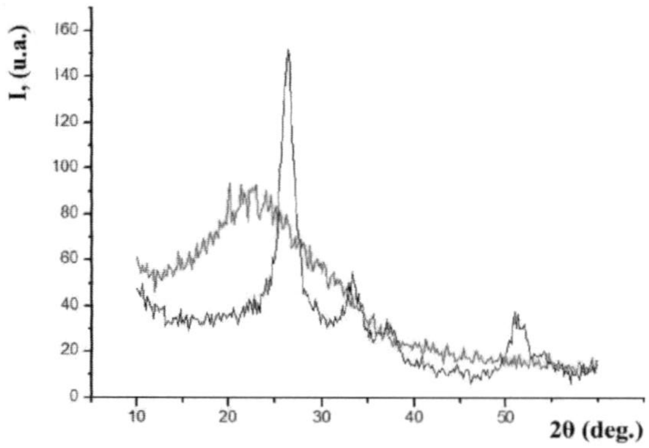

Figure 3.XRD diagram samples of SnO_2

IOP Conf. Series: Materials Science and Engineering **124** (2016) 012148 doi:10.1088/1757-899X/124/1/012148

4. Conclusions

Pure SnO_2 films were deposited at room temperature by reactive magnetron sputtering, followed by Ar ions beam irradiation with energy of 400-600 eV. It allows obtaining the films with 100% of transmittance and satisfactory surface resistance of 100 kOhm/sq. The surface resistance value can be reduced by optimizing the deposition conditions and ion beam parameters. Thereby, transmission increases proportionally with the irradiation time increase, while the surface resistance disproportionately. The crystal structure of the SnO_2 film after ion irradiation is transformed into a quasi-crystalline one.

References

[1] Shamala KS, Murthy L C and NarasimhaRao K 2004 *Bull. Mater. Sci.* **27** 295
[2] McNeil J R, Barron A C, Wilson A Ret al 1984 *Applied Optics***23** 552
[3] McNeil J R, Al-Jumaily G A, Jungling K C et al 1985 *Applied Optics***24** 486
[4] Meng L-J,Gao J, dos Santos M P, Wang X et al 2008*Thin Solid Films***516**1365
[5] Yang Y, Huang Q, Metz A W et al2004*Advanced Materials***16**321
[6] LeeS H,ChoH C, KimH J et al2012*Mol. Cryst. Liq.Cryst.***564** 185
[7] Bae J W,Kim J S, Yeom G Y 2001 *Nuclear Instruments and Methods in Physics Research B***178** 311
[8] Gilo M, Dahan R and Croitoru N 1999 *Opt. Eng.***38(6)** 953
[9] Kılıç C and Zunger A 2002 *Phys. Rev. Lett.***88** 095501
[10] Ginley D SandBright C 2000 *MRS Bull.***25**15
[11] Umnov S and Asainov O 2015 *IOP Conf. Ser.: Mater.Sci. Engineering***81**012001
[12] Asainov O and Umnov S 2012 *7th International Forum on Strategic Technology IFOST 2012*6357554

Principles of construction of ultrasonic tomographs for solution of problems of nondestructive testing in mechanical engineering

I Bolotina[1], A Bulavinov[2], R Pinchuk[2] and Y Salchak[3]
[1] Industrial and Medical Electronics Department, Tomsk Polytechnic University, 30, Lenina ave., Tomsk, 634050, Russia
[2] GmbH 'I-Deal Technologies', 2, Science Pk, Saarbruecken, 66123, Germany
[3] International Scientific Educational Laboratory of Nondestructive Testing, Tomsk Polytechnic University, 30, Lenina ave., Tomsk, 634050, Russia

E-mail: bolotina@tpu.ru

Abstract. The paper considers the problems of ultrasonic nondestructive testing of products intended for mechanical engineering. The functional and electronic circuits of an ultrasonic tomograph are presented. The function of signal radiation from the clocked multielement apparatus is described, the cross-functional flowchart of the prototype of a US tomograph is considered. The development trends of ultrasonic tomography for near-term outlook are demonstrated.

1. Introduction

The risk assessment of the failure of the loaded systems in mechanical engineering is based on precise data on the loading, parameters of the material and possible locations of defects, resulting from manufacturing or operation. Manufacturing of equipment, elements, articles and pieces of equipment, the damage of which is of high risk to people and environment, or can lead to considerable financial losses, should undergo inspection of the presence of possible defects. Therefore, an ultrasonic pulse-echo inspection (USI) has become obligatory for welding joints experiencing excessive pressure, for structural materials used in aircraft industry and etc. The USI advantage is that this method possesses perfect identifiability of flat defects, such as incomplete fusion (welding defect), delamination and cracks in the production volume or in the hard-to-rich areas. Other reasons in favour of USI application are a relatively simple configuration, a possibility of control automation, absence of environmental requirements.

Nevertheless, USI results still present only qualitative information on the presence of the reflectors, the analysis of which was conducted by comparing it with data of the typical reflectors, such as notches, lateral apertures or punts. Thus, standard inspection procedures do not allow quantitative evaluation of the results and identification of the type and parameters of defect.

Consequently, there is an actual necessity of development of an ultrasonic method and monitoring equipment with high resolution and contrast sensitivity to visualize reflectors, which will allow proper evaluation of the type and geometry of defect, and will meet the requirements of the regulatory documents in force and market requirements.

A new technology of US inspection based on application of antenna arrays differs from a traditional one by four features: a way of viewing of space inside the object under inspection and the way of reflection of the inner structure of the material of this object. One more difference is presentation of the inspection results to the operator in the form of images of the inner structure of the material of the object under control. Three-dimensional images, besides the convenience of perception by the operator of the inspection result, which is completely adequate to the actual location of boundaries and discontinuities of the material, allow in many cases measuring actual rather than equivalent sizes of discontinuities.

2. A functional circuit of a US tomograph

All known tomographs (technical and medical ones) are virtually constructed according to a single-type functional diagram (Figure 1) and represent a multichannel flaw detector having M independent channels. In each of the channels there is a programmable delay line and a separate probe pulse generator. In a receiving part of each channel there is a preamplifier, an analog-to-digital converter and a digital programmable delay line. The outputs of all reception channels are combined in the summator, at the output of which there is a signal corresponding to the present angle of incident (in digital form).

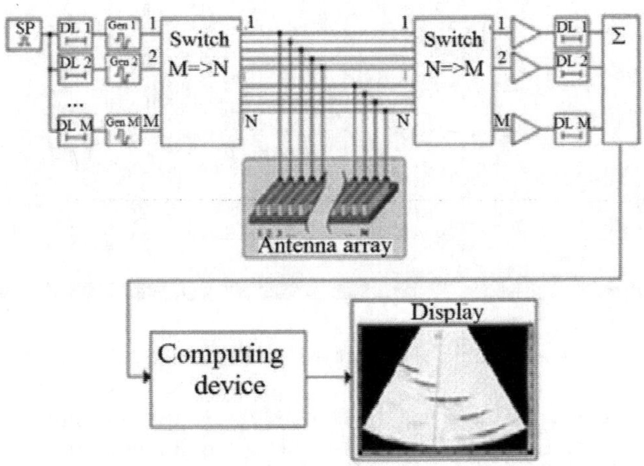

Figure 1. A functional circuit of an ultrasonic tomography.

As the number of array elements generally exceeds the number of tomograph's channels (N>M), the generators' outputs and the receivers' inputs are connected to the elements of the antenna array (AA) through multichannel switches intended for connecting of the required group of antenna elements to the apparatus channels. All electronic circuitries are usually located in the apparatus' case, but the array is connected to it by a multiconductor cable.

The summator output is connected to a computing device, which provides transformation of coordinate systems (from polar into Cartesian) to adequately represent a cross-section image on the display screen. Besides, it implements an operational control over the delay lines and other units of the tomograph.

The most important part of the tomograph is an antenna array (AA). Its properties in many respects define the quality of the cross-section under reconstruction, expressed in terms of noise levels, the extension of the dead space, spurious suppression and others. The development of the construction and adjustment of the manufacturing technique of these devices are a time consuming and expensive procedure demanding special equipment, materials, design decisions and deep understanding of physics of formation and spreading of ultrasonic (US) oscillations.

The main parameters of the array are the mode of the formed oscillations (longitudinal, transversal or multimode ones), frequency characteristics (band frequency, transmission bandwidth, pulse shape) and design parameters (the number and the size of the elements, the width and the length of the antenna aperture) [1].

In order to create a multichannel system it is possible to specify three basic functions performed by it: radiation of a US signal, its reception and image reconstruction (Figure 2).

The first functional group (radiation and reception) is realised by means of digital ultrasonic electronics. Standard components, such as a delay of the signal radiation, preamplification, an analog-digital converter (ADC) are preliminary distinguished as separate functional modules. The system, performing the necessary functions, such as analog filtration, a module of averaging and an interface for communication with a computer, can serve as a prototype of the system for US tomography. Different types of reconstruction, image and signal processing can be additionally realised by means of the software.

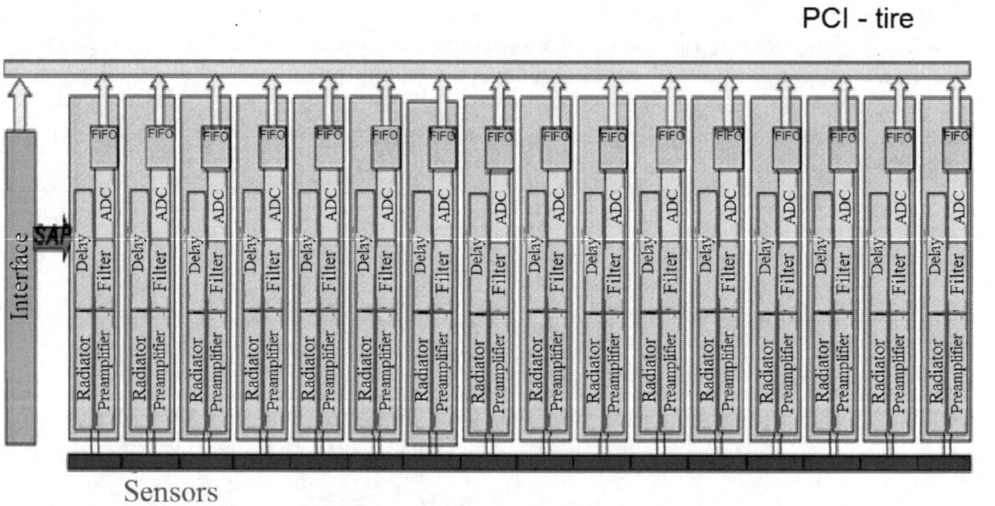

Figure 2. An electronic circuit of the system of US tomography.

3. The function of the signal radiation from the clocked multielement apparatus

From the standpoint of electronics, the function of signal radiation is not connected with the function of reception. The information about the object can be received in an arbitrary sequence, then it is summed and represented in the form of a single ultrasonic signal. In this case the radiating elements of the multielement apparatus are adjusted so that at a definite point of time only one radiator is active, while the rest operate in the reception mode.

Such approach has a number of diagnostic advantages, such as, for example, an opportunity to vary the aperture of each element, thus influencing the resolution capability and allowing adjustment of the sensitivity of the whole system. In case of a smaller aperture of the unit element of the apparatus, it is possible to obtain a higher resolution. In classic multielement apparatuses, the directional radiation pattern of the US signal is determined by means of corresponding time shifts. A wide directional pattern of one of the array elements can be actually useful for obtaining wide radiation angles. If the distance between the elements is less than the length of a half-wave, the directional pattern will not have any peaks, which are necessary for focusing in a definite control area. Therefore, in practice the work is conducted using limited rotation angles of the radiators, which is achieved by applying the radiators with a narrow aperture (Figure 3) [2].

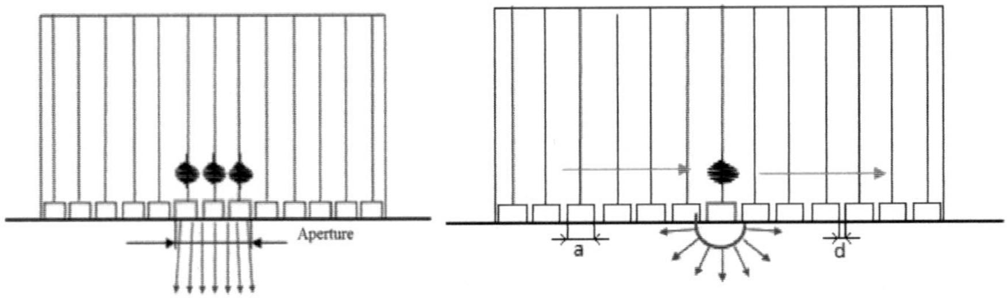

a b

Figure 3. A variation of the radiation aperture: *a* – the front of acoustic radiation from the element having a wide aperture; *b* – focused acoustic radiation from the elements with a narrow aperture.

In case of clocked faceted arrays, it is possible to vary both the size of the element (a) and the distance between the elements (d) (Figure 3a).

If *d* and *a* have large values, it is possible to conduct the analysis faster and with higher sensitivity, but with lower resolution capability. Conversely, if *d* and *a* have low values, it is possible to obtain an improved resolution capability. Due to such flexibility it is possible to realise different analysis strategies, for example, quick scanning of a poor quality, and then slow examination of the dangerous areas using high resolution.

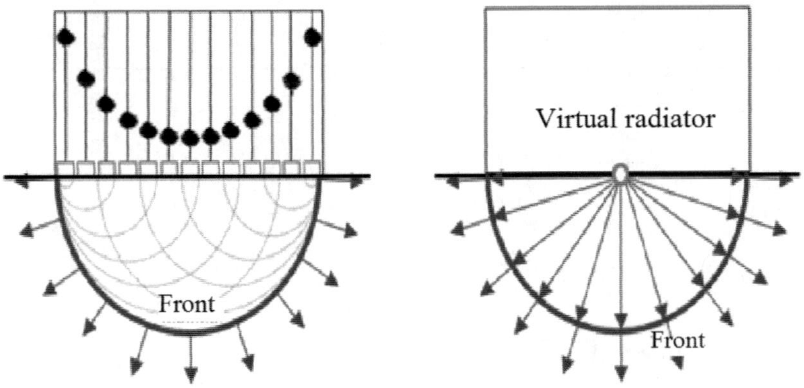

a b

Figure 4. Defocused radiation: *a* – created by means of the group of radiators; *b* – created by means of a unit radiator of high power.

A defocused impulse provides still greater opportunities (Figure 4) [3]. All elements of the array operate in the radiator mode in the same time. It is necessary to select such distance between the elements that would allow the wave front to assume the shape of the sphere surface (Figure 4a). In this case, this group can be considered as a unit radiator of higher power (Figure 4b).

Let us formulate the following advantages of such approach:

- High speed of control. Per one time, it is possible to radiate in all directions;
- The uniform energy distribution over the half space and, thus, uniform sensitivity is obtained independently of the radiation angle;
- A higher sensitivity in comparison to the alternating impulse radiation by the elements.

A part of the processing of the received signal occurs directly in each of the channels of the digital electronic device. It includes amplification of the received signal, analog filtration, analog-digital conversion, averaging and adjustment of the depth. It is necessary to digitize the received ultrasonic signal using corresponding frequency.

Figure 5 represents a cross-functional flowchart of the multielement system. As it was mentioned above, an acoustic beam is controlled independently when sending and receiving a signal. The pulse delay in each of the channels has the accuracy of about 1 ns. Source [4] considers the angle resolution in the following way:

$$\Delta\theta = \frac{\Delta\tau_{\min}[ns]\cdot N\cdot f[MHz]}{12000\cdot\cos\theta} \tag{1}$$

When $\Delta\tau = 1$ ns, a number of elements are $N = 16$, and when the frequency of the analysis is $f = 5$ MHz, the angle amounts to $\Delta\theta = 0.4^\circ$. Hence, electronics meets one of the essential requirements for further reconstruction of a signal and an image.

A receiving electronic module is equipped with ADC, having the resolution beginning from 10 bits, taking into consideration a sign bit (the amplitude value is from –512 to +512). This corresponds to a dynamic area of 54 dB. Under a low ratio between the signal and the noise in separate channels and under digital stacking of the signals, this is an important parameter, influencing the image quality. Thus, the selected electronics provides all conditions for signal and image processing, conducted by the module of reconstruction (Figure 5).

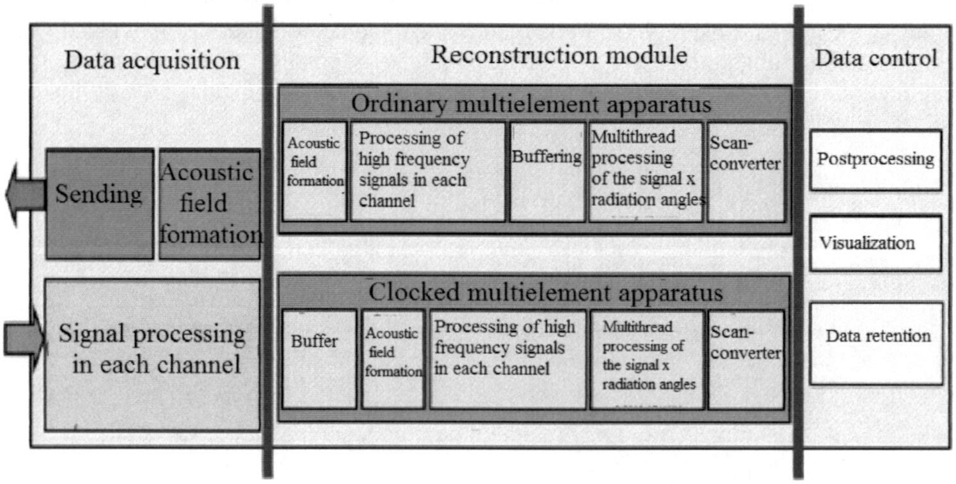

Figure 5. A cross-functional flowchart of the prototype.

In both cases the following functional groups – signal processing algorithms – will be applied:

a) Formation of the acoustic field.

Let us imply a reception as a coherent addition of the received unit signals of the array element, having a corresponding time delay for each radiation angle and focal depth. Unit ultrasonic signals can be projected on the temporal axis; at that, a synthetic focus is achieved in the summary, physically realizable near field. The result of beam formation, in case of using an ordinary multielement apparatus, will be a high-frequency signal for a given radiation angle and, correspondingly, a focal depth. In case of the clocked multielement apparatus, it is necessary to carry out several times of measurements. When applying the clocked multielement apparatus, it is possible to calculate several US signals in the specified angular range using the same data. In both cases (standard faceted arrays or clocked faceted arrays), one or several high frequency US signals are obtained.

b) Processing of high frequency signals of the multielement apparatus for each radiation angle.

The signals, obtained by coherent summation, can be subsequently subjected to processing. First of all, it is a question of filtration. For this purpose it is possible to apply many different methods, for example, Fourier filtration and others. In case of a poor signal/noise ratio, a signal can be averaged. Further, it is possible to logarithmically compact the dynamic field and to establish corresponding thresholds of registration.

c) Multiple-line processing of a signal for a specified angular domain.

It implies processing of several unit signals after coherent summation for a definite radiation angle:

- Automatic adjustment of amplification as a function of the angle to compensate the loss of sensitivity at wide angles;
- Interpolation between two neighbouring signals;
- A comparative analysis of the signals by their properties (an amplitude, a structure, a phase term).

d) A scan-converter.

A scan-converter is a very important module of signal processing. It converts linear ultrasonic signals into two- and three-dimensional images and vice versa, and also calculates two-dimensional projections (B-scan, C-scan, D-scan, TD-scan) from a three-dimensional image. The basis for 3D reconstruction is the use of 2D matrix arrays or a bias of the linear array perpendicular to the direction of rotation.

e) Postprocessing.

Postprocessing is used for the analysis of the signal and images. This implies the following:

- Representation of echo-amplitudes by colour or brightness;
- Adaptive contrast enhancement;
- Automatic recognition of the geometry of the measuring object (automatic search of boundaries).

The above-mentioned functions operate similarly for standard and clocked faceted arrays, independently of the field of application, be it nondestructive testing or medicine [5, 6]. The functional separation is important from the position of the modular composition of the system equipped with the faceted array, where each separate function or functional group is determined by a separate module.

4. Conclusions

The development of ultrasonic tomography for near-term outlook should proceed in the following directions [7]:

- Improvement of the design procedures of residual life, taking into account the parameters of actual defects for resource management. A great number of industrially dangerous objects, subjected to inspection according to a new technology, offer a prospect of creation of a wide range of design procedures of residual life with the aim of managing the resource of these objects.
- Establishment of standards for evaluation of the actual defects for the objects, being in service. On the basis of lifelength of design procedures of residual life it is possible to establish operating standards of the object quality evaluation.
- Provision of defects measurement and residual life calculation in unified combination with the issue of inspection results in the form of operating life immediately after conducting the inspection.
- Creation of measuring equipment with high performance, serviceability and unification of pieces of equipment. It is necessary to develop of a great number of scanning devices, which would provide the inspection of the whole nomenclature of the objects.
- Integration of search and measuring modes into a unique entire measuring mode. Although, in this case, the amount of incoming data significantly increases, the inspection quality will jump sharply owing to improved reliability, especially because the modern machinery develops rapidly and does not create any obstacles for realisation of this mode.
- Automation of processing operations, analysis and documenting of the results. Automation will lower manning requirements and raise the inspection efficiency sharply.

Acknowledgments

The study was financially supported by the Ministry of Education and Science of the Russian Federation within the Federal Target Program entitled 'R&D in Priority Areas of Development of Russia's Scientific and Technological Complex for 2014–2020' (unique project ID: RFMEFI57514X0048).

References

[1] Alekhin S, Bishko A, Dureiko A, Zhukov A, Samokrutov A, Sokolov N and Shevaldykin V Available at: http://acsvs.ru/article/7article id=21

[2] Bulavinov A, Kröning M, Pinchuk R and Reddy K 2009 *Echtzeit-3D-Bildgebung mittels der getakteten Gruppenstrahlertechnik für Ultraschallprüfanwendungen* (Münster: DGZfP)

[3] Nihei K, Nakagawa S, Myer L and Majer E 2002 *Fracture Imaging with Elastic Waves* (Canada: 5th Internat. Workshop on the Applic. of Geophysics in Rock Engineer.) pp 20–24

[4] Schmitz V, Langenberg K, Chakhlov S 2003 *Berechnung hochfrequenter Signale für eine automatische Uitraschallprüfung und Transversalwelleneinschallung mit einem auf der Kirchhoffschen Näherung beruhenden Ray Tracing Verfahren* (Mainz: DGZfP)

[5] Bolotina I, Kryoning K, Kvasnikov K, Sednev D and Sumtsova O 2014 *Advanced Materials Research* **1040** 959–964

[6] Ploskov N, Danilov V, Zuev L, Zavodchikov A, Bolotina I and Orlova D 2012 *Journal of Applied Mechanics and Technical Physics* **53**(4) 611–615

[7] Vopilkin A 2009 *Control. Diagnostics* **4** 37–47

Method of Rotational Turning With Multifaceted Cutters

N Indakov, Y Gordeev and A Binchurov
Siberian Federal University, Polytechnic Institute, 79/10, Svobodniy ave., Krasnoyarsk, 660041, Russia

E-mail: mexanixs@mail.ru

Abstract. The proposed method combines the elements of skew turning (in which a cut layer moves along the cutting edge) and rotary turning (in which the sections of the cutting edge are constantly renewed). This leads to better cooling and, hence, a longer tool life. The required cutting speed is ensured by selection of the required number of revolutions of the cutter; the rotation of the machined shaft corresponds to the rotary feed.

1. Introduction

There are known traditional methods of turning: with vertex and peak less cutters, with rotary cutters with autorotation or forced turning. Each of these methods has its reasonable use. At the same time, during the turning of products from viscous and plastic materials, alloy and heat-resistant steels, the abovementioned methods reveal some drawbacks. Among them area relatively low resistance of the vertex and oblique cutters, the use of expensive coolants, formation of flow chip, limited stroke speed due to turning machines possibilities, and in some cases due to form of work pieces (long, non-rigid and unbalanced shafts) [1-4].

A more effective method of metalworking of such shafts is the method of face milling, in which the optimum stroke speed is provided by an independent actuating device of the tool. Besides, intermittent cutting provides a guaranteed chip control. In turn, intermittent cutting leads to the formation of kinematic waviness and the appearance of shocks and vibrations of the technological system, which significantly reduces the surface quality and eliminates the possibility of using such milling as finish. Moreover, during face milling there are heavy radial forces that lead to shaft deflection [5].

In terms of the mentioned shortcomings, the method of rotational turning with multifaceted cutters, developed at Krasnoyarsk Polytechnic Institute, is one of the most promising methods [6, 7], in which one of two rotational movements is reported to the instrument, and another to the work piece. In this case the rotation axis of the instrument is installed perpendicularly to the axis of a work piece rotation and is placed in the line with the axis of centers. The cutting edge of the cutter is multifaceted and consists of N cutting blades. The cutter makes a forced rotation around its axis, what leads to constant renewal of the cutting edge in the cutting zone.

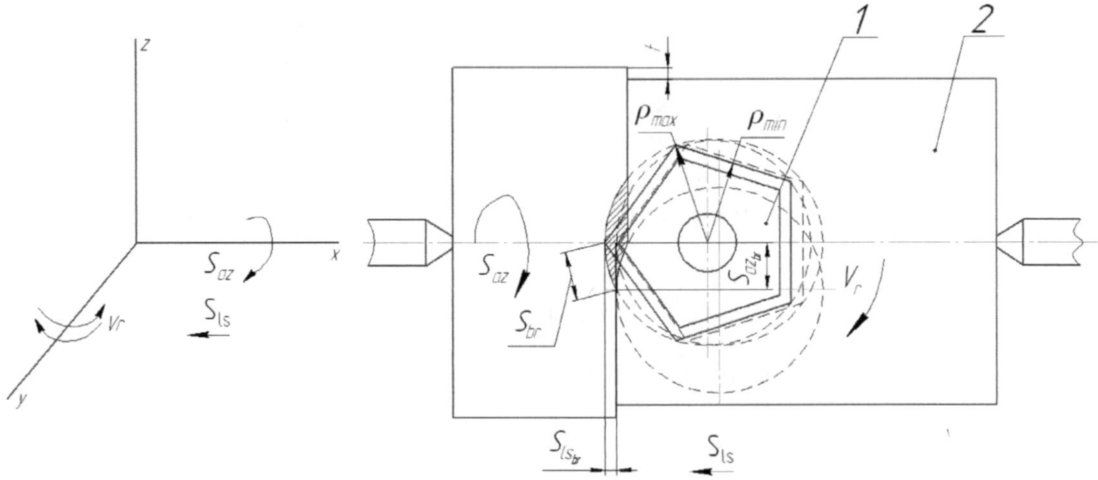

Figure 1. Kinematic scheme of machining with multifaceted rotational cutter.

In terms of novelty, the studied method of turning can be characterized as new, since it is based on new principles of formation of a cut-off material layer and setting of cutting modes.

The presented method enables a simultaneous production of the processed surface and chips in the form of commercial powder particles. This capability makes the method near-waste less.

In the proposed method the rotational cutter is a multifaceted bowl-shaped object with a tool bore and a cutting section consisting of N cutting blades. Each blade has a rear surface and a front surface, which together form a linear cutting edge [7]. The cutting edges formed at each blade in the plane, which is perpendicular to the cutter's axis of rotation, together constitute the cutting polygon [8].

Processing with a rotational cutter is made so that the main direction of the chip deformation occurs along the blade edge, thus the total feed to the cutter profile's face is determined by this formula [8, 9]:

$$S_{br} = \sqrt{S^2_{ls_{br}} + S^2_{az_{br}}}; \tag{1}$$

where: $S_{ls_{br}}$ is line feed to the face of the cutter, mm/rev; $S_{az_{br}}$ is circular feed to the face of the cutter, mm.

Line feed to the face of the cutter is determined by the following formula:

$$S_{ls_{br}} = \frac{S_{ls_{mt}} \cdot n_b}{n_r \cdot N}; \tag{2}$$

where: $S_{ls_{mt}}$ is line feed of the cutter to the turn of the processed shaft, [mm/rev]; n_r is a turning number of a work piece, [rev/min]; n_b is a turning number of the tool, [rev/min];N is a number of cutter profile faces.

The circular feed to the face of the cutter is calculated by the next formula:

$$S_{az_{br}} = \frac{\pi \cdot d_b \cdot n_b}{n_r \cdot N}; \tag{3}$$

where: d_b is the diameter of the blank, mm.

One of the most important characters of the presented method is that the constructive feed appears in the section of the profile with the increasing radius vector.

The constructive feed is determined by this formula:

$$S_c = (\rho_{max} - \rho_{min}) = (\rho_{me} + e) - (\rho_{me} - e) = 2e, \tag{4}$$

where: ρ_{max} is the maximum radius vector of the cutting edge contour, $[mm]$; ρ_{min} is the minimum radius vector of the cutting edge contour, $[mm]$;ρ_{me} is the mid-radius vector of the cutting edge contour, $[mm]$;e is the eccentricity of the tool profile, $[mm]$.

The formation of the treated surface involves three movements: principal movement V_r, linear movement or line feed S_{ls} and circular feed S_{az}.

The principal movement is tool rotation. Its turning speed must be at least by far greater than the turning speed of the work piece.

The cutting speed is found by the formula:

$$V_r = \frac{2\pi \cdot \rho \cdot n_r}{1000}; \tag{5}$$

where: ρis the radius vector of the cutting edge contour, [mm];

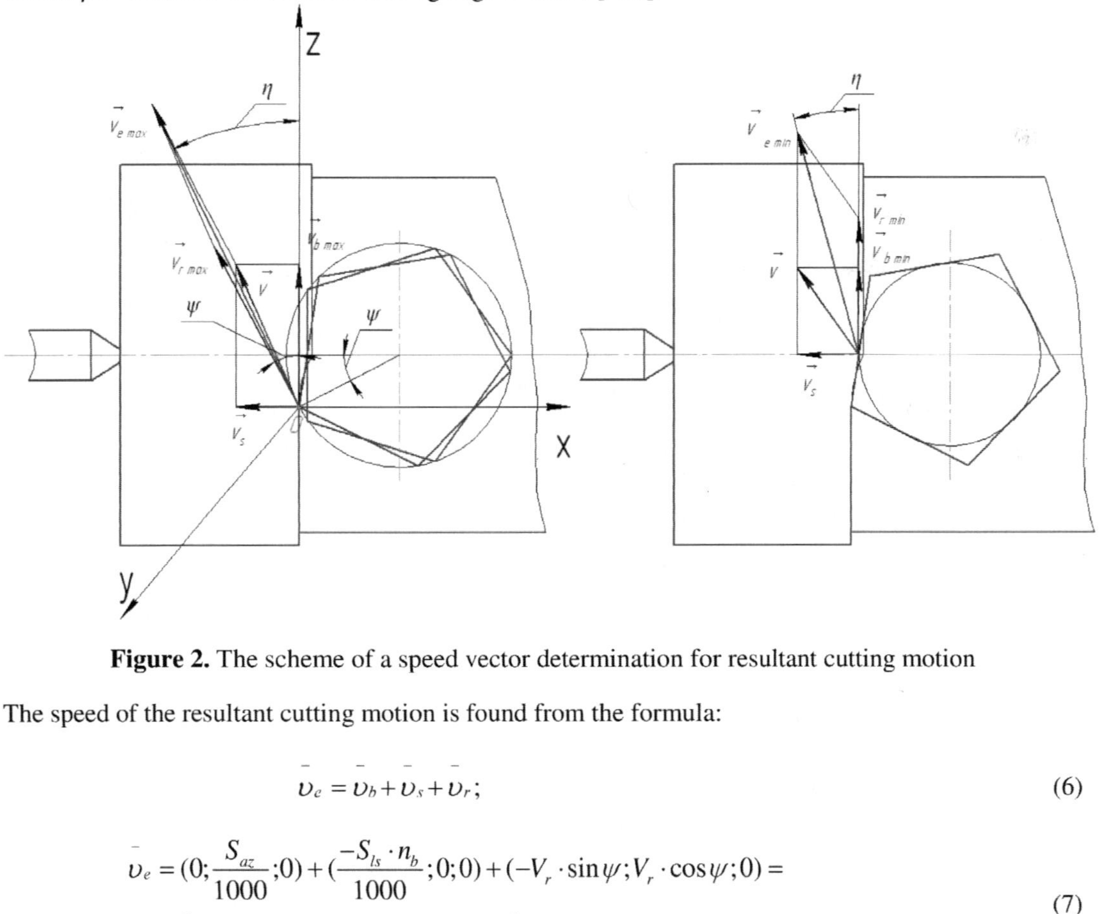

Figure 2. The scheme of a speed vector determination for resultant cutting motion

The speed of the resultant cutting motion is found from the formula:

$$\bar{\upsilon}_e = \bar{\upsilon}_b + \bar{\upsilon}_s + \bar{\upsilon}_r; \tag{6}$$

$$\begin{aligned} \bar{\upsilon}_e &= (0;\frac{S_{az}}{1000};0) + (\frac{-S_{ls} \cdot n_b}{1000};0;0) + (-V_r \cdot \sin\psi; V_r \cdot \cos\psi; 0) = \\ &= (\frac{-S_{ls} \cdot n_b}{1000} - V_r \cdot \sin\psi; V_r \cdot \cos\psi + \frac{S_{az}}{1000}; 0); \end{aligned} \tag{7}$$

$$\left|\bar{\upsilon}_e\right| = \sqrt{(\frac{-S_{ls} \cdot n_b}{1000} - V_r \cdot \sin\psi)^2 + (V_r \cdot \cos\psi + \frac{S_{az}}{1000})^2}; \tag{8}$$

where: $\bar{\upsilon}_b$ is a speed vector of the work piece circular feed, [m/min]; $\bar{\upsilon}_s$ is a speed vector of the tool line feed, [m/min]; $\bar{\upsilon}_r$ is a speed vector of the principal movement (movement of shaping), [m/min]; ψ is a slope angle between the speed vector of principal movement $\bar{\upsilon}_н$ and coordinate axis y.

Slope angle ψ is calculated by the following formula:

$$\sin\psi = \frac{S_{az}}{\rho_{\max}}; \tag{9}$$

The upward angle of the result ant motion's screw trajectory in M point is determined by the formula:

$$\cos\eta = \frac{V_r \cdot \cos\psi + (S_{az}/1000)}{\left|\bar{\upsilon}_e\right|}; \tag{10}$$

The experimental research was carried out on the turning lathe of 16K20 model with the use of the tool drive consisting of an electric spindle, a supporting arm, a mandrel and specially refinished (typical multifaceted hard-metal) plates, modified by ceramics nanoparticles. The final physical and mechanical properties of hard-metal composites, modified by nanoparticles, depend directly on their microstructural parameters, the proportion of volume ratios and the size of carbide and ultrafine particles. Nanoparticles, introduced into a binder, decrease the thickness of the metal interlayer and therefore ensure the strength increase of the binder and composite as a whole. The stable reduction of the carbide grain's average size was found out as well (due to a nanoparticles blocking influence on the recrystallization process). The test results show that modification by nanoparticles enables the improvement of the main properties: service durability – by 1.3…4 fold; wear resistance – by 1.6…2 fold; fracture toughness – by 1.8…2 fold; flexing strength – by 25…50 % fold [10].

Tests were carried on the basis of the aluminum samples with grades A6 and A0. The preliminary range of cutting modes was detected. The cutting depth varied in the range from 0.1 mm to 2 mm. The feed of pieces (circular feed) varied from 0.2 millimeter per revolution to 3 millimeter per revolution. The speed of the tool rotation changes in the range from 12000 to 18000 turns of the electric spindle by changing the frequency of the supply voltage through the inverter Altivar 31.The tool drive is installed in the tool-holder of the turning lathe.

The line feed values are determined by adjusting the feed gear-box of the working machine 16K20, and the circular feed values are determined by corresponding setting of the speed gear-box of the working machine.

The offered method of rotational turning with multifaceted cutters allows increasing the tool life and improving the quality of processing when turning the viscous and plastic materials, alloyed and heat-resistant steels, as well as providing the guaranteed crushing and removal of chips. In this case, there is no need to use toxic and expensive coolants.

Rotational turning with multifaceted cutters allows obtaining chips in the form of particles of small-sized space forms with complex geometry for powder metallurgy and composite materials from various constructional materials, including viscous, plastic metals (zinc, aluminum, copper).

2. Conclusion

Carrying out the corresponding researches allows revealing the optimal processing modes and providing practical recommendations on their choice for various conditions and materials. In general, the conducted researches have confirmed the prospectively of the proposed method.

Acknowledgements
This work was supported by the Russian Fund of Fundamental Investigations (grant RFFI № 14-08-00508).

References

[1] Kozhevnikov D V and Kirsanov S V 2006 *M: Mashinostroenie* 327
[2] Yashcheritsyn P I, Borisenko A V, Drivotin I G and Lebedev V Ya 1987 *(Minsk: Nauka i Tekhnika)* p. 229
[3] Konovalov E G 1961 *(Minsk: Mashinostroenie)* 286
[4] Bobrov V F and Ierusalimskii D E 1972 *(Moscow: Mashinostroenie)* p. 111
[5] Markhasin E L and Petrosyants A Ya 1960 *(Moscow: Mashgiz)* p. 109
[6] Ryzhov E V, Indakov N S, Petrovskiy E A and others 1984 Patent No. USSR 1126375
[7] Indakov N S, Gordeev Yu I and Binchurov A S 2012 Patent No. RU 2463130
[8] Indakov N S and Binchurov A S 2014 *Russian Engineering Research* **34(1)** 52-54
[9] Indakov N S and Binchurov A S 2014 *Russian Engineering Research* **34(2)** 79-82
[10] Gordeev Y I, Abkaryan A K, Binchurov A S and Jasinski V B 2014 *Advanced Materials Research* **1040** 13-18

New combined antifriction treatment of the guides of machines and mechanisms

S G Bishutin
Bryansk State Technical University, 7, 50 years of October st., Bryansk, 241035, Russia

E-mail: nad-bisch@yandex.ru

Abstract. The technology description of the combined antifriction treatment, including three stages, is presented. These stages are the following: abrasive rough-down for surface preparation to coating with antifriction material; surface rubbing with antifriction material; surface plastic deformation of the processed surface area. These stages are executed with one instrument. The article presents the technological modes of this treatment allowing the formation of wear resistant surface layers of guides of machines and mechanisms.

1. Introduction
The extensive friction surfaces of guides, applied in machines, elevators, conveyers and other handling equipment, are exposed to intensive wear. The effective means of longevity increase of such surfaces is forming wear-resistant coatings at antifriction treatment [1, 2]. However the existing technologies of antifriction treatment are executed, as a rule, stage-by-stage on different equipment, which makes them labour intensive and expensive. Nowadays the application of antifriction coating in the narrow extensive area of the cylindrical guide, which contacts the carriage rollers, presents well-known difficulties. In this connection a new technology of the combined antifriction treatment, including three basic stages performed simultaneously, has been developed: abrasive rough-down for preparation of the surface to application of antifriction material; surface rubbing with antifriction material; surface plastic deformation of the processed surface area with hard-tempered spheres.

2. Basic part
For realization of such technology an instrument (Figure 1) that can process an external cylindrical (part of an external cylindrical surface) or flat extensive surface is developed and manufactured. It consists of body 1 that is executed in the disk form with an opportunity of axial rotation from an external drive and has a conical surface, which vertex faces the workpiece. The following elements are body-mounted: processing element 2 in the form of a removable petalous polishing butt-end disk, having a mechanism of the axial throw in the form of springs 4 tightened with bolts 3; rubbing element 5 from antifriction material in the form of the ring, which consists of the ultimate amount of sectors, having the opportunity of self-centering on the processed surface due to deformation of rubber 6; pressing element 10 in the form of hard-tempered spheres from steel ШХ15.

The rubbing elements have a mechanism of the axial moving in the form of adjustment springs 8 tightened with bolts 7 tightening compound disk 9. Burnishing elements also have a mechanism for moving towards a workpiece in the form of adjustment springs 12 tightened with bolts 11, which allows changing the application force of the burnishing element to the processed surface of workpiece 13, thus creating a required hardened surface layer with durable adhesive bonds.

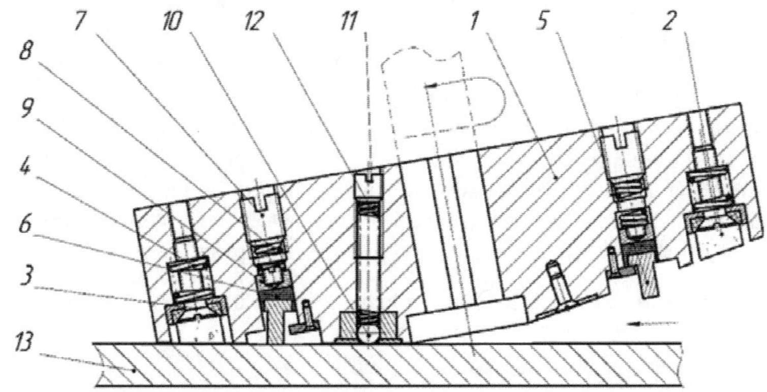

Figure 1. Scheme of the combined antifriction treatment

The working surface of the guide is formed as follows. In the spindle of the machine, for example a vertical milling one, one installs this instrument and brings it to the workpiece so that to provide the simultaneous contact of the workpiece surface with processing, rubbing and burnishing elements. It allows enhancing the quality the applied antifriction layer due to the friction heating of the rubbing element and a subsequent plastic deformation of the coating by burnishing elements. Further, by rotating screws 7 and 11 one provides necessary tightness of rubbing and burnishing elements in the direction of the processed surface. An external drive transmits an axial rotation to an instrument, whereupon one advances it at a tangent to the processed surface of workpiece (Figure 2).

Figure 2. Antifriction sample treatment on the vertical milling machine: 1 – instrument for antifriction treatment; 2 – cylindrical sample; 3 – table of the vertical milling machine

The processing element removes the necessary allowance, and the required surface roughness, which provides the required coating adhesion, is formed. Then, a rubbing element applies a thin antifriction layer, warming up the surface of the workpiece. In future this layer is exposed to surface plastic hardening by means of the burnishing element, which enhances its quality. The construction of the instrument is protected by patent of the Russian Federation №118907 for a useful model.

In order to determine the technological modes of antifriction combined treatment of guide surfaces the number of experiments was conducted using this instrument on vertical milling machine 6T13. During experiments the following processes were modified: the rotation frequency of the spindle (50…200 min^{-1}), longitudinal instrument advance for antifriction treatment (0.2…0.8 mm/rev), pressing effort of the rubbing (250…3000 H) and burnishing (100…500 H) elements against the processed surface, granularity of processing abrasive petals (12…25 according to GOST 3647-80). While experimenting, glycerin was used in order to provide a qualitative antifriction coating. According to the results of experiments the modes of surface treatment of guides, providing stable formation of qualitative antifriction coatings throughout the whole processed area (Table.1), were determined.

Table 1. Technological modes of the combined antifriction treatment of external cylindrical and flat surfaces of guides

Type of the processed surface	Sliding velocity of rubbing (burnishing) elements, m/min	Pressing effort of rubbing elements, H	Pressing effort of burnishing elements, H	Longitudinal device advance, mm/rev	Granularity of abrasive petals according to GOST 3647-80
External cylindrical	60-90 (30-40)	350-400	140-170	0.2-0.3	12-16
Flat	80-100 (40-50)	> 2500	150-200	0.3-0.5	16-20

Notes: 1. Pressing effort depends on the width of the flat processed surface. 2. The indicated granularity of abrasive petals provided the roughness of surface equal to Ra=1.6…2.5 μm. 3. The surface roughness after treatment amounted to Ra=0.65…0.8 μm. 4. The antifriction coating thickness was in the range from 0.01 to 0.05 mm.

The next stage of research was the estimation of wear resistance of the coating layers formed according to the new technology of antifriction treatment. A stand imitating the real external environment of cylindrical elevator guides was created for this purpose. A stand works as follows. A test specimen is fastened to the prism and brought into contact with a roller. The necessary loading on the roller is created by the loads placed on the bowl of the loading device. Later, by switching on the power switch6 the stand is fed with voltage, setting a geared motor into rotation and starting a reciprocating motion of the roller over the cylinder specimen. Then, using the voltage regulator (of rotation frequency) one settles the rotary speed of the geared motor up to a required magnitude, whereupon tribotechnical tests are performed.

During tribotechnical tests the loading on the specimens varied from 100 to 600 H. Loading variation resulted in the change of contact voltage between the roller and the specimen, the values of which were calculated according to a well-known H. Hertz formula.

During tests one fixed the time, after expiration of which the antifriction coating was worn down. The appearance of the base metal of the cylinder specimen in the area of antifriction coating served as a criterion of such wearout. Then the limit number of N-cycles of surface layer loading was determined. Later, one revealed the dependence of the limit number of N-cycles of loading on pressure $p_к$ when the roller contacts the specimen: $N = \alpha\ p_к^{\beta}$, where α, β – are the coefficients determined by a least-square method.

The developed technology of antifriction treatment allows coating of the broken-in and non-broken-in friction surfaces of guides. In this connection antifriction coatings were applied to specimens, which had rolling paths formed by the roller on the stand, and on the specimens without the indicated paths. This circumstance substantially influenced wear resistance of surface layers (Table 2).

Tribotechnical tests of friction specimen surfaces subjected to this antifriction treatment have shown that the service life of the surface layers made of antifriction material amounts to 8000…32000 cycles (from 3 to 12 years of normal operation of cylinder elevator guides). The service life depends on the thickness of antifriction coating, the grade of antifriction material, modes of treatment of the guide and loading at tests.

Table 2. Values of coefficients α and β

Antifriction material	Coating on the non-broken-in friction surface		Coating on the broken-in friction surface	
	α	β	α	β
Bronze	$3.0 \cdot 10^5$	-0.60	$2.1 \cdot 10^5$	-0.65
Copper	$2.5 \cdot 10^5$	-0.45	$1.4 \cdot 10^5$	-0.50
Brass	$1.7 \cdot 10^5$	-0.55	$0.9 \cdot 10^5$	-0.60

Notes: 1.Contact voltages $p_к$ varied in the ranges of 75 to 250 H/mm^2. 2. A significance and adequacy of dependences were checked according to Fisher F-criterion of Fisher with confidence probability of 0.95.

3. Results and Discussion

The analysis of results of experimental researches allowed concluding the following:

At the increase of sliding speed of rubbing elements over the processed surface above 100 m/min there is unevenness of antifriction coating forming. A substantial reduction of sliding speed results in the productivity decline of treatment and thickness of the formed coating, therefore the acceptable value of sliding speed is in the range of 70 - 100 m/min.

The pressing effort of burnishing elements against the processed cylinder surface should not exceed 350…400 H, as local destructions of antifriction surface coating can occur.

At the initial roughness of surface of more than Ra=3.5-4 µm there is a substantial consumption of antifriction material and forming of fragmentary coating on the processed surface. Fragmentary coating is also observed at the tool advance for antifriction treatment of more than 0.5 mm/rev. The most qualitative coating is formed at the initial surface roughness of Ra= 1.6-2.5 µm.

The surface layers containing bronze CuA19Fe3 become more wear-resistant, which is explained by the presence of iron that provides stronger adhesion relations with the specimen metal. The surface layers formed with the use of brass L63 appeared to be less wear resistant.

Antifriction coatings of the non-broken-in friction surfaces possess greater wear resistance (by 1.5-2 times) as compared to the similar coatings applied to broken-in surfaces of specimens. It can be explained by the presence of the considerable defective pre-surface layer of the broken-in surfaces.

The substantial decline (to 3-4 times) of wear resistance of antifriction coatings was observed contact voltages exceeded 130…150 MPa, therefore the accessible voltage in friction pairs with guides, the working surfaces of which are processed according to a new technology should be no more than 100…120 MPa.

4. Conclusion

It is more expedient to apply the described antifriction treatment when producing guides of machines and mechanisms, operating in the fatigue wear conditions. Surface forming by means of the developed instrument will have technological and economic effect, related to upgrading of antifriction coating of frictional unit components, increase of their service life and decrease of repair expenses.

References

[1] Stanski D et al 2015 *Met Sci Heat Tream.* **7(721)** 77-83
[2] Tikhomirov W P and Gorlenko A O 2001 *Problem of engineer and reliability of machines.* **2** 68-76

Preservation of material quality with the use of combined treatment

A A Boldyrev[1], A I Boldyrev[1], V M Skantsev[2] and E V Smolentsev[1]
[1] Voronezh State Technical University, 14, Moskovskiy ave., Voronezh, 394026, Russia
[2] Bryansk State Technical University, 7, 50 years of October st., Bryansk, 241035, Russia

E-mail: alexboldyrev@yandex.ru

Abstract. This research is devoted to some techniques of combined treatment including anodic attack and mechanical impact. The salutary effect of these methods permits one to get the requisite accuracy and quality of work surface. The article contains final results of theoretical and experimental study of metal working for inner apparent surfaces.

1. Introduction

The research of new technologies theory in combined treatment for achievement of the best results was conducted in Voronezh State Technical University and Bryansk State Technical University. The results of it are previously unknown processes that the authors called electromechanical treatment and treatment by unbound particles in neutral salt medium. These processes were successfully used in military industry and are now being used in the production of compressors, pumps, transport facilities. The above mentioned method permits to produce complex parts without finishing operations and with considerable saving.

2. Results and Discussion

It is known that every method of treatment changes material surface coating and affects its mechanical properties, especially the endurance strength of alloys. If the final processing operation is yielding, its influence on surface quality greatly depends on the method and conditions of the previous treatment, that is, such conditions as machine direction stresses, the importance of cutting tool and strain hardening. The only method for removal of previous operations residual factors without new changes of surface coating is electromechanical treatment in electrolyte (aqueous solutions of neutral salts are used as electrolytes).

However it is necessary to achieve the relevant surface strength to guarantee a high quality of mechanical properties of machine parts. For this purpose we need calculated working tolerance of the nominal dimension with a minimum number of mistakes. Therefore in this case it is necessary to have the exact passage form before surface yielding. A separate passage formed by electromechanical precision treatment and plastoelastic strain can guarantee stable results only with total strain hardening of material as long as the passage accuracy changes of electromechanical treatment reach the ISO 9-10 echelon accuracy, but the expected echelon accuracy is ISO 7-8.

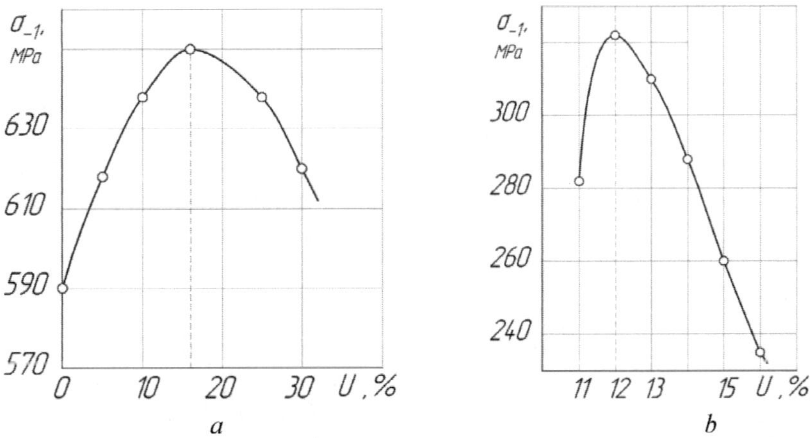

Figure 1. Influence of the surface layer hardening on the endurance limit; *a* –40XHMA steel; *b* –BT3-1 titanium alloy.

The endurance strength of alloys σ_{-1} changes at multicycle tension in accordance with the level of hard-facing *U* is shown in Figure 1.

The general endurance strength of structural steel 40XHMA (Figure 1,a) in the initial state equals 590 MPa and after achievement of full load is 610 MPa. The maximum strength increases up to 650 MPa when hardening level $U = 16$ %. A similar linear connection is considered in titanium alloy BT3-1 (Figure 1,b). Thus some methods of mechanical treatment that are widely used in mechanical engineering industry such as, for example, vibro-strenthening and shot-blast cleaning permit one to achieve the increase of endurance strength in only one part and the reduction of specific quantity of metal in manufactured goods.

We offer the method calculated for achievement of the high passage accuracy with desired strengthening. Figure 2 shows the model of this method.

The control strategies that form the basis of this method are:

- metal removal before the change of a dimension element in proportion to an allowance value along the entire perimeter;
- mechanical properties of material become permanent throughout the depth and they do not affect the force shift of dimensional tool.

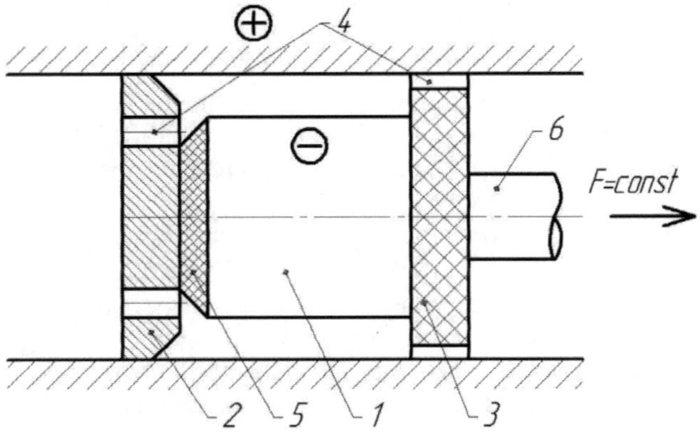

Figure 2. Method of combined treatment:
1 – cathode instrument; 2 – calibrating element; 3 – guide member; 4 – outlets for electrolyte; 5 – dielectric socket; 6 – bar with current lead.

We managed to formulate the method that conducts this process by keeping the calculated force shift of a combined instrument constant.

Cathode instrument 1 (Figure 2) is hold by dielectric rails provided with parallel pits or connected with axes of a tool in the electrolyte pump. Dimensional tool 4 has an external unit holding the contour surface of the passage. It is connected through a dielectric interlayer with a cathode instrument. Expected force F is calculated with the use of the allowance value that is given for optimal strengthening of the obtainable surface. Force F is kept constant regardless of change in the initial passage size. When the allowance exceeds the calculated value, combined instrument stops and does not move until its allowance value reduces, and traces left by the previous treatment are removed by the anodic attack on the cathode. At the time when allowance reaches the calculated value, the feed of the combined instrument is renewed. When the allowance value is unequal throughout the passage perimeter, it is necessary to achieve speeded dissolving in the unit with a thick layer that removes metal. In this case the process of metal removal is intensified by local strengthening.

The design of the combined instrument (Figure 2) helps with this. There is elastic element 5 that is compressed when combined instrument stops in front of the unit with high allowance. It turns off the feeder for some time and bar 6 moves backwards, than forward again, and dimensional element is removed throughout the extended area of the passage. As a result we obtain local strengthening, which permits to intensify the metal dissolution efficiency up to 60 % and quickly equalizes the total allowance.

Feed force is

$$F = \pi \cdot p \cdot z \cdot d \cdot \left(1 + \frac{f}{tg\alpha}\right), \tag{1}$$

where p - regular contact pressure;

z - allowance;

d - dimensional instrument diameter;

f - index of friction;

α - angel of a taper lead in the grade unit.

When F=*const*, random errors of the passage cross section after treatment by the combined instrument include errors caused by nonuniformity of the dimensional unit allowance and a rigidity change. The first group of mistakes is not so important as mistakes caused by the nonuniformity of rough tolerance. The overall error of the passage cross section with the diameter of more than 20 mm at treatment of the parts from structural steel corresponds to a tolerance range with 7 ISO accuracy degree, but in case of treatment of the passages with the diameter of more than 50 mm the achieved accuracy degree is 6 ISO.

We made calculations that permitted the identification of operation conditions for machine-tools with CNC with the combined tool intended for treatment of cylindrical pipes of 37 mm in diameter made from structural steel 40XHMA. Inaccuracy of strain hardening for the rated size was less than 1 %.

The research of a surface quality after combined treatment showed that yielding caused by the dimensional unit improves the quality of surface coating. After finishing the surface is almost free of anodic and pitting attack traces that were typical for previous electrochemical treatment because of defect influence (healding). Moreover the structure of layers united with the surface undergoes the above-mentioned changes. These changes are small distortion in structure formation and a considerable increase of grains. The structure of surface coatings approximates to the structure of amorphous metallic materials.

The measurement of undulation showed the possibility of reaching stability: *Ra* = 0.16-0.08 mym – for steels; *Ra* = 0.32-0.16 mym – for aluminium alloys.

The limit of endurance after combined treatment exceeds the limit of cutting, after which strengthening for most materials followed (Fig.3).

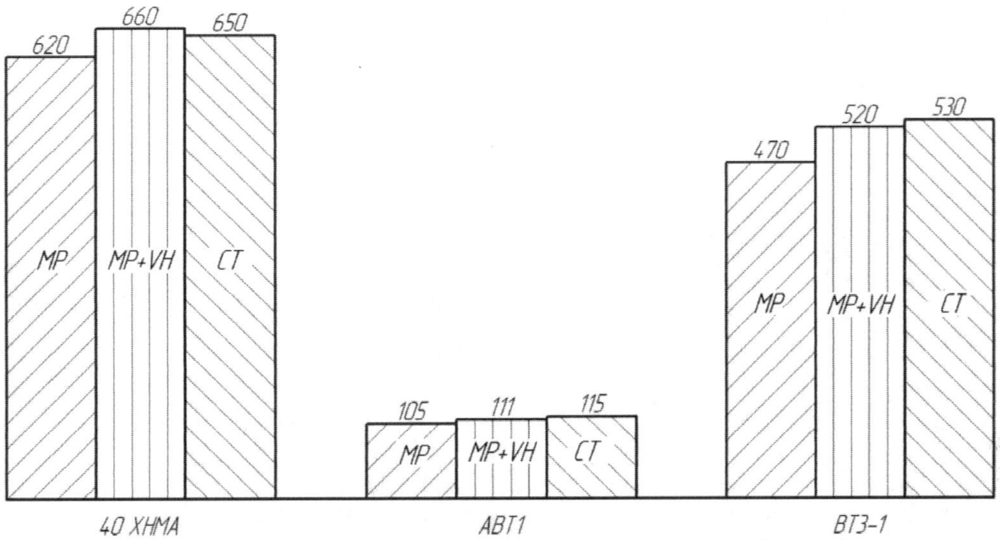

Figure 3. Limit of endurance (MPa) after different treatments: MP – mechanical processing; MP+VH – mechanical processing with vibrohardening; CT – combined treatment.

Thus we recommend the worked out method instead of unrolling, grinding, polishing, etcetera in order to increase the stress state of the surface that is preferable with regard to wearability.

For the complex surfaces formation we suggest a combined technique of electrochemical treatment with the use of a solid current-conducting filler. The basic principle of this method is that spherical filler grains (d = 2-7 mm) are fed to the place of treatment by the electrolyte flow. Figure 4 presents the treatment pattern. The suggested method can be used for treatment of outer and inner surface at a distance of up to 100 mm from the outlet of the nozzle passage.

A hydrodynamic flow regime of the process of the fluid and electromagnetic field distribution in gaps between electrodes is the main factor of a treatment process. Shaping of the workpiece surface coating directly depends on the presence of filler grains in gaps between electrodes. Watching the movement of separate grains and collision with a workpiece it is possible to expect the following consequence of surface coating yielding:

- first of all yielding and microinhomogeneities take place, then the anodic attack starts along tips;
- contact surfaces appear after dissolving and distortion of inhomogeneous cross sections;
- distortion and the anodic attack proceeds in the microprofile basis of the workpiece.

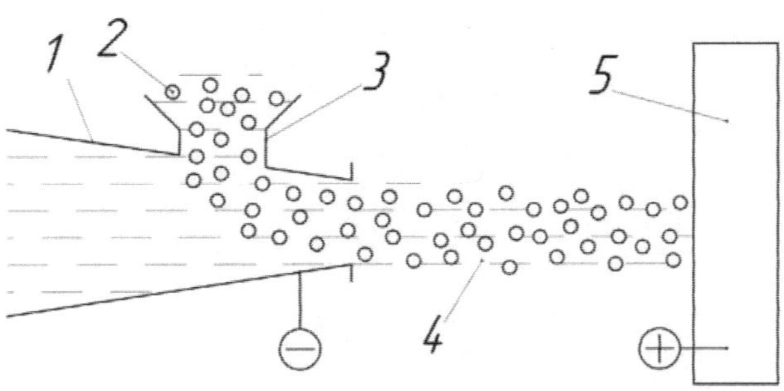

Figure 4. Treatment pattern: 1 – jet; 2 – grains; 3 – doser; 4 – electrolyte; 5 – workpiece.

For qualitative adjectives of surface coating it is necessary to control the speed of filler grain feed supplied to the place of treatment.

A selection of the grain feed speed depends on 2 conditions:

- time of grains movement to a machined surface must be less than time of grain storage;
- speed of grain movement supports the necessary hydrodynamic flow pressure which depends on the surface coating quality of the workpiece.

The linear connection of hydrodynamic pressure with speed can be presented by the following expression:

$$p = \gamma \cdot v^2 \ , \qquad (2)$$

where γ - density of processing medium;

v - speed of grain movement.

The application of this method in treatment of the pump impeller surface decreased its undulation from 5 mym to 0.4-1.2 mym and eliminated the following 3 operations of workpiece treatment that took up to 6 hours.

3. Conclusion

This research presents some engineering opportunities of treatment by combined techniques that are worked out and are being used at some Russian plants with great success. The development of new combined process techniques gives scientists and engineers an opportunity to create new alternative processing techniques ensuring competitiveness of manufactured goods.

References

[1] Baron Y M 1986 *Leningrad: Mashinostroenie* 176
[2] Smolentsev V P 1983 *Moscow: Vyschaya Shkola* **2** 208
[3] Smolentsev E V 2005 *Moscow: Mashinostroenie* p 511
[4] Smolentsev V P 1976 *Moscow: Masinostroenie* **1** 768
[5] Kirichek A V, Solov'ev D L 2008 *Russian Engineering Research* **28(3)** 277–279

IOP Conf. Series: Materials Science and Engineering **124** (2016) 012153 doi:10.1088/1757-899X/124/1/012153

Burnishing tool actuators and their influence on the burnishing force components

Yu Chervach, A Kim and D Dorzhiev
Tomsk Polytechnic University, 30, Lenina ave., Tomsk, 634050, Russia

E-mail: juri@chervach.com

Abstract. The article presents the results of studies on measuring components of force of diamond burnishing of surfaces with variable radius of curvature. The impact of structural elements of burnishing tools of different types on components of burnishing force was analyzed. The impact of different actuators of burnishing tools on components of burnishing force was analyzed. The design of the burnishing tool with an air receiver (bellows with compressed air as an actuator) with a parallelogram-type mounting of the indenter on the four flat posts, which ensures burnishing of profiled and other types of surfaces with constant radial force, was proposed.

1. Introduction

Cold working processes, which are used to increase fatigue strength, contact endurance, wear resistance and corrosion resistance of mechanical components, to increase equipment durability, are the simplest and most effective methods of finishing-hardening treatment [1, 2].

One of the simple processes of cold working, in terms of manufacturing, is diamond burnishing which uses a diamond crystal, held in a special holder, as a deforming element. The diamond indenter is clamped in a special device which is installed in the tool post of a lathe. The pressure of burnishing with elastic contact is usually created by using a calibrated spring, which provides burnishing force P_y.

2. Experimental procedure

It should be noted that a selection of the burnishing parameters – radius of the spherical diamond indenter R_{sp}, feed rate S and burnishing force P_y – depends on the purpose of the workpiece surface treatment, whether it is in ensuring optimal degree of surface layer work hardening, optimal surface roughness or in achieving a compromise between the above-mentioned parameters. This selection is most frequently carried out experimentally using samples of materials from which manufactured components are made. The radius of the diamond indenter is preliminary chosen with respect to the initial hardness of the workpiece material and the desired rate of work hardening. Experimentally found optimal burnishing force P_y should remain constant or have very small deviations during the burnishing process. Slight deviations of burnishing force P_y lead either to a lack of hardening or to over hardening of the surface layer. Also, deviations of burnishing force P_y lead to increased surface roughness compared to the smallest value Ra_{min}, obtained by burnishing with optimal force.

The designs of tools used for diamond burnishing [3, 4] – dynamometric (flat spring is used as an actuator), spring-actuated (coil spring is an actuator), hydraulic (energy of the fluid compressed in the hydraulic cylinder is an actuator), etcetera – have a number of drawbacks which lead to significant fluctuations of burnishing force P_y during the burnishing operation.

IOP Conf. Series: Materials Science and Engineering **124** (2016) 012153 doi:10.1088/1757-899X/124/1/012153

3. Results and Discussion

The use of the dynamometric burnishing tool (Figure 1) is possible with the accurate mounting of the workpiece on the machine tool or in case diamond burnishing is performed immediately after finishing machining in a single setup.

Figure 1. Dynamometric burnishing tool (an actuator is a flat spring with a vibration damper).

Sinusoidal change of the radial burnishing force P_y (Figure 2) shows that the ΔP_y depends only on flat spring constant k (N/m) and radial run-out of the workpiece surface. For comparative experiments, cylindrical samples of 30 mm in diameter made of steel 12H18N10T (0.12% C, 18% Cr, 10% Ni, 1% Ti) GOST 5949-75 were prepared. Radial run during burnishing of the samples was equal to 0.2 mm. Burnishing force components were measured with a KISTLER three-component dynamometer Type 5233A1.

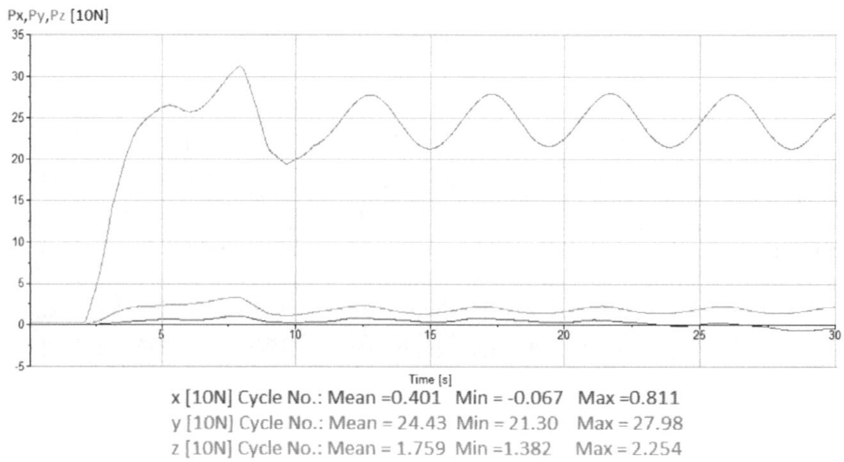

Figure 2. Relationship between burnishing force components P_y, P_x and P_z (kN) and time T (s) during burnishing with a dynamometric tool with a flat spring. Burnishing parameters: radius of the indenter R_{sp}=2 mm; feed rate S=0.05 mm per rev.; burnishing speed v≈0.02 m/s; angular velocity ω=1.3 s^{-1}. ΔP_y≈67 N.

Burnishing with a coil spring tool (Figure 3) of samples mounted with 0.2 mm radial run-out as well as surfaces having substantially smaller radial run-out showed a completely different behavior of the radial force (Figure 4), compared to burnishing with a dynamometric burnishing tool equipped with a flat spring.

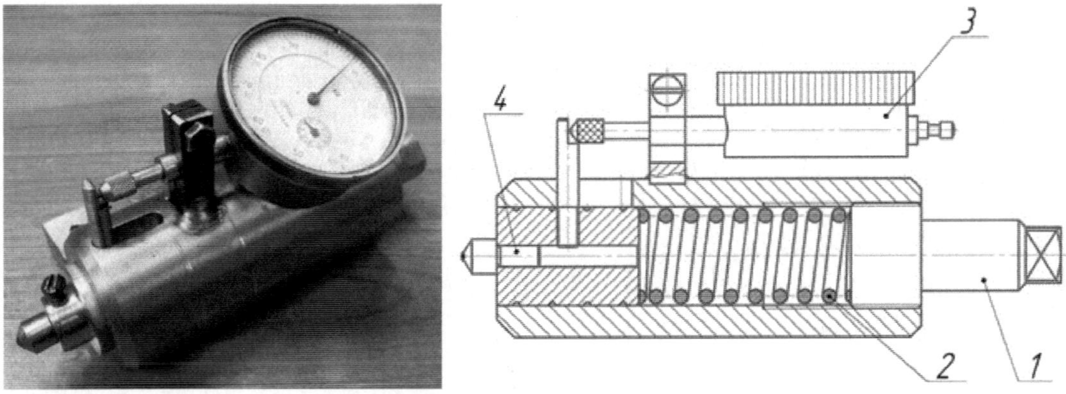

Figure 3. Spring-actuated burnishing tool (coil spring is an actuator): 1 – adjusting screw; 2 – coil spring; 3 – dial indicator; 4 – indenter.

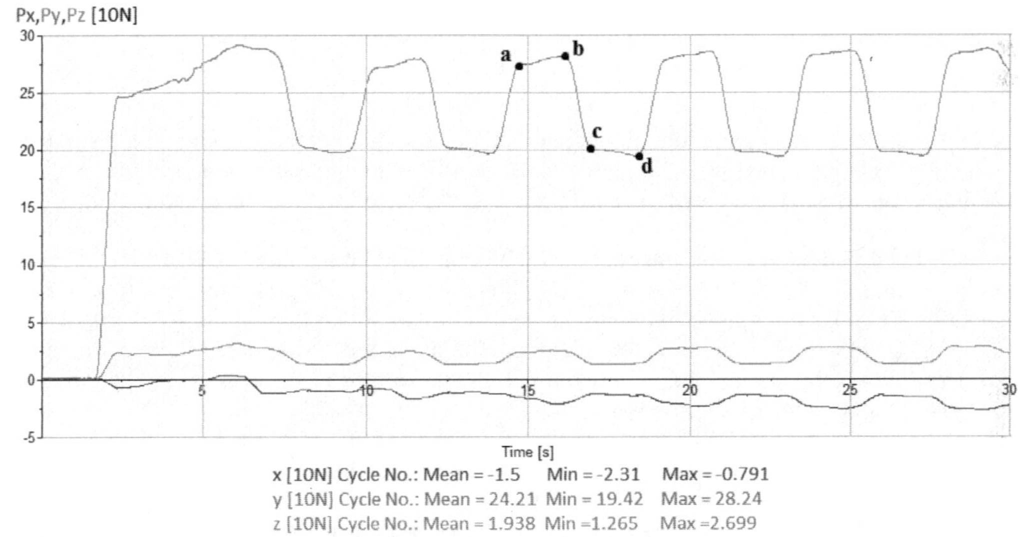

Figure 4. Relationship between burnishing force components P_y, P_x and P_z (kN) and time T (s) during burnishing with a coil spring actuated tool. Burnishing parameters: radius of the indenter R_{sp}=2 mm; feed rate S=0.05 mm per rev.; burnishing speed $v\approx0.02$ m/s; angular velocity ω=1.3 s^{-1}. $\Delta P_y\approx88$ N.

The evident saw-toothed curve for $P_y=f(S)$, as well as $P_x=f(S)$ and $P_z=f(S)$ is, most likely, caused by mechanical sticking (seizure) in the precision reciprocating plunger coupling under dry friction conditions. The use of recommendations to reduce the seizure during the reciprocating movement of the plunger, such as pressure lubrication, rotation of the plunger relative to the cylinder of the tool in the process of burnishing, burnishing with external vibrations of different frequency and amplitude, have not changed the nature of dependence $P_y=f(S)$ and only partially reduced the amplitude of the oscillations of force P_y. The value of ΔP_y depends not only on coil spring constant k (the change in the length of the spring by the amount of radial run-out $\Delta r=r_{max}-r_{min}$), but also on the conditions of the indenter movement over the workpiece surface (Figure 5) and, consequently, on the micro-displacement of the plunger relative to the cylinder of the burnishing tool (motion of the indenter across the surface with the largest radius – from point a to point b and across the surface with the smallest radius – from point *c* to point *d*). If sticking of the reciprocating plunger coupling is eliminated from the process of burnishing with a coil spring burnishing tool, then the dependence of $P_y=f(S)$, shown in Figure 4, takes the form of the curve obtained during burnishing with a flat spring burnishing tool. In these cases, the nature of the force P_y behavior is virtually the same, and depends only on the elastic bending deformations of the flat spring material (Figure 2) or on the elastic torsion deformations of the coil spring. The difference in form of the dependences shown in Figure 2 and 4 is only caused by the process of plunger sticking during micro-displacements.

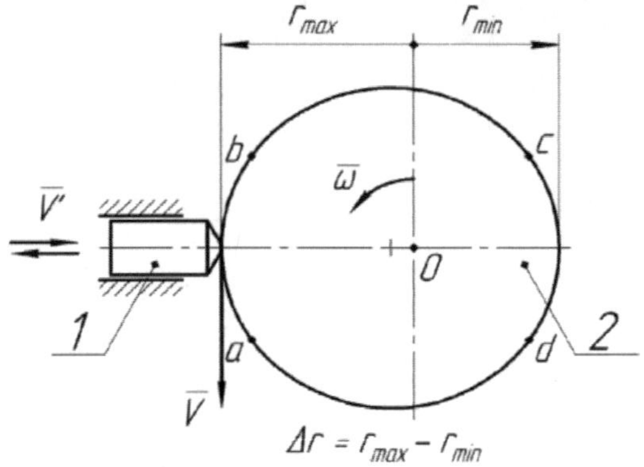

Figure 5. Movement of an indenter on the surface with radial run-out (eccentric surface). 1 – indenter; 2 – surface being burnished. *V* – burnishing speed, m/s; *V'* – sliding speed of an indenter with a plunger relative to the body of the burnisher; ω – angular velocity of the workpiece, s^{-1}.

The analysis of results of burnishing surfaces with radial run-out of up to 0.01...0.02 mm with a coil-spring burnishing tool showed that variation of burnishing force ΔP_y is caused, to a large extent, by sticking in the reciprocating plunger coupling. It can be concluded that the use of spring-actuated burnishing tools with reciprocating sliding elements for plain and profiled burnishing will lead to variable burnishing force P_y, and a non-uniform process of surface layer hardening. The increase in burnishing speed may cause self-oscillation of the system 'indenter-part'.

The task of burnishing force Py stabilization is particularly acute in burnishing complex profiled surfaces of revolution with variable radius of curvature in both the longitudinal and transverse direction relative to the axis of rotation. The tool for hardening rope thread of drill rods [5, 6] (Figure 6) enables diamond burnishing with speed of about 2 m/s without resonance effects inherent to the tools with flat or coil springs. A pneumatic spring, which is a bellows, is used as an actuator in this burnishing tool. The pressure in the bellows is measured with a pressure gauge. A pneumatic burnishing tool is specially designed for burnishing surface of rope thread R32–R38 with constant burnishing force P_y.

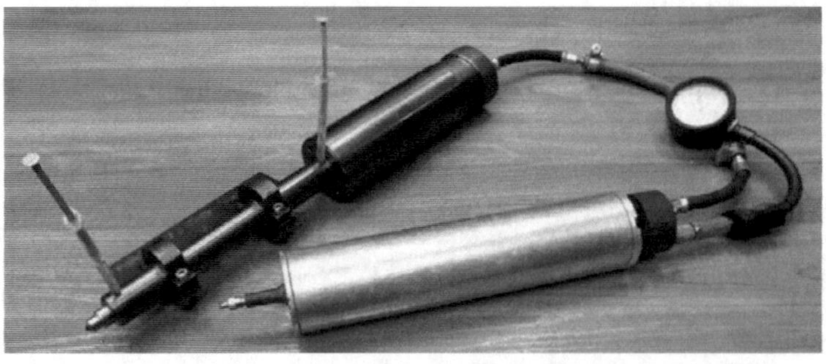

Figure 6. Pneumatic burnishing tool with a receiver (bellows with compressed air is an actuator).

However, the limited volume of the bellows used in the tool leads to the fact that the radial run-out of the surface during burnishing (run-out is approximately 3 mm for thread R38) causes a pressure change in the bellows and, consequently, change in the burnishing force P_y value. In order to stabilize burnishing force, the system of 'indenter-bellows' is equipped with a receiver, which can significantly reduce the pressure fluctuation in the bellows due to the increased volume of compressed air. The volume of a receiver is more than 10 times as big as the volume of the working chamber of the bellows.

The special elements of the pneumatic tool design had to provide constant burnishing force P_y even with large run-out of the workpiece surface, since the volume change of the bellows is small in relation to the volume of the receiver, a change in pressure in the bellows and thus change in burnishing force is also small. A trapezoidal shape of curves *Py=f(S)*, *Px=f(S)* and *Pz=f(S)*, as shown in Figure 7, leads to the conclusion that the change in P_y is directly related only to sticking of the plunger reciprocating pair of friction. It should be noted that burnishing force fluctuation $\Delta P_y \approx 35$ N in burnishing with a pneumatic tool is significantly smaller than the fluctuation during burnishing with flat and coil springs: $\Delta P_y \approx 67$ N and $\Delta P \approx 88$ N respectively.

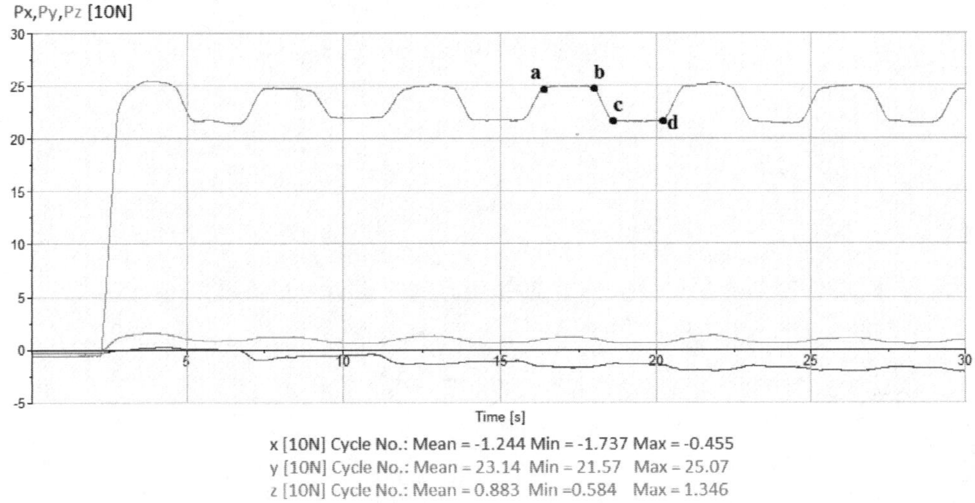

Figure 7. Relationship between burnishing force components P_y, P_x and P_z (kN) and time T (s) during burnishing with a pneumatic tool equipped with a receiver (bellows with compressed air is an actuator). Burnishing parameters: radius of the indenter R_{sp}=2 mm; feed rate *S*=0.05 mm per rev.; burnishing speed $v \approx 0.02$ m/s; angular velocity ω=1.3 s^{-1}. $\Delta P_y \approx 35$ N.

To stabilize force P_y during burnishing of profiled surfaces and surfaces of the components mounted on the machine with a relatively large radial run-out, it was necessary to completely eliminate the reciprocating and rotational pairs of friction, as well as pairs of rolling friction in the design of the burnishing tool. It was supposed to use a well-proven pressure-regulated 'bellows-receiver' system as an actuator. Mounting of the indenter should ensure its free movement in the 'y' direction with respect to the frame, and restrict its movement in orthogonal directions 'x' and 'z' under influence of force components P_x and P_z.

According to these requirements, a pneumatic burnishing tool with an air receiver (actuator is a bellows with compressed air) and a parallelogram-type mounting of the indenter on four flat posts was designed (Figure 8).

The measurement of burnishing force components P_y, P_x and P_z during burnishing of eccentrically mounted parts, showed almost complete absence of variation of force $\Delta P_y \approx 1.9$ N in comparison with other types of burnishing tools (Figure 9a). The burnishing of profiled surfaces, such as rope thread R32, has shown that the design of the burnishing tool ensures a process with constant burnishing force.

Figure 8. Pneumatic burnishing tool with a receiver (actuator is a bellows with compressed air) and a parallelogram-type mounting of the indenter on four flat posts.

(a)

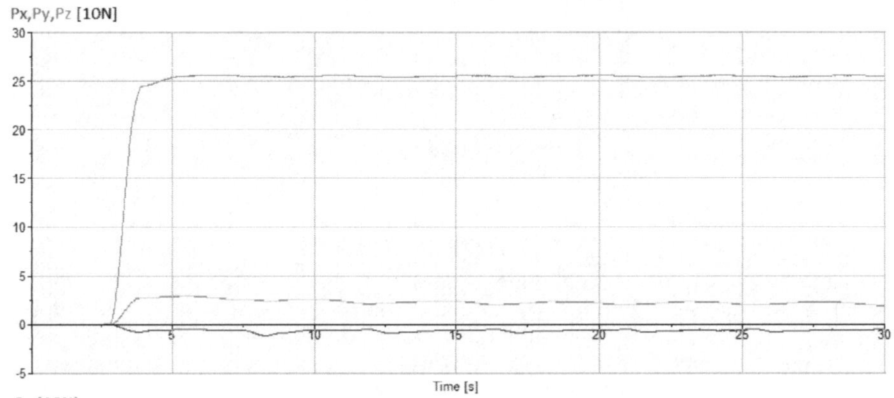

(b)

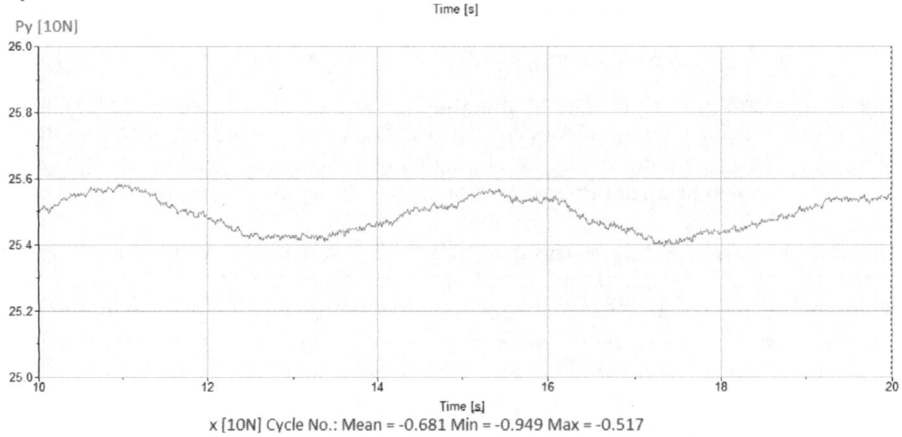

Figure 9. Relationship between burnishing force components Py, Px and Pz (kN) and time T (s) during burnishing with a pneumatic tool equipped with a receiver (bellows with compressed air is an actuator) and a parallelogram-type mounting of the indenter on four flat posts. Burnishing parameters: radius of the indenter Rsp=2 mm; feed rate S=0.05 mm per rev.; burnishing speed v≈0.02 m/s; angular velocity ω=1.3 s^{-1}. ΔPy ≈1.9 N.

Figure 9b shows an enlarged part of dependence *Py=f(S)* at a steady motion: force fluctuation smoothing is about 2 N with $Py \approx 250$ N, which is less than 1%. The shape of the curves almost completely corresponds to the shape of the curve shown for the dynamometric burnishing tool: fluctuation of force *Py* is associated with axial elastic deformation of the bellows as a metal spring, and elastic bending deformations of the four flat posts.

4. Conclusion

The investigations of force dependences of diamond burnishing of parts mounted on machine tools with the radial run-out (or burnishing of profiled surfaces) with burnishing tools equipped with different actuators and special elements show that force components fluctuations depend not only on the rigidity and type of the actuator, but also on non-uniform movement in pairs of friction.

The extensive use of diamond burnishing tools with reciprocating pairs of friction requires a very accurate mounting of workpieces prior to burnishing or implementation of a finishing-hardening operation immediately after a preceding operation in the same setup.

Design features of a pneumatic burnishing tool with a receiver (actuator is compressed air in bellows) and a parallelogram-type mounting of the indenter on four flat posts ensure burnishing of surfaces with an intricate profile and surfaces of parts mounted with the relatively large radial run-out with constant burnishing force P_y. The actuator and special design features of the tool make it possible to perform burnishing at high speeds without causing oscillations in the 'indenter-burnished surface' system.

References

[1] Kuznetsov V P, Smolin I Yu, Dmitriev A I, Smirnov S V, Makarov A V, Kiryakov A E and Yurovskikh A S 2011 *Physical Mesomechanics* **16(6)** 87–97
[2] Shiou F J and Chuang C H 2010 *Prec Eng* 76–84
[3] Yatsenko V K, Zaytsev G Z, Pritchenko V F and Ivshchenko L I 1985 (*Moscow: Mashinostroenie)*
[4] Torbilo V M 1972 (*Moscow: Mashinostroenie*)
[5] Chervach Yu B 2011 *Proceed. of the VI Internat. Science and Engineering Conference* 306–311
[6] Dmitriev A I, Nikonov A Yu and Kuznetsov V P 2011 *Proceed. of the International Confer. on Phys. Mesomech., Comp.-Aided Design and Development of New Materials*

Hardening treatment of friction surfaces of ball journal bearings

A O Gorlenko and S V Davidov
Bryansk State Technical University, 7, 50 years of October st., Bryansk, 241035, Russia

E-mail: bugi12@bk.ru

Abstract. The article presents the technology of finishing plasma hardening by the application of the multi-layer nanocoating Si-O-C-N system to harden the friction surfaces of the ball journal bearings. The authors of the paper have studied the applied wear-resistant anti-friction coating tribological characteristics, which determine the increase in wear resistance of the ball journal bearings.

1. Introduction

One of the new methods of surface hardening, which ensures the application of wear-resistant thin film coatings, is the process of finishing plasma hardening (FPH) based on the application of the plasma jet flowing at atmospheric pressure. The efficiency of this process derives from the compact and economical equipment, which allows for applying the hardening nanocoatings. This method belongs to the additive technologies [1].

The FPH technology fundamental principle of applying a thin-film wear-resistant coating based on the Si-O-C-N system is diffusion of the vapours produced by the liquid organoelemental agents, which are introduced into the plasma chemical reactor of the arc plasma gun, followed by plasma chemical reactions and formation of a coating on work pieces.

Argon is used as plasma-supporting gas as it provides the increased durability and reliability of the plasma gun components in the course of the long-term process. The vapours of volatile liquid reagents are used as the coating-formation materials. They are supplied into the reactor by a special feed-control device. The power for the plasma gun is supplied by a DC inverter with special current-voltage characteristics. The sTable cooling of the reactor and the plasma gun is ensured by a cooler made from a refrigerating unit. The monitoring system of the process ensures monitoring and control of the treatment parameters and defines thickness of the applied coating in the course of its precipitation.

The main technologies of applying the wear-resistant nanocoatings, which are commonly used abroad, are chemical vapour deposition (CVD) and physical vapour deposition (PVD). In the aerated FPH technology, the coating is applied in 30...3 nm layers at typical speeds of the plasma jet movement of 10...100 mm/s.

Unlike the condensed coatings in vacuum performed in PVD and CVD processes, in this method the coating is formed locally, where the plasma jet contacts the base coat, and only for the multilayer coating, which is an important feature of the FPH technology. The FPH cyclic relative movement of the plasma jet and the hardened surface determines obtaining of a layered structure of the coating and allows reducing the thermal effect of the plasma on the base coat to the minimum, thus completely eliminating the softening drawing-back for all steels. The integral temperature of the hardened work pieces during the application of the coating is usually not more than 150 °C. The hardening coating is formed as a transparent film. On the polished surface it looks as an interference pattern with rainbow hues from purple-blue to green-red, depending on the thickness of the coating.

The choice of the material for the coating applied by FPH method is determined based on knowledge of the wear mechanisms experienced by different products, as well as the analysis of available experience of using various compounds as coatings

If we consider fundamentally any tribosystem operating under the adhesive, fatigue, oxidation and abrasive wear, the most perspective thin-film coatings would be non-metallic solids - carbides, nitrides, borides, silicides, oxides, composite and nanocomposite materials based on them, as well as cermets and diamond [2]. In this case, the coating must have the maximum adhesion and thermal expansion coefficient close to the hardened work piece, and its surface properties should comply with the characteristics that increase the product durability, i.e. they must have high hardness, chemical inertness, thermal stability, low thermal conductivity, minimal friction coefficient, etc.

In recent decades, the processes of coating with chemical precipitation widely employ the element-organic (organometallic) compounds; their use in coatings determines the improved level of safety (considering they are non-toxic) and zero explosion risk (considering they are used in the liquid state). Ii is significant that the element-organic compounds can contain all the necessary elements for producing the coatings in a single substance, which improves the effectiveness of the monitoring over the process and reproducibility of the coatings properties.

The X-ray diffraction analysis confirms that after the FPH, the coating is formed in the amorphous state, with no dislocation activity, and the coating exhibits high values of resistance to plastic deformation and elastic recovery. The Si-O-C-N system coating applied with the FPH technology is characterized by high hardness at low elasticity modulus and close values of the coating elasticity modulus and the base coat material, which should objectively result in the improved wear resistance of the surface layer.

2. Findings and discussions

The object of the tribological characteristics study were the journal ball bearings ШС30 (GOST 3635-78, ISO 6125-82). The hardening was performed on the exterior ball surfaces of the inner rings of ШС30 journal ball bearings which are shown in Figure 1.

Figure 1. The ball surfaces of the inner rings of ШС30 journal ball bearings treated by FPH (left) and the bearings manufactured by the factory technology (right).

The factory technology (baseline option) provides the bulk hardening of the inner ring of journal ball bearing made of ШХ15 alloy steel with the following abrasion (grinding) of the ball surface. The FPH technology for the bearings ball surfaces was implemented in NPF Plazmacentr OJSC (St. Petersburg).

The wear-resisting properties of the modified ball journal bearings and those manufactured by the factory technology were tested on a specially designed and produced installation; using the automated research system, the installation allowed determining the tribological performance of the ball journal bearings friction surfaces.

The bearings test plan is presented in Table 1, which shows the numbers of the tested bearings, the method of finishing applied to friction ball surfaces of the inner ring of the bearing, the lubricant used for the test surface preconditioning and the main lubricant

Table 1. Bearings test plan

Bearing No.	Finishing method	Lubricant	
		For preconditioning	Main
1	Baseline option	–	Molykote BR 2 Plus
2	Baseline option	–	Gazpromneft Grease L
3	FPH	Molykote D-321R	Molykote Longterm 2 Plus
4	FPH	Molykote G-Rapid Plus	Molykote Longterm 2 Plus

The bearings were tested in the following conditions: relative sliding velocity of the ball surfaces is $\upsilon = 0.84$ m/s (with ball surfaces diameter $d = 40$ mm and rotating velocity $n = 400$ min^{-1}); normal force loading $N = 2000$H (corresponds to the pressures calculated by Hertz, about 11 MPa); type of lubrication - boundary; a predominant wear mode - fatigue; lubricant - in accordance with the test plan (see Table 1); total time of every bearing test is 6 hours.

During the tests, the sensor system continuously and synchronously recorded the test time, load, friction coefficient and linear wear. Their numerical values were displayed on the PC monitor. The strain gauges were used to measure the friction torque and the load. To ensure the continuous measurement of wear in the course of the test, we developed a special scheme using an inductive sensor, which allowed having the measurement results free from the radial motion variation and thermal deformation of the tested specimen.

The analysis of recorded parameters established the following indicators of tribological properties:

- running-in time t_0 (hrs), which is defined as the time from the start of the test until the time when the wear curve reaches the area of normal wear;
- running-in wear h_0 (micrometers) which is defined as the value of binding at the end of running-in time t_0;
- friction coefficient value at the end of test, f;
- f_0 / f - the ratio of the friction coefficient maximum value during running-in f_0 to its value at the end of the test f;
- the average value of the wear rate during normal wear is $I_h = (h - h_0) / (L - L_0)$, where h (micrometers) is the total value of the sample wear during the test; L (micrometers) is the friction path covered by the sample surface during the test; $L_0 = 3{,}6 \cdot 10^9 \cdot t_0 \cdot \upsilon$ (micrometers) is the friction path covered by the sample surface during the running-in;
- the value of the wear rate during the total test time is $I_{h\Sigma} = h / L$.

The results of the tribology tests of the bearings are presented in Table 2.

Table 2. The results of the journal ball bearings test

Tribotechnical property	Indicator	Indicator value for the bearing			
		1	2	3	4
Running-in	t_0, hrs	2.73	2.23	2.23	2.53
	h_0, micrometers	15.0	14.5	2.0	2.2
	f_0/f	1.71	1.53	2.31	1.62
Antifrictionality	f	0.045	0.059	0.013	0.021
Wear resistance	h, micrometers	22.6	27.5	2.9	3.8
	$I_h \cdot 10^{-10}$	7.69	11.40	0.79	1.52
	$I_{h\Sigma} \cdot 10^{-10}$	12.46	15.16	1.60	2.09

Note. The bearings numbers correspond to those in Table 1.

It should be noted that the applied lubricant never leaves the friction area on the working surfaces of the journal ball bearings for the duration of the test, which is ensured by its properties.

Despite the significant wear of the hardened layer during the test of the journal ball bearings (initial running-in and normal wear), one should bear in mind that the period of running-in and the initial period of normal wear is the time of establishing the basic provisions and patterns of further friction and wear of the workpiece in the continuous service, which primarily affects its wear resistance and durability, along with the other relevant factors.

Also, the FPH coating wear products never leave the friction area and keep being an additional solid lubricant and a means of healing of micro-defects (microcracks, microchips, microcutting effects (abrasive scratches), etc.) at the friction surface.

Having determined that the products of wear of the coating of the Si-O-C-N affect the tribological characteristics of friction pairs and can help to prevent direct transfer of coating material to the indenter, and fill microwave and to gain a foothold in microvascular roughness of the contacting surfaces, which reduces specific pressures and increase the wear resistance of the friction pair.

The micrograph of the wear track in tests on a tribometer Tribometer (CSM, Switzerland) in conditions of dry friction coatings of the Si-O-C-N confirmed the formation of products that are not brought up, but remain at the bottom of the track, providing 'healing' of wear areas (Figure 2).

Figure 3 shows parameters of a strip of the wear coating of Si-O-C-N, with measurements of its width. The antiwear and antifriction action of the wear coating of Si-O-C-N is stored for a long time for the entire test period (the curve of coefficient of friction decreases) of friction pairs and are not associated with the running process.

When compared to the journal ball bearings manufactured by the factory technology, the wear resistance of the FPH-treated journal ball bearings increased by 5...14 times (in the values of wear rate during normal wear), by 6...9 times (in the values of wear rate for total test time).

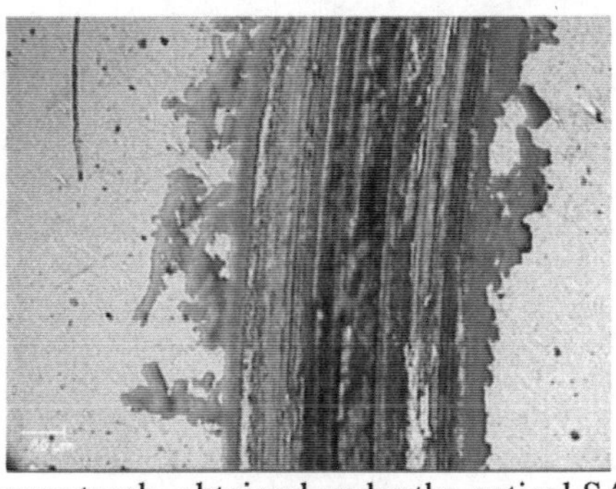

Figure 2. Micrograph of the wear track, obtained under the optical SA25 AXIOVERT microscope (Karl Zeiss, Germany), 100×

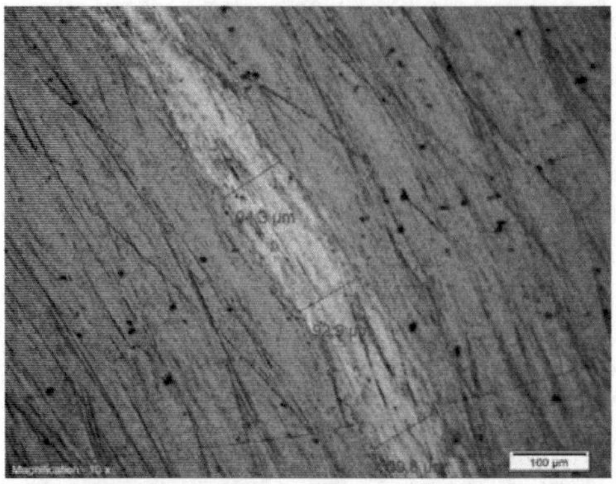

Figure 3. Appearance of the stripe wear coating of Si-O-C-N

The friction coefficient on the working surfaces of the FPH-treated journal ball bearings compared with the journal ball bearings manufactured by the base technology is 2...4 times lower.

3.Conclusion

The analysis of the comparative tests results of durability shows that the best tribological parameters are exhibited by the friction surfaces of the journal ball bearings formed by the technology of their modification with the finishing plasma hardening. The FPH technology can be applied at engineering plants as a highly efficient method of ensuring and improving the operational performance of machine parts at the stage of their manufacturing, in particular, in the manufacture of journal ball bearings.

References

[1] Grigoriev N and Tarasova T 2015 *Met Sci Heat Treat.* **10(724)** 5-10
[2] Stanski D et al 2015 *Met Sci Heat Treat.* **7(721)** 77-83

Features of plastics edge cutting machining

A V Handozhko, A N Shcherbakov, L A Zaharov and T V Gavrilenko
Bryansk State Technical University, 7, 50 years of October st., Bryansk, 241035, Russia

E-mail: msi@tu-bryansk.ru

Abstract. This article describes the features of pieces from thermoplastic materials in the form of electrical insulators cut by a disk edge tool. The problems in question are possible defects arising during machining and technological conditions that reduce their quantity. The necessity of required machining conditions matching substantiated in accordance with a specific grade of the material which is treated. Equipment and machining attachments, developed for experimental studies, determine the rational conditions of plastic electrical insulators machining. As a result of experiments the dependences of cut face quality parameters of plastics are obtained by machining conditions. The obtained results allowed us to make valid conclusions and recommendations.

1. Introduction

Good performance characteristics, high electrical properties, corrosion resistance, low-level noise during operation, high level of labor productivity during parts manufacture and low operating costs of plastics promote their integration into various branches of engineering.

The range of plastics application is wide: they are used as heat-insulating, electroinsulating, radio-technical, frictional, antifrictional, structural materials in units and mechanisms of machinery and devices.

Plastics are non-metallic materials, composition and production technology of which is different from metals composition and production technology. Details of plastics are produced by press forming, casting, molding, stamping. At the same time their size and shape are modified due to material shrinkage at cooling.

Except for polymer the plastic consists of: fillers, plasticizers, stabilizers, hardeners, lubricants, colorants, blowing agents and others. All these components affect the properties of plastics, including technological [1, 2, 3]. In terms of machining the fillers are very important. Fillers are used to improve mechanical properties of plastics, reduce shrinkage, and increase resistance to various types of ambient. The type of the filler, its properties and structure fundamentally affect the machinability of plastic and tool life [3, 4].

Plastics machining is required, widespread and one of important operations in the whole manufacture process for parts made of these materials. The necessity of machining requires the study of theoretical and practical issues of plastics cutting, edge tools progressive structures development, rational cutting conditions and special or specialized high-performance equipment.

Many scientists and nongraduates, such as A. Kobayashi, B.P. Shtuchnyi, V.I. Drozhzhin, N.V. Vyrezub et al., have studied machining of plastic materials. The guidance on practical implementation of different plastics machining, including requirements for cutting tools, have been developed according to these and other studies.

The influence of the tool cutting ability on the formed quality parameters of the workpieces surface layer is sufficiently large. The tool cutting ability changes during machining, which negatively affects the formation of quality parameters. The need to adjust machining conditions becomes inevitable. From this point of view an adaptive cutting process control system use is of great interest. One of them was developed by Russian scientists A.G. Suslov and D.I. Petreshin [5].

New types of plastics with special performance characteristics appearing in recent years are difficult to classify as proposed groups with cutting performance. This requires an additional research and development of new recommendations for their machinability.

A typical example is the manufacture of insulators of electrical connectors (Figure 1).

The main box-shaped workpiece of the electrical connector is an insulator made of thermoplastics of various grades, such as Lexan 3412R; Polyamide 66 structural and glass-filled; Teksan APK-2 black; Penlite GN-3130, Makrolon-8030 and others. Insulators are produced in wide range depending on operational characteristics and standard sizes of connectors.

A group of insulator with a maximum possible number of contacts is obtained by molding of plastics under pressure (Figure 1), which is subsequently cut up into required standard sizes depending on the number of contacts. This leads to a number of issues connected with machined surfaces required accuracy and quality provision (surface roughness, trims size, no chippings, melting).

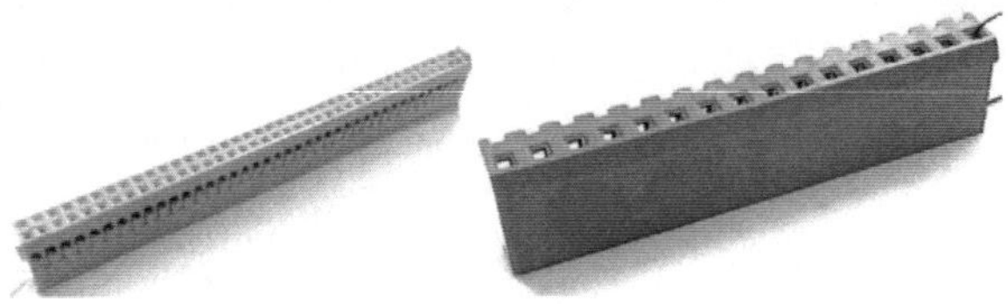

Figure 1. Insulators of electric connectors

Despite the widespread use of plastics in mechanical engineering, their cutting is most often carried out on equipment for metal working. In some cases, companies themselves design and manufacture special machine tools for plastics machining in order to fit their needs.

Disc cutters with a fine tooth of HSS (GOST2679-93) for metal working are used for plastics cutting.

The use of the cutter for metal as a cutter for plastics is a compulsory measure. The plastics, cut by the metal cutter, include thermoplastics (GOST 20324-74) produced only with a large diameter of 315 and 400 mm with thickness of 4 and 5 mm, respectively.

Insulators cutting by means of such technological equipment allows producing relatively reliable parts of required accuracy. However, roughness provision causes difficulties and the burr, which arises, is large in size and is strongly bounded with the part. Burr cutting and, in some cases, surfaces grinding must be done manually.

2. Task description

The following problems must be solved in order to study the cutting process of blanks of thermoplastic materials, such as electrical insulators:

- To develop a methodology for research;
- To design and manufacture the equipment and machining attachments for experimental studies;
- To conduct necessary studies on cutting the items from thermoplastic materials;

Process results will be obtained in order to identify dependencies of machined surfaces quality parameters on machining conditions and make appropriate conclusions and recommendations.

3. Hardware development for experimental studies

A special system based on the horizontal milling machine was developed in order to conduct researches to provide a required accuracy of the surface layer and quality parameters (Figure 2).

Figure 2. A system for experimental studies of plastic insulators cutting

The electric spindle of the SHFV grade with a capacity of 0.6 kW and the shaft with the speed of up to 18,000 min^{-1} were used in order to ensure a wide range of cutting rate control. The rotation of the electric spindle shaft (cutting speed) was changed by a variable frequency driver. The electric spindle is mounted on the support plate, attached on the machine arbor support. The pneumatic pipeline with an air-borne oil circulating system has been placed in order to provide its lubrication or cooling.

The capabilities of the drive of the machine axis were used for cross-feed movement. The cutting was performed by a disc cutter for metal working with increased (up to 15°) back angles.

A special device installed on the working surface of a special dynamometric table was designed and produced for insulators blanks cutting (Figure 3).

Figure 3. System for experimental studies of plastic insulators cutting

The dynamometric table was used in order to measure the cutting force components. Strain-measuring module LTR-212 of crate system L-CARD with appropriate software processed the obtained data.

Burr size was measured by means of a microscope with an output for connection to PC.

The roughness integrator, model 'ABRIS PM7', measured the roughness on the cut surface.

Machining preconditions, namely cutting depth, chosen equal to thickness of the insulator (from 4 to 10 mm), were selected on the basis of available literary recommendations. These recommendations include: the range of cutting speeds is from 750 to 4500 m/min, the feed per tooth is from 0.0005 mm to 0.01 mm, the cutter rake angle is 10 °, the back angle is15 °. The feed during cutting is associated.

The cutting and feed speed varied during experiments. The roughness and burr size were used as output parameters.

4. Results and conclusions

Electrical insulators made of material Lexan 3412R were used as the samples for research implementation.

During the experimental research the following results have been obtained.

A feed increase causes a significant increase of roughness within the feed change range under study. The cutting speed influence on generated surfaces roughness is insignificant (Figure 4.5).\

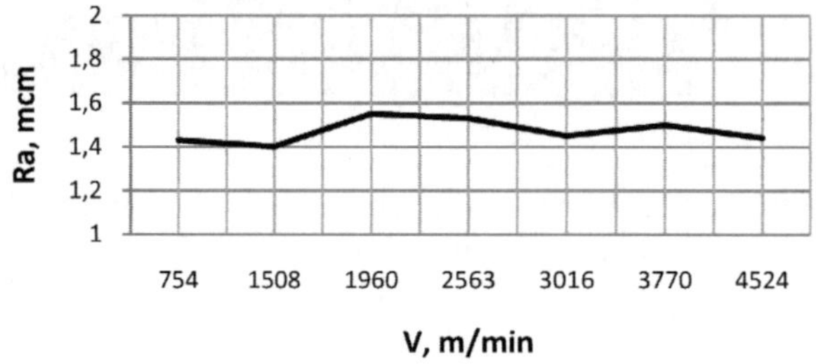

Figure 4. Diagram of the dependence of cut surface roughness on cutting speed Ra= f (V) at Vs = 500 mm/min,\

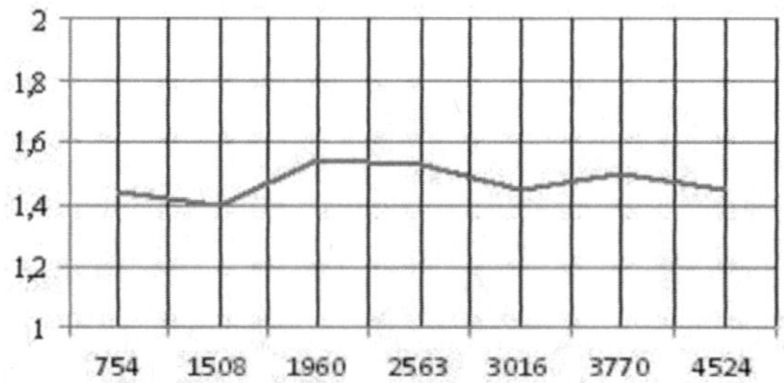

Figure 5. Diagram of the dependence of cut surface roughness on cutting speed Ra= f (V) at Vs = 1000 mm/min,

The burr size was evaluated through thickness (S) of its root when analyzing samples photos made by the microscope. Figure 6 shows the typical burrs arising at cutting.

a) b)

Figure 6. Photos of burrs at insulators cutting:
a) burr thickness is 0.46 mm, b) burr thickness is 0.85 mm

The results of measurements processing are presented in Figure 7.

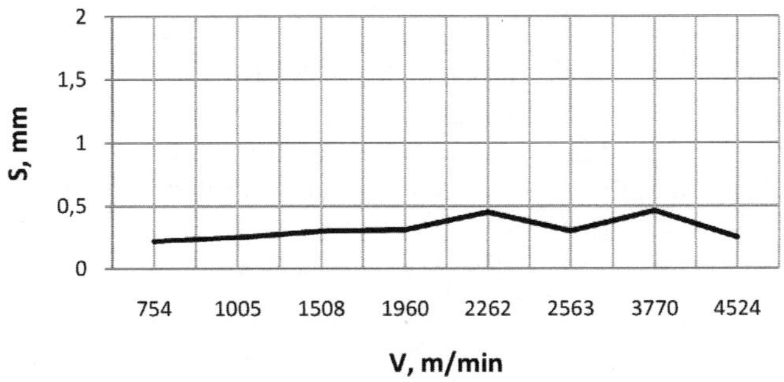

Figure 7. Diagram of the dependence of the burr size on cutting speed s = f (V) at Vs = 500 mm/min.

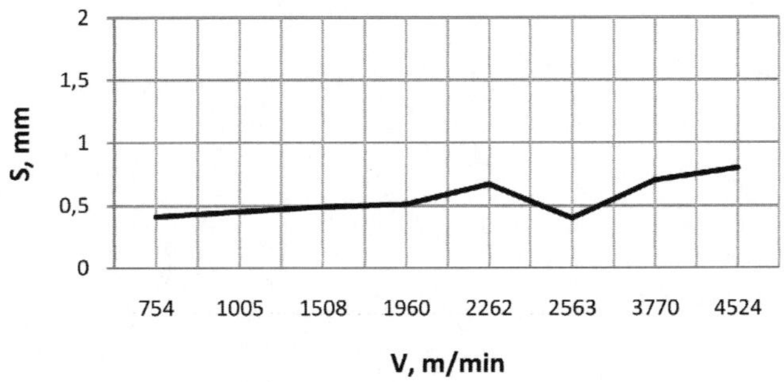

Figure 8. Diagram of the burr size dependence on cutting speed s = f (V) at Vs = 1000 mm/min.

Graphical dependences partly give an indication of cutting conditions influence on the burr size.

Burr bond strength is difficult to assess experimentally with detail. Therefore, this assessment can be considered as conditional. The efforts separating the burr declined with cutting forces increase and increased with feeds growth.

The radius of cut edge rounding and a wear rate along the rear surface greatly influence the cut quality. Cutter blunting leads to deterioration of the cutting quality. This effect decreases with the cutting speed increase.

According to the results of experiments a number of conclusions can be made: 1) high speed machining of thermoplastics allows improving a generated surfaces quality; 2) the cutting speed increase as a whole has a positive effect on cutting process characteristics and machining quality within the range under study; 3) the feed increase within the range under study leads to the cutting forces and roughness increase, in some cases the defects destruction are possible at material melting; 4) in some ranges of spindle rotational speed the resonance phenomena develops, tool and spindle equipment should be precisely balanced, as well as the strict regulation of their beating must be taken into account; 5) under proper selection of machining conditions, as well as the strict control of the cutting tool state the machining can proceed without burr formation.

The experimental data have been obtained relatively only one of the above-mentioned materials, which is Lexan 3412R.

The implementation of further experimental studies is required in order to subject plastics of other grades to DB machining.

The results of these studies were used to develop a special machine for cutting the parts of the electrical plastic insulator with low frequency.

IOP Conf. Series: Materials Science and Engineering **124** (2016) 012155 doi:10.1088/1757-899X/124/1/012155

Reference

[1] Osswald T A, Turng Lih-Shehg and Gramann P J 2002 *Hanser Gardner Publications, Inc., Cincinnati*
[2] Crowder M L, Ogale A M, Moore E R and Dalke B D 1994 *Polym.Eng.Sci.* **34(19)** 1497
[3] Domininghaus H 1993 *Hanser Publishers, Munich*
[4] Schwarz O, Ebeling F W and Furth B 2002 *Vogel Buchverlag*
[5] Suslov A G and Petreshin D I 2010 *Russian Engineering Research* **30(4)** 418-423

Kinematic Analysis of Rotary Deep-Depth Turning Parameters

A M Gurtyakov, A S Babaev and A I Chudinova
Tomsk Polytechnic University, 30, Lenin Avenue, Tomsk, 634050, Russia

E-mail: temkams@mail.ru

Abstract. This article offers a parameterization procedure for deep-depth turning without simplifying assumptions. In this paper the authors will show comparative results of researching parameters for rotary turning according to the developed methodology and according to the methodology of finish-machining conditions. A theoretically found kinematic coefficient of rotary cutting is presented in the paper

1. Introduction

Several investigators have made analytical researching of the parameters of rotary turning. However, not all the authors solved this task in relation to finish machining with the help of simplifying assumptions. If not to doubt the integrity of the obtained results when analyzing “shallow-depth” ($t \leq 1{,}5$ mm) rotary cutting parameters it is logical to suppose that the solution of the mentioned task with the developed methods will be incorrect. Therefore, this work offers researching parameters for rotary deep-depth turning without simplifying assumptions.

2. Methods

Figure 1 shows a scheme for defining the parameters of rotary deep-depth turning and their symbols [1, 3-5]: φ – horizontal cutter of the taper angle; β – vertically dipped cutter angle; h – cutter tool point position in relation to the feed plane going through the axis of the workpiece; B – crown of the cutter; C – crown on circumferential cutting edge of the cutter; A – cutter`s entry point of contacting with the workpiece; M – AB contact of the arc current point of the cutter with the workpiece; Θ – cutter contact angle with the workpiece; Ψ_M – current angle parameter defining the M point position on the contact arc; H – point C position in relation to the plane.

Figure 1 shows two frames: XYZ axes and X´Y´Z´ axes. The frame of axes X,Y,Z is connected to the workpiece and the frame of axes X´Y´Z´ is connected to the cutter [1]. Axis OX coincides with the axis of the workpiece and is directed towards the longitudinal feed. Axis OY is in the base plane. Axis OY is on the basic plane. Point B, the crown of the cutter, is in plane YOZ. Point O, central point, is on the axis of the workpiece.

The central point in the frame of axes X´Y´Z´ is combined with point O', the cutter`s center. A conversion from (one) frame of axes to the other one is done according to the formulas [2]:

$$X = a + a_{11}X' + a_{21}Y' + a_{31}Z';$$
$$Y = b + a_{21}X' + a_{22}Y' + a_{32}Z'; \qquad (1)$$
$$Z = c + a_{13}X' + a_{23}Y' + a_{33}Z'.$$

Where a, b, c are the coordinates of the cutter`s center in system XYZ.

The matrix of linear transformation consists in the rotating coordinate system in space at angles φ and β:

$$c = \begin{vmatrix} a_{11} & a_{21} & a_{31} \\ a_{12} & a_{22} & a_{32} \\ a_{13} & a_{23} & a_{33} \end{vmatrix} \quad (2)$$

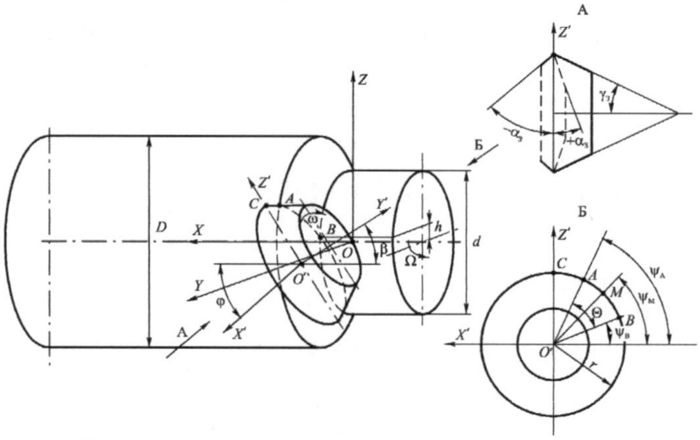

Figure 1. Scheme of rotary deep-depth (t) turning

In our case coefficients a_{ij} are expressed through the setting parameters in the following way:

$$c = \begin{vmatrix} \cos\varphi & -\sin\varphi\cdot\cos\varphi & \sin\varphi\cdot\sin\varphi \\ \sin\varphi & \cos\varphi\cdot\cos\varphi & -\cos\varphi\cdot\sin\varphi \\ 0 & \sin\varphi & \cos\varphi \end{vmatrix} \quad (3)$$

The coordinates of point O in system XYZ will accordingly be:

$$X_B = 0;$$
$$Y_d = \sqrt{\frac{d^2}{4} - h^2}; \quad (4)$$
$$Z_B = h.$$

where d – a diameter of the machined surface. A tangent line vector to the cutting edge at point B has coordinates $\{-X'_B, 0, Z'_B\}$

The equation of the tangent plane to the cylinder of the machined surface at point B:

$$Y_B \cdot Y + Z_B \cdot Z = \frac{d^2}{4}. \quad (5)$$

This is an equation of this plane in system X'Y'Z' taking into account that (4):

$$\sqrt{\frac{d^2}{4} - h^2}\cdot\sin\varphi x' + \left(\sqrt{\frac{d^2}{4} - h^2}\cdot\cos\varphi\cdot\cos\varphi + h\cdot\sin\beta\right)\cdot y' +$$
$$+\left(-\sqrt{\frac{d^2}{4} - h^2}\cdot\cos\varphi\cdot\sin\varphi + h\cdot\cos\beta\right)\cdot Z' + \sqrt{\frac{d^2}{4} - h^2}\cdot b + h\cdot c = 0 \quad (6)$$

The criterion for the parallelism of tangent to the cutting edge and the plane can be written like this:

$$Z'_B\sqrt{\frac{d^2}{4}-h^2}\cdot\sin\varphi+X'_B\left(\sqrt{\frac{d^2}{4}-h^2}\cdot\cos\varphi\cdot\sin\varphi-h\cdot\cos\beta\right)=0 \tag{7}$$

The equation for the cutting edge of the cutter is:

$$\begin{aligned}X'_B+Z'^2_B&=r^2;\\ Y_B&=0\end{aligned} \tag{8}$$

Having solved equations (7) and (8) simultaneously we obtain the coordinates of cutter`s point B crown in system X´Y´Z´:

$$X'_B=\frac{r\cdot\sin\varphi\sqrt{\frac{d^2}{4}-h^2}}{\sqrt{\frac{d^2}{4}-h^2\cdot\sin^2\varphi+\left[\sqrt{\frac{d^2}{4}-h^2}\cdot\cos\varphi\cdot\sin\varphi-h\cdot\cos\beta\right]^2}};. \tag{9}$$

$$Y'_B=\frac{r\cdot\left(\sqrt{\frac{d^2}{4}-h^2}\cdot\cos\varphi\cdot\sin\varphi-h\cdot\cos\beta\right)}{\sqrt{\frac{d^2}{4}-h^2\cdot\sin^2\varphi+\left[\sqrt{\frac{d^2}{4}-h^2}\cdot\cos\varphi\cdot\sin\varphi-h\cdot\cos\beta\right]^2}} \tag{10}$$

$$Z'_B=\frac{r\cdot\left(\sqrt{\frac{d^2}{4}-h^2}\cdot\cos\varphi\cdot\sin\varphi-h\cdot\cos\beta\right)}{\sqrt{\left(\frac{d^2}{4}-h^2\right)\cdot\sin^2\varphi+\left[\sqrt{\frac{d^2}{4}-h^2}\cdot\cos\varphi\cdot\sin\varphi-h\cdot\cos\beta\right]^2}} \tag{11}$$

An angle parameter, which defines the position of point B on the cutting edge, can be calculated like this:

$$tg\psi_B=\frac{Z'_B}{X'_B}=\frac{\sqrt{\frac{d^2}{4}-h^2}\cdot\cos\varphi\cdot\sin\varphi-h\cdot\cos\beta}{\sin\varphi\cdot\sqrt{\frac{d^2}{4}-h^2}} \tag{12}$$

Let us simplify the expression (12) and insymbol

$$tg\zeta=\frac{h}{\sqrt{\frac{d^2}{4}-h^2}}.$$

Thus we will finally have:

$$tg\psi_B=ctg\varphi\cdot\sin\beta-tg\zeta\cdot\frac{\cos\beta}{\sin\varphi} \tag{13}$$

We will find parameters a, b, c from system (1) when substituting values X_B, Y_B, Z_B, X'_B, Y'_B, Z'_B:

$$a=\frac{r\cdot\left[\cos\varphi\cdot\sin\varphi\cdot\sqrt{\frac{d^2}{4}-h^2}-\sin\varphi\cdot\cos\beta(\cos\varphi\cdot\sin\beta-tg\zeta\cdot\cos\beta)\right]}{\sqrt{\sin^2\varphi+(\cos\varphi\cdot\sin\beta-tg\zeta\cdot\cos\beta)^2}} \tag{14}$$

$$b=\sqrt{\frac{d^2}{4}-h^2}+\frac{r\left(\sin^2\varphi+\cos^2\varphi\cdot\sin^2\beta-tg\zeta\cdot\cos\beta\cdot\cos\varphi\cdot\sin\beta\right)}{\sqrt{\sin^2\varphi+(\cos\varphi\cdot\sin\beta-tg\zeta\cdot\cos\beta)^2}} \tag{15}$$

$$c=h\frac{r\cdot\cos\beta(\cos\varphi\cdot\sin\beta-tg\zeta\cdot\cos\beta)}{\sqrt{\sin^2\varphi+(\cos\varphi\cdot\sin\beta-tg\zeta\cdot\cos\beta)^2}} \tag{16}$$

Expressions (14), (15), (16) can be written in a simpler and more practically convenient form:

$$a=r\cdot(\cos\varphi\cdot\cos\psi_B-\sin\psi_B\cdot\sin\varphi\cdot\sin\beta) \tag{17}$$

$$b=r\cdot(\cos\varphi\cdot\cos\psi_B-\sin\psi_B\cdot\sin\varphi\cdot\sin\beta)+\sqrt{\frac{d^2}{4}-h^2} \tag{18}$$

$$c=h-r\cdot\cos\beta\cdot\sin\psi_B \tag{19}$$

It is necessary to know the position of the highest cutter`s point C with known parameters φ, β, h, r, d when designing rotary cutters and making experiments. The coordinates of point C in system X′Y′Z′ will be {0, 0, r}. We obtain the following from equation 3 of simultaneous equation (1):

$$H=c+r\cdot\cos\beta \tag{20}$$

The angle of a cutter`s angle with a workpiece can be defined if we know the values of angles ψ_B and ψ_A. We can calculate angle ψ_A from the conditions that point A of the cutting edge belongs to the cylinder of the surface to be machined:

$$Y^2+Z^2=\left(\frac{d}{2}+t\right)^2, \tag{21}$$

where t – depth of cutting.

Equation (20) will look like this in the system of coordinates X′Y′Z′.

$$(b+a_{12}X'+a_{22}Y'+a_{33}Z')^2+(c+a_{13}X'+a_{23}Y'+a_{33}Z')^2=\left(\frac{d}{2}+t\right)^2 \tag{22}$$

Let us solve equations (8) and (22) simultaneously but in relation to point A:

$$X'_A+Z'_A=r^2;$$

$$Y'_A=0$$

And when substituting coordinate values $X'_A=-r\cdot\cos\psi_A$, $Y'_A=0$, $Z'_A=r\cdot\sin\psi$ into equation (19) we will obtain:

$$(b-r\cdot\sin\varphi\cdot\cos\psi_A-r\cdot\cos\varphi\cdot\sin\beta\cdot\sin\psi_A)^2+(c+r\cdot\cos\beta\cdot\sin\psi_A)^2=\left(\frac{d}{2}+t\right)^2 \tag{23}$$

Transcendental equation (23) in relation to parameter ψ_A can be transformed to the algebraic equation of degree 4 with the help of trigonometric rearranging. It is quite time-consuming to solve the equation of degree 4 directly. Here we use the known methods to simplify the solution of this task. However, equation (23) was solved without any simplifying assumptions in relation to ψ_A and with the use of the computer to avoid inaccuracy when defining the contact arc.

The angle of contact Θ is defined the following way:

$$\Theta=\psi_A-\psi_B \tag{24}$$

3. Results and discussion

Figure 2 shows the dependences of contact angle Θ on the depth of cutting. These dependences are calculated with the simplifying assumptions (curve 1) and without them (curve 2). The dependences imply that the method of approximate solution of equation (22) leads to essential inaccuracies in defining the value of contact angle Θ if we have deep depths of cutting. The bigger is the depth of cutting, the higher is the inaccuracy. So, the inaccuracy can reach 98% if depth of cutting t is 15 mm.

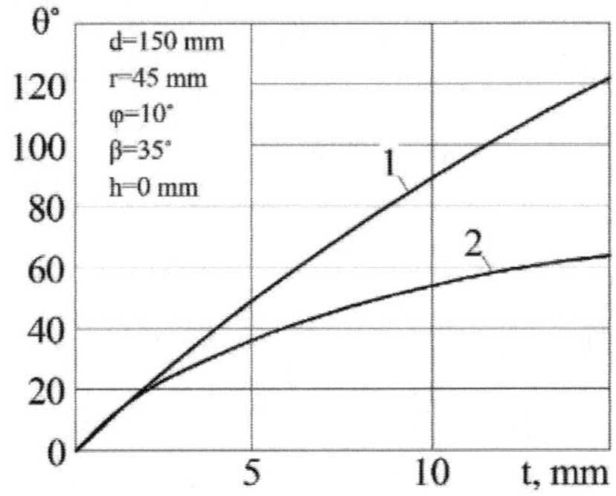

Figure 2. Dependence of depth of cutting t on contact angle θ
(1 – with simplifying assumptions; 2 – without the assumptions)

The inaccuracy in defining the contact angle is higher if setting angle φ increases and setting angle β decreases (Figure 3).

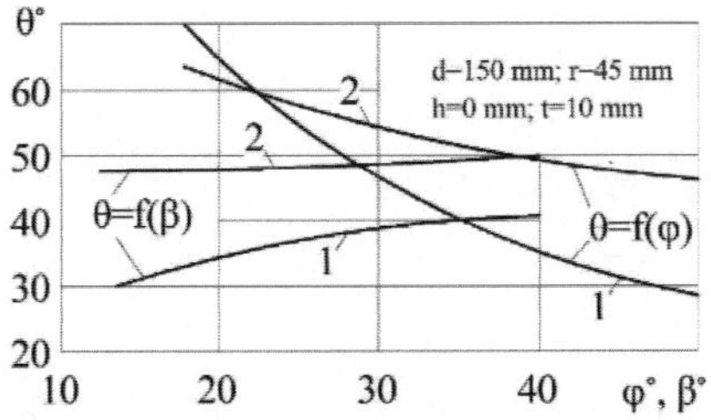

Figure 3. Influence of setting angles β and φ on the inaccuracy when defining contact angle θ
(see symbols in Figure 2)

One of the most essential parameters in the process of rotary turning is the value of ratio between cutter rotary velocities V_{tool} and workpiece to be machined V_w. This ratio is accepted to be called as kinematic coefficient K.

The importance of ratio V_{tool} / V_w is explained by the fact that the ratio predefines power, temperature and wear resistant characteristics of the cutting process as well as the value of sliding velocity of tool`s useful areas in relation to the material to be machined. The kinematic coefficient is defined theoretically without simplifying assumptions:

$$K = \frac{1}{\Theta R_W}\begin{bmatrix} c \cdot \sin\varphi(\cos\psi_A - \cos\psi_B) - r \cdot \cos\beta \cdot \sin\varphi \cdot \Theta + \\ (c \cdot \cos\varphi \cdot \sin\beta + b \cdot \cos\beta)(\sin\psi_A - \sin\psi_B) \end{bmatrix}, \tag{25}$$

where R_w – radius of the workpiece to be machined.

Conclusions

This paper is particularly valuable because it describes the kinematic coefficient that allows one to make theoretical analysis of the rotary turning process and also to design cutters with optimal geometric and setting parameters. The experimental check of the obtained dependences for the analytical research of the main parameters of rotary deep-depth turning proved the reliability of the

IOP Conf. Series: Materials Science and Engineering **124** (2016) 012156 doi:10.1088/1757-899X/124/1/012156

results in case of changing wide-range machining conditions. Disarrangements between experimental data and the data obtained by means of calculation according to the presented dependences did not exceed 5,8%.

The main results and conclusions are summarised as follows:

- Using the methods of researching rotary turning with simplifying assumptions leads to sufficient inaccuracies and cannot be applied in researching parameters of rotary deep-depth turning.
- The developed method allows making computerized analytical researches of all the parameters of rotary turning with high reliability of the obtained results.

References

[1] Marghitu D 2005 *Academic Press* 792
[2] Kadry S 2013 *Elsevier* 144
[3] Olgun U and Budak E 2013 *Procedia CIRP* **8** 81-87
[4] Kossakowska J and Jemielniak K 2012 *Procedia CIRP* **1** 425-430
[5] Suzuki N 2014 *Procedia CIRP* **14** 125-129

Effect of the structure heterogeneously hardened by impact deformation waves upon impact strength of the material

A V Kirichek[1], **D L Soloviev**[2] **and S A Silant'ev**[2]

[1] Bryansk State Technical University, 7, 50 years of October st., Bryansk, 241035, Russia
[2] Murom Institute of Vladimir State University, 23, Orlovskaya st., Murom, 602264, Russia

E-mail: avk.57@yandex.ru

Abstract. This work demonstrates possibilities of development of a heterogeneous hardened structure with alternating hard and plastic areas using impact deformation waves. There is a possibility of preserving impact strength of the metal material during formation process of heterogeneous hardened structure.

1. Introduction

A heterogeneous system is a system consisting of phases with different physical or chemical properties separated by surfaces, where one or several system's properties change rapidly in a step-like manner. If such separation surfaces do not exist, the system is considered to be uniform or homogeneous.

All metal alloys used in machine engineering possess a non-homogeneous polycrystalline structure, and therefore they belong to heterogeneous materials with anisotropic physical and mechanical properties [1, 2]. To meet the requirements of "ideal" metal, the heterogeneous metallic material should have high yield strength and tensile strength together with high fatigue limit. It also should be strong and have plastic and sticky properties at the same time. So, when the material is formed and processed, the heterogeneous properties of the material must be specified.

In general, uniformity of the structure of the metallic material is controlled by introduction of alloying additives at a production stage. Thus, the material obtains required working properties.

In recent times, the heterogeneous hardened structure was successfully formed by gradient hardening treatment, such as coating, surface heat treatment, laser hardening and chemical heat treatment that form hard and plastic areas with a specified alternation in the surface layer.

The purposeful development of advanced technologies on creation of the gradient heterogeneous material with improved structural and functional properties reinforced with areas of high-strength phase surrounded by plastic and viscous layers is discouraged by the stereotype view related to necessity of uniform hardening of the surface layer. Such perception is explained by the complex effect of plastic interlayers on mechanical properties of the heterogeneous material. The transition regions from the solid material to the plastic one formed during the hardening process can become additional stress concentrators and develop fatigue micro cracks and premature failure under the action of cyclic loads.

Therefore, the use of strain hardening, which provides a smooth transition from the high-strength phase with the hardness of 6500 MPa to a plastic interlayer of the hardened material, is quite promising for the development of the heterogeneous gradient hardened structure.

In addition, hardening by surface plastic deformation (SPD) has other advantages, such as machining of parts of any size and configuration, production effectiveness, ease of implementation, mechanization and automation of the process, the possibility of local hardening in a given component representing structural stress concentrators [3-6].

For the first time ever, a strain hardened heterogeneous structure was created by using impulse waves; the method was named as static-pulse treatment [7, 8]. In case of hardening with the static-pulse treatment method, periodic impact force applies the load upon the surface using a striker via a statically loaded waveguide. When the striker hits the waveguide it forms an impulse wave which is transferred to the treated material. Static pre-loading of the waveguide with a tool does not allow the tool to leave the contact with the loaded surface after impact, providing recovery of reflected deformation waves. A geometry of the striker and waveguide is selected in such a way that the impulse waves would transfer maximum energy to the hardened metallic material. This has provided a better way to create a deeper hardened surface layer, than using well-known techniques of strain hardening, such as rolling, flattening or shot peening.

A mechanism that controls the uniformity of the hardened surface layer, depending on the parameters of the hardening with deformation impulse waves is as follows. Impulse waves form plastic impressions on the surface with a shape and a size according to specified numbering determine the depth, extent and uniformity of the surface layer hardening. A coefficient of plastic overlapping impressions is a generalized characteristic that links the size of the impression and numbering of the deformation action.

$$K = 1 - \frac{S}{\delta f 60},$$

where S – conveying speed of the work piece relative to the tool, mm/min; δ – typical size of the impression measured towards a conveying direction, mm; f – impact impulse frequency, Hz.

With $K = 0$, the edge of one impression is adjacent to the edge of the other one; with $0 < K < 1$, the impressions overlap; with $K = 1$ the tool indents into in the same place several times [9, 10].

Such heterogeneous hardened structure works under contact cyclic loads quite effectively. As a result, the study has shown an increase of its durability up to 7 times [11, 12].

Often, machine components (gears, cams, tappets, etc) working under different types of loads at the same time, require an increase of the impact strength in addition to contact fatigue strength. However, it has been established that the hardening with increasing hardness forces the material to reduce its capabilities to withstand impact loads. No studies have been carried out on the effect of the heterogeneous hardened structure upon the change of impact toughness..

2. Results and Discussion

An experimental investigation was undertaken to establish the effect of uniformity of the steel hardened structure created by surface plastic deformation with the use of impulse waves upon impact strength.

The investigation involved a plane sample made of 30HGSA steel (initial hardness of 2800 MPa) of 6 mm thick. The sample was hardened with rod-shaped rolls: roll diameter was 10 mm and width of the roll was $b_r = 20$ mm, 30 mm and 40 mm. The energy of the deformation wave was $A = 150$ J, impact frequency was $f = 10$ Hz. Specific energy of blow $a = A / b_r$ was 7.5 J/mm, 5 J/mm and 3.75 J/mm, accordingly. The heterogeneous structure was formed in the conveying direction with overlapping of plastic impressions (Figure 1). For the purpose of investigation of the impact strength, the sample was cut out across plastic impressions created on the hardened surface (section A-A) and along the impressions (section B-B). This is obvious that the hardened structure in section A-A has greater non-uniformity than in section B-B.

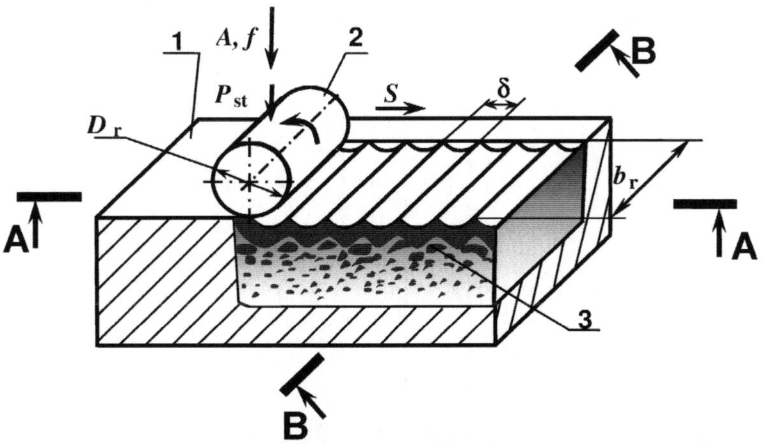

Figure 1. Pattern of impressions made at the flat surface of the work piece carried out with rod-shape roll at hardening with deformation wave:
1 – sample, 2 – rod-shaped roll, 3 – hardened region
A – energy of impulse deformation waves, *f* – impact frequency, P_{st} – static load, *S* – conveying speed of the work piece relative the roll, D_r – roll diameter, b_r – roll width, δ – typical size of the impression, mm

It has been established that with the increase in the specific energy of the impulse wave deformation used for hardening, the hardness of the hardened samples is increased and the impact strength is decreased. The maximum value of hardness for solid regions located at the top of the hardened surface layer was recorded in operation mode a=7.5 J/mm, K=0.66, the measured value was 4100-4200 MPa. The minimum value for solid regions was obtained in operation mode a=3.75 J/mm, K=0.48 the measured value was 3900-4000 MPa, size and number of solid regions are significantly less than those in the soft plastic interlayer. The hardness changed throughout the whole thickness of the hardened samples.

It has been established that with the increase of the overlap ratio, the impact strength is virtually the same. Thus, with a=3.75 J/mm, K=0.48 the impact strength measured across the plastic impressions (section A-A, Figure 1) was 117 J/cm^2, and for K=0.8 it was 116 J/cm^2. With a=7.5 J/mm the impact strength measured across the plastic impressions (section A-A, Figure 1) equals 108 J/cm^2 as it is when K=0.39 and K=0.66 (Figure 2). The explanation implies that the impact strength is more affected by the structure of the hardened material located below the surface, in a subsurface layer at a certain depth. Hardening with overlap factor K=0.48 leads to development of the highly heterogeneous hardened surface layer throughout the depth of the material. Hardening with K=0.8 in the upper part leads to development of the hardened highly uniform surface layer. This layer has little effect on the impact strength as it is supported by the heterogeneous hardened underlayer.

It has been established that a hardened surface has larger impact strength in direction with the heterogeneous structure rather than in direction with a more uniform hardening. Thus, with *a*=3.75 J/mm, *K*=0.48 the impact strength measured across plastic impressions (section A-A, Figure 1) was 117 J/cm^2, and the impact strength measured along plastic impressions (section B-B, Figure 1) was 102 J/cm^2. With *a*=5 J/mm, *K*=0.6 the impact strength measured across plastic impressions (section A-A, Figure 1) was 108 J/cm^2, and the impact strength measured along plastic impressions (section B-B, Figure 1) was 102 J/cm^2. With a=7.5 J/mm, K=0.39 the impact strength measured across plastic impressions on the hardened surface (section A-A, Figure 1) was 108 J/cm^2, and the impact strength measured along plastic impressions (section B-B, Figure 1) was 94 J/cm^2.

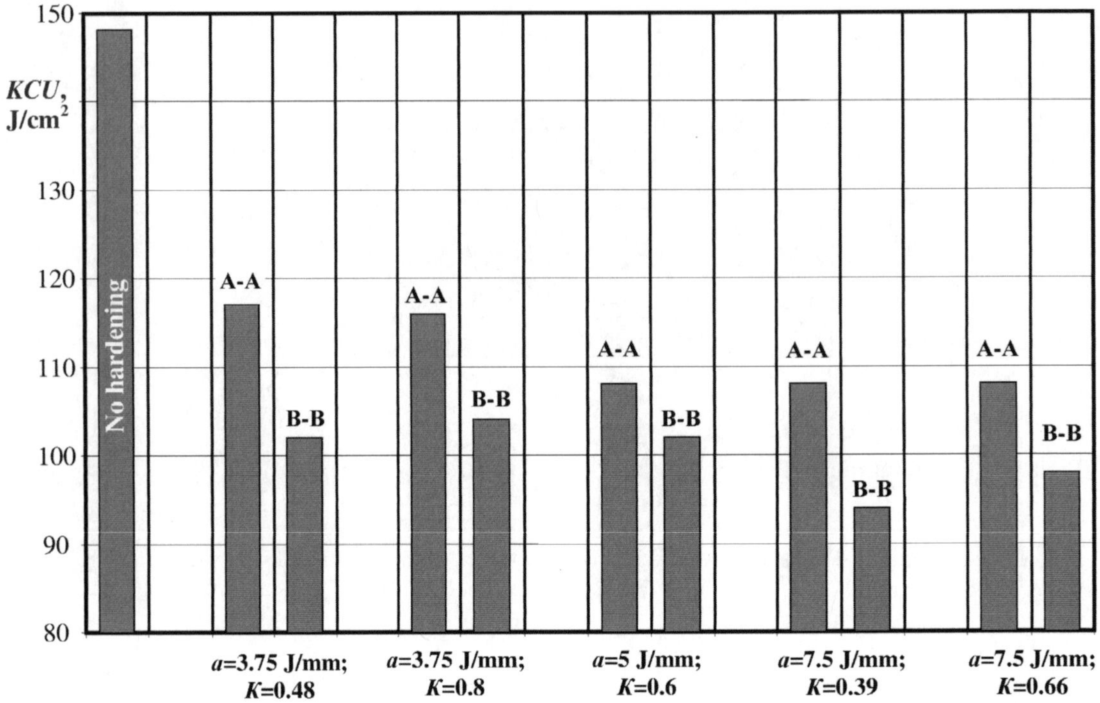

Figure 2. Impact strength of the material after hardening with the impact deformation wave measured in section:
A-A – across plastic impressions on the hardened surface;
B-B – along plastic impressions on the hardened surface

3. Conclusion

Hardening with impulse deformation waves develops a heterogeneous hardened structure in metallic materials; this structure is characterized by high hardness and plasticity.

Hardening of metallic materials by impulse deformation waves contributes to an increase in hardness and size of solid areas related to the heterogeneous hardened structure of the hardened surface layer, and reduces the impact strength as well.

The increase of the impression overlap ratio within investigated range K=0.39-0.8 virtually has no effect upon the impact strength change.

It has been established that a hardened surface has larger impact strength in direction with heterogeneous structure rather than in direction with a more uniform hardening.

Acknowledgement

The study was supported on the base of grant provided by RFBR, 14-08-00112.

References

[1] Dmitriev A I, Zolnikov K P, Psakhie S G, Panin V E and Goldin S V 2005 Low-density layer formation and "lifting force" effect at micro- and meso-scale levels *Theoretical and Applied Fracture Mechanics* **43(3)** 324-334

[2] Panin V E 1998 Overview on mesomechanics of plasticdeformation and fracture of solids *Theor. Appl. Fract. Mech.* **30** 1-11

[3] Odintsov L G 1987 *Strengthening and Parts Finishing by Surface Plastic Forming: Reference Book* (Moscow: Mechanical Engineering) p 328

[4] Smelyansky V M 2002 *Mechanics of Parts Strengthening by Surface Plastic Forming* (Moscow: Mechanical Engineering) p 300

[5] Schneider Yu G 1998 *Techniques for Pressure Finishing: Reference Book* (St. Petersburg: Polytechnic)

[6] Proskuryakov Yu G, Romanov V N and Isaiev A N 1984 *Opening Volumetric Burnishing*, (Moscow: Mechanical Engineering) p 224

[7] Kirichek A V and Solov'ev D L 2001 The methods of dynamic surface strengthening by plastic deformation *Kuznechno-Shtampovochnoe Proizvodstvo (Obrabotka Metallov Davleniem)* **7** 28-32

[8] Kirichek A V, Soloviev D L and Altuhov A Yu 2014 Deformation Wave Hardening of Metallic Materials *Journal of Nano and Electronic Physics* **6(3)** 03069(4pp)

[9] Kirichek A V, Solov'ev D L and Silant'ev S A 2004 Influence of regimes of static-pulse processing on uniformity of superficial layer hardening *Kuznechno-Shtampovochnoe Proizvodstvo (Obrabotka Metallov Davleniem)* **2** 13-17

[10] Kirichek A V and Solov'ev D L 2008 Creating heterogeneous surface structures by static-pulsed treatment *Russian Engineering Research* **28(3)** 277-279

[11] Kirichek A V and Soloviev D L 2013 Properties and Technology for Quasi-Composite Blanket Using Natural Reinforcement of the Metal by Strain Affected Areas *Journal of Nano and Electronic Physics* **5(4)** 04010(5pp)

[12] Kirichek A V, Soloviev D L and Altuhov A Yu 2014 Production of Quasicomposite Surface Layer of a Metal Material by Shock Wave Strain Hardening *Journal of Nano and Electronic Physics* **6(3)** 03070(4pp)

Increasing durability by deformational hardening under the conditions of back-to-back endurance by creating heterogeneous patterns

A V Kirichek[1], S V Barinov[2], A V Yashin[2]

[1] Bryansk State Technical University, 7, 50 years of October st., Bryansk, 241035, Russia
[2] Murom Institute of Vladimir State University, 23, Orlovskaya st., Murom, 602264, Russia

E-mail: box64@rambler.ru

Abstract. A new method of increase the back-to-back endurance of machine parts – the static-pulse processing, which creates a work-hardened surface layer with high hardness on large depth is proposed. An important feature of the static-pulse processing is the possibility to obtain the difference in uniformity of hardened (heterogeneous) surface layer. The researches on influence of parameters of the hardened surface layer obtained by the static-pulse treatment on the durability under the conditions of back-to-back endurance were conducted. As a result, there were the first obtained recommendations on the uniformity of strain hardening of the surface layer, contributing to the increase of the back-to-back endurance of machine parts from steels Ck45, 41Cr4 and 35HGSA.

1. Introduction

The performance capabilities of machines and mechanisms were usually determined by the strength and durability of the most loaded critical parts, the external influences perceived by the surface layer on which are usually the cause of failure. Therefore, in the manufacture of such parts a hardening treatment that forms the surface layer as efficient as possible during operation is used.

For parts subject to high cyclic contact loads, a hardened layer should have a reasonably large depth (1.5...2 mm) to prevent deformation and destruction. Furthermore, as the results of recent researches show, the uniformity of hardening touch on a great impact on performance, including back-to-back endurance. The advantage of surface hardened layer with uneven (heterogeneous) structure, experiencing the effect of cyclic contact stresses, is slowdown by tough and ductile material the brittle crack that is originated in a solid structural component. Today, however, the creation by traditionally used methods of hardening machining the deep hardened surface layer with the requisite uniformity is quite a challenge. For its analysis it is proposed to use a relatively new method of surface plastic deformation (SPD) – static-impulse processing (SIP), which has wide technological capabilities in the formation of the hardened surface layer and allows you to create a greater depth (up to 6...10 mm) and a high degree of hardening (up to 6500 MPas) [1]. During SIP the plastic deformation of the hardened material is carried out by shock pulses, whose shape is fully adapted to the material properties and loading conditions, that increases the EFFICIENCY of the process. The static component of the load is designed for the best use of the pulse component. The SIP technology allows to adjust pretty exactly the uniformity of hardening, creating as homogeneous and heterogeneous hardened structure [2]. The adjustment of the uniformity is achieved by overlapping the plastic imprints, received as the result of influence on the material of the shock pulses. The uniformity of plastic overlapping of imprints is evaluated using the coverage factor *K*. If $K = 0$, the edge of one imprint borders the edge of the other; if $0 < K < 1$, then the imprints are overlapped; if $K = 1$ there is multiple indentation of the tool in the same place [8-10].

2. The study on the possibility of increasing durability by deformational hardening under the conditions of back-to-back endurance by creating heterogeneous patterns

To assess the ability of the SIP to create the heterogeneous structure improving the durability of machine parts under the conditions of back-to-back endurance the samples of steels Ck45, 41Cr4 and 35HGSA with initial microhardness of 2000 – 2200 MPas were taken. The choice of material is due to the fact that the steel Ck45 is the standard in the engineering and steels 41Cr4 and 35HGSA are used to create the heavy-duty machine parts, working in conditions of cyclic contact loads. [11-12] The samples are the plates of 20 mm thick which were reinforced with different intensity of shocks ($5 < a < 7$ J/mm) and with different overlap factor K. The rod rollers with a diameter of 10 mm were used as a tool. The choice of these technological parameters of the SIP is due to the fact that they are significantly affect the shapes, sizes, depths and degrees of hardening of the forming heterogeneous structures [1-3].

On the hardened tracks cut under the samples, the microhardness on the grid with a certain step on depth and in the direction of feed of the SIP tool was measured. According to the results of measurements of micro hardness, the graphs characterizing the distribution of the degree of hardening on the depth and along the surface of the samples were built (fig. 1-3). The analysis of graphs showed that the increase of the overlap factor (from K=0.2 to K=0.5) leads to the formation of an explicit heterogeneous structure of the hardened surface layer, which is characterized by alternating the hard and soft sections of different sizes. [7]

To study the durability the hardened SIP samples were subjected to cyclic contact loading by the components in the form of balls with the same contact load on a specially designed device [4-6]. This allowed to form on the surface of samples at the same time the traces of several raceways, which were studied after testing.

The resistance to contact spalling ΔI is a complex parameter taking into account the change during testing of sizes of the raceways of rolling balls and the area of spallings arised on them, which allows to evaluate the durability of machine parts, working in conditions of contact fatigue.

The results of experimental testing of resistance to contact spalling are presented in form of graphs in figure 4.

The results of the testing of samples with heterogeneous unevenly hardened structure produced by SIP proved the efficacy of use of this technology. So, when studying the effect of overlap factor on ΔI was found that for the steel Ck45 the maximum values of $\Delta I = 4$ to 6 was observed on the value of $K = 0.4$ (Fig. 2, a). The most effective value of ΔI for steels 41Cr4 and 35HGSA were achieved when $K = 0.3$ and were respectively 2 to 2.3 and 2.1 to 2.8 (Fig. 2, b, c). Influence of the value of intensity of shock on resistance to contact spalling on all of the tested steels showed that with increasing of *a* the effective range of values ΔI moves in the direction of smaller values of the overlap factor, which is associated with the increase of the zone of deformable metal under a single imprint.

Conclusions

1. One of the most perspective directions of increase of the back-to-back endurance of machine parts is the creation of surface plastic deformation of heterogeneous hardened surface layer. One of the most effective ways of surface plastic deformation, allowing with high accuracy and in wide range to adjust the parameters of the hardened surface layer is static-pulse processing (SIP).

2. As a result of experimental studies is found that the most important technological factors of SIP that have the greatest impact on the value of resistance to contact spalling of the studied materials

(steels Ck45, 41Cr4 and 35HGSA) with heterogeneous structure, are the energy of the shock pulses, the overlap of the plastic imprints and elastic-plastic properties of material.

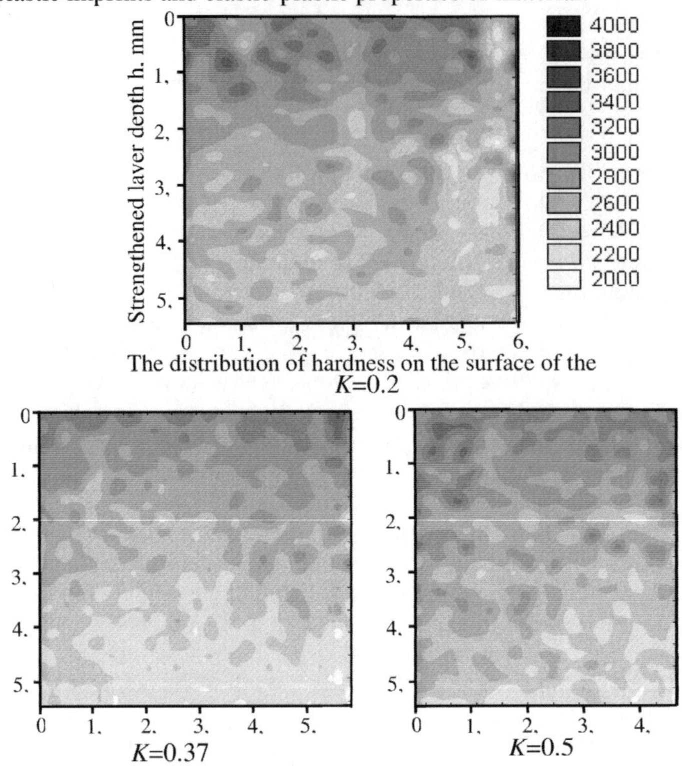

Figure 1. Diagrams of distribution of microhardness in the surface layer of steel Ck45.

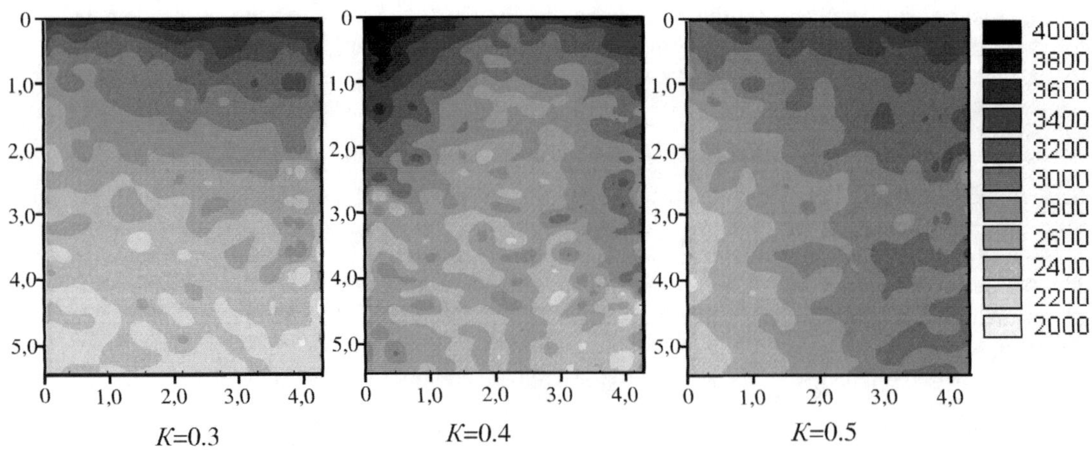

Figure 2. Diagrams of distribution of microhardness in the surface layer of steel 41Cr4.

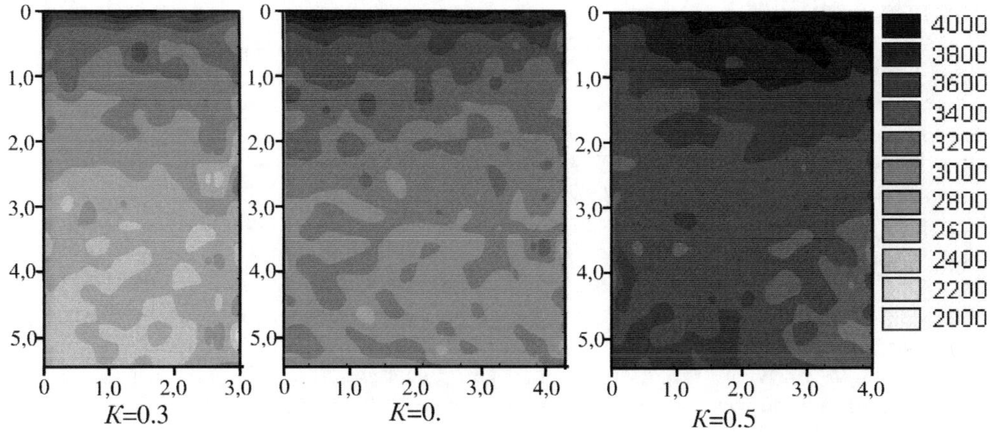

Figure 3. Diagrams of distribution of microhardness in the surface layer of steel 35HGSA.

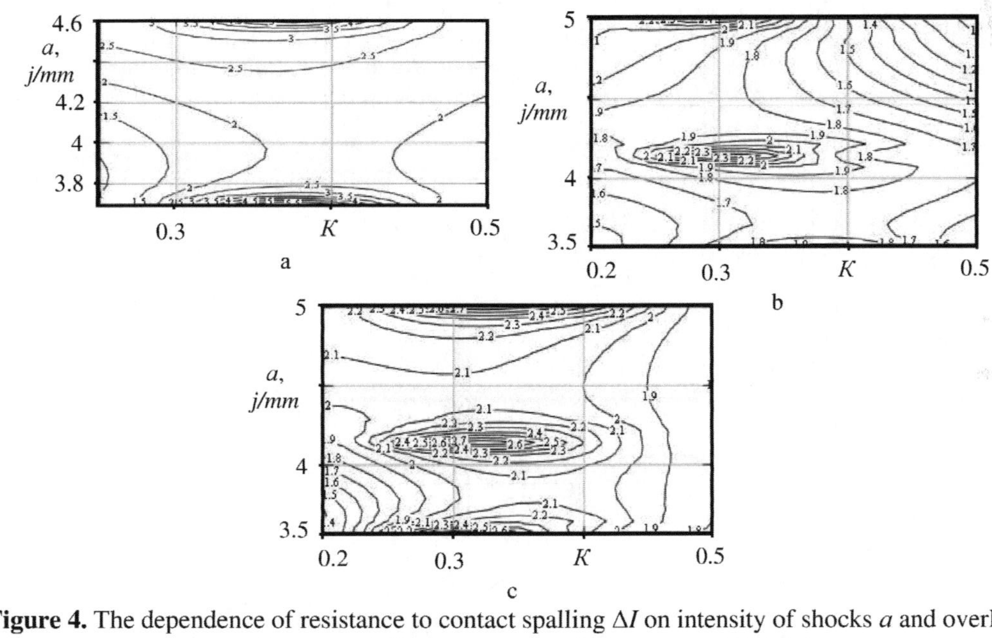

Figure 4. The dependence of resistance to contact spalling ΔI on intensity of shocks a and overlap factor K: a – steel Ck45; b – steel 41Cr4; c – steel 35HGSA.

3. For example for the steel 41Cr4 and 35HGSA, unlike steel Ck45, the maximum range of the resistance to contact spalling *(ΔI)* was achieved by the creation of heterogeneous structures with less uniformity of hardening, forming when the overlap factor of imprints K=0.3 (for the steel CK45 K=0.4). The numerical values of the resistance to contact spalling when creating a heterogeneous structure for steel Ck45 were in 2 to 2.5 times higher than the ΔI, obtained for steels 41Cr4 and 35HGSA.

4. When increasing the intensity of shock when creating a heterogeneous structure of the SIP, the optimal values are shifted towards smaller values of the overlap factors due to the increase of the zone of deformed metal under the single imprint.

Acknowledgement
The researches has been performed with the assistance of the Russian Foundation for Basic Researches No. 14-08-31263 mol_a.

References

[1] Kirichek A V, Soloviyev D L 2013 Nanostructure changes in iron-carbon alloys as a result of impulse deformation wave action, *Journal of Nano- and Electronic Physics.* **5 (4).**
[2] Kirichek A V, Barinov S V 2015 "Study of Methods Relating to Increase of Contact Pitting Resistance in 45, 40H, 35HGSA Steel due to Development of Heterogeneous Structure Involving Mechanical Hardening Technique", *Applied Mechanics and Materials*, **756**, 65-69.
[3] Kirichek A V , Barinov S V 2015 "Development of Parameters Describing Heterogeneous Hardened Structure", *Applied Mechanics and Materials*, **756**, 75-78.
[4] Kirichek A V, Soloviyev D L 2013 Nanostructure changes in iron-carbon alloys as a result of impulse deformation wave action, *Journal of Nano- and Electronic Physics.* **5 (4).**
[5] Kirichek A V, Soloviev D L, Altukhov A Y 2014 Production of quasicomposite surface layer of a metal material by shock wave strain hardening, *Journal of Nano- and Electronic Physics.* **6 (3).**
[6] Kirichek A V, Soloviev D L, Altukhov A Y 2014 *Deformation wave hardening of metallic materials, Journal of Nano- and Electronic Physics.* **6 (3).**
[7] Panin V E and Deryugin E E 2003 Mesomechanics of the formation of banded structures at mesoscopic and macroscopic levels. *The Physics of Metals and Metallography* **96** 2-15.
[8] Schneider Yu G 1998 *Techniques for Pressure Finishing: Reference Book* (St. Petersburg: Polytechnic).
[9] Siegel R W and Fougere G E 1995 *Mechanical properties of nanophase metals Nanostr. Mat.* **6** 1-4.
[10] Smelyansky V M 2002 *Mechanics of Parts Strengthening by Surface Plastic Forming* (Moscow: Mechanical Engineering)
[11] Valiev R Z, Korznikov A V and Mulyukov R R 1993 *Structure and Properties of Ultrafine-Grained Materials Produced by Severe Plastic Deformation Materials Science and Engineering* **168** 141-148.
[12] Valiev R Z, Krasilnikov N A and Tsenev N K Plastic 1991 *Deformation of Alloys with Submicro-Grained Structure Materials Science and Engineering* **137** 35-40.

Strain hardening of metal parts with use of impulse wave

A V Kirichek[1] and D L Soloviev[2]

[1] Bryansk State Technical University, 7, 50 years of October st., Bryansk, 241035, Russia
[2] Murom Institute of Vladimir State University, 23, Orlovskaya st., Murom, 602264, Russia

E-mail: avk.57@yandex.ru

Abstract. This work describes a strain hardening method with the use of impulse waves. This method increases energy transfer to the strained material extending its technological capabilities with development of a deep strengthened layer and allowing formation of a heterogeneous hardened structure using plastic deformation. This structure has specified distribution of the hard and soft (visco-plastic) areas. Due to development of the heterogeneous structure in the surface layer created by strain hardening with impulse wave, durability of parts that suffer contact fatigue loading is significantly increased.

1. Introduction

The current technology progress requires development of new methods that can strengthen metal materials as this helps to provide high physical and mechanical properties.

A hardening process involving surface plastic deformation (SPD) is successfully used for machine components, which operate under severe conditions and mainly under fatigue loads. The following SPD methods are widely used in industry: bead-blasting treatment, knurling and levelling as they form a hardened surface layer of 0.5-2 mm deep [1-3].

However, for the products operating under complicated stress in hard and extreme conditions, the point with the highest load is located at a certain depth under the surface layer and, therefore, a deeper hardened layer is needed to improve its operating properties. This layer should be more than 1-3 mm, and in extreme conditions it should be up to 6 mm deep. Therefore, new process engineering solutions are required to advance capabilities of the cold-work strengthening.

One of the directions for the development of the strain hardening technologies practices is the hardening process with the use of controlled impulse waves. The world scientific and engineering practice shows that due to significant deformations developed by the high pressure in the contact area at relatively low power, the shock loading is very effective for structural damage or a hardening purpose. However, when different impact systems with the same blow kinetic energy are involved, the result can differ. This happens because the impact effect depends on the method how the power is delivered to the deformation region.

The hardening process involving SPD in machine industry uses only short impulses with the high amplitude generated in primitive impact systems without an intermediate member.

A different shape of the impact momentum obtained in various impact systems with the same energy is described by experimental oscilloscope pictures (figure 1), (table 1) [4]. Thus, in case of the ball, the impulse had a triangle shape with the high amplitude, short duration and low energy. When the object was hit with the flat end of the cylinder rod (striker) the impulse was of a trapezoid shape with the lesser amplitude, greater duration and greater energy. When the object was hit with the striker through an intermediate member statically pressed to the loaded surface (a wave guide), the impulse was extended, e.g. it had a leading and a trailing edge. The trailing edge is formed by the energy of deformation reflected waves and it depends on the loaded material properties. Impulse study showed that the impulse formed in case of the striker and the wave guide had the highest energy as the trailing edge prolongs the shock effect upon the loaded material and it intensifies elastoplastic strain and rises up the performance factor of the process.

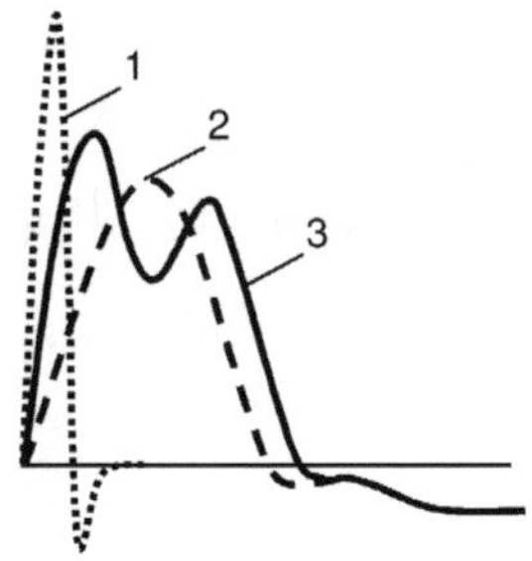

Figure 1. Dependence of the impulse shape formed in a deformation region from the impact system

Table 1. Delivery of hit energy to a deformation region

	Hit with tool		Hit wave guide with a striker
tool	ball	striker	flat end of a wave guide
method	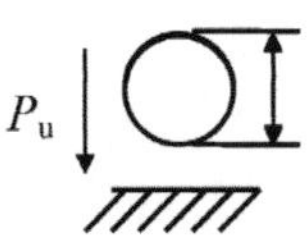 P_u	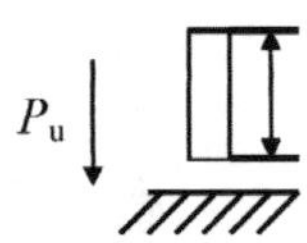 P_u	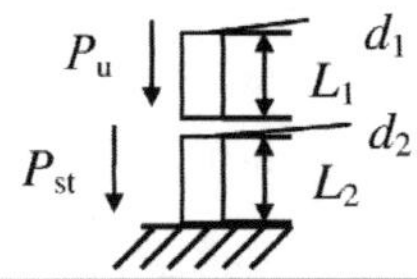 P_u, P_{st}, L_1, L_2, d_1, d_2
Curve number	1	2	3

P_u – impulse load, P_{st} – static load, L_1, L_2 – lengths of a striker and a wave guide, accordingly, d_1, d_2 – cross sections of a striker and a wave guide, accordingly

2. Results and Discussion

Based on the aforesaid, the gradient toughening method was developed. This method implies static and impulse treatment (SIT) that uses impulse waves [5] to develop plastic deformation. In SIT hardening the striker and the wave guide system is used, which generates a plane acoustic wave characterized by the law of the force variation (the deformation wave amplitude) in the course of time, the maximum value of the force, duration of the force (deformation wave duration) and energy of the deformation wave. These features depend on the striker geometry and the wave guide, material properties and impact speed. This wave period is called impact impulse. The form of the impact impulse delivered into the deformation region, for example, a tool contact area with the hardened material, shall define effectiveness of dynamic loading. Static pre-pressing of the wave guide and the tool does not allow the striker to lose contact with the loaded surface after impact providing recuperation of the reflected deformation waves. During hardening, the impulse shape is adapted to the maximum extent to the properties of the material and loading conditions. This adaptation increases a performance factor and extends treatment capabilities providing a deep 6–10 mm hardened layer.

Design facilities for the hardening with impact deformation waves are: an impact impulse generator and production tooling that provides implementation of the treatment process [6].

Design and process-dependent parameters for the hardening with impact deformation waves are: a form and measurements of the deforming tooling and new SPD parameters – a form and measurements of impact system components.

An impact system incorporates a striker and a smooth cylinder wave guide. The system can generate singular square-wave pulses; the amplitude of such pulse defines the force of impact.

The strain hardening tool is fixed at the free end of the wave guide, or the wave guide end is inserted into the required form and works as a tool. The tool can be of a spherical (ball) or roll (rod, disk or profiled rolls) shape. Depending on deformation regions generated at the same time the tool can be a single-contact or multi-contact tool. The tool represents a link transmitting the impact impulse to the loaded metal piece, therefore, the tool shape and measurements define a degree of the pulse distortion delivered through such tool. The shape and measurements of the tool will have an effect upon the configuration and the size of the single plastic impressions.

In the process of hardening with impulse deformation waves the form of the impulse can be adapted to physical and mechanical properties of the loaded material, tool's curvature and the loaded surface, when maximum kinetic energy of the impulse is transferred to the material. For this purpose they use characteristic of structural resistance of the material, which is subjected to indentation of the tool and expressed in terms of the multiplying factor determined by a "force – impression" dependence in the range of $(2.5\ldots7.5)\cdot10^8$ N/m, which is taken in account in calculation of geometrical relationship of the striker and the wave guide [7].

Basic process parameters of the deformation wave hardening technique include energy and frequency of impulses, static pre-pressure of the tool applied to the hardened surface and the depth of advance.

A static part of the loading force should provide the effective use of reflected deformation waves when the tool is impressed into the material; it should equal more than 10% of the impact loading.

The energy of the impact wave is one of the main parameters that defines a plastic flow value. Usually, SPD consumes the energy for plasto-elastic deformation, rubbing friction between tool and a contact spot and the energy of the wave. The energy used for the wave processes can be controlled by changing of the form and measurements of the impact system. Directional deformation waves, generated in the impact system at the moment of impact, develop the best form of the impulse for a given condition. Kinetic energy of the impulse (up to 200 J), determined by striker's speed and weight and its shape and size, determine the amplitude, duration and the form of deformation wave generated in the impact system. These factors define an impulse waveform in the deformation region.

Impulse frequency (5-40 Hz) controls the pulse ratio and a number of the deformation action. The pulse ratio couples the duration of the impact deformation waves and impulse frequency. It represents an important characteristic of impulse systems. A high pulse ratio is used mainly in systems with high frequency and short duration of the impact; and a low pulse ratio is used mainly in systems with low frequency and long duration of the impulse.

Along with impact frequency and the geometry of the impression, the advance controls the amount of the dynamic loading. To increase the hardening performance factor these parameters increase the impact power of the impulse generator to a maximum value and treat the material with maximum possible frequency delivering the required energy. In this case the number is controlled by the change of advance; in general, the advance does not exceed 2000 mm/min.

The above mentioned parameters of hardening with deformation waves shall be regarded as one system that controls surface layer quality characteristics.

The energy of the impact waves spent for plastic deformation of the hardened material is determined by design parameters: the form and measurements of the striker and the wave guide; and process parameters, such as impact kinetic energy and static pre-pressure of the tool applied to the surface to be hardened. The change in these parameters leads to microdisplacement of the tool and development of the impression of a definite size, e.g. – a singular trace of the tool. The shape and the size of the impression determine the pattern of change of material properties along the depth of the surface layer.

The size of single impressions and their relative displacement are determined by the process parameters: impact frequency and advance, adjust overlap of impressions. This overlapping defines uniformity of hardening providing a significant effect upon performance characteristics of the hardened surface layer [8, 9].

Specified deformation wave hardening parameters offer wide technological capabilities to create a hardened surface layer. The obtained results include the following:

– the surface layer with maximum hardness up to 6500 MPa and depth of hardening of 6-10 mm; to ensure specified accuracy and to reduce roughness of the material, subsequent fine machining can be used (a fine or finish turn operation, milling, abrasive machining, SPD finishing and hardening processing); a deep depth of the hardened layer allows selecting the allowance for machining for non-treated steel up to 1-1.5 mm, and for heat-treated steel – up to 0.5 mm;

– the surface layer with maximum hardness up to 6500 MPa and depth of hardening of 2-3 mm and low surface roughness up to Ra = 0.08-0.1 μm.

Adjusting the uniformity of hardening it becomes possible to form [10, 11]:

– a complete unevenly hardened surface layer with a specified change of alternating hard and malleable areas in depth and along the surface as well;

– a complete uniformly hardened surface layer similar to the layer obtained by using known SPD methods (e.g. knurling, levelling, etc);

– a uniformly hardened thin upper layer (up to 1-1.5 mm) and an unevenly hardened layer located under the upper one;

– a uniformly hardened thick upper layer (1.5-3 mm) and an unevenly hardened layer located under the upper one.

High (acoustic) speed of propagation of the wave deformation in the material, the ability to control the intensity and duration of forces acting on fragments of the surface layer suggests this method of treatment to be a method of intensive plastic deformation of the material [12-15]. The study of the metal samples' structure hardened by deformation waves, revealed the presence of nanostructured areas similar to that obtained by intensive plastic deformation. The size of such areas varies from 30 nm to 90 nm. The uniformity of their interchange can be adjusted by the energy of impulse deformation wave, size and the geometry of the tool's contact spot and the deformed surface, the number and a displacement level of emerging deformation regions [16].

Field investigation of the heterogeneous hardened structure obtained by impulse wave deformation showed its durability increase under the contacting fatigue loads up to 7 times.

3. Conclusion

The method of gradient hardening, which uses impulse waves to create plastic deformation, has been developed. This method is characterized by high energy efficiency and it generates a greater depth of the hardened surface layer up to 6-10 mm.

With plastic deformation it becomes possible to form a heterogeneous hardened structure in the surface layer as a result of the multiple action of deformation waves shifted in space and time in relation to each other.

There is determined high efficiency of use of hardening by impulse deformation waves forming a heterogeneous hardened structure to improve performance characteristics of machines' parts experiencing contact fatigue loads.

Acknowledgement

The study was supported on the base of grant provided by RFBR, 14-08-00112.

References

[1] Odintsov L G 1987 *Strengthening and Parts Finishing by Surface Plastic Forming: Reference Book* (Moscow: Mechanical Engineering) p 328

[2] Smelyansky V M 2002 *Mechanics of Parts Strengthening by Surface Plastic Forming* (Moscow: Mechanical Engineering) p 300

[3] Schneider Yu G 1998 *Techniques for Pressure Finishing: Reference Book* (St. Petersburg: Polytechnic)

[4] Kirichek A V and Soloviev D L 2013 Properties and Technology for Quasi-Composite Blanket Using Natural Reinforcement of the Metal by Strain Affected Areas *Journal of Nano and Electronic Physics* **5(4)** 04010(5pp)

[5] Kirichek A V and Solov'ev D L 2001 The methods of dynamic surface strengthening by plastic deformation *Kuznechno-Shtampovochnoe Proizvodstvo (Obrabotka Metallov Davleniem)* **7** 28-32

[6] Kirichek A V, Solov'ev D L and Silant'ev S A 2002 Impact device for static-pulse deformation working *Kuznechno-Shtampovochnoe Proizvodstvo (Obrabotka Metallov Davleniem)* **10** 35-40

[7] Kirichek A V, Soloviev D L and Altuhov A Yu 2014 Deformation Wave Hardening of Metallic Materials *Journal of Nano and Electronic Physics* **6(3)** 03069(4pp)

[8] Panin V E and Deryugin E E 2003 Mesomechanics of the formation of banded structures at mesoscopic and macroscopic levels *The Physics of Metals and Metallography* **96** 2-15

[9] Kirichek A V, Solov'ev D L and Silant'ev S A 2004 Influence of regimes of static-pulse processing on uniformity of superficial layer hardening *Kuznechno-Shtampovochnoe Proizvodstvo (Obrabotka Metallov Davleniem)* **2** 13-17

[10] Kirichek A V and Solov'ev D L 2008 Creating heterogeneous surface structures by static-pulsed treatment *Russian Engineering Research* **28(3)** 277-279

[11] Kirichek A V, Soloviev D L and Altuhov A Yu 2014 Production of Quasicomposite Surface Layer of a Metal Material by Shock Wave Strain Hardening *Journal of Nano and Electronic Physics* **6(3)** 03070(4pp)

[12] Zhu Y T, Huang J Y, Gubicza J, Ungár T, Wang Y M, Ma E and Valiev R Z 2003 Nanostructures in Ti processed by severe plastic deformation *Journal of Materials Research* **18(8)** 1908-1917

[13] Valiev R Z, Krasilnikov N A and Tsenev N K Plastic 1991 Deformation of Alloys with Submicro-Grained Structure *Materials Science and Engineering* **137** 35-40

[14] Valiev R Z, Korznikov A V and Mulyukov R R 1993 Structure and Properties of Ultrafine-Grained Materials Produced by Severe Plastic Deformation *Materials Science and Engineering* **168** 141-148

[15] Siegel R W and Fougere G E 1995 Mechanical properties of nanophase metals *Nanostr. Mat.* **6** 1-4

[16] Kirichek A V and Soloviev D L 2013 Nanostructure Changes in Iron-Carbon Alloys as a Result of Impulse Deformation Wave Action *Journal of Nano and Electronic Physics* **5(4)** 04009(4pp)

Morphological analysis of galvanized coating applied under vibrowave process system conditions

V A Lebedev[1], **V V Ivanov**[1] **and V P Fedorov**[2]
[1] Donskoy State Technical University, 1, Gagarina sq., Rostov-on-Don, 344000, Russia
[2] Bryansk State Technical University, 7, 50 years of October st., Bryansk, 241035, Russia

E-mail: va.lebidev@yandex.ru

Abstract. The article presents the morphological research results of galvanized coating applied to the metal surface in the course of mechanical and chemical synthesis realized under vibrowave process system conditions. The paper reveals the specifics of the coating morphology, its activating role in free-moving indentors formed under the impact of low-frequency vibrations and its positive influence on the operational performance of the part surface layer. The advantages of this galvanized coating application method are presented in comparison with conventional methods.

1. Introduction

Applying galvanized coating to the metal surface under vibrowave process system conditions is a complicated process; it is a result of the combined influence of a mechanical component and a chemical reaction. Applying coating by a mechanical and chemical method helps to achieve one of the surface quality parameters: absence of the hydrogen-charged layer.

Vast technological possibilities of vibrowave process systems present the most efficient means to obtain quality coatings for different functional use as regards operational performance. The research results show that these properties appear due to the morphological specifics of the galvanized coating applied to the metal surface during activating a vibrowave influence of free-moving indentors in the form of spherical bodies.

2. Research results

The coating was applied using a vibration shaker with a 10 l vibration chamber. Porcelain and glass balls of 5…10 mm in diameter served as a working medium. Processing regimes were changed within the following limits: vibration amplitude of 2 to 5 mm, vibration frequency of 16 to 33 Hz. Samples were made of steel grade 3, 20 and 40 with 20 sq.sm sample surface area subjected to coating. The composition of the galvanized coating was the following: 250 g/l of zinc chloride and 100 g/l of zinc powder.

The galvanized coating density applied to the metal surface was defined by the density of plastic impressions formed on the surface as a result of a shock pulse input by working medium indentors. The area of zinc impressions and the distance between them was measured by the МИМ-7 metallographic microscope. For quantity estimation of zinc coating density (N) the following ratio was used:

$$N = So / S,$$

where *So* is a total area of plastic impressions, S is a sample area under investigation.

It has been found that in case of a 50 % load of the work vibrochamber with porcelain indentors of 5 to 10 mm in diameter, the impression area is 1×10^5 sq. μm, an average distance between the impressions does not exceed 250 μm; in case of an 80 % load the impression area is 0.225×10^5 sq. μm and an average distance between the impressions is 50 μm. A 30 minute change in processing timing (Figure 1) results in the increased galvanized coating density. Subsequent 30 minutes of processing contribute to the increase of structured bodies dimensions and elimination of sharp interfaces between plastic impressions.

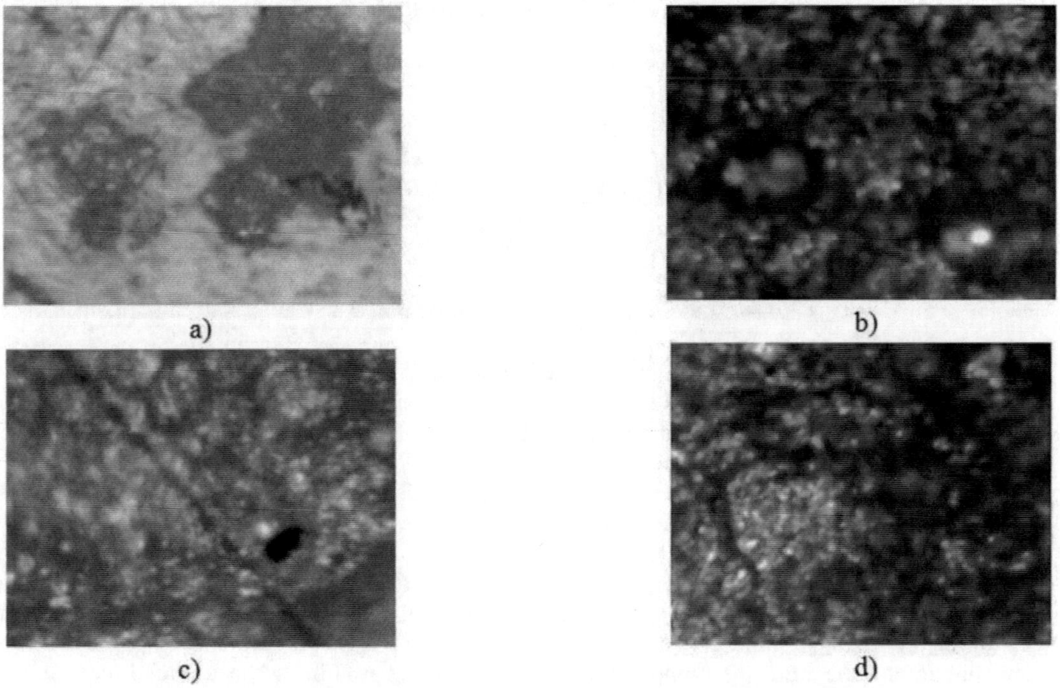

Figure 1. A zinc coating representation depending on the processing time (scale: 100μm):a – 20 min; b – 30 min; c – 40 min; d – 60 min.

Carbon steel samples were used to study a micro/nanostructure of the zinc coating surface. The given area of the coating was investigated layer by layer from top to bottom, which enabled definition of the crystal dimensions, their form and bond type [1]. The research results are mentioned below. The coating has a chaotically oriented surface structure, where the open areas of the galvanized coating represent ≈ 100 μm. (Figure 2a). The scale of 10 μm (Figure 2b) shows that 2…5 μm structures are formed at the interference of large surface areas (marked up area). The surface analysis in a 2 μm scale (Figure 2c) reveals the presence of a transitional area in the structure composed of 1 μm crystals (marked up by an arrow). An increase of this area to a 1μm scale displayed the change of morphology and appearance of bonds on top of the crystals in the form of 0.2..0.5 μm filaments. The further coating surface study in the scale of 200 nm (figure 2d) helped to define the length of intercrystalline bonds and the dimensions of separate crystal elements up to 100 nm. A zinc coating surface morphology illustrated in figure 2f in the scale of 100 nm indicates the type of intercrystalline bonds and separate coating crystals which can be seen on the background in the form of light oval areas.

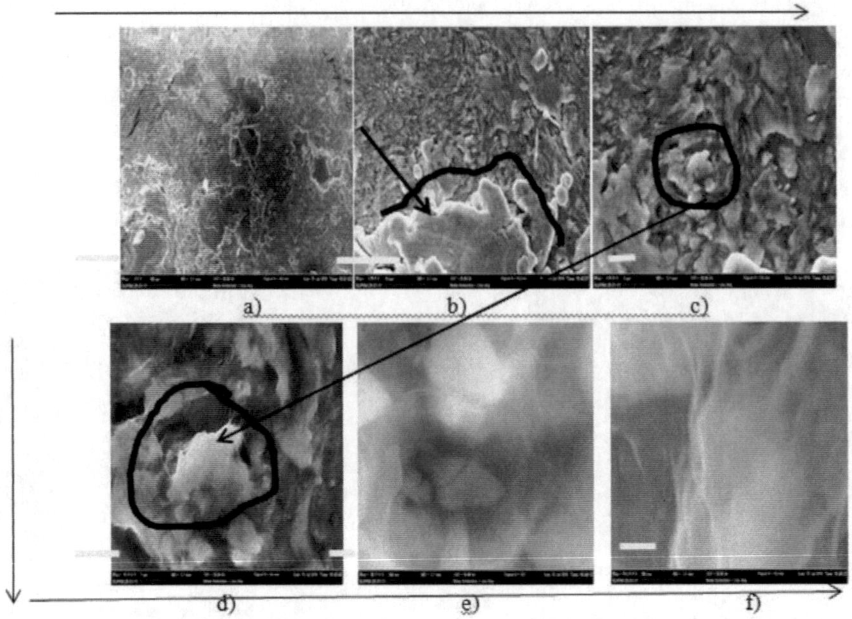

Figure 2. A zinc coating morphology formed under vibration process system conditions (T = 60 min, A = 3 mm.), for different scales of study:
a) 100 μm, b) 10 μm, c) 2 μm, d) 1 μm, e) 200 nm, f) 100 nm.

In order to clarify the dimensions of zinc coating crystals and intercrystalline bonds between them, the specimen stage with a sample was positioned at an angle of 30^0. Figure 3a features coating structural components in the scale of 2 μm representing a combination of 1 to 2 μm crystal contrast peaks and dark cavity areas of up to 2 μm width between them. The biggest crystal is indicated by an arrow in figure 3b in the scale of 1 μm which has an oval shape (2.3 μm in a transversal diameter; 2.9 μm in a horizontal diameter) and represents an upper coating layer. If the scale is enlarged to 200 nm (Figure 3c), the surface morphology can be seen, which was in the background (cavities) in the previous pictures.

The analysis demonstrates that crystal dimensions are within 100 nm and 3 μm; the cavities between them are within the same limit. Crystal bond filaments have a dimension of 10 nm up to 100 nm and are located not only on the crystal surface, but also occupy an intercrystalline space giving additional properties to the coating (corrosion resistance), which implies zinc powder sedimentation in the course of zinc ions reduction.

Figure 3. A zinc coating surface morphology derived along with a sample positioned at an angle of 30^0 and the following scale of magnification: a) 2 μm, b) 1 μm, c) 200 nm.

The following positioning of zinc crystals in the coating can be acquired only as a result of shock pulse input of free-moving indentors in vibrowave process systems. The morphological research illustrates that the mechanical and chemical synthesis coating contains micro/nanostructured zinc particles in the form of circular shape flakes. A plate-like shape of zinc flakes provides the possibility to decrease the thickness of galvanized coating while preserving the adequate protection. The combination of micro and nanoelements comprising the coating gives high operational surface properties.

3. Conclusion

The performed research, as well as the acquired experience of the galvanized coating practical application allow for the following conclusion:

1. The galvanized coating applied under vibrowave process system conditions possesses a number of advantages, such as the absence of a hydrogen-charged layer, low porosity and high adhesion strength.
2. The coating application technology does not require special equipment, facilities, highly qualified specialists and special waste treatment and recovery facilities.
3. Using this way of the galvanized coating application is appropriate for parts lacking deep openings, as well as for protective and decoration solutions.

References

[1] Ivanov V V *et al* 2014 *J. Friction and Wear* **35** 339–342
[2] Babichev A P 1994 Fundamentals of vibration technology (*Rostov-on-Don*) p. 621
[3] Ivanov V V, Popov S N, Marchenko V Yu and Marchenko E V2015 *J. World outside*. **1** 34–38
[4] Babichev A P, Ivanov V V, Burlakova E V and Filipchuk O S 2014 *J. Strengthening Technol. and Coverings*. **7** 24–26
[5] Lebedev V A, Ivanov V V, Stine S Y and Davydova I V 2015 *J. Strengthening Technol. and Coverings*. **8** 19–23

A new approach to application of end milling cutters of compound hard metal

B Ya Mokritsky[1], **A V Morozova**[2] **and D A Pustovalov**[1]
[1] Komsomolsk-on-Amur State Technical University,27, Lenina ave., Komsomolsk-on-Amur, 681013, Russia
[2] Bryansk State Technical University, 7, 50 years of October st., Bryansk, 241035, Russia

E-mail: boris@knastu.ru

Abstract. The results of designing end milling cutters having a hard metal cutting point and a shank made of structural steel are shown. The shank and a cutting point are directed and connected. It allows reducing the requirements in expensive integral hard metal milling cutters and defining the field of rational application of integral and compound mills.

1. Introduction

By the time of hard metal appearance the design of end and key milling cutters changed constantly by means of equipping a mill body made of structural steel with hard metal cutting elements. These milling cutters were equipped with soldered-on blades, with mechanical fastening hard metal cutting blades. Each of these solutions was rational for those times and had its drawbacks. Updating NC (numerical control) metal-working machinery necessitated a machining productivity increase that caused the creation of compound hard metal milling cutters, the entire cutting point of which was made of hard metal and it was jointed with a steel shank either by means of soldering or by welding. Growing requirements to productivity and machining accuracy resulted in the creation of integral hard metal milling cutters. Initially they were expensive and were used only in those cases when the productivity or accuracy of machining was not ensured by compound milling cutters.

After that, due to various reasons, the application of compound milling cutters either decreased, or reduced to zero. It happened first of all because of the decrease in the set of purchased metal-cutting tools.

Even the idea of compound milling cutters with reference to milling cutters of a small diameter disappeared. Thus, for instance, if at the end of GOST (All-Union State Standard) 18372-73 in the part of "Hard Metal End Milling Cutters" there was as yet a record of the possibility of compound milling cutters application, then in its modern substitution GOST 32405-2013 in the part of "Integral Hard Metal End Milling Cutters" there is not any record of compound hard metal milling cutters at all.

Such a situation, in the judgment of the authors, resulted in excess expenditures for metal-cutting tools. Thus, for instance, during machining billets of parts made of aircraft industry material, the application of integral hard metal milling cutters is justified during machining high-test titanium alloys. But, when milling grooves, pockets, projections and so on in parts billets made of less durable titanium alloys, stainless steel or aluminum alloys the application of compound milling cutters is quite possible.

The study in the paper has been carried out with the support of Grant № 9.251.2014/K of the Ministry of Education and Science of the Russian Federation.

2. Methods and materials

One of the ways of the problem solution of the tool costs decrease at the expense of the hard metal decrease is the application of updated compound end milling cutters. Updating is possible due to the fact that general mechanical engineering requirements to accuracy are not so high in case of grooves, projections, pockets and so on. Even in the aircraft industry these requirements are not high. It allows without detriment to a machining quality admitting a divergence for a compound end milling cutter to a larger value, than that for an integral end milling cutter. It is necessary that the optimization should be carried out only for a part of the shank in the length of a milling cutter, reasoning from the required quality of machining and updating of the area of the shank joint with the hard metal point of the milling cutter to ensure a required strength and rigidity of the joint (Figure 1).

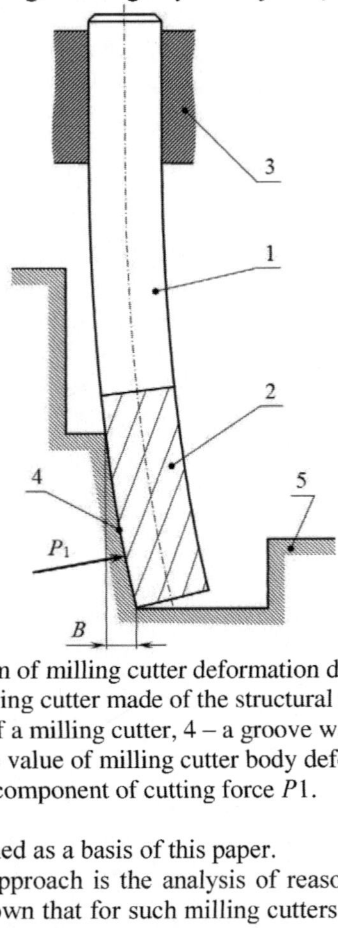

Figure1. The general diagram of milling cutter deformation during groove wall machining: 1 – the shank of a compound milling cutter made of the structural material, 2 – a hard metal point, 3 – a device for fastening the shank of a milling cutter, 4 – a groove wall formed during cutting, 5 – the profile of a billet machined, B – the value of milling cutter body deformation under the impact of the component of cutting force $P1$.

This is the approach that is assumed as a basis of this paper.

As a reason for using such an approach is the analysis of reasons for the failure of integral end milling cutters. The analysis has shown that for such milling cutters operating in alternating changing dynamically milling conditions at high-speed machining geometrically-complex surfaces the cyclical strength of hard metal is insufficient. It is clear even from the statistics of milling cutter failures: milling cutters are withdrawn from the operation in 73…90 cases out of 100 not because of point wear, but, owing to a shank fracture in the integral milling cutter. Such examples are shown in Figure 2 for milling cutters with the 140 mm length and the diameter of 16 mm (three right samples) and the length of 110 mm and the diameter of 12 mm (left sample). The investigation of the surface of failure has shown that it was an endurance failure on a tough-brittle mechanism by means of the multi-cyclic branching growth of intercrystal main cracks.

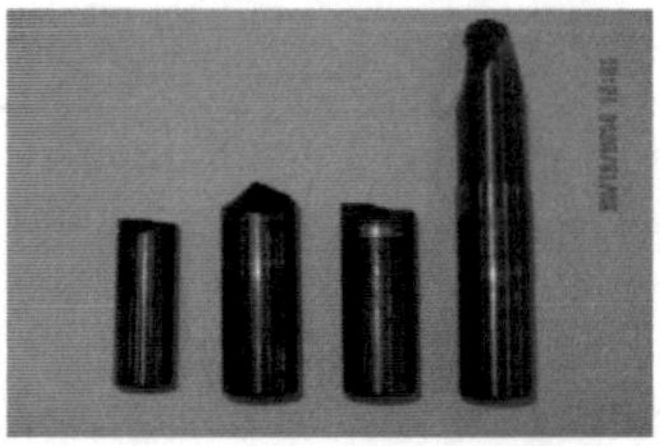

Figure 2. Characteristic examples of the integral hard metal milling cutters failure caused by tough-brittle destruction of the shank (three left cases) or by the area of the shank transition into the point

Dynamometric investigations (simultaneous registration of constituents of cutting force and vibratory displacements in three mutually perpendicular directions) have shown [1], that a normally directed constituent of cutting force changes within the limits of (1600…3000) H under manufacturing conditions of aircraft materials machining with such milling cutters. A small value is obtained during chamfering in aluminum sheets, a greater value can result under complex profile of milling pockets in titanium alloys, intermediate values are calculated when machining aluminum-lithium alloys, stainless steel.

The application of numerical simulation with the use of standard computation programs through the method of finite elements of stress fields, Figure 3, and values of deformations (rod deflection imitating the body of the integral or compound milling cutter), Figure 4, allowed obtaining design charts of milling cutter comparative deformations, Figure 5.

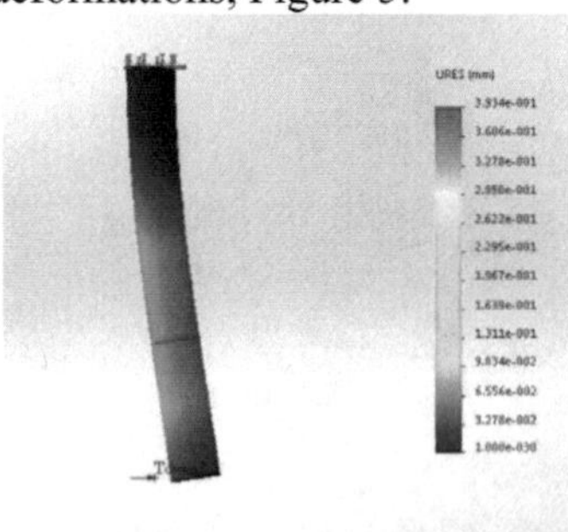

Figure 3. The image of stress fields of the model of a compound hard metal end milling cutter body under the influence of the normal constituent of cutting force.

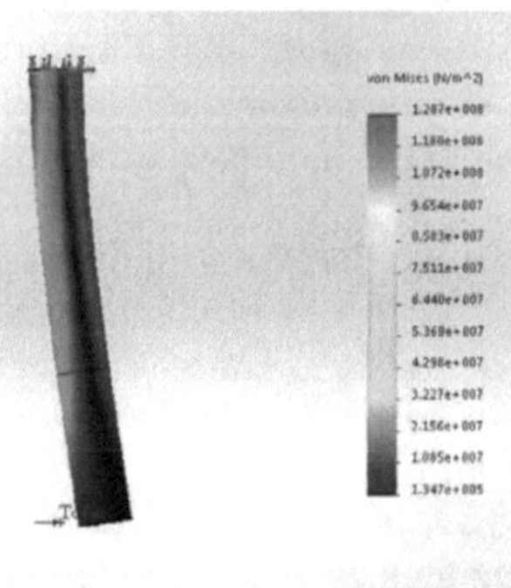

Figure 4. The image of values of area travels (sections) for the body of a compound hard metal end milling cutter under the influence of the normal constituent of cutting force.

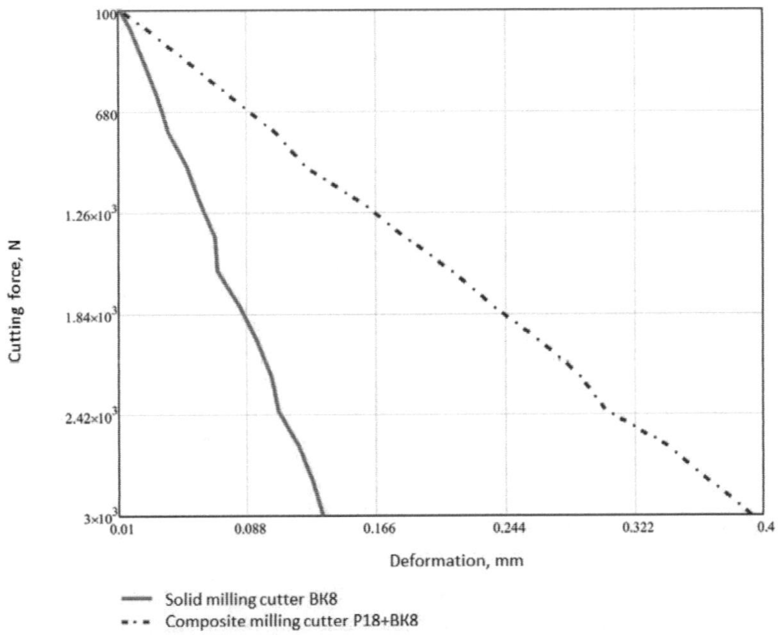

Figure 5. The results of numerical modeling of the deformation value for a milling cutter operation end: the right line – for a compound milling cutter, the shank of which is made of material R18 (high-speed steel (ISO 1.3302, field of application K30 – K40), and the point is made of hard metal TC8 (S6-5-3)(ISO 1.3343); the left line – for an integral hard metal milling cutter made of hard metal TC8 (S6-5-3)(ISO 1.3343).

Figures 3, 4 and 5 are shown only as examples: a milling cutter 140 mm long with the diameter of 16 mm is fixed in the chuck for the shank length of 40 mm, in case of the compound milling cutter a hard metal point is made for the length of 40 mm. in this case the presence of chip grooves at the point and the design of the joint of the shank with the point are not taken into account. The deformation simulation in the rupture-test machine, Figure 6, confirmed the adequacy of design charts of milling cutter comparative deformations.

Figure 6. The plant for simulation of the milling cutter body deformation depending on an effort.

The physical and numerical simulation of the milling cutters deformation allowed differentiating the fields of possible application of integral and compound hard metal milling cutters, revealing the field of preferred application of integral milling cutters, defining the field of integral and compound milling cutters competition. These fields are defined as preliminary for the time being, as the number of considered fulfilled variants of the produced compound milling cutters is still limited. Only saving in hard metal is taken into account, the cost of inputs assuring the shank and point coupling is not assessed. But now it is clear that possible savings should be within 30…60 % of hard alloy.

3. Results

1. Compound hard metal milling cutters are the alternative to integral hard metal milling cutters when they are applied with consideration of general engineering requirements to accuracy of the produced part surfaces. The alternative is more vivid, when the deflection of a milling cutter body is less under the influence of cutting forces, that is, the more partial load modes of cutting are used.

2. Without damage to the quality and machining productivity, compound milling cutters allow saving more than one third of hard metal used for manufacturing of the integral hard metal milling cutter.

3. For integral hard metal milling cutters the preferable field of application is highly productive machining of hard-to machine materials.

4. Conclusion

The authors have borrowed common geometrics of the point of hard metal end milling cutters and considered [2-6] the possibility of manufacturing the shanks of structural metal for milling cutters.

By the example of manufacturing of the shank with the use of tool steel R 18 (as an example of the structural material with high physical-stress-strain properties) and manufacturing of the point with the use of hard metal TC 8 the authors have shown the rationality of compound milling cutters application at least from the position of hard metal savings.

It is useful to proceed with the works for the identification of rational fields of compound milling cutters application taking into account variants of the shank and point coupling, geometrics of chip grooves and so on.

References

[1] Mokritskij B Y, Pustovalov D A, Stupin A, Vereschaka A A, Serebrennikova A G and Altukhova V V 2015 *Applied Mechanics and Materials* **703** 131–136

[2] Mokritskii B Y, Pustovalov D A, Vereschaka A A, Vereschaka A S, Verhoturov A D 2015 *Applied Mechanics and Materials* **719-720** 96–101

[3] Mokritskii B Y, Pustovalov D A, Vereschaka A A, Vereschaka A S and Verhoturov A 2015 *Applied Mechanics and Materials* **719-720** 96–101

[4] Mokritskii B Y, Pustovalov D A and Vereschaka A A 2015 *Applied Mechanics and Materials* **798** 452–456

[5] Vereschaka A A, Mokritskii B Y, Pustovalov D A, Vereschaka A S, Bublikov J I and Oganyan M G 2015 *Applied Mechanics and Materials* **798** 351–356

[6] Vereschaka A A, Mokritskii B Y, Pustovalov D A, Vereschaka A S, Bublikov J I and Litkin 2015 *Applied Mechanics and Materials* **798** 344–350

IOP Conf. Series: Materials Science and Engineering **124** (2016) 012162 doi:10.1088/1757-899X/124/1/012162

Development of methods and equipment for bellows assemblies testing under hydrostatic pressure

E Ol'khovik
Admiral Makarov State University of Maritime and Inland Shipping, 5/7, Dvinskaya st. Saint-Petersburg, 198035, Russia

E-mail: olhovick@gmail.com

Abstract. The paper presents a newbellows assemblies test method for stop and safety valves. To create the real conditions it is suggested testingunder external hydrostatic pressure, which is available in the piping system.A designed test apparatus allows testing of the bellows under the effect of external hydrostatic pressure.The paper describes the design of theapparatus, its specifications and the test procedures.

1. Introduction

Pipeline valves with a bellows seal ofa spindle (stem) is widely used in engineering, shipbuilding and energy, because it has a higher specification for its reliability. However, for a number of structural and technological solutions the bellows unit does not always provide the necessary durability. Major problems occur for the valves operating under a high temperature of coolant, in the modes of above 200 °C. The traditional methodsof bellows fixing on the spindle are provided byits soldering, less seam welding or mechanical crimping. These technologies are well established, fully worked out and tested in practice.

The analysis of strength and stiffness of the welded metal bellows is performed on the basis of theoretical studies, numerical modeling and direct testing.In paper [1] it was found that the main type of the bellows fracture is the stress relaxation. Then the residual capacity is confirmed as a criterion of failure of the bellows. Jha and Diwakar in [2] conducted a detailed metallurgical and chemical analysis of causesof bellow destruction in the weld zone. They established the effect ofthermal stresses from welding as a typical cause of stress corrosion cracking (SCC). Wan and Wang in [3] examined the geometrical parameters of bellows for developing of theempirical relationship between the seal and the axial stiffness.They have proposed a coefficient taking into account the change in thetemperature of elasticity. In paper [4]the metal bellows is modeled with the finite element method (FEM), using commercial software ABAQUS to find the optimal design parameters. The use of the design parameters of the construction allowed obtaining the product with the acceptable quality. Article [5] presents the method of determining theformation oflimit diagramsand limit stress in the bellows assemblies; the data correlate well with the calculations on FEM.Thus, from the literary analysis we can conclude that there is the necessity of the direct tests of bellows assemblies and calculations by means of FEM. Only in this combination we can obtain the adequate data for new compounds of the bellows designs.

Most of the known faultsare associated with the violationof bellowstightness during its operation at the high temperatures. The destruction usually takes place in the region of its connection with the spindle or the upper flange.Itcan be a soldered connection or the break in the weld zone. Moreover, thewidespread cause of destruction of the bellows assembly is the excess of high hydraulic pressure.

One of the major defects that are manifested in the traditional welding of the bellows assembly is skewness or offset of thebellows end with respect to the spindle or the top flange. Even a slight shift from the central axis of the bellowsleads to formation of the eccentric forces of compression and reduction of the critical pressure. The effective area of the bellows, and, hence, the properties of the fatigue resistance remain constant only under the axial tension and compression, the presence of the force skew leads to the constant bending moment of loading of the bellows,which negatively affects its durability. Moreover, deep penetration of the tungsten electrode leads to significant warming of both the bellows and the spindle construction that adds to the internal thermal stresses, which reduce the need of the additional heat treatment. Current technology requires the subsequent machining of the weld that reduces the individual compounds strength.

2. Description of the problem

Most of the standards (ISO 15348: 2002 Pipework –Metal bellows expansion joints - General)intended for the bellows testing provide only a limited set oftest methods. Typically, they include the tests for the uniaxial 'tensile' – 'compression' and internal pressure. In practice, the pipeline stop and safety valves with the bellows seal stem operate under high hydraulic pressure, which exerts the load effecting the bellows. When designing strength and stiffness of bellows assemblies, as a rule,the external hydraulic load is takenas dependent on the area of the pressure action, i.e. the outer area of the bellows. If the assembly is in the stretched condition, the contribution of the external pressure is consideredto be minimal.Most of the methods for calculation of the long-term low-cycle bellows strength with the influence of the high temperature and the residence time under the load are based on the use of the strain-kinetic criteria. It is necessary to solve the problem of the stress-strain state of the bellows with long-term cyclic loading, as well as to obtain the data on mechanical properties of materials under these conditions. However, the analysis of these methods does not allow us to identify the conditions of the effect of the external hydraulic pressure on the strength of the bellows assembly.

3. Description of experimental equipment

A special stand was used in a classical test of bellows assemblies.It allows for the desired mode of cyclic deformation of the bellows under axial tension and compression with a given scale displacement, where in the frequency of loading composes $10 \div 50$ cycles per minute. Additionally, the bellows assembly received the excess internal pressure of 0.1 MPa. Then, the moment of destruction was recorded automatically during the pressure drop due to the leakage of the air through the cracks. Slight pressure has almost no effect on the strain state of the structure and its durability.

Thus, the experimental methods for assessing of the durability of the bellows assemblies under the external hydrostatic pressure have not been identified. The need of performance of such tests has been confirmed by us in the development of the new models of the ship stop valves. The main objective of this work is creation of instruments and methods for the bellows testing in the mode of loading uniaxial tension-compression under the external hydrostatic pressure. While doing this work, we used the existing scientific potential and experience in the development of the test benches (stands) – [6, 7].

In works [8, 9] we offered technologies allowing executing a high-quality connection of the bellows with a spindle or a basic flange due to the use of the precision laser welding and high-energy plasma processing of the surface. These technologies allow achieving a stable quality of the connection and herewith they donot cause residual (superficial) stresses.

The designed test apparatus is shown in Figure 1. Its main purpose is to test the bellows assemblies having the diameter of not more than 100 mm. The experimental test apparatus is simple to use and it does not require a special training. The basis of the experimental setup is a high-pressure chamber (position 1, Figure 1), which locks the top cover (position 5, Figure 1), that establishes an upper support flange bellows assembly. The drive spindle (position 3, Figure1) passes through the upper cover and has mobility in the guide sleeve (position 6, Figure 1). With an adjustable mechanical drive (position 8, Figure 1) through the eccentric shaftit is loaded by an axial force, wherein the axial deformation is controlled by the geometry of the cam-follower (position 7, Figure 1).The hydrostatic pressure on the bellows assembly is fed through the passages in the high pressure chamber; the measurement pressure method is realized by an electronic manometer.The liquid used for hydraulic loading is the industrial oil. We have done a series of tests on the bellows assemblies in order to develop the new models of the ship stop valves. The bellows material is stainless steel 316L (ISO 2604-1 F59), whereas the wall thickness of the bellows for various tests was from 0.5 to 1.0 mm.All tests were successful; the destruction of the bellows was fixed with the sufficient accuracy by the number of cycles and the amount of deformation (sweep).In all experiments the deformation of the bellows was only positive (tensile).It corresponds to a majority of the designs of the stop valves, for which the normal position is open, the bellows are not loaded by the axial force and there is only the external hydrostatic pressure.

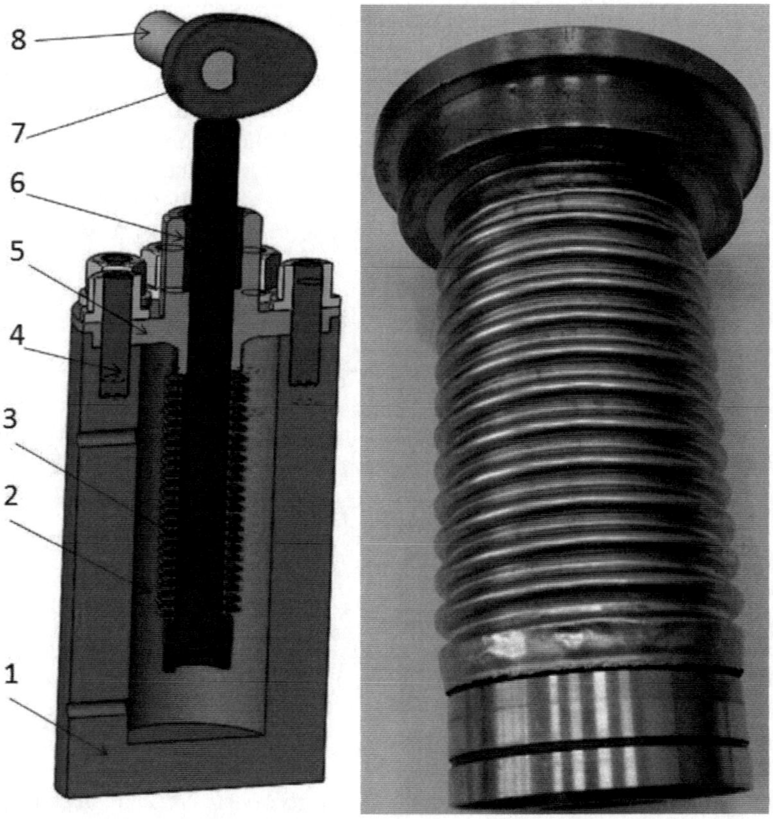

a b

Figure 1. The apparatus of bellows testing for external pressure:
a – Tested bellows assembly (diameter 65 mm, bellows length 120 mm)
b – 1 –the high-pressure chamber; 2 – the bellows; 3 – the spindle; 4 –the stud fastening; 5 –the camera shutter pressure; 6 –the guide bushing; 7 –the cam-follower; 8 – the drive.

4. The results of the experimental studies

In total 15 different tests of bellows were carried out, which is sufficient to assess the quality of the proposed test method and to develop the experimental apparatus. The value of the linear deformation of the bellows is ranged from 5 to 35 mm, which is consistent with real working conditions of the bellows assembly. The external hydrostatic pressure was adjusted from 1.6 MPa to 5 MPa that also corresponds to the type series of pressures for pipes and valves for various purposes.The number of theload cycles per minute was between 10 and 50 cycles, this option is recommended by a variety of standards. A higher loading rate affects critically the research process as it leads to self-heating of the bellows sample.To prevent self-heating of the bellows there should be a possibility to pump hydraulic fluid for cooling.

5. Conclusion

The tested scheme has several advantages compared with the classical ones. Firstly, it is possible to adjust the axial deformation of the bellows by changing the cam.Secondly, due to adjustment of the frequency drive it is possible to create different modes of the fatigue testing from low cyclic to cyclic ones.Thirdly, through the gauge method of the pressure measuring it is possible to control the destruction of the bellows assembly precisely. In case ofthe bellows destruction there is a sharp decrease in hydrostatic pressure, which is measured by an electronic manometer.

A notable result,which is associated with the change in the stress-strain state of the metal bellows using different options of its force loading, has been obtained. At normal (atmospheric) pressure, in most cases a crack is developed in the direction of the inner radii of the bellows, whereas the additional impact of external hydrostatic pressure developed cracks directed at the outer radius of the bellows.This is due to a change in the stress-strain state.Under the influence of hydrostatic pressure,the exterior elements of the bellows are compressed in the axial direction, and the tensile force on the spindle cannot establish equilibrium. Thus, it is possible to make practical recommendations aimed at increasing of the service life of the bellows assembly.

References

[1] Wu J, Li H, Zheng D, Liu W and Xu J 2011 *Reliability, Maintainability and Safety (ICRMS), 9th International Conference on IEEE* 184–188
[2] Jha A K, Diwakar V and Sreekumar K 2003 *Engineering Failure Analysis* **10** 699–709
[3] Wan H and Wang L 2009 *Journal of Mechanical Strength* **31** 787–790
[4] Faraji G, Besharati M K, Mosavi M and Kashanizadeh H 2008 *International Journal of Advanced Manufacturing Technology* **38** 641–648
[5] Hashemi R, Faraji G, Abrinia K and Dizaji A F 2010 *International Journal of Advanced Manufacturing Technology* **46** 551–561
[6] Ol'khovik E O 1997 *State Industrial laboratory* **7** 423–428
[7] Ol'khovik E O, Figovsky O L and Feigin V 1998 *Material Research Innovations* **2** 115
[8] Khmelevskaya V B, Moseyko Y S and Ol'khovik E O 2013 *Bulletin of the Admiral Makarov State University of Maritime and Inland Shipping* **3(19)** 81–87
[9] Ol'khovik E O and Reznik U A 2014 *Bulletin of the Admiral Makarov State University of Maritime and Inland Shipping* **3(25)** 119–122
[10] Ol'khovik E O 2015 *IOP Conference Series: Materials Science and Engineering* **91** 012–022

The Problems of Mechanical Treatment of Leucosapphire Blanks

D I Petreshin[1], A V Handozhko[1], V B Protasev[2], O N Fedonin[1], N A Bizykina[1] and D Y Shukin[1]
[1] Bryansk State Technical University, 7, 50 years of October st., Bryansk, 241035, Russia
[2] Tula state University, 92, Lenina ave., Tula, 300012, Russia

E-mail: msi@tu-bryansk.ru

Abstract. The problems of the technology of mechanical treatment of hard crystal materials, particularly leucosapphires and their possible solutions, are discovered. General questions of the project works, solutions of accurate basic tasks and secure fixation of the blanks or tooling backup of the work technology are considered. The results of experimental studies of the diamond tool dressing and their efficiency evolution results are provided. Some recommendations on the technological process development are worked out.

1. Introduction
Recently the application of leucosapphire as a construction material has become more popular. It is also more often treated mechanically. For some of the products, base plates of semi-conductive devices for example, the technologies are practiced comparatively well, including manufactured equipment, tools, accessories. But the experience of direct transfer of electrical equipment gross production technology to the general machine and instruments building is not always available due to the variety of the parts and series production. The existing experience of metal blanks treatment does not consider the specific features of leucosapphire mentioned below.

Leucosapphire has high hardness and its mechanical treatment can be performed only by a diamond tool. Leucosapphire is characterized by clear properties anisotropy, which is used for the elements design and should be considered, controlled, and maintained during the manufacture. A leucosapphire's friction steel ratio (cast iron) is low, which together with fragility creates problems during blanks fixation. Leucosapphire is very sensitive to thermal shock, which leads to high risk of micro and macro defects.

The methods of the leucosapphire mechanical treatment technology development have not been studied enough, which prevents its distribution despite its unique properties and interesting operational characteristics.

2. Task description
The literature analysis and preliminary studies showed that the leucosapphire cutting process is not very difficult to organize. The problems are to ensure this process stability, productivity, accuracy of the treated form, its position towards its crystal axe and surface quality.

Basic method of monocrystals formation is Kyropoulos method. By means of this method, it is possible to obtain the blanks of an irregular pear-shaped form with the diameter of up to 300 mm and the height of up to 500 mm. Grown crystals are primary blanks. Then, depending on the sort, the blanks for the details are cut outfrom the primary blank. As a rule cutting is realised by diamond circle drilling. The requirements to the form correctness, surfaces arrangements relatively crystallographic axes, dimensions accuracy, gritness are very high. For example, the orientation accuracy of the cylinder blanks for optoelectronics in the C/A/M/R plane is usually taken as $\pm(0.1\div0.2)^{\circ}$; and in a set of cases – as $\pm0.05^{\circ}$ [1].

The arrangement of crystalographic axes in a boule depends on the arrangement of the corresponding inoculating crystals axes and its orientation accuracy and fixation during the growing process. As a rule the orientation of crystalographic axes of the parts does not coincide with the orientation of the optimal direction of the crystal (boule) growth. Thus, the formation of secondary blanks is complicated.

The task of the research was to determine the peculiarities of mechanical treatment of leucosapphire blanks, problems concerning technical preparation and the practical implementation process of the particular operations and their components. A special attention should be paid to the problems concerning the required parameters of the surface layer, which determine operational characteristics of the future product [2]. It is especially important for the production of blanks in optics or electronics industries, as it is associated with the form errors measured in microns, or grits – in nanometres.

Such task description is complex and it demands a larger amount of experiments. Consequently, the tasks of theoretical and experimental research of the basic operations of mechanical treatment and their particular components were set.

The disc cutting, flat and circular sanding and circular drilling by the diamond drill of the crystals were discussed.

During the study of mechanical treatment processes, the questions associated with accurate basing and secure fixation of the blanks, treatment modes designation, diamond tool cutting properties restoration, ensuring the treated surfaces quality arose.

Some questions arise while arranging experimental installations, tools and other technological equipment, necessary for the research.

3. The results

The grown crystal has no accurate surfaces suitable for the restoration of the crystallographic surfaces position. A first approximation provides a very rough direction of plane C; there are specific 'folds' along this axle on the crystal known as cleavage planes, Figure 1.

These planes are clear and in practice their appliance as a preliminary blank installation ensures the orientation with the accuracy of up to 3…5 °. Then, the position of C plane is clarified using different methods, primarily X-Ray and glass plates cut out from the boule. The corresponding metrological supervision is required for the implementation of the mechanical treatment technology with the required accuracy.

The required accuracy and the crystallographic direction of the blanks are mainly ensured by the installation devices, which enable the provision of correct, accurate basing and secure fixation.

Figure 1.A Leucosapphire Crystal.

The errors may be caused by either basing errors or blank shift during the treatment. The installation and secure fixation of the crystal having an incorrect form with the weight of up to 85 kg and more is a complicated task. Additional difficulties are connected with the fragility and a low fraction ratio. The boule basing and fixation are especially complicated. The secondary blanks are already of the correct form, they are more accurate, but they may still be difficult to install.

The existing design methods of the installation of machine accessories were analysed, and their correctness was assessed when developing the equipment for leucosapphire treatment. The fixation of the metallic blanks in the devices is performed mainly by friction forces [3]. Let us determine the possibility of leucosapphire blanks fixation in such a case.

Due to anisotropy the leucosapphire properties are fluctuating, thus the most suitable basic and treated surfaces location values are used for the calculation. The leucosapphire-steel fraction ratio is 0.15, leucosapphire strength is 300 MPa [4].

Having the compression strength of leucosapphire, the maximal force, applicable to the carefully sanded and polished blank during the fixation, should be determined at first approximation. Let us suggest, that the blank and the device are in contact along the whole surface, and fixation force is distributed equally along the whole blank surface (clamp fixation on the top through a completely rigid and perfectly flat plate).

This evaluation is applied for the diamond disc cutting of the prism-shaped seed blank with the cross section area of 10×10 mm and the length of 100 mm. The blank contact surface for the accepted contact scheme is 10^{-3} m^2. The breaking compression force of the blank for the accepted loading scheme is 300 kN. Consequently, the elicited friction force on the blank-device joint, even considering the friction ratio reduction due to the applied lubricating-cooling liquid, will be at least 30 kN. A shifting component of the friction force, measured during the experiment, amounted to 120 N in average with peak values of up to 250 N.

Clearly, that in such conditions the blank is securely fixed due to the friction forces. However, in real conditions, the preliminary cut blankh as spacious defects and high gritness. The studies showed, that due to the disk extraction, cut width spacing of such defects can be 0.1…0.4 mm. The calculations are provided with the consideration of these defects. The following calculation scheme is agreed. The blank is presented as a beam on supports impacted by the equally distributed load. The calculation is executed for the above-mentioned leucosapphire seed blank on the parallelepiped shape with the length of 0.1m and a square cross section of the profile of 0.01m. For such beam let us calculate the breaking force and the deflection during the breaking. For such loading scheme, which is closer to real conditions, the secure fixation due to the cut force is possible only with reservations. Seed blank is broken only if the deflection exceeds 0.16 mm, and the load is around 3 kN. The elicited fraction force is close to the cutting force shift component value. The correctness of the performed calculation estimate was verified experimentally. A blank shift was observed when cutting the blanks with spatial defects exceeding the designed ones and fixed on the device by the scheme close to the designed one by means of the torque wrench (Figure 2). Thus, the design methods of machine accessories associated with treatment of hard large crystals should be corrected.

Figure 2. A leucosapphire blank cut by a diamond disc.

The calculation of fixing forces should be supplemented with the calculation of the blank breaking force considering the existing form errors. If the breaking forces exceed the forces necessary for secure fixation, the device's construction should be changed. For the case of impossibility of secure fixation only by the friction forces the changes to standard design solutions were provided. Particularly, the constructions with additional underwater supporting elements and clamping units, equipped with rigid rubber and polymer shims for preventing the risk of contact breaking at the blanks fixation point were successfully tested.

The blanks fixation security can be sufficiently increased by the use of glue. This method provides secure fixation, but increases material and labour costs. At the same time production standards also increase. Provided studies showed that in case of seed blanks gluing, if no special measures are applied, the difference of the glue layer and other reasons can lead to the planes non-parallelism during flat sanding in the amount of up to 0.05 mm.

To merge a spindle axle with the specified crystallographic direction of the blank, the device should provide a high-accuracy turn of the installed leucosapphire blank in two planes. Some principal engineering solutions of this task were analysed. Few schemes were considered: 1) the device is based on the central spherical joint – the plane turn is provided by the central joint, equipped with the plate with the installed blank: 2) the device, where inclination of the plane with the fixed boule is adjusted by few screws, which support ends, forms a plane inclined to the spindle axle by two axles; 3) the device with two cylindrical joints, where each joint enables the independent turn of each plane. All solutions have benefits and drawbacks. Based on the analysis the recommendations concerning their rational use were provided.

Practically all sapphire treatment operations are diamond-abrasive treatment. It is known that for such operations normal cut is possible either in self-sharpening conditions or during involuntary correction. The study results showed that binder metal tools self-sharpening, recommended for sapphire treatment, is difficult to implement. During the study of cutting, front and peripheral sanding, circular drilling, a self-sharpening mode was available only for the cutting disk for the cut of rotating boule. This can be explained by the fact that with such cutting system the tool and the blank contact along the line. Due to this contact the worn products are removed freely from the cutting zone, and even with the low cutting forces the specific loads are high, which leads to intensive wear of the binder and new seeds discovery. Normally the disc loses its cutting properties when being worn. If such disk is operated the cutting forces and temperature increase. Thermal shocks cause a large number of micro- and then macro-defects or a complete destruction of the blank.

This is especially important for the boules cutting and circular drilling. In such case the tool almost constantly remains in contact with the blank and sanding products; removing from cutting zone is complicated. Even in case of segment tools use the slug remains in the cutting zone and impacts the normal cutting process [6]. Such treatment reminds of the process of cutting by a fine pitch cutter with a large stroke or the case of deep cutting. The chip groove volume in such case should be enough for the placing of the obtained chip. Otherwise, different accidents are possible, including tool destruction or blank rupture. This phenomena is quite specific in case of sanding operations. During the diamond cutting by the leucosapphire disks the tool diameter increases. The investigation of the sanding disk under the microscope showed that the worn products had not only blocked all the pores, but also intruded in the binder. The use of this cutting process was monotonously deteriorating or terminating at some stage. The attempt to continue cutting applying larger forces caused the crystal destruction; thermal shock induced the microcracks net on the surface.

Thus, the trueing of diamond disks on the binder metal is mandatory. Within this work the studies of the basic trueing processes were conducted in order to identify optimal solutions.

Manufacturing plants prefer the diamond disk trueing method. However, for the plants applying these disks in practice the method complicates the efficient use of equipment – for quite a long time the machines are involved in trueing operations. Special staff for trueing and maintenance is required. Commercial equipment is often equipped with the devices of electrical chemical treatment (anode etching). Trueing is merged with the cutting process, which provides high efficiency of the process. However, such equipment is comparatively expensive, the process demands the application of quite aggressive lubricant-cooling agents, which leads to fast machines corrosion, deteriorates labour conditions and ecology.

Electro-erosive trueing is harmless for the machine and relatively green. Loose abrasive trueing is very efficient. The experimental studies of the electro-erosive, electrical chemical and loose abrasive trueing were conducted. All methods enable solution of the described tasks of the cutting properties and the disk shape recovery.

The most efficient and qualitative was the loose abrasive method. Binder removing in such case is performed without causing any harm to diamond seeds. But this method failed its application for diamond disks. The disk breaking corner value and consumption of working trueing medium were exceeded. This process was not completely implemented for the circular diamond drill due to the insufficient cutting rate.

Electro-erosive trueing provides practically the same efficiency. The process is well controlled and relatively easily implemented on the existing and new developed equipment. It was also implemented directly on the processing equipment and using a special device. The important advantage of such trueing type is the developed relief of the operating part of the disk. However, the experiments showed that disks firmness after such trueing was significantly lower (up to 30 %), than in the delivery condition or after chemical trueing. Due to high temperature of the process, the seed tops were found to be graphitized, which was also proved by microscope observations.

An electrical chemical impact was not severe. During anodic dissolution the binder was removed, the seeds were discovered and relief was formed. The seeds are of the light colour, without a defect layer on them. Although disks' relief is not much developed in this case, they operate well. Thus, considering cutting properties, this method can be regarded as the most attractive, although its efficiency is much worse. The reagents selection enables minimization of the corrosive impact on the machine. Owing to this fact the process efficiency is reduced. This quite efficient process demands serious organizational-engineering solutions and costs.

During the experiments one more problem, associated with the cutting process control was also found. Typical for most machines, a constant delivery operational mode is available only with the constant cutting properties of the tool. Comparatively fast glazing and the change of cutting properties complicate a constant delivery operation. Thus, the work with the stable force should be organized, which is typical for the optical industry machines, or proceed to the application of adaptive systems, including self-teaching ones [2].

4. Conclusion

The basic method of leucosapphire treatment is diamond-abrasive treatment. The methods of development of the leucosapphire mechanical treatment technology in general and for particular operations are favourable only for separate product groups and should be studied concerning the improvement of labour efficiency and products quality.

The existing design methods of the metal treatment machines demand a set of changes. The content and sequence of calculation procedures should be corrected, they should include the estimation of acceptable forces, limited by blank strength. Additionally, the auxiliary design solutions are required for accuracy and reliability increasing of the leucosapphire blank installation.

Leucosapphire, which is hard for fragile material treatment has its specific features. The provision and recovery of high-quality cutting properties of the diamond-abrasive instrument is an important issue.

Diamond-abrasive tool trueing is usually a mandatory procedure, a self-sharpening mode for binder metal tools can be implemented only occasionally. In case of operation with worn disks, the slug was identified in the tool binder, which even caused the diameter increase with the drop in cutting properties.

Loose abrasive trueing demonstrated very high results concerning both the efficiency and the quality. But the existing problem of edge effects limits the use of this method for cut off disks and circular diamond drills.

An electro-erosive trueing method of diamond disks negatively affects the cutting properties of diamond seeds in case of the high-efficiency process. Consequently, the tool firmness reduces by 20…30 %.

Electrical chemical trueing of the disks is relatively low-efficient, but provides a high quality. The technology is high-efficient if trueing is merged with treating. Due to this fact the risks of the negative ecological impact of reagents and labour safety risks should be considered.

During the treatment the cutting properties of the tool within its firmness period were significantly changed. This factor requires correction of the treatment process in the performance mode. Ignorance of the specified dynamics may lead to quality deterioration, including irreparable spoilage. This problem can be solved by the adaptive process control or proceeding to the actual cutting force treatment instead of delivery of the actual value.

References

[1] http://www.techsapphire.ru/products/blanks
[2] Suslov A G, Petreshin D I 2010 *J. Russian Engineering Research* **30** 418–423
[3] Durmus Karayel 2009 *J. of Materials Processing Technology.* **209** 3125–37
[4] Handozhko A V 2012 *J. Science intensive technologies inmechanical engineering* **2**
[5] Dobrovinskaya E R, Litvinov L A and Pishik V V 2004 *Sapphire Encyclopedia.* (Kharkov:Monocrystals Institute)
[6] Shaw M C 1996 *Principles of Abrasive Processing* (Oxford: Oxford Science Publications, Clarendon Press)
[7] Quan Y, Liang L and Zhong W 2009 *Advanced Materials Research* **76** 491–96

Pulsed welding plasma source

A Knyaz'kov[1], O Pustovykh[1], A Verevkin[1], V Terekhin[2], A Shachek[1], A Tyasto[1]
[1] National Research Tomsk Polytechnic University, 30, Lenina ave., Tomsk, 634050, Russia
[2] Seversk Technological Institute, Branch of State Autonomous Educational Institution of Higher Professional Education 'National Research Nuclear University 'MEPhl', 65, Communistic ave., Seversk, 636036, Russia

E-mail: bos1983@tpu.ru.

Abstract. It is shown that in order to form the current pulse of a near rectangular shape, which provides conversion of the welding arc into a dynamic mode, it is rational to connect a forming element made on the basis of an artificial forming line in series to the welding DC circuit. The paper presents a diagram of a pulsed device for welding with a non-consumable electrode in argon which was developed using the forming element. The conversion of the arc into the dynamic mode is illustrated by the current and voltage oscillograms of the arc gap and the dynamic characteristic of the arc within the interval of one pulse generation time in the arc gap. The background current travels in the interpulse interval.

1. Introduction

Argon-arc welding with a non-consumable electrode provides a high weld quality when constructing metal structures for critical applications. One of the disadvantages of this method is low melting capacity. Various methods and means to eliminate this disadvantage can be found in literature. The dynamic mode of arc burning is among them. This method allows an increase in the melting capacity by 25…30 % [1] with a more concentrated heat input into the product, in comparison with the conventional welding mode. This improves weld formation and mechanical properties of the welded joint.

It is known [1, 2] that in the dynamic mode of arc burning the pulse duration is to be less than 450 μs. The resumption of the electric discharge occurs through gas molecule ionization in the arc gap. When the arc is powered by short-time current pulses with a steep front and edge, ionization processes leg behind the change in the current, and the arc discharge does not become steady. This process provides a contracted arc column and anode spots adjacent to the column. The arc discharge in the dynamic mode is characterized by a greater density of the energy input into the product and greater melting capacity if compared with those of the arc discharge by a stationary arc with the current equal to the average current of the dynamic mode. Since the concentration of the heat input in the product in the dynamic mode is greater due to the arc energy parameters without auxiliary structures and devices used, the study of this discharge for practical purposes is of great interest.

The lack of technical means for implementation of the dynamic mode of arc burning hinders its wide application.

2. Theoretical analysis

The analysis of the scientific literature showed that the energy storage of the electrical network and its further conversion into the welding current pulses can be provided by chemical elements; a centrifugal mass; magnetic elements; coaxial cables and forming lines.

Electrochemical energy storage devices utilize the energy from chemical reactions for energy storage and regeneration [3]. When using an accumulator, the energy is accumulated due to the chemical reactions between the electrodes immersed in the electrolyte solution. The upper limit of the currents flowing through the accumulator is equal to 300 A. To use the accumulator as the energy storage device, for the dynamic mode of arc welding in particular, is not rational due to the increased heat losses and consequently low efficiency if compared to the stationary mode and other energy storage devices.

The rotating masses are most economical in comparison with capacitors and inductors. The accumulation of energy in the inductive load occurs in two stages: 1) the prime mover of small capacity accelerates the rotor of the impact-excited generator up to the nominal speed; 2) the impact-excited generator converts the stored kinetic energy into the electromagnetic one and transmits it to the inductive load. The losses of the rotating mass energy are insufficient if compared with the transmitted energy, i.e. the efficiency of the charging unit is high, and the weight-size parameters of the system are large.

A homogeneous artificial forming line (*AFL*) is of great importance in creation of the current pulse which provides the dynamic mode of arc burning [4, 5].

The line (Figure 1) consists of L_c and C_c cells with equal inductance values of chokes L_c and capacitances C_c. When analysing the processes occurring in this line, the resistance of the chokes is typically taken into account and the losses in the capacitors are neglected.

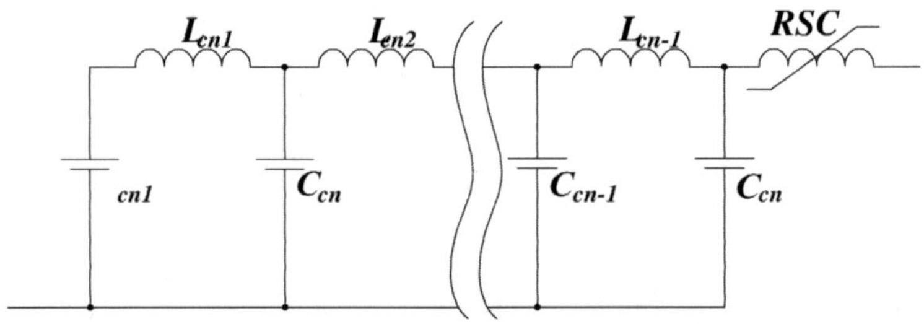

Figure 1. Equivalent circuit of the homogeneous artificial line.

In this system, the capacitor is charged through charging choke L_o and the charge thyristor, and discharging to the arc gap is carried out through the discharge thyristor. The discharge of this line produces rectangular pulses with a steep front and edge. The disadvantage of the system is a limited pulse formation frequency due to a sufficiently long phase of capacitor pre-charging and possible occurrence of the through current which causes failure.

In addition to the *AFL*, coaxial cables or stripline circuits can be used as a device for pulse formation and energy storage [7]. Such lines are used to generate pulses of nanosecond duration. However, their practical application is attended by structural difficulties; limited duration (from 20 ns to 0.5 μs); low operating voltage.

3. Practical implementation

To eliminate the identified shortcomings, the discharge of the forming element (*FE*) and its recharge are to be combined within the time of the pulse current travelling through the arc gap. For this purpose, the *FE* is included in the diagonal of the bridge converter, and the other diagonal of the bridge converter is connected in series with the welding arc to the external circuit of the constant current source.

A functional diagram of the developed experimental model of the power system is presented in Figure 2 [6].

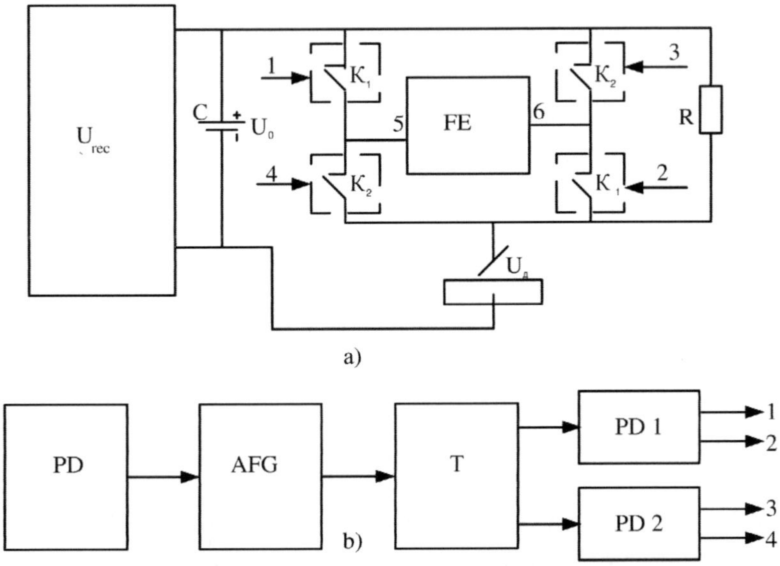

Figure 2. A functional scheme of the device.

The device consists of three-phase rectifier U_{rec} with a rigid external characteristic and regulated open circuit voltage. To reduce the internal resistance to output terminals U_{rec}, the capacitor bank of high capacitance *C* is connected.

The connection of the *FE* to the diagonal of the bridge and presence of capacitor bank C allows for *FE* recharging from the opposite polarity through the arc gap, in the sequential opening of key pairs $K_1 - K_1$and $K_2 - K_2$, without pre-charging from the power supply unit.

To ensure the arc continuity within the inter pulse interval, the interval is shunted by resistor *R* to provide the background current flow through the arc gap. The range of the background current values is limited by a minimum value (requirements for the arc physical stability and protection), and by a maximum value (technological requirements).

The power supply unit (*PSU*) supplies power to the control circuit. The frequency of the device operation is determined by the adjustable frequency generator (*AFG)* capable of varying frequency from a few hertz to several thousand hertz with a corresponding decrease in the welding current pulse width. The sequence of pulses from the generator is supplied to the input of the T-trigger to divide the frequency by two. The pulse sequence frequency is two times less than the adjustable generator frequency, and the pulses shifted relative to each other by the period of the generator pulse frequency are fed to keys $K_1 - K_1$ and $K_2 - K_2$from the two outputs of the T-trigger.

The *AFL* consisting of n-th number of cells is used as a forming element of the *FE* (Figure 3).In turn, each cell consists of series-connected capacitor C_c and choke L_c.

The last cell choke is designed as a rapidly saturated choke (*RSC)* with the core having are ctangular magnetization curve, and the inductance in a saturated state is equal to the inductance of the cell chokes.

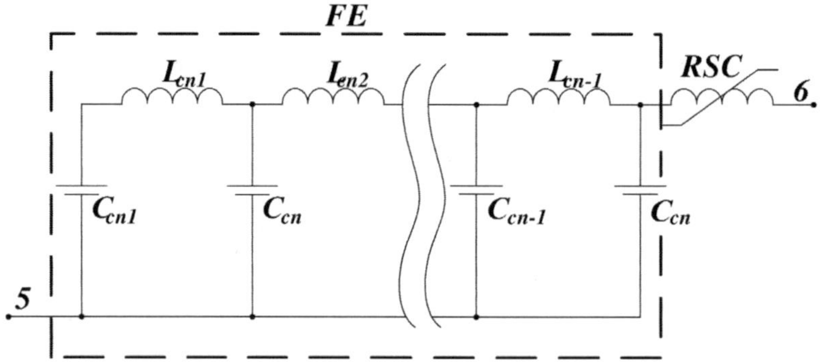

Figure 3.A forming element

The device operates as follows. When feeding the control pulses to keys K_1 – K_1 located in the opposite arms of the bridge, the lower plates of capacitors C_c are negatively charged and the upper plates of the *AFL* capacitor are positively charged. Capacitor bank *C* is always charged from rectifier U_{rec} with the upper plates being positive. When keys K_1 - K_1 are opened under total power supply voltage U_{rec} and the voltage of charged capacitors C_c, the *AFL* is recharged as follows: rectifier U_{rec} (plus) – key K_1 – *FE* – *RSC* – key K_1 – arc gap (between the product and the electrode) – rectifier (minus). When recharging is completed, capacitors C_c change their polarity: the lower plates become positive, and the upper plates become negative.

When keysK_2 – K_2 are opened, the recharging process of the *AFL* is similar to the recharging process with keysK_1 – K_1 opened. In this case, unipolar current pulses travel through the arc gap.

The energy is transferred from the device to the welding arc via the coaxial cable to reduce the welding circuit inductance. The oscillogram of the welding process was recorded by a non-inductive shunt.

Figure 4 shows typical oscillograms of one of the dynamic arc welding recorded under the following parameters: the arc length (distance from the electrode end to the plate surface) was maintained equal to 3.5 mm. The welding conditions were as follows: I_p = 320 A; *Iii* = 16 A; pulse duration was 70 μs; U_p = 32 V. The tungsten electrode had a spherical surface that enhances the operational life and did not violate the spatial stability of the arc column in the dynamic burning mode. The pulse frequency of the welding current was 3.3 kHz.

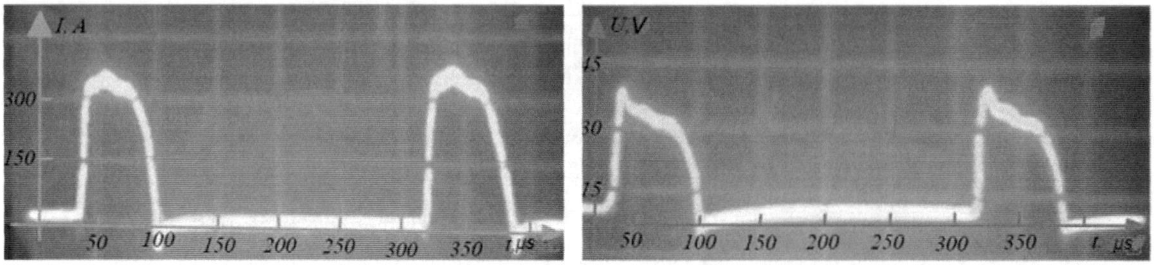

a b

Figure 4. Oscillograms of *a* – arc current pulse, and *b* – arc voltage.

The current pulse amplitude in the welding circuit was determined by [3]:

$$I_p = \frac{U + U_c - U_d}{2\rho};$$

Where U, U_c, U_d are voltages of power supply U_{rec}, the *AFL* charge, and the arc gap, respectively; $\rho=\sqrt{L_c/C_c}$ is the *AFL* impedance; L_c is the choke inductance of the *AFL* cell; C_c is the capacitance of the *AFL* cell capacitor.

The superimposed oscillations occur at the top of the current pulses; the number of oscillations depends on the number (n) of the *AFL* cells. The value of the first maximum spike of the current and voltage oscillations [3] is virtually independent of the number of the *AFL* elements and it is determined by the parameters of the first cell. However, due to severe requirements for the constancy of the pulse peak, the peak is to be corrected. [3]. A rapidly saturated choke is used for the first oscillation spike reduction and smooth transition from the background current to the pulse (at low currents). The active duration of the current pulse to be formed is [3]:

$$t_p = 2,2n\sqrt{L_c \cdot C_c}\ ;$$

where n is the number of cells in the forming line.

The high rate of the current rise and arc voltage assumes that the arc behaves like active resistance within the edge pulse interval. When the current of the pulse peak arc is conventional dc, the voltage drops smoothly due to the development of the ionization processes in the arc column. The properties of the arc column as nonlinear energy storage are observed at the pulse edge. A large number of charge carriers in the arc column facilitate the transition of the arc column voltage to the negative region. This owes to the fact that the number of charge carriers from the power supply required to provide the background current flow should be less than the number of carriers in the arc column volume.

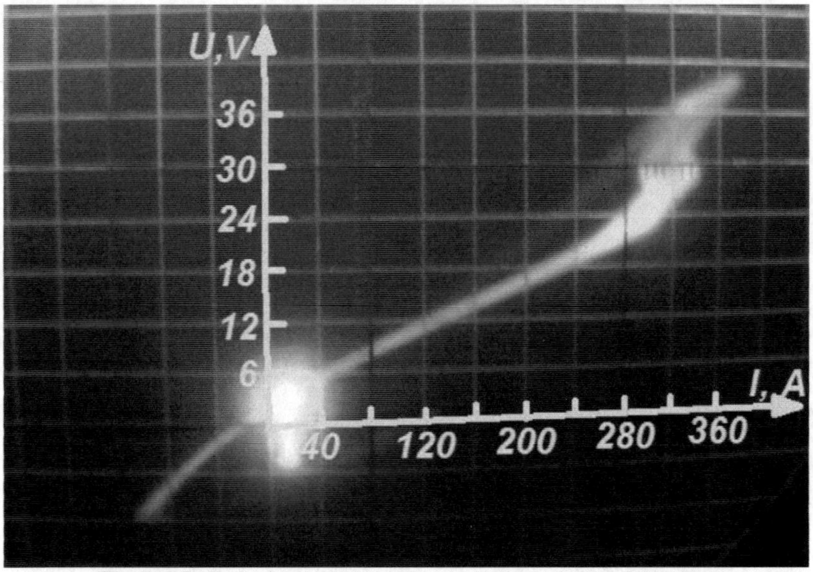

Figure 5. A dynamic volt-ampere characteristic of the arc.

The dynamicvolt-ampere characteristic(DVAC) of the arc presentedin Figure 5indicates a dynamic mode of arc burning, which causes the hysteresis. The upper branch of the DVAC correspondsto thepulse front edge, but the branch cannot be clearly seen on the oscilloscope screen because of the rapidrise of the currentandarc voltage. The lower branch of the DVAC corresponds to the pulse edge.

4. Conclusion

The pulsed device, consisting of a bridge converter with the *AFL*-based forming element connected to one of its diagonals, which is connected in series with the welding circuit, enables generating welding current pulses of a near rectangular shape, provides a dynamic mode of arc burning, and increases its reliability.

The proposed circuit of the power unit increases the frequency of welding current pulse generation through combining the charging and recharging processes of the *AFL* during the welding current pulse.

The proposed circuit, if compared to known circuits, enhances the *AFL* efficiency with the same cell parameters through generating current pulses under the total effect of the power supply and pre-charged capacitors of the homogeneous artificial line to increase the current amplitude.

References

[1] Slavin G, Stolpner E 1974 *Welding production* **2** 3–5
[2] Knyaz'kov A, Biryukova O 2011 *Bullet. of the Tomsk Polytech. Univ: Power Engineer.* **318** 104–107
[3] Nekrasov V, Gavrilov G 1968 *Electricity* **12** 82
[4] Zaitsev A, Knyaz'kov A 1969 *Electricity* **10** 54
[5] Knyaz'kov A, Knyaz'kov S, Lolyu Ya, Pronyaev A Patent No. RU 2294269
[6] Knyaz'kov A, Knyaz'kov V, Biryukova O, Ustinov V Patent No. RU 2343051
[7] Itskhohi Ya, Ovchinnikov N 1972 *Pulsed and digital devices* (Moscow: Izd Soviet radio)
[8] Knyaz'kov A, Krampit N, KrampitA 2008 *Welding International* **22**(8) 534–535
[9] Knyaz'kov A, Knyaz'kov V 2010 *Welding International* **24**(12) 955–957
[10] Knyaz'kov A, Dementsev K, Knyaz'kov V 2013 *Welding International* **27**(2) 147–149

Design Models for Shaping of a Tooth Profile of External Fine-Module Ratchet Teeth

O V Sharkov[1,2], S I Koryagin[1] and N L Velikanov[1]
[1] Immanuel Kant Baltic Federal University, 14 A, Nevskogo st., Kaliningrad, 236041, Russia
[2] Kaliningrad State Technical University, 1, Sovetsky ave., Kaliningrad, 236022, Russia

E-mail: osharkov@kantiana.ru

Abstract. Simulation of the shaping for the fine-module external ratchet teeth at which the contacting surfaces are formed by the straight segments is considered in this paper. The design schemes for shaping of the proposed ratchet teeth by a shaper cutter and a rack are obtained. It is defined that the maximum length of the straight segment of the front edge ratchet teeth will be formed at shaping by a rack cutter. The effect of a module, a gradient angle and a radius of blank circles on the length of the straight segment of the front edge ratchet teeth is investigated.

1. Introduction

One-way clutches are widely used in various machine drives. According to the method of load transfer one-way clutches are classified into two main types. The load transmission in these clutches can occur due to a frictional force (roller one-way clutches) or normal forces (ratchets of one-way clutches) [1-6].

Roller one-way clutches most often are applied in machine drives. However, their load capacity and durability depend on the value of the friction coefficient, which can change in a wide range of f = 0.04…0.20 [2, 3].

Ratchet one-way clutches have a large load capacity and reliability. However during the operation of ratchet engagement there is a significant noise and impact loads [3, 5].

It is possible to lower these deficiencies at the expense of reduction of the size of the module. For example, in the eccentric one-way clutches of a non-friction type the fine-module ratchet teeth with module m_t = 0.3…1.0 mm can be used.

In article [7] determination of geometrical parameters of fine-module ratchet engagement for the eccentric one-way clutches of the non-friction type is considered. The proposed fine-module ratchet teeth of the new profile, which provide their contact in teeth on the flat surfaces.

Modern design schemes used at shaping of involute and ratchet teeth profiles can not be applied for proposed teeth due to the peculiarities of their geometry.

2. Analysis of shaping methods of external fine-module ratchet teeth

Modern production methods of teeth are diverse and include more than 50 types [8-14]. One of the most efficient and widespread methods for shaping of different teeth profiles on cylindrical surfaces is tooth-cutting by continuously indexing method.

Let us analyze the methods used for shaping fine-module ratchet teeth of the described profile on the basis of a procedure for the generating of working surfaces by a continuously indexing method [8-14].

Shaping external teeth by the continuously indexing method is possible in two ways – a shaper cutter or a rack cutter [12, 14].

2.1. Simulation of shaping of the teeth by the shaper cutter

For the preparation of design models let us assume that r_U and r_{U1} – radii of dedendum and addendum circles of teeth of the edge tool; r and r_1 – radii of blank circles, passing through the dedendum of external and internal of the shaped teeth ($r = r_{f1}$ and $r_1 = r_{f2}$).

The shaper cutter with centre O_U and the blank with centre O (Figure 1) are shown in reversed motion with respect to each other with the motionless blank. During forming of point *M* the shaper cutter is in position O_U and a point on the shaper cutter is coinciding with point *M*. Because it is owned by its cutting edge.

There is point M_0 (Figure 1), which is formed by the tooth point of the shaper cutter, located on the circle of radius r_{U1}.

All points located on segment AM_0 do not have conjugate points on the shaper cutter, because they are formed outside the circles of radius r_U and r_{U1}. Consequently, the shape of the cutting teeth will be created at segment $|BM_0| = l_{t11}$ by a straight line and at segment AM_0 by a transition curve formed by the tooth point of the shaper cutter.

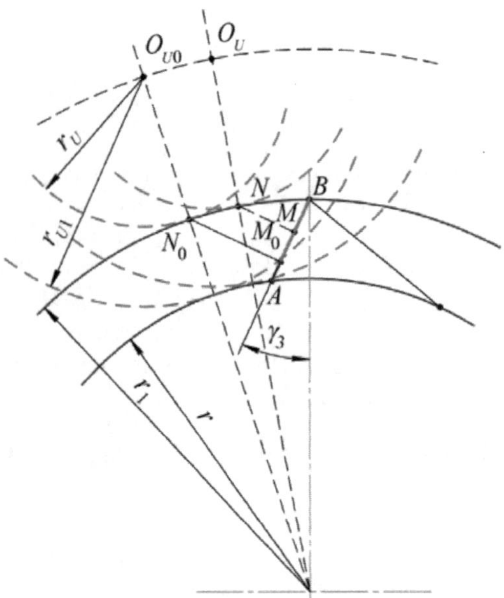

Figure 1. Shaping the external ratchet tooth by the shaper cutter.

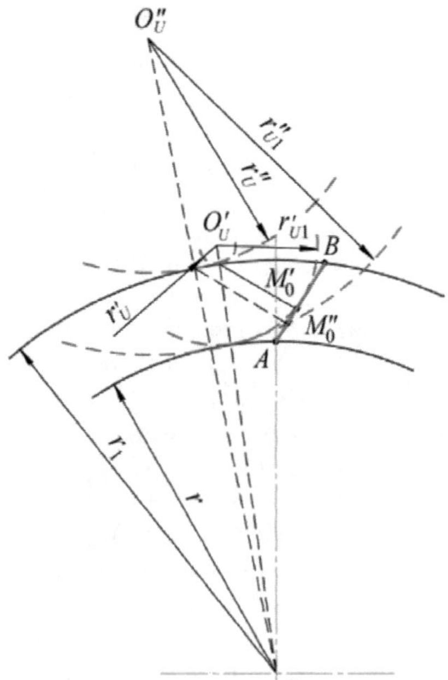

Figure 2. An effect of the shaper cutter diameter on the straight segment length of the tooth front edge.

When selecting the type and the size of the cutting tool for shaping external teeth it is necessary to provide a greater length of the straight segment of the tooth profile in relation to its theoretical length.

Figure 2 shows that the desire to increase the length of straight segment l_{t11} for the same tooth leads to the need of a significant increase of the radius of the shaper cutter. Therefore, for shaping the external teeth it is the most appropriate to use the rack cutter which is $r_U = \infty$.

2.2. *Simulation of shaping of the teeth by the rack cutter*

Figure 3 shows the position of the rack cutter and blanks when the tooth point of the rack cutter is forming extreme point M_0 on the straight segment of tooth edges l_{t11}.

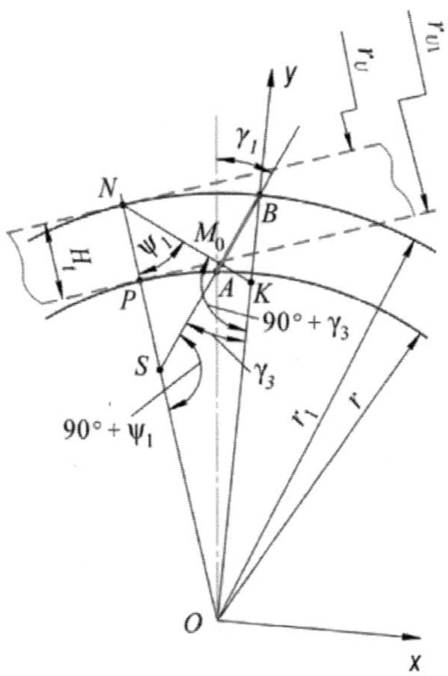

Figure 3. Shaping the external ratchet tooth by the rack cutter.

Using triangle OBS we can write

$$\frac{r_1}{\sin(90+\psi_1)} = \frac{r_1 - H_t/\cos^2\psi_1}{\sin\gamma_3},$$

where $H_t = |PN|$ – theoretical depth of the rack cutter tooth.
Then we obtain expression

$$r_1\cos^2\psi_1 - r_1\sin\gamma_3\cos\psi_1 - H_t = 0,$$

where $\cos\psi_1 = \dfrac{r_1\sin\gamma_3 + \sqrt{r_1^2\sin^2\gamma_3 + 4r_1H_t}}{2r_1}$. (1)

Using triangle ONK we can be write

$$\frac{r_1 - l_{t11}/\cos\gamma_3}{\sin\psi_1} = \frac{r_1}{\sin(90+\gamma_3)}.$$

After the transformation we will have

$$l_{t11} = r_1(\cos\gamma_3 - \sin\psi_1), \quad (2)$$

where $\sin\psi_1 = \sqrt{1-\cos^2\psi_1}$ or $\sin\psi_1 = \sqrt{\dfrac{r_1(2-\sin^2\gamma_3) - \sin\gamma_3\sqrt{r_1^2\sin^2\gamma_3 + 4r_1H_t} - 2H_t}{2r_1}}$.

After the transformation of formula (2) and taking into account $H_t = m_t$ the length of the straight segment on the front edge of the external ratchet tooth can be expressed as

$$l_{t11} = r_1\left[\cos\gamma_3 - \sqrt{\frac{r_1(2-\sin^2\gamma_3) - 2m_t - \sin\gamma_3\sqrt{r_1^2\sin^2\gamma_3 + 4r_1m_t}}{2r_1}}\right]. \quad (3)$$

3. Numerical Results and Discussion

The effect of module m_t, gradient angle γ_1 and radius r_1 on the length of the straight segment of front edge l_{t1} is shown in Figure 4. Calculations are executed by formula (3).

The initial data for the calculations are accepted as: r_1 = 20...100 mm; m_t = 0.4...0.6 mm; γ_1 =0...30°. The value of gradient angle γ_3 is defined by formula [7]

$$\gamma_3 = \arcsin\left(\frac{r_{f1}\sin\gamma_1}{r_{f2}}\right).$$

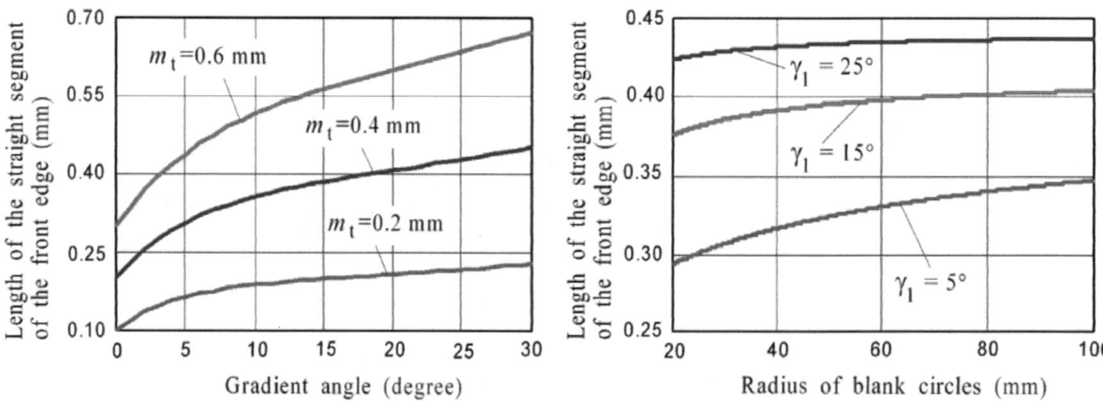

a b

Figure 4. The effect of the module, gradient angle and blank circles radius on the length of the straight segment of the tooth front edge: *a* – r_1 = 30 mm; m_t = 0.4...0.6 mm; *b* – m_t = 0.4 mm; γ_1 = 5...25°.

The calculation results allow one to draw the following conclusions. A 1.5 time increase of module m_t causes a 2.7…3.0 time linear growth of the length of the straight segment of front edge l_{t1}. The increase of gradient angle γ_1 in the range from 0 to 30° causes a 2.20…2.25 time nonlinear growth of the length of the straight segment of the front edge. The increase of radius r_1 in the range from 20 to 100 mm causes a 1.02…1.17 time decrease of the length of the straight segment of the front edge.

4. Conclusion

For shaping ratchet teeth it is recommended to use the continuously indexing method. One can select a shaper cutter or a rack cutter as the cutting tool. The shaper cutter is a universal cutting tool because it allows shaping both external and internal fine-module ratchet teeth.

IOP Conf. Series: Materials Science and Engineering **124** (2016) 012165 doi:10.1088/1757-899X/124/1/012165

References

[1] Aliukov S, Keller A and Alyukov A 2015 *Jour. SAE Technical Paper* 2015-01-1130
[2] Orthwein W C 2004 *Clutches and brakes: design and selection* (New York: Marcel Dekker) p 312
[3] Neale M J 2001 *The tribology handbook* (Oxford: Butterworth-Heinemann) p 640
[4] Feng Z, Gao Z and Lu W 2013 *J. Advanced Mater. Research* **823** 43–46
[5] Bondaletov V P 2008 *Russian Jour. Russian Engineering Research* **28(9)** 845–848
[6] Lee C J, Samie F and Kao C-K 2010 *Proceedings of the ASME Dynamic Systems and Control Conference* DSCC2009 **B** 1517–1521
[7] Sharkov O V, Koryagin S I and Velikanov N L 2015 *J. Appl. Mechan. and Mat.* **770** 264–268
[8] Xia L, Li D and Han J 2014 *J. Key Engineering Materials* **579-580** 300–304
[9] Stadtfeld H J 2011 *American Gear Manufacturers Association Fall Technical Meeting* pp 1–14
[10] Huang C-L, Fong Z-H, Chen S-D and Chang K-R 2009 *J Mechanism and Machine Theory* **44(2)** 401–411
[11] Khurmi R S and Gupta J K 2005 *Theory of Machines* (New Dehli: S.Chand & Co. Ltd.) p 1060
[12] Radzevich S P 2012 *Theory of Gearing: Kinematics, Geometry, and Synthesis* (Boca Raton: CRC Press) p 743
[13] Mott R L 2014 *Machine elements in mechanical design* (Upper Saddle River, New Jersey: Pearson Education) p 816
[14] Youssef H A and El-Hofy H 2008 *Machining Technology: Machine Tools and Operations* (Boca Raton, London, New York: CRC Press) p 672

Selecting a vibration diagnostic defect 'Mechanical weakening of a structure'

A V Ermolaev and E A Sikora
Department of Automation and Robotics in Mechanical Engineering, Tomsk Polytechnic University, 30, Lenina ave., Tomsk, 634050, Russia

E-mail: ermolaevavw@gmail.com

Abstract. The article describes the method of analysis of the levels of vibration signals of a support machine rotor in the radial directions and the results of its application for the diagnosis of mechanical loosening of structures of rotating machines for vibration acceleration peak values. Information about correspondence of the degree of a defect 'mechanical weakening of a structure' to the relative level of acceleration in three frequency bands is given in a table.

1. Introduction

It is known [1, 2] that the information obtained by the sensor of the absolute acceleration installed on poles of the rotor makes it possible to monitor the condition of the bearing units of the unit (Normal, Warning, Alarm), to diagnose the defects at different stages during its operation, and to draw conclusions on such defects as 'mechanical weakening of the structure'.

It should be noted that in most cases the defect appears in conjunction with the features of other types of defects on the unit: imbalance, misalignment, bearing defects, loose fit connecting the coupling halves onto the shaft that occurs for wear and violations of mounting technology, etc.

The origin of the mechanical loosening of rotating equipment is very diverse and is associated with a large number of design parameters, installation and operation.

In general all mechanical weakening can be divided into two major groups [2]:

• mechanical weakening like 'backlash', a manufacturing defect, assembly and installation of rotating equipment: free seating of rotating parts is associated with the presence of non-linearity observed in the support bearings, couplings, in adjacent parts in the support mechanisms and foundations;

• mechanical weakening of the 'mechanical weakening of the structure', which is a result of natural wear of the structural elements of the unit, resulting in the sudden disaster and the destruction of the structural elements; cracks and defects in the design and foundation arising from the operation of controlled equipment.

The division of the causes of mechanical loosening of these two groups is fairly conventional. And in fact, in both cases the characteristics of the defect in the spectrum of the vibration signal are about the same. It should be noted that this diagnostic of similarity is observed in the spectra of the vibration signals, in both qualitative and quantitative ratio.

In electrical machines vibrations occur due to deviation of the shape and arrangement of elements, bearings and other reasons. Interaction of brushes with collector plates is additional sources of vibration in the DC motor. Fan and pump vibrations are caused by the vibration of the motor structural elements and aero-hydromechanical noise. A common feature of the considered devices is the concept of their operations and changes of the technical condition in the most determined elements with rubbing contacts. It should be borne in mind that the technical condition of these elements depends on the dynamics and vibrations of the entire system.

The greatest effect of the application of vibration analysis and prediction takes place when it is used at all stages of the existence of active products, from manufacturing to maintenance. Diagnostic models are implemented in the form of algorithms that allow improving the processes of manufacturing and test systems and defect detection.

2. Analysis of the system of vibration parameters

The changes taking place in any system are linked to external and internal influences. Mechanical stress is generated by the static and dynamic processes, such as the vibration load, determined by the nature of the working process, the relative movement of the elements friction in kinematic pairs. An external temperature field and thermal energy released by friction also have a significant impact on the operation of the devices. It is possible to cite other examples of influences.

The changes occurring in the system may exceed permissible limits and cause decrease of parametric failures. Faults in the system can be divided into gradual and sudden. Degradation failures arise due to changes in parameters such as due processes of wear and tear; sudden faults are associated with the transition of quantitative to qualitative changes. The consequences of failure are also different. Sudden faults leads to the fact that the product cannot perform its function. The consequence of failure is output parameters of the product from the permissible values.

Parametric failures for precision tribological systems are caused by a gradual process of a changing technical condition. The basis of parametric failures is physical laws, but the diversity of factors at these frequencies is regarded as stochastic. The deep knowledge of regularity causes allows solving the problem of forecasting changes in the parameters, properties and condition of materials better. Modern science is studying patterns of changes in the properties and condition of materials at the following levels [3].

A submicroscopic level allows, based on solid-state physics, exploring the imperfections in crystals, non-equilibrium positions of the atoms in the lattice. These studies are the physical basis of solving the basic problems of the analysis of durability of materials.

Microscopic properties of materials are based on the study of the processes occurring in a small area. Primarily, this refers to the interaction of the surface layers of the friction elements in the presence of lubricants and surfactants. A macroscopic level is based on the change of the integral parameters characterizing elements in general (sizes, elastic characteristics, coefficient of friction, etc.). The analysis of processes at the macroscopic level allows one to generalize the physical laws of changes of the technical condition of parts and to investigate the relationship dynamics of systems with the destruction processes.

In the study of the dynamics of systems of particular importance is the analysis of the natural oscillations. The importance of research of vibration parameters of the systems is determined by their technical condition, as well as the change in the parameters of elements containing rubbing contacts (bearings, current-carrying elements, gears, etc.), depending on their natural oscillations. The forces, which excite vibration and noise of machines, appliances and devices, by their nature may be of mechanical, magnetic and aerodynamic origin. In accordance with this division mechanical, magnetic and aerodynamic vibration is produced. The source of mechanical vibration and noise are unbalanced rotating or oscillating parts, bearings, gears, current-carrying nodes, such as a brush-collector, and other elements. Unbalance elements cause oscillations with frequencies that are multiples of the frequency of rotation or stationary traffic. The amplitude of the driving forces is proportional to the square of the frequency and imbalance.

The main reasons for support oscillations are cycled stiffness in rotation or other stationary motion and geometric imperfections of their contacts and the mating surfaces. Parameters of the driving forces depend on the working conditions, dimensions and elements of technological errors. Magnetic vibration and noise occur due to periodic changes in the electromagnetic forces in the air gap of the electromagnetic system. The sources of vibration and noise of the aerodynamic origin are the moving parts of machinery. Vibrations from these factors interact with each other, and as a result vibrations occur in a wide range of frequencies.

More effective methods that provide control of a functioning and allow one to define proper disturbance, are the methods of technical diagnostics, the basic content of which is to examine the rationale and methods of indirect measurements of hidden parameters of devices affecting the nature of its functional behavior. The existing diagnostic methods are generally based on the analysis of the main operating parameters of the object. The object model and signals indicating of the object status as an element of the system does not fully reflect the internal processes and their changes, but only some have conclude a formal link between the input and output object. In this model, it is impossible to diagnose the unobservable internal processes therefore a special diagnostic model of the technical condition of the system and a change of the parameters of the object is required. It is necessary to choose the best information signals to establish the relationship between object as an element of the system and the model for diagnosis.

The main properties of the object as a member of the system are characterized by operator L, which connects input and output signals U_1(t) and U_2(t), and also accounts for dependence U_2(t) of disturbing factor U(t), generated by its own internal processes. The quality of operation depends not only on the design parameters, but also on perturbation U(t), which changes over time and can cause failure of the parametric system.

Changing the technical condition can be controlled by changing the natural oscillations of z(t) (vibration), which are generated by internal processes. In this case, as a rule studies are limited by the establishment of diagnostic models. The disadvantage of this approach is the lack of communication between U(t) and z(t).

It is necessary to create such a model for the diagnosis, which makes it possible to record the changes in internal processes and also establishes a link between the disturbing factor of U(t) and natural oscillations z(t).

As the block diagram model is illustrated in figure 1, the main parameter is r, which binds U(t) and z(t).Parameter r is defined by geometric characteristics deviations from the nominal values, technological uncertainties and other fluctuations. The relationship between U(t) and r, z(t) is set by operator T, and between r and z(t) – by W.

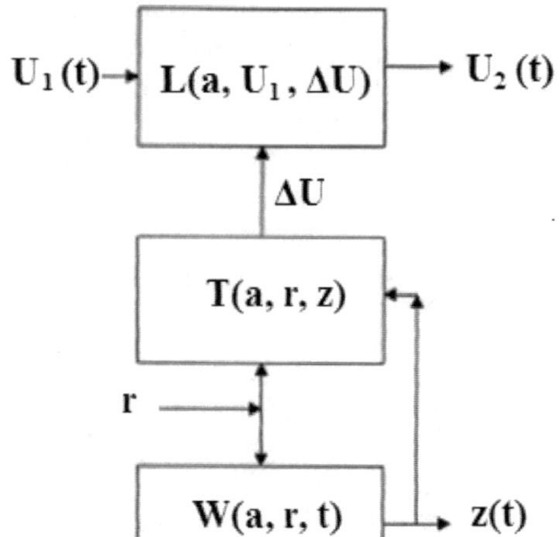

Figure1. A block diagram of a diagnostic device model.

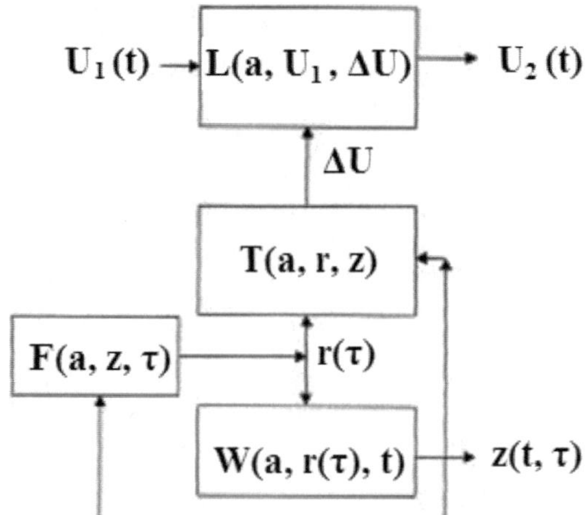

Figure 2. A block diagram of a general diagnostic device model.

This model is not complete, as parameter r operates in long-term system transient. Changing of r is not only aging, but also the dynamic vibration effects; it is slow compared to the vibration and the fluctuation of the main operational indicators; it is a vibrorheological process. The relationship between the change and vibration is set by the operator *F* (Figure 2).

Thus, a generalized model has two types of characteristic processes: a rapid (time *t*) – vibration and fluctuation in performance and a slow of parameters. Fast processes determine the quality of functioning at a given moment of time, and slow – parametric reliability of the system. Selection of diagnostic signal must be carried out so as to be sufficiently informative for evaluation of the vector r, its variations, and hence to estimate $z(t)$ and $U(t)$.

Defects mechanisms eliciting vibrodiagnostic control techniques are presented in table 1.

Table 1. Defects detected by the mechanisms of vibrodiagnostic control methods.

Defect	Explanation
Imbalance	Rotor, impeller, the system 'rotor motor – coupling – the impeller', etc.
Misalignment of shaft	Fracture and mixing of shafts. The curved shaft.
No rigid attachment	Cracks in the frame or body. Loosening of the foundation or mounting bolts. 'Soft foot'.
Uneven defects of motor	Air gap between the stator and the rotor. Damage of the stator windings or insulation. The eccentricity of the rotor. Broken or loose bars in the cage. Looseness stator windings. Voltage unbalance in phases, etc.
Defects of drive sleeve	Weakening of landing on the shaft. Uneven torque of transmission engagement elements.
Defects of compressor, pump, fan cavitation	Surging. Stall. Faulty blades, etc.
Defects of gear	Wear gear tooth. Broken tooth. Shock engagement. Shaft misalignment.
Defects of belt wear.	Easing tension. Misalignment of the pulleys. The eccentricity of the pulley. Resonance of belt etc.
Defects of plain bearings	Wear. Elliptical shaft neck. Forcing the oil film, oscillations, etc.
Defects of rolling bearings	Defects of the rolling elements, the separator, and the inner and outer rings. All defects in manufacture, installation and wear. As a lubricant.

There are dynamic forces in each machine. These forces are a source of the noise and vibration damage mechanism nodes. Vibrational forces acting in units of machines may have very different nature and a variety of frequency components.

The following variants should be highlighted from the mechanical forces of nature[4]:

• centrifugal forces determined by unbalance of rotating parts;

• kinematic forces determined, as a rule, by smooth irregularities of interacting surfaces and above all the friction surfaces;

• parametric forces defined by variable component stiffness rotating assemblies or supports rotation;

• friction force that cannot always be regarded as mechanical, but almost always they are the result of cumulative action of multiple snatching with deformation (elastic) contact of asperities on the surfaces of friction;

• type of impact forces arising from the interaction of individual elements of friction, accompanied by their elastic deformation.

The following variants should be highlighted because of the forces of the electromagnetic origin in electric mechanisms:

• magnetic forces determine the change of magnetic energy in a certain limited space, usually limited by the length of the site of the air gap;

• electrodynamics forces determine the interaction of the magnetic field with an electric current;

• magnetostrictive forces, determine the effect of magnetostriction, i.e. a change in linear dimensions of the magnetic material by the magnetic field.

From the aerodynamic forces the origin should be highlighted: the lifting force, i.e. pressure force on the body, for example, impeller blades moving in a stream or flow are; frictional force at the boundary of the flow and stationary parts of the machine (the inner wall of the pipeline and the like mechanism); pressure pulsations in the flow, determined by its turbulence. The origin of hydrodynamic forces is substantially of the same nature as in the gas atmosphere, but it adds to the pressure pulsations due to cavitation, which in certain conditions can arise in the liquid flow.

The temporary implementation of vibration carries a large amount of information that is invisible to the naked eye. Some of this information may consist of very weak components, the value of which is less than even the thickness of the line graph. To circumvent the limitations of the analysis in the time domain, in practice one usually applies frequency or spectral analysis of the vibration signal. When selecting the vibration frequency domain used for diagnostics of machines and equipment it must respond to different properties of different vibration frequencies. At high frequencies, the vibration becomes of the wave nature of the spectrum of a little line, little (at first glance) information, but small enough power to drive the measured vibration.

The dynamic forces excite vibrations in machines, either directly or forces excite the noise, and the noise – vibration of the body. Vibration, depending on the nature of the forces may be either deterministic (most periodical) or random.

To maximize the dynamic range of the measurement parameter onemust use the vibration (displacement, velocity or acceleration), which provides the most uniform spectrum. Usually this parameter is the vibration velocity, but in general, it depends on the type of machine [5].

In the analysis of the spectrum, there are three groups of vibration components: harmonics, asynchronous components and sub-harmonics. Harmonics are peaks at frequencies that are multiples of the frequency of the action (rotation speed) of a technical complex: they can be used to draw conclusions about the imbalance, misalignment or loose connections [5]. Nonsynchronous components occur at frequencies of no multiple frequency cycle: the analysis of the components of this group allows one to detect defects, such as elements of bearings and belts. Subharmonic components are below the rate of rotation. They can be caused by such phenomena as whirlwinds in the oil bearing wedge, increased friction between elements, defects in the belt drive, the excessive weakening of the compounds or a knock on the car.

The most important are normally one or more harmonics of the fundamental frequencies. The reason is the presence of harmonic distortion of the wave oscillations with the fundamental and non-sinusoidal periodic motion. Harmonic frequencies may coincide with resonant frequencies of certain parts or elements of the designs and significantly increase and form a major source of acoustic noise or dynamic forces imparted by other mechanical elements.

A very popular method of analyzing the vibration activity of the complex in the frequency domain, is the method of the envelope. The spectrum of the vibration signal generated by nonlinear vibrational forces loses a considerable amount of information: it is quite difficult to recognize an amplitude located close in frequency. Moreover, even at a high frequency resolution it is difficult to detect modulation of the strongest components of the vibration at frequencies close to the frequency of rotation. The spectrum of the envelope allows one to select random non-periodic signal components (shock pulses) and therefore control the sliding surface.

3. Results and discussion

From a temporary weakening of the vibration signal when the mechanical contains a significant non-stationary, non-periodic component, it is possible to conclude that even from the turn to turn the waveform can vary. It is quite difficult to diagnose because it is often fickle, there are occasional peaks and shifts.

There is a large number of peaks from the collisions, which have, at first glance, a chaotic character (Figure 3). This form of temporary signal usually accompanies all the defects of the 'backlash' or 'mechanical weakening of the structure' [1].

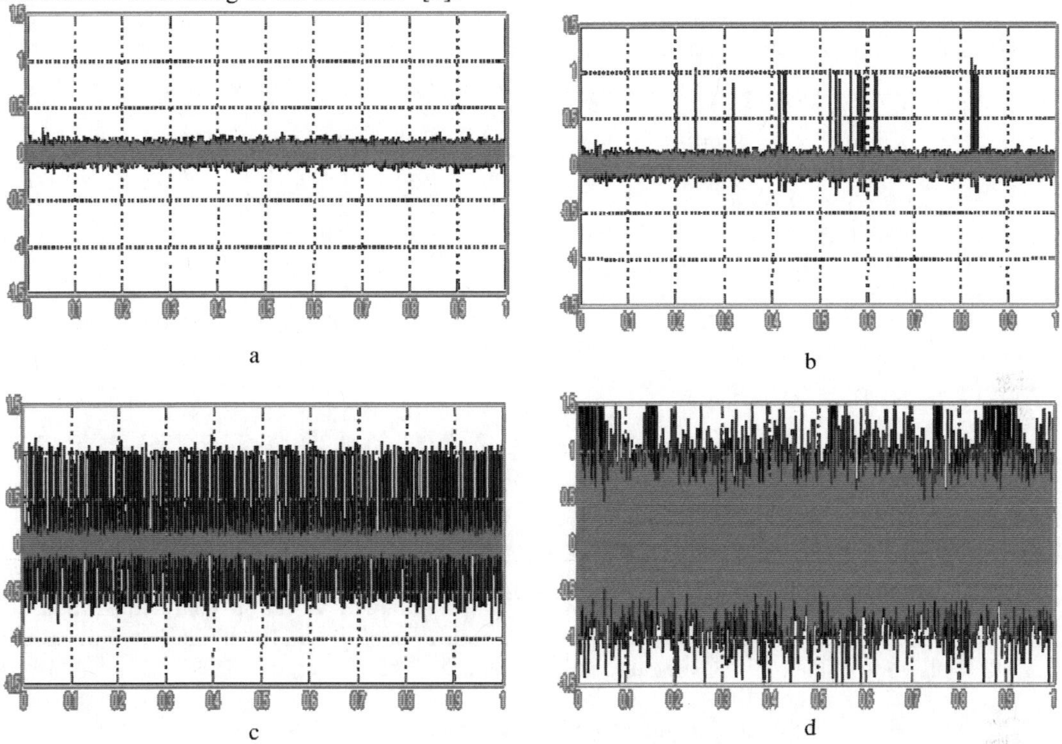

Figure 3. A time signal corresponding to: a – the absence of the defect, b – the emergence of the defect, c – the progression of the defect, d – fixing of the bug.

In order to improve the reliability of defect determination 'Mechanical design facilitation' by the search parameter and the vibration frequency range, sensitive to changes in the stiffness of the rotor poles, we can propose an additional diagnostic feature.

Generally, based on rotor speed *fr*, there are three frequency bands:

- low-frequency (LF) – from 0 to *fr*;
- midrange (MF) – fo *b* from up to 10 • fr;
- radio frequency (RF) – 10 • *fr* to 10 kHz.

Diagnosis Symptoms of separate units of aggregates are often determined by the analysis of the parameters that were previously under the control operations are not regulated and considered to be secondary.

To select the setting of the vibration produced as a diagnostic measure let us define vibration displacement S, vibration velocity V and acceleration W on the two pillars of the rotor with variable stiffness in three frequency bands (LF, MF and HF), in the standard views magnitude, RMS and Peak, respectively. Thus the fob rotor speed was 50 Hz, which corresponds to the most common speed electric drive motor of 3000 rev/min. The analysis of the obtained vibration parameters values is produced by normalization of up to 1: a parameter in each frequency band is divided into maximum (Table 2).

Reducing the stiffness of support reduces the relative value of acceleration in the low and mid ranges (Table 3). The relative values of vibration velocity and vibration displacement are less informative. Diagnostic indication must meet several requirements: uniqueness, accessibility, ease of measurement, information, and manufacturability. To do this, check that the identification of signs of change will maintain the integrity of the rotor imbalance.

Table 2. Sensitivity of vibration parameters to change of the stiffness of supports(I – the first pillar of the rotor, II – the second pillar of the rotor).

Frequency range, Hz	Stiffness of pillar	S_I, Swipe	S_{II}, Swipe	V_I, RMS	V_{II}, RMS	W_I, Peac	W_{II}, Peac
2…50		1.0	1.0	1.0	1.0	0.1	0.1
50…500	maximum	0.1	0.1	0.3	0.3	0.7	0.7
500…10000		0.0	0.0	0.0	0.0	1.0	1.0
2…50		1.0	1.0	1.0	1.0	0.1	0.1
50…500	middle	0.1	0.1	0.3	0.3	0.6	0.6
500...10000		0.0	0.0	0.1	0.1	1.0	1.0
2…50		1.0	1.0	1.0	1.0	0.1	0.1
50…500	minimum	0.1	0.1	0.3	0.3	0.5	0.5
500…10000		0.0	0.0	0.1	0.1	1.0	1.0

To do this, at a constant frequency of the rotor back (fr = 50 Hz) produced by acceleration measurements W for balanced and unbalanced rotor on two pillars with variable stiffness in three frequency bands (LF, MF and HF). The analysis of the obtained vibration parameters values is produced by normalization of up to 1: a parameter in each frequency band is divided into maximum (Table 3). Balancing of the rotor acceleration value is reduced by an average of 15 %.

Table 3. Sensitivity of vibration parameters to change the stiffness of supports (I – the first pillar of the rotor, II – the second pillar of the rotor).

Frequency range, Hz	Stiffness of pillar	Unbalancing		Balancing	
		W_I, Peac	W_{II}, Peac	W_I, Peac	W_{II}, Peac
2…50		0.1	0.1	0.1	0.1
50…500	maximum	0.7	0.7	0.7	0.7
500…10000		1.0	1.0	1.0	1.0
2…50		0.1	0.1	0.1	0.1
50…500	middle	0.6	0.6	0.6	0.6
500...10000		1.0	1.0	1.0	1.0
2…50		0.1	0.1	0.1	0.1
50…500	minimum	0.5	0.5	0.5	0.5
500…10000		1.0	1.0	1.0	1.0

It has been found that the magnitude of the imbalance does not affect the relative values of the acceleration (Table 3).

Conclusion

1. It is advisable to carry out measurements in the low, mid and high frequency bands.

2. It has been empirically found that the reduction in rigidity leads to a reduction of the relative values of acceleration in LF and MF bands.

3. The magnitude of rotor imbalance does not affect the relative values of acceleration on all frequency bands.

References

[1] Rusov V A 2012 *Diagnosis of defects in rotating equipment by vibration signals* (Perm: Vibrocenter)
[2] Belov S V, Barbin F A, Koziakov A F, Pavlikhin G P 1983 *Environmental protection* (Moscow: Higher School)
[3] Peng Z K, Tse P W, Chu F L 2005 *Journal of Sound and Vibration* **286(1)** 187
[4] Antoni J 2007 *Mechanical Systems and Signal Processing* **21(1)** 108
[5] Ahmad M S, Kukrer O, Hocanin A 2011*Digital Signal Processing* **21(4)** 491

IOP Conf. Series: Materials Science and Engineering **124** (2016) 012167 doi:10.1088/1757-899X/124/1/012167

Detection of bearing damage by statistic vibration analysis

E A Sikora
Department of Automation and Robotics in Mechanical Engineering, Tomsk Polytechnic University, 30, Lenina ave., Tomsk, 634050, Russia

E-mail: o-sea@tpu.ru

Abstract. The condition of bearings, which are essential components in mechanisms, is crucial to safety. The analysis of the bearing vibration signal, which is always contaminated by certain types of noise, is a very important standard for mechanical condition diagnosis of the bearing and mechanical failure phenomenon. In this paper the method of rolling bearing fault detection by statistical analysis of vibration is proposed to filter out Gaussian noise contained in a raw vibration signal. The results of experiments show that the vibration signal can be significantly enhanced by application of the proposed method. Besides, the proposed method is used to analyse real acoustic signals of a bearing with inner race and outer race faults, respectively. The values of attributes are determined according to the degree of the fault. The results confirm that the periods between the transients, which represent bearing fault characteristics, can be successfully detected.

1. Introduction

Feature extraction of mechanical fault signals is vital for early fault diagnosis of mechanical equipment. In a rotating machine, a rolling bearing is a key component in various electromechanical devices, and it plays an important role in the entire system. Any defects in bearings, regardless of size, will lead to a series of failures in the parts of the connection. The non-linear and non-stationary characteristics of the fault vibration signals and the interference of the random noise complicate the feature extraction process. Allowing total characteristic of the system state, the signal processing method is generally applied to fault diagnosis [1]. Currently many methods are used to analyze fault vibration signals, such as time-domain synchronous average analysis, an development of the analysis method [2], singular value decomposition (SVD) [3], spectral kurtosis [4-5], wavelet (packet) transformation, fuzzy pattern recognition, neural networks, genetic algorithms, and empirical mode decomposition (EMD) [6-9].

2. The vibration characteristics of rolling bearings

Vibrations measured on the body of the rolling bearing are generated by four major sources [10]:

- the rotation of bearing elements;
- the resonance of bearing elements and their mounting;
- acoustic emission;
- external vibration.

The rotation of bearing elements consists in the following. Each element of bearing (an inner ring, a cage, rolling elements and an outer ring) has a characteristic frequency at which the vibrational energy is excited due to cyclic stresses or periodic attacks on a defect. A set of characteristic frequencies can be defined for each type of the bearing with original geometry. In practice, ideal geometry is a rare phenomenon, and there are usually additional frequency components generated by inaccuracies such as lobing, out-of-roundness of the ring and different sizes of the rolling elements.

The resonances of bearing elements. Each bearing member has a proper frequency that may be varied during assembly and element loading in the presence of a lubricant. Proper fluctuations excite load changes that occur at specific frequencies, and under the influence of irregular contact in defective areas.

Acoustic emission. Acoustic emission is due to vibrations excited by the movement of the material at the atomic level. When the material deforms, dislocations are generated, and their subsequent movement leads to an unsteady stress wave. In rolling bearings significant local stresses occur in the areas of elastic-hydrodynamic loading and this stress can be a source of active radiation of elastic waves.

External vibration. Bearings are usually a single natural connection between rotating and stationary units of machinery and therefore can be considered as a main place of vibration transmission. This vibration may occur on the rotating member under the influence of unbalance, gearing, impeller blades and other bearings. In combination with radiation in a fixed structure, they provide a complete picture of the vibration measured on a bearing housing.

3. Browse vibratory monitoring methods of bearing condition

The main method of the vibration data analysis for diagnosis of bearing damage is to monitor changes in root mean square levels (hereinafter referred to as a RMS level) and the spectral density ofacceleration. The RMS level increaseswith the development of damage [10].

Refinement of a simple measurement of the RMS level is enabled by the analysis of changes in different octave or third-octave frequency bands; more details can be obtained from the narrow-band frequency ranges.

Currently it is a well established fact that impulse vibration can be seen in bearings with fatigue damage and chipping. Changes in the maximum level of vibration in the high frequency range of 10 kHz are a good indicator of incipient damage, and the ratio of maximum vibration to RMS is independent of changes in load and speed (crest factor).

The measured values of the RMS level, the maximum level, the spectral density and a shock pulse depend on the load on the bearing, speed, seating density, lubrication and bearing clearance. Consequently, it is difficult to determine the condition of the bearing by a separate measurement, except when there is a significant amount of additional information. The ratio of the maximum level to RMS provides a more direct assessment of the bearing condition with minimal reference to previously obtained information, because this method is less sensitive to changes in operating conditions.

For evaluation of vibration characteristics of rolling bearings a series of tests has been completed outside the bearing life, in the range from new bearings prior to the mass of fatigue failure.

4. Probability ofthe bearing vibration function

The amplitude response of vibration signal X(t) (which is considered to be a stationary random process) can be expressed in terms of the instantaneous probability density [11]. This function is assessed by the length of stay in each signal of the plurality of amplitude windows, and for a typical window width and amplitude x and width Δx. Instantaneous probability can be written as

$$P\,(x \leq X(t) \leq x + \Delta x = \textstyle\sum_{i=1}^{N} \frac{\Delta t_i}{T}. \tag{1}$$

The solution of equation (1) for all X with small Δx gives the probability density estimate of X. The logarithmic scale change increases the likelihood of low probability, which is important for the detection of bearing damage. At the early stages of testing, when the bearing is damaged, i.e. $0.067LB_{10}$ in (3.35 h), the probability density of an inverted parabola indicates normal or a Gaussian distribution (in a linear scale, it would have a more familiar bell shape). Incipient damage in $1.4LB_{10}$, causes a distinct change in the tail of the distribution curve. This is consistent with the observation that can be made by studying the behavior of the peak-factor, where the measured maximum acceleration level increases, but the RMS level remains relatively unchanged. The tail of the distribution curve initially expands with increasing operating time and the development of a damage. This characteristic can be enhanced by calculating the integral of the probability density, which gives the probability of event $X(t) > X_0$ (probability of exceeding a predetermined level):

$$P(X(t) \geq X_0) = 1 - \int_{-\infty}^{X_0} p(x)dx. \tag{2}$$

The result is that the information on the probability signal exceeds a specific value of amplitude X_0. An obvious measure of the bearing condition is obtained by comparing the probabilities for specific levels of amplitude. Levels above 3σ provide the most relevant information.

5. Statistical vibration acceleration moments of a bearing housing

The shape of the probability density is described by consistency of statistical moments, and these moments are determined by generalized integral

$$M_n = \int_{-\infty}^{+\infty} x^n p(x)dx,\ n = 1, 2, 3, \ldots, \tag{3}$$

the first and the second of which are well known: $n = 1$ gives a mathematical expectation, $n = 2$ – dispersion. They are similar to the first and the second mechanical moments, i.e. the coordinate of a central axis and the moment of inertia of a plane figure correspond to the shape of the probability density.

Odd moments (i.e. $n = 1, 3, 5, \ldots$) provide information about the position of the maximum density with respect to the median value. Even moments (i.e. $n = 2, 4, 6, \ldots$) indicate the spread of the distribution.

Moments ($n > 2$) can be normalized by eliminating expectation $\bar{x}$ and dividing it by standard deviation σ, raised to the power of the order of time, i.e. by σ^n. For example, the third central moment is:

$$\sqrt{\beta_1} = \frac{\int_{-\infty}^{+\infty} x^3 p(x)dx}{\sigma^3}, \tag{4}$$

$$\beta_2 = \frac{\int_{-\infty}^{+\infty} x^4 p(x)dx}{\sigma^4}, \tag{5}$$

where β_1 is defined as asymmetry, β_2 – excess.

The first six moments of the measured accelerations housing units have been studied for testing of bearing durability. They are evaluated using less precise, but a more practically convenient operator of averaging over time

$$M_n = \frac{1}{T}\int (x(t) - \bar{x})^n dt. \tag{6}$$

Odd moments are close to zero, indicated on the distribution symmetry of the acceleration amplitudes, while higher even moments are very sensitive to impulsive signals associated with damage to the bearing. The normalized fourth moment (or excess) is a measure of compromise between slightly lower sensitive moments and hypersensitive higher moments. For the intact bearing it is close to 3 (± 8 %), RMS and maximum levels vary respectively between ± 50 and ± 65 % over the same range of load and speed.

Some common waveforms have a specific, but ambiguous excess value [12]:

- 1.0 – square wave;
- 1.5 – sine wave;
- 3.0 – wave with a Gaussian distribution.

Thus, the measured excess value for the intact bearing is equal to three points for the Gaussian distribution of acceleration amplitudes.

Changes in excess, RMS and maximum levels of acceleration in the process of durability testing during $1.6LB_{10}$ showed that the value of the excess increases markedly above three immediately after $1.23LB_{10}$, and this increase is consistent with the change of the maximum level, but far ahead of the change of RMS. Maximum is reached in the vicinity of $1.47LB_{10}$, and at that time the RMS level increases significantly.

6. A change of excess in different frequency bands with damage development

The value of excess increases rapidly near 9 LB_{10}, reaching at 9.6 LB_{10} the maximum value of 6.7. At this point the lesion is located on the inner ring and has the form of a fatigue crack.

At first, a subsequent rotation of excess declined, but then it rose again to a maximum of 6.7 at 11.3 LB_{10}. Then the change has become quite chaotic, and excess has found a tendency to return to a level equal to three. Appropriate acceleration versus time clearly shows that at an early stage of development of damage in 9.1 LB_{10} (454 h), transient vibrations are generated by packets (one for each revolution of the inner ring). This pattern persists until 13 LB_{10} (653 h), followed by a quiet period between packets, then it begins to fade, and the packages become less distinct. The amplitude of the acceleration is normalized with respect to the RMS level, and it is important to note that between $3LB_{10}$ (149 h) and 13.15 LB_{10} (657 h) the RMS level increases from 0.53 g to 19.1 g.

The disappearance of a clearly discernible pattern of unsteady shock vibration occurs with increasing damage. Initially the fatigue crack length is much less than the circumferential distance between two rollers, whereby the system is subjected to discrete shock loading. The shock loading becomes more continuous, and consequently bursts become a seeming continuous signal due to the damage extent and exceeds the distance between adjacent rollers. When damage captures more than 60 % of the inner ring, it is always in the loading zone, which leads to continuous shock loading and vibration bursts with a set of the indistinguishable pattern. It is at this stage that excess is reduced to 3 (or the value inherent in undamaged bearings), so at first glance it seems that this circumstance prevents the application of the proposed method for the defects detection. However, information about the extent of the defect can be extracted by the observing of excess changes in selected frequency bands. This conclusion is illustrated by examining the value of excess in four frequency bands at certain moments of the durability test. Excess equals 3 in all frequency bands when a defect is absent for 0.06 LB_{10} (3 h), i.e. the instantaneous distribution of acceleration amplitude is Gaussian. A marked increase of excess occurs in the lower frequency band with defect nucleation at 9.1 LB_{10}. With increasing operating time (amplification defect) the excess value in the low band drops again to 3 and the maximum value is moved to the higher frequency band. At 13.15 $LB_{10}$the damaged area exceeds 60 % of the inner ring and captures a greater number of rollers and a maximum of the selected frequency bands available. The importance of these changes can be assessed by analyzing the acceleration versus time for each band at a particular point in the test. In the frequency range of 3 Hz to 5 kHz, it is difficult to identify any specific time-dependent pattern, but in the band range of 5...10 kHz one can see the frequency of bursts corresponding to the frequency of passage of the rollers on the inner ring. At higher frequencies, the quiet periods between bursts of consecutive packets appear once again during each revolution of the inner ring.

Thus it can be seen that in this method the initial lesion causes changes mainly in the low frequency band, while more advanced damage to the greatest extent affects the high frequencies. However, the degree of damage can be defined only by monitoring the excess distribution in selected frequency bands. This situation provides a unique opportunity to determine the condition of the bearing of a set of measurements of excess. If all values are close to 3, the bearings are not damaged. Deviation from three indicates a damage, the dimensions of which can be specified from the shift of the distribution excess maximum towards lower or higher frequencies [13].

7. The proposed mechanism for transfer of discrete frequency pulses with increasing bearing damage

The vibration acceleration of a bearing housing is a result of a complex interaction between the compelling forces and a transfer characteristics design. Building an accurate model, which would allow one to study the development of defect from the transfer of frequencies, requires a full understanding of the driving forces at different stages of bearing damage, as well as the dynamic characteristics of the structure [14]. The system simulation is necessary using the idealized characteristics of the driving forces and structures, because this full information is absent.

A condition of a 'new' bearing is characterized by compelling force that gradually decreases in amplitude with increasing frequency, and is a combination of the dynamic characteristics of the bearing and a machine. Small surface cracks occurring under generation of fatigue damage dramatically change the distribution of loads by contacting a loading area that can be modeled in the form of a pulse power sine half-wave. If such a waveform summarizes the background noise, the characteristics of its impulse will occur only at low frequencies. Consequently, the time-dependent excitation of the structure will occur only in this region, it indicates the excess above 3.Broadband background noise still dominates at high frequencies so the excess value is equal to three.

The damage is distributed by deepening and expanding the initial crack in conjunction with formation of additional cracks and surface spalling. The transition to an intermediate state is simulated by sine half-waves, but with increased amplitude and recurrence frequency. It causes an increase at the level of the spectrum at all frequencies. Now impulsive characteristics are distinguished above the background noise in the higher frequency scale, and the value of excess increases in this area. With increasing damage the frequency of impact repetition reaches the value when the time between beats is not sufficient for the natural damping of the structure reaction. The resulting interaction between successive bursts has random character, and the resulting signal has the Gaussian distribution of amplitudes in the low frequency band. However, the impact of energy at high frequencies will have a relatively low repetition rate and therefore generates more discrete reaction designs, which can be identified by analyzing a signal with an appropriate band pass filter. Therefore, the development of damage to the value of the low-frequency excess returns to 3, and the value of the high excess remains high.

The basic premise of this model is to ensure that:

a) the relative levels of shock loading are mixed in background noise;

b) the frequency of shock repetition is comparable to the time decay response of the structure.

Consequently, the area in which discrete bursts with defect nucleation appear for the first time depends on the excitation spectrum of the background and the nature of damage.

8. The classification of the damage extent

The terms 'nascent' and 'development' damages have been used earlier, but their definitions have not been given.

The nascent damage refers to damage at the macroscale, which in the case of fatigue is a primary surface crack, and not previous microscale movement of dislocations within the material. The rate of transition in the development of damages defined as a state, in which the damage is continuously in contact with rollers in a loading area, depends mainly on the particular nature of damage to an element and a load. If damaged, the outer ring of continuous contact will occur when the extent of damage exceeds the distance between successive rollers and the inner ring, for continuous contact occurs when damage covers 60...70 % of the circumference (for pure radial load). Damage to a single element alone rarely increases and the continuous contact can be an interaction result of elements damage. It should highlight a fact that the absolute area of contact should not be a major factor in predicting failure rates. The intensity of the spread depends to a large extent on the intensity of the impact and frequency of their recurrence. And since these factors result in transfer of frequency corresponding to the maximum excess, then it is obvious that a predisposition to failure can be predicted by the instantaneous values of excess aggregate.

Conclusion

It is shown that values of excess in selected bands are potentially a powerful tool for quantitative evaluation of the bearing condition. The main advantages of this parameter are insensitiveness to changes in speed and load bearing and an ability to indicate damage extent and the tendency to damage spread. 'Maximum operating time of failure' cannot be determined by this method. However, the obtained information allows one to estimate a bearing condition. This evaluation together with such factors as production schedules, the effects of secondary injury, availability of spare parts, allows one to take a decision to replace the bearing.

The method is applicable to all types of bearings, which is confirmed by laboratory and field tests. Particular difficulties arise in explaining the results of split bearings and bearings operating under conditions close to cavitation. However, by revising the frequency bands and interpretation of the measurements it is still possible to determine the condition of the bearing, but with less sensitivity and reliability.

The results of this study are the beginning of the solution of the contradiction of the 'best' frequency ranges for detection of bearing damage. Discrete bursts are normally associated with a defect and occur in different frequency bands with different degrees of damage. This feature can be explained qualitatively by considering the relative levels of damage associated with the disturbance and background noise, so the efficiency of detection of damage depends on the level of background noise. Simple rules for evaluating detection effectiveness depending on the position of the sensor can be formulated, but in the meantime placing the sensor as close as possible to the bearing can be recommended.

The size of damage is determined beforehand on the basis of the absolute spalling area. A more reasonable interpretation, reflecting the rate of spread of damage, should take into account this area due to the frequency of its contacting loading area.

References

[1] Wen C, Dong L, Jin X 2015 *Journal of Shanghai Jiaotong University (Science)* **20(1)** 56
[2] Tse P W, Peng Y H, Yam R 2001 *Journal of Vibration and Acoustics* **123** 303
[3] Yang W X, Tse P W 2003 *NDT & E International* **36(6)** 419
[4] Antoni J 2006 *Mechanical Systems and Signal Processing* **20(2)** 282
[5] Antoni J 2007 *Mechanical Systems and Signal Processing* **21(1)** 108
[6] Ricci R,Pennacchi P 2011 *Mechanical Systems and Signal Processing* **25(3)** 821
[7] Xue Y J, Cao J X, Tian R F 2012 *Internat. J. of Digital Content Technol. and Its Appl.* **6(4)** 150
[8] Yu D J, Cheng J S, Yang Y 2005 *Mechanical Systems and Signal Processing* **19(2)** 259
[9] Peng Z K, Tse P W, Chu F L 2005 *Journal of Sound and Vibration* **286(1)** 187
[10] Rusov V A 2012 Diagnosis of defects in rotating equipment by vibration signals (Perm: Vibrocenter)
[11] Ahmad M S, Kukrer O, Hocanin A 2011 *Digital Signal Processing* **21(4)** 491
[12] Sikora E A, Krauinsh P Y 2008 Modern Problems of Mechanical Engineering (Tomsk: TPU) pp 450–454
[13] Jiang G Z, Jin L H, Li Q and Han Y L 2010 *Beading* **12** 27
[14] Shen Z J, He Z J, Chen X F, Sun C, Liu Z W 2012 *Sensors* **12** 10109

IOP Conf. Series: Materials Science and Engineering **124** (2016) 012168 doi:10.1088/1757-899X/124/1/012168

Stampless fabrication of sheet bars using disposable templates

V P Smolentsev[1], S V Safonov[2], E V Smolentsev[3] and O N Fedonin[4]
[1] Voronezh State Technical University, 14, Moskovskiy ave., Voronezh, 394026, Russia
[2] Bryansk State Technical University, 7, 50 years of October st., Bryansk, 241035, Russia

E-mail: vsmolen@inbox.ru

Abstract. The article is devoted to the new method of small-scale fabrication of sheet bars. The procedure is performed by using disposable overlay templates, or those associated with a sheet, which parameters are obtained directly from the drawing. The proposed method used as a substitution of die cutting enables to intensify the preparatory technological process, which is particularly effective when launching the market-oriented items into production. It significantly increases the competitiveness of mechanical engineering and creates the conditions for technical support of present-day flexible production systems.

1. Introduction

The cost of a cutting die is economically justified if the production program assumes the fabrication of tens or hundreds of thousands of identical thin sheet bars. The modern flexible production systems require a limited number of bars, and their fabrication carried out by using cutting dies is sometimes unprofitable, taking much time for manufacture and finalization. The available technological methods of fabricating small lots of sheet bars are quite labor-intensive and applicable only for manufacturing products with simple contours and limited sheet thickness (primarily, from 0.1 to 1.0 mm).

To separate the sheets of less than 1.0 mm in thickness, the new method of shape creation has been devised which involves the use of disposable templates with contours of various complexity. To implement this method, it was necessary to work out the scientific principles and technical means of their application by the control of local processing, based on unilateral (or bilateral) disposable dielectric and metal templates, regulating the parameters of primary and ancillary electric fields. As a result, new templates have been designed, including those having conductive layers connected to voltage source [1-3]. By this method it is possible to obtain the bars of required accuracy during separation of sheet materials having constant (variable) thickness, or various workability of their segments. Such products include the components obtained from sheets (rather than by stamping) of various thickness, and also the parts made from sheets having areas of various workability (e.g. by soldering or welding of homogeneous or heterogeneous materials). An acute need for such components is caused by the present-day development of flexible production systems, especially, small enterprises manufacturing diversified products.

2. Technological schemes of material separation using the disposable templates

Figure 1 shows the initial state of the system, preceding the material separation, with a unilateral (Figure 1a) or bilateral (Figure 1b) location of templates, one of which is a dielectric one.

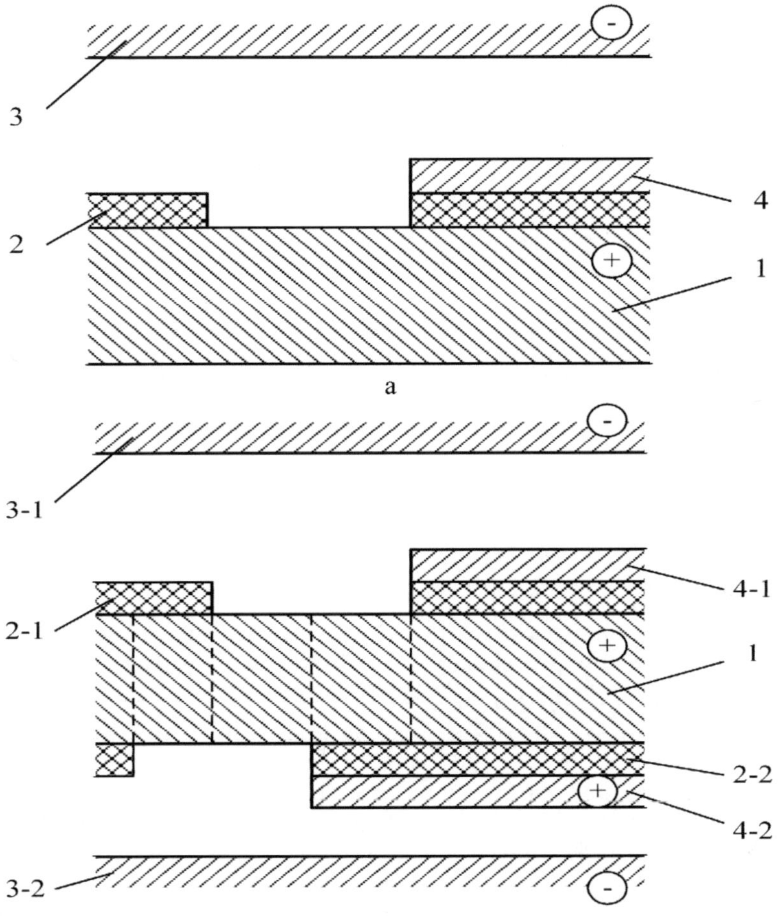

Figure 1. A scheme of material separation using the disposable templates: *a* – unilateral processing; *b* – bilateral processing; 1 – a bar; 2 – a dielectric template; 3 – an EDM cathode; 4 – a metal template.

The bilateral location of templates (2-1; 2-2 and 4-1; 4-2) makes it possible to double the thickness of a sheet, available for separation, increasing it to 1.0…1.2 mm. This requires the use of two EDM cathodes (3-1 and 3-2). The displacement of templates, as shown in Figure 1b, enables matching of the skewed slots, formed on both sides of a sheet bar, and helps to control the position of the separation plane by energizing the metal templates (4, as shown in Figure 1a and 4-1; 4-2, as shown in Figure 1b). This provides the possibility of obtaining skewed side walls and openings with a curvilinear axis, passing through the bar and having one-way outlet openings

3. Technological modes

Table 1 presents various technological modes for fabricating sheet metal bars, using disposable dielectric templates, shown in Figure 1.

Table 1.Technological modes.

Length of the contour to be cut, mm	Thickness of the dielectric template, mm	Initial inter-electrode gap, mm	Electrode voltage, V	Speed of electrolyte pumping, m/s	Note
up to 20	up to 0.2	0.15...0.2 (h_2=0, h_3=0)	8...10	1...3	Unilateral processing
	0.2...0.4	0.15...0.3 (h_2=0, h_3=0)	8...10 10...12	1...3 2...4	Bilateral processing
	0.4...1.0	0.5...0.6	10...12	2...4	
20-50	up to 0.2	0.2...0.25 (h_2=0, h_3=0)	10...12	2...3	Unilateral processing
	0.2...0.4	0.2...0.3 (h_2=0, h_3=0)	10...12	2...3	Bilateral processing
	0.4...1.0	0.5...0.7	12...15	2...4	
over 50	up to 0.2	0.25...0.3 (h_2=0, h_3=0)	10...12	2...3	Unilateral processing
	0.2...0.4	0.25...0.3 (h_2=0, h_3=0)	12...15	22...4	Bilateral processing
	0.4...1.0	0.7...0.8	12...15	3...4	

If the anode process requires the separation of sheets having constituent segments of various workability, it is necessary to set the technological modes, providing equal time of anode dissolution for all constituent materials. It is most feasible in cases whenever the sufficiently precise data related to electrochemical equivalents of alloys are known, and the initial constancy of inter-electrode clearance value is available. In these cases, it is possible to accelerate or retard the process of local anode dissolution depending on specified parameters for constituent fragments of bars, as well as to obtain the high-precision bars composed of heterogeneous materials.

The detailed recommendations on the selection of parameters for various technological modes are presented in [4, 5].

The results of the estimated technological modes help to design the procedure for the technological process, which includes:

- selection of the medium suitable for anode dissolution of all the constituent materials of the bar;
- setting of the medium pumping speed;
- calculation of the medium pressure at the inlet of the inter-electrode gap;
- electrode voltage selection for types of materials subjected to processing, in consideration of the properties of all bar segments;
- selection of the value of voltage loss from manuals;
- selection of the electrochemical equivalent ;
- calculation of voltage at bar segments;
- creation of the technological scheme of processing performed by still electrodes (i.e., unilateral and bilateral types of processing), selection of the template pattern, its type of manufacture, position and connection to power supply.

4. The area of the surveyed method application in fabricating sheet bars

Table 2 presents the results of feasibility study obtained from implementation of the surveyed method for manufacturing the batches of bars (up to 100 items) from metal sheets (up to 1 mm in thickness).

Table 2. Technical and economic parameters of electrochemical template-based sheet separation [4].

Designation	Original sheet		Production time, min	Tolerance, mm	Roughness of the cut, R_a, μm	Edge finishing
	Material	Thickness, mm				
Screen filters with openings, 0.02×0.02 mm	Heat-and rust-resistant steel	0.08	1...2	0.02	0.32	Not required
Slats	Brass	0.15	1 ... 2	0.02	0,63	Not required
Stator plates	Transformer steel	0.35	2 ... 3	0.03 ... 0.04	0.63 ... 1.25	The smoothing of the rounding radius of the exit edge
Edge strips	Constructional steel	0.5	4 ... 5	0.05 ... 0.1	0.63	Not required
Inserts (bilateral processing using the dielectric and metal templates)	Alloy steel	0.85	6 ... 7	0.7... 0.8 (0.02...0.04 (after calibration)	0.63	Calibration of openings (up to the 5th degree of precision)

In recent years, the extensive study dealing with electrochemical and combined separation of sheet metal materials has been conducted in Russia. The scientific works touch upon processing of bars with variable thickness. The new methods and devices (mainly developed by Voronezh scientific school) (patents [1-3]) provide the basis for advancement in research and solution of problems related to effective separation of metal materials with the most common sheet thickness (up to 1.0…1.2 mm) in small-scale bar fabrication. Patent 2257981 [1] solves the task by means of the proposed new method of electrochemical sheet bar processing using the dielectric templates. In this procedure, the processing is performed in flowing electrolyte using the tools, connected to the source of DC power supply, with a bar acting as an anode. The novelty of the method consists in the following. In the first part of the procedure, one of the EDM electrodes is connected to the negative pole of power supply, and the electrochemical processing is implemented through the stencil located on the bar (namely, at the EDM electrode side), up to a depth of 0.75…0.8 of the bar thickness. Further, at the second stage, the electrochemical processing is performed with disconnecting of the first EDM electrode and connecting of the second one, until obtaining the separating cavity of the size equal to the sheet thickness.

To conduct the electrochemical processing, the stencils are installed coaxially relative to the position of the cavity being processed. The template being used in processing by the first EDM electrode has the operating contour resembling that of the bar. In turn, the second stencil has the operating contour copying that of the cavity, and corresponding to the pattern, specified by the drawing. In this case, the thickness of the first template is half that of the second one.

The separation scheme is presented in Figure 2. Prior to processing, conductive sheet bar 1 (as shown in Figure 2), having thickness t, is supplied with the dielectric template (thickness – t_{T1}, contour size – H_{T1}), and stencil 3 (thickness – t_{T2}, contour size – H_{T2}), equal to preset width of the separation cavity. Thickness t_{T1} of the first template is half the corresponding thickness t_{T2} of the second stencil. EDM electrodes 4 and 5 are located at a distance of S_0 from the bar. This is the initial position for the bar, the templates and the EDM electrodes.

Figure 2 b represents the position and the dimensions of the cavity after completing the first stage of processing. At the end of the first processing stage, the cavity depth of bar 1, processed with electrode 4, equals (0.75…0.8) h. The size of the cavity contour is H_{T2}, due to broadening ΔH.

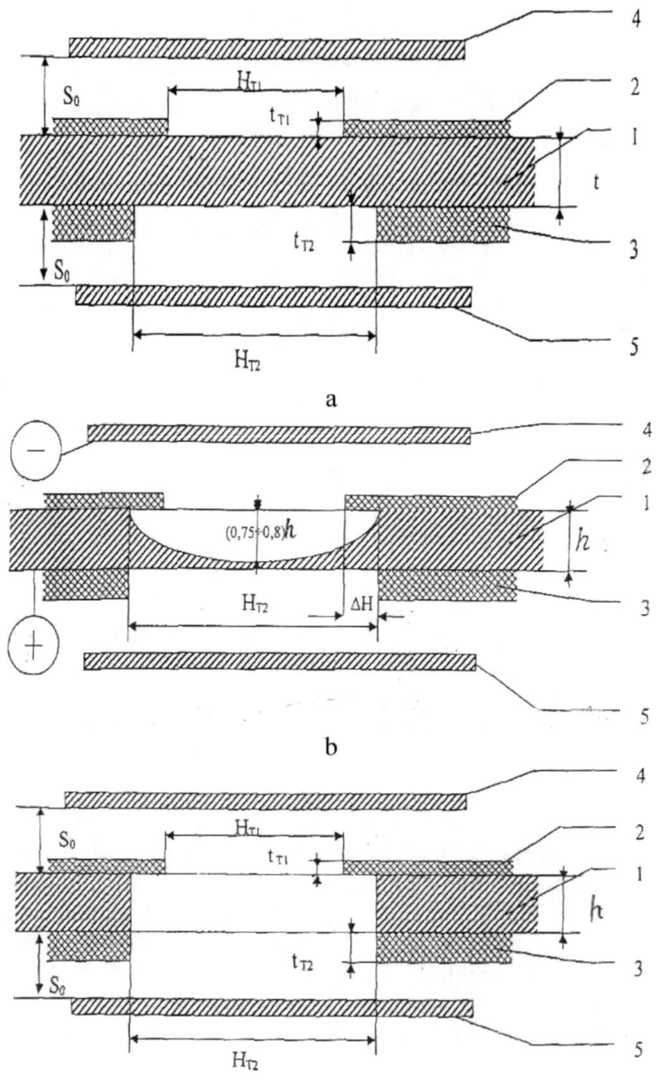

Figure 2. A scheme of obtainment of precision sheet bars using the dielectric templates: a – the adjustment of electrodes and templates before processing; b – the position and size of the cavity at the end of the 1st processing stage; c – the position of tools after finishing the separation zone processing. 1 – sheet bar; 2,3 – dielectric templates; 4,5 – EDM electrodes.

Figure 2 demonstrates the completion of the second stage of processing, performed by connecting EDM electrode 5 to the negative pole of power supply, provided that EDM electrode 4 is disconnected. In view of the small thickness of the cavity part, unprocessed after the first stage of piercing, and the doubled thickness of the second stencil, as compared with the first one, the size of the opening contour (as viewed from the side of template 3) is identical to size H_{T2} of the opening, predefined by the drawing, as well as to the size of the processed fragment (as viewed from the side of template 2).

The time of the first (τ_1) and the second (τ_2) processing stages is calculated according to the following formula:

$$\tau_1 = \frac{\left(\left[(0.75 \div 0.8)h + S_0\right]^2 - S_0^2\right)\gamma}{2\alpha\eta\chi(U - \Delta U)},$$

$$\tau_2 = \frac{\left[(0.2h + S_0)^2 - S_0^2\right]\gamma}{2\alpha\ \alpha\ \eta - \chi\Delta(U)},$$

where h – bar thickness; S_0 – initial gap between the processed bar surface and the EDM electrode; α – electrochemical equivalent of the material to be processed; η – current output; χ – electrolyte conductivity; γ – bar material density; U – electrode voltage; ΔU – voltage loss.

According to patent 2275279 [3], the goal of the invention is to obtain the vertical or special cutting profile, which increases the accuracy of sheet material separation by local template-based anode dissolution. To achieve this goal, the amount of electricity, applied to the metal template for creating the vertical cutting profile, is determined by the following dependence:

$$Q = \gamma h L \Delta H / 2\alpha,$$

where Q – the amount of electricity; h – the thickness of the bar; L – the length of the bar contour; ΔH – inverse broadening.

The proposed method is implemented according to the scheme, presented in Figure 3. To obtain a precision part with the desired inclination of the side wall (e.g. vertical or curvilinear), dielectric template 2 is placed on metal sheet material 3.

On the side of the backward shift of the material, it is covered with metal template 6, for example, the titanium one, passivated against anode dissolution.

The edge of the separation area shifts relative to dielectric template 2 by the value, equal to inverse broadening of the cutting profile.

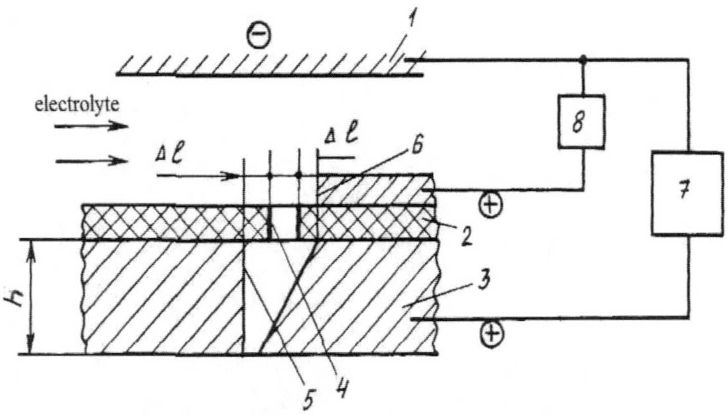

Figure 3. A scheme of creating the area of material separation with controlled position of the lateral surface/1 – EDM cathode, 2 – dielectric template; 3 – processed bar. 4 – slot in the dielectric template, 5 – part to be cut out; 6 – slot in the metal template, 7 – the first power supply, 8 – the second power supply; h – specified bar thickness; l – size of the part to be cut out.

Material 3 and EDM cathode 1 are connected to the first power supply 7. Metal template 6 and EDM cathode 1 are connected to the second power supply 8. The electrolyte is conducted along the template (as marked with arrows in figure 3), by applying constant voltage 7 and the variable voltage, produced by the power supply.

5. Conclusion

The analysis, taking into account table 3, shows that the method of the electrochemical phototemplate-based separation, used in fabrication of a batch of sheet bars (20…100 items) has clear advantages with regards to per-unit costs. Besides, the process of sheet bar fabrication can take several hours for manufacturing products of every size, which is unattainable for any other method of small-scale fabrication of sheet bars.

References

[1] Zakirova A R, Sadykov Z B, Smolentsev V P and Gazizullin K M 2005 The method of electrochemical processing. Patent No. RU 2257981

[2] Zakirova A R, Sadykov Z B, Smolentsev V P, Gazizullin K M and Odintsov I A 2006 The method of electrochemical processing of a sheet bar and the device of its implementation. Patent No. RU 2275994

[3] Smolentsev M G, Smolentsev E V, Ryabova S A and Koptev I T 2006 The method of electrochemical separation of sheet materials. Patent No. RU 2275279

[4] Smolentsev M G, Smolentsev V P 2004 *Metalloobrabotka* **4** 6–8

[5] Smolentsev V P, Smolentsev G P and Sadykov Z B 1983 *Electrochemical marking of components* (Moscow: Mashinostroenie) p 72

IOP Conf. Series: Materials Science and Engineering **124** (2016) 012169 doi:10.1088/1757-899X/124/1/012169

Theoretical bases of the surface layer formation in the finishing and hardening treatment of details by SPD in flexible granular environment

M A Tamarkin[1], E E Tishchenko[1] and V P Fedorov[2]

[1] Donskoy State Technical University, 1, Gagarina sq., Rostov-on-Don, 344000, Russia
[2] Bryansk State Technical University, 7, 50 years of October st., Bryansk, 241035, Russia

E-mail: tehn_rostov@mail.ru

Abstract. The article presents results of theoretical studies of the surface layer formation during finishing and hardening treatment of details by SPD in flexible granular environment. The dependencies are fixed for determining the surface roughness, processing time, the depth of the hardened layer and the degree of hardening for different methods of treatment by SPD in flexible granular environment. The process of residual stresses formation is researched.

1. Introduction

Methods of processing by the surface plastic deformation (SPD) in the flexible granular working environments are increasingly used in various industries at the finishing stages. They have fairly broad technological capabilities that allow processing the details of different shapes, ensuring high quality of the surface layer and achieving high processing performance.

It should be noted that the processing based on the plastic deformation of a thin surface layer (finishing and hardening treatment) using flexible environment has several advantages compared to other SPD finishing treatment methods: the metal fibers integrity is retained and a fine structure is formed in the surface layer; a treated surface charging with particles of abrasive is absent (if processing is performed in abrasive environment); no thermal defects; provides a stable surface quality; it is possible to achieve a minimum altitude of the surface roughness for crude steel, non-ferrous alloys and high-strength materials; it creates a favorable form of asperities with a high share of the supporting area at the level of roughness; favorable residual compressive stresses in the surface layer are created; surface microhardness increases; usually finishing and strengthening treatment with flexible granular medium (FST FGM) does not require the use of complex equipment and tooling; the FST FGM may use a variety of details of different shapes and sizes made of different materials [1-3]. On the basis of the studies, a generalized theory of FST FGM was created for finishing and hardening vibration treatment (FHVT), decorative and reinforcing centrifugal rotary processing (DRCRP), fraction processing (FP) and other similar methods.

2. Research results

In the study of the basic FST FGM methods of technological parameters (process performance and the quality of the treated surface) one of the most important issues is a question of theoretical modelling of the interaction process between the environment particles and the workpiece surface. The development of theoretical dependencies describing the shape and the size of the processing marks allows further movement to probability theory describing the distribution of the marks on the workpiece surface, which in turn makes it possible to develop a model of the roughness profile and physico-mechanical properties of the workpiece surface.

Deformation of irregularities in the FST FGM is as follows: a particle of environment interacts with protrusions of the workpiece surface microprofile resulting in plastic deformation causing the metal flow (in microvolumes), with plastic deformation occurring in the surface layer of the workpiece. This is due to a significant excess of working bodies hardness compared to the hardness of the processed material.

At the beginning of the interaction, the environment particle contacts the tops of the workpiece irregularities as an arc. The length of the contact segment increases with the introduction of the particle into the surface up to partial or total deformation of the asperities. Frictional force at the contact surface prevents asperities deformation in the circumferential direction. Shear stresses resulting from the interaction are maximal at the contact surface and decrease in the material with depth. The roughness increases at the contact of the surface with the environment particle and causes plastic flow of the metal towards the least resistance, i.e. in the direction of free surface - microroughness cavity. As a result of the efforts redistribution, the cavities 'raise' up to the contact with the processing body surface. This process is more intensive if the deformable protrusion is close to the zone of maximum pressures [1, 2].

The depth of the environment particle (mostly a metal ball) penetration can be determined by

$$h_{\max} = 2 \cdot V_{ef} \cdot R \cdot \sin\alpha \cdot \sqrt{\frac{\rho_c}{3\kappa_s \cdot c \cdot \sigma_s}},$$

where V_{ef} is effective speed of the particle (determined by the characteristics of a particular treatment type), R is the radius of the particle, α is the angle of particle falling to the workpiece surface, ρ_c is the density of the particle material; k_s is a coefficient taking into account the effect of the workpiece roughness; c is a coefficient of the contact surface bearing capacity; σ_s is the workpiece material flow ability limit.

For the fraction processing, this dependence is as follows:

$$h_{\max} = 1.6 \cdot K_L \cdot R \cdot \sqrt{\frac{P_{din} \cdot \rho_c}{k_s \cdot c \cdot \sigma_s \cdot \rho_{cm}}},$$

where P_{din} is dynamic pressure; K_L is loss ratio, which takes into account the distance to the workpiece and the balls flux density (determined during experimental studies); ρ_{cm} is the density of the mixture.

The volume of the surface layer deformation is largely determined by the physico-mechanical properties of the workpiece material and the number of interactions relating to each point of the treated surface. The contact stresses in the treatment zone under the influence of the environment particles depend not only on the force of impact, but also on the particle size and the height of the original protrusions.

Due to its dependence on the angle of impact and the shape of the initial roughness, the contact surface has a complex spatial form. In the case of processing with polished steel balls, ellipsoidal shape of the nominal contact surface seems to be possible. This greatly simplifies the further theoretical calculations.

The point of the particle contact with the surface is an ellipse, semiaxis sizes of which are

$$b=\sqrt{R^2-(R-h_{\max})^2},\ a=\frac{\pi}{2}\cdot(ctg\alpha-f)\cdot h_{\max}+b,$$

where f is a coefficient of the particle friction on the workpiece surface.

Plastic deformation of the workpiece surface occurs with repeated interaction of the environment particles with it. Individual wells with a transverse roughness are formed depending on the parameters of the original profile. The microrelief of the surface is formed by overlapping and crossing of individual tracks (holes).

If the process of treatment is stopped before the steady roughness formation, an increase of the profile bearing surface compared with the original one will be observed. Continuation of the process will lead to redeformation of the original roughness if the parts originally did not have very high hardness and significant initial roughness.

On the workpiece surface, a new specific micro-relief is formed, which is characterized by uniform properties in all directions. Under continued processing, it is permanently reproduced and its parameters will not change during a certain period of time (until the surface lapping). Parameters of the established roughness will be determined only by technological conditions and the size of environmental particles.

During the FST FGM the average arithmetic roughness deviation of the profile can be determined

$$Ra_{\text{уст}}=k_R\cdot\sqrt{\frac{h_{\max}\cdot a\cdot b\cdot l_{ed}}{R^2}},$$

where l_{ed} is the length of the unit, k_R is an empirical coefficient.

For the fraction treatment, this dependence is as follows:

$$Ra_{us}=0.009\cdot\frac{b}{R}\sqrt{h_{\max}\cdot l_{ed}}$$

Reduction of the initial roughness height parameters during the FST FGM is complex. Therefore, obtaining a rigorous theoretical model to determine the processing time is not yet possible.

According to many researchers, the change in surface roughness is exponential. The speed of the new relief formation is reduced from its maximum in the initial period to values slightly different from zero at the time of the steady roughness formation, which indicates the formation of continuously replicating surface topography. Each combination of technological parameters corresponds to a specific density of interactions and the level of the environmental particles energy determining both the length of time and altitude parameters of the established relief, and the latter depend on the initial roughness.

The above conditions are met with

$$Ra=\left(Ra_{is}-Ra_{us}\right)e^{-k_i t}+Ra_{us},$$

where Ra_{is} is arithmetic average deviation of the original surface roughness, k_i is coefficient of roughness intensity reduction, t is processing time, Ra_{us} is set surface roughness.

Expressing the processing time from the resulting dependence we obtain the following formula:

$$t_{zad}=-\frac{1}{ki}\cdot\ln\frac{Ra_{zad}-Ra_{us}}{Ra_{is}-Ra_{us}},$$

where k_u is a coefficient of roughness intensity changes; $Ra_{\text{исх}}$, $Ra_{\text{з}}$, $Ra_{\text{уст}}$ are original, set and specified surface roughness respectively.

For solving the FST FGM technological problems, analytical calculation of the expected value of the hardened layer depth and the degree of hardening is important. Many operational properties of the workpiece, such as fatigue strength, depend on the thickness of the hardened layer. Value $h_{\text{н}}$ designates a zone of the surface layer, which has residual deformation and an increase of the crystal lattice dislocation density formed by the application of external loads. Analytical determination of the hardened layer thickness and the degree of hardening depending on the physical and mechanical properties of the material and the process parameters is a very difficult task. Existing mathematical relationships were based on the theory of elasticity and plasticity after the adoption of numerous simplifications and assumptions. Significant researches in this area were conducted by I V Kudryavtsev, V P Pshibylski, E G Papshev and others. Using their dependencies, the FST FGM depth of the hardened layer and the degree of hardening can be determined by formulas:

$$h = 3k\sqrt{a\cdot b}\,,\ \varepsilon = \sqrt{\frac{h_{\max}}{R}\cdot\left(2-\frac{h_{\max}}{R}\right)}\,.$$

Formation of residual stresses in the FST FGM is of great research interest. In the works by R B Heywood [2] and V T Troshchenko, it is noted that the value of the endurance limit is affected by the average value of the cycle stress, the influence of which, in turn, can be compensated with the value of residual stresses.

Using the analysis of the cycle stress average value impact on the endurance limit for cyclically stable steels and aluminum alloys, the following equation is proposed for determining the limiting cycle stress:

$$\sigma_a(h) = \sigma_b\left(1-\frac{\sigma_{cr}(h)}{\sigma_b}\right)\left[F+\gamma(1-F)\right],$$

where σ_b is the sample material strength limit; σ_{cr} – residual stresses; F and γ are parameters depending on the sample material.

For the aluminium alloys samples

$$\gamma = \frac{\sigma_{cr}}{\sigma_b\left[1+\frac{(\lg N_p)^4}{225}\right]};\ F = \left(1+\frac{0,0031(\lg N_p)^4}{1+0,0031(\lg N_p)^4}\right)\Bigg/\left(1+0,0031(\lg N_p)^4\right)$$

For steel samples

$$\gamma = \frac{\sigma_{cr}(2+\sigma_{cr}/\sigma_{в})}{3\sigma_b};\ F = \frac{1+0,0038(\lg N_p)^4}{1+0,008(\lg N_p)^4},$$

where N_p is the number of cycles before destruction.

3. Conclusions

The results of studying surface residual stresses formed after various kinds of FST FGM allows making a conclusion that compressive residual stresses are formed in the workpiece surface layer at a depth of 0.2…0.8 mm of 200…300 MPa that allows one to predict the improvement of the workpieces operational properties.

The generalized theory has passed a comprehensive experimental verification, and can be used to design the FST FGM processes that enhance the quality of the workpieces surface layer. On its basis, a method of the FST FGM technological processes optimization is created.

References

[1] Tamarkin M 2005 *Internat. Sympos.'Advances in Abrasive Technology-YIII' ISAAT* 46–49
[2] Tishchenko E, Korolkov Yu, Tamarkin M, Rozhnenko O. 2009 *Russian Engineer. Res.* **5** 71–76
[3] Tishchenko E Tamarkin M, , Lebedenko V 2010 *Russian Engineering Research* **2** 88–93
[4] Heywood R 1962 *Designing against fatigue* (London: Chapman and Hall Ltd.) pp. 242

Capturing action of cooling lubricants in grinding and evaluation of its effectiveness

A V Tyukhta[1], A V Khandozhko[2], Yu V Vasilenko[1]
[1] Prioksky State University, 29, Naugorskoe shosse, Orel, 302016, Russia
[2] Bryansk State Technical University, 7, 50 years of October st., Bryansk, 241035, Russia

E-mail: vyuv-post@yandex.ru

Abstract. Methods of coolant supply are considered to be aiming at capturing and neutralization of the flare of grinding waste and noxious coolant fumes during flat grinding with a wheel periphery. Study and evaluation of the new coolant functional property – capturing – are presented.

1. Introduction

Recently, in Russia and abroad active researches aimed at improving the methods of lubricating and cooling fluid (coolant) delivery and the use of different types of energy impact on it have been conducted [1]. The features of the coolant application at abrasive processing are related to the specific design of the instrument, mode and method of processing [2]. The main functions of the coolant are: lubricating, cooling, cleaning, dispersant and anti-corrosive. Today, there are more than 2.000 different ways of coolant supplying. Systematizing them, 30 major ways as well as a significant number of their varieties and combinations, have been allocated.

2. A barrage method of coolant supply

In 2002, a new main way of the coolant supply – a barrage one – was designed to fight the waste flare moving from the cutting area, which is a major source of pollution in the grinder's working area (Figure 1).

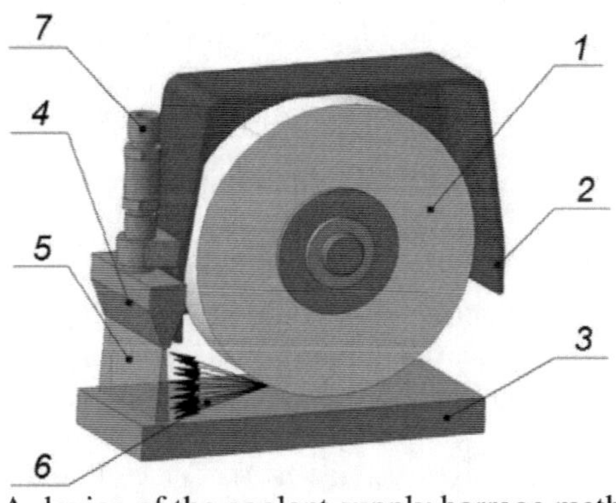

Figure 1. A device of the coolant supply barrage method:
1 – grinder, 2 – protective cover, 3 – workpiece, 4 – nozzle, 5 – coolant jet, 6 – grinding waste torch.

In accordance with GOST 12.1.005-88, the working area is defined as the space bounded by the height of 2 m above the floor or a platform, which is a place of permanent (more than 50 % of the time or more than 2 h continuously) stay of the worker [3].

In the barrage method of coolant supply, the grinding torch capture is of primary importance. The grinding torch capture is resulted from the vertical arrangement of nozzles for coolant supply behind the machining area so that the path of the torch is barraged by a curtain of liquid providing its efficient capture.

However, the barrage method does not ensure the effective implementation by the coolant of its known action that paved the way for the development of the combined technique of coolant supply using the barrage method in 2012.

3. A combined method of coolant supply

By the hierarchy analysis method, the main ways of coolant supply were analyzed and the methods, by means of which the process fluid carries out its functions most effectively, were identified: freely falling jet, pressure jet, hydroaerodynamics in the coolant environment. The most important ones are: freely falling jet and the coolant environment (Figure 2a, b), which formed the basis for a new combined method of coolant supply (Figure 3).

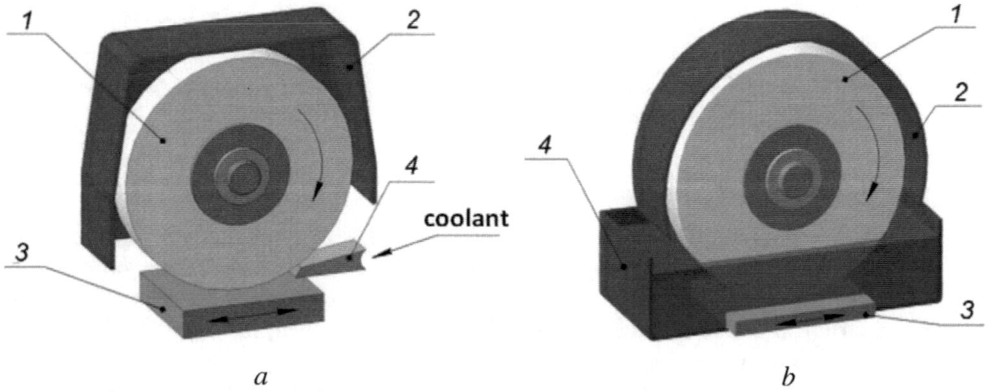

Figure 2. The most important ways of coolant supply during flat grinding:
a – irrigation, *b* – grinding in the coolant environment
1 – grinding wheel, 2 – protective cover, 3 – workpiece, 4 – coolant nozzle

The essence of the method is that the treatment zone perimeter is completely surrounded by the coolant curtains, and due to inclination of these curtains, a bath of processing liquid is created in the cutting area. Consequently, the entire sludge departing from the cutting area is fully captured by protective curtains of the coolant, and in the cutting zone there is a process liquid bath. The captured sludge particles are washed away to the coolant cleaning system of the machine.

4. New device of combined coolant supply

As shown by the experimental operation, the device of combined coolant supply presented above has several disadvantages:

- increased consumption of the coolant;
- low efficiency in the grinding spatial detail;
- increased energy consumption by the machine coolant system.

To eliminate these drawbacks, the method is amended taking into account the application of a hydroaerodynamic way of coolant supply, and a new device of combined coolant supply is developed (Figure 4). The device is a piece fixed to protective cover 2 of the machine; it is joined by pipes 1

IOP Conf. Series: Materials Science and Engineering **124** (2016) 012170 doi:10.1088/1757-899X/124/1/012170

providing coolant supply. The liquid itself forms protecting circuit 5 of a smaller perimeter than in the prior device.

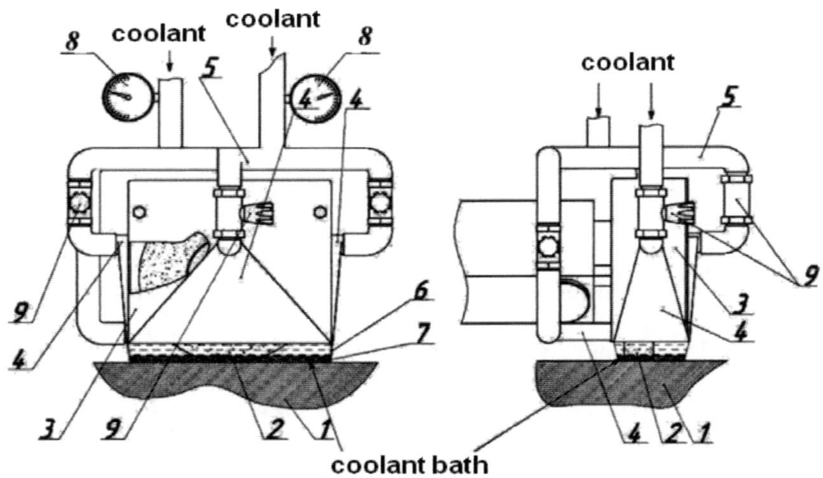

Figure 3. A device of combined coolant supply:
1 – workpiece, 2 – grinding wheel, 3 – protective cover, 4 – nozzles, 5 – pipe, 6 – coolant curtain, 7 – coolant bath, 8 – manometers, 9 – valves.

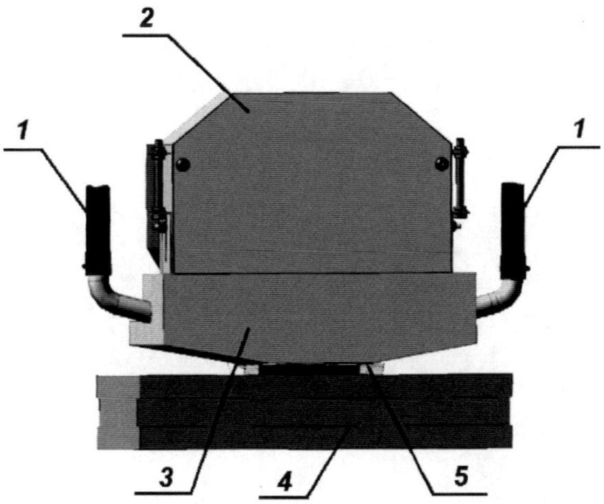

Figure 4. A new device of combined coolant supply for flat grinding with the wheel periphery:
1 – pipeline, 2 – protective cover of the machine, 3 – device of combined coolant supply, 4 – magnetic table, 5 – coolant curtain.

At the bottom, the device is provided with a number of non-regulated nozzles 1…7 positioned as shown in Figure 5, due to which a continuous barrage of the coolant curtain is formed around the cutting zone capturing all grinding waste and the coolant fume. Due to a combination of at least six major ways of supply (treatment zone irrigation, irrigation of the wheel periphery and the workpiece beyond the treatment zone, grinding in the coolant environment, hydroaerodynamic and barrage methods) a negative influence of peripheral air flows on the coolant action is neutralized. During the contact of the wheel with the workpiece, a running dynamic spatial bath of processing liquid is instantaneously formed in the cutting zone.

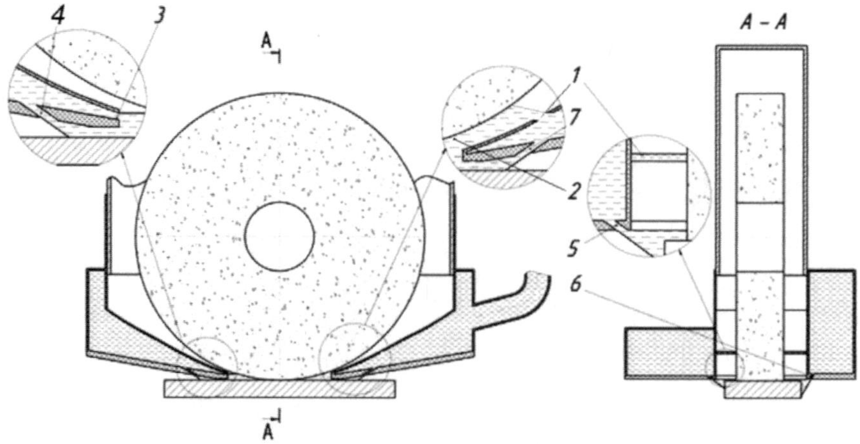

Figure 5. Design features of the new device of combined coolant supply:
1 – rear nozzle, 2 – constructive nozzle, 3 – frontal nozzle, 4…7 – contour nozzles.

The device is equipped with screws 1 (Figure 6) providing the adjustment of the device position when the grinding wheel is worn.

It is generally known that grinding is accompanied by dust emission, the intensity of which depends on the processed material, abrasive tools used and techniques of the coolant supply.

5. Background and evaluation of the coolant capturing effectiveness

The grinding dust is an aerodisperse system, in which the dispersion environment is the air of the working area, and the dispersed phase consists of suspended solids of the material being treated, abrasive and ligaments of the grinding wheel and the atomized coolant.

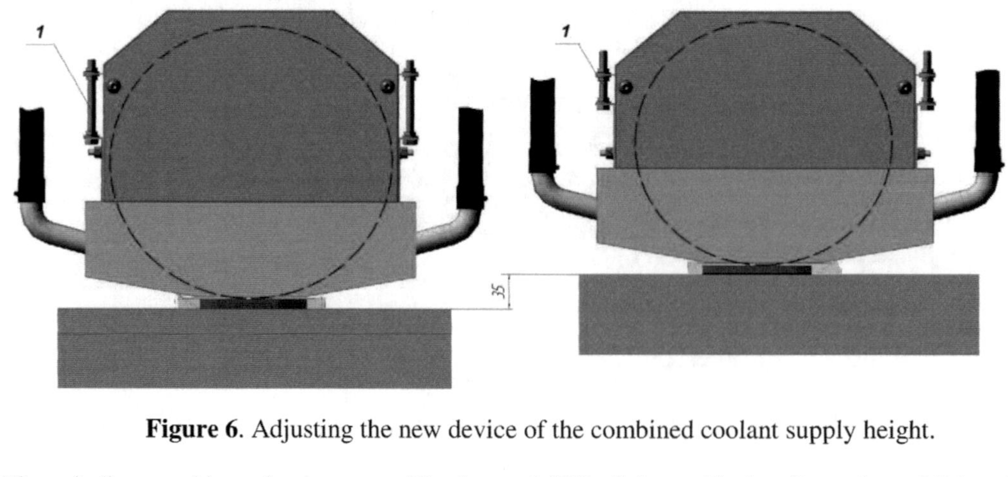

Figure 6. Adjusting the new device of the combined coolant supply height.

The grinding machine releases up to 30 g/h per 1 kW of dust with the dispersion of 0.1 to 200 microns. Due to the relatively rapid subsidence of large dust particles (10 mcm or more), the air of production areas contains mainly dust particles of up to 10 microns, 70 ÷ 90 % of which consist of particles larger than 5 microns.

One of the most common hazardous substances released during grinding is silicon dioxide (SiO_2) coming from the cutting zone in the form of condensation and disintegration aerosols and amounting to 70 % of the total weight of the dust. Silica refers to the third class of danger; its MPC is 2 mg/m^3.

On the basis of the coolant supply technology, we may speak of allocating a new functional action (property) of the coolant – the capturing one, which is to reduce the concentration of hazardous substances in the working area of the machine operator by neutralizing the flare of grinding waste and disintegration and condensation aerosols in the cutting zone by their physical capture and removal by the process fluid.

Numerically the capturing effect of the coolant can be evaluated through dimensionless capture coefficient μ by the following formula:

$$\mu = \frac{K}{MPC}, \tag{1}$$

where K is the concentration of harmful emissions in the working area, mg/m^3; MPC is maximum permissible concentration of harmful substances in the working area, mg/m^3.

The capturing action of the coolant can be considered satisfactory if coefficient $\mu \le 1$, and the less is this value, the better is the coolant capturing effect. It should be taken into account that the hot waste intensive penetration to the coolant in the cutting zone intensifies evaporation and heating being able to dramatically alter its composition and structure, so in each case the capturing process requires a balanced approach.

The results of discussed ways of the coolant supply test are presented in Figure 7.

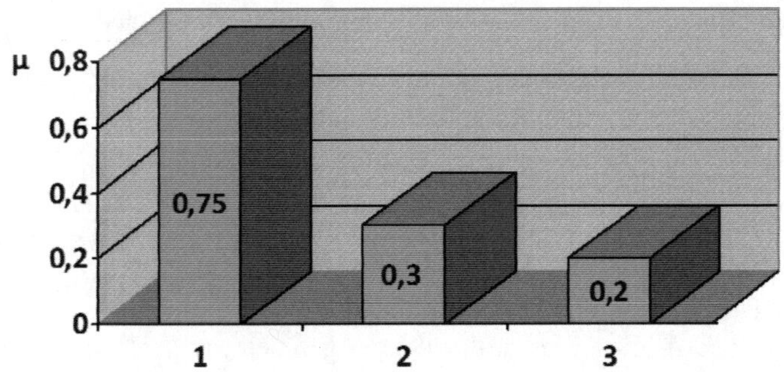

Figure 7. Evaluation of the coolant capturing action for the developed methods: 1 – barrage, 2 – combined, 3 – new combined (predicted value).

6. Conclusion

Analyzing the data obtained it can be concluded that the most effective device in terms of capturing and neutralization of the grinding waste flare, disintegration and condensation aerosols in the cutting zone is a recently developed device of the combined coolant supply. This device also provides the cutting zone with a dynamic spatial bath of the coolant that allows processing flat workpiece surfaces of any form factor.

References

[1] Stepanov Y S 2009 *Russian Engineering Research* **29(8)** 778–781
[2] D'yakonov A A 2015 *Russian Engineering Research* **34(12)** 778–780
[3] D'yakonov A A 2015 *Russian Engineering Research* **34(12)** 814–816

About creation of machines for rock destruction with formation of apertures of various cross-sections

I A Zhukov[1], L T Dvornikov[1] and S M Nikitenko[2]

[1] Siberian State Industrial University, 42, Kirov st., Novokuznetsk, 654007, Russia
[2] Federal Research Centre of Coal and Coal Chemistry at the Siberian Branch of the Russian Academy of Sciences, Leningradskij ave., Kemerovo, 650065, Russia

E-mail: tmmiok@yandex.ru

Abstract. The article presents the results of the experimental research of the high strength rock destruction by a bladeless tool. Rational circuit designs of disposing of indenters in the impact part of the drill bits and a diamond tool are justified. New constructive solutions of reinforcing bladeless drill bits, which allow drilling blast-holes of the various cross-section, are shown.

1. Statement of the problem

Elaboration of new and improvement of the known systems of development of the minerals are connected with wide application of drilling operations of different functions [1-5]. The primary way of destruction of high strength rocks is a percussive method. When striking blows on the tool, they can be transmitted to the destroyed object with the highest energy per time unit. Drilling without rotation of the tool round its axis [6], except for the rotation mechanism, allows essentially simplifying the drilling machine design and drilling blast-holes of any geometrical form. The percussive method of destruction can be realized by application of a bladeless tool – non-coring bits with indenters [7].

Commercial tests showed considerable advantages of indenters over the tool equipped with the blade in terms of endurance, the drilling rate, the specific charge of the firm alloy, expenses for drilling. Application of indenters in equipment of bits allows one to create almost an unlimited quantity of alternatives of crown designs using simple means. Changing, for example, their form, quantity and a disposition of circuit design for specified mountain-technological conditions of the deposit, it is possible to choose the most optimal equipment design. Now there are no well-founded recommendations for choosing the size, form, quantity of indenters and circuit designs and their optimal disposition on the chisel tool.

The project was financially supported by the Ministry of Education and Science of the Russian Federation within the Federal target program 'Research and development of priority directions of scientific-technological complex development of Russia for 2014-2020' (agreement No. 14.607.21.0028 from 05.06.2014, unique ID RFMEFI60714X0028).

2. Problem solution

For the purpose of acknowledging the possibility of rocks drilling without axial rotation of the tool, as well as revealing features of the penetration of bladeless indenters, experimental research [8] was conducted. During the experiment indenters were introduced into rocks by the universal testing machine 'IK-500.01' (Figure 1).

Drill bits of a cylinder-spherical form reinforced by indenters and placed on four circuit designs were used as the prototypes. They included a single indenter (Figure 2a), two indenters (Figure 2b), three indenters located at apexes of the equilateral triangle (Figure 2c), four indenters placed at apexes of the square (Figure 2d). The rock blocks with the sizes of 40×36×15 cm were used for destruction at the installation. The rock is presented by fine-grain and tight granite with strength factor f = 18 on the scale of Protodjakonov.

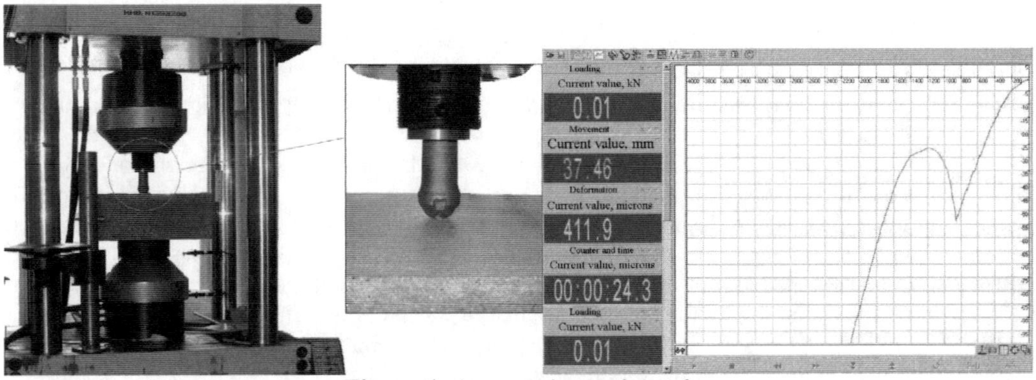

Figure 1. An experimental stand.

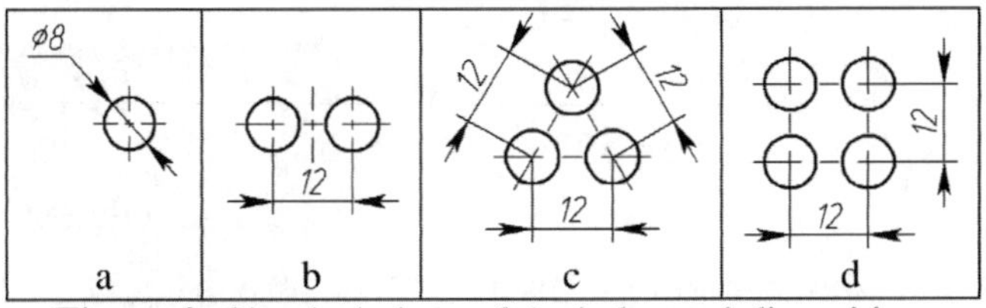

Figure 2. Circuit designs of the indenters' disposition.

The experimental technique consisted in the following. The rock block was installed on the lower platform of the stand. A vertically oriented non-coring bit was fixed in the overhead part of the stand and so that it could rotate in the course of the test. With appropriate setting of the hydraulic station the speed and the maximum value of loading was established. After turning of the stand the contact between the bit and the granite block was installed. Then the crown loading with speed V_H = 1 kN/s was shifted. The computer control system recorded a diagram of dependence of the crown penetration depth into the granite on the magnitude of the loading (a 'force – penetration' diagram). The first cycle of the test proceeded until the muck was destroyed and the tool penetration formed a characteristic hole under indenters. The subsequent test was carried out after cleaning a hole from the destroyed muck and installation of the non-coring bit into the same hole. The loading proceeded until applied force magnitude F_{max} = 100 kN was achieved. After each test the analysis of the broken rock was made.

The above-mentioned diagram 'force-introduction' allow us to study the mechanism of rock destruction at dynamics by means of the tool. The diagram 'force-introduction' allows identifying the various stages of the process of tool introduction into the rock. The first section of the chart until the reduction of the load indicates the accumulation of elastic deformation resulting in the brittle fracture of the rock. The next section shows a sharp drop in the loading at a small depth of penetration. Then, under the almost constant loading, the depth of the tool penetration increases. This step corresponds to brittle fracture of the rock. Then the cycle is repeated.

3. Analysis of the results

During the experiment the following results have been obtained.

As the result of the statistical analysis of various functional relationships the diagram 'force – introduction' allowed us to determine that the hyperbolic function is the closest to the truth:

$$P_z(h) = k_1 + k_2 \cdot \sqrt{k_3^2 + h^2} \ ,$$

where P – value of the loading, h – depth of tools' penetration, k_i – empirical coefficients that characterize the object of destruction, $i = 1,2,3$.

The physical meaning of coefficients k_i is the compliance of the object of destruction with the tool. The resulting function describes adequately the experimental results and allows one to judge the effectiveness of the tool.

During penetration of the single indenter under the bonding pad a tightly pressed and quite thin bowl-shaped layer of the rock is formed. Its diameter is slightly less (1…2 mm) than the diameter of the hole. The other part of the destroyed muck represents less pressed shattered layers of the muck removed without considerable efforts by a sharp object.

During simultaneous penetration of two, three, or four indenters (Figure 3) the muck part located between indenters is destroyed by the large-scale spalling. At an optimum distance between three next indenters l_i = 1…2.5 d_i, located at the apexes of the equilateral triangle, there is a 1.5…2.5-fold increase in the total volume of destruction on the average.

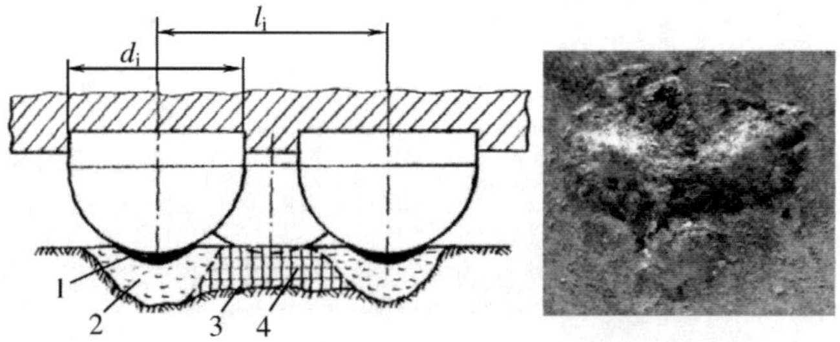

1 – a strongly compacted layer of the muck, 2 – a zone of the shattered muck, 3 – a contour joint of the hole, 4 – a zone of the large-scale spalling

Figure 3. The hole resulted in during simultaneous penetration of three indenters.

According to the analysis results of the obtained diagram «force - penetration », the distribution of the magnitude of the applied forces on the number of simultaneously introduced indenters is set. If the force per one indenter is related to the volume of fracture (Figure 4), the most optimum scheme from the point of view of energy consumption will be that in which three indenters are located at the apexes of the equilateral triangle.

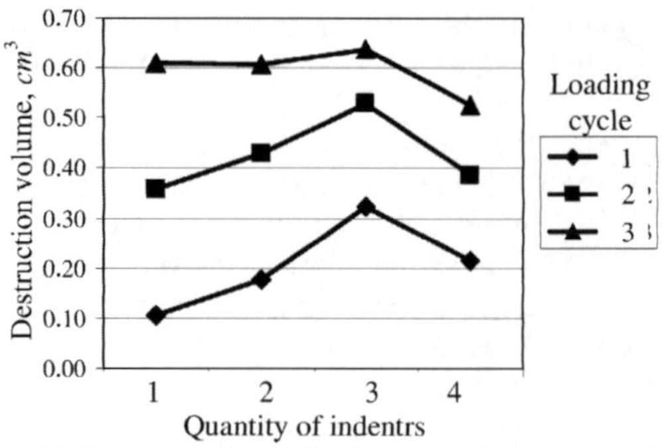

Figure 4. The estimation of volume of the destroyed muck.

Thus, the experimental verification of the interference effect of closely spaced and, at the same time, introduced indenters provides the effect of simultaneity. Assuming a particular value of the diameter of the blast-hole, it is possible to find such mutual bracing of three indenters which allows drilling the aperture holes without the tool rotation around its geometrical axis. This type of drilling mode provides an opportunity of obtaining apertures of a non-circular cross-section. It is also necessary to note that the figure formed by the next indenters should not necessarily be a precise equilateral triangle; the angle between the sides can be in the range of 60 ± 12°. Instances of the recommended circuit designs of indenters' disposition on the working part of non-coring bits are shown in Figure 5.

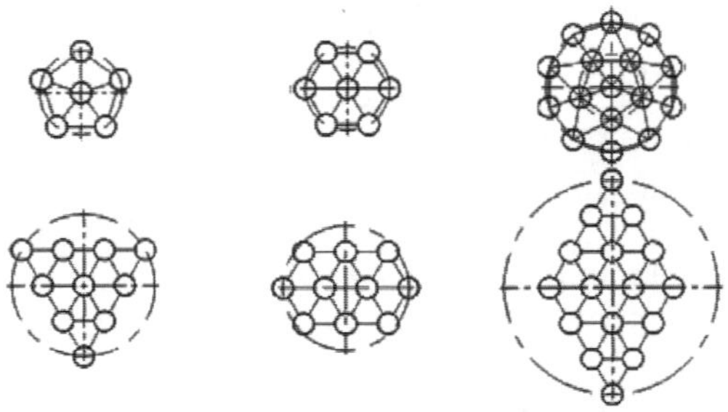

Figure 5. Rational circuit designs of indenters' dispositions.

In Figure 6 the bit with the rational disposition of indenters, by means of which churn drilling with the participation of the author, a blast-hole of the rhombic cross-section is obtained.

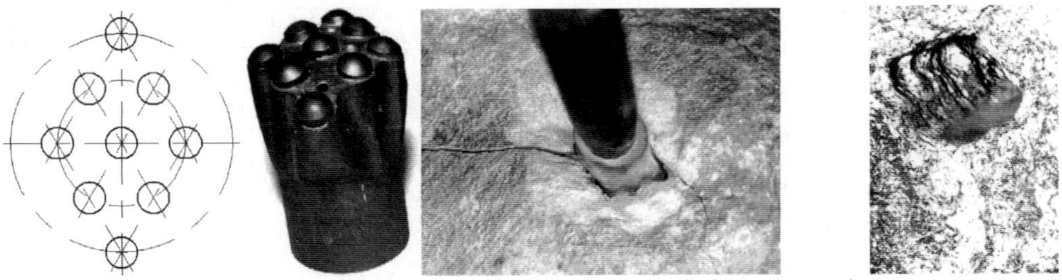

Figure 6. Drilling of a blast-hole of the rhomboid cross-section.

4. Conclusion

New aspects of the drill bits and a diamond tool provide a possibility of the short-hole drilling of the various cross-section including a non-circular one, when in the sharp corners of the drilled blast-hole, stress concentrators are created and there are possibilities of purposeful influence on the array with an essential decrease in drill footage.

References

[1] Lipin A A 2005 *J. Physical-technical problems of development of mineral resources* **2** 74
[2] Smolyanitsky B N, Tishchenko I V, Chervov V V, etc. 2009 *J. Physical-technical problems of development of mineral resources* **4** 65
[3] Yungmeister D A, Sudenkov J V and Pivnev V A 2011 *J. Mountain Inform.-Analyt. Bullet.* **8** 288
[4] Saruev L A, Pashkov E N, Ziyakaev G R and Kuznetcov I V 2013 *J. Mountain Inform.-Analyt. Bullet.* **S4(1)** 482
[5] Zhukov I A, Molchanov V V 2014 *J. Advanced Materials Research* **1040** 699
[6] Dvornikov L T, Zhukov I A 2007 *J. Mountain Inform.-Analyt. Bullet.* **9** 269
[7] Zhukov I A, Dvornikov L T 2009 *J. Mining Equipment and Electromecanics* **2** 23
[8] Zhukov I A, Tsvigun V N 2009 *J. MachineStructure* **19** 125

A method of definition of life-cycle resources of electromechanical equipment

Y Zhukovskiy[1] and N Koteleva[2]
[1] Electrical Power Engineeringand Electromechanics Department, National Mineral Resources University, 2, 21 Line, St. Petersburg, 199106, Russia
[2] Technological Process Automation and Production Department, National Mineral Resources University, 2, 21 Line, St. Petersburg, 199106, Russia

Email: spmi_energo@mail.ru

Abstract. The problems of developing a diagnostic system of electromechanical equipment of mining industry are described in this article. This system is a platform for changing the strategy from periodic maintenance in the direction of predictive maintenance. The researches of different methods of malfunctions detection have been carried out and described. These researches have helped to make a conclusion that implementation of a safety diagnostic system without determination of many interdependent diagnostic parameters is not possible. The results of these researches will help to develop the system of monitoring, diagnostic and definition of life-cycle resources of electromechanical equipment. This system will able to take into account the differences of technological processes and will base on the subsystems of simulation and prediction of malfunctions. In addition, development of a subsystem of monitoring, diagnostic and definition of life-cycle resources of electromechanical equipment has been described. This subsystem will evaluate the life-cycle resources in different periods of timeand make the machine control possible, using the information about the fault values and availability of spare parts, tools and equipment in addition.

1. Introduction

The safety of technological processes is the main tasks of factories in mining industry. Electromechanical equipment plays a main role in safety, especially when electromechanical equipment is used in hazardous and harmful areas. To ensure protection against faults, periodic maintenance is used. During the periodic maintenance faults are detected by traditional methods. Consequently the cost of maintenance is increased because the time between two maintenances is less than the time that is really required for safe use of mechanical equipment [2]. Moreover, sometimes the electromechanical equipment operates in the hard-to-get-at places and specialists have to halt technological processes during maintenances. Instead of using typical methods, condition-based predictive methods are more reasonable to use, especially for electromechanical equipment of mining industry [2, 4]. They allow evaluating the life-cycle of equipment,decreasing the cost of maintenances and increasing the economic efficiency of production.

So, the definition of life-cycle resources of electromechanical equipment is an important topic of scientific researches.

2. A traditional approach to fault detection of electromechanical equipment

There are a few parameters that are measured by a traditional approach of fault detection of electromechanical equipment: vibrations, audible noise, heating of individual elements of a machine, a thermal imaging survey, electrical parameters, including deviations of currents and voltages from the nominal values (amplitude, frequency, phase), an origin of high-harmonic in the spectrum of currents and voltages [3]. Each of the measured parameters can be correlated with one or more diagnostic elements of the object. Therefore, each of the traditional approaches of fault detection has some disadvantages thatdo not allow any possibilities to detect fault atan early stage of fault emergence, but let usidentify the indirect causes of fault (table 1). There are approaches that can help detect fault, but cannotassess the extent of the fault, hence, we are not able to predict machinemalfunction.The reason for this is that the defective parts of a machine influencethe parts that are faultless. Besides,these parts have close electromagnetic, mechanical and technical relationships.If the physical fault appears in one of the machine parts, other parts will have an indirect fault.

If the physical fault appears in the several parts of the machine, the task of finding machine fault parts will be difficult.The complex methods of measuring the diagnostic parameters are needed in this case. In addition to the above-mentioned, plenty of diagnostic parameters can be measured at the same time, so all of the measuring devices must be time-triggered.

Table 1. A traditional approach to fault detection of the induction motor.

A method of fault detection	Diagnostic parts of the induction motor							
	Isolation winding relative to the housing	Phase barrier	Winding Insulation	The stator winding	The rotor winding	Bearings	Rotor	Active steel stator
Vibration Analysis	0[a]	0	0	K	K	X	X	K
Acoustic analysis	0	0	X	K	K	X	X	K
Spectral analysis of the currents	0	0	X	X	X	K	K	B
Measurement of an external electromagnetic field	0	0	0	X	B	K	K	0
Temperature, thermal analysis	0	K[b]	B[d]	B	0	B	0	0
Leakage currents	X[c]	X	0	0	0	0	0	0
The level of dielectric strength	X	X	X	0	0	0	0	0
Multivariate analysis of electrical parameters (wattmeter graphics)	B	B	X	X	X	K	K	B

[a] The type and values of fault cannot be detected.
[b] The type and values of fault can be detected indirectly.
[c] The type of fault can be detected.
[d] The type and values of fault can be detected.

Most diagnostic systems based on traditional approaches to respond to one of its parts at a moment when the motor is already out of order or it suffered from internal damage or exceeded the setting of one of the monitored parameters (current, temperature, vibration), the cause of the incident is not defined [4]. However, the main types of damage to the electro-mechanical equipment, such as violations of the electric and magnetic symmetry of the rotor, develop gradually, so monitoring of the

IOP Conf. Series: Materials Science and Engineering **124** (2016) 012172 doi:10.1088/1757-899X/124/1/012172

accumulation of the diagnostic information in the time domain is a very important factor that allows one to predict the failure of the engine at a certain moment beforehand on the basis of the modeling state.

3. A new approach of fault detection of electromechanical equipment

The complex systems of monitoring and protection along with the systems of prediction and detection of the life-cycle stage of electromechanical equipmentfunctioningtogether ina diagnostic center, can help to solve the problem of traditional methods [4].

3.1. The module information analysis systemof fault detection of electromechanical equipment

The module information analysis system (IASFD) consists of several modules. Each of them is developed specifically for each type equipment. In summary, it has a typical structure, links and connections. A scheme of this system is shown in figure 1. It consists of subsystems of obtaining data: on-line and off-line parameters, subsystems of data verification, subsystem of fault simulation, subsystem of prediction and detection of the life-cycle stage of electromechanical equipment

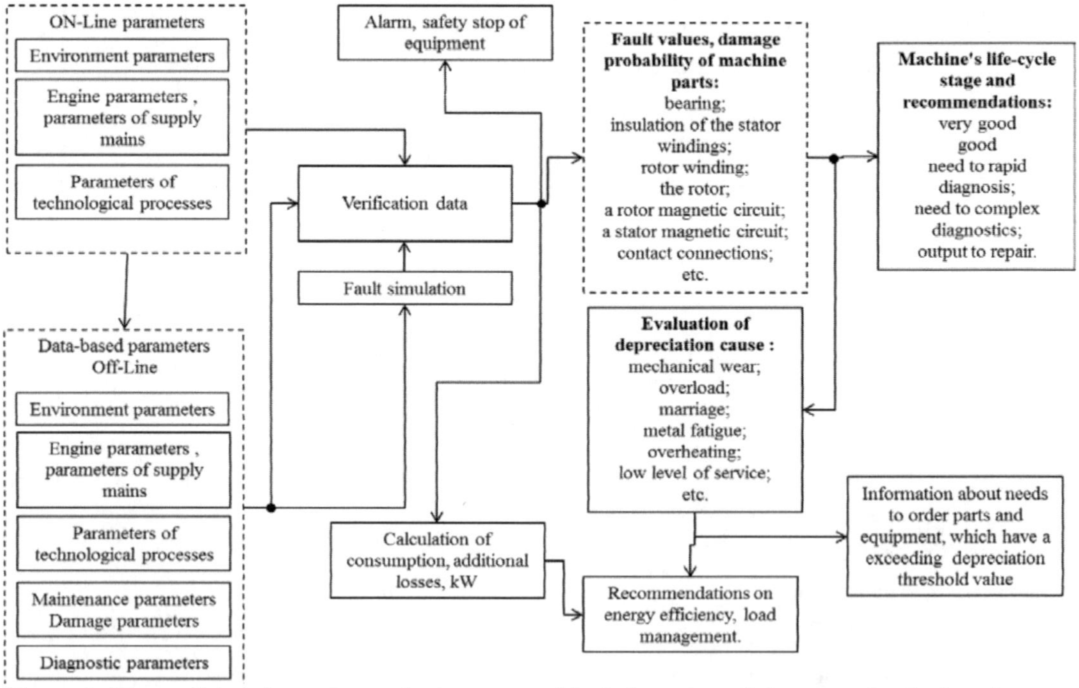

Figure 1. The module information analysis system of fault detection of electromechanical equipment.

ON-Line parameters are required for operational monitoring and prediction in the short period of time, which suffice to satisfy the continuity of the process and the possibility of early detection of the accident. Based on data,IASFDwill advance to signal a possible accident and halt the equipment if necessary.

Off-Line parameters are required for the subsequent modeling and forecasting for a short or long period of time, which, in itsturn, will organize optimally scheduled maintenance, decommissioning work and ordering spare parts. In addition to the basic parameters, which are removed to the On-Line mode for the system requires data on the repairs, accident data, the data obtained from the comprehensive and rapid diagnostics. On the basis of the aggregate input IASFD generates a report about the deterioration of the individual electric nodes and the likelihood of the release of their failure. At the next stage, the causes of wear and their contribution to the deterioration of a certain motor unitare considered. Based on this knowledge, we can make a decision about how to manage stress and ordering of spare parts, and the wear exceeding a threshold value. A separate module of IASFDcalculates the additional losses that arise in the motor resulting in considerable wear and tear, inadequately supplied voltage, overheating or electric motor optimum performance.These data are a proposition on energy saving.

A distinctive feature of the proposed system is ability to inform a manager about the impending threat of a failure of the equipment. To do this on the basis of modeling it is necessary to select the appropriate time intervals based on a survey to obtain data on the probability of occurrence of a specific defect. This in turn not only allows avoiding serious accidents, but also managing the loading equipment on the basis of its projected state and the availability of spare parts. In the case of absence of certain parts it is possible to decide on the removal of the load on the production site so that the failure of the equipment does not lead to a lengthy downtime of the enterprise or an individual plant [1].

3.2. The subsystemof fault prediction

The main module of IASFD is a module of fault prediction. The module implements a constant or periodic monitoring of equipment condition and determines the remaining service life of the equipment [1, 3]. In this case, decisions on the repair and renovation plan is based on information about the technical condition of the monitored objects formed with the help of the module, including intelligent processing of diagnostic parameters.

In developing the structure and algorithm of the module one first has to allocate an additional subsystem allowing one to solve the issues of information failure. For the problem considered it is advisable to carry out the development of the following subsystems:

1. The subsystem of life-cycle assessment of the electromechanical equipment.
2. The subsystem of forecasting the development of the defect.
3. The intelligent subsystem determines the parameters of the equipment in the period between diagnoses (a subsystem solving a problem of information insufficiency).
4. The subsystem forming a part of the information field of the information database of the enterprise.

To solve analytical problems and a diagnostic module based on the selected subsystems it is possible touse a system block diagram, which is shown in Figure 2.

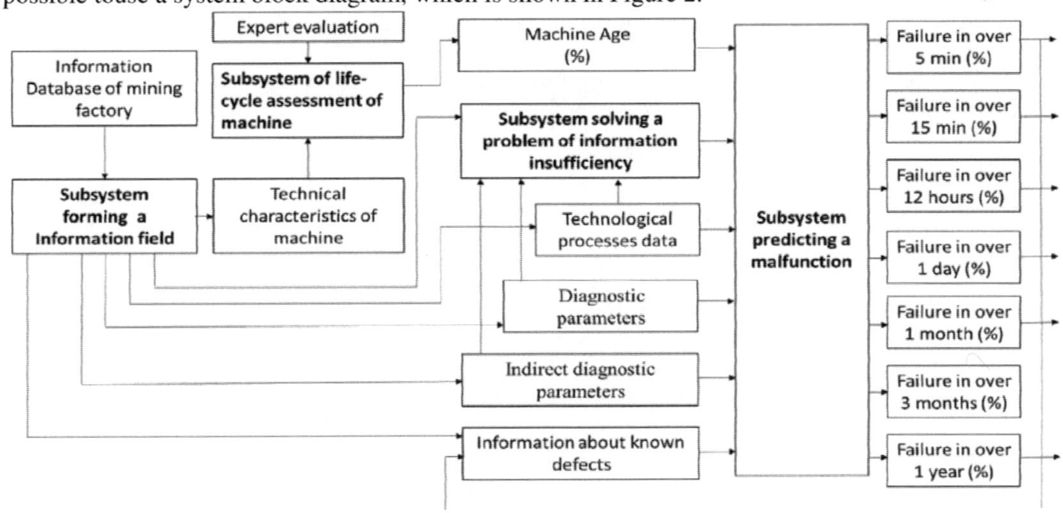

Figure 2. Asubsystemof fault prediction.

For the operation and design of analysis modules of the system we can use various principles, among which it is advisable to use neuro-fuzzy algorithms of neural networks and genetic algorithms [3, 4]. The efficiency of these algorithms depends on the quality of the information that was the basis for their development. Therefore, upon receipt of unsatisfactory results in the process of testing these algorithms on the performance it is necessary to make adjustments in the information space of the data generated for the system operation.

3.3. The prospect of the research

The formation of the information field is the next step of this research. In future the main task is the search of calibrated values of the parameters of electromechanical equipment, especially the values thatcannot be found using traditional methods. These values help to predict the stages of electromechanical equipment more correctly because they will use a special knowledge about working parameters of equipment.

4. Conclusions

Currently, a transition to the system of condition-based predictive maintenance is blocked by deficiency of methods of definition of life-cycle resources of electromechanical equipment. In the existing systems there are no simulation modules and fault prediction modules and they are unable to provide a high level of safety for technological processes, therefore many companies are unwilling to change the maintenance strategy of electromechanical equipment. The results of this research help to solve this problem and allow developing the system, which will be able to define life-cycle resources of electromechanical equipment on the base of the simulation, taking into account the differences of technological processes that will be widely used in technological practices. These systems will allowevaluatingthe current state of electromechanical equipment without halting the processes and losing the profit of mining industry companies.

References

[1] Xu X, Chen T and Wang S 2009 *The 4th World Congr. on Engineer. Asset Manag.* 614–619

[2] Schreiner Z, Bizjak J and Middleton A 2008 *The 42nd Internat. Confer. on Large High Voltage Electric Syst.* **5** 213

[3] Aguiar E, Nogueira F, Amaral R, Fabri D, Rossignoli S, Ferreira J and Ribeiro M 2015 *Neural Computing and Applications*1–15

[4] Lili G, Kebao X and Zhaoxue L 2015 *Open Electr. and Electr. Engineer. J. Appl. Phys.* **9** 166–169

Distribution of contact loads over the flank-land of the cutter with a rounded cutting edge

V Kozlov[1], A Gerasimov[2] and A Kim[1]
[1] Tomsk Polytechnic University, 30, Lenina ave., Tomsk, 634050, Russia
[2] Tomsk State University, 36, Lenina ave., Tomsk, 634050, Russia

E-mail: kozlov-viktor@bk.ru

Abstract: In this paper, contact conditions between a tool and a workpiece material for wear-simulating turning by a cutter with a sharp-cornered edge and with a rounded cutting edge are analysed. The results of the experimental study of specific contact load distribution over the artificial flank wear-land of the cutter in free orthogonal turning of the disk from titanium alloy (Ti6Al2Mo2Cr), ductile (63Cu) and brittle (57Cu1Al3Mn) brasses are described. Investigations were carried out by the method of 'split cutter' and by the method of the artificial flank-land of variable width. The experiments with a variable feed rate and a cutting speed show that in titanium alloy machining with a sharp-cornered cutting edge the highest normal contact load ($\sigma_{h\,max}$ = 3400…2200 MPa) is observed immediately at the cutting edge, and the curve has a horizontal region with the length of 0.2…0.6 mm. At a distance from the cutting edge, the value of specific normal contact load is dramatically reduced to 1100…500 MPa. The character of normal contact load for a rounded cutting edge is different – it is uniform, and its value is approximately 2 times smaller compared to machining with a sharp-cornered cutting edge. In author's opinion it is connected with generation of a seizure zone in a chip formation region and explains the capacity of highly worn-out cutting tools for titanium alloys machining. The paper analyses the distribution of tangential contact loads over the flank land, which pattern differs considerably for machining with a sharp-cornered edge and with a rounded cutting edge.

Abbreviation and symbols: m/s – meter per second (cutting speed v); mm/r – millimeter per revolution (feed rate f); MPa – mega Pascal (specific contact load as a stress σ or τ); h_f – the width of the flank wear land (chamfer) of the cutting tool, flank wear land can be natural or artificial like the one in this paper [mm]; x_h – distance from the cutting edge on the surface of the flank-land [mm]; σ_h – normal specific contact load on the flank land [MPa]; τ_h – tangential (shear) specific contact load on the flank land [MPa]; HSS – high speed steel (material of cutting tool); P_y – radial component of cutting force [N]; $P_{y\,r}$ – radial component of cutting force on the rake face [N]; P_z – tangential component of cutting force [N]; γ – rake angle of the cutting tool [°]; α – clearance angle of the sharp cutting tool [°]; α_h – clearance angle of the flank wear land [°]; ρ – rounding off radius of the cutting edge [mm]; b – width of the machined disk [mm].

1. Introduction

Cutting tool wear causes a change in its geometry – a wear-land with width h_f forms on the flank surface, the cutting edge rounds with radius ρ, a wear crater forms on the tool face. In machining of titanium alloys, which are widely used in aerospace and chemical industry, shipbuilding, etc., very intensive wear occurs on the flank surface with rounding of the cutting edge. Contact loads vary on the face and flank surfaces of the cutting tool, as well as the temperature in the chip formation zone.

In machining of difficult-to-machine titanium alloys the width of wear-land h_f reaches 2.5…5 mm, and cutting edge rounding radius reaches $\rho = 0.3...5$ mm [1, 2], but even with this big value, cutting tool is still capable of working, which is obscure. It is necessary to know distribution of contact loads and temperature over the surfaces of the cutting tool for estimation of cutting wedge strength at various stages of the wear process.

The investigation of specific contact loads distribution over the work surfaces of the cutting tool is a difficult task due to large values of contact loads, especially in machining of difficult-to-machine materials, to which the widely used in aerospace industry titanium alloy BT3-1 belongs.

2. Research Methods and Preparation for Experimental Study of Contact Load Distribution

The distribution of specific normal σ_h and tangential τ_h contact loads over the wear-land of the flank surface was studied by the method of 'split cutter', and by the method of the artificial flank-land of variable width near the cutting edge [1].

The experiments were carried out in free orthogonal turning of a disk made from the above-mentioned materials, i.e. with radial feed f of a cutter. The ductile brass (63Cu), which forms continuous chip, the brittle brass (57Cu1Al3Mn), which forms discontinuous chip, the difficult-to-machine titanium alloy BT3-1 (Ti6Al3Mo2Cr), which forms discontinuous chip, were used as a workpiece material.

Brass was selected to fine-tune the method of a split cutter and to compare its results with the results received by the method of the artificial flank-land of the variable width. The latter method was used to study contact load distribution over the artificial flank-land near the cutting edge in machining of titanium alloy BT3-1, since the split cutter, which is made from cemented carbide BK8 (WC8Co), does not have sufficient strength when a split between its parts approaches to a cutting edge at a distance smaller than 0.6 mm. The comparison of the results received by both methods in brass machining, has displayed good coincidence, the deviation did not exceed ±15 %.

The wear was simulated by sharpening a chamfer with width h_f on a flank surface ground with clearance angle $\alpha_h = 0$ °. To study influence of the cutting edge rounding, artificial rounding of a cutting edge stimulated the wear to required radius ρ. The width of artificial flank-land h_f was measured by means of a toolmaker's microscope. The length of the artificial flank wear land was accepted equal to the width of disk b, since ordinary wear of a cutting tool occurs along the cutting edge and its length is usually bigger than its width.

The change in length of a flank-land insignificantly affects distribution of specific contact loads over the face of a cutting tool – only through the change of cutting temperature. It allows us to use the data on contact load distribution over the rake surface of the cutting tool, received in cutting by a sharp cutting tool. Therefore, the focus was given to an experimental research of the contact loads distribution over a flat section of an artificial flank-land, which was used to simulate flank wear.

3. Results of experiments

In the ductile brass (63Cu) machining with a sharp cutter an artificial flank-land (width $h_f = 2.4$ mm, clearance angle $\alpha_h = 0$ °) extreme pattern of σ_h curves is observed, i.e. the highest normal contact loads $\sigma_{h\ max}$ are at some distance from the cutting edge (Figure 1 and 2). In the brittle brass (57Cu1Al3Mn) machining by a cutter with the same geometry and cutting speed, the highest contact loads $\sigma_{h\ max}$ are near the cutting edge (Figure 3).

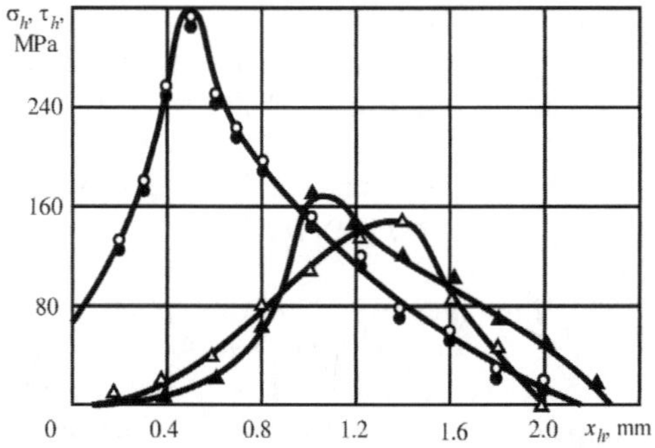

Figure 1. Distribution of contact loads over the flank-land of the cutter in ductile brass (63Cu) machining. $\gamma = 0$ °, $\alpha = 18$ °, $\alpha_h = 0$ °, v =1.7 m/s, ○, ● – f = 0.06 mm/r; Δ, ▲ – f = 0.21 mm/r; ○, Δ – normal contact load σ_h; ●, ▲ – tangential contact load τ_h. Ordinate – normal σ_h [MPa], and tangential τ_h [MPa] specific contact load over the flank-land; abscissa – distance from the cutting edge on the surface of the flank-land x_h [mm]

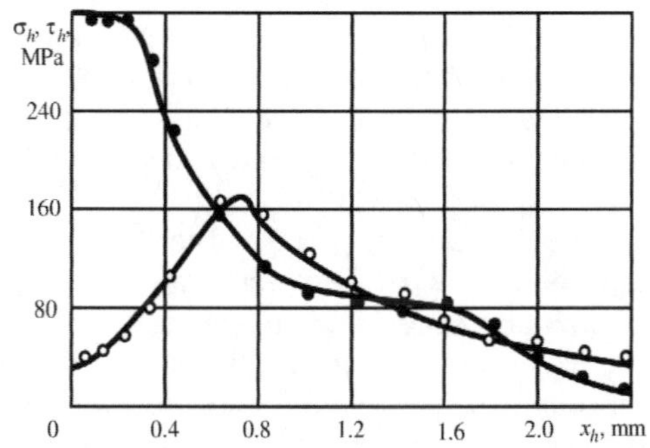

Figure 2. Distribution of contact loads over the flank-land of the cutter in cutting of the ductile brass (63Cu). $\gamma = 0$ °, $\alpha = 18$ °, $\alpha_h = 0$ °, v =3.6 m/s, f = 0.21mm/r; ○ – normal contact load σ_h; ● – tangential contact load τ_h. Ordinate – normal σ_h [MPa], and tangential τ_h [MPa] specific contact load over the flank-land; abscissa – distance from the cutting edge on the surface of the flank-land x_h [mm]

The results of investigations of contact load distribution over a flat section of the flank-land of the cutter with artificial rounding of the cutting edge in machining of brass are presented in Figure 4, and in machining of titanium alloy BT3-1 - in Figure 5.

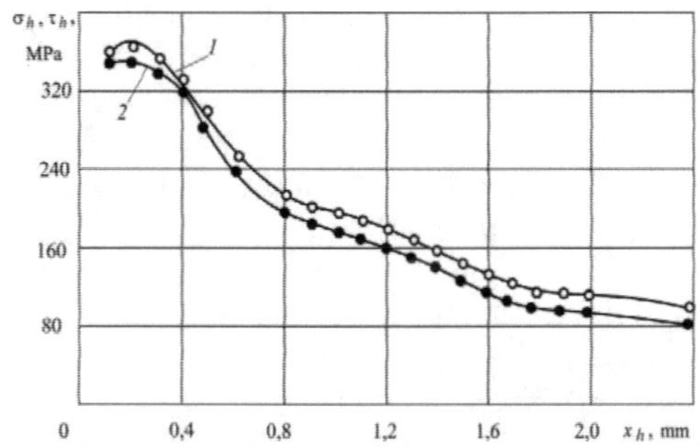

Figure 3. Distribution of normal σ_h (○) and tangential τ_h (●) contact loads over the flank-land of the cutter in cutting of the brittle brass (57Cu1Al3Mn). $\gamma = 0$ °, $\alpha = 18$ °, $\alpha_h = 0$ °, ν=1.7 m/s, f =0.41 mm/r. Ordinate – normal σ_h [MPa], and tangential τ_h [MPa] specific contact load on the flank-land; abscissa – distance from the cutting edge on the surface of flank-land x_h [mm]

IOP Conf. Series: Materials Science and Engineering **124** (2016) 012173 doi:10.1088/1757-899X/124/1/012173

4. Discussion of experiments results

The extreme pattern of the curves of normal σ_h and tangential τ_h specific contact loads in machining of ductile materials was also mentioned by other researches [2], however authors have not discovered explanations of this phenomenon.

In our opinion, the highest normal contact loads σ_h in cutting with continuous chip are at some distance from a cutting edge due to a sag of the transient surface under the radial component of the cutting force on rake surface $P_{y\,r}$ [1].

In continuous chip formation the influence of the radial component of cutting force $P_{y\,r}$ on the face is stable [3], a sag of the transient surface is constant, therefore, pressure from the elastic recovering transient surface on the flank-land is higher at some distance from the cutting edge [1].

During discontinuous chip formation at the moment when formed chip elements are separated from the workpiece, the radial component of the cutting force on face $P_{y\,r}$ quickly decreases (sometimes to zero) [3], which leads to elastic recovery of the transient surface and its pressure upon the cutting tool flank surface. To a greater degree this pressure acts near the cutting edge [1], therefore the highest normal contact loads $\sigma_{h\,max}$ are observed near the cutting edge, which is confirmed by the results of experiments in machining of the brittle brass (57Cu1Al3Mn), which forms discontinuous chip (Figure 3).

In machining of brass, tangential contact loads τ_h are equal to normal ones σ_h, except for the case of the ductile brass (63Cu) machining at elevated cutting speed v =3.62 m/s, when the highest value of $\tau_{h\,max}$ is observed near the cutting edge, where it is not equal to normal contact loads σ_h (Figure 3). The equality of tangential and normal contact loads is associated, in our opinion, with a plastic character of the contact on the flank-land. The high coefficient of friction $\mu = \tau_h / \sigma_h \approx 1$ does not correspond to usual external friction, when coefficient of friction is equal to 0.15 … 0.1. In case of plastic contact, tangential contact loads τ_h cannot be calculated by formula $\tau_h = \sigma_h \times \mu$, as they will be equal to shear strength of a material τ_{max} at the operating temperature in the contact zone ($\tau_h = \tau_{max}$).

At elevated cutting speed v =1.17 m/s (Figure 2), the higher temperature of cutting promotes adhesion of contact surfaces, thus strong seizure takes place even at insignificant pressure ($\sigma_h \approx 40$ MPa), what was observed near the cutting edge. A high value of $\tau_{h\,max} = 320$ MPa is explained, in our opinion, by increased strain rate and hardening of the work material. Softening of the work material due to influence of temperature at increased cutting speed does not have time to occur [5].

In case of the rounded cutting edge, the workpiece material hardens as it moves under the rounded section, therefore there is an increase in tangential contact load τ_h (Figure 4) compared to cutting with a sharp cutting edge (Figure 1 and 2).

In machining of plastic brass the cutting edge rounding leads to change in the distribution pattern of contact loads and their value (Figure 4). When a cutting edge is rounded, the volume of the work material, pressed under the rounded section of the cutting edge, is increased, therefore the sag of the transient surface affects pressure on a flank surface to a lesser degree. Normal contact load σ_h on a flat section of the flank land near the cutting edge is not equal to zero in machining of the plastic brass (63Cu), and the distribution pattern of contact loads becomes uniform (Figure 4). Normal and tangent contact loads are equal in magnitude ($\sigma_h \approx \tau_h \approx 130$ MPa), which indicates plastic character of contact on the flank-land. The magnitude of these loads is a bit higher than the magnitude of the highest loads on the flank-land of the rounded cutting edge (Figure 1, 2), this is associated with dragging of additional material under the rounded section of the cutting edge and with an increase of elastic deformation of the work material in the chip forming zone.

The work material is work-hardened when it moves under a rounded section, thus an increase of tangential contact load τ_h is observed (Figure 4) compared to cutting with a sharp cutter (Figure 1 and 2).

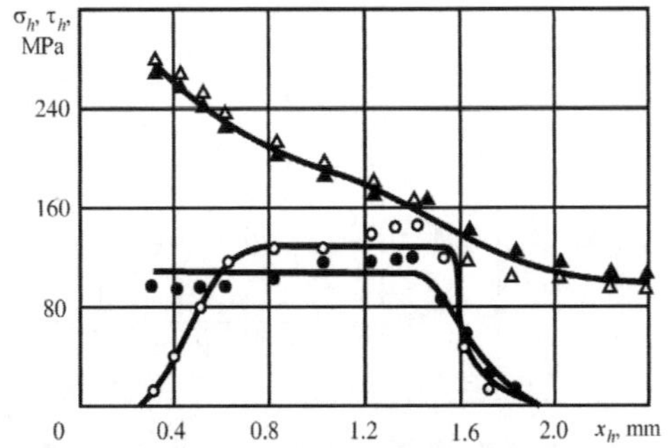

Figure 4. Distribution of contact loads over the flank-land of the cutter in brass machining by the cutter with rounding off cutting edge with radius $\rho = 0.3$ mm. $\gamma = 0$ °, $\alpha = 18$ °, $\alpha_h = 0$ °, v =1.7 m/s, ○, ● – ductile brass (63Cu), f = 0.21 mm/r; Δ, ▲ – brittle brass (57Cu1Al3Mn); f = 0.41 mm/r; ○, Δ – normal contact load σ_h; ●, ▲ – tangential contact load τ_h

The distribution pattern of the contact loads does not change for the brittle brass machining with a rounded cutting edge (Figure 4), and the magnitude of the highest contact load even decreases ($\sigma_{h\ max} \approx \tau_{h\ max} = 280$ MPa) compared to cutting with a sharp cutting edge ($\sigma_{h\ max} \approx \tau_{h\ max} = 360$ MPa) (Figure 3). The formation of a seizure zone in the rounding region, which reduces contact of the transient surface with the flank-land, explains this fact, and this is especially true for conditions of discontinuous chip formation. The presence of the scratch marks on the rounded portion, left during sharpening of the cutting tool, proves this indirectly.

The reduction of tangential contact load τ_h, at a distance from the cutting edge, is caused, in our opinion, by softening of the work material in the contact zone due to temperature increase. The plastic character of the contact on the flank land is indirectly confirmed by a considerable amount of brass sticked to the flank-land after turning.

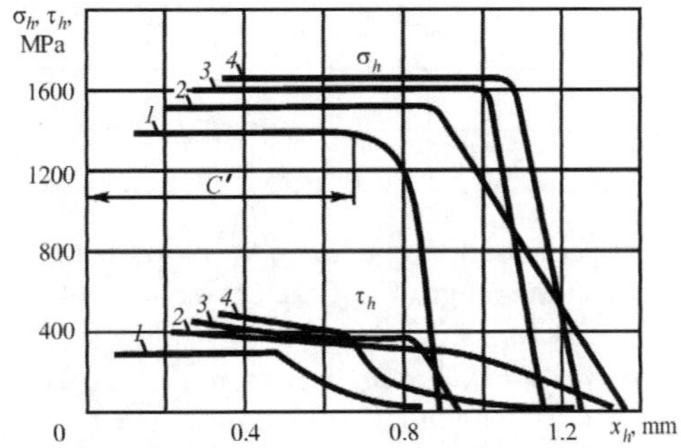

Figure 5. Distribution of contact loads over the flank-land of the cutter in machining of titanium alloy BT3-1 by the cutter with rounding off cutting edge with radius $\rho = 0.3$ mm. BT3-1 – BK8, $\gamma = 0$ °, $\alpha = 10$ °, $\alpha_h = 0$ °, V =1 m/s, f = 0.21 mm/r; 1 – $\rho = 0.07$ mm; 2 – $\rho = 0.2$ mm; 3 – $\rho = 0.28$ mm; 4 – $\rho = 0.35$ mm; σ_h – normal contact load on the surface of the flank-land, τ_h – tangential contact load on the surface of the flank-land

The machining of titanium alloy BT3-1, which forms discontinuous chip, by a cutter with a rounded cutting edge (Figure 5), shows almost a 2 times reduction of normal contact loads compared to cutting without cutting edge rounding (Figure 6). We explain this paradoxical phenomenon by formation of a seizure zone on the cutter face in the region of rounding, which reduces the contact of the transient surface with the flank land. The presence of the scratch marks on the rounded part, left during sharpening of the cutting tool, proves this indirectly.

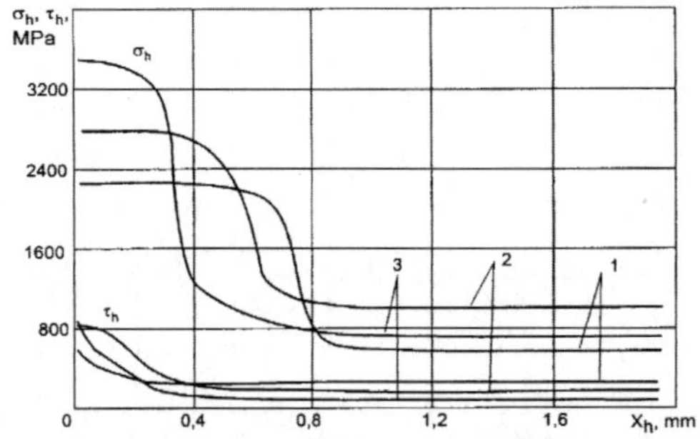

Figure 6. Distribution of normal σ_h and tangential τ_h contact loads over the flank-land of the cutter in titanium alloy cutting without rounding off cutting edge. BT3-1 – BK8, γ=0 °, α_h = 0 °, α = 10 °, v=1 [m/s]: 1 - f=0.11 mm/r; 2 - f=0.21 mm/r; 3 - f=0.41 mm/r. Ordinate – normal σ_h [MPa] and tangential τ_h [MPa] specific contact load on the flank-land; abscissa – distance from the cutting edge on the surface of flank-land x_h [mm].

The tangential contact loads, in the area that is more than 0.3 mm away from the cutting edge, are a bit higher in magnitude for cutting with a rounded cutting edge (τ_h = 350 … 420 MPa) (Figure 5) than the tangential contact loads in cutting with a sharp cutting edge (τ_h = 300 … 200 MPa) (Figure 6). The increase of τ_h for a rounded cutting edge is explained by work-hardening of the work material during its travel under the rounded section of the cutting edge. The absence of influence of normal contact loads on the tangential contact loads indicates plastic character of the contact on the flank-land, which is indirectly confirmed by presence of titanium alloy adhered to the flank-land after turning.

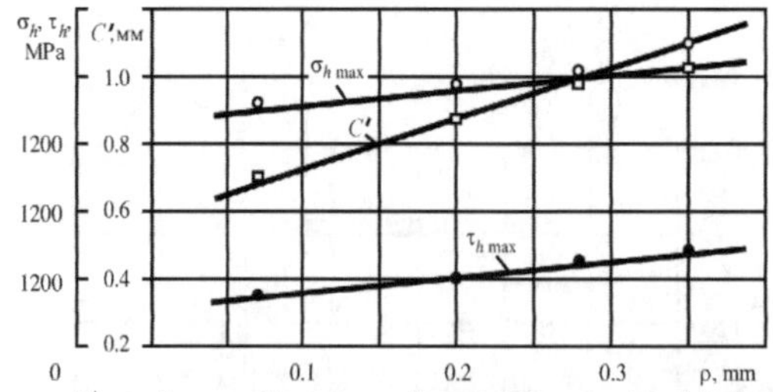

Figure 7. Influence of the cutting edge radius rounding off ρ on the highest contact loads magnitude on the lank-land $\sigma_{h\ max}$, $\tau_{h\ max}$ and on the length C′ of the horizontal region of the curve σ_h in titanium alloy BT3-1 cutting BT3-1 – BK8, γ = 0 °, α = 10 °, α_h = 0 °, v =1 m/s, f = 0.21 mm/rev; 1 – ρ = 0.07 mm; 2 – ρ = 0.2 mm; 3 – ρ = 0.28 mm; 4 – ρ = 0.35 mm; σ_h – normal contact load, τ_h – tangential contact load

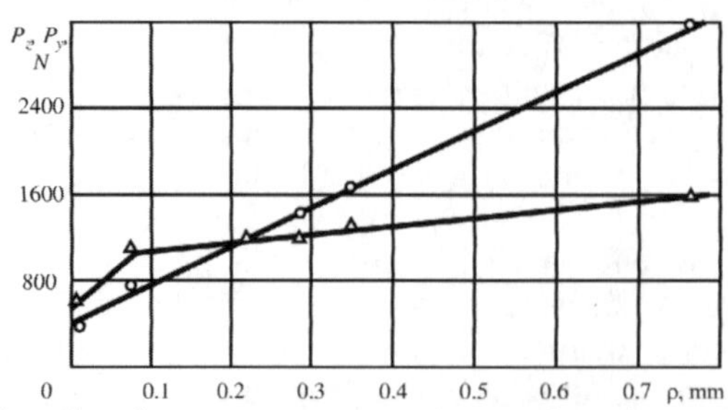

Figure 8. Dependence of cutting force component acting on the rake surface $P_{Y\,r}$ и $P_{Z\,r}$ on radius of cutting edge rounding ρ in titanium alloy BT3-1 machining. BT3-1 – BK8, γ = 0 °, α = 10 °, v =1 m/s, f = 0.21 mm/r; ○ – $P_{Y\,r}$ [N], Δ – $P_{Z\,r}$ [N]; machined disk width b=2.3 mm

The magnitude of the highest contact loads $\sigma_{h\,max}$, $\tau_{h\,max}$ and length C′ of the horizontal region of curve σ_h are directly proportional to rounding radius value ρ (Figure 7) that indicates increase of the material volume, pressed down under the cutting edge.

There is also an increase of the elastic deformation of the work material in a chip formation zone, which is proven by direct proportional dependence of the radial component of the cutting force acting on face $P_{y\,r}$ on rounding radius ρ (Figure 8). The increase of cutting force components, when the seizure zone is formed, is explained by the small sizes of this zone, which cannot cope with resistance increase to chip formation from the rake surface. In order to change normal contact loads even a minor change of the seizure zone is enough due to the rigidity of the contact on the flank surface.

5. Conclusion

In machining of the ductile brass (63Cu), which forms continuous chip, the cutting edge rounding leads to pattern change of the contact loads distribution over the flank-land – it becomes uniform. It is explained by dragging of additional material under the rounded section of the cutting edge, which leads to a smaller influence of the transient surface sag.

In machining of ductile brass L63 with a rounded cutting edge, the magnitude of contact loads is a bit more than the highest loads on the flank-land in cutting with a sharp cutting edge. It is explained by dragging of additional material under the rounded section of the cutting edge.

In machining of ductile brass L63 with a cutting tool with rounded cutting edge, normal and tangential contact loads on the flank-land are equal in magnitude ($\sigma_h \approx \tau_h \approx 130$ MPa), which indicates a plastic character of the contact on the flank-land.

In machining of the brittle brass (57Cu1Al3Mn), which gives discontinuous chip, a cutting edge rounding does not lead to change in the pattern of contact loads distribution over the flank-land, but the magnitude of the highest contact load is smaller compared to cutting with a sharp cutting edge. It is explained by formation of a seizure zone on the rounded section of the cutting edge. It reduces the contact of the transient surface with the flank-land, and to a greater degree it is true for discontinuous chip.

The reduction of the tangential loads on the flank land during machining of the brittle brass (57Cu1Al3Mn), in case the hypothesis about the plastic character of the contact on the flank land is accepted, is explained, in our opinion, by the softening of the work material in the contact zone due to temperature increase.

The magnitude of normal contact loads on the flank land when cutting titanium alloy BT3-1 with a rounded cutting edge is two times as small as in cutting with a sharp cutting edge. It is explained by formation of the seizure zone on the rounded section of the cutting edge.

The two-times reduction of the normal contact loads magnitude on the flank-land with a rounded cutting edge explains the working capacity of a considerably worn cutting tool to machine titanium alloys.

The character of the contact on the flank land is plastic in machining of titanium alloy.

The increase in tangential contact loads on the flank-land in cutting of titanium alloy BT3-1 with a rounded cutting edge is explained by work-hardening of the work material due to high plastic deformations.

References

[1] Kozlov V 2012 *Proceedings of the 7th International Forum on Strategic Technology* 6357713
[2] Ostafjiev V *et al* 1976 *Kiev: Visha shkola* 136
[3] Afonasov A and Lasukov A 2014 *Russian Engineering Research* **3** 152-155
[4] Narutaki N *et al* 1983 *CIRP Annals - Manufacturing Technology* **32(1)** 65-69
[5] Wang B *et al* 2013 *International journal of Machine Tools & Manufacture* **73** 1-8
[6] Hu J and Chou Y 2007 *Wear* **263(7-12)** 1454-1458
[7] Trent E M and Wright P K 2000 *Fourth ed. Butterworth-Heinemann, Boston*
[8] Kozlov V and Li X 2015 *Applied Mechanics and Materials: Scientific Journal* **756** 126-131

Strength of inserts in titanium alloy machining

V Kozlov, Z Huang and J Zhang
Tomsk Polytechnic University, 30, Lenina ave., Tomsk, 634050, Russia

E-mail: kozlov-viktor@bk.ru

Abstract: In this paper, a stressed state of a non-worn cutting wedge in a machined titanium alloy (Ti6Al2Mo2Cr) is analyzed. The distribution of contact loads on the face of a cutting tool was obtained experimentally with the use of a 'split cutting tool'. Calculation of internal stresses in the indexable insert made from cemented carbide (WC8Co) was carried out with the help of ANSYS 14.0 software. Investigations showed that a small thickness of the cutting insert leads to extremely high compressive stresses near the cutting edge, stresses that exceed the ultimate compressive strength of cemented carbide. The face and the base of the insert experience high tensile stresses, which approach the ultimate tensile strength of cemented carbide and increase a probability of cutting insert destruction. If the thickness of the cutting insert is bigger than 5 mm, compressive stresses near the cutting edge decrease, and tensile stresses on the face and base decrease to zero. The dependences of the greatest normal and tangential stresses on thickness of the cutting insert were found.

Abbreviation and symbols: m/s – meter per second (cutting speed v); mm/r – millimeter per revolution (feed rate f); MPa – mega Pascal (dimension of specific contact loads and stresses); γ – rake angle of the cutting tool [°]; α – clearance angle of the sharp cutting tool [°].

1. Introduction
In machining of difficult-to-machine titanium alloys, brittle fracture of the cutting wedge occurs in the form of chipping and spalling, which is especially dangerous for indexable inserts made from cemented carbides commonly used in industry. The working capacity of inserts increases if their thickness increases, as well as their cost. The investigations of stresses distribution were carried out to determine the optimal thickness of indexable inserts made from cemented carbide BK8 (WC8Co).

2. Research methods
A finite element method along with ANSYS 14.0 software was used to calculate stresses in free orthogonal cutting of difficult-to-machine titanium alloy BT3-1 (Ti6Al2Mo2Cr). The distribution of specific contact loads on the face and on the flank surfaces was found experimentally. It was found by the methods of a 'split cutting tool' and the artificial flank-land of the variable width in free orthogonal cutting of the difficult-to-machine titanium alloy (Ti6Al2Mo2Cr), having a constant geometry of the cutting part and with the following cutting conditions: (rake angle $\gamma = 0°$, cutting speed $v = 1$m/s, radial feed $f = 0.21$ mm/r, wear on the flank surface (artificial flank-land) $h_f = 0.2$ mm) [1]. Thickness h of the insert varied from 2 to 20 mm (Figure 1, 2).

3. Research methods. Results of experiments

Small thickness of the cutting insert h = 2 mm leads to extremely high (σ_3 = –6866 MPa) compressive stresses near the cutting edge, stresses that exceed the ultimate compressive strength of cemented carbide (WC8Co) (σ_{UCS} = –4200 MPa). It is practically proved by brittle fracture of cutting inserts when its thickness is less than 2 mm [2-5].

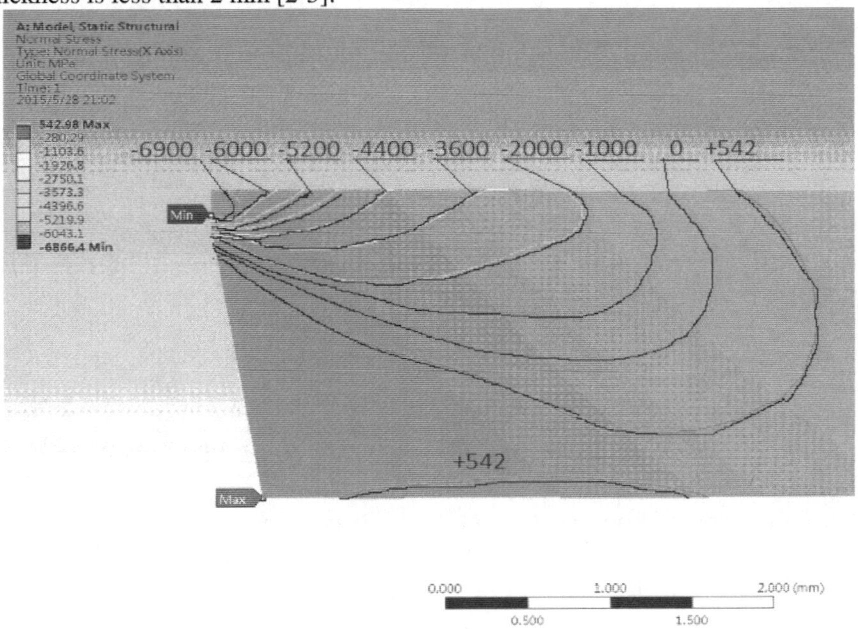

Figure 1. Distribution of normal stresses in the cutting wedge for a cutting insert with thickness of 2 mm. Width of the wear-land h_f = 0.2 mm

Tensile stresses (σ_1 = +542 MПа) are observed on the face surface at a distance equal to approximately four lengths of contact of a chip with a rake surface. The tensile stress value approaches the ultimate tensile strength of cemented carbide (WC8Co) (σ_{UTS} = +750 MPa).

High tensile stresses (σ_1 = +542 MPa) are observed on the base of the cutting insert, which value approaches the ultimate tensile strength of cemented carbide BK8. In practice, the results of calculations are justified by brittle fracture of the base of the insert with small thickness [6, 7].

Tensile stresses at the bottom of the base of the insert are not observed (Figure 2b, the base is not displayed, since there are no lines that correspond to $|\sigma| < 100$ MPa).

Shear stresses τ in the cutting edge were calculated under similar condition (Figure 3). The shear stress near the cutting edge of the cutting insert with small thickness (h=2 mm) exceeds the ultimate shear strength of cemented carbide (Figure 3a), which leads to destruction of a cutting part.

The maximum value of the shear stress does not exceed the ultimate shear strength of cemented carbide when thickness of the cutting insert is bigger than 5 mm (Figure 3b).

The influence of the cutting insert thickness on the maximum value of normal stresses σ_{max} near the cutting edge is represented in Figure 4. The greatest influence of the cutting insert thickness is observed for h ≤ 7 мм (σ_3 = –3000 MPa). This influence reduces significantly with a further increase in the insert thickness, and for h > 12 mm thickness does not influence the magnitude of the normal stress near the cutting edge (σ_3 = –1300…–1177 MPa).

The influence of the cutting insert thickness on maximum value of the shear stress τ_{max} has a similar character (Figure 5), but its influence is sufficiently great for h > 12 mm.

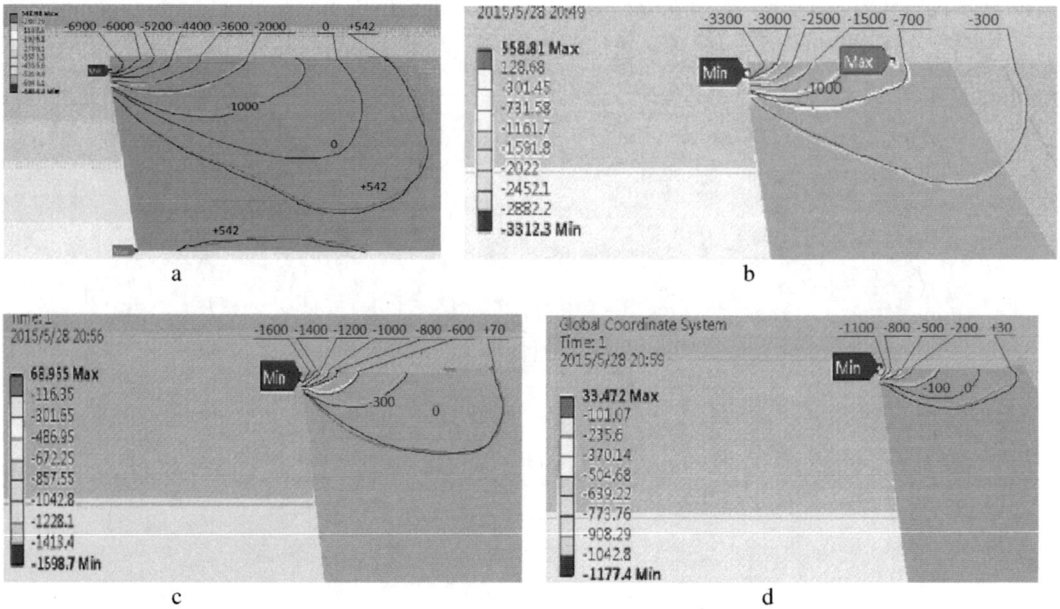

Figure 2. Distribution of normal stresses σ [MPa] in the cutting wedge for a cutting insert with thickness of h: *a* – h = 2 mm; *b* – h = 6 mm; *c* – h = 12 mm; *d* – h = 20 mm.

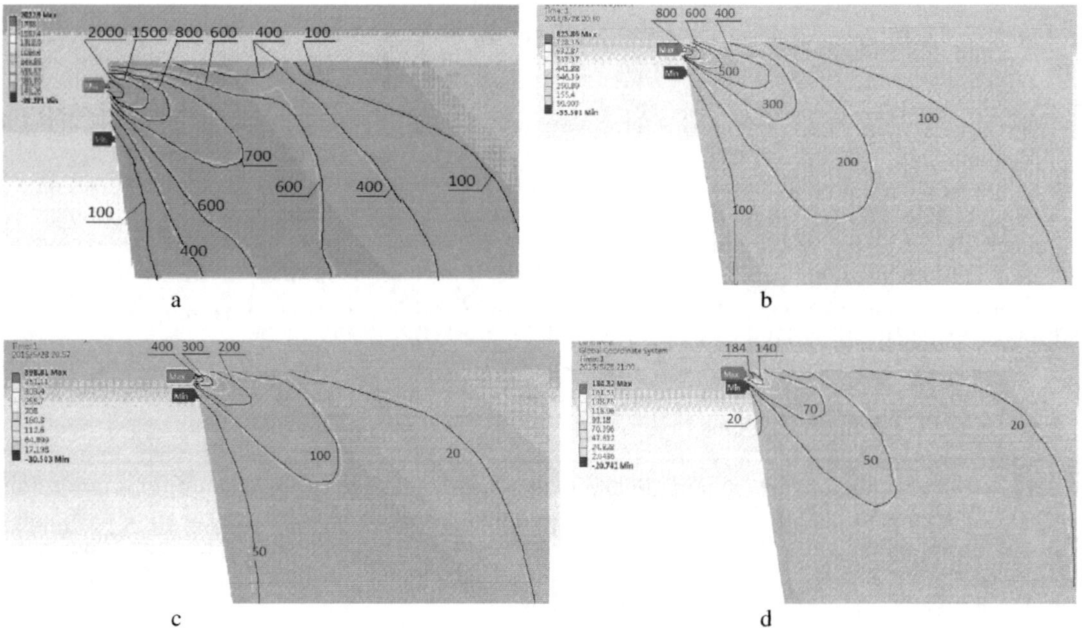

Figure 3. Distribution of shear stresses τ [MPa] in the cutting edge for a cutting insert with thickness of h: *a* – h = 2 mm; *b* – h = 6 mm; *c* – h = 12 mm; *d* – h = 20 mm.

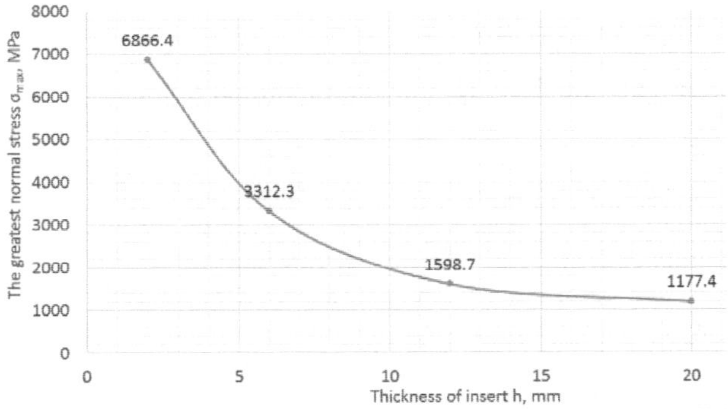

Figure 4. Influence of the cutting insert thickness on the maximum value of the principle normal stress in the cutting wedge

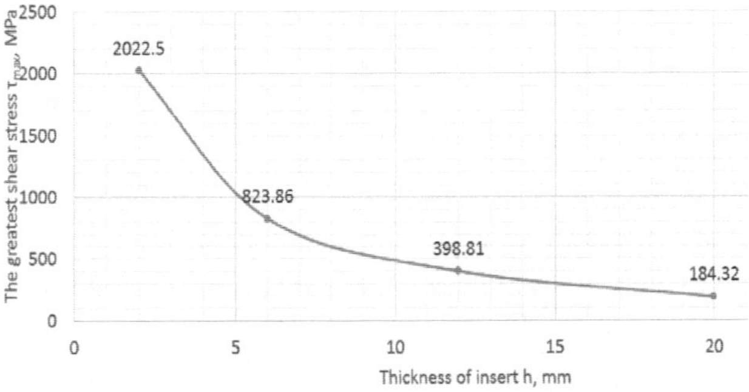

Figure 5. Influence of cutting insert thickness on the maximum value of the tangential stress in the cutting wedge

Relatively low, but elevated stresses ($-3200 > |\sigma_3| \geq -2000$ MPa) influence the probability of the insert destruction, since the presence of small imperfections in a zone of these stresses can lead to development of a macro-crack with subsequent destruction of the cutting insert. The diagram of relationship between the location of a zone with increased normal stresses ($|\sigma_3| \geq -2000$ MPa) and insert thickness is presented in Figure 6. The zone with increased stresses is negligibly small for insert thickness h > 12 mm.

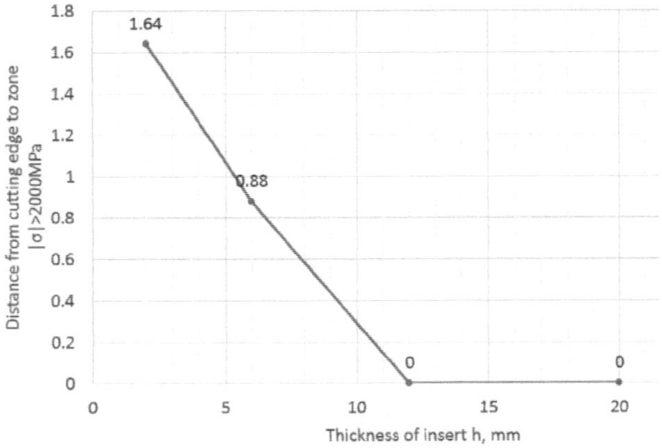

Figure 6. Influence of cutting insert thickness on the distance from the cutting edge to the zone with increased normal stress ($\sigma \geq 2000$ MPa) in the cutting wedge

4. Conclusion

In machining of titanium alloy (Ti6Al2Mo2Cr) with an insert having thickness equal to or less than 5 mm, high compressive stresses, which exceed the ultimate compressive strength of cemented carbide (WC8Co), arise near the cutting edge, which leads to chipping of the major cutting edge.

In machining of titanium alloy (Ti6Al2Mo2Cr) with cutting insert thickness equal to or less than 5 mm, tensile stresses, which approach the ultimate tensile strength of cemented carbide, arise on the face and base surfaces of the cutting insert, which increases the probability of cutting insert destruction.

In machining of titanium alloy (Ti6Al2Mo2Cr) with a cutting insert having thickness equal to or less than 6 mm high shear stresses arise near the cutting edge, exceeding the ultimate shear strength of cemented carbide and leading to destruction of the major cutting edge.

The application of cutting inserts with thickness of more than 6 mm in machining of titanium alloy helps to avoid destruction of easily worn inserts, which are common for finish and semifinish machining [4, 6, 7].

References

[1] Kozlov V 2012 *Proceedings 7th International Forum on Strategic Technology, IFOST* 6357713
[2] Shaw M 2003 *Sadhana* **28** 875–896
[3] Anderson T 1991 *CRC Press, Florida*
[4] Narutaki N *et al* 1983 *CIRP Annals - Manufacturing Technology* **32(1)** 65–69
[5] Wang B *et al* 2013 *International journal of Machine Tools & Manufacture* **73** 1–8
[6] Hu J and Chou Y 2007 *Wear* **263(7-12)** 1454–1458
[7] Kozlov V and Li X 2015 *Applied Mechanics and Materials: Scientific Journal* **756** 126–131

AUTHOR INDEX

AUTHOR INDEX

AUTHOR INDEX

AUTHOR INDEX

AUTHOR INDEX